A Textbook of Machine Drawing

IN FIRST ANGLE PROJECTION

[According to the Bureau of Indian Standards (B.I.S.) SP: 46-1988 & IS: 696-1972]

for

The Students of B.E./B.Tech.

Dr. R.K. Dhawan
Ph.D., M.I.E., M.E., M.I.S.T.E.
Principal
Ramgarhia Institute of Engineering & Technology
Satnampura, Phagwara
(Punjab Technical University, Jalandhar)

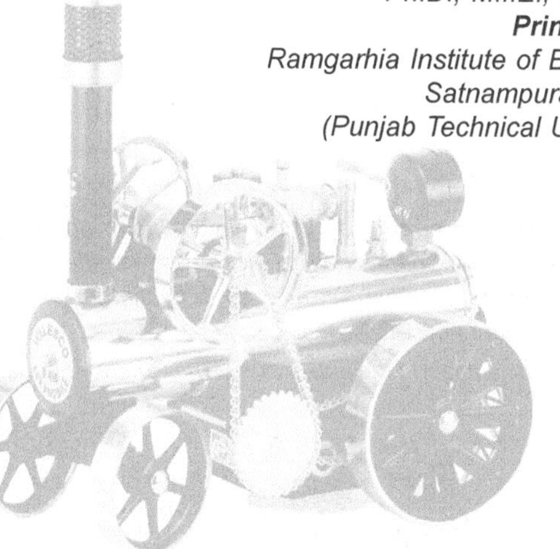

S Chand And Company Limited

(ISO 9001 Certified Company)

S Chand And Company Limited

(ISO 9001 Certified Company)

Head Office: Block B-1, House No. D-1, Ground Floor, Mohan Co-operative Industrial Estate, New Delhi – 110 044 | Phone: 011-66672000

Registered Office: A-27, 2nd Floor, Mohan Co-operative Industrial Estate, New Delhi – 110 044 Phone: 011-49731800

www.**schandpublishing.com**; e-mail: **info@schandpublishing.com**

Branches

Chennai	:	Ph: 23632120; chennai@schandpublishing.com
Guwahati	:	Ph: 2738811, 2735640; guwahati@schandpublishing.com
Hyderabad	:	Ph: 40186018; hyderabad@schandpublishing.com
Jalandhar	:	Ph: 4645630; jalandhar@schandpublishing.com
Kolkata	:	Ph: 23357458, 23353914; kolkata@schandpublishing.com
Lucknow	:	Ph: 4003633; lucknow@schandpublishing.com
Mumbai	:	Ph: 25000297; mumbai@schandpublishing.com
Patna	:	Ph: 2260011; patna@schandpublishing.com

First Edition 1996
Subsequent Editions and Reprints 1998, 2001, 2006, 2007, 2008, 2009, 2010, 2011, 2014
Reprint with Corrections 2016

ISBN : 978-93-856-7649-9

PRINTED IN INDIA

By Vikas Publishing House Private Limited, Plot 20/4, Site-IV, Industrial Area Sahibabad, Ghaziabad – 201 010 and Published by S Chand And Company Limited, A-27, 2nd Floor, Mohan Co-operative Industrial Estate, New Delhi – 110 044.

PREFACE TO THE REVISED EDITION

This revised edition of book takes special care to include the new topics of machine drawing as per the new syllabi of different universities. Twelve Model Test Papers, given at the end of the book are useful for the students to practice before examination.

All chapters, enchance the content value and give the students an idea of which they will be dealing in reality to abridge the gap between theory and actual practice. I hope the readers will find this book more useful.

Three test papers have been added for Self Assessment of the students. These papers contain MCQs and Objective Type Questions.

The author is thankful to Prof. Harvinder Lal, Prof. Jagjot Singh, Prof. Naresh Dr. Naveen Dhillo and Prof. Gurmeet Singh Gahir for their valuable suggestions.

Suggestions for further improvement of the book from the colleagues and learners will be greatly acknowledged and incorporated in the subsequent edition.

Dr. R.K. DHAWAN

PREFACE TO THE FIRST EDITION

PREFACE

A Textbook of Machine Drawing has been prepared to meet the requirements of the students preparing for B.Sc. Engineering, B.E., B.Tech., A.M.I.E. (India), Diploma in Mechanical Engineering, Production Engineering, Automobile Engineering and Textile Engineering, I.T.I. (Draftsman Course in Mechanical Engineering), C.T.I. and other Engineering Examinations. The book is in First Angle Projection as recommended by Bureau of Indian Standards, SP: 46-1988. However, the third angle projection has also been included in it.

The book is divided into three sections, each containing different chapters covering questions for self examination (viva-voce), objective type questions, fill in the blanks, problems for practice, thus covering the full syllabus of all Indian Universities and boards. The book is equally useful for the students of different Technical Universities and Polytechnics of India.

Simple language, easy to follow descriptions of subject matter and clear diagrams are the characteristic features of the book.

Acknowledgement is due to Bureau of Indian standards and other authorties on the subject, whose works author has Consulted for writing the book.

Authors warmest thanks to M/s S. Chand & Co. for bringing out the book in such short period with good get-up, fine printing and reasonable price.

The author is thankful to Prof. Rattan Bhandari, Prof. S.K. Aggarawal, Dr. G.D. Bansal, Prof. L.D. Garg, Prof. S.P. Thapar, Prof. Rakesh Chandera, Prof. J.S. Gill and Prof. S. Charanjet Singh.

Suggestions for the improvement of the book will be acknowledged with thanks.

Dr. R.K. DHAWAN

CONTENTS

CONTENTS

Section – I

Section – II

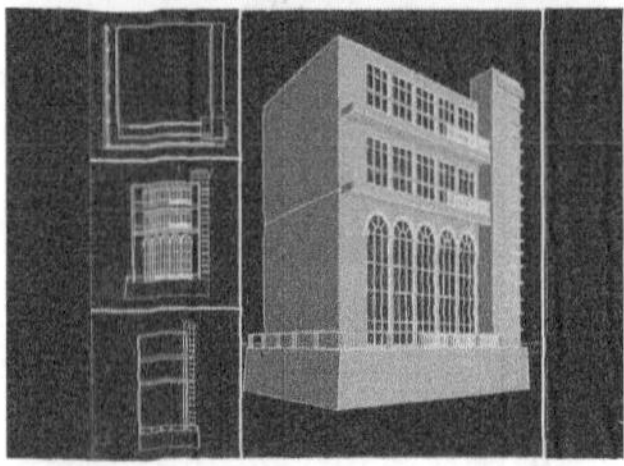

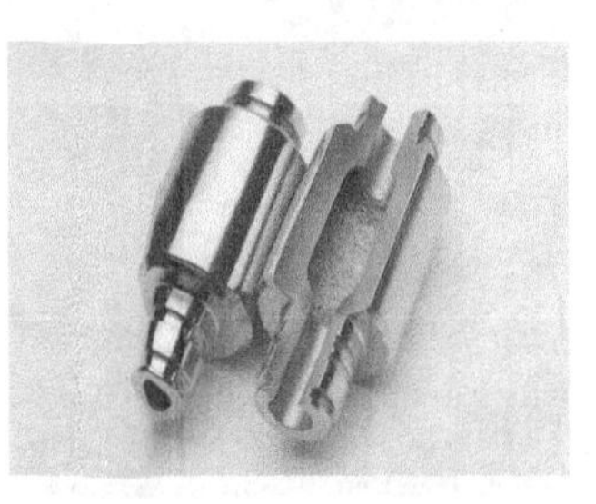

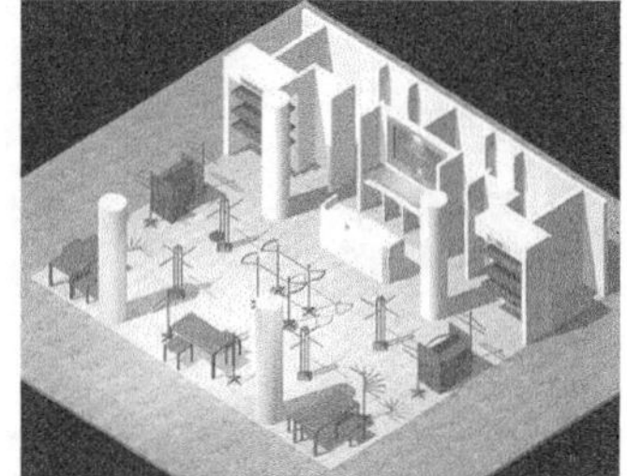

Section – III

SECTION - I

1. Introduction and Drawing Instruments
2. Layout of Drawing Sheet
3. Conventions
4. Lettering
5. Dimensioning
6. Scales

This Page has been intentionally left blank

CHAPTER

1

Introduction and Drawing Instruments

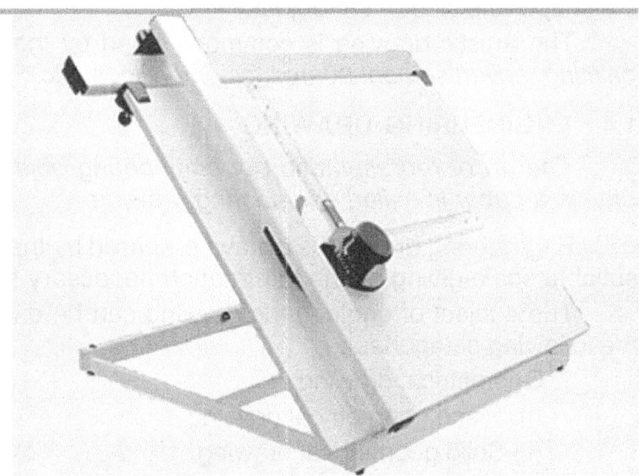

INTRODUCTION

In this fast developing society, an engineer plays a vital role. He is rightly called "The Creator" ... a man who puts his imagination into actual practice. He thinks of the problems in his mind and conveys them to others through the language of systematic lines. It is this language of systematic lines which is called engineering drawing. Therefore, an engineer must have knowledge of this language to project his ideas correctly on the paper and then execute the job efficiently and effectively with the help of this drawing.

Since, the modern research work in engineering depends mainly upon engineering drawing, it is, therefore, necessary for an engineer to acquire a good working knowledge about the subject in order to express and record the shape, size and other informations necessary for the construction of various objects such as buildings, roads, bridges, structures, machines, etc.

In this chapter, we will deal with the study of introduction of drawing, engineering drawing, drawing instruments, their uses and cares in handling them along with modern instruments for speedy and accurate work.

1.1. DRAWING

The art of representation of an object by systematic lines on a paper is called drawing.

1.2. CLASSIFICATION OF DRAWING

The drawing may be classified into the following two main groups:

1. Artistic drawing (Free hand or Model drawing)
2. Engineering drawing (Instrument drawing)

1.3. ARTISTIC DRAWING

The art of representation of an object such as painting, cinema slide, advertisement boards, etc. by the artist by his imagination or by keeping the object before him is known as artistic drawing.

The artist tries to produce the object in the shape of a picture by giving various shades and colours. Dimensions and other details are not given in it, however, one can appreciate the shape and size of the object.

The artistic drawing is commonly used by the artists for commercial purposes, *i.e., in painting, advertisement boards, cinema slides, etc.*

1.4. ENGINEERING DRAWING

The art of representation of engineering objects such as buildings, roads, machines, etc. on a paper is called engineering drawing.

Engineering drawing is always prepared by the drawing instruments according to some sutiable scale giving all the information necessary for the construction of an object.

The subject of engineering drawing can be divided from engineering point of view into the following categories :

1. Geometrical drawing
   (i) Plane geometrical drawing
   (ii) Solid geometrical drawing
2. Mechanical engineering drawing
3. Civil engineering drawing
4. Electrical and Electronic engineering drawing, etc.

1.5. GEOMETRICAL DRAWING

The art of representation of geometrical objects such as rectangle, square, cube, cone, cylinder, sphere etc. on a paper is called geometrical drawing.

1.5.1. Plane Geometrical Drawing : The art of representation of objects having two dimensions, i.e., length and breadth such as square, rectangle, triangle, etc. on a paper is called *plane geometrical drawing.*

1.5.2. Solid Geometrical Drawing : The art of representation of objects having three dimensions i.e. length, breadth and thickness such as cube, prism, cylinder, sphare etc. on a paper is called solid geometrical drawing. It is also called *descriptive or practical solid geometrical drawing.*

1.6. MECHANICAL ENGINEERING DRAWING OR MACHINE DRAWING

The art of representation of mechanical engineering objects such as machines, machine parts, etc. on a paper is called mechanical engineering drawing or machine drawing.

The mechanical engineering drawing is commonly used by mechanical engineers to express mechanical engineering works and projects for actual execution.

1.7. CIVIL ENGINEERING DRAWING

The art of representation of civil engineering objects such as roads, buildings, bridges, dams, etc. on a paper is called civil engineering drawing.

The civil engineering drawing is commonly used by civil engineers to express civil engineering works and projects for actual execution.

1.8. ELECTRICAL AND ELECTRONIC ENGINEERING DRAWING

The art of representation of electrical objects such as motors, generators, poles, towers, transformers, wiring diagrams, etc. on a paper is called electrical engineering drawing or electrical drawing.

The electrical engineering drawing is commonly used by electrical engineers to express electrical engineering works and projects for actual execution.

The art of representation of electronic circuits of T.V., V.C.R., calculators, computers, etc. on a paper is called electronic engineering drawing or electronic drawing.

This drawing is commonly used by electronic engineers to express electronic engineering works and projects for actual execution.

1.9. SELECTION OF DRAWING INSTRUMENTS

For the preparation of any drawing work, it is quite essential to use instruments correctly and accurately, in order to get the required degree of accuracy. But the accuracy of drawing depends largely on the quality of instruments to be used. It is, therefore, essential for the students to purchase good quality instruments. If, however, it is not possible to have costly instruments box, then it will always be better to purchase some required pieces of instruments of good quality.

1.10. DRAWING INSTRUMENTS AND OTHER DRAWING MATERIALS

Following is the list of drawing instruments and other drawing materials required for the preparation of drawing work :

1. Drawing board
2. T-square
3. Set-squares
4. Instrument box
5. Scales
6. Pencil and sand paper block
7. Protractor
8. Clinograph
9. Rubber or Eraser
10. Erasing shield
11. Drawing pins or Cello-tape
12. Irregular or French curves
13. Duster or handkerchief
14. Drawing ink
15. Tracing paper and tracing cloth

16. Drawing paper
17. Sketch-book

(Modern Instruments for accurate and speedy work)

1. Drafting machine
2. Parallel ruling straight edge machine

Important Note: *The students are advised to purchase good quality of instruments in order to get proper degree of accuracy in drawing work.*

1.11. DRAWING BOARD

A first class engineer's drawing board is made of 4 to 6 strips of well-seasoned soft wood such as pine, fir, oak or kail about 18 mm thick. The wooden strips are cleated at the back by two battens by mean of screws (see Fig. 1.1). On the left hand edge of the board, a straight ebony strip is fitted against which the stock of the T-square moves.

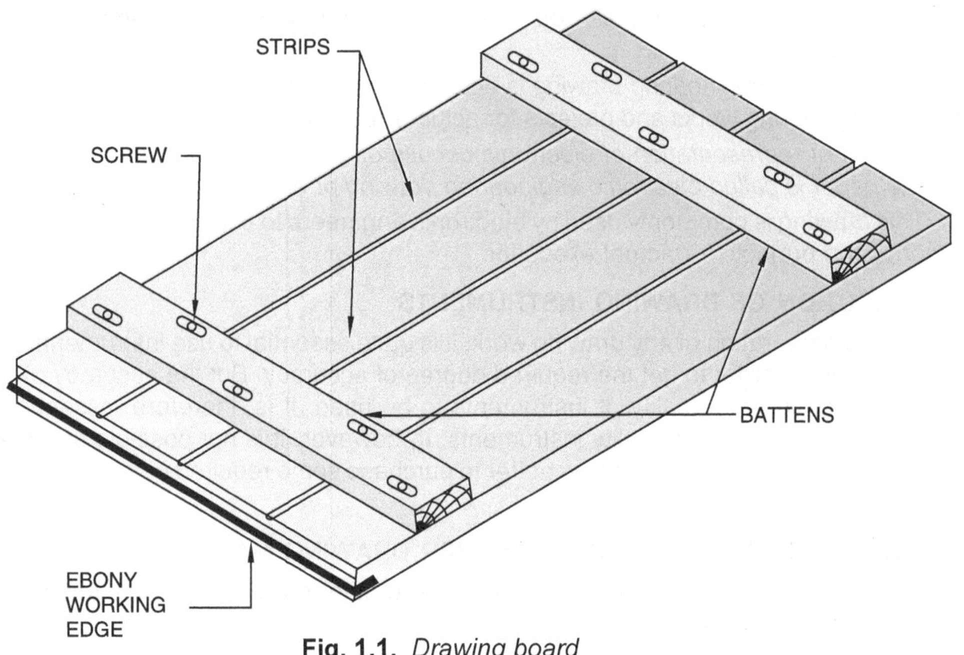

Fig. 1.1. *Drawing board*

The standard sizes of drawing boards according to Indian Standards Institution (I.S.I.) are as given below :

S.NO.	DESIGNATION	SIZE IN MM LENGTH × WIDTH × THICKNESS	TO BE USED WITH SHEET SIZES
1.	D_0	1500 × 1000 × 25	A_0
2.	D_1	1000 × 700 × 25	A_1
3.	D_2	700 × 500 × 15	A_2
4.	D_3	500 × 350 × 15	A_3

Important Note: *For drawing exceeding above mentioned sizes, special drawing boards may be used, but the width of drawing board should not exceed 1000 mm.*

1.11.1. Uses of Drawing Board : The following are the uses of drawing Board :—

1. The flat surface of the board is used to hold the drawing sheet while the drawing is being made.
2. The ebony edge of the board is used as a guide for the stock of T-square.

1.11.2. Cares in Handling of Drawing Board : The following are the cares for handling the drawing board :

1. Handle the drawing board carefully, so that the top flat surface of the board should not be spoiled.
2. The ebony working edge of the board must be straight against which the stock of T-square moves.
3. Fasten an extra sheet of paper on the board to keep the surface clean.

1.12. T-SQUARE

T-square is made of hard quality wood such as teak or mahogany, etc. There are two essential parts of T-square, namely the stock and the blade. The blade is fitted with an ebony or plastic piece to form working edge of T-square. The two parts being held securely together at right angles to each other by means of screws or dowel pins in order to form a straight edge of the blade (see Fig. 1.2). The working length of T-square is equal to the length of drawing board.

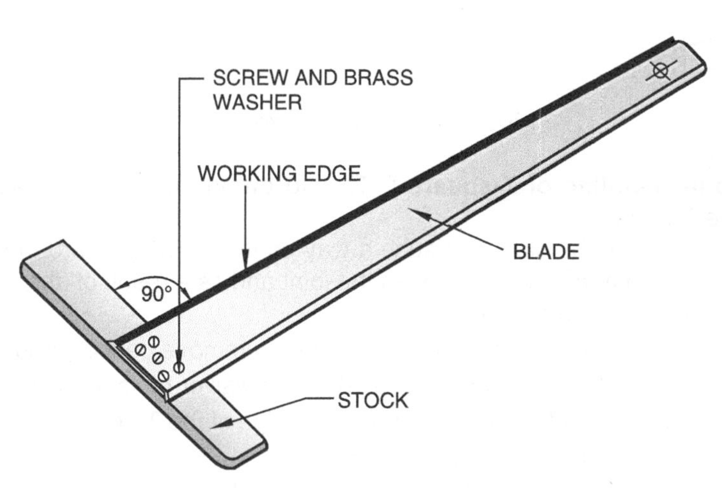

Fig. 1.2. *T-square*

The standard sizes of T-square according to I.S.I. are given below :

S.NO.	DESIGNATION	LENGTH OF WORKING EDGE IN mm
1.	T_0	1500 ± 10
2.	T_1	1000 ± 10
3.	T_2	700 ± 5
4.	T_3	500 ± 5

1.12.1. Uses of T-square: The following are the uses of T-squares :—

1. The T-square is mainly used for drawing horizontal lines.

Fig. 1.3 illustrates the way of drawing horizontal lines along with the position of hands.

Important Note: *While drawing horizontal lines, the pencil should be slightly inclined towards the edge of the T-square from left to right so that the lines can be drawn nearly coincident with the edge of the T-square.*

---

2. T-square is used as a base for drawing various angles with the help of set squares as shown in Fig. 1.7, Fig. 1.8 and Fig. 1.9.

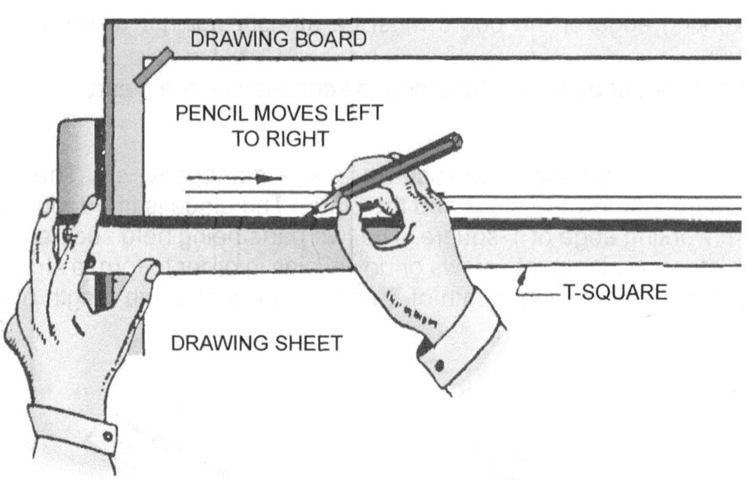

Fig. 1.3. *Drawing horizontal lines*

1.12.2. Cares in Handling of T-square : The following are the precautions for careful handling of the T-square :

1. T-square should be placed in such a way on the drawing board that it may not drop on the floor, otherwise it will loosen the joint and as a result of that, the T-square will no longer draw true parallel lines.
2. Clean the blade of T-square with moist cloth to remove pencil graphite lead.
3. T-square should be hung on a nail against the wall or table when it is not being used.
4. Do not draw horizontal lines with the lower edge of the T-square.
5. Do not use T-square as a hammer.
6. Do not use the edge of the T-square as a guide for cutting the paper with a knife.

1.12.3. Testing the T-square : The following points should be checked while testing the T-square :—

1. Check all screw heads and tighen if necessary [see Fig. 1.4 (i)]
2. In order to check the T-square, first of all draw a horizontal line [see Fig. 1.4 (ii)]. Now reverse the T-square and again draw a horizontal line with working edge. If both the lines coincide with each other, then the working edge of T-square is alright. If there is any difference in two lines, then working edge is not correct and the line gives twice the error of the working edge. This error should be rectified by scraping the edge with a scraper or a sharp knife.

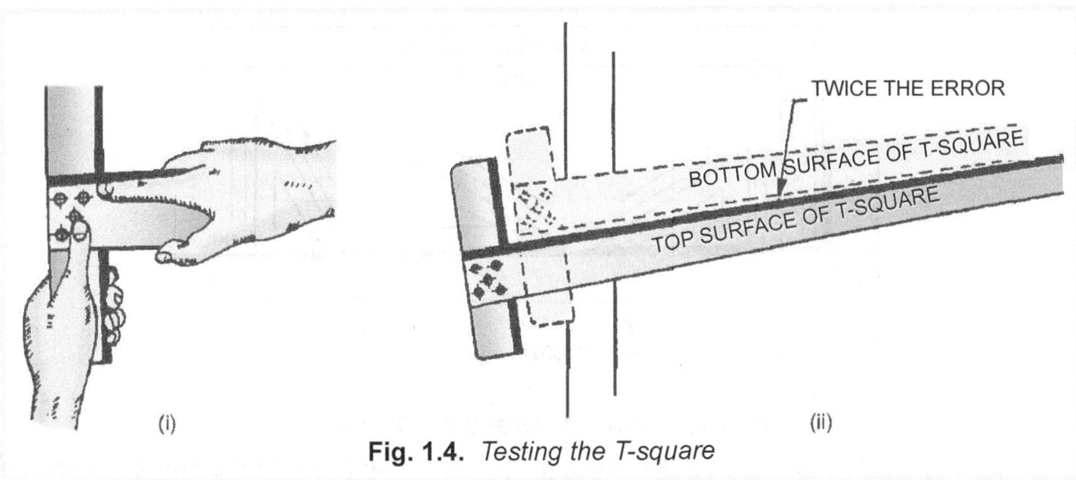

Fig. 1.4. *Testing the T-square*

1.13. SET SQUARE

The set-square is made of transparent celluloid or plastic materials, etc. The set-square made of transparent celluloid is most satisfactory as the line underneath the set-square can be seen quite easily and this often prevents another line being drawn in the wrong place.

1.13.1. Types of Set-squares: The sets-squares are of following two types :

1. Thirty-sixty degree (30°-60°) set square
2. Forty five degree (45°) set square

The 30°-60° set-square has three edges, one of which forms 90° and other edges forming angles of 30° and 60° with the other sides respectively (see Fig. 1.5)

The 45° set-square is similar to the 30°-60° set-square, but its edges form an isosceles triangle in which two of the angles are of 45° each and other one's 90° (see Fig. 1.6).

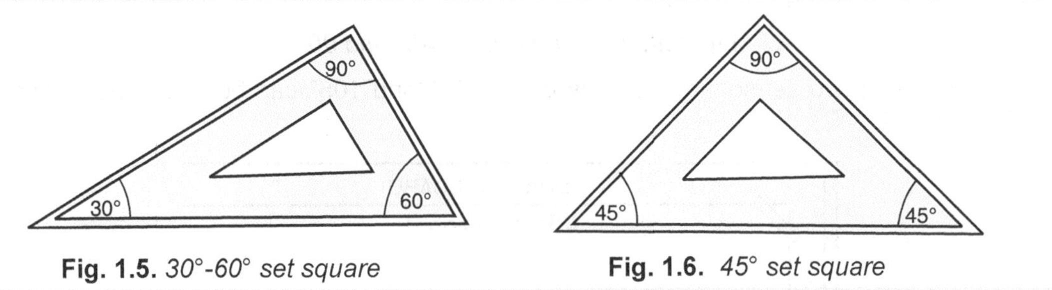

Fig. 1.5. *30°-60° set square*

Fig. 1.6. *45° set square*

1.13.2. Sizes of Set-squares : The set-squares can be available in different suitable lengths, but 30°-60° set squares of 250 mm and 45° set-square of 200 mm are more suitable for general work.

1.13.3. Uses of Set-squares : The following are the important uses of set-squares :—

1. The set-squares are used for drawing all straight lines except the horizontal lines which are usually drawn with T-square.
2. The perpendicular lines or the lines at 30°, 60° and 90° to the horizontal can be drawn by using the set-squares as shown in Fig. 1.7.

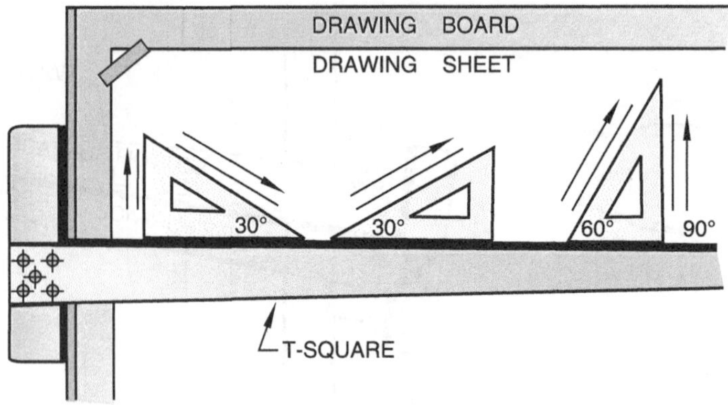

Fig. 1.7. *Drawing lines at 30°, 60° and 90°*

3. The perpendicular lines or the lines inclined at 45° and 90° to the horizontal can be drawn by using the set-squares as shown in Fig. 1.8.

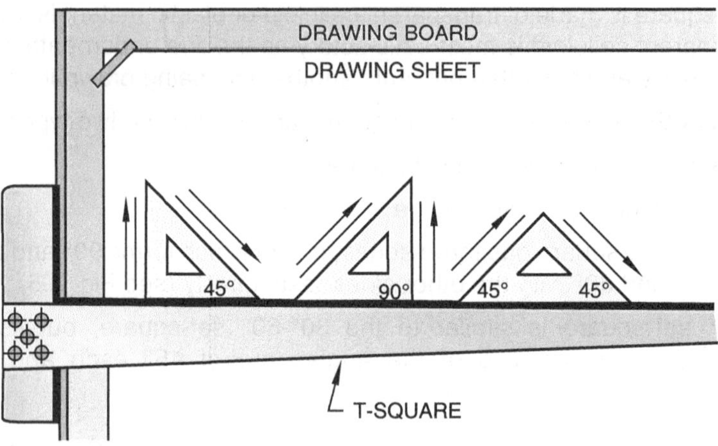

Fig. 1.8. *Drawing lines at 45° and 90°*

4. By using two set-squares, angles of 15°, 75° and 105° can be drawn as shown in Fig.1.9.

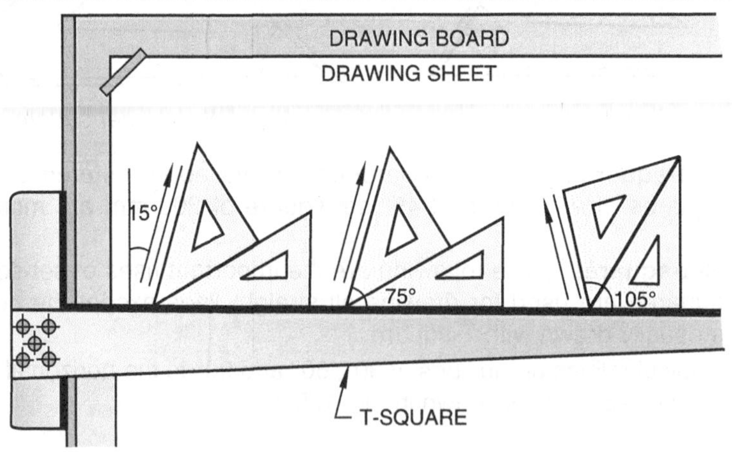

Fig. 1.9. *Drawing lines at 15°, 75°, and 105°*

In general, set-squares are used to draw angles of 15°, 30°, 45°, 60°, 75°, 90°, i.e, any multiples of 15°, as shown in Fig. 1-10.

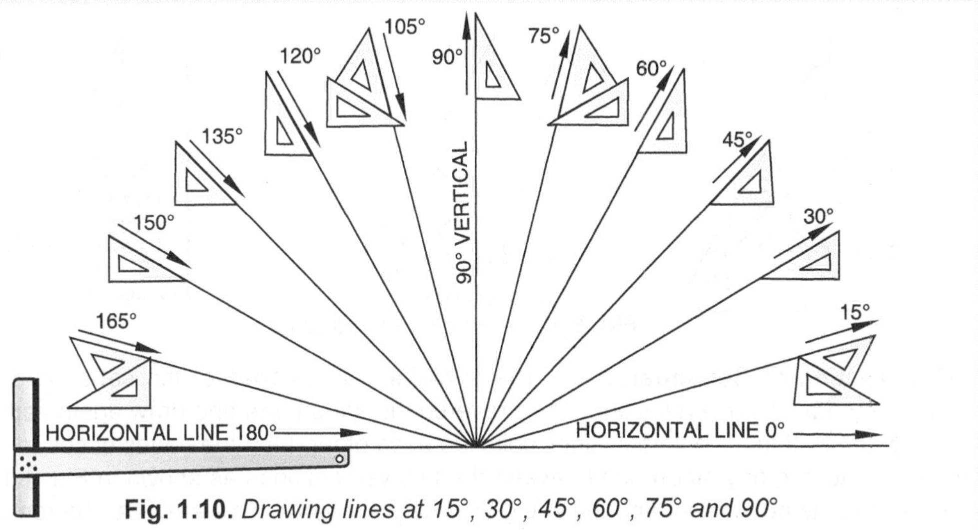

Fig. 1.10. *Drawing lines at 15°, 30°, 45°, 60°, 75° and 90°*

5. The set-squares can also be used for drawing parallel and perpendicular lines to any given line as shown in Fig. 1.11.

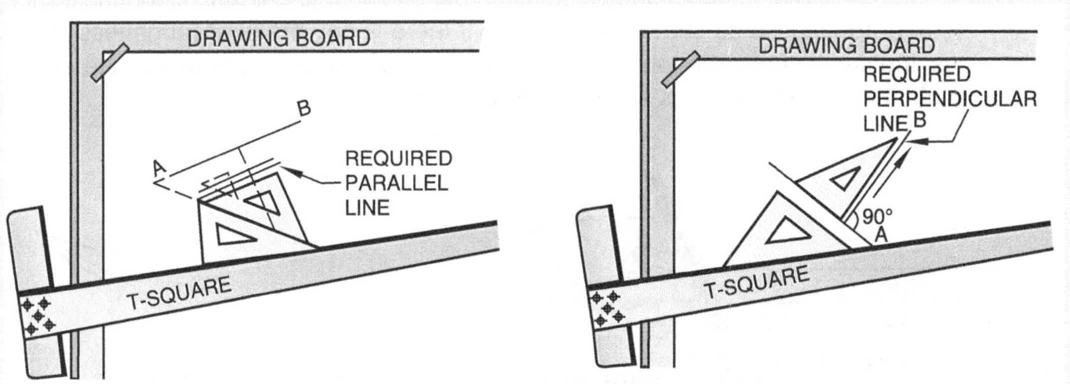

Fig. 1.11. *Drawing parallel and perpendicular lines with set squares*

6. Set-squares are also used for drawing a line between any two given points, say A and B. For this, first of all, place a set square to meet a point A. Then, swing the set square to meet the point B. Finally, draw the required line AB by pencil as shown in Fig. 1.12.

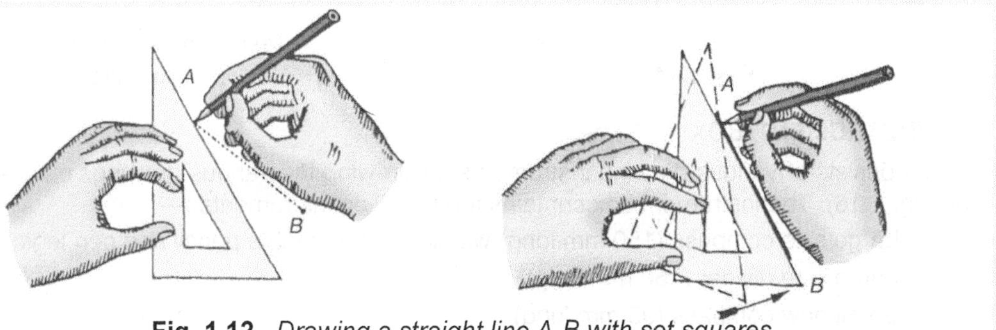

Fig. 1.12. *Drawing a straight line A-B with set squares*

Fig. 1.13, shows adjustable set-squares which are used to draw parallel, perpendicular and inclined lines with speed and accuracy. They are helpful especially in making inclined printing.

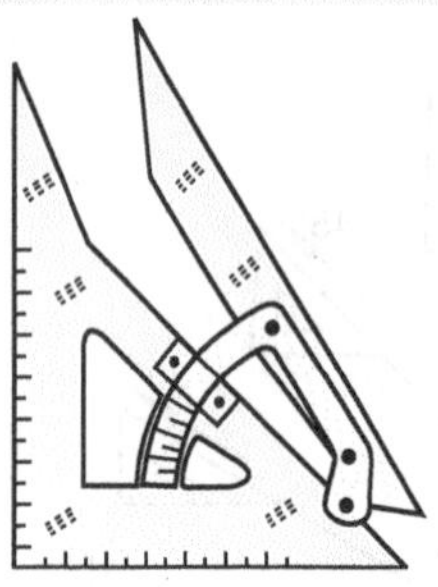
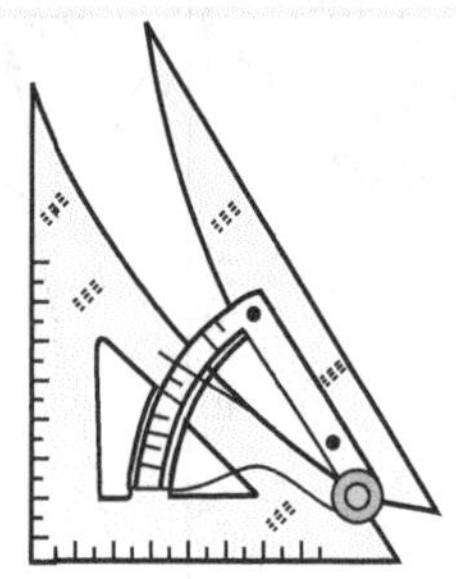
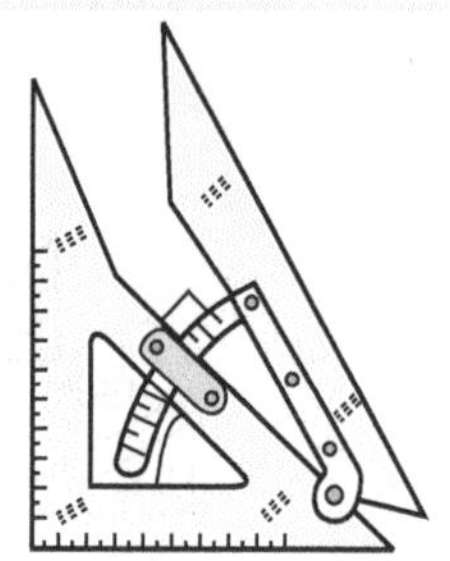

Fig. 1.13. *Adjustable set-squares*

1.13.4. Testing of Set-squares : The straightness of edges of the set-square can be checked by drawing a vertical line. Then reverse the set-square and draw again vertical line (see Fig. 1.14). If the two vertical lines coincide each other, then the edges of set-square are correct. If there is any difference between the two vertical lines as shown in Fig. 1.14, then working edge is not correct and the line gives the twice error. This error can be removed by straightening the edges by means of a scraper or sand paper. If there is too much difference, then set-square should be replaced by new one.

The edges of the set squares should checked by running a finger nail, along the edges for any break or roughness as shown in Fig. 1.15. If there is any break or roughness in the set squares, the same should be removed by scraper or sand paper. If, however, the edges are too rough, then the set-squares should be replaced by new one.

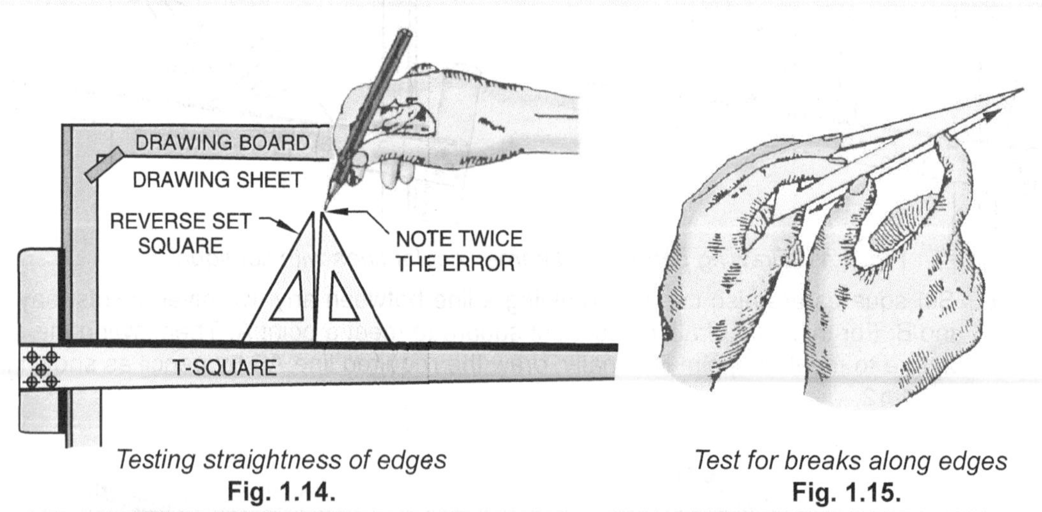

Testing straightness of edges
Fig. 1.14.

Test for breaks along edges
Fig. 1.15.

1.14. INSTRUMENT BOX

It consists of various drawing instruments for drawing the various drawings on the sheet (see Fig. 1.16). The instrument box contains the following instruments :—

1. Large size compass (150 mm long) with inter-changeable pencil and pen legs
2. Large size divider (150 mm long)
3. Small bow compass (95 mm long)

4. Small ink bow compass (95 mm long)
5. Small bow divider (95 mm long)
6. Lengthening bar
7. Pin point
8. Ink point
9. Extension bar
10. Ruling pen or liner
11. Holder croquill
12. Lead case.

Note : *All the instruments are made of silver coated with nickel.* The parts like *divider points, ruling pen nibs, screws, springs, etc. are made of hard steel.*

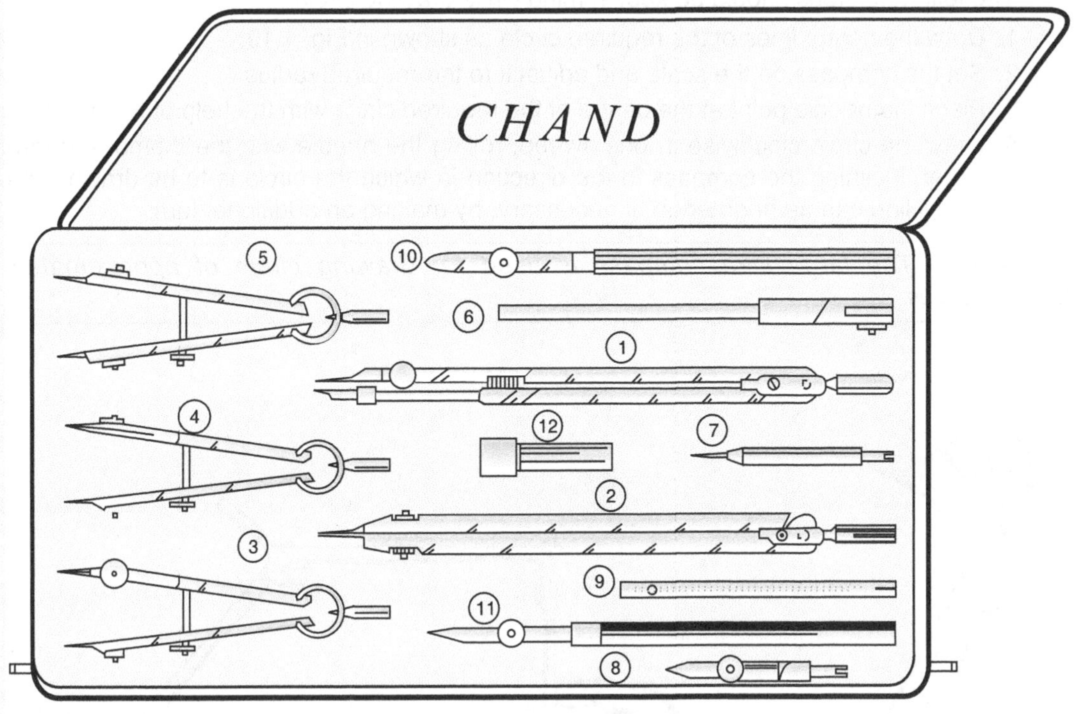

Fig. 1.16. *Instrument box*

1.14.1. Large Size Compass: This compass is used for drawing circles and arcs of circles of required sizes. It consists of two metal legs hinged together at its upper end by means of a joint known as knee joint. An adjustable needle is fitted on to end of one of the two legs and on the other leg is provided an attachment which can be fitted with either on metal-leg carrying pencil lead (see Fig. 1.17) or an inking device.

Important Note : *For better working of compass, the leg containing the lead pencil should be slightly shorter (about 1 mm) than the needle point of the leg (see Fig. 1.21).*

1.14.2. Uses of Compass : The compass is used for drawing circles and arcs of circles of required sizes.

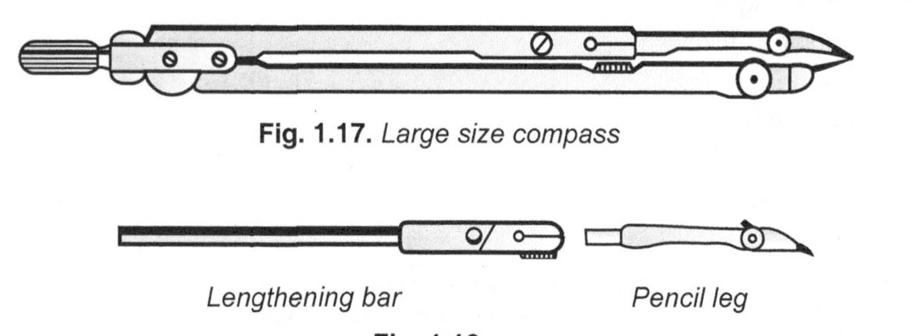

Fig. 1.17. *Large size compass*

Lengthening bar *Pencil leg*

Fig. 1.18.

The following steps should be kept in mind while drawing circle :-

1. Draw the centre lines of the required circle as shown in Fig. 1.19.
2. Set the compass on the scale and adjust it to the required radius.
3. Place the needle point at the centre of the required circle with the help of left hand.
4. Draw the circle clockwise in one sweep, rolling the handle with the thumb and fore finger, inclining the compass in the direction in which the circle is to be drawn. The pencil line can be brightened, if necessary, by making an additional turn.

Note : *The large size compass is used for drawing circle of approximately 120 mm radius.*

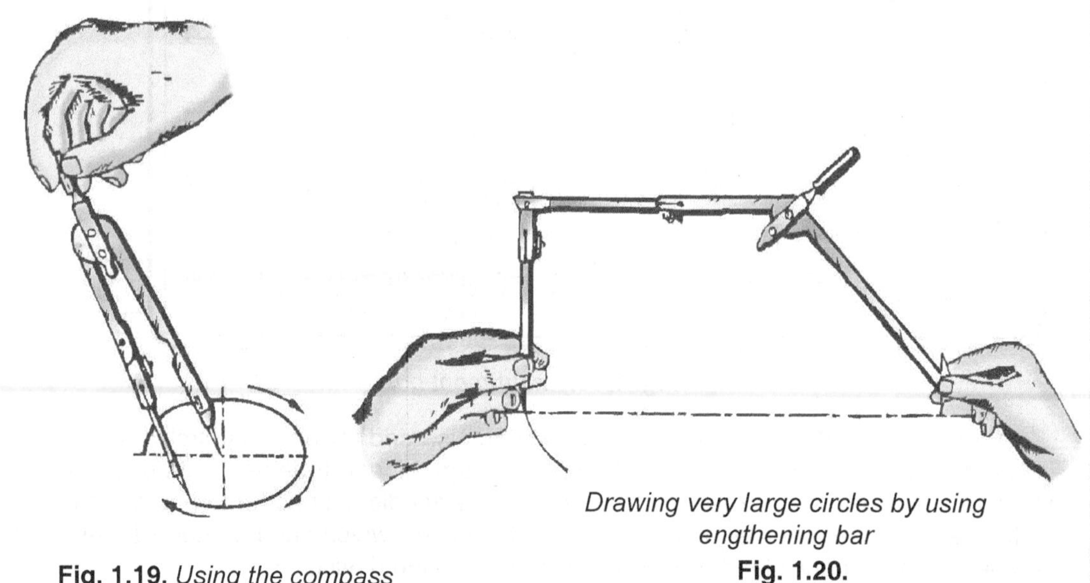

Fig. 1.19. *Using the compass*

Drawing very large circles by using engthening bar

Fig. 1.20.

Very large circles are drawn by using the lengthening bar (see Fig. 1.18) as illustrated in Fig. 1.20. For this, detach the pencil leg and insert the lengthening bar in its place. Connect the detached part with the lengthening bar and finally draw the required circle.

1.14.3. The Compass Lead : In compass, use lead about one degree softer than the pencil lead. This is necessary to maintain the uniform darkness in all the lines and put less pressure on the compass lead than on the pencil.

As the position of the needle is required to be inserted slightly inside the paper, so that it should be kept slightly, about 1mm, longer than the lead point. The adjustment of the pencil-lead with respect to the needle point and shape to which the lead should be ground are shown in Fig. 1.21.

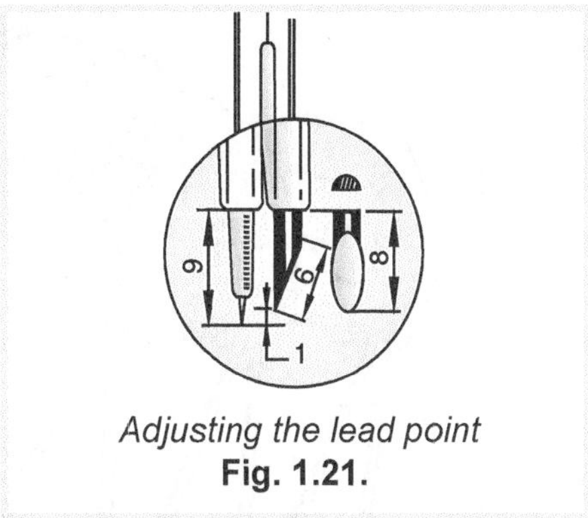

Adjusting the lead point
Fig. 1.21.

Important Note : *The lead of the compass should be used one degree softer than the pencil used in order to maintain the uniform brightness of the drawing.*

1.14.4. Cares in Handling of Compass : The following are important precautions for careful handling of compass :

1. In large-size compass, test the tightness of joint and see that it is tight and firm, yet reasonably easy to set with one hand.
2. Do not oil the joints of the compasses.

1.14.5. Large-size Divider : The divider is used for dividing straight or curved lines into desired number of equal parts. This instrument also consists of two metal legs as in the compass, except for the fact that two steel points are provided instead of pencil point as shown in Fig. 1.22.

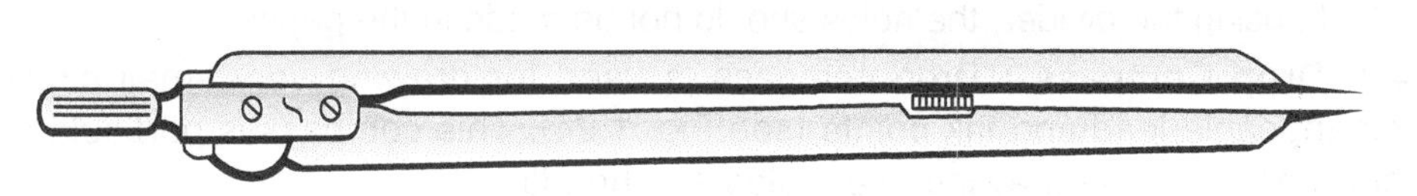

Fig. 1.22. *Large size divider*

In dividing a line into desired number of equal parts, open the divider until the distance between the points is estimated to be equal to the length of a division and step of the line as shown in Fig. 1.23. If the last division falls short or goes beyond, increase or decrease the setting of the divider points and step off the line again. Repeat this procedure until the divider is properly set for dividing the required line into exact number of equal parts.

Similarly, any arc or circle can be divided into any number of equal divisions.

1.14.6. Uses of Divider : The following are the uses of divider :

1. It is used to divide straight or curved lines into desired number of equal parts.
2. It is used to transfer dimensions from one part to another part of the drawing.
3. It is used to set-off given distances from the scale to the drawing.

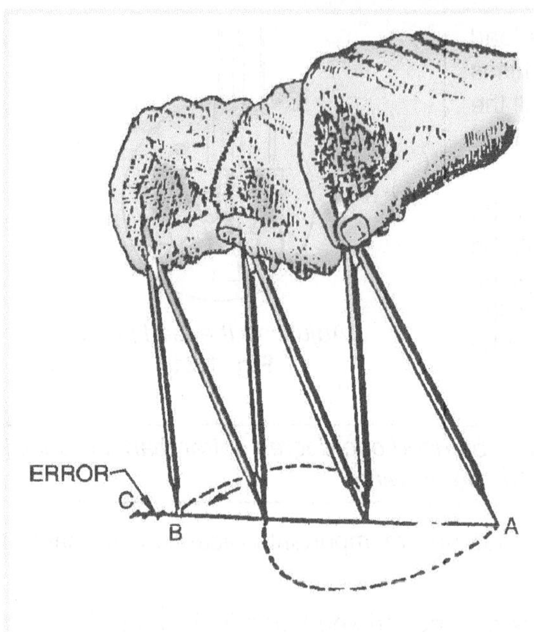

Fig. 1.23. *Dividing a line with divider*

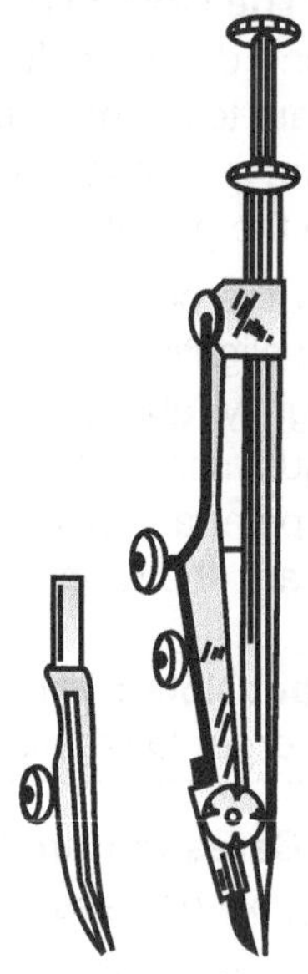

Fig. 1.24. *Drop compass*

1.14.7. Cares in Handling of Divider : Following are the precautions for careful handling the divider :—

1. Test the tension of the joint by moving its legs. If the legs are loose or tight, adjust the legs by means of screw with the help of a screw driver.
2. Keep both the needle points at the same length.
3. In using the divider, the holes should not be made in the paper.

1.14.8. Drop Compass : Drop compass is used for drawing very small circles. It has interchangeable lead and ink points (see Fig. 1.24). This compass is particularly used by structure engineers in drawing rivet holes and heads.

1.15 INK BOW COMPASS, SMALL BOW COMPASS AND SMALL SIZE DIVIDER

1. Small ink bow compass : The small ink bow compass is used to draw small circles in ink. It consists an inking needle instead of lead needle as in case of small bow compass [see Fig. 1.25 (i)].

2. Small bow compass : The small bow compass is used to draw small circles and arcs of about 18 mm radius and if set, circle as small as about 1mm radius can be drawn. This compass can be either wheel type [see Fig. 1.25 (ii)] or side wheel type. The use of small bow compass is similar to large compass.

3. Small size divider : The small size divider is similar to large divider [see Fig. 1.25 (iii)].

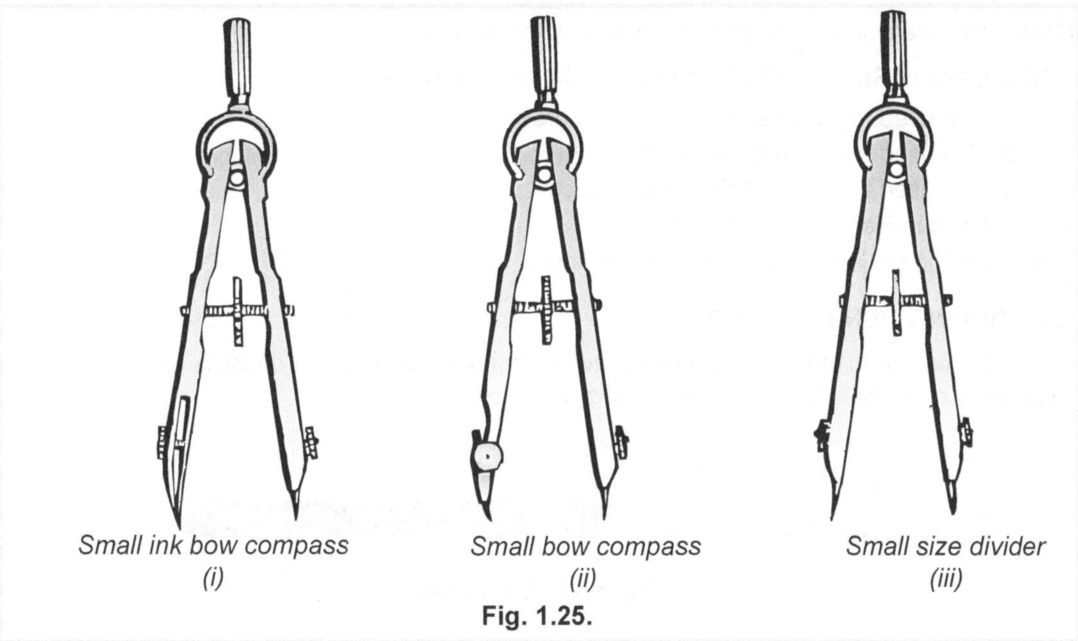

Small ink bow compass (i) Small bow compass (ii) Small size divider (iii)

Fig. 1.25.

1.16. SCALES

All measurements of lengths or distances on a drawing are made with the scales. The scale is a measuring stick, graduated with different divisions to represent the corresponding actual distances of ground according to some fixed proportion, thereby facilitating rapid in marking off distances on drawing. The scale are either flat or triangular and the materials used in their construction may be wood, celluloid, metal, card-board, etc. The Fig. 1.26 shows the scale which is graduated on both the edges with figures 1 :1, called full-size scale and 1 : 2, called half full-size scale.

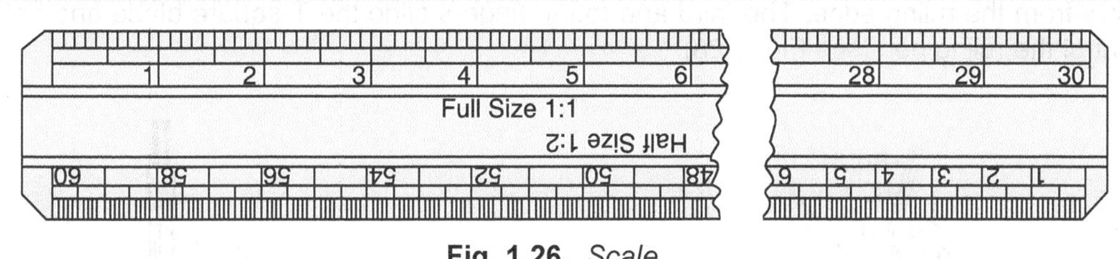

Fig. 1.26. *Scale*

In the scale, the figures such as 1 : 1, 1 : 2, 1 : 5, 1 :10, 1 : 15, 1 : 20, 1 : 50, 1 : 75, 1 : 100, 1 : 200, 1 : 500, 1 : 2000, 1 : 3000 .... 1 : 100000, 1 : 250000, 1 : 500000, etc. are written. Thus, the proportion 1 : 5 means that if 1 centimetre is to be shown on drawing, it will represent its corresponding actual distance of 5 centimeters on ground.

The scales used in engineering practice, according to I.S.I are :

FULL SIZE SCALE	REDUCING SCALES		ENLARGING SCALES
1 : 1	1 : 2	1 : 25	10 : 1
	1 : 5	1 : 50	5 : 1
	1 : 10	1 : 100	2 : 1
	1 : 20		

Note : *The scale should be graduated on both of its edges.*

1.16.1. Uses of Scales : The following are the uses of scales :

1. It is used to prepare or enlarge size drawings.
2. It is used to set off dimensions.
3. It is used to measure distances directly.

1.16.2. Care in Handling of Scale :

Never use the scale as a ruler for drawing lines.

1.17. RULING PEN OR LINER

The ruling pen or liner is used for drawing straight lines in ink. It consists of steel nib fitted to an ivory or metal holder as shown in Fig. 1.27.

Fig. 1.27. *Ruling pen*

The ruling pen is charged with ink by inserting the quill, fitted with the cork of ink bottle (for this purpose), between the two nibs and allowing a small quantity of ink to flow into the gap between the two nibs. The required thickness of the line is controlled by means of a screw provided on the nibs as shown in Fig. 1.27. The inking pen is always used in connection with a guide edge of the T-square or bevelled edge of the set-square.

1.17.1. Using a Ruling Pen : When drawing a line, the ruling pen should be held in a vertical plane and inclined at 60° to 70°, in the direction of movement. It should be held by the thumb and forefinger with the blade against the second finger and adjusting screw on the outside away from the ruling edge. The third and fourth fingers slide the T-square blade and help to control the ruling pen (see Fig. 1.28).

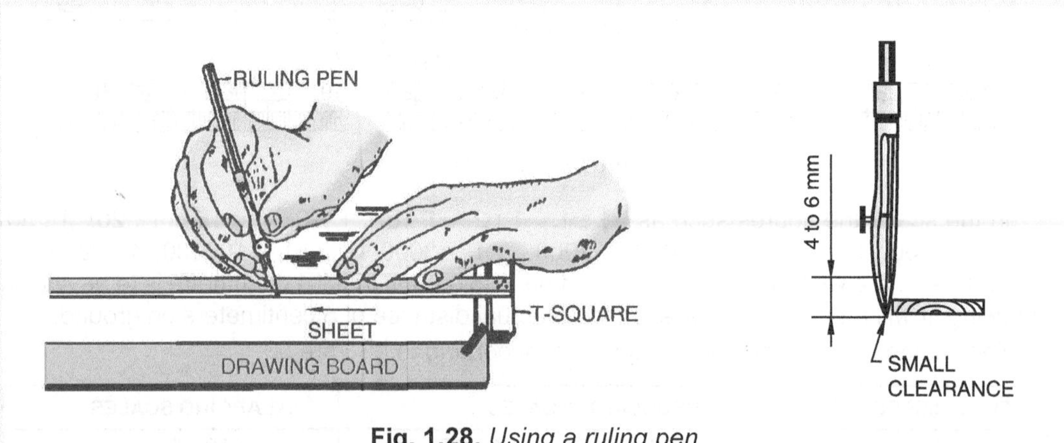

Fig. 1.28. *Using a ruling pen*

Care must be taken to get and keep the correct position of the pen, since only a slight deviation is required to bring disastrous result.

Fig. 1.29 shows the correct and incorrect results from such right and wrong holding of ruling pen as discussed below :—

1. If the pen point is out and off the perpendicular plane, it will rub against blade of T-square and make the line ragged.
2. If the nibs of the pen pressed tightly against the T-square it would close the nibs and thus reduced the thickness of the lines.
3. If the pen is turned in from the perpendicular, the ink is likely to run under the edge of the guide and cause a blot.
4. If the ink refuses to flow, it may be that ink has dried in the extreme point of the pen. If by pinching the blade slightly or touching the pen on the finger, the ink does not start, the pen should immediately be wiped out and fresh ink be supplied.

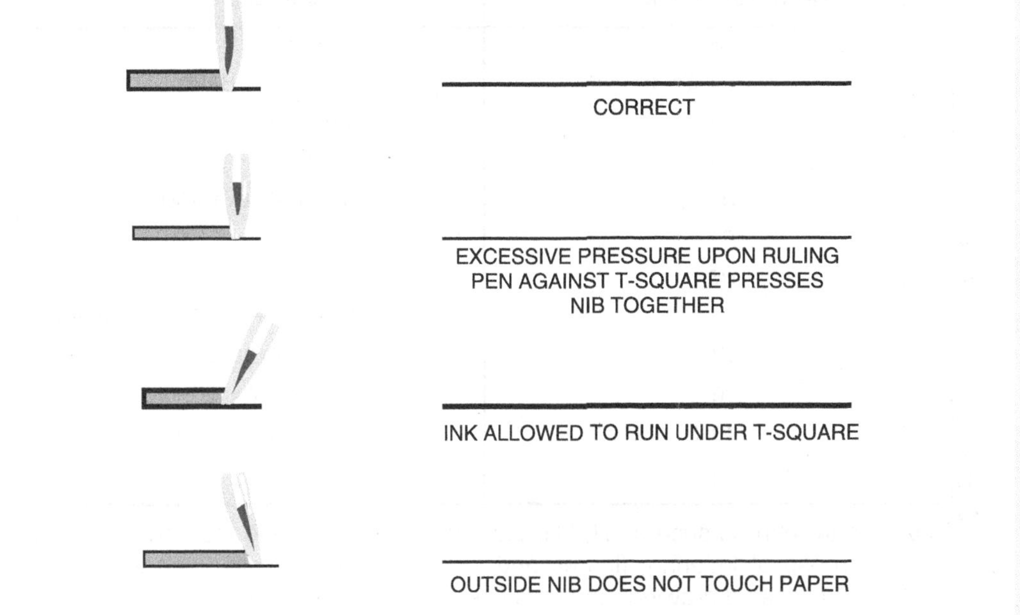

Fig. 1.29. *Correct and incorrect results while holding a ruling pen*

1.17.2. Cares in Handling of Ruling Pen :

1. Keep the small metal part of the pen bright and clean.
2. Clean the ruling pen with a cloth.
3. Under no circumstances, the pen should be dipped into the ink bottle.

Notes :

(1) For all circles and arcs, the small size ink bow compass should be used.

(2) The large circles and the circular arcs should be drawn by using inking attachment in place of pencil leg in the compass

(3) Do not fold the drawing paper.

Fig. 1.29 (i), show the modern development of double nib pens. It is arranged so that the pen barrel contains a reservoir of ink similar to any ordinary fountain pen.

This pen can be used or laid aside for long periods without refilling and is convenient to handle.

Fig. 1.29 (i). *Radiograph style (Ruling pen)*

1.18. PENCIL

The pencils are used for preparing the drawings on the sheets. The accuracy and appearance of drawing depend upon the quality of the pencil used. Pencils are of various grades easily recognised by the letters marked on pencils.

The description of different grades of drawing pencils according to I.S.I. are given below :

GRADE OF PENCIL	HARDNESS
9H	HARDEST
6H, 5H, 4H	EXTREMELY HARD
3H	VERY HARD
2H	HARD
H	MODERATELY HARD
F	FIRM
HB	MEDIUM
B	MODERATELY SOFT AND BLACK
2B	SOFT AND BLACK
3B	VERY SOFT AND BLACK
4B, 5B AND 6B	VERY SOFT AND VERY BLACK
7B	SOFTEST

Usually hard pencils such as H, 2H, etc. are used for making the engineering drawing, but for the purpose of lettering, figures and sketching, soft pencils such as HB or H etc. are used. As complicated drawing demands fine lines and minute details, harder pencils such as of 4H, 5H and 6H grades should be used for this purpose. Fig. 1.30 shows different grades of pencils.

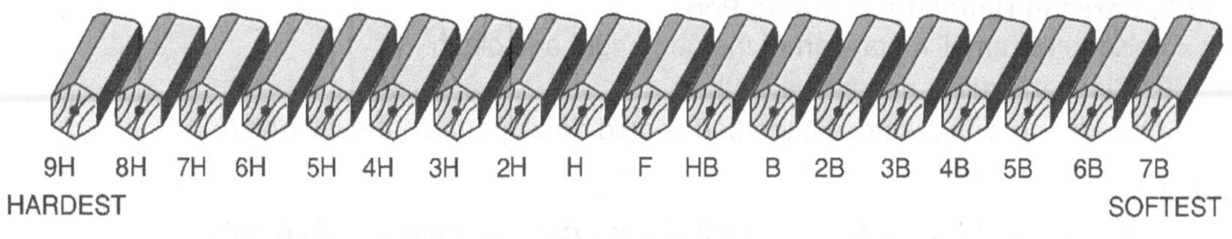

Fig. 1.30. *Different grades of pencils*

1.18.1. Ways of Mending Pencils :

The following are the two types of pencils according to the way of mending, for good and accurate work :

1. Chisel edge pencil
2. Conical or round point pencil

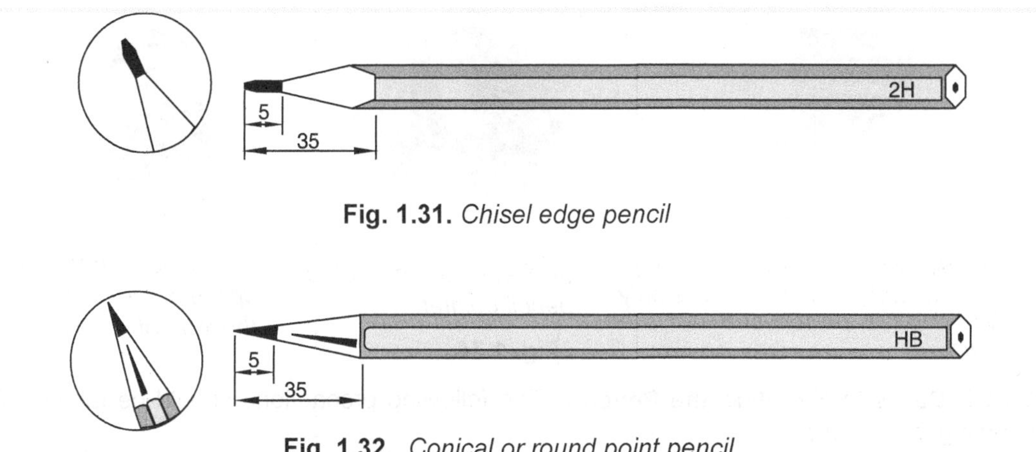

Fig. 1.31. *Chisel edge pencil*

Fig. 1.32. *Conical or round point pencil*

The chisel pencil has a chisel edge, flat on each side which remains fine for a long period and is generally used for drawing straight lines. 2H, 3H etc. pencils are generally mended to the chisel shape as shown in Fig. 1.31.

Medium grade pencils such as HB, etc. are sharpened to a conical or round point as shown in Fig. 1.32 and are meant for free hand work.

1.18.2. Sharpening the pencil : After some drawing work, the point of the pencil becomes dull and it needs sharpening. For this, a small piece of sand paper of zero grade, pasted upon a piece of wood will be very useful for keeping the point in good condition.

Important Note : *Pull and roll your pencil on sand paper block as shown in Fig. 1.33 to sharpen the pencil point.*

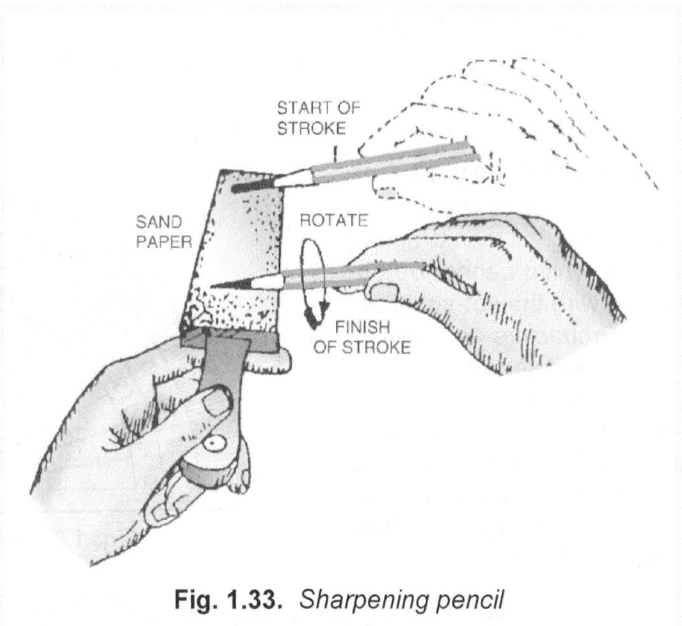

Fig. 1.33. *Sharpening pencil*

Fig. 1.32 shows the position of lead point in relation to the straight edge for normal and very accurate work.

1.18.3. Pencil Sharpeners : To simplify the process of sharpening a pencil, mechanical sharpeners equipped with cutters are used. This sharpener removes the wood only, leaving the lead exposed so that it may be pointed to any desired shape [see Fig. 1.34(i)]. Fig. 1.34(ii) shows an electrically operated pencil pointers. Fig. 1.34 (iii) shows pencil pointer which is used to shape the lead to a conical point after wood is removed.

(i) Sharpener equipped with special cutter which removes only wood on a pencil and not the lead.

(ii) Electrically operated pencil pointer.

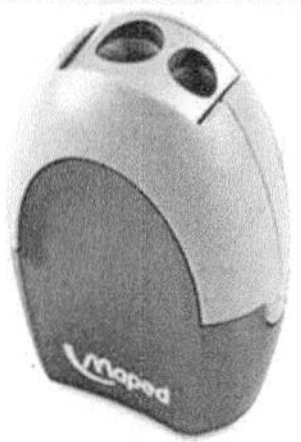

(iii) A pencil pointer speeds up the process of shaping the lead of pencil.

Fig. 1.34.

1.18.4. Cares in Handling the Pencil : The following precautions should be taken while handling the pencils :

1. For drawing vertical lines, the pencil should be slightly inclined towards the edge of T-square from left to right so that the line can be drawn nearly coincident with the edge of T-square.
2. For drawing horizontal lines, the pencil should be slightly inclined towards the edge of T-square from left to right so that the line can be drawn nearly coincident with the edge of the T-square.
3. While sharpening the pencil, do not allow the lead graphite to fall on the drawing sheet.
4. Do not put either end of the pencil into mouth.

1.19. PROTRACTORS

Protractors are used for measuring or constructing angles which cannot be obtained with the set-squares.

Protractors can be flat, circular and semicircular. The circular and semicircular protractors are usually made of boxwood or ivory. Figs. 1.35 and 1.36 show the semicircular and rectangular protractors.

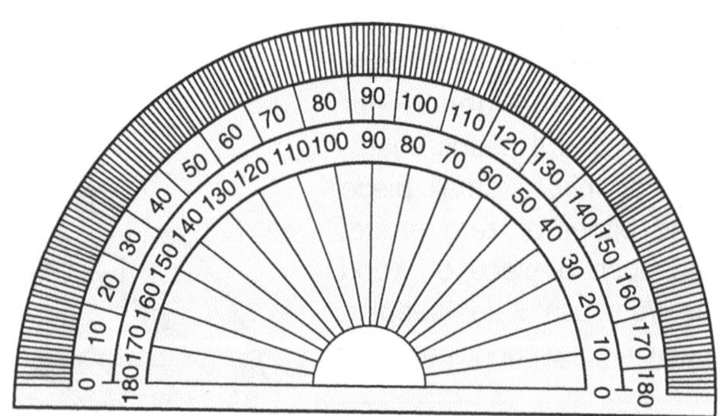

Fig. 1.35. *Semicircular protractor*

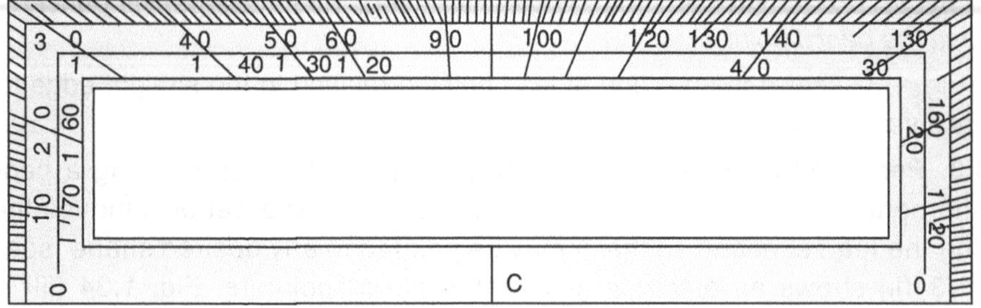

Fig. 1.36. *Rectangular protractor*

1.19.1. Care in Handling : Handle the protractor carefully and wash it with soap and water before using.

1.20. CLINOGRAPH

Clinograph is an adjustable set-square and is used to draw parallel lines at any inclination. The two sides of clinograph are fixed at 90° and the third side can be adjusted at any desired angle (see Fig. 1.37).

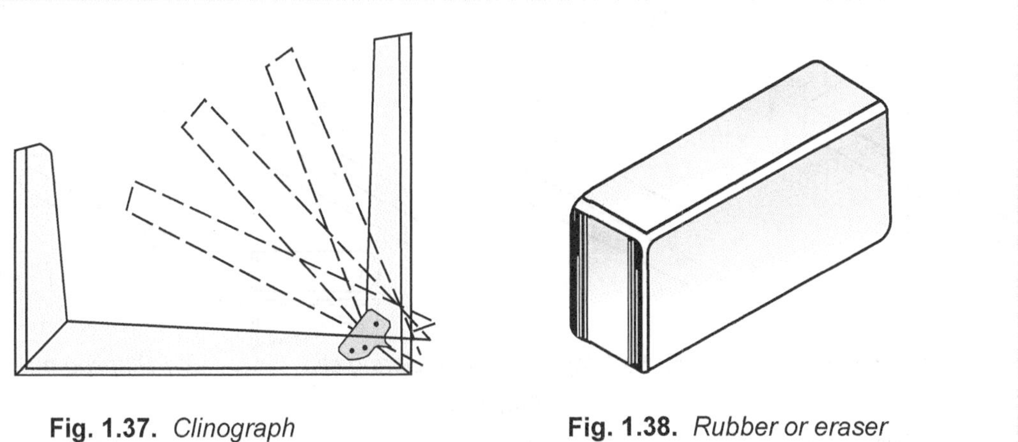

Fig. 1.37. *Clinograph* **Fig. 1.38.** *Rubber or eraser*

1.21. RUBBER OR ERASER

The most suitable eraser for pencil work is soft eraser as shown in Fig. 1.38.

1.21.1. Use of Rubber : The rubber is used for erasing extra pencil lines.

1.21.2. Cares in Handling :

1. Frequent use of rubber should be avoided.
2. Rubber crumbs should be swept away with a duster and should not be brushed off with hands.

1.22. ERASING SHIELD

An erasing shield is used to protect the adjacent lines on the drawing when some part of a line is being erased. It is usually made of thin metal sheet in which gaps of different width, curves, small circles, arcs, etc. are cut according to the lines to be erased (see Fig. 1.39).

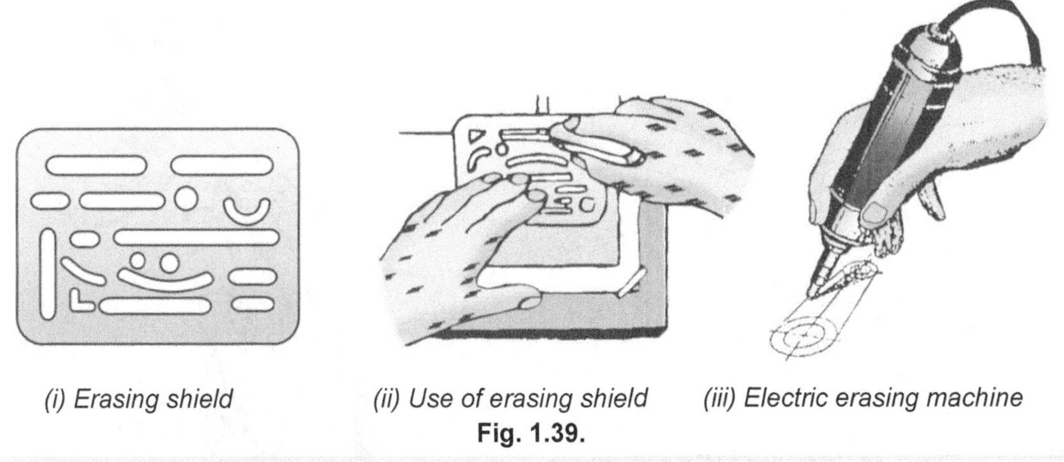

(i) Erasing shield *(ii) Use of erasing shield* *(iii) Electric erasing machine*

Fig. 1.39.

For speedy erasing of lines and to save time, electric erasing machine is used as shown in Fig. 1.39.(iii).

1.23. DRAWING PINS, SELLOTAPE AND CLIPS

Drawing pins are used for fixing the drawing sheet on the drawing board. These are best furnished with a steel pin and a brass bevelled top of about 15mm diameter (see Fig. 1.40).

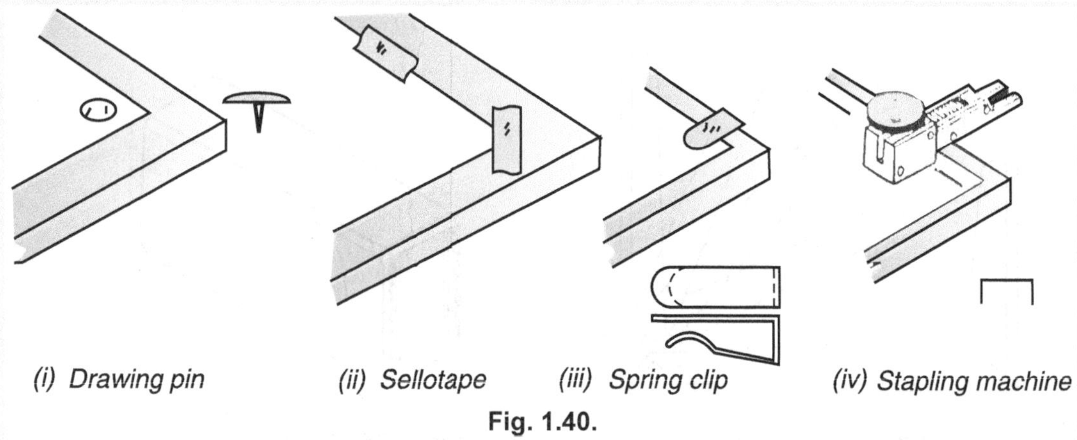

(i) Drawing pin *(ii) Sellotape* *(iii) Spring clip* *(iv) Stapling machine*

Fig. 1.40.

Nowadays, sellotapes are used in place of drawing pins for its practical convenience for the drafting machine. T-square and set-square can be moved very easily over the tape. It is transparent and is available in rolls varying in width from 13 mm to 35 mm.

Sometimes spring clips or staples are also used in fixing the sheet on the drawing board (see Fig. 1.40).

Note : *Paper tapes can also be used for fixing the sheet on the drawing board where economy is the main consideration.*

1.24. IRREGULAR OR FRENCH CURVES

Curved rules, called irregular curves or french curves are used for drawing curved lines and circular arcs. These are patterns of templates made of flexible transparent sheet material such as celluloid and having a series of different shaped curved edges as shown in Fig. 1.41.

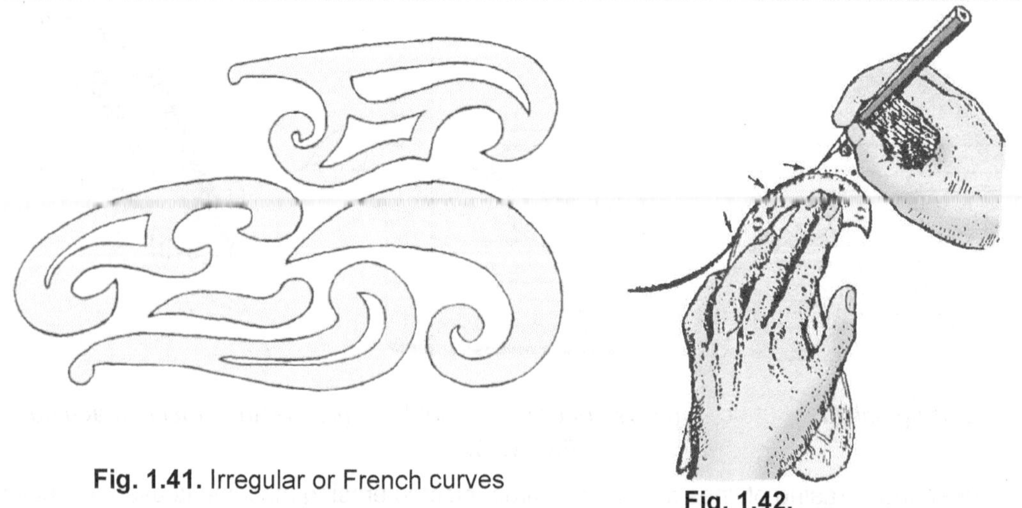

Fig. 1.41. Irregular or French curves

Fig. 1.42.

Fig. 1.42 shows the use of irregular or French curves for drawing a smooth curve between various points.

1.24.1. Flexible or Adjustable Curves or Splines: This is an other form of irregular or French curves used to draw long curves between irregular points owing to the great length of a spline. It is impossible to hold it in place by hand, then lead weights called ducks are used to hold the spline in desired position (see Fig. 1.43).

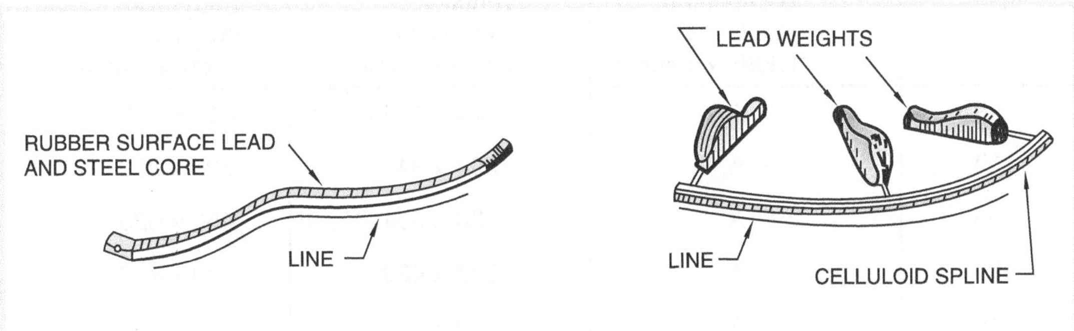

Fig. 1.43. *Flexible or adjustable curves or splines*

1.25. DUSTER OR HANDKERCHIEF

Duster is used for cleaning drawing instruments and other drawing materials. It should preferably be cleaned with a handkerchief or a towel cloth. It is also used to sweep away the crumps formed after the use of rubber on the drawing paper.

1.26. DRAWING INK

The drawing ink is used for preparing drawings in ink on tracing paper or tracing cloth as on drawing sheet. It is composed of carbon in colloidal and gum. It is water proof and gives deep and black lustre on drying.

1.26.1. Precautions :

The ink bottle should not be left uncorked otherwise by evaporation, the ink will get thickened. If, however, the ink becomes thick, then it can be thinned by adding, a few drops of solution containing four parts of aqua-ammonia to one part of distilled water.

1.27. DRAWING PAPER OR DRAWING SHEET

There are different qualities of drawing papers. The quality of paper to be used for a drawing depends upon the nature of drawing. The drawing paper should be uniform in thickness and of such a quality that the erasing effect should not be there. In addition to this it should be of such a quality that the ink should not spread out.

The drawing papers are obtained either in rolls or in sheets. The rolls available usually vary form 750 mm to 1800 mm in width and generally 20 meters in length. They are designated by their weight per double crown ream.

One of the sides of the drawing paper is usually rough and the other smooth. The smooth surface is the proper side for drawing work.

1.28. SIZES OF DRAWING SHEETS

The standard sizes of trimmed and untrimmed drawing sheets according to I.S.I. are given below:

STANDARD SIZES OF DRAWING SHEETS

S.No.	Designation size in mm Width × Length	Trimmed size in mm Width × length	Untrimmed size in mm Width × length
(i)	A_0	841 × 1189	880 × 1230
(ii)	A_1	594 × 841	625 × 880
(iii)	A_2	420 × 594	450 × 625
(iv)	A_3	297 × 420	330 × 450
(v)	A_4	210 × 297	240 × 330
(vi)	A_5	148 × 210	165 × 240

Note : *Whenever necessary size of sheets with length 1189 mm may be further extended by steps of 210 mm in length only.*

Fig. 1.44 shows the various sizes of trimmed drawing sheets such as A_0, A_1, A_2, A_3, A_4 and A_5 according to Indian Standards Institution (I.S.I.).

In arriving at the various trimmed sizes of the drawing sheets, the following basic principles have been taken into consideration :

1. Halving or Doubling of Sheet: Two successive sizes of the series are obtained by halving or doubling. Consequently the surface area of the two successive sizes are in the ratio of 1 : 2 (see Fig. 1-45).

2. Ratio of Sides and Areas : The formats or forms are geometrically similar to one another. The sides of each size being in the ratio of 1 : 2 (see Fig. 1-45).

3. Surface Area : The surface areas of the basic size of A_o is one square metre.

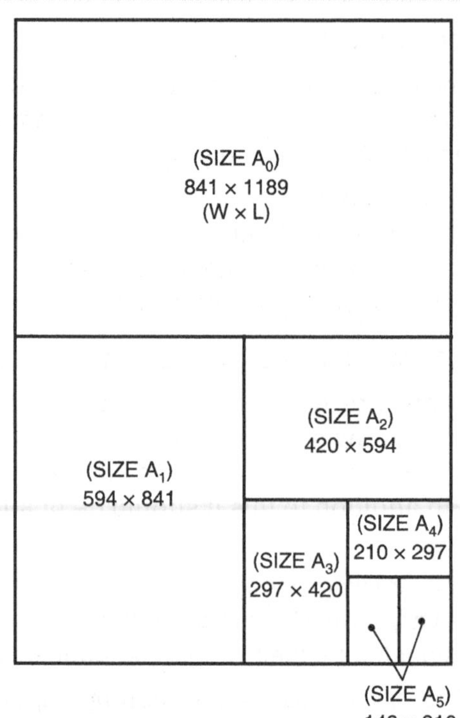

Standard sizes of drawing sheets according to I.S.I.

Fig. 1.44.

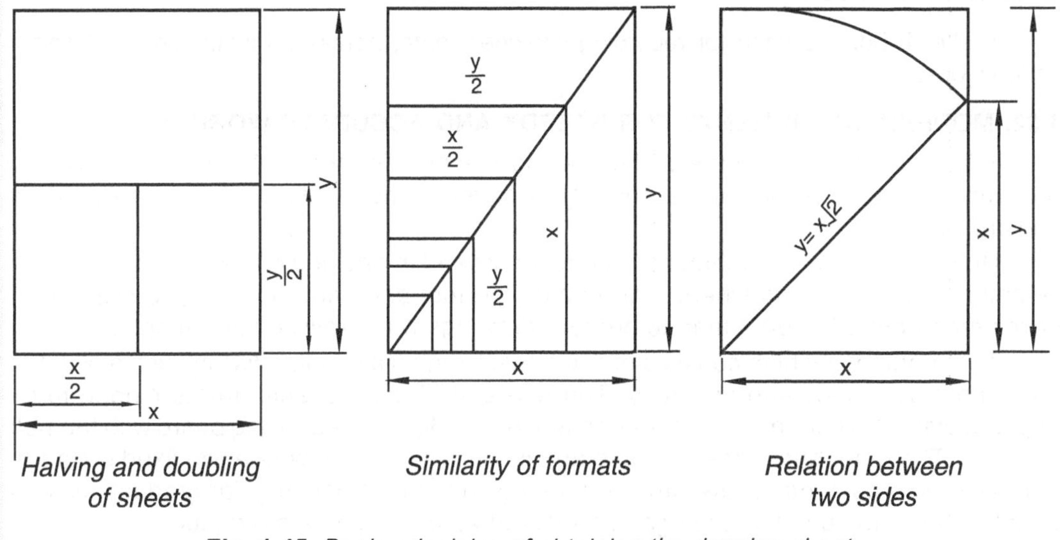

Fig. 1.45. *Basic principles of obtaining the drawing sheet*

Important Note : *Drawing sheet of size 420 x 594, i.e., A_2 size known as half imperial sheet is generally used by engineering students as it is very handy and easy for drawing work in the class.*

Fig. 1.46 shows the untrimmed size (or uncut size) and trimmed size (cut size) of drawing sheet.

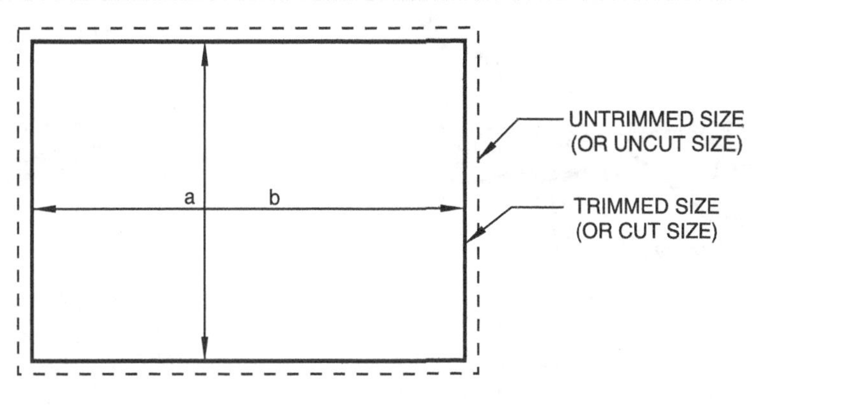

Fig. 1.46. *Trimmed and untrimmed sheet*

1.29. TRACING PAPER

Tracing paper is a thin transparent paper on which drawings are traced in ink or pencil. From the traced drawings blue prints are prepared.

1.30. TRACING CLOTH

Tracing cloth is a transparentized fabric and is used when the original tracing has to be preserved for a longer period. It is available in either white or blue tinted colour with one side dull and the other glazed.

1.31. SKETCH BOOK

A sketch book is used for recording sketches, data, notes and instructions relating to various jobs.

1.32. MODERN INSTRUMENTS FOR SPEEDY AND ACCURATE WORK

1.32.1. Drafting Machine : The drafting machine is the most important device for preparing drawings quickly and accurately. It performs the functions of T-squares, scales and protractor.

Now a days, drafting machine is commonly used in drawing offices of industries and commercial departments where high quality, performance and accuracy is required. It saves time from 25 to 50% than to prepare drawings with drawing instruments.

The drafting machine consists of two arms, fixed on a circular disc in such a way that they can move freely and smoothly. The one end of the movable arm is hinged to the gripping plate. This gripping plate is attached to the edge of the drawing board with the help of screw. The other end of the movable arm is hinged to a disc where two fixed plated (at right angles to each other) are provided. Along-with the scales angular graduations from 0° to 360° or 0° to 180°, left and right are marked which acts as a protractor.

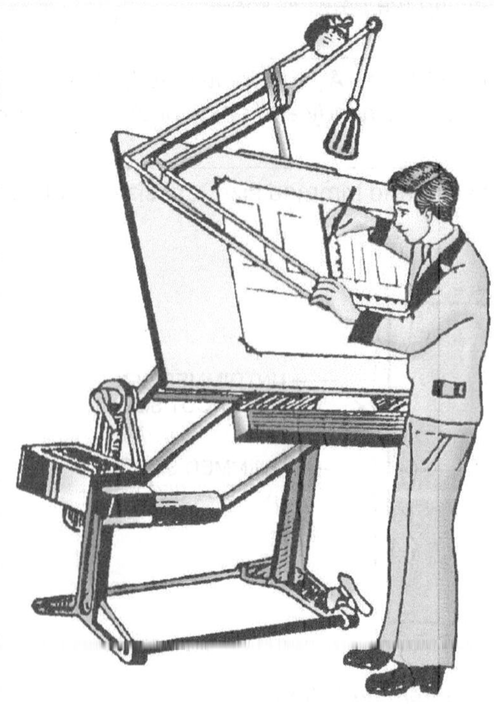

Drafting machine fitted on pedestal drawing table

Fig. 1.47.

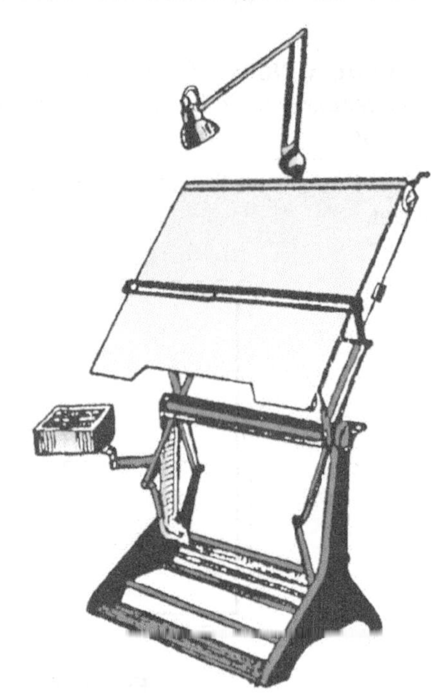

Parallel ruling machine fixed on a drawing board

Fig. 1.48.

Two screws are provided on the drafting machine, one on the gripping plate and other on the scale to loosen or tighten the arms.

Fig. 1.47. shows the pictorial view of a drafting machine fitted on pedstal drawing board.

The board can be fixed to any position and may be adjusted to any height.

1.32.2. Size of Drafting Machine : The size of drafting machine depends upon the size of the drawing board. Generally, the following drafting machines are available in the market.

S.No.	DESIGNATION	SIZES
1.	Imperial size	575 × 800 mm
2.	Double elephant size	750 × 1090 mm
3.	Antiquariam size	825 × 1350 mm
4.	Emperor size	1000 × 1500 mm

Note: *The Imperial size (Mini drafting machine) is generally used by the engineering students.*

1.32.3. Parallel Ruling Straight Edge Machine

The parallel ruling straight-edge machine is used to draw horizontal lines with great speed and accuracy. It serves the purpose of T-square (see Fig. 1.48).

This machine consists of a parallel straight edge with guide cords attached to the ends of the straight edge of drafting table or on the drawing board.

Since, the parallel straight edge is supported at both ends, its advantage over the T-square is that it maintain parallel motion automatically and may be moved up or down with slight pressure.

QUESTIONS FOR SELF EXAMINATION

1. Define engineering drawing. Why drawing is called the universal language of engineers?
2. What is the difference between electrical, civil and machine drawings and where each is used ?
3. Name different types of drawing instruments.
4. While drawing horizontal lines, in what direction the pencil should be inclined?
5. Why set squares of transparent materials are used nowadays?
6. Why pencil is rotated in finger while drawing a long line?
7. Why the needle point is slightly more in length than the lead point?
8. What are the standard sizes of drawing sheets according to I.S.I. and which is suitable for drawing work by engineering students?
9. What are the ways of sharpening a pencil for good and accurate work and which type of pencil is more suitable for drawing work?
10. Why sellotape is used instead of drawing pins, nowadays?
11. How you will test the set-square and T-square?
12. Name the modern instruments which are replacing other drawing instruments.
13. Fill in the blanks:
    (*i*) The art of representation of an object by systemic lines on a paper is called ....
    (*ii*) The working edge of the drawing board should be on ...... side of the draughtman.
    (*iii*) T-square is used to draw ..... lines ..... to each other.
    (*iv*) .... is used for marking off short distances.
    (*v*) To remove a particular spot on drawing .... is used.
    (*vi*) To draw circle of about 80 mm radius, ... is used.
    (*vii*) A scale should not be used as a ..... for drawing lines.
    (*viii*) The hardness of the pencil lead .... as numeral letter H increases.
    (*ix*) .... is designed to performs the functions of T- square, set squares, scale and protractor.

(*x*) The artificial light on the drawing board is to be provided from the ..... side of the draughtman.

Ans. (*i*) drawing (*ii*) left (*iii*) horizontal, parallel (*iv*) Bow compass (*v*) erasing shield (*vi*) large compass (*vii*) ruler (*viii*) increases (*ix*) Drafting machine (*x*) left

14. Choose the correct answer :

(*i*) The needle point of the compass should be (*a*) equal to (*b*) longer than (*c*) shorter than the lead point.

(*ii*) The lead becomes softer as the numeral letter H on pencil (*a*) increases (*b*) decreases (*c*) remains same.

(*iii*) Chisel pencil are used for drawing (*a*) straight lines (*b*) curved lines (*c*) free hand line work.

(*iv*) Parallel ruling straight edge machine is used to draw *(a)* horizontal lines (*b*) vertical lines *(c)* inclined lines.

Ans. *(i) (b), (ii) (b), (iii) (a), (iv) (a).*

15. Match the statements of column A with the corresponding ones in column B

Column A	Column B
(*i*) French curves are used to draw	(*a*) Irregular curve lines.
(*ii*) Splines are used to draw	(*b*) 420 x 594 mm.
(*iii*) Engineering student use imperial size drafting machine of	(*c*) 575 x 800 mm.
(*iv*) Half imperial sheet is generally used by engineering student of	(*d*) Long curve lines.

Ans. 15. (*i*) (*a*) ; (*ii*) (*d*); (*iii*) (*c*) *and* (*iv*) (*b*)

PROBLEMS FOR PRACTICE

1. Copy Fig. 1.49. by using T-square and set square or drafting machine only.

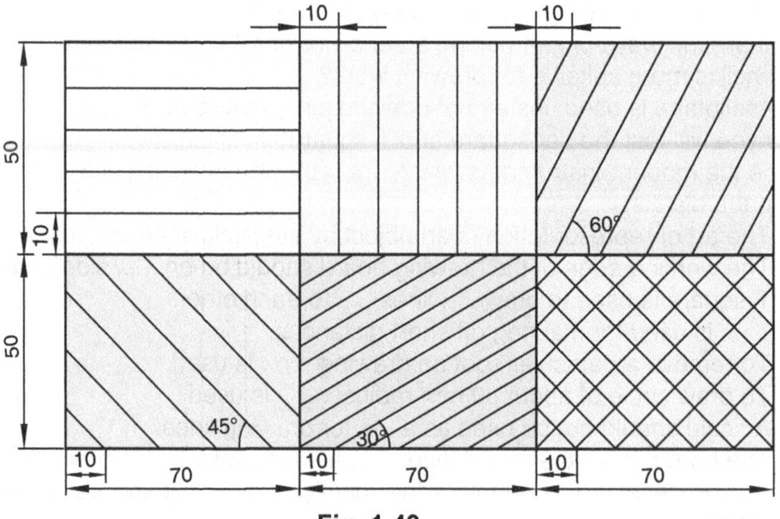

Fig. 1.49.

2. Draw a line 100 mm long and divide it into 8 equal parts by means of a divider.
3. Draw a circle of 60 mm diameter and divide it into (*i*) six, (*ii*) eight, (*iii*) twelve and (*iv*) twenty equal parts by using T-square and set squares only.
4. Draw the following figure (Fig. 1.50) with a horizontal line by using protractor :

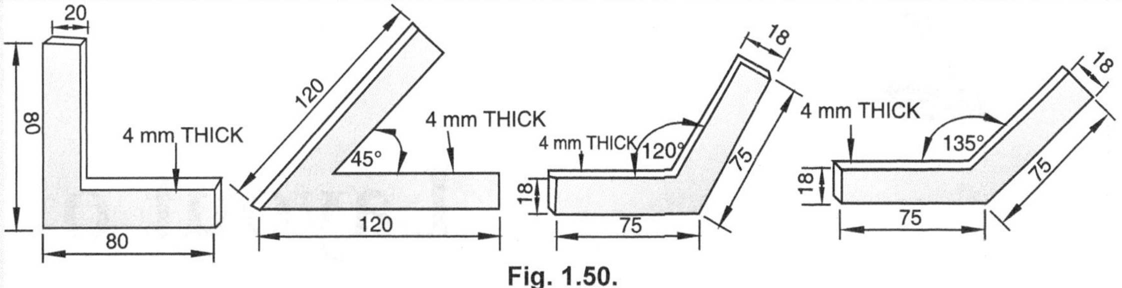

Fig. 1.50.

5. Draw the Fig. 1.51 by using various drawing instruments.

Fig. 1.51.

6. Copy Fig. 1.52 by using various types drawing instruments with circle 120mm dia.

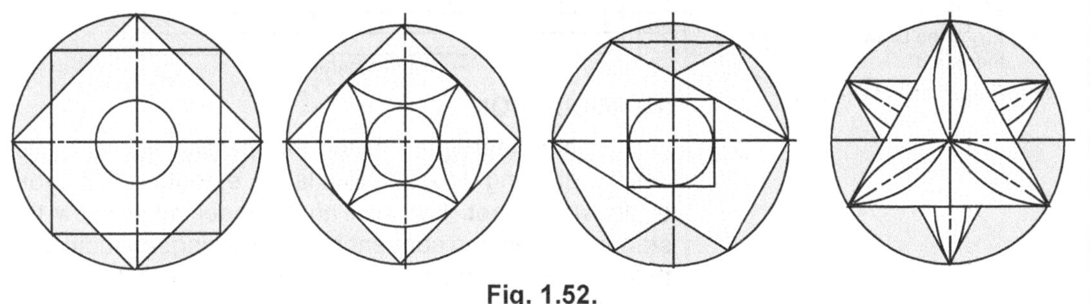

Fig. 1.52.

CHAPTER 2

Layout of Drawing Sheet

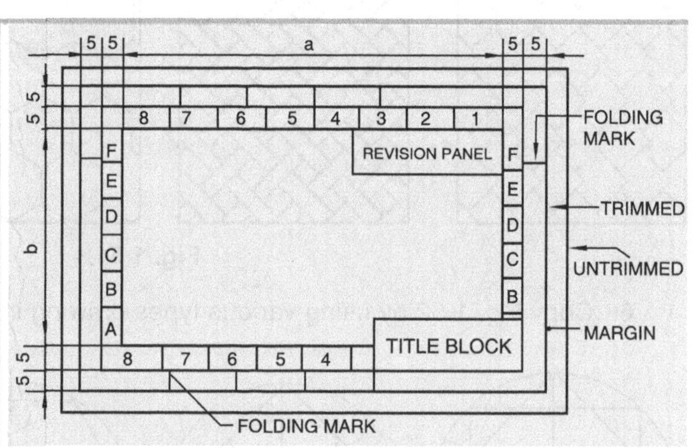

INTRODUCTION

In engineering drawing, the preparation of a successful drawing depends to a large extent on the layout of drawing sheet. For this, an engineer must know the standard rules and conventions of drawing. In addition to this, he has to keep in his mind the other aspects such as margin, title block, parts list, revision panel, zone system, folding marks, etc., so as to facilitate the reading and interpretation of the drawing.

In this chapter we will deal with the study of layout of drawing sheet, numbering of sheets, folding marks and folding of sheets, etc.

2.1. LAYOUT OF DRAWING SHEET

The selection of suitable scale and allotment of proper space for margin, title block, revision panel, folding marks, etc. on the drawing sheet is known as layout of drawing sheet.

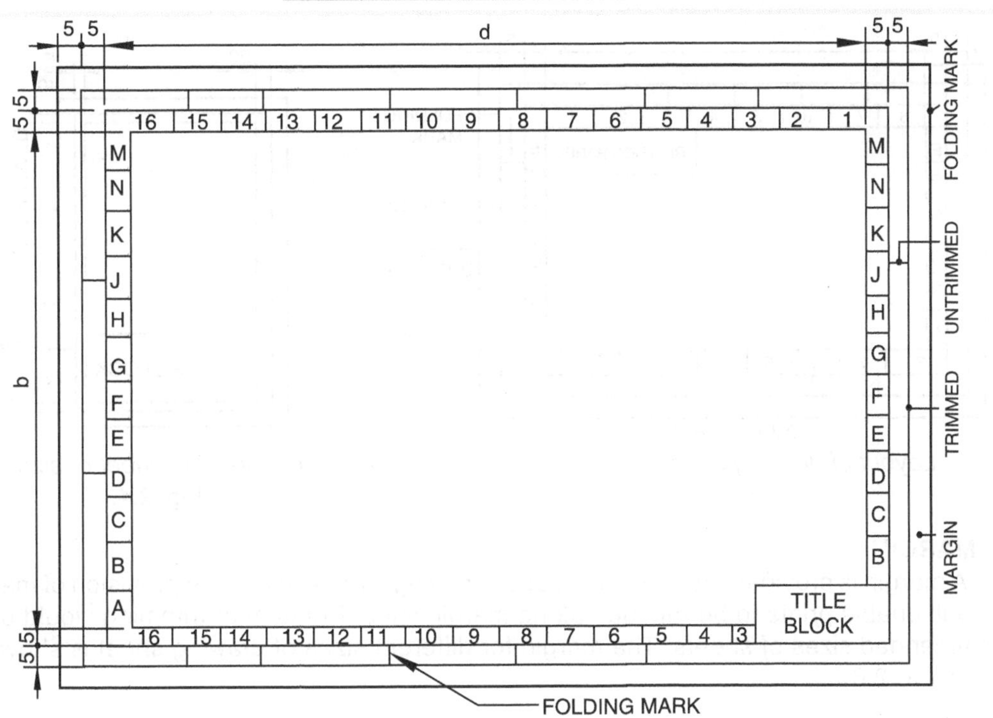

Fig. 2.1. *Layout of drawing sheet* A_o *(841 × 1189) according to I.S.I.*

Figs. 2.1 to 2.6 show typical layout of drawing sheets A_o, A_1, A_2, A_3, A_4 and A_5 as specified by I.S.I.: 696-1972. The various aspect such as margin, boarder line, title block, list of parts, revision pannel, zone system, folding marks, numbering of sheet, etc. are discussed in the next pages.

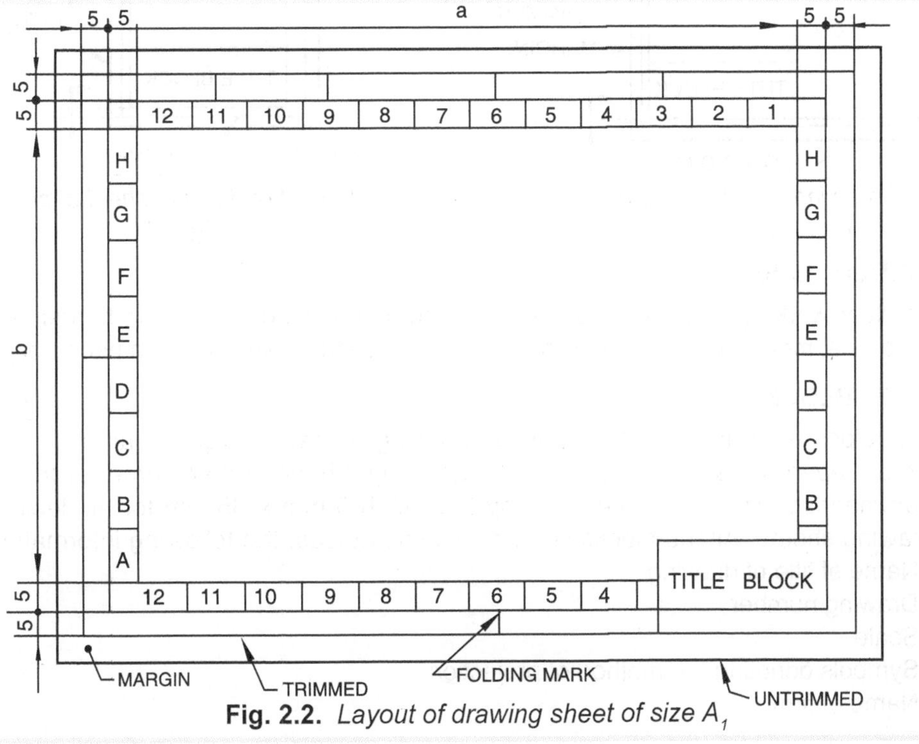

Fig. 2.2. *Layout of drawing sheet of size* A_1

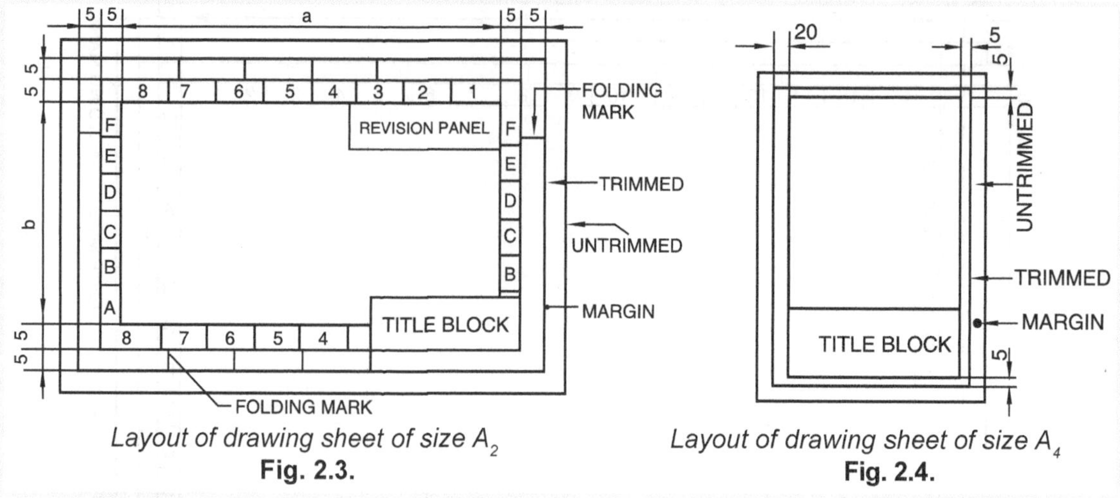

Layout of drawing sheet of size A_2

Fig. 2.3.

Layout of drawing sheet of size A_4

Fig. 2.4.

2.2. MARGIN

A margin is provided around the sheet by drawing margin lines. The provision of margin lines will enable prints to be trimmed along margin lines. Prints after trimming would be of recommended sizes of sheets. The margin for different sizes of drawing sheet is shown in Figs. 2.1 to 2.6.

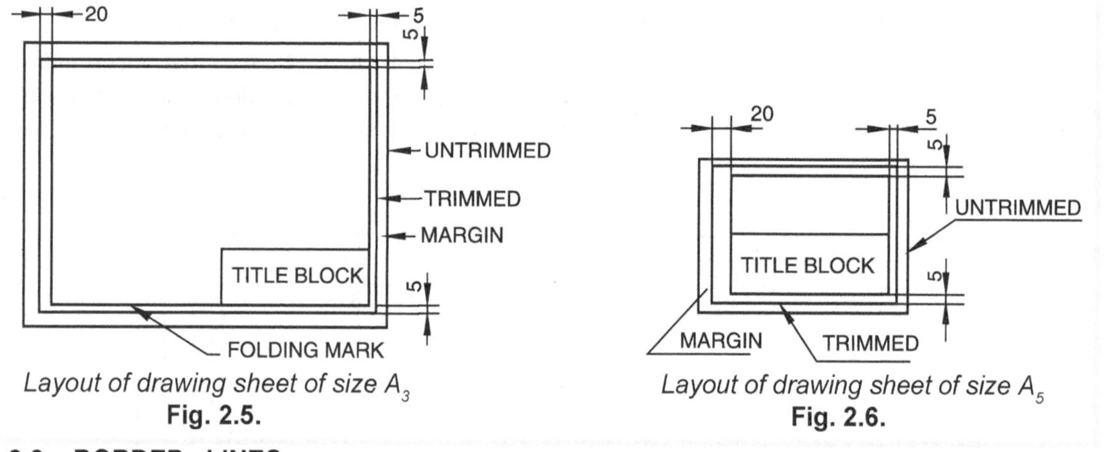

Layout of drawing sheet of size A_3

Fig. 2.5.

Layout of drawing sheet of size A_5

Fig. 2.6.

2.3. BORDER LINES

The clear working space on the drawing sheet is obtained by drawing border lines. In general practice, more space is kept on the left side for filing or binding when necessary.

2.4. TITLE BLOCK

Title block is an important feature in drawing, because it gives all the informations of the prepared drawing. It is provided at the right hand bottom corner of the sheet. The recommended size of the title block by I.S.I. is 185 mm × 65 mm for all designation of the drawing sheets. All the blocks should contains at least the following informations :

1. Name of title of drawing
2. Drawing number
3. Scale
4. Symbols denoting the method of projection
5. Name of firm

6. Initials with dates, of staff who have designed, drawn, checked, standards and approved the drawing.

Figs. 2.7 and 2.8 show the details of the side title block used by the industries according to I.S.I.

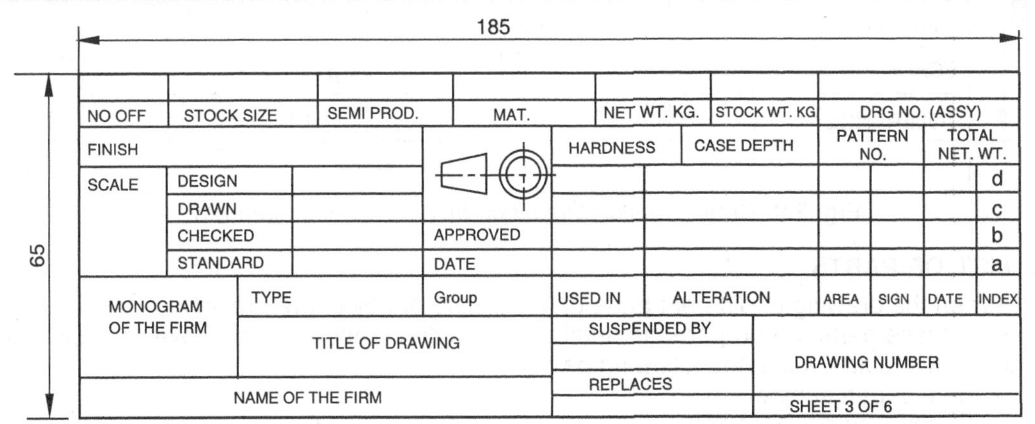

Fig. 2.7. *Details of side title block according to I.S.I.*

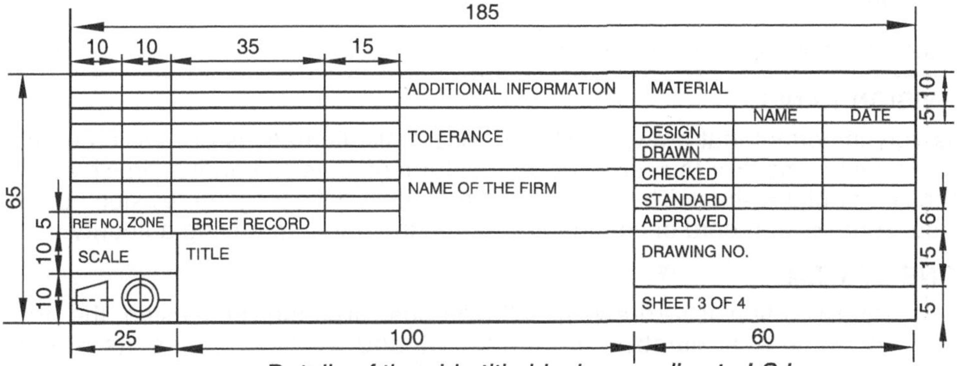

185						
10	10	35	15	ADDITIONAL INFORMATION	MATERIAL	
				TOLERANCE		NAME / DATE
					DESIGN	
					DRAWN	
				NAME OF THE FIRM	CHECKED	
					STANDARD	
REF NO.	ZONE	BRIEF RECORD			APPROVED	
SCALE		TITLE			DRAWING NO.	
					SHEET 3 OF 4	
25		100			60	

Dimensions: 65; 5, 10, 10; 10, 5, 6, 15, 5

Details of the side title block according to I.S.I

185			
NAME OF THE FIRM	20	25 NAME	10 DATE
	DESIGNED		
	DRAWN		
	CHECKED		
	STANDARD		
	APPROVED		
SCALE / TITLE	DRAWING NO.		
	SHEET 1 OF 4		
25 / 105	55		

Dimensions: 65; 10; 10, 7, 7, 7, 7, 7, 20

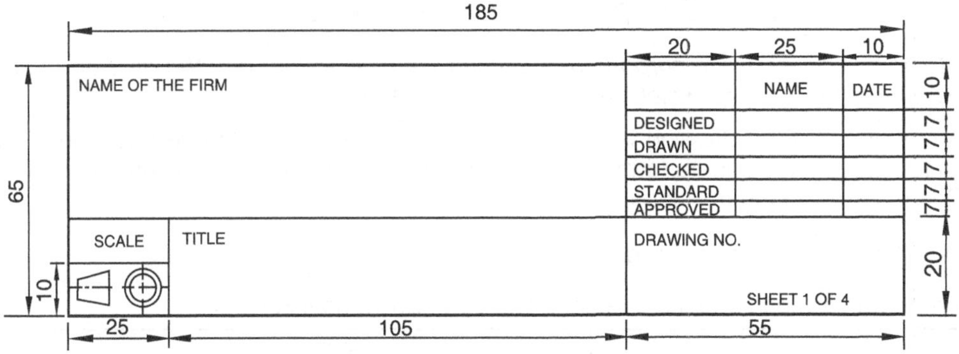

Fig. 2.8. *Details of the side title block according to I.S.I.*

Fig. 2.9 shows the title block which is suitable for use by engineering students.

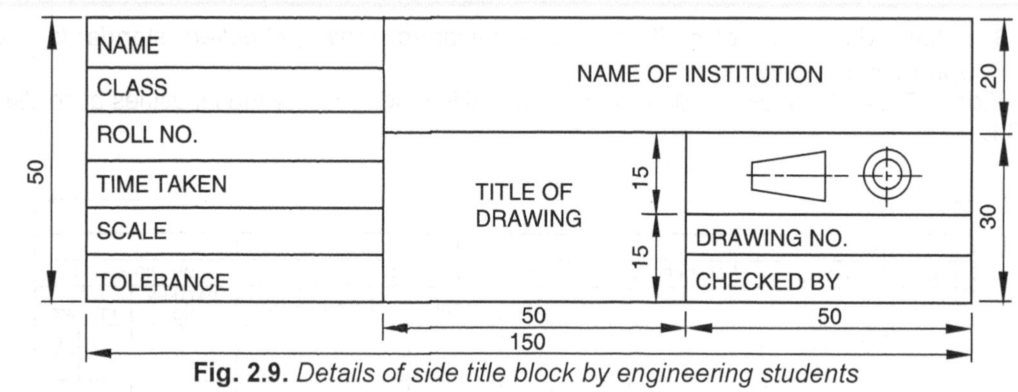

Fig. 2.9. *Details of side title block by engineering students*

2.5. LIST OF PARTS

When the drawing of different parts of a machine or object are drawn in a single sheet, the list of these parts are prepared in another rectangle in tabular form above or beside the title block. It should contain the following particulars :

1. Item or part number
2. Description of title of part
3. No. off, i.e., quantity required
4. Cross references to other detail drawings and
5. Material specification.

Note : *To facilitate extention, entries in the table should begin from bottom to upwards.*

2.6. REVISION PANEL

It is very important that drawings should record all alternations or revision which are made from time to time. For this, one convenient form of revision panel is drawn at the top right hand corner of the sheet. The revision are recorded in it giving the revision number or letter, date, zone, brief record of change and signature of the approving authority.

2.7. ZONE SYSTEM

For locating a portion of the drawing for the purpose of change or revision, the drawing sheet is divided into a number of horizontal and vertical zones as shown in Fig. 2.10. The lengthwise devisions are numbered as 1, 2, 3, 4, etc. from right to left and widthwise divisions as A, B, C, D, etc. from bottom to top.

Note : *In general, the purpose of providing zones is to facilitate the division of sheet for the figures to be drawn which can be helpful to the beginers.*

Any revision in the drawing is located by the referring to the square it falls in. Thus, a revision following in the square formed by the intersection of the horizontal zone 8 and vertical zone D is recorded in the revision panel as D8.

Its full description can be given as : $\frac{\text{X}}{\text{3DB}}$

where X is given in sheet 3, horizontal zone 8 and vertical zone D.

DIVISION		SHEET SIZES		
		A_0	A_1	A_2
NO. OF ZONES	a	16	12	8
	b	12	8	6

Table 2.1. *Zones of sheets A_0, A_1 and A_2 according to I.S.I.*

The number of zones suggested by I.S.I. for A_o, A_1 and A_2 sizes of sheets along the length a are 16, 12 and 8 respectively, while those along the width b are 12,8 and 6 respectively (see Table 2.1).

2.8. NUMBERING OF SHEETS

When there is more than one sheet is required for a drawing or part or assembly and a particular sheet is in from one such drawing, the numbering should show the numbers of the particular sheet. The total number of sheets of the drawing is indicated as below :

Sheet 1 of 4

This entry should be made below the drawing number in the title block.

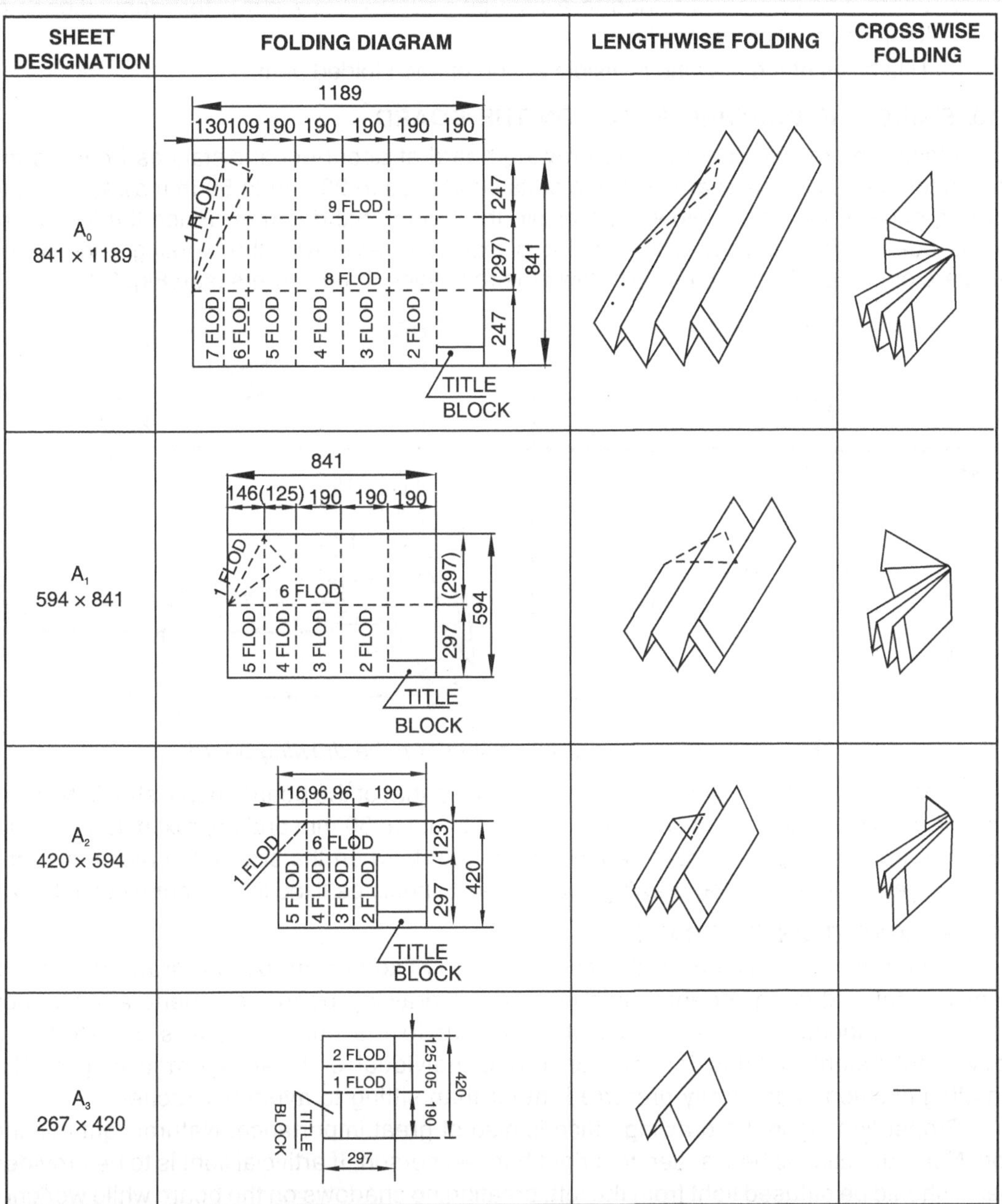

Fig. 2.10. *Folding marks and folding of sheets according to I.S.I. : 696-1972*

2.9. FOLDING MARKS AND FOLDING OF SHEETS

Folding marks are made on the sheet to facilitate folding of prints for the purposes of filing and binding in the proper and easy manner (see Fig. 2.10).

Fig. 2.10 shows the folding diagrams for folding the A_0, A_1, A_2 and A_3 sheets as recommended by I.S.I.: 696-1972.

The method of folding has the following features :

1. Prints can be folded and unfolded (when attached to the other papers) without the removal from the file.
2. Large size prints are folded to a final size (210 × 2197 mm) for convenience of keeping in office record files.
3. Title block of the drawing is visible on top of each folded print.

2.10. FIXING THE DRAWING SHEET ON THE BOARD

Place the drawing sheet on the drawing board at about equal distances from the top and bottom edges of the board with one of its shorter edge 25 mm to 50 mm away from the working edge of the board. Now, tape or pin the top left hand corner. Place the T- square on the board and swing the sheet until it is exactly in level with the working edge of the T-square or drafter. Then, tape or pin down the remaining three corners (see Fig. 2.11).

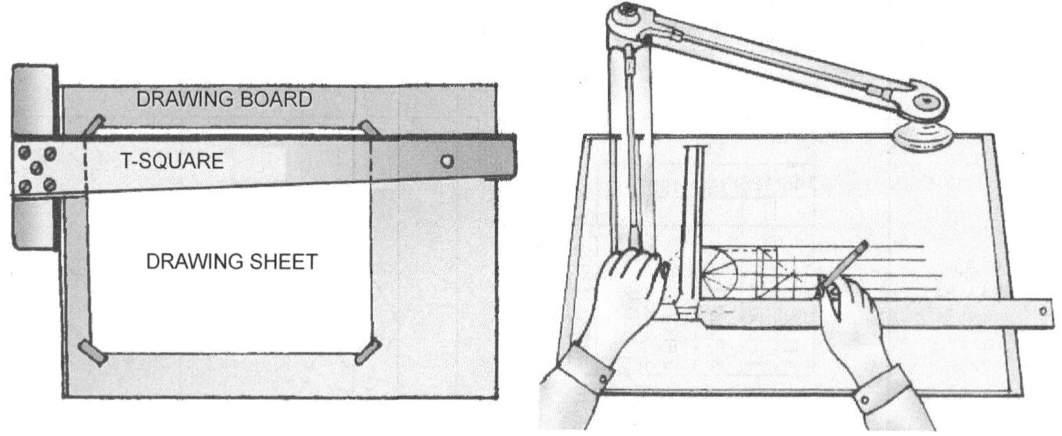

Fig. 2.11. *Fixing the drawing sheet on the drawing board*

When the size of the sheet is much smaller than that of the board, then it should be fixed near the working edge of the board varying from 25 mm to 50 mm and approximately at equal distances from the top and bottom edges of the board. When the size of the paper is nearly of the same size as that of the drawing board, then it should be fixed in the centre of the board.

2.11. PREPARATION OF DRAWING

Preparation of drawing on the sheet which is fixed on horizontal drawing board is inconvenient and tiring. Keeping this in view the drawing boards are generally titled by different arrangements, to an angle of about 20° to the horizontal. It is adjusted to a convenient height suitable for the concerned individual or to an average height. The standing position is generally preferred rather than sitting to avoid backache.

Proper lighting in the drawing office is also of great importance. Natural light coming from North direction is best suited to a right handed person. If artificial light is to be provided then it should be diffused light from the left, creating no shadows on the board while working.

The drawing instruments should be placed suitably with in the reach of the person. These instruments must be well cleaned before starting the drawing work.

QUESTIONS FOR SELF EXAMINATION

1. Why is the layout of sheet necessary?
2. What are the standard sizes of drawing sheet according to I.S.I.?
3. List out contents of title block and material list.
4. What do you understand by revision panel and zone system?
5. What is the necessity of folding a drawing print?
6. Fill in the blanks:
   (*i*) The preparation of a successful drawing depends upon the .... of drawing sheet.
   (*ii*) The size of title block of all sizes of drawing sheet is .... x .... mm.
   (*iii*) For A_1 size sheet the number of zones suggested by I.S.I. along the length are...... while those along the width are .....
   (*iv*) For locating a portion of a drawing, the sheet is divided into a number of.....
   (*v*) The zones along the length of the sheet are designated by ......, while along its width by the .....

Ans. (*i*) layout (*ii*) 185,65 (*iii*) 12,8 (*iv*) zones (*v*) numerals, letters

7. Choose the correct answer :
   (*i*) To facilitate extension, entries in the table should begin from (*a*) bottom to upwards (*b*) top to downwards (*c*) horizontal
   (*ii*) The artificial light on the drawing board should be diffused light from the (*a*) right (*b*) left (*c*) top.

Ans. (*i*) *a* (*ii*) *b*

8. Match the statements of column A with the corresponding ones in column B.

Column A	Column B
(*i*) Natural North light on the sheet is best suited to a	(*a*) 185 x 65 mm
(*ii*) The size of the title block for all sizes of drawing sheet is	(*b*) always on the top
(*iii*) The drawing sheet is so folded that title block is	(*c*) right handed draughtsman
(*iv*) The purpose of providing zones is	(*d*) to facilitate the extension
(*v*) Large size prints are folded to a final size of	(*e*) to facilitate the division of sheet.
(*vi*) The entries in the table should begin from bottom to top	(*f*) 210 x 297 mm

Ans. (*i*) (*c*), (*ii*) (*a*), (*iii*) (*b*), (*iv*) (*e*), (*v*) (*f*), (*vi*) (*d*).

PROBLEMS FOR PRACTICE

1. Make a complete layout of A_2 drawing sheet showing title, material list, folding marks and revision panels.
2. Prepare a specimen title block (*i*) as recommended by I.S.I. and (*ii*) for use in class room by engineering students.
3. Fold the drawing sheet A_4 for filing purposes.

CHAPTER

3

Conventions

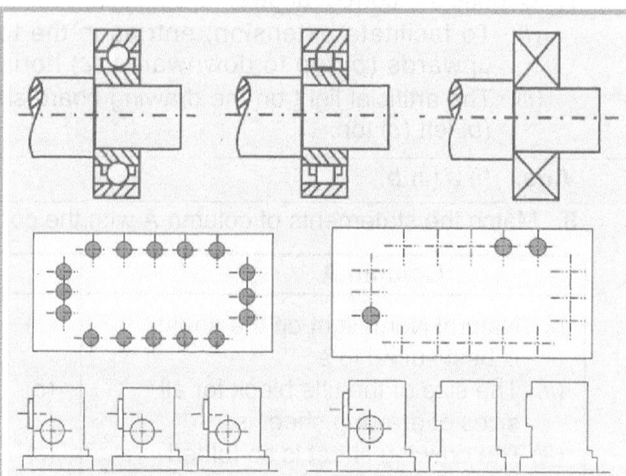

INTRODUCTION

In english textbook, we use correct words for making correct sentences. Similarly, in engineering drawing, the details of various objects are drawn by different types of lines. Each line has a definite purpose and sense to convey. For this, different convention of lines are used to represent the details of the object accurately on the drawing.

The conventions are also used to differentiate the different types of materials used for manufacturing various parts of a machine on the drawing.

Sometimes, long members of uniform cross-section such as bars, rods, etc. are shown by their conventional breaks to accommodate their views on the drawing sheet.

In this chapter, we will deal with the study of convention, convention for lines and materials, conventional breaks, convention for common features and convention for springs etc. used in engineering practice according to I.S.I.

3.1. CONVENTION OR CODE

The representation of any matter by some sign or mark on the drawing is known as convention or code. The conventions make the drawing simple and easy to draw.

3.2. CONVENTION FOR LINES

Since, engineering drawing is the systematic combination of different types of lines, it is, therefore, essential for the students to have clearly in mind the difference between the various types of lines. After following the difference between several types of lines, there will be little difficulty in making further interpretation of engineering drawing. Various types of lines as given by Indian Standards Institution (I.S.I.), are illustrate in Fig. 3.1.

DESCRIPTION AND USES OF VARIOUS LINES [ACCORDING TO I.S.I. 1972]

S. NO.	TYPE OF LINE	ILLUSTRATION	APPLICATION
A	CONTINUOUS THICK		VISIBLE OUTLINES
B	CONTINUOUS THIN		DIMENSION LINES, LEADER LINES, EXTENSION LINES, CONSTRUCTION LINES, OUTLINES OF ADJACENT PARTS, HATCHING AND REVOLVED SECTION
C	CONTINUOUS THIN-WAVY		IRREGULAR BOUNDARY LINES, SHORT BREAK LINES
D	SHORT DASHES MEDIUM	2 to 3 mm APPROX 1mm APPROX	HIDDEN OUTLINES AND EDGES
E	LONG CHAIN THIN	2 to 3 mm APPROX 1 mm APPROX 15 to 30	CENTRE LINES, LOCUS LINES, EXTREME POSITIONS OF THE MOVEABLE PARTS, PITCH CIRCLES AND PARTS SITUATED INFRONT OF THE CUTTING PLANES
F	LONG CHAIN THICK AT ENDS & THIN ELSEWHERE	2 to 3 mm APPROX 1 mm APPROX 15 to 30	CUTTING PLANE LINES
G	LONG CHAIN THICK	2 to 3 mm APPROX 1 mm APPROX 15 to 30	TO INDICATE SURFACES WHICH ARE TO RECEIVE ADDITIONAL TREATMENT
H	RULED LINE AND SHORT ZIG-ZAG THICK		LONG BREAK LINES

Fig. 3.1. *Various types of lines*

3.3. DETAILED DESCRIPTION AND USES OF VARIOUS LINES

The following are the detailed description and uses of various types of lines :—

1. Visible outline or object line : The outline or object line is represented by thick line and is used to show the outer visible feature of the object in the drawing (see Fig. 3.1). Every edge or surface that is visible is represented by these lines.

2. Section line or hatching line : It is a thin continuous line and is used for the purpose of sectioning an object.

Note : *The section lines are drawn at an angle of 45° to the horizontal line and are spaced uniformly from 2 to 4 mm apart depending upon the size of the object (see Fig. 3.3).*

3. Centre line, locus line, pitch circles, extreme position of moveable parts and parts situated infront of cutting plane : These lines are represented by long and short dashes in proportion ranging from 6 : 1 to 4 : 1, closely and evenly spaced in any drawing. The proportion once selected should be maintained through the drawing (see Fig. 3.1). These are used to show the centre and location of cylindrical, conical and spherical object.

The following rules should be kept in mind while drawing centre lines of various objects :

(i) The centre lines should not end at out line representing surfaces but should extend approximately from 2 to 5 mm beyond the out lines of the object.

(ii) Where centre lines cross, the short dashes should intersect at symmetrically. In case of very small circles, the short dashes should be neglected while drawing centre lines.

4. Hidden line : The hidden line is represented by closely and evenly spaced short dashes, It is used to show the invisible or hidden parts on the drawing (see Fig. 3.1).

5. Construction line : It is a thin continuous line and is used for constructing an object (see Fig 3.1)

These lines do not appear in finished drawing except in geometrical drawings where these are not removed.

6. Dimension line : It is thin continuous line for giving dimensions. This line terminates in arrow head where the dimension lines meet the extension lines (see Fig. 3.1).

7. Extension line : It is a thin continuous line used for dimensioning an object (see Fig. 3.1).

8. Projectors line : It is a thin continuation of outlines and is used for drawing projector (see Fig. 3.1).

9. Cutting plane line : The cutting plane line is represented by thick long line at the ends with thin (long and short) lines at the centre. It is used to show the edge of the cutting plane.

10. Short break line : The short break line is represented by a thin free-hand and line is used to show the break of an object for a short length (see Fig. 3.1). It results in a saving in space and time used for drawing without loss of any details.

11. Long break line : The long break line is represented by thin ruled line provided with freehand zig-zags at suitable intervals and is used to show the break for a considerable length of the object (see Fig. 3.1).

CONVENTIONS FOR VARIOUS LINES [ACCORDING TO B.I.S. S.P : 46 - 1988]

LINE	DESCRIPTION	GENERAL APPLICATION
A 	CONTINUOUS THICK	A1 VISIBLE OUTLINE A2 VISIBLE EDGES
B	CONTINUOUS THIN STRAIGHT OR CURVED	B1 IMAGINARY LINES OF INTERSECTION B2 DIMENSION LINES B3 PROJECTION LINES B4 LEADER LINES B5 HATCHING B6 OUTLINES OF REVOLVED SECTIONS IN PLACE B7 SHORT CENTRE LINE
C 	CONTINUOUS THIN FREEHAND	C1 LIMITS OF PARTIAL OR INTERRUPTED VIEWS AND SECTIONS, IF THE LIMIT IS NOT A CHAIN THIN
D	CONTINUOUS THIN (STRAIGHT WITH ZIGZAGS	D1 LINE
E 	DASHED THICK	E1 HIDDEN OUTLINES E2 HIDDEN EDGES
F	DASHED THIN	F1 HIDDEN OUTLINES F1 HIDDEN EDGES
G 	CHAIN THIN	G1 CENTRE LINES G2 LINES OF SYMMETRY G3 TRAJECTORIES
H 	CHAIN THIN, THICK AT ENDS AND CHANGES OF DIRECTION	H1 CUTTING PLANES
J 	CHAIN THICK	J1 INDICATION OF LINES OR SURFACES TO WHICH A SPECIAL REQUIREMENT APPLIES
K	CHAIN THIN DOUBLE-DASHED	K1 INDICATION OF LINES OR SURFACES K2 ALTERNATIVE AND EXTREME POSITIONS OF MOVABLE PARTS K3 CENTROIDAL LINES K4 INITIAL OUTLINES PRIOR TO FORMING K5 PARTS SITUATED IN FRONT OF THE CUTTING PLANE

Fig. 3.2

The various types of lines as given by Beaureo of Indian Standards B.I.S., SP: 46-1988, are illustrated in Fig. 3.2.

Fig. 3.3 shows the application of various types of lines according to B.I.S. SP:46-1988.

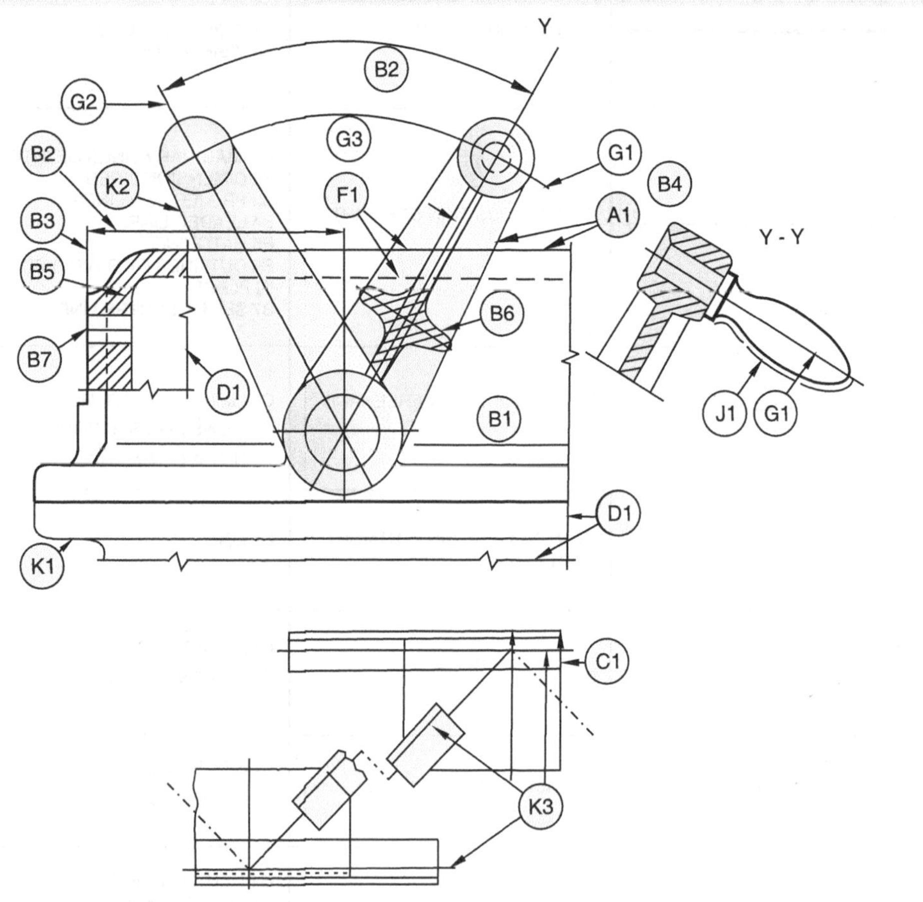

Fig. 3.3. *Application of various types of lines according to B.I.S. SP:46-1988*

3.4. THICKNESS OF LINES

There are three distinct thickness of lines used in engineering drawing. These lines are specified as thick, medium and thin lines. The line specified as thick is usually 3 times thicker and the line specified as medium is 2 times thicker than a thin line.

The thickness of lines should be chosen according to the size and type of the line group used on a drawing. It is identified from the thickest line, i.e., visible outlines. For a given view or section, the line used should be chosen from one of the line groups as shown in Fig. 3.4.

Two thicknesses of lines are used. The ratio of thick to thin line shall not be less than 2 : 1. Grading of lines is in 2 increments.

Note : *The thickness of lines should be chosen according to the size and the type of the drawing.*

Fig. 3.4 shows the comparative thickness of line groups according to I.S.I.: 692-1972.

COMPARATIVE THICKNESS OF VARIOUS TYPES OF LINE GROUPS

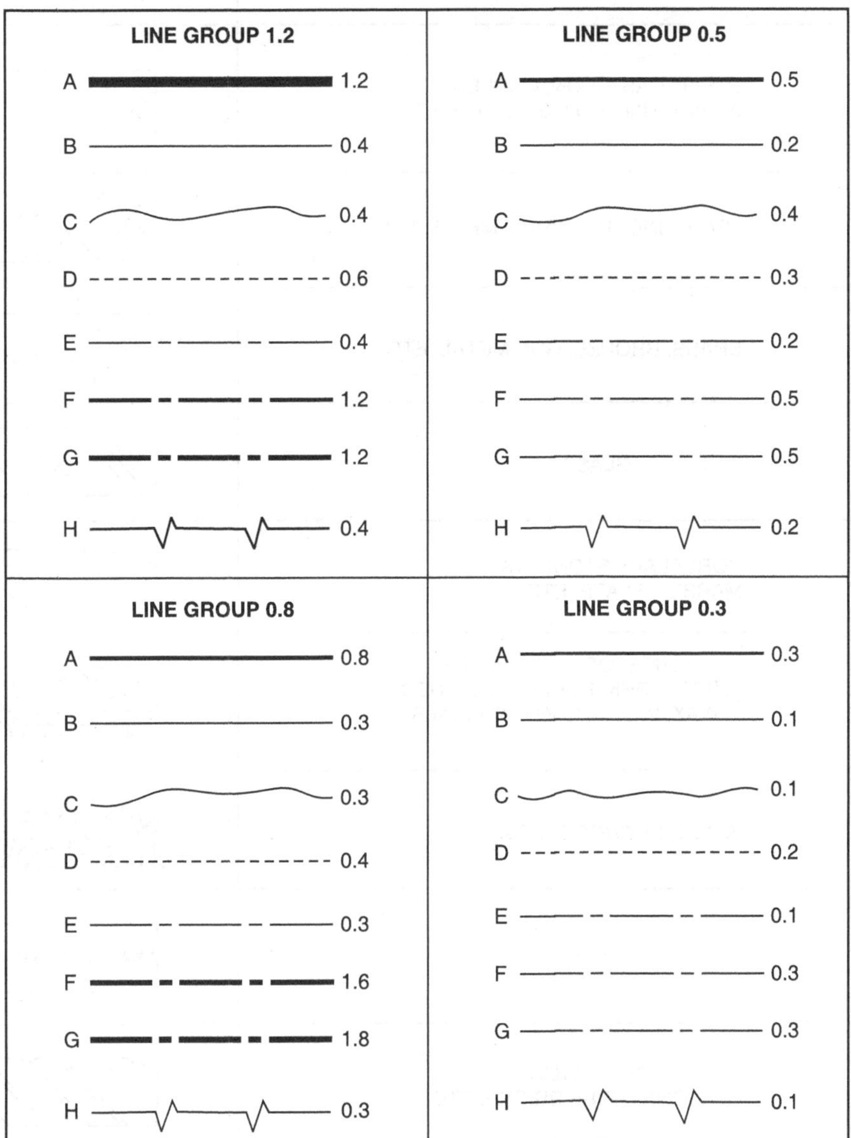

Fig. 3.4. *Comparative thickness of various types of line groups*

3.5. CONVENTIONS FOR VARIOUS MATERIALS

In engineering practice, there are different types of materials used for manufacturing the various parts of a machine. It is, therefore, desirable that different conventions should be adopted to differentiate various materials for convenience on the drawing. The conventions of materials thus save time and labour of the drawing work.

The I.S.I. has recommended the convention for various types of materials as per I.S.I.: 696-1972 (revised) as shown in Fig. 3.5, when these are sectioned.

CONVENTION FOR VARIOUS MATERIALS

S. NO.	MATERIALS	CONVENTION
1.	STEEL, CAST IRON, COPPER ALUMINIUM AND ITS ALLOYS, ETC.	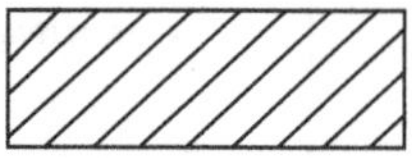
2.	LEAD, ZINC, TIN, WHITE METAL, ETC.	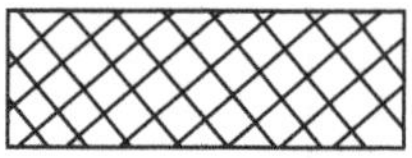
3.	BRASS, BRONZE, GUN METAL, ETC.	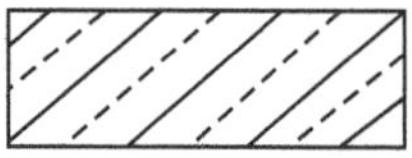
4.	GLASS	
5.	PORCELAIN, STONE WARE, MARBLE, SLATE, ETC.	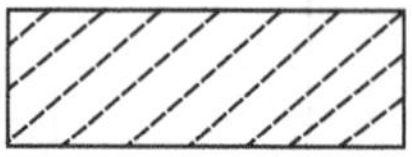
6.	ASBESTOS, FELT, PAPER, MICA, CORK, RUBBER, LEATHER WAX, INSULATING MATERIALS	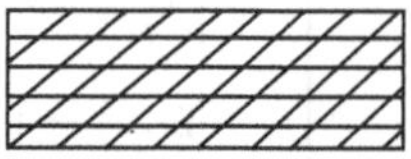
7.	WOOD, PLYWOOD, ETC.	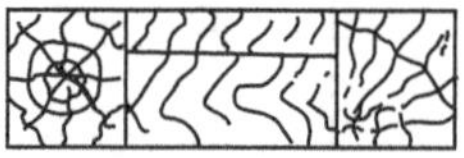
8.	EARTH	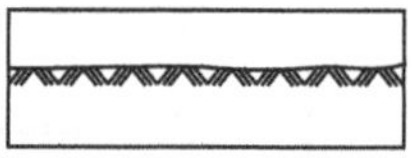
9.	BRICK WORK, MASONRY, FIRE BRICKS, ETC.	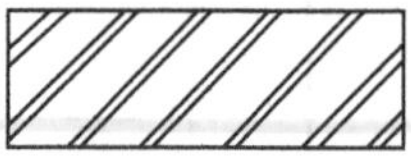
10.	CONCRETE	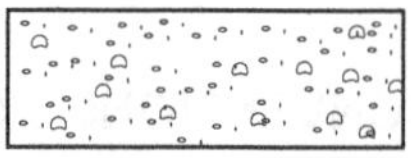
11.	WATER, OIL, PETROL, KEROSENE, ETC.	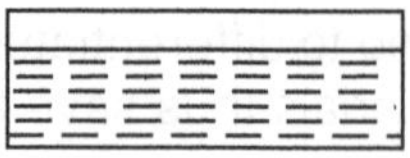

Fig. 3.5. *Conventions for various materials*

3.6. CONVENTIONAL BREAKS

Long members of uniform cross sections such as rods, shafts, pipes, etc. are generally shown in the middle by the conventional breaks so as to accommodate their view of whole length on the drawing sheet without reducing the scale. The exact length of the member is shown by a dimension.

Fig. 3.6 shows the important types of conventional breaks as given by I.S.I.

S. NO.	OBJECT	CONVENTION
1.	RECTANGULAR SECTION	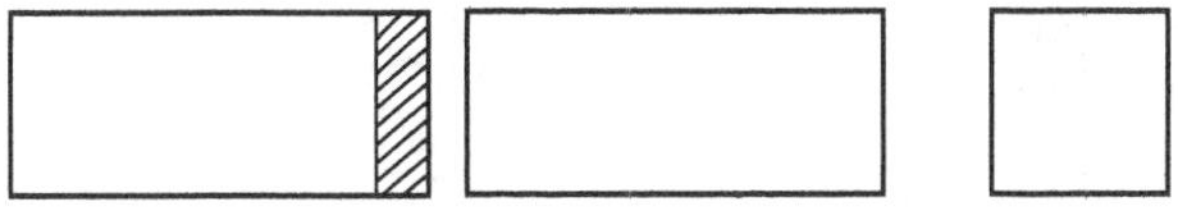
2.	ROUND SECTION	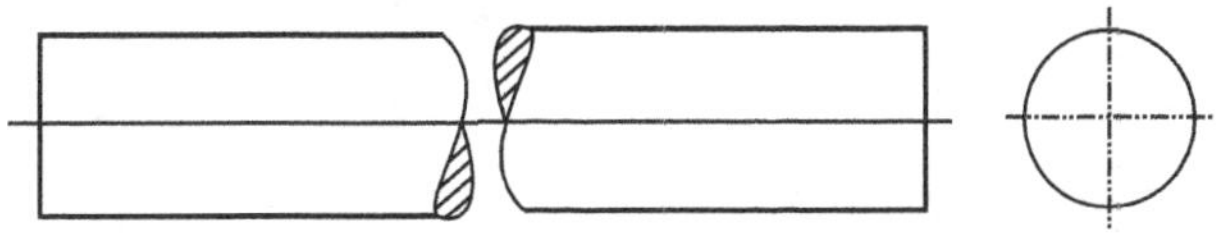
3.	PIPE OR TUBING	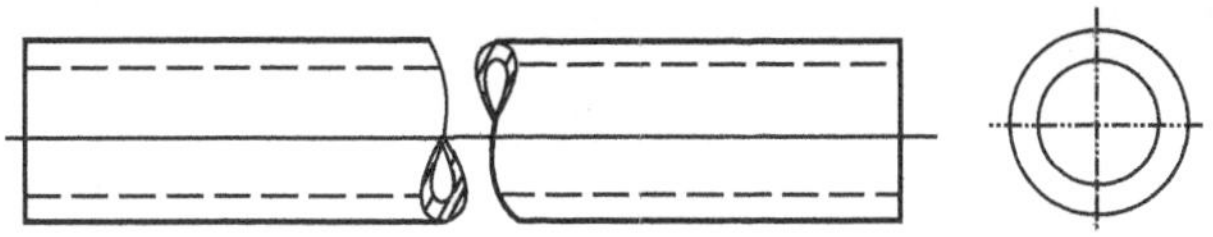
4.	PIPE OR TUBING	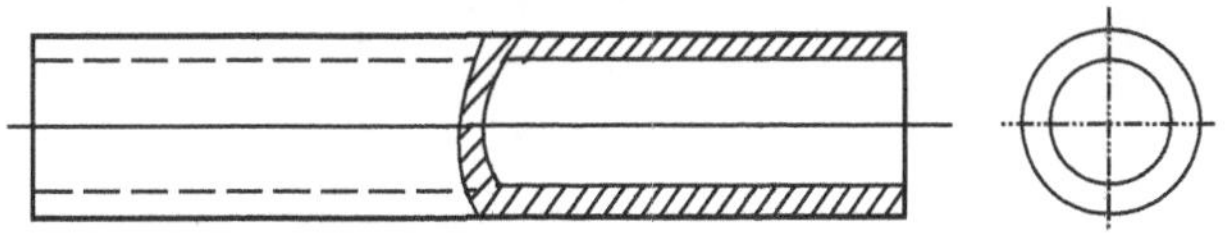
5.	WOOD RECTANGULAR SECTION	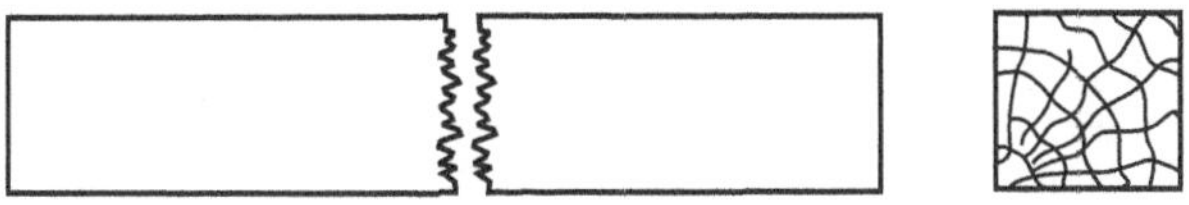
6.	ROLLED SECTION	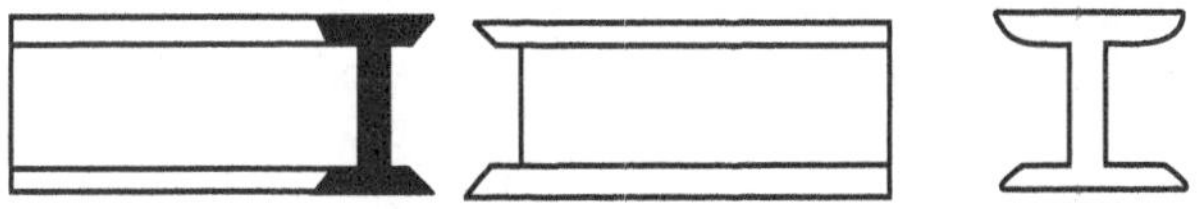
7.	CHANNEL SECTION	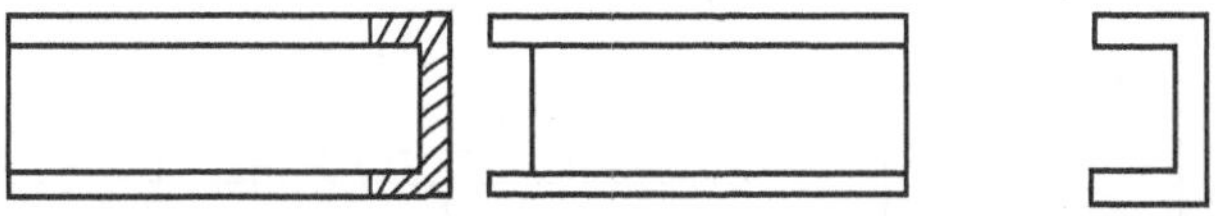

Fig. 3.6. *Conventional breaks as given by I.S.I.*

3.7. CONVENTIONAL REPRESENTATION OF COMMON FEATURES

Conventional representation is adopted in cases where complete description of the machine component would involve unnecessary time or space on the drawing.

Fig. 3.7 shows the conventional representation of common features given by I.S.I.

TITLE	ACTUAL PROJECTION/SECTION	CONVENTION
RATCHET AND PINION		
BEARINGS		
STRAIGHT KNURLING		
DIAMOND KNURLING		
HOLES ON A LINEAR PITCH		
HOLES ON CIRCULAR PITCH		
REPEATED PARTS		

Fig. 3.7. *Convention representation of common features*

Fig. 3.8 shows the conventional representation of springs as per I.S.I.

TITLE	ACTUAL PROJECTION/SECTION	CONVENTION
EXTERNAL THREADS	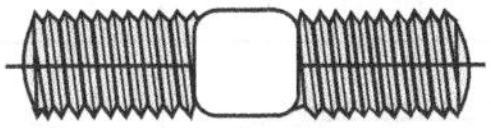	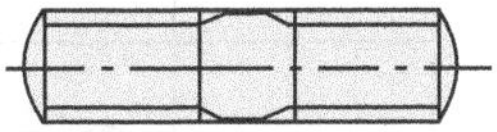
INTERNAL THREADS	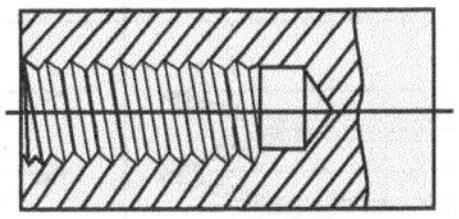	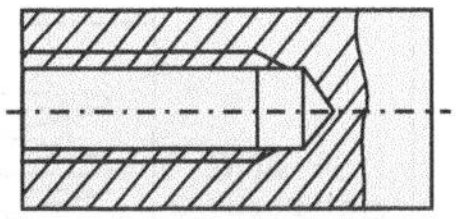
SLOTTED HEAD	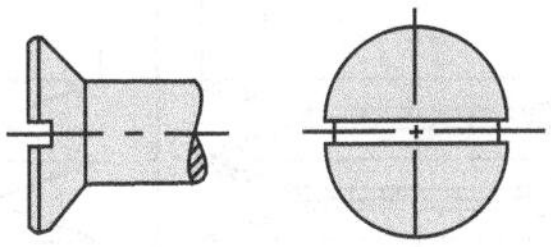	 TO BE DRAWN AT 45°
SQUARE END AND FLAT	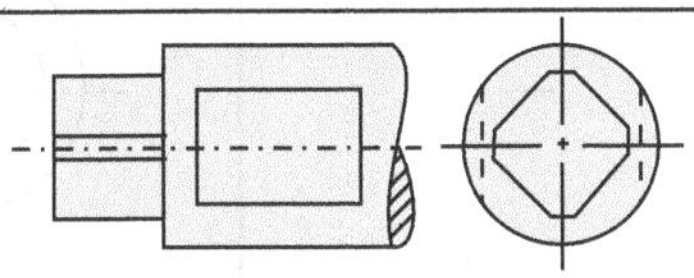	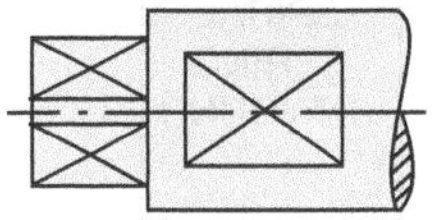
RADIAL RIBS	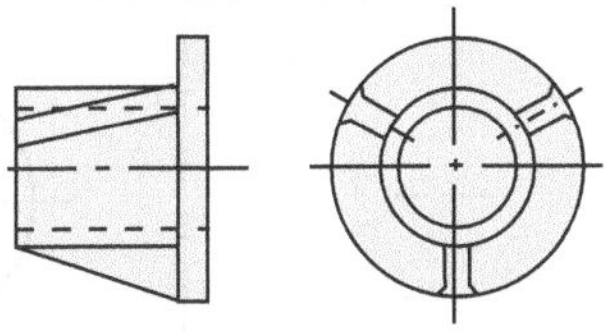	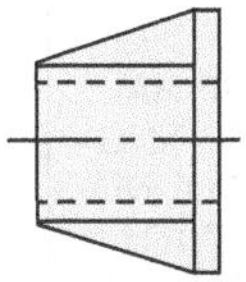
SERRATED SHAFT	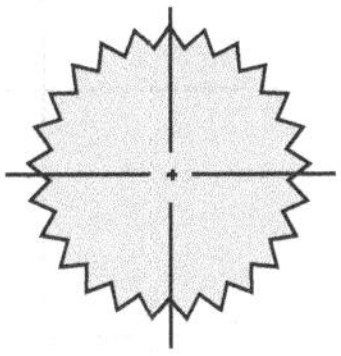	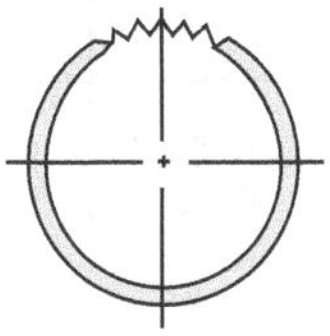
SPLINED SHAFT	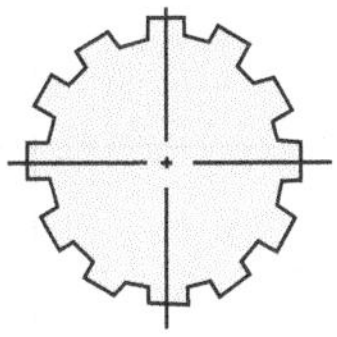	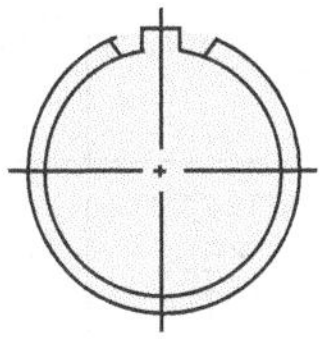
CHAIN WHEEL	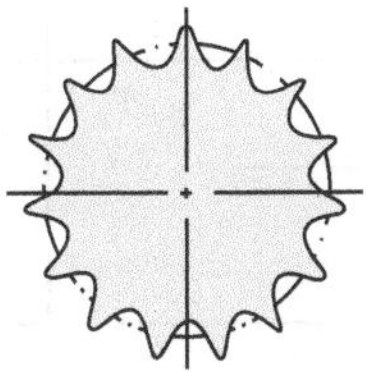	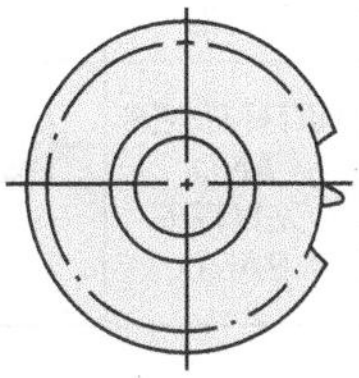

Fig. 3.8. *Conventions for springs*

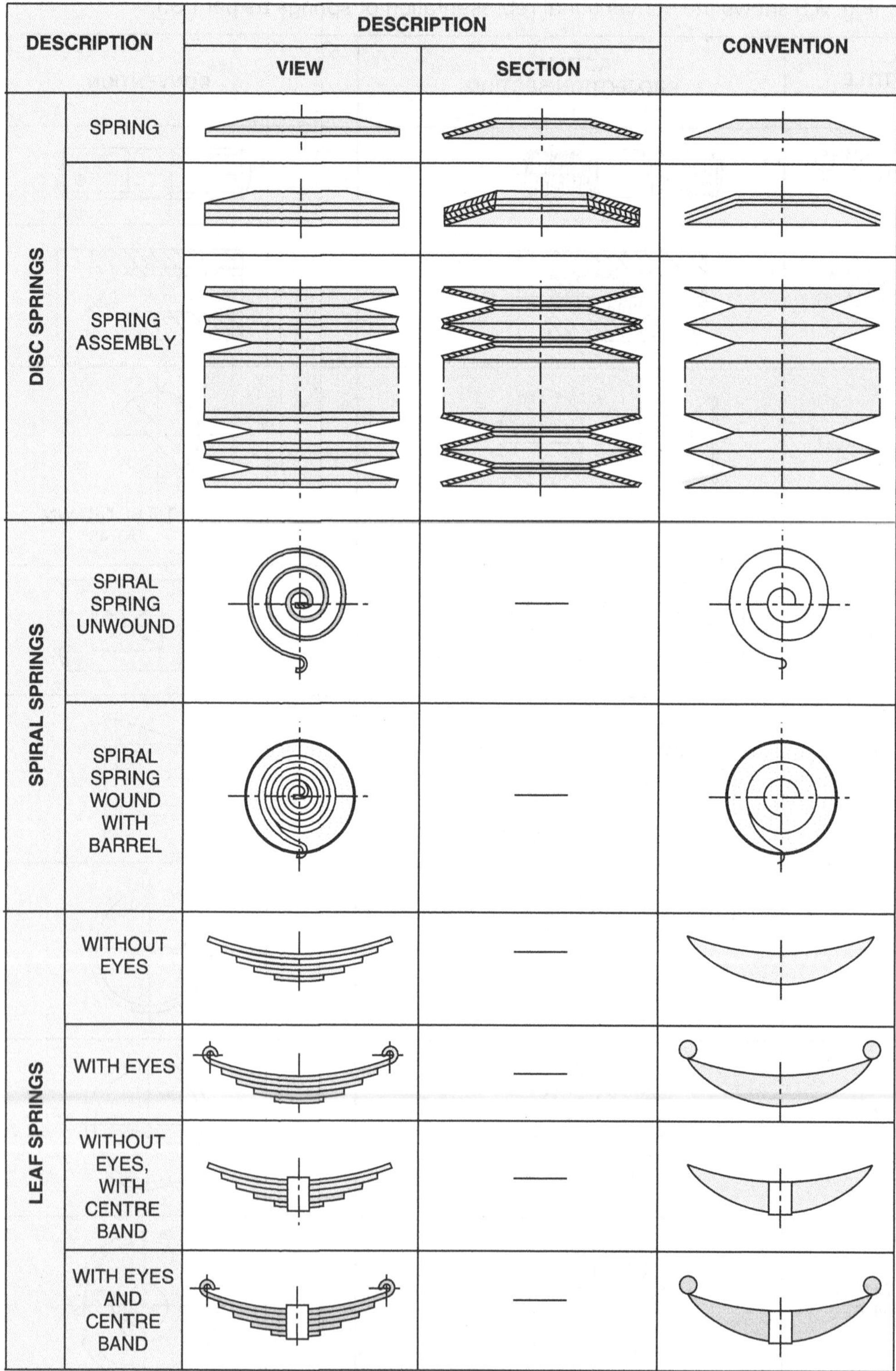

Fig. 3.9. *Conventions for springs*

QUESTIONS FOR SELF EXAMINATION

1. What do you mean by convention?
2. What do you understand by thickness of lines and various line groups?
3. Where and why a cutting plane is drawn in a drawing?
4. What is the necessity of convention breaks and convention of materials?
5. Why the conventional representation of common features are adopted in drawing?

PROBLEMS FOR PRACTICE

1. Draw the conventional signs of different types of lines used in engineering practice as per B.I.S.
2. Draw the conventional signs of different materials in section commonly used in engineering practice as given by I.S.I.
3. Draw the conventional break of a solid bar, a rectangular rod, wooden log and R.S.J.
4. Show the conventional representation of the following :
   (*i*) Plywood, Wax, Slate, Kerosene.
   (*ii*) Steel, Cast iron, Copper, Aluminium, Lead and white metal.
5. Match the statement of column A with the corresponding one in column B

Column A	Column B
(*a*) Visible outlines are drawn as	(1) Chain thin lines
(*b*) Dimension lines, hatching and projection lines are drawn as	(2) Continuous thin lines.
(*c*) Centre lines, locus lines are drawn as	(3) Continuous thick lines.
(*d*) Cutting plane lines are shown by	(4) Continuous thin lines with zig-zags
(*e*) Long breaks are shown by	(5) Continuous thin and wavy lines
(*f*) Irregular boundary lines, short break lines are drawn as	(6) Thick and chain thin lines, thick at ends.

Ans. *(a) (3); (b) (2); (c) (1); (d) (6); (e) (4) and (f) (5).*

6. Fill in the blanks :
   (*i*) The representation of any matter by some sign or mark on the drawing is known as......
   (*ii*) The conventions make the drawing......and......to draw.
   (*iii*) The section lines are drawn at .... to the horizontal line.
   (*iv*) The line specified as thick is usually......times thicker and line specified as medium is.........times thicker than a thin line.
   (*v*) Long members are generally shown in the.......by the convention breaks.

Ans. (*i*) convention (*ii*) simple, easy (*iii*) 45° (*iv*) 3, 2 (*v*) middle

CHAPTER

4

Lettering

INTRODUCTION

In engineering drawing, lettering plays an important role as it explains those parts of the object which cannot be shown by lines. On the other hand a poor lettering not only mars the appearance of drawing but sometimes leads to wrong result which causes a wastage of time and labour. Thus, the lettering is the talk of drawing and therefore it should be legible, uniform in appearance, simple and easy for rapid writing.

In this chapter, we will deal with the study of lettering, general proportions and different types of letters, etc. used in engineering practice.

4.1. LETTERING

The art of writing the alphabets A, B, C, ... Z and numbers such as 1, 2, 3, ... 0, etc. is known as lettering.

It is an important part of drawing and is used to write letters, dimensions, notes and other necessary information required for the complete execution of a machine or structure, etc.

4.2. REQUIREMENTS OF GOOD LETTERING

In engineering, drawing a good lettering must fulfill the following purposes :

1. The knowledge of the shape and proportion of each letter.
2. The knowledge of the order and direction of the strokes used in making letters.
3. The knowledge of the general composition of letters.
4. The knowledge of the rules for combining letters into words and words into sentences.
5. The knowledge of writing the letters in plain and simple styles so that the lettering can be done freehand and speedly.

4.3. GENERAL PROPORTIONS OF LETTERING

The general proportion of lettering means the relationship between the height, width and spacing of each letter. Although there is no fixed standards for the proportions of each letter, yet all letters in a given type of alphabets have a general relation to each other which allows only small variations. In general, the following are the important proportions of lettering :

1. Normal lettering
2. Condensed lettering
3. Extended lettering

4.3.1. Normal Lettering : The normal lettering have normal height and width and are used for general purposes (i) (see Fig. 4.1).

Note : *The width of the normal letter is about 0.67 times of the height of the letter*

4.3.2. Condensed Lettering : The condensed lettering is written in the narrow space. These are used when the space is limited (see Fig. 4.1).

Note : *The width of the condensed letters is less than height.*

4.3.3. Extended Lettering : The extended lettering is wider than normal lettering but of the same height (see Fig. 4.1.)

(i) ENGINEERING
NORMAL LETTERING

(ii) ENGINEERING
CONDENSED LETTERING

(iii) E N G I N E E R I N G
EXTENDED LETTERING

Fig. 4.1. *Proportions of Lettering*

Important Note : *It is often desirable to increase or decrease the width of letters in order to make them neat and pleasing to the eyes.*

4.4. COMPOSITION OF LETTERS

The composition means the composing of letters into words and words into sentences. The letters are so arranged that the open area between two letters of a word appears equal to the eye judgement.

The difference between good composition and poor composition can be well understood by observing the arrangement of different letters. The example of good as well as poor composition is shown in Fig. 4.2.

Fig. 4.2. *Composition of letters*

4.5. SPACING OF LETTERS

The spacing means the distance which is to be left between the two adjacent letters in all types of lettering. Good spacing is as important as good lettering. An effort should be made to equalize the white space area between the letters while composing. The spacing should be judged by observation and not by measurement.

In Fig. 4.2 the word "SPACING" is written in which stress is made to provide equal white space area between the individual letters for good appearance. The reasonable ratio of 1:4 between space and adjacent letters is generally followed in double stroke lettering. The ratio 1:4 means that 1 square is to be left after covering 4 squares leaving few letters, as shown in Fig. 4.3, such as R,A, I, L,W,A and Y alphabetes i.e. in the word "RAILWAY".

Fig. 4.3.

Fig. 4.3 shows the setting of the letters in the word "RAILWAY" in which the letters or alphabets are arranged so as to leave approximately the same amount of space between the letters.

Notes :

(i) *The space between each words should be kept equal to height of letter [see Fig. 4.4(i)]*

(ii) *The space between the two lines should be left equal to twice the height of the letter.*

(iii) *The space between two lines should be kept not less than half or more than one and a half times the height of letter.*

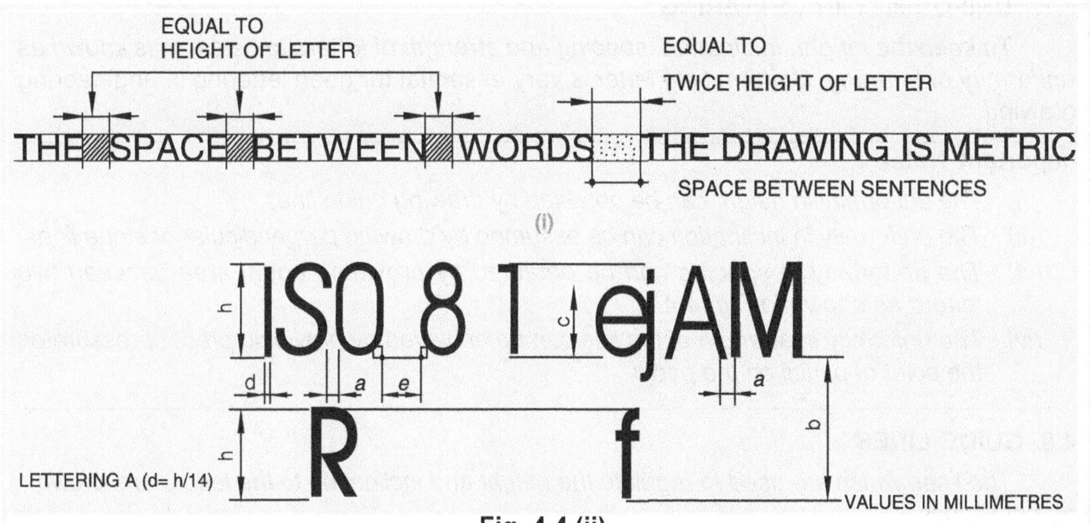

Fig. 4.4 (ii)

CHARACTERISTIC	RATIO	DIMENSION						
LETTERING HEIGHT HEIGHT OF CAPITALS *h*	(14/14) *h*	2.5	3.5	5	7	10	14	20
HEIGHT OF LOWER-CASE LETTERS (WITHOUT STEM OR TAIL) *c*	(10/14) *h*	-	2.5	3.5	5	7	10	14
SPACING BETWEEN CHARACTERS *a*	(12/14) *h*	0.35	0.5	0.7	1	1.4	2	2.8
MINIMUM SPACING OF BASE LINES *b*	(20/14) *h*	3.5	5	7	10	14	20	28
MINIMUM SPACING BETWEEN WORDS *e*	e (6/14) *h*	1.05	1.5	2.1	3	4.2	6	8.4
THICKNESS OF LINES *d*	(1/14) *h*	0.18	0.25	0.35	0.5	0.7	1	1.4

According to B.I.S., the height *h* of capital letters is taken as the base of dimensioning [see Fig 4.4(ii). The two standard ratio for *d*/*h*; 1/14 and 1/10 are most economical (d represents line thickness) as they result in a minimum number of thickness as shown in above table.

The range of standard height *h* for lettering is also shown in table, which involves a ratio of 2, which is derived from the dimensions from the dimensions for drawing sheet sizes.

4.6. SIZE OF LETTERS

In engineering practice, the size of letters means the height of the letters.

The standard sizes of the alphabets recommended by the Indian Standards Institute (I.S.I) are as follows :

SIZES OF ALPHABETS FOR DRAWING

S.NO.	PURPOSE	SIZE OF ALPHABETS IN mm
1.	MAIN TITLE AND DRAWING NO.	6, 8,10 AND 12
2.	SUB-TITLE AND HEADINGS	3, 4, 5 AND 6
3.	NOTES, SUCH AS LEGENDS, SCHEDULES, MATERIALS AND DIMENSIONS	2, 3, 4 AND 5

4.7. UNIFORMITY IN LETTERING

To keep the height, inclination, spacing and strength of letters to be same is known as uniformity of lettering. Uniformity in letter is very essential for good lettering in engineering drawing.

Important Notes :

(i) The uniformity in height can be obtained by drawing guide lines.

(ii) The uniformity in inclination can be assumed by drawing perpendicular or slope lines.

(iii) The uniformity in spacing can be obtained by providing equal area between two letters as shown in Fig. 4.4.

(iv) The unformity in strength of the line can be achieved by providing proper pressure on the point of pencil on the paper.

4.8. GUIDE LINES

The lines which are used to regulate the height and inclination to the letters are known as guide lines.

The guide lines are used to regulate the letters uniform.

Note : *The guide lines should be drawn with 4H to 6H pencil, so lightly that the lines do not need any erasing.*

4.9. GUIDE LINE DEVICES

In order to simplify the process to draw guide lines, many special guide line devices are used. The following are the most common guide line devices used in practice :

1. Guide line triangle
2. Guide line instrument

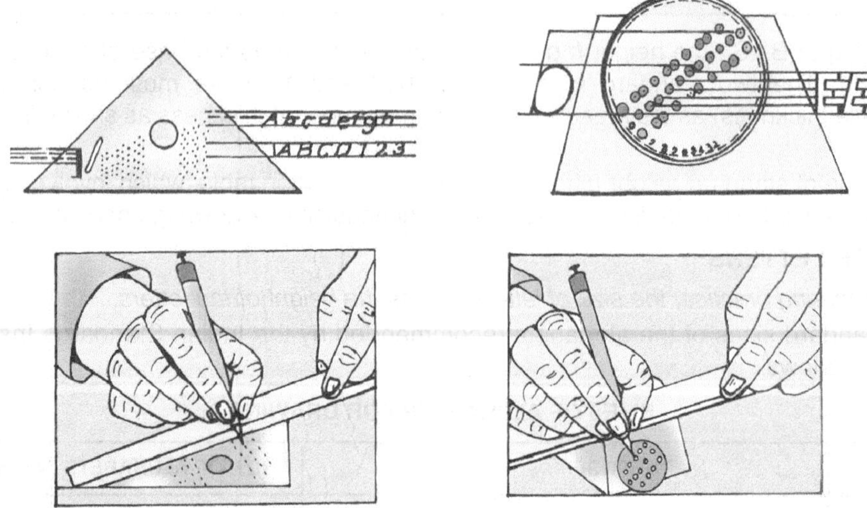

Guide line triangle *Guide line instrument*

Fig. 4.5.

4.9.1. Guide Line Triangle : It is just like a set-square with a series of holes arranged in such a way to provide guide lines for lettering and dimensioned figures (see Fig. 4.5).

To use the triangle, the pencil point is inserted in the proper hole and the triangle moved back and forth along the T-square.

4.9.2. Guide Line Instrument : It is also used for drawing guide lines for letters of various heights (see Fig. 4.5).

4.10. ORDER AND DIRECTION OF THE STROKE OF LETTERS

Fig. 4.7 shows the order and direction of the stroke that forms the letters. Small variation are permissible while combing the letters to form words, but great care must be exercised to preserve the general proportion and direction, order and direction so that the letters will look well together.

4.11. CLASSIFICATION OF LETTERING

Lettering can be classified into the following three basic groups :

1. Gothic lettering
2. Roman lettering
3. Freehand lettering

4.12. GOTHIC LETTERING

The lettering in which all the alphabets are of uniform width or thickness is known as gothic lettering.

The gothic lettering may be further divided into the following groups:

(a) Vertifical or Upright gothic lettering	(b) Inclined or Italic gothic lettering
1. Single stroke vertical gothic lettering	1. Single stroke italic gothic lettering
2. Double stroke vertical gothic lettering	2. Double stroke italic gothic lettering
3. Lower case vertical gothic letterings.	3. Lower case italic gothic letterings.

4.12.1 Vertical or Upright Gothic Lettering

The lettering in which the direction of alphabets is vertically upward is known as vertical or upright gothic lettering. Fig. 4.6 shows the method of constructing the vertical or upright gothic lettering according to I.S.I.

Fig. 4.6. *Vertical or Upright gothic lettering*

1. Single-stroke vertical gothic lettering : *The lettering in which the alphabets are of the same thickness (as the single-stroke of the pencil or pen) is called single-stroke vertical gothic lettering.*

Fig. 4.7 shows the single-stroke vertical gothic lettering and numbers in the ratio of 7 : 4 and 6 : 5 drawn with the help of instruments.

Single-stroke vertical instrumental gothic lettering [ratio of 7:4]

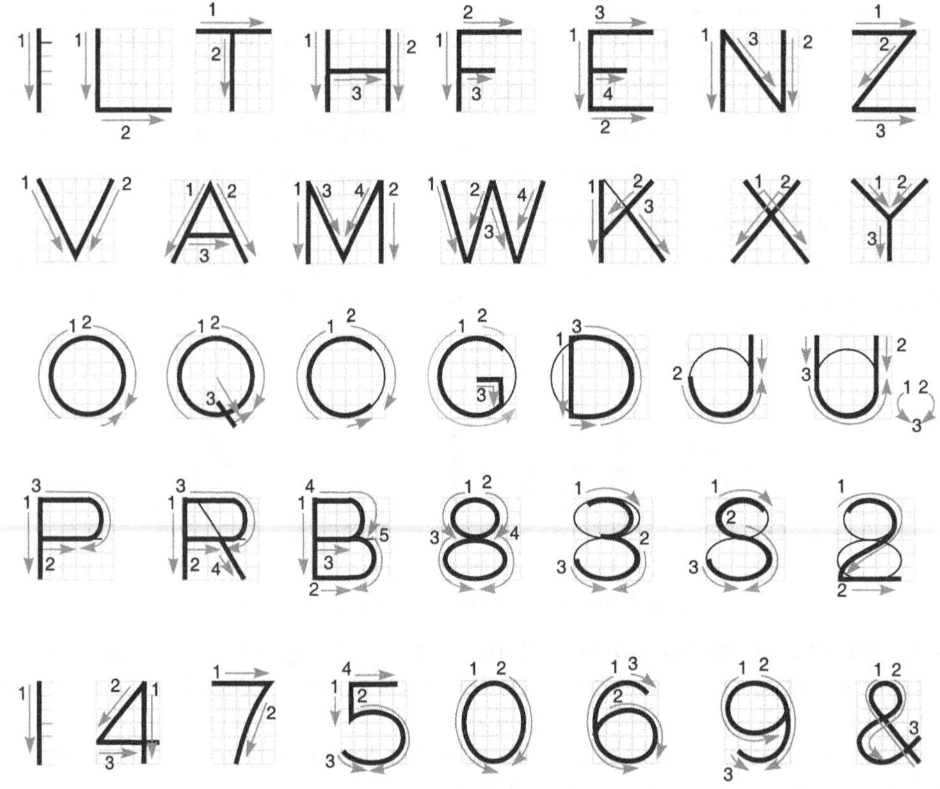

Single-stroke vertical gothic lettering [ratio of 6 : 5]

Fig. 4.7.

ABCDEFGHIJ
KLMNOPQR
STUVWXYZ
1234567890

RATIO - 7 : 4

ABCDEFGHIJ
KLMNOPQR
STUVWXYZ
1234567890

RATIO - 5 : 4

Fig. 4.8. *Double-stroke vertical gothic lettering*

2. Double-stroke vertical gothic lettering : The lettering in which the alphabets are written by double-stroke of the pencil or pen with a uniform spacing in between the strokes is called double-stroke vertical gothic lettering.

Fig. 4.8 shows the double stroke vertical gothic letters in the ratio of 7 : 4 and 5 : 4 drawn with the help of instruments.

Note : *Instrumental lettering means, lettering drawn with the help of instrument.*

3. Lower-case vertical gothic lettering : *The lettering in which the alphabets are of small letters such as a, b, c,... z, etc. is called lower-case vertical gothic lettering.*

This type of lettering is generally used in maps or architectural drawings and very seldom used in engineering drawing.

Fig. 4.9 shows vertical lower-case gothic lettering in which the drawing of alphabets alongwith the various new terms such as cap line, drop line, base line, ascender space and decender space, etc.

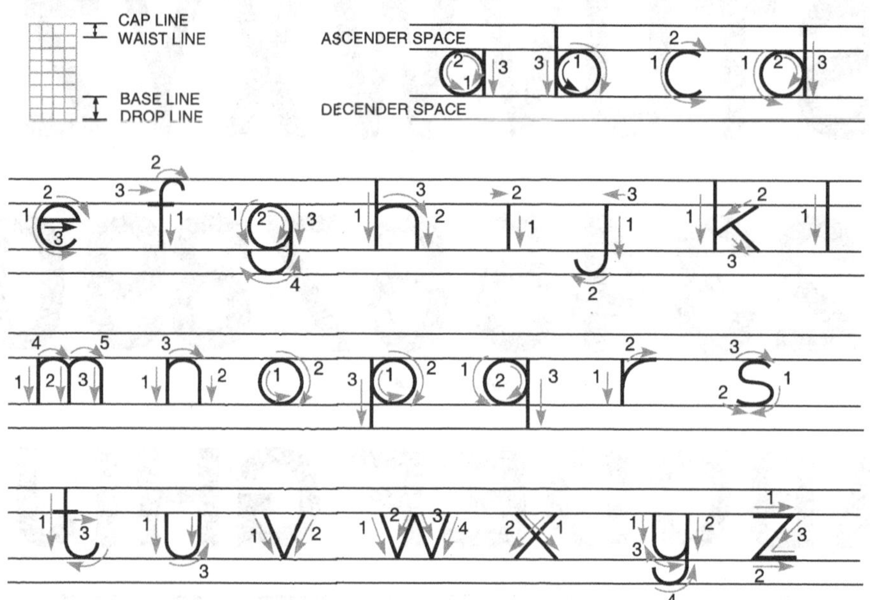

Fig. 4.9. *Lower-case vertical gothic lettering*

4.12.2. Inclined or Italic Gothic Lettering

The lettering in which the direction of alphabets is inclined to the horizontal line is known as inclined or Italic lettering. An inclination of about 75° is recommended, from the right towards the left according to Indian Standards Institute (I.S.I.).

Inclined or Italic gothic lettering

Fig. 4.10. *Inclined or Italic gothic lettering*

Fig. 4.10 shows the method of constructing the inclined or italic gothic lettering as per I.S.I.

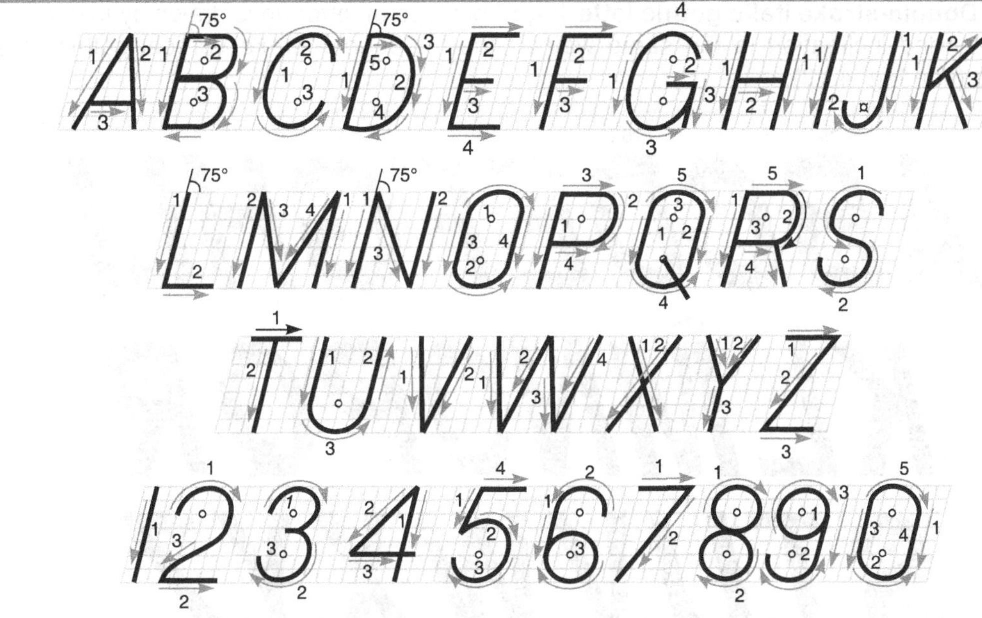

Fig. 4.11. *Single-stroke italic gothic lettering ratio 7 : 4*

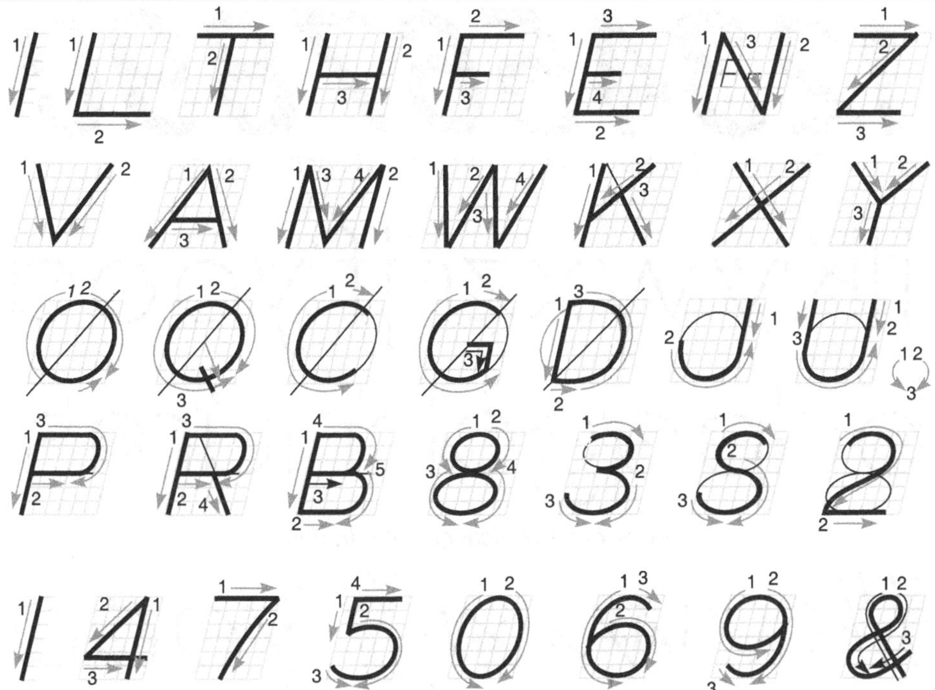

Fig. 4.12. *Single-stroke italic gothic letters ratio 6 : 5*

1. Single stroke italic gothic lettering : This type of lettering is drawn by taking the recommended inclination in the same way as the single-stroke vertical gothic type lettering (see Fig. 4.11 and 4.12).

2. Double-stroke italic gothic lettering : This type of lettering is drawn by taking the recommended inclination in the same way as that of double-stroke vertical gothic type lettering (see Fig. 4.13).

75° 75°
ABCDEFGHIJ
75° 75°
KLMNOPQR
STUVWXYZ
1234567890

Fig. 4.13. *Double-stroke instruments italic gothic lettering (7:4)*

Lower-case vertical gothic lettering

Fig. 4.14. *Single-stroke lower case italic lettering*

3. Lower case italic gothic lettering : The inclined lower case lettering can be drawn by giving recommended inclination of 75° from right towards left (see Fig. 4.14).

Fig. 4.15 shows the specimen of vertical and inclined letters and numerals given by B.I.S., S.P. : 46–1988.

ABCDEFGHIJKLMNOP

QRSTUVWXYZ

Specimen of vertical letters [B.I.S.-1988]

aabcdefghijklmnopq

rstuvwxyz

[(!?:;'' — = + × : · % &)] ϕ

0123456778 9IVX

Fig. 4.15. *Specimen of lower-case letters and numerals [B.I.S.-1988].*

4.13. ROMAN LETTERING

The lettering in which all the alphabets are composed of thick and thin elements is known as roman lettering and can either be vertical or inclined.

This type of lettering is drawn either by means of a chisel-pointed medium soft pencil or ink by various nibs known as speed-balls. Fig. 4.16 shows one of the Roman type lettering made with D-3 type speed balls.

Note : *Roman type lettering is used by the architects because of their beautiful form.*

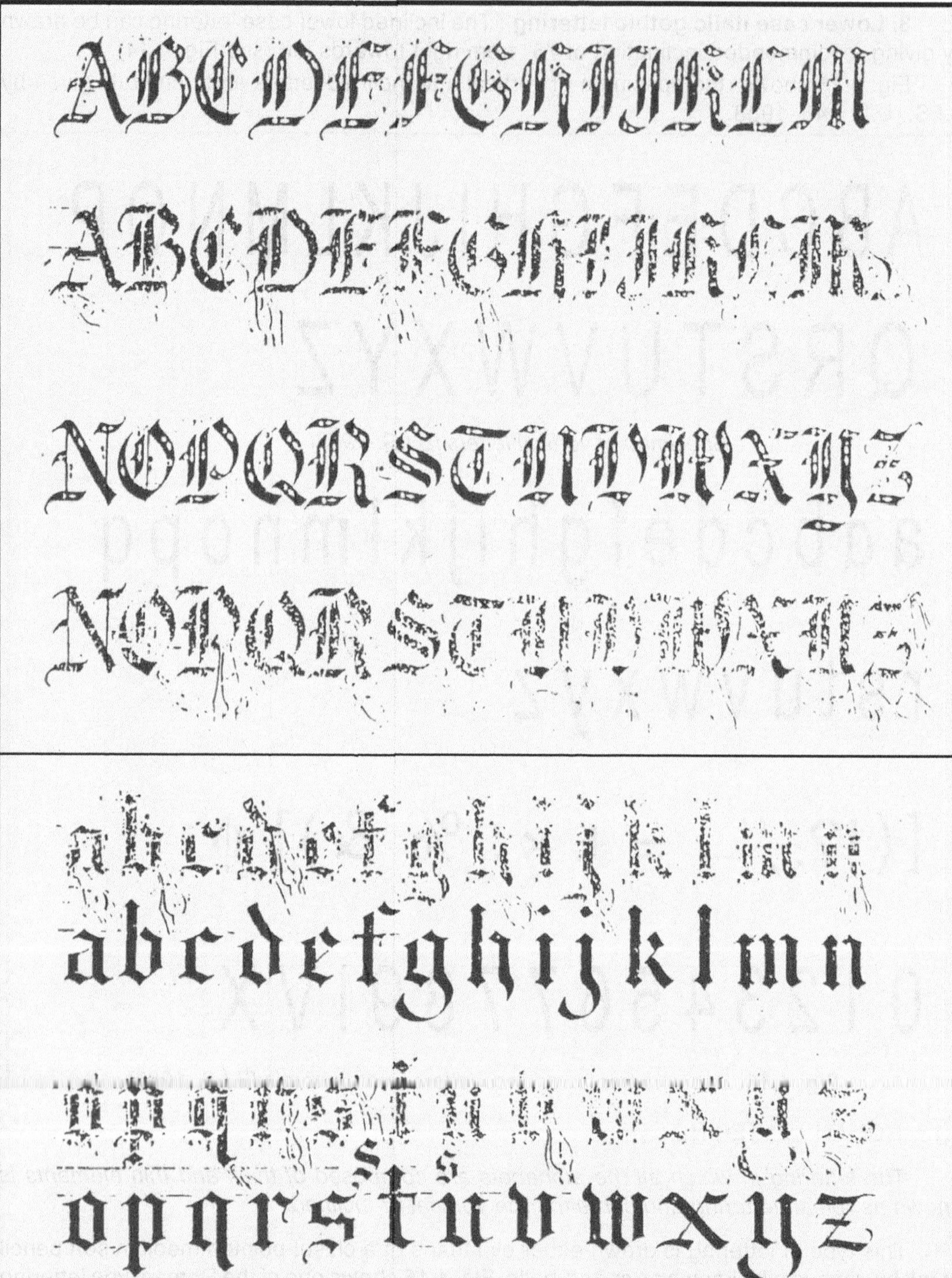

Fig. 4.16. *Roman lettering*

4.14. FREEHAND LETTERING

The art of writing the alphabets without the use of drawing instruments is called freehand lettering. The freehand lettering is of the following types :

(a) Vertical or Upright freehand gothic lettering

1. Single-stroke vertical freehand gothic lettering
2. Lower case vertical freehand gothic lettering

(b) Inclined or Italic freehand gothic lettering

1. Single-stroke italic freehand gothic lettering
2. Lower case italic freehand gothic lettering

4.14.1. Vertical or Upright Freehand Gothic Lettering

The lettering which is written in a vertical upward direction without the use of drawing instruments is called a vertical or upright freehand gothic lettering.

1. Single-stroke, freehand gothic lettering : The lettering in which the alphabets are written vertically with a single-stroke of pencil or pen without the help of drawing instruments is called a single-stroke freehand gothic lettering (see Fig. 4.17).

90°

A B C D E F G H I J K L M N O P

Q R S T U V W X Y Z

1 2 3 4 5 6 7 8 9 0

Fig. 4.17. *Single-stroke vertical freehand lettering (Height = 4mm)*

Fig. 4.17 shows the single-stroke vertical freehand gothic lettering of height = 4mm. Similarly, single-stroke freehand vertical gothic lettering of 4 mm, 5 mm, 8 mm, 12 mm, etc. can be written (see Fig. 4.18).

ENGINEERING DRAWING IS THE SYSTEMATIC COMBINATION OF DIFFERENT TYPES OF LINES

Fig. 4.18. *Single-stroke vertical freehand gothic lettering (Height = 5 mm)*

2. Lower case vertical freehand gothic lettering : The lettering in which the vertical small size alphabets such as a, b, c,... z, etc. are written freehand is called lower case vertical freehand lettering (see Fig. 4.19).

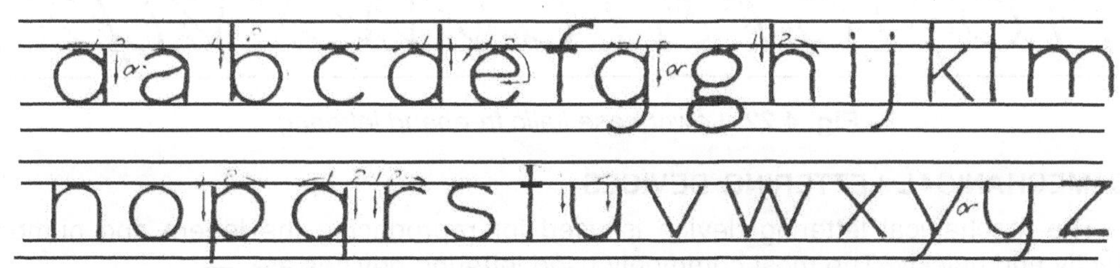

Fig. 4.19. *Lower case vertical gothic lettering*

4.14.2. Inclined or Italic Freehand Gothic Lettering

The lettering which is written with some inclination to the horizontal line without the use of drawing instruments is called inclined or italic freehand gothic lettering. An inclination of about 75° from right towards left is recommended by I.S.I.

1. Single-stroke freehand italic gothic lettering : The lettering in which the alphabets are written inclined with a single stroke of pencil or pen without the help of drawing instruments is called a single-stroke freehand italic lettering.

Fig. 4.20. *Single-stroke italic freehand lettering (Height = 4 mm)*

Fig 4.20 shows the single-stroke italic freehand lettering of height equal to 4 mm. Similarly, the single stroke italic freehand gothic lettering of various heights such as 4 mm, 5 mm and 12 mm can be drawn.

Fig 4.21 shows single stroke italic printing of height equal to 5mm.

ENGINEER SAYS, POOR LETTERING MARS
THE APPEARANCE OF PREPARED DRAWING
ON SHEET

Fig. 4.21. *Single-stroke Italic freehand lettering (Height = 5 mm)*

2. Lower case italic freehand lettering : The lettering in which the small size italic alphabets such as a,b,c,....z, etc. written freehand is called lower case italic freehand lettering (see Fig. 4.22).

Fig. 4.22. *Lower case italic freehand lettering*

4.15. MECHANICAL LETTERING DEVICES

The mechanical lettering device is used for reproducing the letters and numbers perfectly and quickly. The most commonly used lettering devices are :—

1. Lettering guide
2. Lettering instrument.

1. Lettering guide : The lettering guide consists of a plastic stencil containing outlines of letters and numerals. The guides are available with letters of various heights.

The letters and numerals are available by placing the lettering guide over the portion of the paper on which the lettering is to be done and trace the outline with a pencil. The guide is moved back and forth along the edge of the T-square (see Fig. 4.23).

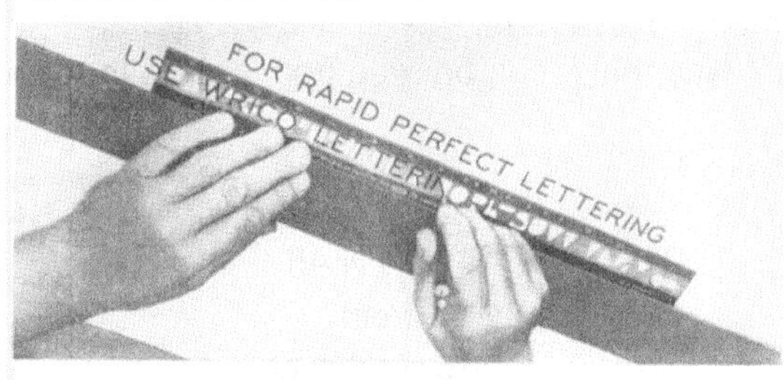

Fig. 4.23. *Lettering guide produces letters when used with pencil*

Fig. 4.24. *Lettering instrument forms letters in ink*

For inking, a special lettering pen is provided. When the pen is moved in contact with the sides of T-guide, letters and numbers are quickly formed without the danger of smearing.

2. Lettering instrument : The lettering instrument is basically an inking lettering device. It consists of a template or guide with groved letters and numbers. The scriber is equipped with a tracer pin that follows the grooved letters on the template and a pen that forms the letters (see Fig. 4.24). The scriber is either of the fixed type for reproducing only vertical letters or adjustable for inclined and vertical letters.

4.16. PENCILS FOR LETTERING

The most important part in lettering is to draw a good black line. This depends upon the selection of a pencil that has proper hardness and on keeping it well sharpened.

HB and H grade pencils sharpened to a conical point should be used for lettering. To keep the stroke of the letters uniform, the pencils should be rotated between the thumb and fingers while lettering. Hard pencils such as 2H or 3H should be used to draw guide lines.

QUESTIONS FOR SELF EXAMINATION

1. What are the main requirements of lettering?
2. What do you mean by composition of letters?
3. What do you understand by uniformity in lettering?
4. What do you mean by normal, compressed and extended lettering?
5. What are guide lines and why they are necessary in lettering?
6. What do you mean by single stroke letters?
7. Which lettering is in universally used throughout the technical world now-a-days?
8. What is the gothic and roman lettering?
9. What do you mean by mechanical lettering devices?
10. What should be the grade of pencil used for lettering?
11. Fill in the blanks with appropriate words selected from list B
    (*a*) The art of writing of title, notes etc on a drawing is called ......
    (*b*) A good lettering is achieved by continuous ....
    (*c*) The lines which are used to regulate the height of letters are known as ....
    (*d*) All letters should be uniform in ......, ..., ..., and .....

(*e*) The size of letter means the .... of the letter.
(*f*) Lettering is generally done in ...... letters.
(*g*) The lettering in which the direction of alphabets is ....., called upright gothic lettering.
(*h*) The lettering in which the direction of alphabets is at .... is called italic lettering.
(*i*) Two types of single stroke letters are ..... and .....
(*j*) Main title of drawing is generally written in ..... letters.
(*k*) Lettering should be written in such a way that it can be read from .... with main title ......
(*l*) Roman lettering is used in .... drawings.

List B

1. vertical	7. height	13. inclined
2. capital	8. shape	14. front
3. practice	9. lettering	15. Architect
4. guide lines	10. size	16. Gothic
5. 75°	11. slope	17. horizontal
6. vertical	12. spacing	

Ans (*a*) 9, (*b*) 3, (*c*) 4, (*d*) 10,8,11,12, (*e*) 7, (*f*) 2, (*g*) 1, (*h*) 5, (*i*) 6,13, (*j*) 16, (*k*) 14,17 (*l*) 15.

PROBLEMS FOR PRACTICE

1. In guide lines, write the following sentences in single stroke vertical letters taking height 30 mm and using ratio 7 : 4 and 6 : 5 :

   ENGINEERS SAY; "POOR LETTERING MARS THE APPEARANCE OF DRAWING".
2. Write following sentence in freehand, single stroke, inclined, upper case letters in 8mm height.
   "INDIAN ATHLETES MUST GIVE AN EXEMPLARY PERFORMANCE IN ASIAN GAMES"
3. Write in double stroke vertical and inclined style, the following statements using ratio 7 : 4 and 5 : 4 **(*i*) WORK IS WORSHIP**
   (*ii*) BEAUREO OF INDIAN STANDARDS
   (*iii*) INDIAN STANDARDS INSTITUTE.
4. Write freehand in single stroke capital letters and inclined letters, using the ratio (*i*) 6 : 5 (*ii*) 7 : 5 and height of letters 10mm, the following statements:-
   (*a*) Drawing is the graphic language of engineers
   (*b*) Laboratory is a temple where search for truth is made.
5. Write in 12 mm hight upper case upright single stroke, instrument letters **"GOVERNOR"**
6. Print freehand in vertical and inclined lower case letters, taking height of letters as (*i*) 10 mm and (*ii*) 6 mm, the statements :

   Importance of Technical Education in India
7. Print vertical and inclined numerals, taking height of numerals as (*i*) 10 mm, (*ii*) 6 mm and (*iii*) 3 mm.

CHAPTER

5

Dimensioning

INTRODUCTION

The main aim of drawing is to represent the correct size of the object to be manufactured or constructed. It is, therefore, necessary that the drawing must carry the proper dimensions and other information of the various parts of the object. The correct dimensioning facilitates the technician in manufacturing the part, whereas, the wrong dimensioning may cause confusion and lead to a great loss of time, labour and material. Therefore, dimensions should be done in such a way which is located simply and easily understood by the technician.

In this chapter, we will deal with the technique of dimensioning the drawing in engineering practice.

5.1. DIMENSIONING

The art of writing the various sizes or measurements on the finished drawing of an object is known as dimensioning.

Dimensioning expresses all the sizes and other information necessary to define the object completely. It must be done with due regard to manufacturing processes and inspection requirements. The dimensioning also includes expression of tolerances necessary for the correct functioning of the part given to be assembled.

5.2. NOTATION OF DIMENSIONING

The notation of dimensioning consists of dimension lines, extension lines, arrow heads, dimension figures, notes, symbols, etc. These notations are explained below :

1. Dimension line : Dimension line is a thin continuous line used to indicate the measurement of an object. It is shown by figure in a space above the dimension line or a space left in the dimension line (see Fig. 5.1).

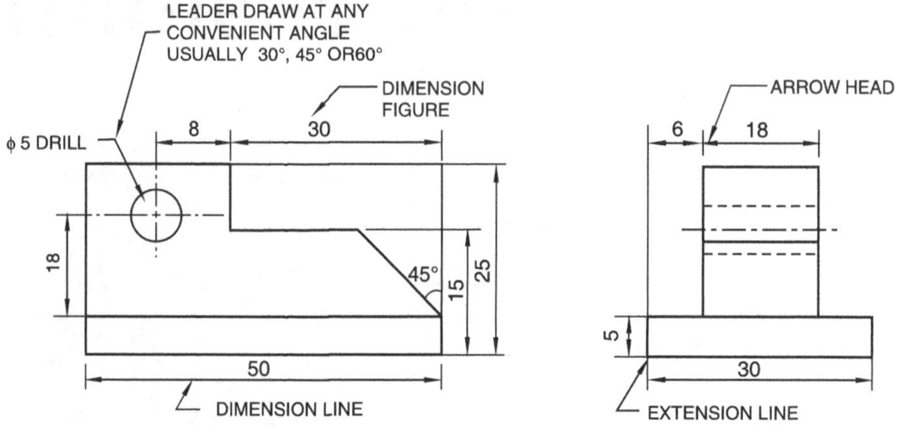

Fig. 5.1. *Notation of dimensioning*

2. Extension line : Extension line is also a thin continuous line extending beyond the outline of the object. It should extend about 3 mm beyond the dimension line (see Fig. 5.1).

3. Arrow head : Arrow heads are used to terminate dimension lines. These touch the extension lines and indicates the extent of a dimension. The length of the arrow-head is about three times its width. The space in the arrow-heads should be filled in (see Fig. 5.2). The size of the arrow-heads should be proportionate to the thickness of the lines of the drawing (see Fig. 5.3).

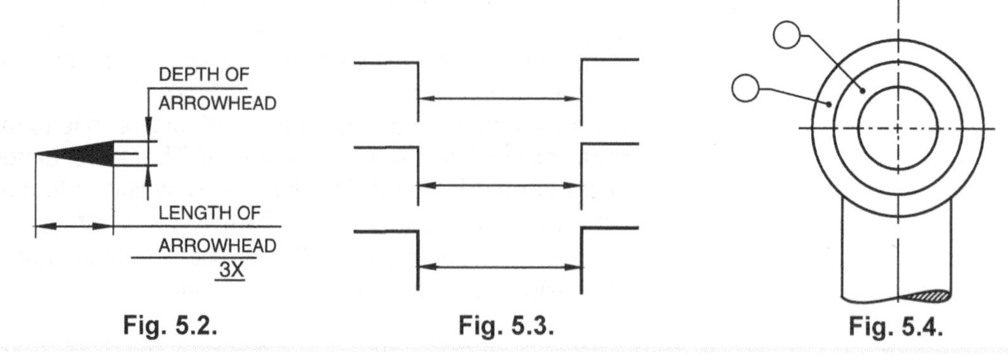

Fig. 5.2. **Fig. 5.3.** **Fig. 5.4.**

Note : *Where the space is limited, adjacent arrow-heads may be replaced by dots.*

4. Dimension figure : A numeral that indicates the size of a particular feature of an object is called dimension figure (see Fig. 5.1).

5. Leader (pointer line) : A leader is a thin continuous line drawn from note of the figure to show where it applies. It is terminated by an arrow-head or a dot (see Fig. 5.4). The arrow-head touches the outline, whereas the dot is placed within the outline of the object.

The leader is generally drawn at any convenient angle, usually 30°, 45°, and 60° but of not less than 30°.

Note : *The use of long leaders should be avoided even if it involves repetition dimensions or notes.*

6. Notes : A note on drawing gives complete information regarding specific operation relating to a feature. It is generally placed outside a view and read in such a way that the drawing is viewed from the bottom edge (see Fig. 5.1).

7. Symbol : A symbol is the representation of any object by some mark on the drawing. It is used to save time and labour of drawing work.

5.3. THEORY OF DIMENSIONING

An object may be considered to be made up of a number of geometrical shapes such as prism, cylinder, pyramid, cone, sphere etc. It then becomes very simple to dimension these geometric forms in a manner that will show their individual sizes and location to each other. Thus, the following two types of dimensions are commonly used in engineering drawing :

1. Size dimensions
2. Location dimensions

1. Size dimensions : The dimensions which indicate the various sizes of the object such as length, breadth, diameter, etc. are known as size dimensions. These dimensions are represented by the letter S' as shown in Fig. 5.5

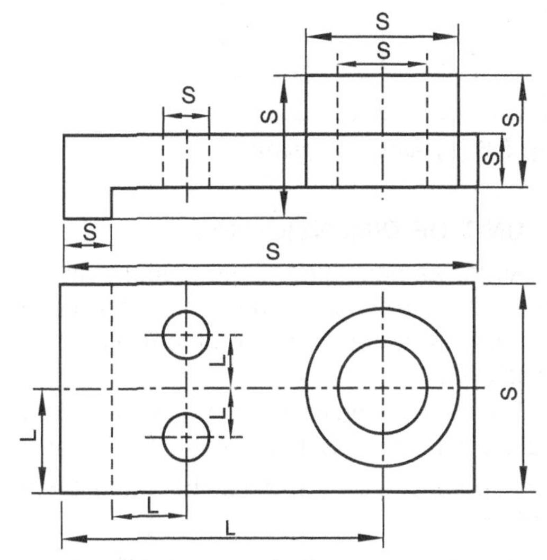

Fig. 5.5. *Size and location dimensions*

2. Location dimensions : The dimensions which locate the position of one feature with respect to the other feature are known as location dimensions. Distances between the centre lines of the holes from the edges of features are given by location dimensions. These dimensions are marked by letter 'L' as shown in Fig. 5.5.

5.4. SYSTEM OF PLACING DIMENSIONS

The following are the two recommended systems of placing dimensions on a finished drawing according to I.S.I. as given by :

1. Aligned system
2. Unidirectional system.

1. Aligned system : In this system, all dimensions are so placed that they may be read from the bottom or the right hand edges of the drawing sheet [see Fig. 5.6 (a)]. Here all the dimensions are placed normal and above the dimension lines.

Aligned system of placing dimensions is commonly used in engineering drawing.

2. Unidirectional system : In this system, all dimensions are so placed that they may be read from bottom edge of the drawing sheet [see Fig. 5.6 (b)]. Note, there is no restriction on controlling the direction of the dimension lines.

Unidirectional system of placing dimensions are used on large drawings as of air-crafts, automobiles, etc., where it is convenient to read dimensions from right hand side.

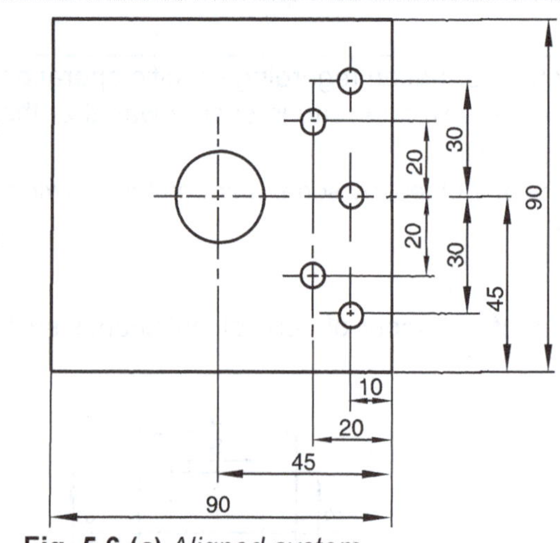

Fig. 5.6.(*a*) *Aligned system*

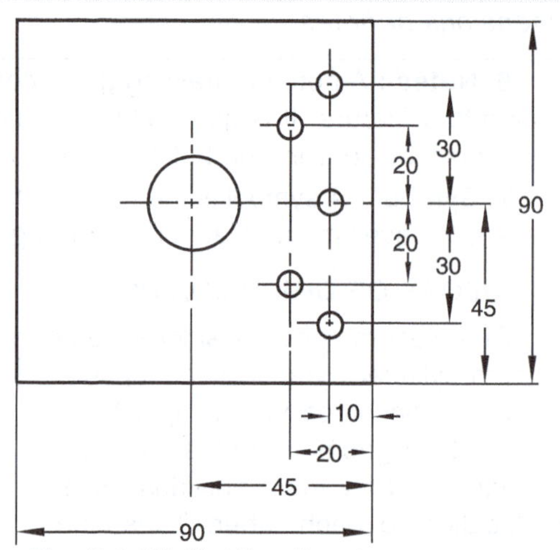

Fig. 5.6.(*b*) *Unidirectional system*

5.5. UNIT OF DIMENSIONING

As far as possible, all dimensions should be given preferably in one unit only, i.e., in millimetres. The symbol of unit "mm" can, therefore, be omitted while writing each dimensions but a note is added in a prominent place near the title block that "All dimensions are in millimetres".

The decimal point in a dimension should be in line with the middle of the figure. When the dimension is less than unity, I.S.I. has recommended that zero should preceed the decimal point e.g. 0·72 and may be omitted when the decimal point is preceeded by a symbol i.e. 45.56 + .76.

5.6. GENERAL RULES FOR DIMENSIONING

The following general rules should be followed for dimensoning a drawing :

1. All the dimensions necessary for the correct functioning of the part should be expressed directly on the drawing.
2. Dimensions should be given on the view which shows the relevant features most clearly.
3. Dimensions marked in one view need not be repeated in another view except for reference purposes.

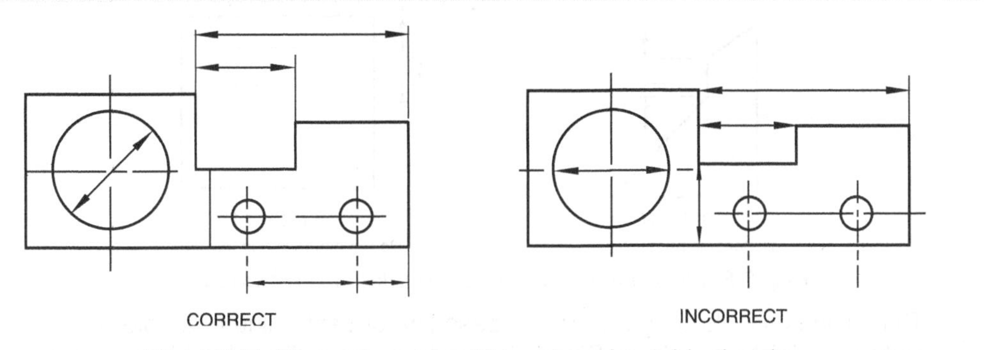

Fig. 5.7.(i) *Dimensions should be placed outside the view*

4. Dimensions should be placed outside the view, as far as possible (see Fig. 5.7). Fig. 5.7 (ii) shows the correct and incorrect ways of dimensioning.

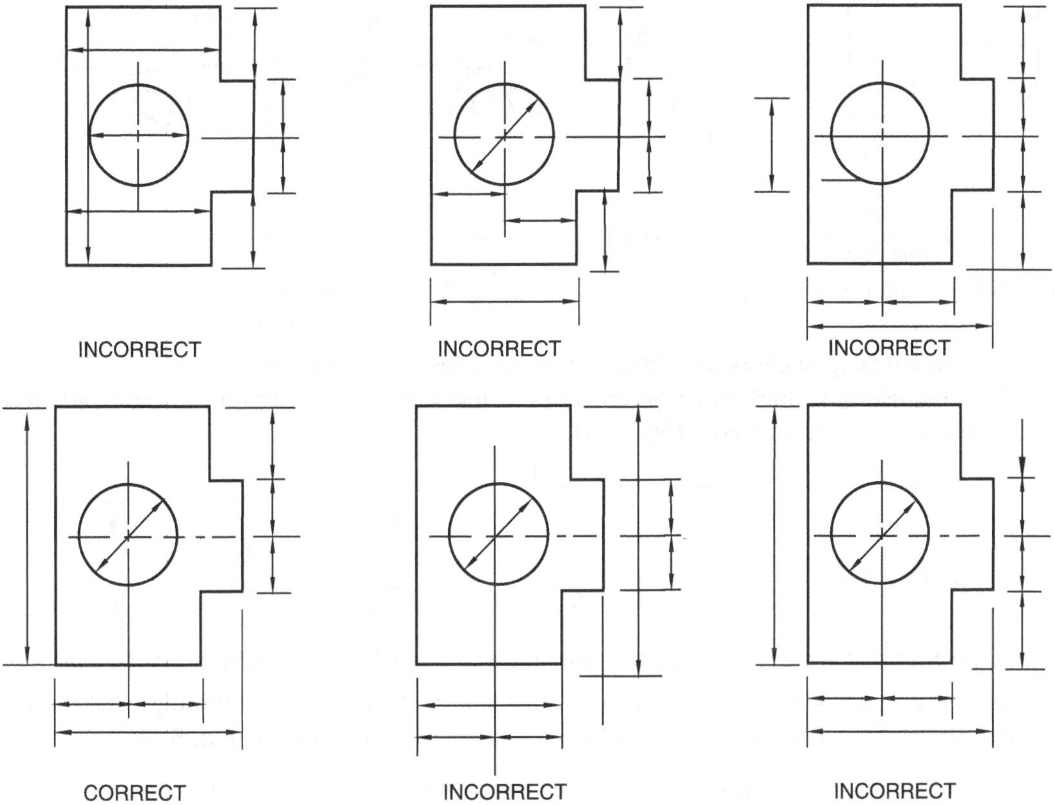

INCORRECT INCORRECT INCORRECT

CORRECT INCORRECT INCORRECT

Fig. 5.7.(ii)

5. Dimensions should be taken from visible outlines rather than from hidden lines (see Fig. 5.8).

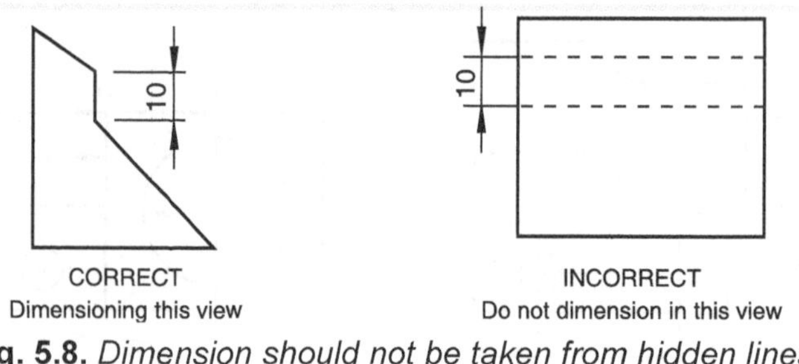

Fig. 5.8. *Dimension should not be taken from hidden lines*

6. Dimensions should be given from a base line or centre line of a hole or an important hole or a finished surface. Dimensioning to a centre line should be avoided, except when the centre line passes through the centre of a hole (see Figs. 5.9 and 5.10).

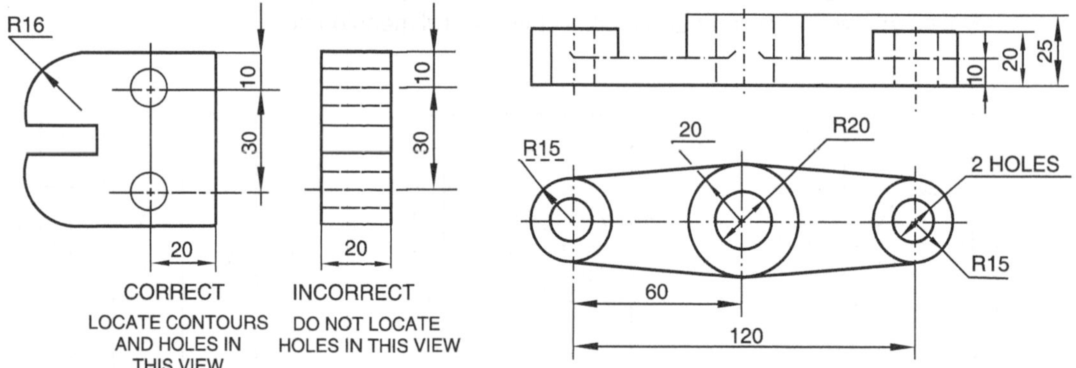

Fig. 5.9. *Location of position holes*

Fig. 5.10. *Dimensioning to the centre line of a hole is permitted*

7. The crossing of dimension lines should be avoided, if possible.
8. Dimensions should never be crowded. If the space is insufficient, arrow-heads may be replaced by dots (see Fig. 5.11).

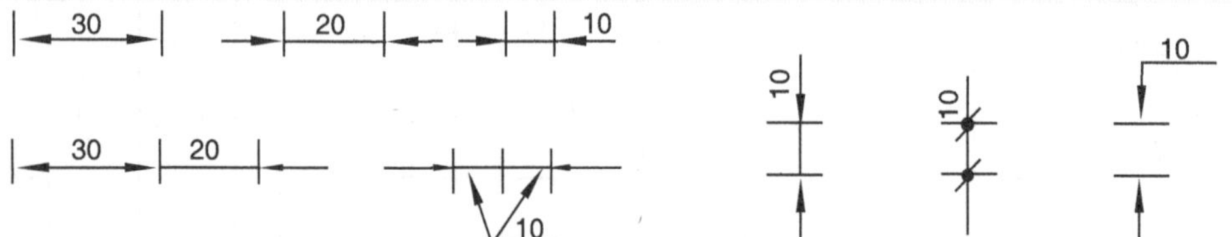

Fig. 5.11. *Different ways of dimensioning distances when the space is insufficient*

9. As far as possible, dimensions should be given in one unit only, perferably in millimetres.
10. Circles of different sizes should be dimensioned as illustrated in Fig. 5.12

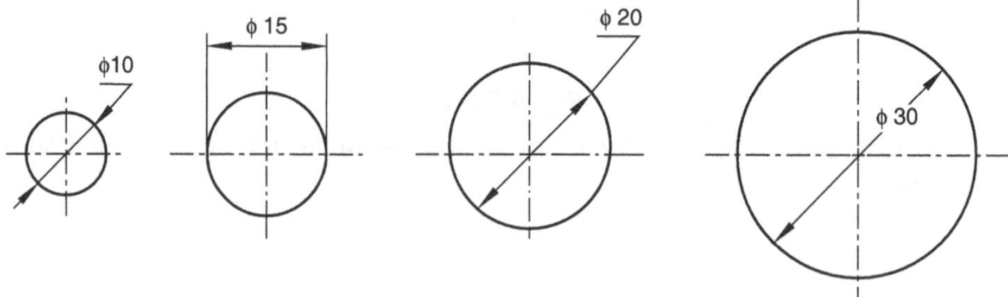

Fig. 5.12. *Dimensioning of circles of different sizes*

When the space for dimensioning is restricted, one of the methods (see Fig 5.13 and Fig. 5.14) should be used.

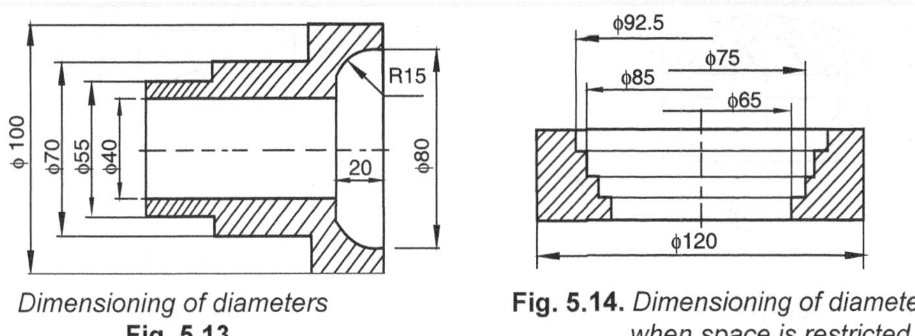

Dimensioning of diameters
Fig. 5.13.

Fig. 5.14. *Dimensioning of diameters when space is restricted*

11. Arcs of circles should be dimensioned by their respective radii (see Fig. 5.15).

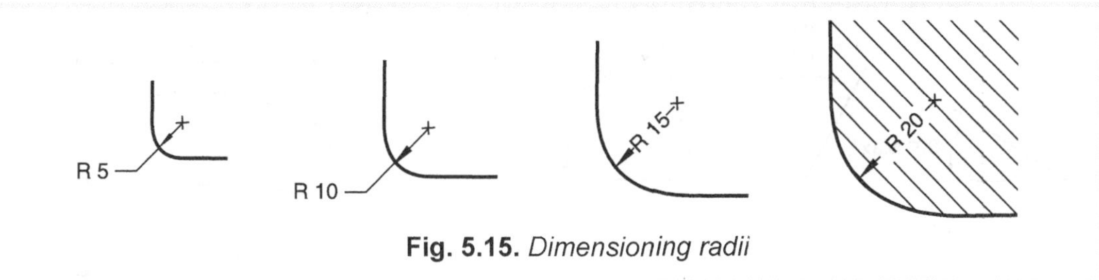

Fig. 5.15. *Dimensioning radii*

While dimensioning small radii, the arrow should be reversed as shown in Fig. 5.16.

Radii of arcs, the centres of which need not be located, should be dimensioned as shown in Fig. 5.17.

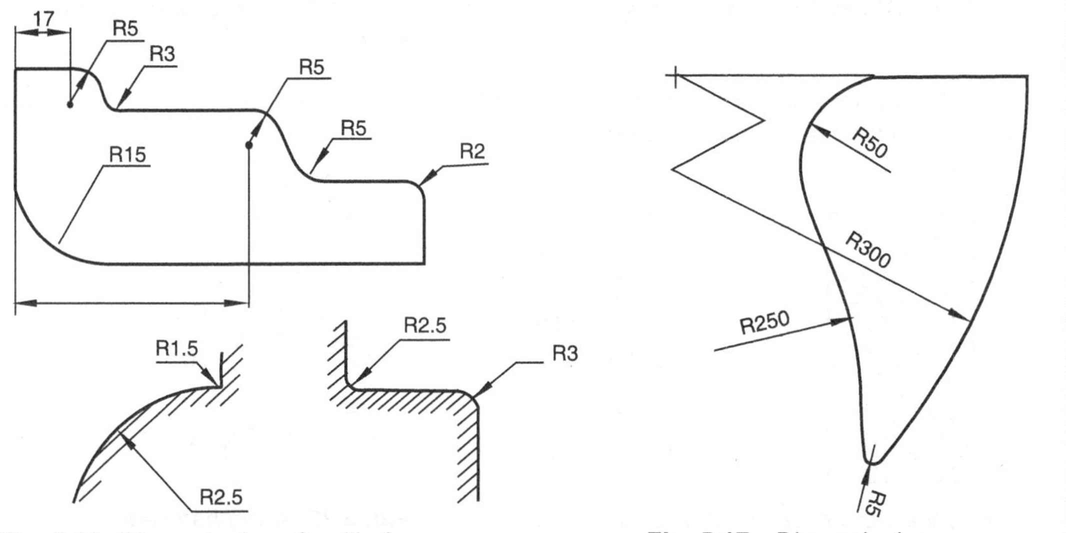

Fig. 5.16. *Dimensioning of radii of arc, the centre of which need not be located*

Fig. 5.17. *Dimensioning of radii of arcs*

12. Radii of a spherical surfaces should be given as shown in Fig. 5.18.

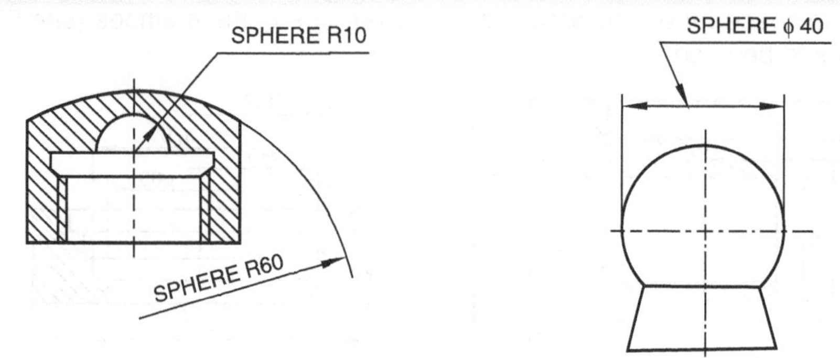

Fig. 5.18. *Dimensioning of diameter of spherical object parts*

13. Dimensions of angles, chords and arcs should be given as shown in Fig. 5.19 and Fig. 5.20.

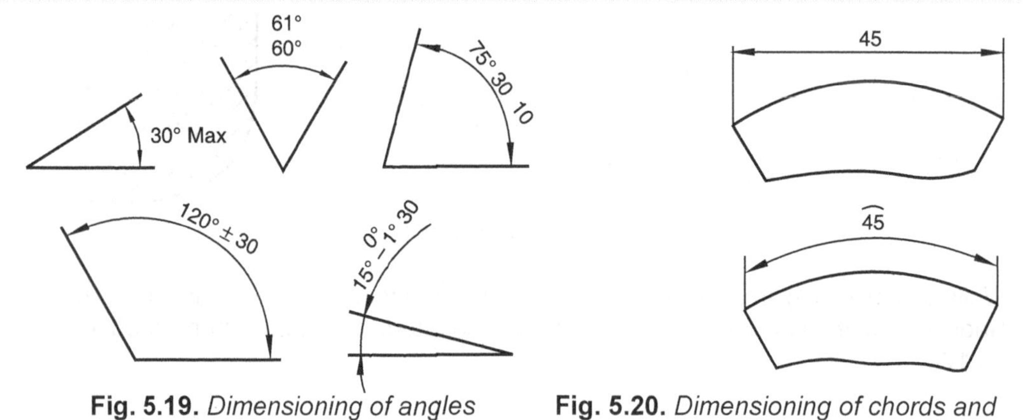

Fig. 5.19. *Dimensioning of angles* **Fig. 5.20.** *Dimensioning of chords and arcs*

14. Dimensions on a sectioned view should be given as shown in Fig. 5.21. A small portion around the figure should be left unsectioned.

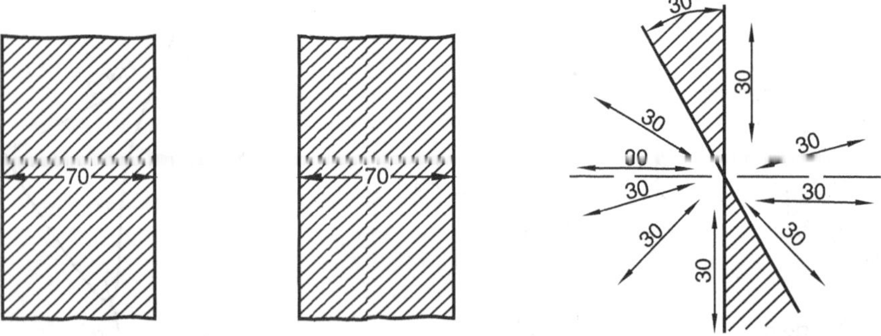

Fig. 5.21. *Dimensions on a sectional view* **Fig. 5.22.** *Avoid hatched area for dimensioning*

15. Avoid hatched area while giving dimensions to circles (see Fig. 5.22).
16. Dimensions of parts which are not drawn to scale should be underlined. However, when it is required to draw the whole drawing not to scale, the abbreciation 'NTS' may be used or the column of scale be scored off.

17. While dimensioning an object, the use of long leaders should be avoided (see Fig. 5.23).

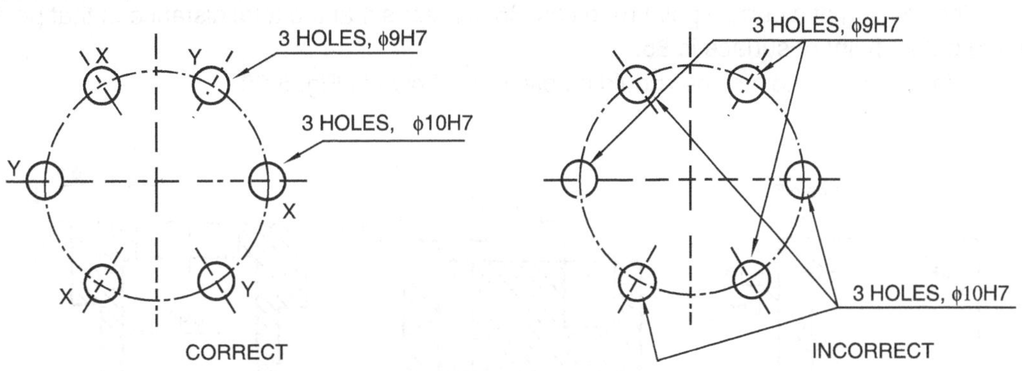

Fig. 5.23.

18. Dimensions in a series may be given on the views in any one of the following arrangements :

(i) Chain dimensioning : In this system, dimensions are arranged in a straight line (see Fig. 5.24). When overall dimension is given, one of the smaller or least important dimension is generally omitted.

(ii) Parallel dimensioning : In this arrangement, all the dimensions are given from a common base line. The smaller dimensions are placed nearer the view and the larger further away, so that the extension lines do not cross dimension lines (see Fig. 5.25).

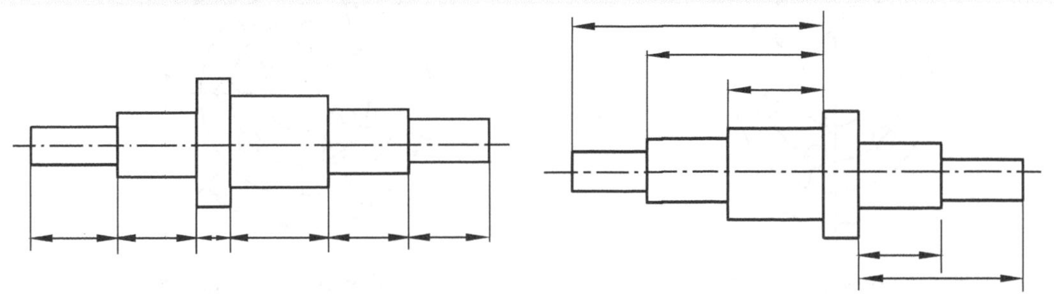

Fig. 5.24. *Chain dimensioning* **Fig. 5.25.** *Parallel dimensioning*

(iii) Combined dimensioning : Combined dimensioning is the result of the simulataneous use of chain and parallel dimensioning (see Fig. 5.26).

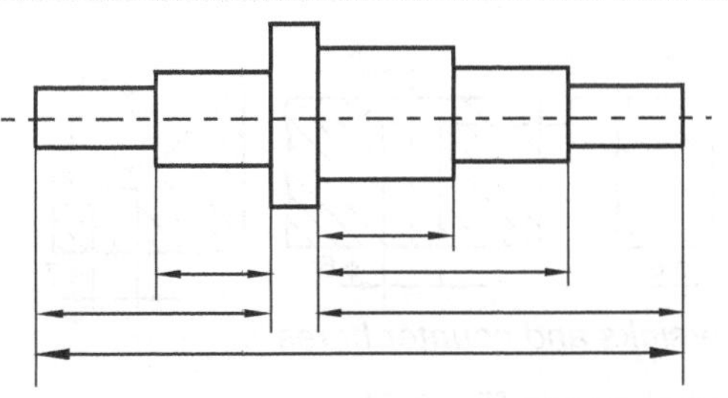

Fig. 5.26. *Combined dimensioning*

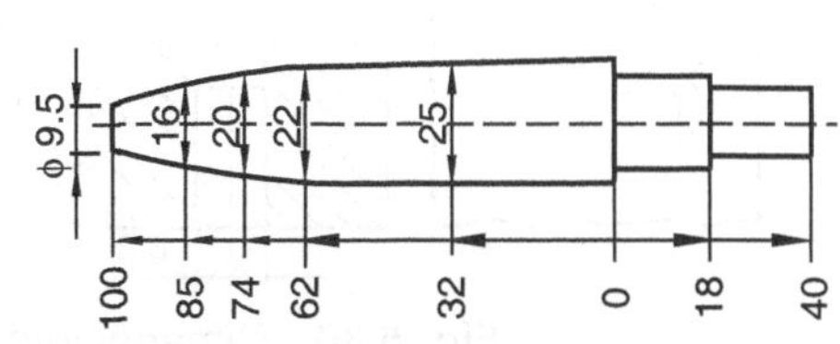

Fig. 5.27. *Progressive dimensioning*

(iv) Progressive dimensioning : In this arrangement, one datum point or surface is selected which reads as zero. All the dimensions are referred to that point or surface (see Fig. 5.27). If the reading at a certain point reads as 85, it means that the total distance of that point from the datum point or surface is 85.

19. Dimensions of hole sizes should be given as shown in Fig. 5.28.

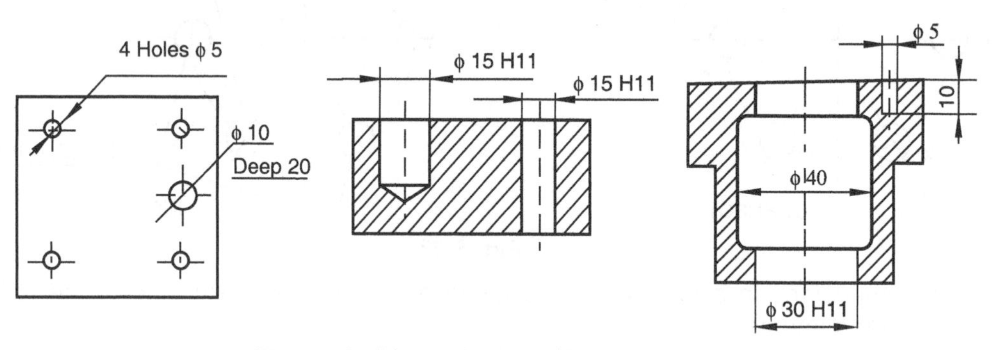

Fig. 5.28. *Dimensioning of hole sizes*

20. The position of holes should be dimensioned as shown in Fig. 5.29.

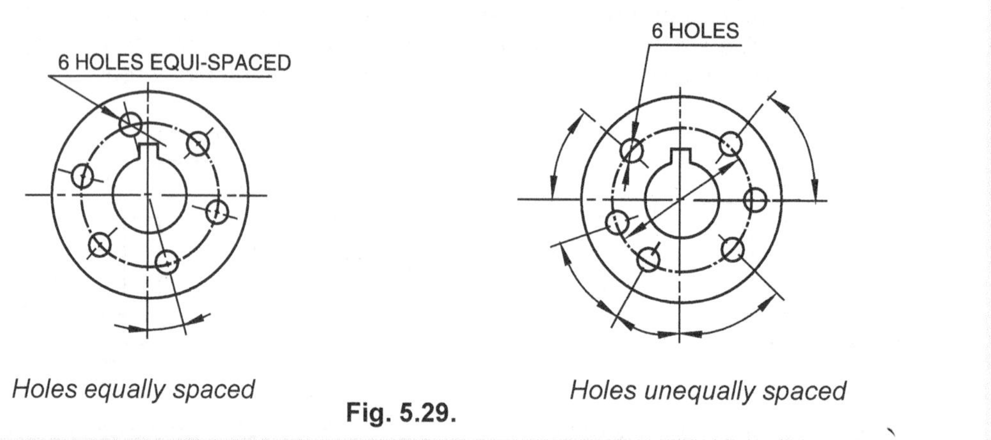

Fig. 5.29.

21. The counter sinks and counter bores should be dimensioned by giving maximum diameter and the included angle as shown in Fig. 5.30.

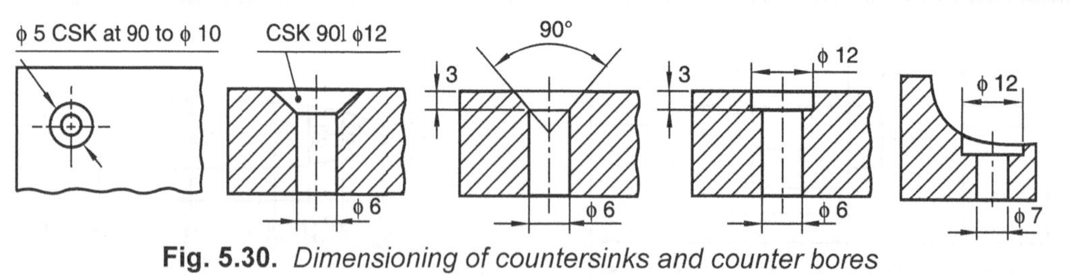

Fig. 5.30. *Dimensioning of countersinks and counter bores*

22. The bent up parts should be dimensioned as shown in Fig. 5.31.

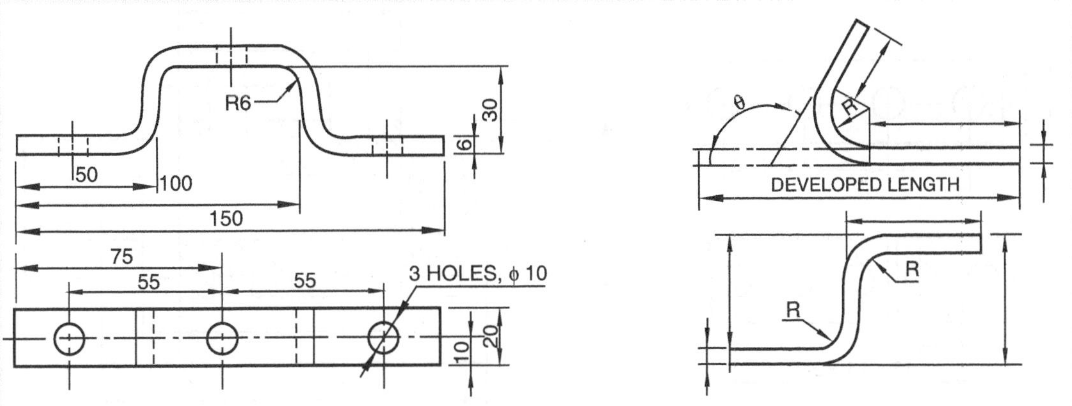

Fig. 5.31. *Dimensioning of bent up parts*

23. The chamfer should be dimensioned by the length and angle of chamfer (see Fig.5.32).

Fig. 5.32. *Dimensioning of chamfer* **Fig. 5.33.** *Addition of letters and symbols*

24. The addition of letters and symbols should be used before dimensioning as shown in Fig. 5.33.
25. Dimensions for specifying size and form of tapered features should be given as shown in Fig. 5.34.

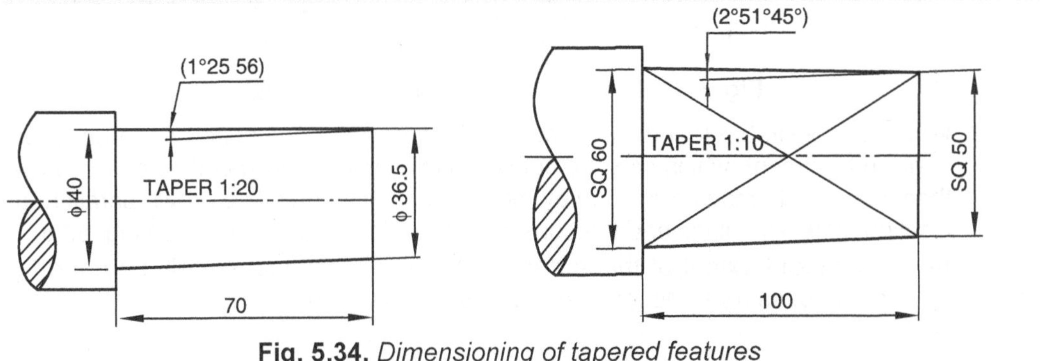

Fig. 5.34. *Dimensioning of tapered features*

26. When equidistance or regularly arranged elements appear on a drawing, it should be dimensioned as shown in Fig. 5.35.
27. When several parts are drawn in assembly, the groups of dimensions related to each other part should be kept separately as shown in Fig. 5.36.

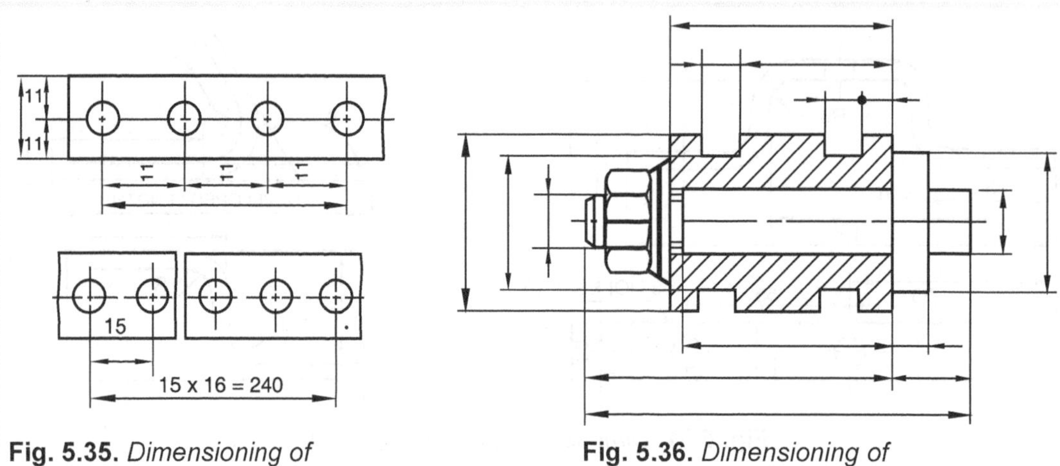

Fig. 5.35. *Dimensioning of equidistance features*

Fig. 5.36. *Dimensioning of assembly parts*

28. The method of dimensioning geometrical shapes are shown in Fig. 5.37.

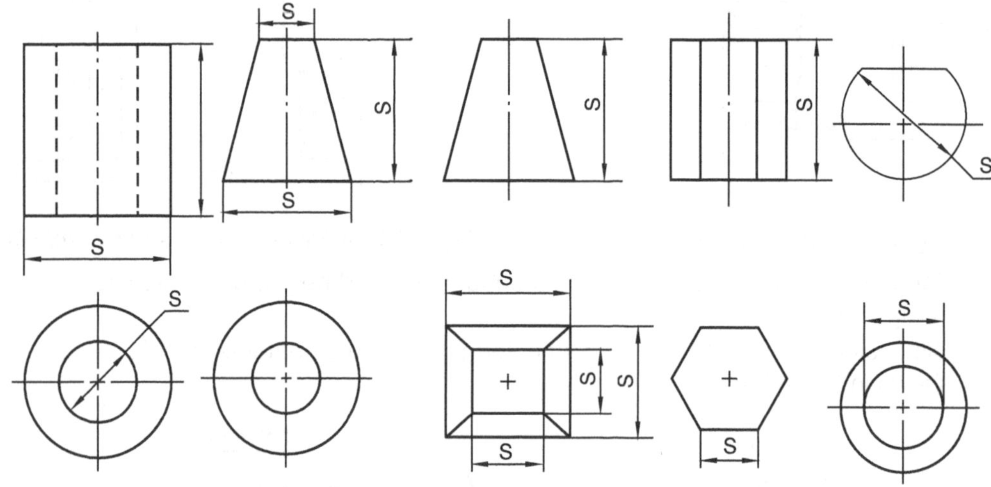

Fig. 5.37. *Dimensioning geometrical shapes*

29. When it is desirable to indicate that a surface or a surface zone has to be given an addition treatment, which shall be applied within limits to be specified on the drawing, then these limits may be defined by means of a thick chain line drawn paralled to the locating dimension lines and the corresponding dimension added (see Fig. 5.38).

If the location and extent of the surface to be treated appear clearly from the figure drawing, it is not necessary to dimension them (see Fig. 5.39).

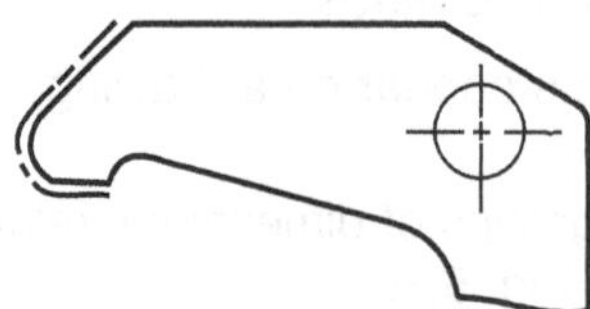

Fig. 5.38. *Surface required additional treatment – Dimensions required*

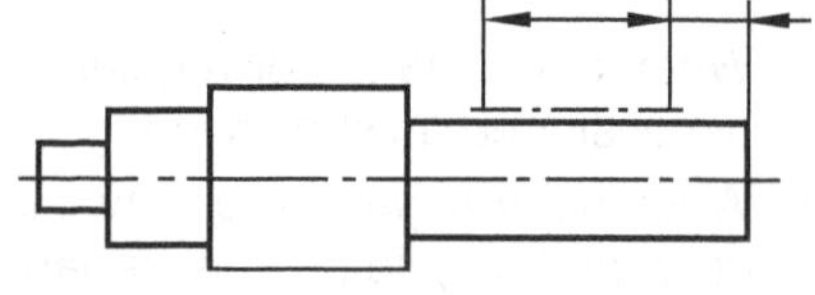

Fig. 5.39. *Surface required additional treatment – No dimension required*

Problem 1: Complete the symmetrical view of the template shown in Fig. 5.40 and dimensions on the template fully.

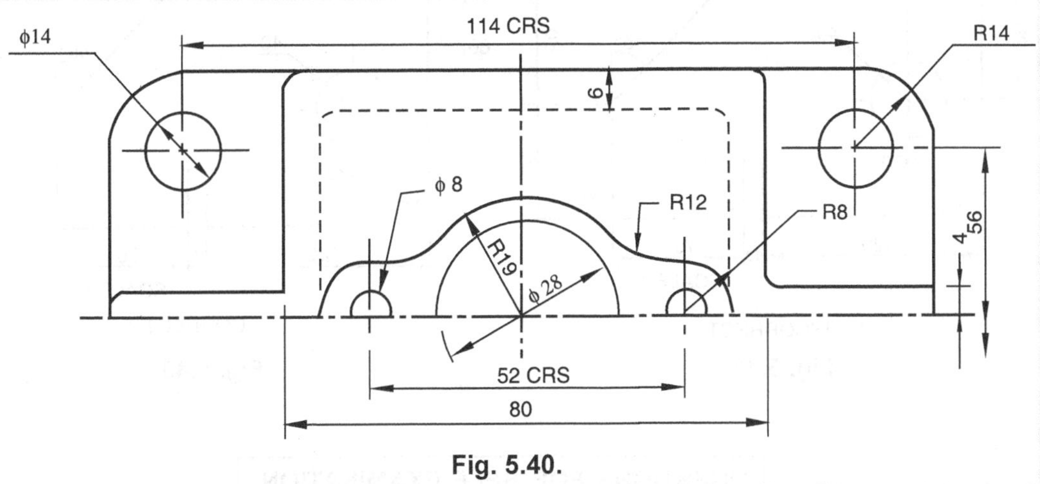

Fig. 5.40.

Solution : For its solution, see Fig. 5.41.

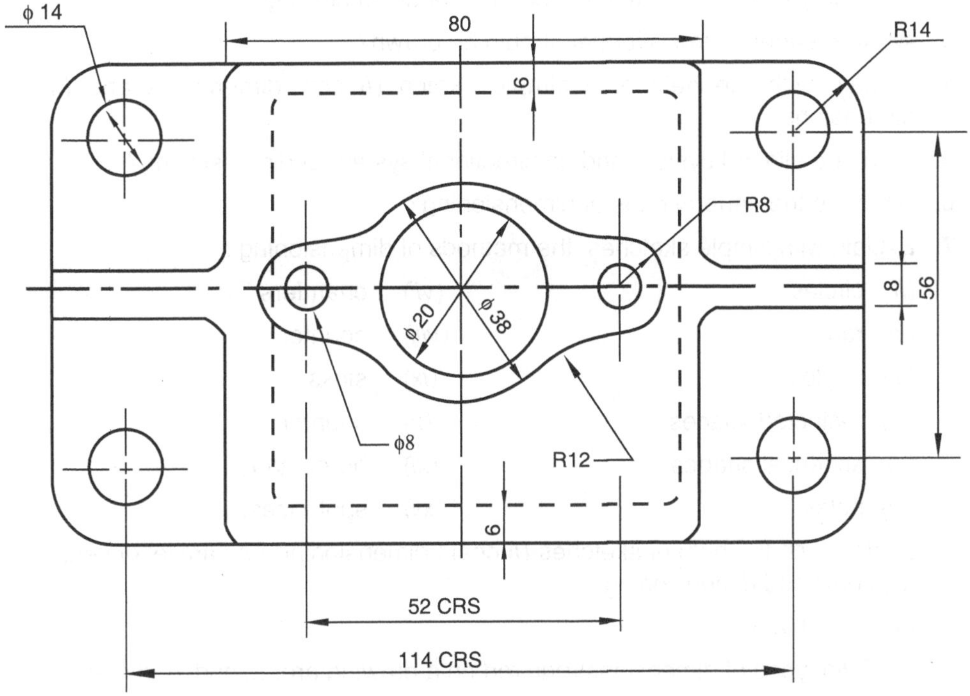

Fig. 5.41.

Problem 2: Read the dimensioned drawing as shown in Fig. 5.42. Redraw the figure and dimension it in correct way as per Indian Standard.

Solution : For its solution, see Fig. 5.43.

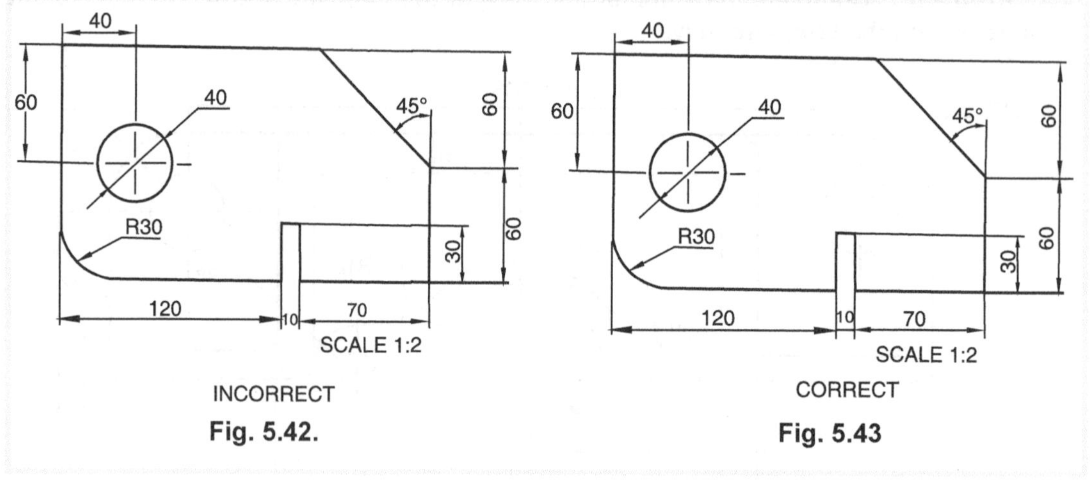

INCORRECT
Fig. 5.42.

CORRECT
Fig. 5.43

QUESTIONS FOR SELF EXAMINATION

1. What is the importance of dimensioning?
2. What do you mean by the term 'notation of dimensioning' ?
3. What is leader ? How a leader should be drawn?
4. Explain with the help of a simple sketch (*i*) size dimensions and (*ii*) location dimensions?
5. What are aligned system and unidirectional system of dimensioning?
6. What are the general rules of dimensioning?
7. Explain with simple sketches, the methods of dimensioning :

(*i*)	circles	(*vii*)	chamfers
(*ii*)	radii	(*viii*)	counter
(*iii*)	angles	(*ix*)	sinks
(*iv*)	restricted spaces	(*x*)	counter
(*v*)	spherical shapes	(*xi*)	bores and
(*vi*)	holes	(*xii*)	spot faces.

8. Explain with the help of sketches (*i*) chain dimensioning (*ii*) parallel dimensioning and (*iii*) combined dimensioning.
9. Fill in the blanks :

   (*i*) Two types of dimensions required on a drawing are..... and .... dimensions.

   (*ii*) Two systems of placing dimensions on a drawing work are .... and ..... systems.

   (*iii*) As far as possible, dimensions should be given in.... unit only, perferably in.....

   (*iv*) Dimensions should be taken from visible.... rather than from hidden lines.

   (*v*) Dimension lines should not ... each other.

   (*vi*) The extension lines should extend about .... mm beyond......

(*vii*) Dimension lines should be drawn about 8 mm away from the ... and from ......

(*viii*) The line connecting a view to note is called a ....

Ans. (*i*) *size,* location (*ii*) aligned, unidirectional (*iii*) one, mm (*iv*) outlines (*v*) cross (*vi*) 3, dimension line (*vii*) outlines, each other (*viii*) leader.

PROBLEMS FOR PRACTICE

1. Fig. 5.44 shows the front view of a object. Copy the front view only and give all necessary dimensions.

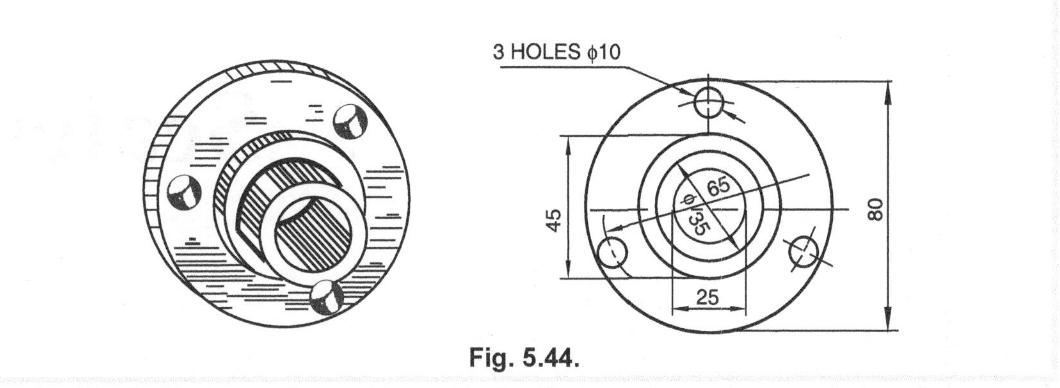

Fig. 5.44.

2. Fig. 5.45 shows views of two objects. Redraw these views and give all necessary dimensions by actual measurement.

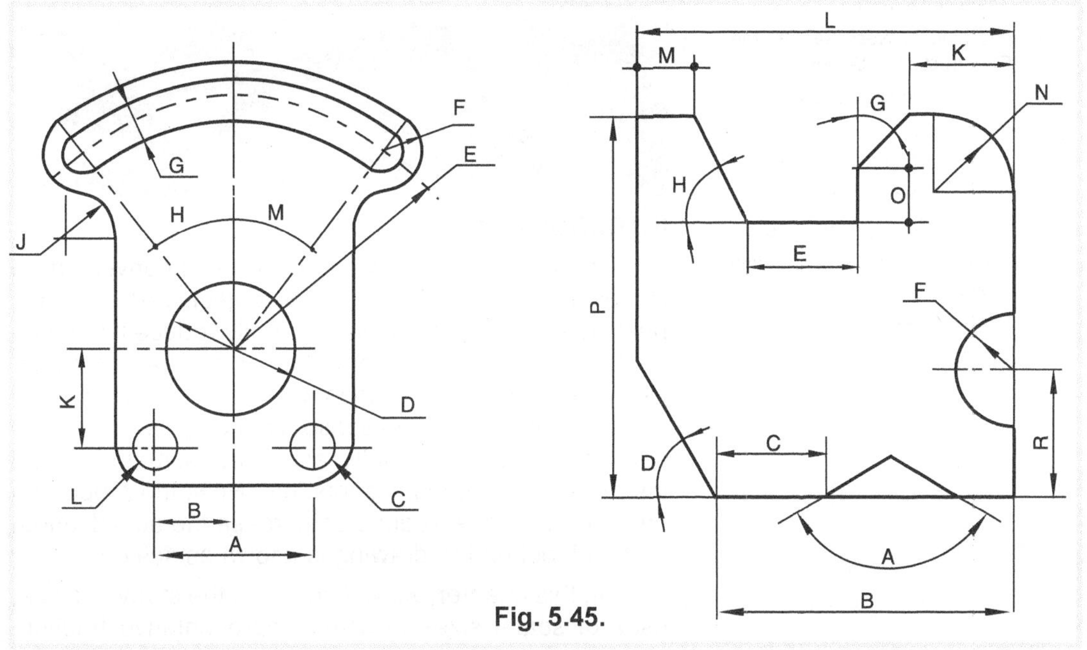

Fig. 5.45.

CHAPTER

6

Scales

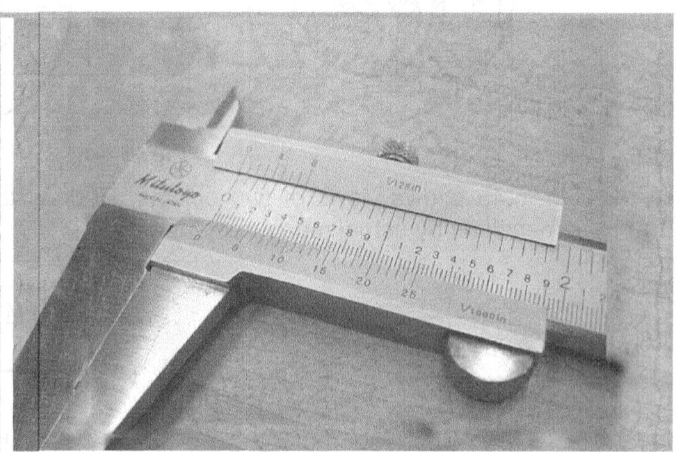

INTRODUCTION

In engineering drawing, it is often inconvenient to represent the actual size of the object. For example, building on the drawing, we generally reduce it to some proportion in order to accomodate on the drawing sheet. However, sometime, it becomes essential to increase the actual size of the object so as to give clear conception, as in the case of small machine parts, mathematical instruments, watches, insects, etc. Now, the proportion by which we either reduce or increase the actual length of the object on the drawing is known as scale.

In this chapter, we will deal with the study of scale, uses of scale, sizes of scale, representative fraction, classification of scale, etc.

6.1. SCALE

The proportion by which we either reduce or increase the actual size of the object on a drawing is known as drawing to scale or simply scale.

The scale is actually a measuring stick, graduated with different divisions to represent the corresponding actual distances according to some proportion, thus giving rapidity in marking off distances on drawing. Numerically scales indicate the relation between the dimensions on the drawing and actual dimensions of the object.

The scales are either flat or triangular and the material used in their manufacturing may be wood, celluloid, metal, etc. In drawing, scale should not be selected arbitrarily, but standard recommended scales should be adopted as far as possible.

6.2. USES OF SCALE

The following are the main uses of scale in engineering practice :

1. The scales are used to prepare reduced or enlarged size drawings.
2. The scales are used to set off dimensions.
3. The scales are used to measure distances directly.

6.3. SIZES OF SCALE

The following are the sizes of scale which are used in engineering practice :

1. Full size scale
2. Reducing scale
3. Enlarging scale

1. Full size scale : *The scale in which the actual measurements of the object are drawn to same size on the drawing is known as full size scale.* It is written on the stick as under:

1 : 1 — drawing made to actual size

2. Reducing scale : *The scale in which the actual measurements of the object are reduced to some proportion is known as reducing scale.* The standard reducing proportion are:

1 : 2	—	drawing made to one half of the actual size
1 : 5	—	drawing made to one fifth of the actual size
1 : 10	—	drawing made to one tenth of the actual size
1 : 20	—	drawing made to one twentieth of the actual size
1 : 50	—	drawing made to one fiftieth of the actual size
1 : 100	—	drawing made to one hundreth of the actual size.

3. Enlarging scale : *The scale in which the actual measurements of the object are increased to some proportion is known as enlarging scale.* The standard proportions are :

2 : 1	—	drawing made to twice the actual size
5 : 1	—	drawing made to five times the actual size
10 : 1	—	drawing made to ten times the actual size.

6.4. UNITS OF MEASUREMENT

The Indian Standard Institute has introduced the metric system of measurements in place of British measures for its practical convenience and other reasons. As in this system of measurements, every successive unit is ten time the proceding one. The following are the basic units of measurements alongwith their symbols :

METRIC MEASURES

(I) Linear Measures

10 millimetres (mm)	=	1 centimetre (cm)
10 centimetres (cm)	=	1 decimetre (dm)
10 decimetres (dm)	=	1 metre (m)
10 metres (m)	=	1 decametre (dam)
10 decametres (dam)	=	1 hectometre (hm)
10 hactometres (hm)	=	1 Kilometre (km)

(II) Square Measures

100 square millimetres (mm^2)	=	1 square centimetre (cm^2)
100 square centimetres (cm^2)	=	1 square decimetre (dm^2)
100 square decimetres (dm^2)	=	1 square metre (m^2)
100 square metres (m^2)	=	1 acre (a)
100 square acres (a)	=	1 hectare = 10^4 square metres
100 hectares	=	1 square kilometre = 10^6 sq. metres

(III) Cubic Measures

1000 cubic millimetres (mm^3)	=	1 cubic centimetre (cm^3)
1000 cubic centimetres (cm^3)	=	1 cubic decimetre (dm^3)
1000 cubic decimetres (dm^3)	=	1 cubic metre (m^3)

BRITISH MEASURES

(I) Square Measures

144 square inches	=	1 square foot
9 square feet	=	1 square yard
4840 square yards	=	1 acre
640 acres	=	1 square mile

(II) Cubic Measures

1728 cubic inches	=	1 cubic foot
27 cubic feet	=	1 cubic yard

(III) Linear Measures

12 inches	=	1 foot
3 feet	=	1 yard
220 yards	=	1 furlong
8 furlong	=	1 mile

6.5. REPRESENTATIVE FRACTION (R.F.) OR SCALE FACTOR (S.F.)

The ratio of the distance on the drawing sheet of an object to the corresponding actual distance of the object is known as representative fraction or scale factors.

$$\text{R. F.} = \frac{\text{Distance of the object on drawing sheet}}{\text{Corresponding actual distance of the object}} \quad \text{[in same units]}$$

For examples :

1. If 1 centimetre represents 1 metre, than $\text{R.F.} = \frac{1\text{ cm}}{1\text{ m}} = \frac{1}{1 \times 10 \times 10} = \frac{1}{100}$ (same units).

The scale used now is written as 1 cm = 1 metre (m) or 1:100 and indicates that each line on the drawing is 1/100 of its actual length.

2. If 1 centimetre represents 1 kilometre, then

$$\text{R.F.} = \frac{1\text{ cm}}{1\text{ km}} = \frac{1}{1 \times 10 \times 10 \times 10 \times 10 \times 10} = \frac{1}{1{,}00{,}000} \quad \text{or } 1:1{,}00,$$

3. If 2.5 centimetres represent 5 hectometres, then

$$\text{R.F.} = \frac{2.5\text{ cm}}{5\text{ hectometres}} = \frac{2.5}{5 \times 10 \times 10 \times 10 \times 10} = \frac{1}{20{,}000} \quad \text{or } 1:20{,}000$$

6.6. SCALES ON DRAWING

In order to prepare a drawing of any size, a scale is essential. After the construction of the figure, a scale must be stated as scale 1 cm = 1 metre (m) or 1:100 or full size.

6.7. NECESSARY INFORMATION FOR THE CONSTRUCTION OF A SCALE

When out of a set of recommended scales, the required scale is not available, it is then constructed on the drawing sheet. For its construction, the following informations are required :—

1. The representative fraction (R.F.) of the scale
2. The units to be presented either in Metric or British measures
3. The maximum length of the scale.

6.8. CLASSIFICATION OF SCALES

1. Plain or simple scales
2. Diagonal scales
3. Comparative or Corresponding scales
4. Vernier scales
5. Chord scales or Scale of chords.

6.9. PLAIN SCALE

A line which is divided into suitable number of equal parts or units, the first part of which is further sub-divided into small parts or sub-units of main unit is known as plain scale.

The plain scales are used to represent either two units or one unit and its fraction such as kilometers, hectometres, decimetres or metres and 1/10th of metres, etc.

6.10. POINTS TO DRAW SCALE

In every scale, the following key points should be remembered while constructing a scale :—

1. Find the representative fraction (R.F.), if not given.
2. Find the length of the scale = R.F. × Actual length of the object.

Note : *If the length of the scale is not mentioned in the problem, then take it about 15 cm or 150 mm or 6 inches].*

3. The mark 0 (zero) should be placed at the end of the first main division.
4. The main units should be numbered to the right and its sub-units to the left from the zero mark.
5. The scale or its R.F. should be mentioned along with the figure.
6. The name of main unit and its sub-units should be mentioned either below or at the respective ends of the scale.

Problem 1 : Construct a plain or simple scale to show metres and decimetres when one metre is represented by 2.5 centimetres and long enough to measure upto 6 metres. [Fig. 6.1]

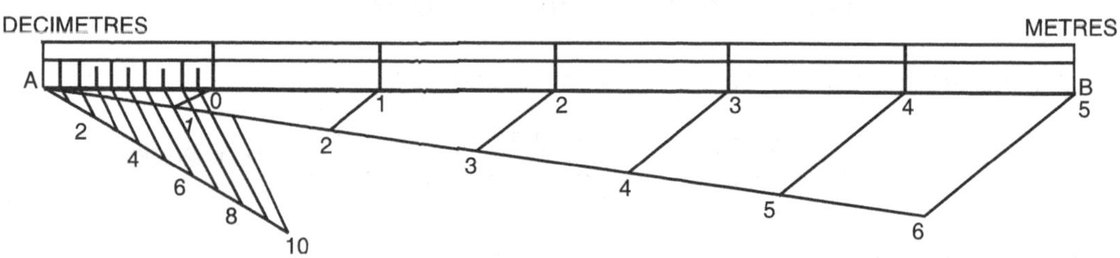

Fig. 6.1. *Scale 1 :40*

Solution : $\text{R. F.} = \frac{\text{Distance on sheet}}{\text{Distance of object}} = \frac{2.5\,\text{cm}}{1\text{m}} = \frac{2.5}{1\ 100} = \frac{1}{40}$ or 1:40

Length of the scale = R.F. × maximum length

$= \frac{1}{40} \times 6 \times 100 = 15\text{ cm}$

Now, take a line AB = 15 cm as shown in Fig. 6.1.

Divide the line AB into 6 equal parts thereby each representing 1 metre

Mark 0 (zero) at the end of 1st main division and 1,2 and 3 metres etc. at the end of the subsequent divisions towards right.

Divide the 1st division into 10 equal parts to represents single decimetre.

Mark decimetres to the left of 0 (zero) as shown in Fig. 6.1.

Write the names of units and sub-units either to the respective ends or below the scale.

Problem 2 : To construct a plain scale to show kilometres and hectometres when 2.5 centimetres are equal to 1 kilometre and long enough to measure upto 6 kilometres. Find R.F. and indicate a distance 4 kilometres and 5 hectometres on the scale. [Fig. 6.2]

Solution : $\text{R.F.} = \frac{2.5\,\text{cm}}{1\text{km}} = \frac{2.5}{1 \times 10 \times 10 \times 10 \times 10} = \frac{1}{40{,}000}$ or $1 : 40{,}000$

Length of scale = R.F. × maximum length

$= \frac{1}{40{,}000} \times 6 \times 1000 \times 10 \times 10 = 15\text{ cm}$

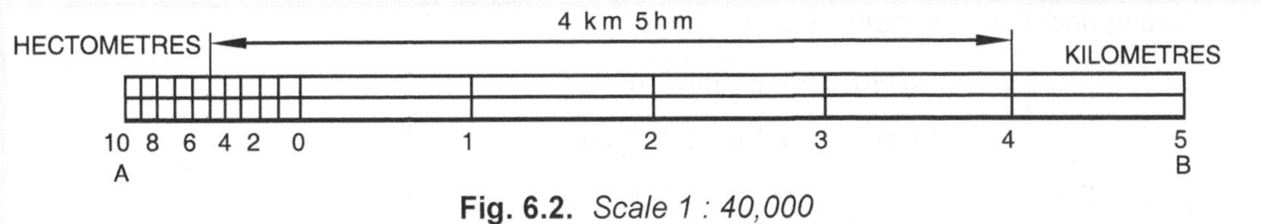

Fig. 6.2. *Scale 1 : 40,000*

Take a horizontal line AB = 15 cm to represent 6 kilometers and follow the same method of its construction as in problem 1.

Indicate on scale, the given distance, i.e., 4 kilometres and 5 hectometres (see Fig. 6.2).

Note : *Always take the length of the scale in centimetres.*

Problem 3 : Construct a scale of 1 cm = 0.4 metre to show the metres and decimetre and large enough to measure upto 5 metres. Show a distance of 4 metres and 6 decimetres on it. [Fig. 6.3]

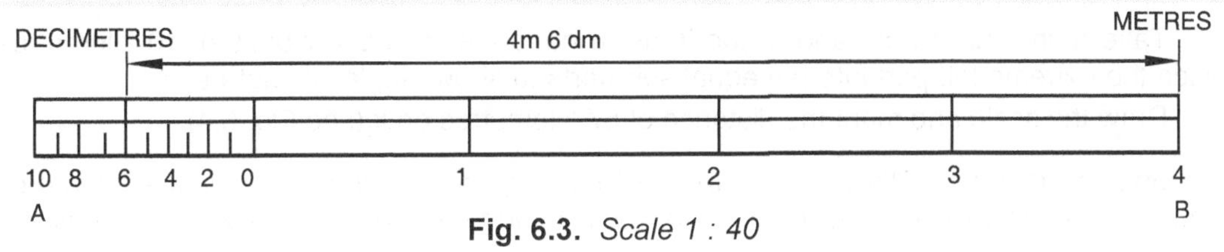

Fig. 6.3. *Scale 1 : 40*

Solution : $\text{R.F.} = \dfrac{1\text{cm}}{0.4\text{m}} = \dfrac{1}{0.4 \times 100} = \dfrac{1}{40}$

Length of scale $= \text{R.F.} \times \text{actual length of object}$

$$= \frac{1}{40} \times 5 \times 100 = 12.5 \text{ cm.}$$

Take a line AB =12.5 cm and divide it into 5 equal parts so that each part represents 1 metre.

Construct the scale as shown in Fig. 6.3. and indicate the given distance i.e. 4 m 6 dm.

Problem 4 : A rectangular plot of land of area 16 square kilometres is represented on a certain map by area 1 sq. cm. Draw a plain or simple scale to show units of 10 kilometres and single kilometre. Find R.F. and mark on it a distance of 57 kilometres. [Fig. 6.4]

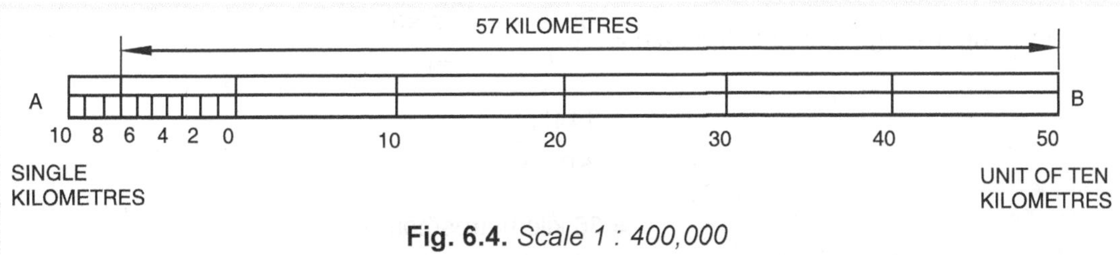

Fig. 6.4. *Scale 1 : 400,000*

Solution : Since the area of the plot 1 sq.cm = 16 sq. kilometres

Taking under root on both sides, we have

$$\sqrt{1\text{sq. cm}} = \sqrt{16\,\text{sq.km}}$$

$$1\text{ cm} \quad = 4\text{ km (kilometre)}$$

$$\text{R. F.} = \frac{\text{Distance on sheet}}{\text{Distance of object}} = \frac{1\text{ cm}}{4\text{km}} = \frac{1\text{ cm}}{4 \quad 100{,}000\text{cm}} = \frac{1}{4{,}00{,}0}$$

Let the length of the scale be equal to 15 centimetre as not given

$$\text{The maximum length of an object} = \frac{\text{Length of scale}}{\text{R.F.}}$$

$$= \frac{15}{1{,}00{,}000} \times \frac{4{,}00{,}000}{1}$$

$$= 60 \text{ kilometres}$$

Within which a given distance of 57 kilometres can be measured.

Take a line AB=15 cm and divide it into 6 equal parts to show units of 10 kilometres. Divide the extreme left part into ten equal sub-parts to show single kilometres.

Draw the scale and mark the distance of 57 kilometres on it (see Fig. 6.4).

Important Note : *While solving the problems on area, change the unit of area into the corresponding linear measures in order to mark the length of the scale by linear measuring instrument.*

Problem 5 : The distance between two stations A and B is 144 kilometres and it is covered by Janta train in 4 hours. Draw a plain scale to measure the time upto single minute. The R.F. of the scale is 1 / 240,000. Calculate the distance covered by the train in 45 minutes and show minutes on the scale. [Fig. 6.5]

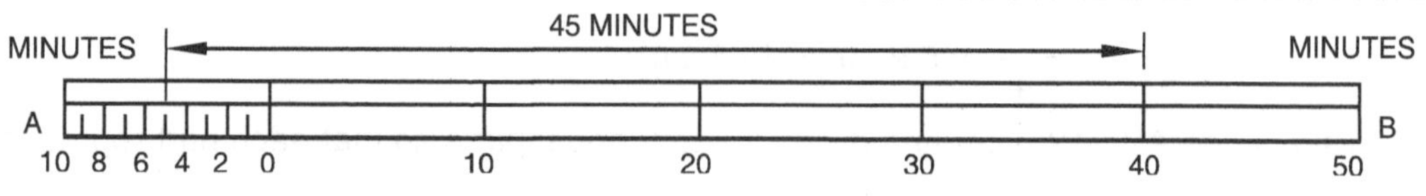

Fig. 6.5. *Scale 1 : 2,40,000*

Solution : Take the length of the scale equal to 15 cm [assume]

As 1 cm represents the actual distance of 2,40,000 cm, therefore, 15 cm will represent the actual distance = 2,40,000 × 15 cm = 36 kilometres.

$$\text{Now, the average speed of the train} = \frac{\text{Distance covered}}{\text{Time taken}}$$

$$= \frac{144\text{km}}{4\text{hour}}$$

$$= 36 \text{ kilometres/hour.}$$

It means that 36 kilometres are covered in 1 hour or 60 minutes.

Note : *The scale is to be drawn to show single minutes.*

Draw a straight line AB equal to 15 cm to represent 36 kilometres covered in 60 minutes (see Fig. 6.5).

Divide AB into six equal parts so that each part represent 10 minutes.

Divide the first main part into 10 sub-parts so that each sub-part should represent single minute which one is required.

The distance covered in kilometres by the train in 45 minutes (as shown) is given as,

i.e., $= \frac{36}{60} \times 45 = 27$ kilometres.

Problem 6 : A car is moving at a speed of 60 km per hour. On a scale one cm representing one third of a km. Show the distance travelled by the car in 3 minutes and 30 seconds. What is R.F. of scale. [Fig. 6.6]

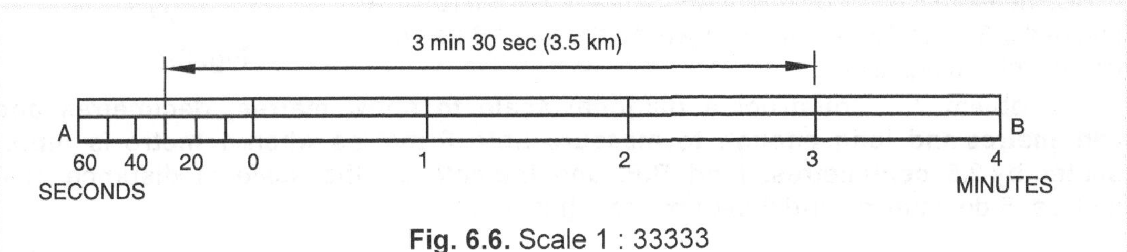

Fig. 6.6. Scale 1 : 33333

Solution: R.F. $= \frac{\text{Distance on drawing}}{\text{Actual distance}}$ (in same units)

$$= \frac{1\text{cm}}{1/3\text{km}} = \frac{1\text{cm}}{0.33333\text{km}} = \frac{1\text{cm}}{33333\text{cm}} = \frac{1}{33333}$$

Speed of the car $= 60\frac{\text{km}}{\text{hr}} = \frac{60\text{km}}{60\text{min}} = \frac{1\text{ km}}{1\text{ min}}$

or $1\text{km} = 1\text{min}$ .....(i)

Also $1/3\text{ km} = 1\text{cm}$ (given)

$\therefore$ $1\text{km} = 3\text{ cm}$ ..... (ii)

From equations (i) and (ii), we get,

$$1\text{min} = 3\text{ cm}$$

Take a line AB = 15 cm (assumed) which will represent 5 min or 5 km.

Divide AB into 5 equal parts, each representing single minute or single km.

Divide the first part into 6 equal parts each representing 10 seconds (see Fig. 6.6).

Show on the scale, the given distance travelled in 3 minutes and 30 seconds, i.e., 3.5 km = 3 km and 50 hectometres.

6.11. DIAGONAL SCALE

The scale in which small divisions of short lines are obtained by following the principle of diagonal divisions is known as diagonal scale.

Diagonal scales are used to represent either three units of measurements such as metres, decimetres and centimetres or two units and a fraction of its second unit.

6.12. PRINCIPLE OF DIAGONAL SCALE

The principle of diagonal scale is to divide a short line into any number of equal parts by following the diagonal division's method of construction which is as follows :

Let the given short line be AB which is required to be and divided into 10 equal parts.

Draw a perpendicular BC of suitable length and divide it into ten equal parts.

Join AC and draw horizontal line 1-1', 2-2', 3-3', etc. parallel to AB.

It is obvious from Fig. 6.7 that the triangles C-1-1', C-2-2', C-3-3' .... CBA are all similar.

Now consider the triangle ABC and 5'-5-C. Since 5B is half of BC, therefore 5'-5 will be half of AB. Similarly 1-1' =1/10 of AB and 3-3' = 3/10 of AB, etc. In other words, the line AB has been divided into 10 equal parts.

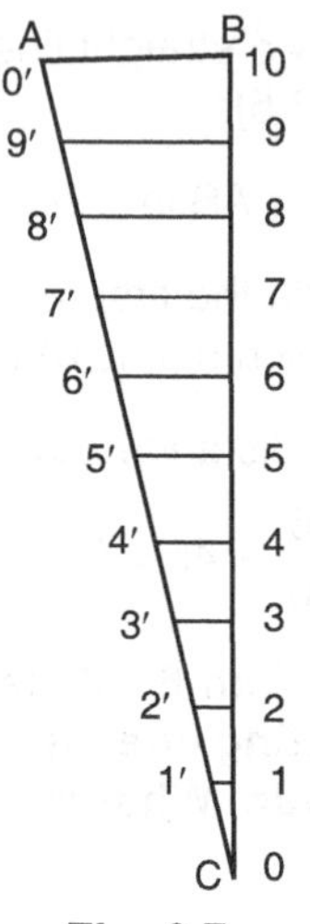

Fig. 6.7.

Problem 7 : Construct a diagonal scale to show metres, decimetres and centimetres and long enough to measure upto 6 metres when 1 metre is represented by 2.5 centimetres. Find R.F. and indicate on the scale, a distance of 4 metres, 5 decimetres and 6 centimetres. [Fig. 6.8]

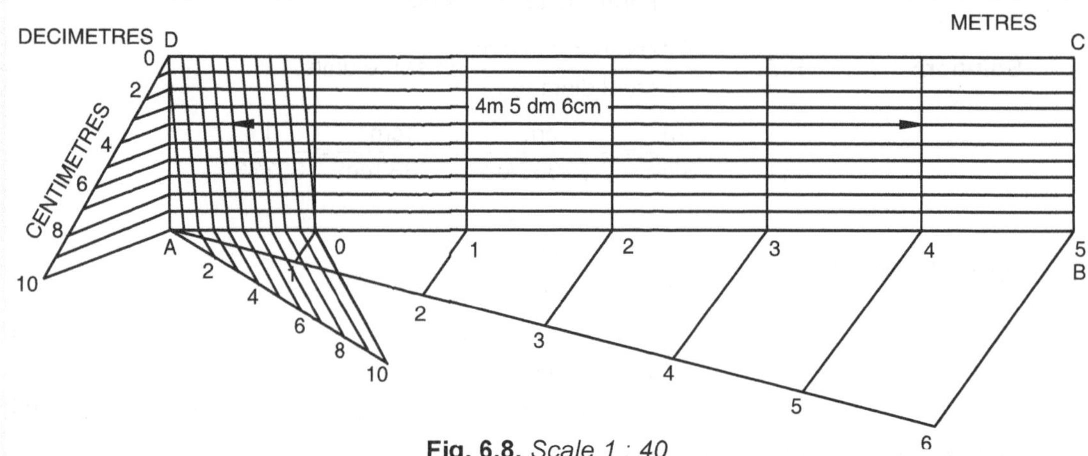

Fig. 6.8. *Scale 1 : 40*

Solution : R.F. $= \dfrac{2.5\,\text{cm}}{1\,\text{m}} = \dfrac{1}{40}$ or 1 : 40

Length of scale = R.F. × actual length

$$= \frac{1}{40} \times 6\text{ m} \times 10 \times 10 = 15\text{ cm}$$

Draw a horizontal line AB = 15 cm (assume) as shown in Fig. 6.8.

Divide AB into 6 equal parts so that each part represents 1 metre.

Divide the first part into 10 equal sub-parts so as to show single decimetre.

Draw a line AD, perpendicular to AB of suitable length and complete the rectangle ABCD.

Show diagonally by drawing ten equidistant parallel lines and show on the scale, the given distance, i.e., 4 metres 5 decimetres and 6 centimetres as shown in Fig. 6.8.

Problem 8 : Construct a diagonal scale to read metres, decimetres and centimetres for a R.F. of 1/50 and long enough to measure upto 5 metres. [Fig. 6.9]

Show on it a length of 2.89 metres, 3.67 metres and 4.44 metres.

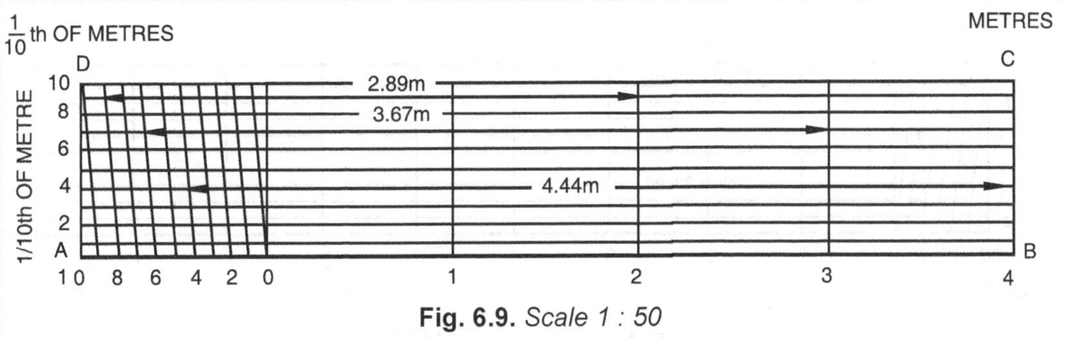

Fig. 6.9. *Scale 1 : 50*

Solution : The length of scale = R.F. × actual length of an object

$$= \frac{1}{50} \times 5 \times 100 = 10 \text{ cm}$$

Draw a line AB = 10 cm as shown in Fig. 6.9.

Divide AB into 5 equal parts so that each part should represent 1 metre.

Divide the first part (A-0) into 10 equal parts so as to show single division.

Draw a line AD, perpendicular to AB of suitable length and complete the rectangle ABCD. Show diagonally by drawing ten equidistant parallel lines.

Show on the scale, the given distance, i.e., 2.89 metres, 3.67 metres and 4.44 metres as shown in Fig. 6.9.

Problem 9 : To construct a diagonal scale of R.F. = 1/5000 or 1:5000 to show single metres and long enough to measure upto 600 metres. On the scale show a distance of 457 metres. [Fig. 6.10]

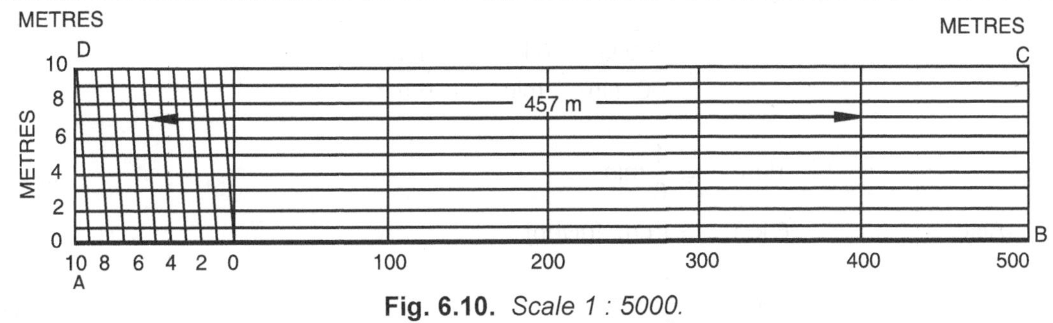

Fig. 6.10. *Scale 1 : 5000.*

Solution : The length of the scale = R.F.× actual length of an object

$$= \frac{1}{5000} \times 600 \text{ m} \times 100$$

$$= 12 \text{ cm}$$

Draw a line AB = 12 cm and for its complete construction, proceed as of Problem 7 and show the given distance of 457 metres (see Fig. 6.10).

Problem 10 : An aeroplane is flying at a speed of 360 km/hr. Draw a diagonal scale to represent 6 km by 1 cm and to show a distance up to 60 km. Find the R.F. of the scale.

Find from the scale the distance covered by the aeroplane in (i) 3 minutes 22 seconds (ii) 5 minutes 36 seconds and (iii) 7 minutes 49 seconds. [Fig. 6.11]

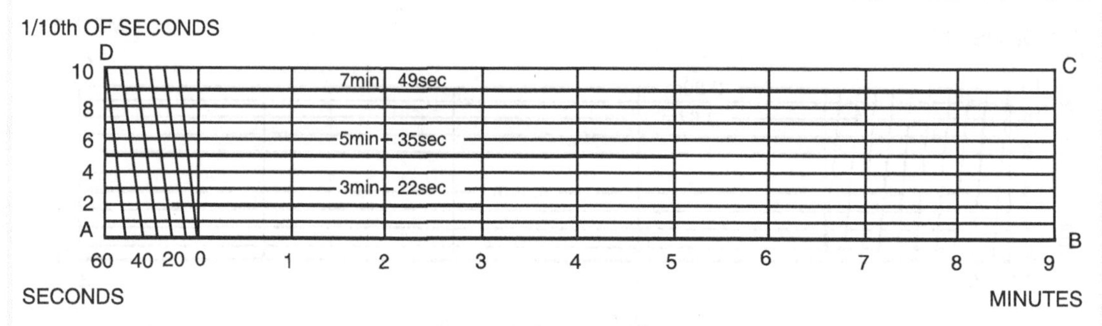

Fig. 6.11. *R.F. 1: 600,000*

Solution :

$$\text{R.F.} = \frac{\text{Distance on drawing}}{\text{Actual distance}} \text{ (in same unit)}$$

$$= \frac{1\text{ cm}}{6\text{ km}} = \frac{1\text{ cm}}{6 \times 1{,}000 \times 10 \times 10\text{cm}} = \frac{1}{6{,}00{,}000}$$

Length of scale = R.F. × total distance

$$= \frac{1}{6{,}00{,}000} \times 60\text{km}$$

$$= \frac{1}{6{,}00{,}000} \times 60 \times 1{,}000 \times 10 \times 10\text{cm} = 10\text{cm}$$

Speed of aeroplane = 360 km/hr

$$= \frac{360\text{ km}}{60\text{ min}} = \frac{6\text{ km}}{\text{min}} = \frac{1\text{ km}}{1/6\text{ mir}}$$

$\therefore$ 1km $= \frac{1}{6}$ min ......(i)

Also 6 km = 1 cm [given]

$\therefore$ 1 km $= \frac{1}{6}$ min ......(ii)

From equations (i) and (ii) we get,

$$\frac{1}{6}\text{ min} = \frac{1}{6}\text{ cm}$$

$\therefore$ 1min = 1cm

Take a line AB = 10 cm which will represent 10 minutes or 60 km.

Divide AB into 10 parts so that each part represents 1 minute or 1 km.

Divide the first part (A-0) into 6 equal parts each representing 10 seconds divisions.

Draw a line AD, perpendicular to AB of suitable height and complete the rectangle ABCD Show diagonally by drawing 10 equidistance parallel lines (see Fig. 6.11).

Show on the scale, the given distances, i.e.,

(i) 3 minutes 22 seconds, i.e., 20.2 km

(ii) 5 minutes 35 seconds, i.e., 33.6 km [∵ 1 min = 6 km]

(iii) 7 minutes 49 seconds, i.e., 46.9 km

Problem 11 : The distance between two stations A and B is 100 kilometres and its equivalent distance on railway map measures 2.5 centimetres. What is the R.F.?

Draw a diagonal scale showing single kilometres and indicate on the scale, the following distances :

(i) 577 kilometres

(ii) 455 kilometres

(iii) 333 kilometres

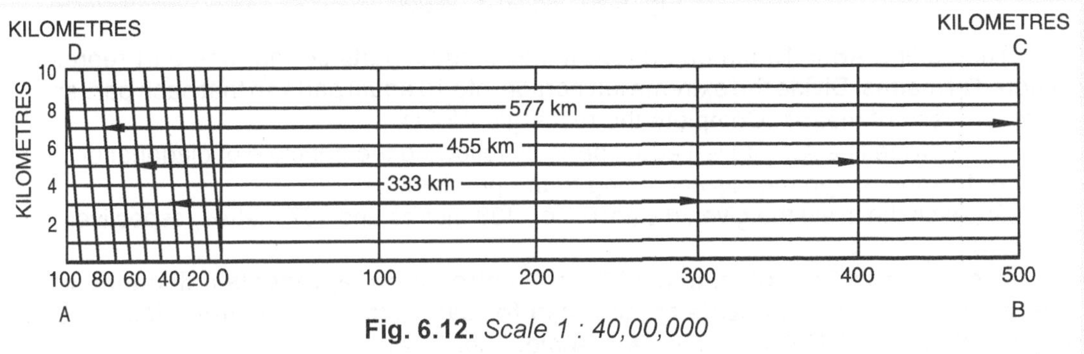

Fig. 6.12. *Scale 1 : 40,00,000*

Solution : R.F. = $\frac{\text{Distance on drawing}}{\text{Actual distance}}$ (in same unit)

$$= \frac{2.5\,\text{cm}}{100\,\text{km}} = \frac{2.5\,\text{cm}}{100{,}00{,}000\,\text{cm}} = \frac{1}{40{,}00{,}000}$$

Take a line AB = 15 cm (assume) to represent the actual distance = $\frac{100 \times 15}{2.5}$ = 600 kilometres within which a maximum given distance of 577 kilometres can be measured.

Proceed for its complete construction in similar way as in problem 8 and indicate the given distance on the scale (see Fig. 6.12).

Important Note : *The students are advised to use the method of dividing any straight line into desired number of equal parts (in every problem).*

Problem 12 : On a certain drawing 40 metres are represented by 1 centimetre. Draw a suitable scale to read hundreds of metres, tens of metres and metres. Find R.F. and mark the following distances on the scale:—

(i) 433 metres (ii) 467 metres

Solution : R.F.= $\frac{1\text{ cm}}{40\text{m}} = \frac{1\text{ cm}}{40 \times 10 \times 10\text{cm}} = \frac{1}{4,000}$ or 1 : 4,000

As the length of the scale is not given, so assume it to be equal to 15 centimetres. Also we know that 1 cm = 40 metres

∴ 15 cm = 15 × 40 = 600 metres (within which given maximum distance of 467 metres can be measured).

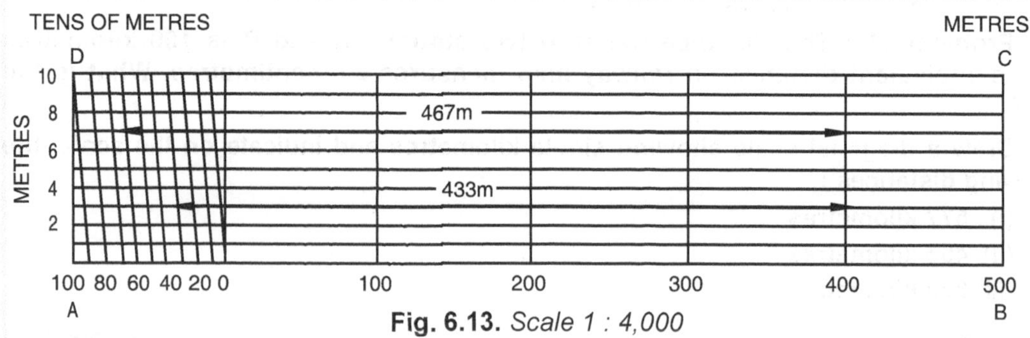

Fig. 6.13. *Scale 1 : 4,000*

Take a line AB = 15 cm and divide it into 6 equal parts so that one part represents hundred of metres. Divide the extreme left portion into 10 equal parts so that each part should represent tens of metres. Complete the rectangle ABCD.

Now, divide line AD into ten equal parts by following the principle of diagonal divisions in order to show single metres according to the problem.

Show on the scale the given distance, i.e., 433 metres and 467 metres respectively (see Fig. 6.13).

Problem 13 : On a plan, a line 22 centimetres long represents a distance of 440 metres. Draw a diagonal scale for the plan to read upto single metres. Measure and mark a distance of 187 metres on the scale. [Fig. 6.14]

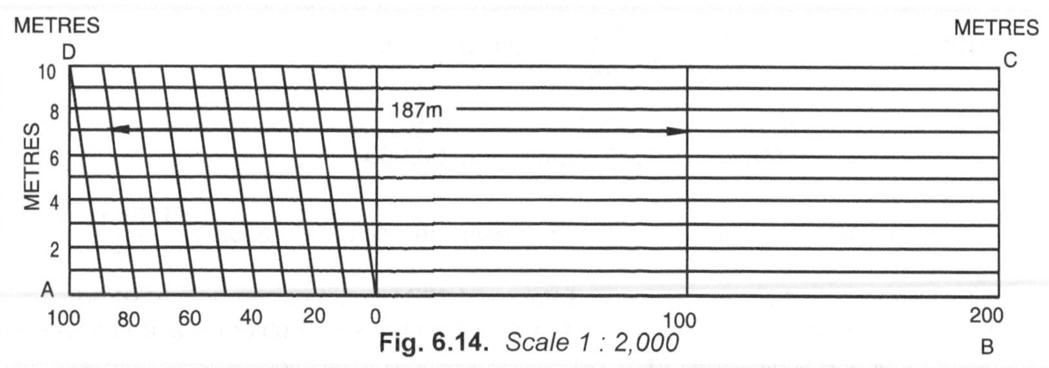

Fig. 6.14. *Scale 1 : 2,000*

Solution : R.F. = $\frac{22\text{cm}}{440\text{m}} = \frac{22}{440 \times 10 \times 10} = \frac{1}{2,000}$ or 1 : 2,000

Assuming the length of scale = 15 cm as it is not given in the problem.

Since 22 cm = 440 metres from the given data.

∴ 1 cm = 20 metres

∴ 15 cm will represent 300 metres which can measure a given distance of 187 metres. Draw a line AB = 15 cm and follow the same method of construction as in problem 6 and indicate the required distance of 187 metres (see Fig. 6.14).

Second possible solution of problem 13.

We know that

22 cm = 440 metres (given)

or 1 cm = 20 metres

or 10 cm = 200 metres (within which a given distance of 187 metres can be measured). Take a line AB = 10 cm and divide it into two parts of 100 metres each and show in a similar way the given distance of 187 meters (see Fig. 6.14).

Problem 14 : The distance between two cities A and B is 300 kilometres. Its equivalent distance on the map measures only 6 centimetres. What is the R.F.?

Draw a diagonal scale to show hundreds of kilometres, tens of kilometres and kilometres. Indicate on the scale, the following distances :—

(i) 525 kilometres

(ii) 313 kilometres

(iii) 277 kilometres.

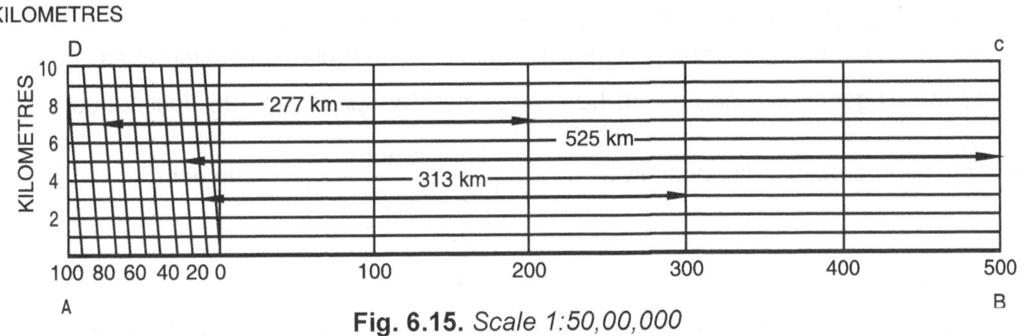

Fig. 6.15. *Scale 1:50,00,000*

Solution : R.F. = $\dfrac{6\,\text{cm}}{300\,\text{km}} = \dfrac{6\,\text{cm}}{300\,\text{km} \times 10 \times 10 \times 10 \times 10 \times 10} = \dfrac{1}{5{,}000{,}000}$

or scale 1 : 50,00,000

We know that

6 cm = 300 kilometres

or 1 cm = 50 kilometres (But we require a maximum distance 525 kilometres)

Now let the length of the scale be 12 cm (Assume).

12 cm = 600 kilometres (Within which a distance of 525 kilometres can be measured)

Draw a line AB = 12 cm and divide it into 6 equal parts and complete the figure as shown in Fig. 6-15 which is itself a self explanatory.

Problem 15 : Construct a diagonal scale to measure kilometre, hectometres and spaces of 125 decimetres when a distance of 1 kilometre is represented by 3 centimetres and long enough to measure upto 5 kilometres. Indicate on the scale, a distance of 3 kilometres, 5 hectometres and 5 decametres (or 500 decimetres or 4 spaces of 125 decimetres). [Fig. 6.16]

Solution : R.F. = $\dfrac{\text{Distance on drawing}}{\text{Actual distance}} = \dfrac{3\,\text{cm}}{1\,\text{km}} = \dfrac{3\,\text{cm}}{1{,}00{,}000\,\text{cm}}$ (Same unit)

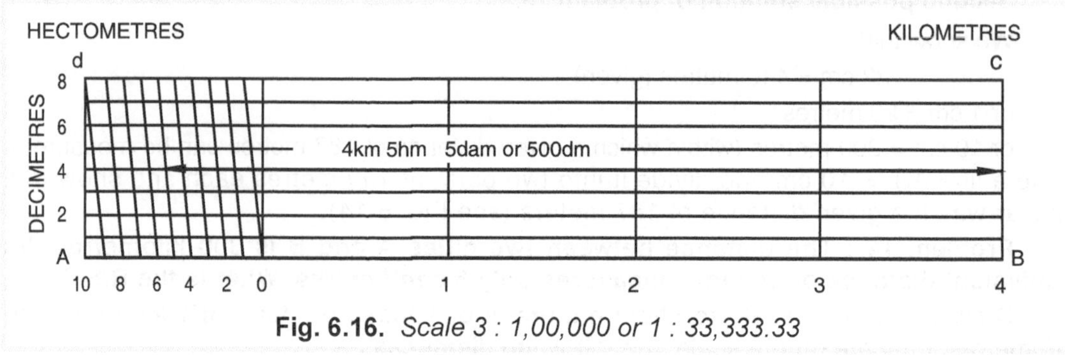

Fig. 6.16. *Scale 3 : 1,00,000 or 1 : 33,333.33*

Length of scale = R.F. × actual distance

$$= \frac{3}{1,00,000} \times 5 \text{ km} \times 1000 \times 10 \times 10 = 15 \text{ cm.}$$

Draw a line AB = 15 and divide it into 5 equal parts to show kilometres (see Fig. 6.16). Divide the extreme left hand part into 10 equal parts to represent hectometres.

As the diagonal divisions are to be represented by spaces of 125 decimetres, so the

number of short parts, by diagonal division $= \frac{1 \text{ hectometre}}{125 \text{ decimetres}}$ (in same unit)

$$= \frac{1000}{125} = 8$$

So divide the line AD into 8 equal parts so that each short part by diagonal divisions should represent a space of 125 decimetres which is required in the problem.

Problem 16 : A rectangular plot of land area 0.45 hectare is represented on a map by a similar rectangle of 5 square centimetres. Calculate the R.F. of the scale of the map. Also draw a scale to read upto single metre from the map. The scale should be long enough to measure upto 400 metres. 1 hectare = 10, 000 square metres. [Fig. 6.17]

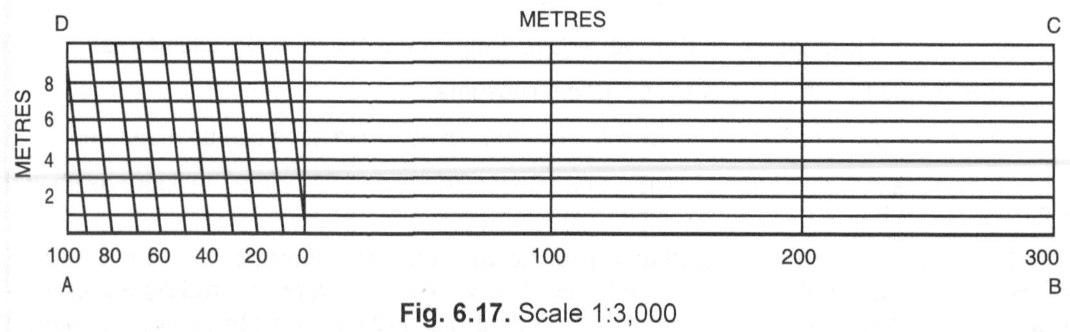

Fig. 6.17. Scale 1:3,000

Solution : Since 5 square centimetres = 0.45 hectare

$$1 \text{ square centimetre} = \frac{0.45}{5} \text{ hectare} = \frac{0.45 \times 10,000}{5} \text{ square metres.}$$

(Since 1 hectare = 10,000 square metres)

Important Note : *While solving problems on area, first change the units of areas into linear units in order to make the length of the scale by linear measuring instrument on the sheet.*

Taking the under-root on both the sides, we have

$$1 \text{ centimetre} = \frac{\sqrt{0.45 \times 10{,}000}}{5} = 30 \text{ metres}$$

Now 400 metres will represent its corresponding length on drawing = $\frac{400}{30}$ = 13.33 cm.

Take a line AB = 13.33 centimetres. Divide AB into four equal parts so that each part should represent 100 metres.

For its complete construction, see Fig. 6.17 which is self explanatory.

6.13. COMPARATIVE OR CORRESPONDING SCALES

When the given scale of a plan reads a certain measure and if it is required to construct a new scale for the same plan to read in some other measure, then the new scale is called comparative or corresponding scale.

The important point which should be kept in mind is that the representative fractions of both the scales must be the same.

The comparative scales may either be plain or diagonal and may be constructed separately or one above the other.

Important Note : *Comparative scales can only be drawn by considering two different* units of different measures, so in above problem another unit of British measure i.e. inch unit has *been taken along with the metric measure for its comparison.*

Problem 17 : A drawing is drawn in inch unit to a scale 3/8 full size. Draw the scale showing 1/8 of an inch division and to measure upto 16 inches. Construct a comparative scale showing centimetres and millimetres and long enough to measure upto 40 centimetres. [Fig. 6.18]

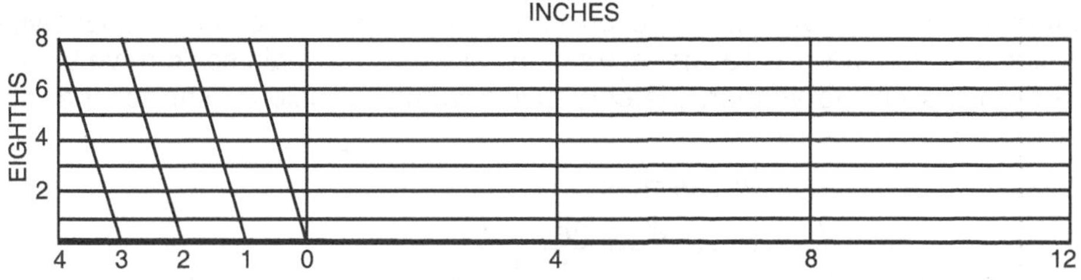

Fig. 6.18 (i) *Inch scale Scale 3 : 8 or 1 : 2.667*

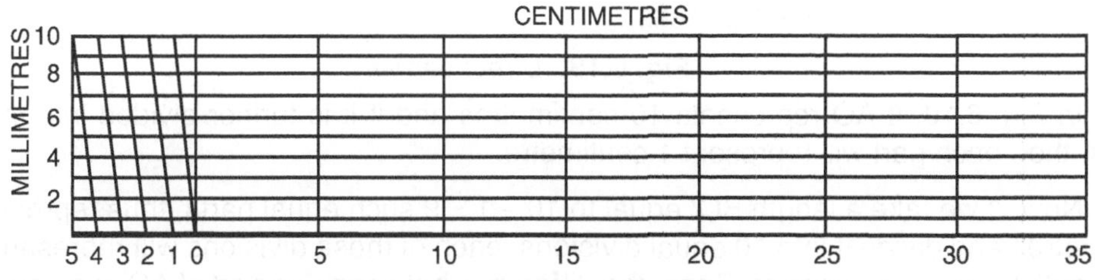

Fig. 6.18 (ii) *Comparative scale 3 : 8 or 1 : 2.667*

Solution :

Inch Scale : We know that the length of the Scale = R.F × Maximum length

$$= \frac{3}{8} \times 16$$

= 6 inches.

Take a line AB = 6 inches and construct a diagonal scale [see Fig. 6-18 (i)].

Comparative Scale : Here the length of the scale $= \frac{3}{8} \times 40 = 15$ centimetres.

Draw a line AB = 15 centimetres and construct a comparative scale [see Fig. 6-18 (ii)].

6.14. VERNIER SCALE

The vernier scale is a device used for measuring the fractional parts of the smallest division of the main scale. The vernier scale consists of two parts, a *fixed scale* (called the *main scale*) and a *movable graduate scale* (called the *vernier*). The vernier with its graduate edge moves along the graduate edge of a long fixed scale. The vernier carries an index mark which represents the zero of the vernier divisions and is shown by an arrow. The vernier scale may either be straight or curved one.

These scales, like diagonal scales, are used to read to a very small units with great accuracy.

Practical Applications : *These scales are used in surveying and astronomical instruments for the determination of angular measurements. These are also used in slide-calliper, screw-gauge and other similar instruments for recording linear measurements.*

The vernier are classified as under :

1. Direct vernier
2. Rectrograde vernier

6.14.1. Direct Vernier : In this, the principle on which the vernier scale is constructed is based upon the fact that if it is required to read 1/4th part of the smallest division on the main scale, then (n-1) main divisions are taken which are further divided into equal divisions on the vernier as shown in Fig. 6.19. In this, 9 small divisions of the main scale have been divided into 10 divisions on the vernier, thereby, giving vernier divisions shorter than the main divisions.

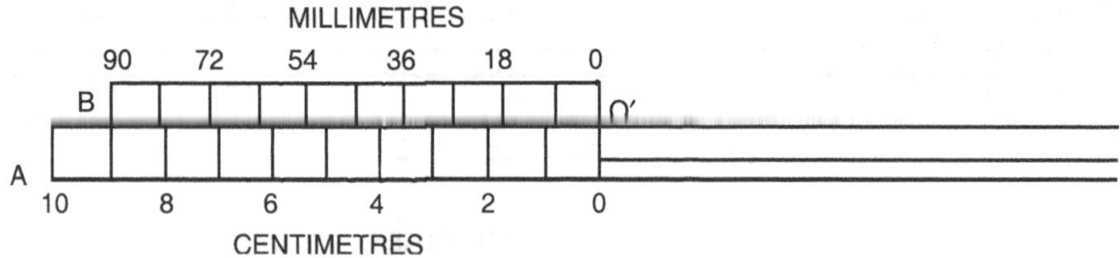

Fig. 6.19. *Direct vernier*

In Fig. 6.19, if AO represents 10 centimetres and if it is further divided into ten equal parts, then each part will represent 1 centimetre.

Now, if we take a length BO′ equal to 10 – 1 = 9 such equal parts, thus representing 9 cetimetres and divide it into 10 equal divisions, each of these divisions will represent 9/10 = 0.9 centimetre or 9 milimetres. Then, the difference between one part of AO and one division of BO will be equal to 0.1 centimetre or 1 millimetre. Similarly, the difference between two parts of each will be 2 millimetres. Here, the upper scale BO′ represents the vernier and the combination of main scale and the vernier, represent the vernier scale.

6.14.2. Rectrograde Vernier : This type of vernier is also based upon the same principle as direct vernier, but 11 smallest divisions of the main scale are taken to divide it into 10 divisions on the vernier, thus giving vernier divisions longer than the one part of BO′ and AO, is again equal to 0.1 centimetre or 1 millimetre (see Fig.6.20).

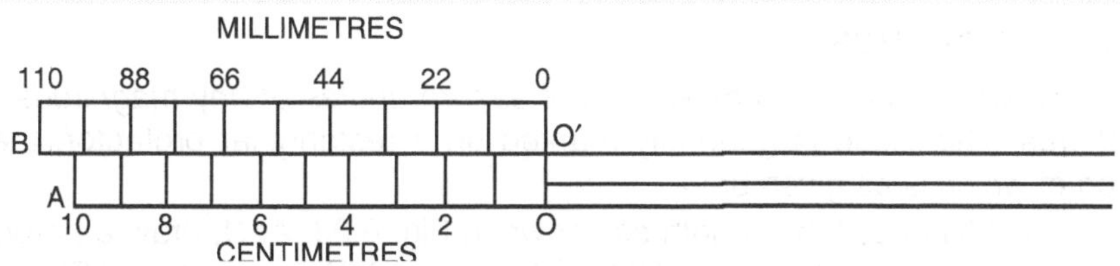

Fig. 6.20. *Rectrograde vernier*

Problem 18 : Construct a vernier scale to read metres, decimetres and long enough to measure upto 6 metres when 1 metre is represented by 2.5 centimetres. Find R.F. and show on it, a distance of 4.33 metres. [Fig. 6.21]

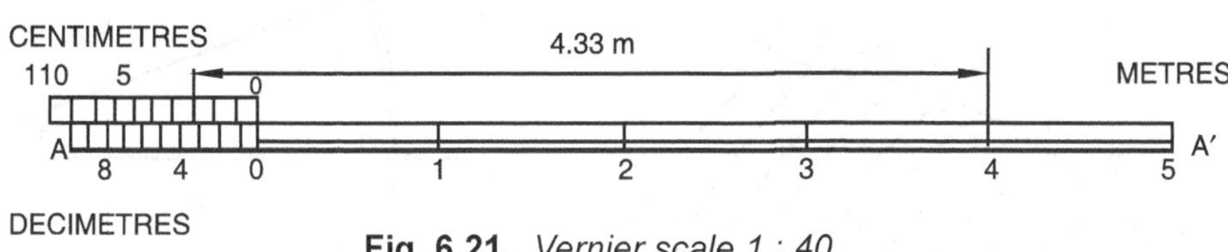

Fig. 6.21. *Vernier scale 1 : 40*

Solution : R.F. $= \frac{2.5\,\text{cm}}{1\,\text{m}} = \frac{2.5}{1 \times 100} = \frac{1}{40}$ or 1:40

The length of scale = R.F × maximum distance

$$= \frac{1}{40} \times 6 \times 100 = 15 \text{ cm.}$$

Take a horizontal line AA' = 15 cm and divide it into six equal parts so that each part should represent 1 metre.

Now, to read decimetres, divide the extreme left part into ten equal sub-parts.

From the zero (0) mark, measure a length equal to 11 parts to the left (Rectrograde vernier). Mark the divisions and place the vernier as shown in Fig. 6.20

Show on the scale, a given distance of 4.33 metres.

Problem 19 : Construct a vernier scale to read tenth (or 0.1) and hundredth (or 0.01) of one centimetre and show on it, a distance of 12.22 centimetres. [Fig. 6.22]

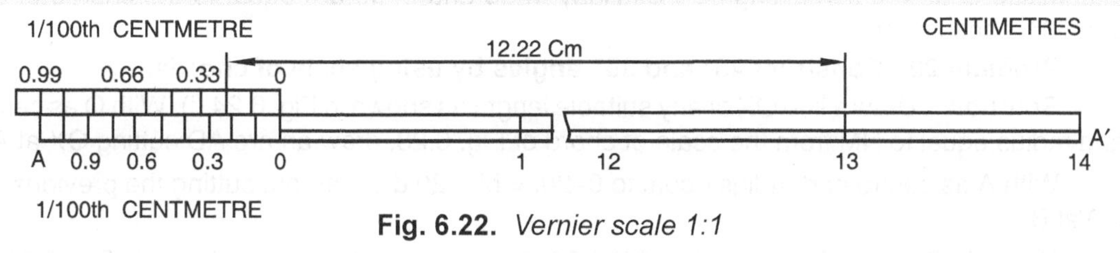

Fig. 6.22. *Vernier scale 1:1*

Solution : Let a line AA' be equal to 15 cm as it is not given. Divide it into 15 equal divisions so that each division should represent 1 cm.

Now to read tenth of a centimetre, divide extreme left divisions into ten equal subdivisions.

From the zero (0) mark, measure a length equal to 11 sub-divisions to the left and divide it into ten equal parts (Rectrograde vernier). Mark the divisions and place the vernier as shown in fig. 6.22. Show on a scale, a given distance of 12.22 centimetres.

6.15. SCALE OF CHORDS

The scale which is used to make or to measure angles of any magnitude is known as scale of chords. This scale is generally marked on a rectangular protactor, the method of construction of which is as given below :

Draw a line MN of suitable length as shown in Fig. 6.23. At N, draw a perpendicular NT to MN. With N as centre and radius = NM, strike an arc MP, thus cutting NT at P. Then, the arc MP or chord MP subtends at an angle of 90° at the centre N.

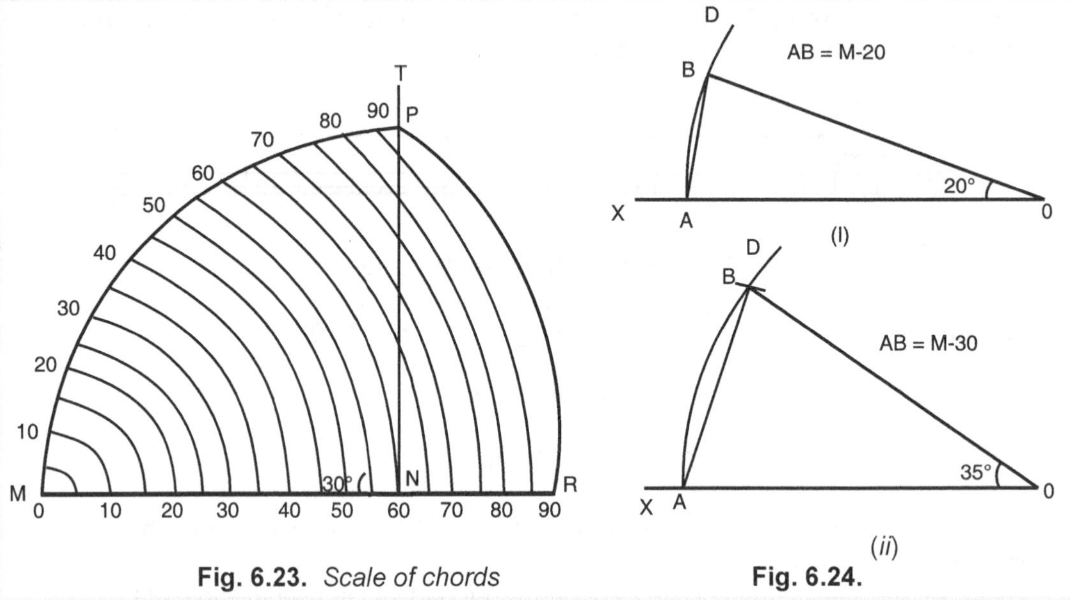

Fig. 6.23. *Scale of chords* **Fig. 6.24.**

Divide the arc MP into 18 equal parts, each part therefore, subtends at N an angle of 90/18 = 5°, as shown. Now with M as centre, turn down the divisions to the line MR and finish the scale as shown in Fig. 6.23.

Then MR is the required scale of chords to read each part equal to 5°.

Refering to Fig. 6.23, the horizontal line MR is equal to the chord of 90° which joins the extremities of an arc containing 90°. Similarly M-30 and M-60 are equal to the chords of 30° and 60° respectively.

Problem 20 : Construct 20° and 35° angles by using scale of chords.

Solution : Draw a line OX of any suitable length as shown in Fig. 6.24 (i). With O as centre and radius equal to NM from the scale of chord of Fig. 6.23, draw an arc AD cutting OX at A.

With A as centre and radius equal to 0–20, = M – 20 draw an arc cutting the previous arc AD at B.

Now join O with B. Then, angle AOB = 20° is the required angle as shown in Fig. 6.24 (i).

Similarly, make another angle, i.e., 35° in the same manner as shown in Fig. 6.24(ii).

Problem 21 : Construct an angle of 105° by scale of chords. [Fig. 6.25]

Solution : Since the angle to be constructed is greater than 90°, so it should be taken into two parts, i.e., the sum of the two angles which must be equal to the given angle 105° (say, 60° + 45° = 105°).

For its construction, proceed in the same manner as in the last problem (see Fig. 6.25).

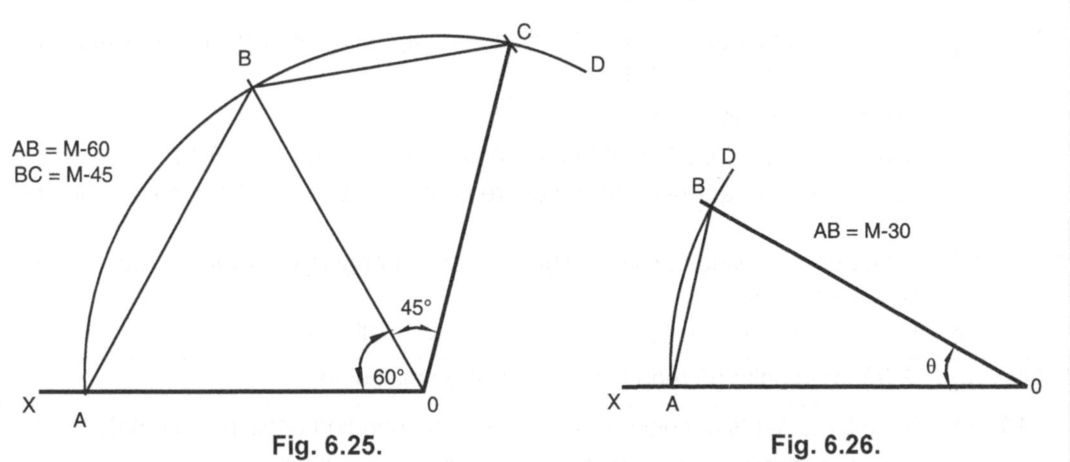

Fig. 6.25.

Fig. 6.26.

Problem 22 : Measure the given angle of Fig. 6.26 by using the scale of chords. [Fig. 6.26]

Solution : From the given figure, take O as centre and radius equal to MN from the scale of chords of Fig. 6.23. Join A with B. The chord distance AB to the scale of chords gives an angle, θ = 30°, in this case (see Fig. 6.26).

QUESTIONS FOR SELF EXAMINATION

1. What is scale?
2. What is the difference between reducing and increasing scale?
3. What is representative fraction (R.F.) or scale factor (S.F.)?
4. What are the main uses of scale?
5. What are the necessary informations for the construction of scale?
6. Give the name of different types of scales.
7. What is the difference between a plain scale and a diagonal scale?
8. What is the principle of a diagonal scale?
9. What is the difference between direct and rectrograde verniers?
10. What is the difference between vernier scale and scale of chords?
11. Fill in the blanks :
    (*a*) When a drawing is made to the same size of the object, the name of the scale is ....
    (*b*) For drawing of small instruments, e.g., watch parts, ... scale is used.
    (*c*) Plain scales are used to represent .... units.
    (*d*) When measurements are desired in three units, ... scale is used.
    (*e*) The relative values of the R.F. (S.F.) of enlarged, full size and reducing scales are ...., ..., and .... respectively.

(*f*) The ratio of distance on the drawing sheet and corresponding actual distance of the object is known as ....

(*g*) Length of scale = R.F. × ....

(*h*) Vernier scales are used in surveying and astronomical instruments for .... measurements.

(*i*) Scale of chords are used to measure ....

Ans. (*a*) full size, (*b*) enlarged, (*c*) two, (*d*) diagonal, (*e*) >1, 1, <1,(*f*) R.F. (*g*) actual length of object. (*h*) angular (*i*) angles.

12. Choose the correct answers :

(*a*) Drawing made to one half of the actual size is (*i*) 2:1 (*ii*) 1:1 (*iii*) 1:2

(*b*) Diagonal scales are used for measurement of (*i*) one unit (*ii*) two units (*iii*) three units

(*c*) Scales usually used in civil engineering are (*i*) englarged scale (*ii*) reduced scale (*iii*) full size scale

(*d*) Vernier scales are to read (*i*) angles (*ii*) very small units (*iii*) two units,

Ans. (*a*) 1:2 (*b*) three units (*c*) reduce scale (*d*) very small units.

13. Match the statements of column A with the corresponding ones in column B.

Column A	Column B
(*i*) Plain scale	(*a*) Three units
(*ii*) Diagonal scale	(*b*) Dissimilar units
(*iii*) Scale of chords	(*c*) Two units
(*iv*) Comparative scale	(*d*) Angle measurement

Ans. (*i*) *c*, (*ii*) *a*, (*iii*) *d*, (*iv*) *b*

PROBLEMS FOR PRACTICE

1. Construct a plain scale to show metres and decimetres, when 3 centimetres are equal to 2 metres and long enough to measure upto 5 metres.

2. Construct a plain scale to show kilometres and hectometres when 25 centimetres are equal to 1 kilometre and long enough to measure upto 6 kilometres. Find R.F. and indicate, a distance of 5 kilometres and 6 hectometre on the scale.

3. Construct a scale of R.F. = 1 : 60 to show metres and decimetres and long enough to measure upto 5 metres. Show on the scale, a distance of 4 metres and 7 decimetres.

4. The distance between two stations A and B is 180 kilometres and it is covered by Janta train in 6 hours. Draw a plain scale to measure the time upto single minute. The R.F. of the scale is 1/2,00,000. Calculate the distance covered by the train in 37 minutes and show these minutes on the scale.

5. Construct a plain scale to compute time in minutes and distance covered by a train in km., when the train passes between two stations 240 km apart in four hours. The scale should have R.F. 1/4,00,000. Show the distance covered in 45 minutes on the scale.
6. Construct a diagonal to read metres, decimetres and centimetre, and long enough to measure upto 5 metres when one metre is represented by 3 centimetre. Find R.F. and indicate on the scale, a distance of 4 metres, 7 decimetres and 6 centimetres.
7. Construct a diagonal scale to measure 0.01 and 0.1 of a metre and long enough to measure upto 5 measure when 1 metre is represented by 2.5 centimetres. Find R.F. and indicate on the scale, the following distance :

   (*i*) 5.55 metres (*ii*) 4.44 metres (*iii*) 3.33 metres
8. Construct a scale of R.F. = 1:50 to show metres, decimetres and centimetres to measure upto 6 measures. Show on this scale the following length :

   4 metres, 3.7 metres, 2.58 metres, 6 decimetres and 9 cm.
9. Draw a diagonal scale of 1 : 2 5 showing centimetres and millimetres and long enough to measure upto 20 centimetres. Show a distance of 134 cm on it.
10. Construct a diagonal scale of R.F. = 1/50,000 to show single metres and long enough to measure upto 500 metres. On the scale, indicate a distance of 467 metres.
11. On a building plan a line 15 cm long represents a distance of 300 metres. Calculate its R.F. Construct a diagonal scale to read upto 300 metres showing single metre by diagonal divisions. Indicate the following distances on this scale: 245, 160, 70 and 8 metres.
12. Construct a diagonal scale to read upto 1/100 of kilometres having given the value of R.F. = 1/50,000 and to measures upto 8 kilometres. Indicate on the scale, a distance of 6 .76 kilometres.
13. Construct a diagonal scale in kilometres, hectometres and spaces of 125 decimetres when a distance of 1 kilometre is represented by 2.5 centimetres and long enough to measure upto 6 kilometres. Indicate on the scale, a distance of 5 kilometres, 6 hectometres and 5 decametres.
14. A map 500 cm × 50 cm wide, represents an area of 6250 sq. kilometres. Construct a diagonal scale to measure kilometres, hectometres and decametres. Indicate on the scale a distance of 5.56 kilometres. Find R.F. of the scale.
15. Construct a vernier scale to read metres, decametres and centimetres and long enough to measure upto 5 metres when 1 metre is represented by 3 centimetres. Find R.F. and show on it, a distance of 5.33 metres.
16. Construct a vernier scale to read tenth and hundreth of one centimetre and indicate on it, a distance of 13.66 centimetres.
17. Construct 25° and 55° angles by using scale of chords.
18. Using the scale of chords, construct an angle of 135°.

This Page has been intentionally left blank

SECTION - II

1. Theory of Projection and Orthographic Projection
2. Orthographic Reading or Interpretation of Views
3. Identification of Surfaces
4. Missing Lines and Views
5. Sectional Views
6. Isometric Projections
7. Auxiliary Views
8. Freehand Sketching
9. Sections of Solids

This Page has been intentionally left blank

CHAPTER

1

Theory of Projection and Orthographic Projection

INTRODUCTION

Engineering drawing is the universal graphic language by means of which the shape and size of an object can be specified on a plane of paper. In order to represent the true shape and size of the object, different straight lines are drawn from the various points on the contour of the object on to the plane of paper. The image or figure, thus, formed on the paper by joining different points in correct sequence is known as projection of that object. It is therefore, necessary for an engineer to acquire a good working knowledge of projections to express and record the shape and size of the object.

In this chapter, we will deal with the study of projections, types of projections, orthographic projections, ways of projecting them on drawing, etc.

1.1. PROJECTION

Any kind of representation of an object on a paper, screen or similar surface by drawing or by photography is called the projection of that object.

If different straight lines are drawn from the various points on the contour of an object to meet a plane, the figure, thus, formed by joining these points in correct sequence is called the projection of that object [see Fig. 1.4].

1.2. TYPES OF PROJECTION

In engineering practice, the following are the common types of projections :—

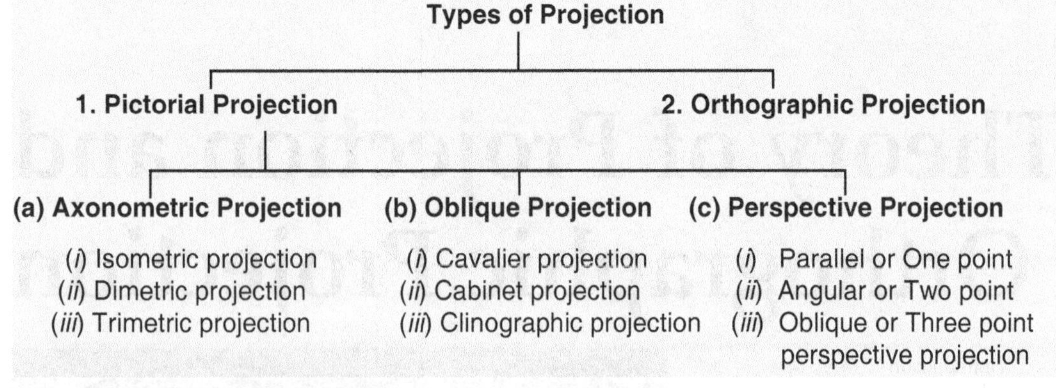

1.3. PICTORIAL PROJECTION

The projection in which the length, breadth and height of an object is shown in one view is known as pictorial projection.

This type of projection has the advantage of conveying an immediate impression of the general shape of the object, but does not necessarily show the exact dimensions. The pictorial projection are (a) Axonometric (b) Oblique and (c) Perspective projections (see Figs. 1.1 to 1.3).

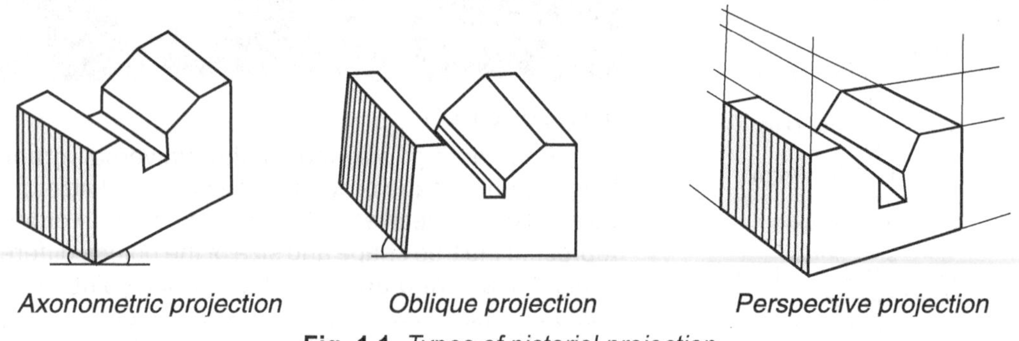

Axonometric projection *Oblique projection* *Perspective projection*

Fig. 1.1. *Types of pictorial projection*

1.3.1. (*a*) **Axonometric Projection :** *The projection obtained on a plane of paper when the projectors are perpendicular to the plane and parallel to each other is known as axonometric projection.* The axonometric projections can be :

(*i*) **Isometric projection :** The projection obtained on a plane of paper when the projectors are parallel but inclined at angle of 30° to the plane of projection is known as isometric projection (see Fig. 1.2).

(*ii*) **Dimetric projection :** In dimetric projection, two planes are equally fore shortened and the two axes are equally spaced (see Fig. 1.2).

(*iii*) **Trimetric projection** : In trimetric projection, all the three faces are unequally fore shortened and the three axes are also unequally spaced (see Fig. 1.2).

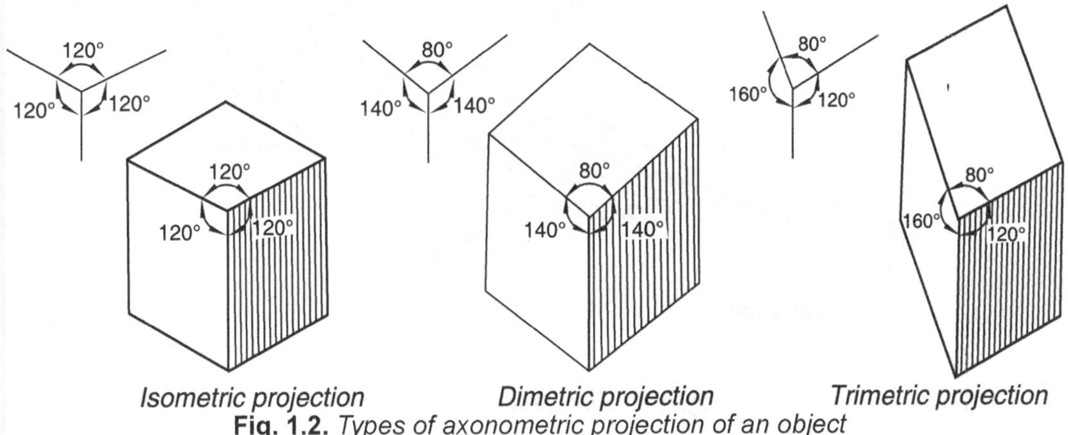

Fig. 1.2. *Types of axonometric projection of an object*

1.3.2. Oblique projection : *The projection of an object on a plane of projection when one face of the object is parallel but the other adjacent face is inclined at an angle of 45° to the plane of projection is known as oblique projection (see Fig. 1.3).*

The *oblique projections* are further discussed as under :—

(*i*) **Cavalier projection :** When the projection lines make angle of 45° with the plane of projection, the projection is called a *cavalier projection* (see Fig. 1.3).

(*ii*) **Cabinet projection :** When the angle that the projecting lines make with the plane of projection is such that the scale on the receding axis in the drawing is about one half as long as the two axes, the result is called a *cabinet projection* (see Fig. 1.3).

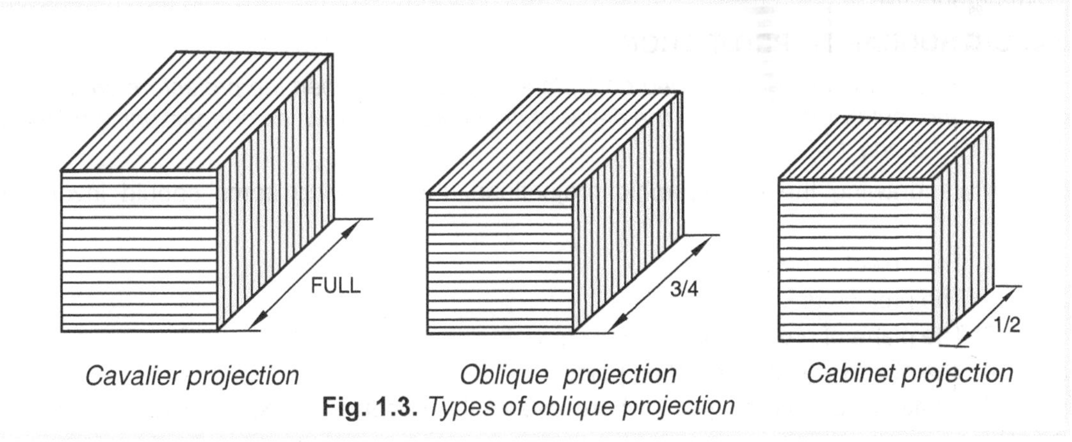

Fig. 1.3. *Types of oblique projection*

(*iii*) **Clinographic projection :** In cavalier and cabinet projections the principal face of the object is made parallel to plane of projection. For some cases it may be desirable to turn the object at an angle w.r.t. the plane of projection and is known as *clinographic projection.*

1.3.3. (c) Perspective Projection : *The projection obtained on a plane when the projectors converge to a point is known as perspective projection (see Fig. 1.4).*

It does not represent the true size of the object.

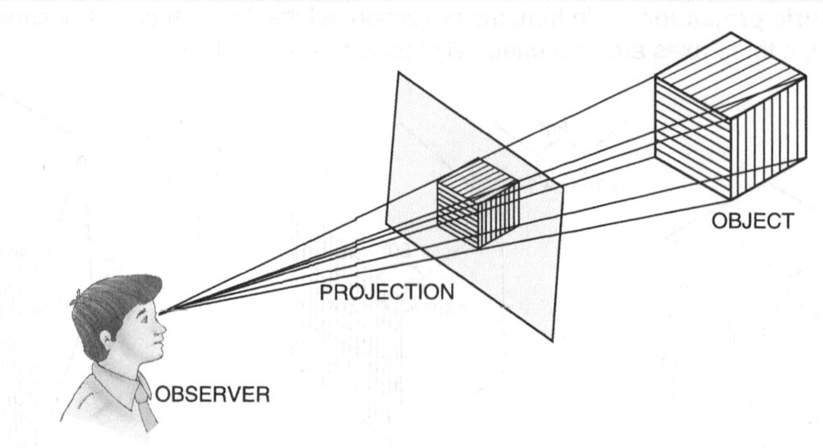

Fig. 1.4. *Perspective projection*

Three kinds of perspective projection may be drawn to serve different purposes as :—

(i) Parallel or One point perspective : If the principal face of the object is parallel to the plane of projection and there is only one vanishing point, the projection is known as a parallel or one point perspective.

(ii) Angular or Two point perspective : When the two faces of the object are at an angle with the plane of projection, where the third face is perpendicular to it, the two principal vanishing points occur and the projection is known as angular or two point perspective.

(iii) Oblique or Three point perspective : If the three principal faces of the object are inclined to the plane of projection, the projection obtained is known as an oblique or three point perspective.

1.4. ORTHOGRAPHIC PROJECTION

The projection or view obtained on a plane of projection when the projectors are parallel to each other but perpendicular to the plane of projection, is known as orthographic projection (see Figs. 1.5 and 1.6).

While drawing the orthographic projections, the following items should invariable exist :—

(1) The object to be projected

(2) The projectors

(3) The plane of projections

(4) The observer's eye or station point

The station point is the point where eye of the observer is located while seeing the object.

1.5. HOW TO GET THE ORTHOGRAPHIC PROJECTION

In order to obtain an orthographic projection of an object proceed as follows :—

Suppose that the plane of projection (transparent plane) is assumed to be situated in front of the object [see Fig. 1.5(i)]. If the imaginary lines, called projectors are drawn to the transparent plane from the various points on the contour of the object, the shaped figure, thus formed by joining the different points in correct sequence will give the projection of the front surface of the object in the true shape and proportion.

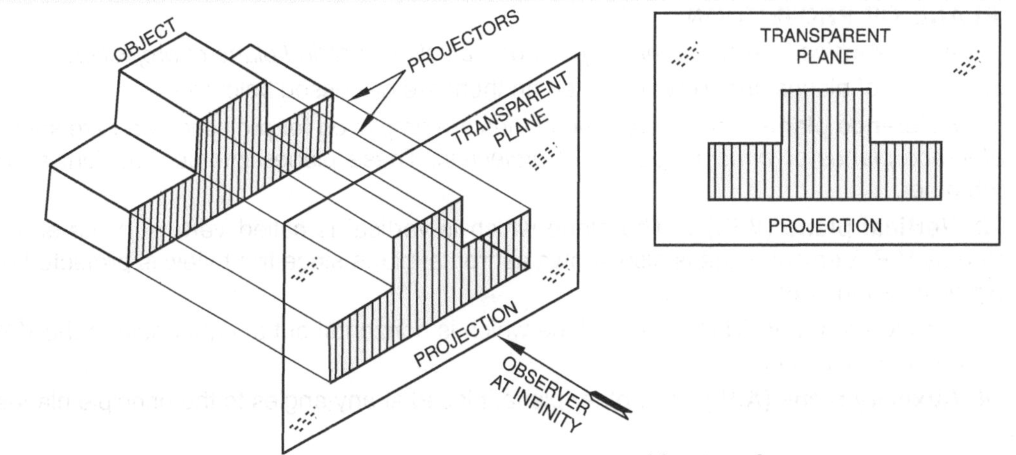

Fig. 1.5. *Orthographic projection of front surface*

Fig. 1.5 (ii) shows the front surface projection which represents only two dimensions of the object viz., the length and height. Since this single view does not show the breadth of the object so one or more additional projections are necessary to complete its description. In general, two projections are usually sufficient to describe the simple objects, but sometimes, however, three or more projections are required for complicated objects.

Important Note : *In case of orthographic projection, either imaginary lines should be considered which should be mutually parallel and perpendicular to plane of projection or the observer should look at the object from an infinite distance so that the rays of sight from his eyes become mutually parallel to each other and perpendicular to the plane of projection.*

Now, to show the third required dimension, consider the transparent plane of projection to be horizontal and draw the projection or image on it. Fig. 1.6 (i) shows the projection of the top surface of the object when the direction of observation is from top, which is marked by an arrow head. For its corresponding orthographic projection see Fig. 1.6 (ii).

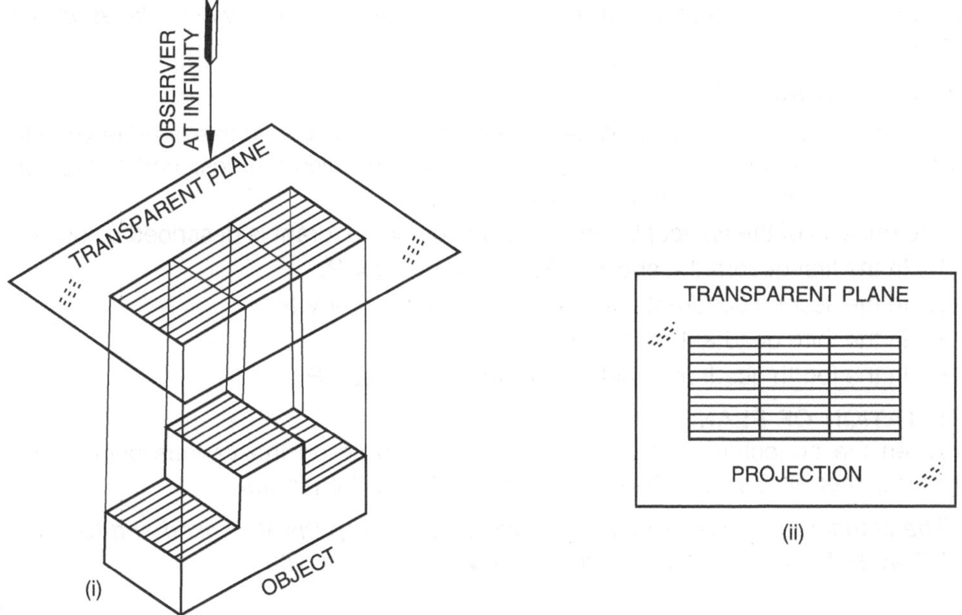

Fig. 1.6. *Orthographic projection of top surface*

1.6. PLANE OF PROJECTION

The plane which is used for the purpose of projection is called plane of projection. Different types of planes and views obtained on them are discussed below :—

1. Reference plane : In general, two planes are employed for projection and are known as *reference planes* or *principal planes of projection.* These planes intersect at right angle to each other.

2. Vertical plane (V.P.) : The plane which is vertical is called *vertical plane* and is denoted by V.P. Vertical plane is also known as frontal plane since front view is projected on this plane (see Fig. 1.6).

3. Horizontal plane (H.P.) : The plane which is horizontal but at right angle to the V.P. is called *horizontal plane.*

4. Auxiliary plane (A.P.) : Any other plane, placed at any angles to the principle planes, is called *auxiliary plane.*

5. Profile plane (P.P.) : The plane which is at right angles to the two principle planes is called *auxiliary vertical plane* (A.V.P.) or profile plane (P.P.).

6. Ground line : The line of intersection of two principle planes of projections, i.e., V.P. and H.P. is called *reference or intersection or ground line* and is denoted by x-y line.

7. Front view or Elevation : The projection of the object on vertical plane (V.P.) is known as *front view or elevation.*

8. Top view or plan : The projection of the object on the horizontal plane is known as *top view or plan.*

9. Side view or side elevation or Profile view : The projection of the object on an Auxiliary Plane (A.V.P.) or profile plane is known as *side view* or *side elevation* or *profile view or end view.*

10. Auxiliary view : The projection of the object on an auxiliary plane is known as *auxiliary view.*

Note : The *front view and top view are called the projections of the object. When we speak about the projections of an object we mean the front view (elevation) and top view (plan).*

1.7. FOUR QUADRANTS

If the horizontal and vertical planes of projection are assumed to extend beyond the line of interaction, the four dihedral (see Fig. 1.7) are formed which are designed as first, second, third and fourth angles or fourth quadrants.

The position of the object placed in any one of the quadrant is described below :—

1. In the first quadrant – above H.P. and infront of V.P.
2. In the second quadrant – above H.P. and behind of V.P.
3. In the third quadrant – below H.P. and behind V.P.
4. In the fourth quadrant – below H.P. and infront of V.P.

1.8. ROTATION OF PLANES

When the projection of an object have been made on the various planes, they are brought together on a single sheet of paper by rotating the planes.

The standard practice of rotation of planes is to keep the V.P. fixed and to rotate H.P. and P.P. away from the object so that they may come in line with V.P.

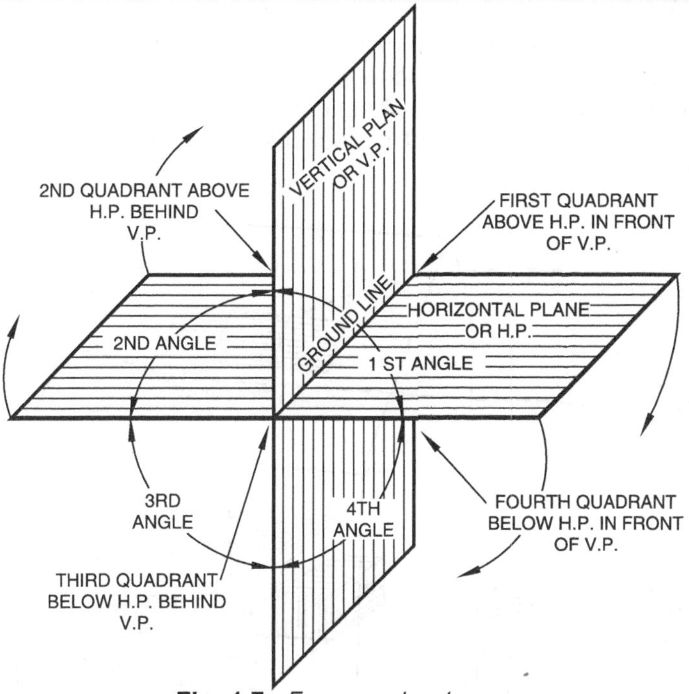

Fig. 1.7. *Four quadrant*

1.9. TYPES OF ORTHOGRAPHIC PROJECTION

The following two types of orthographic projection are used in engineering practice : —

(1) First Angle Projection (2) Third Angle Projection

1.10. FIRST ANGLE PROJECTION

In this type of projection, the object is assumed to be situated in first quadrant. The position of the object can be situated on the H.P. and infront of V.P.

Here, each view is so placed that it represents the side of the object remote (or far from it in adjacent view (see Fig 1.8).

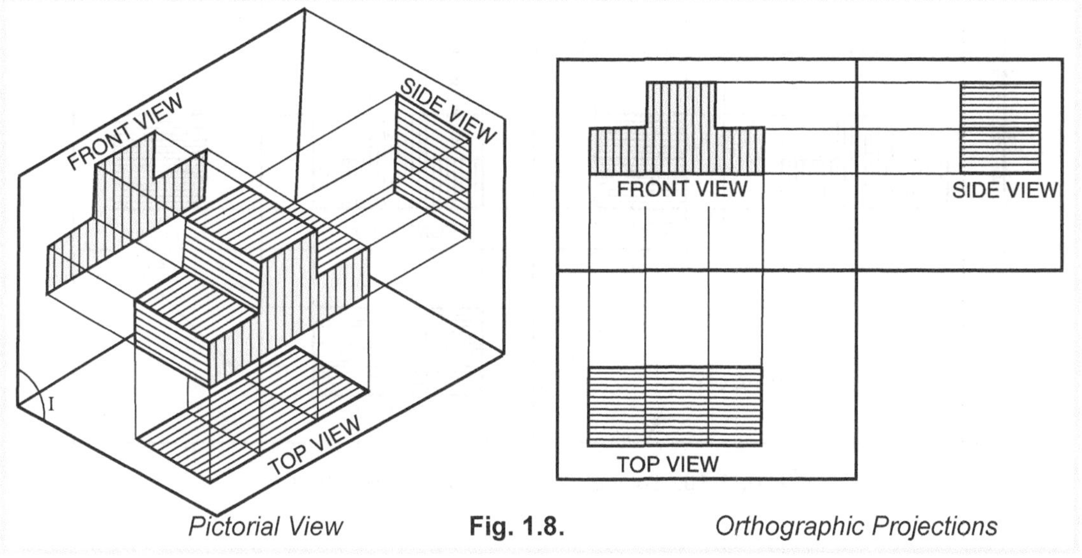

Pictorial View **Fig. 1.8.** *Orthographic Projections*

1.11. SIX VIEWS

If the same object is assumed to be situate in a transparent box (see Fig. 1.9) in which the object itself appears to be suspended in air. Then, the required projections of all the six respective planes of the object on the sheet of paper can be drawn by unfolding the various planes. Fig. 1.10 shows six different projections, along with their names, on a sheet of drawing paper.

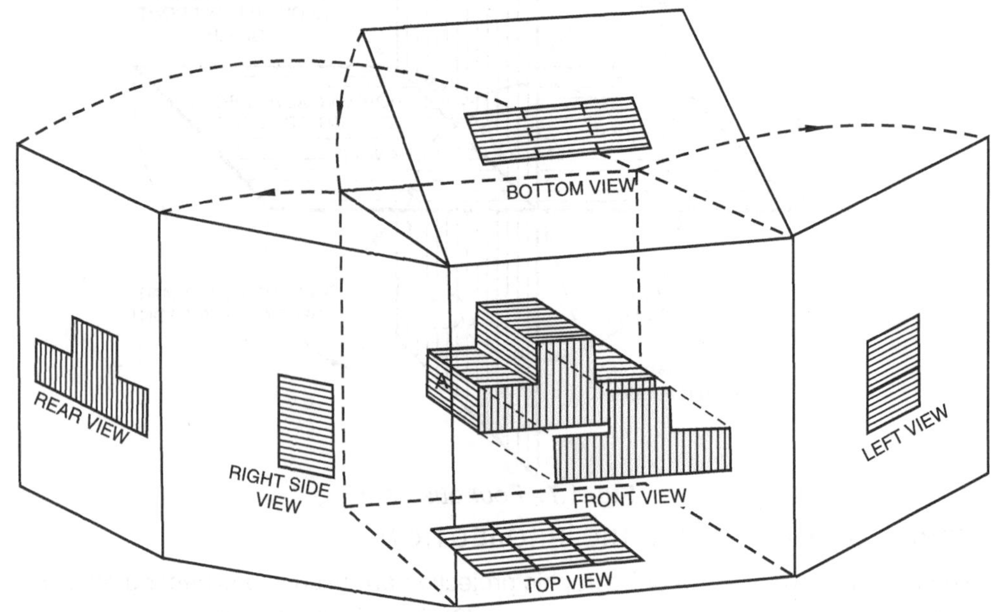

Fig. 1.9. *Orthographic projections of unfolded transparent box*

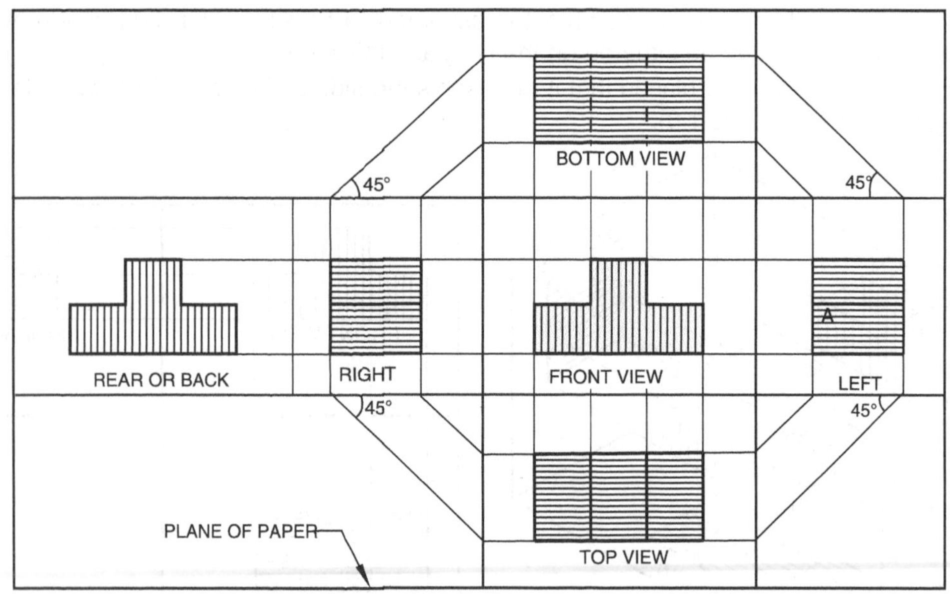

Fig. 1.10. *Six views in first angle projection*

In the first quadrant, with reference to the front view or elevation, the other views are arranged as follows :—

(i) View from below (bottom view) placed above of front view
(ii) View from above (top view) placed below of front view
(iii) View from left (left side view) placed on the right of front view
(iv) View from right (right side view) placed on the left of front view
(v) View from the rear (rear view) may be placed on the left or on the right of the front view as found convenient.

1.12. SYMBOL FOR FIRST ANGLE PROJECTION

According to I.S.I. Fig. 1.11 shows conventional representation symbol for first angle projection.

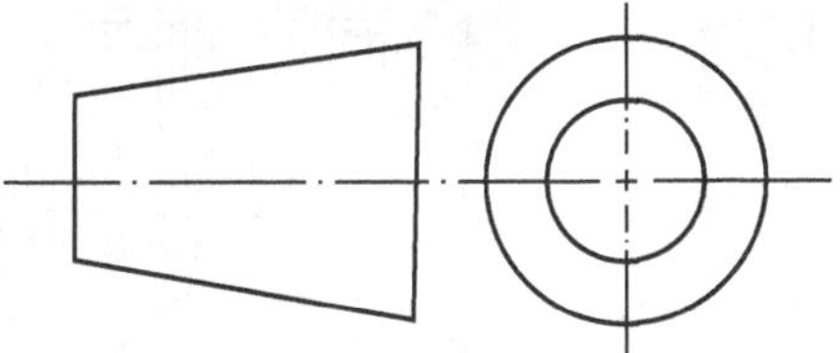

Fig. 1.11. *Symbol for first angle projection*

1.13. THIRD ANGLE PROJECTION

In this type of projection, the object is assumed to be situated in the third quadrant. The situation of the object is said to be below H.P. and behind V.P. Each view is so placed that it represents the side of the object near to it in the adjacent view (see Figs. 1.12 and 1.13).

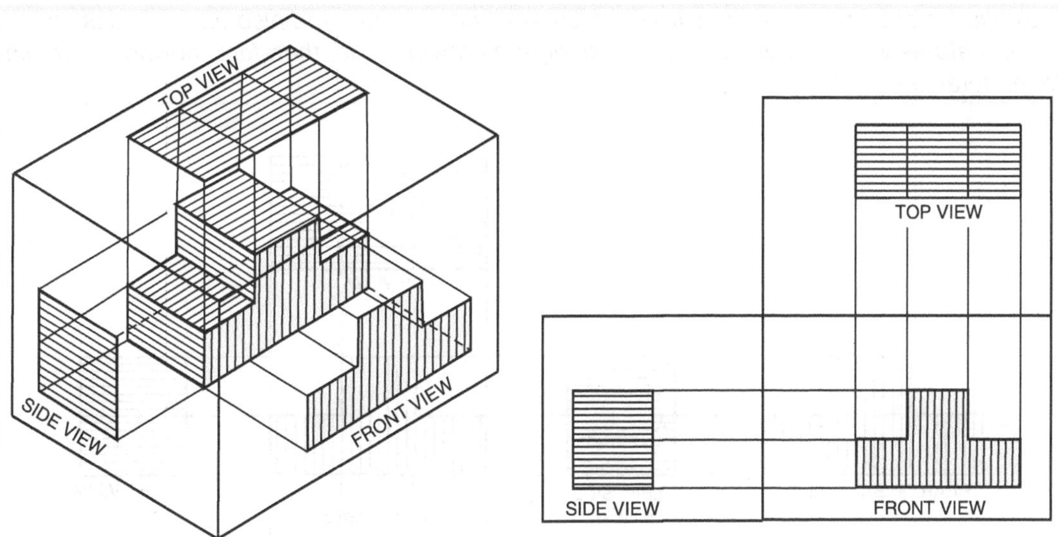

Fig. 1.12. *Pictorial view* **Fig. 1.13.** *Orthographic projections*

1.14. SIX VIEWS

If the same above object is assumed to be situated in a transparent box (see Fig. 1.14) in which the object itself appears to be suspended in air, then the required projections or images of all the six respective planes on a sheet or paper can be drawn by unfolding the various planes. Fig. 1.15 shows six different projections along with their names on a sheet of drawing paper.

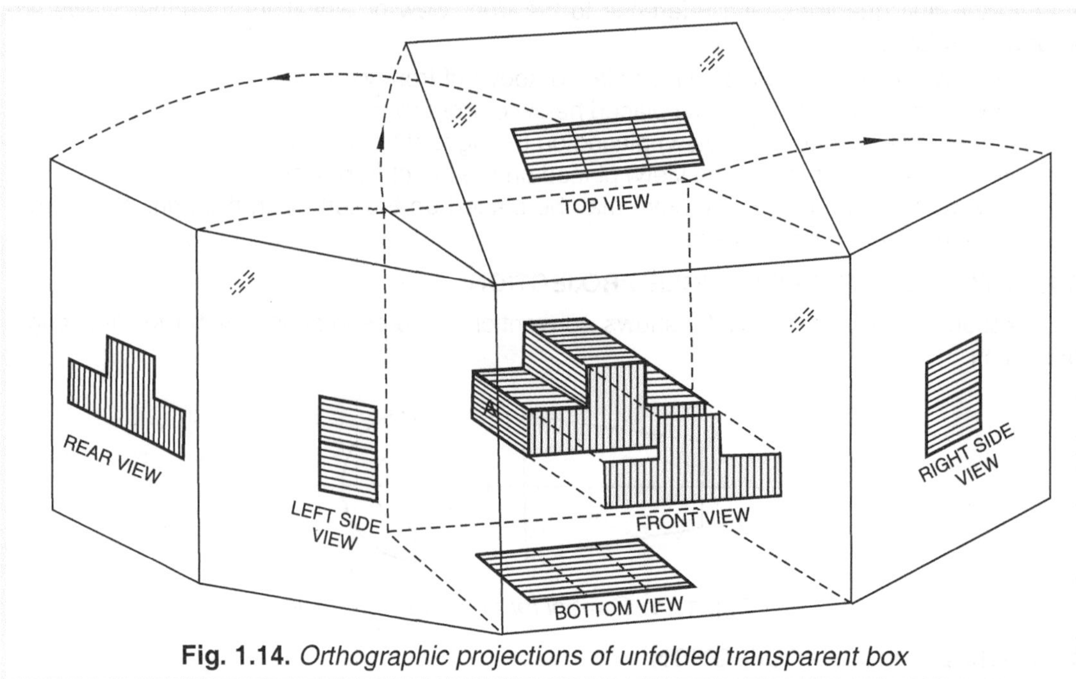

Fig. 1.14. *Orthographic projections of unfolded transparent box*

Note : *This method has an important advantage that the features of adjacent view are in-just a position (or placed nearer in actual contact), thus it is easier than the first angle projection in projecting one view from the other when drawing: and also easier in assignating these features when dimensioning or reading the drawing. It is, therefore, nowadays coming for general use.*

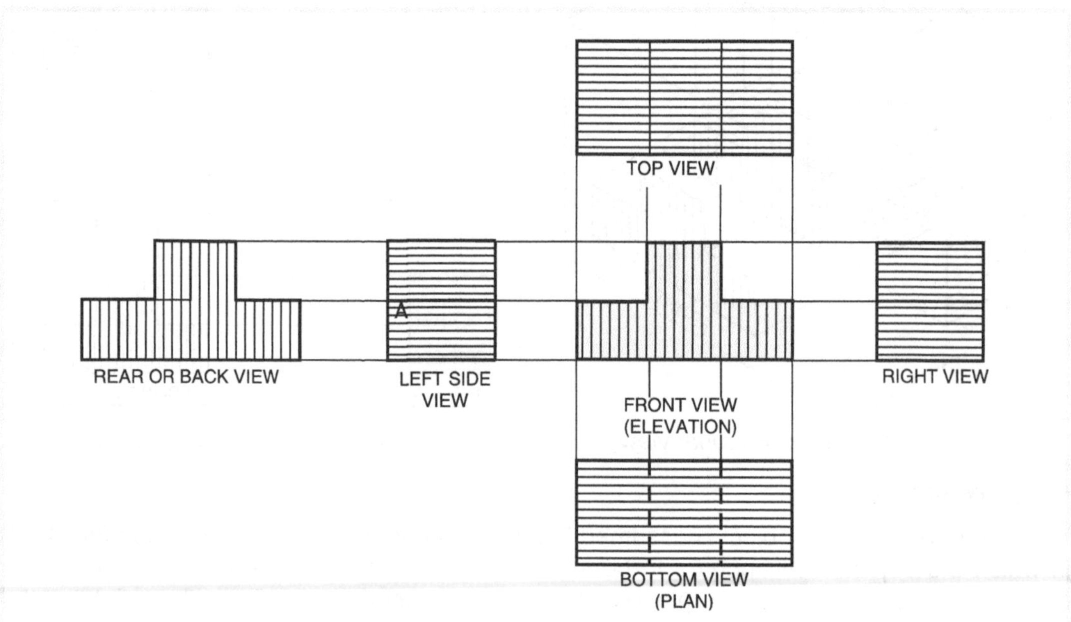

Fig. 1.15. *Six views in third angle projection*

In third quadrant, with reference to the front view or elevation, the other views are arranged as follows :—

(i) View from above (top view) placed above of front view.

(ii) View from below (bottom view) placed below or underneath of front view.

(iii) View from left (left side view) placed on the left of front view.

(iv) View from right (right side view) placed on the right of front view.

(v) View from the rear (rear view) may be placed on the left or on the right of front view as found convenient.

1.15. SYMBOL FOR THIRD ANGLE PROJECTION

According to I.S.I. Fig. 1.16 shows the conventional representation symbol of third angle projection.

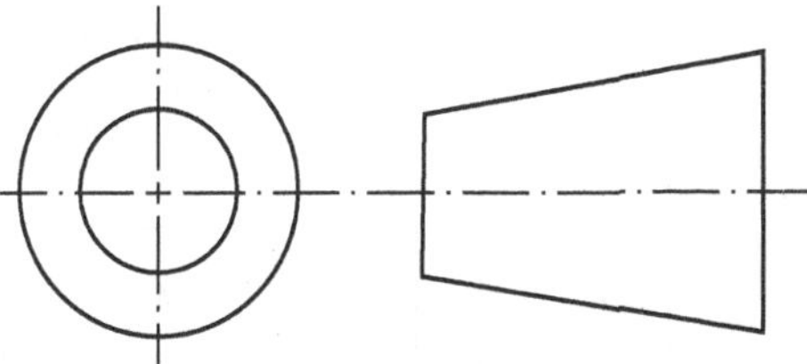

Fig. 1.16. *Symbol for third angle projection*

COMPARISON OF FIRST AND THIRD ANGLE PROJECTIONS

FIRST ANGLE PROJECTION	THIRD ANGLE PROJECTION
1. The object is assumed to be situated in the first quadrant.	1. The object is assumed to be situated in the third quadrant.
2. The object lies in between the observer and the plane of projection.	2. The plane of projection lies in between the observer and the object.
3. View from above (top view) is drawn below of front view.	3. View from above (top view) is drawn above of front view.
4. View from below (bottom view) is drawn above of front view.	4. View from below (bottom view) is drawn below of front view.
5. View from left (left side view) is drawn on the right of front view.	5. View from left (left side view) is drawn on the left of front view.
6. View from right (right side view) is drawn on the left of front view.	6. View from right (right side view) is drawn on the right of front view.
7. View from the rear (rear view) is drawn on left or right of front view.	7. View from the rear (rear view) is drawn on right or left of front view.

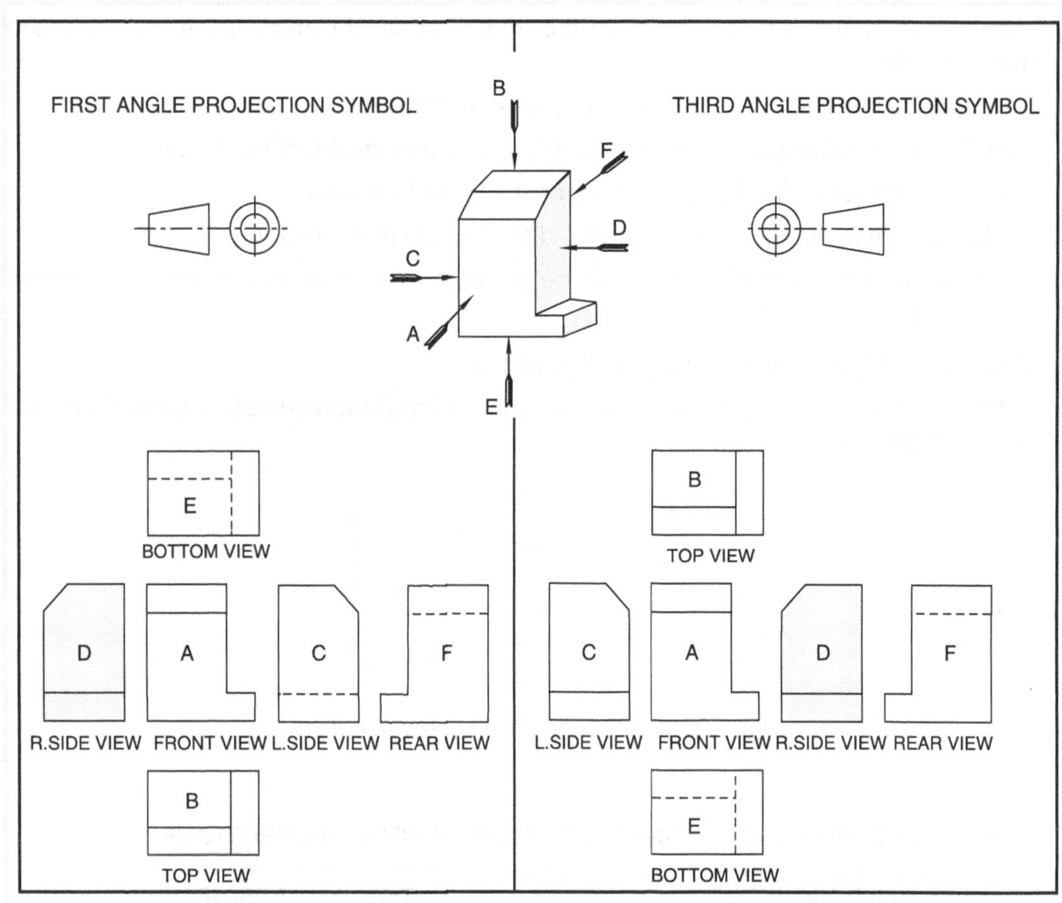

Note : *The first angle projection is almost universally used in Britain as standard practice. Third angle projection is the standard practice followed in America and in the continent of Europe. In our country, the first angle projection was the standard practice, but the Indian Standards Institute has recommended the adoption of third angle projection method since 1960 as a standard practice. In December 1973, the committee responsible for its preparation left the option of selecting first or third angle projection to the user. In September, 1986, the committee again reviewed the situation and finally recommended for implementation of first angle projection method by the year 1991 in our country.*

1.16. SELECTION OF VIEWS

The proper selection of views is of most importance in orthographic drawing. In general, the students should select the least number of views for clear and complete description of an object. Unnecessary or poorly choosen views should be avoided, since they are confusing and waste the valuable time. The proper selection of views is a skill that comes only with practice. In general, the following slection should be followed while selecting the views :—

1. One view drawing : In general, atleast, two views are required to describe the complete shape of a simple object. However, in some cases one view is sufficient, provided a note is given on the drawing. Such drawings are called one view drawings.

For example, *cylindrical objects* can be adequately represented by one view, if the necessary dimensions are indicated by dimeters [see Fig. .1.17 (*i*)]. Similarly one view of the *shim* with a note indicating the thickness is sufficient to describe the shim [see Fig. 1.7(*ii*)]. Also *bolts, screws, shafts* and similar parts may be represented by single view.

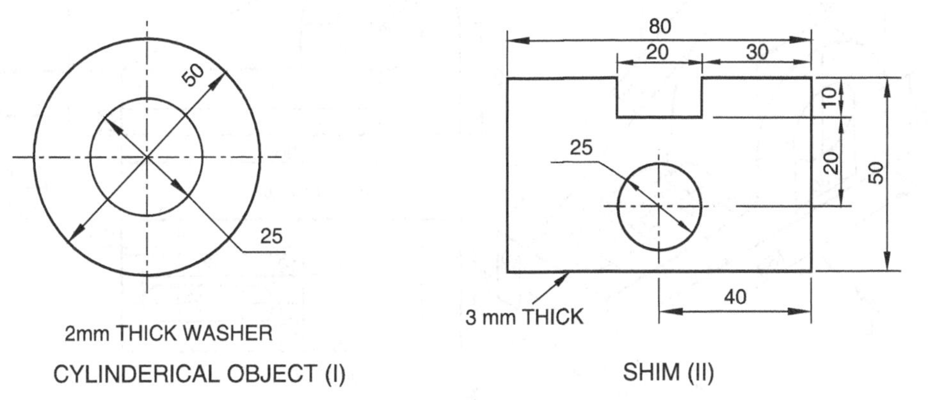

Fig. 1.17. *One view drawing*

2. Two view drawing : Many objects can be described completely if atleast two properly choosen views are drawn. The choice of the views is determined by descriptive value of the view.

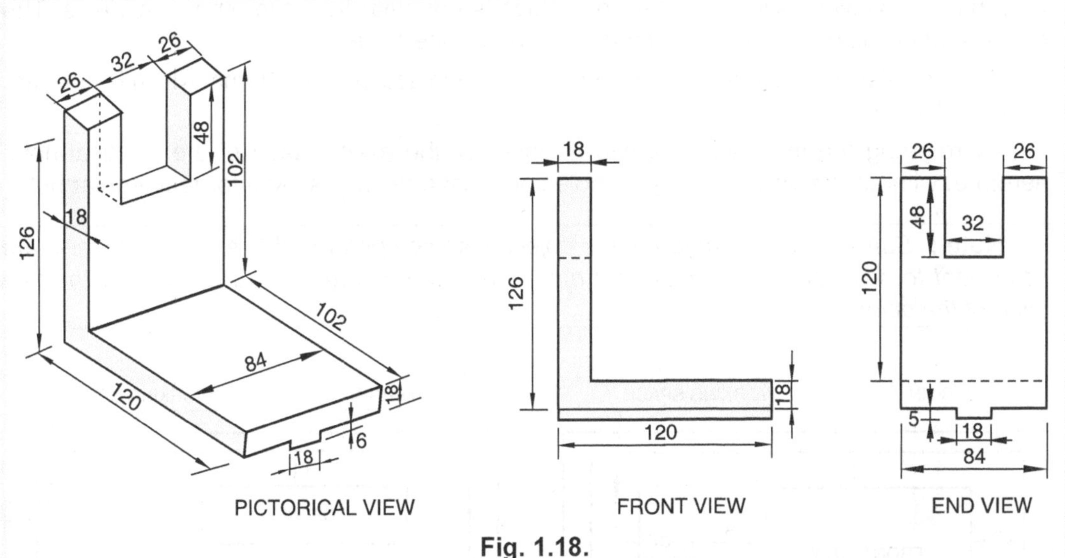

Fig. 1.18.

Fig. 1.18 shows two views of a cast iron block. Here, the front view shows the length and height of the base. The dotted line in the front view also indicates that there is a slot in the vertical portion. But without the side view it would not be possible to tell any thing about the shape of this slot. By referring to the side view, it is evident that the slot is of rectangular shape 48 × 32 mm size. The side view also shows that the base has a projecting section of 18 × 6 mm size.

3. Three view drawing : When the three views are required to describe the shape of an object, the drawing is known as three view drawing. The most common combination of views is that of front view, top view and side view.

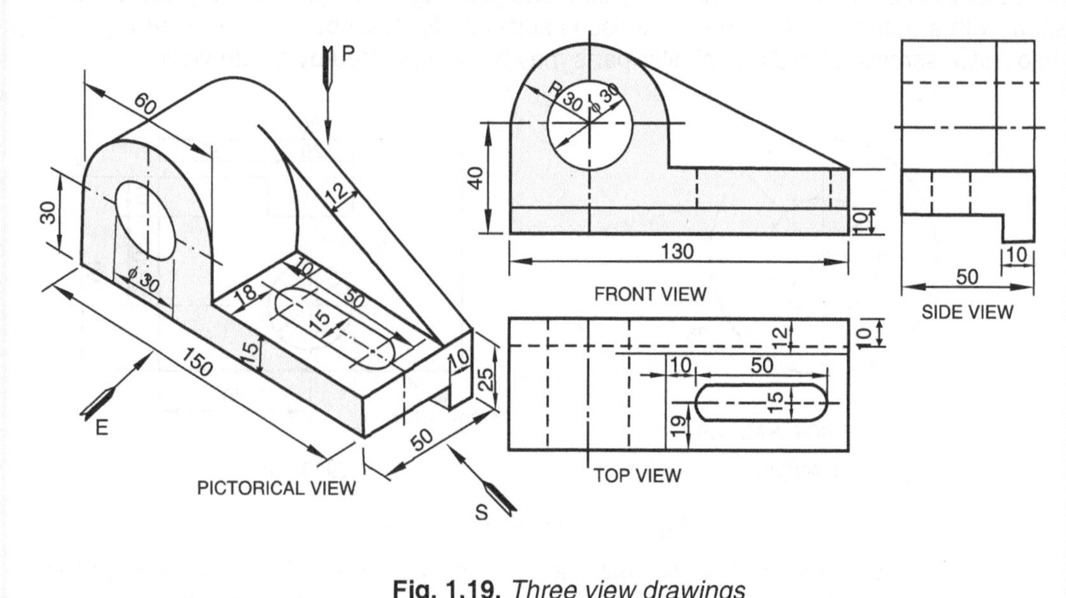

Fig. 1.19. *Three view drawings*

Fig. 1.19 illustrates a typical example of an object which requires atleast three views. Here, the front view shows a circular hole which is not clear from the other two views. The front view also shows the contour of the rib as an inclined line.

The side view shows that the base has a projecting section 10 x 10 mm which is not clear from the other views.

By referring the top view, one can at once tell the exact shape of the elliptical hole. Therefore, without this view, the elliptial hole could have been mistaken for rectangular hole.

Note : *Sometimes, the shape of the object is so complicated that even three views are in-sufficient to describe it completely. So more than three views are drawn to show the true shape of the object.*

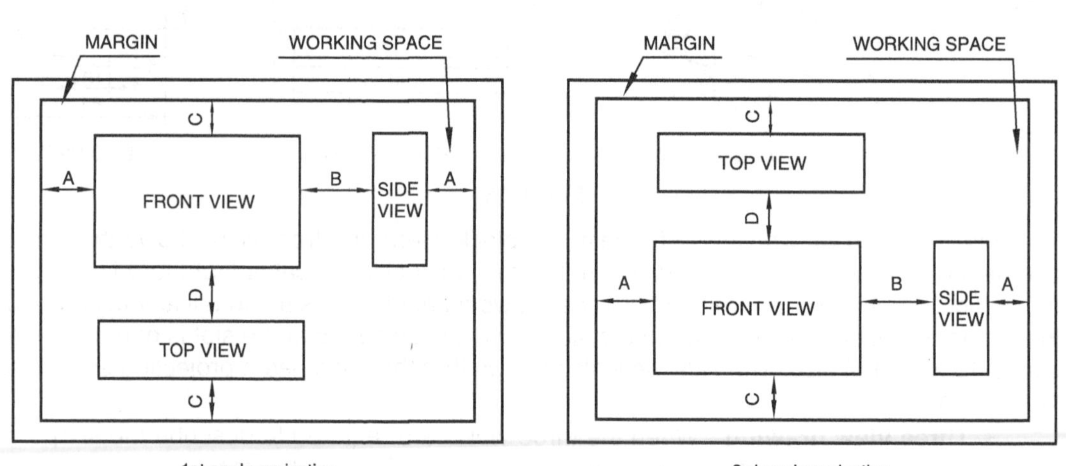

Fig. 1.20. *Spacing of three view drawing*

1.17. SPACING OF VIEWS

Before commencing a drawing, it is of most importance to obtain an idea of the space required for different views on the drawing sheet. When there is only one view, it should be drawn in the centre of the working space. For more than one view, the space should be divided into suitable rectangles and each view should be drawn in the centre of its respective rectangle. In Fig. 1.20, the space A and A should be kept almost equal. Space B should be equal to or slightly less than either of space A.

Space C and C should be nearly equal and space D should be equal to or slightly less than the space C. Further, adjustment for giving dimensions, notes, etc. on the particular drawing should be made accordingly.

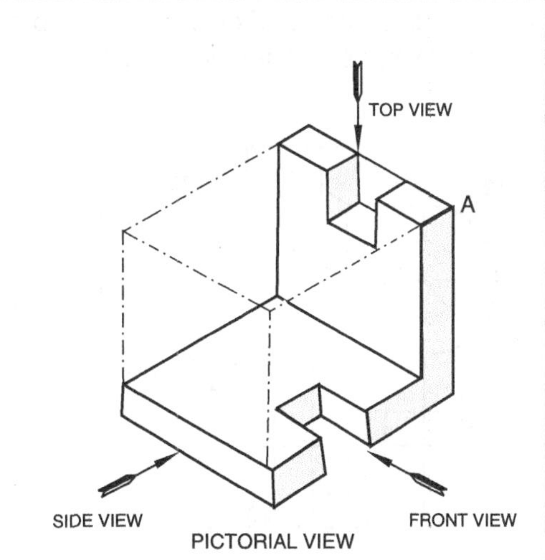

1.18. WAYS OF PROJECTING THREE VIEW DRAWINGS

Fig. 1.21 illustrates the ways of projecting three views from one view to other. In Fig. 1.21 (i) depth dimensions from the top view to the side view and vice-versa are transferred by means of a 45° line. The right side view can be drawn to the left or the top view can be drawn downward by shifting the 45° line accordingly.

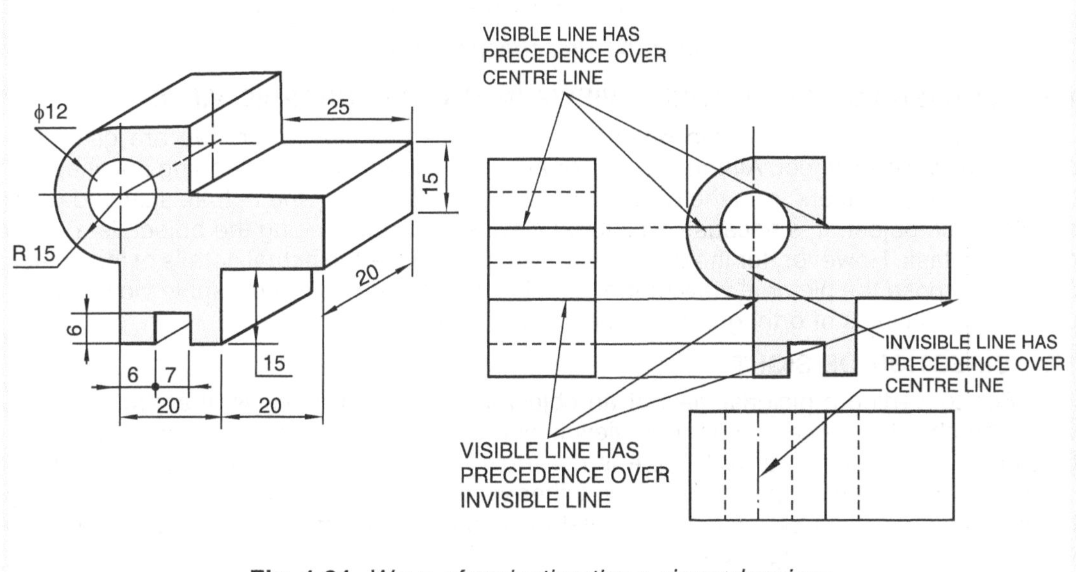

Fig. 1.21. *Ways of projecting three views drawings*

In actual pratice, the depth dimension from top to the side view or vice versa are transferred by using the compass as shown in Fig. 1.21 (ii) This method of transferring dimensions is quicker and more accurate than the 45° line mehod.

1.19. PRECEDENCE OF LINES

In a drawing, solid lines, centre lines and dotted lines often coincide and so a definite precedence has to be set to avoid any confusion. Order of precedence of lines should be as.

(i) When a centre line coincides with a solid line, the solid line will take precedence over dotted line (see Fig. 1.22).

(ii) When a dotted line coincides with a solid line, the solid line will take precedence over center line. The centre line is then indicated by extending at both ends.

(iii) When a centre line coincides with cutting plane line, the one which is more important for the understanding of drawing, takes precedence over the other.

(iv) Dimension lines and extension lines take precedence over section lines.

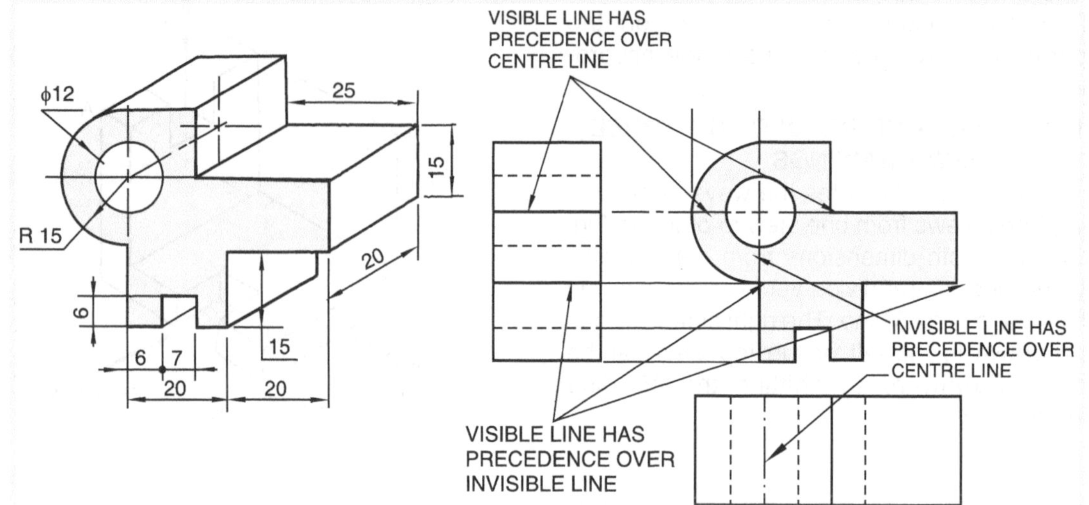

Fig. 1.22. *Precedence of lines*

1.20. CONVERSION OF PICTORIAL VIEWS INTO ORTHOGRAPHIC VIEWS

In engineering drawing, we often come across a situation where we are given the pictorial view of the object. Although, a pictorial view helps us in understanding the shape of the object, yet, it suffers from the drawback that it fails to convey the actual, size and inner details of the object. It is because a pictorial view is drawn by seeing the object, is a three directional task. However, for the design purpose, we require the actual details of the object. For this purpose the pictorial view of the object is converted into orthographic views by applying the principles of orthographic projections.

1.21. DIRECTION OF SIGHT

For converting a pictorial view of an object into othographic views, the direction from which the object is viewed for its front view is generally indicated by arrow. The arrow must be parallel to the sloping axis. If there is no arrow, the direction for the front view may be decided to give the most prominent view. Other views (top and side views) are obtained by looking in the direction perpendicular to first direction and are drawn in correct relationship with the front view (see Fig. 1.23).

1.22. ORTHOGRAPHIC VIEWS OF SIMPLE OBJECTS

Fig. 1.23 shows the pictorial view of a rectangular plate. Its front view when seen in the direction of arrow X, side view in the direction of arrow y and top view in the direction of arrow z are shown in Fig. 1.24.

When the position of the same plate is changed with the longest edge vertical (see Fig. 1.25), its front view when seen in the direction of arrow 'X', side view in the direction of arrow 'Y' and top view in the direction of arrow 'Z' are shown in Fig. 1.26.

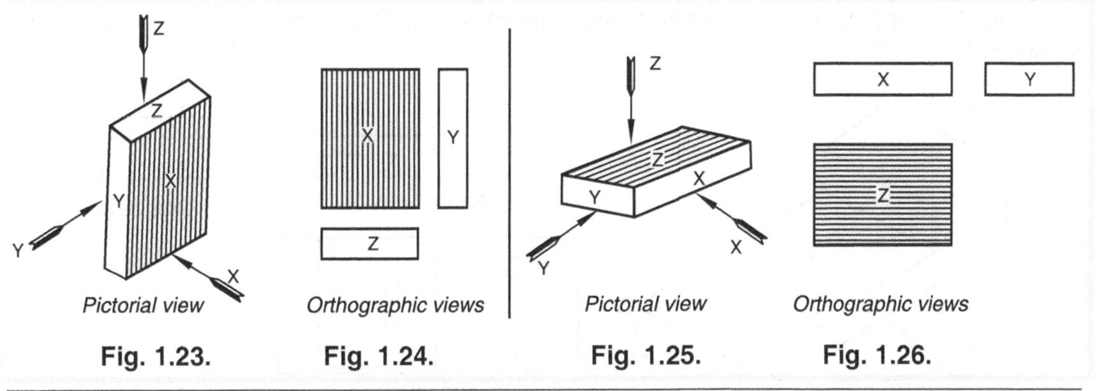

Pictorial view — Orthographic views — Pictorial view — Orthographic views

Fig. 1.23. **Fig. 1.24.** **Fig. 1.25.** **Fig. 1.26.**

Note : *By changing the position of plate, only position of views have been changed.*

Fig. 1.27 shows a plate cut into three different ways. The front and side views in each case are the same. The top view in each case will show a different shape (see Fig. 1.28).

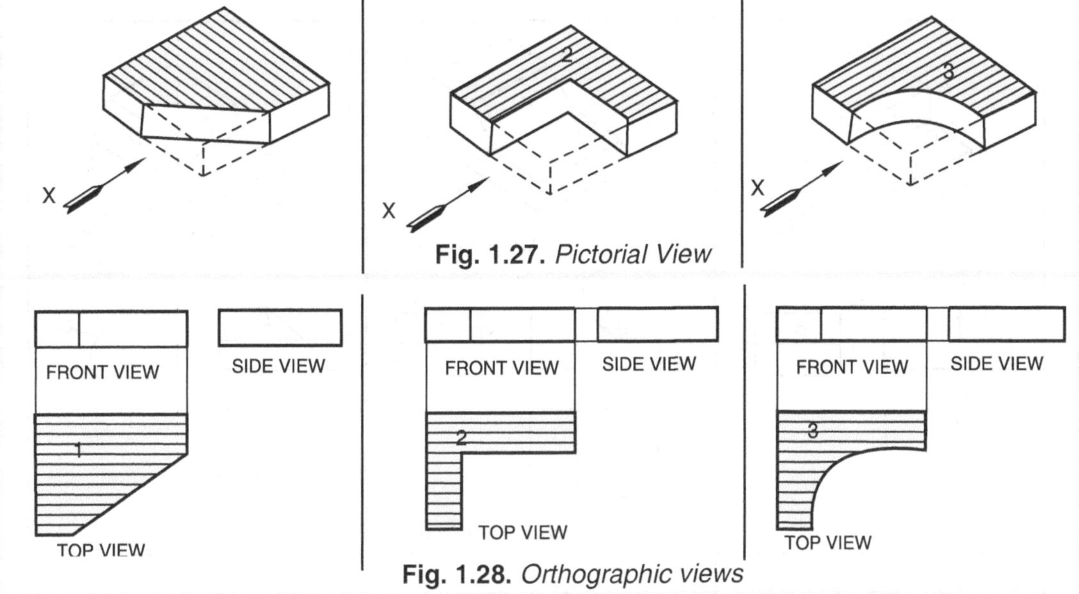

Fig. 1.27. *Pictorial View*

Fig. 1.28. *Orthographic views*

Fig. 1.29 shows a plate with a triangular, circular and square holes. The front view, side view and top view along with hidden lines of holes are shown in Fig. 1.30.

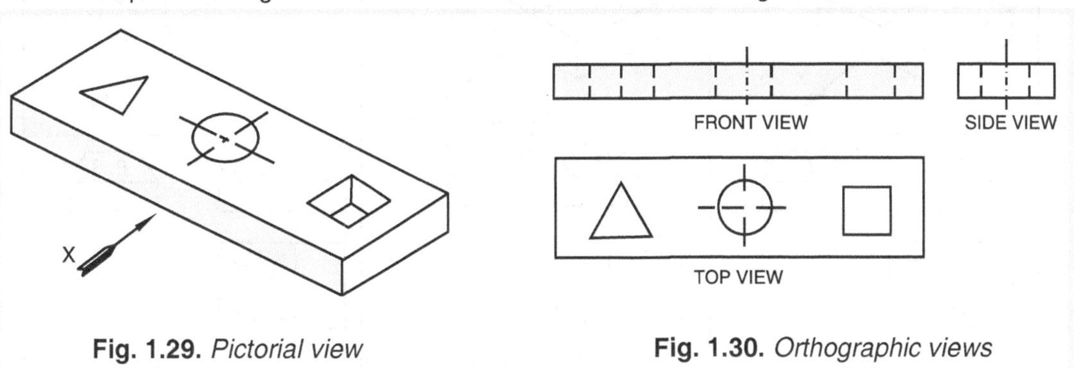

Fig. 1.29. *Pictorial view*

Fig. 1.30. *Orthographic views*

Fig. 1.31 shows a plate cut into three different ways. The front, top and side views are shown in Fig. 1.32.

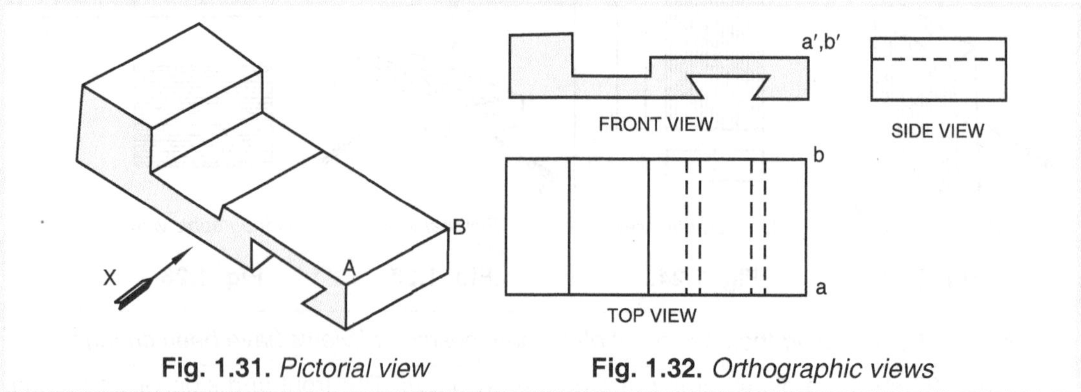

Fig. 1.31. *Pictorial view* **Fig. 1.32.** *Orthographic views*

Figs. 1.33 and 1.35 show pictorial views of different shapes of the blocks in the form of steps. Their front, side and top views along with inclined and straight lines are shown in Figs. 1.33 and 1.36.

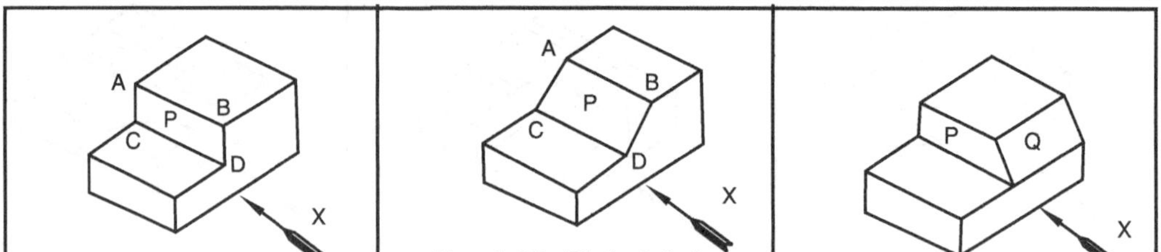

Fig. 1.33. *Pictorial view*

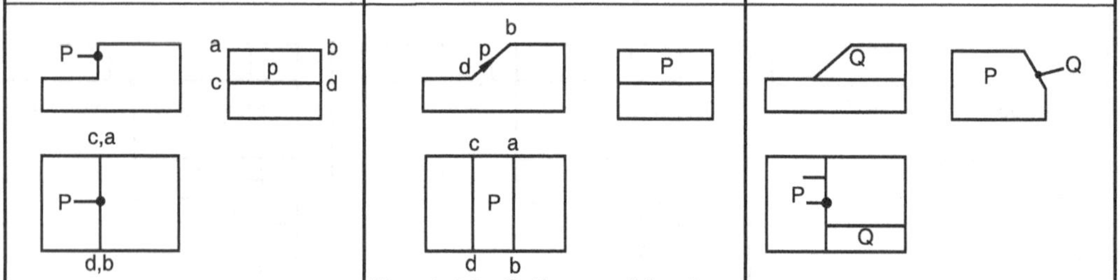

Fig. 1.34. *Orthographic views*

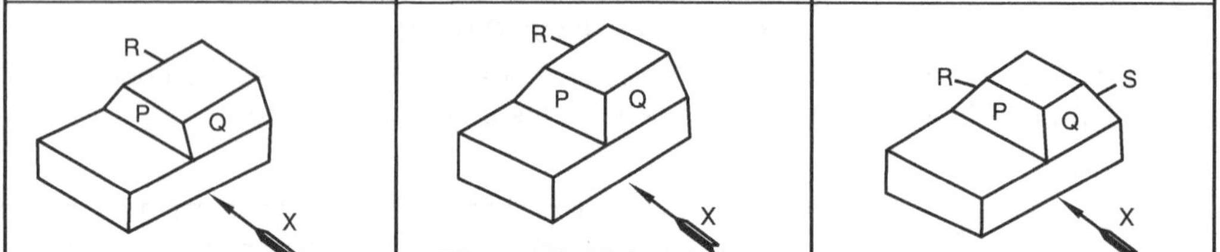

Fig. 1.35. *Pictorial view*

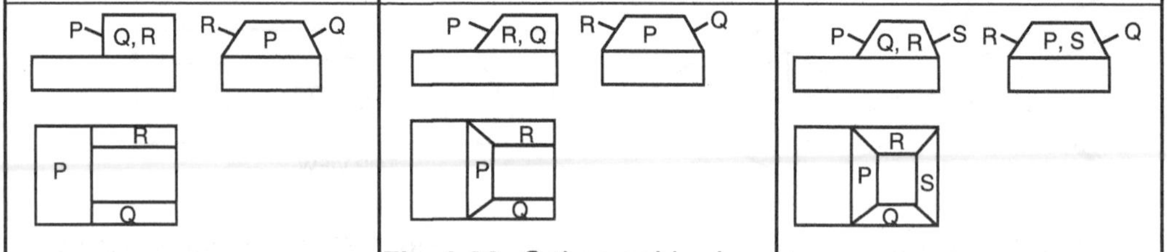

Fig. 1.36. *Orthographic views*

1.23. INVISIBLE LINES IN ORTHOGRAPHIC VIEWS

The parts of the object not visible to the observer are represented by dotted lines on the drawing.

The dotted lines should be carefully drawn, otherwise the drawing would be difficult to read. Fig. 1.37 shows the correct and incorrect ways of drawing the dotted lines. The following points should be kept in mind while drawing the invisible and visible lines :—

1. An invisible line should intersect a visible line with a dash in contact.

2. An invisible line should intersect another invisible line at crossing point of two dashes in contact.

3. Invisible lines meeting at a corner should have two dashes at the centre.

4. Three invisible lines meeting at a corner should have three dashes intersecting at a corner.

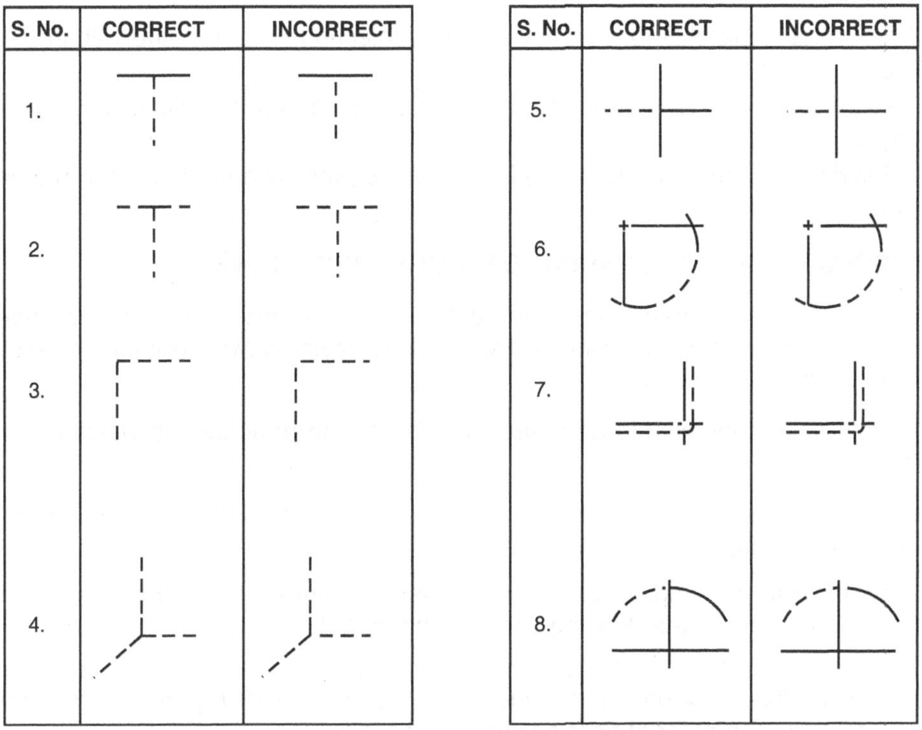

S. No.	CORRECT	INCORRECT
1.		
2.		
3.		
4.		

S. No.	CORRECT	INCORRECT
5.		
6.		
7.		
8.		

Correct and incorrect way of drawing dotted lines

Fig. 1.37.

5. An invisible line can be shown as a continuous line of a visible line, the invisible line begins with a space.

6. Invisible arcs should begin with a dash.

7. When arcs are too small to be made dashed, they should be made solid.

8. When two invisible arcs meet, the intersection at the point of tangency should be located with a dash on each arc.

1.24. GENERAL PRINCIPLES OF ORTHOGRAPHIC DRAWING

The following principles should be thoroughly understood before drawing orthographic projections (Fig. 1.41) :—

(i) The top view (plan) and front view (elevation) always should be vertical.

(ii) The front view (elevation) and side view (end elevation) should always be horizontal.

(iii) The length of the top view should always the same as the length of the front view.

(iv) The breadth of the top view should always the same as the breadth of the side view.

(v) The height of the side view should always the same as the height of the front view.

(vi) If a line is parallel to a plane of projection, then it will show its true length on that plane.

(vii) If a line is perpendicular to the plane of projection, then it will represent shorter length than true length on the plane of projection.

(viii) If a line is perpendicular to a plane of projection, then its projection on that plane will be a point.

(ix) If a surface is parallel to a plane, then its projection on that plane will show its true shape and size.

(x) If a surface is inclined to a plane, then the projection on that plane will be foreshortened.

(xi) If a surface is perpendicular to a plane, then its projection on that plane will be a line simply.

1.25. PROCEDURE FOR PREPARING ORTHOGRAPHIC VIEWS

The procedure for preparing orthographic views is very important, as speed and accuracy largely depend upon the method to be used in laying down the lines. The systematic working of drawing saves time.

The following steps should be noted while preparing orthographic views of an object (see Fig. 1.38).

(1) Determine overall dimensions of the required views and select a suitable scale so as to accommodate the required views on drawing sheet.

(2) Draw different rectangles for the views, keeping suitable space between them and from the borders of the drawing sheet (see Fig. 1.39).

(3) Draw the centre lines in all the views. When any cylindrical portion or hole is seen as rectangle, draw only *one centre line*, but if the circle is visible, *draw two centre lines* intersecting each other at right angles at its centre.

(4) Draw simultaneously the required details in different rectangles in the following, i.e.,

  (i) the circles and arcs of circles.

  (ii) the straight lines for proper shape of the object.

  (iii) the straight lines, the small curves, etc. for minor details.

(5) After completion of the required views, rub off all unnecessary lines except the centre lines and mark the outlines so as to give good appearance.

(6) Give the dimensions, the scale and print the title along with the other required particulars such as notes, etc.

(7) Check the drawing carefully and see that it is complete in all respect.

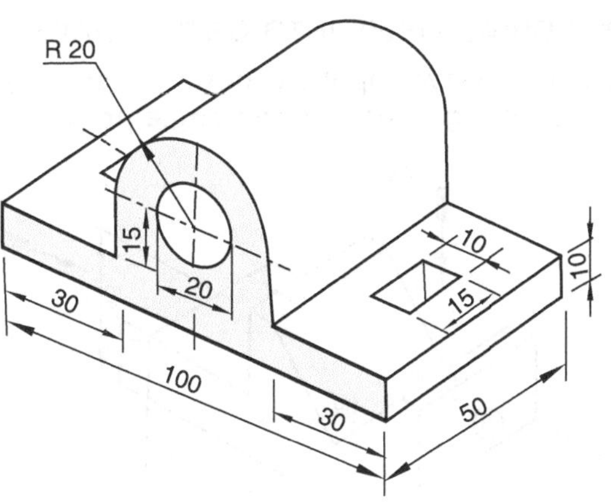

Fig. 1.38. *Pictorial view*

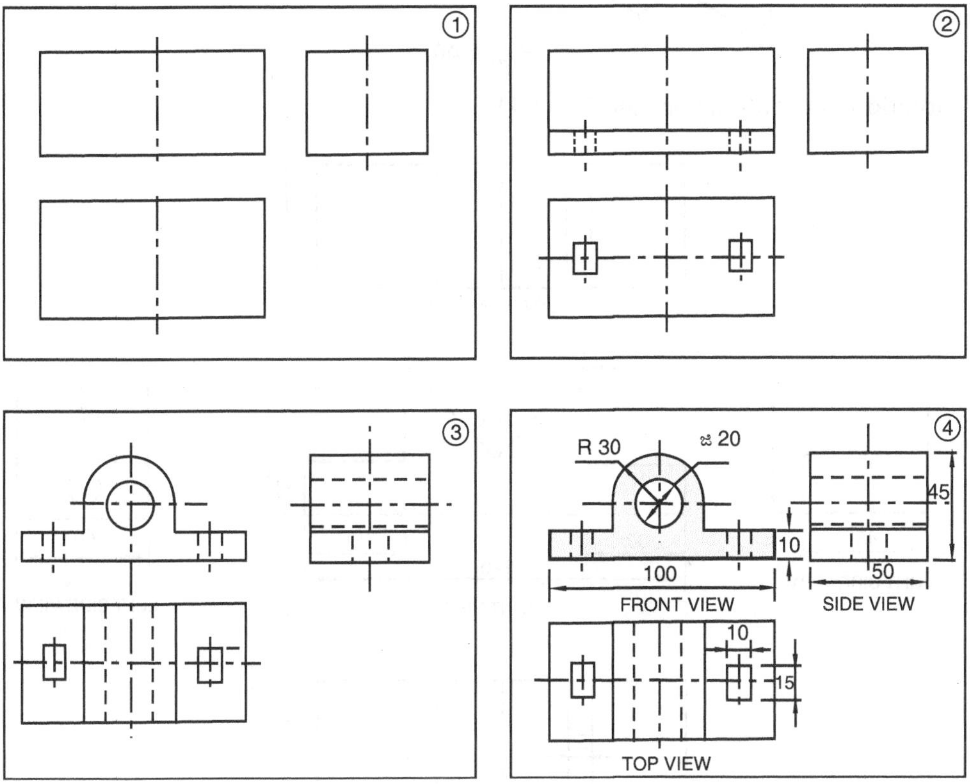

Fig. 1.39. *Steps for preparing orthographic views*

PROBLEMS ON

CONVERSION OF PICTORIAL VIEWS INTO ORTHOGRAPHIC PROJECTIONS

Problem 1 : Draw the front view, top view, bottom view, left side view and right side view of a given object (see Fig. 1.40) by first angle projection method.

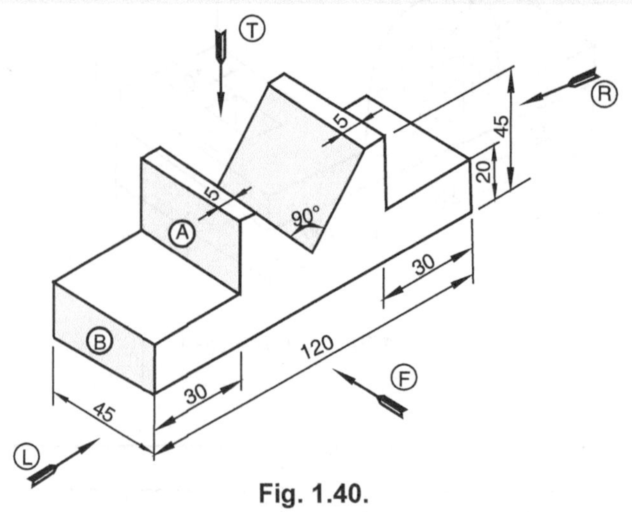

Fig. 1.40.

Solution : For its solution, see Fig. 1.41.

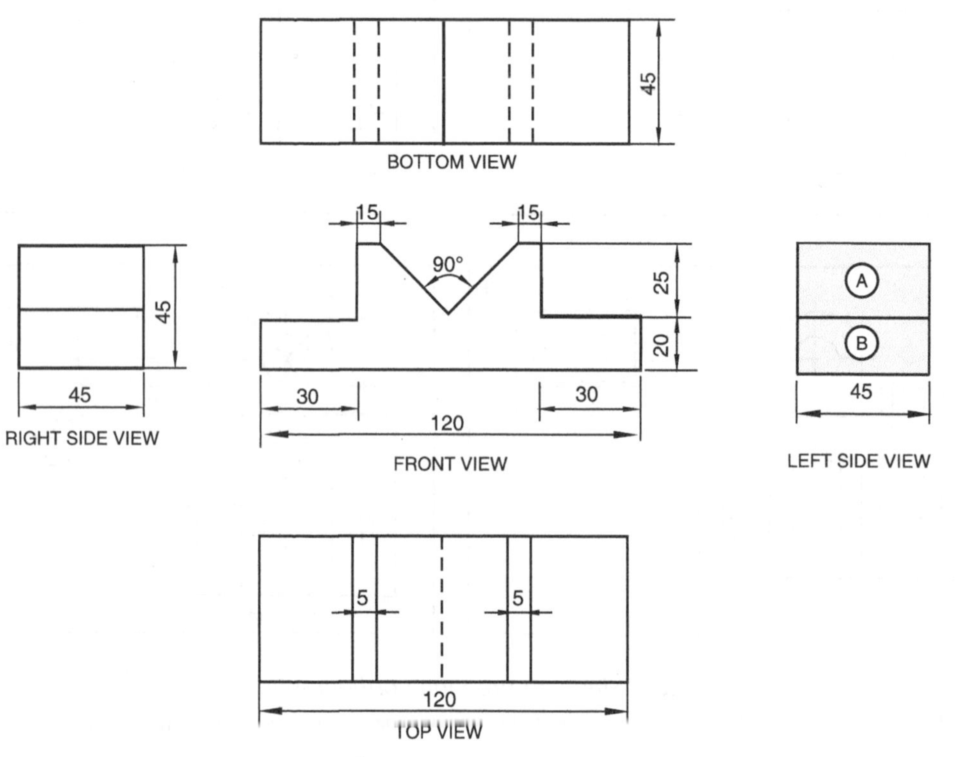

First angle projection

Fig. 1.41.

Problem 2 : Draw the front view, top view, side views and bottom view of a given object (see Fig. 1.40) by third angle projection method. [Fig. 1.42]

Solution : For its solution, see Fig. 1.42.

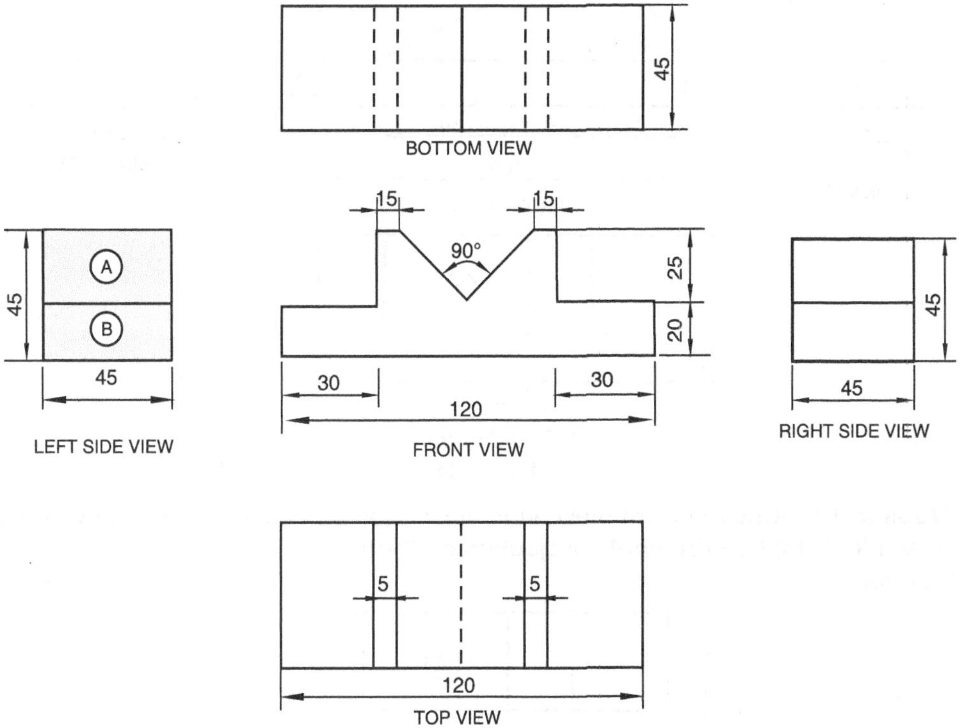

Fig. 1.42. *Third angle projection*

Problem 3 : Draw the front view, top view and side views of a given object (see Fig. 1.43) by first angle projection method.

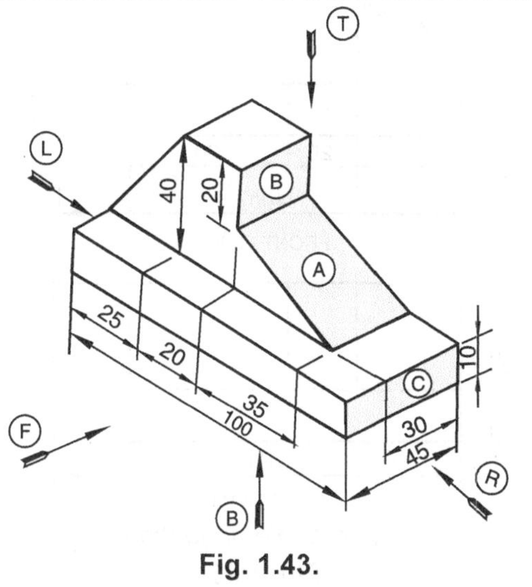

Fig. 1.43.

Solution : For its solution, see Fig. 1.44.

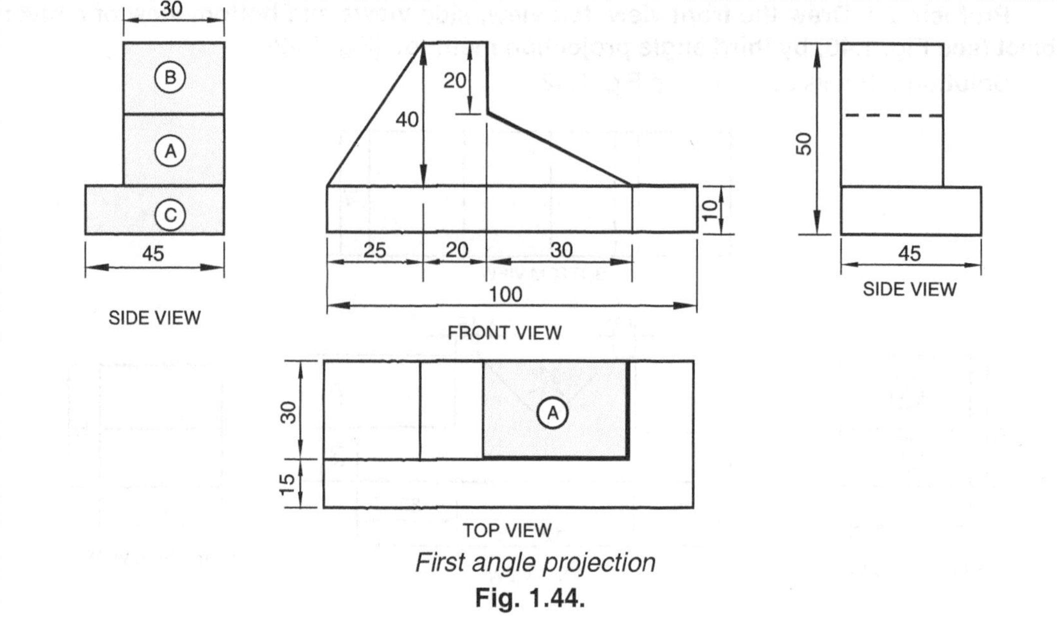

First angle projection

Fig. 1.44.

Problem 4 : Draw the front view, side views, top view and bottom view of a given block (see Fig. 1.43) by third angle projection method.

Solution :

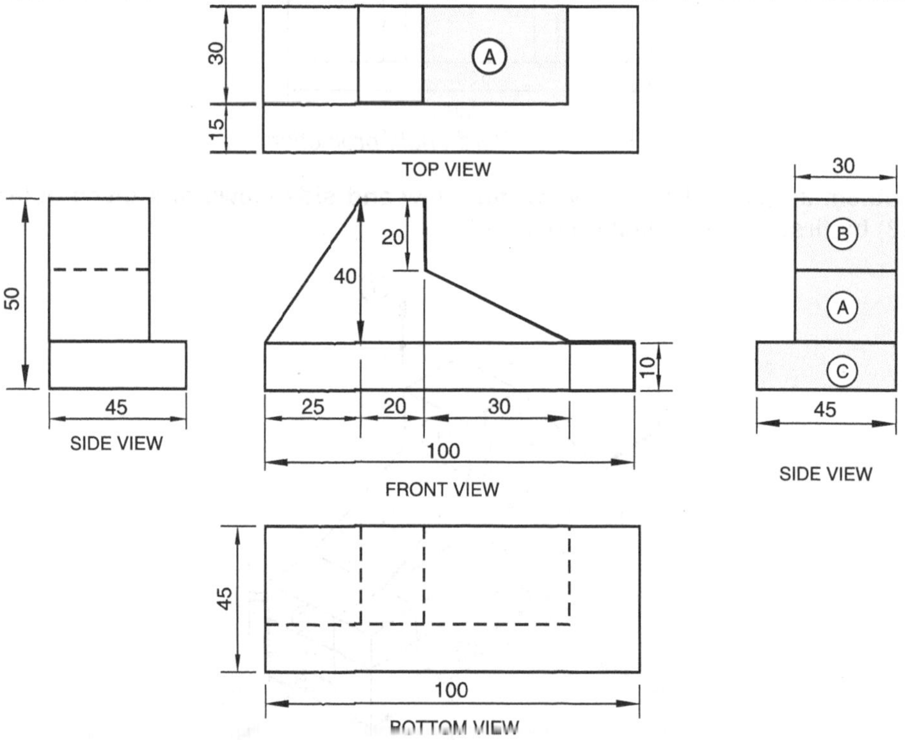

Third angle projection

Fig. 1.45.

Solution : For its solution, see Fig. 1.45.

Problem 5 : Fig. 1.46 shows pictorial view of an object. Draw the following views in first angle projection.

(i) Front view in the direction of F

(ii) Right-side view and (iii) Top view

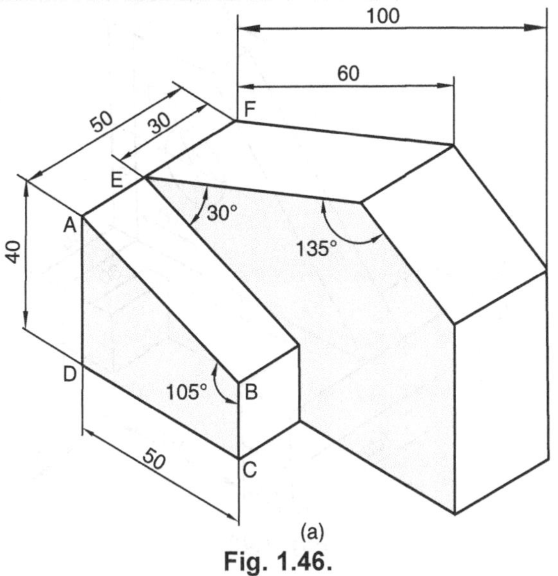

Fig. 1.46.

Solution : For its solution, see Fig. 1.47.

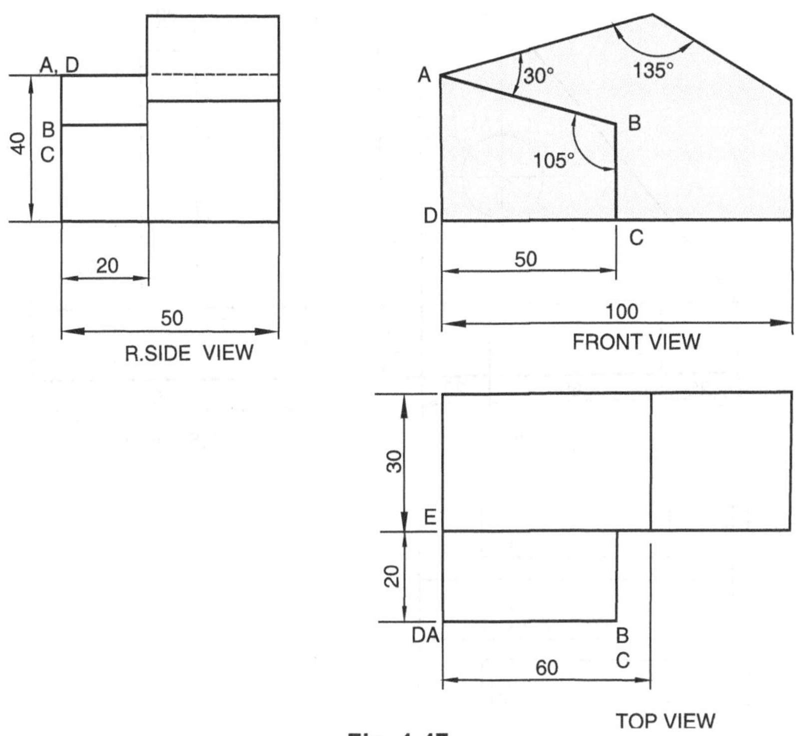

Fig. 1.47.

Problem 6 : Fig. 1.48 shows isometric view of an object. Draw the following views.

(i) Front view in the direction in the directions of A.

(ii) Side view and (iii) Top view

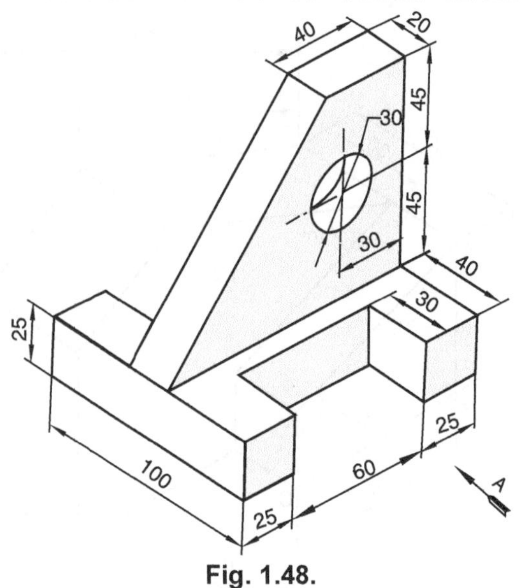

Fig. 1.48.

Solution : For its solution, see Fig. 1.49.

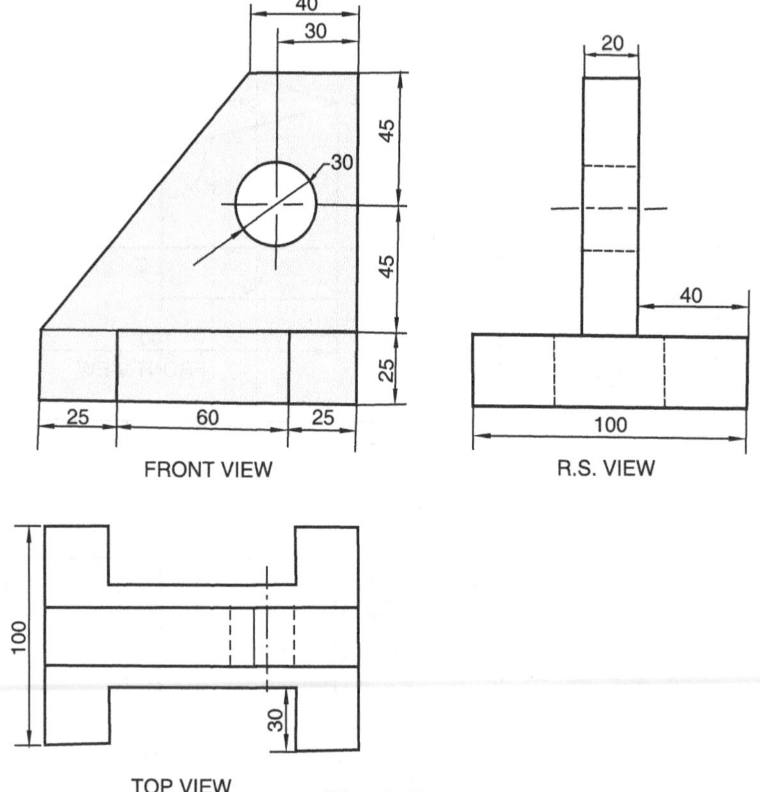

Fig. 1.49.

Problem 7 : Draw the top view, front view and side view of the given object (see Fig. 1.50) by following third angle projection.

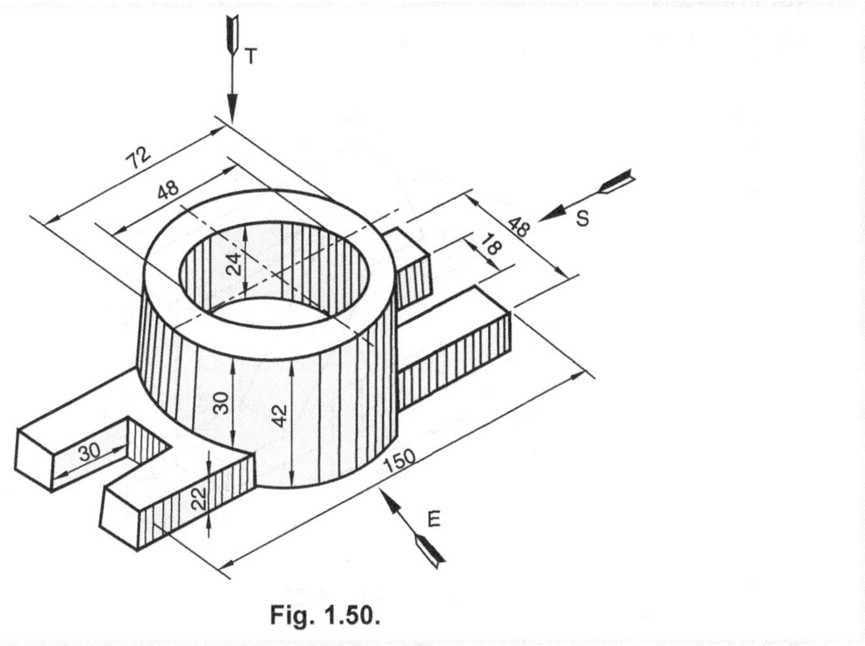

Fig. 1.50.

Solution : For its solution, see Fig. 1.51.

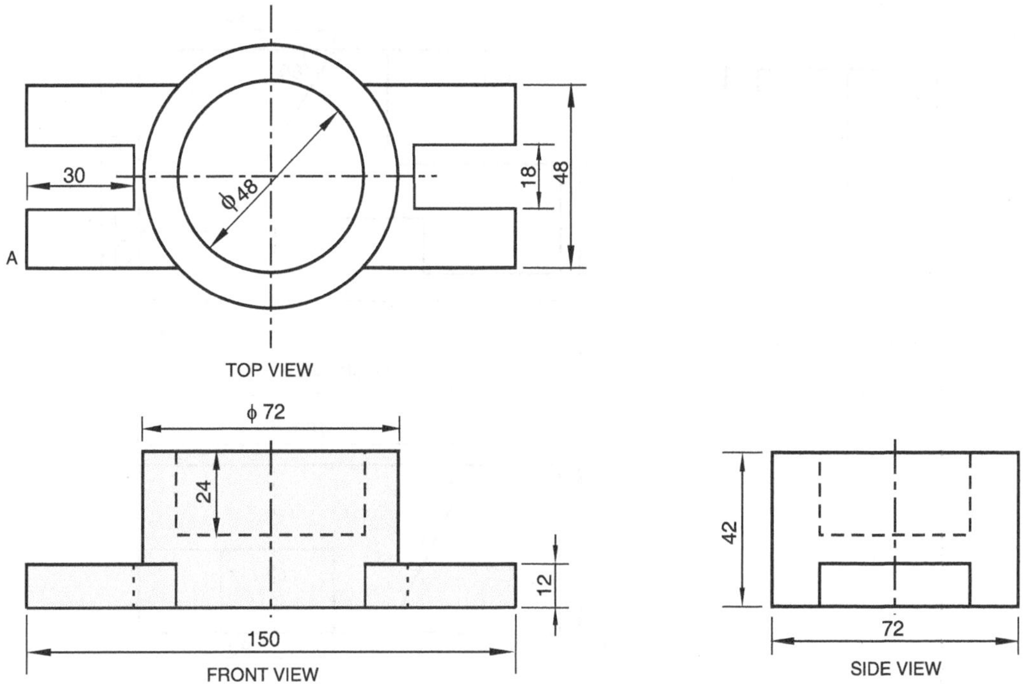

TOP VIEW

FRONT VIEW

SIDE VIEW

Fig. 1.51.

Problem 8 : Fig. 1.52 shows isometric view of an object. Draw the following views:

(a) Front view

(b) End view

(c) Top view

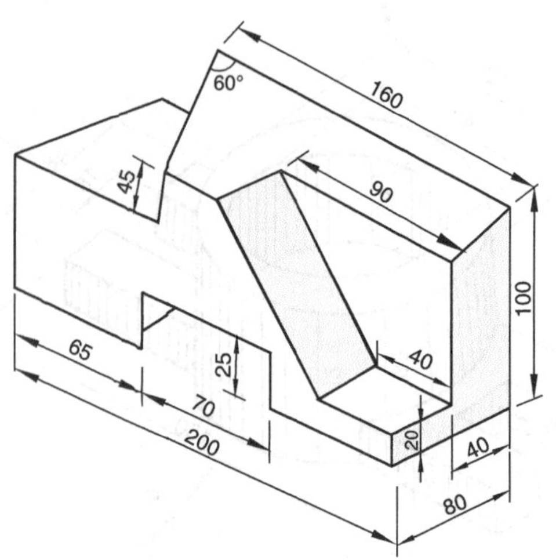

Fig. 1.52.

Solution : For its solution, see Fig. 1.53

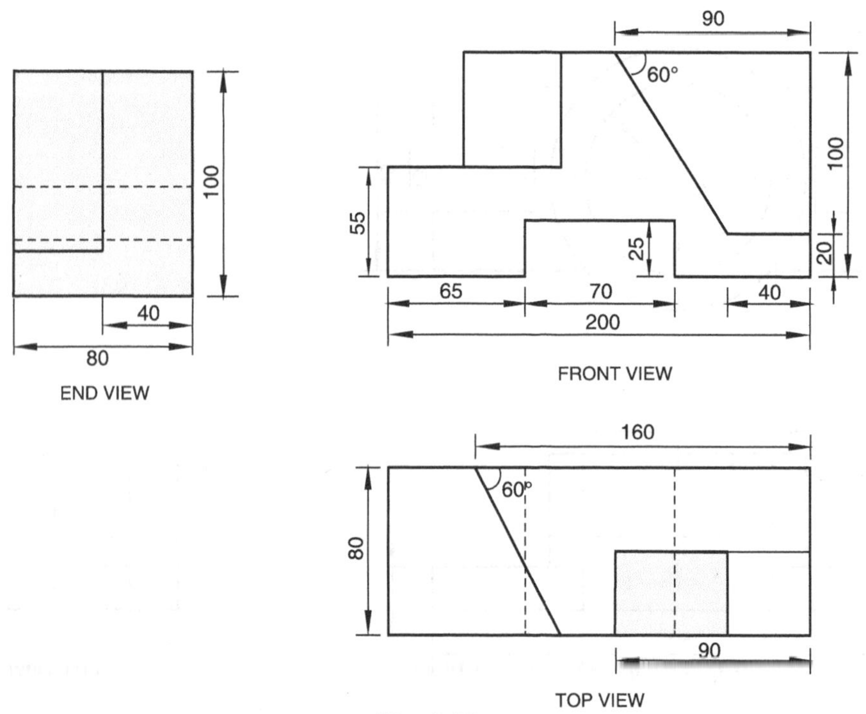

Fig. 1.53.

Problem 9 : Fig. 1.54 shows a sketch of an angle block: Draw the following views:
(a) Front view (Elevation) **(b) Side view (End view)**
(c) Top view (Plan)

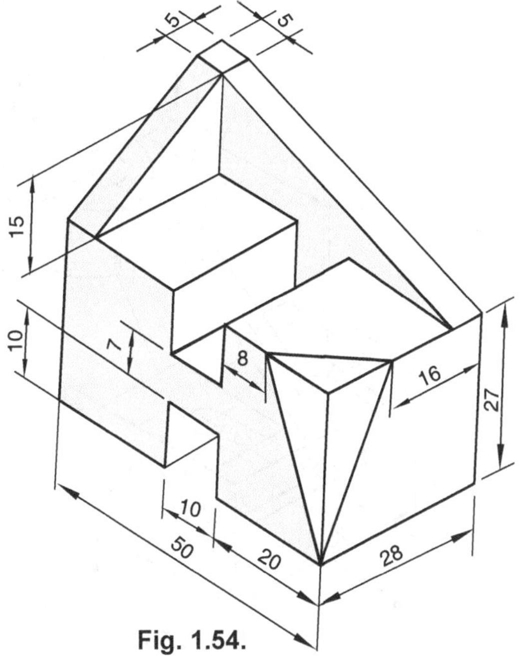

Fig. 1.54.

Solution : For its solution, see Fig. 1.55.

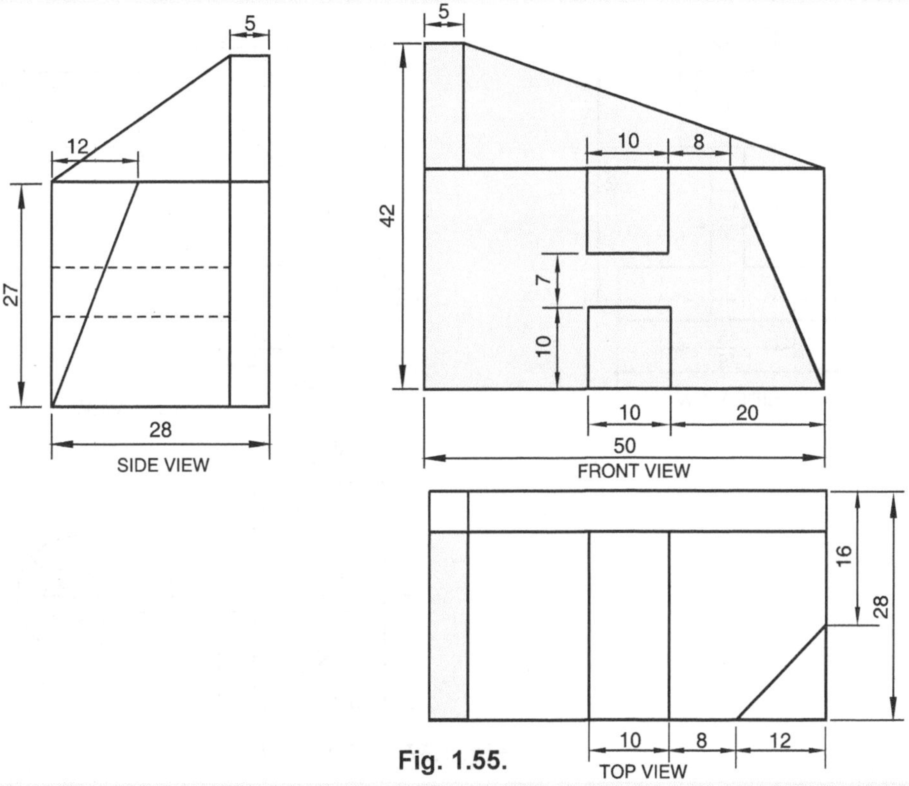

SIDE VIEW

FRONT VIEW

TOP VIEW

Fig. 1.55.

Problem 10 : Fig. 1.60 shows a sketch of a casting. Draw the following views :—

(a) Front view in the direction of S **(b) Side view in the direction F**

(c) Top view.

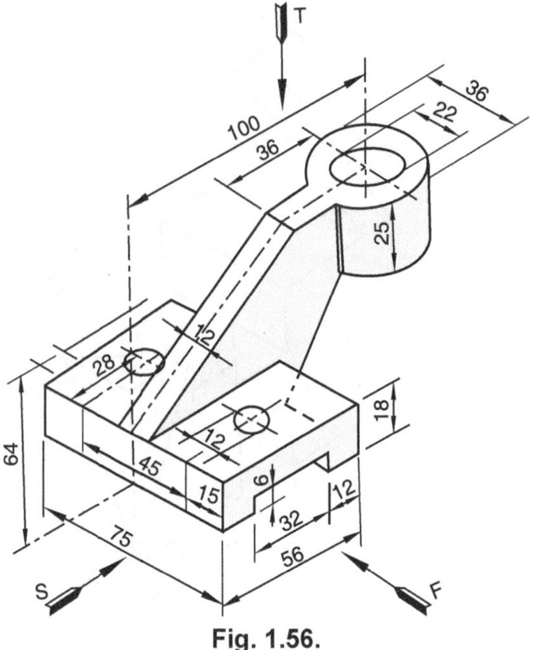

Fig. 1.56.

Solution : For its solution, see Fig. 1.57.

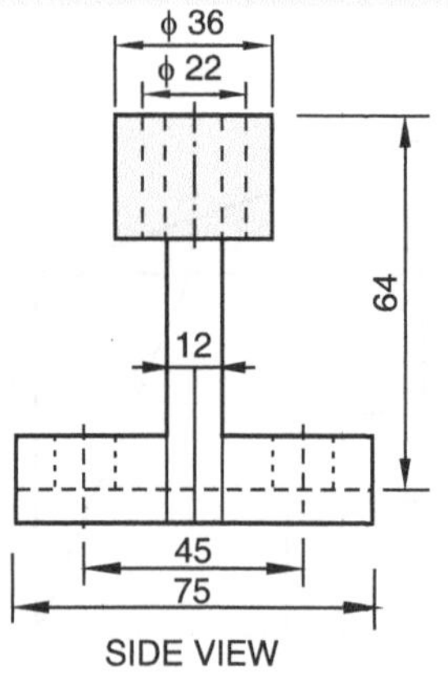

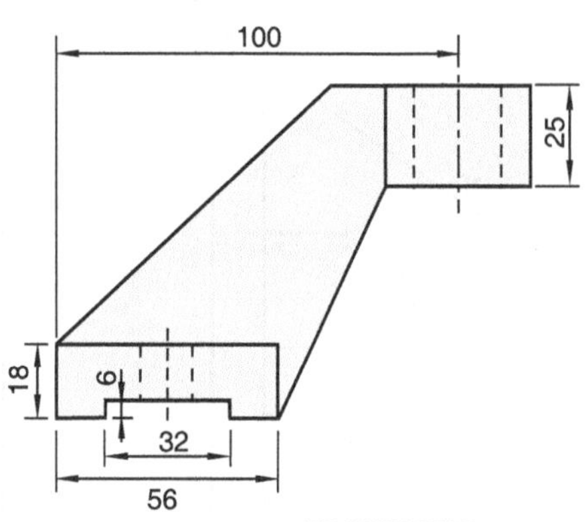

28

ϕ 36

ϕ 12

Fig. 1.57. TOP VIEW

Problem 11 : Fig. 1.58 shows a sketch of an assembling block. Draw to full size the following views by third angle projection method :—
(a) Front view through E (b) Side view through S (c) Top view (plan) through P

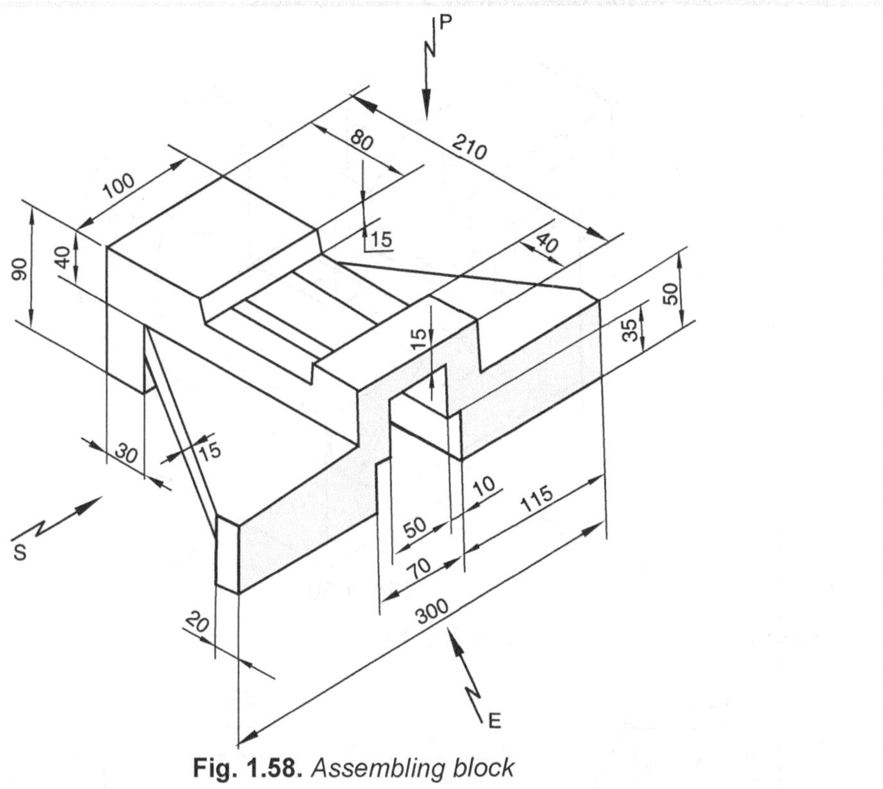

Fig. 1.58. *Assembling block*

Solution : For its solution, see Fig. 1.59.

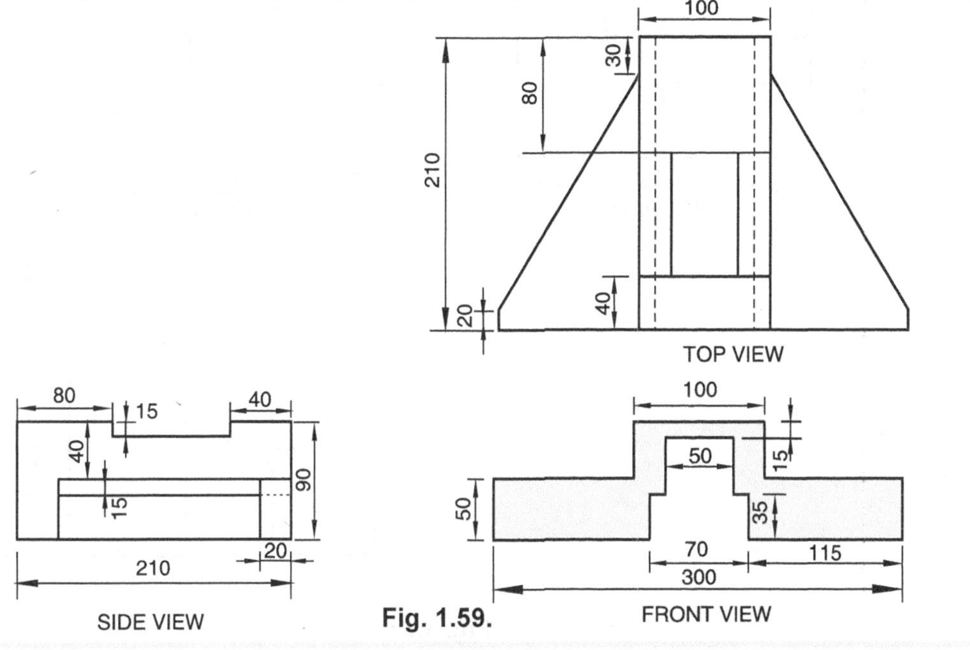

Fig. 1.59.

Problem 12 : Fig. 1.60 shows pictorial view of an object. Draw the following views in 1st angle pojection:

(a) Front view in the directions of F

(b) Side view in the direction of S (c) Top view

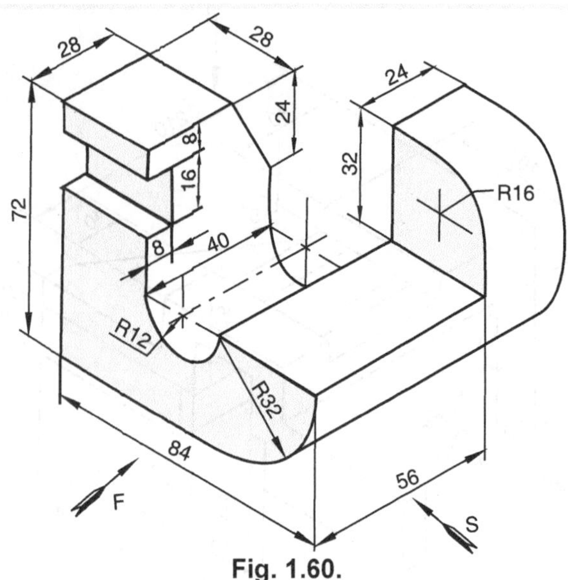

Fig. 1.60.

Solution : For its solution, see Fig. 1.61.

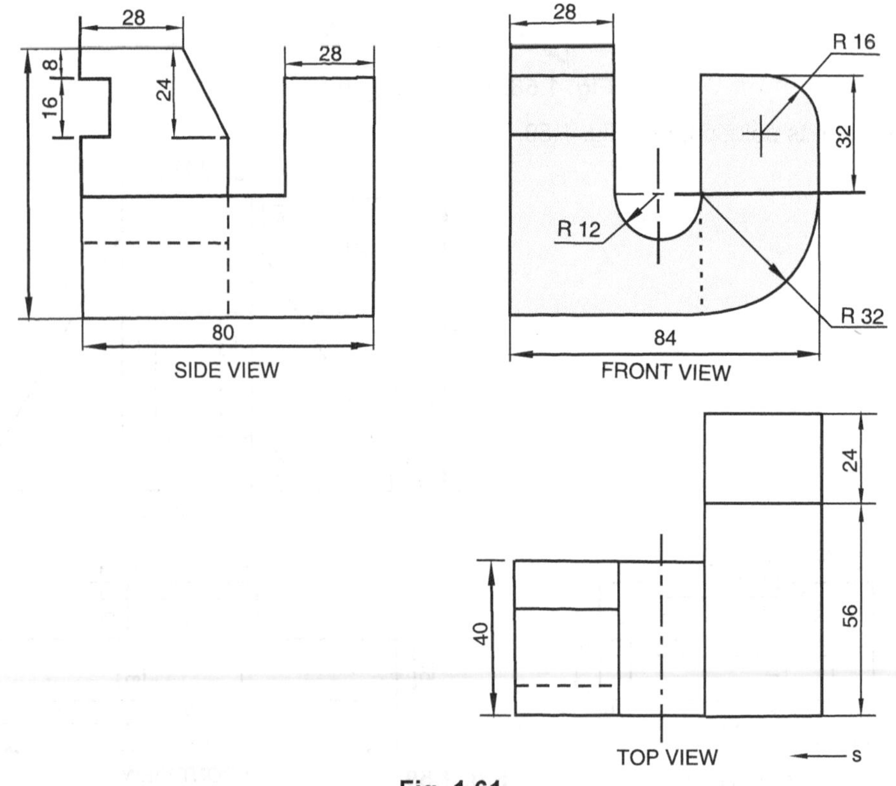

Fig. 1.61

Problem 13 : Fig. 1.62 shows a sketch of a block. Draw to full size the following views by third angle projection method :—

(a) Front view through-A (b) End view through-B (c) Top view through-C.

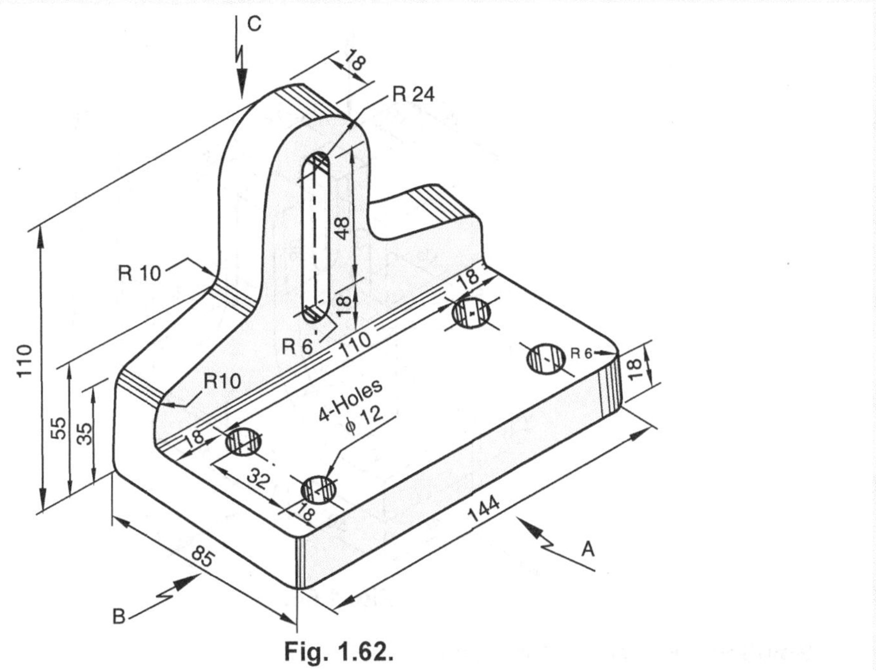

Fig. 1.62.

Solution : For its solution, see Fig. 1.63.

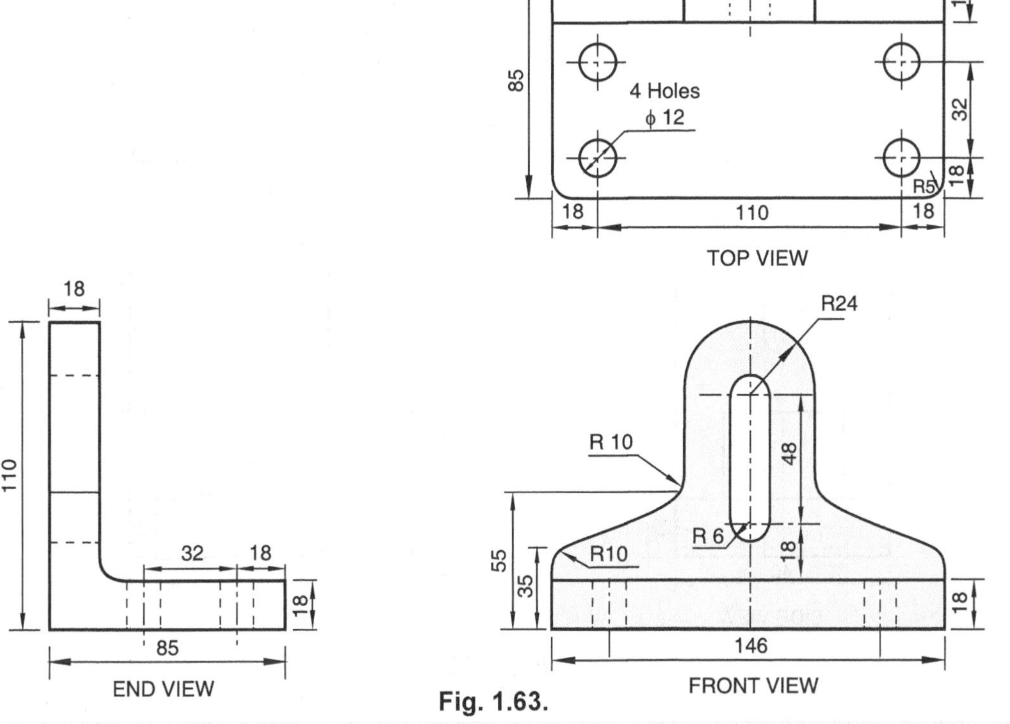

Fig. 1.63.

Problem 14 : Fig. 1.64 shows a sketch of the adjuster. Draw to full size the following views by third angle projection method :

(a) Front view (elevation)-F (b) Side view (end view)-S (c) Top view plan-T

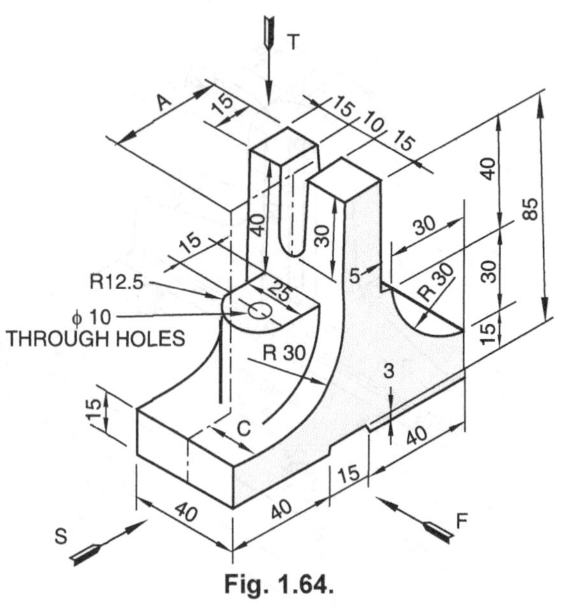

Fig. 1.64.

Solution : For its solution, see Fig. 1.65

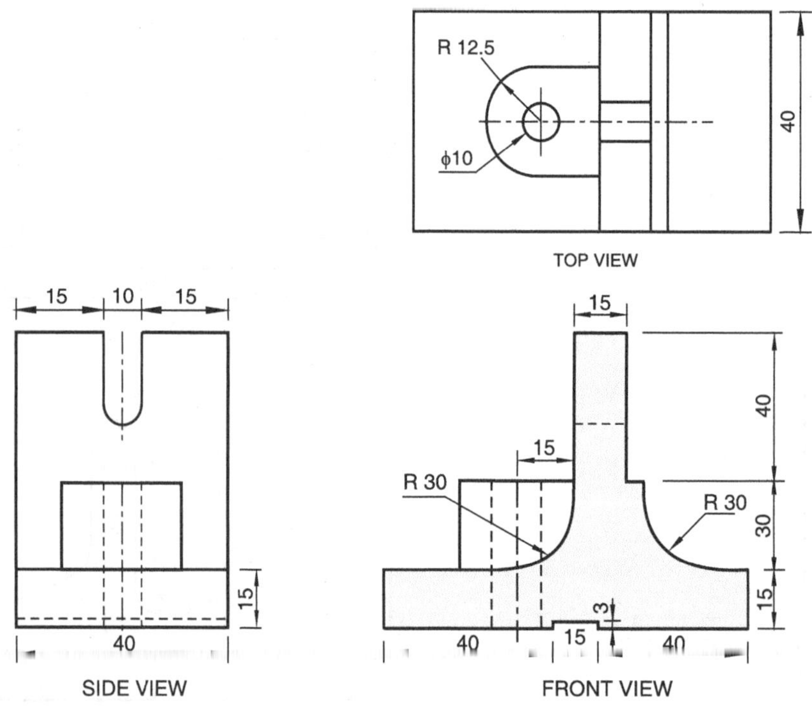

Fig. 1.65.

Problem 15 : Fig. 1.66 shows isometric view of an object. Draw the following views.

(a) Front view in direction of 'F' **(b) Side view**

(c) Top view

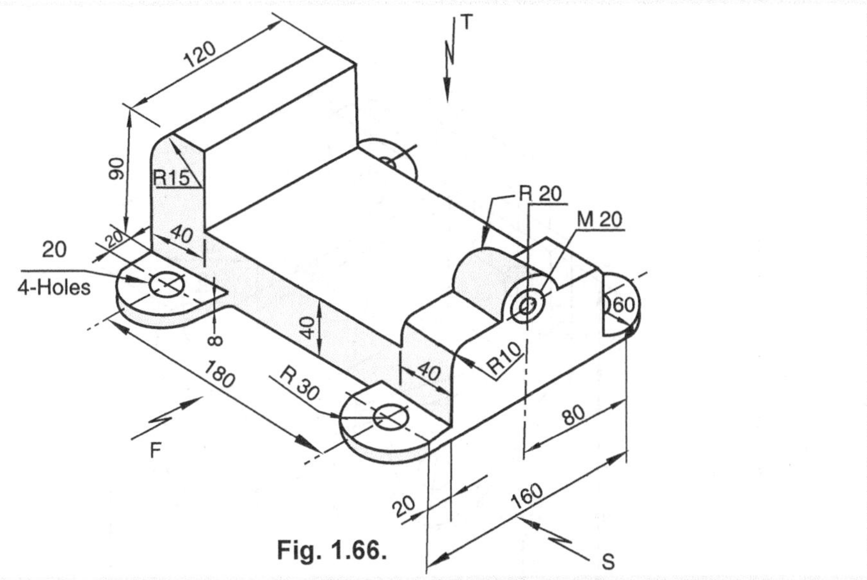

Fig. 1.66.

Solution : For its solution, see Fig. 1.67.

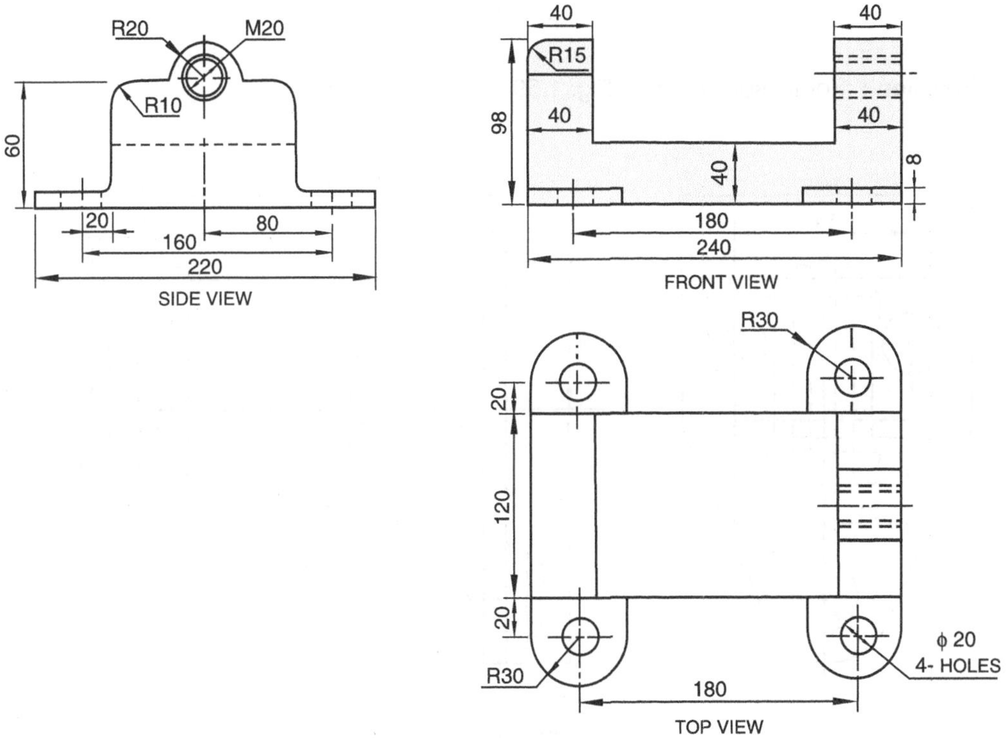

Fig. 1.67.

Problem 16 : Fig. 1.68 shows the pictorial view of a cast iron block. Draw the following views

(a) Front view

(b) Side view

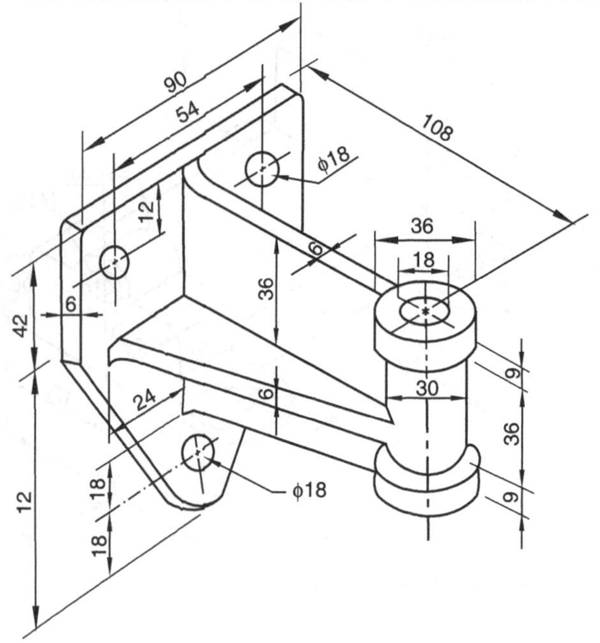

Fig. 1.68.

Solution : For its solution, see Fig. 1.69.

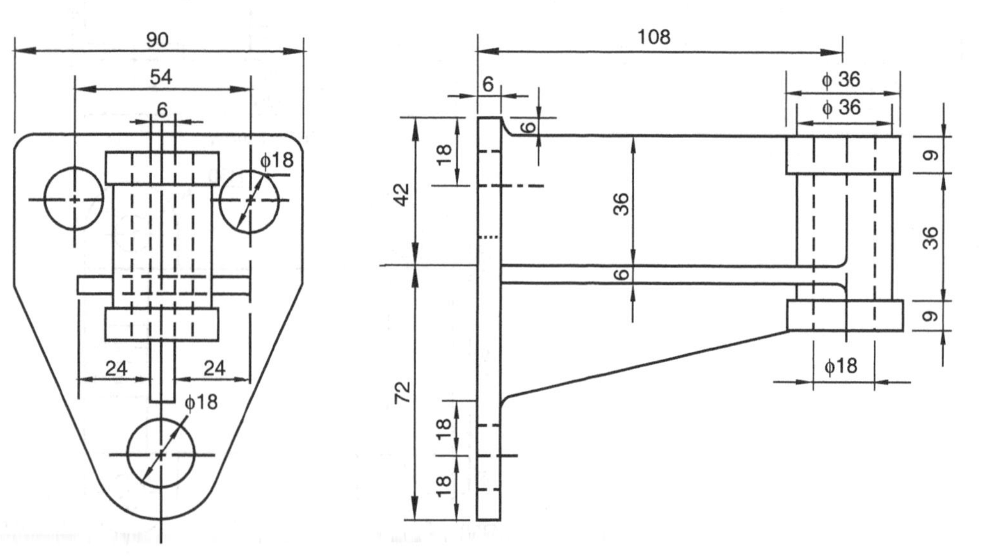

Fig. 1.69.

PROBLEMS ON CONVERSION OF ORTHOGRAPHIC PROJECTIONS INTO PICTORIAL VIEW

Problem 17 : Fig. 1.70 shows two views of an object. Draw the pictorial view, i.e., front face as front view (as orthographic), and depth at 45° receding axes. [Fig. 1.71]

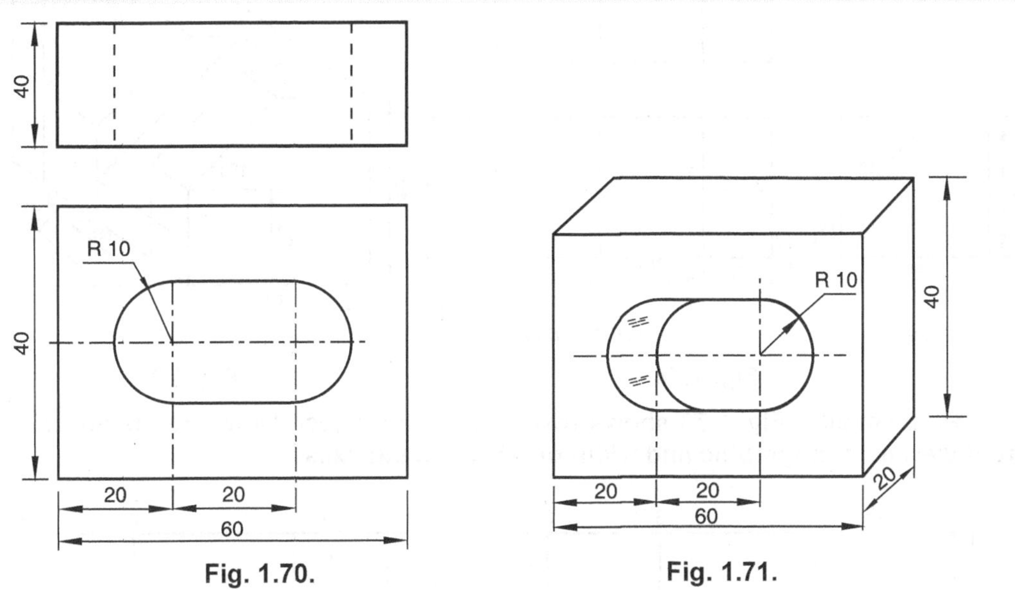

Fig. 1.70. Fig. 1.71.

Solution : For its solution see Fig. 1.71.

Problem 18 : Fig. 1.72 shows front view and side view of an object. Draw the following in scale 1:1:

(a) Copy the given views

(b) An oblique view (pictorial sketch)

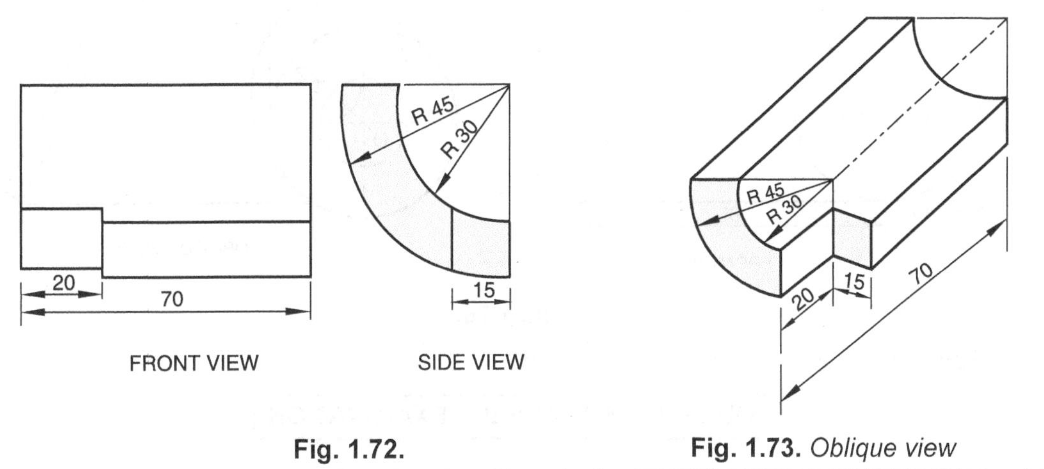

Fig. 1.72. Fig. 1.73. *Oblique view*

Solution : For its solution, see Fig. 1.73

Problem 19 : Make a pictorial view of the casted block shown in orthograpic projection in Fig. 1.74.

Solution : For its solution see Fig. 1.75.

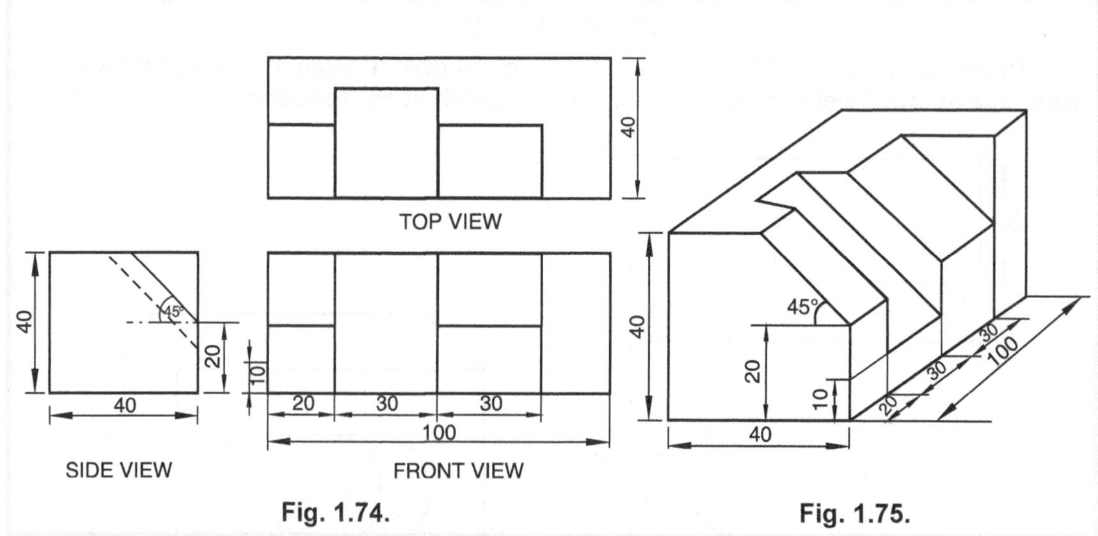

Fig. 1.74. Fig. 1.75.

Problem 20 : Fig. 1.76 shows two views of an object. Draw the oblique view, i.e, front view as orthographic and width on 45° receding axes.

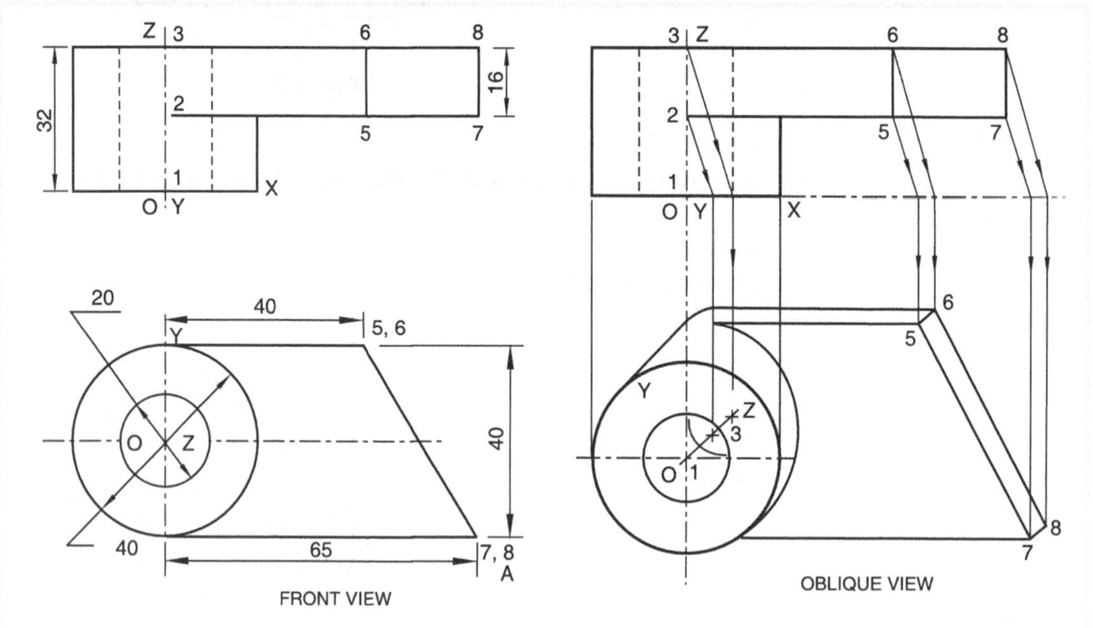

Fig. 1.76.

Solution : For its solution, see Fig. 1.76.

QUESTIONS FOR SELF EXAMINATION

1. Define projection and give the different types of projections.
2. What are different types of pictorial, oblique and perspective projections?
3. Define axonometric projection and its various types.
4. Name the principal planes of projections.

5. Give the name of three coordinate planes.
6. What is the difference between first angle projection and third angle projection?
7. Which types of projection (whether third or first angle projection) is followed now-a-days in India, Britain and America?
8. Explain one view drawings. For what type of objects these are used?
9. Explain the different ways of projecting three view drawing.
10. What do you mean by spacing of views?
11. What are the different ways of projecting three view drawings?
12. What do you understand by precedence of lines?
13. Fill in the blanks :
    (*a*) In orthographic projection, the lines of sight are .... to the plane of projection.
    (*b*) A surface of an object appears in its true shape when it is .... to the plane of projection.
    (*c*) In orthographic projection, the ...... are perpendicular to the .... of projection.
    (*d*) In first angle projection, the .... comes between the ...... and the plane.
    (*e*) In third angle projection, the plane comes between the .... and the ....
    (*f*) In first angle projection:
        (*i*) The view from above is placed .... of front view.
        (*ii*) The view from left is placed on the .... of front view.
    (*g*) To make a one view drawing complete ...... must be added.
    (*h*) When a dotted line coincides with a solid line, the .... line takes preference.
    (*i*) When a centre line coincide with a solid line, the .... line takes preference.
    (*j*) The three principle planes of projection are ...., ... and .... planes.

Ans. (*a*) perpendicular (*b*) parallel (*c*) projectors, plane (*d*) object, observer (*e*) observer, object (*f*) (*i*) below, (*ii*) right (*g*) a note (*h*) solid (*i*) solid (*j*) horizontal, vertical, profile.

14. Choose the correct answer :—
    (*i*) In orthographic projection, the projectors to the plane of projection are
    (*a*) *parallel* (*b*) *perpendicular* (*c*) *inclined.*
    (*ii*) The number of mutually perpendicular planes that can be surround an object in space is
    (*a*) *three* (*b*) *four* (*c*) *six.*
    (*iii*) Clinographic projection belongs to the major class of projection
    (*a*) *orthographic* (*b*) *oblique* (*c*) *perspective.*
    (*iv*) In first angle projection, the object is imagined to be placed.
    (*a*) *above H.P. and behind V.P.*
    (*b*) *above H.P. and infront of V.P.*
    (*c*) *below H.P. and behind V.P.*
    (*v*) In first angle projection, to obtain right side view, the A.V.P. is assumed to be on
    (*a*) *right side of object* (*b*) *left side of object* (*c*) *top of object.*
    (*vi*) In third angle projection, if the Auxiliary V.P. is placed to the right of the object, the view obtained is called
    (*a*) *left side view* (*b*) *right side view* (*c*) *bottom view.*
    (*vii*) An invisible line should intersect a visible line with
    (*a*) *a dash* (*b*) *no dash* (*c*) *a visible line in contact.*

Ans. (*i*) *b*, (*ii*) *c*, (*iii*) *b*, (*iv*) *b*, (*v*) *a*, (*vi*) *b*, (*vii*) *a*.

15. Match the statements of column A with the corresponding ones in column B.

Column A	Column B
(*i*) In orthographic projection, the projectors are	(*a*) the observer and plane of projection
(*ii*) In first angle projection, the object is placed between	(*b*) perpendicular to plane of projection
(*iii*) According to B.I.S. the symbol for 1st angle projection is	(*c*)
(*iv*) As per B.I.S. symbol used for the 3rd angle projection is	(*d*)
(*v*) In first quadarant, view from right is	(*e*) drawn to top of front view
(*vi*) In third quadrant, view from top is	(*f*) drawn to left of front view
(*vii*) Dimetric	(*g*) perspective projection
(*viii*) Cavalier	(*h*) orthographic projection
(*ix*) Third angle	(*i*) oblique projection
(*x*) Three points	(*j*) axonometric projection

Ans. (*i*) *b*, (*ii*) *a*, (*iii*) *d*, (*iv*) *c*, (*v*) *f*, (*vi*) *e*, (*vii*) *j*, (*viii*) *i*, (*ix*) *h*, (*x*) *g*,

PROBLEMS FOR PRACTICE

1. Draw the front view, side view and top view of the object as shown in Fig. 1.77 by following first angle and third angle projections.

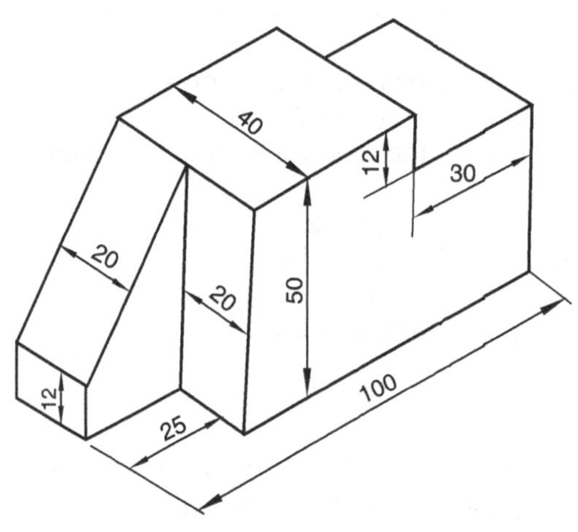

Fig. 1.77.

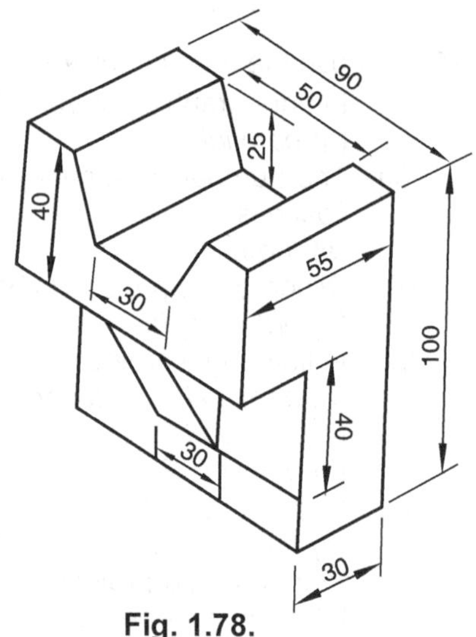

Fig. 1.78.

2. Draw the elevation, plan, right end view and left end view of a given block (see Fig. 1.78) by first angle projection.

3. Draw the front view, top view, right side view and left side view of a given block (see Fig. 1.79) by third angle projection.

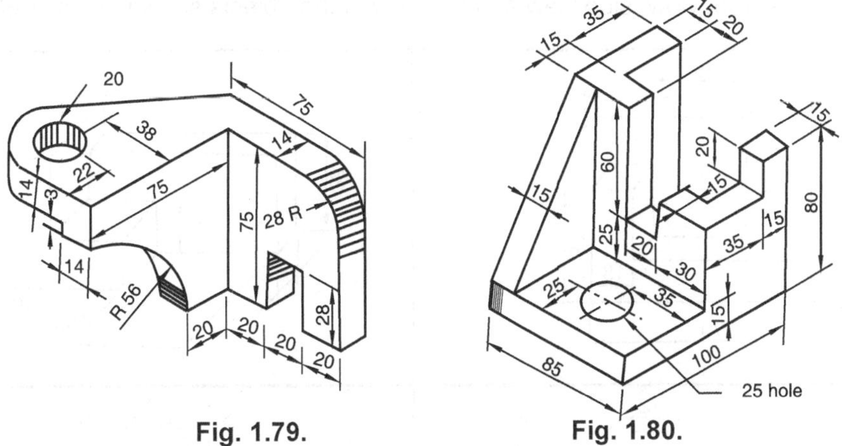

Fig. 1.79. Fig. 1.80.

4. Draw the top view, front right, side view and left side view of a given block (see Fig. 1.80) by following first angle projection.

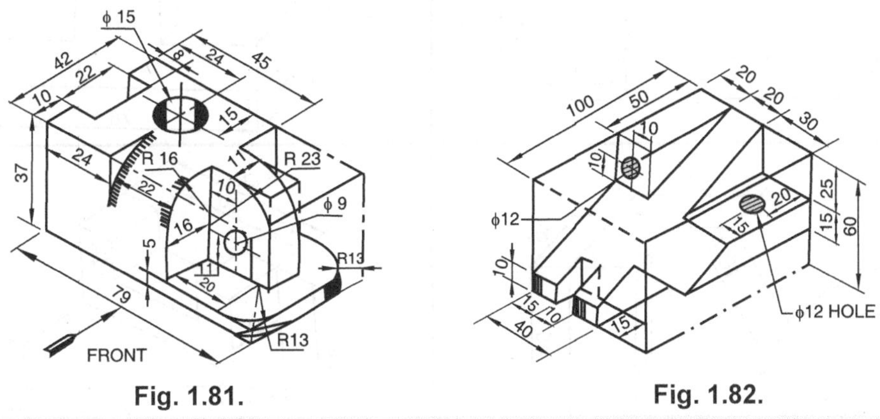

Fig. 1.81. Fig. 1.82.

5. Draw the front view, side view of the objects shown in Figs. 1.81 and 1.82.
6. Draw the elevation, end view and plan of the objects as shown in Figs. 1.83 and 1.84.

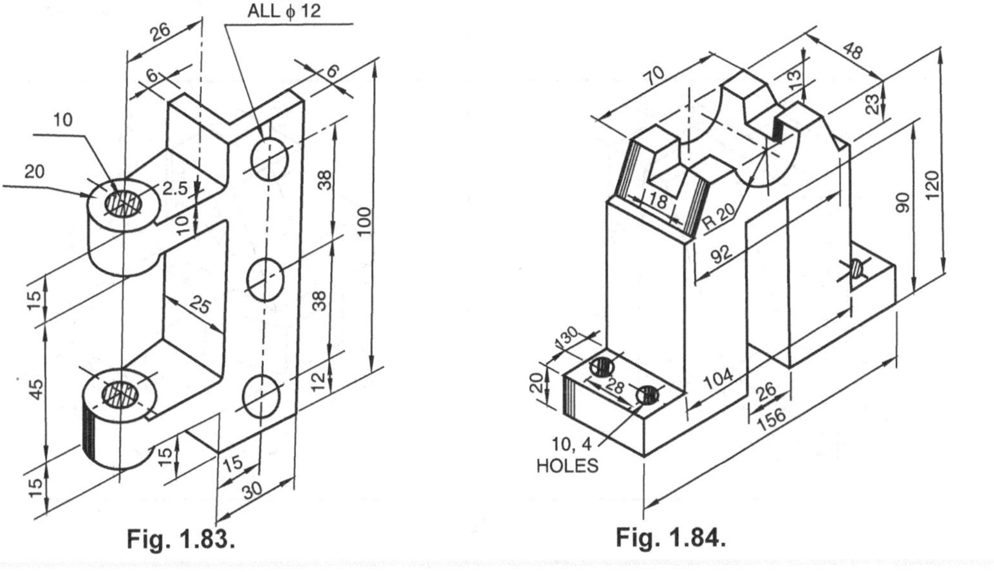

Fig. 1.83. Fig. 1.84.

7. Draw the front view, side view and top view of the objects shown in Fig. 1.85.

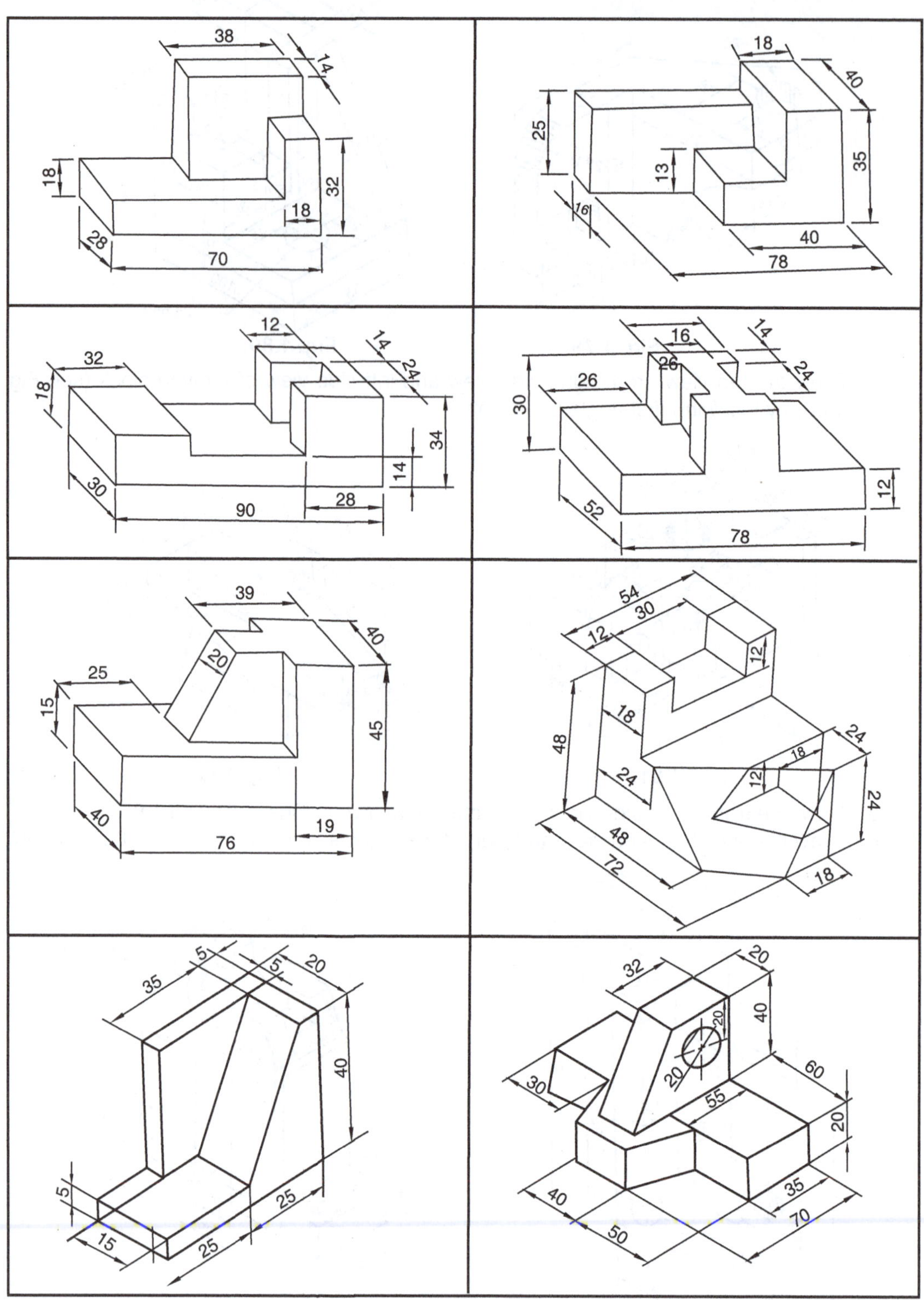

Fig. 1.85.

8. Draw the fronts, in the direction of arrow side and top views of the objects given in Fig. 1.86.

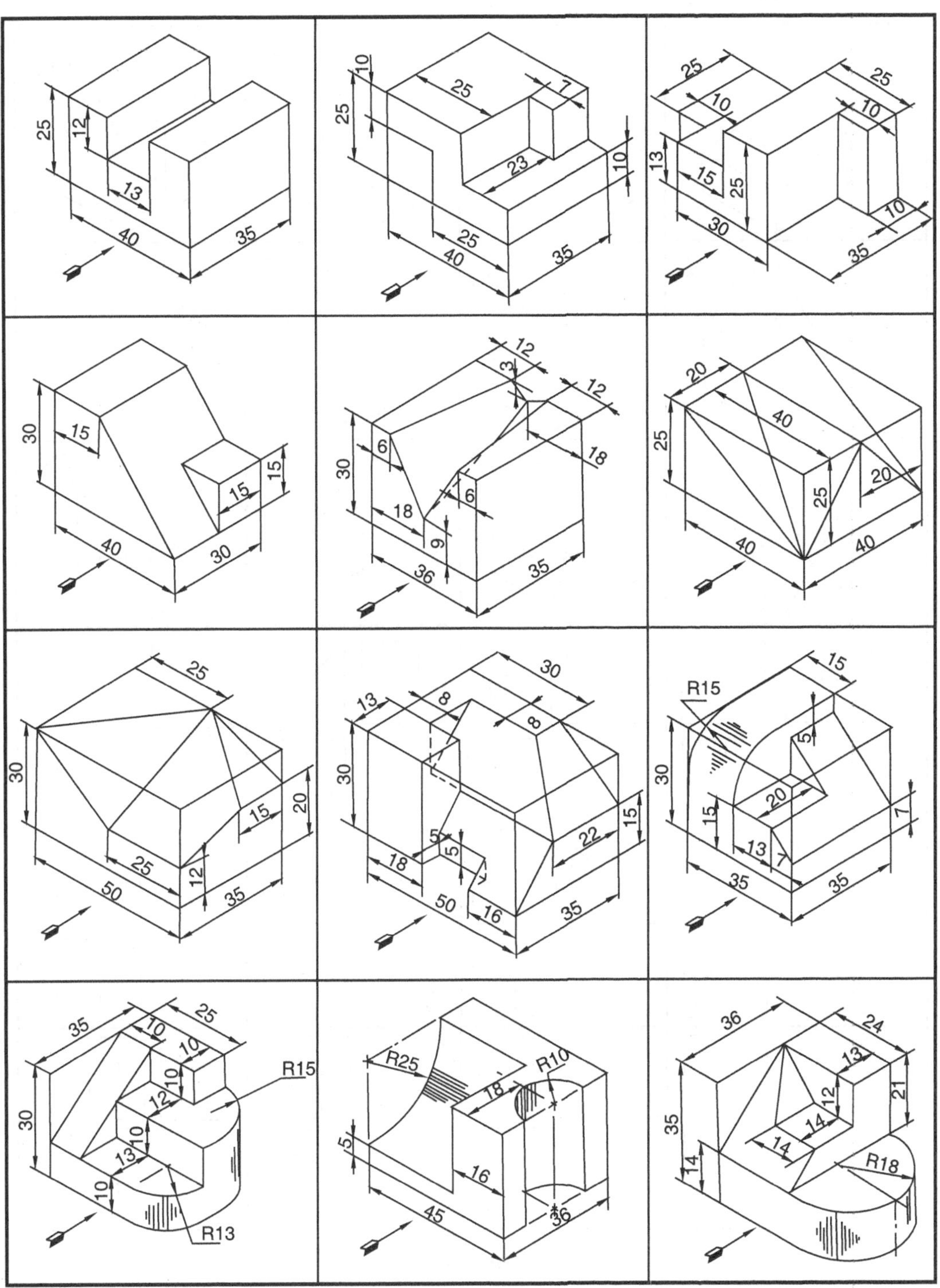

Fig. 1.86.

9. Fig. 1.87 shows the pictorial view of the objects. Match the orthographic views with the respective objects.

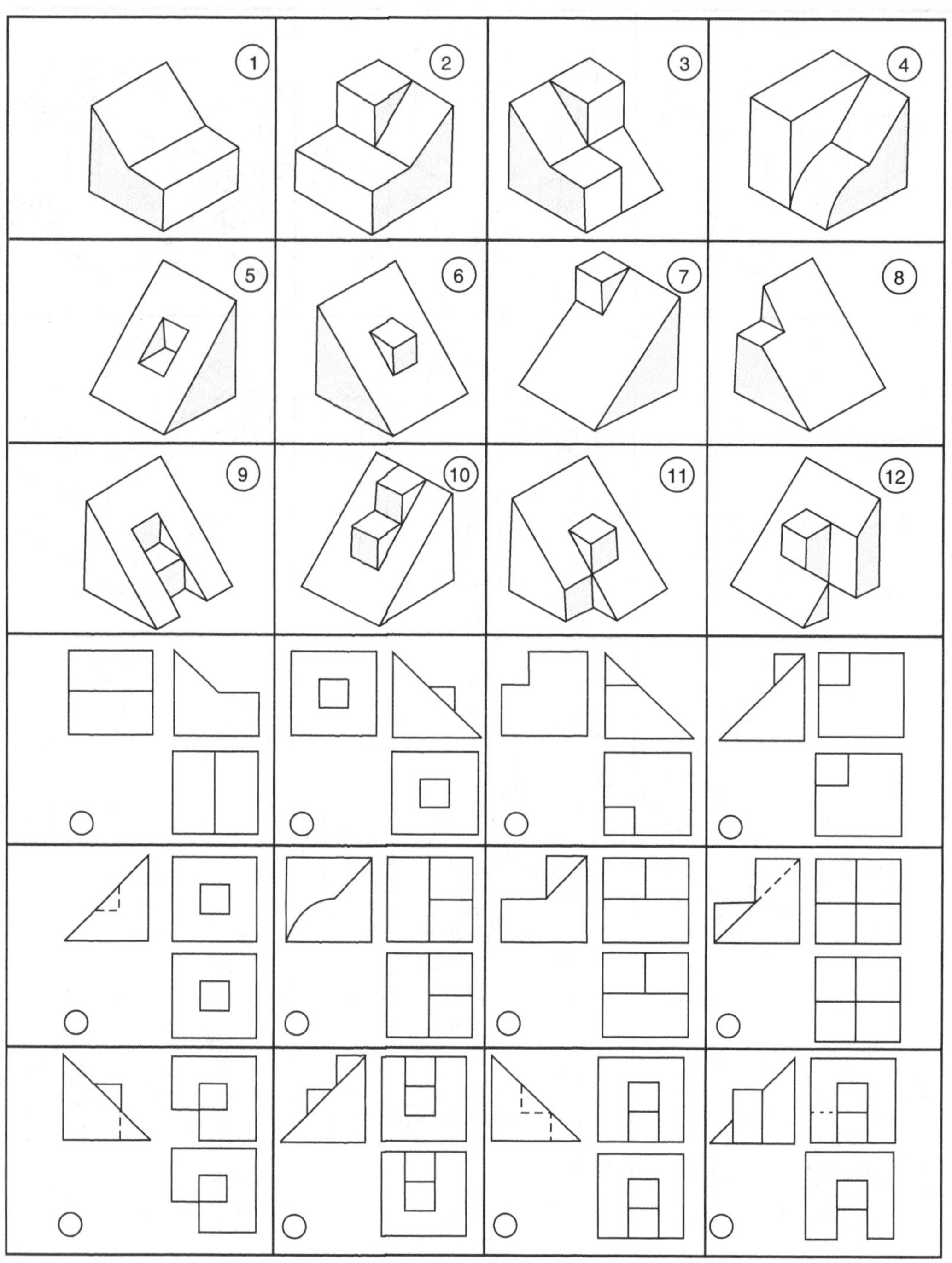

Fig 1.87

CHAPTER

2

Orthographic Reading or Interpretation of Views

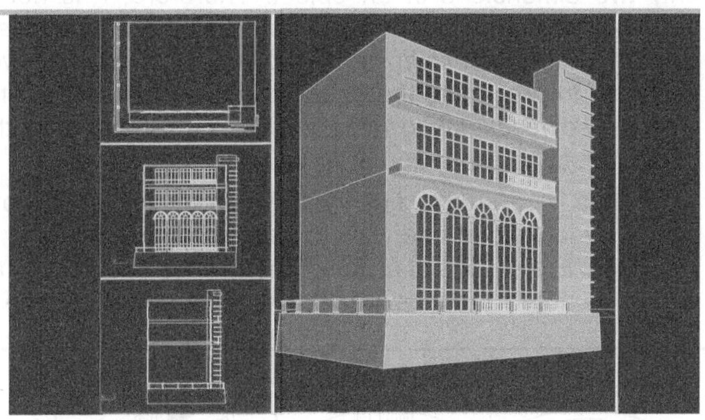

INTRODUCTION

In engineering, an object is represented on the drawing by its orthographic views. These views give the complete external details of the object. The ability to visualise the shape of the object from its orthographic drawings is known as *orthographic reading*. Often the orthographic drawings are prepared in the design office and these have to be interpreted by the engineers and technicians concerned with the work of manufacturing. An engineer can make an object correctly, only if he has thorough knowledge of reading or interpreting the orthographic views. Therefore, the key to successful orthographic reading is the possession of sound knowledge of the principles of orthographic projection.

In this chapter, it will be explained that how the knowledge of orthographic projection can help us to read and interpret a given drawing.

2.1. ORTHOGRAPHIC READING OR INTERPRETATION OF VIEWS

The ability to visualize the shape of the object from its orthographic views is known as orthographic reading or interpretation of views.

Reading the orthographic views consists of visualizing the views of the object to find out the purpose and meaning of each line and then to develop mentally the shape and location of various surfaces of the object. In this way, a mental picture of the object is obtained giving its shape, size and other details.

Thus, reading the orthographic view is the reverse process of making the drawing and therefore, sound knowledge of the principles of orthographic projection is quite essential to read the orthographic drawing.

Note : *In practice, one of the best aids in visulization is the use of an actual model of the object. Such a model can be made to any scale using any convenient material such as wood, modelling clay, laundry soap, etc.*

2.2. HOW TO READ ORTHOGRAPHIC VIEWS

As already stated that one view in orthographic projection gives only two dimensions of an object. Therefore, it is not possible to visualise the shape of the object from a single view. The second view gives the third dimension of the object. Thus, atleast two views are essential to determine the shape of the object. Some times, a third view is also necessary to visualize the exact shape and other features of the object having irregular or complicated shape.

Thus, one should not try to visualise the shape of the object simply by looking at one view but should simultaneously examine all the views. Projections of all the corners, lines, surfaces in all the views should be noted and thus meaning of the same should be interpreted.

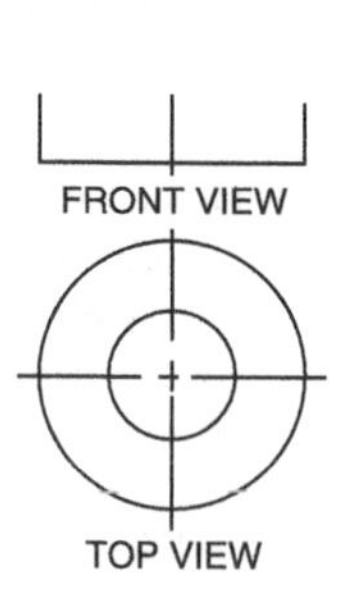

Fig. 2.1.

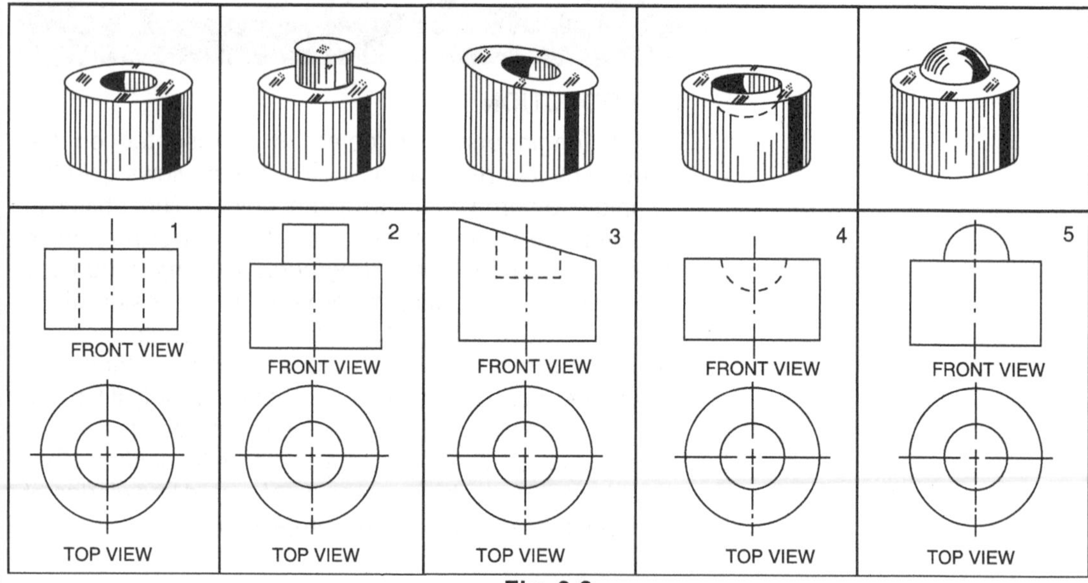

Fig. 2.2.

Fig. 2.1 shows top view (one view) of a simple object. The top view does not describe the exact shape of the object. Thus, atleast two views are necessary to determine the shape of the object. Fig. 2.2 shows the second view of the objects, but their top views remain the same. From front view and top view, i.e., two views of various objects, we can visualise the shape of the objects. The corresponding objects are shown by pictorial views at the top.

Fig. 2.3 shows front view (one view) of a simple object. The front view does not describe the exact shape of the object. Thus, atleast two views are necessary to determine the shape of the object. Fig. 2.4 shows the second view of the objects, but their front view remain the same. From front views and side view (i.e., two views), we can visualize the shape of the objects. The corresponding objects are shown by pictorial views.

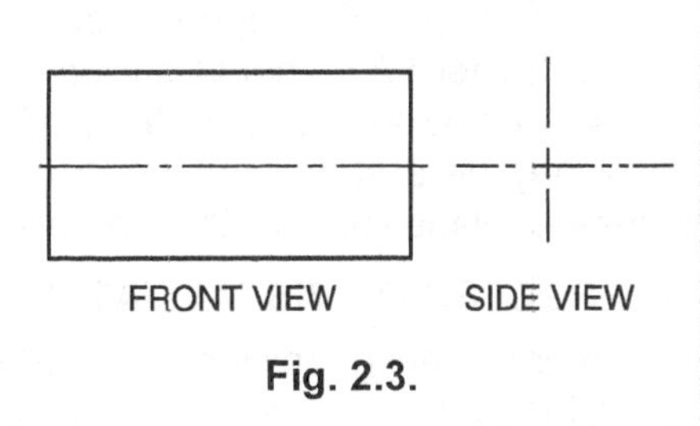

Fig. 2.3.

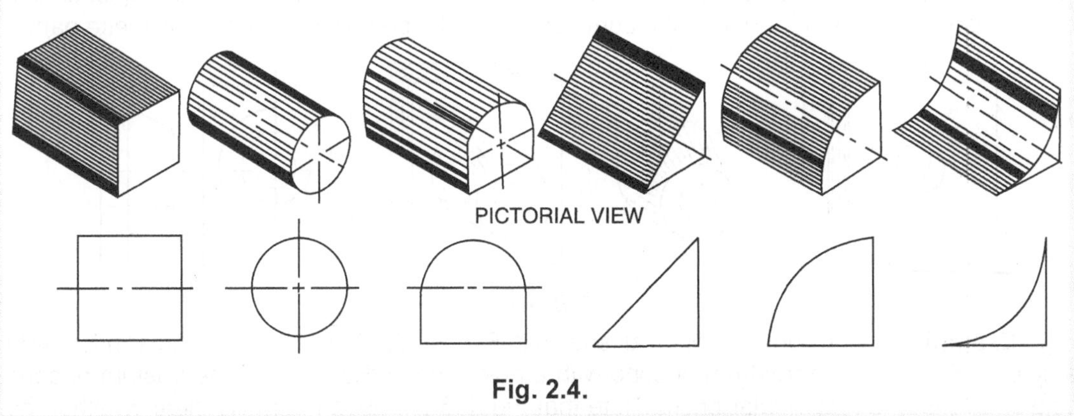

Fig. 2.4.

2.3. SIMPLE FACTS OF READING ORTHOGRAPHIC PROJECTIONS

The following points should be kept in mind while reading orthographic drawing :

1. Surface : (i) A surface AEFHG is represented by a line AEFHG in two views (front and side views) and a closed figure in the third view (see Fig. 2.5).

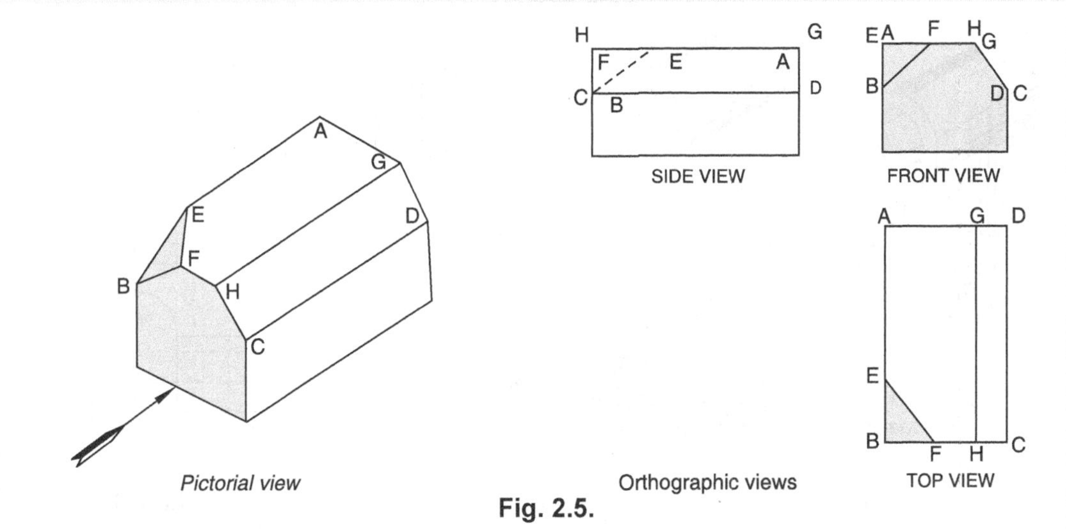

Fig. 2.5.

(ii) A surface CDGH is represented by a line in one view (side view) and a closed figure in the remaining two views (see Fig. 2.5).

(iii) A surface BEF is represented by a closed figure in all the three views BEF (see Fig. 2.5).

2. Edge : (i) An edge AE is represented by a point (AE) in one view (front view) and a line AE in other two views (see Fig. 2.5).

(ii) An edge EF is represented by a line EF in all the three views (see Fig. 2.5).

3. Corner : A corner C is represented by a point C in all the three views (see Fig. 2.5).

4. Area bounded by outlines : An area bounded by outline represents continuous surface which may be a straight plane or curved or combination of both e.g. in top view AEFHG represents plane surface, BEF represents bounded area and GHCD represents plane surface.

2.4. PROCESS OF VISUALIZATION

In engineering drawing, every object may be imagined to be built up of a number of parts having forms of simple solids such as rectangular plates, prisms, cylinders, cones, pyramids, etc. Addition or subtraction or both of these parts make the object. Additions are seen in the form of projections and substractions as holes, grooves, cavities, etc. Thus, the object should be first visualized by part and then the complete object by combining together all these parts.

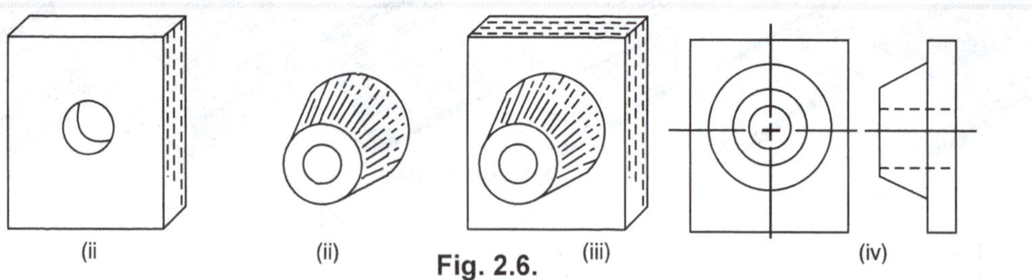

Fig. 2.6.

Example 1 : Consider an object as shown in Fig. 2.6 (i), there is a rectangular prism with a hole; at Fig. 2.6 (ii), a frustrum of cone with a hole and at Fig. 2.6 (iii), the frustum of cone combined with the rectangular prism. Note that Fig. 2.6 (iii) does not make clear whether the hole is blind or through. In Fig. 2.6 (iv) hole is seen by dotted lines or the combination of lines which describe the views clearly in a simple way

Example 2 : Fig. 2.7 shows an object. The shape of the object may be visualized by imagining it to be broken up in three rectangles and one cylinder. Three views of the object are drawn in Fig. 2.8 after visualizing it thoroughly.

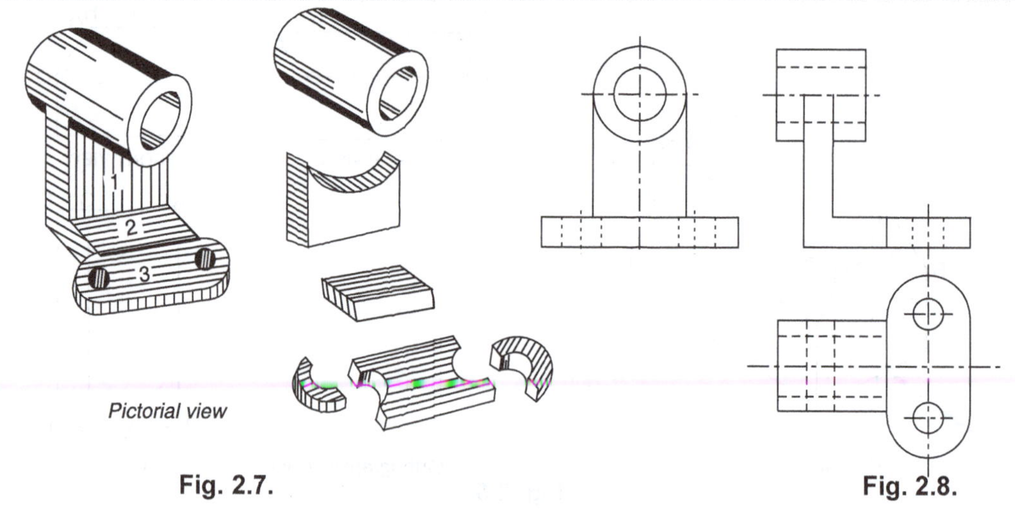

Fig. 2.7. **Fig. 2.8.**

Example 3 : Fig. 2.9 shows a pictorial view of a block. The shape of the block may be visualized by imagining it to be split up into five parts, i.e., A,B,C,D and E. After referring to split up pictorial views it can be visualized that the part 'A' is in the form of a rectangular prism with 2 holes. The part B is in the form of a hollow cylinder. The part C and D are in the form of triangular block. The part E is in the form of rectangular prism with curved portion at the top.

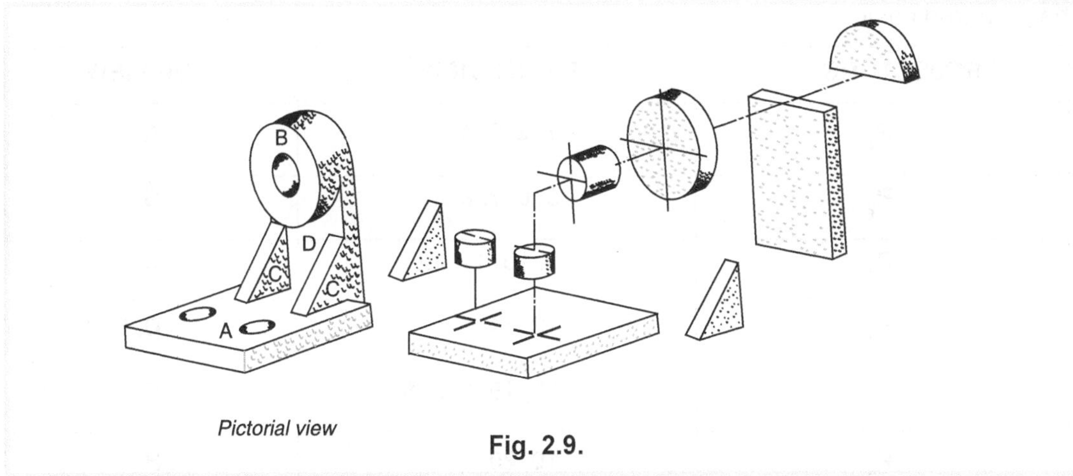

Fig. 2.9.

Example 4 : Fig. 2.10 shows the front view and top view of a block. To find out the relations between points and lines in front view and top view, few projections are drawn. The position of these projections in front view and top view are marked by various points as given in the table.

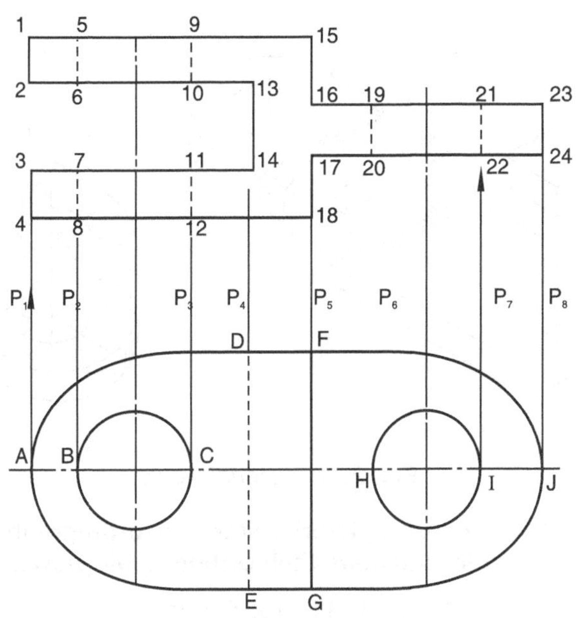

Fig. 2.10. *Orthographic view*

After noting the corresponding points from front view and top view, the given views can be at once imagined to be split up in convenient parts as shown in Fig. 2.11.

Each part is a geometrical form and can be easily visualized as shown in Fig. 2.11. Finally the complete object can then be visualized as one piece built up of all those elementary parts. In net sheel, once a student has grasped this process, actual drawing can be prepared easily.

PROJECTORS	FRONT VIEW	TOP VIEW
P_1	1, 2, 3, 4	A
P_2	5, 6, 7, 8	B
P_3	9, 10, 11, 12	C
P_4	13, 14	D, E
P_5	15, 16, 17, 18	F, G
P_6	19, 20	H
P_7	21, 22	I
P_8	23, 24	J

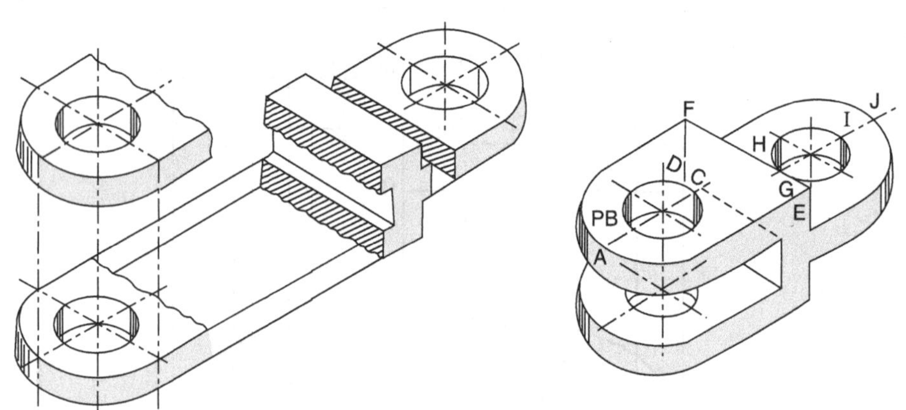

Fig. 2.11. *Pictorial views*

Problem 1 : Fig. 2.12 shows the pictorial views and orthographic views of 12 objects. Visualize the objects carefully and mark their orthographic views.

Solution : 1, 6, 8, 7, 5, 4, 2, 3, 11, 10, 9 and 12.

VISUALIZE THE OBJECT AND MARK THEIR ORTHOGRAPHIC VIEWS

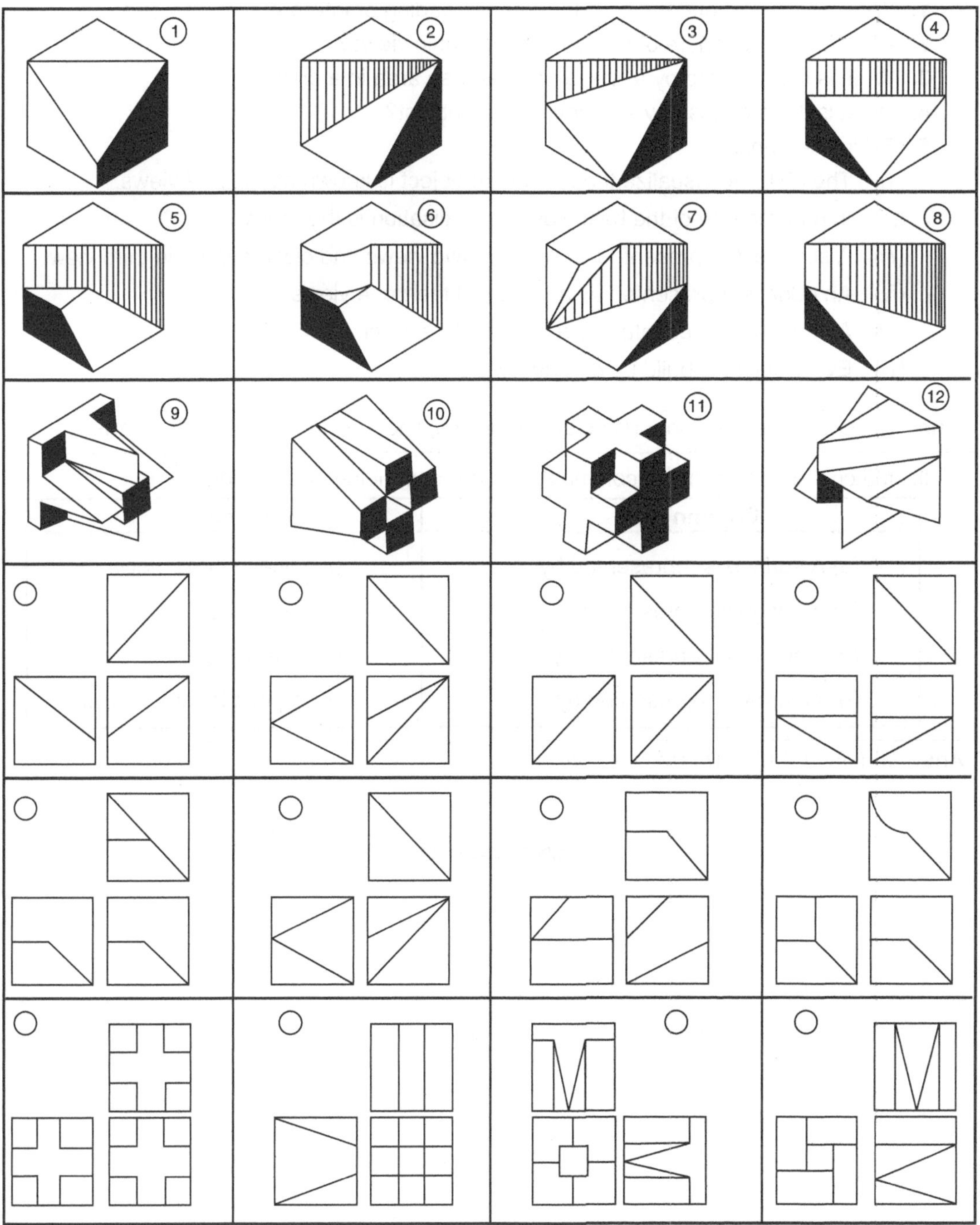

Fig. 2.12.

QUESTIONS FOR SELF EXAMINATION

1. What do you understand by orthographic reading or interpretation of views?
2. What is the procedure of reading orthographic views?
3. What is represented by a line, a point and an area in a drawing?
4. What is the process of visualization of an object?
5. Fill in the blanks :
  (*i*) The ability to visualize the ...... of the object is known as .... of views.
  (*ii*) In practice one of the best aids in visualisation is the use of ....
  (*iii*) A surface is represented by a ...... in two views and closed figure in third view.
  (*iv*) An edge is represented by a ...... in all the three views.
  (*v*) A corner is represented by a ...... in all the three views.
  (*vi*) Every object is build up of a number of ....

Ans. (*i*) shape, interpretation (*ii*) actual model (*iii*) line (*iv*) line (*v*) point (*vi*) parts

6. Match the statements of column A with corresponding ones in column B.

Column A	Column B
(*a*) A surface is represented by	(*i*) a line in three views
(*b*) An edge is represented by	(*ii*) continuous surface
(*c*) A corner is represented by	(*iii*) closed figure
(*d*) An area is represented by	(*iv*) a point in three views

Ans. (a) (*iii*), (*b*) (*i*), (*c*) (*iv*), (*d*) (*ii*).

CHAPTER

3

Identification of Surfaces

INTRODUCTION

In engineering drawing, we come across with such objects having simple to complex surfaces. When orthographic views of these objects are drawn on the drawing paper, then it is very important to identify the different surfaces of the object in orthographic views. For this, surfaces of the object and their corresponding surfaces in the orthographic views are marked by alphabets such as A,B, C, ... or 1, 2, 3, etc.

The main object in identification of surfaces is to develop the imagination for recognising the different surfaces of an object in order to give clear understanding of the projections.

In this chapter, we will deal with the study of a large number of problems in order to give the perfect understanding regarding the identification of surfaces.

3.1. IDENTIFICATION OF SURFACES

The art of marking the corresponding surfaces of an object from pictorial views to orthographic views and vice-versa is known as the identification of surfaces.

The identification of surfaces of an object can be had from :

(i) Pictorial view to orthographic views

(ii) Orthographic views to pictorial view

3.2. IDENTIFICATION OF SURFACES FROM PICTORIAL TO ORTHOGRAPHIC VIEWS

Fig. 3.1 shows the pictorial view of an object in which the different surfaces are marked by alphabets such as A, B, C, D, E, F and G.

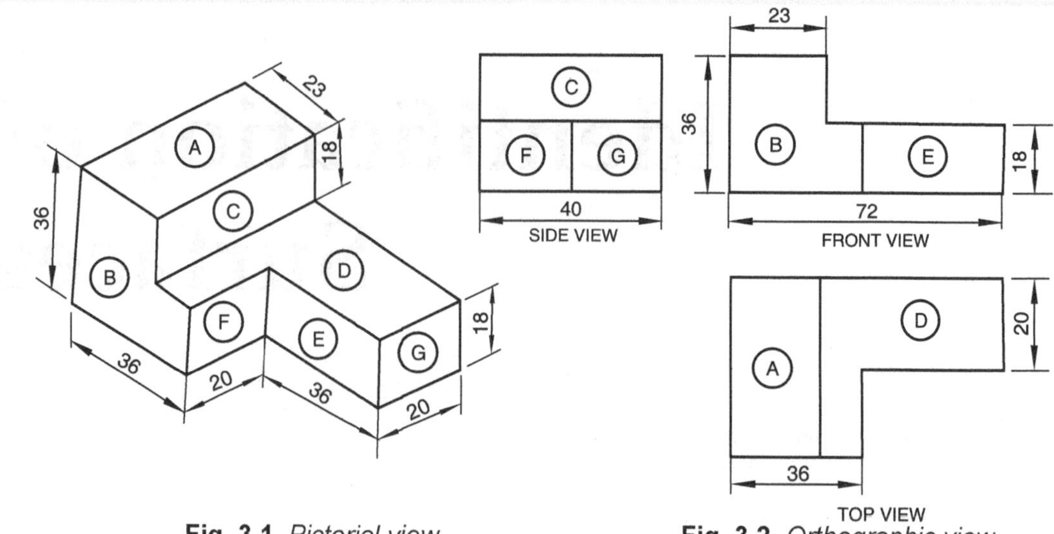

Fig. 3.1. *Pictorial view*

Fig. 3.2. *Orthographic view*

Fig. 3.2 shows the orthographic views alongwith the marking of the corresponding surfaces in front view, top view and side view. In front view, the surface A is shown. In top view, the surfaces such as B,C,D,E and F are shown. In the side-view the surfaces such as E and G are shown.

Problem 1 : Fig. 3.3 shows the pictorial view of an object in which the various surfaces are marked by different alphabets. Identify and mark the various surfaces from the pictorial view to the orthographic views. [Fig. 3.4]

Solution : For the identification of various surfaces, see Fig. 3.4.

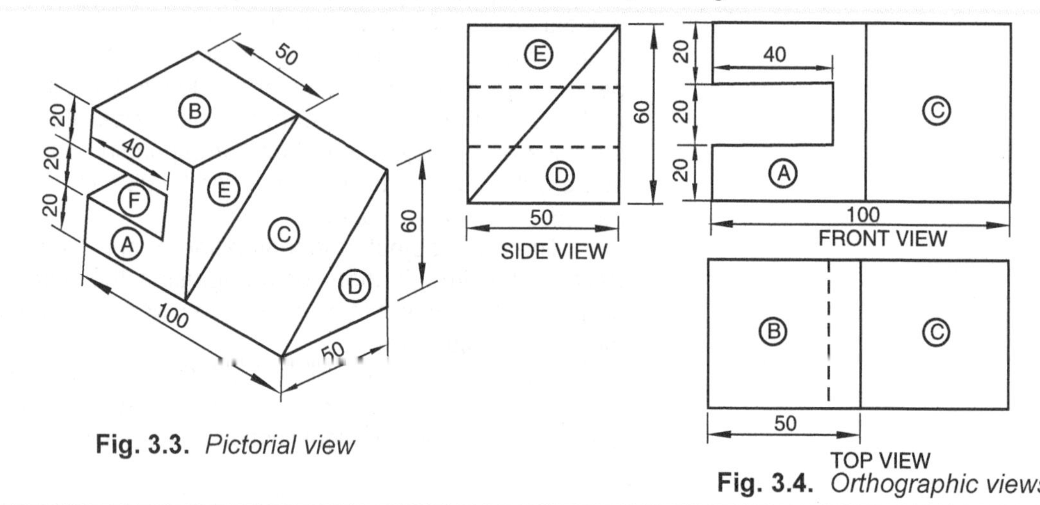

Fig. 3.3. *Pictorial view*

Fig. 3.4. *Orthographic views*

Problem 2 : Fig. 3.5 shows the pictorial view of an object in which the various surfaces are marked by different alphabets. Identify and mark various surfaces from the pictorial view to the orthographic projection. [Fig. 3.6]

Solution : For identification of various surfaces, see Fig. 3.6.

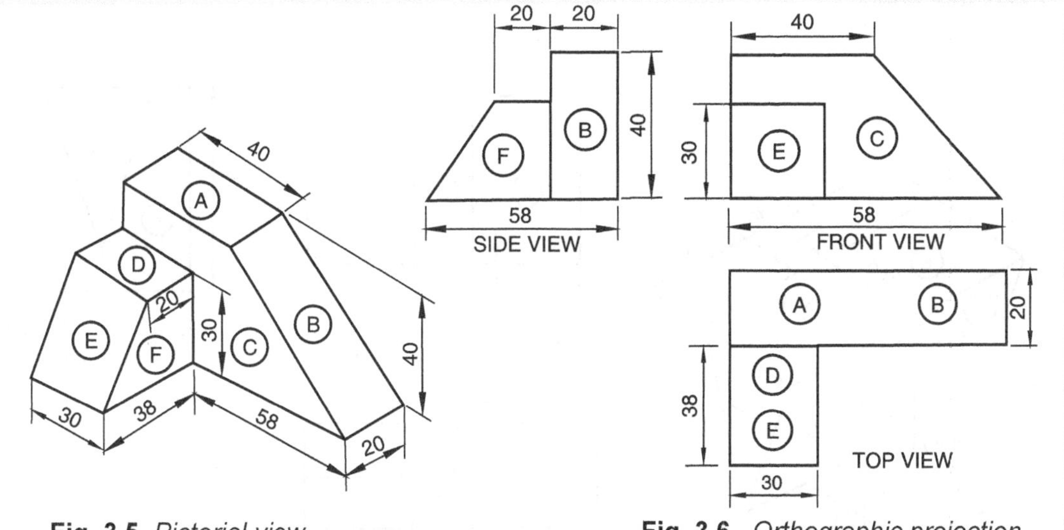

Fig. 3.5. *Pictorial view* **Fig. 3.6.** *Orthographic projection*

Problem 3 : Fig. 3.7 shows the pictorial view of an object, in which the various surfaces are marked by different alphabets. Identify and mark various surfaces from the pictorial view of the object to the orthographic projection. [Fig. 3.8]

Solution : For the identification of surfaces, see Fig. 3.8 which is itself a self explanatory sketch.

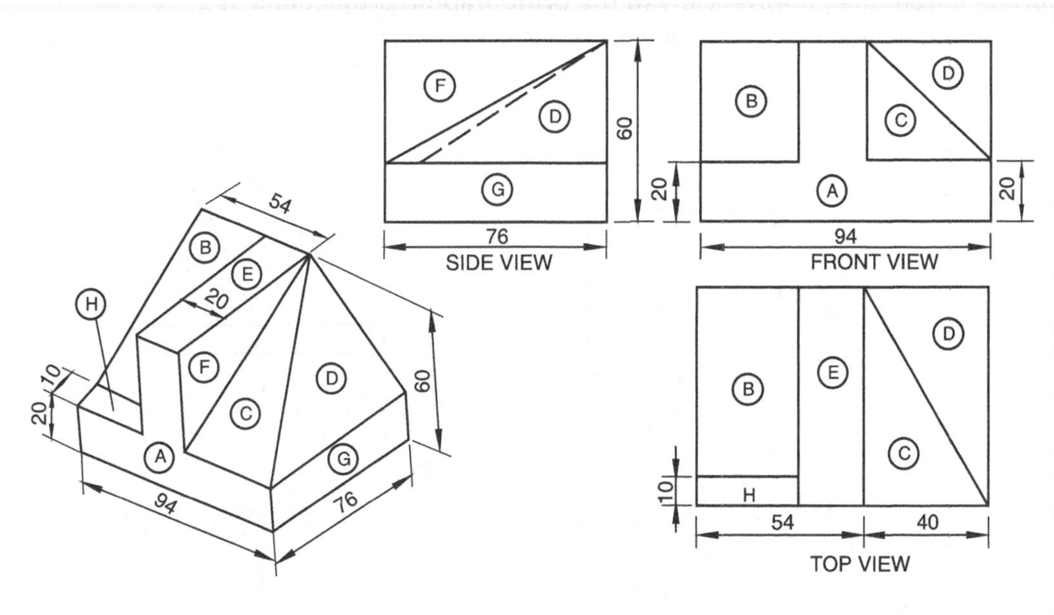

Fig. 3.7. *Pictorial view* **Fig. 3.8.** *Orthographic views*

Problem 4 : Fig. 3.9 shows the pictorial view of an object, in which the various surfaces are marked by different alphabets. Identify and mark various surfaces from the pictorial view to the orthographic projection. [Fig. 3.10]

Solution: For the identification of various surfaces, see Fig. 3.10 which is itself a self explanatory sketch.

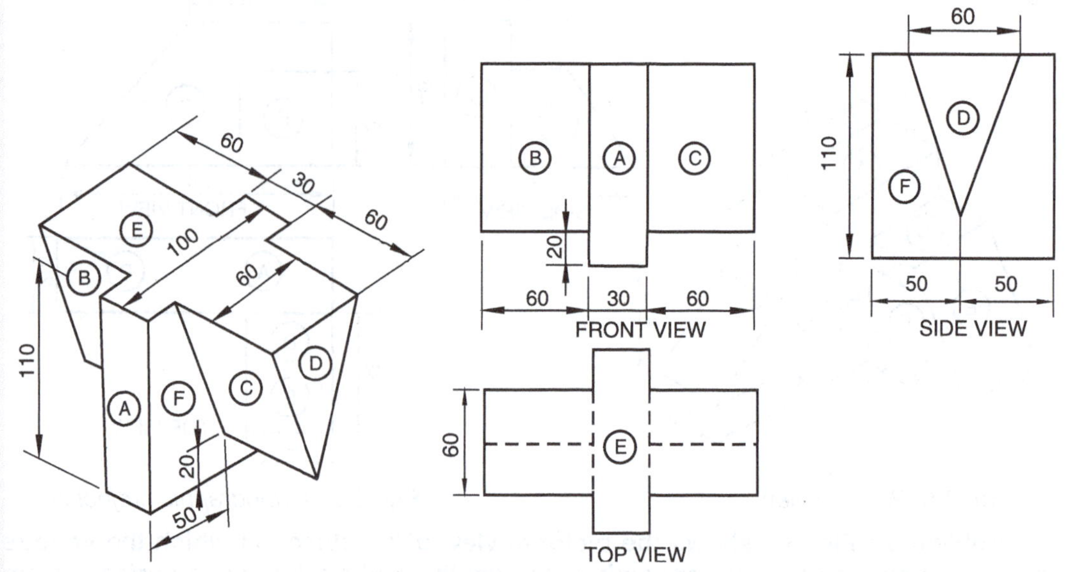

Fig. 3.9. *Pictorial view* **Fig. 3.10.** *Orthographic projection*

Problem 5 : Fig. 3.11 shows the pictorial view of an object in which the various surfaces are marked by different alphabets. Identify and mark the corresponding surfaces from the pictorial view to the orthographic projections. [Fig. 3.12]

Solution : For the identification of various surfaces, see Fig. 3.12.

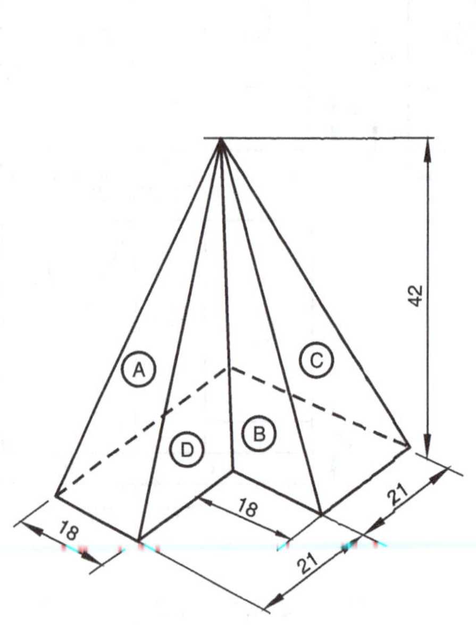

Fig. 3.11. *Pictorial view*

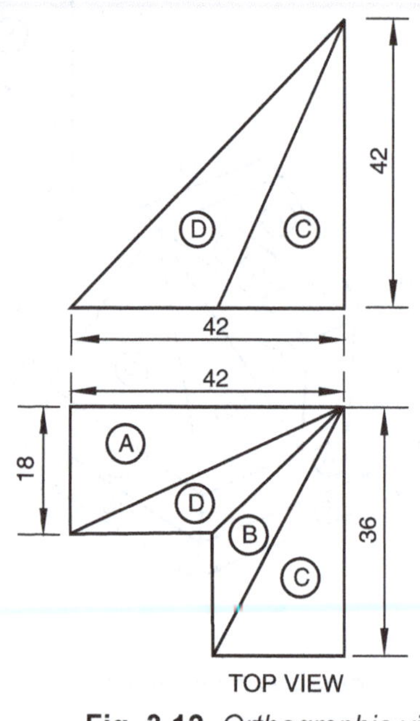

Fig. 3.12. *Orthographic views*

Problem 6 : Fig. 3.13 shows the pictorial view of an object in which the various surfaces are marked by different alphabets. Identify and mark various surfaces from the pictorial view to the orthographic projections. [Fig. 3.14]

Solution : For identification of various surfaces, see Fig. 3.14.

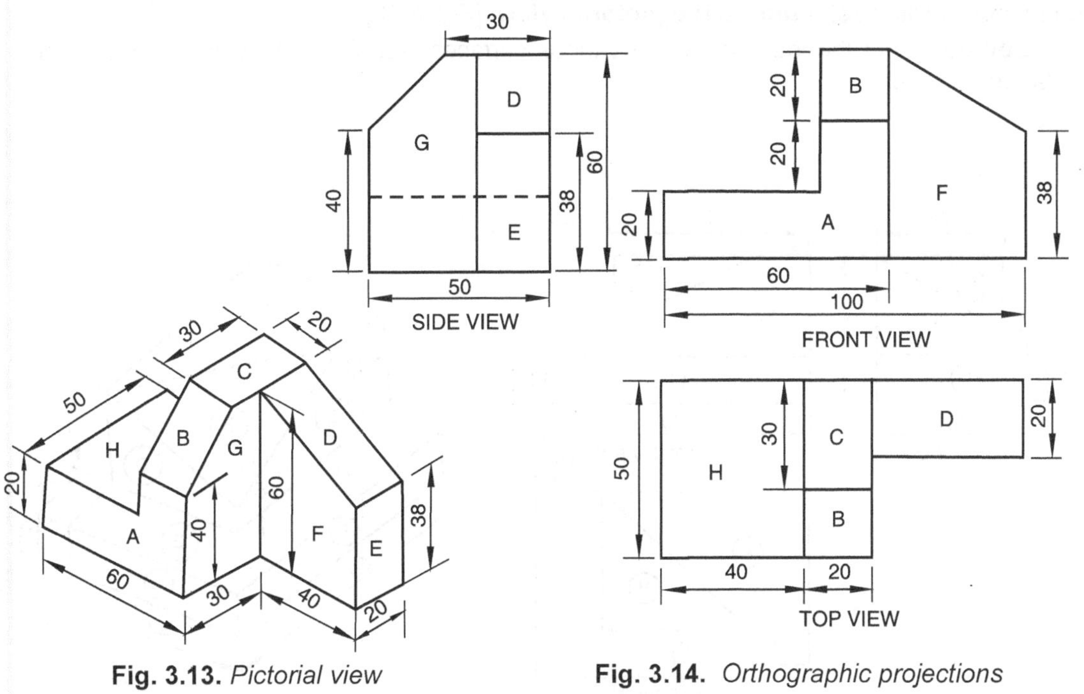

Fig. 3.13. *Pictorial view*

Fig. 3.14. *Orthographic projections*

3.3. IDENTIFICATION OF SURFACES FROM ORTHOGRAPHIC PROJECTIONS TO PICTORIAL VIEWS

Fig. 3.15 shows the orthographic projections of an object in which different surfaces are marked by the alphabets such as A, B, C, D, F, G and H.

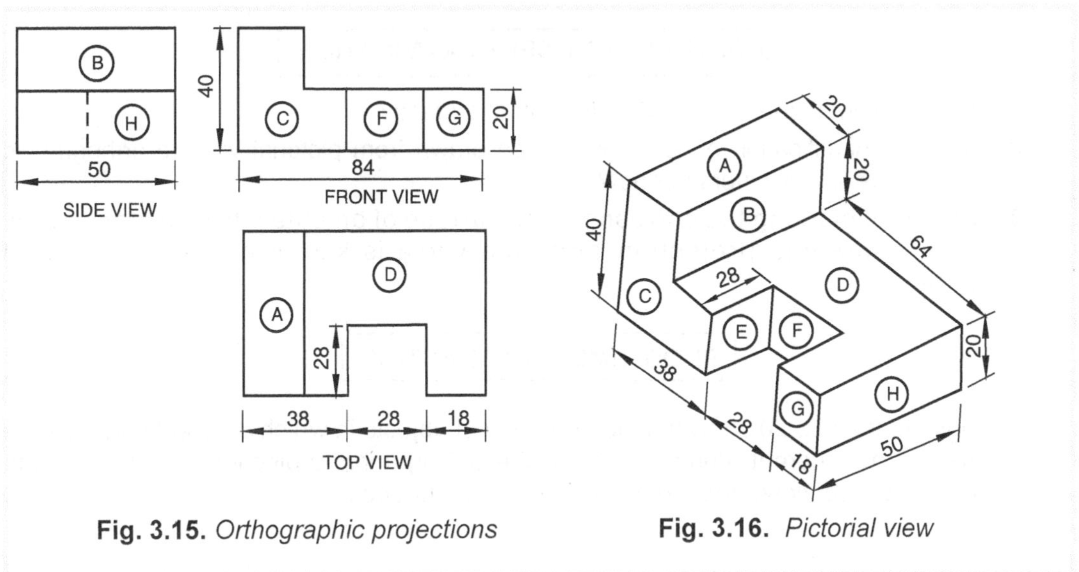

Fig. 3.15. *Orthographic projections*

Fig. 3.16. *Pictorial view*

Fig. 3.16 shows the pictorial view of the object (see Fig. 3.15) in which the corresponding surfaces are marked by the alphabets such as A, B, C, D, F, G and H.

Problem 7 : Fig. 3.17 shows orthographic views of an object in which the various surfaces are marked by different alphabets. Identify and mark various surfaces from the orthographic projection to the pictorial view. [Fig. 3.18]

Solution : For the identification of various surfaces, see Fig. 3.18 which is itself a self explanatory sketch.

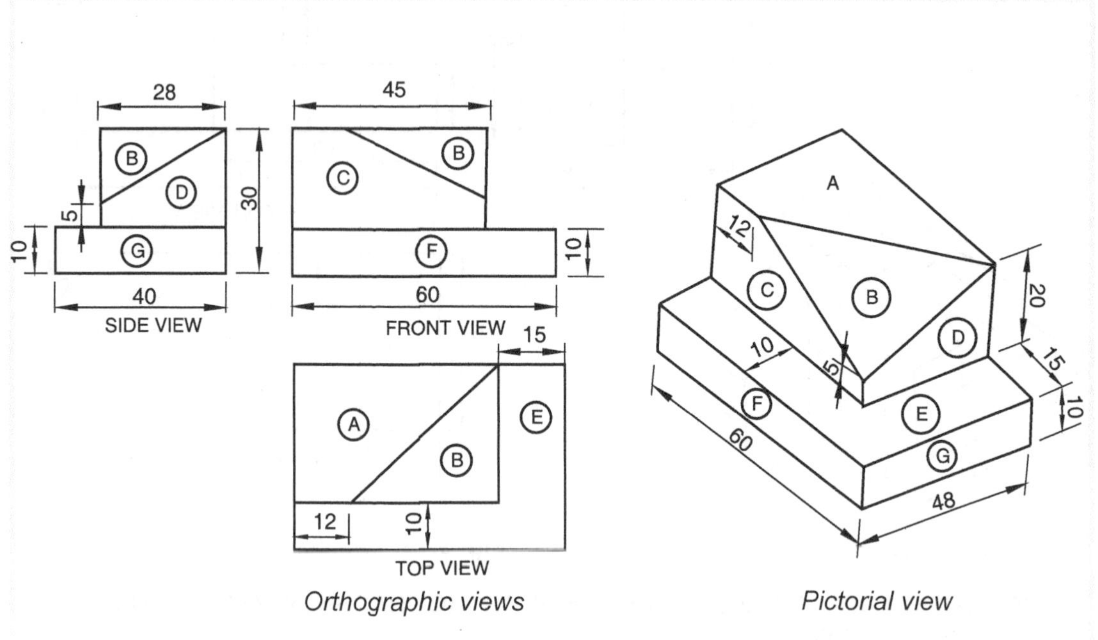

Orthographic views *Pictorial view*

Fig. 3.17. **Fig. 3.18.**

QUESTIONS FOR SELF EXAMINATION

1. What do you understand by identification of surface?
2. Whether identification of surfaces can be drawn from pictorial view to orthographic projections and vice versa or not?
3. The art of marking the corresponding surface of an object from pictorial view to orthographic projection and vice versa is known as the .... of surfaces.

PROBLEMS FOR PRACTICE

1. Figs. 3.19 and 3.20 show the pictorial view of objects in which different surfaces are marked by different alphabets. Draw the orthographic projections (front view, top view and side view) and identify the different surfaces.

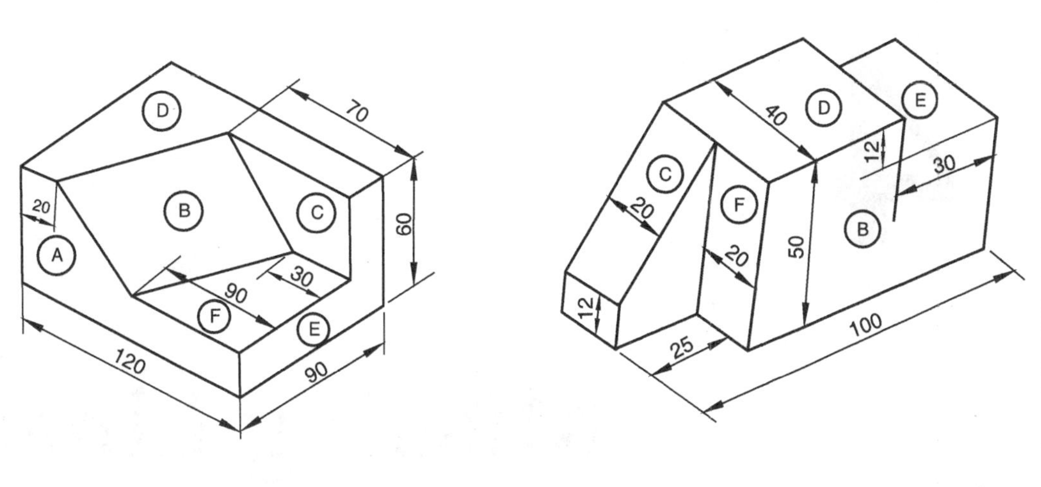

Fig. 3.19. Fig. 3.20.

2. Figs. 3.21 and 3.22 show the pictorial view of objects in which different surfaces are marked by different alphabets. Draw the three orthographic projections and identify the different surfaces.

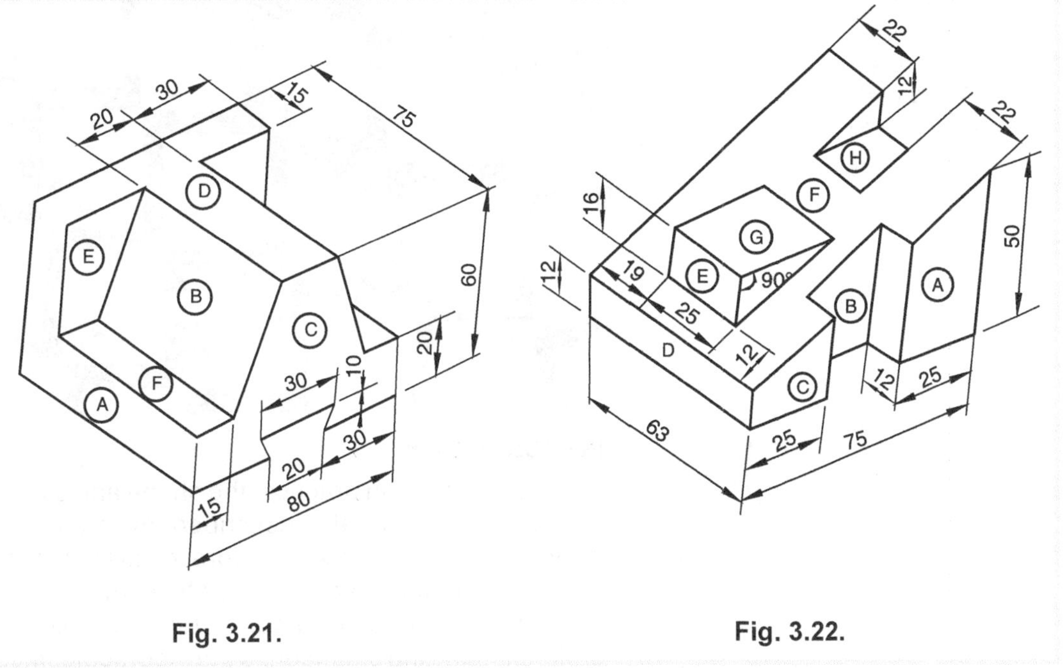

Fig. 3.21. Fig. 3.22.

CHAPTER 4

Missing Lines and Views

INTRODUCTION

In engineering practice, the shape and size of an object is generally described by means of orthographic views. It is, therefore, desired that the students should have a complete knowledge of orthographic views. In order to test the ability in understanding the engineering drawing, sometimes, one or more *lines or views* of the object are intentionally ignored in the orthographic projection. The ignored lines or views which are added in the given orthographic views are known as missing lines and views.

In this chapter, we will study missing lines, missing views and procedures for drawing the missing lines and views.

4.1. MISSING LINES

The lines which are added in the given orthographic projection in order to complete the drawing of an object are called missing lines.

4.2. PROCEDURE FOR DRAWING THE MISSING LINES

The following points should be kept in mind while drawing the missing lines of various objects:

1. First of all, draw the given orthographic projections of the object with missing lines (see Fig. 4.1).

2. If the object is simple, then draw the missing lines directly without drawing the rough proportionate pictorial view.

3. If the object is complicated, then first of all, draw the rough proportionate pictorial view without dimensions in order to understand the shape of the object.

4. After completing the pictorial view, draw the required missing lines on the orthographic projection.

Problem 1: Fig. 4.1 shows an incomplete orthographic projections of an object. Draw the missing lines and complete the orthographic projections. [Fig. 4.2]

Solution : For its solution, see Fig. 4.2 which is itself a explanatory sketch.

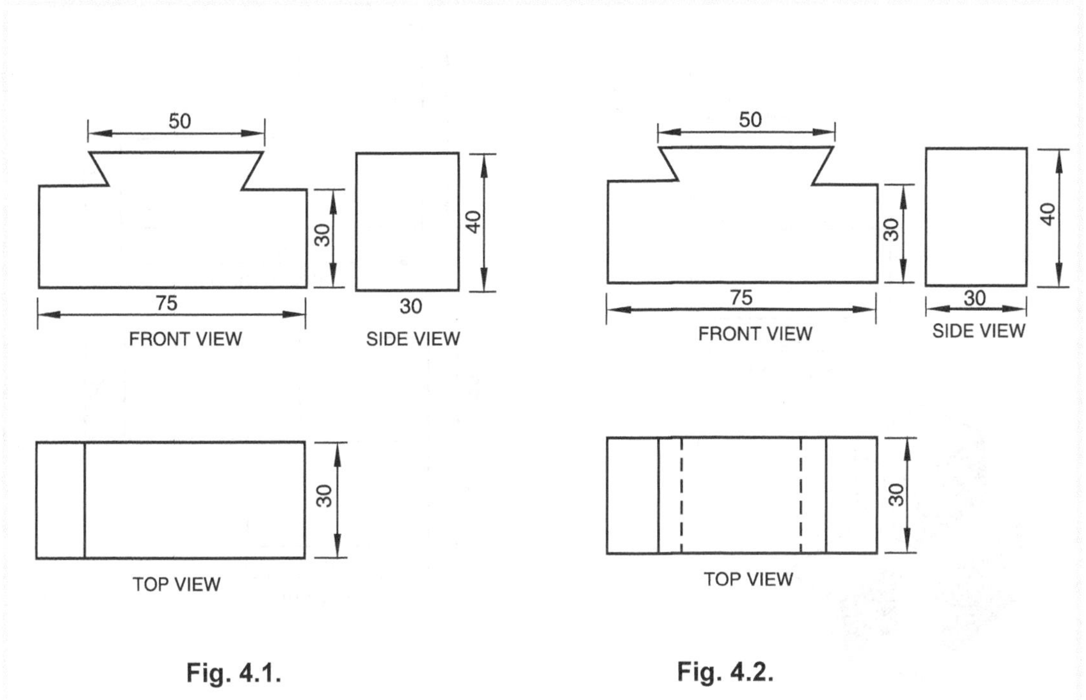

Fig. 4.1. **Fig. 4.2.**

Problem 2 : Fig. 4.3 shows a set of orthographic drawings. Redraw plan, elevation and complete the side view. [Fig. 4.4]

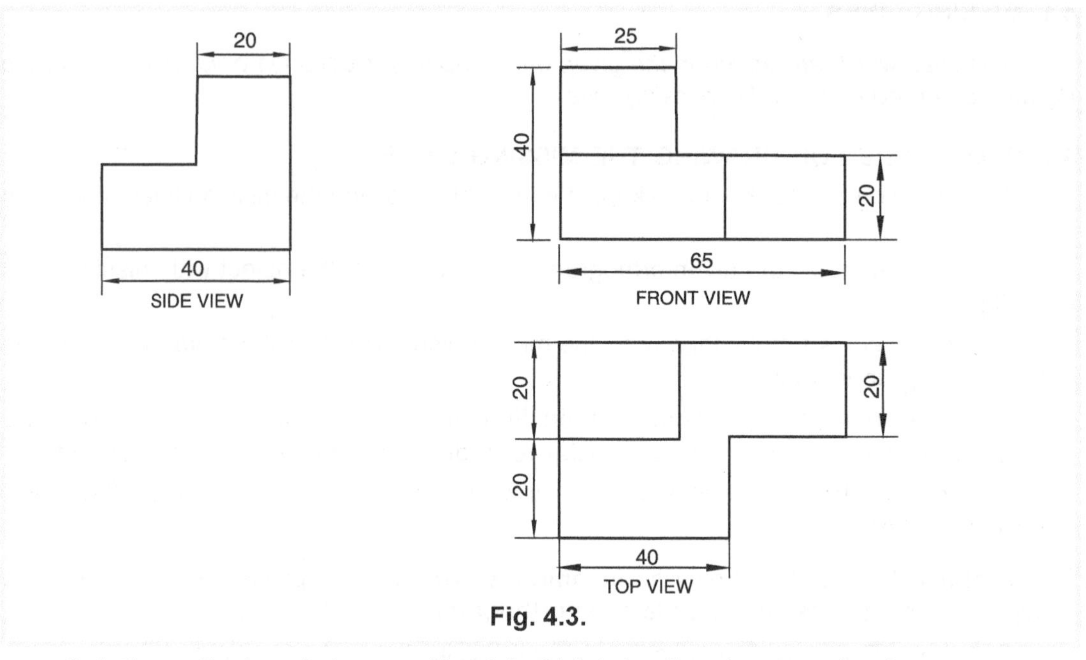

Fig. 4.3.

Solution : For its solution, see Fig. 4.4 which is itself explanatory sketch.

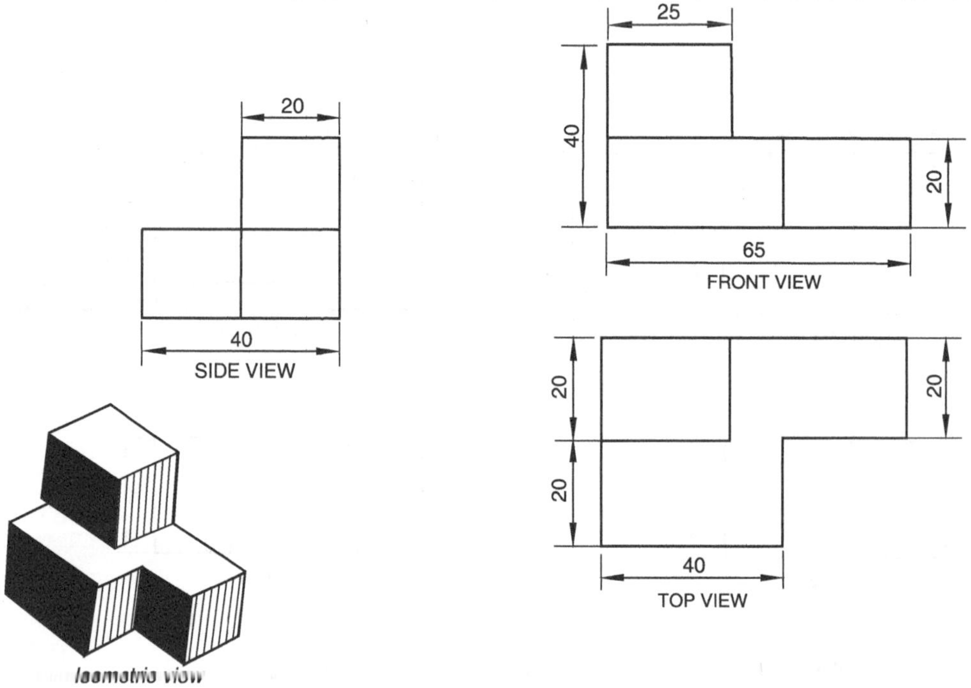

Isometric view

Fig. 4.4.

Problem 3 : Fig. 4.5 shows the incomplete orthographic projection of an object. Draw the missing lines and complete the orthographic projections. [Fig. 4.6]

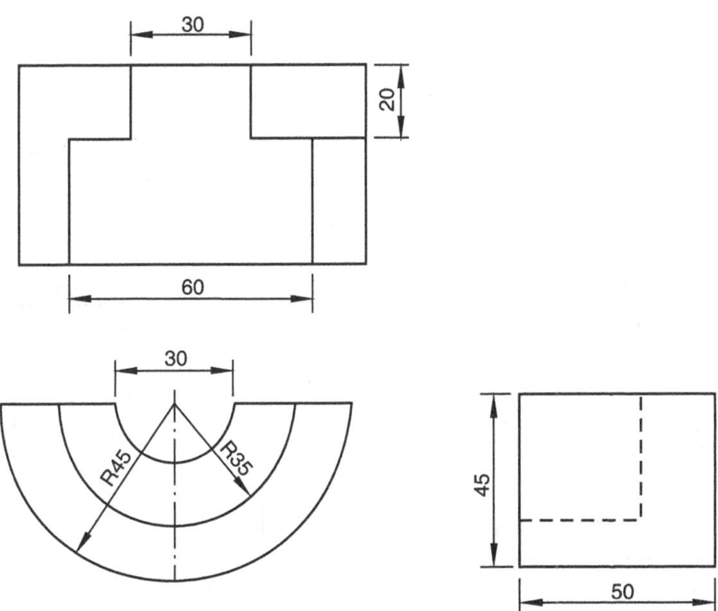

Fig. 4.5.

Solution : For its solution, see Fig. 4.6 which is itself a self explanatory sketch.

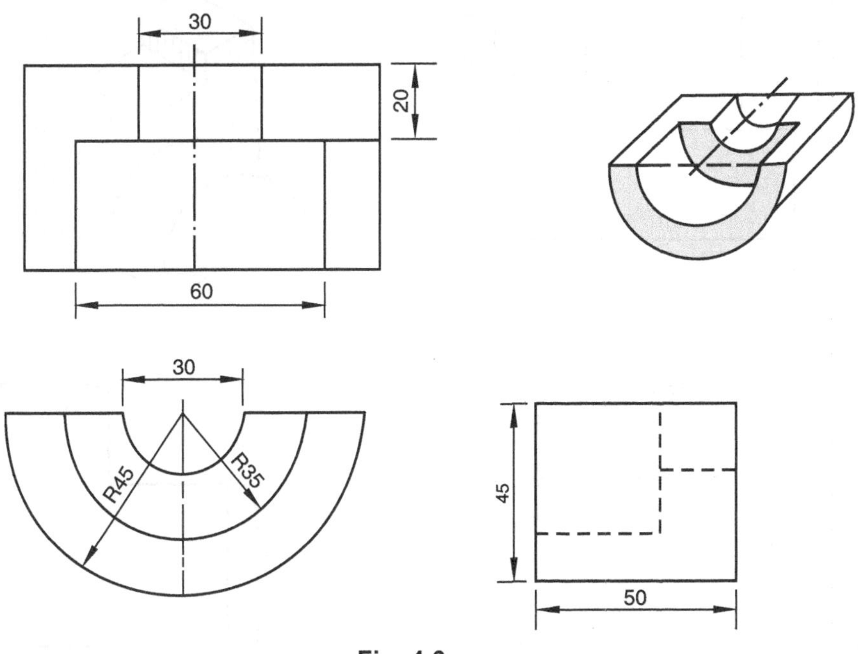

Fig. 4.6.

Note : *The students should not draw the isometric view in the examination.*

Problem 4 : Fig. 4.7 shows the incomplete orthographic projection of an object. Draw the missing lines and complete the orthographic projections. [Fig. 4.8]

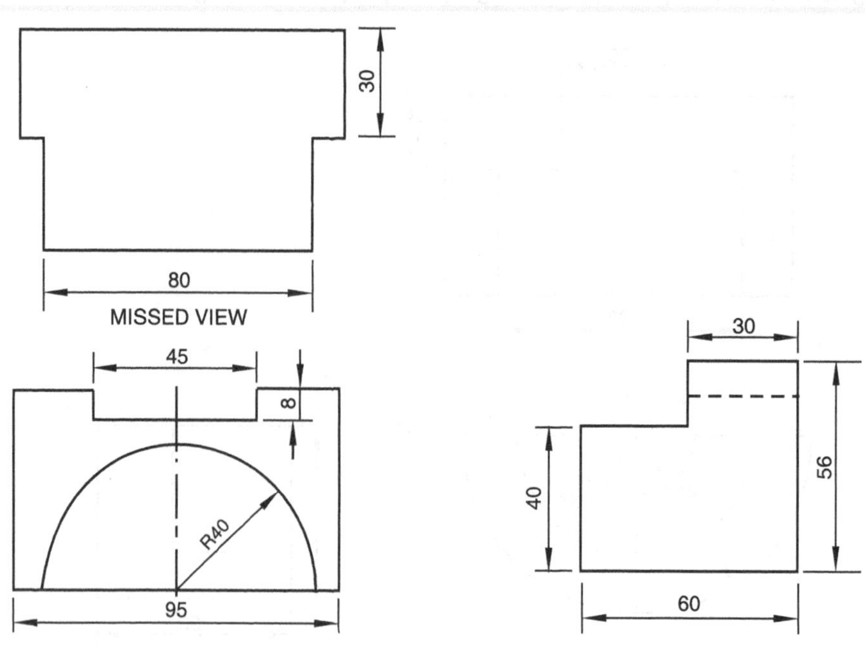

Fig. 4.7.

Solution : For its solution, see Fig. 4.8 which is itself a self explanatory sketch.

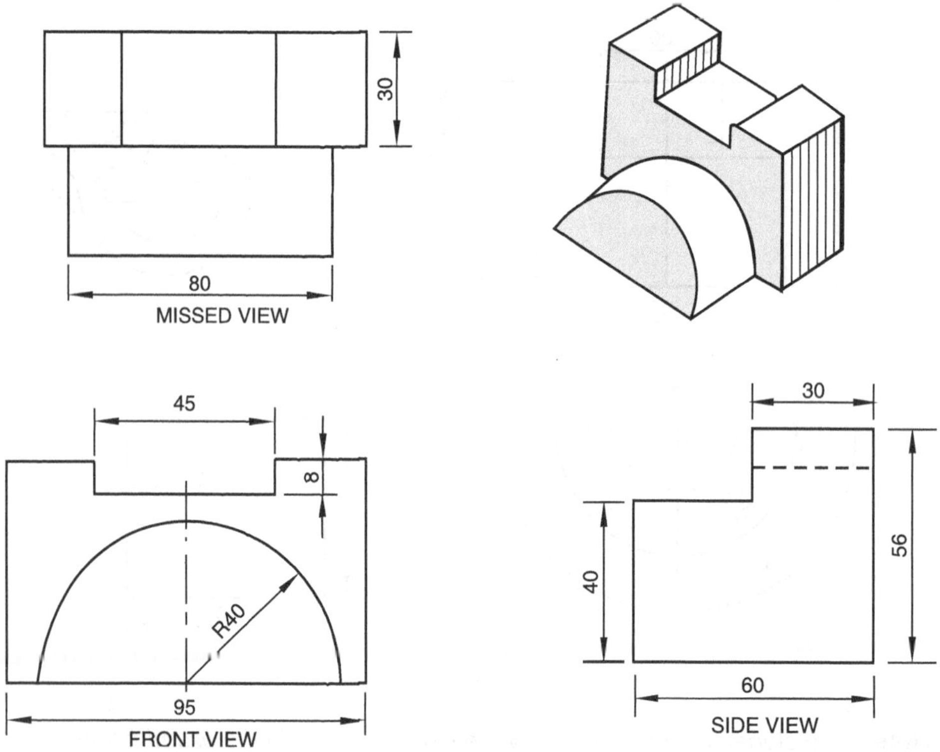

Fig. 4.8.

4.3. MISSING VIEWS

The view which is added in the given orthographic projections in order to complete the drawing of an object is called missing view.

The following points should be kept in mind while drawing the missing views of various object :—

1. First of all, draw the given orthographic projections of the object with missing views.
2. If the object is simple, then draw the missing view directly without drawing the rough proportionate pictorial view.
3. If the object is complicated, then first of all, draw the rough proportionate pictorial view without dimensions in order to understand the shape of the object.
4. After completing the pictorial view, draw the required missing view of the object.

Problem 5 : Fig. 4.9 shows the incomplete orthographic projections of an object. Draw the missing view. [Fig. 4.10]

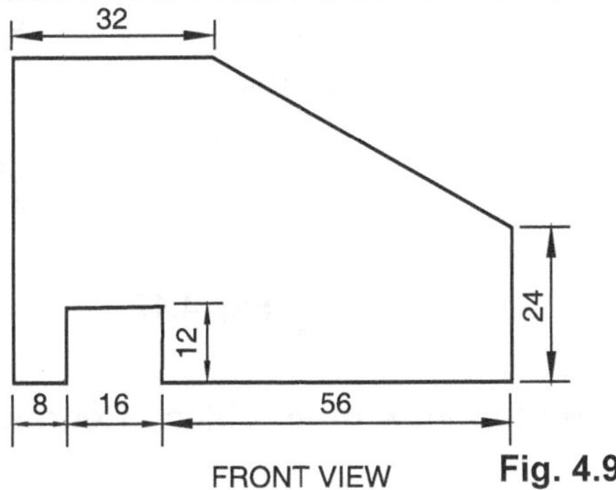

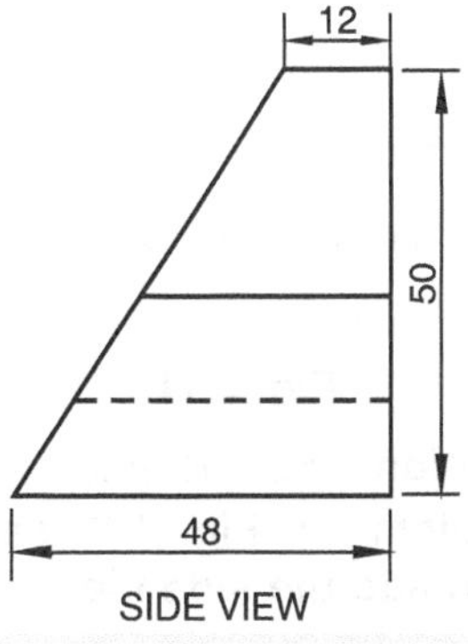

Fig. 4.9.

Solution : For its solution, see Fig. 4.10 which is itself a self explanatory sketch.

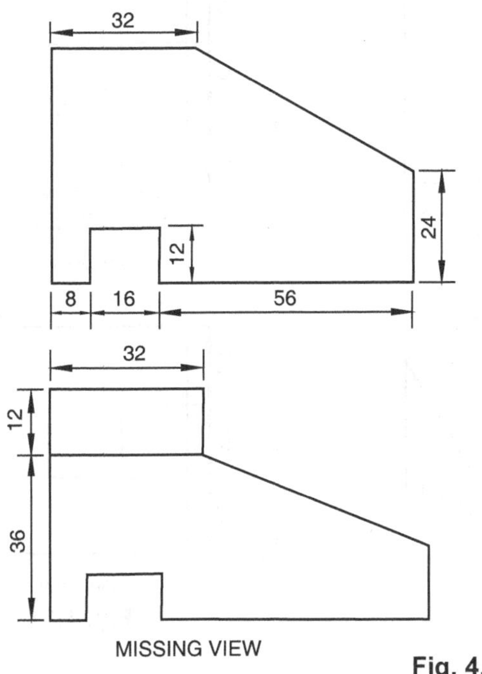

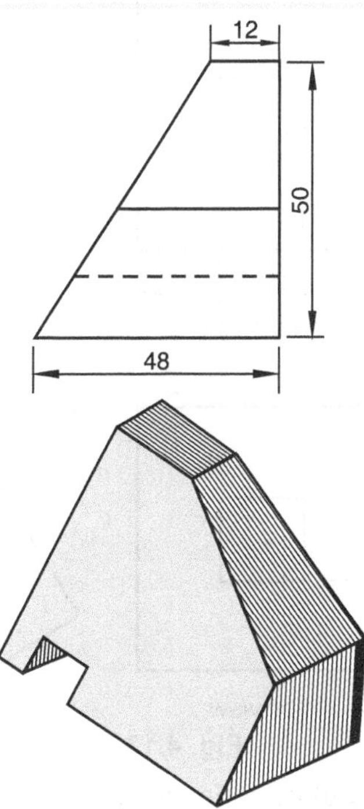

Fig. 4.10.

Problem 6 : 4.11 shows two views of an object. Draw the given views to a full size scale and add the side view. [Fig. 4.12]

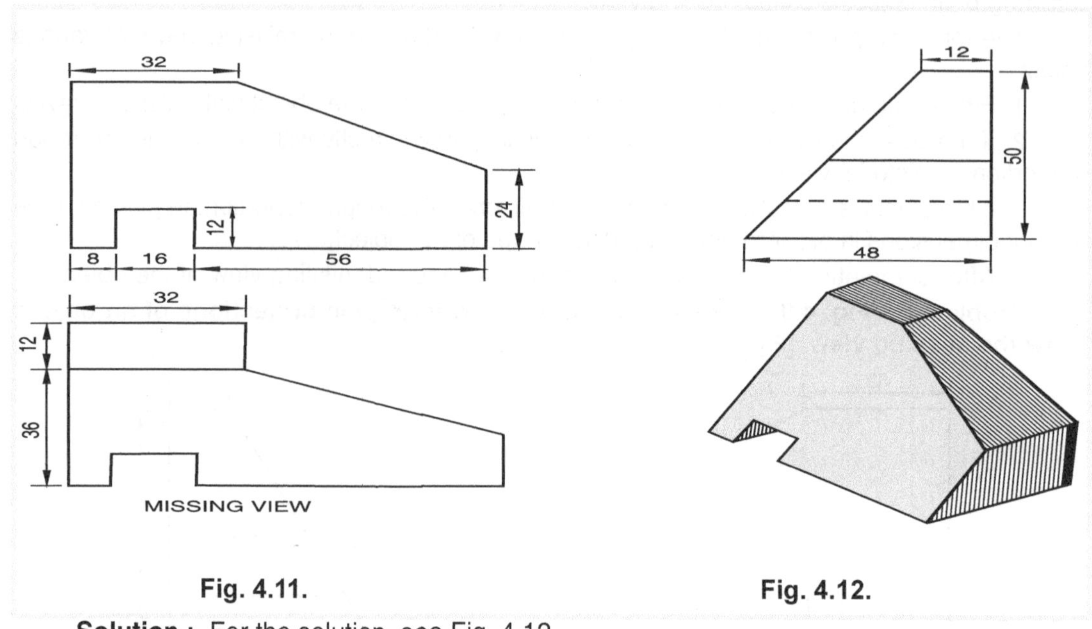

Fig. 4.11. **Fig. 4.12.**

Solution : For the solution, see Fig. 4.12.

Problem 7 : Fig. 4.13 shows front and top view of an object. Redraw the given views and add the side view. [Fig. 4.14]

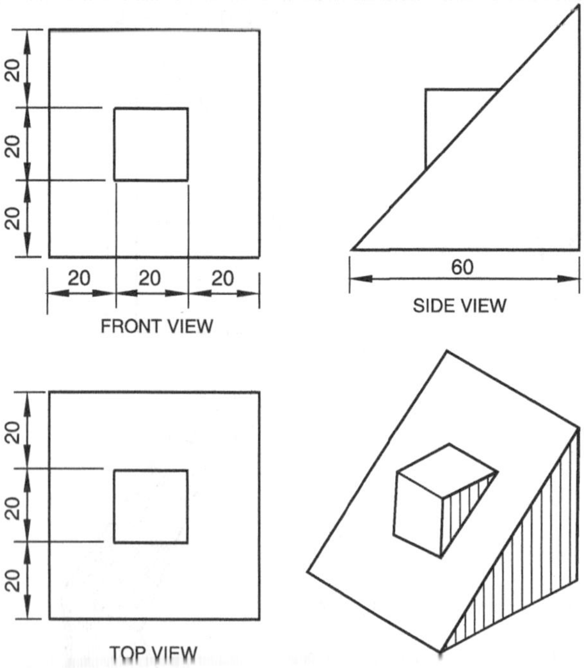

20 20 20

20 20 20

FRONT VIEW

60

SIDE VIEW

20 20 20

20 20 20

FRONT VIEW

20 20 20

TOP VIEW

20 20 20

TOP VIEW

Fig. 4.13. 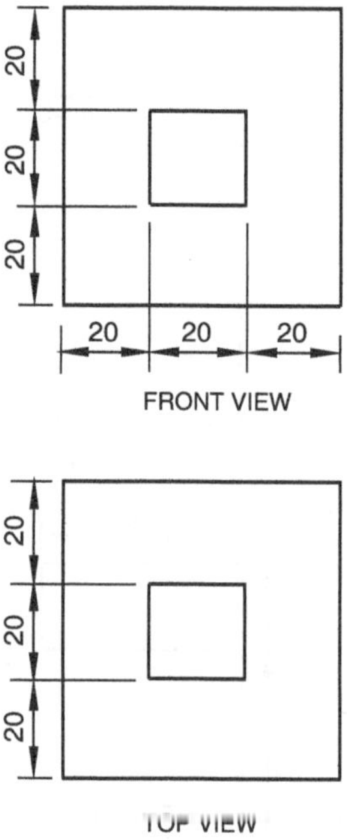 **Fig. 4.14.**

Solution : For its solution, see Fig. 4.14.

Problem 8 : Fig. 4.15 shows the incomplete orthographic projection. Draw the given views and add missing view.

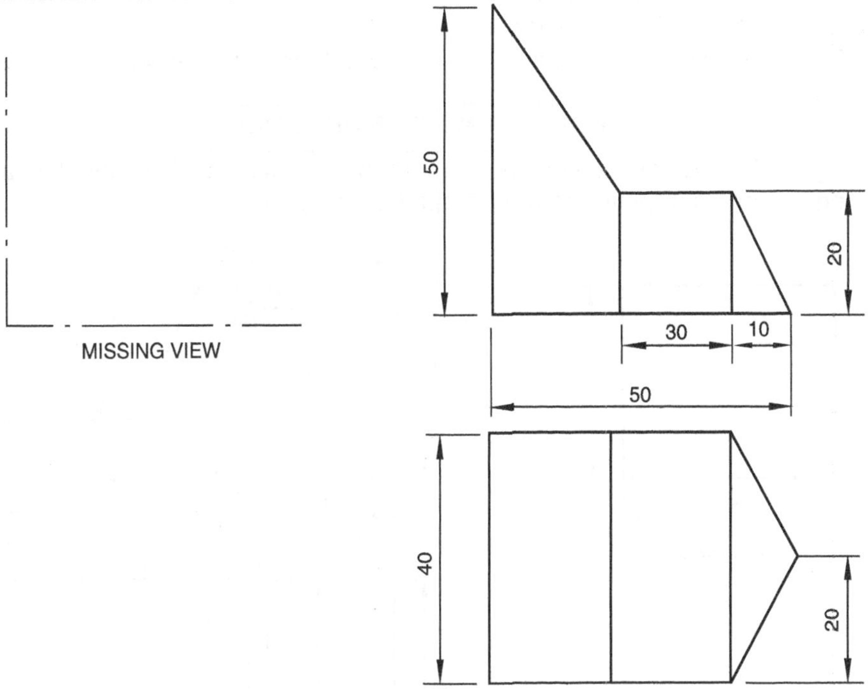

MISSING VIEW

Fig. 4.15.

Solution : For solution, see Fig. 4.16, which is itself a self explanatory sketch.

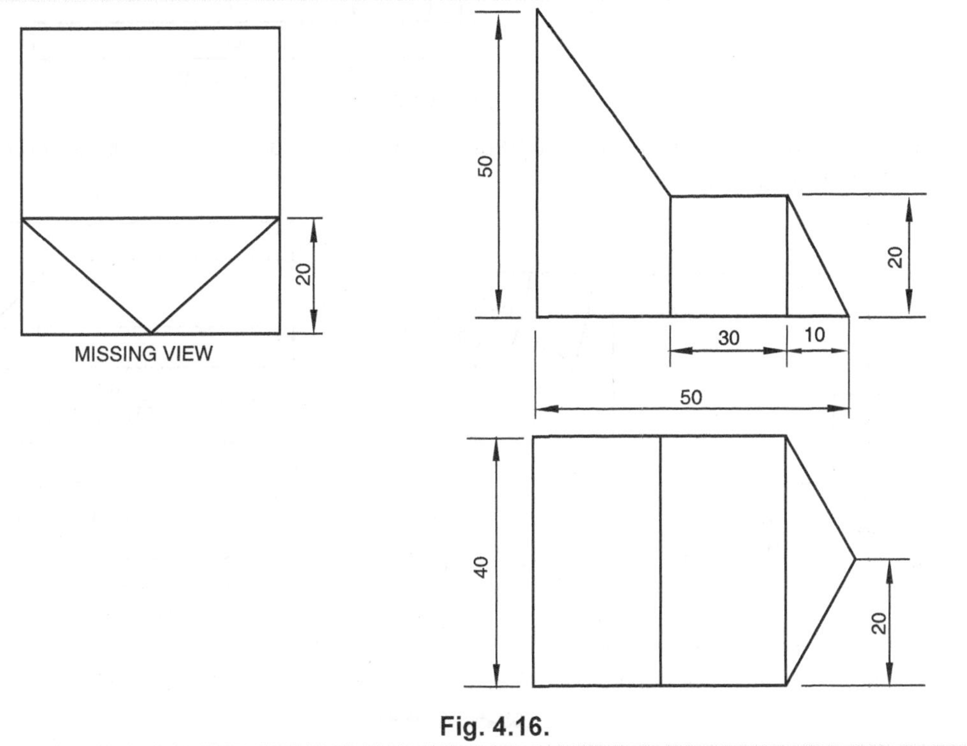

Fig. 4.16.

QUESTIONS FOR SELF EXAMINATION

1. What do you understand by missing lines?
2. What is the need of drawing missing view?
3. The lines/views which are .... in the orthographic views are known as .... lines/views.

PROBLEMS FOR PRACTICE

1. Fig. 4.17 shows an incomplete orthographic projections of an object. Draw the missing line and complete the orthographic projections.

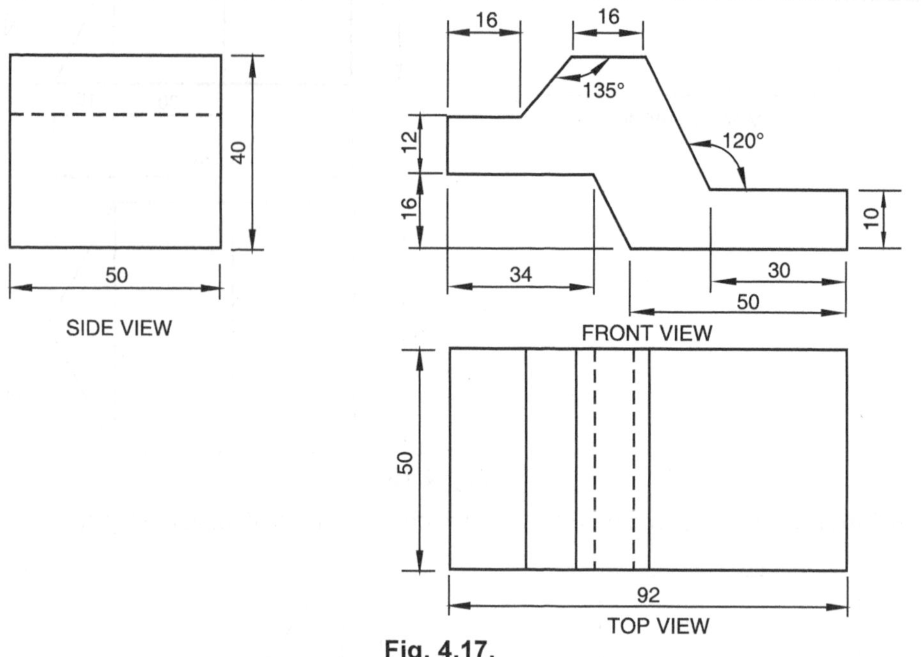

Fig. 4.17.

2. Fig. 4.18 shows a set of orthographic drawings. Redraw the drawings and complete the missing view.

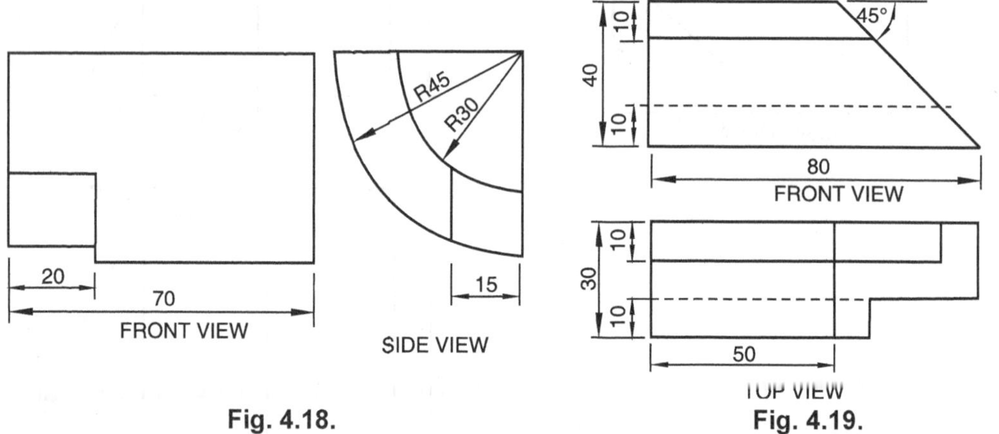

Fig. 4.18. **Fig. 4.19.**

3. Fig. 4.19 shows the incomplete orthographic projections of an object. Draw the missing view.

CHAPTER

5

Sectional Views

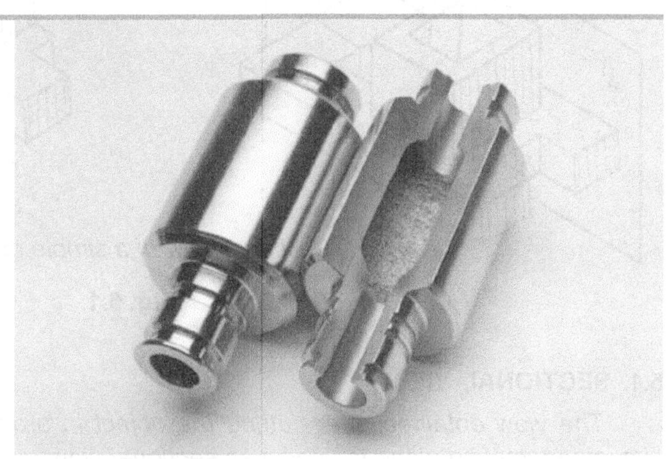

INTRODUCTION

In engineering practice, it is often required to make the drawing showing the interior details of the object, which are not visible to the observer from outside. These interior details are shown on the drawing by the dotted lines or hidden lines. If the object is simple in its interior construction, then there are few hidden lines on the drawing. On the other hand, if the object is complicated in its interior, then there are large number of hidden lines, which make the drawing unnecessarily complicated and confusing to interpret. To overcome this difficulty, complicated objects are assumed to be cut by an imaginary cutting plane. The part of the object between the imaginary cutting plane and the observer is assumed to be removed and the orthographic view of the cut object then obtained is known as *sectional view* or *view in section* (see Fig. 5.1).

In this chapter, we will deal with the study of sectional views, cutting planes, section lines, types of sectional views and some important sections.

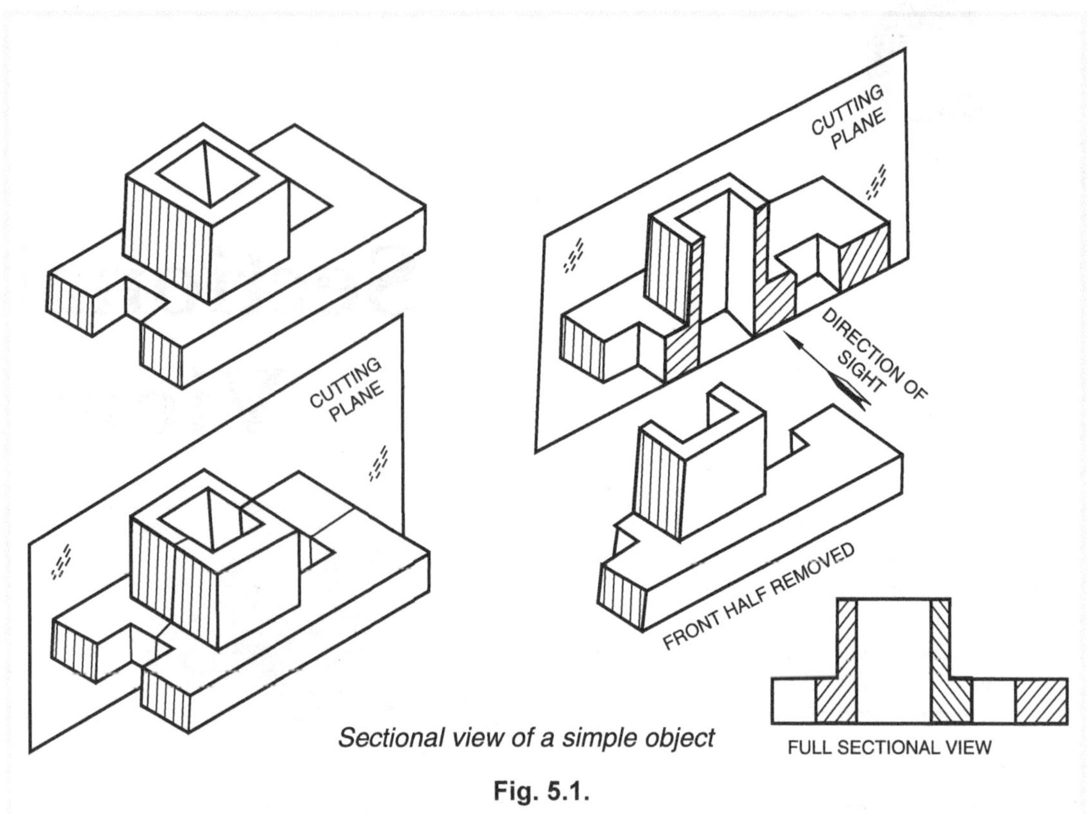

Sectional view of a simple object

Fig. 5.1.

5.1. SECTIONAL VIEW

The view obtained after cutting the object in order to show the inner details by an imaginary cutting plane is known as sectional view.

The sectional view shows the shape of the section as well as all the visible edges and contours of the object behind the cutting plane. The hidden details of the object behind the section are generally omitted from the sectional view, unless it is absolutely essential for further imaginary cutting plane to emphasize the contour of the interior.

When a section is assumed in one view, it does not affect the other views of the object. The other views are drawn as if the entire object exists as a whole. Only a cutting line is included to show the location of the section and the direction in which the object is viewed.

The sectional views are very important in engineering drawing, as they help in manufacturing and explaining the construction of complicated machines and their parts.

5.2. CUTTING PLANE OR SECTIONAL PLANE

The imaginary plane by which the object is assumed to be cut is called the cutting plane or sectional plane (see Fig. 5.2).

Fig. 5.2. *Cutting plane line*

The cutting plane used in sectioning is indicated by a line in a view adjacent to sectional view. This line is called cutting plane line. It is thick long chain at the ends and thin long and short lines at the centre. Arrow heads indicate the direction in which the cut away object is viewed i.e., direction of sight. At the end of the cutting plane line, capital letters such as X-Y or A-B are often marked to identify the cutting plane line with the corresponding section.

Note : 1. The cutting plane line is omitted in case of symmetrical object, unless it is needed for clarity.

2. *When the cutting plane line coincides with a centre line, the cutting plane line take precedence.*

5.3. SECTION LINES

The lines used to represent the material which has been cut by the cutting plane are called section lines. They are also called hatchings and crosshatchings.

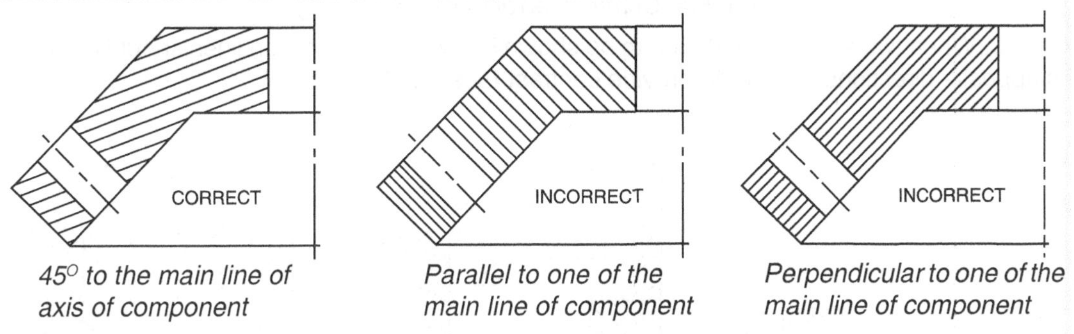

Fig. 5.3. (i) *Correct and incorrect section lining*

Section lines are thin, equally spaced, parallel and inclined lines drawn in those area of the view where the cutting plane cuts the material of the object. The following points should be followed while drawing section lines :—

1. Section lines are generally drawn at an angle of 45° to the horizontal line or outline or axis of the section. Fig. 5.3 (i) shows the correct and incorrect method of sectioning an object.

Note : *For better appearance the section lines should not be parallel to any of the main boundery lines of area [see Fig. 5.3 (i)].*

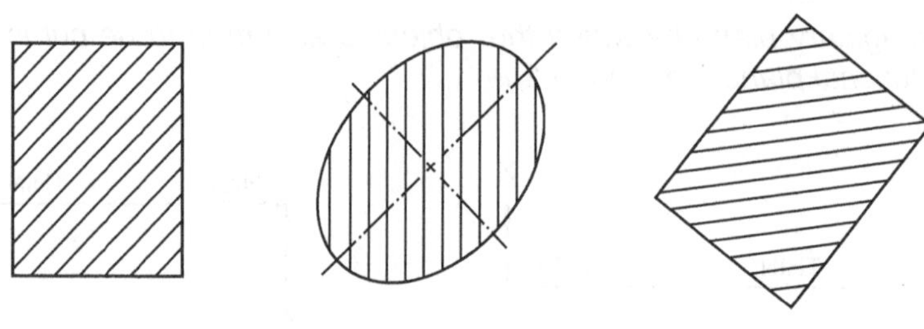

Fig. 5.3. (*ii*) *Correct angles of section lining*

2. The spacing between the section lines should be chosen in proportion to the size of the area to be sectioned. For ordinary sectioning the spacing varies from 1 mm to 3 mm. Fig. 5.4 shows the common errors in sectioning the object.

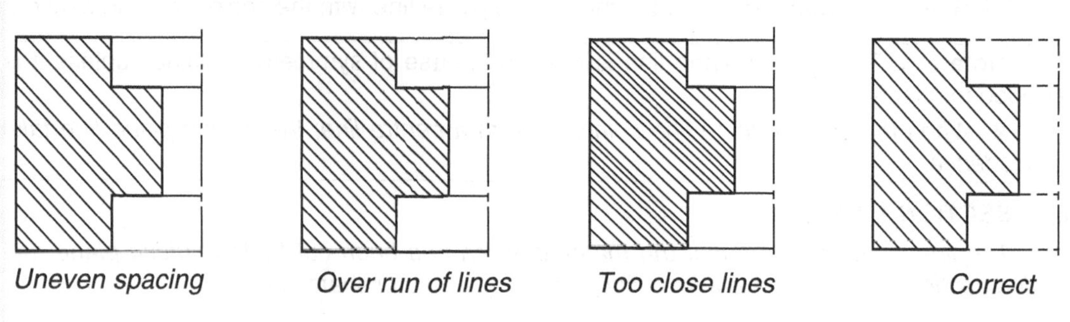

Fig. 5.4. *Common errors in section lining*

3. When two adjacent parts are to be shown in section, the section lines should be drawn in opposite directions in order to provide contrast (see Fig. 5.5).

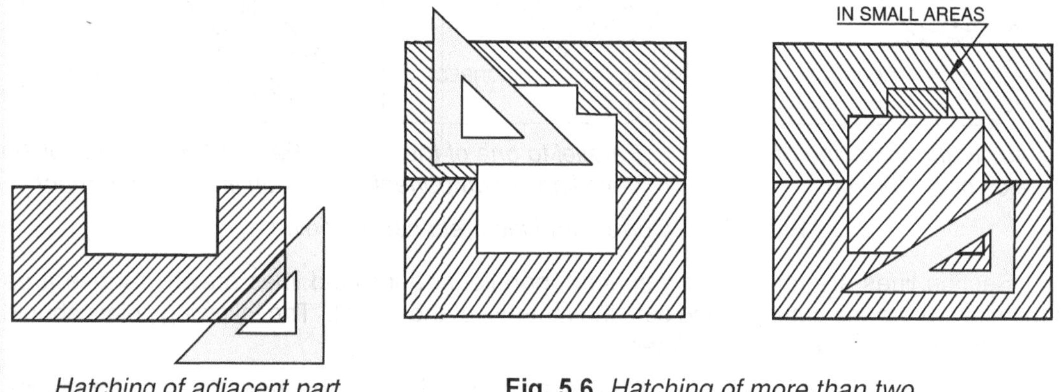

Hatching of adjacent part
Fig. 5.5.

Fig. 5.6. *Hatching of more than two adjacent parts*

4. A third part adjacent to the first two parts should be sectioned at 30^0 or 60^0 with the horizontal (see Fig. 5.6)

Note : *Sometime different materials are identified by varying the types of section lines. Section lines for various materials are shown in the table on Page 41, Chapter 3 (Section-I).*

5. For large areas, the section lines should be limited to a small portion (see Fig. 5.7)

6. In case of the adjacent section of the same parts in different plane, section lines should be drawn as shown in Fig. 5.8.

7. If a dimension need be given in any sectioned area, the section line should be interrupted for giving dimension (see Fig. 5.9)

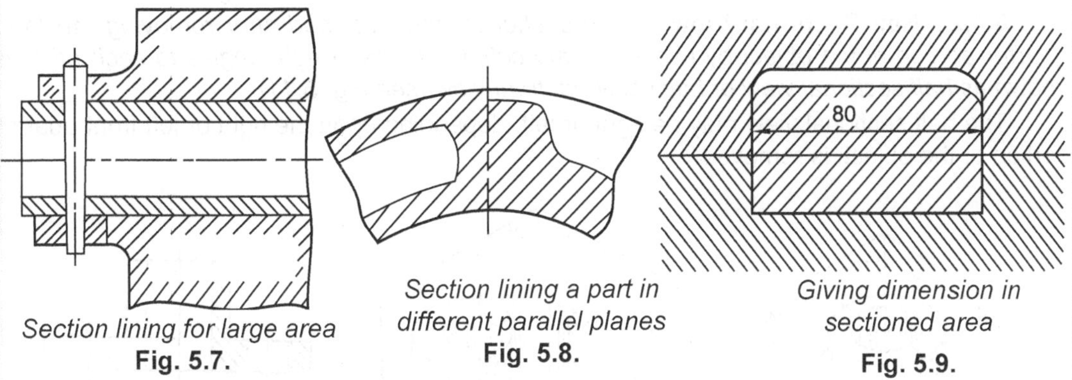

Section lining for large area
Fig. 5.7.

Section lining a part in different parallel planes
Fig. 5.8.

Giving dimension in sectioned area
Fig. 5.9.

5.4. TYPES OF SECTIONAL VIEWS

The following types of sectional views are commonly used in engineering practice :—

1. Full sectional view, and
2. Half sectional view.

5.4.1. Full Sectional View : *The sectional view obtained after removing the one half portion of the object through its centre line by an imaginay cutting plane is known as full sectional view (see Fig. 5.10).*

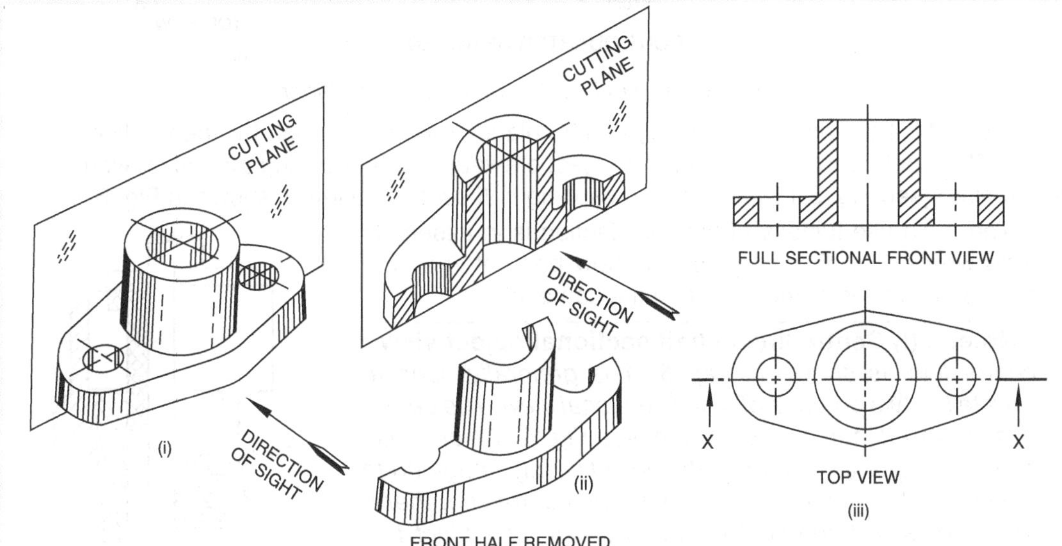

Fig. 5.10. *Full sectional front view of an object*

Fig. 5.10 (i) shows an imaginary cutting plane passing entirely through the centre line of the object. The half portion of the object in front of cutting plane is assumed to be removed exposing the internal construction of the object [see Fig. 5.10 (ii)]. The observer views the object in the direction of arrow and the resulting full sectional front view is obtained as shown in Fig. 5.10 (iii).

The following points should be kept in mind while making a full sectional view :—

(i) Invisible lines behind the cutting plane should be omitted.

(ii) Visible lines behind the cutting plane should be shown.

(iii) The parts which are actually cut by cutting plane should be hatched.

(iv) The position of the cutting plane should be shown on final drawing.

5.4.2. Half Sectional View : *The sectional view obtained after removing the front quarter portion of the object by two imaginary cutting planes at right angles to each other is known as half sectional view or half sectional front view (see Fig. 5.11).*

The half sectional view may be right or left, depending upon the right or left front quarter portion of the object, is removed.

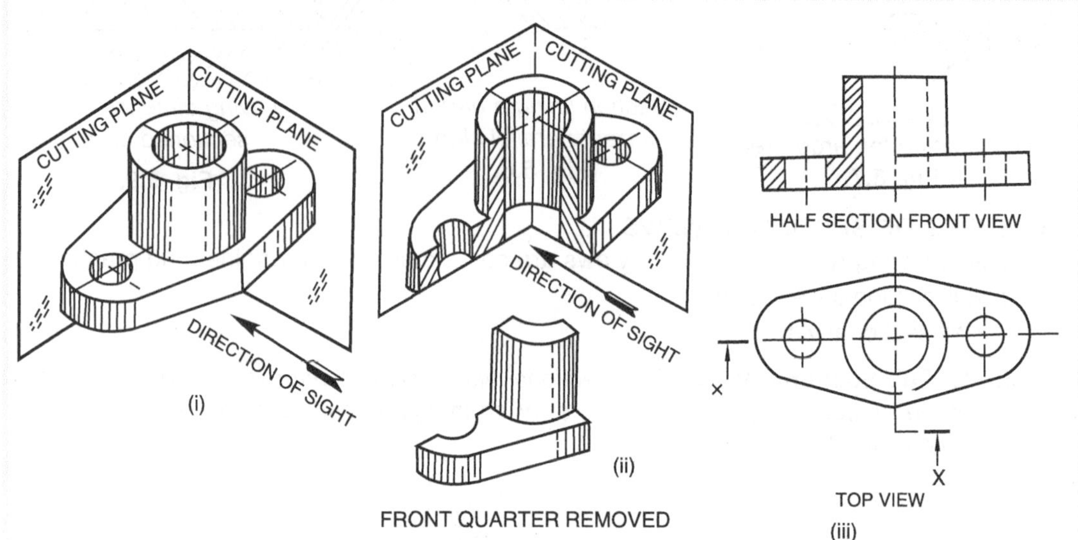

Fig. 5.11. *Half sectional view of an object*

Fig. 5.11(i) shows two imaginary cutting planes passing through two centre lines of the object. The front quarter portion of the object between the two cutting planes is assumed to be removed [see Fig. 5.11 (ii), and the resulting half sectional view is shown in Fig. 5.11 (iii).

The main use fulness of the half section is in assembly drawing where it is often required to show both internal and external construction of the object on the same view.

Notes : (i) Centre line in half sectional object view : *A centre line as shown in Fig. 5.11 is generally used to separate the two portions of a half sectional view. The centre line is preferred than the object line (see Fig. 5.12), because the removed quarter portion of the object is only imagined to be removed and not actually cut out. The centre line, therefore, denotes a theoretical edge, whereas, an object line would indicate an edge. That is why a centre line is generally preferred.*

(ii) Hidden lines in sectional object : *It is a standard practice to omit all lines for hidden details in the sectioned part of a view.*

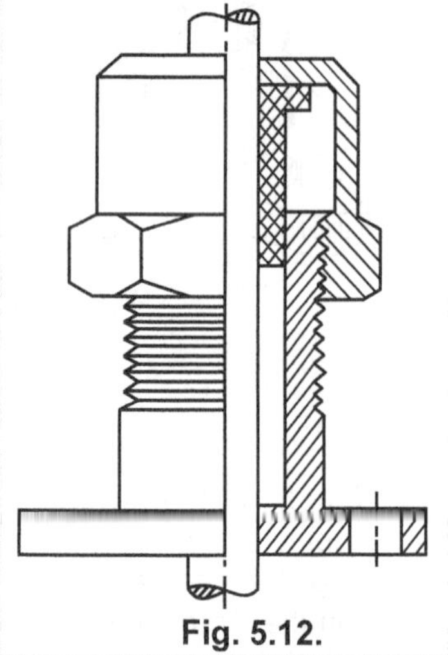

Fig. 5.12.

In making a half sectional assembly, the invisible lines are usually omitted from the unsectioned half as well as from the sectioned half so that the sectioned side shows the internal construction and unsectioned side represents outer appearance (see Fig. 5.12). In making details of a single part, the invisible lines may be shown on the unsectioned half if they are needed for dimensioning.

5.5. SOME IMPORTANT SECTIONS

The following are the important sections which are mostly used in engineering drawing:—

1. Partial or Broken out section
2. Offset section
3. Revolved section
4. Removed section
5. Outline section
6. Phantom section
7. Thin material in section
8. Spokes of wheel in section
9. Web in section
10. Auxiliary section
11. Aligned section

1. Partial or Broken out section : Sometimes, only a particular hidden detail of the object is required to be shown instead of full section or a half section. In such a case, only a partial section suffices which is limited by short break line and is called a partial or broken out section (see Fig. 5.13).

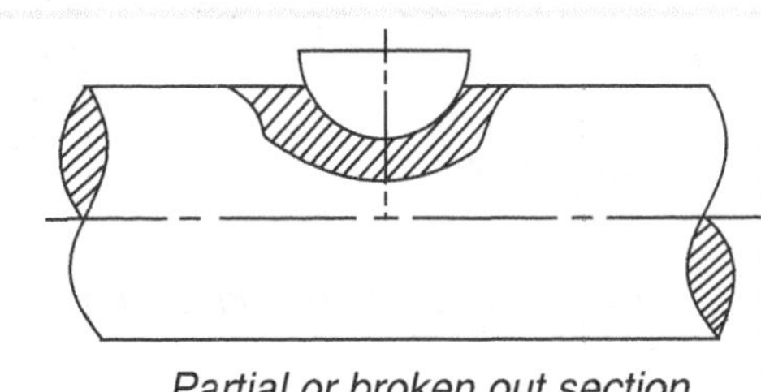

Partial or broken out section

Fig. 5.13.

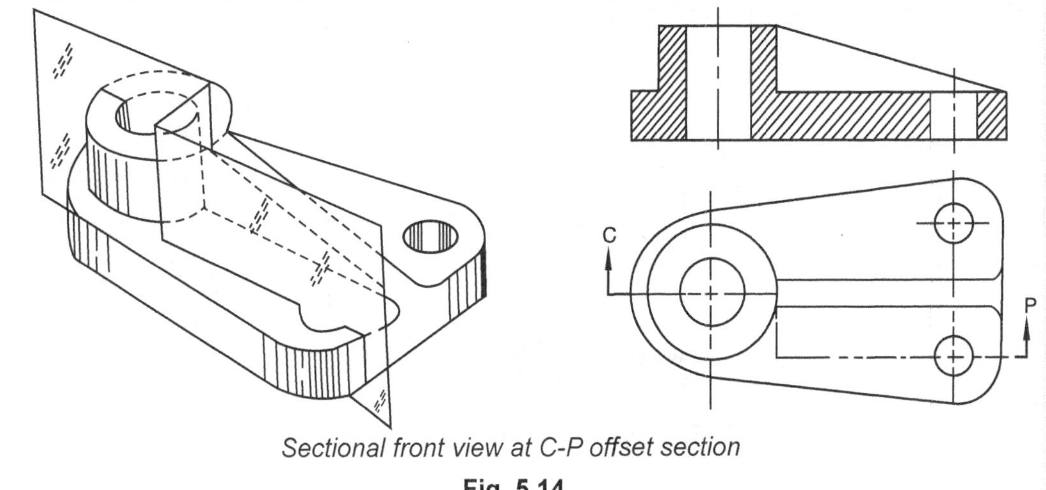

Sectional front view at C-P offset section

Fig. 5.14.

2. Offset section : When an object is cut by an offset section plane (i.e. two or more planes) in order to show the maximum details, the view is shown in off-set section (see Fig. 5.14).

When an off-set section is drawn, the edges of the off-set section plane are not shown in sectional view, but the position or path of the off-set section is shown in the related view by means of a cutting plane line (see Fig. 5.14).

3. Revolved section : The cross-section of elongated objects such as bars, arms, spoke or ribs are drawn on its longitudinal view by means of a revolved section. To obtain such a section, an imaginary cutting plane is made to pass perpendicular to the centre line of the object as shown in Fig. 5.15. Then, the resulting section is revolved at right angles to its axis and shown there. The section shown in this manner is known as revolved section.

The revolved section not only saves an additional view but the dimensions can be frequently placed on such section.

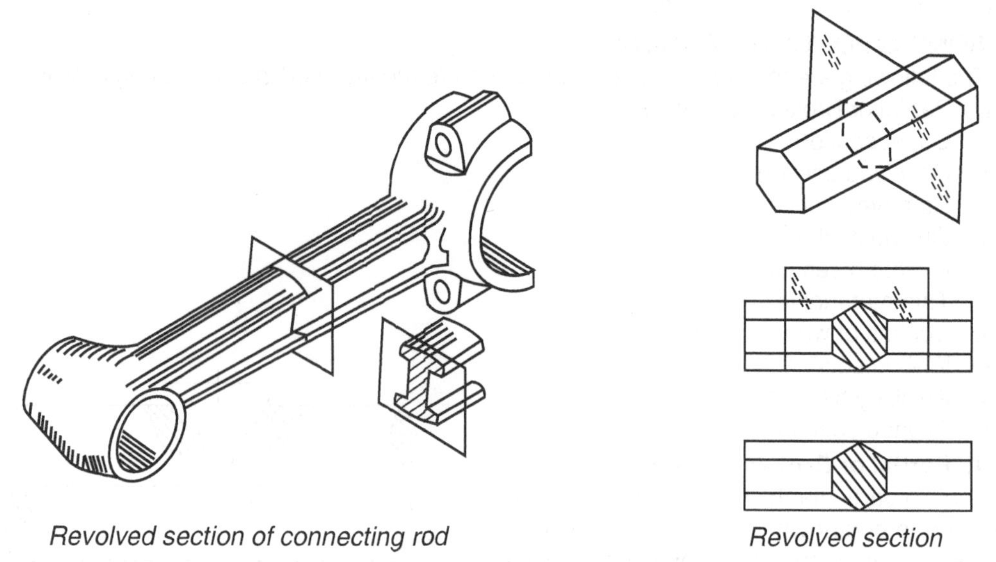

Revolved section of connecting rod *Revolved section*

Fig. 5.15.

4. Removed section : A removed section is obtained in the same manner as the revolved section, but is drawn separately outside the view, i.e., generally around the extension of the cutting plane line (see Fig. 5.16).

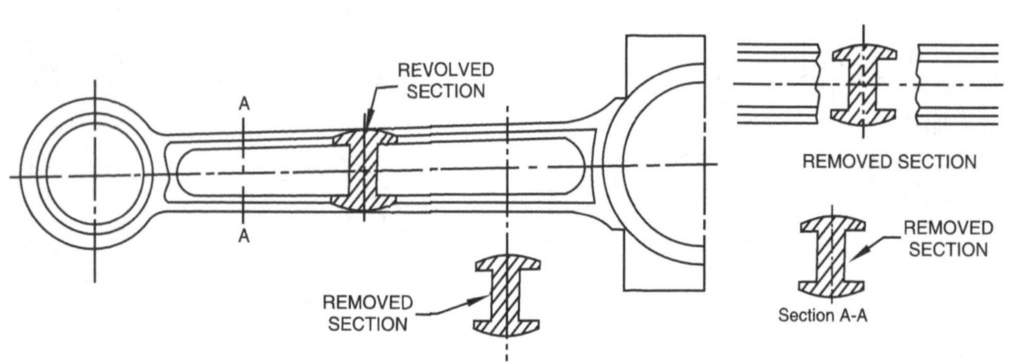

Fig. 5.16. *Revolved section and removed section*

This section is also drawn at some suitable place where sufficient space is available for its enlargement and the section is identified by a note e.g., section at A-A (see Fig. 5.17).

5. Outline section : Large surfaces are sectioned around the edges only as shown in Fig. 5-17.

6. Phantom section : The object which are not completely symmetrical, a hidden or phantom section may be drawn instead of full or half section as shown in Fig. 5.18.

7. Thin material in section : The cross-sectional views of thin parts such as steel structures, packing, gasket, sheet metal etc, are shown by solid block. In such cases, a thin white space is left between adjacent sections as shown in Fig. 5-19.

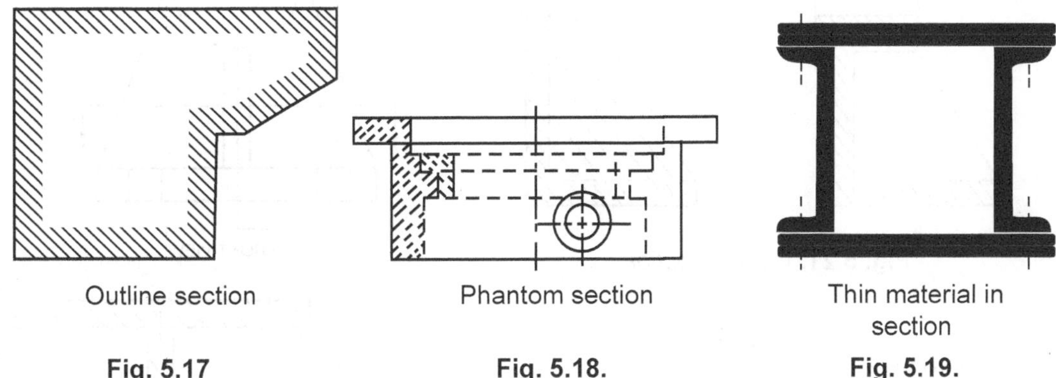

Fig. 5.17 **Fig. 5.18.** **Fig. 5.19.**

8. Spokes of wheel in section : Spokes of wheel are not sectioned even though the cutting plane passes through them. This method not only saves the time but also gives a method of distinguishing between a wheel with spokes as well as a wheel with solid web in the sectioned view (see Fig. 5.20).

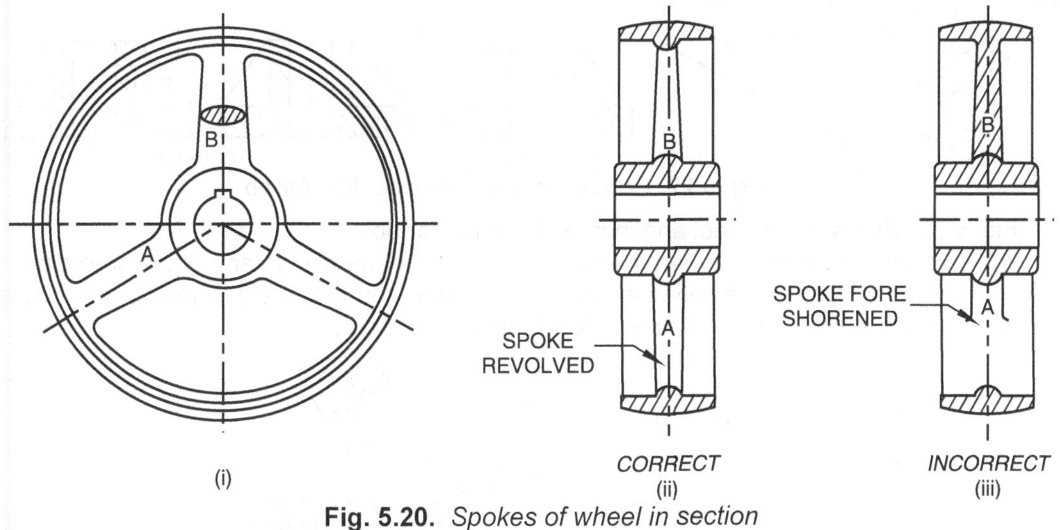

Fig. 5.20. *Spokes of wheel in section*

Fig. 5.20 (i) shows the front view of a wheel having an odd number of spokes. In conventional practice the spoke A is assumed to be revolved and brought in line with the spoke B and then its sectional side view is drawn [see Fig. 5.20 (ii)].

If the spoke A is not revolved, it will appear for-shortened as shown in Fig. 5.20 (iii).

9. Web in section : A web is a thin part of the object which is used as bracing for adding strength to the object. When the cutting plane passes through web parallel to its larger dimension as shown in Fig. 5.21, the most common practice is to omit the section lines on the web with the body of the object [see Fig. 5.21(i)].

Another practice is to show the web by means of dotted lines and the section lines are drawn on it, omitting alternative line [see Fig. 5.21(ii)].

When a section plane cuts a web perpendicular to its large dimension, the web is shown in section (see Fig. 5.21 for sectional top view).

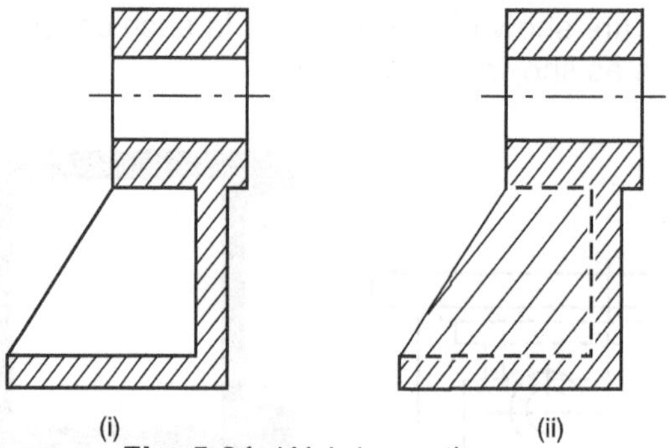

(i) (ii)

Fig. 5.21. *Web in section*

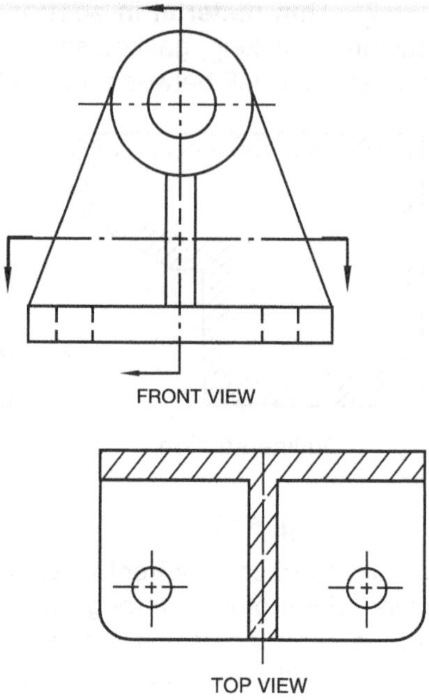

FRONT VIEW

TOP VIEW

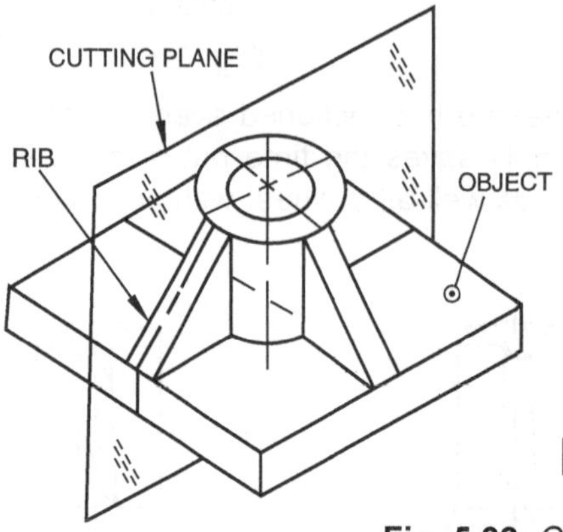

CUTTING PLANE

RIB

OBJECT

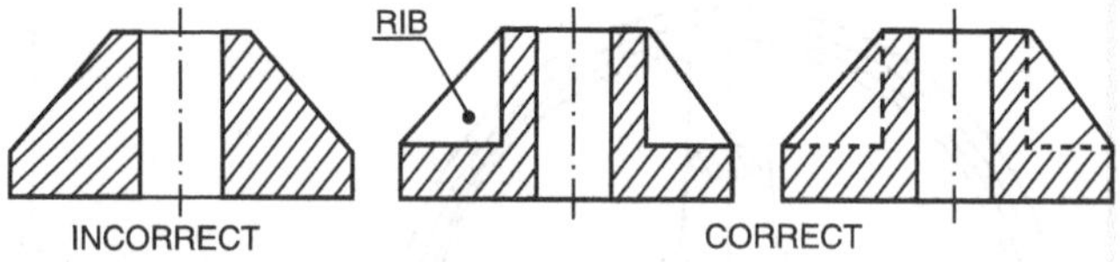

INCORRECT

CORRECT

Fig. 5.22. *Correct and incorrect section for rib*

Fig. 5.22. shows the correct and incorrect section for rib.

10. Auxiliary section : When the cutting plane is not parallel to any of the co-ordinate planes, the section obtained is called an auxiliary section (see Fig. 5.23). These sections are used to show the interior details of the inclined features of the object.

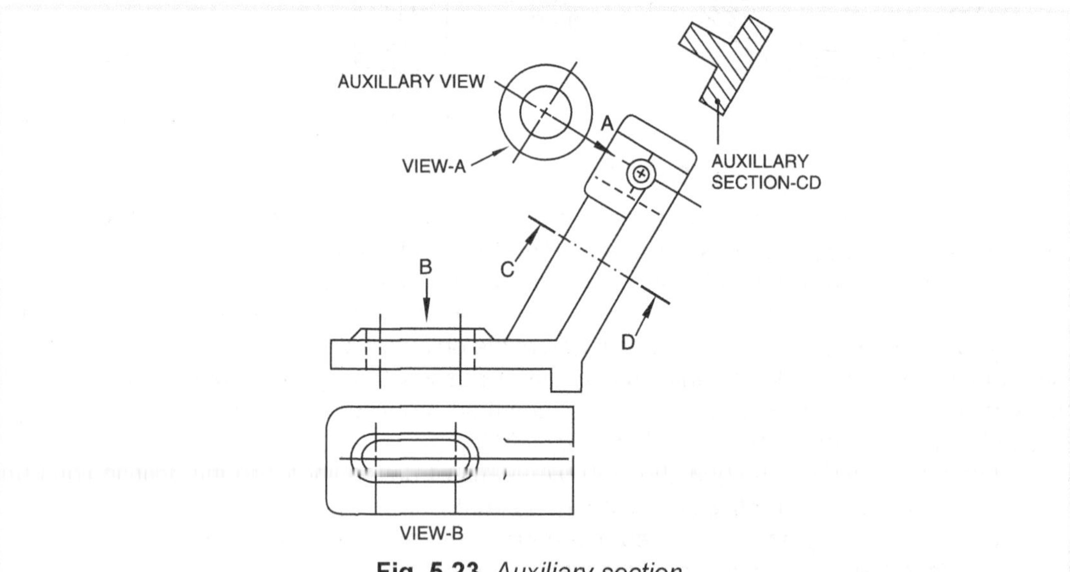

Fig. 5.23. *Auxiliary section*

11. Aligned section : Certain objects give unsymmetrical and confusing section if drawn according to the rules of projection. In sectioning such objects, more details can be shown if the cutting plane is to pass through the required features of the object.

While drawing the aligned sectional view, the various segments of the cutting plane are imagined to be revolved until they are parallel to the plane of projection (see Fig. 5.24).

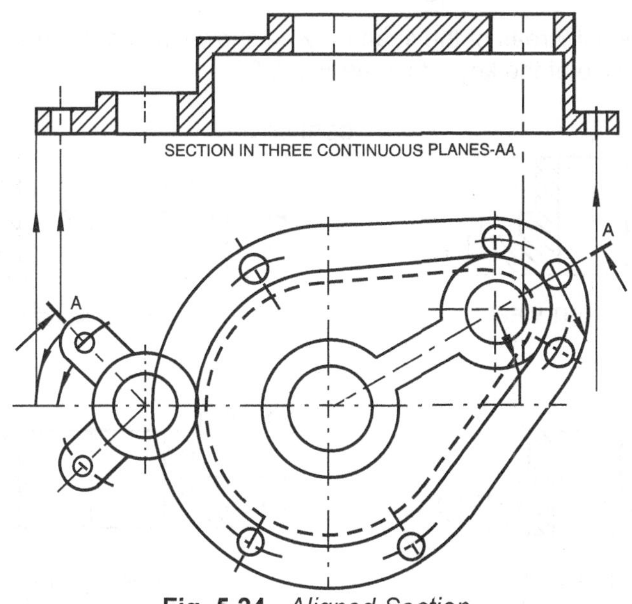

Fig. 5.24. *Aligned Section*

5.6. PARTS NOT SECTIONED

The main idea of sectioning is to show the internal details of the parts. Some parts such as bolts, nuts, rivets, pins, gibs, spokes, ribs, webs, shafts, rods, livers, links, axles, gear teeth, balls, rollers, etc. have no internal details and more easily recognised by their exterior

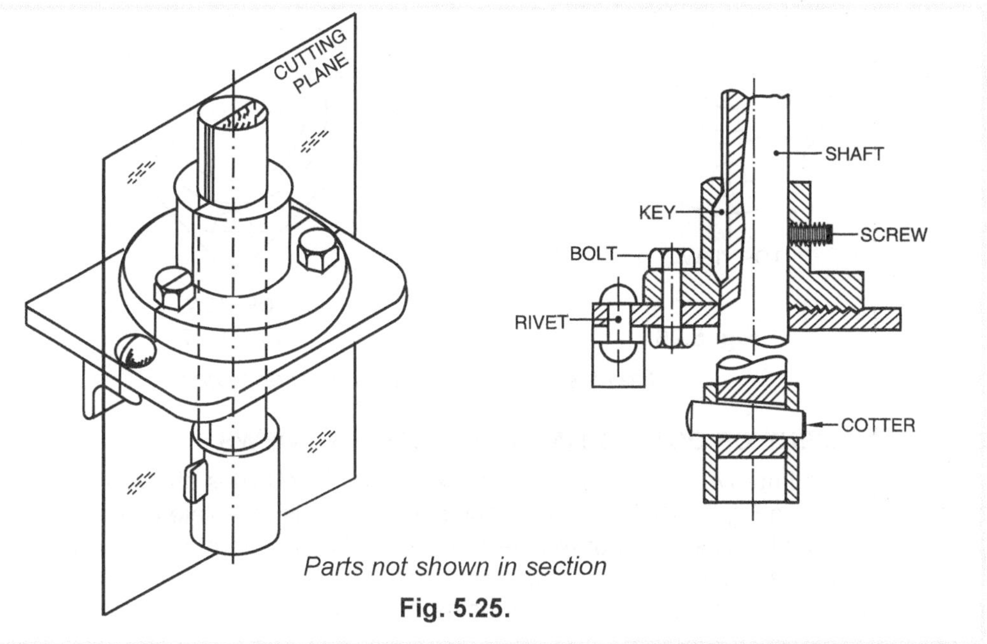

Parts not shown in section

Fig. 5.25.

views. These parts are not usually shown in section when cut longitudinally, because there is no details inside to show dimension. But if they are sectioned, they are more difficult to read because their identifying features are removed. Thus, features of this kind should be left in full view and not sectioned (see Fig. 5.25).

These parts are, however, cut and shown in section when the cutting plane passes at right angles to their axes [Refer to Fig. 5.21 (top view)].

Occasionally, small areas of the shaft may be shown in section to serve some particular purpose as to show size of the key, etc. (see Fig. 5.25).

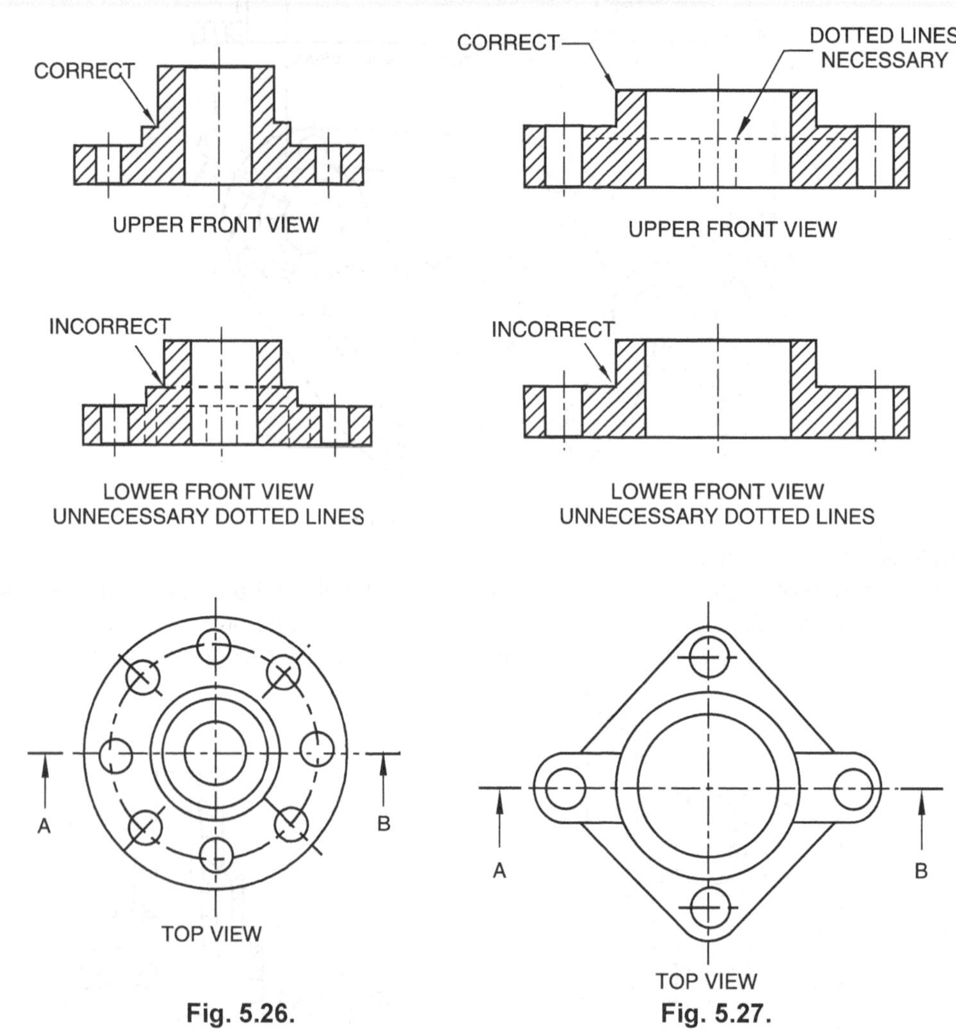

Fig. 5.26. **Fig. 5.27.**

Treatment of dotted lines in sectional view

5.7. TREATMENT OF DOTTED LINES IN SECTIONAL VIEWS

In order to improve the clarity of the drawing, the dotted lines should be omitted in the sectional view (see Fig. 5.26). Here, the shape of the object is completely clear from the upper front view and dotted lines are not required as shown in lower front view. While at Fig. 5.27 if the dotted lines are not omitted, the upper front view does not shows detail clearly and also the lower front view without dotted lines remains incomplete.

PROBLEMS ON SECTIONAL VIEWS

Problem 1 : Fig. 5.28 shows a pictorial view of a cast iron block. Draw the following view to full size scale :—

(a) Sectional front view

(b) Half sectional side view

(c) Top view.

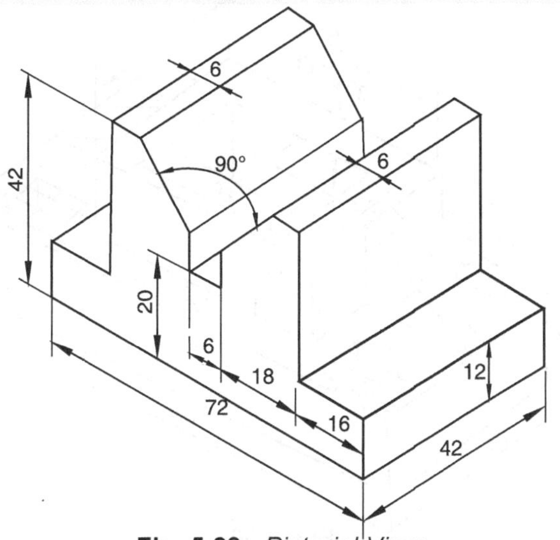

Fig. 5.28. *Pictorial View*

Solution : For its solution, see Fig. 5.29.

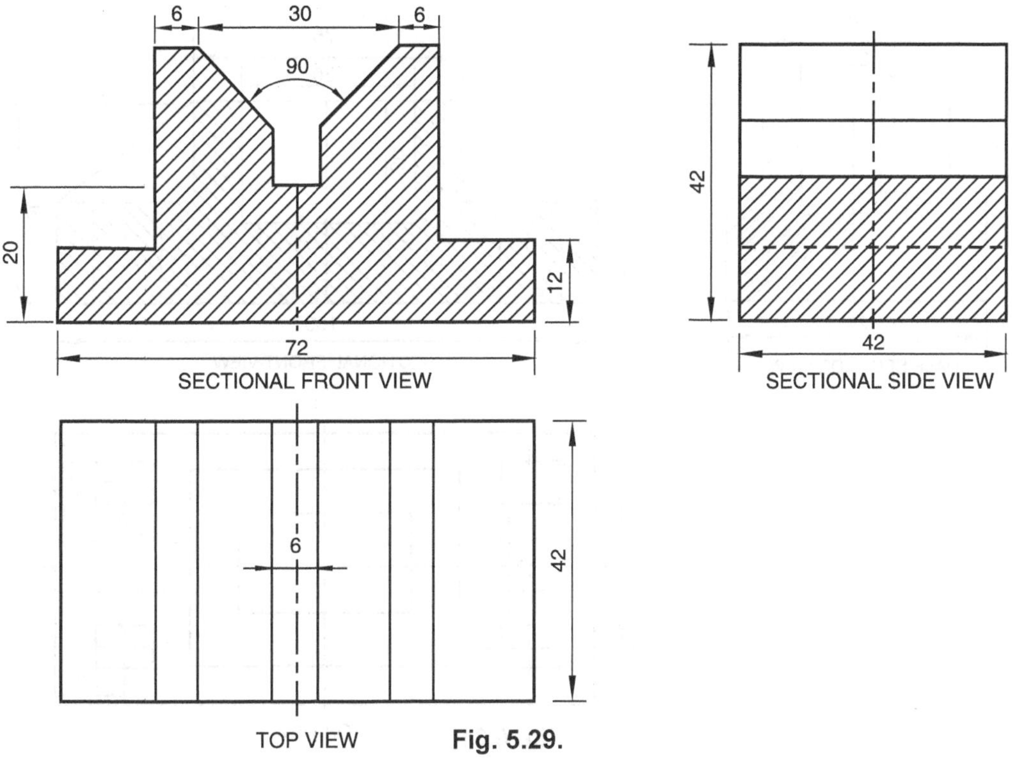

6
30
6
90
20
12
72
42
42
SECTIONAL FRONT VIEW
SECTIONAL SIDE VIEW
6
42
TOP VIEW

Fig. 5.29.

Problem 2 : Fig. 5.30. a shows a cast iron block. Draw the following views to a full size scale :—

(a) Sectional front view

(b) Half sectional side view

(c) Top view

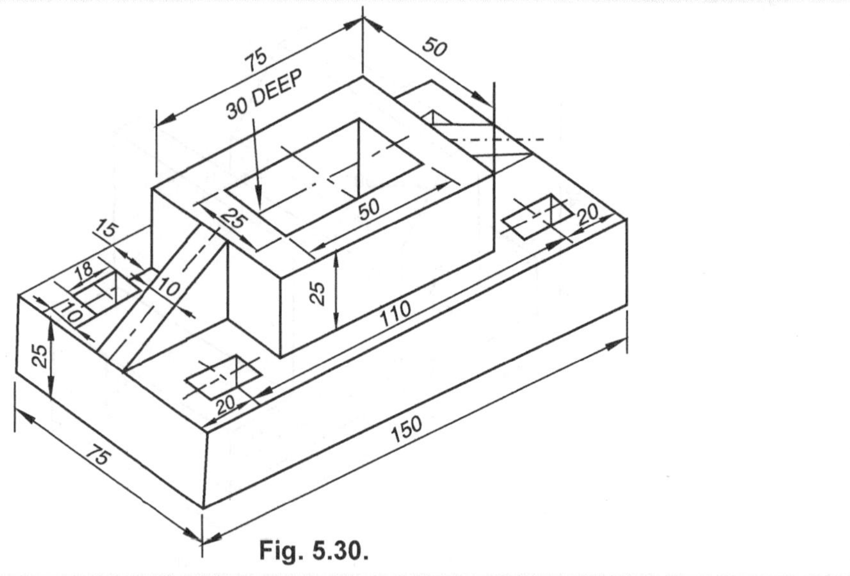

Fig. 5.30.

Solution : For its solution, see Fig. 5.31.

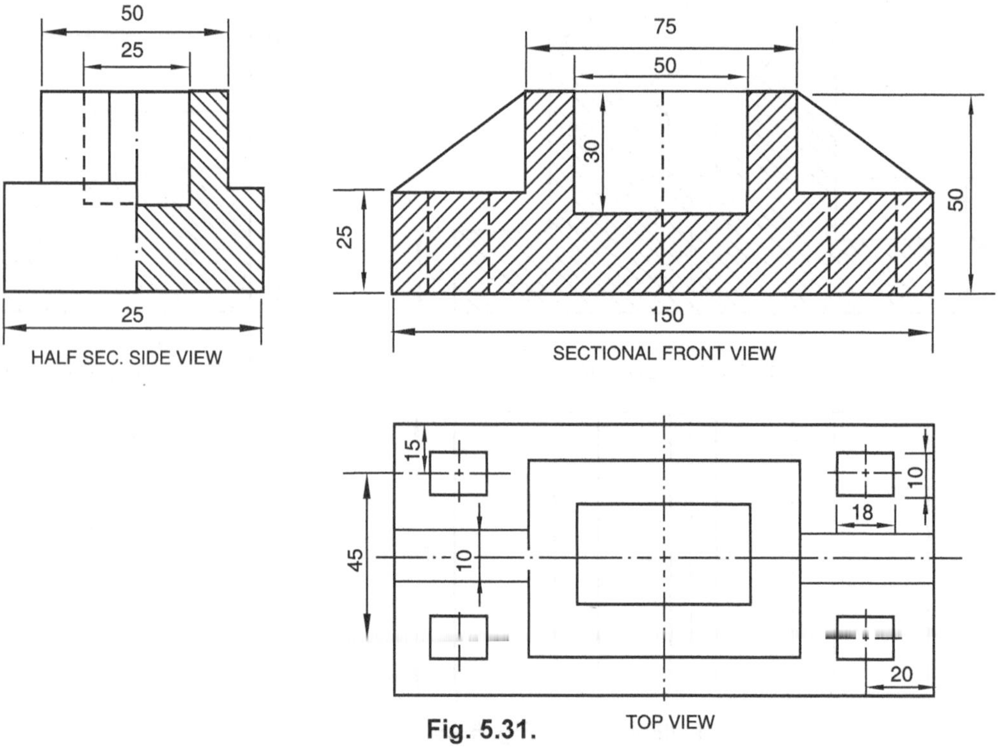

HALF SEC. SIDE VIEW

SECTIONAL FRONT VIEW

TOP VIEW

Fig. 5.31.

Problem 3 : Fig. 5.32 shows the pictorial view of a block. Draw to a scale 1:1 the following views :—

(a) Sectional front view

(b) Half sectional side view

(c) Top view

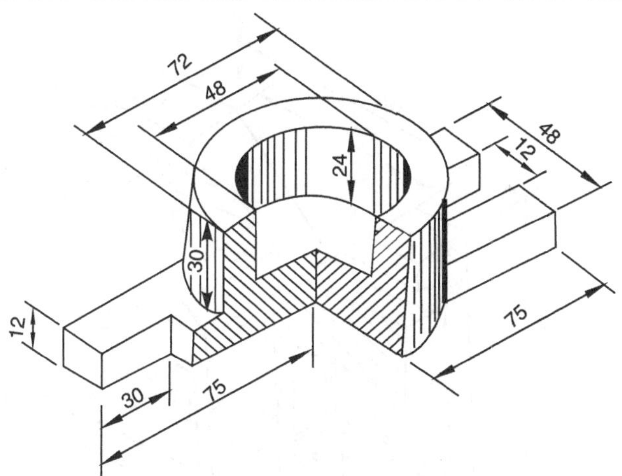

Fig. 5.32. *Pictoral view*

Solution : For its solution, see Fig. 5.33.

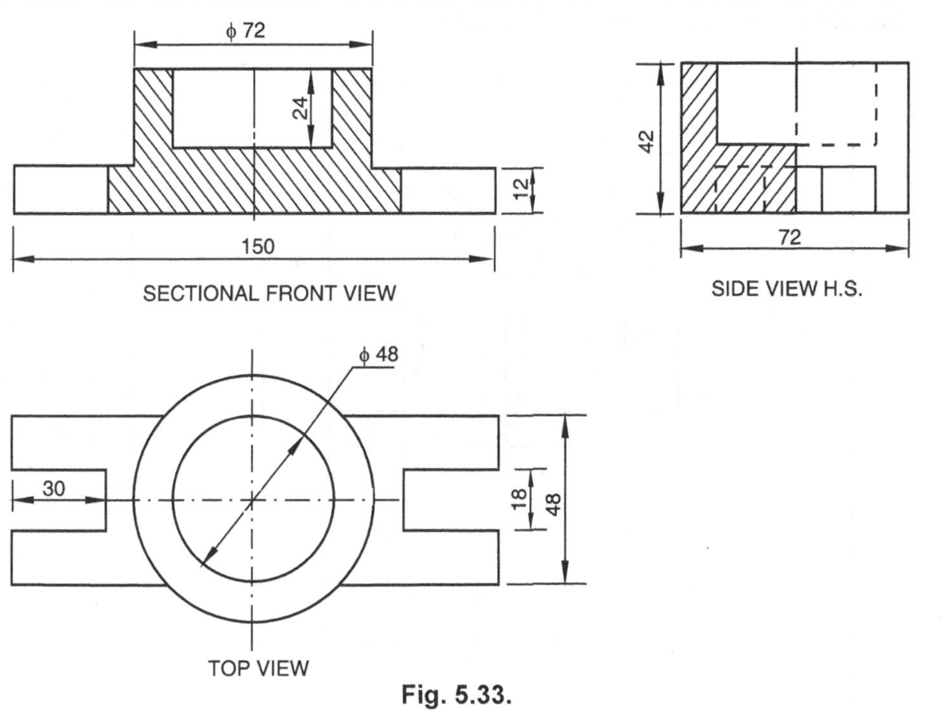

Fig. 5.33.

Problem 4 : Fig. 5.34 shows a casting of a V-block with quarter portion removed. Draw the following views to a full size scale :—

(a) Half-sectional front view, looking in the direction of arrow X.

(b) Half-sectional side view, looking in the direction of arrow Y.

(c) Top view through arrow Z.

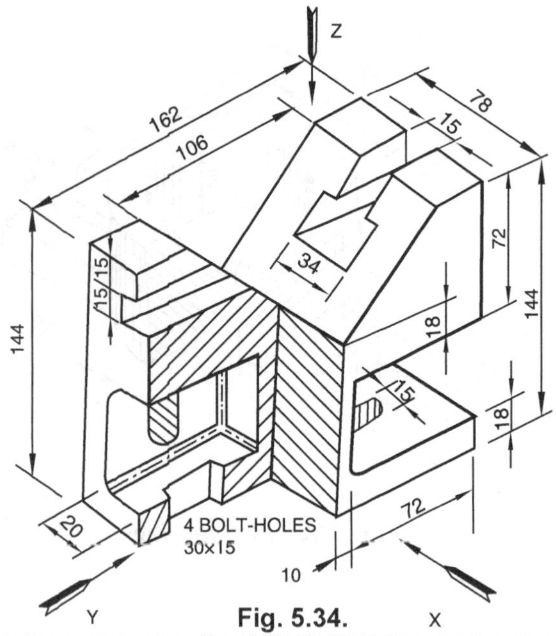

Fig. 5.34.

Solution : For its solution, see Fig. 5.35.

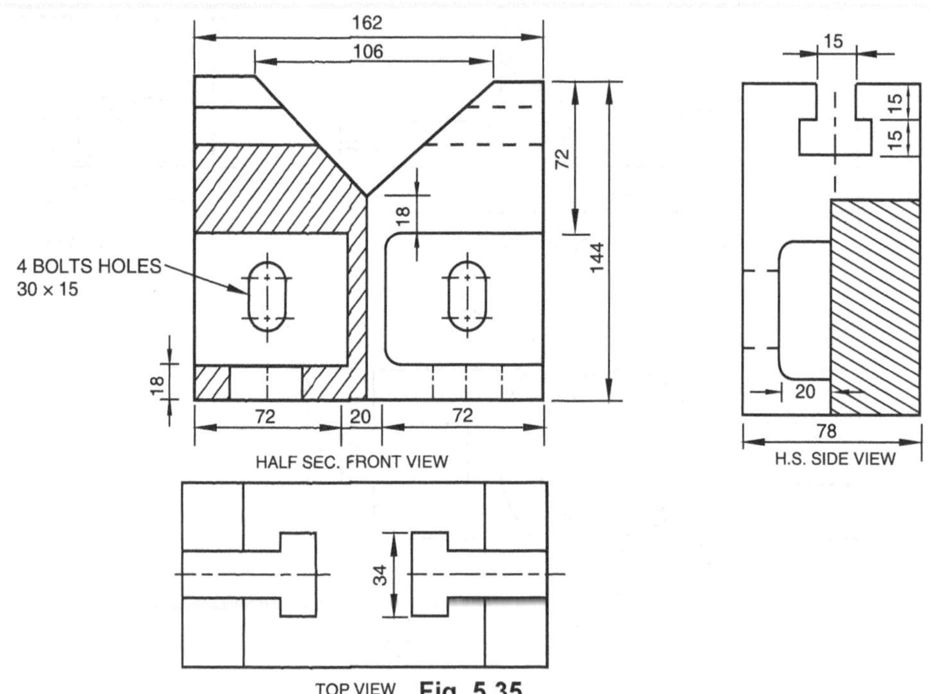

Fig. 5.35.

Problem 5 : Fig. 5.36 shows casting of an object with quater postion removed. Draw the following views to 1 : 1 scale, the following views :

(a) Half sec. front view in the direction of arrow X

(b) Half sectional side view

(c) Top view

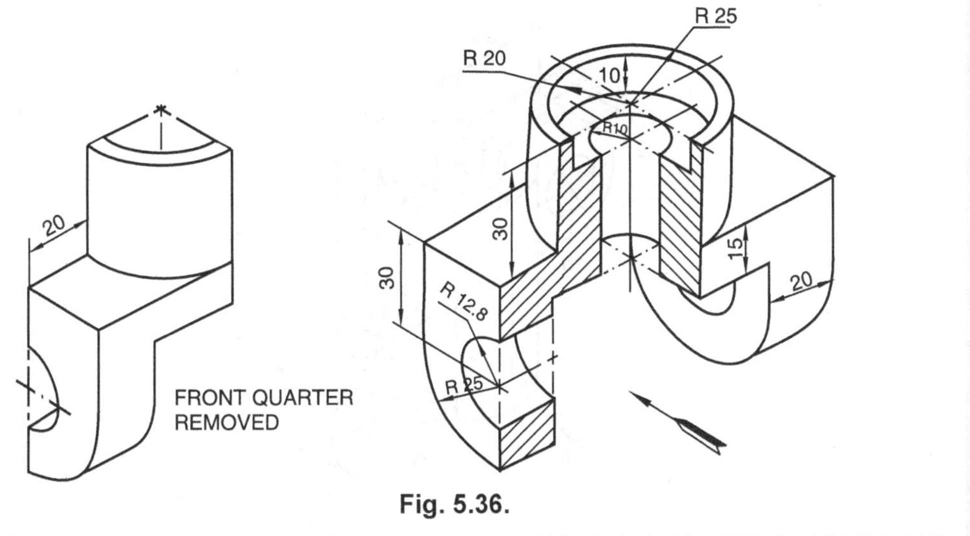

Fig. 5.36.

Solution : For its solution, see Fig. 5.36.

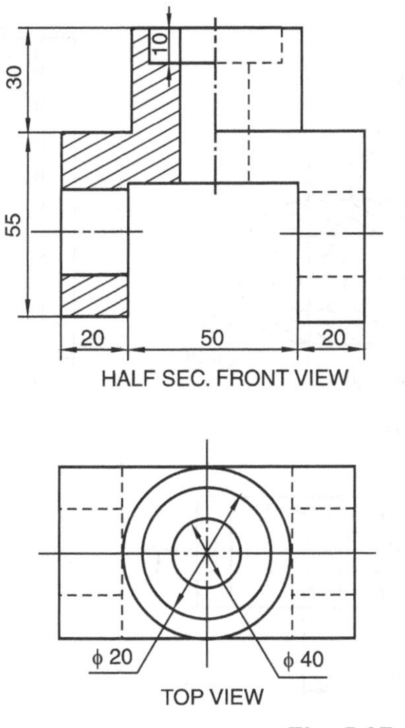

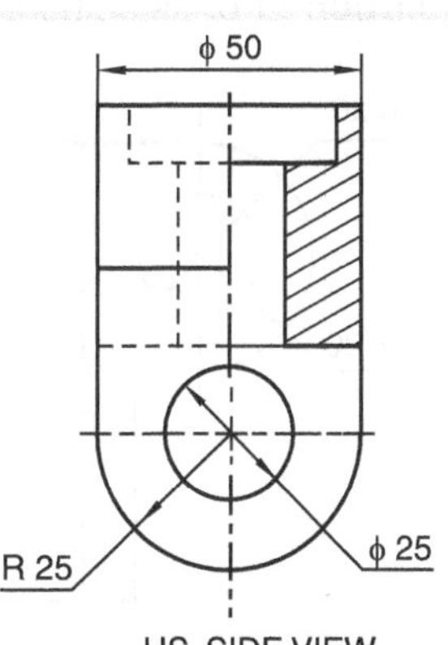

Fig. 5.37.

Problem 6 : Fig. 5.38 shows pictorial view of an object. Draw the following views, to scale 1 : 1 :

(a) Sectional front view in the direction of A

(b) Side view

(c) Top view.

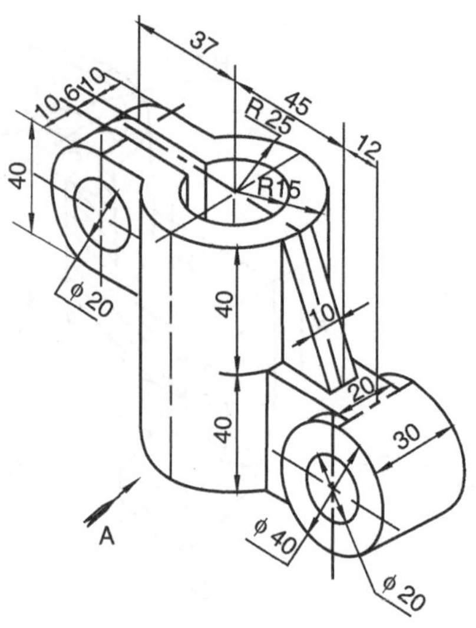

Fig. 5.38.

Solution : For its solution, see Fig. 5.39

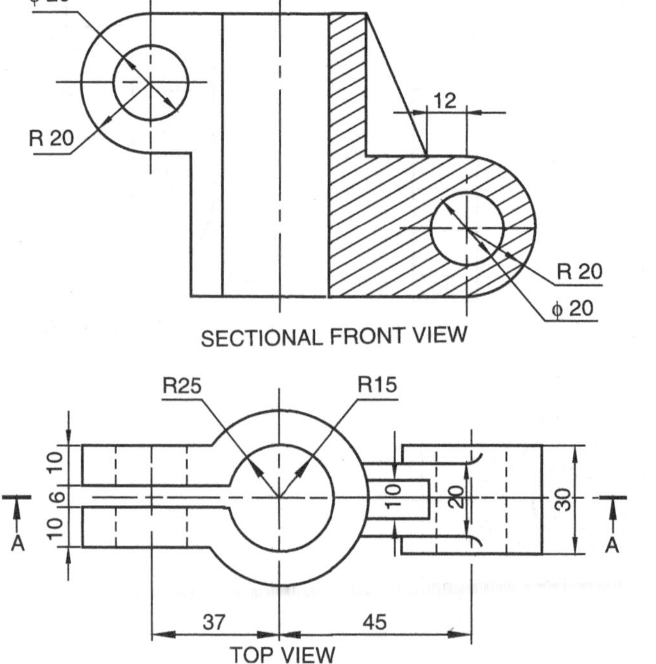

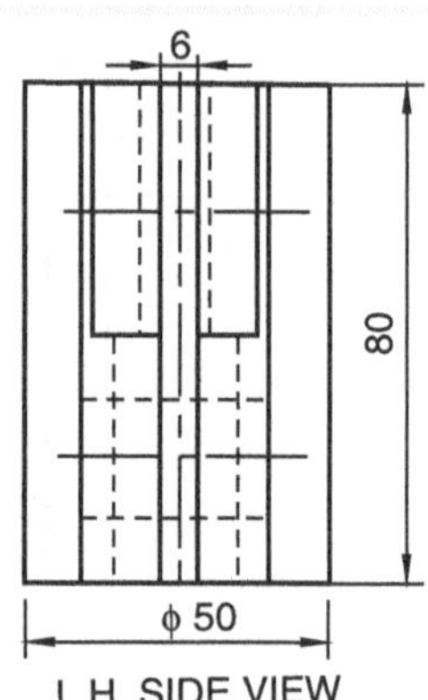

Fig. 5.39.

Problem 7 : Fig. 5.40 shows the pictorial view of a casting. Draw the following views by third angle projection method on a full size scale :—

(a) Sectional front view from A

(b) Top view

(c) Side view

Solution : For its solution, see Fig. 5.41

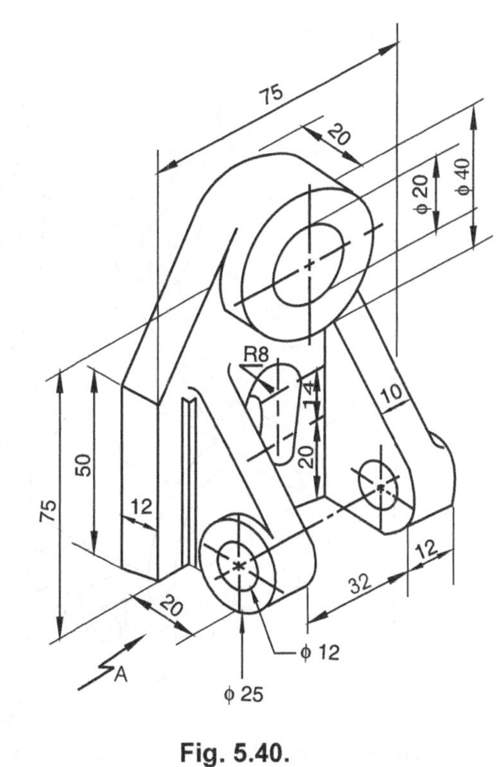

Fig. 5.40.

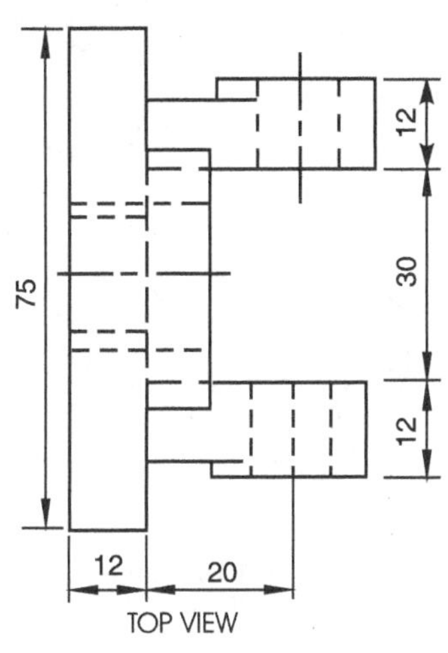

TOP VIEW

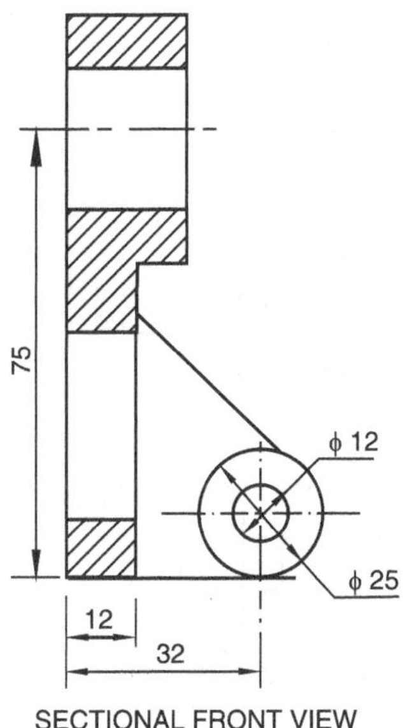

SECTIONAL FRONT VIEW

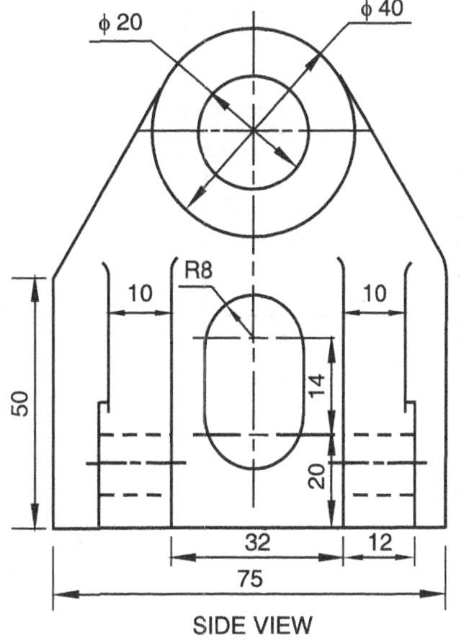

SIDE VIEW

Fig. 5.41.

Problem 8 : Fig. 5.42 shows the pictorial view of a casting. Draw the following views by third angle projection method on a full size scale :

(a) Front view.

(b) Sectional side view, section along A-A

(c) Top view.

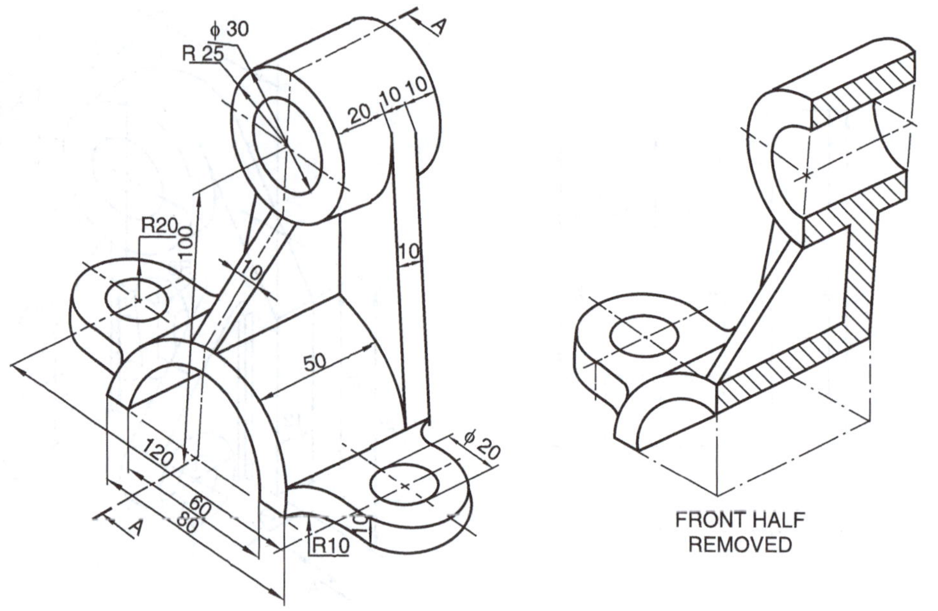

Fig. 5.42.

Solution : For its solution, see Fig. 5.43

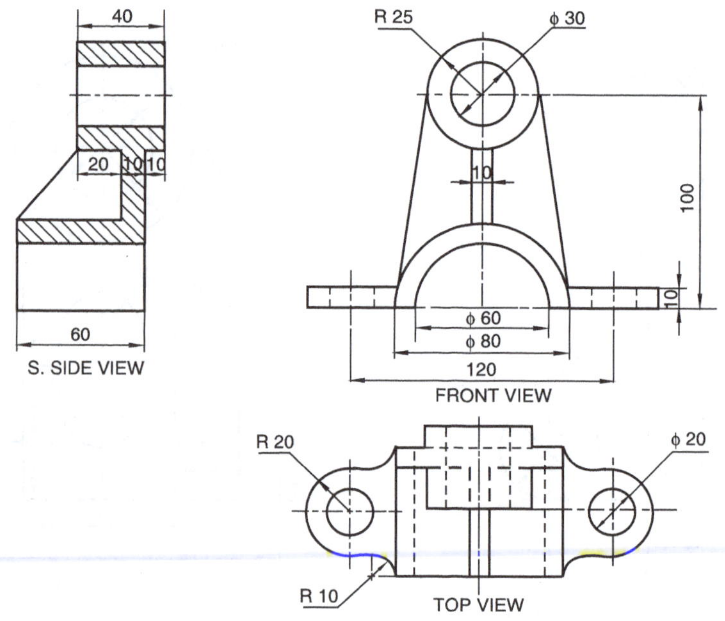

Fig. 5.43.

Problem 9 : Fig. 5.44 shows the pictorial view of an object. Using the first angle projection, draw

(a) Front view full in section the direction arrow

(b) Side view,

(c) Top view.

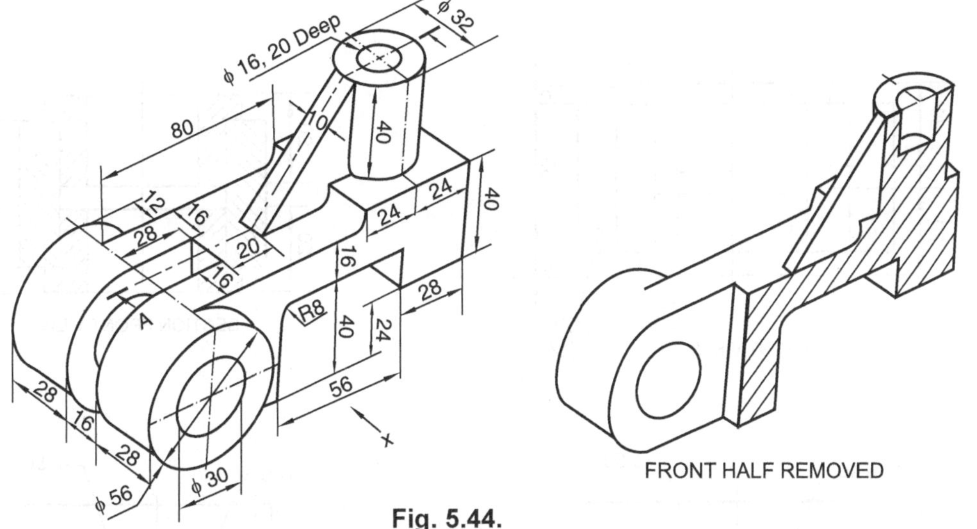

Fig. 5.44.

Solution : For its solution, see Fig. 5.45.

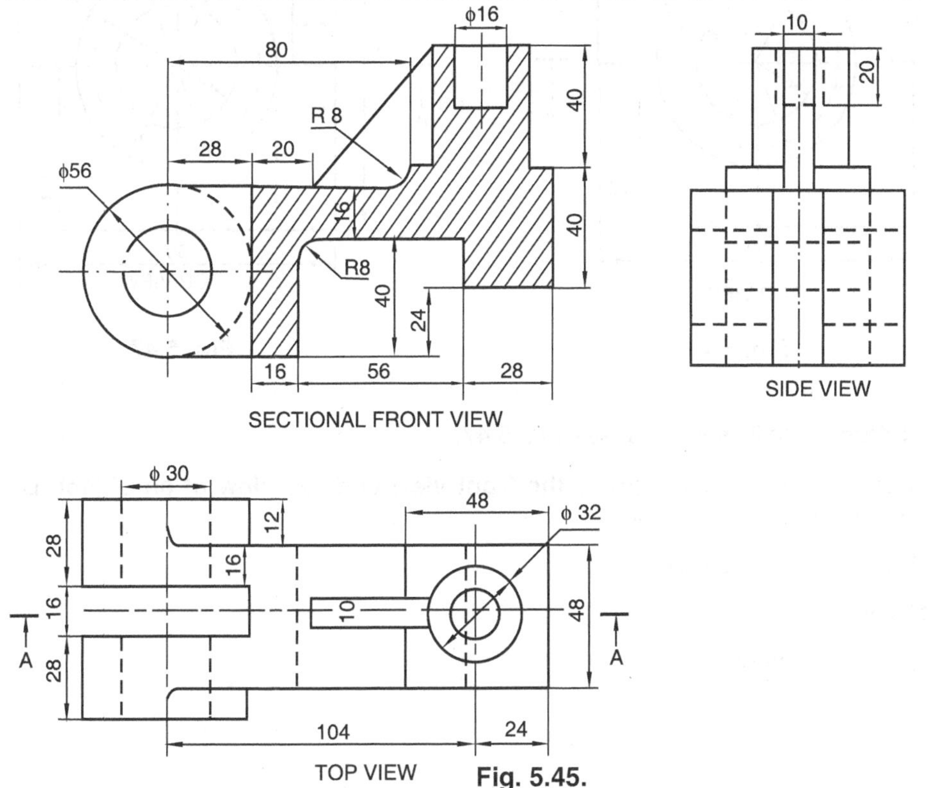

Fig. 5.45.

Problem 10 : Fig. 5.46 shows the front and top views of a component (numbered 1) in which two bushes (numbered 2 and 3) are fitted. Draw the following views :—

(a) Top view as shown

(b) Sectional front view by the plane A-A.

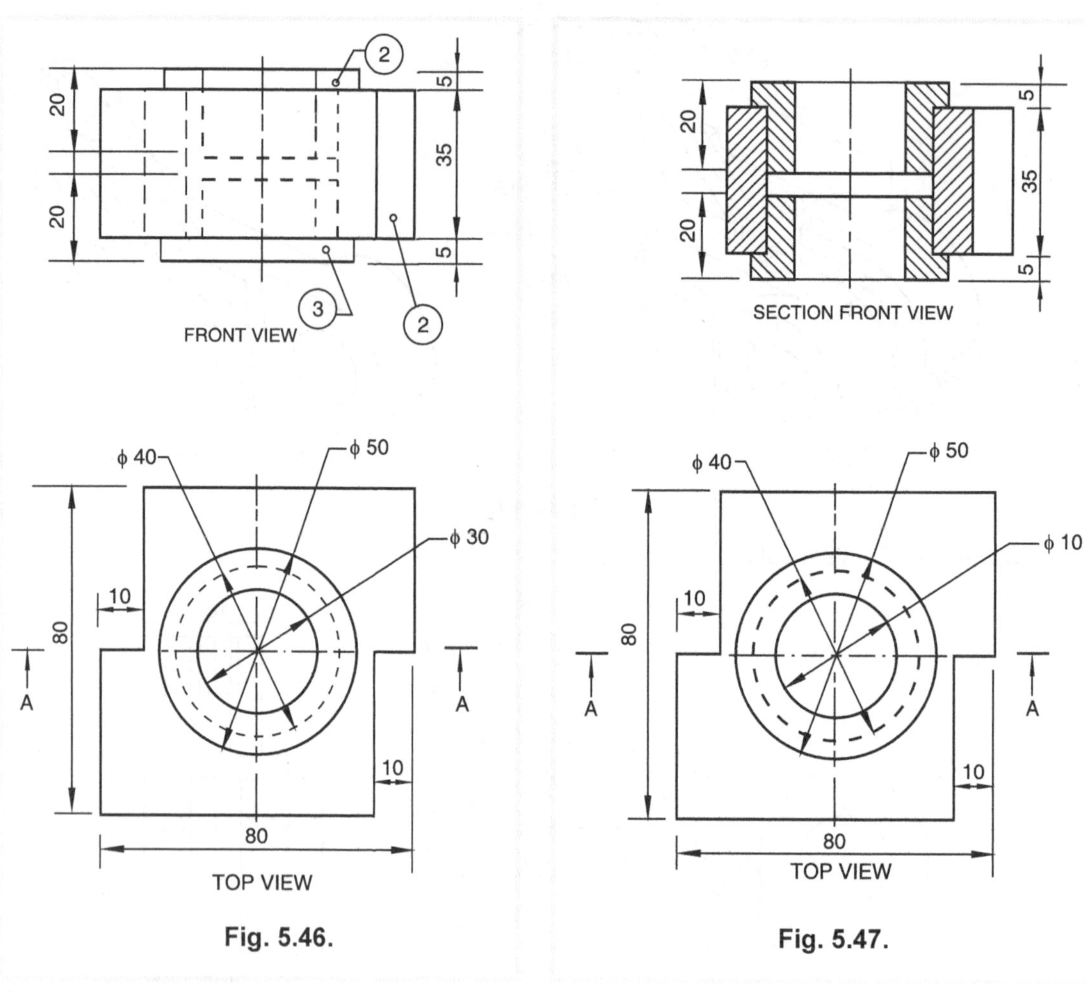

Fig. 5.46. **Fig. 5.47.**

Solution : For its solutions, see Fig. 5.47.

Problem 11 : Fig. 5.48 shows the front view and top view of an object. Draw the following view to scale 1 : 1 :—

(a) Sectional front view.

(b) Top view

(c) R.H. side view

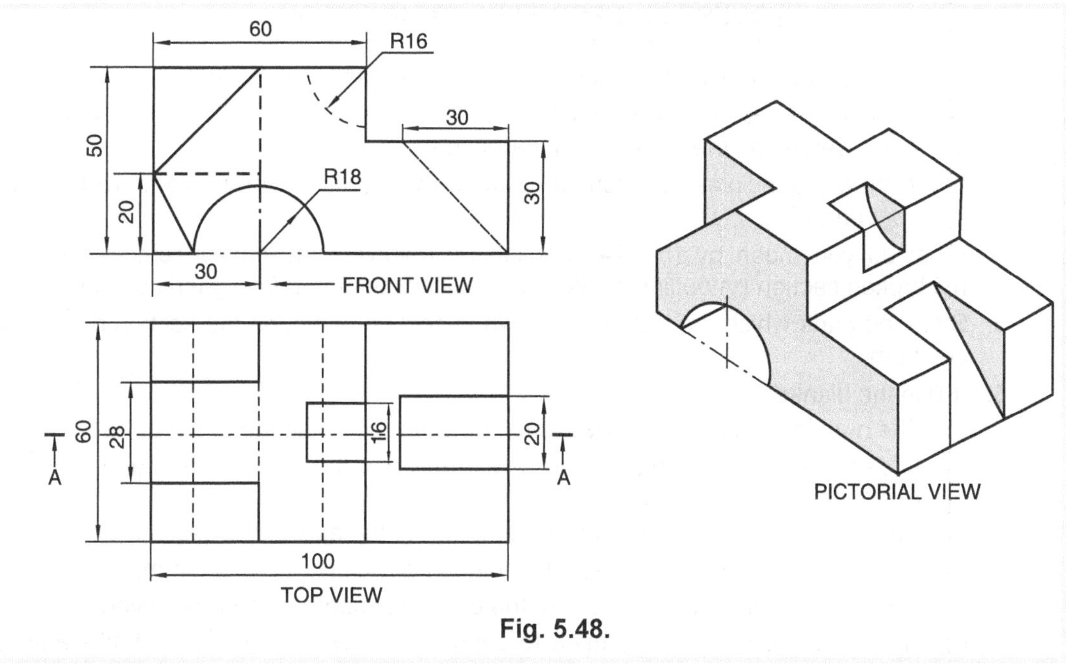

Fig. 5.48.

Solution : For its solutions, see Fig. 5.49.

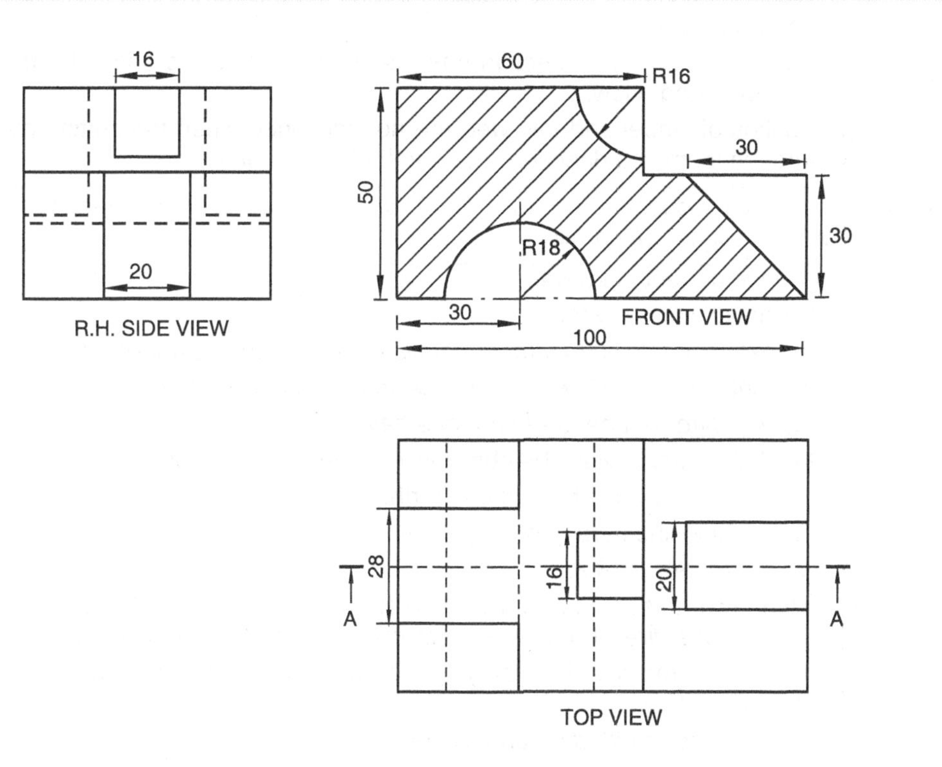

Fig. 5.49.

QUESTIONS FOR SELF EXAMINATION

1. What is a sectional view? Why sectional views are used in drawing?
2. Which section is used to show interior and exterior features of an object? What type of the line serves as the boundary separating them?
3. What are the principles of hatching? How do you hatch if there are more than two adjacent parts?
4. What do you mean by (*i*) offset section (*ii*) revolved section (*iii*) removed section (*iv*) broken section (*v*) outline section (*vi*) auxiliary section (*vii*) aligned section?
5. State the parts which are not to be shown in section, although the section plane may cut them.
6. Fill in the blanks :
   (*a*) The purpose of sectional views is to show the .... shape of the object.
   (*b*) In sectional view, the part of the object between the cutting plane and .... is assumed to be removed.
   (*c*) Section lines are generally drawn at .... to the axis of the section.
   (*d*) For ordinary sectioning the spacing varies from .... to .... mm.
   (*e*) In making a full section .... of the object is imagined to be removed.
   (*f*) A .... line is generally used to separate the two portions of a half sectional view.
   (*g*) In the sectioned view, all hidden detail lines are......
   (*h*) When the cutting plane passes through the axis of the solid shafts, bolts, rivets and screws, are...........
   (*i*) An .... section has the section rotated to show the true shape in the sectioned view.
   (*j*) Spokes of wheel and thin web are not sectioned when the cutting plane passes through them in a direction .... to their larger dimension.

Ans. (*a*) internal (*b*) observer (*c*) 45° (*d*) 1,3 (*e*) half (*f*) centre (*g*) omitted (*h*) not shown in section (*i*) aligned (*j*) parallel

7. Choose the correct answers :
   (*i*) Sectional views reveal
   (a) external details (b) interior details (c) length and height of the object.
   (*ii*) In a half sectional view the object is imagined to be cut off
   (a) one third (b) one fourth (c) one half.
   (*iii*) In a full sectional view the object is imagined to be cut off
   (*a*) one third (*b*) one half (*c*) one forth.
   (*iv*) The section lines are generally drawn to the horizontal lines at angle of
   (a) 45° (b) 30° (c) 60°.
   (*v*) When two adjacent parts are to be shown in section, the lines should be drawn
   (a) opposite direction (b) same direction (c) parallel to horizontal.
   (*vi*) In order to improve the clarity of the drawing, the dotted lines in the sectioned view should be
   (a) drawn (b) omitted (c) partially drawn.

Ans. (*i*) (*b*), (*ii*) (*b*), (*iii*) (*b*), (*iv*) (*a*), (*v*) (*a*), (*vi*) (*b*).

PROBLEMS FOR PRACTICE

1. Fig 5.50 shows some objects. Draw to a full size scale the following views :
   (a) Sectional front view in direction F
   (b) Side view
   (c) Top view

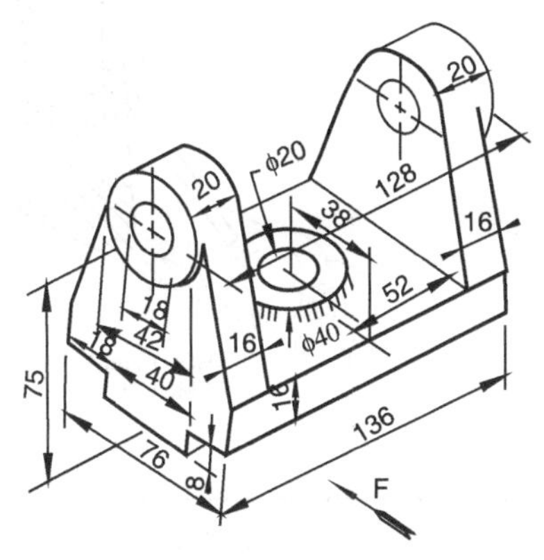

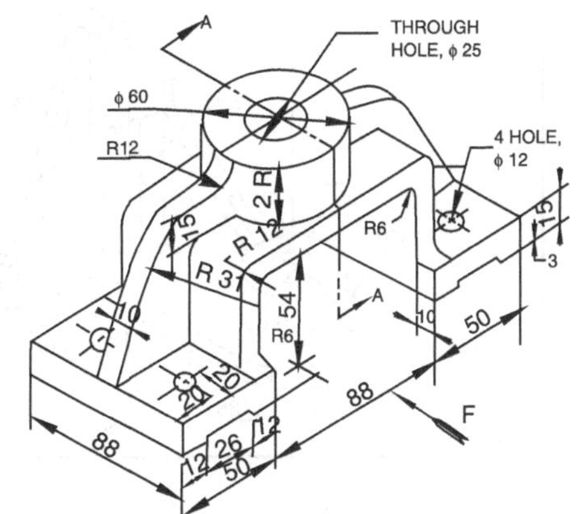

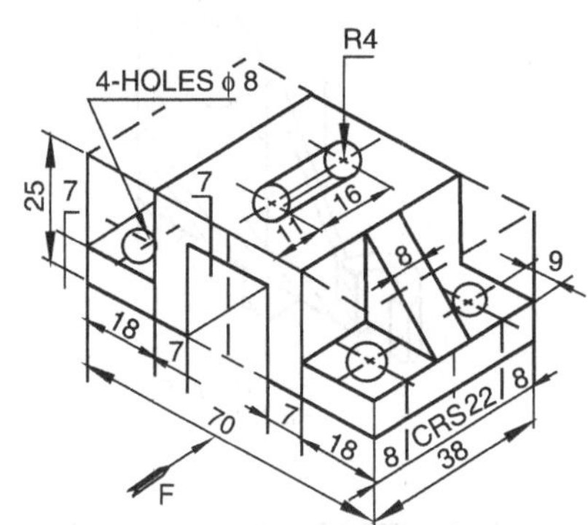

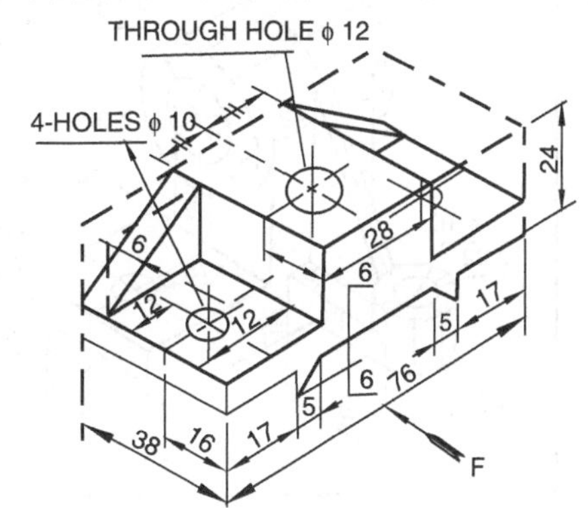

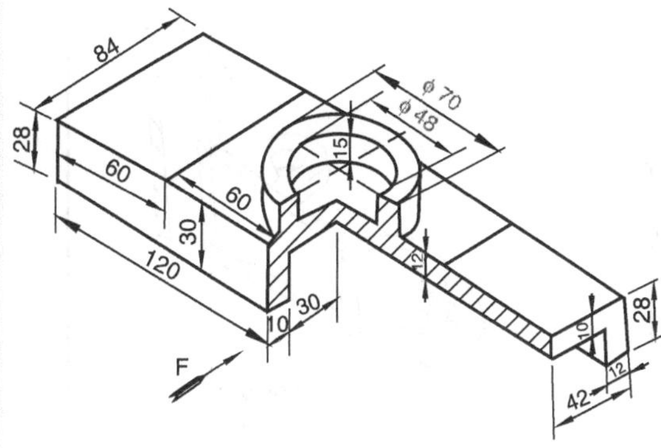

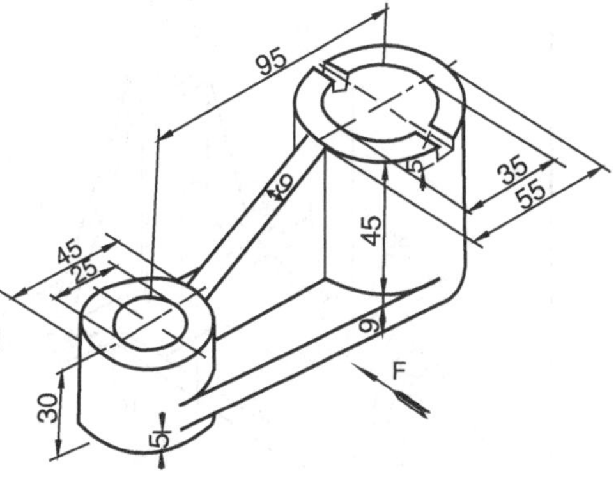

Fig. 5.50.

2. Draw the sectional front view, side view and top view of the objects given in the Fig. 5.51.

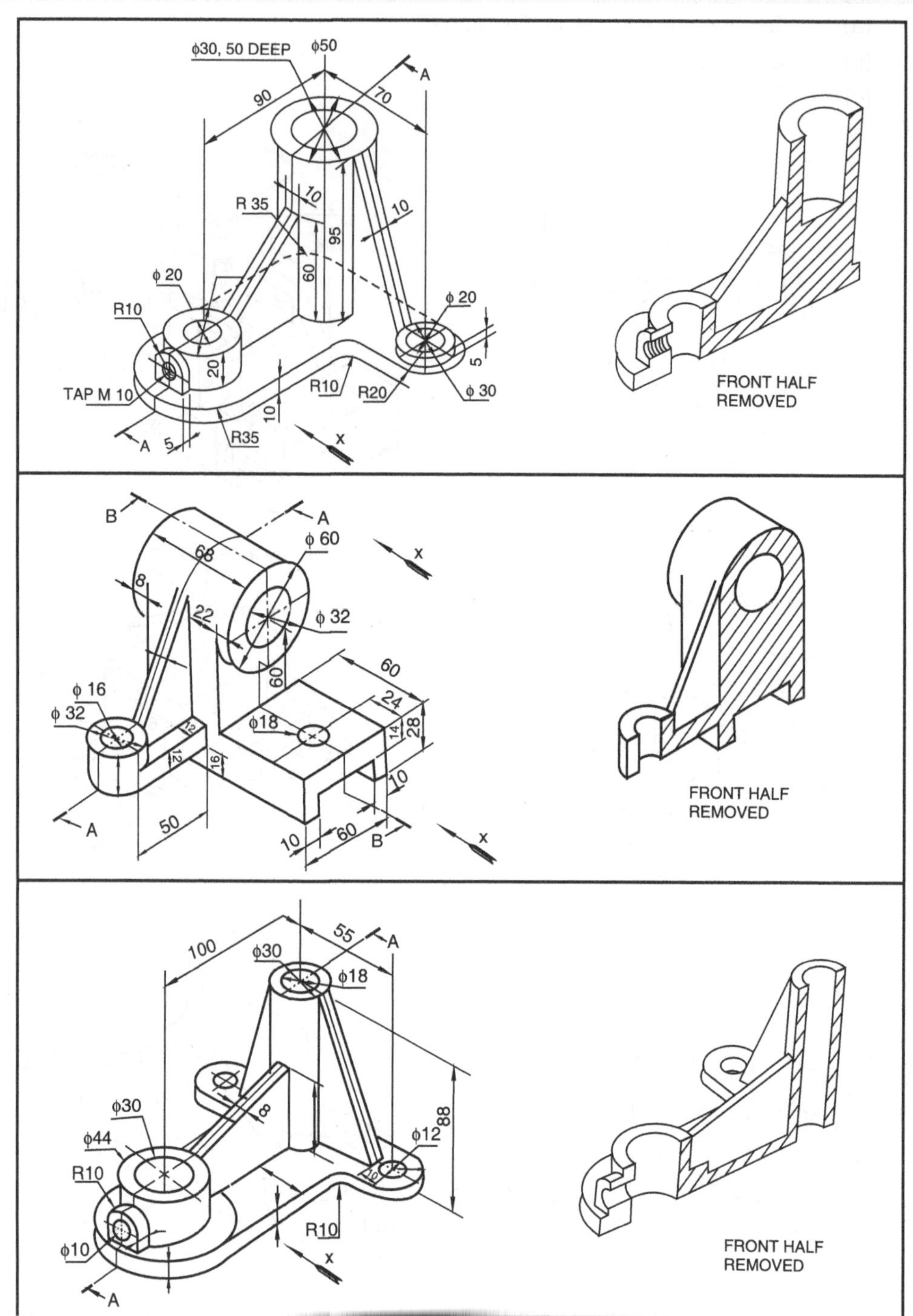

Fig. 5.51.

3. Draw the sectional front view, side view and top view of the objects given in the Fig. 1.52.

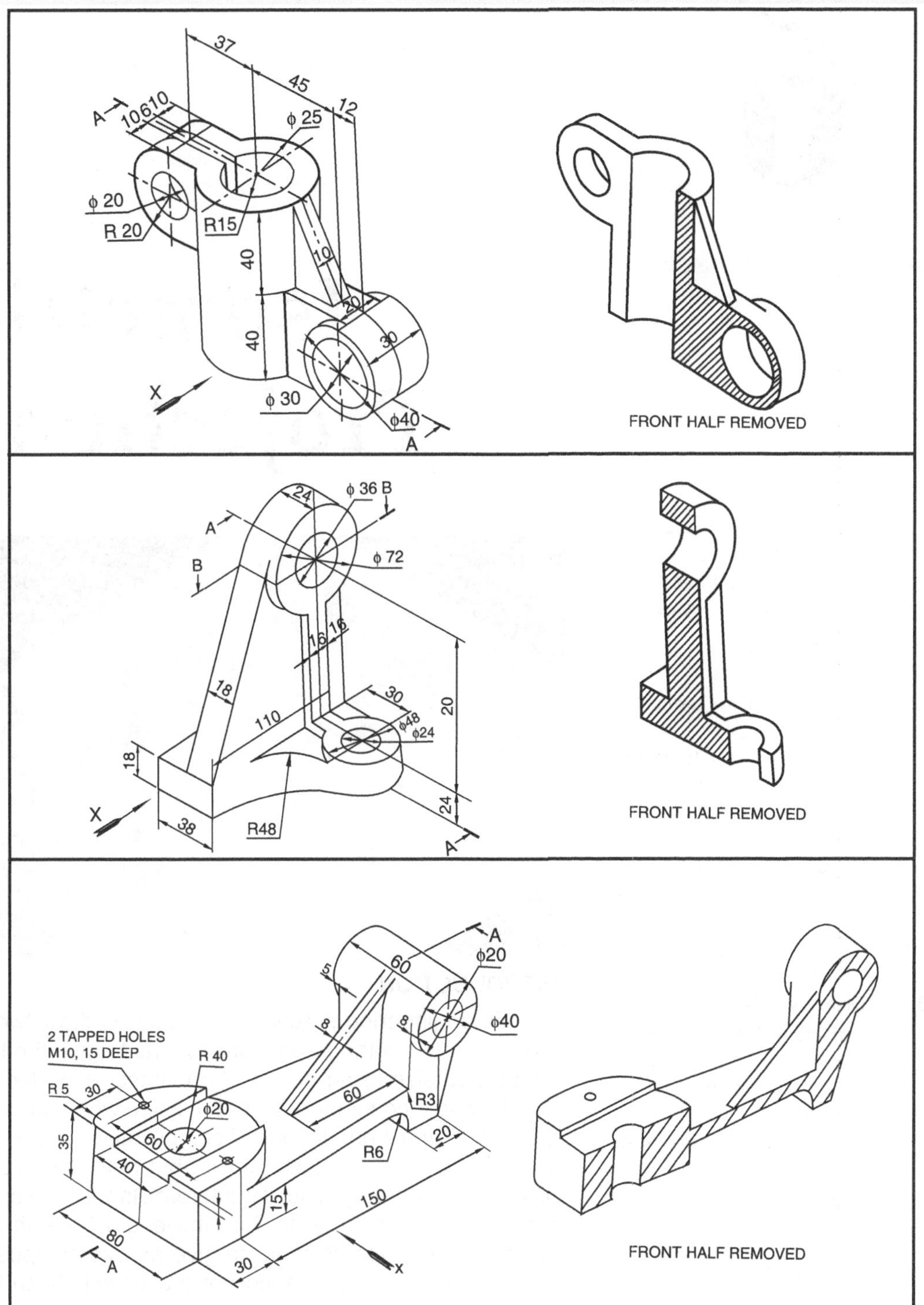

Fig. 5.52.

CHAPTER

6

Isometric Projections

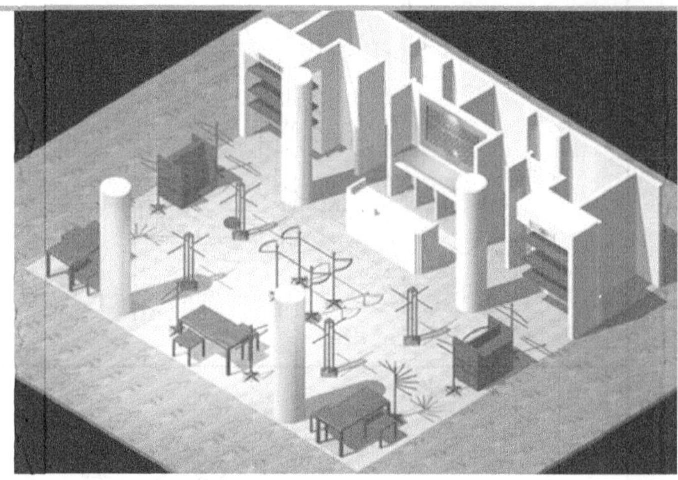

INTRODUCTION

In engineering practice, it is usual to draw two or more than two orthographic projections to give the true shape and size of an object. Sometimes, even technician having long experience get puzzled when studying the orthographic projections of complicated parts which might have been correctly drawn, but may have some details which are very difficult to interpret. In such cases an extra projection is added to show the length, breadth and height of the object in order to give an overall appearance for the clear conception of the object. Such like projections which give the overall appearance of the object are known as *isometric projections.*

In this chapter, we will deal with the study of isometric view, theory of isometric projection, isometric length, isometric scale, methods to draw isometric projections, etc.

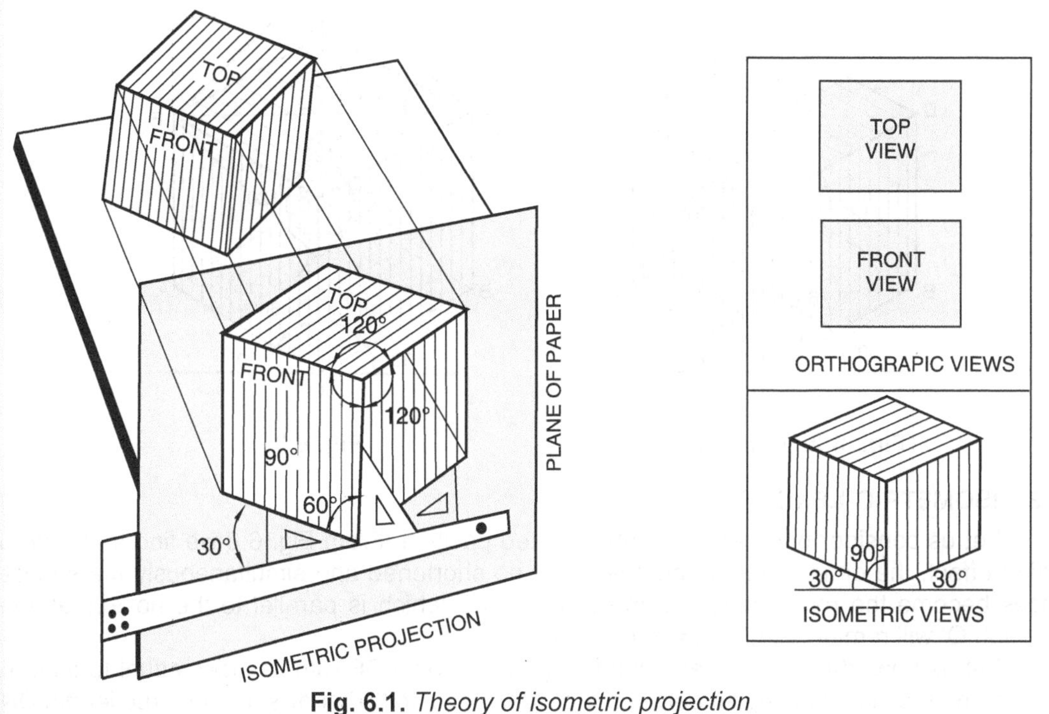

Fig. 6.1. *Theory of isometric projection*

6.1. ISOMETRIC VIEW OR PROJECTION

The view or projection obtained on a plane when the object is so placed that all the three axes make equal angle with the plane of projection is called an isometric view or projection.

This type of projection is pleasing to the eyes than oblique or perspective projections, as it is easier to draw, because all the edges are fore-shortened equally for which only one scale is required.

6.2. THEORY OF ISOMETRIC PROJECTION

Fig. 6.1 shows how an isometric projection is obtained on a plane. Let, a cube be revolved through an angle of 45° about a verticle axis and then tilted in the forward direction so that one of its corners is nearer to the plane of projection. The three edges meeting at the nearer corner are equally inclined to the plane of projection so that their projections are of equal length. Any line so inclined is known as *Isometric line* and the projection thus obtained on the plane will be *isometric projection.* Since three edges meeting at a point are equally inclined to each other, therefore, the angle between any two edges will be equal to 360/3 = 120°.

If we consider these three lines represented by the three edges of a cube and rest of the sides parallel to these three lines as shown in Fig. 6.2, then the view thus obtained will be the isometric view or projection.

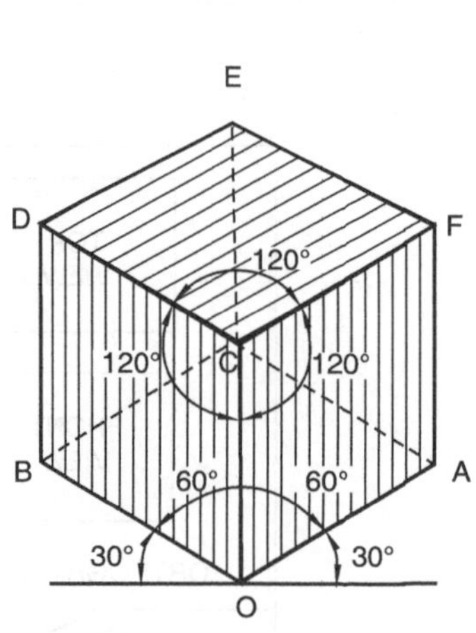

Fig. 6.2. *Isometric view*

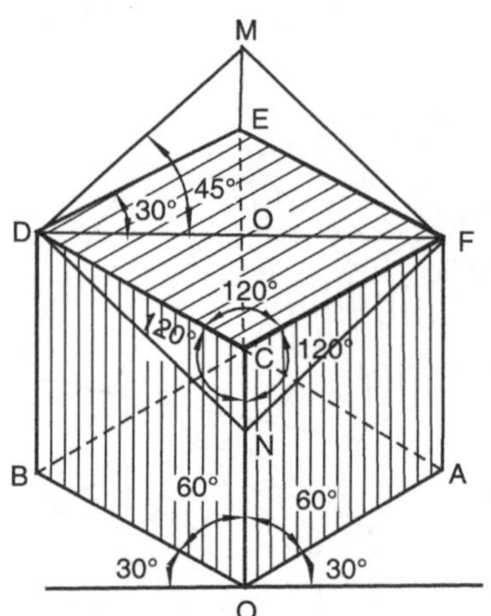

Fig. 6.3.

6.3. ISOMETRIC LENGTH

Let us consider once again a cube in tilted position. From Fig. 6.3 we find that due to tilting effect, the dimensions of the object will be shortened and simultaneously the square faces become the rhombuses, but the diagonal DF which is parallel to the horizontal line through O' will remain equal to its true length.

Let us now, draw the square DMFN on the diagonal DF with its sides equal to its true length and the included angles between the sides equal to 90° shows that the true length DN has fore-shortened and becomes equal to DC. The ratio to which these lengths are fore-shortened may be calculated as under :

Consider two triangles DO'E and DO'M.

In triangle DO'E, $\frac{O'D}{DE} = \cos 30^{\circ\circ}$ or $\frac{DE}{O'D} = \frac{1}{\cos 30^\circ} = \frac{1}{\sqrt{3}/2} = \frac{2}{\sqrt{3}}$

In triangle DO'M, $\frac{O'D}{DM} = \cos 45^\circ$ or $\frac{DM}{O'D} = \frac{1}{\cos 45^\circ} = \frac{1}{1/\sqrt{2}} = \sqrt{2}$

$$\frac{DE}{O'D} \Big/ \frac{DM}{O'D} = \frac{2}{\sqrt{3}} \Big/ \frac{\sqrt{2}}{1} \text{ or } \frac{DE}{O'D} = \frac{\sqrt{2}}{\sqrt{3}} = 0.815$$

The ratio, $\frac{\text{Isometric length}}{\text{True length}} = \frac{DE}{DM} = 0.815 = \text{or } \frac{9}{11}$ approximately

Hence, the isometric projections are reduced in the ratio of $\sqrt{2} : \sqrt{3}$.

i.e., isometric lengths are 0.815 of the true lengths. For example,

(i) A square changes to a rhombus

(ii) A circle changes to an ellipse.

6.4. ISOMETRIC SCALE

The proportion by which the actual distances are reduced to isometric distances are known as isometric scale.

6.5. METHOD TO DRAW ISOMETRIC SCALE

To obtain an accuracy in isometric projection, it is necessary to convert true lengths into isometric lengths for measuring and marking the sizes. This is done easily by constructing and making the use of an isometric scale. The method of this construction is as follows :

1. Draw a base line OA.
2. Through O, draw two lines OB and OC, making angles of 30° and 45° respectively with the line OA such that AOB = 30° and AOC = 45°.

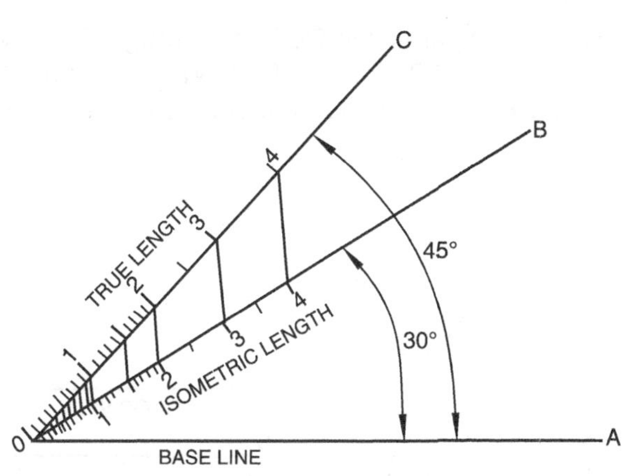

Fig. 6.4. *Isometric scale*

3. On the line OC, construct the original scale (full size) and divide the major and minor divisions such as 1, 2, 3, 4 cm, etc.
4. Through these divisions, draw lines perpendicular to OA thereby cutting the line OB at points 1, 2, 3, 4 etc. on the line OB which gives the isometric scale.

Note : *If the isometric projection of an object is to be drawn on an enlarged or reduced scale then the enlarged or reduced distances should be drawn on the line OC which in turn will give the corresponding isometric scale on the line OB.*

6.6. DIFFERENCE BETWEEN ISOMETRIC VIEW AND ISOMETRIC PROJECTION

The method of drawing the isometric view and isometric projection is absolutly the same, but there is only difference in size, i.e., isometric view is bigger in size (of actual scale) while the isometric projection is smaller (of shortened size) as shown in Fig. 6.5.

6.7. METHOD TO DRAW ISOMETRIC VIEW OR PROJECTION

For obtaining isometric view of an object say a rectangular block proceed as follow :

(1) Draw the isometric scale (see Fig. 6.5).

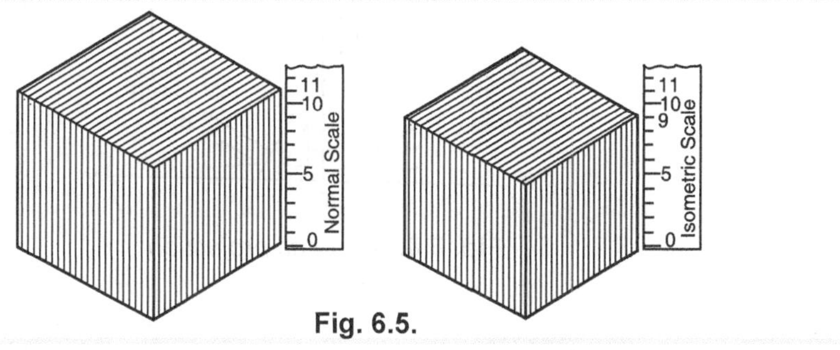

Fig. 6.5.

(2) Draw the orthographic projections of the given block by using isometric scale (see Fig. 6.6)

Now for isometric view or projection :—

(3) Draw a horizontal line xy and take any point O (see Fig. 6.7). Through O draw the three isometric axes, OA', OB', and OC' with the help of set-square such that; OC' is perpendicular to xy line, $\angle$ B'OY = $\angle$ A'OX = 30°

4. Cut off OA, OB and OC the length, breadth and height of the object along the axes OA', OB' and OC' respectively.

5. Through A, draw a line AD parallel to OC′ and through C, draw a line CD parallel to OA' so as to meet each other at a point D. Through B draw a line BF parallel to OC' and through C, draw a line CF parallel to OB' so as to intersect each other at a point F.

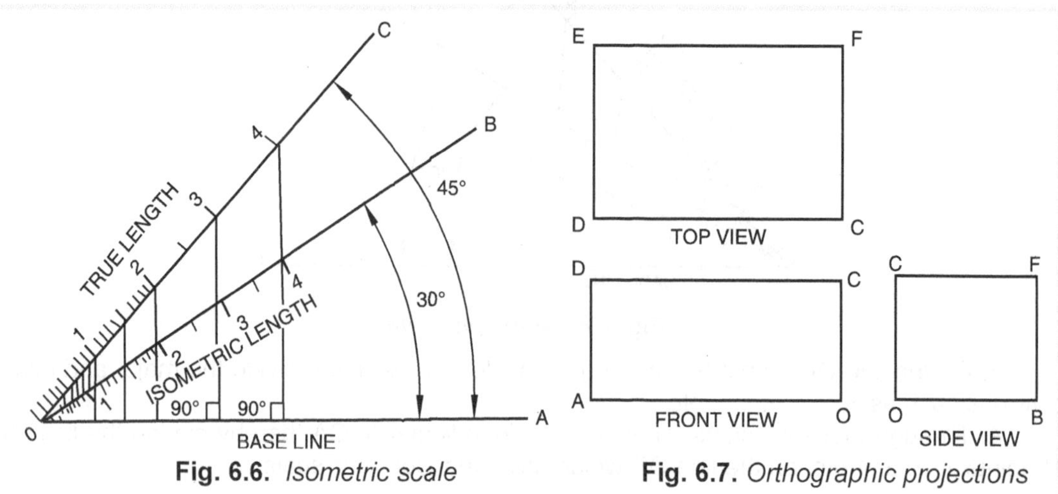

Fig. 6.6. *Isometric scale* **Fig. 6.7.** *Orthographic projections*

Similarly, through D draw a line DE parallel to OB' and through F, draw a line FE parallel to OA' so as to meet at E. Now complete the rectangular block after showing the position of dotted lines as shown in Fig. 6.8 which is itself a self explanatory sketch.

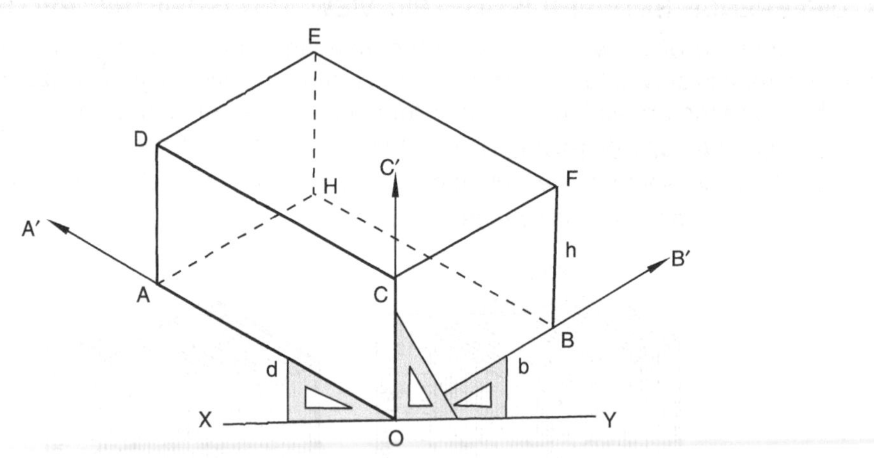

Fig. 6.8. *Isometric projection*

6.8. SOME IMPORTANT TERMS

The following are some important terms used in the isometric views :—

(1) Isometric axes : The three lines OA,OB and OC meeting at a point and making 120° angles with each other are termed as *isometric axes.*

(2) Isometric lines : The lines parallel to the isometric axes are called *isometric lines.*

(3) Non-isometric lines : The lines which are not parallel to isometric axes are called *non-isometric lines.*

(4) Isometric planes : The planes representing the faces of the cube as well as other planes parallel to these planes are called *isometric planes.*

6.9. VARIOUS POSITIONS OF ISOMETRIC AXES

In order to show the different sides of an object, the object is revolved about its three isometric axes. The angles between the axes should not be changed, however, even though their position on the drawing sheet are varied. These different positions are used to show clearly the required faces. The figures from 6.9 to 6.11 show the various positions for the isometric axes.

Fig. 6.9 shows the top surface of square clear with one axis OC vertical.

Fig. 6.10 shows the bottom surface of square clear with one axis OC vertical.

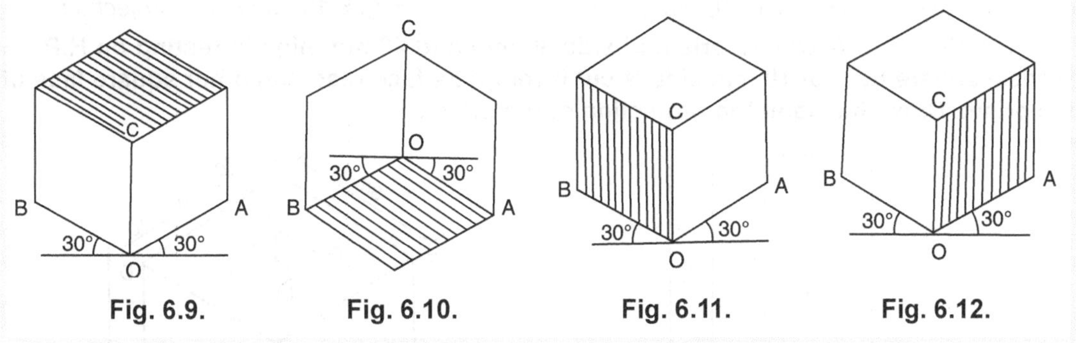

Fig. 6.9. **Fig. 6.10.** **Fig. 6.11.** **Fig. 6.12.**

Fig. 6.11 shows the left face of the square clear with one axis OC vertical.

Fig. 6.12 shows the right face of the square clear with one axis OC vertical.

After going through the above given positions of the isometric axes, it will be easy to show clearly the required faces.

Note : *As it is easier to draw an isometric projection from the lower front corner, therefore, the axes with first position is generally used.*

6.10. ISOMETRIC VIEWS OF VARIOUS OBJECTS

Objects may be divided into the following types from isometric projection point of view :—

(1) Objects with isometric lines

(2) Objects with non-isometric lines

(3) Objects with curved surfaces.

6.11. OBJECTS WITH ISOMETRIC LINES

The objects in which all the edges or lines are parallel to the isometric lines are called objects with isometric lines.

This type includes the objects such as cubes, rectangular prisms and combination of these.

PROBLEMS ON OBJECTS WITH ISOMETRIC LINES

Problem 1 : A cube of 40 mm side rests centrally on a square block of 60 mm edges and 20 mm thick. Draw the isometric projection of the two objects with the edges of the two blocks mutually parallel to each other.

Solution : Draw two lines TA and TB equal to 60 mm each, making an angle of 30° with the horizontal line (see Fig. 6.13).

Draw TC = 20 mm perpendicular to the horizontal line and complete the square prism as shown in Fig. 6.13.

Now, locate the centres of CD and DE which will represent the centre lines of the cube. Construct the cube over the square block so that the edges C'D', D'E', etc. should be parallel to CD, DE, etc. respectively and equal to 40 mm each.

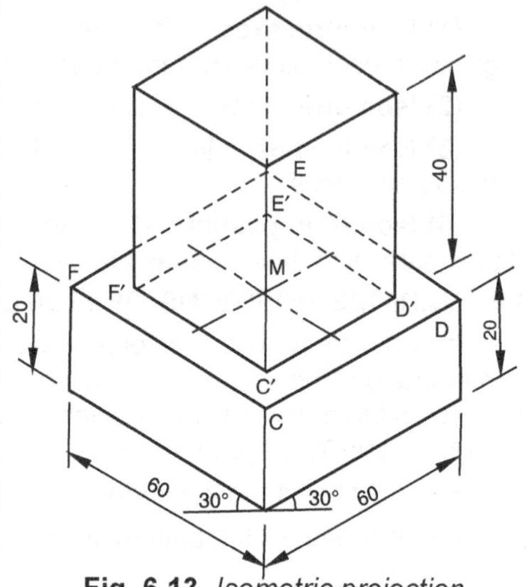

Fig. 6.13. *Isometric projection*

Problem 2 : A square prism of side 30 mm and 40 mm high is resting on H.P. A vertical square bore of 10 mm side is cut through its face reaching other square face of the prism. Draw the isometric projection of the prism.

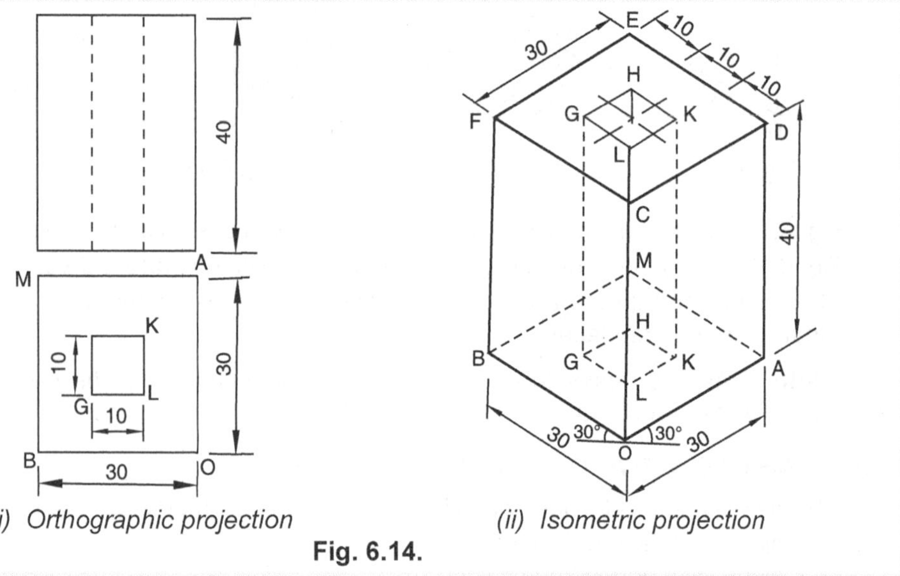

(i) Orthographic projection *(ii) Isometric projection*

Fig. 6.14.

Solution : Draw the orthographic projections of the prism by using the isometric scale [see Fig. 6.14 (i)].

Take three lines OA = 30 mm, OB = 30 mm and OC = 40 mm, the part of the isometric axes so as to represent the length, breadth and height as shown in Fig. 6.14 (ii) and complete the square prism.

Construct a rhombus LKHG in the centre of bottom by taking care that the centre point Z of the rhombus CDEF is also the same as the centre of the rhombus LKHG.

Complete the smaller prism showing hidden edges by dotted lines construction.

Problem 3 : Draw the isometric projection of the three bricks of size 200 mm x 200 mm x 100 mm from the given top and front views. [Fig. 6.15 (i)]

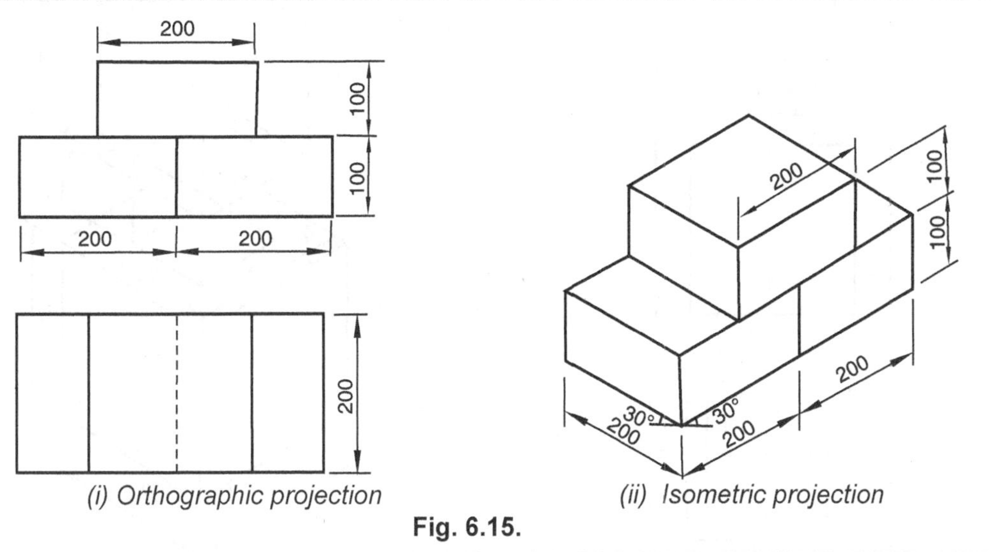

(i) Orthographic projection *(ii) Isometric projection*

Fig. 6.15.

Solution : For its isometric projection, see Fig. 6.15 (ii).

Problem 4: Fig. 6.16 shows two orthographic views of an object. Draw the isometric projection of the object.

Solution : For its solution, see Fig. 6.17.

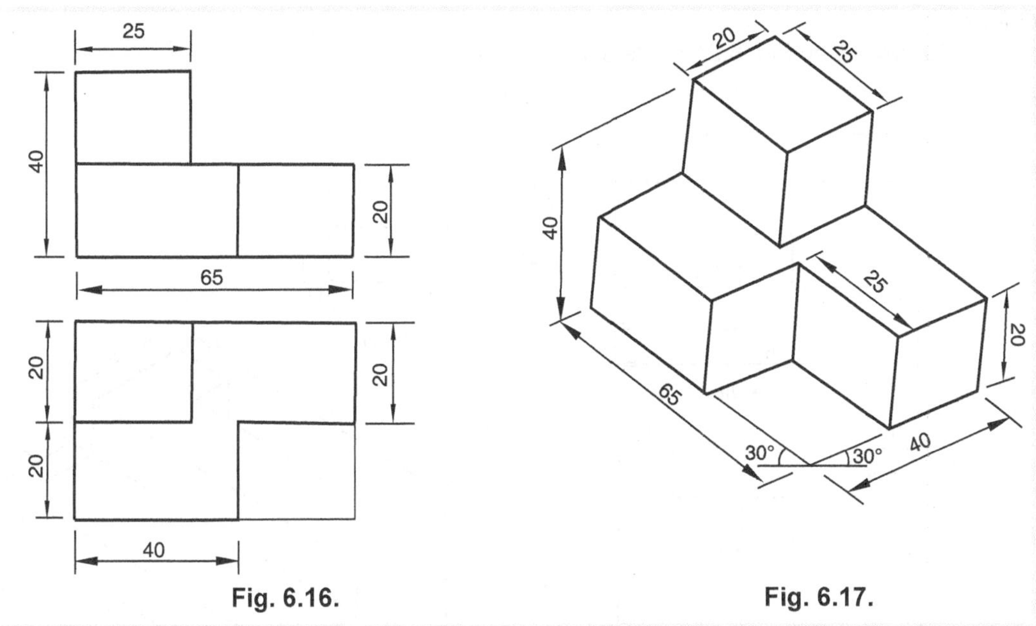

Fig. 6.16. **Fig. 6.17.**

Problem 5 : Three cubes of sides 30 mm, 20 mm and 40 mm respectively are resting one upon another such that the vertical axes of all the three cubes are in the same straight line. From the given orthographic projections shown in Fig. 6.18, draw the isometric projection.

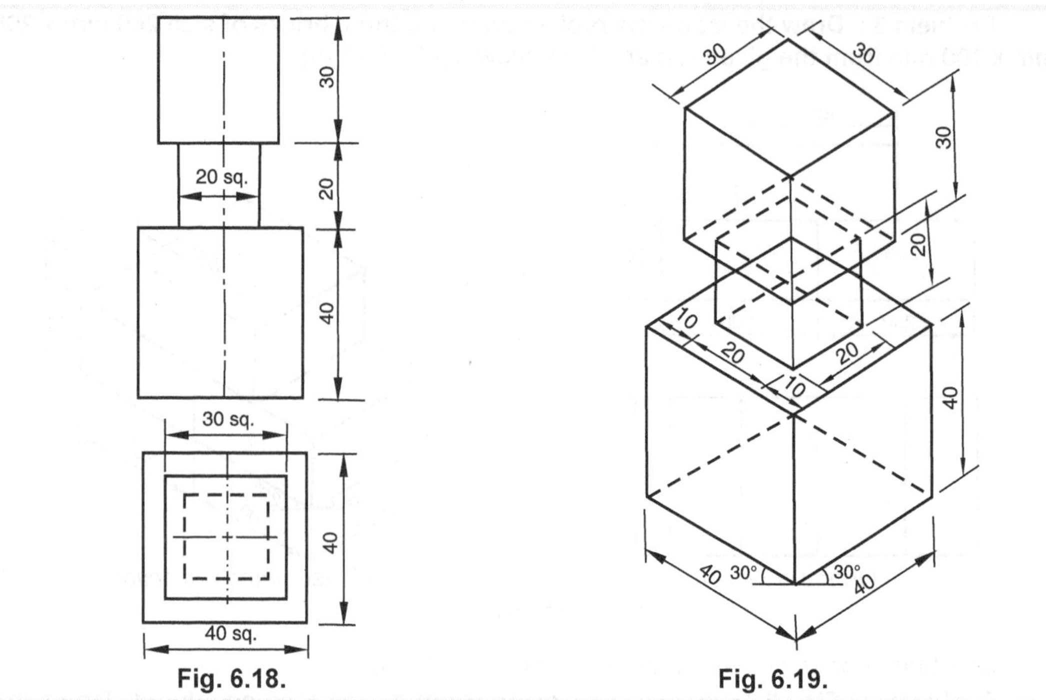

30
20 sq.
20
40
30 sq.
40
40 sq.

Fig. 6.18.

30
30
30
20
10
20
20
10
40
40
30°
30°
40

Fig. 6.19.

Solution : Follow the same method as for Problem 1 and see Fig. 6.19 for its complete solution.

Problem 6 : Fig. 6.20 shows three views of an object. Draw the isometric view of the object.

Solution : See Fig. 6.21 for its solution.

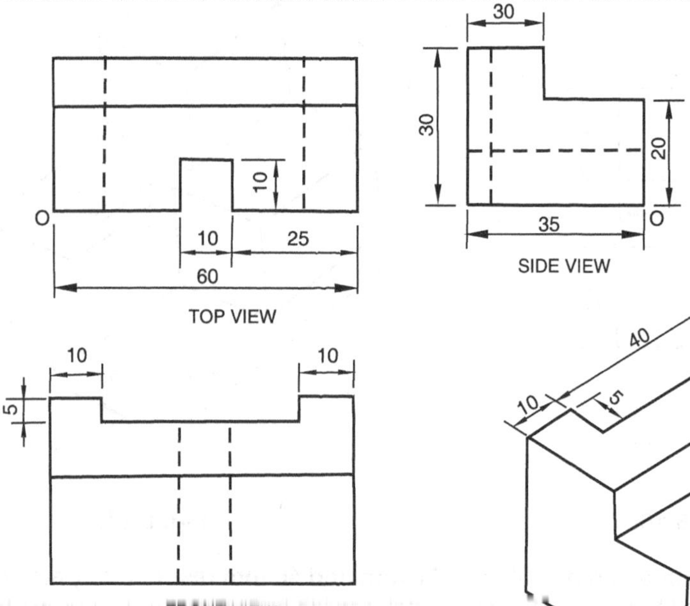

O
10
10
25
60
TOP VIEW
30
30
20
35
O
SIDE VIEW
10
10
5
FRONT VIEW

Fig. 6.20.

10
15
40
10
10
5
20
20
10
25
10
30°
30°
O
ISOMETRIC VIEW

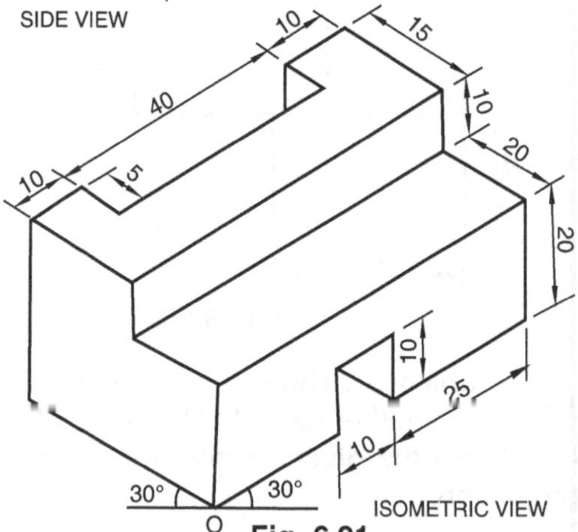

Fig. 6.21.

Problem 7 : Fig. 6.22 shows three views of an object. Draw its isometric view.

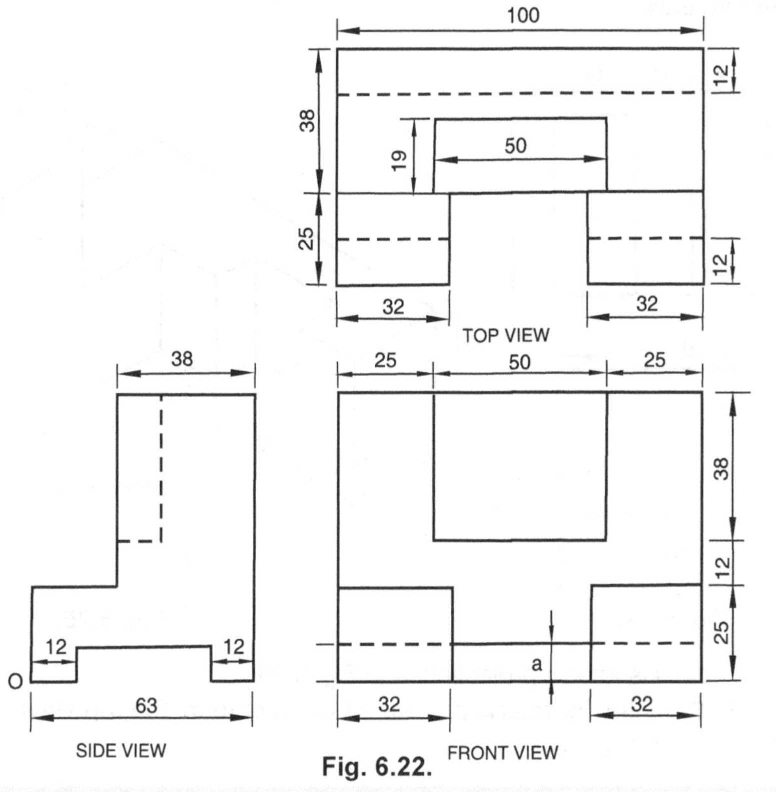

Fig. 6.22.

Solution : For its solution, see Fig. 6.23.

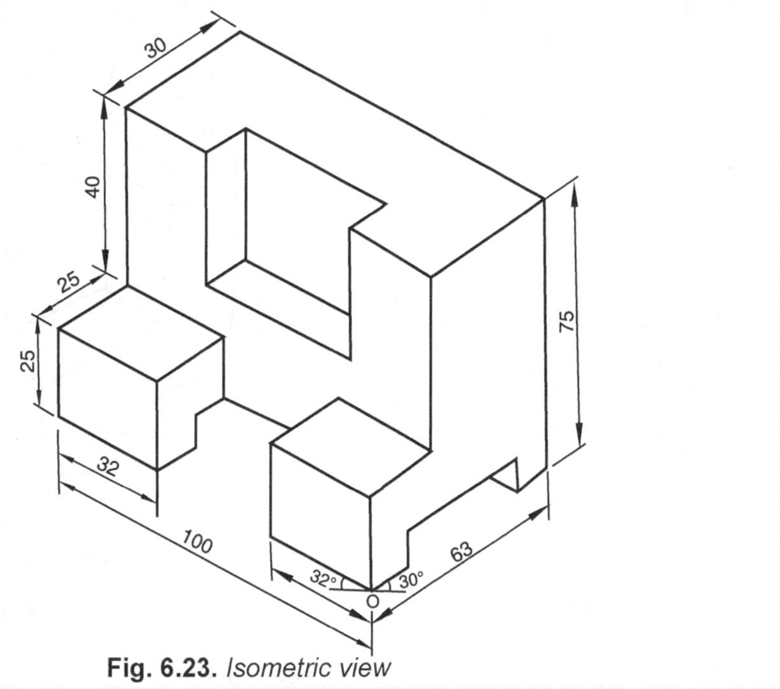

Fig. 6.23. *Isometric view*

Problem 8 : Draw the isometric projection of an object from the top view and front view given in Fig. 6.24.

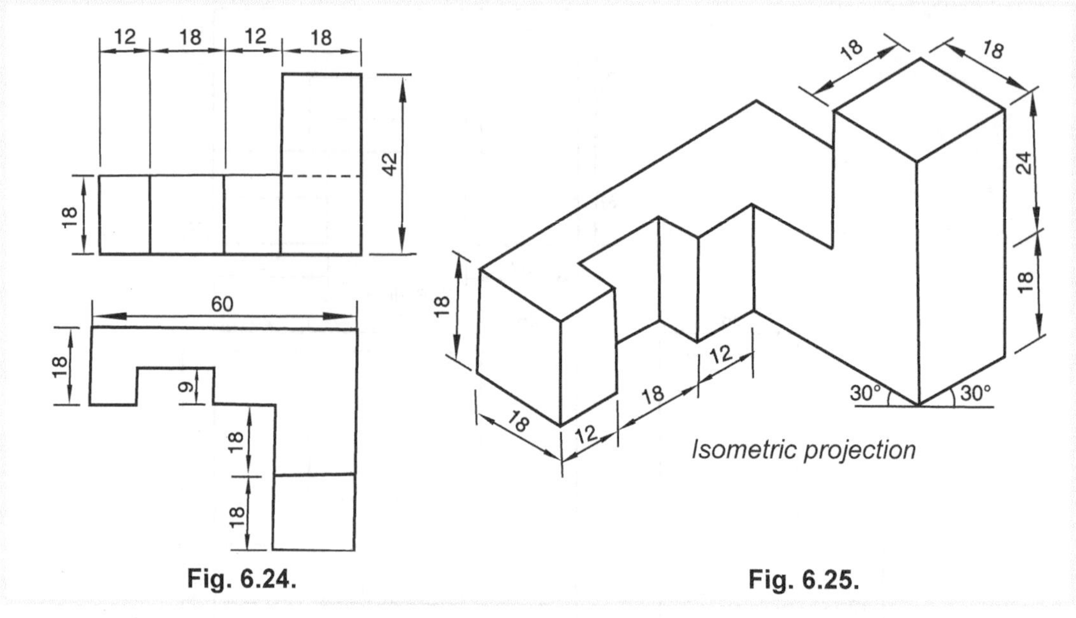

Fig. 6.24. Fig. 6.25.

Solution : For its isometric projection, see Fig. 6.25.

Problem 9 : Draw the isometric projection of an object. The top view and front view of which are shown in Fig. 6.26.

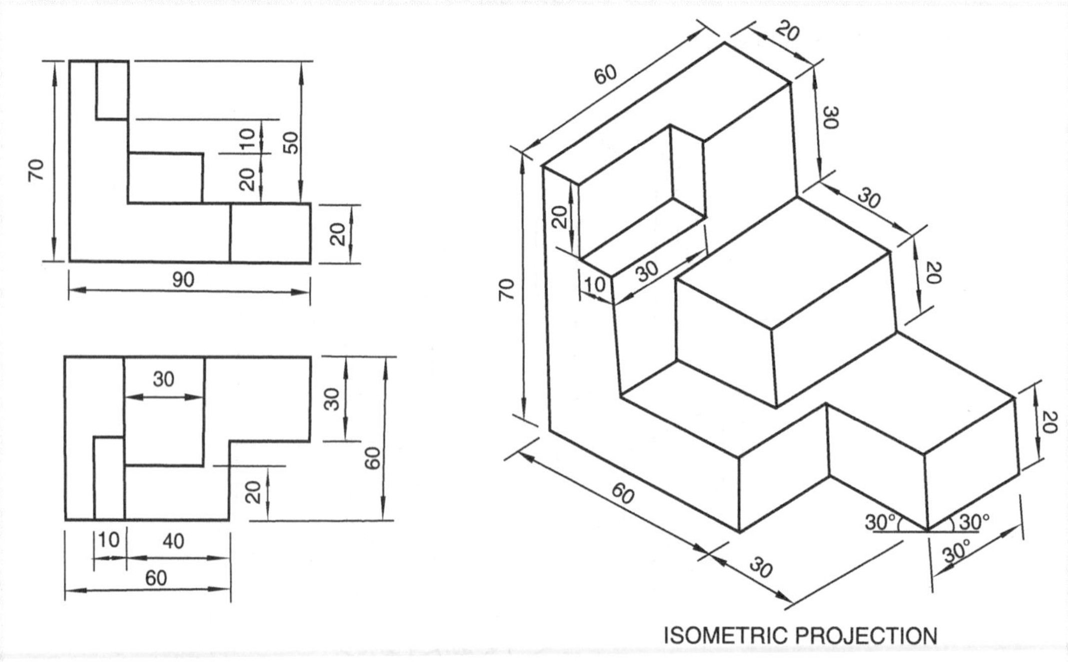

Fig. 6.26. Fig. 6.27.

Solution : For isometric projection, see Fig. 6.27.

6.12. OBJECTS WITH NON-ISOMETRIC LINES

The objects in which the edges or lines are not parallel to the isometric axes are called the objects with non-isometric lines.

This type of problems include the objects such as *pyramids, frustum of pyramids, etc.*

The best and easy method for solving such problems is to enclose the objects in squares or rectangles so as to form the isometric lines by means of which the isometric projections can be easily drawn. In some cases, the inclinations are converted into horizontal and vertical components so as to complete the isometric projections easily.

PROBLEMS ON OBJECTS WITH NON-ISOMETRIC LINES

Problem 10 : A hexagonal pyramid of 20 mm side and 50 mm height is resting on a horizontal plane. Draw the isometric projection of the pyramid.

Solution : Draw the top view and front view (by using isometric scale) of the hexagonal pyramid and enclose the top view in a rectangle *klmn* as shown in Fig. 6.28 (i). Mark the points *abcde* and *f* in the rectangle *klmn* so as to show the corners of the pyarmid.

Show isometric projection of the top view *klmn* as KLMN by taking K as a tilting point on the horizontal line and mark the points A, B, C, D, E and F as shown in Fig. 6.28 (ii). Through centre S, draw SO, the axis and cut it equal to the height of the pyramid.

Join AB, BC, CD, EF and FA which form the base of the pyramid. Also join the point O (the apex) to A, B, C, D, E and F keeping in mind that visible adges are shown in continuous lines and hidden edges with dotted lines.

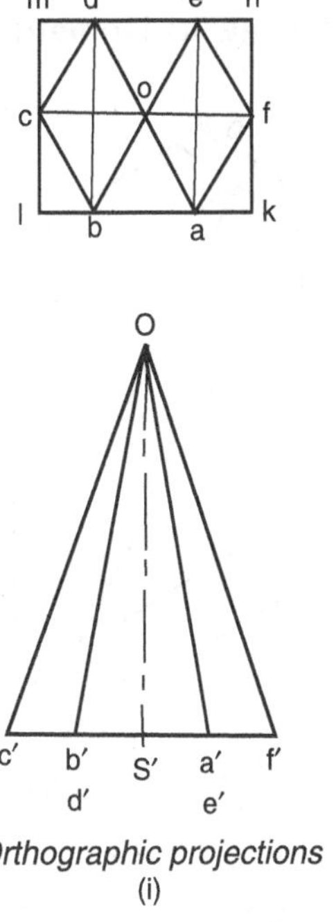

Orthographic projections
(i)

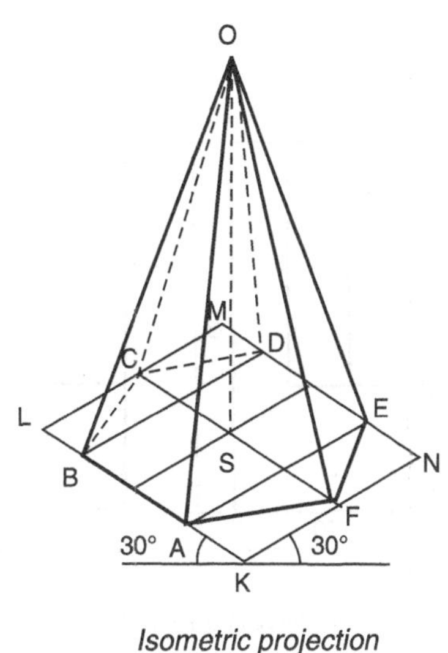

Isometric projection
(ii)

Fig. 6.28.

Problem 11 : The top and front views of a triangular pyramid are shown in Fig. 6.29. Draw its isometric projection.

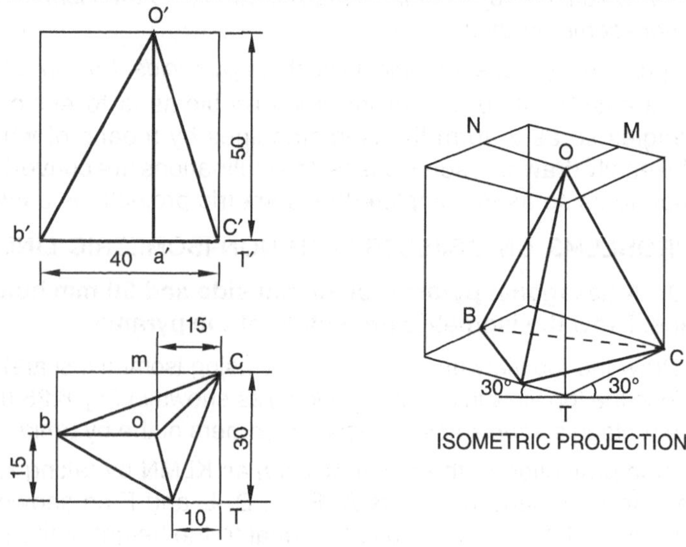

Fig. 6.29. Fig. 6.30.

Solution : Draw the given top view and front view by using isometric scale and enclose them in rectangles as shown in Fig. 6.29.

For its isometric projection, see Fig. 6.30 which is itself a self explanatory sketch.

Problem 12 : Fig. 6.31 shows two views of an object. Draw its isometric projection.

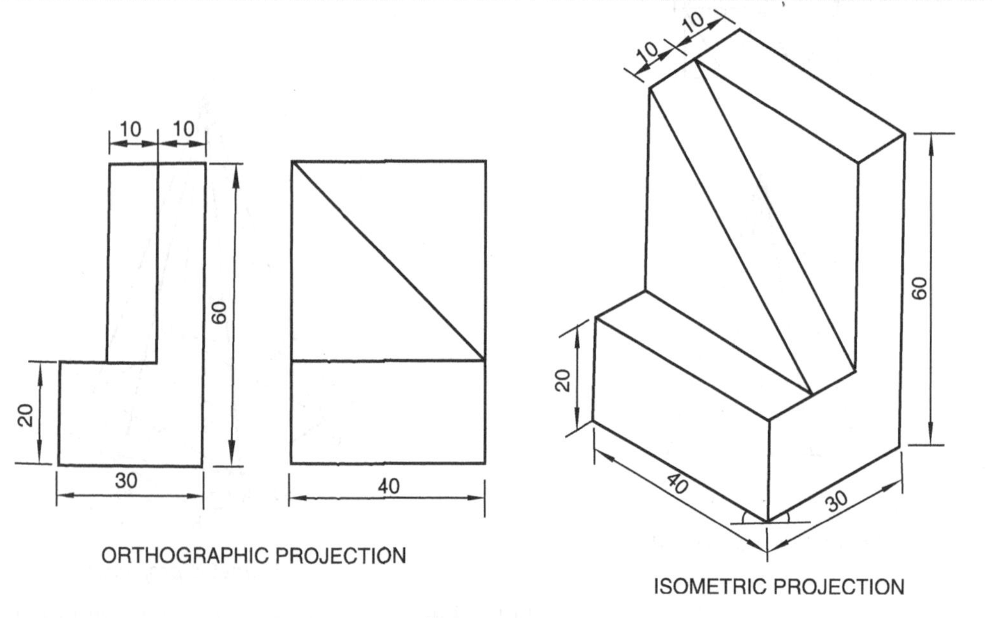

Fig. 6.31. Fig. 6.32.

Solution : For its solution, see Fig. 6.32.

Problem 13 : Draw the isometric projection of the frustrum of the hexagonal pyramid when two orthographic projections are given (see Fig. 6.33).

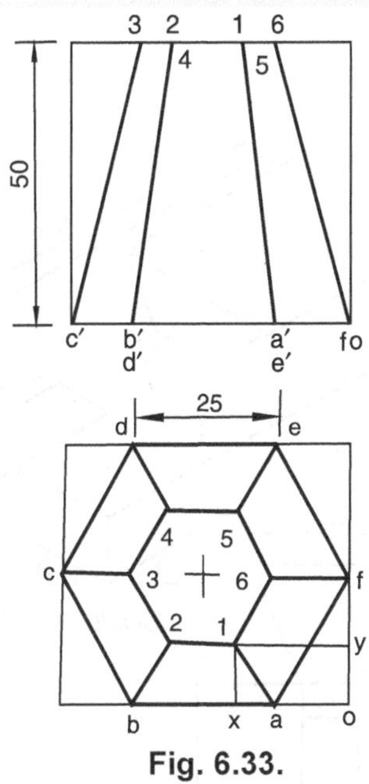

Fig. 6.33.

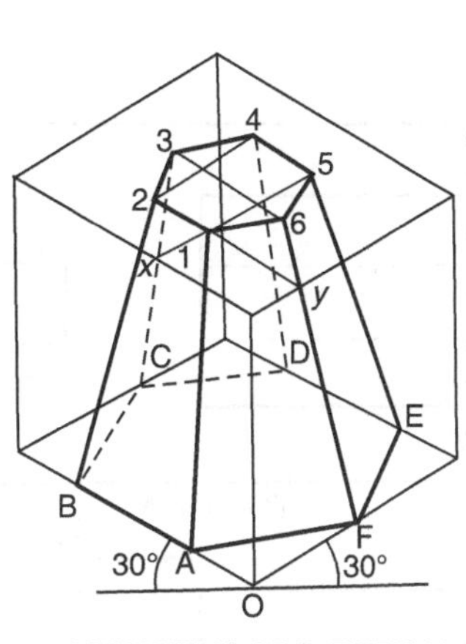

Fig. 6.34.

Solution : For its isometric projection, see Fig. 6.34.

Problem 14 : Fig. 6.35 shows the top and front view of an object. Draw the isometric projection of the given object.

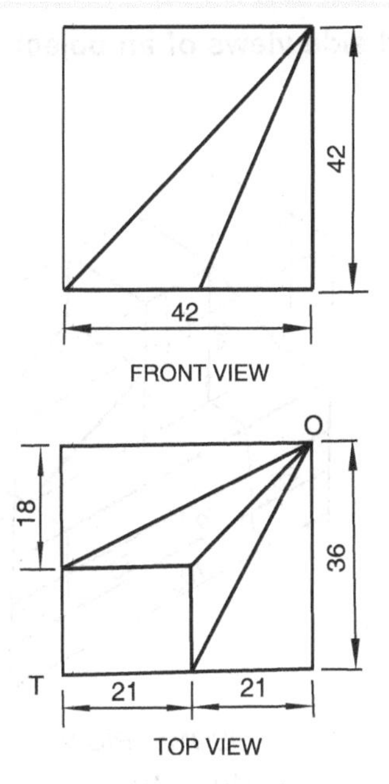

Fig. 6.35.

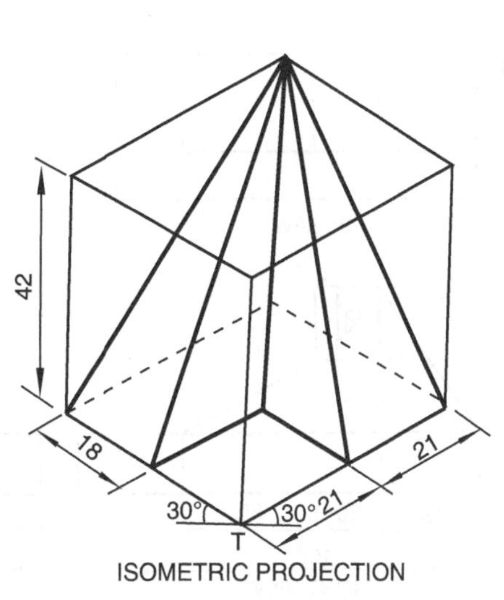

Fig. 6.36.

Solution : For its isometric projection, see Fig. 6.36.

Problem 15 : Draw the isometric projection of an object whose orthographic projections are given in Fig. 6.37.

Solution : For its solution see Fig. 6.38.

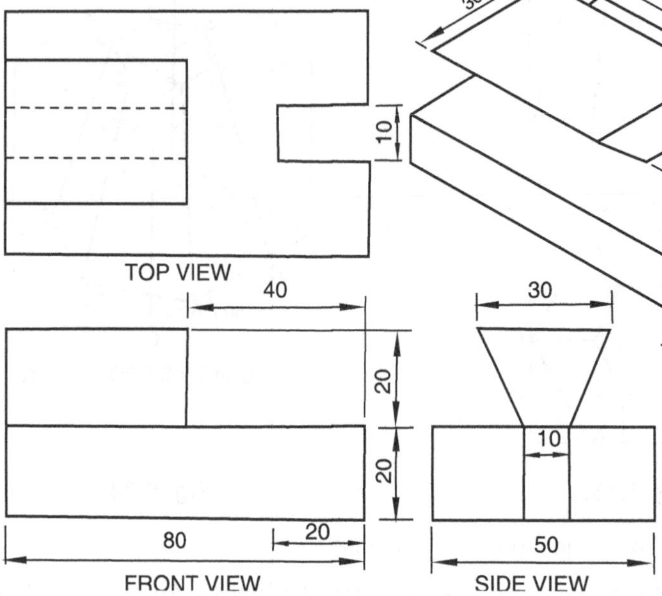

TOP VIEW

FRONT VIEW

SIDE VIEW

Fig. 6.37.

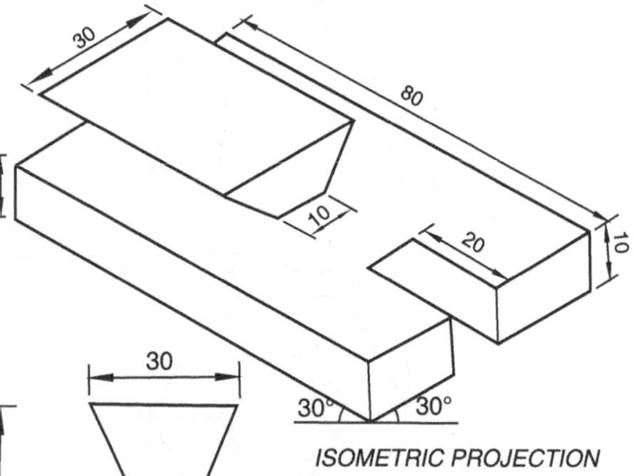

ISOMETRIC PROJECTION

Fig. 6.38.

Problem 16 : Fig. 6.39 shows the top, front and side views of an object. Draw the isometric view of the object.

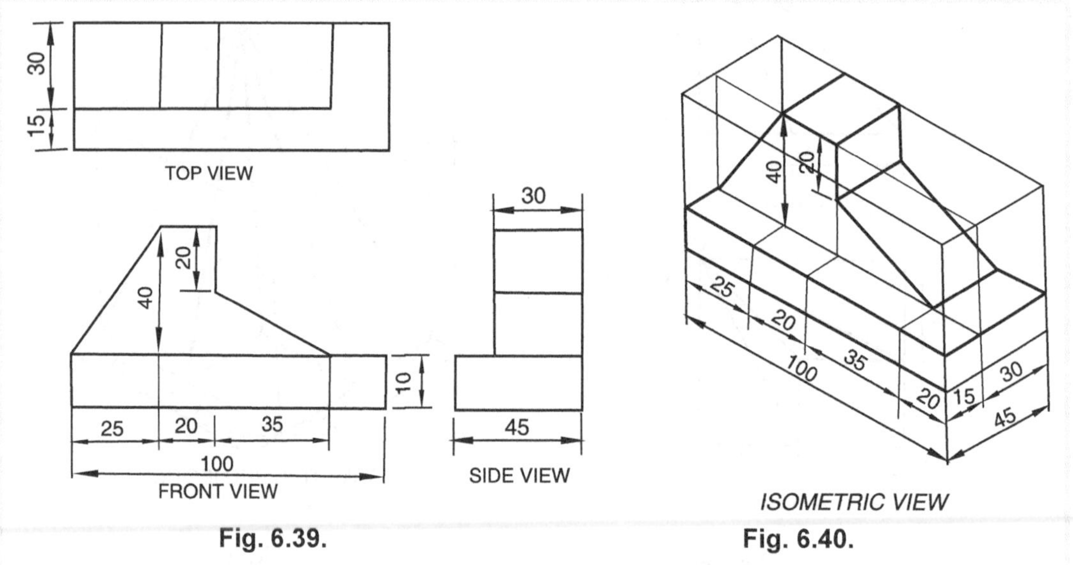

Fig. 6.39.

Fig. 6.40.

Solution : For its isometric view, see Fig. 6.40.

Problem 17 : Fig. 6.41 shows two views of an object. Do not copy the given views but draw an isometric view.

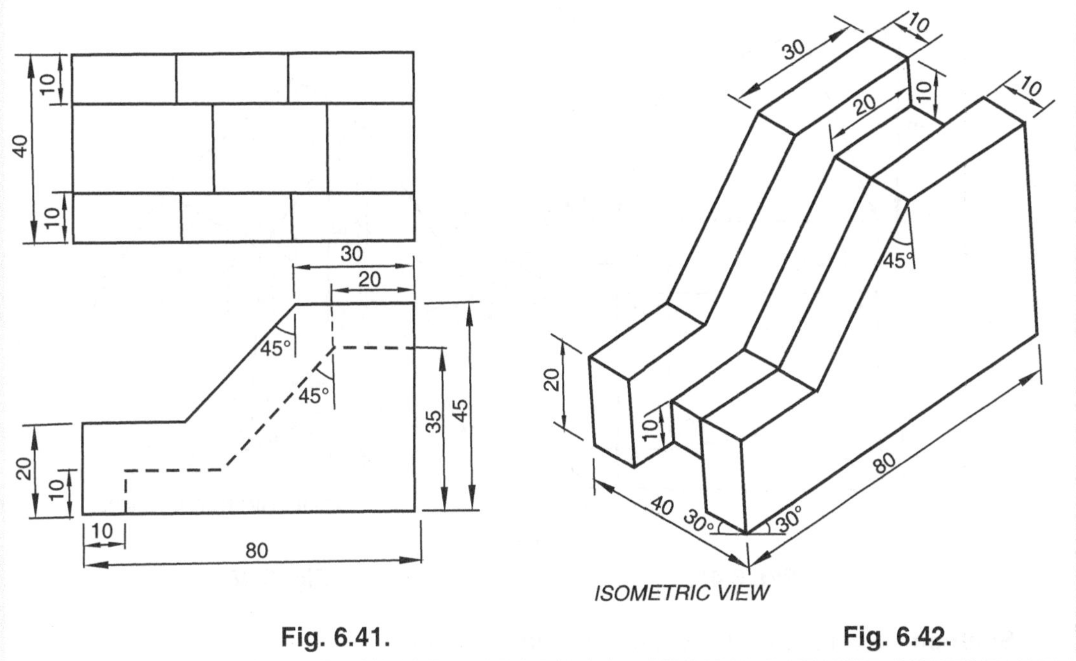

Fig. 6.41. **Fig. 6.42.**

Solution : For its isometric view, see Fig. 6.42.

Problem 18 : Draw an isometric projection of the cross as shown in Fig. 6.43.

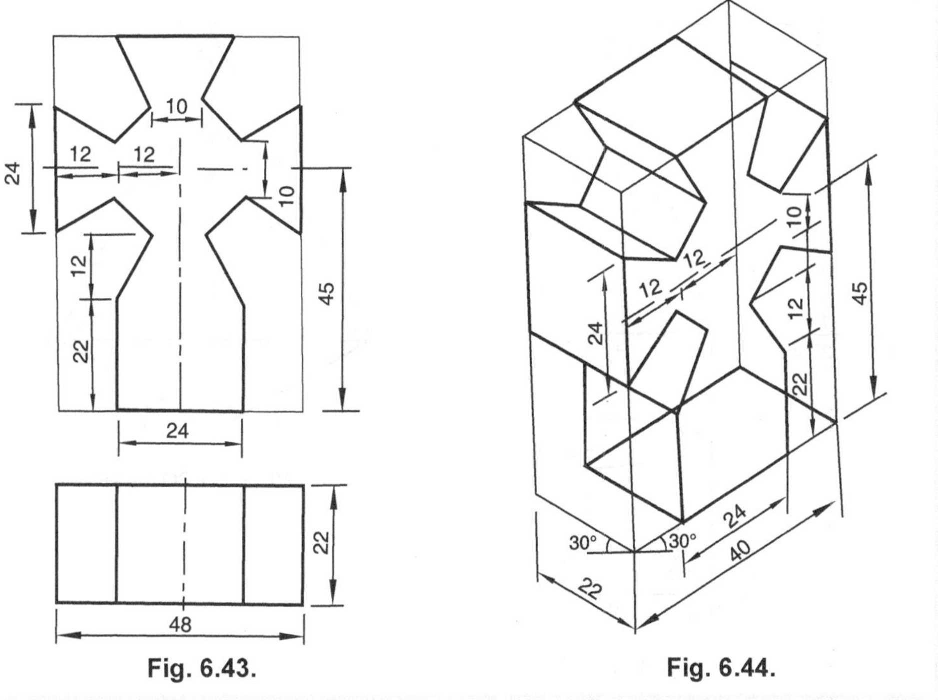

Fig. 6.43. **Fig. 6.44.**

Solution : For its isometric projection, see Fig. 6.44.

Problem 19 : Fig. 6.45 shows three views of an object. Draw its isometric projection.

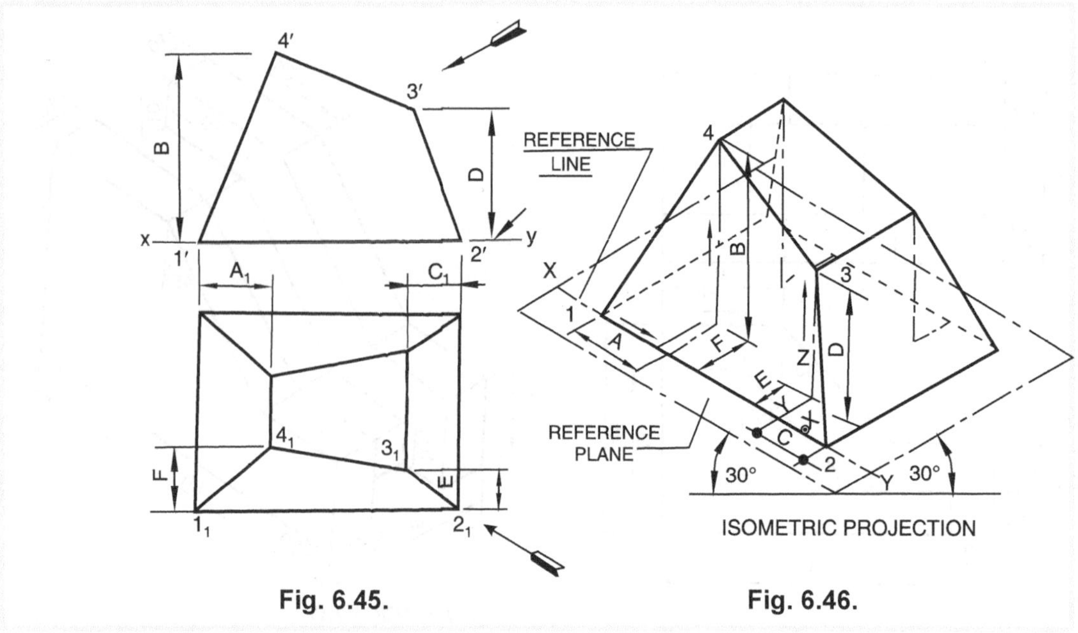

Fig. 6.45. Fig. 6.46.

Solution : For its isometric projection, see Fig. 6.46.

Problem 20 : Fig. 6.47 shows the two orthographic projection. Draw its isometric projection.

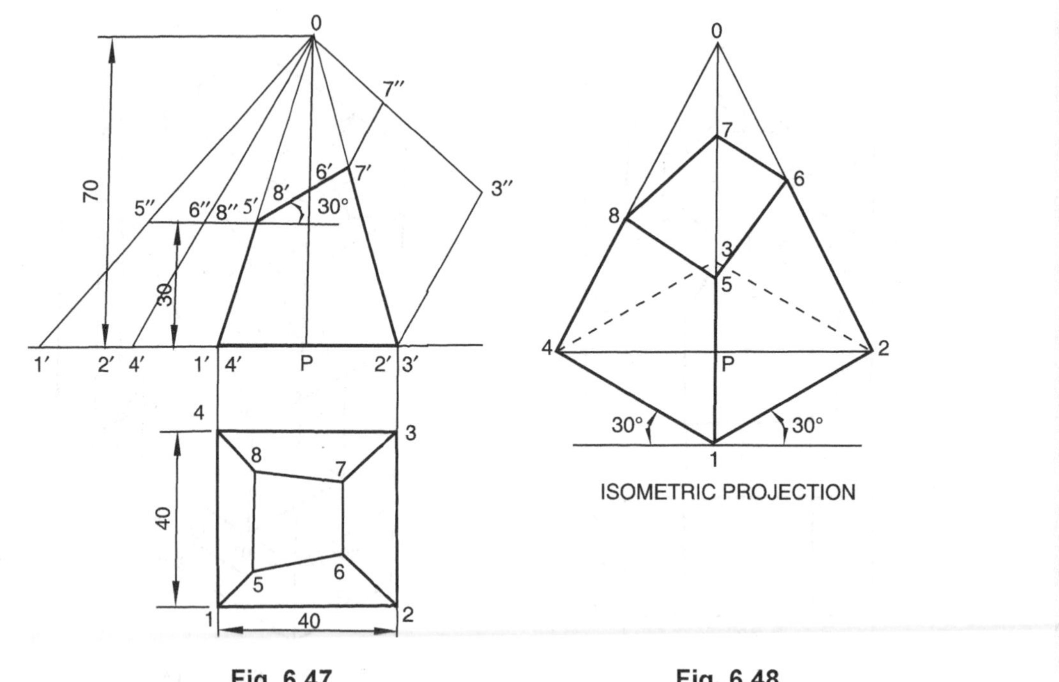

Fig. 6.47. Fig. 6.48.

Solution : For its isometric projection, see Fig. 6.48.

Problem 21 : A plate of 10 mm thickness is so cut that its elevation and plan appear as shown in orthographic projections in Fig. 6.49. Draw its isometric projection.

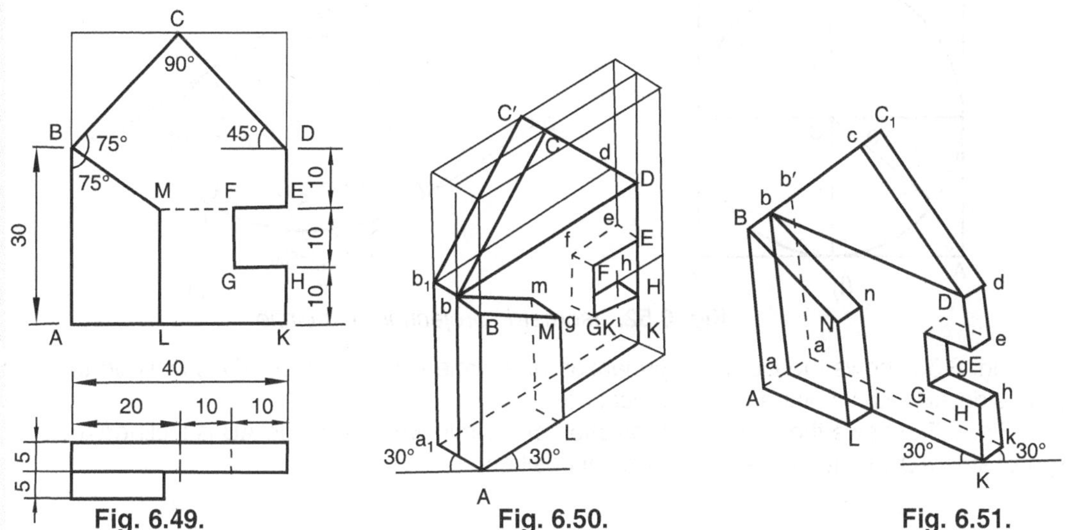

Fig. 6.49. Fig. 6.50. Fig. 6.51.

Solution : A close study of the elevation gives an idea that the figure can be bounded by a rectangle of thickness 10 mm and breadth can be measured from the elevation by producing the lines.

Note : *Find unknown distances from the orthographic projections corresponding to the given examples.*

For isometric projection, see Fig. 6.50.

Note : *Thick lines show the object lines whereas the thin are enclosing lines and if not required should be removed after the completion of work.*

An additional isometric projection of the same object can be drawn when looking in the direction of K (see Fig. 6.51).

6.13. OBJECTS WITH CIRCULAR OR CURVED SURFACES

The objects in which the different surfaces are of circular shapes and not parallel to the isometric axes are called objects with circular or curved surfaces.

This type of surfacs includes, the *circles, irregular curves, circular discs, slots, sphere etc.*

The isometric projection of the curved surfaces have been dealt as under.

6.14. ISOMETRIC PROJECTION OF A CIRCLE

The following steps should be followed while drawing the isometric projection of a circle:

1. Enclose the given circle in a square ABCD [see Fig. 6.52(i)].

2. Draw the isometric projection of the square which will be a rhombus ABCD, by tilting the sides AB and AD at 30° to the horizontal [see Fig. 6.52 (ii)].

3. Mark the mid-points of the sides of rhombus such as E, F, G and H. Join the mid-points E and F to C and similarly G and H to A, the opposite lines thereby, intersecting at C_1 and C_2.

4. With C_2 and C_1 centres and radius $R_1 = C_2F = C_2G = C_1E = C_1H$, draw arcs FG and EH. Similarly, with C and A as centres and radius $R_2 = CF = CE = AG = AH$, draw arcs EF and GH. Now the enclosing curve EFGH is the required isometric projection of the given circle.

Note : *The isometric view of a circle is an ellipse.*

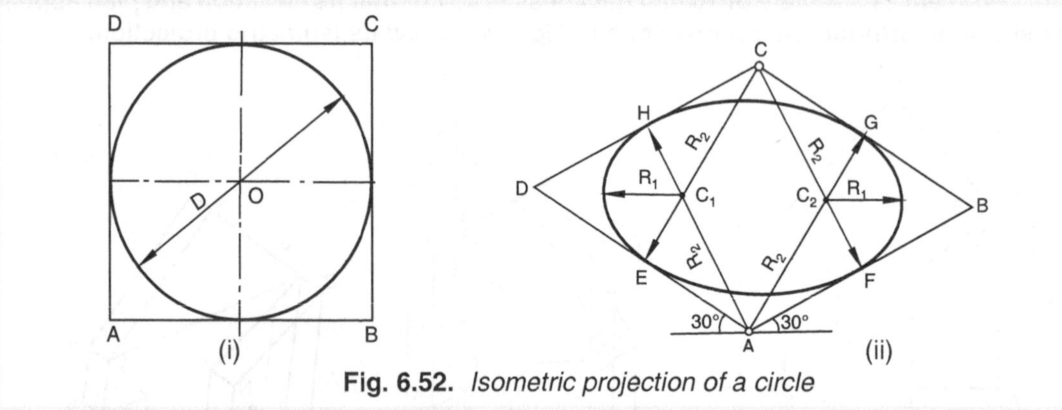

Fig. 6.52. *Isometric projection of a circle*

Fig. 6.53 shows the isometric projection of a circle when the side AD is inclined towards left of side AB with its plane in the vertical position.

Fig. 6.54 shows the isometric projection of a circle when the side AB is inclined towards right of side AD with its plane in the vertical position.

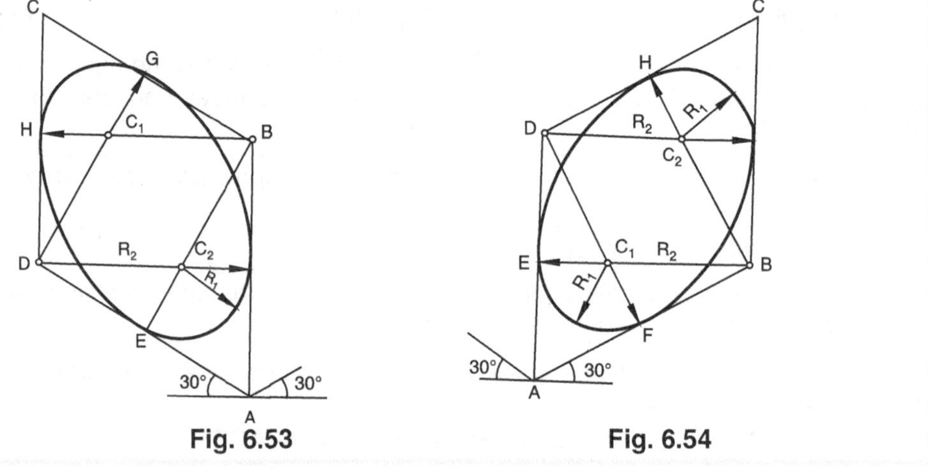

Fig. 6.53 **Fig. 6.54**

Fig. 6.55 shows the isometric projection of concentric circles which is drawn as described above.

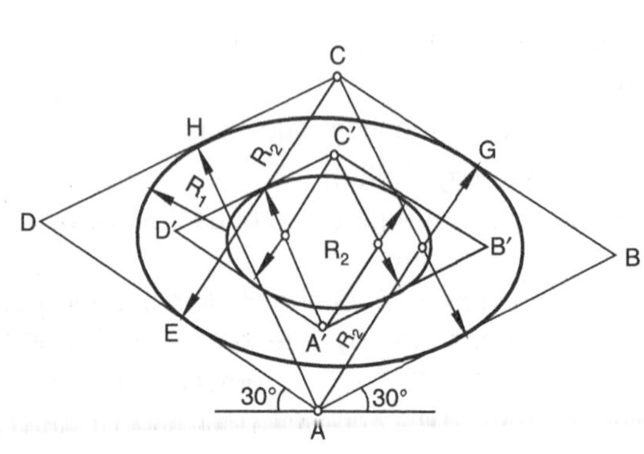

Fig. 6.55. *Isometric projection of concentric circles*

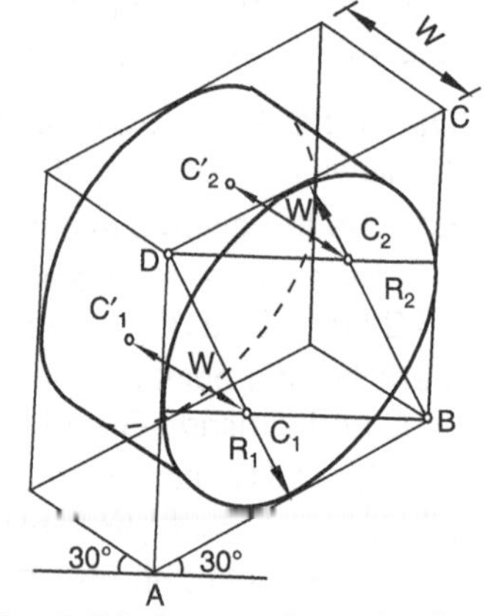

Fig. 6.56. *Isometric projection of a circular disc or shaft*

6.14.1. Isometric Projection of a Circular Disc or Shaft : The isometric projection of a circular disc or shaft is shown in a similar way as circle, but in this case two isometric circles at a certain width W are taken and joined as shown in Fig. 6.56.

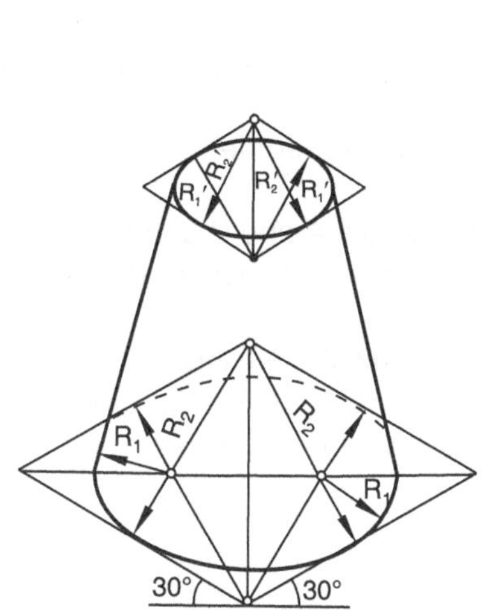

Fig. 6.57. *Isometric projection of a tapered shaft*

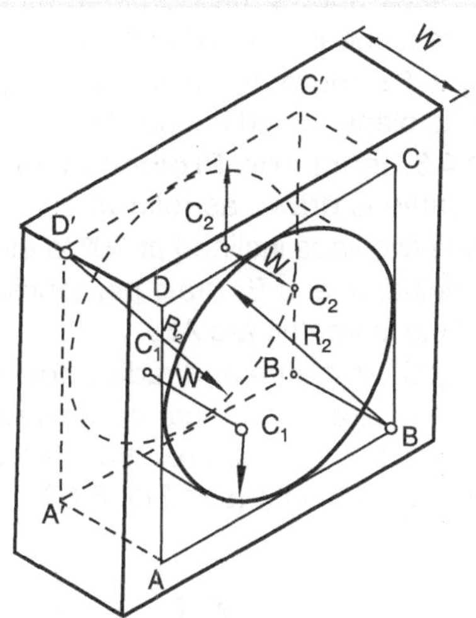

Fig. 6.58. *Isometric projection of a circular hole in a block*

6.14.2. Isometric Projection of a Tapered Shaft : The isometric projection of a tapered shaft is drawn in a similar way as a simple shaft, but in this case two different isometric circles are drawn first of all (see Fig. 6.57).

6.14.3. Isometric Projection of a Circular Hole in a Rectangular Block : The isometric projection of a circular hole in a rectangular block is drawn in a similar way as a circular shaft (see Fig. 6.58).

6.14.4. Isometric Projection of Irregular Curves : The isometric projection of irregular curves are drawn graphically as under:

Imagine the given curve to be enclosed by two mutually perpendicular lines OX and OY and further assume some points A, B, C etc. on the curve (see Fig. 6.59). Find the co-ordinates of these points and let these be x_1y_1, x_2y_2, x_3y_3 etc.

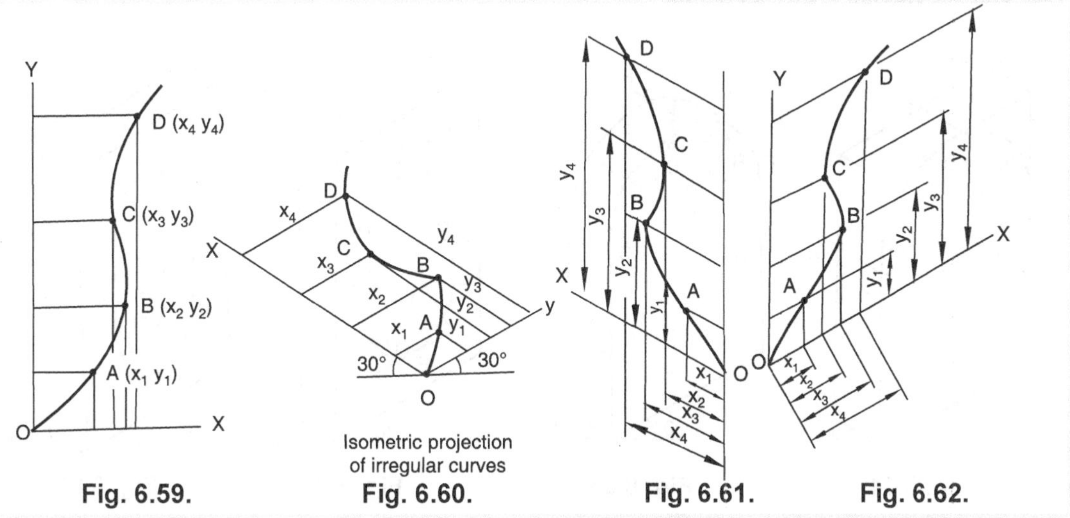

Fig. 6.59. **Fig. 6.60.** **Fig. 6.61.** **Fig. 6.62.**

Now, draw OX and OY as the isometric axes and plot the co-ordinates of points A, B, C etc. as shown in Fig. 6.60. A smooth curve passing through points will be the isometric projection of the given curve.

Fig. 6.61 shows the isometric projections when the axis OX is inclined towards left of OY with its plane in vertical position.

Fig. 6.62 shows the isometric projection when the axis OX is inclined towards right of OY with its plane in vertical position.

6.14.5. Isometric Projection of a Circular Curve : The isometric projection of a circular curve is drawn as follows :

Draw two lines inclined at 30° to either side and fix the points A and B on them.

At points A and B, draw perpendicular lines so as to intersect C_1 which is the required centre of the isometric arc AB.

Take C_1 as centre and radius equal to R = C_1A = C_1B. Draw the arc AB.

From C_2 draw a vertical downward line C_1C_2 such that C_1C_2 = H, Height of the circular disc. Now, with C_2 as centre, draw the arc A'B'. Thus, the arcs AB and A'B' are the circular arcs at required height (see Fig. 6.63).

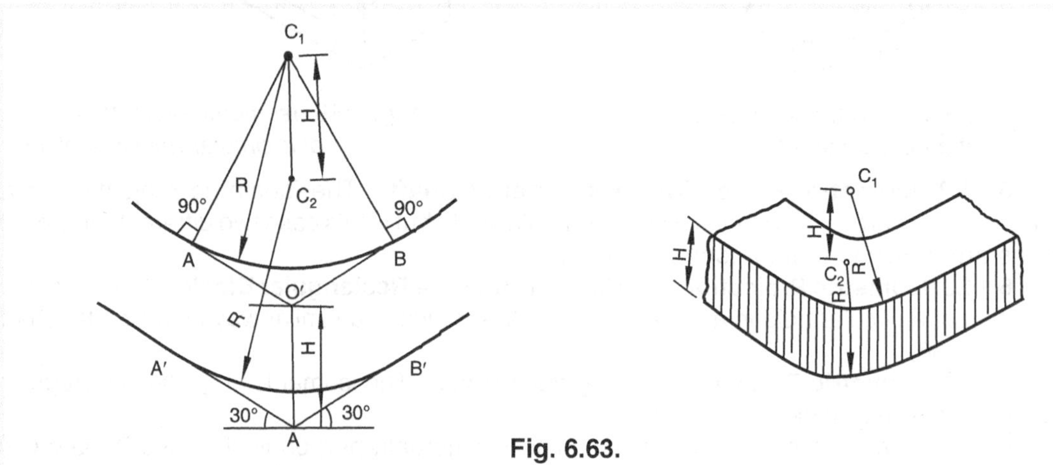

Fig. 6.63.

6.14.6. Isometric Projection of a Round End Object : The isometric projection of a round end object is drawn as under :

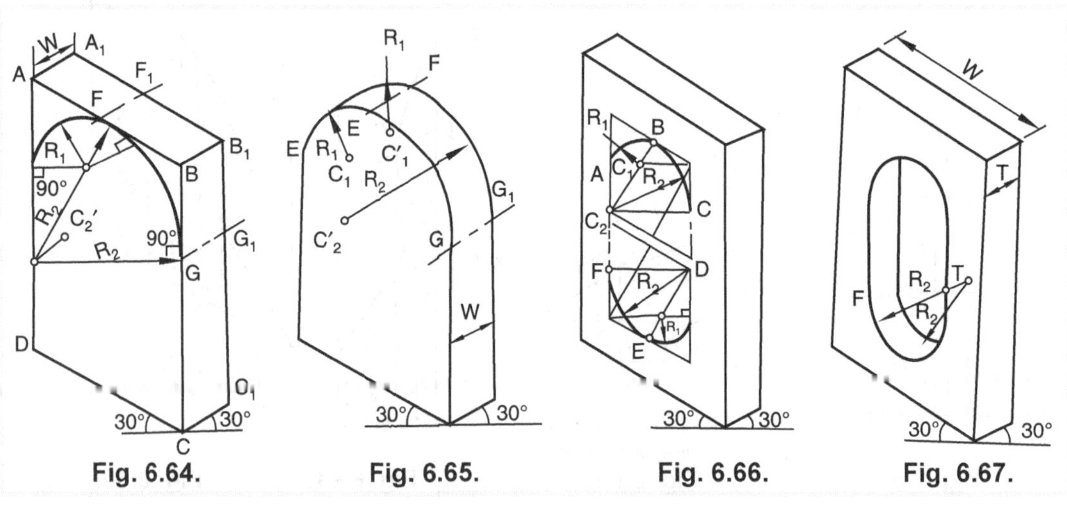

Fig. 6.64. Fig. 6.65. Fig. 6.66. Fig. 6.67.

First of all draw the isometric projections of rectangular prisms ABCD and $A_1B_1C_1D_1$ with given dimensions (see Fig. 6.64). Mark the midpoint F on the AB and F_1 on A_1B_1.

For arc EF :

From E and F erect two perpendiculars to intersect at C_1. With C_1 as centre and radius equal to $R_1 = C_1E = C_1F$ draw the required arc EF.

For arc FG :

From F and G erect two perpendiculars to interest at C_2. With C_2 as centre and radius equal to $R_2 = C_2G = C_2F$ draw an arc FG which completes the front portion of the object.

For arc E_1F_1 :

From C_1 draw a line C_1C_1' parallel to BB_1 and equal to width W of the object. With C'_1 as centre and radius equal to C_1' F_1 draw E_1F_1.

For arc F_1G_1 :

From C_2 draw a line C_2C_2' parallel to BB_1 and equal to width. With C_2' as centre and radius equal to $C_2'F_2$ draw the arc F_1G_1 which completes the rear portion of the object (see Fig. 6.65).

6.14.7. Isometric Projection of a Round End Slot : *The isometric projection of a round end slot is drawn in similar way as isometric projection of a round end object (see Figs. 6.66 & 6.67).*

6.14.8. Isometric Projection of a Sphere : Draw the front view of a sphere of diameter D by using true lengths and enclose it in a square as shown in Fig. 6.68 (i). In front view, O represents the centre of the sphere.

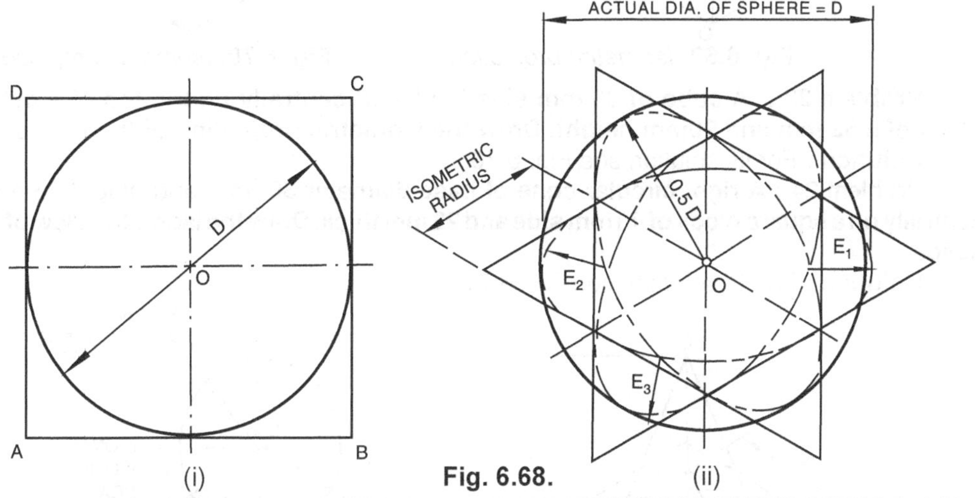

Fig. 6.68.

Consider a vertical sectional plane through the centre of the sphere which will be a circle of diameter D. The isometric projection of this circle is shown in Fig. 6.68 (ii) by ellipses E_1 and E_2, drawn in two different vertical positions around the same centre O. In this case, the length of the major axis in each case is equal to D and its distance from the point of contact, to the point O is equal to the isometric radius of the sphere.

Now, again consider a horizontal sectional plane passing through the centre of the sphere. The isometric projection of this circle is shown by an ellipse E_3, drawn horizontally around the point of contact. In this case, also, the distance of the outermost points on ellipse from the centre is equal to 0.5D. Thus, it is clear from the isometric projection that the distances of all the points on the surface of the sphere from its centre are equal to the radius of the sphere.

Note : *The isometric projection of the sphere is a circle with its diameter equal to the true diameter of the sphere and also the distance of its centre from the point of contact is equal to the isometric radius of the sphere.*

PROBLEMS ON OBJECTS WITH CURVED SURFACES

Problem 22 : A cylinder having its length and diameter equal to 60 mm and 30 mm respectively is lying on ground with its circular faces parallel to the V.P. Draw its isometric projection.

Solution : Draw the three isometric axes OX, OY and OZ. Cut of the length OO' = 60 mm, breadth OD and height OB = 30 mm and complete the square prism with two end faces OBCD and O'B'C'D' as explained earlier. Join HH' and EE' and complete the cylinder (Fig. 6.69).

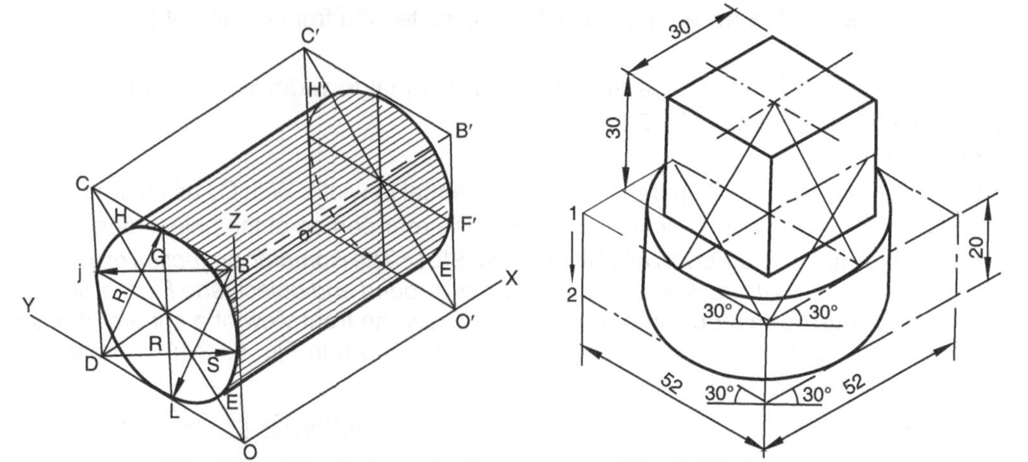

Fig. 6.69. *Isometric projection* **Fig. 6.70.** *Isometric projection*

Problem 23 : A cube of 30 mm side is placed centrally on the top of a cylindrical block of ϕ 52 mm and 20mm height. Draw the isometric projection of the solids.

Solution : For its solution see Fig. 6.70.

Problem 24 : A right circular cone of base diameter 30 mm and height 36 mm rests centrally on a square block of 48 mm side and 22 mm thick. Draw the isometric view of the two solids.

Solution : For its solution, see Fig. 6.71.

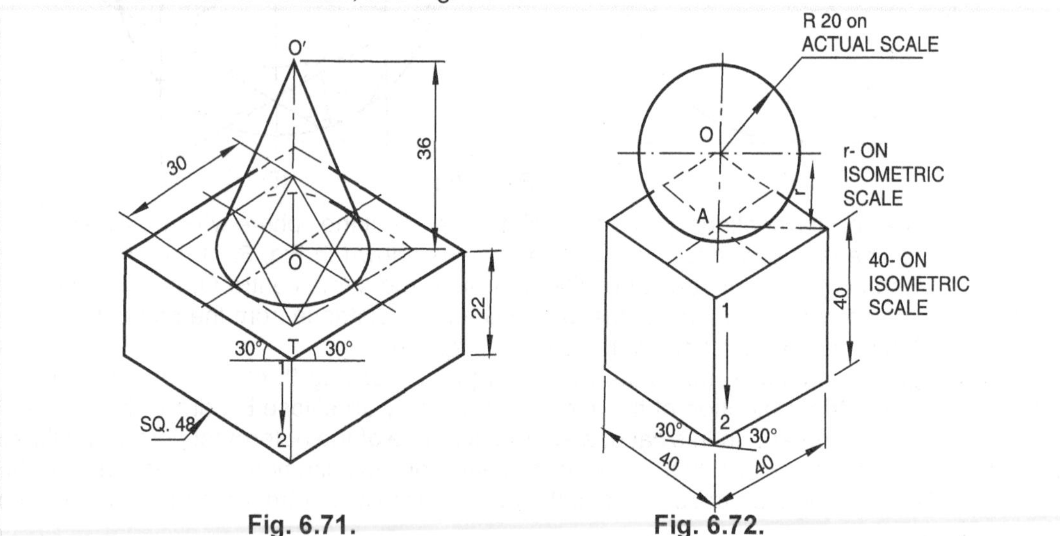

Fig. 6.71. **Fig. 6.72.**

Problem 25 : Draw an isometric projection of a sphere of 40 mm diameter, resting centrally on a square prism of 45 mm edges and 20 mm thick.

Solution : Draw the isometric projection of the square prism with isometric scale. Mark the centre point C on its top surface (see Fig. 6.72).

Draw a vertical line through C and mark the point O on it in such a way that CO = isometric radius of the sphere.

With O as centre and radius equal to the true radius of the sphere, draw a circle which will be the required isometric projection of the sphere.

Problem 26 : A hemisphere of ϕ 40 mm rests on its circular base on the top of a cube of 40 mm side. Draw the isometric projection of the solids.

Solution : For its isometric projection, see Fig. 6.73. Here the ellipse of hemisphere is drawn on isometric scale. Whereas, the arc of spherical portion of hemisphere is drawn with radius, R = 40/2 = 20 mm on actual scale.

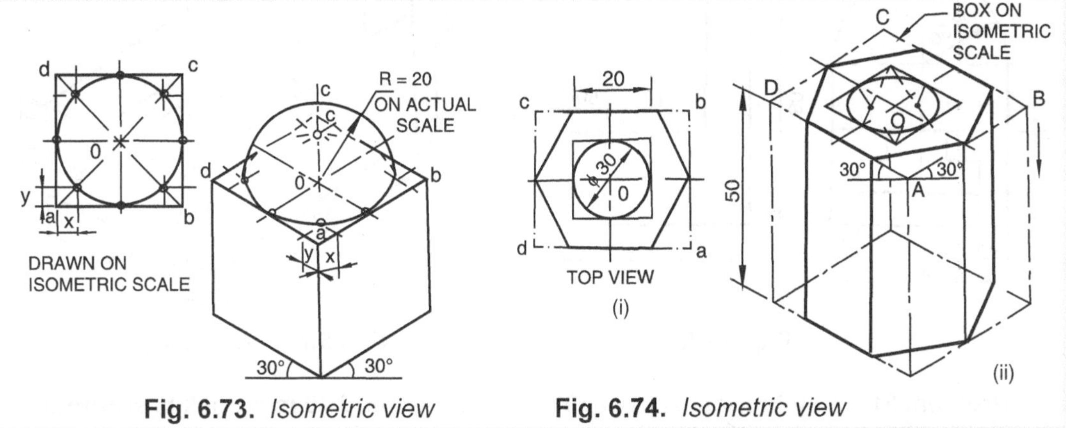

Fig. 6.73. *Isometric view* **Fig. 6.74.** *Isometric view*

Problem 27 : Fig. 6.74 (i) shows the orthographic view of an object. Draw its isometric projection.

Solution : For its isometric projection, see Fig. 6.74 (ii).

Problem 28 : A hexagonal prism lies horizontally on one flat face and on the parallel top face, a cylinder stands vertically as shown in Fig. 6.75 (i). Draw to full-size, an isometric projection of the arrangement.

Solution : For its solution, see Fig. 6.75.

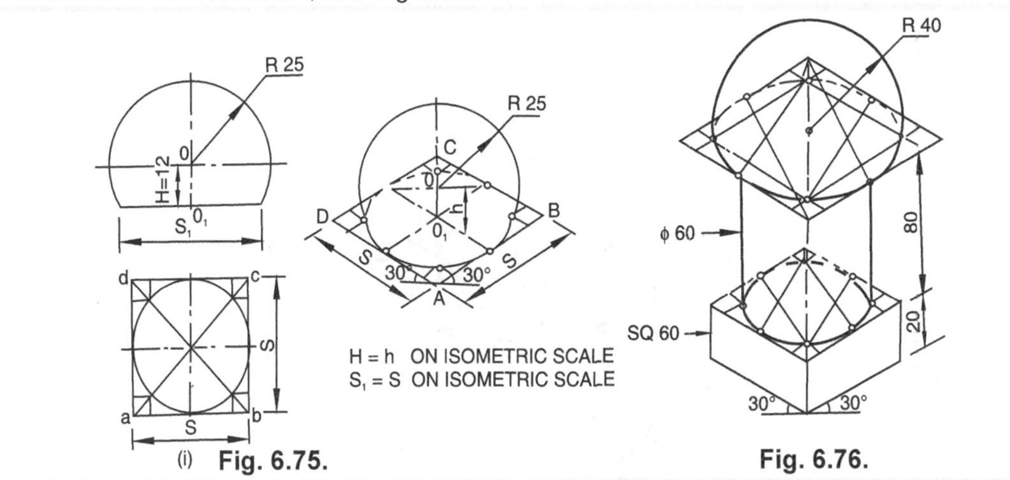

(i) **Fig. 6.75.** **Fig. 6.76.**

Problem 29 : An object consists of a hemispherical vessel of 80 mm diameter which is placed centrally over a cylinder of 60 mm diameter and 60 mm height. The cylinder in turn is placed centrally over a square prism of 60 mm base side and 20 mm height. Draw the isometric projection of the object.

Solution : For its isometric projection, see Fig. 6.76.

Problem 30 : Fig. 6.77 shows the plan and elevation of an object. Draw the isometric projection of the given object.

Solution : For its isometric projection, see Fig. 6.78.

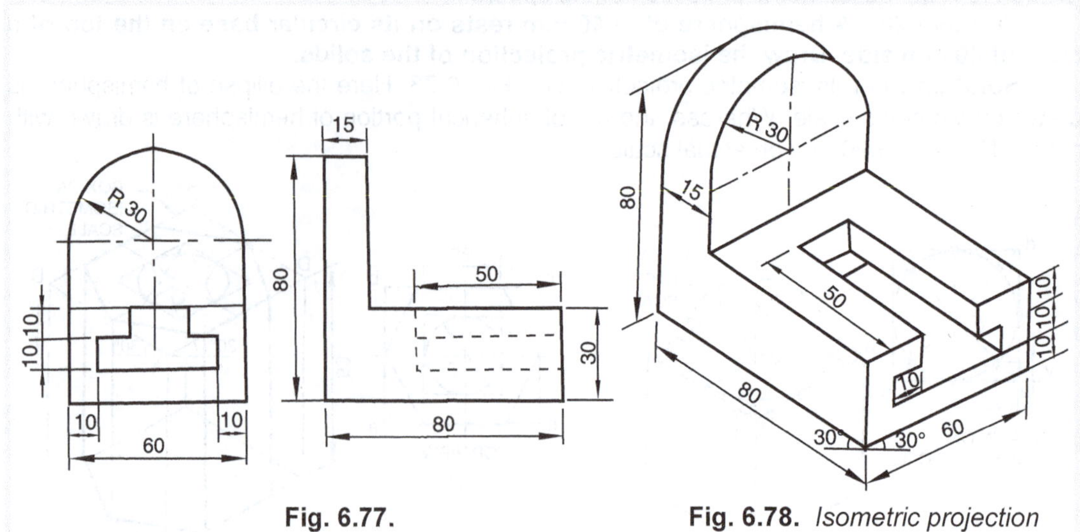

Fig. 6.77. **Fig. 6.78.** *Isometric projection*

Problem 31 : A cylindrical slab of 70 mm diameter and 40 mm thick is surmounted by a cube of 35 mm edge. On the top of a cube, rests a square pyramid, altitude 35 mm and side of base 20 mm. The axes of the solids are in the same straight line. Draw isometric projection of solids.

Solution : For its solution, see Fig. 6.79.

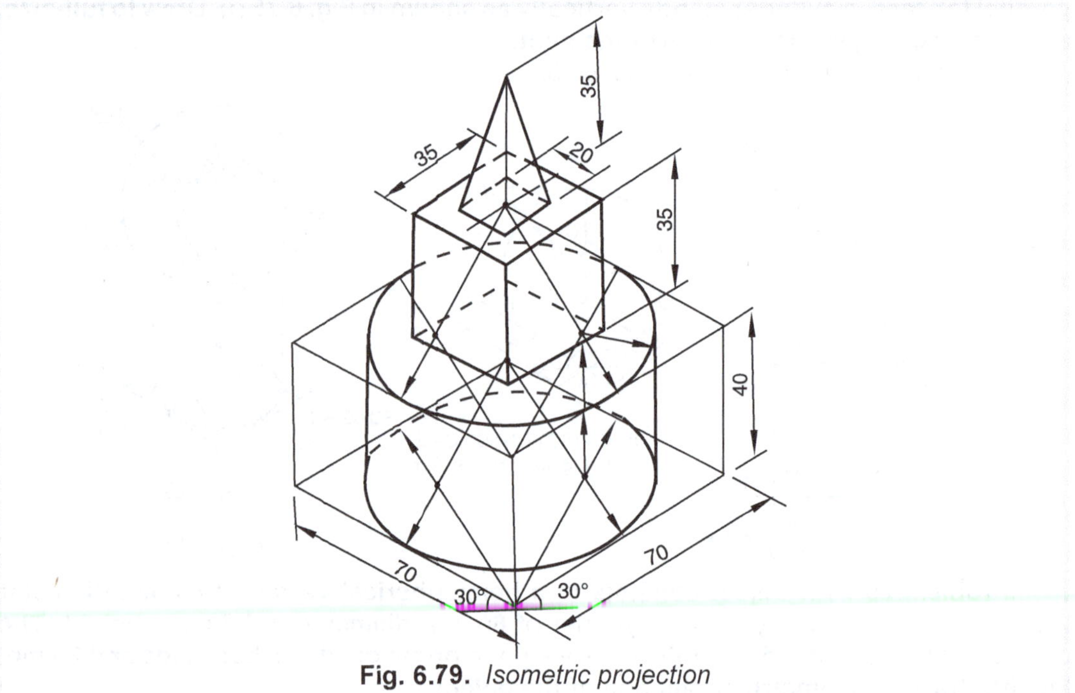

Fig. 6.79. *Isometric projection*

Problem 32 : Fig. 6.79 shows one view of an object. Draw its isometric projection.

Solution : For its solution, see Fig. 6.80.

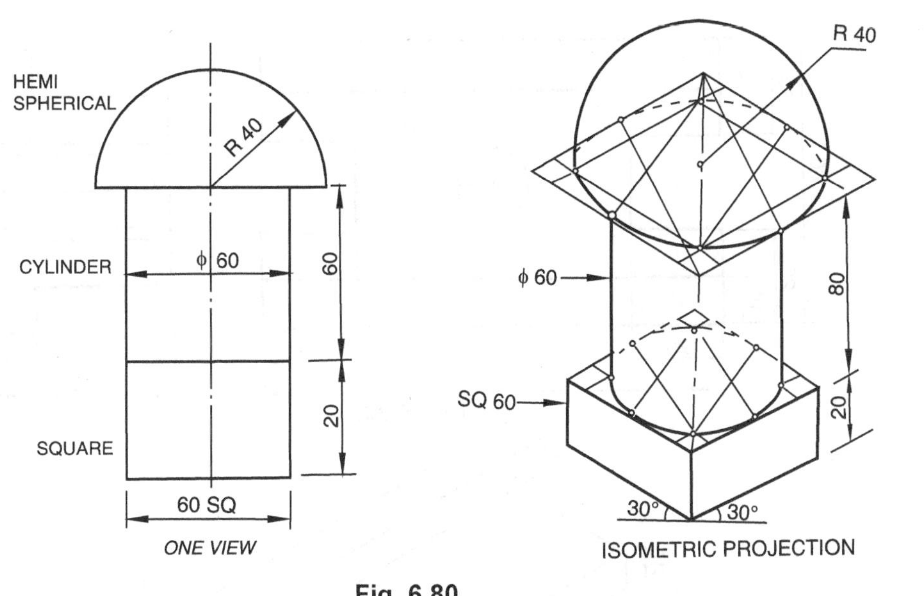

Fig. 6.80.

Problem 33 : Fig. 6.81 shows two views of an object. Draw isometric view of this object.

Solution : For its solution, see Fig. 6.82.

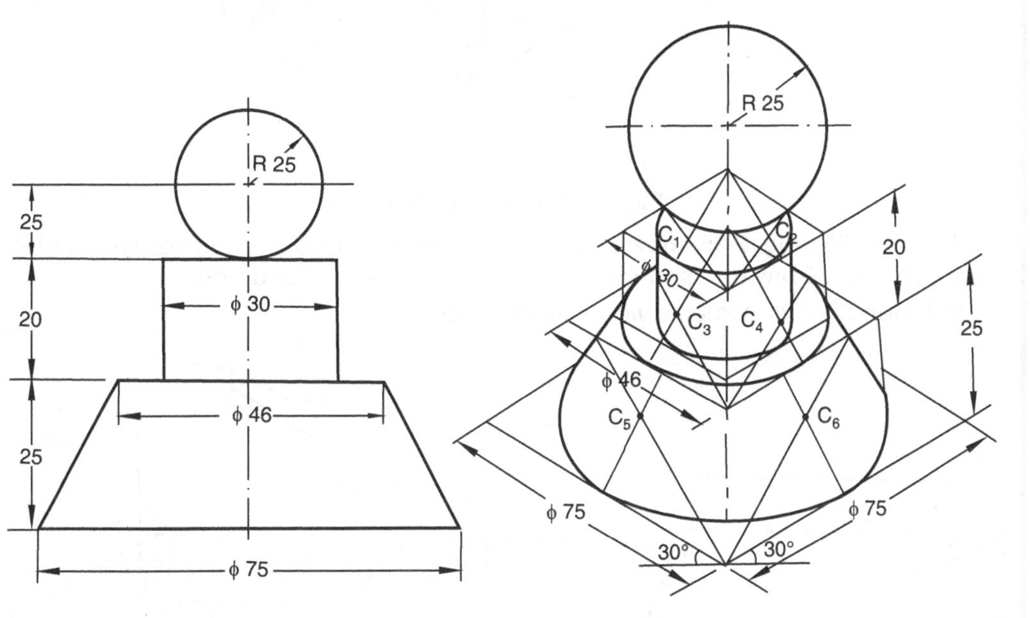

Fig. 6.81.

Fig. 6.82. *Isometric view*

Problem 34 : Fig. 6.83 shows front and side view of a cast iron block. Draw its isometric projection.

Solution : For the solution of problem 13, see Fig. 6.84.

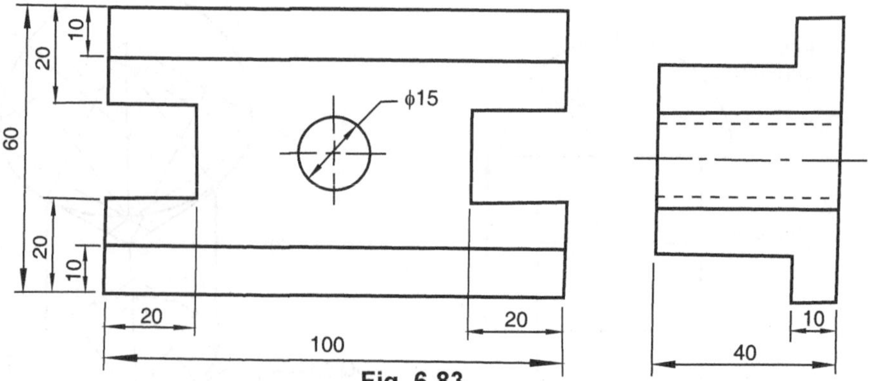

Fig. 6.83.

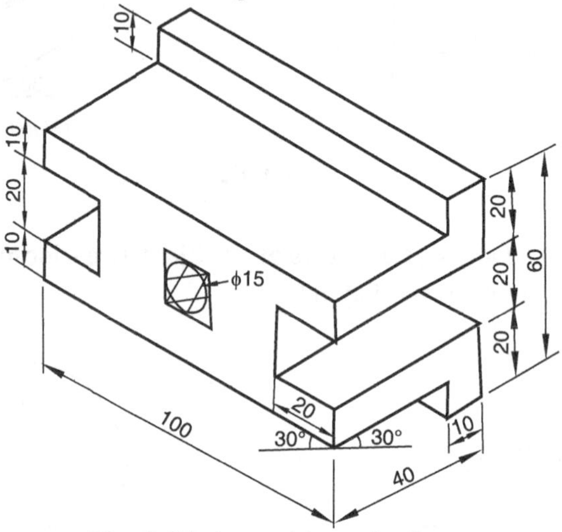

Fig. 6.84. *Isometric projection*

Problem 35 : Fig. 6.85 shows two orthographic projections of a square headed bolt 20 mm diameter. Draw the isometric view of the square headed bolt.

Solution : For its isometric view, see Fig. 6.86.

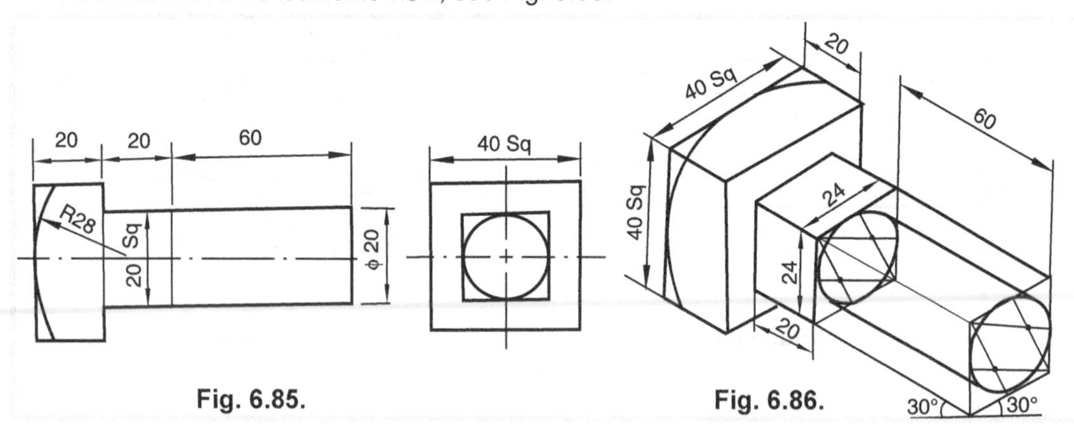

Fig. 6.85. **Fig. 6.86.**

Problem 36 : Fig. 6.87 shows three orthographic projections of an object. Draw the isometric projection of the object.

Solution : For its solution see Fig. 6.88.

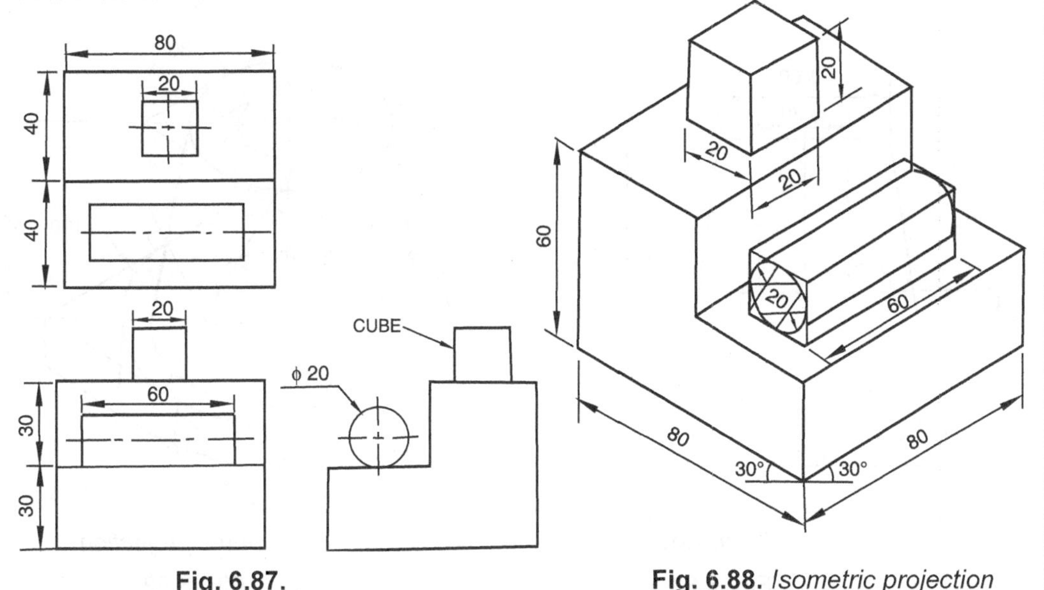

Fig. 6.87.

Fig. 6.88. *Isometric projection*

Problem 37 : The front view of an object is shown in Fig. 6.89. Draw the isometric projection of the object.

Solution : For its isometric projection, see Fig. 6.90.

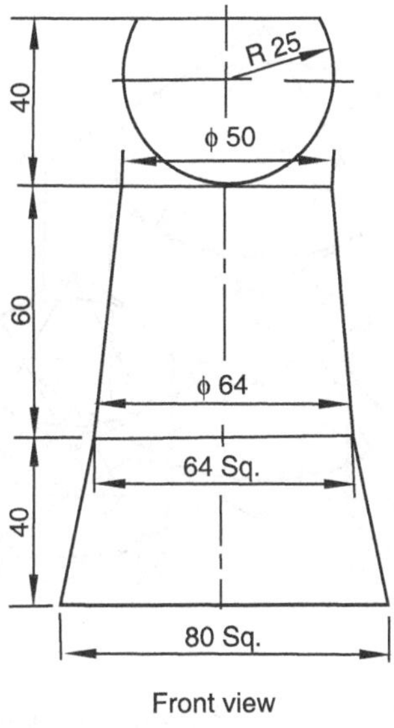

Front view

Fig. 6.89.

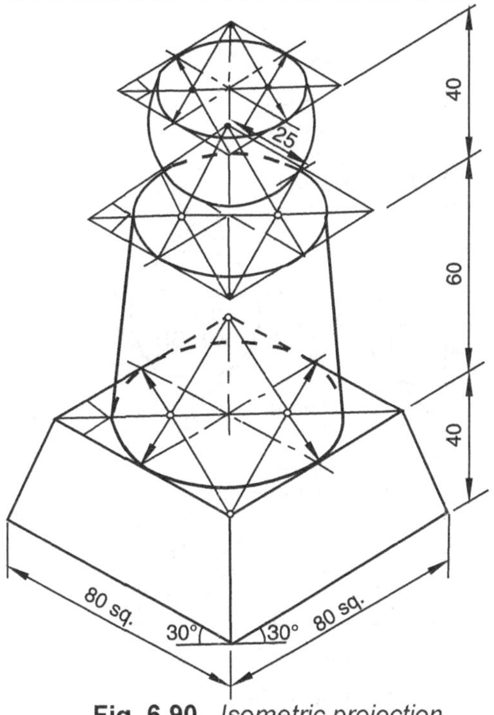

Fig. 6.90. *Isometric projection*

Problem 38 : Fig. 6.91 shows the front view of a cone bearing. Draw the isometric view of the bearing.

Solution : For its isometric projection, see Fig. 6.92.

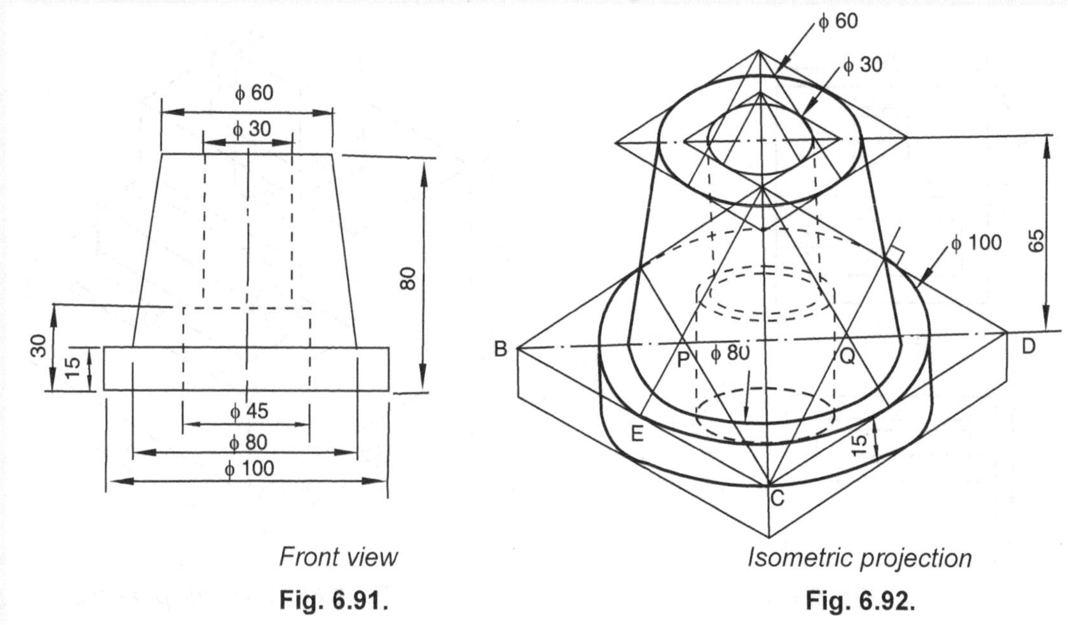

Front view

Fig. 6.91.

Isometric projection

Fig. 6.92.

Problem 39 : Fig. 6.93 shows the two views of three cylinders. Draw the isometric projection of the given cylinders.

Solution : For its isometric projection, see Fig. 6.94.

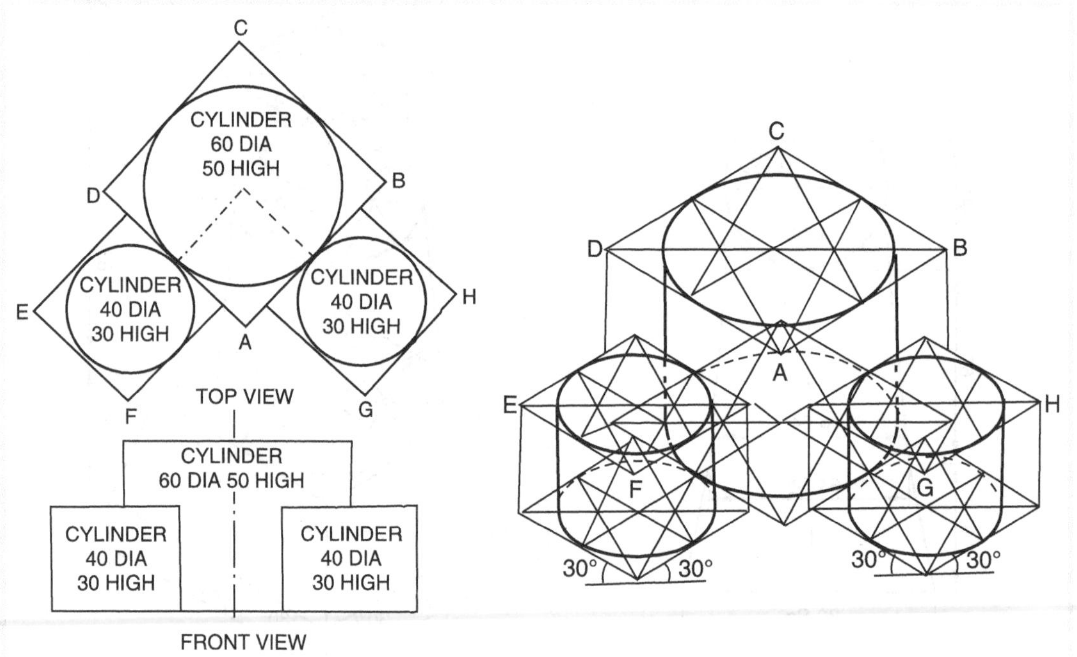

Fig. 6.93.

Fig. 6.94. *Isometric projection*

Problem 40 : Fig. 6.95 shows two views of an object. Draw the isometric projection of the given object.

Solution : For its solution, see Fig. 6.96.

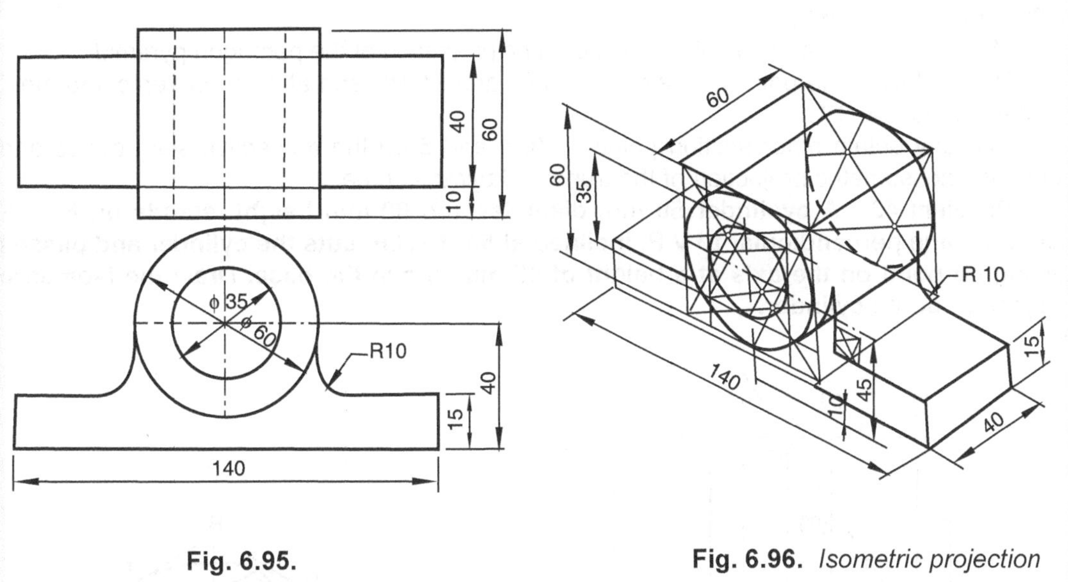

Fig. 6.95. **Fig. 6.96.** *Isometric projection*

Problem 41 : A pentagonal pyramid, 30 mm edge of base and 65 mm height, strands on H.P. such that an edge of base is parallel to V.P. and the nearer to it. A section plane perpendicular to V.P. and inclined at 30° to H.P. cuts the pyramid passing through a point on the axis at a height of 35 mm from the base. Draw the isometric projection of the truncated pyramid. [Fig. 6.97]

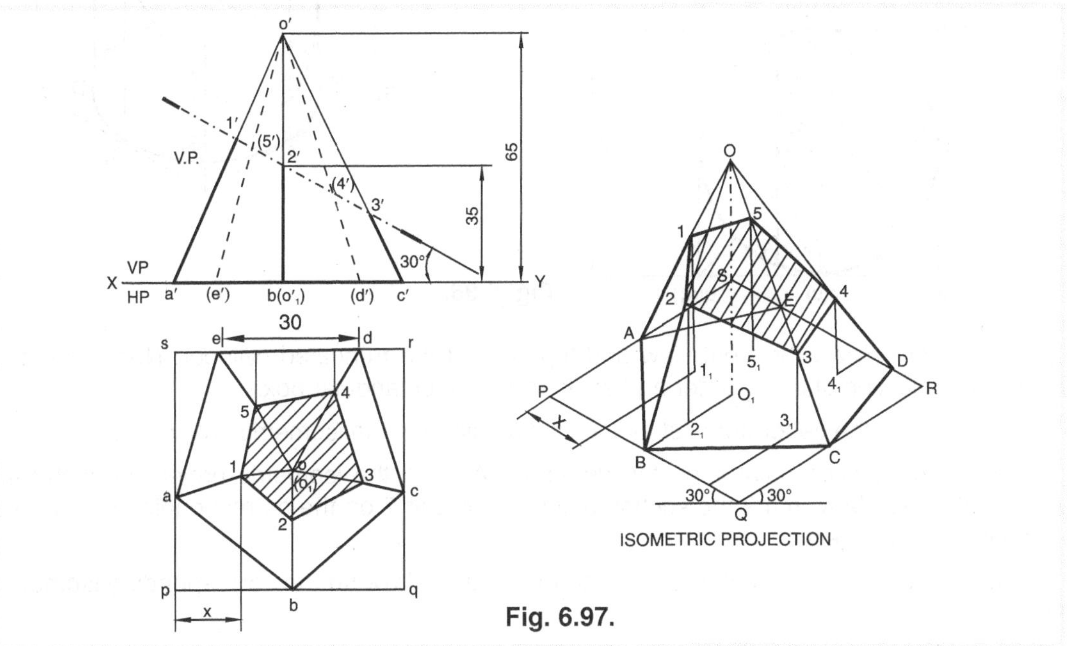

Fig. 6.97.

Solution : Draw the front view and top view of the truncated pyramid as per given condition.

Enclose the pentagon in top view as rectangle pqrs (see Fig. 6.97 (i)]

Draw the isometric projection of the pentagonal base pqrs in the parallelogram PQRS at 30° both sides (see Fig (ii)].

Mark the apex O and complete the isometric projection of the pentagon pyramid.

Now to locate the point 1 obtain 1_1 in the plane PQRS and at 1_1 erect vertical to meet the slant edge OA at 1.

Similarly obtain other section points 2, 3, 4 and 5 on the respective slant edges and complete the isometric projection of the truncated pyramid (see

Problem 42 : A cyclinder 50 mm diameter and 60 mm height, stands on H.P. A section plane perpendicular to V.P. inclined at 55° to H.P. cuts the cylinder and passes through a point on the axis at a height of 45 mm above the base. Draw the isometric projection of the cylinder.

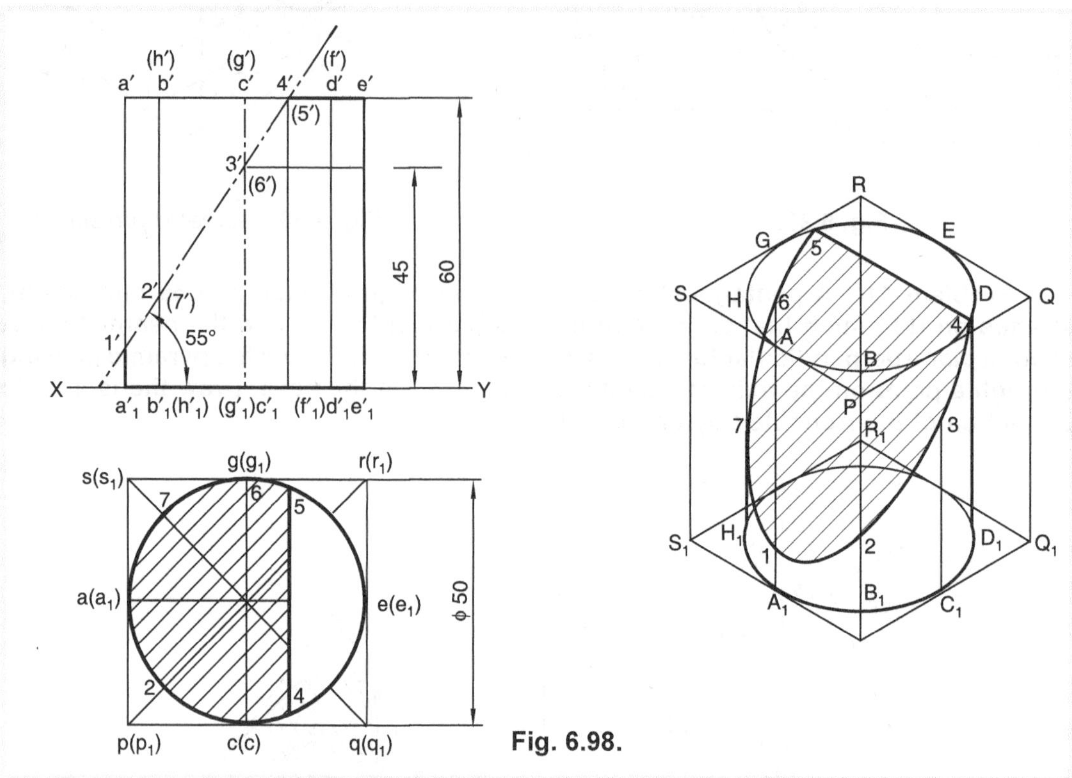

Fig. 6.98.

Solution : Draw the front view and top view of the truncated cylinder. Now, draw the isometric projection of the cylinder and enclose inside rectangular box.

Now, show the generators cut by the section plane in the isometric projection.

Mark the section point 1 on the generator A_1 A in the isometric projection such that A_1 1 = a'_1 1, similarly, mark the section point 2, 3, 6 and 7 on the corresponding generators in the isometric projection.

Mark the section point 4 and 5 on the top face by drawing the corresponding isometric line.

Join the points 4, 3, 2, 1, 7, 6, 5 by a smooth curve and complete the isometric projection of truncated cylinder.

Problem 43 : A hexagonal prism, side of base 25 mm and axis 50 mm long rests on H.P. and one of the edges of its base is parallel to V.P. A section plane perpendicular to V.P and inclued at 50° to H.P. bisect the axis of the prism. Draw the 150 metric projection of the truncated prism.

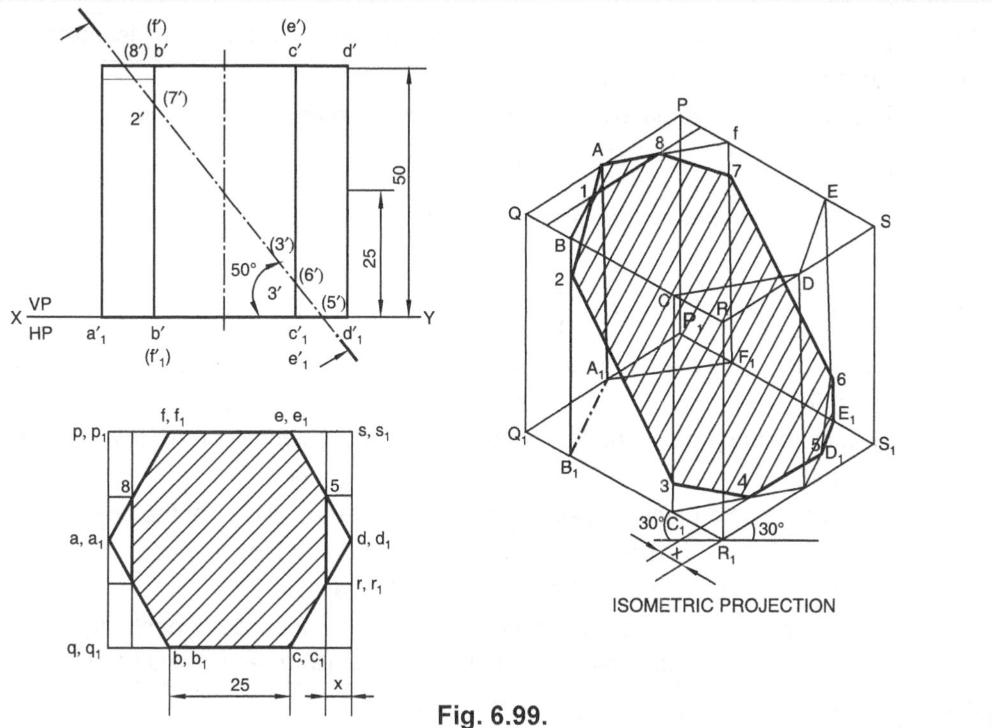

Fig. 6.99.

Solution : Draw the front view and top view of the truncated hexagonal prism. Enclose the top view of hexagon in a rectangle pqrs. Draw the isometric projection of unit hexagonal base and top face in parallelogram PQRS and complete the hexagonal prism unit. Mark the section point 3 and 6 on the vertical edges C_1C and E_1 E such that C_1 3 = c'_1 3′ and E_1 6 = e'_1 6′, in isometric projection.

Similarly mark section points 2 and 7 on vertical edges 13 B_1 and EF_1 respectively.

Locate the section points 4 and 5 on the respective edges of the base by drawing the corresponding isometric line.

Mark the section points 1 and 8 on the respective edges of the top face.

Join 1, 2, 3, ... 8 in the correct sequence and obtain the cut surface.

Join the visible edges of the remaining portion of truncated prism as thick lines.

QUESTIONS FOR SELF EXAMINATION

1. What do you mean by isometric projections ?
2. Define isometric axes, isometric line and isometric planes.
3. Define isometric scale. How is it constructed ?
4. Give the various positions of isometric axes.
5. Fill in the blanks :—
   (*a*) The isometric view is in the ratio of .... of the true length.
   (*b*) The isometric length is about .... of the true length.

(*c*) In isometric projection, dimension lines are drawn parallel to ....

(*d*) A circle in isometric projection appears as ....

(*e*) Isometric projection of a sphere is a circle having a diameter .... to that of sphere.

(*f*) The three forms of axonometric projection are dimetric, trimetric and .... projection

Ans. (*a*) $\sqrt{2}:\sqrt{3}$ *3* (*b*) 82% (*c*) isometric axes (*d*) ellipse (*e*) equal (*f*) isometric.

6. Choose the correct answer :—

(*i*) Compared to actual diameter, isometric diameter of a sphere is

(*a*) *greater* (*b*) *smaller* (*c*) *equal.*

(*ii*) The ratio between the isometric and true length is

(*a*) $\frac{2}{\sqrt{3}}$ (*b*) $\frac{\sqrt{2}}{3}$ (*c*) $\frac{\sqrt{2}}{\sqrt{3}}$

(*iii*) The angle between isometric axes is

(*a*) 90° (*b*) 120° (*c*) 60°

(*iv*) In isometric projection, the receding lines are drawn with the horizontal at

(*a*) 45° (*b*) 30° (*c*) 60°

(*v*) In isometric projection, circles are represented as ellipse of the object in

(*a*) *all faces* (*b*) *only the front face* (*c*) *only the top face.*

(*vi*) The receding lines in an isometric projection are drawn

(*a*) *parallel* (*b*) *perpendicular* (*c*) *inclined to each other.*

(*vii*) An isometric projection is pictorial drawing of (*a*) *three* (*b*) *two* (*c*) *one view*

Ans. (*i*) (*c*) (*ii*) (*c*) (*iii*) (*b*) (*iv*) (*b*) (*v*) (*a*) (*vi*) (*a*) (*vii*) (*c*)

7. State whether the following statements are true or false :

(*a*) In isometric projection, various lines are represented by their true lengths.

(*b*) In isometric view, the three principal axes of the object will be equally foreshortened.

(*c*) A circular arc will appear as a part of a parabola in an isometric view of sphere.

(*d*) In an isometric view, vertical lines on the object appear vertical in the drawing.

(*e*) Isometric scale must be used while drawing isometric view of sphere.

(*f*) The isometric length is about 0.816 of the true length.

Ans. (*a*) False (*b*) True (*c*) True (*d*) True (*e*) False (*f*) True.

PROBLEMS FOR PRACTICE

1. A cube of 36 mm side rests centrally on a square block of 56 mm edges and 10 mm thick. Draw the isometric projection of the two objects with the edges of the two blocks mutually parallel to each other.

2. A cube of 30 mm sides rests on the top of a cylindrical slab of 60 mm diameter and 25 mm thick. The axes of the solids are in same straight line. Draw an isometric projection of the solid.

3. Draw the isometric projection of the cylindrical axes of various diameters as given in Fig. 6.100 by assuming the suitable lengths.

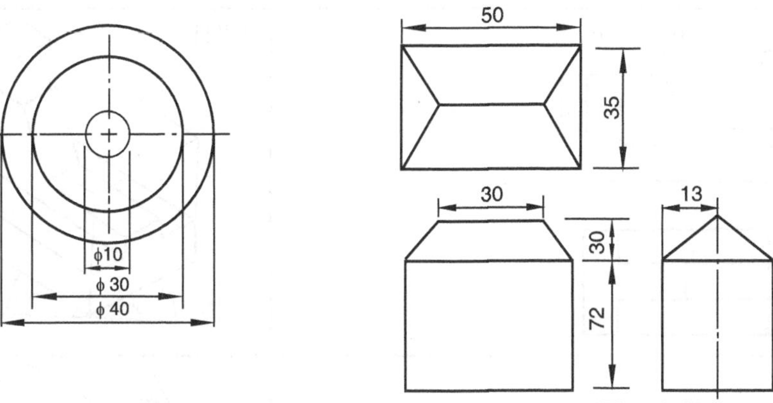

Fig. 6.100. Fig. 6.101.

4. Draw the isometric projection of a solid whose three orthographic projections are given in Fig. 6.101.

5. Fig. 6.102 show the two views of an object. Draw the isometric views.

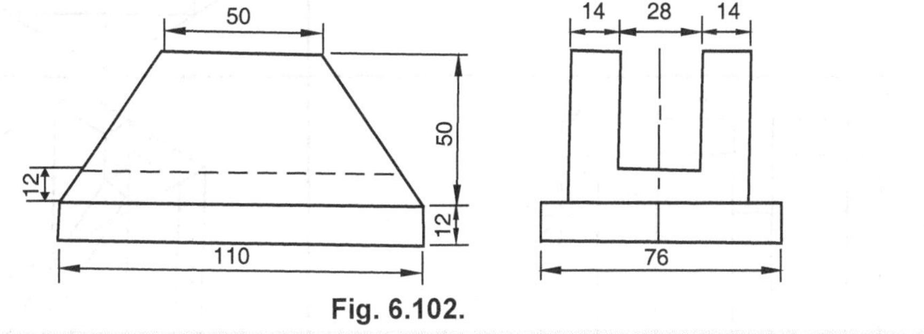

Fig. 6.102.

6. Draw the isometric projection of the views, shown in Figs. 6.103 and 6.104

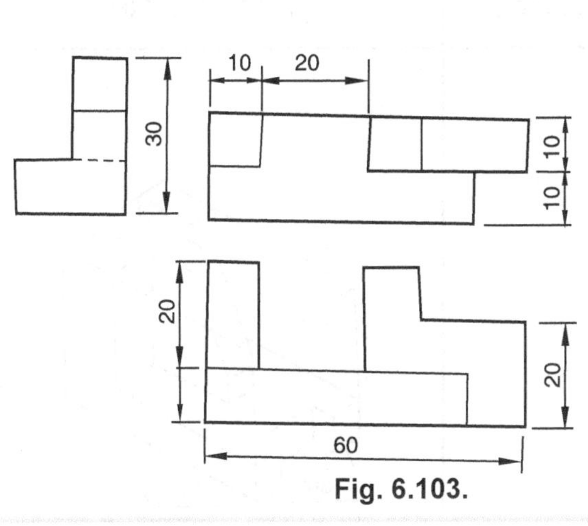

Fig. 6.103.

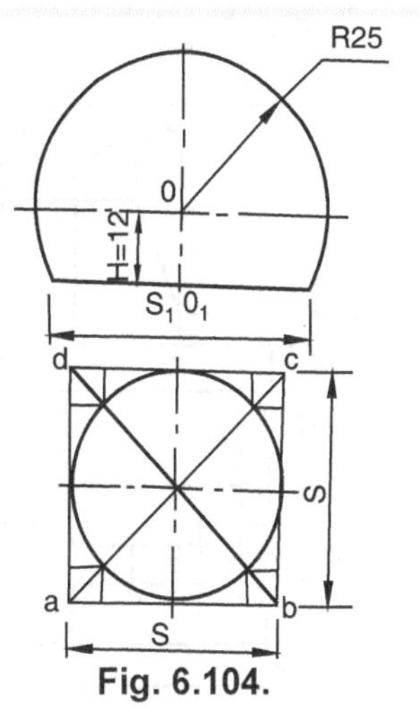

Fig. 6.104.

7. Two orthographic views of the objects are given. Draw the isometric views with complete dimensions.

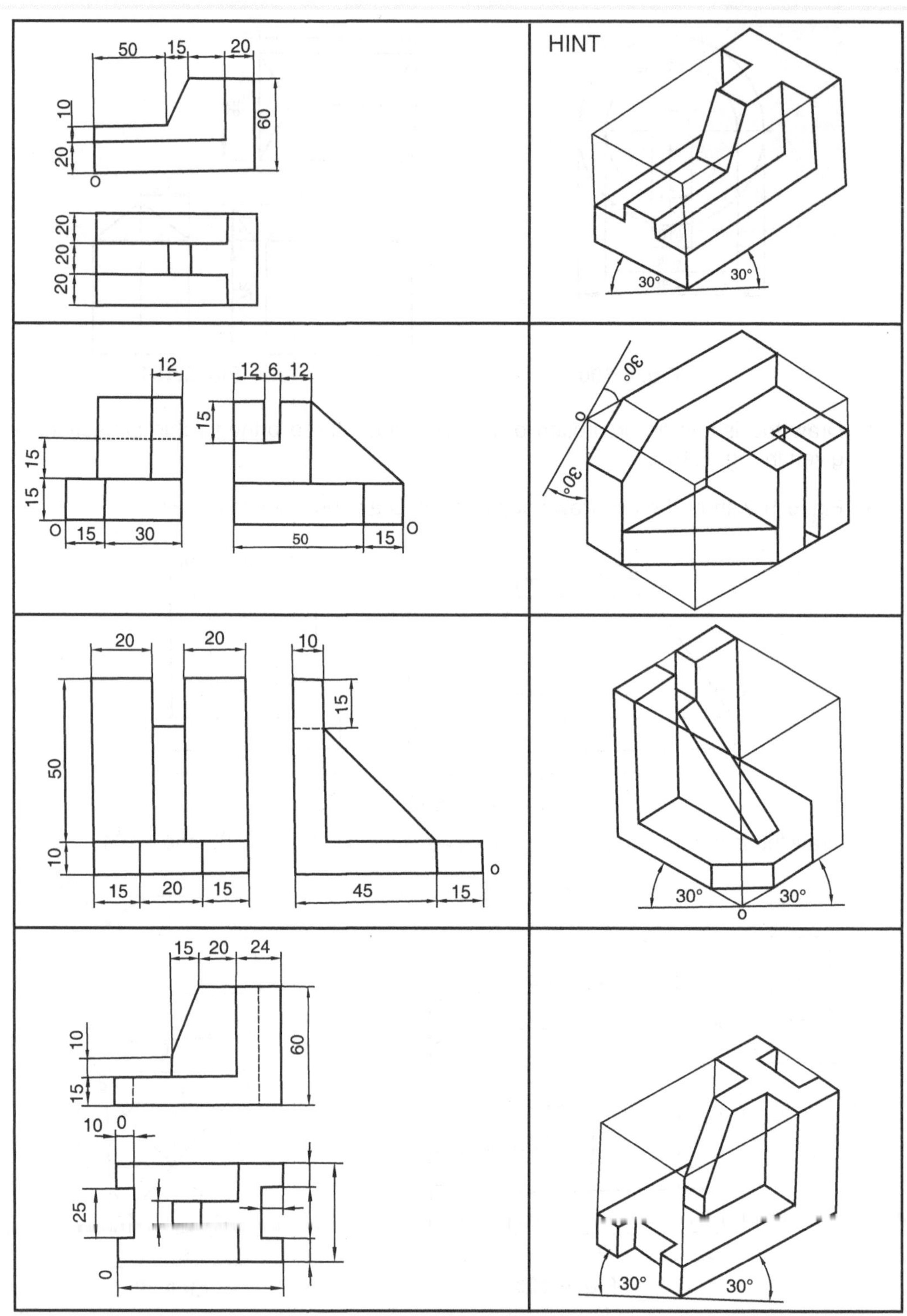

Fig. 6.105.

8. Draw the front view, side view and top view of a the objects shown in Fig. 6.106. in first angle projection to scale 1 : 1.

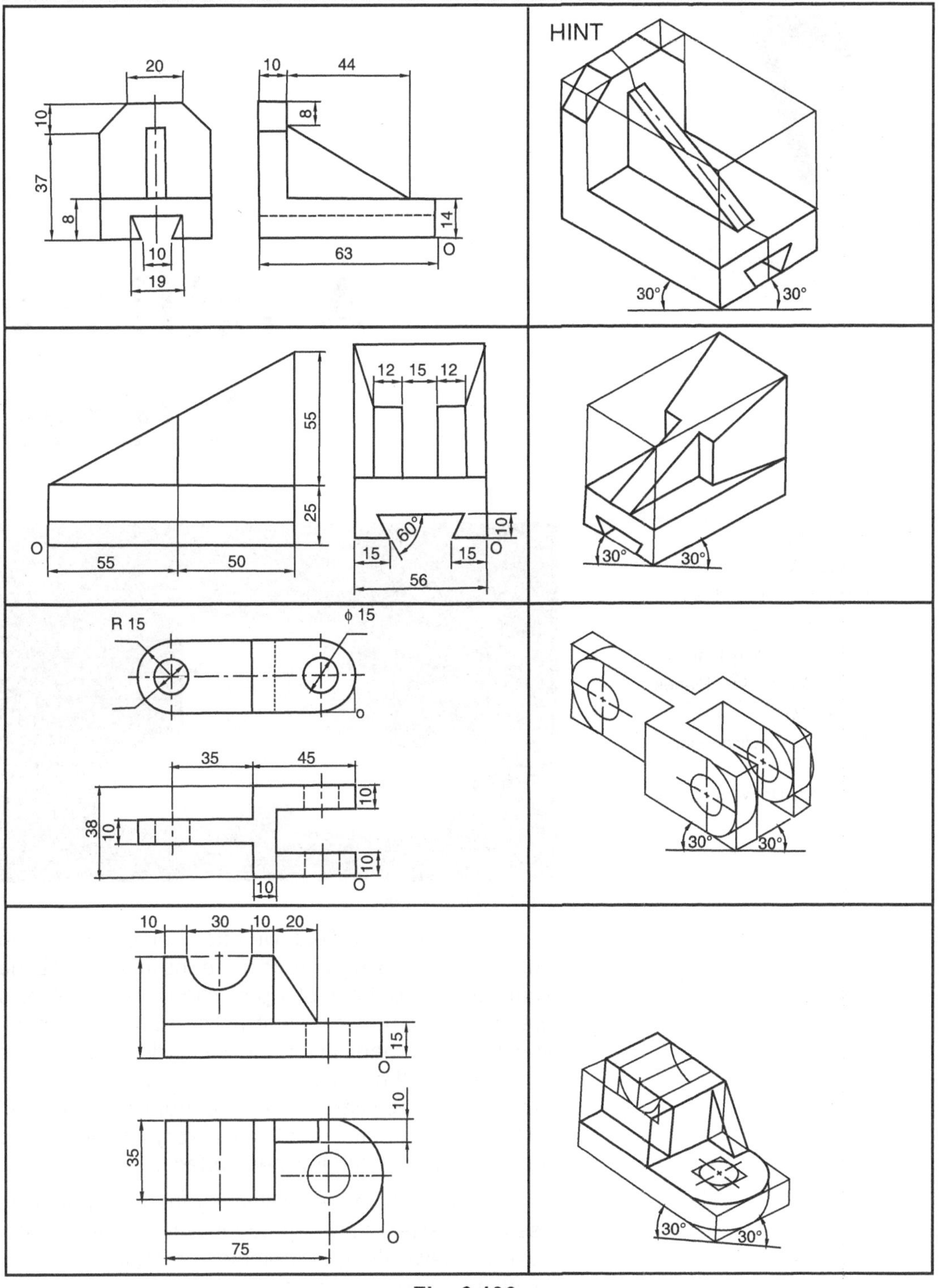

Fig. 6.106.

Auxiliary Views

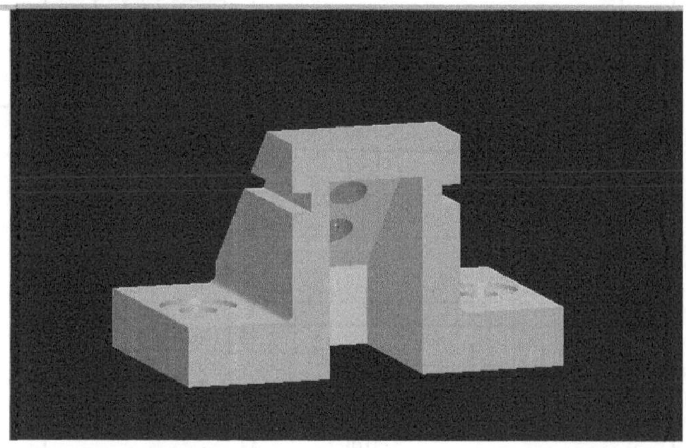

INTRODUCTION

In engineering field, we often come across with such objects having one or more inclined or oblique surfaces. If the orthographic projections of these objects are drawn, the projections of inclined surfaces, do not show the true shape in any view. In order to obtain the true shape of these inclined surfaces, the projections are drawn on auxiliary plane. The projection obtained on the auxiliary plane is called auxiliary view and gives the true shape of the inclined surface.

In this chapter, we will study the auxiliary view, uses of auxiliary views, procedure for projecting the auxiliary views along with simple to complex problems used in engineering practice.

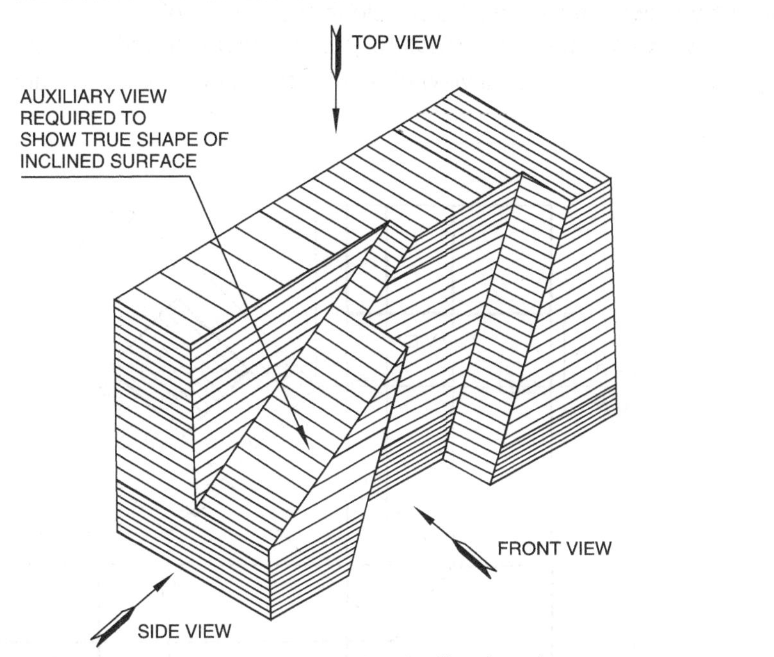

Fig. 7.1. *Pictorial view showing inclined surfaces*

7.1. AUXILIARY VIEW

The view obtained on the auxiliary plane which is parallel to the inclined surface of an object is called auxiliary view.

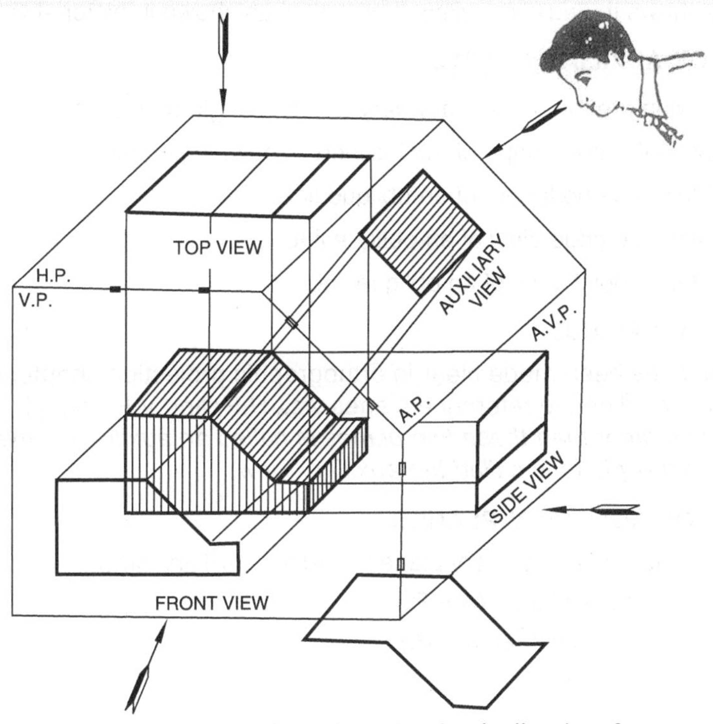

Fig. 7.2. *Pictorial view of an object having inclined surface*

Auxiliary views play an important role in the design of objects having inclined surfaces. The auxiliary views can be *primary auxiliary view* and *secondary auxiliary view.*

Fig. 7.2 shows the pictorial view of an object having inclined surface for auxiliary view.

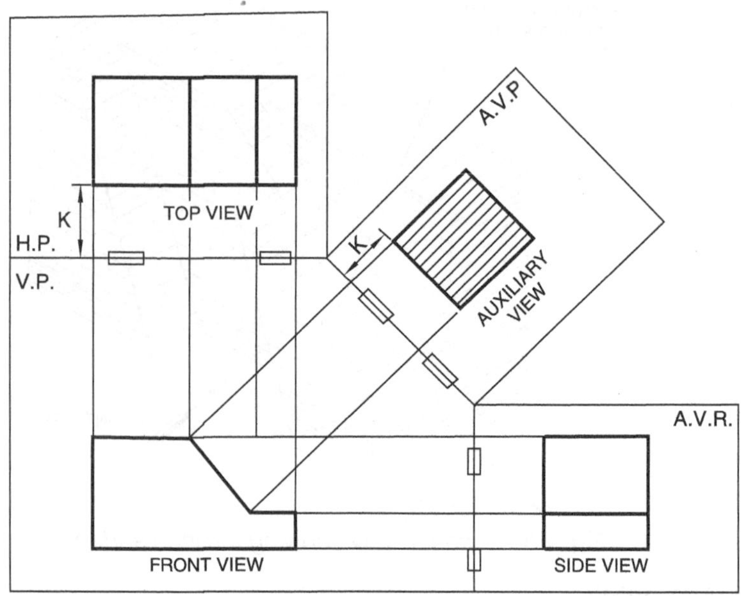

Fig. 7.3. *Auxiliary view of an object*

Fig. 7.3 shows the auxiliary view of an object as shown in pictorial view.

7.2. USES OF AUXILIARY VIEWS

The following are the uses of auxiliary views in engineering practice :—

1. To obtain the true shape of the inclined or oblique surface.
2. To obtain the true length of an oblique line.
3. To obtain the edge view of an oblique line.
4. To obtain a point view of an oblique line.

7.3. AUXILIARY PLANE

It has already been made clear in orthographic projection chapter that two views can always be projected on the reference planes viz., H.P. and V.P. that give top view and front view. *Any plane other than these two planes is called an auxiliary plane and the view thus obtained upon this plane is called the auxiliary view.*

7.4. TYPES OF AUXILIARY PLANES

The following are the auxiliary planes used for auxiliary views :—

1. Auxiliary vertical planes (A.V.P.)
2. Auxiliary inclined planes (A.I.P.)

1. Auxiliary vertical planes (A.V.P.) : *The planes inclined to the V.P. and perpendicular to the H.P. are called auxiliary vertical planes. The projection on these planes gives an auxiliary front view.*

2. Auxiliary inclined planes (A.I.P.) : *The planes inclined to H.P. and perpendicular to V.P. are called auxiliary inclined planes. The projection on these planes will give an auxiliary top view.*

7.5. METHOD TO DRAW AUXILIARY VIEW

Fig. 7.4 shows the front and top views alongwith pictorial view of an object having an inclined surface whose auxiliary view is to be drawn.

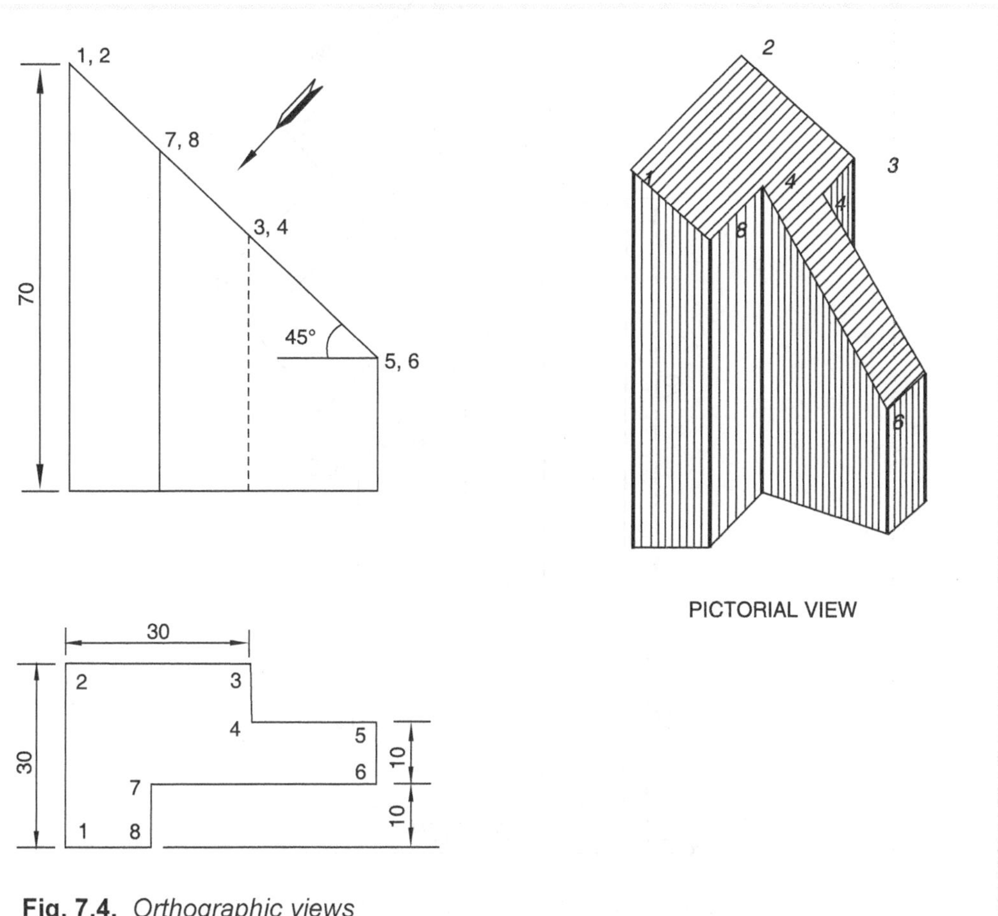

Fig. 7.4. *Orthographic views*

The following steps should be followed for projecting the auxiliary view of inclined surface of an object.

1. Draw the two given views of the object such as front and top views and show the X-Y line below top view (see Fig. 7.5).

2. Draw the auxiliary plane line X-Y parallel to the inclined surface of front view. The X-Y line should be at such a convenient distance from the view so that auxiliary view falls in a clear space on the drawing.

3. Draw projections perpendicular to X-Y line from the various points of the inclined surface.

4. With divider or compass mark the point on the projections by keeping the heights of the points above X-Y line equal to the height of their respective points from X-Y line, like K distance.

5. Complete the required auxiliary view by joining various points in correct sequence (see Fig. 7.5).

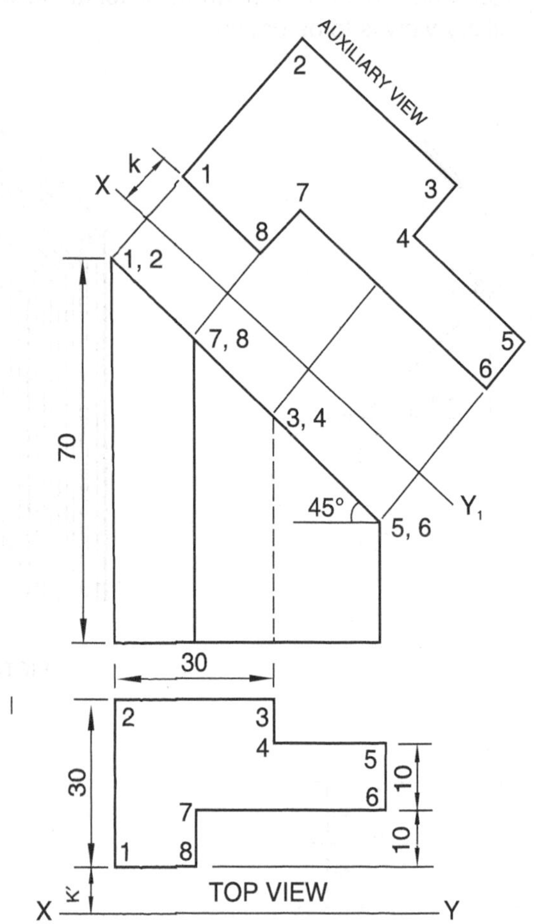

Auxiliary view of the inclined surface of an object
Fig. 7.5.

Problem 1 : Fig. 7.6 shows the front view and top view of an object. Draw to scale 1:1, the given views and add the auxiliary view in the direction of arrow. [Fig. 7.7]

Solution : For its solution, see Fig. 7.7 which is itself a self explanatory sketch.

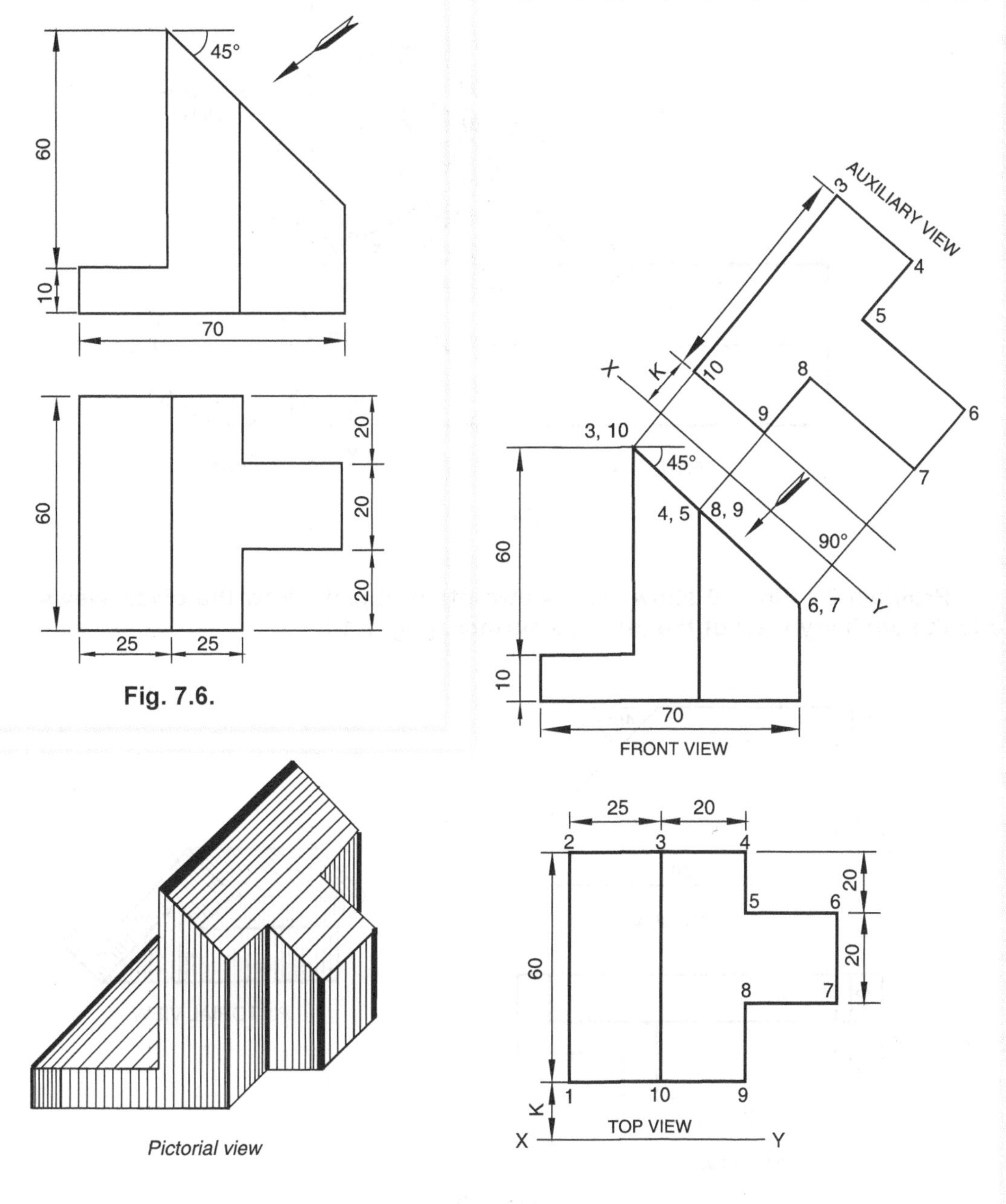

Fig. 7.6.

Pictorial view

Fig. 7.7.

Problem 2 : Fig. 7.8 shows two views of the cut cylinder. Draw the given views and add the auxiliary view of the inclined surface. [Fig. 7.8]

Solution : For its solution, see Fig. 7.8.

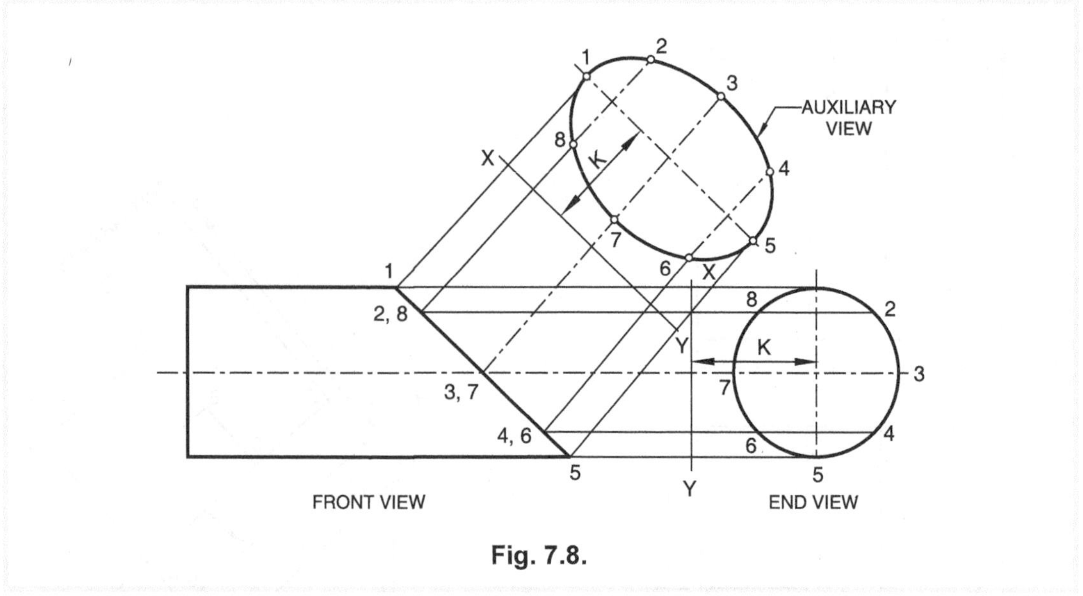

Fig. 7.8.

Problem 3 : Fig. 7.9 shows two views of an object. Draw the given views and add the auxiliary view of the inclined surface. [Fig. 7.10]

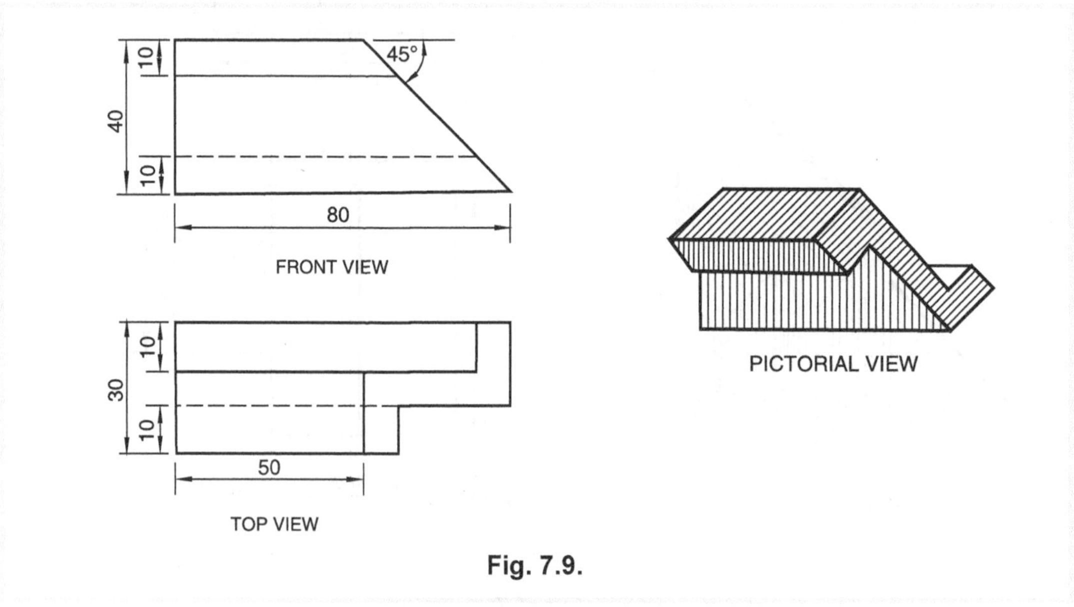

Fig. 7.9.

Solution : For its solution, see Fig. 7.10 which is itself a self explanatory sketch.

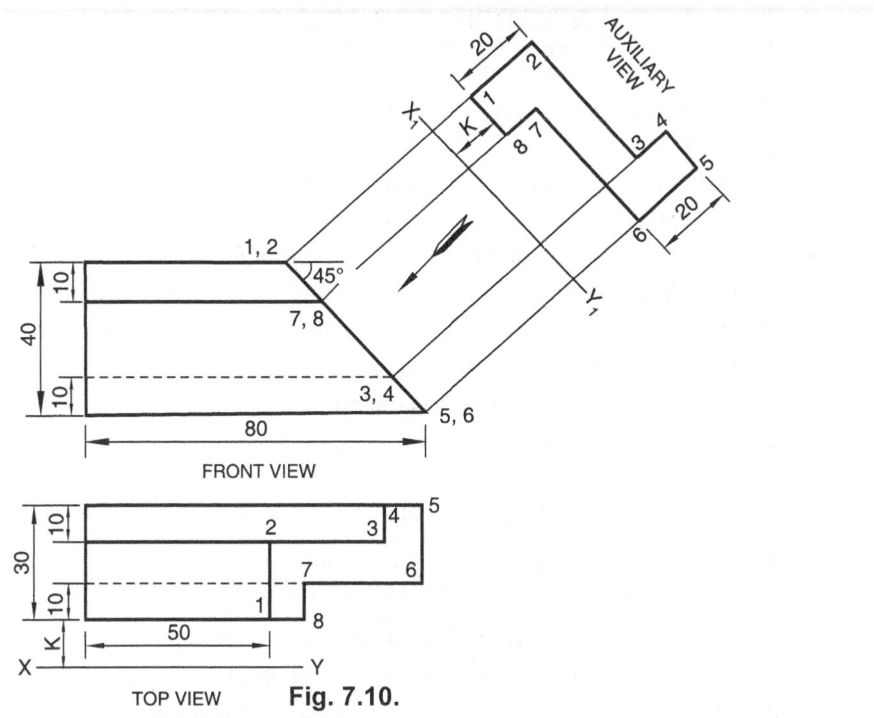

Fig. 7.10.

Problem 4 : Fig. 7.11 shows front and top views of an object with inclined surface. Draw
(a) Primary auxiliary view showing true shape of the inclined surface
(b) Secondary auxiliary view

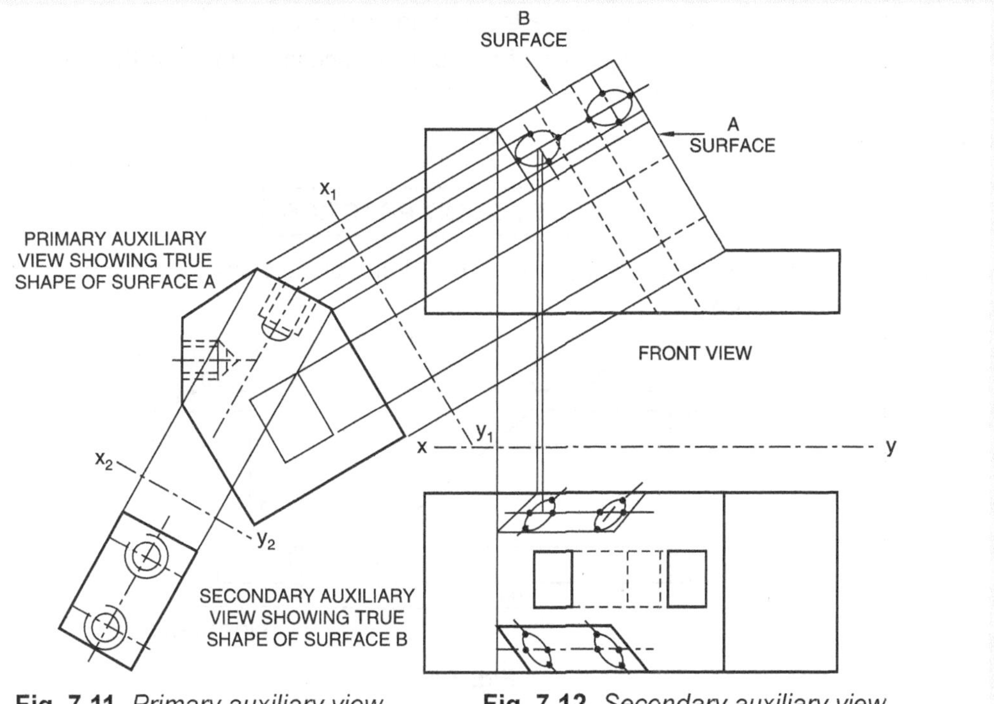

Fig. 7.11. *Primary auxiliary view* **Fig. 7.12.** *Secondary auxiliary view*

Solution : For its solution see Fig. 7.12.

QUESTIONS FOR SELF EXAMINATION

1. What do you mean by auxiliary view?
2. What is the necessity of auxiliary view?
3. Fill in the blanks :
   (*a*) The view obtained on the auxiliary plane which is .... to the inclined surface of an object is called auxiliary view.
   (*b*) The auxiliary view gives the .... shape of the oblique surface.
   (*c*) The planes which are .... to V.P. but .... to H.P. are called A.V.P.
   (*d*) The planes which are .... to V.P. but .... to V.P. are called A.I.P.
   (*e*) The projection on an A.V.P. is called ....
   (*f*) The projection on an A.I.P. is called ....
   (*g*) When the auxiliary plane is perpendicular to a principle of the projection, the view obtained is called .... auxiliary view.
   (*h*) Auxiliary view can be .... auxiliary view and .... auxiliary view.

Ans. (*a*) Parallel (*b*) true (*c*) inclined, perpendicular (*d*) inclined, perpendicular (*e*) auxiliary front view (*f*) auxiliary top view (*g*) primary (*h*) primary, secondary.

4. State whether the following statements are true or false:
   (*a*) Projection on the auxiliary plane reveals the true shape of the inclined surface.
   (*b*) To get a front auxiliary view, the auxiliary plane is parallel to V.P. and perpendicular to H.P.
   (*c*) To get the top auxiliary plane, the auxiliary plane is inclined to H.P. but perpendicular to V.P.
   (*d*) A side auxiliary view is projected from the side view
   (*e*) To find the point projection of a line, it is necessary to have the auxiliary plane must be set up perpendicular to the line.

Ans. (*a*) True, (*b*) False, (*c*) True, (*d*) True, (*e*) True.

PROBLEMS FOR PRACTICE

1. Fig. 7.13 shows front view and side view of an object. Copy the given views and add the auxiliary view.

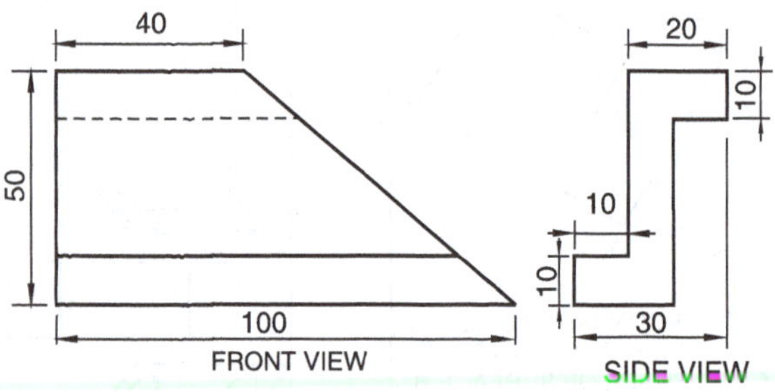

Fig. 7.13.

2. Fig. 7.14 shows front and top views of an object. Copy the given views and add the auxiliary view in the direction of arrow.

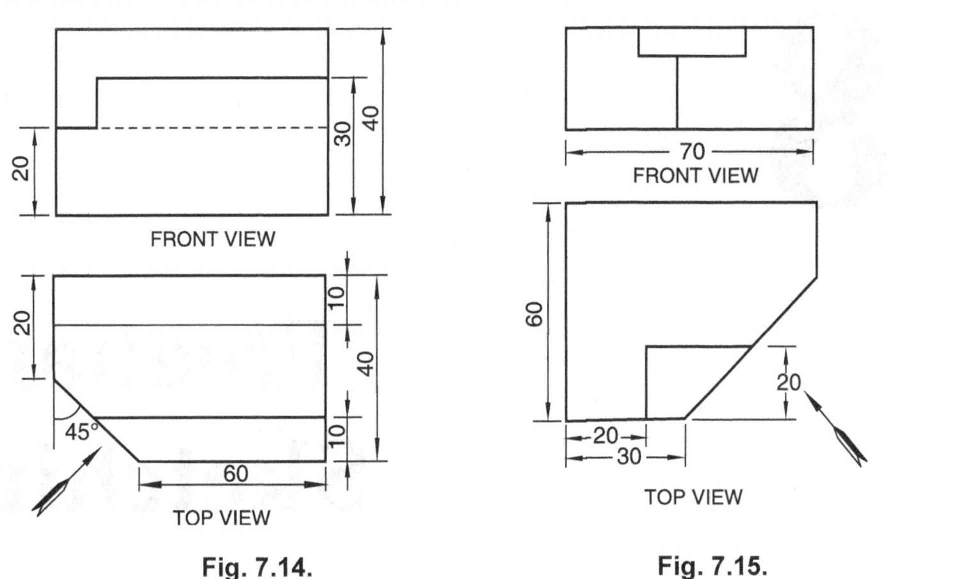

Fig. 7.14.

Fig. 7.15.

3. Fig. 7.15 shows two views of an object. Draw the given views and add the auxiliary view in the direction of arrow.

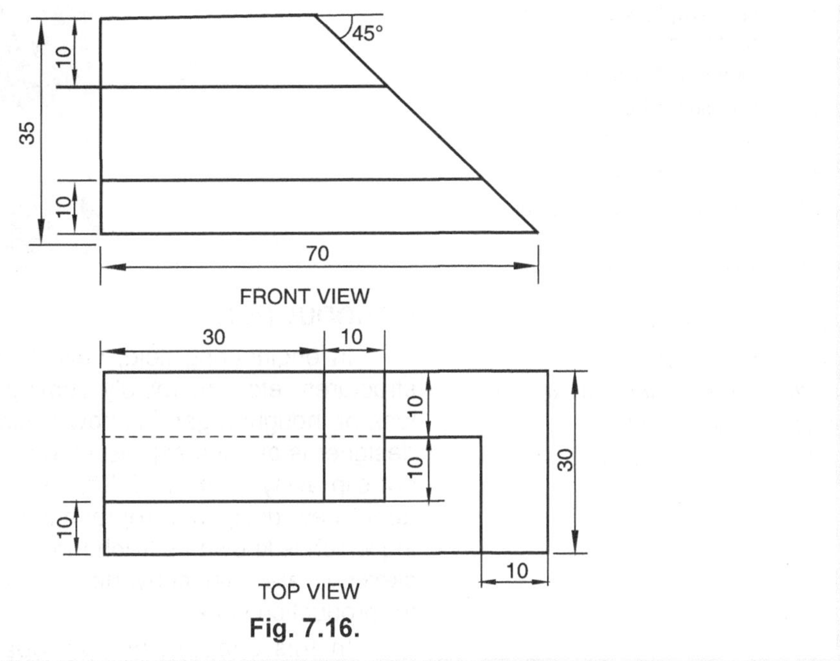

Fig. 7.16.

4. Fig. 7.16 shows the two views of an object. Draw the given views and add the auxiliary view of inclined surface.

CHAPTER

8

Freehand Sketching

INTRODUCTION

In engineering field, new designs of machines or structures, etc. are rapidly coming up day-to-day. The idea or thought regarding new design by the engineer or designer is at once expressed on a paper so that it may not slip away from mind. This technical idea or thought about new designs is represented in the form of drawing which is known as freehand sketching. The freehand sketches are later converted into actual scale drawing for production work.

In this chapter, we will deal with the study of freehand sketching, uses of sketching, sketching materials and various methods of sketching different shapes of objects.

8.1. FREEHAND SKETCHING

The drawing prepared by a pencil without the use of drawing instruments is known as freehand sketching or simply sketching.

8.2. USES FREEHAND OF SKETCHING IN ENGINEERING

In engineering practice, the freehand sketches are used for :—

1. formulating, expressing and recording new ideas in the technical drawing and project,
2. the production of temporary fixtures,
3. designing and working up new fixtures, and
4. showing the different arrangements for making actual drawings of various objects.

8.3. IMPORTANT POINTS FOR GOOD SKETCHING

The following important points should be kept in view for getting good sketching :—

1. Good practice
2. Proper patience
3. Proper proportions
4. Proper proficiency.

8.4. SKETCHING MATERIALS

The following materials are required for making good sketching :—

1. Paper
2. Pencil
3. Rubber.

The paper in the form of sketch book should be used frequently for various sketches in the workshop or field. But in drawing class, plain or graph paper may be used for making freehand sketches. However, it is recommended by I.S.I. that the engineer should use graph, as it is convenient to draw sketch in correct proportions on it.

A soft pencil preferably H.B. or F sharpened to conical point should be used for good sketching. The grade of pencil depends upon the quality of paper to be used. *In general, H.B. pencils are used on hard surface paper whereas F pencils for soft surface paper.*

A good quality of soft rubber is used for erasing the extra lines drawn in the sketch work.

8.5. HOLDING THE PENCIL

The pencil should be held by the thumb and first finger in such a way that it should rest on the second finger at a distance of about 30 mm from the pencil point.

8.6. SKETCHING STRAIGHT LINES

Straight lines are drawn by making a succession of short distances or points or short straight lines. The gap is then filled up from one to another dashes or points or lines and at the same time the direction of the lines are corrected (see Fig. 8.1).

Fig. 8.1. *Sketching straight lines*

Notes : (*i*) *The whole line should not be drawn in a single stroke.*
(*ii*) *Always keep your eye in the direction of drawing the line but not on the pencil point.*

8.6.1. Sketching Horizontal Lines : The horizontal lines are drawn with the motion of wrist and fore arm, left to right with series of overlapping strokes (see Fig. 8.2).

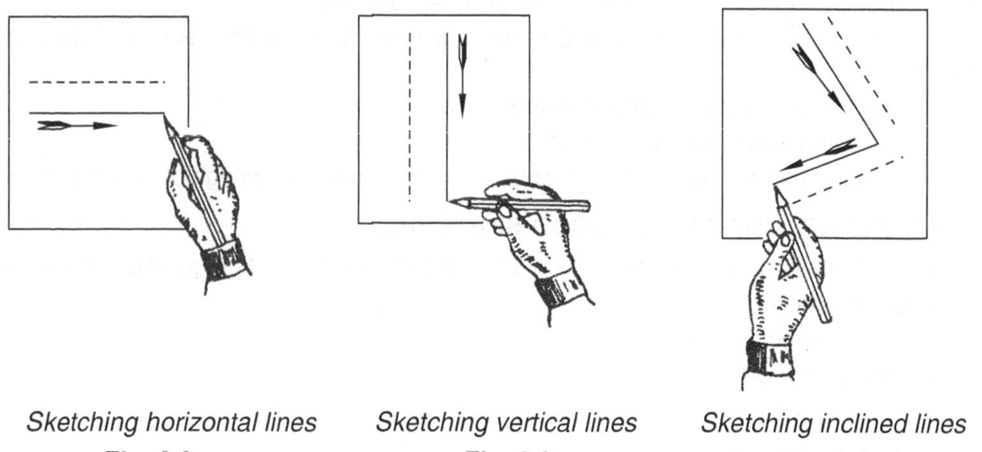

Sketching horizontal lines **Fig. 8.2.** *Sketching vertical lines* **Fig. 8.3.** *Sketching inclined lines* **Fig. 8.4.**

8.6.2. Sketching Vertical Lines : The vertical lines are always drawn downward, i.e., from top to bottom with the movement of finger and wrist (see Fig. 8.3).

8.6.3. Sketching Inclined Lines : The inclined lines are drawn from left to right when they are nearly horizontal and should be drawn downward when they are nearly vertical (see Fig. 8.4).

Fig. 8.5 shows the steps for sketching plane figure on the drawing paper.

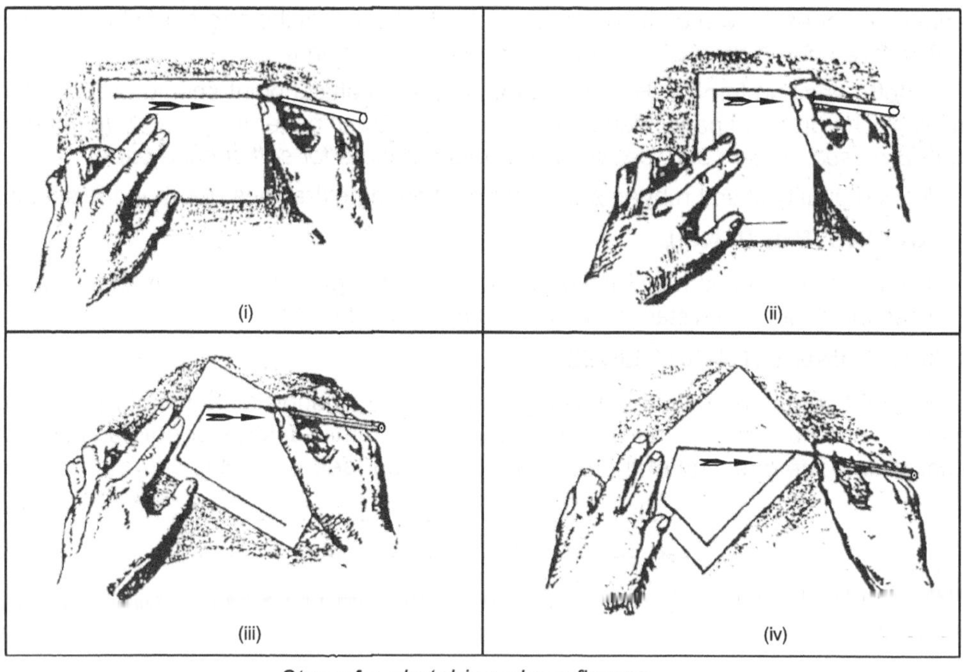

Steps for sketching plane figures
Fig. 8.5.

8.7. SKETCHING CIRCLES

There are different methods for sketching circles, depending upon the convenience and practice of the sketches.

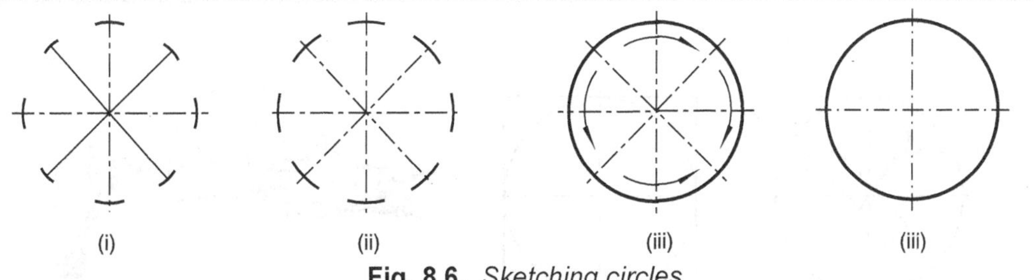

Fig. 8.6. *Sketching circles*

In general, small circles are sketched very easily in one or two strokes, but the large circles are sketched in a number of strokes. The following methods are generally adopted for sketching the circles.

8.8. METHODS FOR DRAWING SMALL CIRCLES

First method :

First of all, draw two centre lines about right angle to each other and add two diagonals in addition to the centre lines [see Fig. 8.6 (i)].

Now take the radius of the circle, make arcs from the centres [see Fig. 8.6 (ii)] and complete the circle as shown in Fig. 8.6 (iii).

After completing the circles, rub off all the unrequired lines and diagonals [see Fig. 8.6 (iv)].

Second method :

Firstly, draw a square equal to the diameter of the circle and mark the mid-point of the sides [see Fig. 8.7 (i)].

Draw diagonals and mark points at a distance equal to the radius of a circle from the centre [see Fig. 8.7 (ii)].

Now complete the circle by drawing short dashes on the marked points [see Fig. 8.7 (iii)].

After completing the circle, rub all the unrequired lines [see Fig. 8.7 (iv)].

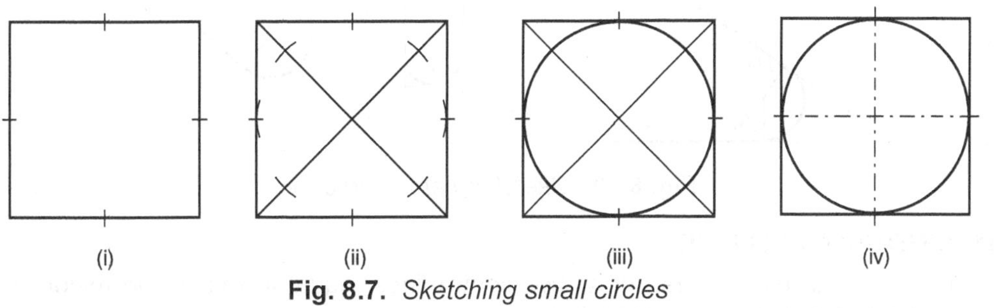

Fig. 8.7. *Sketching small circles*

8.9. METHODS FOR DRAWING LARGE CIRCLES

First method :

For sketching a very large circle, first draw centre lines of the circle. Take piece of paper called trammel and mark the radius of the required circle on it. Fix one end of the trammel by a pin or hand at the centre and take the pencil point on the marked point and complete the circle as shown in Fig. 8.8.

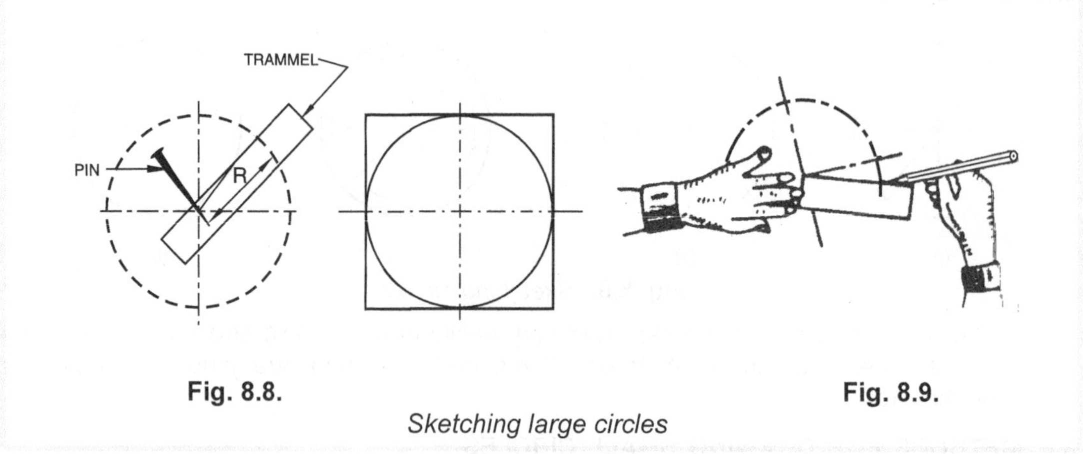

Fig. 8.8. **Fig. 8.9.**

Sketching large circles

Second method :

Fig. 8.9 shows the second method of sketching a circle on the loose paper. It is an efficient quick method of sketching a circle in which, hand is used as a compass. In this, the little finger is used as a point at the centre and the pencil is held stationary at the required radius from the centre. Rotate the paper under the hand and pencil and complete the circle as shown in Fig. 8.9.

8.10. SKETCHING ARCS AND CURVES

Arcs and curves are sketched by the same method as dealt with for sketching circles, Fig. 8.10 shows the different methods of sketching various types of arcs and curves.

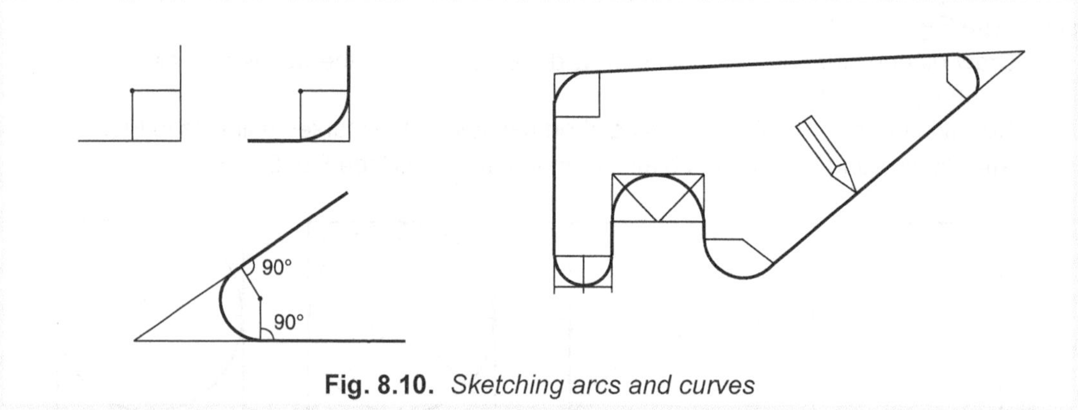

Fig. 8.10. *Sketching arcs and curves*

8.11. SKETCHING ELLIPSE

To sketch an ellipse, first, draw two centre lines, equal to major and minor axis and enclose the rectangle [see Fig. 8.11 (ii)].

(The large diameter of the ellipse is called the major axis and the shorter diameter is called minor axis).

Sketch one-fourth of the ellipse by drawing arc [see Fig. 8.11 (iii)] and finally complete the ellipse as shown in Fig. 8.11 (iii).

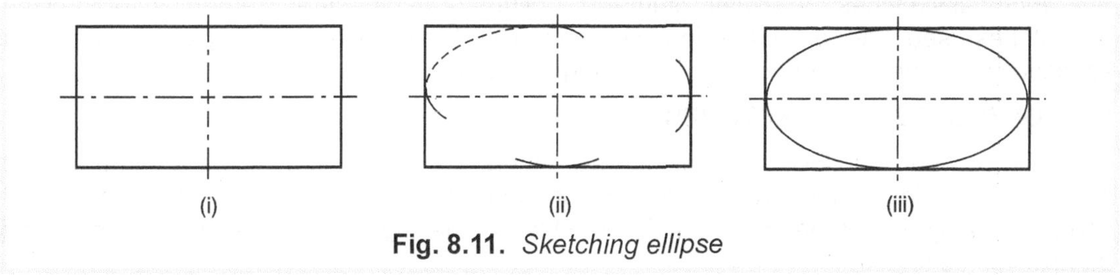

(i) (ii) (iii)

Fig. 8.11. *Sketching ellipse*

8.12. TYPES OF FREEHAND SKETCHES

The freehand sketches may be classified as under depending upon the function for which they are intended :-

(i) Orthographic sketches

(ii) Pictorial sketches.

In the orthographic sketches, all the necessary principel views are drawn and each detail is dimensioned so as to present a complete shape and size description of the object [see Fig. 8.13(iii)].

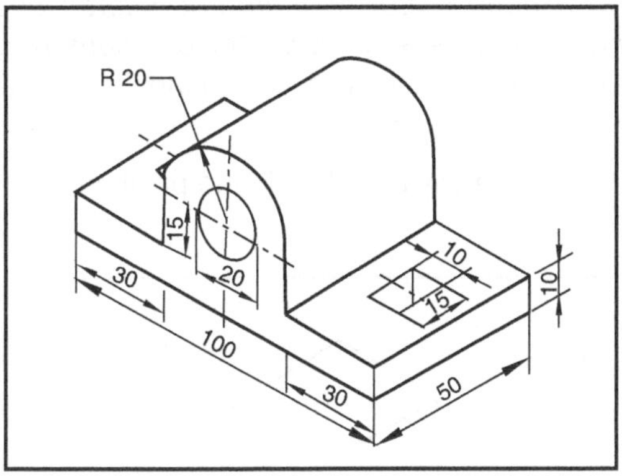

R 20
15
20
30
100
30
50
10
15
10

Fig. 8.12

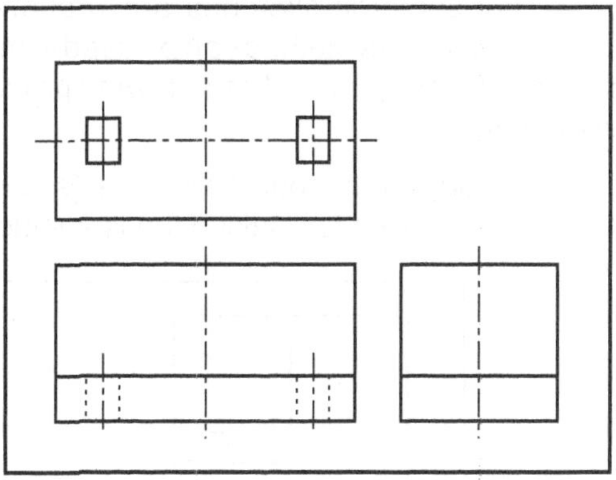

(i)

(ii)

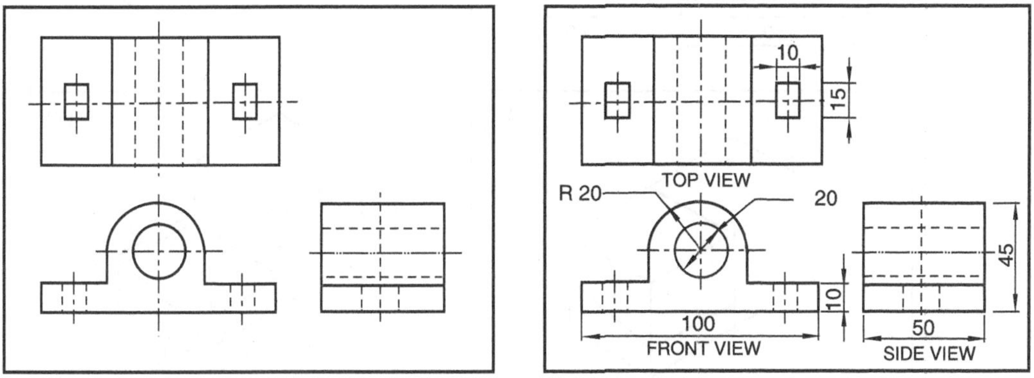

10
15
TOP VIEW
R 20
20
45
10
100
FRONT VIEW
50
SIDE VIEW

(iii)

Fig. 8.13. *Orthographic sketches*

Pictorial sketches are usually isometric, oblique or perspective. Regardless of the type of sketches used, they are made in accordance with the rules governing pictorial drawings as discussed in detail in the chapter of orthographic projections.

8.13. ORTHOGRAPHIC SKETCHES

The procedure followed in making an orthographic sketch is almost the same as that in drawing with instrument. Though the sketch is not made to scale with instruments, yet it should be fairly proportionate. The different steps in making an orthographic sketch should be followed as under :—

1. Study the object (see Fig. 8.12) carefully and decide the views which describe the shape of the object best.

2. Draw the rectangles or squares in which the view are to be sketched and sketch the outline of every feature by firm lines [see Fig. 8.13 (i)].

3. Sketch in the required dotted lines for hidden features of the object [see Fig. 8.13 (ii)].

4. Finish all the lines of the views and rub out lines which are not required. Add the necessary dimension and note [see Fig. 8.13 (iii)].

8.14. ISOMETRIC SKETCHES

The isometric sketches are prepared on *a plain paper* or *isometric sketch paper.* Though any of the method can be used in making isometric sketches, yet the box construction method is mostly used. The following procedure should be followed for making the isometric sketches:

1. Study the orthographic views [see Fig. 8.14] carefully and decide the position in which it should be placed for making the isometric sketches.

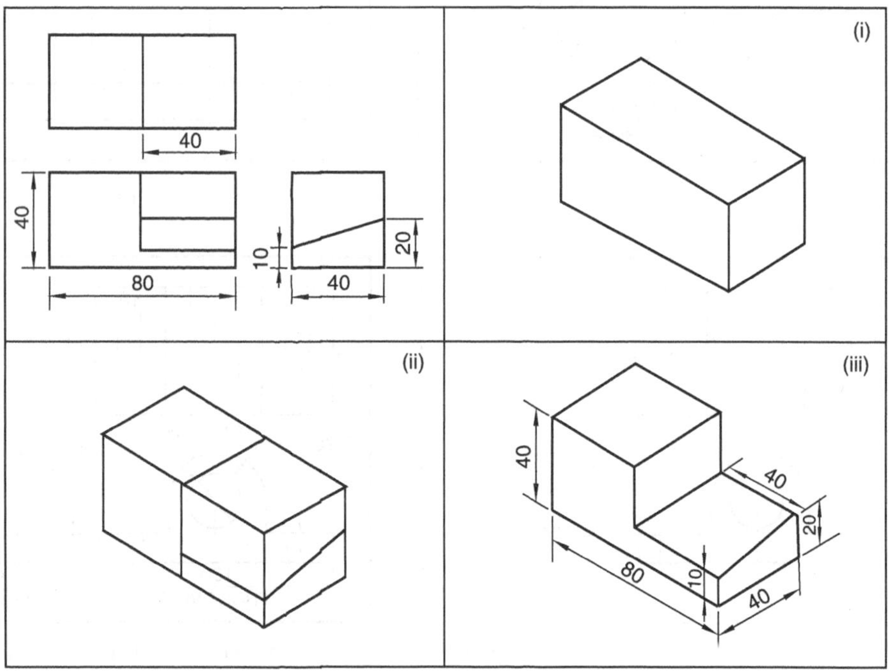

Isometic sketches

Fig. 8.14

2. Enclose the orthographic view in a box as shown in Fig. 8.14 (i) to give length, breadth and height.

3. Lay off the isometric lines and dark the required lines [see Fig. 8.14 (ii)].

4. Sketch in the required hidden lines.

5. Finish all the lines and rub out whatever is not required.

6. Add the necessary dimensions, notes, etc. [see Fig. 8.14 (iii)].

Problem 1 : Sketch a hall showing the black board, door, ventilators and any other details visible to you in the drawing hall. [Fig. 8.15]

Solution : For its solution, see Fig. 8.15 which is a self explanatory sketch.

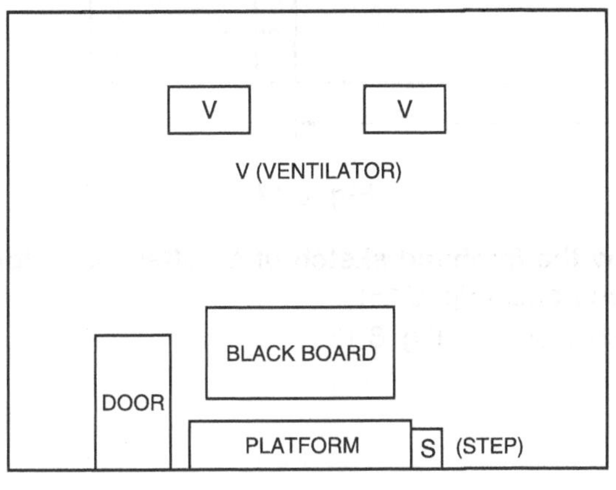

Fig. 8.15.

Problem 2 : Sketch a left-side wall showing the doors, windows, ventilators and any other details visible to you in the drawing hall. [Fig. 8.16]

Solution : For its solution, see Fig. 8.16 which is itself a self explanatory sketch.

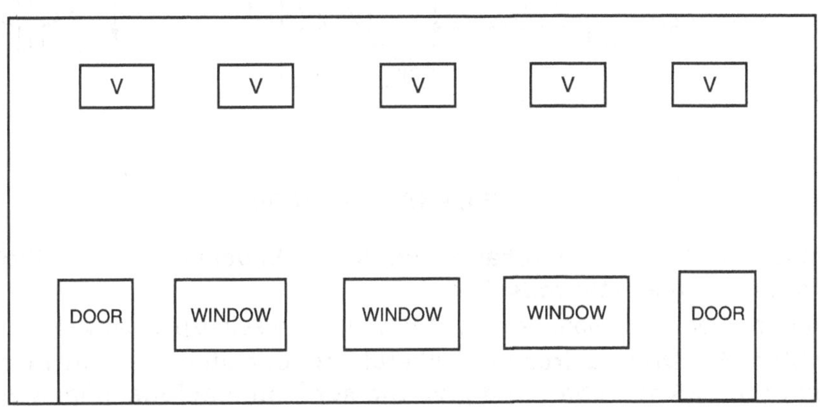

Fig. 8.16.

Problem 3 : Draw a freehand sketch of a V-block which you have seen in the work.

Solution : For its solution, see Fig. 8.17.

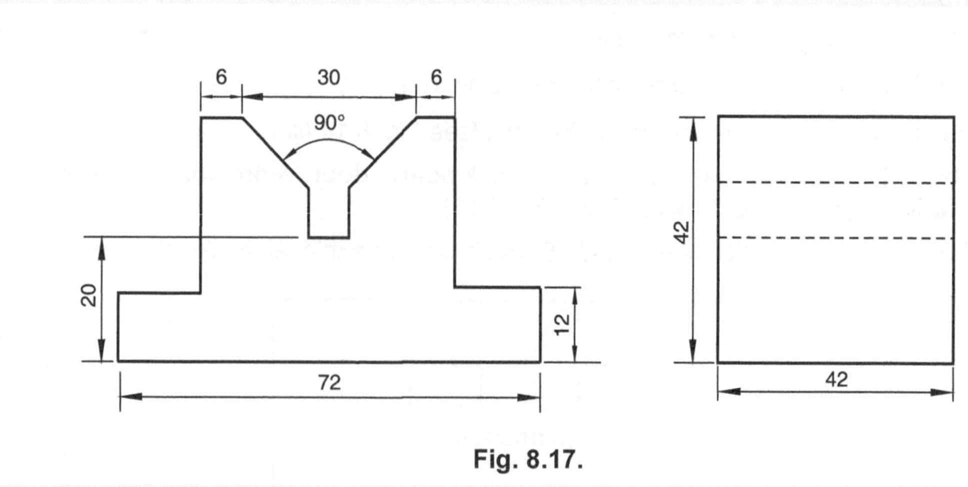

Fig. 8.17.

Problem 4 : Draw the freehand sketch of a cotter joints for rods of ϕ 40 mm and give the principle dimensions. [Fig. 8.18]

Solution : For its solution, see Fig. 8.18.

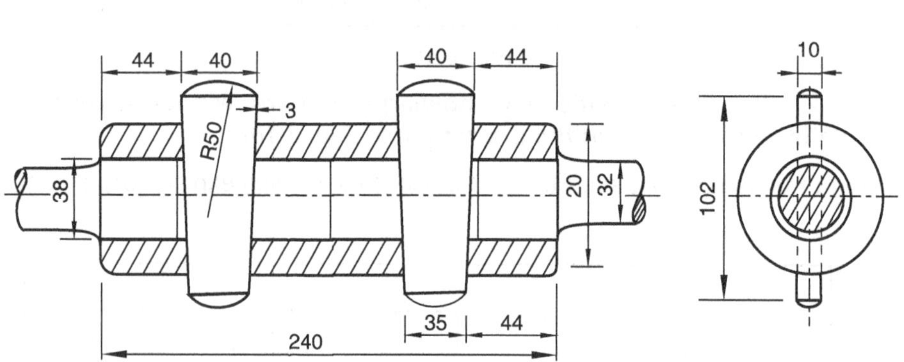

Fig. 8.18. *Cotter joint*

Problem 5 : Draw the freehand sketch of a knuckle joint in orthographic projection for 24 mm diameter rods. [Fig. 8.19]

Solution : For its solution, see Fig. 8.19 which is a self explanatory sketch.

Problem 6 : Draw a freehand sketch of foot step bearing in orthographic projection for ϕ 25 mm shaft and give the semi dimensions. [Fig. 8.20]

Solution : For its solution, see Fig. 8.20.

Note : *Students are required to keep the jobs prepared by them in the workshop and take the measurements with the scale and draw them free hand on the sheet.*

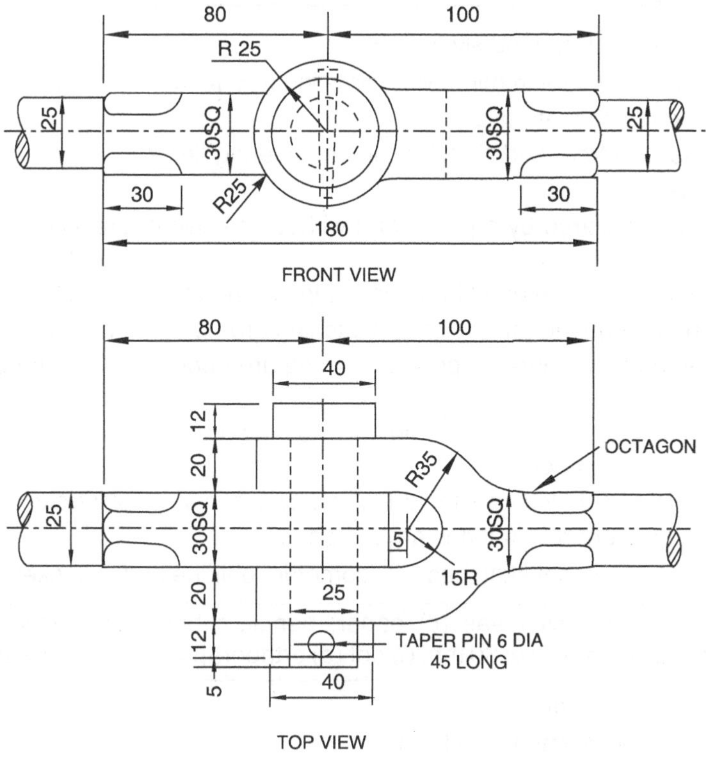

Fig. 8.19. *Knuckle joint*

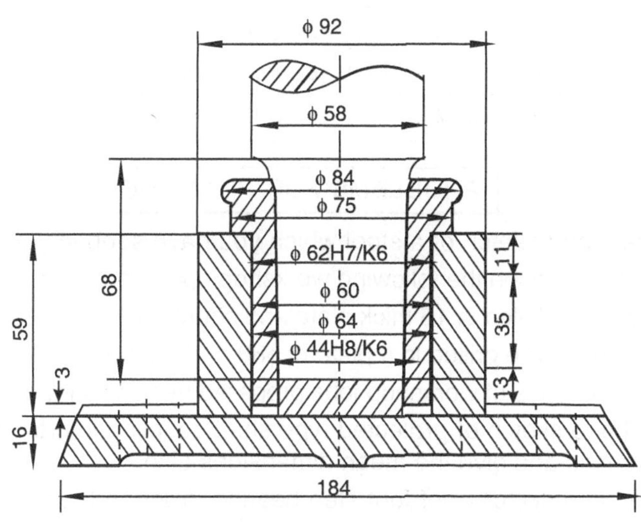

Fig. 8.20. *Foot step bearing*

QUESTIONS FOR SELF EXAMINATION

1. What do you mean by freehand sketching?

2. What points should be kept in mind while drawing freehand sketching?

3. What are the sketching materials?

4. What are the uses of freehand sketching in engineering field?

5. Fill in the blanks :—

(*a*) The drawing prepared by a pencil and without the use of drawing instruments is called ....

(*b*) Freehand sketches are used for expressing and recording ....

(*c*) The materials required for freehand sketching are ......, ... and ....

(*d*) The four important points for good sketching are good ......, ... proper....... and proper........

(*e*) The pencil should be held by thumb and first finger that it resets on second finger at about ..... from pencil point.

(*f*) The whole line in sketching should .... be drawn in a single-stroke.

(*g*) Two types of freehand sketches are...... and ....

(*h*) Three types of pictorial sketches are isometric, oblique and ...... sketches.

Ans. (*a*) sketching (*b*) new ideas (*c*) paper, pencil, rubber (*d*) practice, patience, proportion, proficiency (*e*) 30 mm (*f*) not (*g*) orthographic, pictorial (*h*) perspective.

6. Choose the correct answers :

(*i*) The horizontal lines are drawn from

(*a*) *left to right* (*b*) *right to left* (*c*) *up to down*

(*ii*) The vertical lines are always drawn from

(*a*) *down wards* (*b*) *top to bottom* (*c*) *upwards*

(*iii*) The inclined lines are drawn from left to right when they are nearly

(*a*) *horizontal* (*b*) *vertical* (*c*) *circular*

(*iv*) Small circles are sketched very easily in

(*a*) *one stroke* (*b*) *no. of strokes* (*c*) *no. of arcs.*

Ans. (*i*) *a;* (*ii*) *b;* (*iii*) *a;* (*iv*) *a.*

PROBLEMS FOR PRACTICE

1. Draw a freehand sketch of a stool which you have seen in your drawing room.

2. Sketch and dimension the following workshop operation as per I.S.I.:—

(*a*) A hole of ϕ5 drilled in 15 thick plate and countersunk 4 deep and at angle of 90°.

(*b*) 6 holes of ϕ8 are drilled in a plate of ϕ80 and on a pitch circle of ϕ60.

3. Draw any four types of workshop jobs prepared by you. Show mean dimensions also.

4. Draw freehand sketch of protected type shaft coupling in orthographic projection for ϕ 50 mm shafts and give the mean dimensions also.

5. Draw free hand sketches of foot step bearing (front and top views). Show the mean dimensions also.

6. Draw the free hand front view and top view of pipe joint for ϕ50 mm. Give the mean dimensions also.

CHAPTER 9

Sections of Solids

INTRODUCTION

In engineering practice, it is often required to make the drawing showing the interior details of the object, which are not visible to the observer from outside. If the object is simple in its construction, the interior portion of the object can easily be interpreted by dotted lines in the orthographic projection but in cases where the interior construction of the object is complicated, it cannot be clearly interpreted by the dotted lines. In such cases, views can be drawn by cutting the object apart by an imaginary cutting plane and the portion between the cutting plane and the observer is assumed to be removed so as to show the internal constructional details of the invisible features. Such views are known as *sectional views* or *views in section* (see Fig. 9.1).

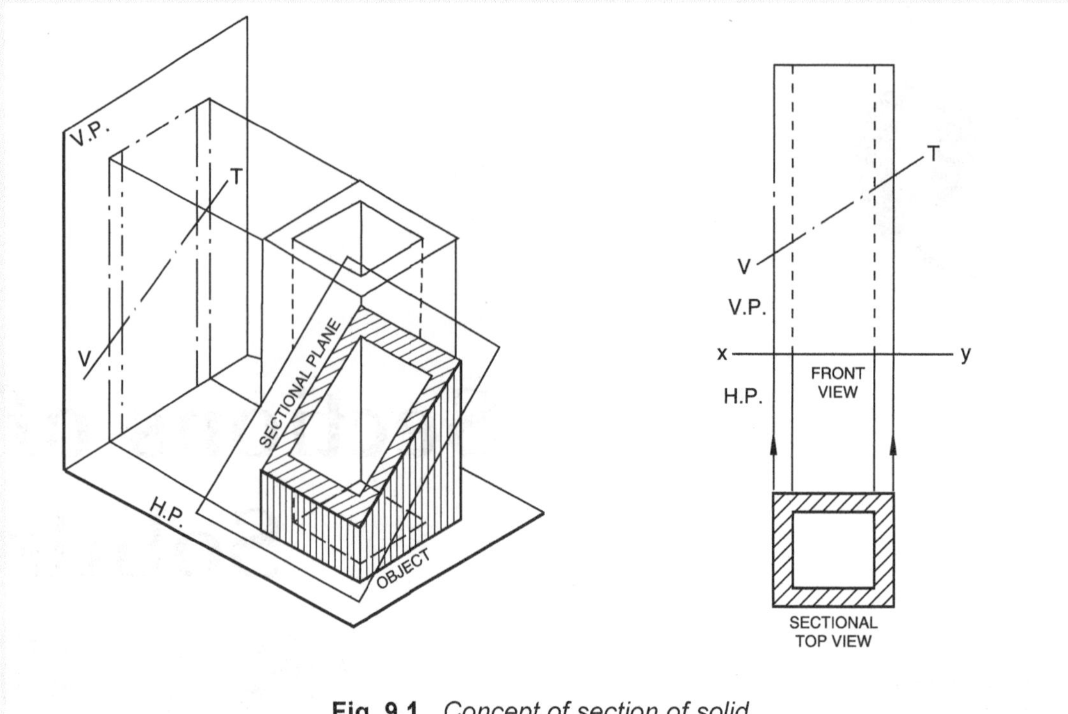

Fig. 9.1. *Concept of section of solid*

The section surfaces are indicated by section lines, evenly spaced and inclined at 45° to the horizontal. Suitable spacing to these lines is kept from 3 mm to 5 mm and 2 mm for small parts. Different parts are denoted by changing the direction of section lines even if the parts are made of the same material.

In this chapter, wherever found necessary, different pictorial views have been added to enable the students to visualize easily the positions of various solids along with the section planes.

9.1. SECTIONS OF SOLIDS

The solids which are cut by the section planes to visualise the internal constructional details of the invisible features are known as sections of solids.

9.2. TERMS USED IN SECTIONS OF SOLIDS

The following are the important terms used in the sections of solids :–

1. Section plane or cutting plane : The imaginary plane by which the object is assumed to be sectioned or cut, is known as section plane or cutting plane. It is represented by its traces on H.P. and V.P. (see Fig. 9.1).

A section plane may have the following positions with respect to the reference planes in space shown by its trace as given below :–

(i) Section plane perpendicular to V.P. and parallel to H.P. : When the section plane is perpendicular to V.P. and parallel to H.P., then its V.T. will be a straight line parallel to xy line (see Fig. 9.2).

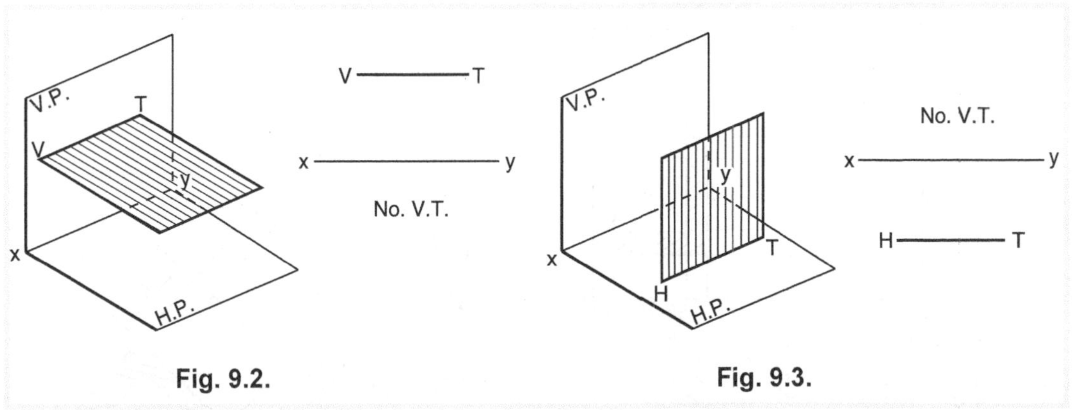

Fig. 9.2. Fig. 9.3.

(ii) Section plane perpendicular to H.P. and parallel to V.P. : When the section plane is perpendicular to H.P. and parallel to V.P., then its H.T. will be a straight line parallel to xy line (see Fig. 9.3).

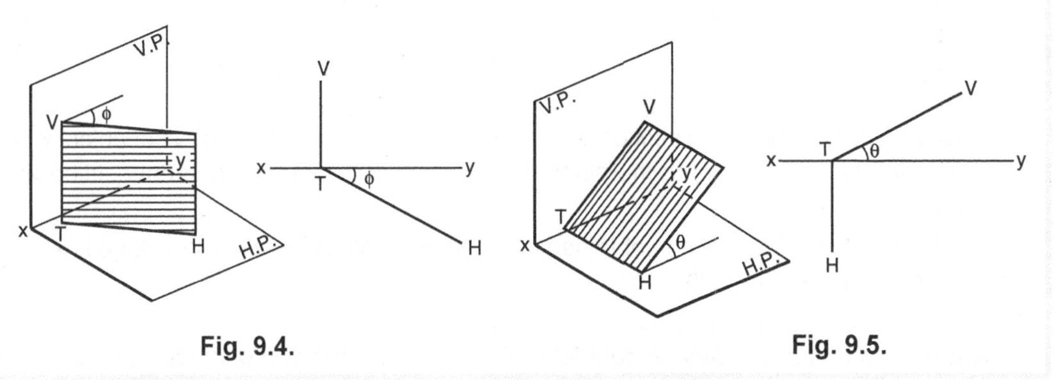

Fig. 9.4. Fig. 9.5.

(iii) Section plane perpendicular to H.P. and inclined to V.P. : When the section plane is perpendicular to H.P., and inclined at an angle ϕ to V.P., of its traces are represented by straight lines V.T. and H.T. as shown in Fig. 9.4.

Important Note : *In general, the perpendicular line is omitted as it serves no useful purpose. The inclined plane, therefore, is represented by its H.T. only and is shown in the top view by means of a straight line as H.T. (see Fig. 9.4).*

(iv) Section plane perpendicular to V.P. and inclined to H.P. : When the section plane is perpendicular to V.P., and inclined at an angle θ to H.P., then its traces are represented by straight lines V.T. and H.T. as shown in Fig. 9.5.

2. Sectional view : The projection obtained on a plane of projection of a cut object is called the section or sectional view.

3. Sectional top view : The projection obtained on a horizontal plane (H.P.) of a cut object, is called the sectional top view (see Fig. 9.6).

4. Sectional front view : The projection obtained on a vertical plane (V.P.) of a cut object when viewed horizontally forward, is called the sectional front view (see Fig. 9.6).

5. Sectional side view : The projection obtained on a profile plane (P.P.) of a cut object when viewed perpendicular to profile plane, is called the sectional side view (see Fig. 9.6).

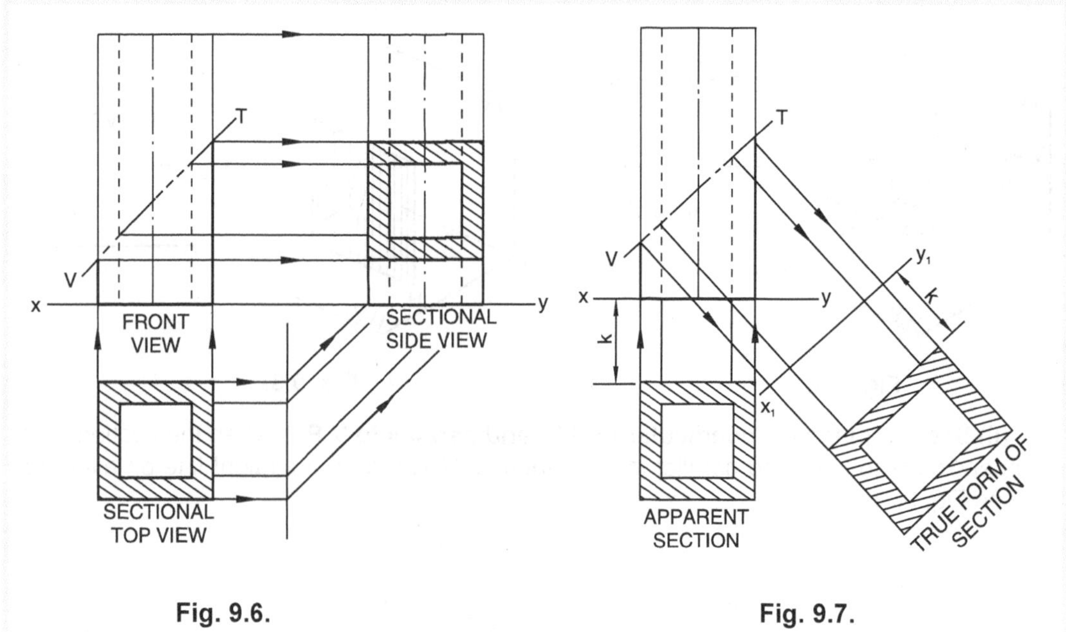

Fig. 9.6.

Fig. 9.7.

6. Apparent section : The section obtained on a plane of projection of a cut object when the section plane is not parallel to the plane of projection is called the apparent section (see Fig. 9.7).

7. True section : The section obtained on a plane parallel to the section plane is known as the true section (see Fig. 9.7).

Notes : *If the section plane is parallel to the H.P., then the sectional top view will show the true shape of the section.*

If the section plane is parallel to the V.P., then the sectional front view will show the true shape of the section.

But when the sectional plane is inclined to H.P., and perpendicular to V.P., the section is to be projected parallel to the plane of section in order to obtain the true section (see Fig. 9.7).

8. Full sectional view : It is a view drawn by a cutting plane which is assumed to be cut half of the object.

9. Half sectional view : It is a view drawn by a cutting plane which is assumed to be cut one fourth of the object stopping at the axis or the centre line.

PROBLEMS ON SECTIONS OF SOLIDS

Problem 1 : A square pyramid, base edge 25 mm and height 45 mm, is resting on its base on the H.P. in such a way that one of its base edges makes an angle of 30° with the V.P. It is cut by a section plane parallel to H.P. and passing at a distance 25 mm from the base along the axis. Draw the front view and sectional top view. [Fig. 9.8]

Solution : Draw the top view and front view of the square pyramid in the required position (see Fig. 9.8).

Draw a line V.T. for the section plane parallel to H.P. and passing at a distance 25 mm from the base along the axis. We see from Fig. 9.8 that the four slant edges o′a′, o′b′, o′c′ and o′d′ are cut by the section plane V.T. Name the four points as 1′, 2′, 3′ and 4′ in the front view. Project these points on the corresponding slant edges in the top view to meet oa at 1; ob at 3; oc at 4 and od at 2. Join these points and draw the section lines in it. it will give the required sectional top view.

Note *: As the section plane is parallel to H.P., therefore, the figure in the top view 1-2-4-3 gives the true shape of the section.*

Important Note : *The marking of the intersection points of the slant edges and the cutting plane lines can also be done this way, i.e., 1′ on o′a′, 2′ on o′b′, 3′ on o′c′ and 4′ on o′d′ in an alphabetical order. The students are suggested to use either this way of marking the points on section plane V.T. or the other method which is given in the Fig. 9.8.*

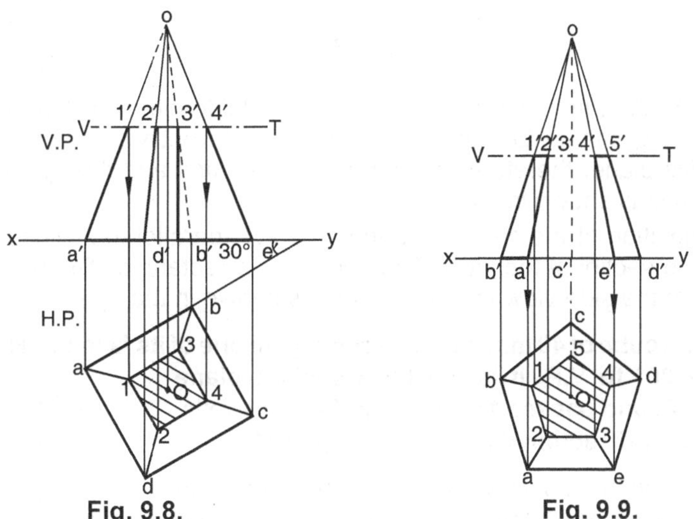

Fig. 9.8. Fig. 9.9.

Problem 2 : A pentagonal pyramid, edge of base 25 mm and height of axis 55 mm, has its base on the horizontal plane and an edge of the base parallel to V.P. It is cut by a horizontal section plane and passes at a distance of 30 mm above the base along the axis. Draw its front view and sectional top view. [Fig. 9.9]

Solution : Draw the projections of the pentagonal pyramid in the required position.

Show a line V.T. for the section plane parallel to the H.P. and at a distance of 32 mm above H.P., i.e., xy line (see Fig. 9.9).

As all the five slant edges of the pyramid are cut by the section plane, name the five points in front view and project them on the corresponding edges in the top view. Join these points. Draw section lines inside the figure, which gives the required sectional top view.

Note *: The point 5 cannot be projected directly in the top view as the slant edge o′c′ in the front view and oc in the top view is on vertical line. To locate point 5 in top view, draw a horizontal line through 5′ to cut o′d′ at 4′. From 4′, draw a vertical projector to cut od in top view at the point 4. With o as centre as o4 as radius, draw an arc to cut oc line at 5, which is the projection of the point 5′.*

Problem 3 : A triangular prism, side of base 45 mm and length of axis 75 mm is lying on one of its rectangular faces on H.P. Such that its axis is parallel to both H.P. and V.P. It is cut by a section plane parallel to the H.P. at a distance of 22 mm from the H.P. Draw its front view and sectional top view. [Fig. 9.10]

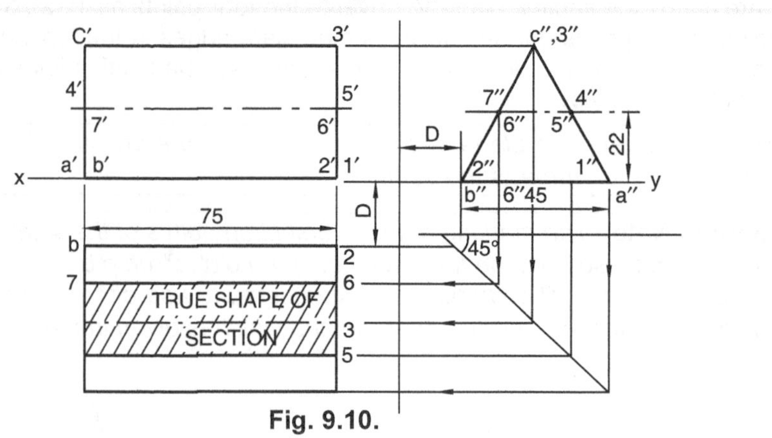

Fig. 9.10.

Solution : As the triangular end face is not parallel to either H.P. and V.P. in the given position, so begining cannot be start from the top and front views. Since the end face is parallel to a profile plane, therefore, draw the side view first and then project the front view and top view from it as shown in Fig. 9.10.

Show the position of the section plane by its V.T. parallel to xy line and 22 mm away from it. Project the sectional top view. The section is a rectangle. The depth dimension (D) is transferred from the side view to the top view as shown in the figure.

Problem 4 : A cube of 40 mm edges, is resting on one of its faces on H.P. with a vertical face inclined at 30° to V.P. It is cut by a section plane parallel to the V.P. and passes 15 mm away from the axis. Draw its top view and sectional front view. [Fig. 9.11]

Solution : Draw the top view and front view of the cube in the required position. Mark the centre o for the axis in the top view (see Fig. 9.11).

Since the section plane is parallel to V.P., it will be represented by its H.T. Draw a line H.T. for the section plane parallel to xy line and at a distance of 15 mm from the centre o. Number the points as 1, 2, 3 and 4 as shown in the top view. From these points, draw projectors and locate the corresponding points in front view. Join these points in proper order and draw section lines in it. It will give the required sectional front view.

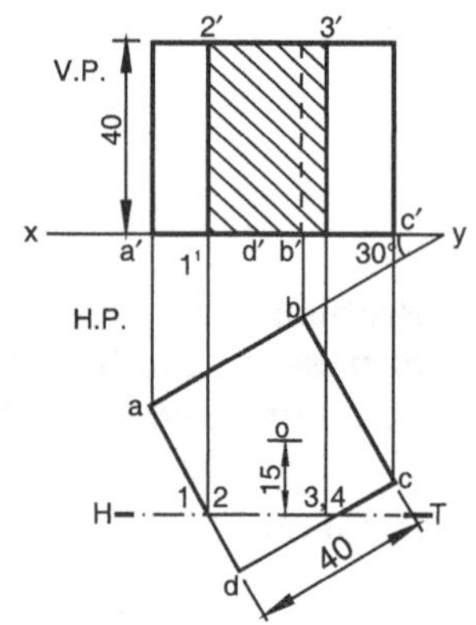

Fig. 9.11.

Note : *As the section plane is parallel to V.P., therefore, the figure 1′2′3′4′ shows the true shape of the section in front view.*

Problem 5 : A right regular pentagonal pyramid, base 30 mm, axis 65 mm long, has its base on the horizontal plane and an edge of the base parallel to V.P. An auxiliary inclined plane perpendicular to the H.P. and inclined at 45° to the V.P., cuts the solid at a distance of 12 mm from the axis. Draw its sectional front view and true shape of the section. [Fig. 9.12]

Solution : Draw the projections (top view and front view) of the pentagonal pyramid in the required position (see Fig 9.12).

As it is cut by an auxiliary inclined plane which is at 45° to the V.P. and passing 12 mm away from the axis of the pyramid in top view. To locate the position of the cutting plane, draw a circle, with o as centre and radius equal to 12 mm. Show a line H-T for the section plane tangential to this circle and inclined at 45° to xy line. The cutting plane cuts the pyramid at points 1, 2, 3 and 4 in top view. From these points, draw the projectors and locate the corresponding points in the front view. Join these points and obtain the required sectional front view as shown in Fig. 9.12.

In order to obtain the true shape of the section, draw a line x_1y_1 parallel to the cutting plane H-T at some convenient distance. Draw projectors from 1, 2, 3 and 4 perpendicular to x_1y_1 line. Along these projectors, set off distances of the respective points from the sectional front view to the distance below x_1y_1 line, i.e., distances of 1_1, 2_1, 3_1 and 4_1 above x_1y_1 is equal to the distances of 1′, 2′, 3′ and 4′ respectively measured above xy line.

Join 1_1, 2_1, 3_1, and 4_1 points and obtain the required true shape of the section as shown in Fig. 9.12.

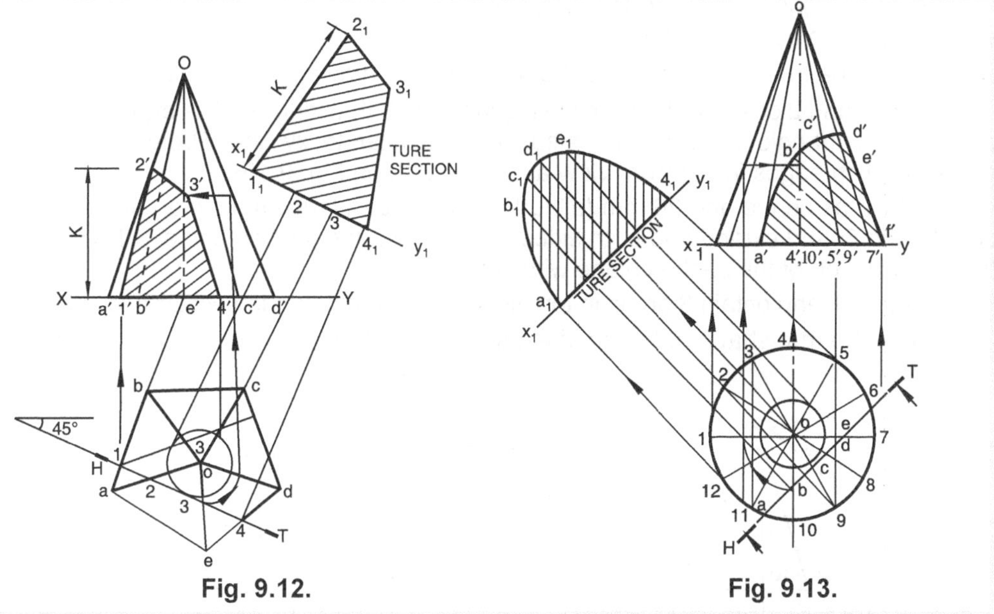

Fig. 9.12.

Fig. 9.13.

Problem 6 : A right circular cone 45 mm diameter, axis 65 mm long, is resting on its base on horizontal plane. It is cut by a plane, the horizontal trace (H.T.) of which makes an angle 45° with the V.P. and is passing 15 mm from the top view axis. Draw the sectional front view and true shape of the section. [Fig. 9.13]

Solution : Draw the top view and front view of the cone in the required position. Divide the top view into 12 equal parts and join them with the centre point o so as to represesnt the generators. Project them in the front view.

To locate the position of the cutting plane, draw a circle of 15 mm radius with o as centre in top view. Draw a line H-T for the cutting plane, tangential to the circle and inclined at 45° to xy line as shown in Fig. 9.13.

The section plane cuts the generators at points b, c, d, e and the base at a and f. Project these points on the corresponding lines in front view. Join these points and obtain the sectional front view.

A true section is projected on a line x_1y_1 parallel to cutting plane H-T [see Fig. 9.13].

Problem 7 : A cube of 55 mm edge, is resting on one of its faces on horizontal plane with an edge of its base making an angle of 30° to the V.P. A section plane inclined at 60° to the H.P., cuts the cube and passing through a point 25 mm from the base along the axis. Draw the apparent and the true sections of the solid. [Fig. 9.14]

Solution : Draw the top view and front view of the pyramid in the required position.

Show a line V-T for the section plane, inclined at 60° to xy line and at a distance of 25 mm from the base.

Name in proper sequence, the points at which the edges are cut. Project them on the corresponding edges in top view and complete the apparent section as shown in Fig. 9.14.

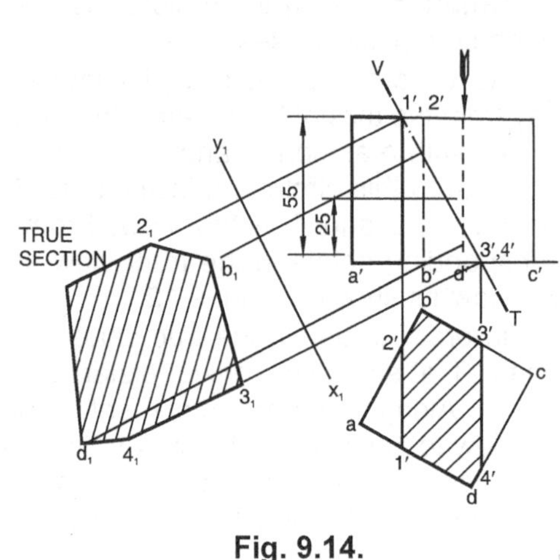

Fig. 9.14.

Obtain the true section of the cube on a line x_1y_1 drawn parallel to V-T.

Mark 1 and 2 on oa and ob lines in top view such that length 1-2 is equal to 50 mm (base of isosceles triangle 50 mm) and also perpendicular to xy line. Project 1 and 2 on the corresponding edges in the front view as 1′ and 2′ (see Fig. 9.15).

With 1′ as centre and 36 mm (altitude of triangle) as radius draw an arc to cut o′c′ at 3′. Join 1′ with 3′ and extend it to both sides to represent the section plane V.T. Measure angle θ, the inclination of the section plane with H.P. as 42°.

Project 3′ and obtain 3 on oc in top view. Then complete the sectional top view.

Draw the true shape of the section as usual which is the required isosceles triangle.

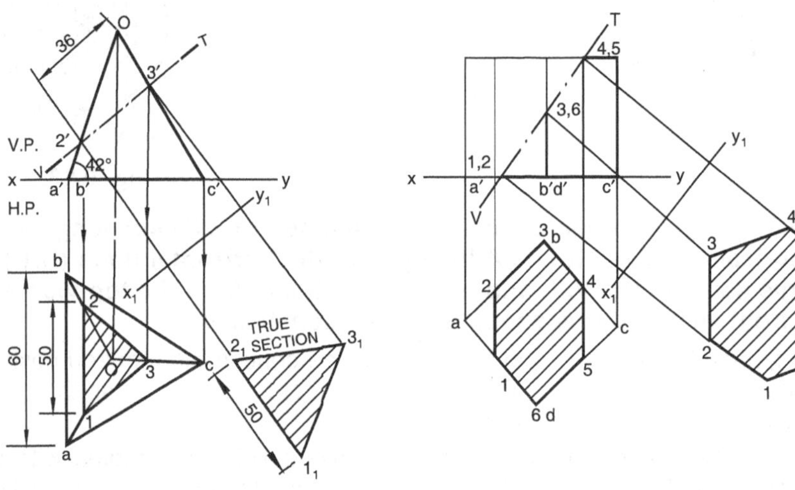

Fig. 9.15. Fig. 9.16

Problem 8 : A cube of 50 mm edges, is resting on one of its faces on H.P. It is cut by a plane in such a way that the true section available is a regular hexagon. Obtain the apparent and true section of the cube. Find the inclination of the sectional top view with H.P. and V.P. [Fig. 9.16]

Solution : The true section will be a six-sided figure only, when the six edges of the cube are cut by the section plane.

The cube, therefore, must be kept in such a way that its vertical faces make equal angles (45°) with the V.P. Also if the section is to be a regular hexagon, all the six edges of the cube in top view must be cut at their mid points as shwon in Fig. 9.16.

The cutting plane now passes trough the points 1, 2, 4 and 5 which are the mid points of edges in top view. Project these points in the front view to get the section plane 1′2′-3′6′-4′5′.

Project the true shape of section on a line x_1y_1 parallel to V-T.

QUESTIONS FOR SELF EXAMINATION

1. Why the solids are sectioned?
2. What do you mean by sectional view?
3. What is cutting plane or section plane?
4. What is the difference between apparent section and true section?
5. How the half and full section of solids are obtained?
6. What do you understand by V.T. and H.T. of section plane?
7. Fill in the blanks :
   (*i*) The sectional views are used to visualise the .... details of the objects.
   (*ii*) The section plane are represented by its .... on H.P. and V.P.
   (*iii*) The projection obtained on a H.P. of a cut object is called sectional ....
   (*iv*) The section obtained on a plane when the section plane is not parallel to plane is called .... section.
   (*v*) The true shape of the section is a ...., when a cylinder is cut by a section plane inclined to the axis.
   (*vi*) When a sphere is cut by a section plane, the true shape of the section is ....

Ans : (*i*) internal (*ii*) traces (*iii*) top view (*iv*) apparent (*v*) eclipse (*vi*) eclipse.

PROBLEMS FOR PRACTICE

1. A cube of 40 mm edge, is resting on one of its base edges on the H.P. in such a way that the face containing that edge is inclined at 45° to H.P. The cube is cut by a horizontal section plane V.T. at a distance of 15 mm away from the axis. Draw its front view and sectional top view.
2. A pentagonal pyramid, base edge 30 mm and axis 55 mm long, has its base on the H.P. with an edge of its base inclined at 30° to V.P. It is cut by a horizontal section plane at a distance of 35 mm above the base. Draw the front view and sectional top view.
3. A hexagonal prism, side of base 35 mm and axis 70 mm long, is resting on one of its corners on the H.P. with a longer edge containing that corner inclined at 60° to the H.P. and a rectangular face parallel to the V.P. It is cut by a horizontal plane which bisects its axis. Obtain the sectional top view and true shape of the section.
4. A triangular pyramid, base 32 mm side and axis 65 mm long, is lying on one of its triangular faces on the H.P. A section plane normal to H.P., and parallel to V.P. cuts the pyramid at a distance of 12 mm from the axis. Draw the sectional front view.

5. A tetrahedron of 40 mm sides, is resting with one of its edges on the H.P., so that one of its triangular faces is parallel to and 10 mm from V.P. The tetrahedron is cut by a vertical plane parallel to V.P. and 15 mm from the V.P. Draw the sectional front view and top view of the tetahedron.
6. A cube of 45 mm edges stands on one of iits faces on the H.P. with an edge of its base inclined at 30° to V.P. A section plane inclined at 60° to the H.P. cuts the cube and passes through a point 25 mm from the base along the axis. Draw the apparent and true sections of the cube.
7. A right circular cone of base diameter 50 mm and 65 mm height of the axis, is resting on its base on H.P. It is cut by a plane, the H.T. of which makes an angle of 45° with the V.P. and 10 mm away from the top view of the axis. Draw the sectional front view and true shape of the section.
8. A cone, diameter of base 60 mm and axis 80 mm long, is resting on its base on the H.P., is cut by a section plane perpendicular to V.P. inclined at 45° to H.P. and bisecting the axis. Draw the sectional top view, true shape of the section and the sectional end view.
9. A right circular cylinder of diameter 36 mm and 60 mm long, is resting on its base. An auxiliary inclined plane inclined at 45° cuts the axis of the solid at 15 mm from the top. Draw the sectional top view and true shape of the section.

SECTION - III

1. Production Drawings
2. Limits, Fits and Machining Symbols
3. Rivets and Riveted Joints
4. Welding
5. Screw Threads
6. Fastenings
7. Keys, Cotters and Joints
8. Shaft Couplings
9. Bearings
10. Brackets
11. Pulleys
12. Pipe Joints
13. Steam Engine Parts
14. I.C. Engine Parts
15. Valves
16. Gears
17. Cams
18. Jigs and Fixtures
19. Miscellaneous Drawings
20. Computer Aided Drafting

CHAPTER

1

Production Drawings

INTRODUCTION

In this fast developing engineering world a large number of simple to complex components are being manufactured. The shape and size of these components are decided by the designer. The basic idea of the designer is usually expressed through freehand drawings based on which production drawings are made. Therefore, the study of detail and assembly drawing is very important to understand the concept for the production of components.

In this chapter, we shall deal with the study of production drawing, detail and assembly drawings in cases of machines, machine parts, structure, wooden joints, etc.

1.1. PRODUCTION DRAWING

A drawing which provides complete information for the production of a machine or structure is called production drawing. It is also known as working drawing.

Fig. 1.22 shows the production drawing of a connector used in engineering practice.

Working drawing is also known as production drawing, because it completely describes the shape, size and specifications for the kinds of material, the method of finish, the accuracy required for the fabrication of a product.

A complete knowledge for the production of a machine or structure is given by a set of production drawings, conveying all the facts fully so that further instructions are not required. Fig. 1.40 shows a set of working drawing of a machine vice.

1.2. ENGINEERING PROCEDURE FOR MAKING A PRODUCTION DRAWING

When a new machine or structure is designed, the first drawings are usually made in the form of freehand sketches on which the original ideas and designs are worked out. These drawings are followed by calculations to prove the suitability of the design.

Working from the free hand sketches and calculations, the design department produces a *design assembly known* as a *design drawing or design layout.* This is the preliminary production drawing on which more details of the design are worked out. It is accurately made with drawing instruments showing the shape and position of various parts. The designer's general specifications for materials, clearances, heat-treatments, finishes, etc. are given in the from of notes on drawing.

This drawing is then passed to the draftsman who carefully works out the various detail drawings. All the views necessary for complete shape description of a part are provided and the necessary dimensions and manufacturing directions are given in the detail drawing. A set of drawings is completed with the addition of a assembly drawing and a part list or bill of materials. These drawing are then sent to the workshops for the execution of production.

A *set of production drawings [see Fig.1.23] will include, in general,the following classes of drawing:*

1.3. DETAIL DRAWING

The drawing which provides complete information for the production of each part of a machine or structure is called detail drawing.

A successful detail drawing provides the following informations :—

(i) Shape of the part
(ii) Size of the part
(iii) Material
(iv) Grade of finish
(v) Necessary shop operations
(vi) Limits of accuracy
(vii) No. of parts required
(viii) No. of view and sections required to describe exact shape of the part.

Fig. 1.1 shows the detail drawing of a C-clamp used in engineering practice.

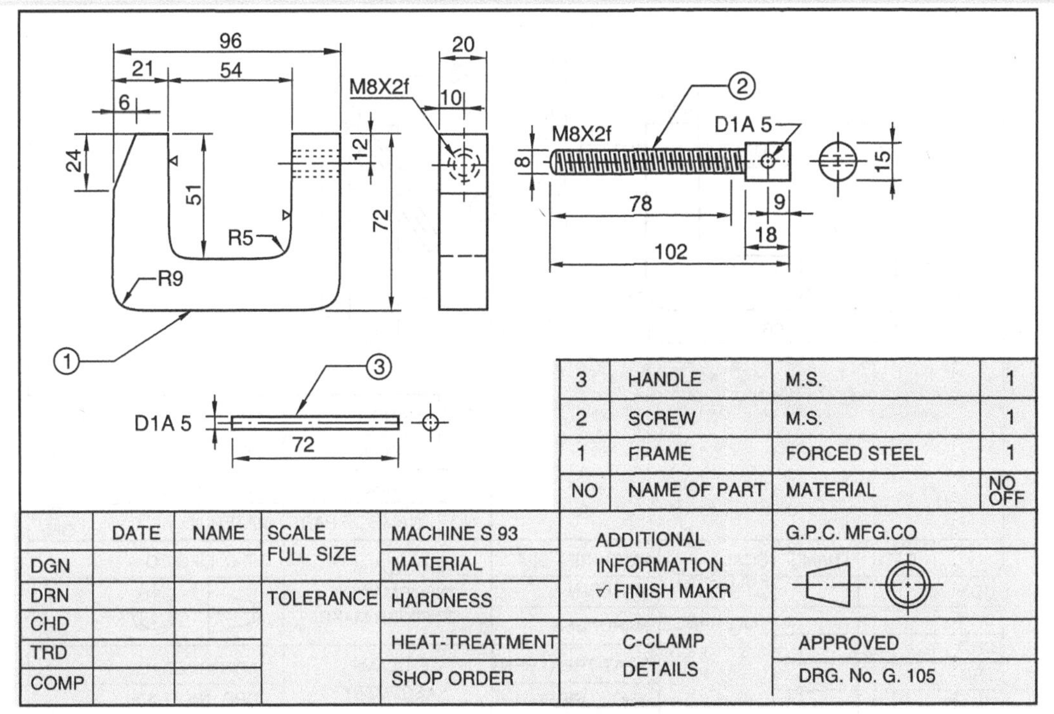

Fig. 1.1. *Detail Drawing of a C-clamp*

Sometimes, separate detail drawings are made for the use of different workmen such as the pattern-maker, hammer smith, machinist, welder, etc. Such drawings have only dimensions and informations needed by the workmen for whom the drawing is made.

When a large number of machines are to be manufactured, it is usual practice to make a detail drawing for each part on a separate sheet, especially when some parts are manufactured on different machines. When several parts are used on single machine, it is a common practice to draw a number of detail drawings on a large sheet.

In general, the detailing practice varies somewhat according to the industry and the requirements of the shop system. For example, in structure work, details are grouped together on one sheet, while in modern mechanical practice, a separate sheet is used for each part.

1.4. ASSEMBLY DRAWING

The drawing of a complete machine or structure showing all the parts assembled in their functional positions is called the assembly drawing.

The assembly drawing furnishes the following informatio in engineering practice :—

1. One main view to show the best assembly
2. Selected overall dimensions and important centre to centre distance
3. Identification of different parts on assembly drawing
4. Necessary sections
5. Parts list, notes, titles etc.

Fig. 1.2 shows the assembly drawing of a C-clamp used in engineering field.

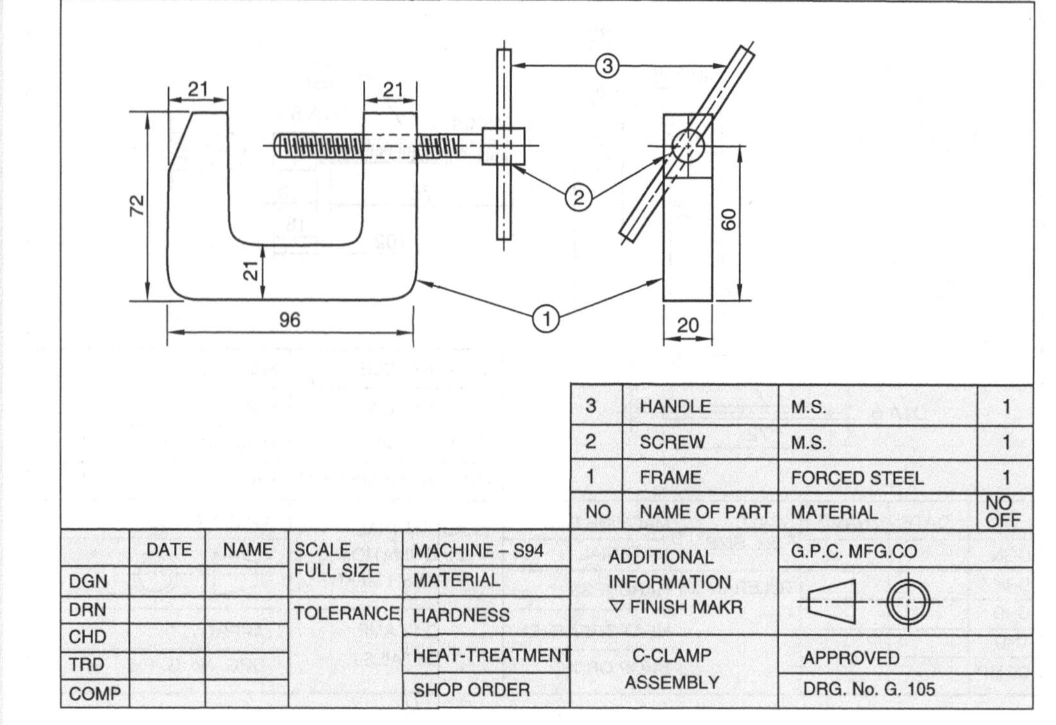

Fig. 1.2. *Assembly drawing of a C-clamp*

1.5. TYPES OF ASSEMBLY DRAWINGS

The following are the main types of assembly drawings according to their use in particular industry :

1.5.1. Design assembly drawings : *The drawings made at the time of developing any machine i.e. at the design stage are called design assembly drawing.*

The design assembly drawings only show how various parts of a machine are to be assembled together. These drawings are generally made to a large scale so that the designers can attempt further improvements keeping in view their functional requirements and appearance.

1.5.2. Working assembly drawings : *The drawings made for simple machines comprising a relatively smaller number of parts are called working assembly drawings.*

Here, as each part is completely dimensioned so there is no need for separate part drawings to fabricate the machine.

1.5.3. Group assembly drawings : *The drawings of a group of related parts that form a part in a more complicated machine are called group or unit or sub-assembly drawings.*

A set of unit drawings are required to make a complete machine. It is difficult to give the complete working assembly drawing of a entire machine but working parts or group assembly drawings are necessary for each unit.

Thus, in the drawing of a lathe, there would be included unit assemblies of group of parts such as the head stock, tail stock, gear box, etc.

1.5.4. Installation assembly drawings : *The drawings which show how to install or erect a machine or structure are known as installation assembly drawings.*

These drawings show only the out lines and relationship of exterior surfaces, that is why these are often called *outline drawings.*

In these drawings dimensions of a few important parts and overall dimensions of assembled unit are indicated.

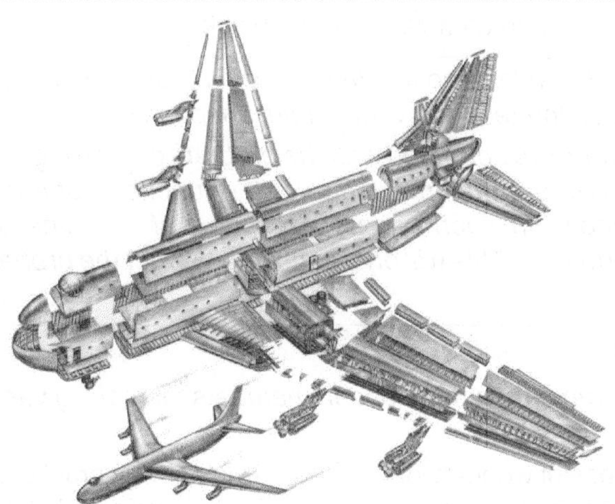

Fig. 1.3. *Exploded pictorial drawing of an aeroplane*

1.5.5. Exploded pictorial assembly drawings : *The drawings used in the parts-list of company, catalogues and instruction manual in exploded shapes are called exploded pictorial drawings.*

Fig. 1.3 shows the exploded pictorial drawing of different parts of an aeroplane.

1.5.6. Check assembly drawings : *The drawings made for checking the correctness of the details and their relationship in assembly are known as check assembly drawings.*

Such drawings are drawn accurately to scale. After the check assembly has served its purpose, it may be converted into a general assembly drawing. Thus, the assembly drawing may be used to check against detail drawing.

Since Machine Drawing depends upon the application of geometrical drawing for the representation of machines and machine parts, it is therefore, necessary for the beginners to acquire a good working knowledge of geometerical drawings so as to represent with its applications, the correct shape, size and construction of various machines and machine parts.

1.6. PROCEDURE FOR MAKING ASSEMBLY DRAWING FROM DETAIL

To make assembly drawing from details, the relative positions of all the parts are visualised and then the required views to be drawn are decided. In the begining, the students find it difficult to visualise the views of the assembled unit. It will be advisible for them to prepare the assembly drawing part by part.

The following steps should be followed to make an assembly drawing from detail drawing:-

1. Decide the relative positions of various parts depending upon their function, shape and size.
2. Examine the external and internal features of each part.
3. Determine overall dimensions of the assembly views and select a suitable scale.
4. Draw different rectangles and axes of symmetry for all the views.
5. While drawing the main view *i.e.* front view, draw main part of the machine (*i.e.* body) first and then draw the remaining parts in the sequence of assembly.

6. Project the other views from the main view *i.e.* front view.
7. Mark the overall dimensions and part numbers on the drawing.
8. Prepare the parts list and complete the title block.

Notes : *1. It is not advisable to complete one view before commencing the other views. The better way is to draw all the required views simultaneously.*

2. Some times students are confused with the problem of drawing an assembly drawing of a unknown machine. Without the knowledge of the functions of each parts one cannot decide about the assembly. In such cases, each part of machine should be carefully visualised for its shape and size. This will give some idea about the probable fitting positions of various parts.

1.3. WOODEN JOINTS

The joining of two or more than two wooden members for making various fittings is known as wooden joints.

There are various types of wooden joints used in engineering practice, but some important joints, whose detail and assembly drawings given in the following problems are as follows :—

(i) Halving joint
(ii) Cross halving joint
(iii) Tee halving joint
(iv) Dovetailed halving joint
(v) Open mortice and tenon joint
(vi) Bridle joint
(vii) Cogged joint
(viii) Notched joint etc.

Problem 1 : Fig. 1.4 (i) shows the assembly of a Corner Bridle wooden joint. Draw the detail drawings.

Solution : For its solution, see Fig. 1.4 (ii).

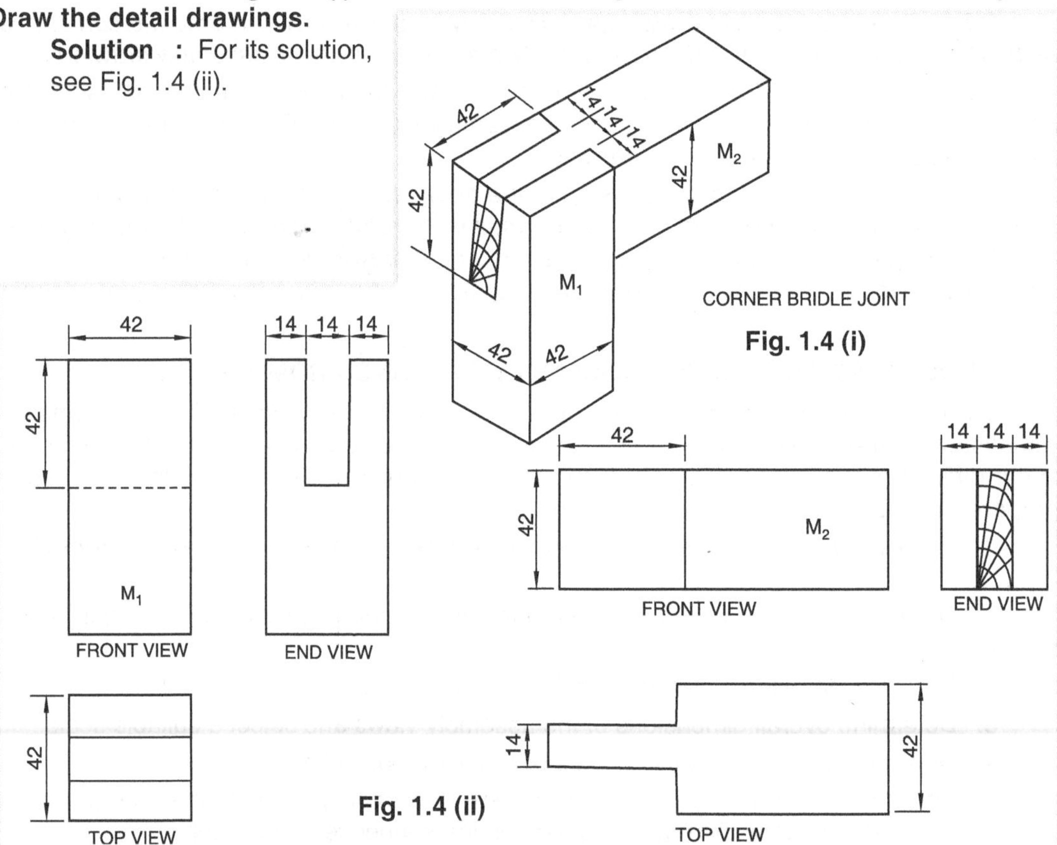

Fig. 1.4 (i)

Fig. 1.4 (ii)

Problem 2 : Fig. 1.5 shows the detail drawings of a Tee Halving joint, Assemble the parts and draw the following views :—

(i) Front view

(ii) Side view

(iii) Top view

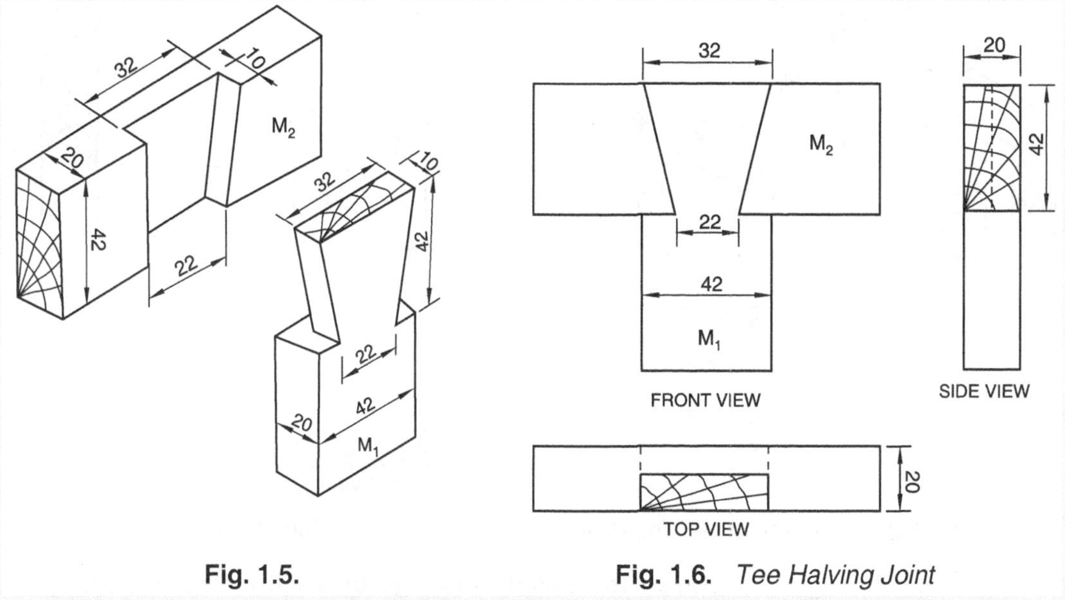

Fig. 1.5. **Fig. 1.6.** *Tee Halving Joint*

Solution : For its solution, see Fig. 1.6.

Problem 3 : Fig. 1.7 shows the detail drawings of a Dove-tailed Bridle wooden joint. Assemble the parts and draw the following views :—

(i) Front view (ii) Side view (iii) Top view.

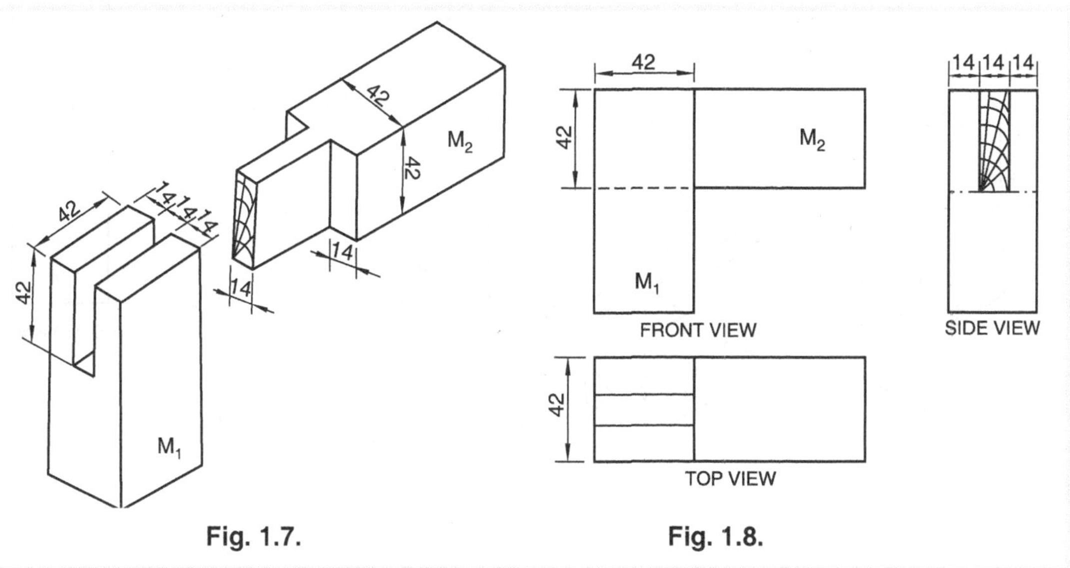

Fig. 1.7. **Fig. 1.8.**

Solution : For its solution, see Fig. 1.8.

Notes : *M_1 stands for wooden No. 1 and M_2 stands for wooden No. 2.*

Problem 4 : Fig. 1.9 shows the detail drawings of a Dove-tailed Halving joint. Assemble the parts and draw the following views :—

(i) Front view (ii) Side view (iii) Top view

Solution : For its solution see Fig. 1.10.

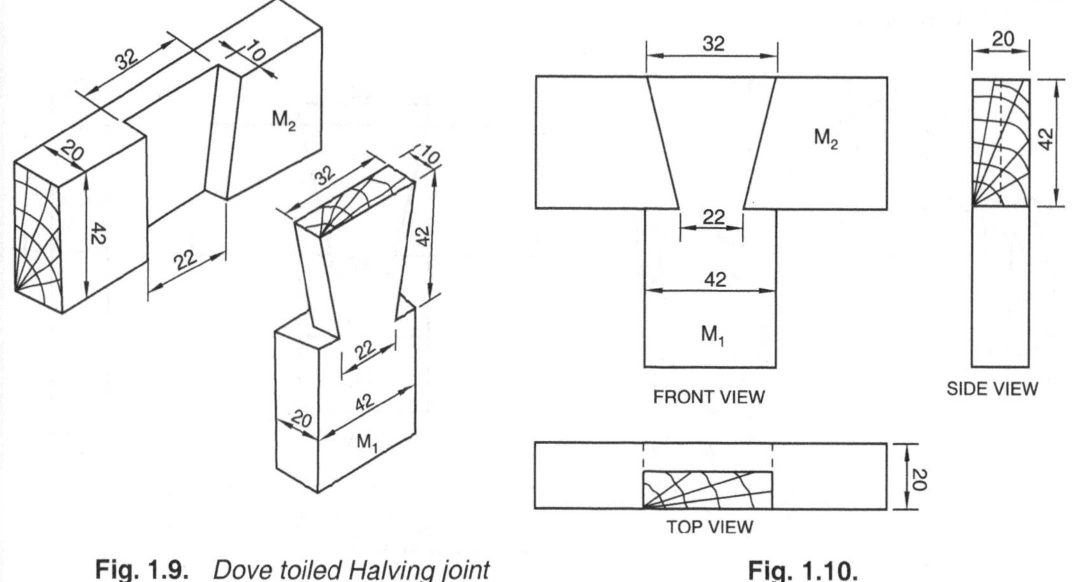

Fig. 1.9. *Dove toiled Halving joint* **Fig. 1.10.**

Problem 5 : Fig. 1.11 shows the detail drawings of a Mitre Faced bridle joint. Assemble the parts and draw the following views :—

(i) Front view (ii) Side view (iii) Top view.

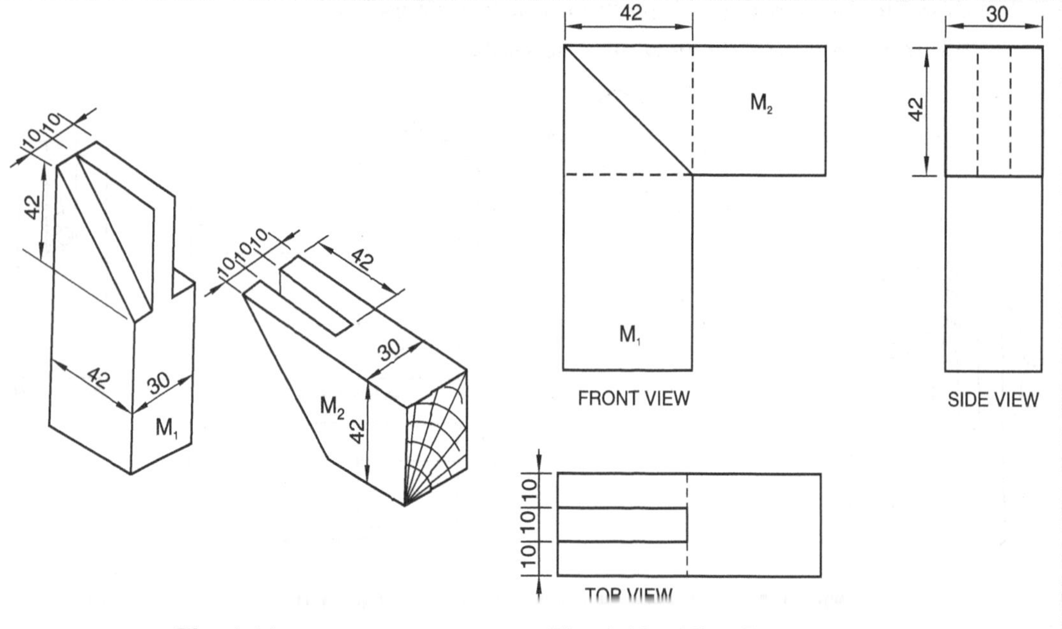

Fig. 1.11. **Fig. 1.12.** *Mitre Faced bridle Joint*

Solution : For its solution, see Fig. 1.12.

Problem 6 : Fig 1.13 shows the detail drawings of a Bevelled halving joint. Assemble the parts and draw the following views :—

(i) Front view (ii) Side view (iii) Top view.

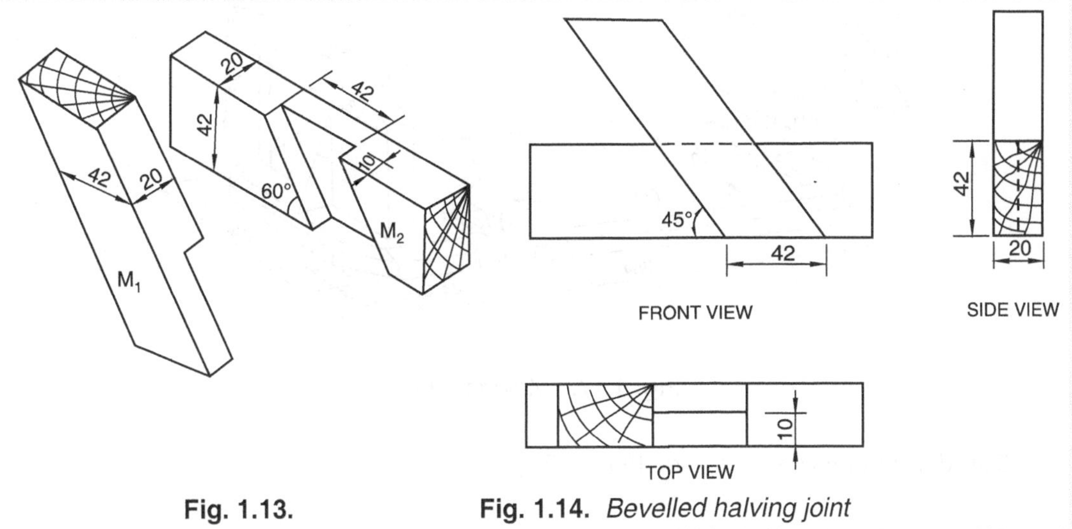

Fig. 1.13. **Fig. 1.14.** *Bevelled halving joint*

Solution : For its solution, see Fig. 1.14.

Problem 7 : Fig. 1.15 shows the pictorial views of a wooden joint. Assemble the parts and draw the following views :—

(i) Front view (ii) Side view (iii) Top view.

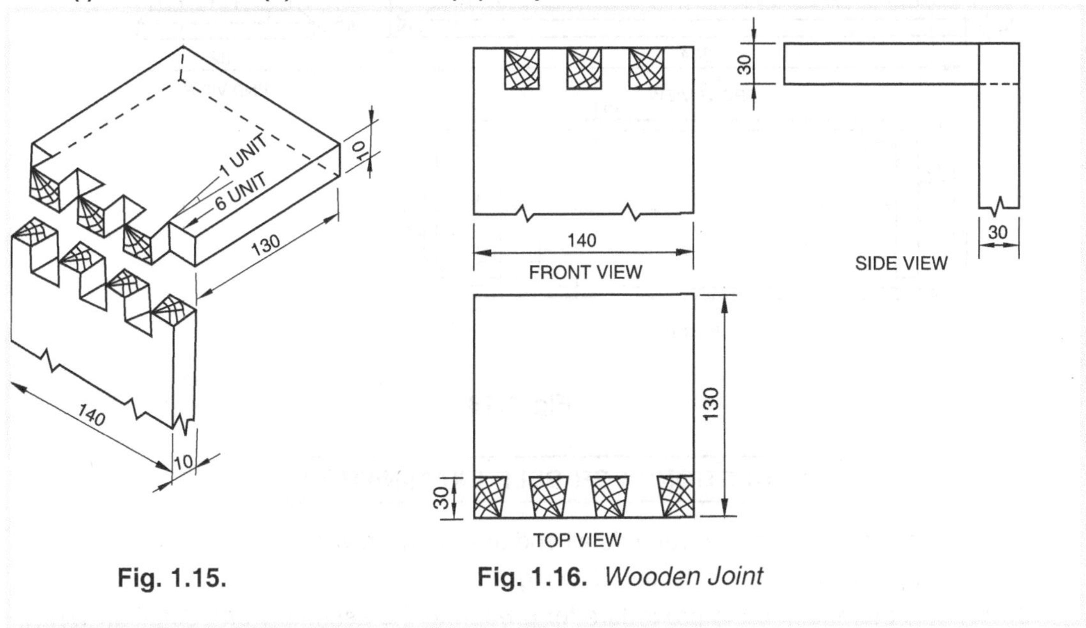

Fig. 1.15. **Fig. 1.16.** *Wooden Joint*

Solution : For its solution, see Fig. 1.16.

Problem 8 : Fig. 1.17 shows the pictorial views of a wooden joint. Draw (i) Front view (ii) End view (iii) Top view

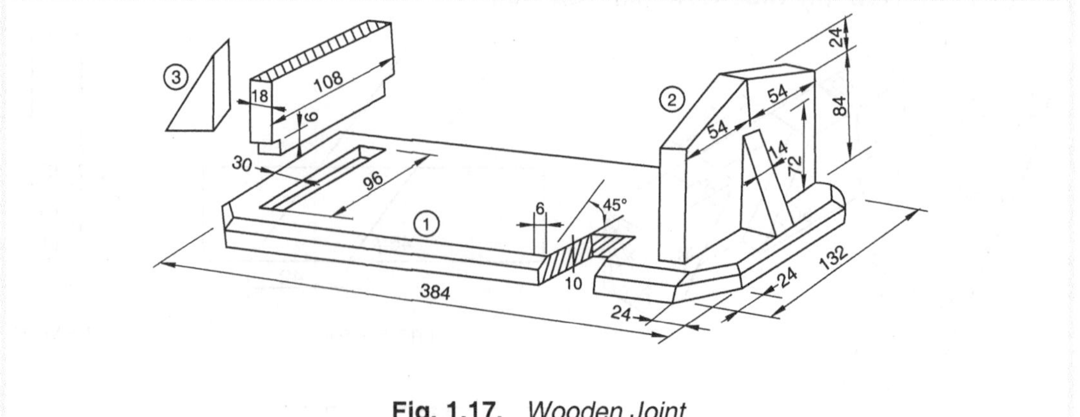

Fig. 1.17. *Wooden Joint*

Solution : For its solution, see Fig. 1.18

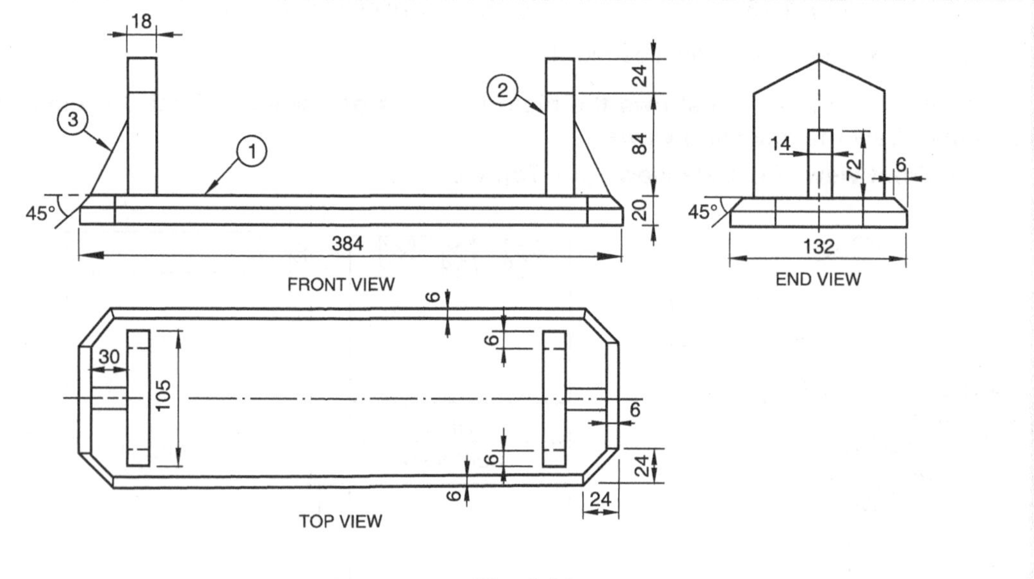

Fig. 1.18.

QUESTIONS FOR SELF EXAMINATION

1. What is the difference between detail and assembly drawing ?
2. what do you mean by production drawing ?
3. What is the engineering procedure for preparing the assembly and detail drawing?
4. Define wooden joints. Name any five wooden joints.

PROBLEMS FOR PRACTICE

1. Fig. 1.19 shows the detail of a Bridle Joint. Assemble the parts and draw the following views :—

   (i) Front view (ii) Side view (iii) Top view.

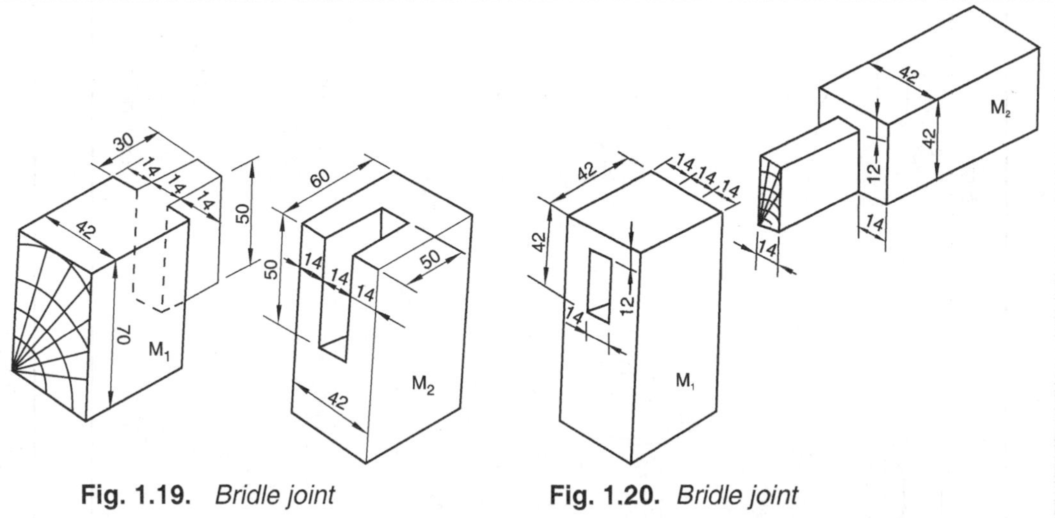

Fig. 1.19. *Bridle joint* **Fig. 1.20.** *Bridle joint*

2. Fig. 1.20 shows the detail of a bridle joint. Assemble the parts and draw the following views to some suitable scale :—

   (i) Front View (ii) Side view (iii) Top view.

3. Fig. 1.21 shows the assembly of dovetailed joint. Draw the details of its different parts.

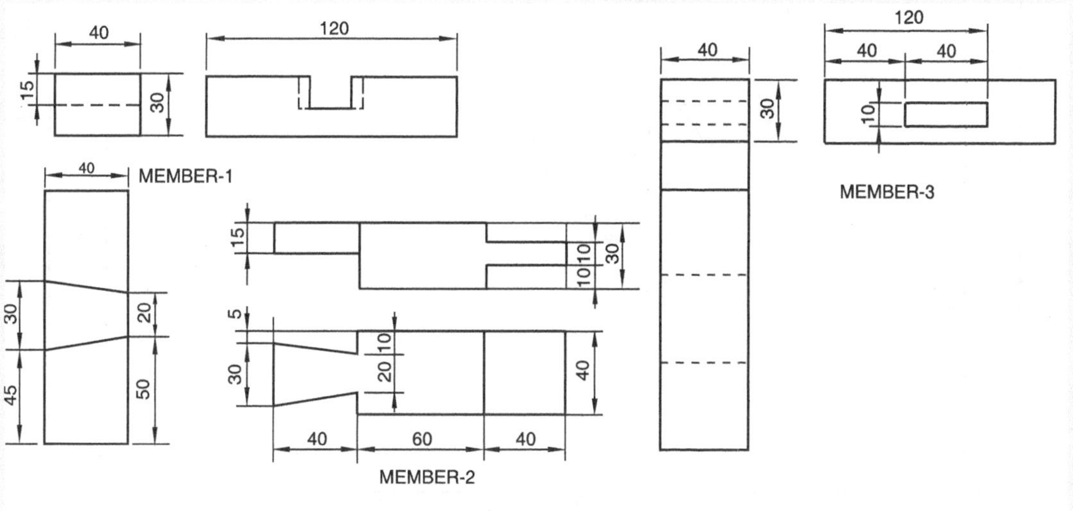

Fig. 1.21. *Dovetailed Joint*

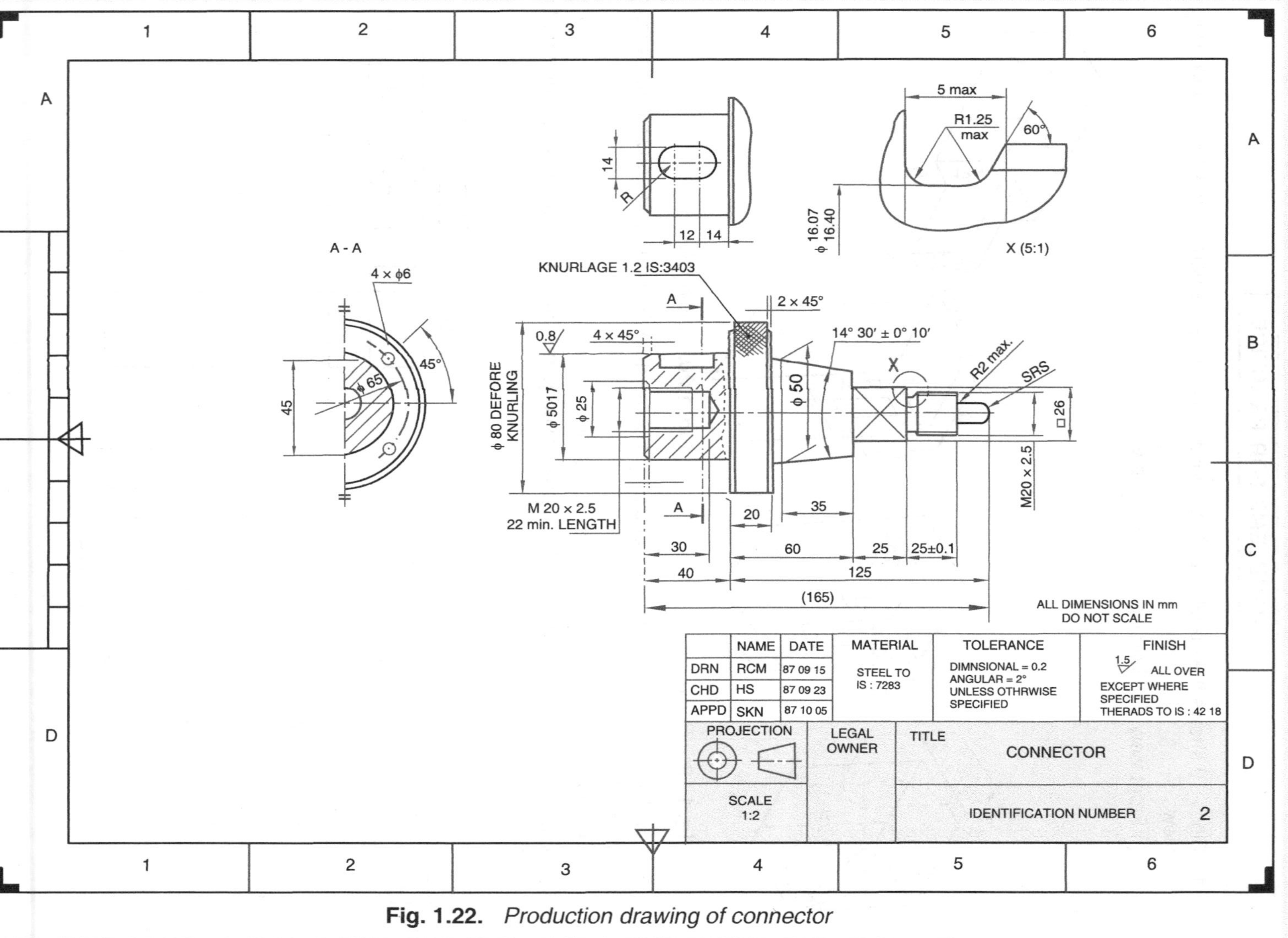

Fig. 1.22. *Production drawing of connector*

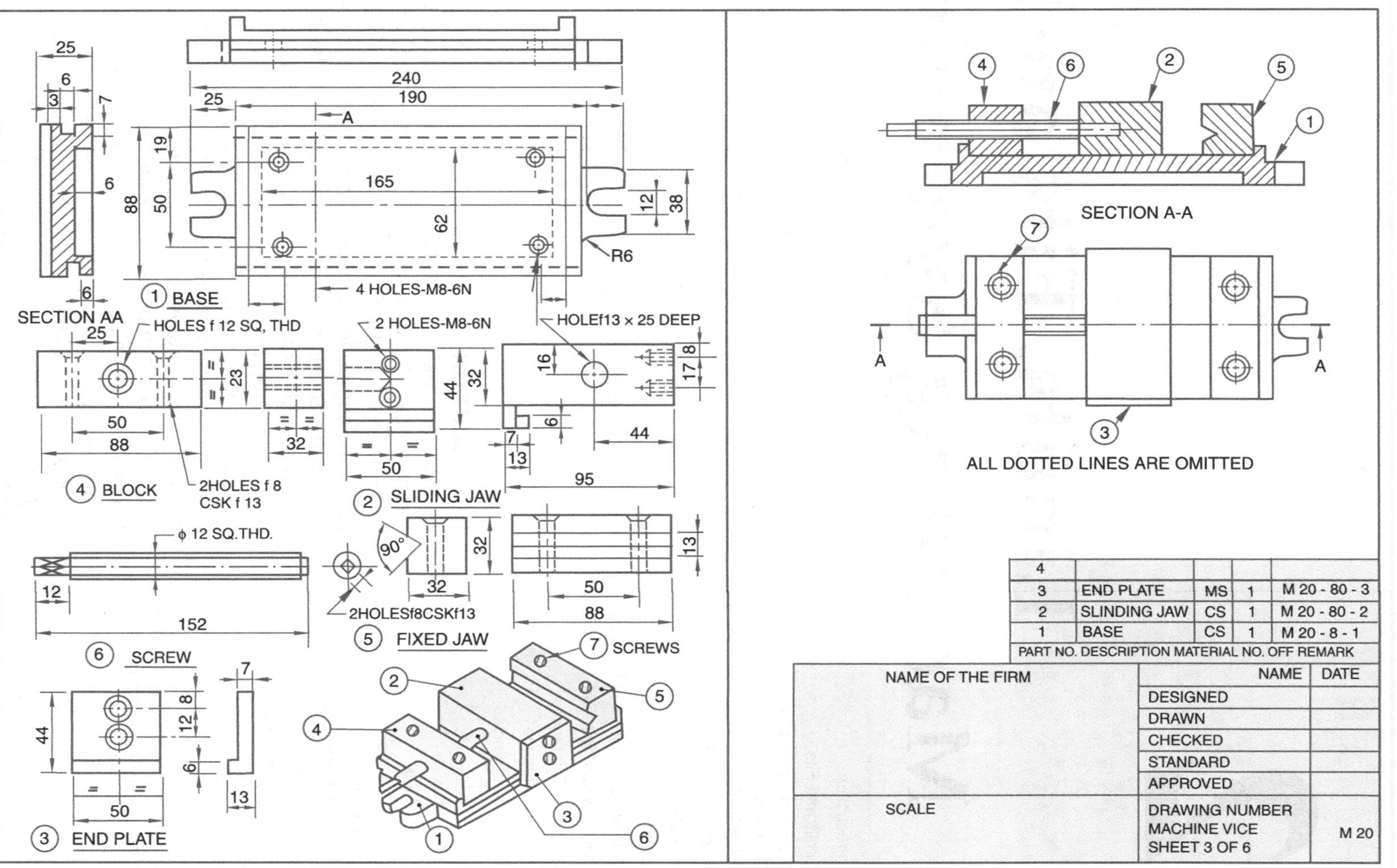

Fig. 1.23. *A set of production drawing of a machine vice*

CHAPTER

2

Limits, Fits and Machining Symbols

INTRODUCTION

In this fast developing engineering world a large number of simple to complex components are being manufactured. One of the basic requirments is that the components must be manufactured to the specified size. However, it is practically impossible to achieve this exactness due to many reasons such as human error, high cost and lack of sophisticated measuring instruments. Therefore, in practice, the variations which can be tolerated in the size of a component are always given. This permitted variation in the size of a component is called tolerance. The tolerance in the size of the components depends upon the place where these are to be used. *Thus, a component to be used in jet-plane requires a very high accuracy and hence very small tolerance is permitted.* *On the other hand, a component required for Bi-cycle may be permitted to have more tolerance in size i.e. more tolerance. Therefore, the tolerance of any component varies according to the degree of accuracy necessary for that particular work. The closer the tolerance, the higher would be the cost and there are greater chances of rejection of components during manufacturing stage.*

The purpose of this chapter is to study the various aspects of limits and fits, tolerances, machining symbols and surface finish with special emphasis on its increasing utility in the modern industry according to B.I.S.

2.1. LIMITS AND FITS

Two extreme permissible sizes of a part between which the actual size is contained are called limits. Whereas the relationship existing between two parts which are to be assembled with respect to the difference in their sizes before assembly is called a fit.

Thus there are two limits in the dimensions, a maximum limit and a minimum limit in any basic size.

It is a common experience that every machinery big or small consists of a large number of components assembled together properly. Due to mass production and specialisation the components of a machinery are manufactured by different organisations. The result is that the manufacturer of a particular component does not have knowledge about the other compont of the machinery under reference. This necessitates that each component should be manufactured with specified accuracy so that they can be properly fitted into the machinery. Use is, therefore, made of standardized limits of variations and fits from the nominal dimension.

2.2. NEED OF LIMITS AND FITS

The growing need of limits and fits is due to the following reasons:–

1. Mass production and specialisation: Due to numerous advantages of mass production and specialisation, it is not possible for a single industrial unit to manufacture all the components of a machinery. Therefore, with the permitted variation in the size, a component must be limited so that it can be properly fitted into the machinery. This calls for the need of maximum and minimum limits of the components.

2. Standardisation : The growing technology of advancement have introduced elements of standardisation i.e. specifying the size of a particular component nationally and if possible internationally in order to enjoy the benefit of large scale production and specialisation. This calls for the maximum and minimum size of the component so that it can be used in any part of the country and if possible of the world.

Use of standardisation : The following are the use of standardisation :-

1. It reduces the manufacturing costs and labour requirements.
2. It helps in improving the quality of products.
3. It simplifies the repair and maintenance of machines.
4. It reduces the time and effort needed to make new machines.

3. Interchangeability : Due to standardisation, interchangeability has become an essential feature in industry. It ensures the possiblity of assembling a unit or mahcine or replacing a worn out compnent without resorting to the extra machining or fitting operation. So if, various machines are provided with interchangeable spare parts, they can be dismantled for replacement of worn out parts in service condition. This necessitates that the components must be manufactured within the permitted variation in size so that they can be accepted without any change. Therefore, the interchangeablitity of parts, is a major per-requiste for economic production, operation and maintenance of machinery, mechanism and instruments.

2.3. TOLERANCE

Due to the inevitable accuracy of manufacturing methods, it is not possible to make a part precisely to a given dimension. Therefore, in practice, the variations which can be permitted in the size of a part are given. The amount of variation permitted in the size of a part is called tolerance and lies between two limits, upper and lower.

Thus, the difference between the upper and lower limit on a dimension of a part is called the tolerance.

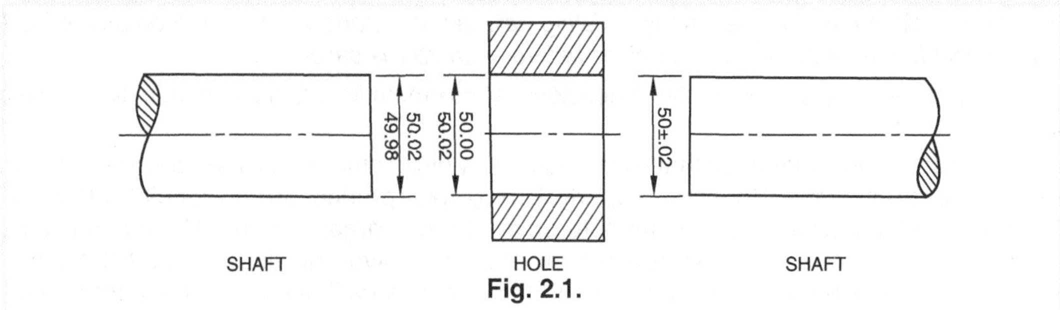

Fig. 2.1.

If the diameter of a shaft is written in mm as $50 \pm .02$ or $\frac{50.02}{49.98}$, the tolerance is 0.04 mm on 50 mm diameter (see Fig. 2.1)

The following examples will make tolerance more clear :–

1. $50 \pm .02$ — Upper limit 50.02, Lower limit 49.98 — $\therefore$ Tolerance = 50.02 – 49.98 = 0.04

or

$50\,^{+.02}_{-.02}$ — Upper limit 50.02, lower Limit 49.98 — $\therefore$ Tolerance = 50.02 – 49.93 = 0.04

2. $50\,^{+.025}_{+.005}$ — Upper limit 50.025, Lower limit 50.005 — $\therefore$ Tolerance = 50.025 – 50.005 = .020

3. $50\,^{-.025}_{-.040}$ — Upper limit 49.975, Lower Limit 49.960 — $\therefore$ Tolerace = 49.975 – 49.960 = .015

It is seen from above that a tolerance can be the difference between

(1) A positive uper limit and negative lower limit.

(2) A positive upper limit and a positive lower limit.

(3) A negative upper limit and a negative lower limit.

2.4 ALLOWANCE

The difference between the dimension of two mating parts is called the allowance or It is difference between the smallest hole and largest shaft as shown in Fig. 2.2.

$\therefore$ Allowance = Smallest hole – largest shaft

= 50.00 – 49.96

= 0.04

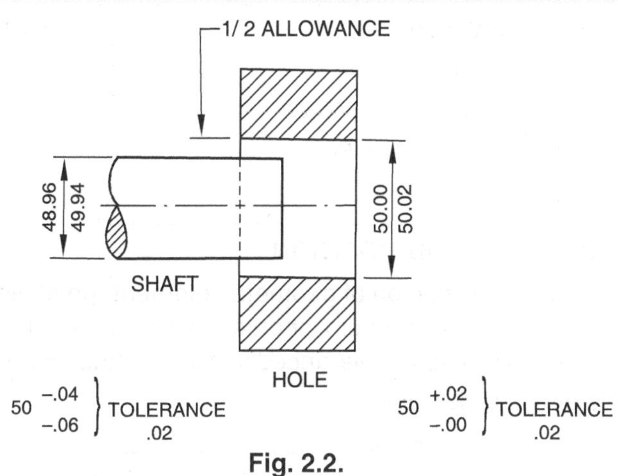

Fig. 2.2.

Note : *The terms tolerance and allowance represent different quantities, as tolerance is the variation providing for reasonable error in work-manship and allowance is necessary in the sizes of two mattng parts.*

2.5. APPLICATION OF TOLERANCES

Tolerance should be specified for all requirements critical to functioning and interchangeability wherever it is doubtful that ordinary or established workshop technique and equipmnt may be relied upon to achieve a satisfactory standard of accuracy. Tolerance should be specified to indicate where unusually wide variatitons are permissible. Tolerance should be indicated either by a general note, or notes assigning uniform or graded tolerances to specified classes of dimensions.

2.6. SELECTED ASSEMBLY

There are many instances when it is desirable to maintain almost an identical difference in size between a shaft and a hole produced in quantity without the expense of the extremely small tolerances that would make this possible. A free fit class with larger tolerance can be used if the parts are paired e.g., a small piston with a medium-sized cylinder bore, and a large piston is paired with a large cylinder bore. The difference in size between each pair is fairly constant.

2.7 GENERAL TOLERANCES

It is found that the tolerancing is indicated either by a general note assigning uniform or graded tolerances to specific classes of dimensions or by tolerances assigned to individual dimensions. In a general practice, a general tolerance note simplifies the drawing and saves time. The use of the typical notes on the drawing are given in the following typical examples.

Example 1: Tolerance except where otherwise stated ± . 02 mm

Example 2: Tolerance for dimensions 150 mm and below ± .5 mm and above 150±.1 mm.

Example 3: Tolerance on cast thickness ± 12.5 percent.

Example 4: Tolerance except where otherwise stated;

SIZE (mm)	TOLERANCE (mm)
UP TO AND INCLUDING 75	± . 02
OVER 75 UP TO AND INCLUDING 300	± . 40
OVER 300 UPTO AND INCLUDING 600	± . 50
OVER 600	± . 75

2.8. TOLERANCES OF FORM AND POSITION

The tolerance of form and position of geometric element (point, line, surface or plane) defines the one within which the element should be contained. The tolerance zone is according to the characterstics which is toleranced and the manner in which it is dimensioned;

(a) either a circle or a cylinder, or

(b) the distance between two straight lines, or

(c) the space between two surfaces or two planes, or

(d) the space in a parallelopiped.

If the tolerance zone is a circular or cylinderical, its width is in the direction of the arrow terminating the line joining the symbols frame to the feature which is toleranced. If the tolerance zone is circular or cylinderical, the sign is placed after the tolerance value in the symbol frame as shown in Fig. 2.3.

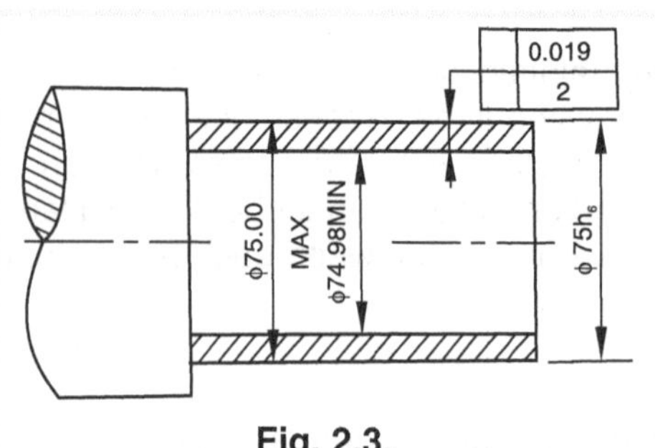

Fig. 2.3.

	CHARACTERISTIC TO BE TOLERANCED	SYMBOL
TOLERANCE OF FORM	FLATNESS	
	STRAIGHTNESS	—
	CIRCULARITY	○
	ACCURACY OF ANY PROFILE	⌒
	ACCURACY OF ANY SURFACE	
TOLERANCE OF POSITION	PARALLELISM	//
	PERPENDICULARITY	⊥
	ANGULARITY	∠
	POSITION	⌖
	CONCENTRICITY OR COXIALITY	◎
	SYMMETYRY	⌯

Symbols for tolerance of forms and position

Fig. 2.4.

The symbols for tolerance of form and position are shown in Fig. 2-4. These symbols represent the types of characteristics to be controlled by the tolerance. Tolerance of form and position should be specified in addition to the dimensional tolerances only on functional ground.

The necessary indications are written in a rectangular frame which is divided into two components or some times into three components. These sections are filled in from left to right in the following order :-

//	0.1	A

Fig. 2.5.

(a) Tolerance symbol. (see Fig. 2.5)

(b) Tolerance value in the unit used for length dimensioning.

2.9. UNILATERAL AND BILATERAL TOLERANCES

On the drawing, the limits on a dimension can be specified in two ways as :–

1. Unilateral (one way)
2. Bilateral (two ways)

2.9.1. Unilateral Tolerance : A unilateral tolarance is one in which the variation is permitted in one direction i.e, either plus or minus, from the design size (see Fig. 2.6).

For exampple 50 $^{+.000}_{-.035}$

Here, 50 mm is the design size

Upper limit = 50.000

Lower limit = 49.965

∴ Tolerance = 50.000-49.965=.053 mm

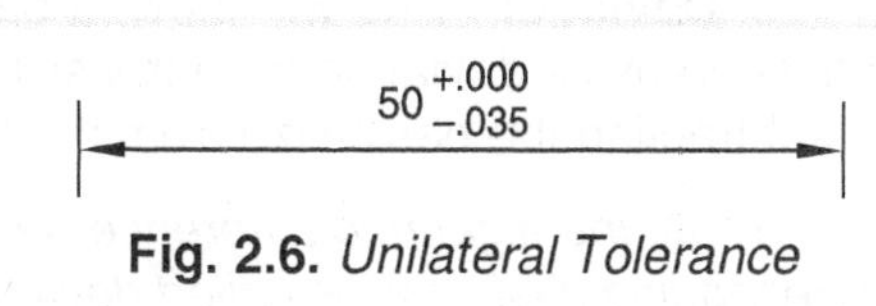

Fig. 2.6. *Unilateral Tolerance*

In this, the tolerance .053 is all in one direction towards smaller size.

2.9.2. Bilateral Tolerance : A bilateral tolerance is one in which variation is permitted in both directions i.e. plus and minus from the design size (see Fig. 2.7)

For example 50 $^{+.050}_{-.020}$

Here, 50 mm is the design size

Upper limit = 50.050

Lower limit = 49.980

∴ Tolerance 50.050 - 49.980 = .070 mm

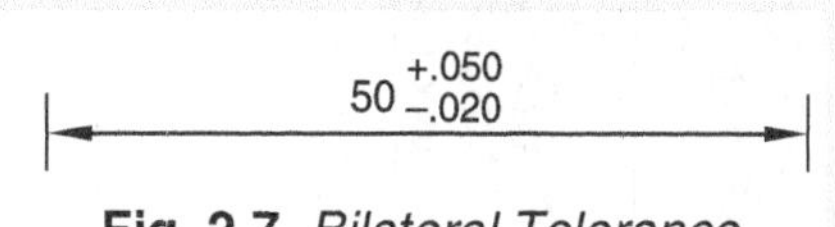

Fig. 2.7. *Bilateral Tolerance*

If the tolerance .070 specifies equal variations in both directions, then the bilateral tolerance is written as ± .035.

2.10. GRADE OF TOLERANCE

In a standardized system of limits and fits, grade of tolerance considered is corresponding to the same level of accuracy for all the basic sizes.

2.11. SIZE OF TOLERANCE

1. Size : A number expressing the numerical value of a length in a particular unit, is called size.

2. Actual size : The size of a part as may be found by measurement is called actual size.

3. Limit of size : The two exterme permissible possible sizes between which the actual size contained is called limit of size (see Fig. 2.1).

4. Maximum limit of size : The greaterof the two limits of size is called maximum limit of size.

5. Minimum limit of size : The smaller of the two limits of size is called minimum limit of size.

6. Nominal size : The size refered to as a matter of a convenience is called nominal size. In Fig. 2.1 the norminal size of shaft is 50 mm.

7. Basic size : The basic size is the size with referance to which the limit of size is fixed (see Fig. 2.8).

In other words, we can say that the basic size of a dimension or part is the size in relation to which all limits of variations are determined. The basic size is same for both members of a fit.

The basic size is expressed in decimal equivalent of the nominal size. If the nominal size of a shaft is 50½ mm, the basic should be written as 50.5 mm.

8. Deviation : As we know, a basic size is fixed to the part and each of the two limits is defined by its deviation from the size. The magnitude and sign of the deviation is obtained by subtracting the basic size from the limit problem.

Thus, the algebraic difference between a size (actual, maximum, minimum, etc) and the correspoding basic size is called deviation.

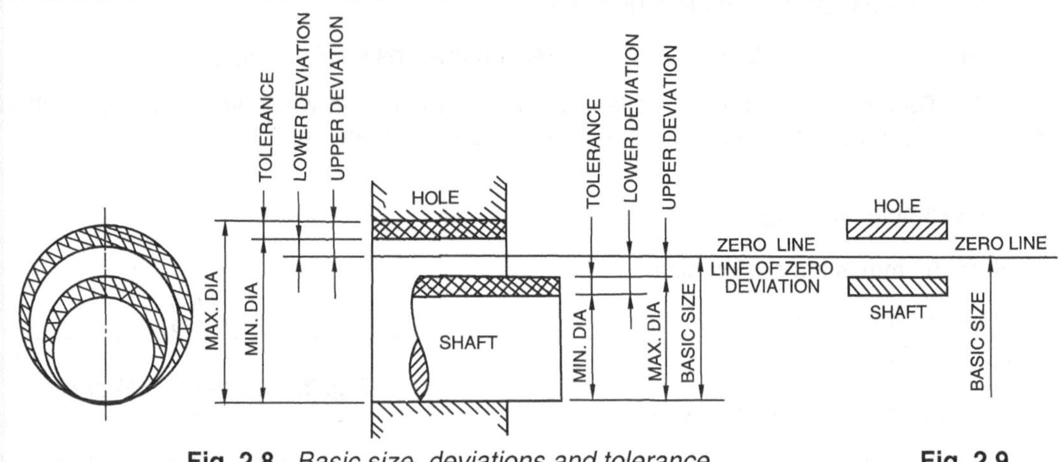

Fig. 2.8. *Basic size, deviations and tolerance* **Fig. 2.9.**

Fig. 2.8 illustrating the basic size; deviation and tolerances. The axis of the part is to be maintained below the diagram. Fig. 2.9 shows the simplified diagram conventionally used.

9. Actual deviation : The algebraic difference between the actual size and the corresponding basic size is called actual deviation

10. Upper deviation : The algebric difference between the maximum limit of size and the corresponding basic size is called upper deviation (see Fig. 2.9).

Symbols: *Upper deviation of the hole* - Es (E'cart superior)------High

Upper deviation of a shaft - es -------high

11. Lower deviation : The algebraic difference between the minimum limit of size and the corresponding basic size is called lower deviation (see Fig. 2.11).

Symbol: *Lower deviation of a hole* –E (E'cart inferior)-----Low

Lower deviation of a hole –ei ------Low

Fig. 2.11.Shows the basic size and its deviation.

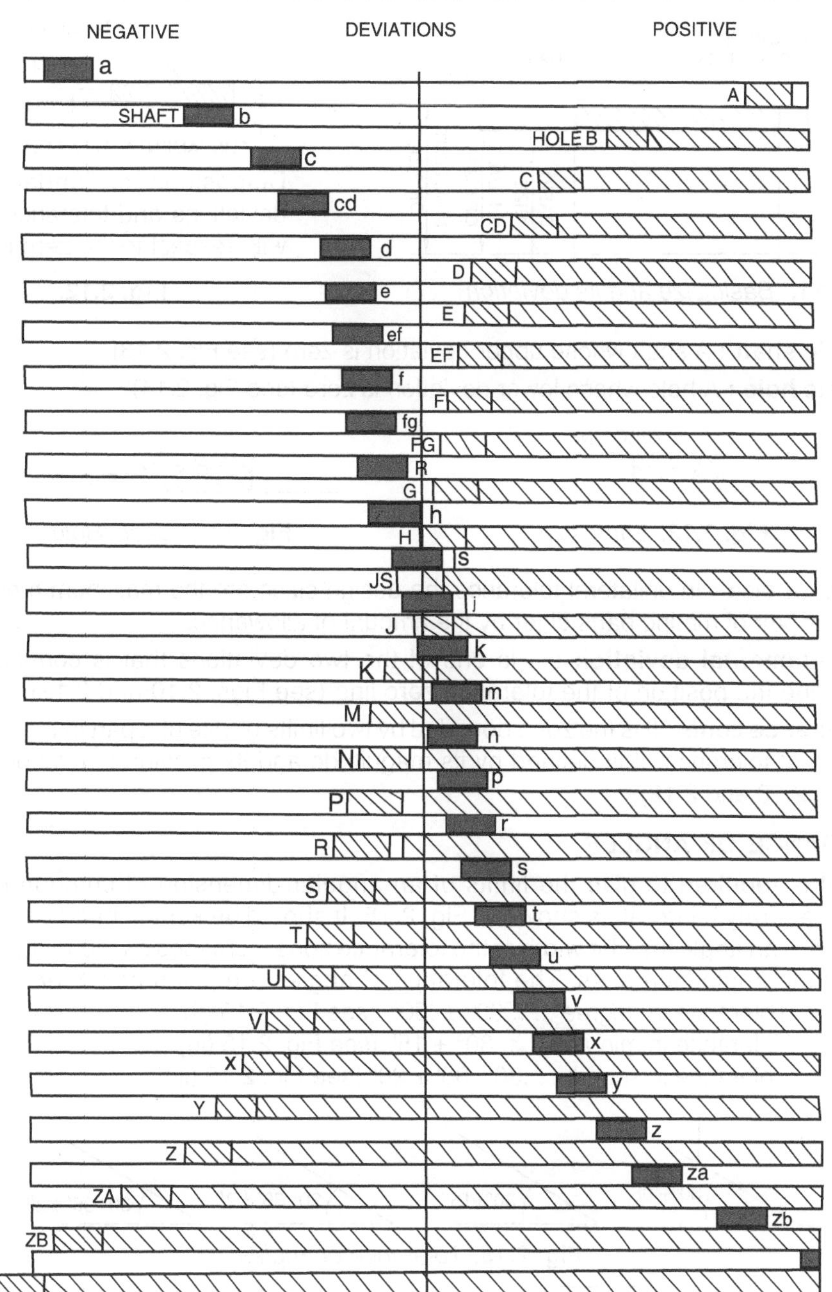

Chart showing fundamental deviation, tolerance zone and basic size for holes A to ZC and for shaft a to zc

Fig. 2.10.

12. Zero line : In a graphical representation of limitis, a straight line to which the deviations are referred is called zero line. The zero line is the line of zero deviation and represents the basic size (see Fig. 2.12). By convention, when the zero line is drawn horizontally, positive deviations are shown above and the negative deviations below it.

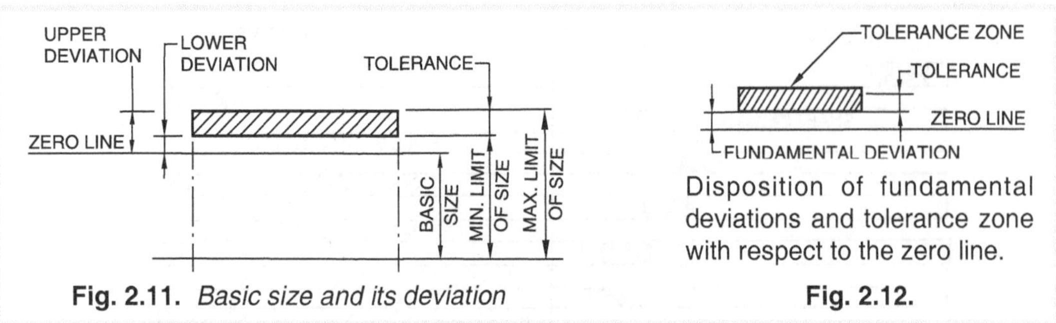

Fig. 2.11. *Basic size and its deviation*

Disposition of fundamental deviations and tolerance zone with respect to the zero line.

Fig. 2.12.

13. Basic shaft : A shaft whose upper deviation is zero (see Fig. 2.13)

14. Basic hole : A hole whose lower deviation is zero (see Fig. 2.14)

Fig. 2.13. *Shaft size* **Fig. 2.14** *Basic Hole*

15. Design size : In unilateral tolerance the design sizes are the maximum metal limits and those for a pair of mating parts differ by the amount of allowance.

16. Fundamental deviation : It is one of the two deviations that is conventionally chosen to define the position of the tolerance zero line (see Figs. 2.10 and 2.12).

17. Tolerance zone : It is the zone bounded by two limits of size of a part in the graphical representation of tolerance. It is defined by its maguitude and its position in relation to zero line (see Figs. 2.10 and 2.12).

2.12. TOLERANCE OF ANGLES

When it is required to give the limits of an angular dimension of components, the tolerance is generaly bilateral as shown in Fig. 2.15. It should be kept in mind when giving the tolerances of an angle that the width of the tolerance zone increases as the distance from the vertex of the angle increases. The tolerance of angles are given as under:

Tolerance in degree ... ...30° ± .50° [see Fig. 2.15 (i)]
Tolerance in minutes ... ...30° ± 15' [see Fig. 2.15 (ii)]
Tolerance in scond... ...30° .35' ± 20" [see Fig. 2.15 (iii)]

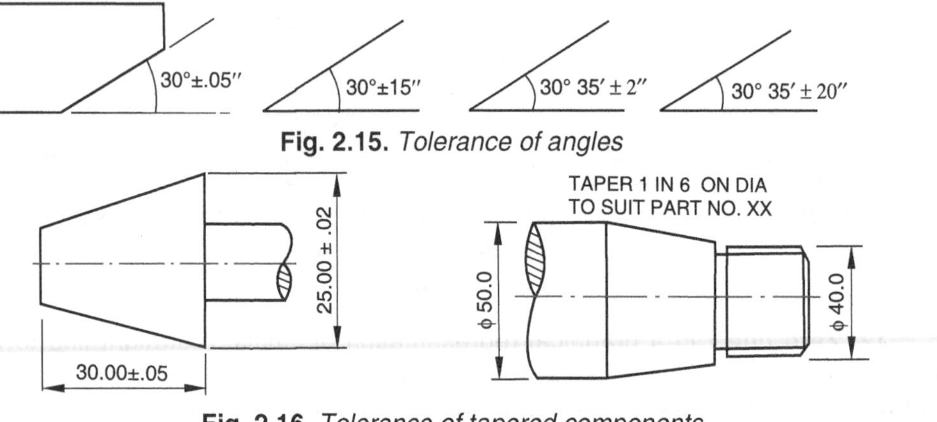

Fig. 2.15. *Tolerance of angles*

Fig. 2.16. *Tolerance of tapered components*

2.13. CLEARANCE

In a fit a positive difference between the size of the hole and the shaft (the hole being greater than the shaft) allowing relative movement between them is called a clearance (see Fig. 2.17).

2.13.1 Maximum clearance : The positive difference between the maximum size of a hole and the minimum size of a shaft is known as maximum clearance (see Fig. 2.18).

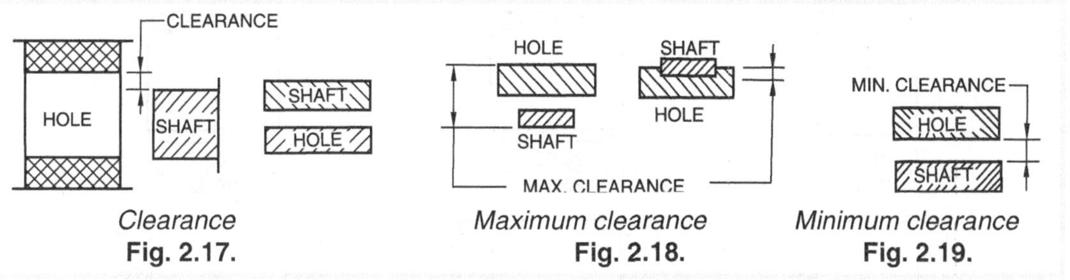

Clearance
Fig. 2.17.

Maximum clearance
Fig. 2.18.

Minimum clearance
Fig. 2.19.

2.13.2 Minimum clearance : The positive difference between the minimum size of hole and the maximum size of shaft is known as minimum clearance (see Fig. 2.19).

The arithmetical mean of the maximum and minimum clearance is known as mean clearance (see Fig. 2.19).

2.14. INTERFERENCES

In a fit, a negative difference between the sizes of the hole and the shaft (the shaft being greater than the hole) is called an interference (see Fig. 2.20).

1.14.1 Maximum interference : The negatilve difference between the maximum size of the hole and the shaft and the minimum size of the hole is known as maximum interference (Fig. 2.21).

2.14.2. Minimum interference : The negative difference between the minimum size of the shaft and the maximum size of the hole is known as minimum interference (see Fig. 2.22).

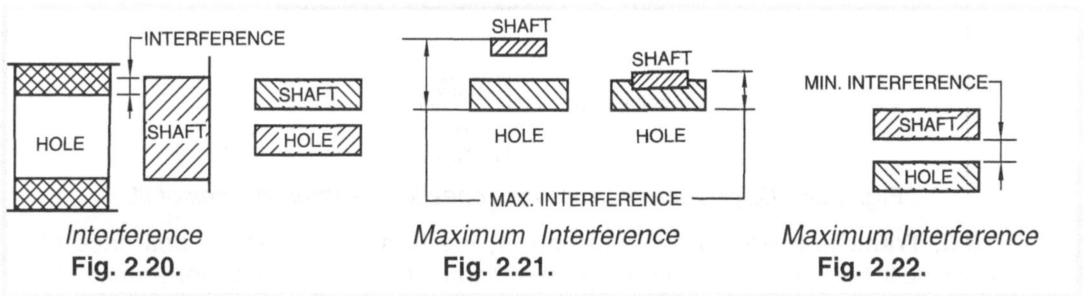

Interference
Fig. 2.20.

Maximum Interference
Fig. 2.21.

Maximum Interference
Fig. 2.22.

2.15. TYPES OF FITS

Depending upon the actual limits of the hole or shaft, the fit may be divided into three main classes as follows:

(i) Clearance fits

(ii) Interference fits

(iii) Transition fits

2.15.1 Clearance fits : In a clearance fit there is always a positive allowance between the largest possible shaft and the smallest possible hole. In this the shaft is smaller than hole.

Such type of fits give loose joint *i.e.* there must be some degree of freedom between shaft and a hole (see Fig. 2.23). Clearance fits may be (1) slide fit (2) easy slide fit, (3) running fit (4) slab running fit, (5) loose running fit.

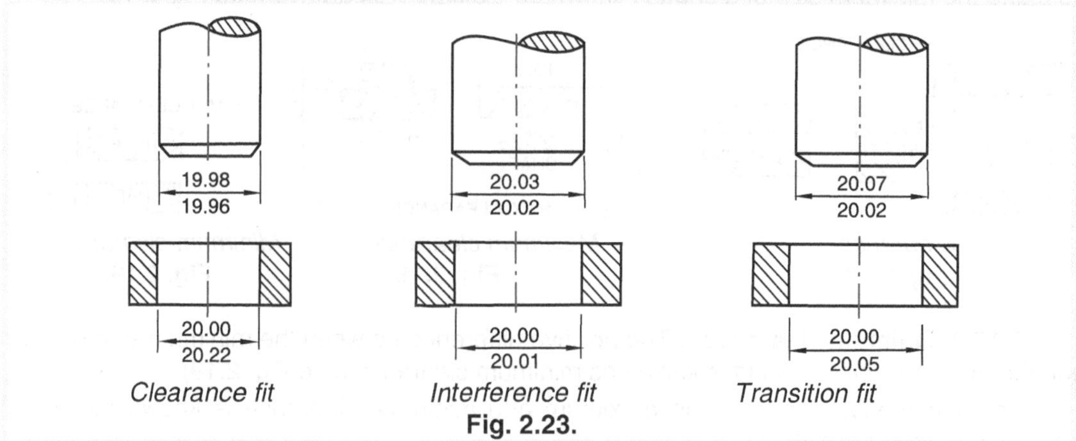

Fig. 2.23.

2.15.2. Interference fits : In a interference fit there is always a negative allowance between the largest hole and the smallest shaft (see Fig. 2.23). In this, the shaft is larger than hole. Interferences fits may be (1) shrink fit (2) heavy drive fit and (3) light drive fit.

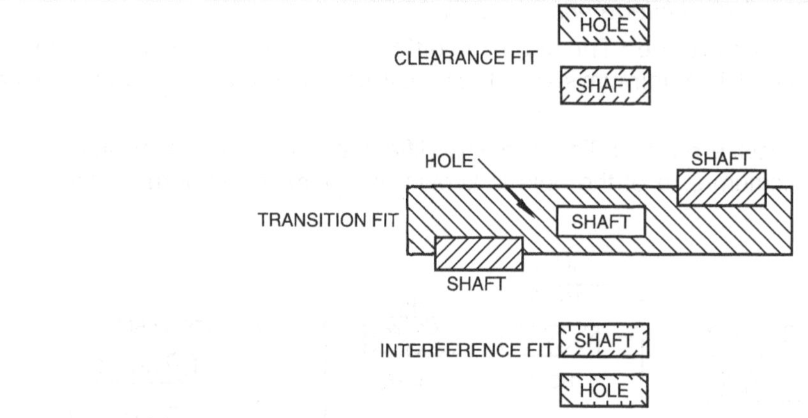

Fig. 2.24. *Disposition of tolerance zones for the three classes of fits*

2.15.3. Transition fits : In a transition fit, the fit might be either clearance fit or interference fit. In this, any pair of parts mating with transition may fit with interference, while another pair with the same fit may be a clearance fit (see Fig. 2.23 and 2.24).

Transition fit may be (i) force fit (ii) tight fit (iii) driving fit and (iv) push fit.

In ordinary machine construction the following classes of fits are commonly used:-

(1) Running fit (2) Push fit (3) Driving fit (4) Forced fit or shrink fit.

1. Running fit : In a running fit, one part can be assembled into the other so as to rotate or slide freely *e.g., shaft freely rotating in a bearing.*

2. Push filt : In push fit, one part can be assembled into other with light hand pressure and there being no sufficient clearance to allow shaft to rotate, as in locating plugs, dowels etc.

3. Driving fit : In a driving fit, one part can be assembled into the other with a hand hammer or by medium pressure e.g., a pulley fitted on a shaft with a key.

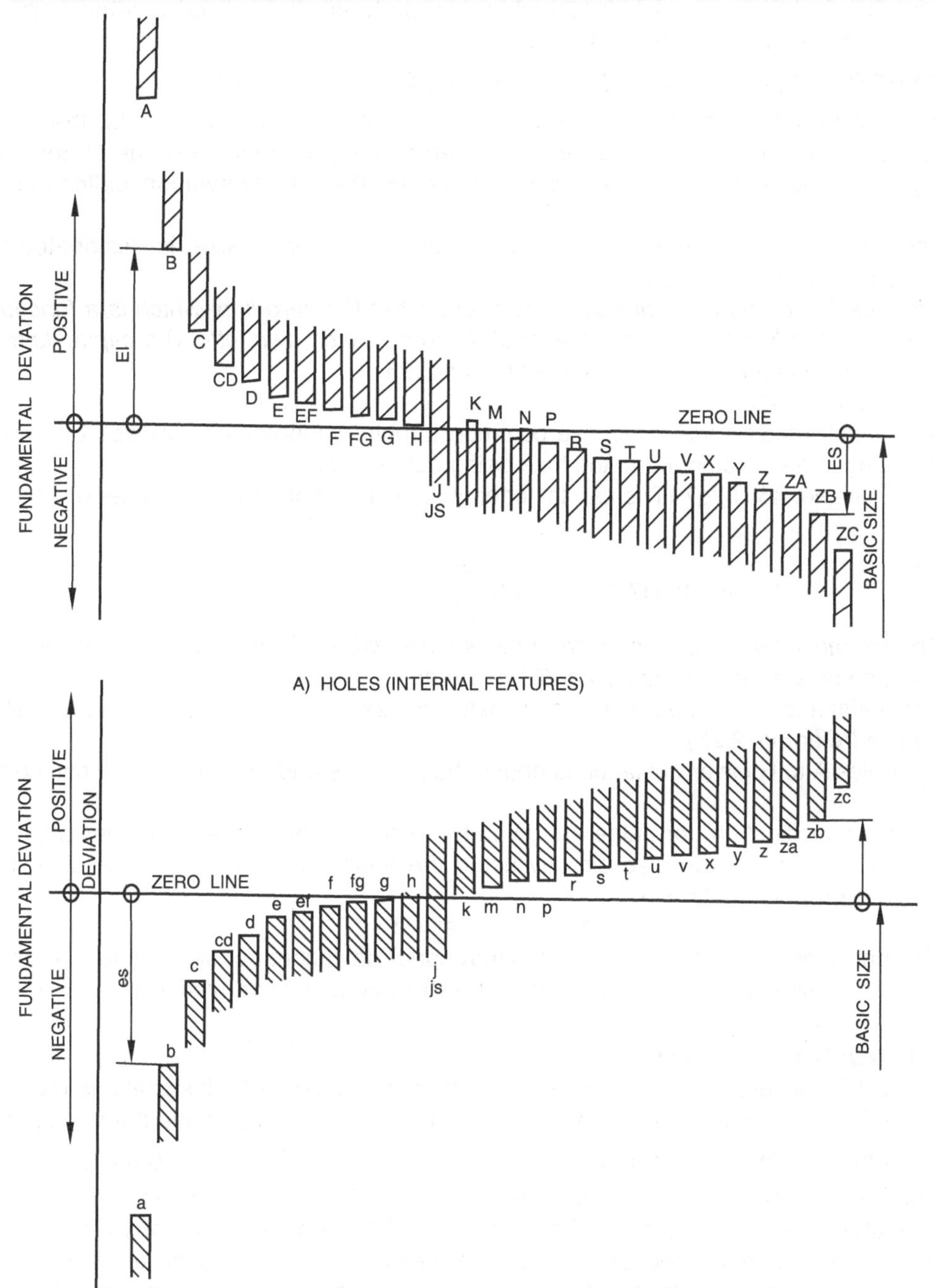

Fig. 2.25. *Letter symbols for Tolerance (examples for diameter 30 to 40 mm)*

Fundamental deviations for shafts j and k and for holes j to ZC slightly vary or different grades or sets of grades for the same diameter step and letter symbol. Therefore, Grade 7 for shafts j,k and holes J,K and grade above 7 for holes M to SC have been represented in those cases to avoid confusion in the example.

4. Forced fit or shrink fit : In force fit or shrink fit, one part (say shaft) is assembled into the other (hole) either with great pressure or the hole is expanded by heating, so as to shrink on the shaft. It is specially used when two parts are to be fixed together *e.g., cart wheels, railway and tram car wheels are fitted.*

2.16. SYMBOLS FOR TOLERANCES DEVIATIONS AND FOR FITS

In order to satisfy the usual requrements both of individual part of fits, the system provides for the given basic size, a range of tolerance together with a range of deviations defining the position of these tolerances with respect to the line of deviation, called the zero line.

The tolerance, the number which is a function of the basic size, is designated by a number symbol, called the grade.

The position of the tolerance zone with respect to the zero line, which is a function of the basic size is indicated by a letter symbol (in some cases, two letters) a capital letter for holes and a small letter for shaft as shown in Fig. 2.25.

For example : 45g6

A fit is indicated by the basic size common to both compoments, followed by symbols correspanding to each component, the hole being quoted first.

For Example: 45 H7/g7 (The capital letter H stands for hole and smal letter g stands for shaft)

$$\text{or } 45\ H7\text{-}g6 \qquad \text{or } 45\ \frac{H7}{g6}$$

The toleranced size is, thus, defined by its basic value followed by a symbol composed of a letter (in some cases, two letter) and a numeral.

For shafts a to h the upper deviation is below the zero line and for shaft j to ZC is above the zero line (refer Fig. 2.25).

For holes A to H lower deviation is above the zero line and for j to ZC is below the zero line.

For the specification, formulae are given to determine the fundamental deviation. The other deviations may be derived directly using the absolute value of the tolerance IT by means of the graphic relationship.

$$e_i = e_s - IT \qquad e_s = e_i + IT$$

The deviation given in the table 2-3 is that upper deviations for shafts a to h and lower deviation e_i for shafts J to Zc. For holes, the deviations are derived from these of the corresponding shafts i.e., $EI = e_s$

2.17. STANDARD TOLERANCES

The B.I.S. has provided, 18 grades of standard tolerances with designations ITO1, ITO and IT1 to IT 16. The numerical values of these standard tolerances have been determined in terms of the standard tolerance unit i. i.e. $i = 0.45\sqrt[3]{D} + 0.001D$

where i = standard tolerance unit in microns D = Diameter in mm

The above formula has been empirically calculated on the basis of former national standards of other countries and taking account of the fact that in most usual cass the tolerance varies more or less parabolically in terms of diameter for the same manufacturing conditions. The relative magnitude to each grade is given as under for the grade 5 to 6 in terms of standard tolerance unit 'i'.

For grade IT7, the value is 16i and above it the tolerance magnitude is multiplied by 10 to each fifth step.

GRADE	IT5	IT6	IT7	IT8	IT9	IT10	IT11	IT12	IT13	IT14	IT15	IT16
VALUES	7i	10i	16i	25i	40i	64i	00i	106i	200i	400i	640i	1000i

The value of standard tolerances corresponding to grade ITOI, ITO, ITI are:

The values of IT2 to IT4 are scaled approximately geometrically between the values of ITI and I+5

VALUES IN MICRONS	ITOI	ITO	ITI
FOR D IN MM	0.1+.008	0.5+.012D	0.8+.020D

The most commonly used shafts and holes in the industries are given in Table 2.1 where Table 2.2 gives the values of tolerances of grades from ITI to ITI6 for diameter steps upto 500 mm

TABLE 2.1

SHAFTS		HOLES	
LETTER SYMBOL	GRADES	LETTER SYMBOL	GRADES
a	9,11	A	9,11
b	8,9,11	B	8,9,11
c	8,9,11	B	8,9,11
d	5 to 11	C	6 to 11
e	5 to 9	D	5 to 10
f	4 to 9	E	5 to 9
g	4 to 7	F	5 to 7
h	1 to 16	G	1 to 16
j	5 to 7	H	6 to 8
js	1 to 16	js	1 to 16
k	4 to 7	K	5 to 8
m	4 to 7	M	5 to 8
n	4 to 7	N	5 to 11
p	4 to 7	P	5 to 9
r	4 to 7	R	5 to 8
s	4 to 7	S	5 to 7
t	5 to 7	T	5 to 7
u	5 to 8	U	6 to 7
v	5 to 7	V	6 to 7
x	5 to 8	X	6 to 7
y	6 to 7	Y	7
+z	6 to 7	Z	7 to 8
za	6 to 7	ZA	7 to 8
zb	7 to 8	ZB	8 to 9
zc	7 to 8	ZC	8 to 9

TABLE 2.2

	FUNDAMENTAL TOLERANCES OF GRADES ITI TO IT16															
DIAMETER STEPS	TOERANCE GRADES, MICRONS										MICRON or MICROMETER = 0.001 MM					
IN MM	IT1	IT2	IT3	IT4	IT5	IT6	IT 7	IT8	IT9	IT10	IT11	IT12	IT13	IT14	IT15	IT16
OVER 1 TO AND *INC. 3	1.5	2	3	4	5	7	9	14	25	40	60	90	140	250	400	600
OVER 3 TO AND INC 3	1.5	2	3	4	5	8	12	18	30	48	75	120	180	300	480	750
OVER 6 TO AND INC.10	1.5	2	3	4	6	9	15	22	36	58	90	150	220	360	580	900
OVER 10 TO AND INC.18	1.5	2	3	5	8	11	18	27	43	70	110	180	270	430	700	1100
OVER 18 TO AND INC.30	1.5	2	4	5	9	13	21	33	52	84	130	210	330	520	840	1300
OVER 30 TO AND INC.50	2	3	4	7	11	16	25	39	62	100	160	250	350	620	1000	1600
OVER 50 TO AND INC.80	2	3	5	8	13	19	30	46	74	120	190	300	460	740	1200	1900
OVER 80 TO AND INC120	3	4	6	10	15	22	35	54	87	140	220	350	540	870	1400	2200
OVER 120 TO AND INC180	4	5	8	12	18	25	40	63	100	160	250	400	630	1000	1600	2500
OVER 180 TO AND INC.250	5	7	10	14	20	29	46	72	115	185	290	460	720	1150	1800	2900
OVER 250 TO AND INC.315 3200	6	8	12	16	23		32	52	81	130	210	320	520	810	1300	2100
OVER 315 TO AND INC.400	7	9	13	18	25	36	57	89	140	230	360	570	810	1400	2300	3600
OVER 400 TO AND INC.500	8	10	15	20	27	40	63	97	155	250	400	630	970	1550	2500	4000

*INC.=INCLUDING

2.18. LIMIT GAUGES

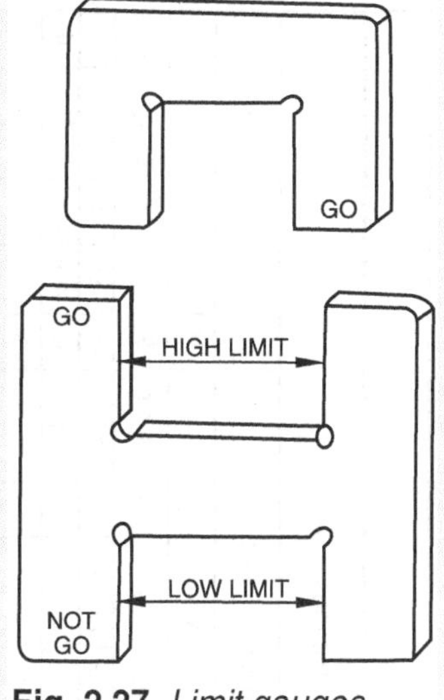

Fig. 2.27. *Limit gauges*

In the production of interchangeable parts, two sets of limit gauges are necessary for checking the size, shape and relative positions of various parts.

One set is necessary during the course of production and other set for inspection purpose. There are two gauges in one set named as Go limit gauge for Go and not Go gauge for Not go limit.

2.18.1 Go limit : The Go limit applied to that of the two limits of size which corresponds to the maximum material condition, that is ;

(a) The upper limit of a shaft, and

(b) The lower limit of a hole.

When limit gauages are used, this is the lilmit of size checked by the `Go' gauge.

2.18.2 Not Go Limit : The Not go limit applied to that of the two limits of size which corresponds to the minimum material condition, that is ;

(a) the lower limit of a shaft, and

(b) the upper limit of a hole.

When limit gauges are used, this is the limit of size checked by the Not Go gauge.

Example : To find upper and lower limits of 50 H7/g6. [Refer Table 2.3]

OVER TO		HOLE H7	SHAFT g6
50 80	GO	+ . 000	- . 009
	NOT GO	+ . 025	- . 025

In go limit	+ . 000 lower limit of hole - . 009 upper limit of shaft	$\therefore$ 50 H7 = $50^{+.025}_{+.000}$
in not go limit	+ . 025 upper limit of hole - . 025 lower limit of shaft	$\therefore$ 50 g6 = $50^{-.009}_{-.025}$

2.19. HOLE BASIS AND SHAFT BASIS SYSTEMS

In a general limit system it is necessary to decide on what basis the limit are to be taken to give the desired fit. The following two systems are used for varying the size of parts.

1. Hole basis system ;

2. Shaft basis system.

2.19.1. Hole basis system : In the hole basis system the hole is the constant member and different fits are obtained by varying the size of the shaft (see Fig. 2.28 (i)). Here the high and low limits of the hole are constant (or fixed) for all fits of the same accuracy grade and for the same basis size.

UNIT HOLE SYSTEM I.S.O.

			DIFERENCE IN MEASUREMENT OF NOMINAL DIAMETER IN MICRONS																																
			VERY ACCURATE WORK							ACCURATE WORK												ORDINARY WORK										ROUGH WORK					
			HOLE							HOLE	SHAFT											HOLE	SHAFT									HOLE	SHAFT				
				CLEAR. FIT		TRANS. FIT			INT. FIT		CLEAR. FIT					TRANS. FIT				INT. FIT		CLEAR. FIT					TRANS. FIT					CLEAR. FIT				
OVER	TO		H6	g5	h5	J5	mS	mS	nS	H7	ds	eS	f7	g6	h6	j6	k6	m6	n6	P6	H8	d10	e9	f8	h8	h7	j7	k7	m7	n7	HII	all	bll	cll	dll	hll
1	3	GO	0	−3	0	+4	–	+7	+11	0	−20	−14	−7	−13	0	+6	–	+9	+3	+16	0	−20	−14	−7	0	0	+7	–	–	+15	0	−270	−140	−60	−20	0
		NOT GO	7	−8	−5	−1	–	+2	+6	+0	−45	−28	−16	−10	−7	−1	–	+2	+6	+9	+14	−60	−39	−21	−14	−9	−2	–	–	+6	+60	−330	−200	−120	−80	−60
3	6	GO	0	−4	0	+4	–	+9	+13	0	−30	−20	−10		0	+7	–	+12	+16	+20	0	−30	−20	−10	0	0	+5	–	–	+20	0	−270	−140	−70	−30	0
		NOT GO	8	−3	−5	−1	–	+4	+8	−12	−60	−38	−22	−12	−8	−1	–	+4	+8	+12	+18	−78	−50	−28	−18	−12	−3	–	–	+8	+75	−345	−215	−145	−105	−75
6	10	GO	0	−5	0	+4	+7	+12	+16	0	−40	−25	−13		0	+7	+10	+15	+19	+24	0	−40	−25	−13	0	0	+10	+16	+21	+25	0	−280	−150	−180	−40	0
		NOT GO	9	−11	−6	−2	+1	+6	+10	−15	−76	−47	−28	−14	−9	−2	+1	+6	+10	+15	+22	−58	−61	−35	−22	−15	−5	+1	+6	+10	+90	−370	−240	−170	−130	−90
10	18	GO	0	−6	0	+5	+9	+15	+20	0	−50	−32	−16	−6	0	+8	+12	+18	+23	+29	0	−50	−32	−16	0	0	+12	+19	+25	+30	0	−290	−150	−95	−50	0
		NOT GO	11	−14	−8	−3	+1	+7	+12	−18	−93	−59	−34	−17	−11	−3	+1	+7	+12	+18	+27	−120	−75	−43	−27	−18	−6	+1	+7	+12	+110	−400	−260	−205	−150	−110
18	30	GO	0	−7	0	+5	+11	+17	+24	0	−65	−40	−20	−7	0	+9	+15	+21	+28	+35	0	−65	−40	−20	0	0	+13	+23	+29	+36	0	−300	−160	−110	−65	0
		NOT GO	13	−16	−9	−4	+2	+8	+15	−21	−117	−73	−41	−20	−13	−4	+2	+8	+15	+22	+33	−149	−92	−53	−33	−21	−8	+2	+6	+15	+130	−430	−290	−240	−195	−130
30	50	GO	0	−9	0	+6	+13	+20	+28	0	−60	−50	−25	−9	0	+11	+18	+25	+33	+42	0	−80	−50	−25	0	0	+15	+27	+34	+42	0	−310	−170	−120	−80	0
		NOT GO	16	−20	−11	−5	+2	+9	+17	−25	−142	−69	−50	−25	−16	−5	+2	+9	+17	+26	+39	−180	−112	−84	−39	−25	−10	+2	+9	+17	+160	−400	−340	−290	−240	−160
50	80	GO	0	−10	0	+6	+15	+24	+23	0	−100	−50	−30	−10	0	+12	+21	+30	+39	+51	0	−100	−60	−30	0	0	+18	+32	+41	+50	0	−340	−130	−140	−100	0
		NOT GO	+19	−23	−13	−7	+2	+11	+20	−30	−174	−108	−60	−29	−18	−7	+2	+11	+20	+32	+46	−220	−134	−76	−46	−30	−12	+2	+11	+20	+190	−550	−390	−340	−290	−190
80	120	GO	0	−12	0	+6	+18	+28	+38	0	−120	−72	−36	−12	0	+13	+25	+35	+45	+59	0	−120	−72	−38	0	0	+20	+38	+48	+58	0	−380	−220	−170	−120	0
		NOT GO	+22	−27	−15	−9	+3	+13	+23	−35	−20	−126	−71	−34	−22	−9	+3	+13	+23	+37	+54	−260	−159	−90	−54	−35	−14	+3	+13	+23	+220	−630	−400	−400	−340	−220
120	180	GO	0	−14	0	+7	+21	+33	+45	0	−145	−85	−43	−14	0	+14	+28	+40	+52	+68	0	−145	−85	−43	0	0	+22	+43	+56	+67	0	−460	−260	−200	−145	0
		NOT GO	+25	−32	−18	−11	+3	+15	+27	−40	−245	−148	−83	−33	−25	−11	+3	+15	+27	+43	+63	−305	−185	−106	−63	−40	−19	+3	+15	+27	+250	−830	−560	−460	−395	−250
180	250	GO	0	−15	0	+7	+24	+37	+51	0	−170	−100	−50	−15	0	+16	+33	+46	+60	+79	0	−170	−100	−15	0	0	+25	+50	+63	+77	0	−660	−340	−240	−170	0
		NOT GO	+29	−35	−20	−13	+4	+17	+31	46	−285	−172	−96	−44	−29	−13	+4	+17	+31	+50	+72	−355	−215	−122	−72	−46	−21	+4	+17	+31	+290	−120	−710	−570	−460	−290
250	315	GO	0	−17	0	+7	+27	+43	+57	0	−190	−110	−56	−17	0	+16	+36	+52	+66	+88	0	−190	−110	−56	0	0	+26	+56	+72	+86	0	−920	−480	−300	−190	0
		NOT GO	+32	−40	−23	−16	+4	+20	+34	−52	−320	−191	−108	−49	−32	−16	+4	+20	+34	+56	+81	−400	−240	−137	−81	−52	−26	+4	+21	+34	+320	−1370	−560	−650	−510	−320
315	400	GO	0	−18	0	+7	+29	+48	+52	0	−210	−125	−52	−18	0	+18	+40	+57	+73	+98	0	−210	−125	−62	0	0	+29	+61	+78	+94	0	−1200	−600	−360	−210	0
		NOT GO	+36	−43	−25	−18	+4	+21	+37	−57	−350	−214	−119	−54	−36	−18	+4	+21	+37	+62	+89	−440	−265	−151	−89	−57	−28	+4	+21	+37	+360	−1710	−1040	−760	−570	−360
400	500	GO	0	−20	0	+7	+32	+50	+57	0	−230	−135	−68	−20	0	+20	+45	+63	+60	+108	0	−230	−135	−68	0	0	+31	+68	+86	+103	0	−1500	−760	−440	−230	0
		NOT GO	+40	−47	−27	−20	+5	+23	+40	−63	−385	−232	−131	−60	−40	−20	+5	+23	+40	+68	+97	−480	−290	−165	−97	−63	−32	+5	+21	+40	+400	−2050	−1240	−880	−630	−400

CLEAR. — CLEARANCE TRANS. — TRANSITION INT. — INTERFERENCE

TABLE 2.4

GUIDE TO THE SELECTION OF FITS

H11-c11 SLACK RUNNING FIT	USED TO GIVE FLEXIBILITY UNDER LOAD, EASY ASSEMBLY OR A CLOSE FIT AT ELEVATED WORKING TEMP.	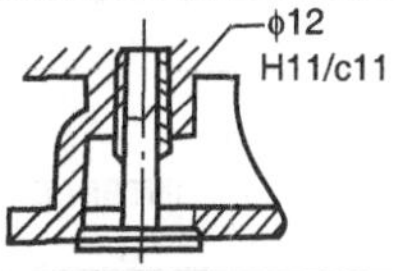 ϕ12 H11/c11	EXAMPLES I.C. ENGINE EXHAUST VALUE IN GUIDE
H9-d10 LOOSE RUNNING FIT	USED FOR GLAND SEALS PULLEYS AND VERY LARGE BEARING	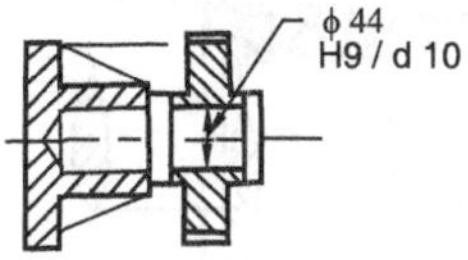 ϕ 44 H9 / d 10	DLER GEAR ON SPINDLE
H9-c9 EASY RUNNING FIT	USED FOR WIDELY SEPARATED BEARINGS OR SEVERAL BEARINGS IN LINE	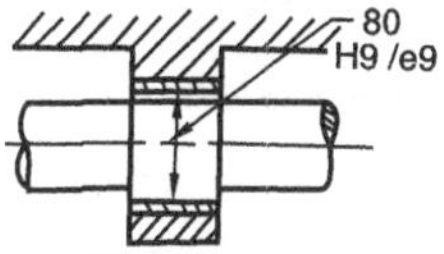 80 H9 /e9	CAMSHAFT IN BEARING
H8-f7 NORMAL RUNNING FIT	SUITABLE FOR APPLICATIONS REQUIRING A GOOD QUALITY FIT THAT IS EASY TO PRODUCE	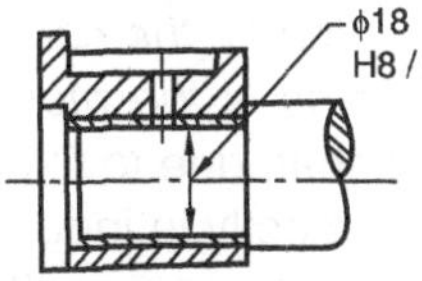 ϕ18 H8 / f7	GEARBOX SPINDLE IN BEARING
H7 - g6 SLIDING AND LOCATION FIT	NOT NORMALLY USED FOR CONTINUOUSLY RUNNING BEARINGS UNLESS LOAD IS LIGHT, SUITABLE FOR PRECISION SLIDING AND LOCATION.	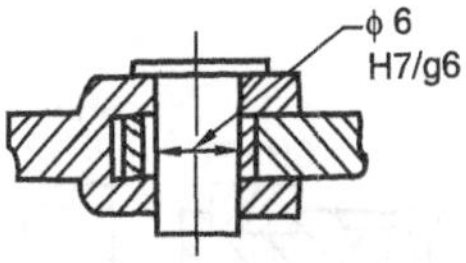 ϕ 6 H7/g6	VALAVE MECHANISM LINK
H7- h6 LOCATION FIT	SUITABLE FOR MANY NON-RUNNING ASSEMBLIES.	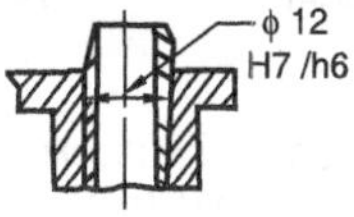 ϕ 12 H7 /h6	VALVE GUIDE IN HEAD
H7 - K6 PUSH FIT	USED FOR LOCATION FITS WHEN A SLIGHT INTERFERENCE, WHICH ELIMINATES MOVEMENTS OF ONE PART RELATIVE TO THE OTHER, IS AN ADVANTAGE	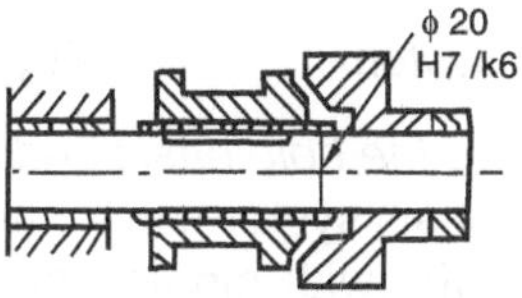 ϕ 20 H7 /k6	CALUTCH MEMBER KEYED TO SHAFT
H7 - n6 TIGHT ASSEMBLY FIT	USED WHEN DEGREE OF CLEARANCE THAT CAN RESULT FROM H7-k6 FIT IS NOT ACCEPTABLE.	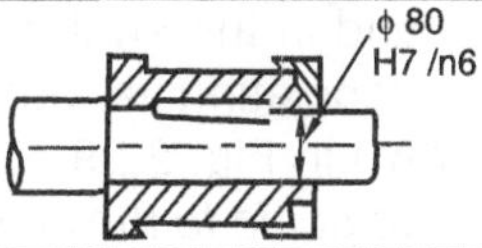 ϕ 80 H7 /n6	CAMMUTATOR SHELL ON SHAFT
H7 - n6 PRESS FIT	FERROUS PARTS ARE NOT OVER STRAINED DURING ASSEMBLY OR DISMANTLING.	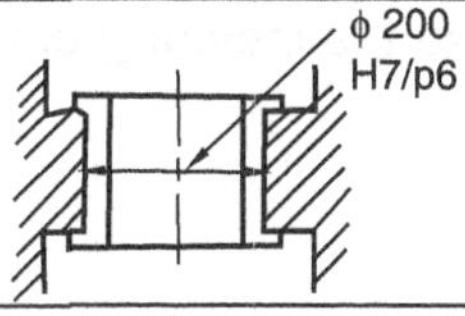 ϕ 200 H7/p6	SPLIT JORUNAL BEARING
H7 - s6 HEAVY FIT	MAINLY USED FOR PERMANENT ASSEMBLIES	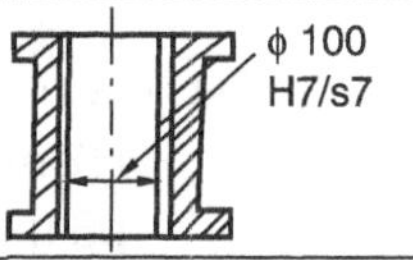 ϕ 100 H7/s7	CYLINDER LINER IN BLOCK

2.19.2. Shaft basis system : In the shaft basis system the shaft is constant member and different fits are obtained by varaing the size of the shaft (see Fig.2.28 (ii)). Here, the high and low limits of the shaft are constant for all fits of the same accuracy grade and for the same basis size.

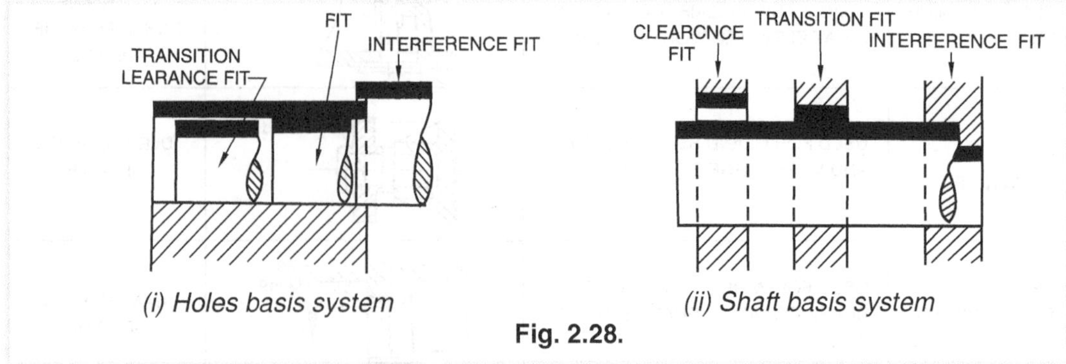

(i) Holes basis system *(ii) Shaft basis system*

Fig. 2.28.

The application of either system depends on many conditions such as the nature of the product, manufacturing methods, the condition of the raw material etc.

The hole basis systems is the most extensive in use. This is because a hole is more difficult to produce than a shaft due to fixed character of hole producing tools. The shaft basis system, however, is used mostly in industries using semifinished shafting in raw materials.

The designer should decide on the adoption of either system to secure general interchangeability.

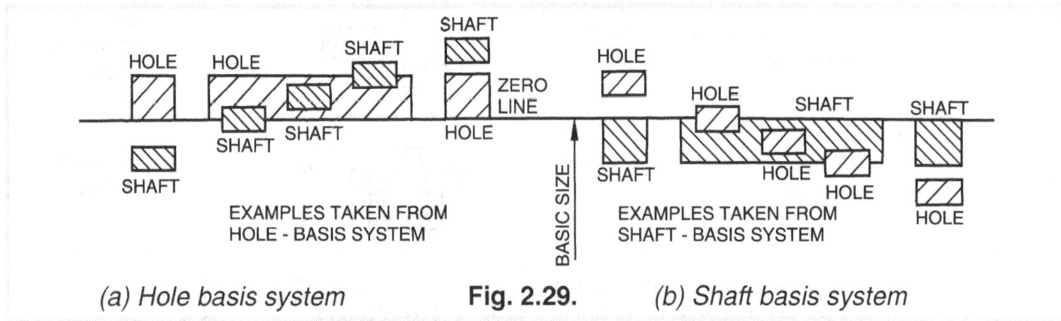

(a) Hole basis system **Fig. 2.29.** *(b) Shaft basis system*

Fig. 2.29 *illustrating the hole basis and shaft basis systems.*

Thus, in the hole basis system, the different clearance and interference are obtained in associating various shafts with single hole, whose lower deviation is zero (symbol H) as shown in Fig. 2.29 *(a)* and in the shaft basis system, the different clearances and interferences are obtained in associating various holes with a single shaft whose upper deviation is zero (symbol h) as shown in Fig. 2.29 *(b)*.

Table 2.4 shows the guide to the selection of fits generally used in the industries.

Problem 1: Give the meaning of the following:

(i) ϕ 48H7/p6

(ii) 46g6

(iii) $50^{+.015}_{+.000}$

Solution : (i) Basic size 48 mm diameter

H7 means an H hole to tolerance grade 7

p6 means an p shaft to tolerance grade 6

(ii) Basic size 46 mm

g6 means g shaft to tolerance grade 6

(iii) A tolerance of 0.015 mm on 50 mm diameter

i.e., Tolerance = 50.015 – 50.000 = .015 mm

Problem 2 : Find the basic size, type of fit and tolerance of a 45 mm diameter shaft rotating normal in a bearing.

Solution : (i) Basic size : 45 mm

(ii) Type of fits : Normal running fit

So tolerance for the shaft and bearing is 45 H8 / f7 [Refer table 2-4 of `Guide' to the selection of fits']

H8 means an 'H' hole to tolerance grade 8

f 7 means a "f" shaft to tolerance grade 7

(iii) **Tolerance** : To find tolerance of 45 H8/f7, proceed as under;

(*Refer table 2.3 of unit hole system I.S.O.*)

Method 1 : As 45 mm lies in diameter step of 30 and 50 mm. Therefore tolerance table for this can be prepared as under:

OVER TO		HOLE H8	SHAFT f8
30 – 50	GO	+ . 000	– . 025
	NOT GO	+ . 039	– . 064

In Go limit	– . 025 upper limit of shaft + . 000 lower limit of hole	$\therefore$ 45 H8 = 45 $^{+.039}_{+.000}$
In not go limit	+ . 039 upper limit of hole – . 064 lower limit of shaft	$\therefore$ 45 f8 = 45 $^{-.025}_{-.064}$

Method 2 : The 45 mm lies in diameter step 30 to 50 mm

$$\therefore D = \sqrt{30 \times 50} \quad \sqrt{1500} = 38.67 \text{ mm}$$

Value of tolerance unit i = 0.45 $\sqrt[3]{D}$ + 0.001 D

$$=0.45\sqrt[3]{38.67} + 0.001 \times 38.67$$

$$=1.521+0.0386$$

$$=1.5596 \text{ microns}$$

$$=0.00156 \text{ mm}$$ [1 mm=1000 microns]

For hole of grade H8 :

Fundamental tolerance IT8 = 25*i* [Refer table 2.2 of fundamental tolerance]

= 25 × 0.00156

= 0.039 mm –This value can be read directly from Table 2.4

Fundamental deviation = 0

Hole limits $45\ ^{+.000}_{+.039}$

For shaft of grade f8 :

Fundamental tolerance = 25 *i*

= .25 × 0.00156

= – 0.39 mm

Problem 3 : Fix the limits of tolerance and allowance for a 25 mm diameter shaft and hole pair designated H8/d9. Comment on the application of this type of fit represented.

Solution : To find the limits of tolerance, Refer table 2.4 of unit hole system ISO.

Method 1 : As 25 mm lies in diameter step of 18 and 30 mm. Therefore, tolerance table for this can be prepared as under :

OVER TO		HOLE H8	SHAFT d9
18 – 30	GO	+ . 000	– . 117
	NOT GO	+ . 033	– . 065

In go limit	– . 117 lower limit of shaft + . 000 lower limit of hole	∴ 25 H8 = $25\ ^{+.033}_{+.000}$
In not go limit	+ . 033 upper limit of hole – . 065 upper limit of shaft	∴ 25 d9 = $25\ ^{-.065}_{-.117}$

Method 2 : To find the tolerance for shaft and hole pair designated 25 H8/d9. Here 25 mm lies in step 18 to 30 mm (refer table 2-2).

$$D = \sqrt{18 \times 30} = 540 = 23.2 \text{ mm}$$

∴ Value of tolerance unit i = $0.45\ \sqrt[3]{D} + .001$

$= 0.45\ \sqrt[3]{23.3} + .001 \times 23.2$

1.30 micronons

= 0.001 mm

For hole of quality 8:

Fundamental tolerance IT8 = 33 *i* [Refer table 2.2 of fundamental deviations]

= 33 × 0.001

=.033

Fundamental deviation = 0

∴ Hole limits are 25 H8 = $25\ ^{+.033}_{+.000}$

For shaft quality 9

Fundamental tolerance IT9 = - 52 microns [Refer table]

Fundamental deviation $= -16\ D^{0.44}$

$= -16 \times (23.2)^{0.44}$

$\therefore\ D = \sqrt{18 \times 30}$ limits

$= 23.2$ mm

$= 16 \times (23.2)^{0.44}$

$= -65$ microns

$\therefore$ The limits for shaft 25 d9 are 25 – 0.065 = 24.935 mm and

25 – . 065 = 24.833 mm

i.e., $25\ d9 = 25^{-.56}_{-.117}$

$\therefore$ Tolerance = 24.935 – 24.833 = 0.052 mm

By refering the selection of fits table, we see that designation H8/d9 shows a fit which is suitable for loose running fits e.g. plummer block, loose pulley etc.

2.20. MACHINING SYMBOLS AND SURFACE FINISH

In engineering practice, some surfaces of the casting are required to machined whereas the others are left unmachined. The surfaces to be machined are to be indicated on the drawing so that the pattern maker may leave the machining allowance (see Fig. 2.30). similarly, the grade of surface finish is required to be indicated to enable the machinist to carry out the job. Thus, on production drawings where it is necessary to indicate that the surfaces are to be machined or finished in some way, certain specific symbols are used.

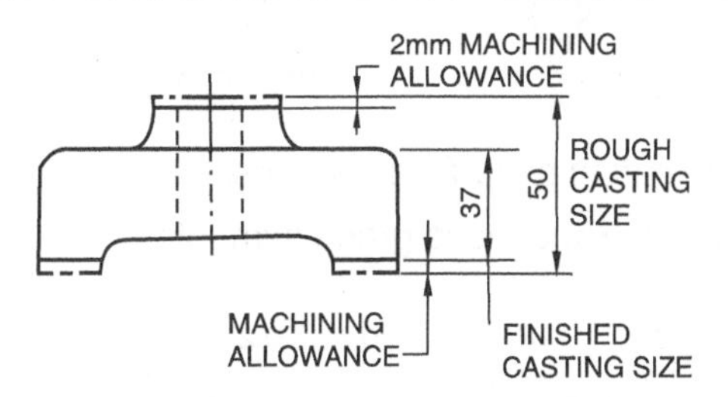

Casting with extra metal allowed for machining

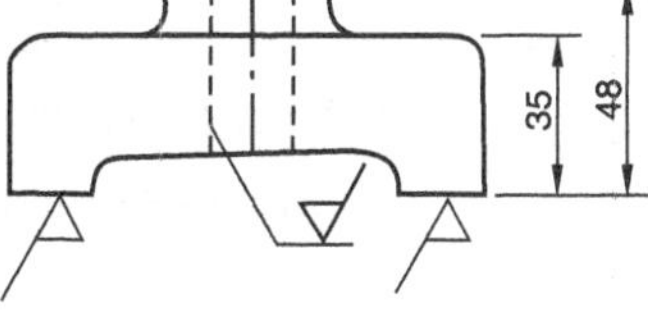

Finished cast part

Fig. 2.30. *Allowance for machining*

Note : *2 to 4 mm allowance are allowed on small casting and forgings for surface requiring machining.*

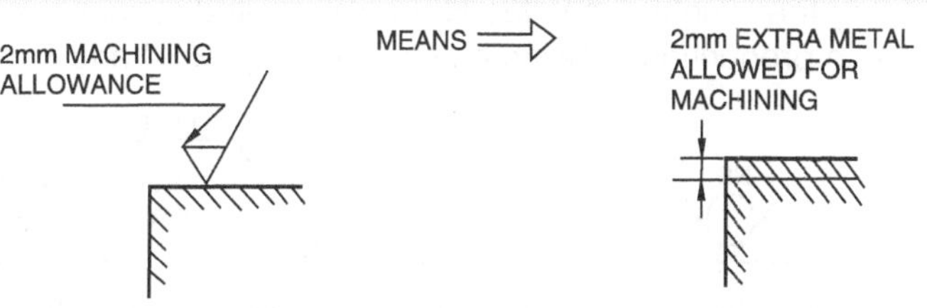

Fig. 2.31. *Indication of machinig allowance*

When the value of the machining allowance is specified on the drawing, it is indicated to the left of the system as shown in Fig. 2.31. The value of allowance is expressed in millimetres.

The basic symbol used for indication of surface roughness consists of two legs of unequal length inclined at 60° to the line representing the surface under consideration (see Fig. 2.32). It may only be used alone when its meaning is expressed by a note.

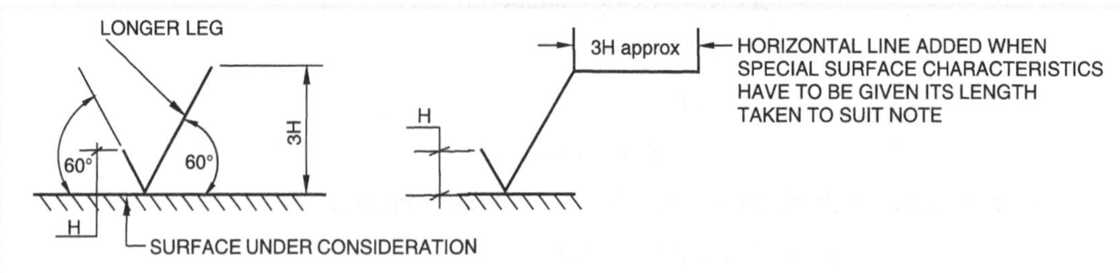

Fig. 2.32. *Basic symbol for indication of surface roughness.*

(a) Thus when the surface is produced by any method , it is indicated as by the shown in Fig. 2.33 (a).

(b) When the removal of material by machining is required, a bar is added to the basic symbol [see Fig. 2.33 (b)]

(c) Wherever the romoval of material is not permitted, a circle is added to the basic symbol [see Fig. 2.33(c)]

(d) When some special surface characteristics are to be indicated (say a milled surface) a line is added to the longer leg of the basic symbol [see Fig. 2.33 (d)]

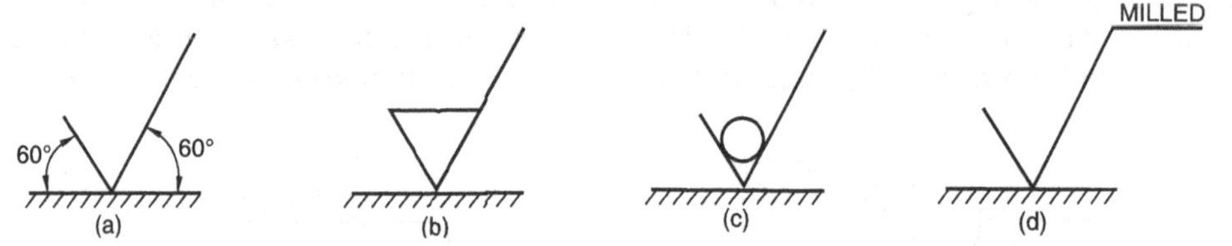

Fig. 2.33. *Symbols used for indication of surface roughness*

Roughness value R_a (μ m) and Roughness grade symbols

ROUGHNESS VALUE R_a	50	25	12.5	6.3	3.2	1.6	0.8	0.4	0.2	0.10	.050	.025
ROUGHNESS GRADE SYMBOLS	N12	N11	N10	N9	N8	N7	N6	N5	N4	N3	N2	N1

2.21. INDICATION OF SURFACE ROUGHNESS

The value defining the roughness value R_a in micron and roughness grade symbols as given above are given in production drawing as shown in Fig. 2.34 and Fig.2.35.

Fig. 2.34. *Indication of surface roughness in micrometres or by roughness grade symbol*

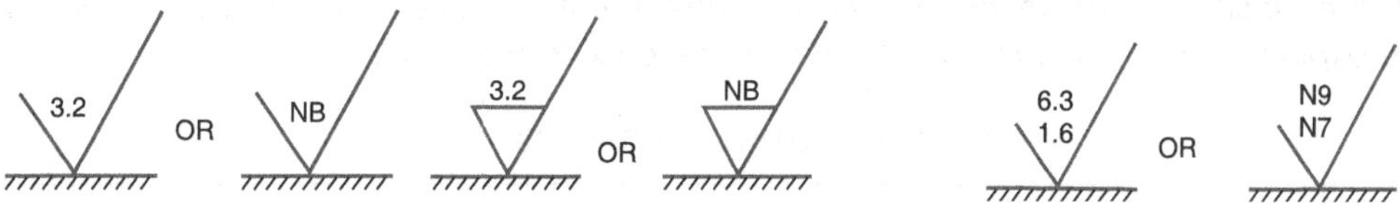

Fig. 2.35 *Indication of surface roughess in micrometres or by roughness grade symbol.*

Fig. 2.36. *Indication of the maximum & minimum limits of surface roughness.*

Notes : 1. *When it is necessary to specify the maximum and minimum limits of the surface roughness, both the values or grades should be given as shown in Fig. 2.36.*

2. *If it is necessary to indicate the sampling length, it is shown adjacent to the symbol [see Fig. 2.37 (a)]*

3. *If it is necessary to control direction of lay or the direction of the predominant surface patterns, it is indicated by a corresponding symbol added to the surface roughness symbol [see Fig. 2.37(b)]*

4. *Whenever, it necessary to specify the value of machining allowance, it is indicated on the left of the symobl [see Fig. 2.37(c)] . This value is generally expressed in millimeters.*

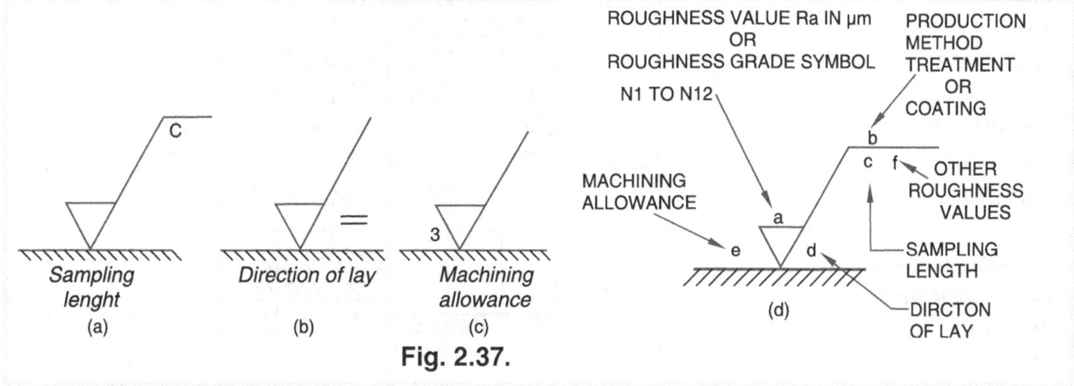

Fig. 2.37.

[*Refer table 2.5 for direction of lay*]

Thus, combining the above points, we can establish that the specification of surface roughness should be placed relative to the symbol as shown Fig. 2.37 (d)].

Where, a = Roughness value R_a in micrometres

or Roughness grade symbol N1 to N12

b = Production method, treatment or coating to be used

c = Sampling length

d = Direction of lay

e = Manufacturing allowance

f = Other roughness value in bracket

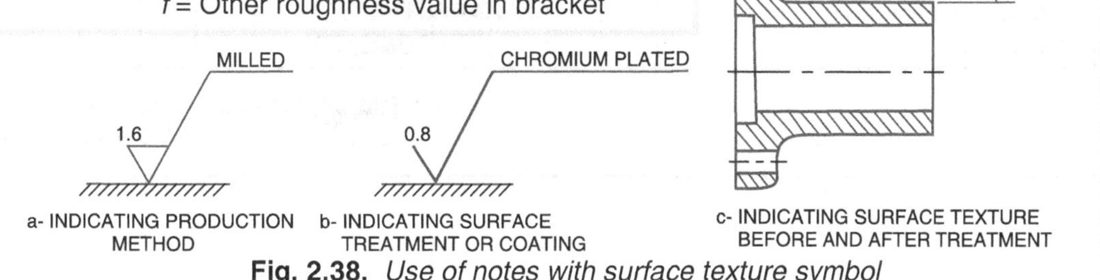

a- INDICATING PRODUCTION METHOD

b- INDICATING SURFACE TREATMENT OR COATING

c- INDICATING SURFACE TEXTURE BEFORE AND AFTER TREATMENT

Fig. 2.38. *Use of notes with surface texture symbol*

5. *If it is necessary to define surface roughness both before and after treatment, this should be explained in a suitable note or in accordance with Fig. 2.38.*

SURFACE TEXTURE ROUGHNESS

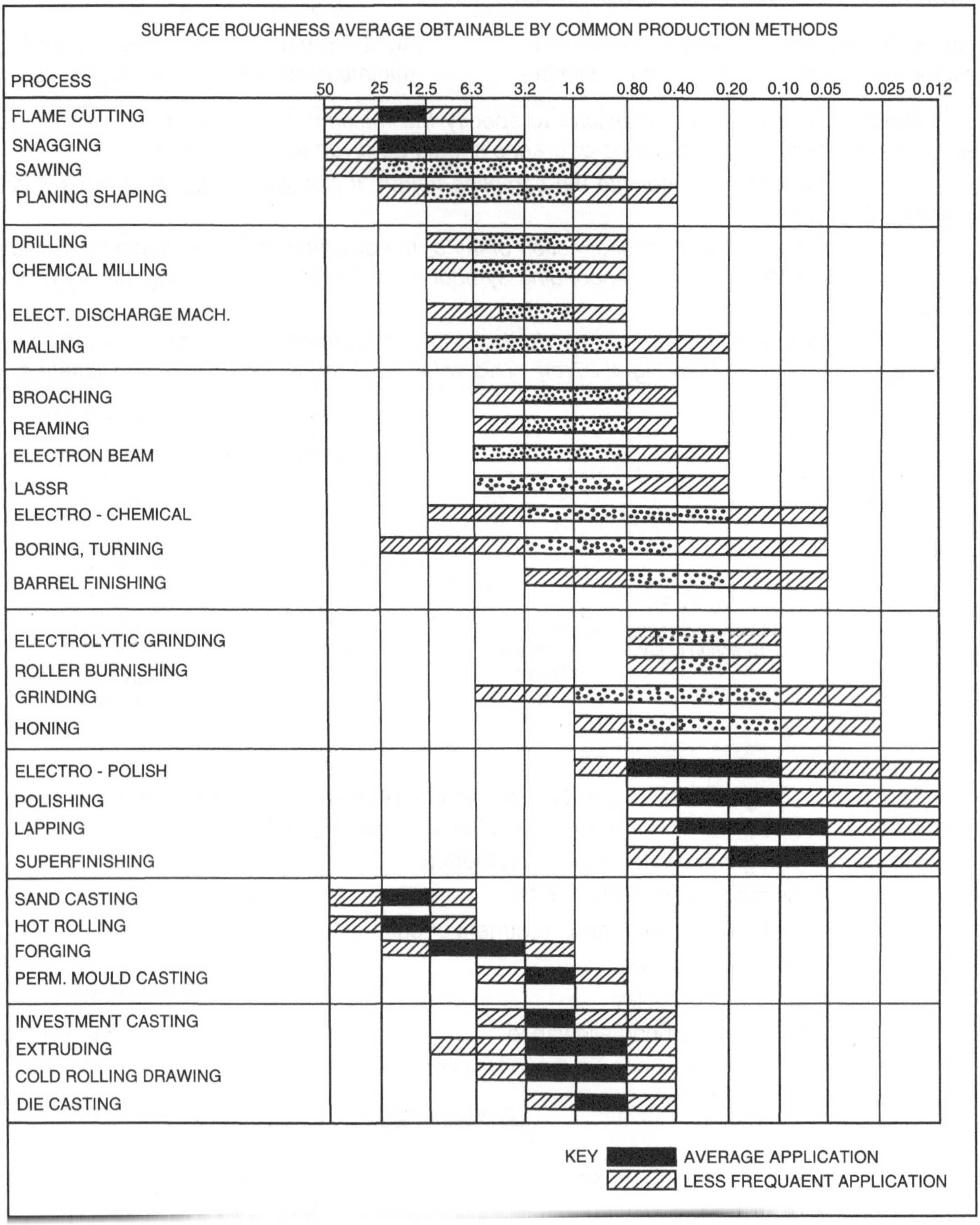

SURFACE ROUGHNESS AVERAGE OBTAINABLE BY COMMON PRODUCTION METHODS

PROCESS	50	25	12.5	6.3	3.2	1.6	0.80	0.40	0.20	0.10	0.05	0.025	0.012
FLAME CUTTING													
SNAGGING													
SAWING													
PLANING SHAPING													
DRILLING													
CHEMICAL MILLING													
ELECT. DISCHARGE MACH.													
MALLING													
BROACHING													
REAMING													
ELECTRON BEAM													
LASSR													
ELECTRO - CHEMICAL													
BORING, TURNING													
BARREL FINISHING													
ELECTROLYTIC GRINDING													
ROLLER BURNISHING													
GRINDING													
HONING													
ELECTRO - POLISH													
POLISHING													
LAPPING													
SUPERFINISHING													
SAND CASTING													
HOT ROLLING													
FORGING													
PERM. MOULD CASTING													
INVESTMENT CASTING													
EXTRUDING													
COLD ROLLING DRAWING													
DIE CASTING													

KEY
AVERAGE APPLICATION
LESS FREQUAENT APPLICATION

TABLE 2.5

SYMBOLS FOR DIRECTION OF LAY

SYMBOLS	INTERPRETATION	
=	PARALLE TO THE PLANE OF PROJECTION OF THE VIEW IN WHICH THE SYMBOL IS USED.	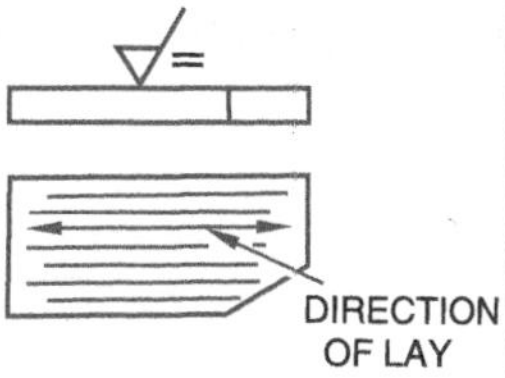 DIRECTION OF LAY
⊥	PERPENDICULAR TO THE PLANE OF PROJECTION OF THE VIEW IN WHICH THE SYMBOL IS USED.	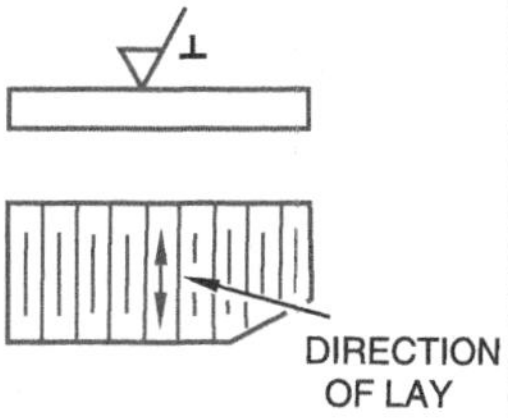 DIRECTION OF LAY
X	CROSSED IN TWO SLANT DIRECTIONS RELATIVE TO THE PLANE OF PROJECTION OF THE VIEW IN WHICH THE SYMBOL IS USED.	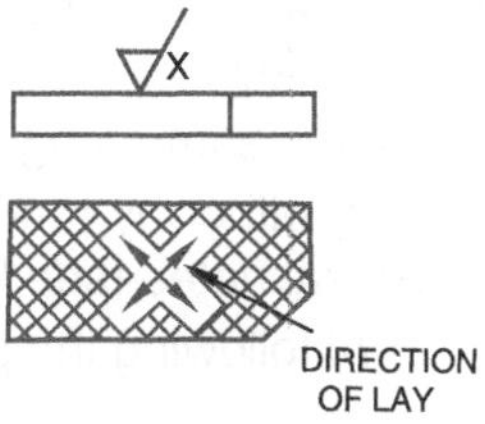 DIRECTION OF LAY
M	MULTI-DIRECTIONAL	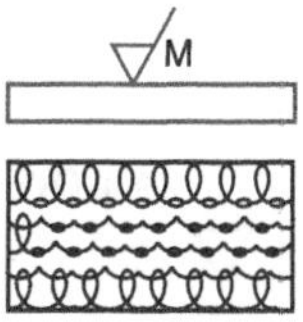 M
C	APPROXIMATELY CIRCULAR RELATIVE TO THE CENTRE OF THE SURFACE TO WHICH THE SYMBOL IS APPLIED	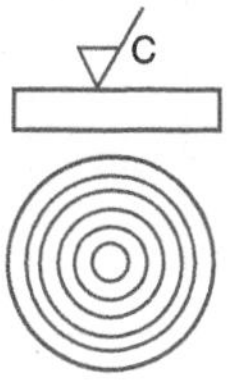 C
R	APPROXIMATELY RADIAL RELATIVE TO THE CENTRE OF THE SURFACE TO WHICH THE SYMBOL IS APPLIED	 R

2.22. METHODS OF PLACING MACHINING SYMBOLS ON ORTHOGRAPHIC VIEWS

The machining symbol should be placed on orthographic views in such a way that along with the inscription, they may be read from the bottom or the right hand side of the view as shown in Fig. 2.39. If required, the symbol may be connected to the surface by a leader line terminating in an arrow. The symbol or the arrow should point from outside the material of the component either to the line representing the surface or to an extension of it.

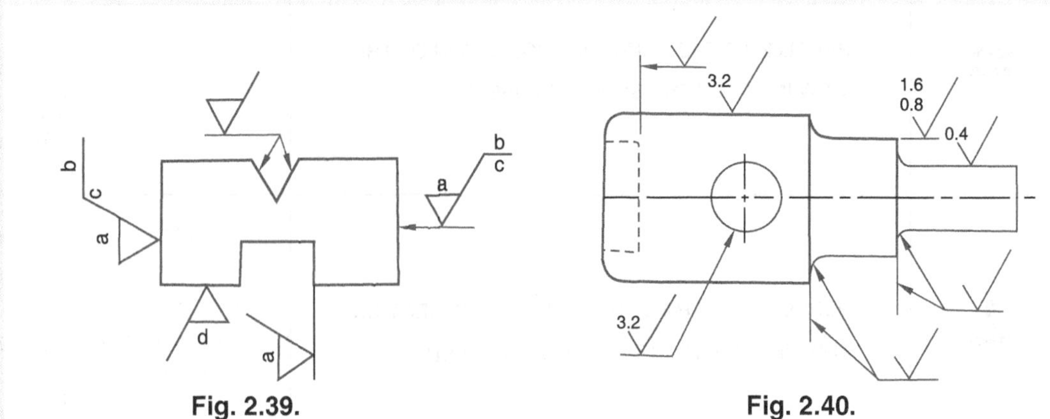

Fig. 2.39. Fig. 2.40.

In accordance with the general principles of dimensioning, the symbol should be used once for a given surface and, if possible on the view which carries the dimension defining the size of position of the surface (see Fig. 2.40).

If the same roughness is required on all the surfaces of a part, it is specified:

(a) either by a note near a view of the part, near the title block or in the space devoted to general notes (see Fig. 2.41).

(b) following the part number on the drawing (see Fig. 2.42).

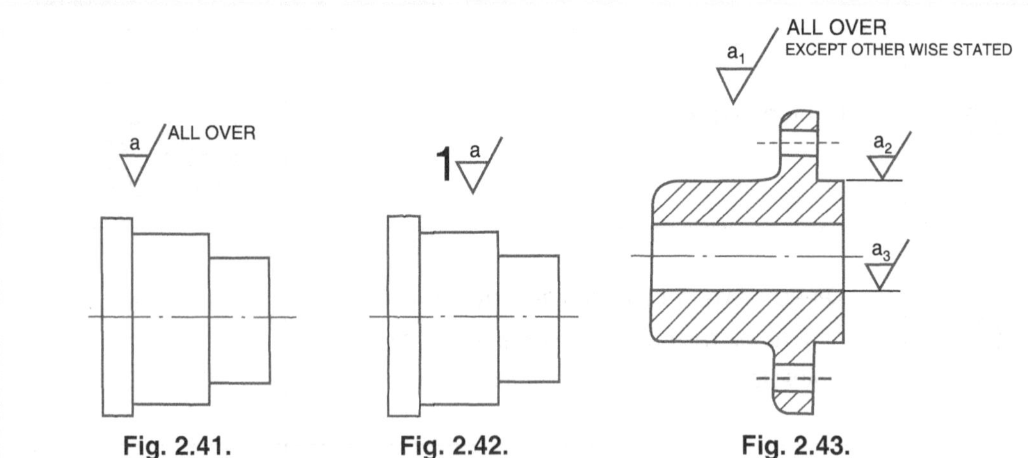

Fig. 2.41. Fig. 2.42. Fig. 2.43.

If the same surface roughness is required on the majority of the surfaces of a part, it is specified as above with the addition of :

(a) the notation except where otherwise stated (see Fig. 2.43).

(b) a basic symbol (in brackets) without anyother indication (see Fig. 2.44).

(c) the symbol or symbols (in brackets) of the special surface roughness or roughness (see Fig. 2.45).

The symbols for the surface roughness which are exceptions to the general symbols are indicated on the corresponding surfaces. [see Figs 2.46 and 2.47].

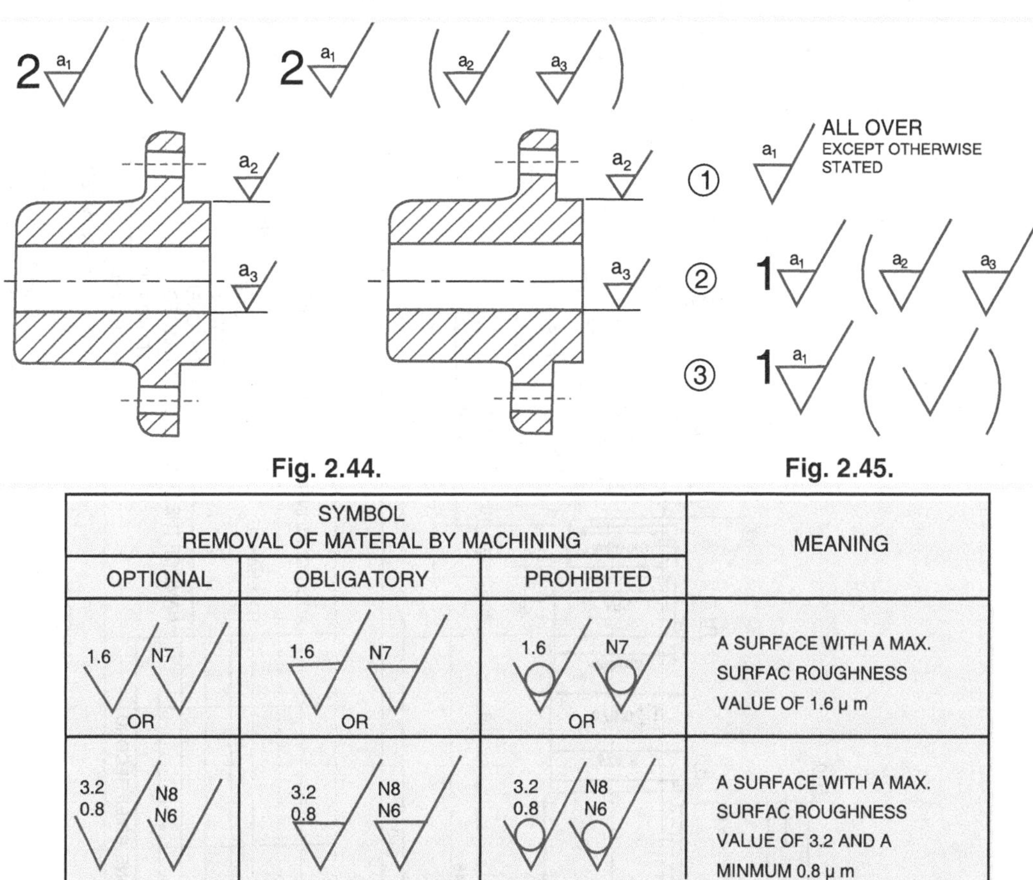

Fig. 2.44.

Fig. 2.45.

SYMBOL REMOVAL OF MATERAL BY MACHINING			MEANING
OPTIONAL	OBLIGATORY	PROHIBITED	
1.6 OR N7	1.6 OR N7	1.6 OR N7	A SURFACE WITH A MAX. SURFAC ROUGHNESS VALUE OF 1.6 µ m
3.2 0.8 N8 N6	3.2 0.8 N8 N6	3.2 0.8 N8 N6	A SURFACE WITH A MAX. SURFAC ROUGHNESS VALUE OF 3.2 AND A MINMUM 0.8 µ m

Fig. 2.46. *Symbols for removal of material by machining and their meaning*

ROUGHNESSS HEIGHT µ m	SURFACE DESCRIPTION	MACHINE PROCESSES
25.0	VERY ROUGH	SAW OR TORCH CUTTING, FORGING OR SAND CASTING
12.5	ROUGH MACHINING	COARSE FEEDS AND HEAVY CUTS IN MACHING
6.3	COARSE	COARSE SURFACE GRIND MEDIUM FEED AND AVERAGE CUTS IN MACHINING
3.2	MEDIUM	SHARP TOOLS, LIGHT CUTS, FINE FEEDS, HIGH SPEEDS, WITH MACHINING
1.6	GOOD FINISH	SHARP TOOLS, LIGHT CUTS, EXTRA FINE FEEDS, WITH MACHINING
0.8	HIGH GRADE FINISH	VERY FINE FEEDS, VERY SHARP TOOLS, AND FINE CUTS
0.4	HIGHER GRADE FINISH	SURFACE GRINDING, COARSE HONING, COARSE LAPPING
0.2	VERY FINE MACHINE FINISH	FINE HONING AND FINE LAPPING
0.05	EXTREMELY SMOOTH MACHINE FINISH	EXTRA FINE HONING AND LAPPING

Fig. 2.47. *Surface finishes obtained by different machining processes.*

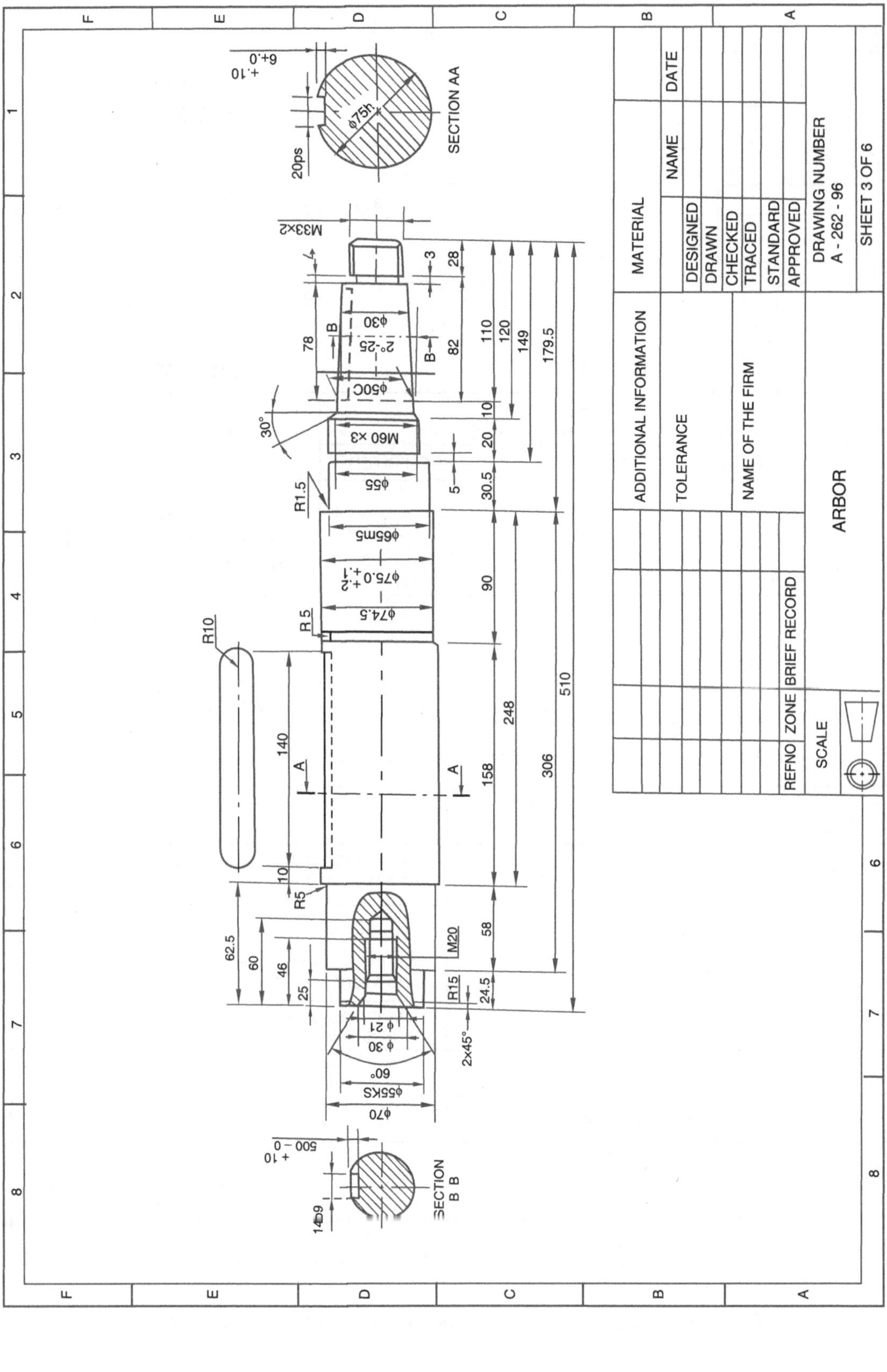

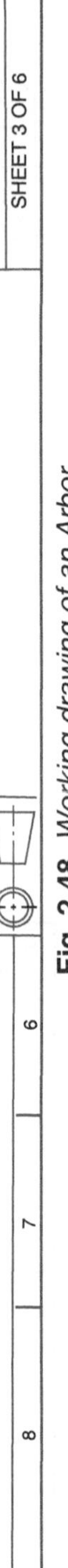

Fig. 2.48. *Working drawing of an Arbor*

Note : *Some times machining symbols as given in Fig. 2.49 are used to indicate surface roughness followed in industries.*

After studying the limits, fits and machining symbols, we come to their applications to the machine parts. Fig. 2.48 shows the working drawing of an Arbor as per I.S.

In this tolerances on dimensions, surface finish, method of dimensioning, title block, etc are shown. Study the drawing carefully and reproduce it on the drawing sheet exactly in the same way as given.

ROUGHNESS GRADE R_aµm	ROUGHNESS NUMBER	ROUGHNESS SYMBOL
50	N12	
25 12.5	N11 N10	▽
6.3 3.2 1.6	N9 N8 N7	▽▽
0.8 0.4 0.2	N6 N5 N4	▽▽▽
0.1 0.05 0.025	N3 N2 N1	▽▽▽▽

Fig. 2.49. *Machining symbols used in industry.*

QUESTIONS FOR SELF EXAMINATION

1. What is the importance of interchangeability of parts?
2. Define the following terms:
   (i) Tolerance. (ii) Limits and Fits. (iii) Allowance and Clearance.
3. Explain the unilateral and bilateral tolerance.
4. Sketch a figure illustrating the following terms as per B.I.S. :
   (i) Basic-size and Zero line. (ii) Upper and lower deviation. (iii) Tolerance zone.
   (iv) High and low limits.
5. Explain the following terms as per I.S. 919:
   (i) Deviation. (ii) Fundamental deviation. (iii) Grade of tolerance.
   (iv) Basic shaft and basic hole. (v) Go and not Go Limit.
6. Explain the following terms with a neat sketch:
   (a) (i) Clearance.
   (ii) Interference.
   (b) (i) Clearance, transition and interference fit.
   (ii) Minimum and maximum clearance.
   (iii) Pulley and Shaft
   (iv) Shaft in bearing
7. Why the machining symbols are used? Draw the symbols used for indication of surface roughness.

CHAPTER

3

Rivets and Riveted Joints

INTRODUCTION

In engineering practice, we often come across such situations when two or more parts of a machine or structure are joined or fastened together by means of various devices. The devices which are used for joining these parts are known as *fasteners* and the process of joining these parts is known as *fastening.* The fasteners generally employed for joining the parts are rivets, bolts and nuts, screws, keys etc.

In this chapter, we shall deal with the study of fastening, fasteners, types of fastenings, rivets, riveted joints, connections of plates, rivet symbols etc.

3.1. FASTENING

The process of joining two or more parts of a machine or structure by means of various devices is known as fastening.

3.2. FASTENERS

The devices which are used for joining two or more parts of a machine or structure are known as fasteners e.g. rivets, bolts and nuts, screws, etc.

3.3. TYPES OF FASTENINGS

The following are the two types of fastenings used in engineering practice :–

1. Permanent Fastenings
2. Temporary Fastenings

3.3.1. Permanent Fastenings : Permanent fastenings are those by which the parts are joined or fastened together permanently and cannot be separated easily. If, it is required to separate, the fastenings are to be broken.

Examples of Permanent fastenings: *Soldering, brazing, riveting and welding are the processes in order of strength by which permanent fastenings are made.*

3.3.2. Temporary Fastenings : Temporary fastenings are those by which the parts are fastened together temporarily and can be separated easily without breaking any part of fastener.

Examples of temporary fasteners: *Bolts and nuts, studs and nuts, keys, pins, etc.*

They are used to fasten parts in such a way as to permit their ready separation as in shaft couplings, various machines parts and temporary structures.

3.4. RIVET

The round bar of steel or wrought iron with a head on one end and tail on the other end is known as rivet.

Fig. 3.1 shows the different parts of a rivet before it is being used.

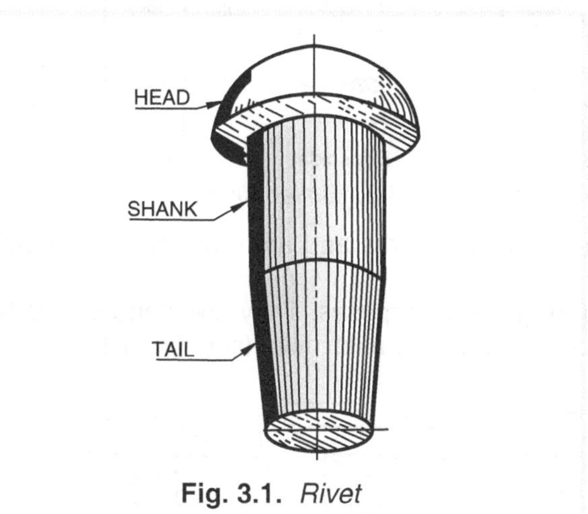

Fig. 3.1. *Rivet*

3.5. PRACTICAL APPLICATIONS OF RIVETS

Rivets are widely used in structural works like roof trusses, bridges, ship & aircraft building and in pressure vessels like boilers, air receivers and other engineering works.

3.6. TYPES OF RIVETS

In general the following types of rivets are used for different works :-

1. Structural rivets (12 to 45 mm dia.)
2. Boiler rivets (12 to 50 mm dia.)
3. Small rivets (2 to 10 mm dia.)

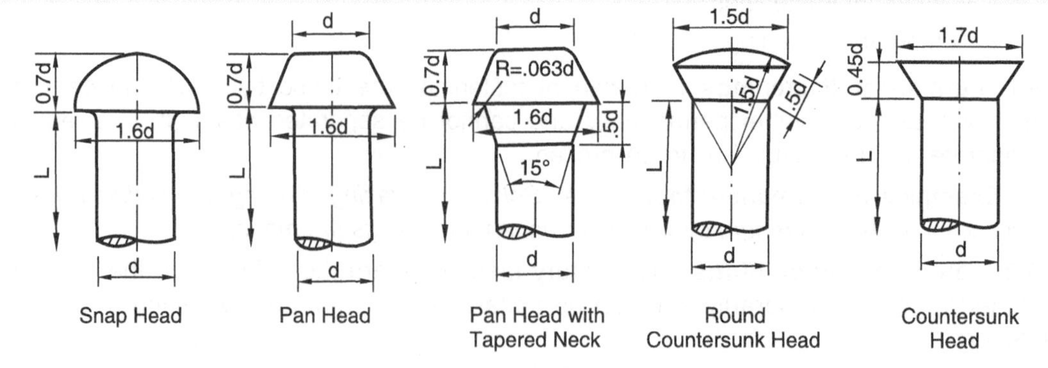

Fig. 3.2. *Structural Rivets*

Structural rivets are used for structure at works with diameters varying from 12 to 45 mm (see Fig. 3.2).

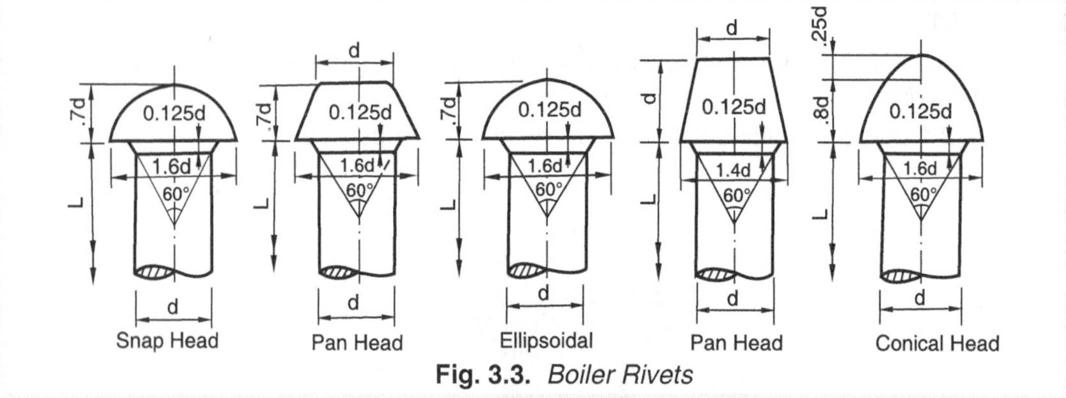

Fig. 3.3. *Boiler Rivets*

Boiler rivets are used for boiler works and are standeradised from 12 to 50 mm in diameter (see Fig. 3.3).

Small rivets are used for general purposes. They are standardised below 2 to 10 mm diameters and are of steel, copper, brass or aluminium (see Fig. 3.4).

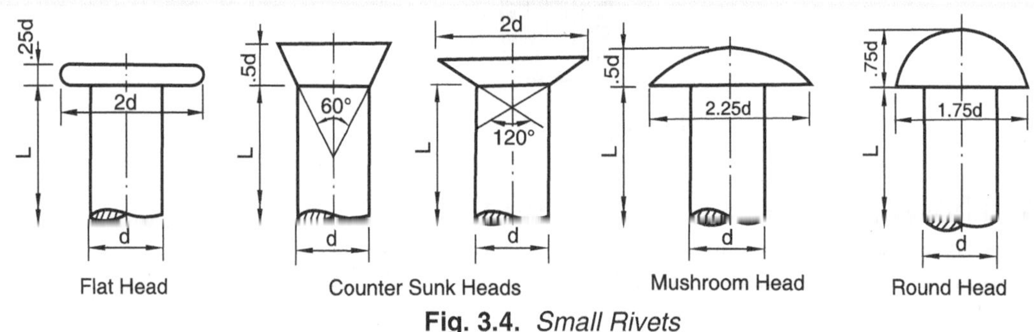

Fig. 3.4. *Small Rivets*

3.7. RIVETS SYMBOLS

In case of large drawings where numbers of rivets are used, it is not necessary to draw all the rivets. For this, the gauge lines are drawn and location of the rivets are shown by line symbol. The various rivet symbols according to I.S.I. are shown in Fig. 3.5 (i).

RIVETS SYMBOLS

SR. NO.	OBJECT	SYMBOL
1.	SHOP SNAP HEADED RIVETS	
2.	SHOP CSK (NEAR SIDE) RIVETS	
3.	SHOP CSK (FAR SIDE) RIVETS	
4.	SHOP CSK (BOTH SIDES) RIVETS	
5.	SITE SNAP HEADED RIVETS	
6.	SITE CSK (NEAR SIDE) RIVETS	
7.	SITE CSK (FAR SIDE) RIVETS	
8.	SITE CSK (BOTH SIDE) RIVETS	
9.	OPEN HOLE	

Fig. 3.5. *(i) Rivet Symbols as per I.S.I.*

BOLT OR RIVET	SYMBOL FOR BOLT OR RIVET TO FIT HOLE			SYMBOL FOR RIVET TO FIT IN HOLE CSK ON BOTH SIDES
	WITHOUT COUNTER SINK	CSK ON NEAR SIDE	COUNTER SINK ON FAR SIDE	
FITTED IN THE WORK SHOP				
FITTED ON SITE				
FITTED ON SITE AND HOLE DRILLED ON SITE				

Fig. 3.5. *(ii) Rivet symbols as per B.I.S.*

3.8.1. CONVENTIONAL REPRESENTATION OF HOLE AND RIVETS (B.I.S.)

In case of large drawings where numbers of rivets are used, it is not necessary to draw all the holes and rivets [also for bolts]. For this, the gauge lines are drawn and location of rivets and holes are shown by line symbols. The various symbols according to B.I.S. S.P.: 46-1988 are given in Fig. 3.5 (iii) :

HOLE	SYMBOL FOR HOLE			
	WITHOUT COUNTER SINKING	COUNTER SUNK ON NEAR SIDE	COUNTER SUNK ON BOTH SIDES	COUNTER SUNK ON BOTH SIDES
DRILLED IN THE WORKSHOP				
DRILLED ON SITE				

HOLE	SYMBOL FOR HOLE		
	WITHOUT COUNTER SINKING	COUNTER SUNK ON ONE SIDE ONLY	COUNTER SUNK ON BOTH SIDES
DRILLED IN THE WORKSHOP			
DRILLED ON SITE			

BOLT OR RIVET	SYMBOL FOR BOLT OR RIVET TO FIT IN HOLE		SYMBOL FOR RIVET TO FIT IN HOLE COUNTER SUNK ON BOTH SIDES	SYMBOL FOR BOLT WITH DESIGNATED NUT POSITION
	WITHOUT COUNTER SINKING	COUNTER SUNK ON ONE SIDE ONLY		
FITTED IN WORKSHOP				
FITTED ON SITE				
FITTED ON SITE AND HOLE DRILLED ON SITE				

Fig. 3.5. *(iii)*

3.8. RIVETING

The process of joining two or more plates by means of rivets is called riveting.

For this process, see Fig. 3.6 in which the rivet is shown by dotted lines in its orginal shape and second head is formed after the riveting process.

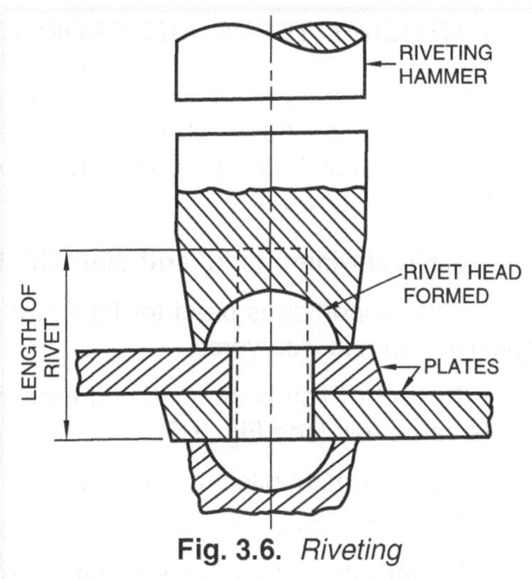

Fig. 3.6. *Riveting*

In riveting process, the diameter of the hole is kept slightly larger (about 1 mm to 1.5 mm) than the rivet diameter. After punching or drilling, any burr formed at the edges of the hole is removed by a little countersinking. The rivet is heated to a red-hot condition in a charcoal furnace and then inserted by a little hammering into hole of the plates. The other head of the rivet is then formed by means of a tool called "dolly." *In modern practice, the machine riveting is employed when the work is to be done speedily or on a large scale.*

When the rivets are of small size and for rivets made of brass, copper, other soft metal or alloy, the riveting is done in cold condition by hand hammer as in the case of rivets used in buckets.

3.9. CAULKING AND FULLERING OPERATIONS

To prevent the leakage of steam or other pressure fluid through the riveted joints, the plates are firmly faced together by caulking and fullering operations.

3.9.1. Caulking : *Caulking is an operation of burring down the edges of the plates and rivet heads by caulking tools so as to prevent the leakage of steam or fluid through riveted joints.*

Fig. 3.7 shows caulking tools like stiff chisel which is generally hammered either by hand or compressed air.

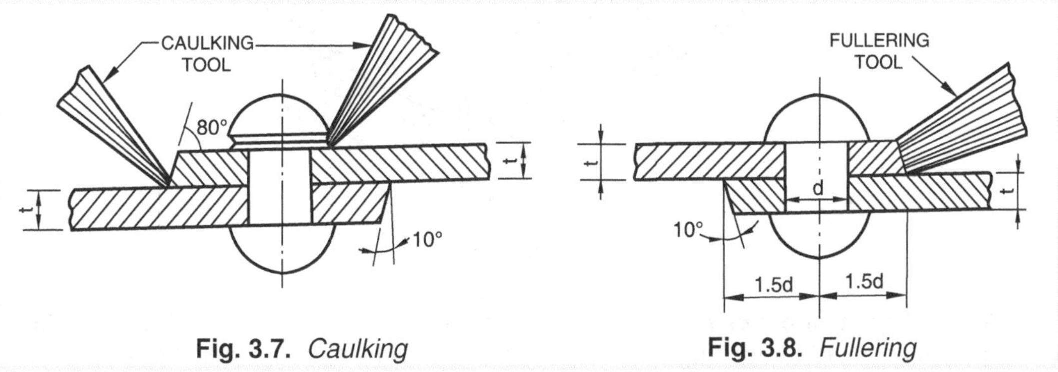

Fig. 3.7. *Caulking* **Fig. 3.8.** *Fullering*

3.9.2. Fullering : *Fullering is the operation of burring down the whole surface of the edges of the plate by means of fullering tool so as to prevent leakage of steam, fluid etc, through the riveted joints (see Fig. 3.8).*

When the tool is about as thick as the plate, it is called a **fullering tool**.

Note: *The caulking operation is done on the edges of the plates as well as on the edges of the rivet heads, but fullering operation is done on the edges of the plates only. To facilitate these operations the edge of the plate are bevelled at about 80°.*

3.10. GENERAL TERMS AND PROPORTIONS OF RIVETED JOINTS

In engineering practice, the riveted joints are carefully designed to prevent failure. The design should be taken about with suitable values for rivet diameter, distance of holes, thickness of plates, etc. Some useful emperical general terms and proportions of the riveted joints are as follows :–

1. Rivet Diameter 'd' and plate thickness 't': According to I.S. : $d = 6\sqrt{t}$

This proportion is used for finding the diameter '**d**' when the thickness '**t**' of the plate is given in mm and vice-versa.

2. Pitch: Pitch is the distance from the centre of one rivet to the centre of the next rivet in the same row (see Fig. 3-9).

The maximum pitch may be taken as **P = 3d**

and minimum pitch **P = d + 30** mm.

3. Margin: Margin is the distance from the centre of the rivet hole to the nearest edge of the plate.

Margin, m=1.5d

4. Perpendicular distance between rows of rivets: In chain type riveting, the perpendicular distance between rows of rivets is **2d + 6** mm while in the zig zag riveting it is **2d** only.

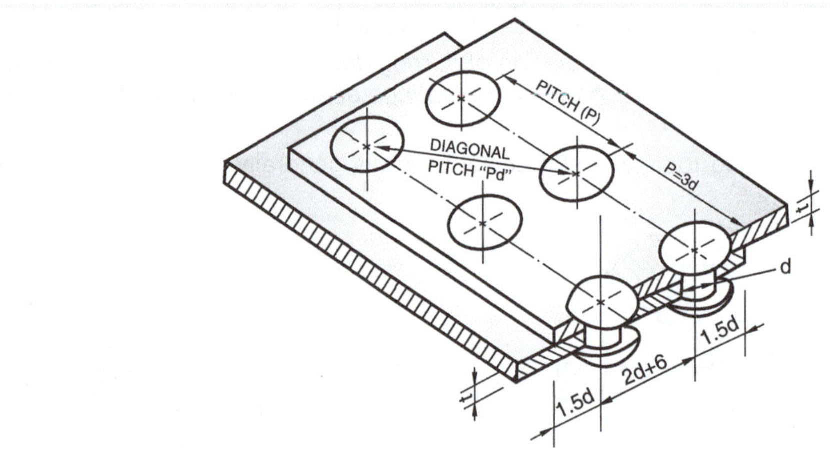

Fig. 3.9.

5. Thickness of cover plate strap: Thickness of cover plate, t_1 **= 0.625 t** for double cover plates and **1.125 t** for single cover plate (in case of butt joint) should be taken.

6. Diagonal pitch: It is the distance between the centres of two rivets in adjacent rows and is denoted by "p_d.".

Diagonal pitch, "p_d" = **2p+d/3**

where **p** = max. pitch

d = diameter of rivet.

3.11. TYPES OF RIVETED JOINTS

The following are two basic types of riveted joints which are commonly used is engineering practice:-

1. Lap joints.
2. Butt joints.

3.10.1. Lap joint : In lap joint, one plate overlaps the other plate and the rivets pass through both the plates :–

Laps joints are further sub-divided as under:-

(a) Single riveted lap joint: In single riveted lap joint, only one row of rivets is used to join two overlapping plates (see Fig. 3.10).

(b) Double riveted lap joint: In double riveted lap joint, two rows of rivets are used to join two overlapping plates (see Fig. 3.11).

Important Notes. *(1) In all problems, the sectional front view have been drawn corresponding to the section plane C-P in top view.*

(2) The rivets should not be shown in section according to conventional practice.

Problem 1 : Draw the top view and sectional front view of a single riveted lap joint. Take the diameter of rivet = 24 mm *[Fig. 3.10].*

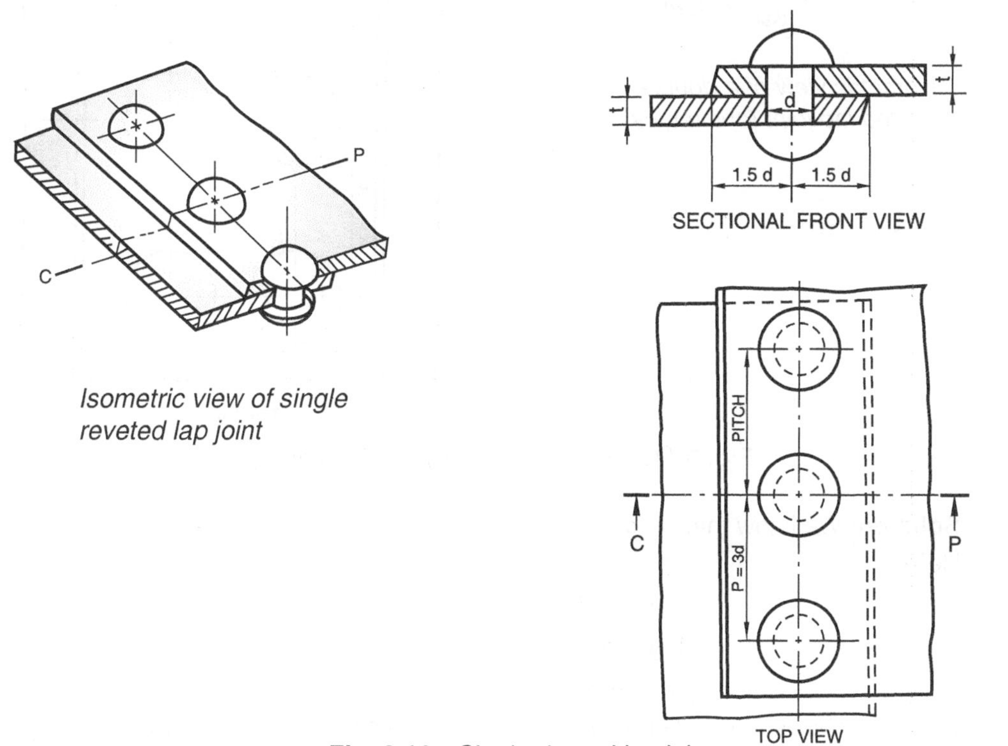

Fig. 3.10. *Single riveted lap joint*

Solution: We know that,

$$d = 6\sqrt{t}$$

Diameter of rivet, d = 24 mm

$$\text{or } 24 = 6\sqrt{t}$$

$\therefore$ t = 16 mm

Knowing the value of t and other proportions, the top view and sectional front view can be drawn. Fig. 3.10 shows the top view and sectional front view alongwith other proportions.

Important Note: *Do not draw the isometric view while giving the solution of the problem in the examination.*

Problem 2 : Draw the sectional front view and top view of a doubled riveted lap joint (Chain type). Take the diameter of rivet = 24 mm (Fig. 3.11).

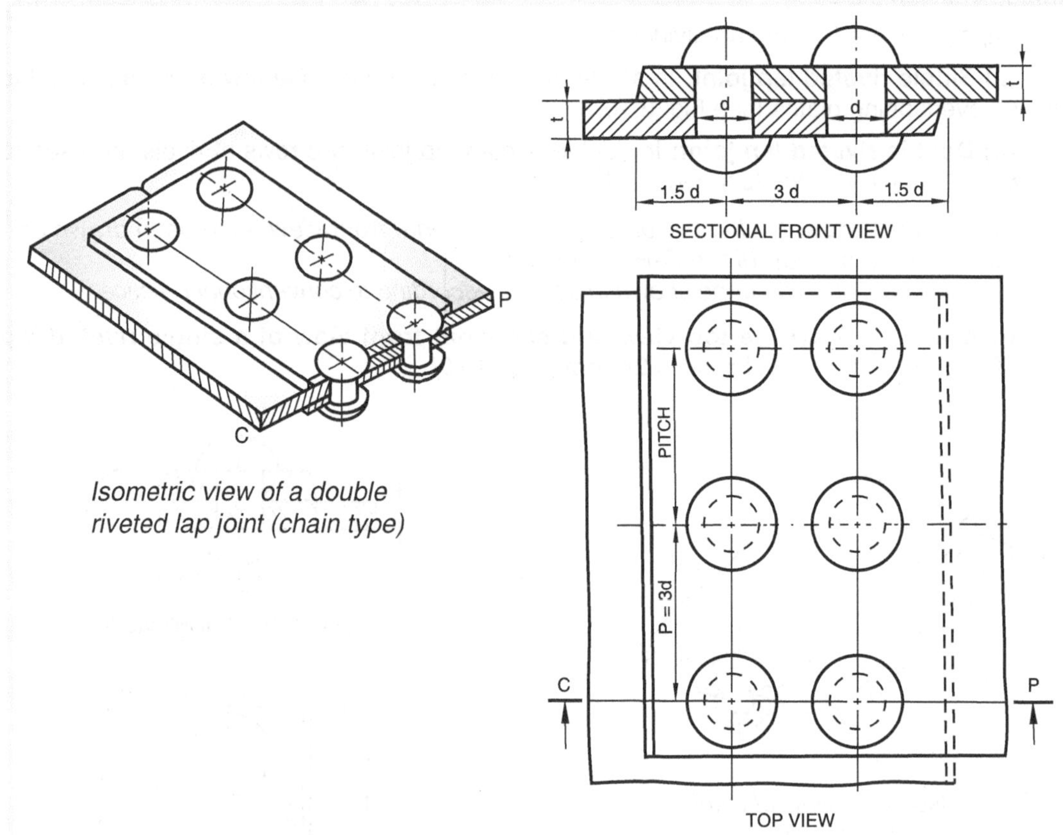

Fig. 3.11. *Double riveted lap joint (chain type)*

Solution: We know that $d = 6\sqrt{t}$

Here,

$$d = 24 \text{ mm}$$

$$\text{or } d = 24 = 6\sqrt{t}$$

$$\therefore \quad t = 16 \text{ mm.}$$

Fig. 3.11. shows the sectional front view and top view of a double riveted Chain type lap joint.

Note : *Chain type riveting is that in which rivets are arranged parallel and directly opposite to the adjoining row (see Fig. 3.11).*

Problem 3 : Draw the sectional front view and top view of a double riveted lap joint (Zig-zag type). Take the diameter of rivet = 24 mm [Fig. 3.12].

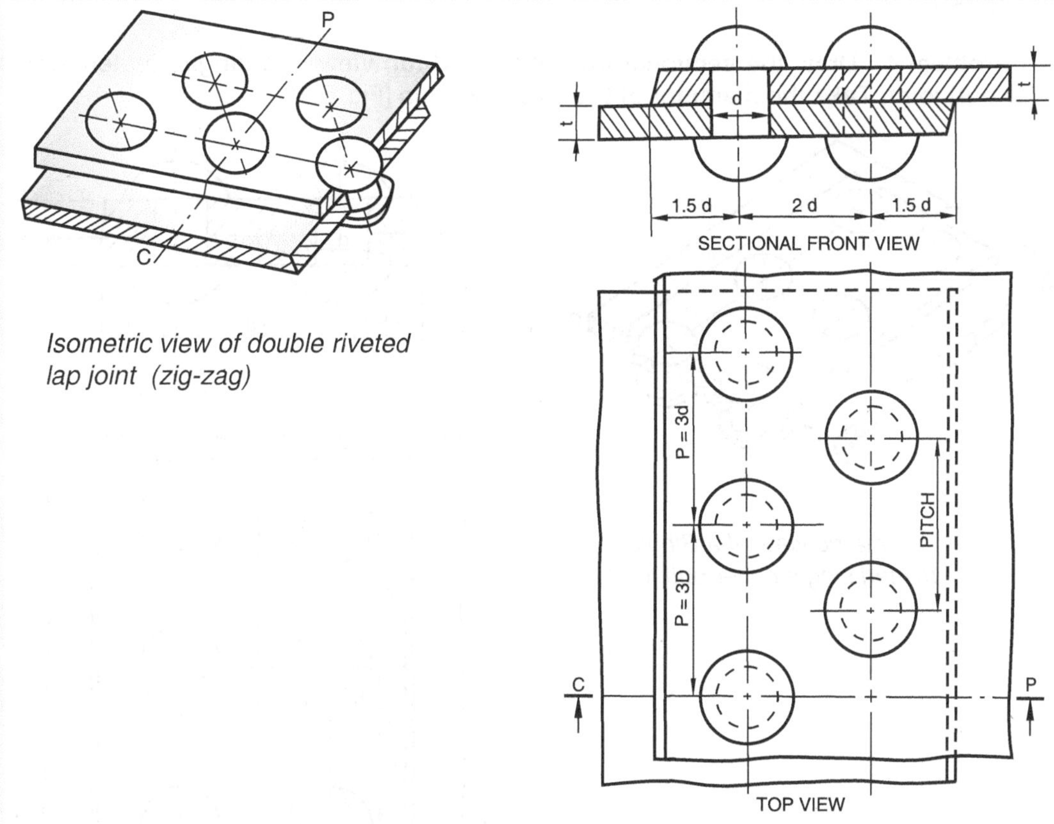

Fig. 3.12. *Double riveted lap joint (zig-zag)*

Solution: We know that,

$$d = 6\sqrt{t}$$

Putting the value of D 24 mm,

we have, $24 = 6\sqrt{t}$

$\therefore$ $t = 16$ mm

Fig. 3.12 shows the sectional front view and top view of a double riveted lap joint along with other required proportions.

Note : *Zig-zag type riveting is that inwhich the rivets are staggered ina zig-zag formation (see Fig. 3.12).*

3.12. BUTT JOINT

In but joint, the edges of the connecting plates butt against each other and covered with cover plate one or both sides. At least two rows of rivets, one on each of the connected plates being joined, are necessary to make a butt joint.

But joints are further sub-divided into the following parts :–

(a) Single riveted single cover (or strap) but joint: In this, joint, only one cover plate and two rows of rivets, one on each side of the connected plates are required to form the joint (see Fig. 3.13).

(b) Double riveted double cover (or strap) butt Joint: In this joint, two cover plates and two rows of rivets one on each side of the connected plate are required to form the joint (see Fig. 3.14).

Problem 4 : Draw the sectional front view and top view of a single riveted single cover butt joint. Take the diameter of the rivet = 24 mm [Fig. 3.13].

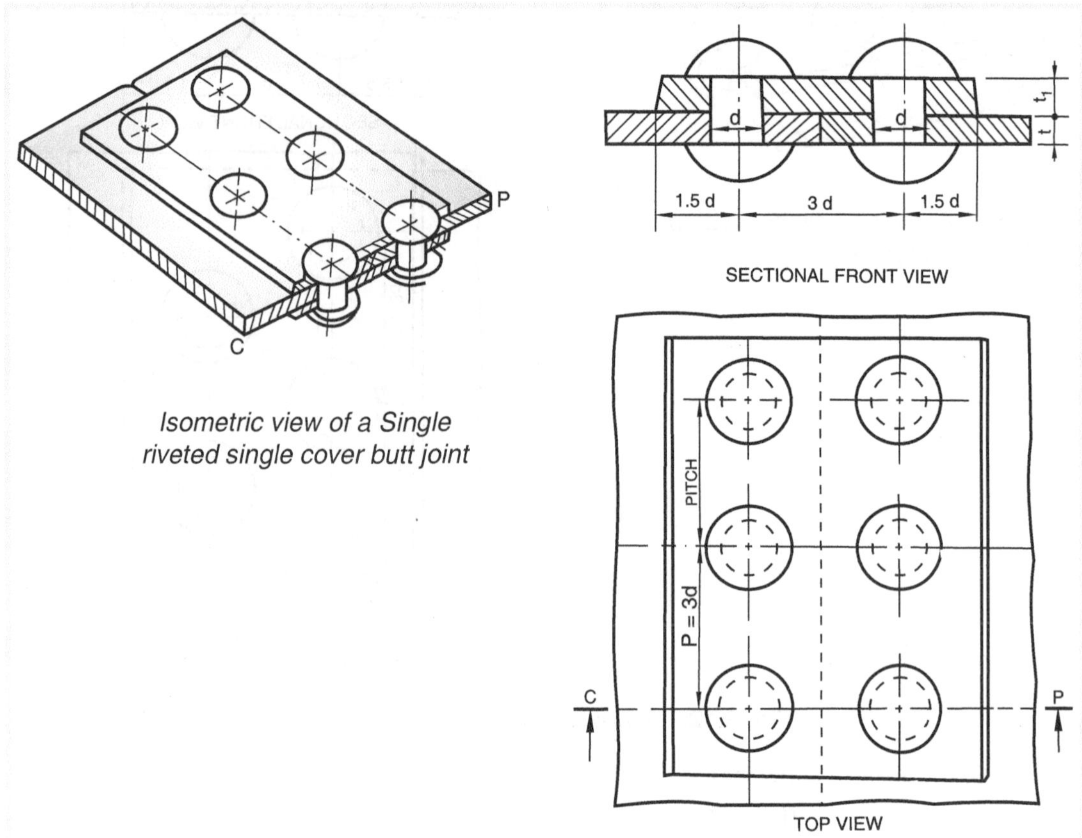

Fig. 3.13. *Single riveted single cover butt joint*

Solution : We know that,

$$d = 6\sqrt{t}$$

Putting the value of d=24 mm

$$24 = 6\sqrt{t}$$

$$\therefore \quad t = 16 \text{ mm}$$

The thickness of cover plate $t_1 = 1.125\,t$

$$= 1.125 \times 16$$

$$= 18 \text{ mm}$$

Knowing the values of t_1 alongwith other proportions, the sectional front view and top view can be drawn (see Fig. 3.13 for its solution).

Problem 5 : Draw the top view and sectional front view of a single riveted double cover butt joint. Take the diameter of the rivet = 24 mm [Fig. 3.14].

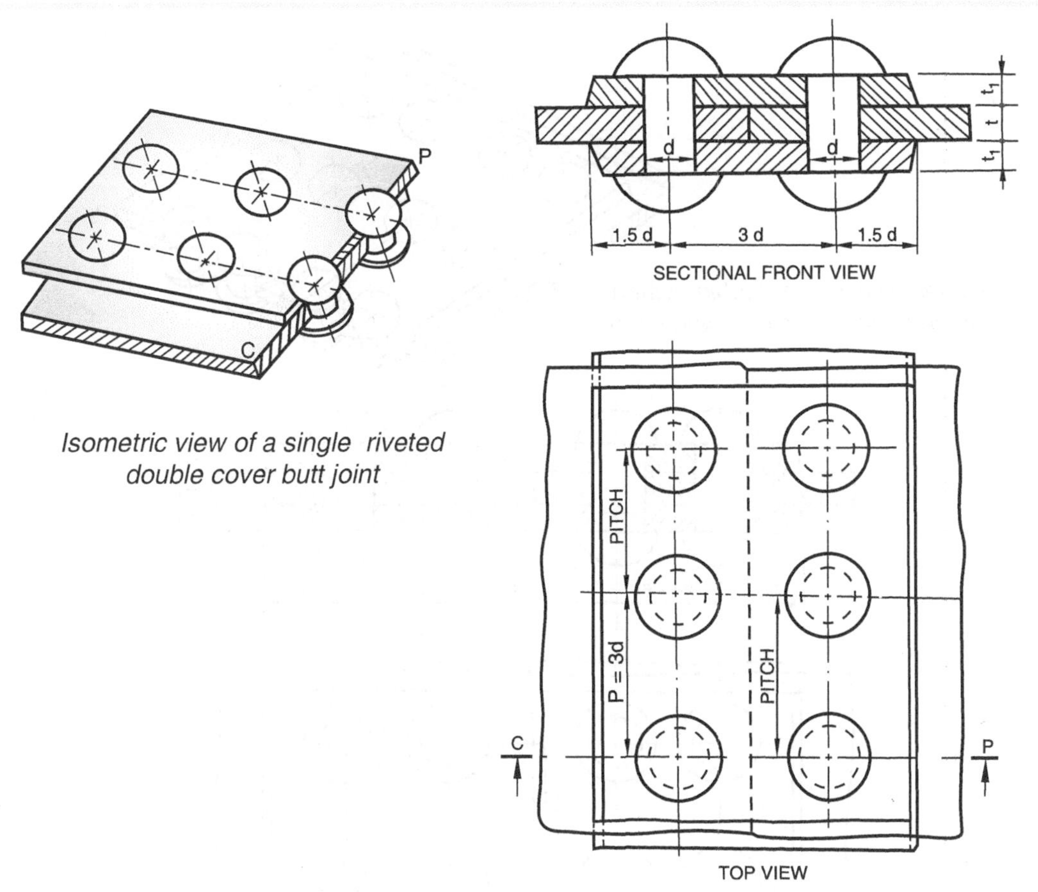

Fig. 3.14. *Single riveted double cover butt joint*

Solution: we know that, $d = 6\sqrt{t}$

Putting the value of d = 24 mm

We have $24 = 6\sqrt{t}$

$\therefore \quad t = 16$ mm

The thickness of the cover plate in case of double cover, $t_1 = 0.625\, t$

$= 0.625 \times 16$

$= 10$ mm

Knowing the value of t and t_1 alongwith other proportions, the top view and sectional front view can be drawn. For its complete solution, see Fig. 3.14.

Problem 6: draw the sectional front view and top view of a double riveted double strap butt joint (Chain type). Take thickness of main plates = 8 mm, cover plates 5 mm, diameter of rivet = 12 mm and pitch = 36 mm.

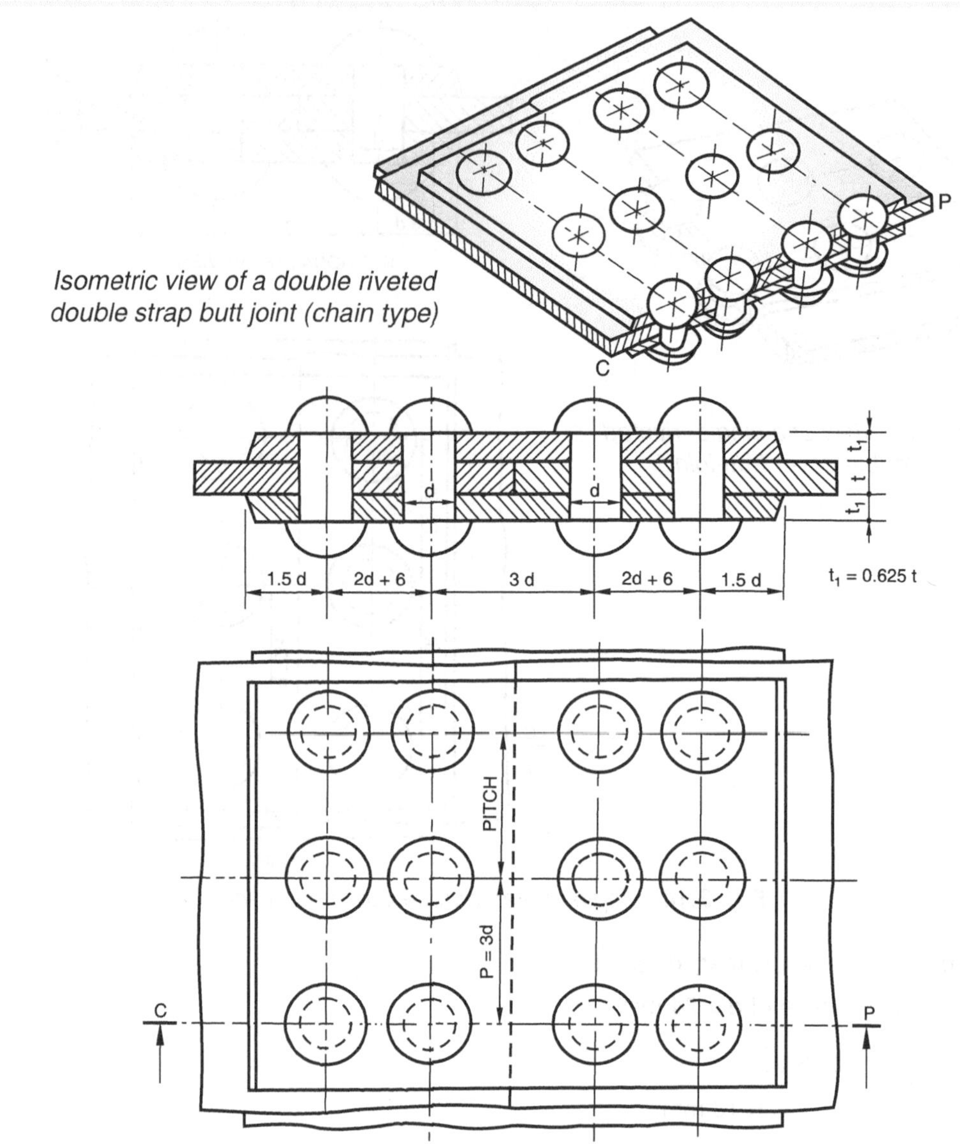

Fig. 3.15. *Double riveted double strap butt joint (chain type)*

Solution: From the given data

Diameter of rivet, d=12 mm

Pitch, P=36 mm

Thickness of main plate, t = 8 mm

Thickness of cover plate, t_1 =5 mm

From the above values alongwith other proportions, draw the sectional front view and top view as shown in Fig. 3.15.

Problem 7 : Draw the top view and sectional front view of a double riveted double strap zig-zag butt joint. Take the thickness of main plates = 10 mm. Assuming pitch of rivets as three times the rivet diameter [Fig. 3.16].

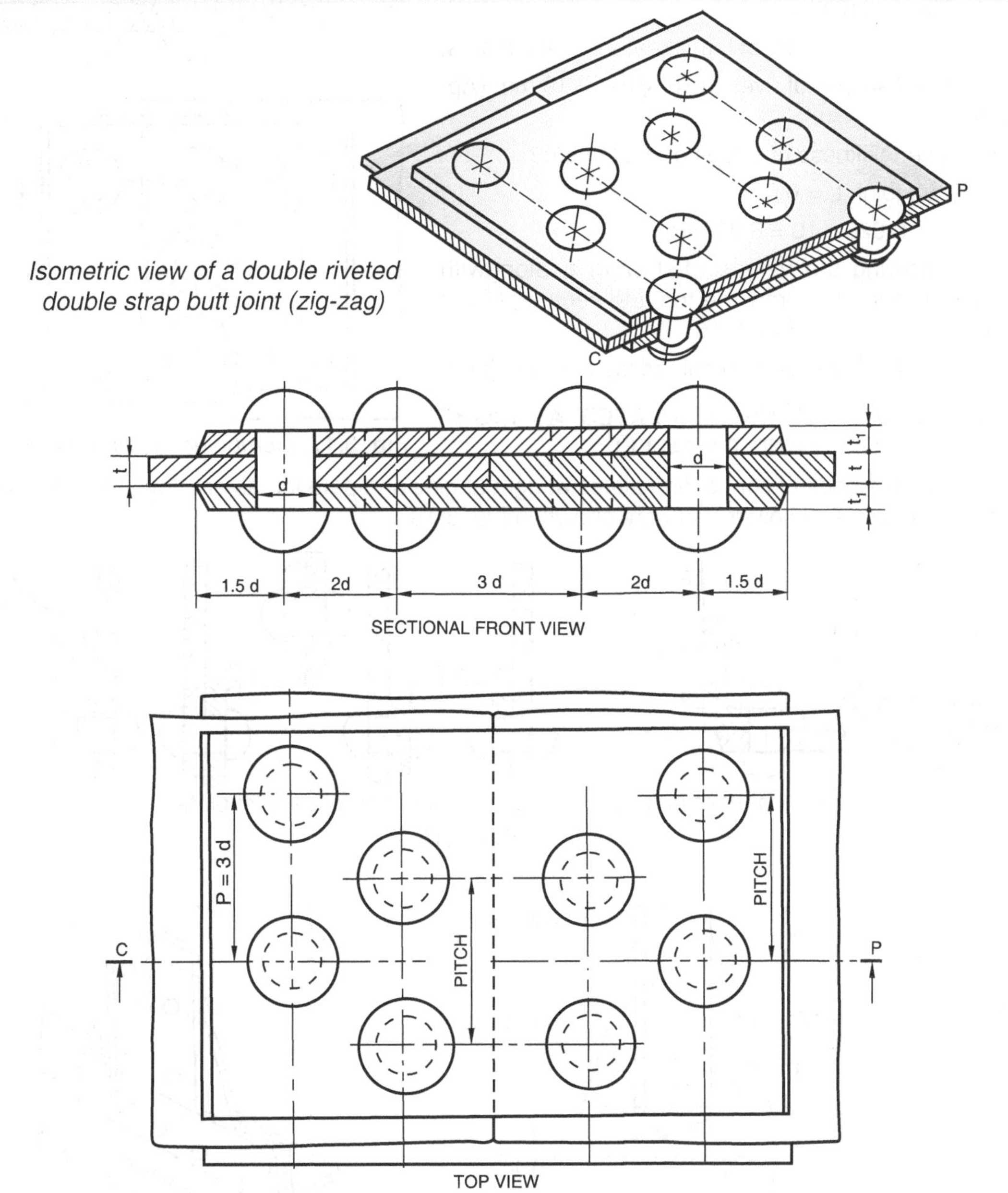

Fig. 3.16. *Double riveted double strap butt joint (zig-zag type)*

Solution: We know that

$$d = 6\sqrt{t}$$

Here, t=10 mm

Putting the value of t, thickness of main plates, we have diameter of rivet, $d = 6\sqrt{10} = 19$ mm (Approx.)

The thickness of straps of double strap butt joint, $t_1 = 0.625\ t$

$t_1 = 0.625 \times 10 = 6.25$ mm

Knowing the values of d sand t_1 alongwith other proportions, draw the top and sectional front views as shown in Fig. 3.16.

Fig. 3.17 shows a combined lap and butt joint

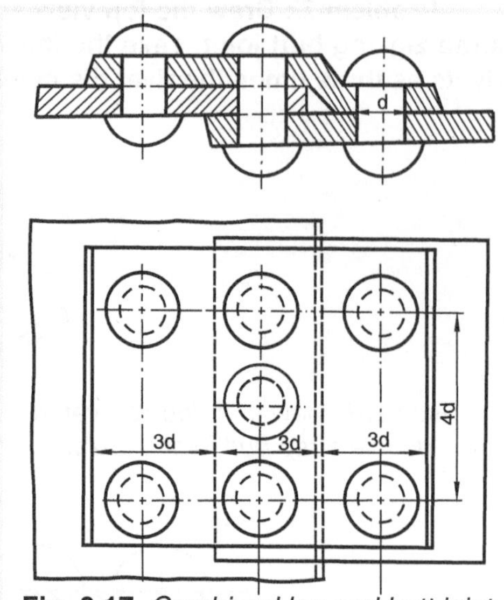

Fig. 3.17. *Combined lap and butt joint*

3.13. CONNECTIONS OF PLATES AT RIGHT ANGLES AND PARALLEL

The methods of connecting plates at right angles and paralled as employed in fire boxes of locomotives and boilers work are shown in Fig. 3.18.

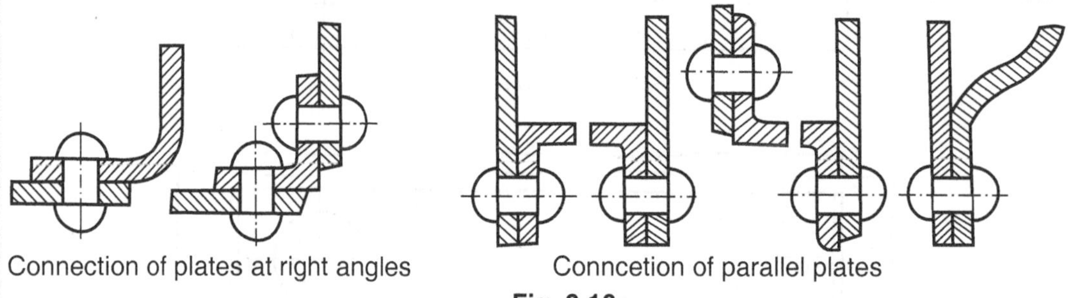

Connection of plates at right angles Conncetion of parallel plates

Fig. 3.18.

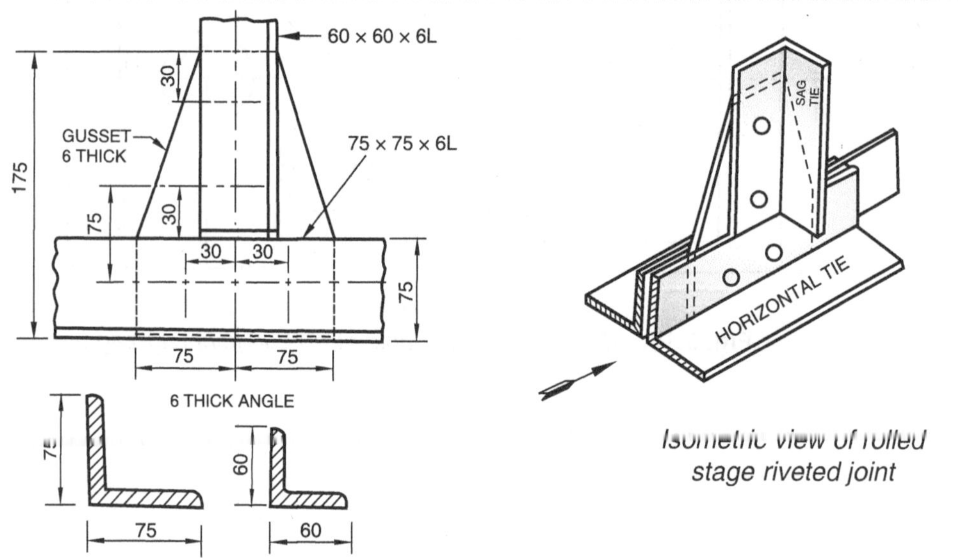

Isometric view of rolled stage riveted joint

Fig. 3.19. *shows a rolled steel riveted joint used in engineering practice*

Fig. 3.20 shows the application of rivet symbols in engineering practice. In long seams, a few rivets should be shown at each end. The intermediate rivets should be omitted and the centre lines of the rivet rows in the direction of seams be shown (see Fig. 3.21). A series of equal pitches should be dimensioned by writing no. of pitches, the pitch and total distance.

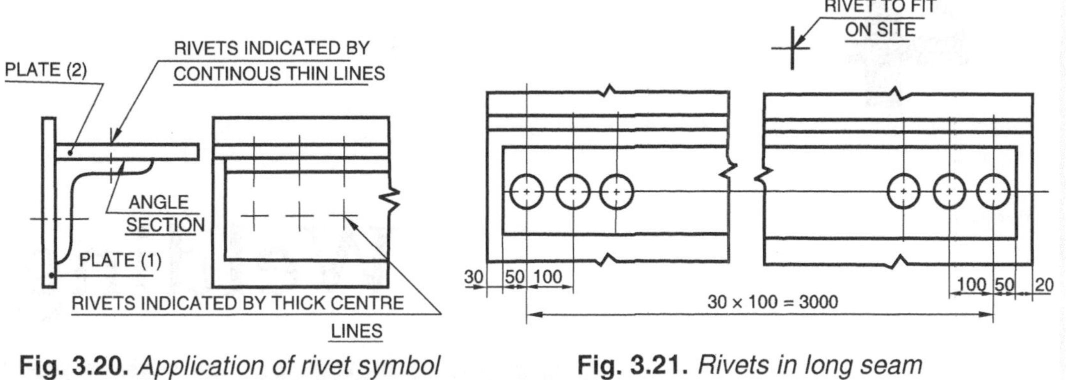

Fig. 3.20. *Application of rivet symbol*

Fig. 3.21. *Rivets in long seam*

QUESTIONS FOR SELF EXAMINATION

1. What is a fastening ? State its types.
2. What is the difference between temporary and permanent fastenings? Give examples.
3. Define: Pitch, diagonal pitch and margin.
4. What do you understand by terms, Caulking and Fullering?
5. What is the difference between lap and butt joint?
6. Why the ends of the overlapping plates are bevelled ? State the angle of bevelling.

PROBLEMS FOR PRACTICE

1. Draw different types of rivet heads.
2. Draw the sectional front view and top view of the following riveted joints, take thickness of plate 13 mm and diameter of rivet 18 mm.
   (a) Single riveted lap joint.
   (b) Double riveted lap joint, chain riveting.
   (c) Double riveted lap joint, zig-zag riveting.
3. Draw the single riveted butt-joint with two cover plates. Take t = 8 mm and d = 13 mm.
4. Draw the double riveted chain type butt-joint with two cover plates, Take thickness of main plates = 8 mm, thickness of cover plates = 5 mm, diameter of rivet = 12 mm and pitch = 36 mm.
5. Draw the top view and sectional front view of a double-riveted zig-zag butt-joint. Take thickness of plates 15 mm, thickness of cover plates 10 mm and diameter of rivet 15 mm. Assume other dimensions.
6. Sketch two method of connecting parallel plates.

CHAPTER

4

Welding

INTRODUCTION

In this modern world, welding has come to play as most important branch of engineering. It is the least expensive process for joining metal parts permanently and widely used now-a-day in engineering field. It not only saves the time and labour but also gives maximum strength and durability to the jointed parts.

In this chapter, we shall deal with the study of welding, applications of welding, types of welding, welded joints and welding symbols used in engineering practice.

4.1. WELDING

The process of joining two or more metal parts permanently by the application of heat is known as welding.

4.2. APPLICATIONS OF WELDING

In industry welding process is used in the *manufacture of automobile parts, aircrafts, road cars, machine frames, structural work and other general repair works.*

In engineering practice, welding process is *used for fabrication of cylindrical works such as pipe lines, containers, ducts, etc.*

4.3. TYPES OF WELDING

Welding may be classified under two main headings in engineering field : –

1. Plastic welding or pressure welding
2. Fusion welding or Non-pressure welding

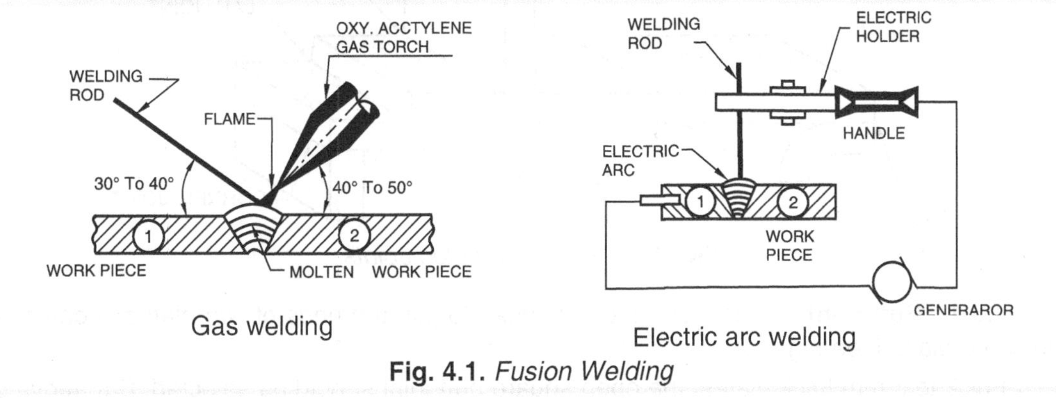

Fig. 4.1. *Fusion Welding*

4.3.1. Plastic welding or pressure welding : In this type of welding the parts of metal to be welded together are heated to a plastic state and then joined together by some external pressure.

4.3.2. Fusion welding or Non-pressure welding : In fusion welding, the parts of matels are heated to a molten state and the space between the two parts is filled with molten material which is supplied by a suitable welding rod. The parts are then allowed to solidify (see Fig. 4.1).

Fusion welding may be (1) *Gas welding* (2) *Electric arc welding* (3) *Thermit welding* (4) *Flash welding.*

4.4. WELDED JOINTS

The joints which are prepared by welding processes are known as welded joints.

4.5. TYPES OF WELDED JOINTS

The fusion type of welded joint depends upon the relative position of the two parts to the joined. But there are five basic types of joints which are:–

(1) Butt joint

(2) Lap joint

(3) Corner joint

(4) Edge joint

(5) T-joint

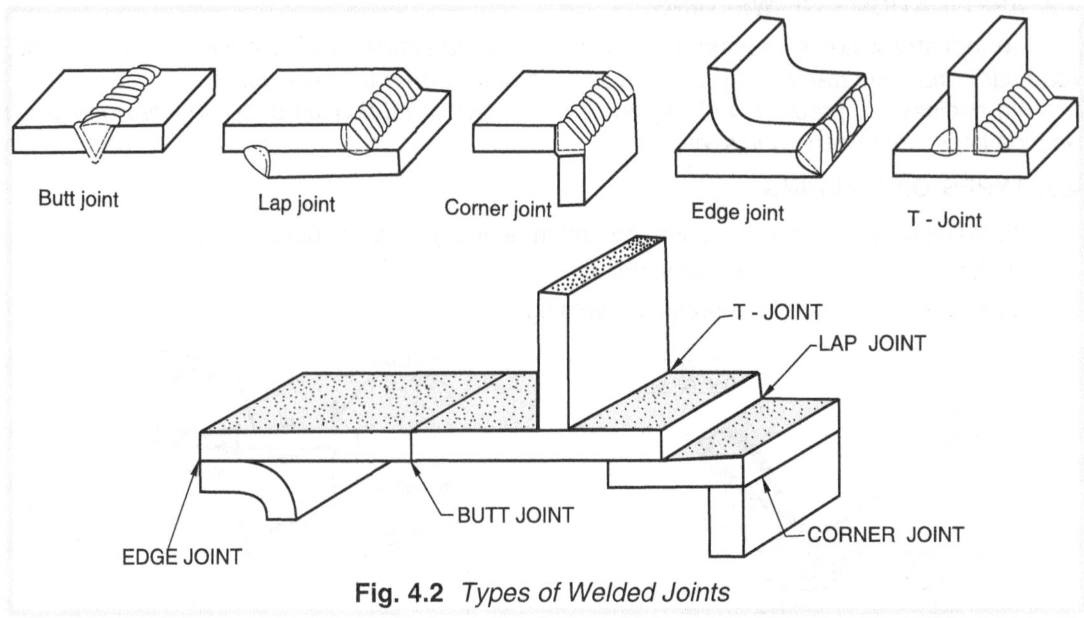

Fig. 4.2 *Types of Welded Joints*

4.5.1. Butt joint : This type of joint is used to join the ends of two plates, located in the same plane [see Fig. 4.2].

For 2 to 5 mm thick plates, the open square butt joint should be selected. But above 5 mm thickness, the joint with edge preparation on one or both sides may be recommended.

4.5.2. Lap joint : It is used to join two overlapping plates such that the edge of each plate is welded to the surface of the other [see Fig. 4.2].

This type of joint is suitable upto 3 mm thick plates.

4.5.3. Corner joint : It is used to weld the edges of two plates. This is suitable for both heavy and light gauges [see Fig. 4.2].

This type of joint is commonly used in the construction of boxes, tanks, frames and other similar items.

4.5.4. Edge joint : The edge joint is used to join two parallel plates. This is generally used in sheet metal works [see Fig. 4.2].

4.5.5. T-joint : T-joint is used to join two plates, the surfaces of which are at right angle to each other [see Fig. 4.2].

It is employed for thickness of plates upto 3 mm and is widely used in thin walled structure.

4.6. EDGE PREPARATION

The following are the different types of edges preparation used in fusion welding processes for welding butt joints :—

1. Plain or Square bult joint
2. Single – V butt joint
3. Double – V butt joint
4. Single – U butt joint
5. Double – U butt joint

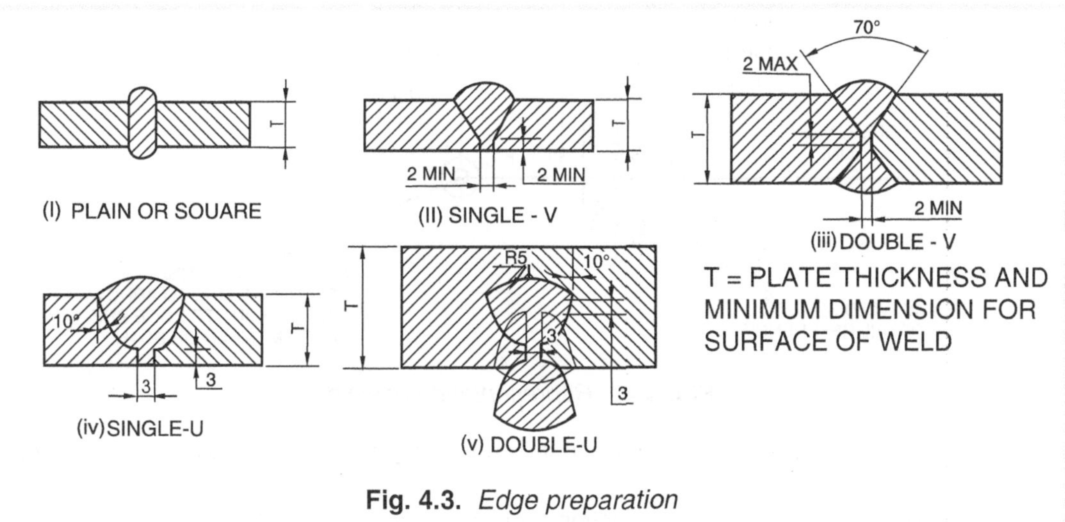

Fig. 4.3. *Edge preparation*

Note : *The preparation of edge depends upon the thickness of metal being welded.*

4.6.1. Plain or Square butt joints are used for thickness of plates from 3 to 8 mm. Before welding the edges, plates are spaced approx 3 mm apart [see Fig. 4.3 *(i)*].

4.6.2. Single-V butt joints are used for joining plates from 8 to 16 mm thick. The edge of the plates forming the joint are bevelled to form an included angle of 70° to 90°, depending upon the welding technique to be used [see Fig.4.3*(ii)*].

4.6.3. Double-V butt joints are used on plates over 16 mm thick. This welding is performed on both sides of the plates [see Fig. 4.3*(iii)*].

4.6.4. Single-U butt joints are used for plates of 16 mm to 20 mm thickness [see Fig. 4-3(iv)].

4.6.5. Double-U butt joints are used on plates over 20 mm thick. They are more satisfactory and require less filler rod but are difficult to prepare [see Fig. 4.3*(v)*].

Note : *The plain or square joints are used to 3 mm thick plates. The single-U or V-joints are used upto 12 mm thick plates and double-U or V-joints are used for plates having thickness above 12 mm.*

4.7. REPRESENTATION OF WELD

The scheme adopted by I.S.I. for representation of weld on drawings requires the use of the following elements:–

1. A basic symbol used to specify the type of weld.
2. An arrow and reference line to indicate the location of the weld.

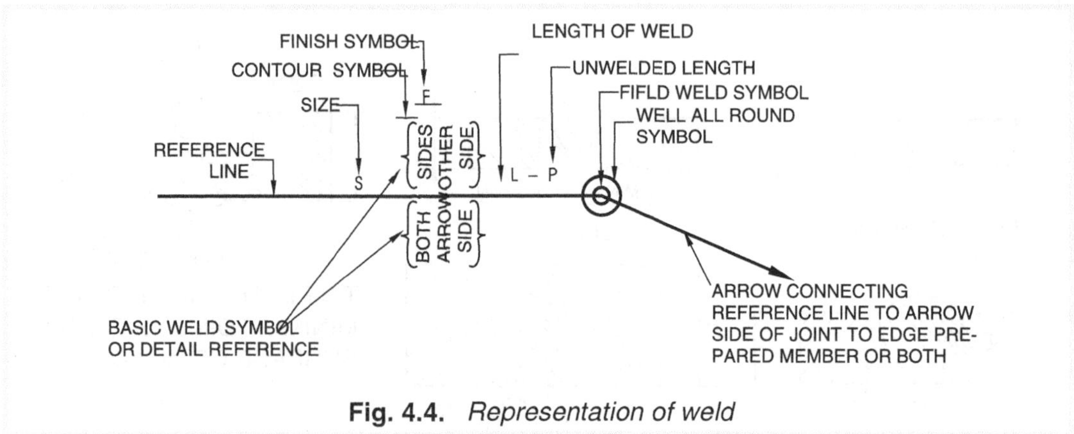

Fig. 4.4. *Representation of weld*

3. Supplementary symbol required to indicate special instructions such as finish, weld-all round, etc.

4. Dimensions of weld in cross section and length, where required.

The elements for the represention of weld shall have specific location with respect to each other as shown in Fig. 4.4.

4.8. ELEMENTARY AND SUPPLEMENTARY WELDING SYMBOLS ACCORDING TO B.I.S.

In practice elementry welding symbols may be completed by a symbol characterizing the shape of the external surface of the weld. But the absence of a supplementry symbol means that the shape of the weld surface does not need to be indicated precisely. Table 4.3 shows the recommended supplementry symbols as per Bureau of Indian Standards (B.I.S.)

TABLE 4.3

DESIGNATION	ILLUSTRATION	SYMBOL
FLAT (FLUSH) SINGLE-V BUTT WELD		
CONVEX DOUBLE-V BUTT WELD		
CONVEX FILLET WELD		
FLAT (FLUSH) SINGLE-V WELD WITH FLAT (FLUSH) BACKING RUN		

Table 4.4 shows the examples of combinations of elementry and supplementary symbols.

Note : *Though it is not forbidden to associate several symbols, it is better to represent the weld on a separate diagram when symbolization become too difficult.*

TABLE 4.4

SHAPE OF WELD SURFACE	SYMBOL
a) FLAT (USUALLY FLUSH)	
b) CONVEX	
c) CONCAVE	

TABLE 4.1
SCHEDULE OF BASIC WELDING SYMBOLS AS PER B.I.S.

S.NO.	DESIGNATION	ILLUSTRATION	SYMBOL
1.	BUTT WELD BETWEEN PLATES WITH RAISED EDGES		
2.	SQUARE BUTT WELD		 ||
3.	SINGLE-V BUTT WELD		 V
4.	SINGLE-BEVEL BUTT		
5.	SINGLE-V BUTT WELD WITH BROAD ROOT FACE		 Y
6.	SINGLE BEVEL BUTT WELD WITH BRAOD ROOT FACE		
7.	SINGLE-U BUTT PARALLEL OR SLOPING SIDES		
8.	SINGLE-J BUTT WELD		
9.	BACKING RUN BACK BACKING WELD USA		
10.	FILLET WELD	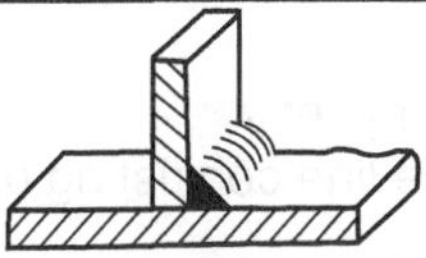	
11.	PLUG OR SLOT WELD		
12.	SPOT WELD	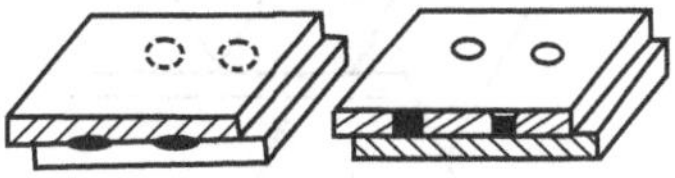	 ○
13.	SEAM WELD	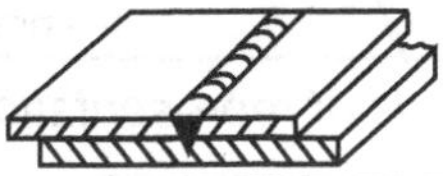	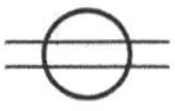

TABLE 4.2

SCHEDULE OF BASIC WELDING SYMBOLS

S.No.	OBJECT	DRAWING REPRESENTATION	SYMBOL
1	WELD ALL ROUND		○
2	SITE WELD (ASSEMBLY WELD)	A	A ○
3	SITE WELD (ERECTION WELD)	E	E ●
4	FLUSH CONTOUR		—
5	CONVEX CONTOUR		⌒
6	CONCAVE CONTOUR		(
7	GRINDING FINISH	G	G
8	MACHINING FINISH	M	M
9	CHIPPING FINISH	C	C

4.9. POSITION OF THE SYMBOLS ON DRAWING AS PER B.I.S.

According to latest B.I.S. SP: 46-1988, the symbols covered form only, part of a complete method of representation (see Fig. 4.5) which comprises in addition to the symbol itself.

1. an arrow line per joint
2. A dual reference line consisting of two parallel lines, one continuous and one dashed line.
3. A certain number of dimensions and conventional signs.

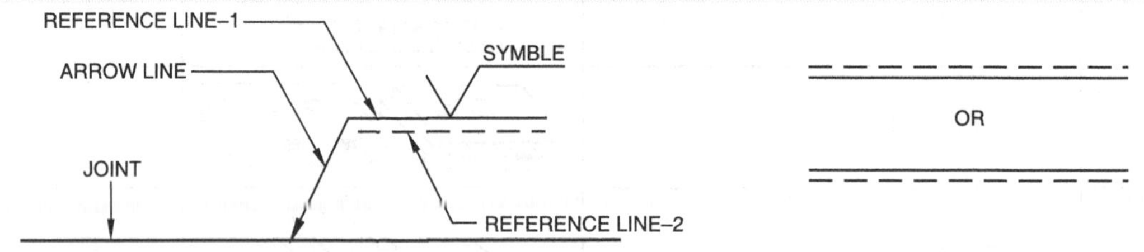

Fig. 4.5. *Methods of representation of symbols on drawing*

Fig. 4.6. *Dashed line*

Notes : *1. The dashed line can be drawn either above or beneath the continuous line (see Fig. 4.6).*

2. For symmetrical welds the dashed line is unnecessary and should be omitted.

3. The thickness of line or arrow line, reference line, symbol and lettering shall be in accordance with the thickness of line for dimensioning.

Following rules define the location of welds by specifying:

(a) the position of the arrow line

(b) the position of the reference line, and

(c) the position of the symbol.

4.10. RELATION BETWEEN THE ARROW LINE AND THE WELD JOINT

Figs. 4.7 and 4.8 explain the meaning of the terms:–

1. Arrow side of the joint, and
2. Other side of the joint

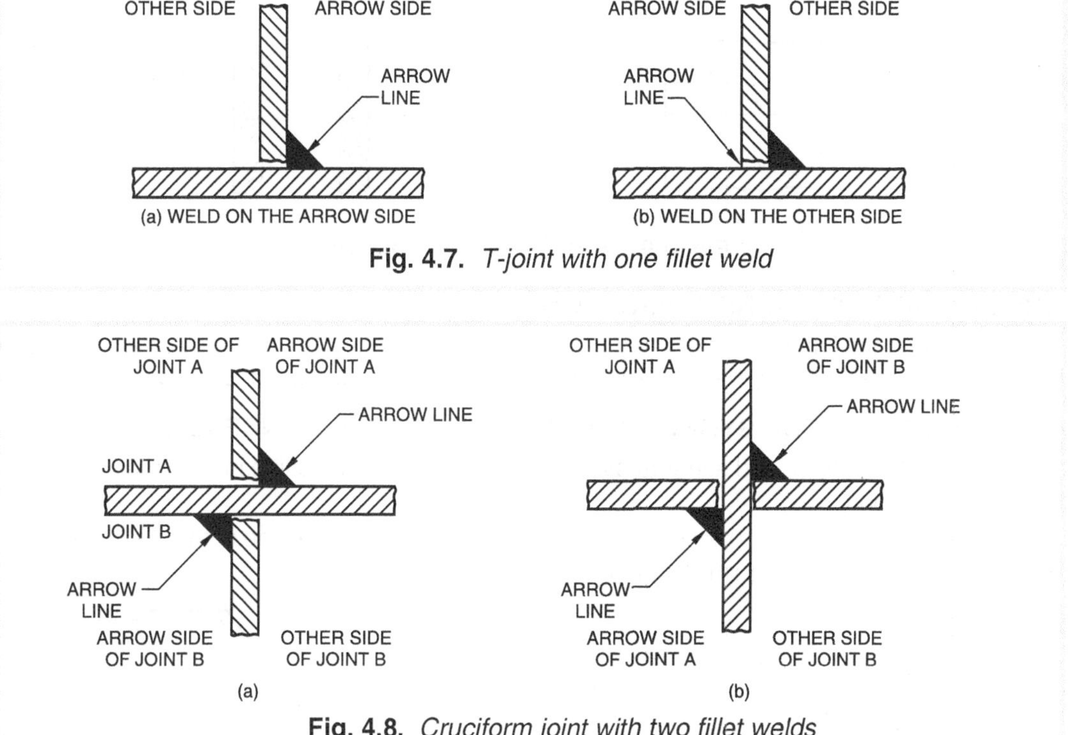

Fig. 4.7. *T-joint with one fillet weld*

Fig. 4.8. *Cruciform joint with two fillet welds*

Note : *The position of the arrow in these figures is chosen for purpose of clarity. Normally it would be placed immediately adjacent to the joint.*

4.11. POSITION OF THE ARROW LINE

The position of the arrow line with respect to the weld is generally of no special significance [see Figs. 4.9 (i) and (ii)]. However in the case of weld which requires only one plate to be prepared [see table 4.1); the arrow line shall point towards the plate which is prepared [see Figs 4.9 *(iii)* and *(iv)*]. The arrow line

1. joins one end of the continuous reference line such as that it forms an angle with it, and
2. shall be completed by an arrow head.

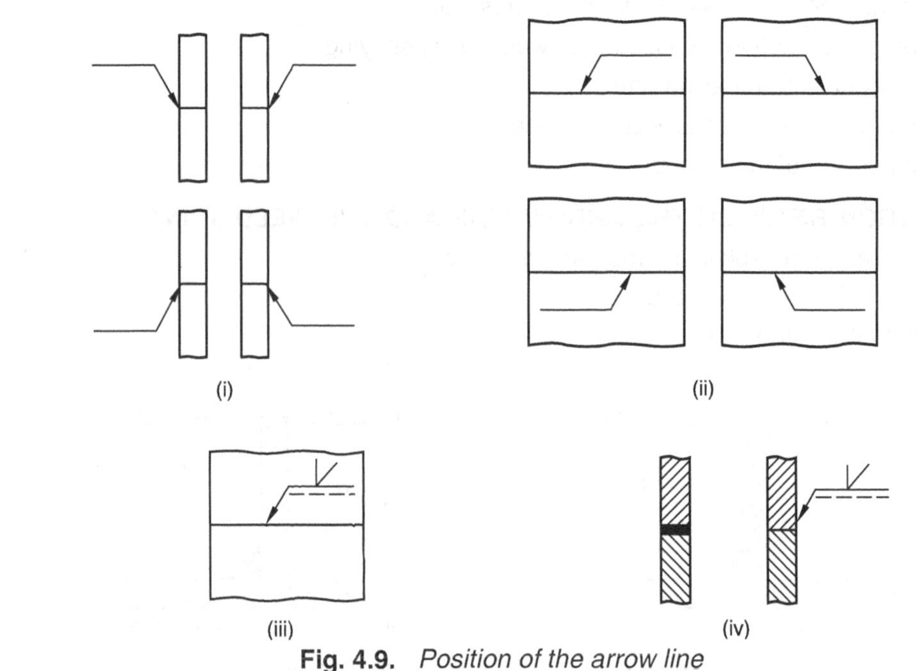

Fig. 4.9. *Position of the arrow line*

4.12. POSITION OF THE REFERENCE LINE

The reference line shall preferably be drawn parallel to the bottom edge of the drawing, or if it is not possible then it is drawn perpendicular to the bottom edge of the drawing.

4.13. POSITION OF THE SYMBOL WITH REGARD TO REFERENCE LINE

The symbol is to placed either above or beneath the reference line in accordance with the following regulations:–

1. The symbol is placed on continuous line side of the reference line of the weld. Weld face is on the arrow side of the joint [see Fig. 4.10 (i)].

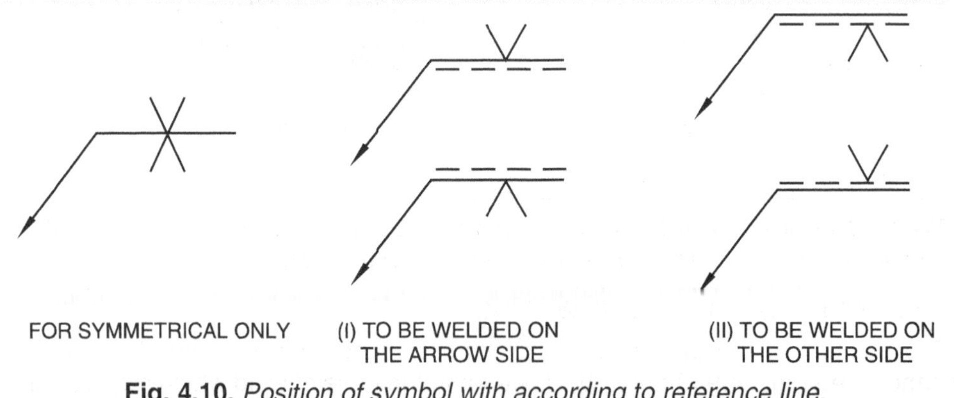

Fig. 4.10. *Position of symbol with according to reference line.*

2. The symbol is placed on the dashed line side of the weld. Weld face is on the other side of the joint [see Fig. 4.10 *(ii)*].

4.14. DIMENSIONING OF WELDS

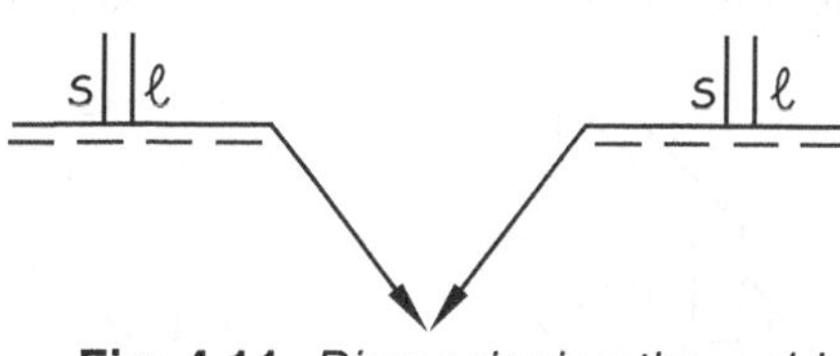

Fig. 4.11. *Dimensioning the weld*

General rules : 1. Each weld symbol may be accompanied by a certain number of dimensions. These dimensions are written in accordance with Fig. 4.11 as follows :–

(i) The main dimensions relative to the cross section of weld are written on the left hand side (i.e. before) of the symbol.

(ii) Longitudinal dimensions are written on the right-hand side (i.e. after) of the symbol.

The method of indicating the main dimensions is defined in table 4.1. Whereas, the rules for setting down these dimension are also given in this table.

(iii) Other dimensions of less importance may be indicated, if necessary.

2. Main dimensions to be shown :

(i) The dimension that locates the weld in relation to the edge of the sheet shall not appear in symbolisation but on the drawing.

(ii) In the absence of any indication following the symbol signifies that the weld is to be continuous over the total length of the work piece.

(iii) In the absence of any indication, butt welds are to have complete penetration.

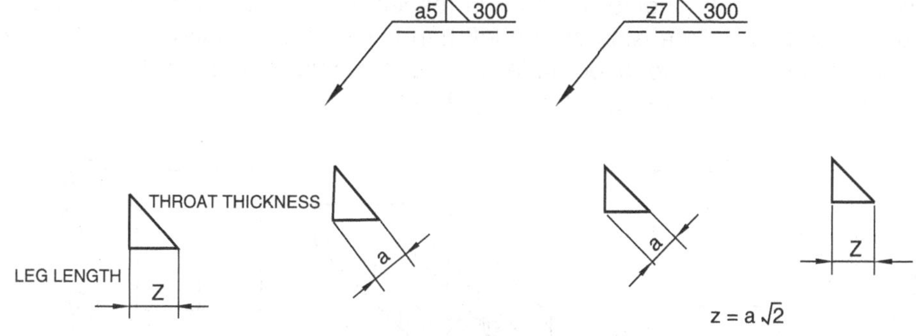

Fig. 4.12. *Dimensioning for the fillet welds*

(iv) In case of fillet welds, there are two methods to indicate dimensions (see Fig. 4.12). Therefore the letters a or z shall always be placed in front of the value of the corresponding dimension.

(v) For plug or slot welds with bevelled edges, it is the dimension at the bottom of the hole which shall be taken into consideration.

According to I.S., the dimension of weld lengths, weld sizes and spacing should be indicated in mm as discussed below:–

Length of welds : All welds should be continuous along the length of the joint unless shown otherwise [see Figs. 4.13 (i) and (ii)].

Note : *The length of each weld should be indicated on the right hand side of the symbol.*

In case where regular intermittent welding is required and the joint commences with a weld, the length of each weld should be indicated as shown in fig. 4.13 (iii).

Where regular intermittent welding is required and the joint commences with an unwelded length, there it should be indicated as shown in Fig. 4.13 *(iv)*.

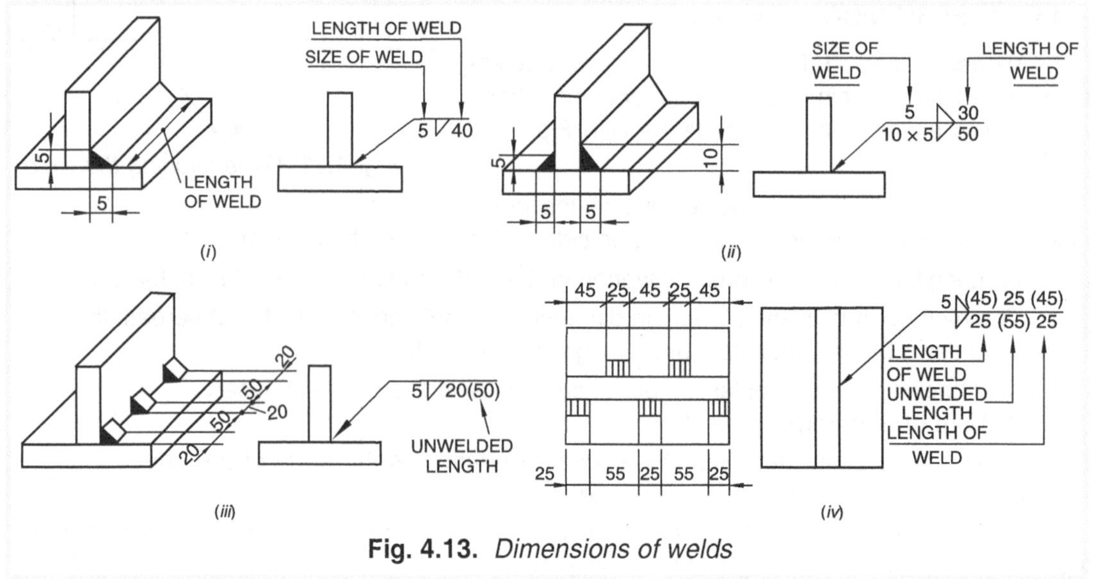

Fig. 4.13. *Dimensions of welds*

4.15. SIZE OF WELD

The size of the weld should be given in respect of the fillet weld. The size should be indicated by inserting a dimension on the left hand side of the symbol. The size should be indicated against each symbol used, unless there is a general note on the drawing stating that all welds are of a particular size (see Fig. 4.14).

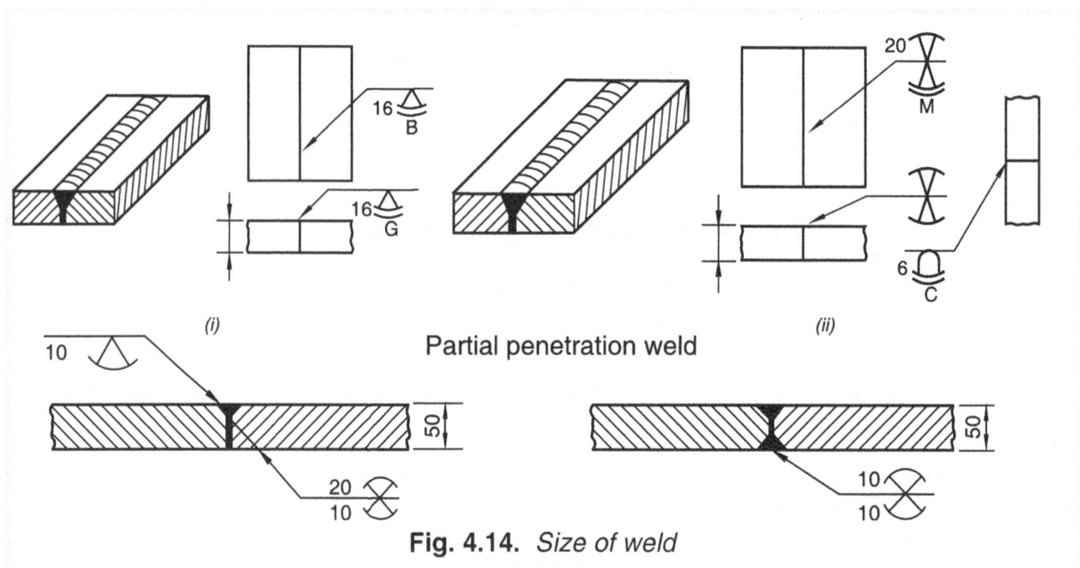

Fig. 4.14. *Size of weld*

The single V, single U, single J or bevel butt welds should be indicated by a dimension as shown in Fig. 4.14 *(i)*.

In case of double-V, double-J or double bevel butt welds, where only partial penetration is required, the same method as given above should be followed. But, the dimension should be added above or below the reference line. Fig. 4.14 (ii) shows the joint for partially penetrated welds.

4.16. SURFACE CONTOUR AND FINISH OF WELDS

Surface contour (i.e. flush, convex and concave) and finish of weld (chipping C, machining M and griding G) are indicated by symbols as shown in Fig. 4.15.

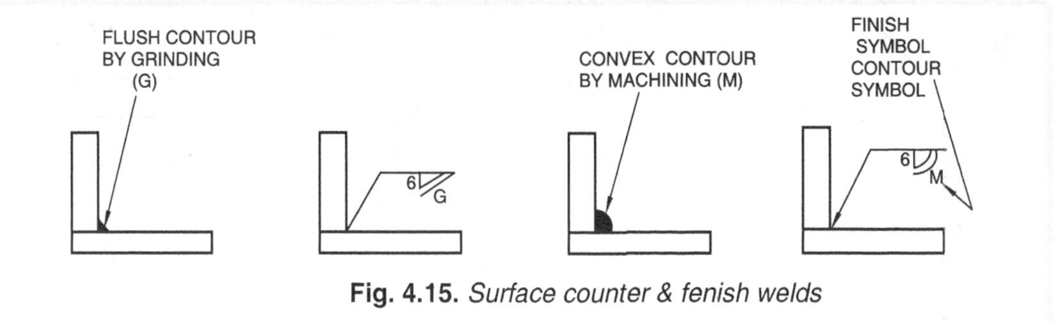

Fig. 4.15. *Surface counter & fenish welds*

4.17. WELD ALL AROUND

When a weld is to be made all around the joint, a circle be made at the elbow of line connecting the reference line as shown in Fig. 4.16.

4.18. SITE WELD

When a weld is to be made in a particular welded structure on site, during assembly or erection, a filled-in-circle be made at the elbow of line connecting the arrow to the reference line as shown in Fig. 4.16.

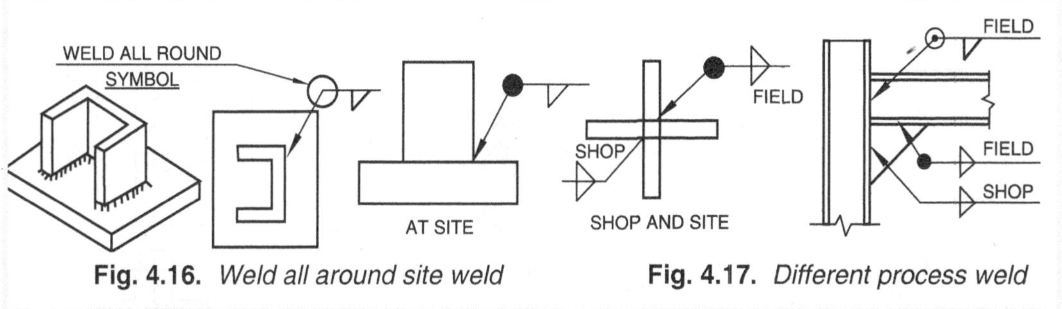

Fig. 4.16. *Weld all around site weld*

Fig. 4.17. *Different process weld*

4.19. DIFFERENT PROCESSES OF WELDING

When different processes are to be used for different positions of the same work piece, the drawing is made as shown in Fig. 4.17.

4.20. COMPOUND WELD

The compound weld as composed of two different weld forms e.g. a fillet weld super imposed on a J-weld is indicated by symbol as shown in Fig. 4.18.

4.21. BACKING STRIP WELD

When backing strip is used to become permanent part of a welded joint, it is indicated by means of symbols shown in Fig. 4.19.

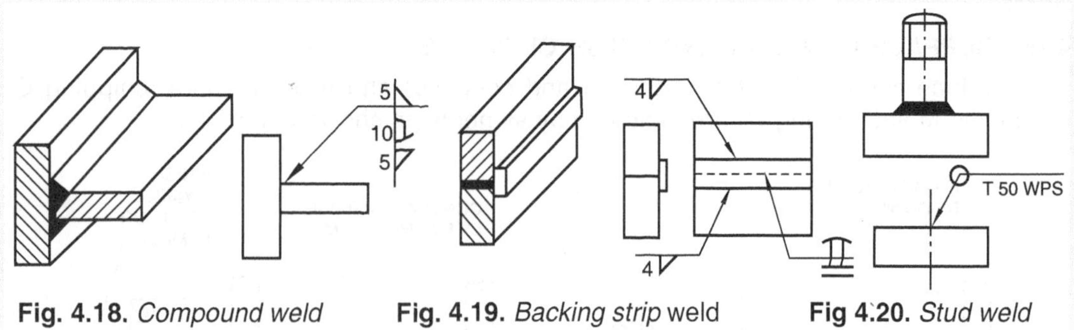

Fig. 4.18. *Compound weld* **Fig. 4.19.** *Backing strip* weld **Fig 4.20.** *Stud weld*

4.22. STUD WELD

The length, size and other features of the stud weld is indicated by symbol shown in Fig. 4.20.

4.23. SPOT WELD

The spot weld is indicated by symbol as centred on the reference line. The dimensions may be shown on either side of the reference line. The location of the first spot, distance from edge and distance between rows of spots is dimensioned, as shown in Fig. 4.21.

4.24. SEAM WELD

The seam weld is indicated by symbol as centred on the reference line [see Fig. 4.22]. The dimension may be shown on either side of the reference line. The width i.e. size of the seam weld is dimensioned to the left of the weld symbol.

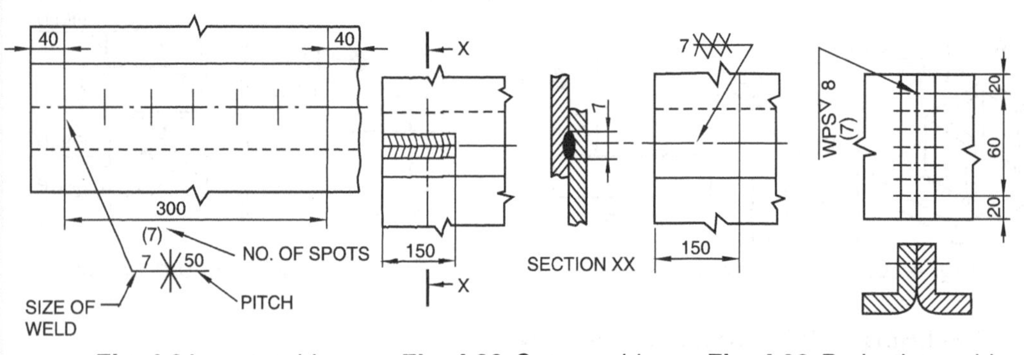

Fig. 4.21 *spot weld* **Fig. 4.22** *Seam weld* **Fig. 4.23** *Projection weld*

4.25. PROJECTION WELD

The projection weld is indicated by symbol either on the drawing or in the welding procedure sheet (W.P.S.) as shown in Fig. 4.23. The spacing between the welds is measured from centre to centre and on the right hand side of the symbol.

QUESTIONS FOR SELF EXAMINATION

1. What do you mean by welding?
2. What are the various types of welding?
3. Why the edges of welding parts are prepared?
4. Why the welding symbols are used in drawing the various joints?
5. What is the significance of the position of arrow head and reference line?
6. What are the various methods by which welded joints are dimensioned?

PROBLEMS FOR PRACTICE

1. Draw the neat sketches of the following welded joints :—
   (a) Butt joint (b) Lap joint (c) T-joint
   (d) Corner joint (e) Edge joint.
2. Draw the following edge preparation which are used in welding processes for welding butt joint. Give the preparation dimensions also.
   (a) Plain face (b) Single-V (c) Double-V
   (d) Single-U (e) Double-U
3. Draw a neat sketch of "Standard location of elements of a welding symbol" according to I.S.I.
4. Draw the "sectional representation" and "appropriate symbol" for the following from of weld :—
   (a) Fillet
   (b) Square butt
   (c) Single-V butt
   (d) Double-V butt
   (e) Single bevel butt
   (f) Double-U butt
   (g) Stud
   (h) Bead (Edge or Seal)
   (i) Plug or slot
   (j) Stitch
   (k) Flash
   (l) Butt resistance.
5. Draw the various supplimentary symbols for weld finish.
6. Draw the following weld joints with reference to the arrow.
7. Draw a proportionate sketch of butt joint showing the various methods for dimensioning.

CHAPTER

5

Screw Threads

INTRODUCTION

In engineering field, it is usual to join different parts of a machine, structure or other engineering products temporarily by means of temporary fasteners such as bolts and nuts, machine screws, keys, cotters-pins etc. By temporary fasteners, the joined parts can be separated easily without breaking them. The most important operating element of temporary fastener is screw thread. Screw threads occur in one form or another on pratically all engineering products.

In this chapter, we shall study screw thread, practical application of screw threads, terms used in screw threads, froms of screw threads, etc.

5.1. SCREW THREADS

The helical grooves cut on the surface of a cylindrical piece for fastening parts or power transmission is known as screw threads.

5.2. PRACTICAL APPLICATION OF SCREW THREADS

The screw threads are used on *bolts, nuts, machine screws, studs, foundation bolts* and other similar parts employed for fastening, power transmission or adjustments.

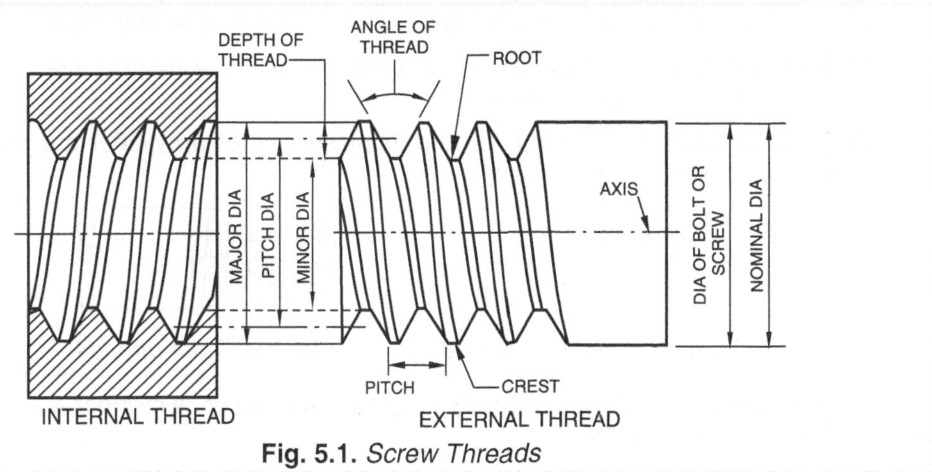

Fig. 5.1. *Screw Threads*

5.3. TERMS USED IN SCREW THREADS

Fig. 5.1 shows different portions of internal and external screw threads. The various terms used in connection with the screw threads are as follows :—

1. Pitch : It is the distance from a point on a screw thread to a corresponding point on the next thread measured parallel to the axis e.g. *the distance from crest to crest or root to root.*

Pitch is the reciprocal of the number of threads per mm.

$$\text{Pitch} = \frac{1}{\text{number of threads per mm}}$$

$$\therefore \quad p = \frac{1}{\text{t.p.mm}}$$

2. Lead : It is the distance moved by a nut or bolt in the axial direction in one complete revolution. In single start threads, the lead is equal to the pitch. In multistart threads lead is equal to number of starts multiplied by pitch.

Thus, *Lead = No. of starts ×Pitch*

3. Crest : It is the outermost portion which joins the two sides of a thread.

4. Root : It is the innermost portion which joins the adjacent sides of a thread.

5. Angle of thread : It is the angle between the two sides of the thread measured in an axial plane.

6. Depth of the thread : It is the distance between the crest and root of a thread measured perpendicular to the axis.

7. Nominal diameter : It is the diameter of the cylindrical piece on which the threads are cut. *The screw is specified by this diameter.*

8. Major diameter : It is the largest diameter of a screw thread. It is also known as outside or crest diameter.

9. Minor diameter : It is the smallest diameter of a screw thread. It is also known as root or core diameter.

10. Pitch diameter : It is the diameter of an imaginary cylinder passing through the width of the threads at which the threads and the width of the gap are equal.

11. Slope : It is the axial distance through which a thread is tilted on the front view. The slope of a thread is equal to half the pitch or lead on single start threads.

12. External threads : Threads on the outer surface of a cylinder are called external threads, such as threads in case of bolt.

13. Internal threads : Threads on the inner surface of a cylinder are called internal threads, *such as threads in case of nut.*

Fig. 5.2. *Straight threads*

Fig. 5.3 *Taper threads*

14. Straight or cylinderical threads : Threads formed on a cylinder are called straight or cylinderical threads (see Fig. 5.2).

15. Taper threads : Threads formed on a cone shaped rod are called tapered threads (see Fig. 5.3).

16. Right hand threads : A right hand thread is one which advances into the nut when turned in a clock-wise or if the direction of thread slopes up towards the left, when the screw is placed horizontally (see Fig. 5.4).

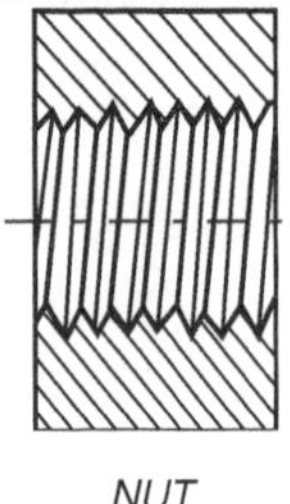

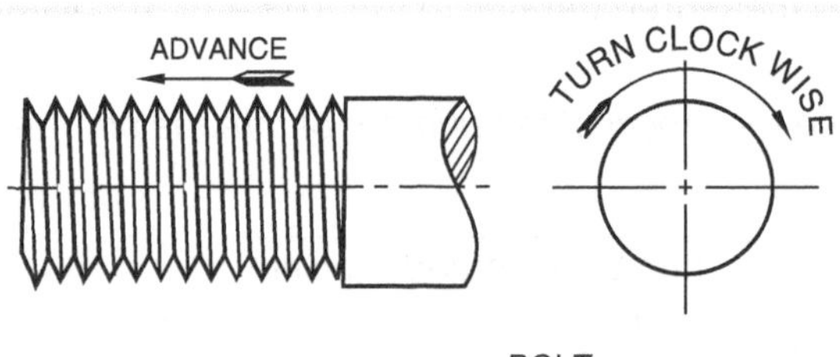

Fig. 5.4. *Right hand threads*

17. Left hand threads : A left hand thread is one which advance into the nut when turned in an anticlock or if the direction of the thread slopes up towards the right, when the threads are placed horizontally (see Fig. 5.5).

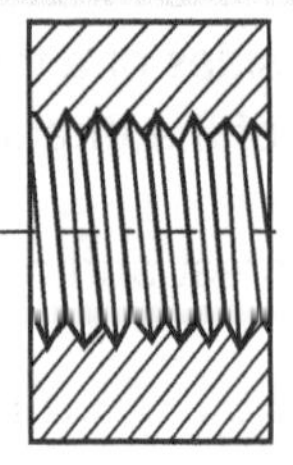

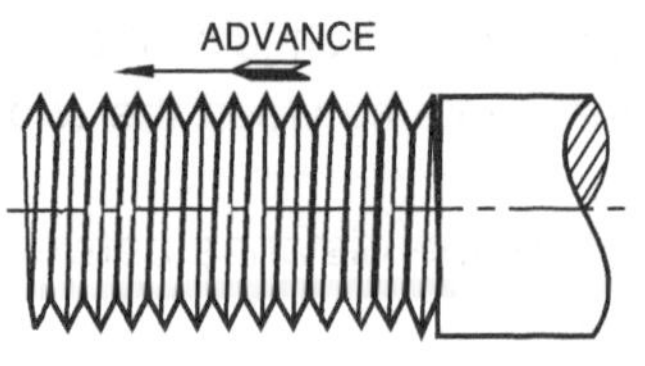

Fig. 5.5. *Left hand threads*

18. Single start threads : When only one helix forming the threads, run on a surface is called a single start thread. It is also called single start threads, because only one starting point is seen on the beginning of the threaded portion. In a single start threads, the pitch is equal to the lead. Threads are always single start unless or otherwise specified [see Figs. 5.6 (i) and 5.7 (i)]

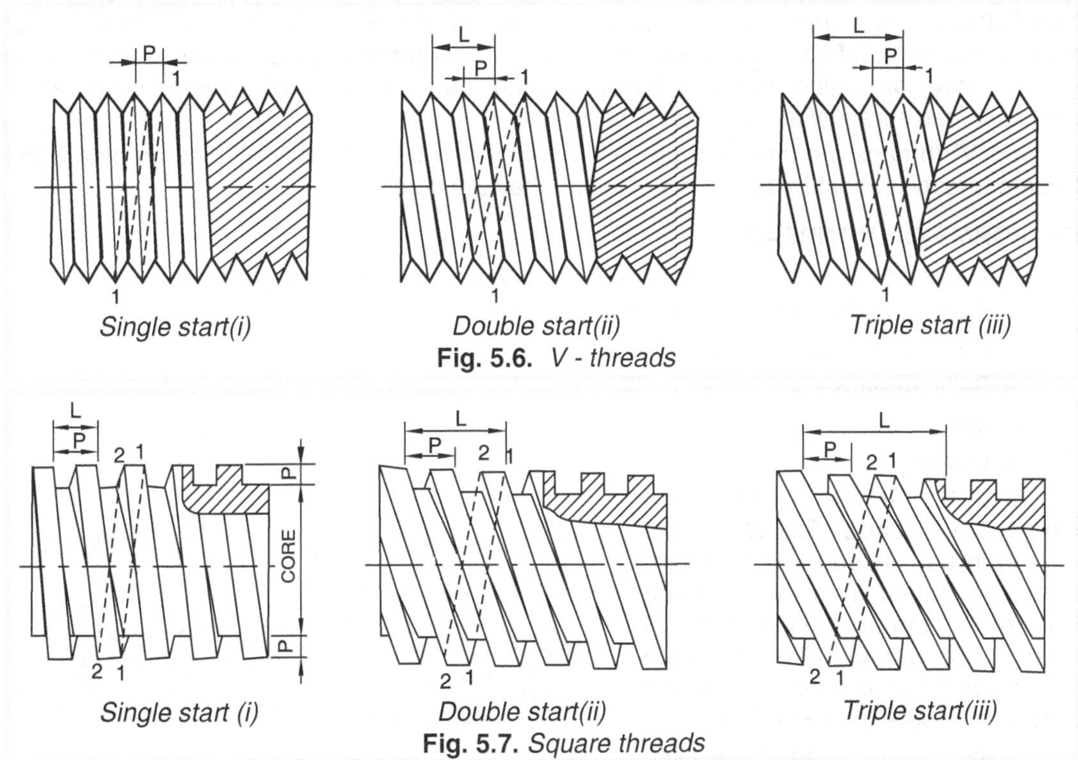

Fig. 5.6. *V - threads*

Fig. 5.7. *Square threads*

19. Multiple-start threads: When two or more helices forming the threads run side by side on the cylindrical surface, it is called multiple-start threads.

It is also called a *multiple start thread,* because two or more than two starting points are seen on the beginning of the threaded portions [see Figs. 5.6 (*i*) and 5.7 (*ii*)].

In double start threads, the lead is twice the pitch; in triple start threads, lead is three times the pitch and so n on [see Figs. 5.6 (*ii*) and 5.7 (*iii*)].

Uses of Multiple Threads: Multiple threads are used wherever quick motion is required and application of great force is not allowed e.g. on *fountain pens, bottles, tooth-paste caps, valves etc.* to impart quick action in opening and closing.

5.4. FORMS OF SCREW THREADS

As stated already the screw threads are used as temporary fasteners on devices for making adjustments or for the transmission of power from one part to another. For these different purposes, different forms of screw threads are used as given below :-

1. Triangular or "V" form threads.
2. Square form threads.

Other forms are either modified from of "V" threads or a combination of above two forms.

5.5. COMPARISON OF 'V' AND SQUARE THREADS

1. "V" threads are stronger than the square threads. As for the same depth the "V" threads have twice the amount of shearing action at the root of the thread than a square threads.

2. "V" threads offer greater frictional resistance to motion than square threads and are thus better suited for fastening purposes as in bolts and nuts, studs and nuts, tap bolts, set screws, etc. For low friction resistance characteristic, sequare threads are used for transmission of power or motion from one part to another as in the case of a *bench vices, valve spindles, lead screw of a lathe, screw jacks, etc.*

3. "V" threads are cheap to produce as they can be cut easily by dies on machines. Square threads cannot by cut by dies, hence they are costly.

5.6. FORMS OF V-THREADS

The following are different forms of V-threads used in engineering practice:-

1. British standard whitworth threads (B.S.W.)
2. British Association threads (B.A.)
3. Seller threads
4. Metric threads
5. Unified thread.

5.6.1. British standard whitworth threads (B.S.W.) : This form of thread is used as a standard thread in Britain. It is the modified from of "V" thread having angle of threads 55° and 0.167 of the depth of the full 'V' is rounded off at the crests and roots (see Fig. 5.8)

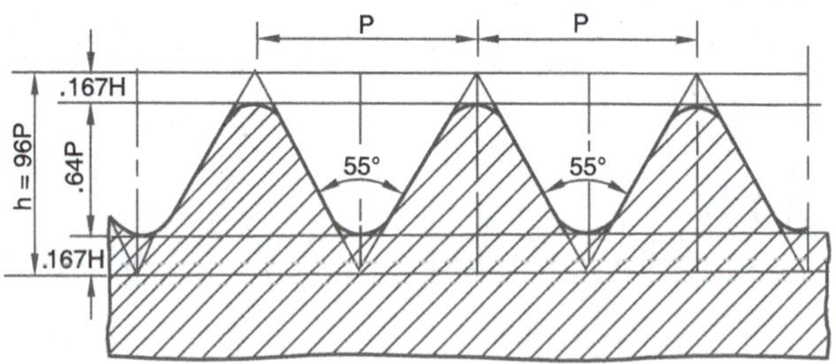

Fig. 5.8. *B.S.W. Threads*

Uses: *These threads are used on bolts, nuts and scerews used for general purpose fastening.*

5.6.2. British Association threads (B.A.) : This form of thread is recommended by British Standard Institution. The angle of threads is 47.5° and 0.236 of the depth is rounded off at crests and roots (see Fig. 5.9).

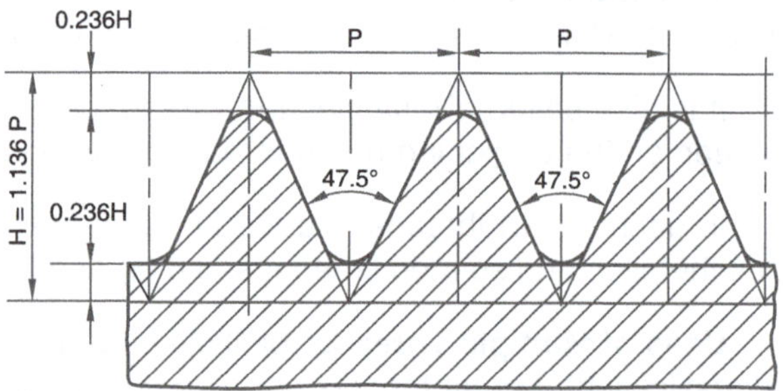

Fig. 5.9. *B.A. Threads*

Uses: *It is generallyl used on fine work as in mechanical instruments, aircarfts construction etc.*

5.6.3. Seller threads(American National thread) : This form of thread is adopted as a standard form in America. The angle between the sides of threads is 60°. The crests and roots are flattened up to 0.125 of the depth (Fig. 5.10)

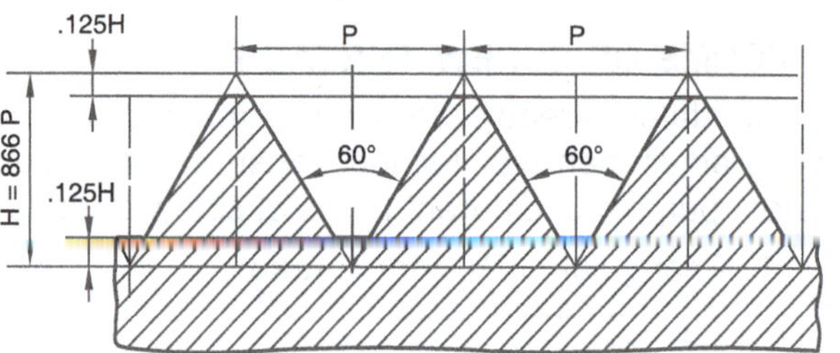

Fig. 5.10. *Seller threads*

Uses: *It is mostly used on fasteners for making adjustment.*

5.6.4. Metric threads (M): The Bureau of Indian Standards has recommended the adoption of the unified screw thread based on metric system as a standard form used in India and has designated it as metric screw thread with I.S.O. profile. In this system instead of fixing the number of threads per unit length, the pitch of the threads is fixed (see Fig. 5.11).

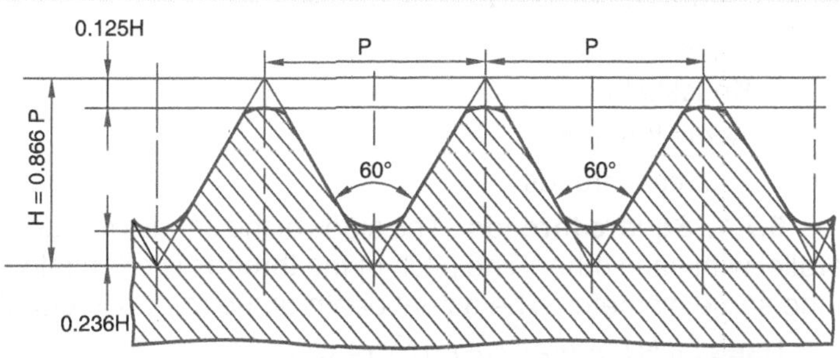

Metric Threads with crests and roots in rounded form.

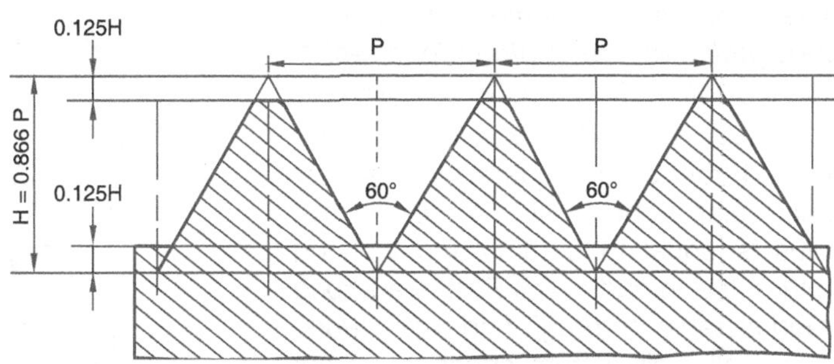

Metric Threads with crests and roots in flatted form.

Fig. 5.11.

5.6.5. Unified threads or international standard organi-sation threads: The International organisation for standardization (I.S.O.) of which India, U.S.A., U.K., Canada and other countries are members have agreed to have a common form of thread to facilitate the interchangeability of their product.

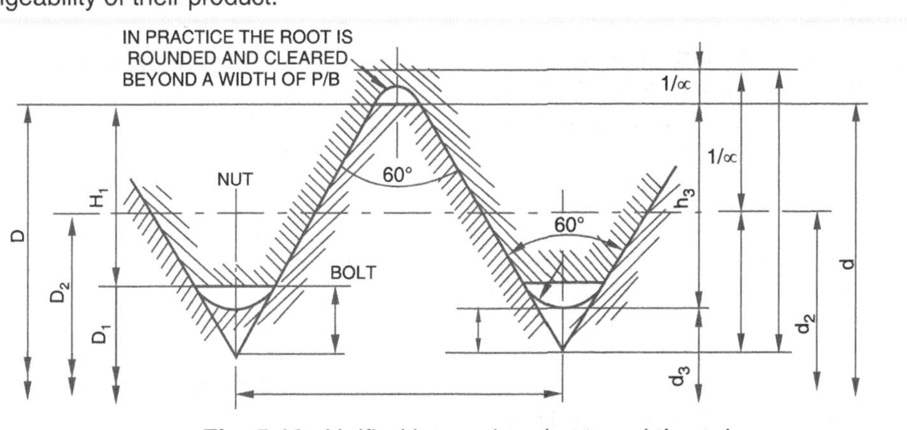

Fig. 5.12. *Unified internal and external threads*

Theoretical depth, H = 0.866025 P

Actual depth, $H_1 = \dfrac{D - D_1}{2} = \dfrac{5}{8}H$

$= 054127$

$$H_2 = \frac{d - d_1}{2} = \frac{17}{24} = 0.61343 \text{ P}$$

Major or nominal diameter D = d

Pitch diameter, $D_2 = d_2$

5.7. FORMS OF SQUARE THREADS

1. Square threads
2. Knuckle threads
3. Acme threads
4. Buttress threads

5.7.1. Square threads : The sides of the square threads are normal to the axis and hence parallel with each other. The pitch of the threads is often taken as twice that of B.S.W. threads of the same diameter. The actual depth of the thread is equal to half the pitch (see Fig. 5.13)

Uses : *It is generally used for power transmission on feed mechanism of machine tools, valve spindles, vice screw, screw jacks, etc. as it offers less resistance to motion than the "V" threads.*

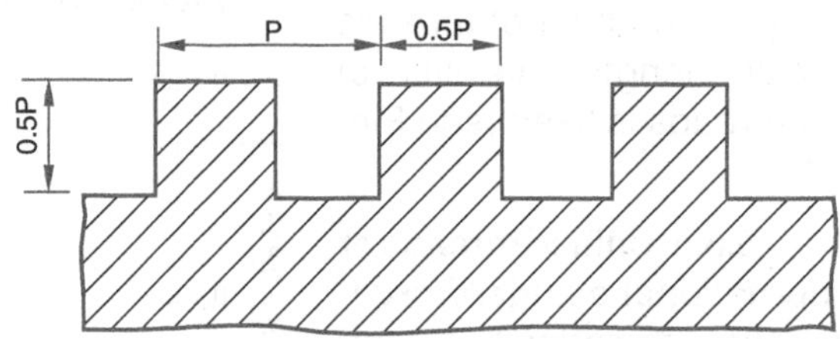

Fig. 5.13. *Square threads*

5.7.2. Knuckle threads : It is the modified from of the square threads. It is semicircular at the crest and root. In this, radius is equal to 0.25 P and working depth of the thread is 0.5 P. It is a strong thread suitable for rough and ready work. It is usually rolled from sheet metal but sometimes casted (see Fig. 5.14).

Uses : *Knuckle threads are used in electric bulbs, bottles, fire hydrants, railway couplers etc.*

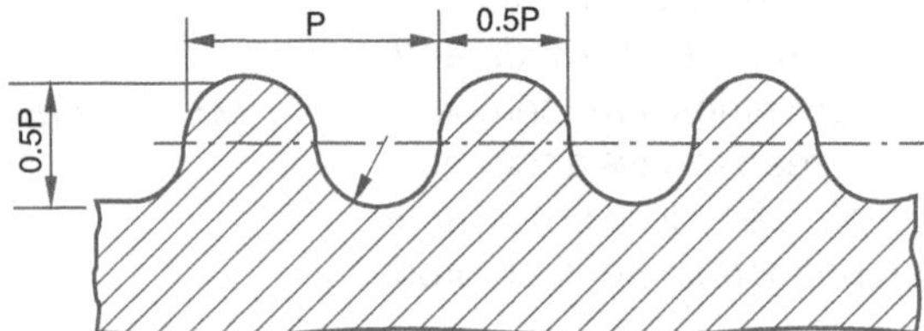

Fig. 5.14. *Knuckle threads*

5.7.3. Acme thread : It is another modified from of square thread. The threaded angles is 29°. The depth of the thread = 0.5 P + 0.25. The width of the thread at the crest is equal to 0.3707 P (see. Fig. 5-15).

Uses: *It is particular used in such cases where the process of engagement and disengagement of threads are frequent as in the lead screw shaft of the lathe.*

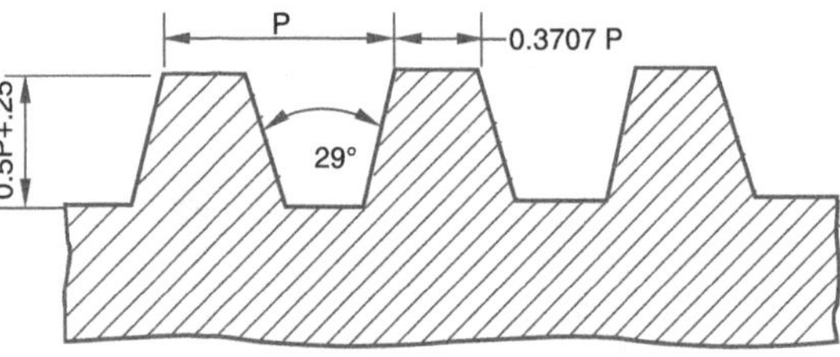

Fig. 5.15. *Acme threads*

5.7.4. Buttress thread : It is a combined form of V and square threads. One side of the thread is perpendicular to the axis of the thread and the other inclined at 45°. Sharp edges at crest and the root are flatened to 0.125 of the depth (see Fig. 5.16).

Uses: *It is used for power transmission in one direction only as in breachlocks of large guns, jacks, in air plane propellers, carpenter's vices, etc.*

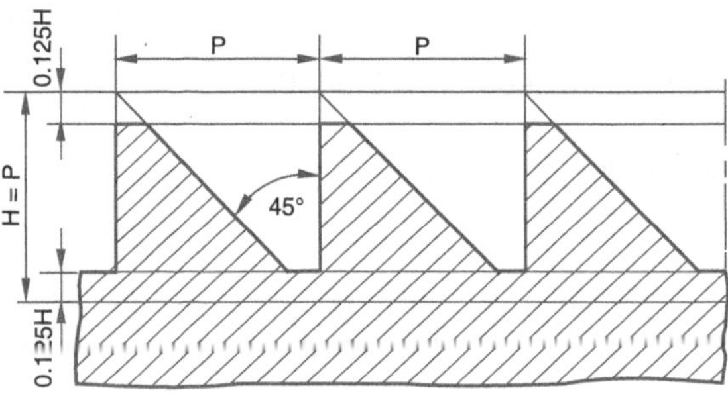

Fig. 5.16. *Buttress threads*

5.8. THREAD DESIGNATION

The size of screw threads is designated by letter M followed by diameter and the pitch, the two being separated by the 'X' e.g. M 16 ×2, where 16 is the diameter of the screw in mm and 2 is the pitch of the screw. Where there is no indication of pitch, it should be considered as coare pitch.

When specifying the length of thread an 'X' is used to separate the length from the rest of the designation e.g. M 16×2×24 where 24 is the length of threads in mm.

5.9. CONVENTIONAL REPRESENTATION OF SCREW THREADS

The true form of a screw threads is a helical curve and it would take considerable time and labour to draw its actual helical form on the drawing. A screw so drawn naturally looks nice, but serves no special purpose. To overcome this difficulty, screw threads are generally represented by standarized conventions, by which the threads can be easily and quickly drawn. Notes are given at the bottom of the threads stating the particulars *e.g.* thread form, pitch, diameter, left hand or right hand threads. The conventional representation of screw threads are shown in fig. 5.17 and 5.18.

Fig. 5.17 show the conventional symbol as recommended by the Bureau of Indian standards (B.S.I.) for external and internal screw threads.

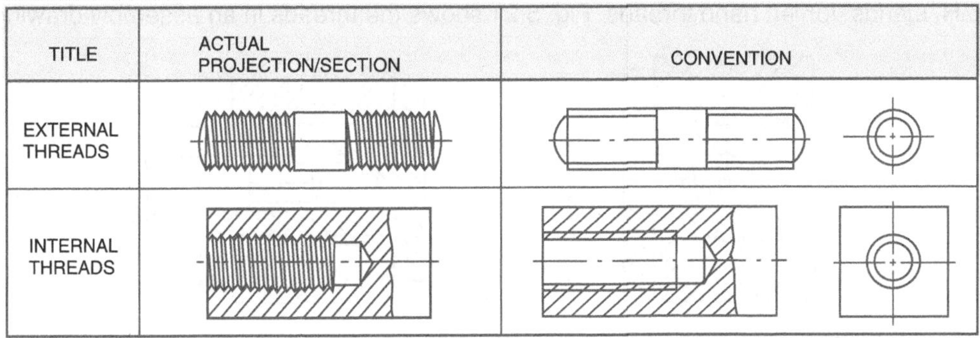

Fig. 5.17. *Conventional symbol for representation of screw Thread.*

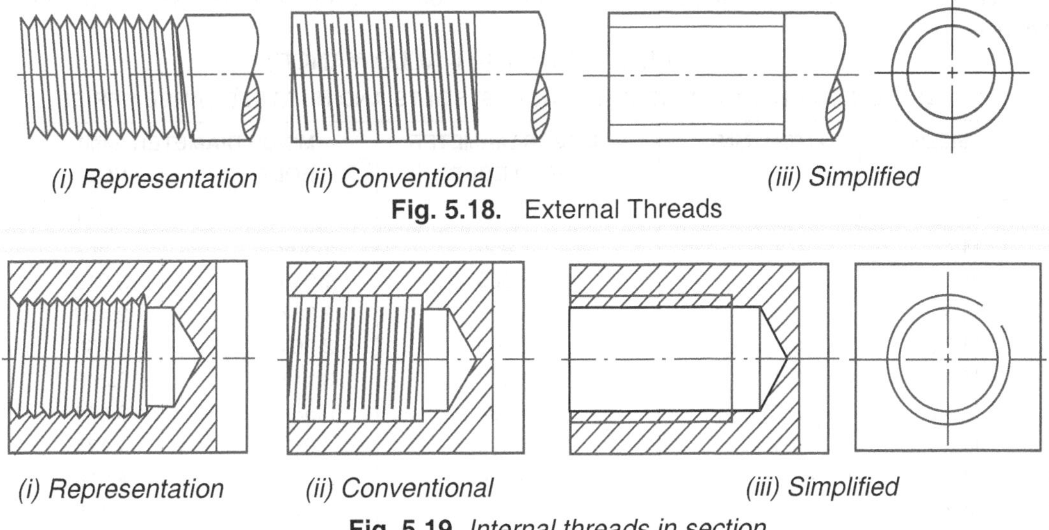

(i) Representation *(ii) Conventional* *(iii) Simplified*

Fig. 5.18. External Threads

(i) Representation *(ii) Conventional* *(iii) Simplified*

Fig. 5.19. *Internal threads in section*

Fig. 5.18. (i), (ii), and (*iii*) shows representation, conventional and simplified methods of representing external threads. In Fig. 5.18 (iii), the threaded portion is shown by two thin line, one half pitch apart.

Fig. 5.19 (i), (ii) and (iii) shows the representation, conventional and simplified methods of representing internal threads in section. The direction of the root and crest lines should slope in a direction opposite to those of the visible threads lines as these are shown in section.

Fig. 5.20 (i) shows single start metric threads indicated by M 20 L H where M stands for metric thread, 20 is the diameter of the screw in mm and L.H. means left hand threads.

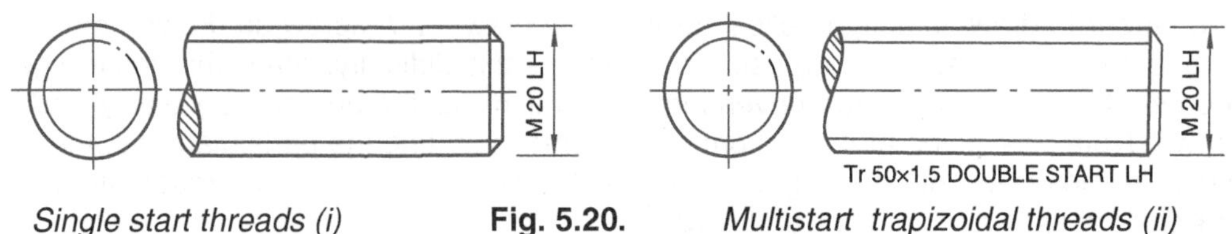

Single start threads (i) **Fig. 5.20.** *Multistart trapizoidal threads (ii)*

Fig. 5.20 (ii) shows multistart threads designated by Tr 20 x 1.5 double start L.H. In this Tr stands for Trapezoidal thread, 20 is the diameter of the screw, 1.5 is the pitch double start and L.H. stands for left hand threads. Fig. 5.21 shows the threads in an assembly drawing.

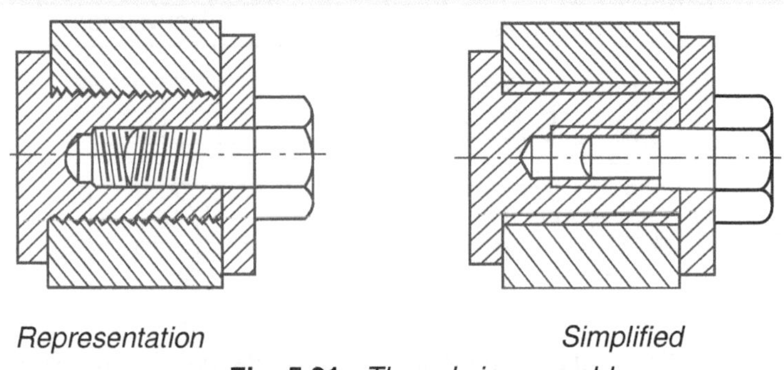

Representation *Simplified*

Fig. 5.21. *Threads in assembly*

Table 5.1 shows dimensions of screw threads, bolts and nut in coarse series without allowance.

TABLE 5.1 (WITHOUT ALLOWANCE)

DESIGN DIMENSION OF SCREW THREADS, BOLTS AND NUTS IN COARSE SERIES

SIZES	PITCH (MM)	MAJOR DIAMETER (MM) BOLT	MINOR DIAMETER (MM)	
			BOLT	NUT
M 1.6	0.35	1.600	1.71	1.221
M 2	0.4	2.000	1.509	1.567
M 2.5	0.45	2.500	1.948	2.013
M 3	0.5	3.000	2.387	2.459
M 3.5	0.5	3.500	2.764	2.850
M 4	0.6	4.000	3.141	3.242
M 4.5	0.75	4.500	3.580	3.688
M 5	0.8	5.000	4.019	4.134
M 6	1	6.000	4.773	4.918
M 7	1	7.000	5.773	5.198

SIZES	PITCH (MM)	MAJOR DIAMETER (MM) BOLT	MINOR DIAMETER (MM)	
			BOLT	NUT
M 8	1.25	8.000	6.446	6.647
M 10	1.5	10.00	8.160	8.376
M 12	1.75	12.000	9.853	10.106
M 14	2	14.000	11.546	11.835
M 16	2	16.000	13.546	13.835
M 18	2.2	18.000	14.933	15.294
M 20	2.5	20.000	16.933	17.294
M 22	2.5	22.000	18.933	19.294
M 24	3	24.000	20.320	20.752

Table 5.2 shows design dimensions of screw threads for screw, bolts and nuts, in fine series without allowance.

TABLE 5.2 (WITHOUT ALLOWANCE)

DESIGN DIMENSION OF SCREW THREADS FOR SCREW, BOLTS AND NUTS IN FINE SERIES

SIZE	PITCH (MM)	MAJOR DIAMETER (MM) BOLT	MINOR DAMETER (MM)	
			BOLT	NUT
M 8X1	1	8.000	6.773	6.918
M 10X1.125	1.25	10.000	8.446	8.647
M 12X1.25	1.25	12.000	10.446	10.647
M 14X1.5	1.5	14.000	12.160	12.376
M 16X1.5	1.5	16.000	14.160	14.376
M 18X1.5	1.5	18.000	16.160	16.376
M 20X1.5	1.5	20.000	18.160	18.376
M 22X1.5	1.5	22.000	20.160	20.376
M 24X2	2	24.000	21.546	21.805

Problem 1: Draw a right handed "V" threads conventionally by thick and thin lines method on a rod of 24 mm diameter and 50 mm long.

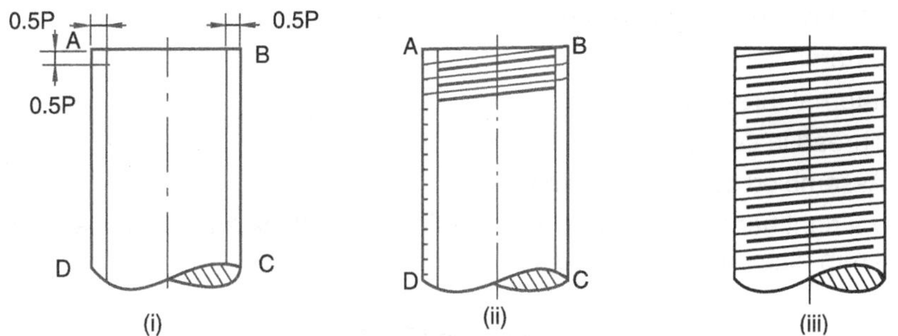

Fig. 5.22 *Method of drawing threads*

Solution : 1. Draw a rectangle ABCD of 24 mmx50 mm as shown in fig. 5.22 (i)

2. On AD, mark off distances equal to 0.5 P, where P = Pitch = 1/t. p. mm. (From the thread table, take pitch for 24 mm diameter).

3. On AB, mark two points at a distance of 0.5 P away from A and B respectively and from these points, draw very thin lines parallel to AD and BC.

4. Now join the first division point on AD with B by a thin line. Then, draw a thick line from the second division point (*between the vertical lines*) parallel to the first thin line. Similarly, draw alternately thin and thick lines through the other points as shown in Fig. 5.22 (*ii*)

5. Erase all construction lines and finalise the view as shown in Fig. 5.22 (iii).

QUESTIONS FOR SELF EXAMINATION

1. What do you mean by screw threads and give its practical application?

2. What is the difference between:
   (i) normal and major diameter,
   (ii) Pitch and lead,
   (iii) right hand and left hand threads,
   (iv) single and multi-start threads,
   (v) V and square threads.

3. What are multi-start threads? Where these are used and why?

4. What is the necessity of conventional representation of screw threads?

PROBLEMS FOR PRACTICE

1. Draw the following threads with pitch equal to 15 mm :
   (a) (i) B.S.W. threads
   (ii) BA threads
   (iii) Seller threads
   (iv) Metric threads
   (b) (i) Square threads
   (ii) Knuckle threads
   (iii) Acme threads
   (iv) Buttress threads

2. Sketch the conventional representation of the following :
   (i) External threads
   (ii) Internal threads
   (iii) Thread in section
   (iv) Assembled thread in section

3. Draw a right handed V Threads conventionally by thick and thin lines method on a rod of 24 mm diameter rod and 40 mm long.

CHAPTER

6

Fastenings

INTRODUCTION

In engineering field, various structures and other engineering products are joined together either by means of temporary or permanent fastenings. Screwed fastening is that type of temporary fastening which is joined together by means of screws known as threads. The important types of screwed fastenings are bolts, nuts, etc. on which threads are cut. A nut and bolt comprise a screw pair (see Fig. 6.1). They are used to join two or more parts together temporarily and can be easily separated by screwing off the unit.

In this chapter we shall study nuts and bolts, their forms and practical applications in engineering field.

6.1. FASTENING

The process of joining two or more parts of a machine or structure by means of various devices e.g. nut and bolt is known as fastening.

6.2. NUT

A device used with a bolt or a stud to join two or more parts together temporarily is known as nut.

A nut alone cannot be used to fasten the parts. It screws on the threaded end of the bolt and draws the parts together by tightening it.

Nut consists of a prism of some particular shape, having a hole through it, the axis of which is coincident with that of the prism. The hole is threaded to be screwed on or off the bolt or stud.

Fig. 6.1. *Nut, bolt and washer*

6.3. FORMS OF NUTS

The following are the important forms of nuts used in engineering practice:

1. Hexagonal nut
2. Square nut
3. Flanged nut
4. Cap nut
5. Dome nut
6. Cylindrical or Capstan nut
7. Wing nut.

6.3.1. Hexagonal Nut : This is the most common form of nut and is used widely in machines and other engineering works. The upper corner of this nut are rounded off or chamferred at an angle of 30° or 45° w.r.t. base of the nut, but 30° conical chamferred is generally preferred. Due to rounding or chamferring, an arc is formed on each vertical face and a circle is thus formed on the top of nut which facilitates the drawing of the nut (see Fig. 6.2)

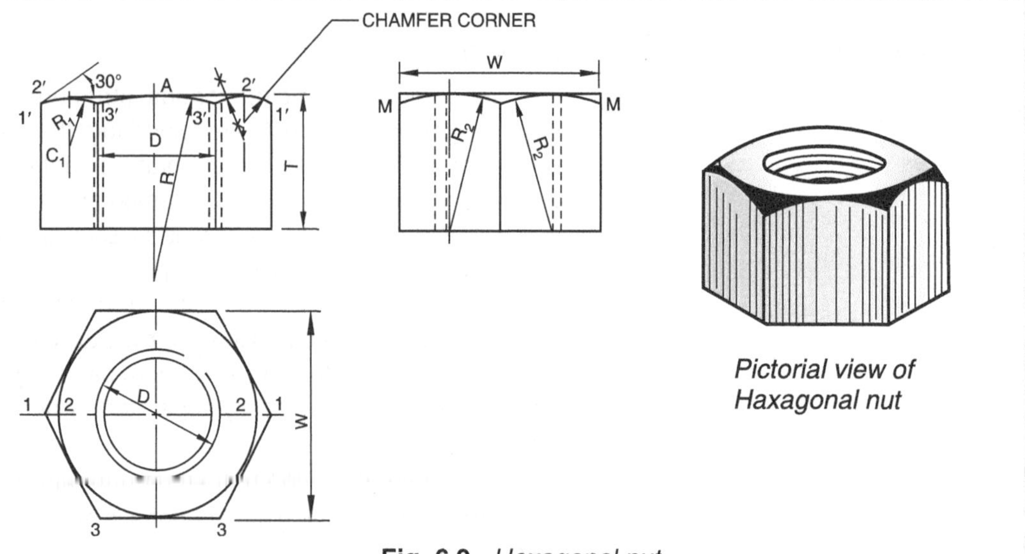

Pictorial view of Haxagonal nut

Fig. 6.2. *Hexagonal nut*

Proportations of the Hexagonal nut : If D, be the nominal diameter of the nut then according to I.S.,

Width across flats, W = 1.5 D + 3 mm [For bolt dia.>12 mm]

Thickness of the nut T = 0.9 D to D

Angle of chamfer = 30° to the base of the nut

Radius of chamfer, R = 1.5 D

The size across corners can be obtained by construction. For drawing purposes, it may be taken as 2 D in which 'size across flat' will come out to be slighly more than the value given above. The thickness and radius of the chamfer can be taken as D and 1.5 D respectively.

Method to draw a hexagonal nut : After rounding or chamfering, the upper corners of the nut form a circle. The top view, therefore, consists of a circle commonly referred as a chamfer circle.

Now, draw a circle of diameter equal to the distance across flats i.e. W = 1.5D + 3mm and circumscribe a regular hexagon about this circle with its two sides being horizontal. This regular hexagon can be drawn very easily with a 30°-60° set square (see Fig. 6.2).

Draw the front view, taking height or thickness of the hexagonal prism equal to T.

Project the marked points from the top view to the top surface of the front view and mark the respective points. Through the point 1', draw line 1'- 2' inclined at 30° to the top face. Draw a horizontal line (not shown) through 1' and 1' to cuts the edges of the central face at point 3'-3'.

With C as centre along the centre line in front view and radius equal 1.5 D, draw an arc passing through the points 3'- 3' and touching the line 2'- 2' at A. After finding C_1, draw arc R_1 as radius through the points 3'-1' and touching the line 2'-2:

Project the right side view in which only two faces of the nut will be visible and the distance between the other edges will be equal to W. Project the line 1'-1' horizontally and draw the arcs as shown in fig. 6.2.

Draw two circles for outer and inner portion of threaded nut in top view and simutaneously, with the dotted lines and the screwed hole in the front view and side view (see Fig. 6.2).

Important Note : *When three faces are visible in front view, the upper outer corners should be shown chamfered whereas when two faces are visible, the corner must be shown straight as shown in right side view of fig. 6.2.*

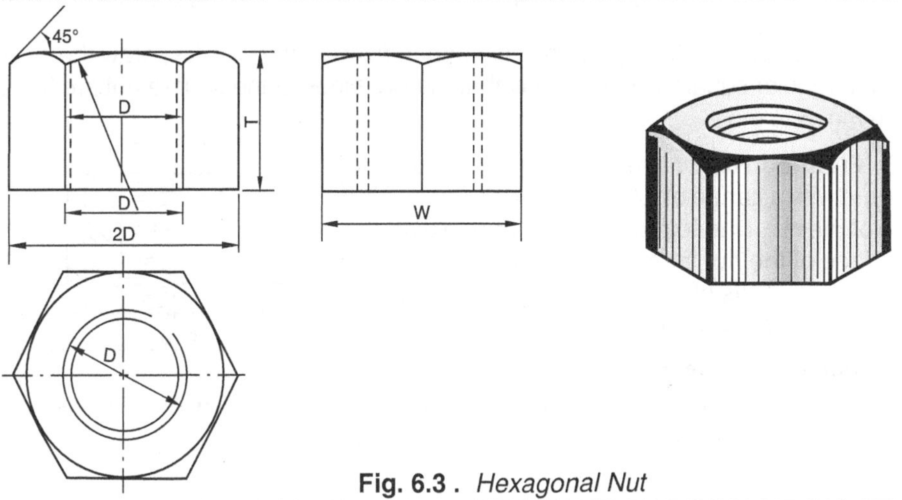

Fig. 6.3 . *Hexagonal Nut*

Simple method of drawing hexagonal nut : In this method the size across corners is taken as 2 D. The thickness and radius of chamber can be taken as D and 1.5 D respectively. Fig. 6-3 shows the three views of nut drawn by simple method.

6.3.2. Square Nut : This type of nut is usually used in conjunction with the square headed bolt. It has four sides instead of six sides as in the case of hexagonal nut. The corners of the square nut are also chamfered in the same way as for hexagonal nut (see fig. 6.4).

Proportions of the square nut : If D, be the nominal diameter of the bolt then according to I.S., the

Width across flats, W = 1.5 +3 mm

Thickness of the nut, T = 0.9 D to D

Angle of chamfer = 30° to the base of the nut

Radius of chamfer arc, R = 2D

Method to draw square nut : *(i)* **When two faces are equally seen in the front view.**

Draw the circle of diameter equal to W = 1.5D + 3 mm and circumscribe a regular square with all its side equally inclined to the horizontal as shown in Fig. 6.4.

Project the two faces in front view and complete it as shown in Fig. 6.4.

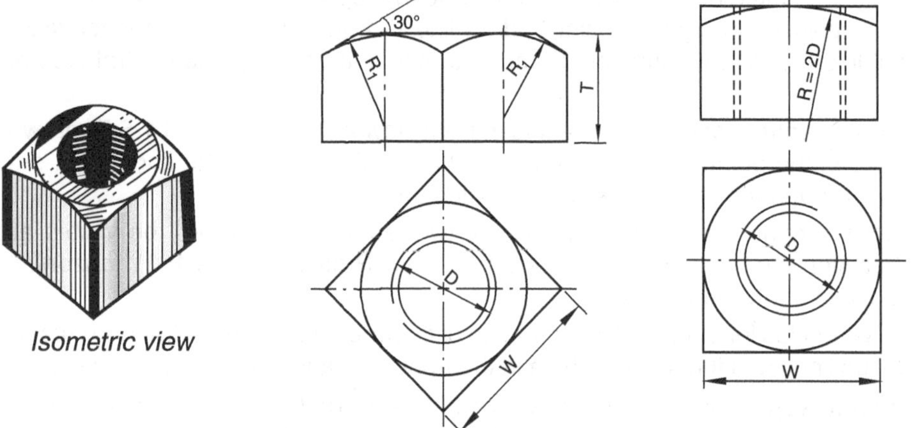

Fig. 6.4. Square nut with two faces visible

Fig. 6.5. Square nut with one face visible

(ii) **When one face of the square nut is seen in front view :** In this, draw the circle and circumscribe a regular square in such a way that the two sides of the square nut are horizontal as shown in Fig. 6.5.

6.3.3 Flanged Nut : The flanged nut is just like hexagonal nut with a flange or a flat circular disc or washer at the base of the nut. Due to provision of flanged disc, the bearing surface of the nut increases for which these nuts are used where the bolts of large diameter are required to pass through the required hole (see Fig. 6.6).

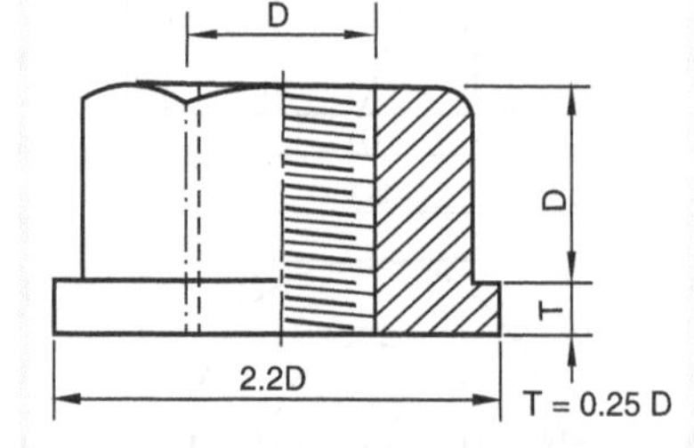

Fig. 6.6. *Flanged nut*

Proportions of the flanged nut ;

Let D, the outside diameter of the bolt, then the

Width across flat, W = 1.5 D + 3 mm

Thickness of the nut = D

Thickness of the disc, T = 0.25 D

6.3.4. CAP NUT : Cap nut made of bronze are commonly used in smoke boxes of locomotives on the main steam pipe and connection etc. These are special nuts, generally of hexagonal form, provided with an integral cylindrical cap at the top of the nut to protect the end of the bolt for corrosion. It also prevents the leakage through threads (see Fig. 6.7).

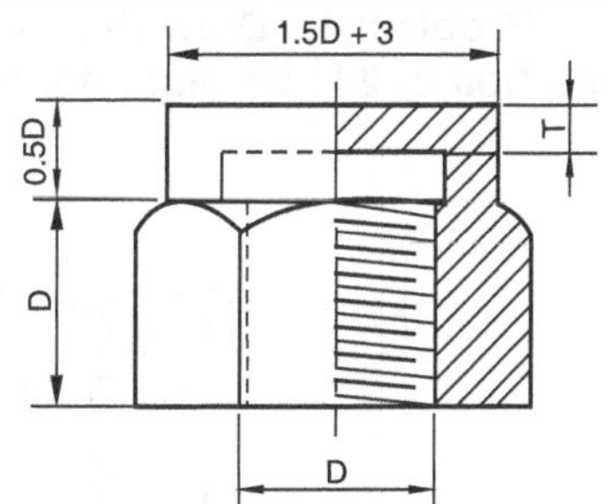

Fig. 6.7. *Cap nut*

Proportions of the Cap nut : Let D the nominal diameter of the bolt, then the

Thickness of the nut = D

Width of the cylindrical cap = 1.5 D + 3mm

Thickness of cylindrical cap, T = 0.5 D

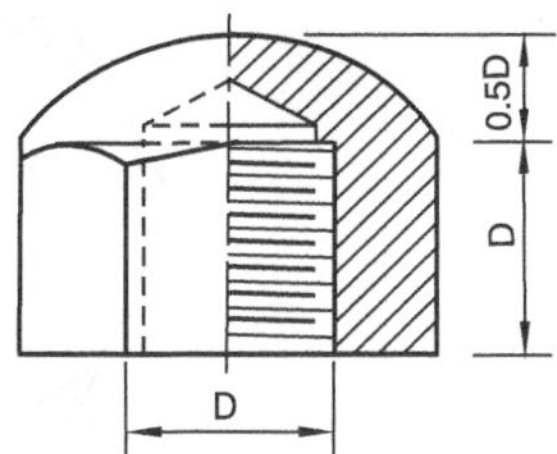

Fig. 6.8. *Dome nut*

6.3.5. Dome Nut : The dome nut serves the same purposes as the cap nut. In this, spherical portion is provided at the top surface of hexagonal nut.

The various proportion of this nut is shown in Fig. 6.8.

6.3.6. Cylindrical or Capston Nut : The nut which secures a crane hook in position, is generally of this type (see Fig. 6.9). It is a circular nut provided with circular holes in the curved surface for turning it with a tommy bar. These holes can also be drilled in the top surface of the nut for which a fine spanner is used to turn the nut.

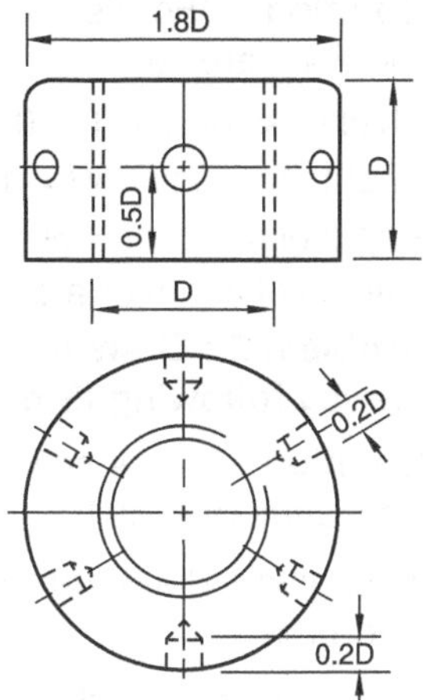

Fig. 6.9. *Capston nut*

Proportions of the Cylindrical or Capston Nut: Let D, the nominal diameter of the bolt, then the nominal diameter of the cylindrical or capston nut = 1.5 D.

Thickness of the nut = D

The diameter of the drilled hole = 0.2 D

6.2.7. Wing Nut : These nuts are designed for operation by thum and finger and are frequently used for securing adjustable fittings such as in the wind screen of a motor car. It is a concial nut having two wings attached with its slant surface (see Fig. 6.10).

The various proportions with respect to the nominal diameter D of the bolt is shown in Fig. 6.10.

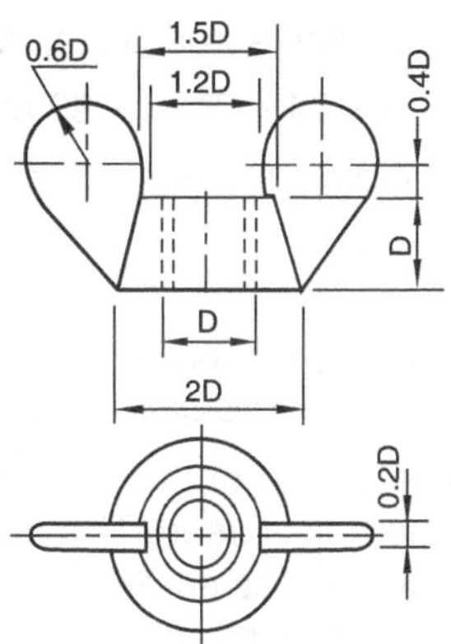

Fig. 6.10. *wing nut*

Problem 1 : Draw the top view, front view and right side view of a haxagonal nut for a bolt of 24 mm diameter by following the I.S.I. proportions [Fig. 6.11].

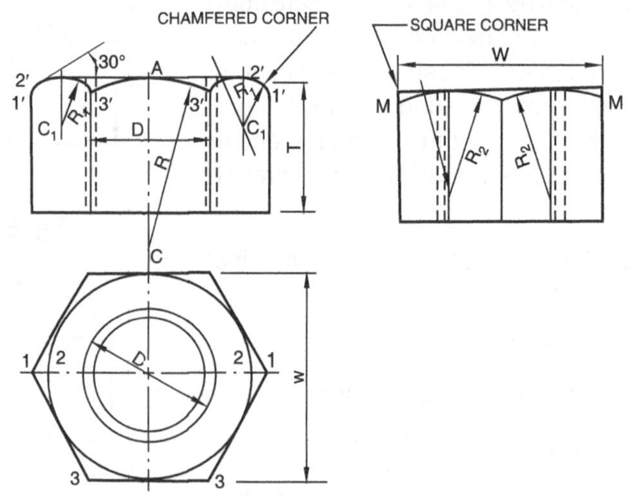

Fig. 6.11. *Haxagonal nut*

Solution : For its solution, see Fig. 6.11. The given diameter of a bolt is 24 mm, then its proportions are i.e.

Thickness of nut, T = 0.9D = 0.9 × 24 = 21.6 mm say 22 mm

Width across flat, W = 1.5 D + 3 = 1.5 × 24 + 3 = 39 mm

Angle of chamfer = 30°

Radius of chamfer arc, R =1.5D = 1.5 x 24 = 36 mm

Problem 2 : Draw the top view and front view of a square nut for a bolt of 24 mm diameter by following IS proportions [Fig. 6.12].

Solution : For its solution, see Fig. 6.12 which is itself a self explanatory sketch.

The given diameter of the bolt is 24 mm, then its proportions are

i.e. Thickness of the nut, T=0.9 D = 0 . 9 × 24=21.6 mm (Say 22 mm)

The width across flats, W = 1.5 D + 3 mm

=1.5 × 24 + 3mm = 39 mm

Angle of chamfer = 30°

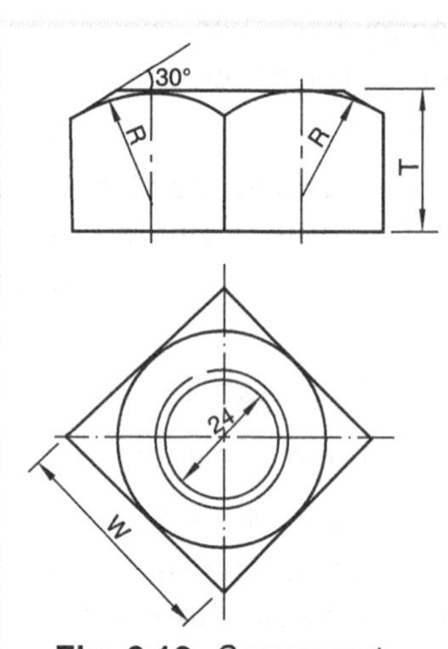

Fig. 6.12. *Square nut*

6.4. BOLT

A cylindrical piece with a head on one end and threads on the other end is called a bolt.

The shape of the head depends upon the purpose for which the bolt is used. The bolt is passed through the clearance holes in two or more aligned parts and a nut is screwed on the threaded side to tighten the parts together. The bolt head is chamfered generally to 30° to remove the sharp corners which may hurt the workman.

The bolts are used to connect two or more parts temporarily.

6.5 FORMS OF BOLTS

The following are the important types of bolts used in engineering practice according to the shape of the head:-

1. Hexagonal headed bolt
2. Square headed bolt
3. Cylindrical or Cheese headed bolt
4. Cup headed bolt
5. T-headed bolt
6. Counter sunk headed bolt
7. Hook bolt
8. Headless tapered bolt
9. Eye bolt
10. Shackle bolt

6.5.1. Hexagonal Headed bolt : Hexagonal headed bolt is the most common form of bolt and widely used in machines and the engineering works to tighten two or more parts together (see Fig. 8-13). In this, the shape of the head is hexagonal and that is why the name given to this type of bolt is hexagonal headed bolt. The hexagonal head is chamfered at its upper end at an angle of 30° to its base. To prevent the rotation of the bolt while screwing the nut on or off, the bolt is held by a spanner.

Proportions of the hexagonal bolt : Let D,be the nominal diameter of the bolt, then,

Width across flat, W = 1.5 D + 3 mm

Thickness of head, T = 0.8 D

Angle of chamfer = 30° to the base of the head

Radius of chamfer arc, R = 1.5D

Length of the bolt, L = 3 D to 20 D

Length of threaded portion, L_1 = 2 D + 5 mm upto 80 mm dia.

or = 2 D + 10 mm for 81 to 200 mm dia

or = 2 D + 20 mm above 200 mm dia.

The size across corners can be obtained by construction.

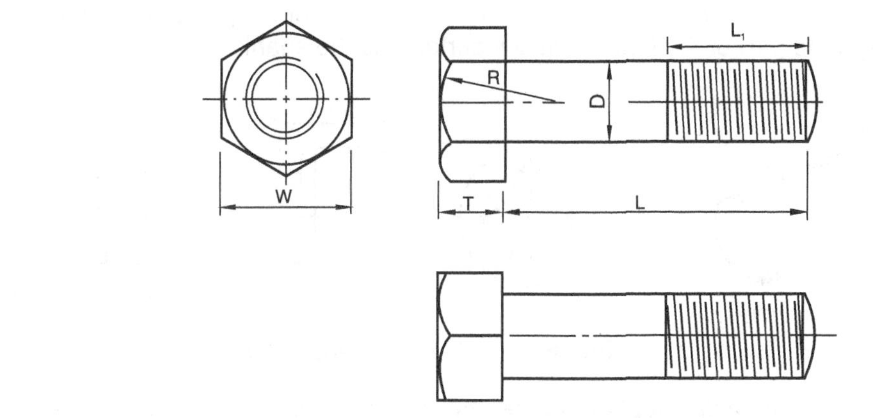

Fig. 6.13 *Hexagonal headed bolt*

Important Note : *While considering the length of the bolt, the thickness of the head should not be taken into consideration.*

6.5.2 Square headed bolt : Square headed bo lt is also a common form of the bolt and is generally used where the head of the bolt is to be accommodated in a recess. The recess is also made of square shape so that the bolt is prevented from turning when the nut is screwed on or off. When the square head of the bolt projects outside the parts to be joined, it is provided with a neck of square cross-section. This neck prevents the bolt from turning when it is tightened.

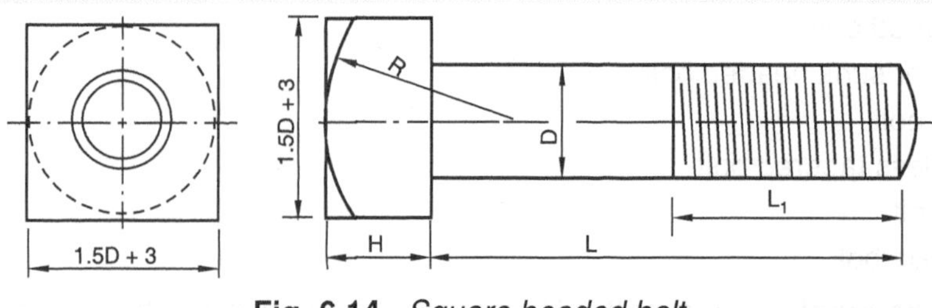

Fig. 6.14. *Square headed bolt*

Proportions of the square headed bolt : Let D,be the nominal diameter of the bolt, then according to I.S. the,

Width across flat, W = 1.5 D + 3mm

Thickness of the head, H = 0.7 D

Angle of chamfer = 30° to the base of the head

Radius of the chamfer arc, R = 2 D

Length of bolt, L = 3 D to 2 0 D

Length of threaded portion, L_1 = 3 D + 5 mm upto 80 mm dia.

or = 2D + 10 mm from 81 mm to 200 mm dia.

or = 2D +20 mm above 200 mm dia.

6.5.3. Cylindrical or Cheese headed bolt : Cylindrical or cheese headed bolt is used when the space for accommodating the bolt-head is comparatively limited or where the use of a spanner for holding and at the same time screwing off or on is to be avoided. The *cross-head connecting rod, eccentric,* etc. are provided with this types of bolt. The rotation of this type of bolt is prevented by means of a pin, called a sung inserted into the shank below the head. The hole in which the bolt is to be inserted has a corresponding recess for the sung. The sung is a separate part of the bolt which is inserted into the shank.

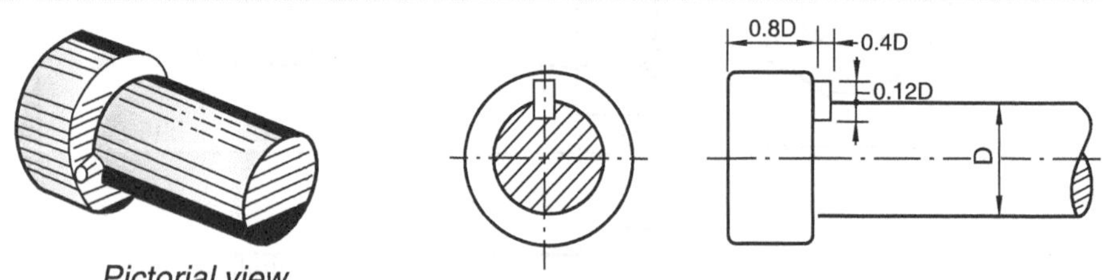

Pictorial view

Fig. 6.15. *Cheese headed bolt*

Proporations of the cylindrical or Cheese headed bolt : Let D, the nominal diameter of the bolt, then the

Diameter of the head = 1.5 D

Thickness of the head = 0.8 D

Diameter of the pin = 0.4 D

Length of the pin = 0.12 D

6.5.4 Cup headed or round headed bolt : The cup headed or round headed bolts are *used largely in tank construction and in certain parts of locomotives constructions.* It is provided with a sung on the shank. The sung is fixed into a corresponding recess in the adjacent piece to prevent rotation of the bolt. Sometimes, the cup headed bolt is provided with a square neck (see Fig. 6.16).

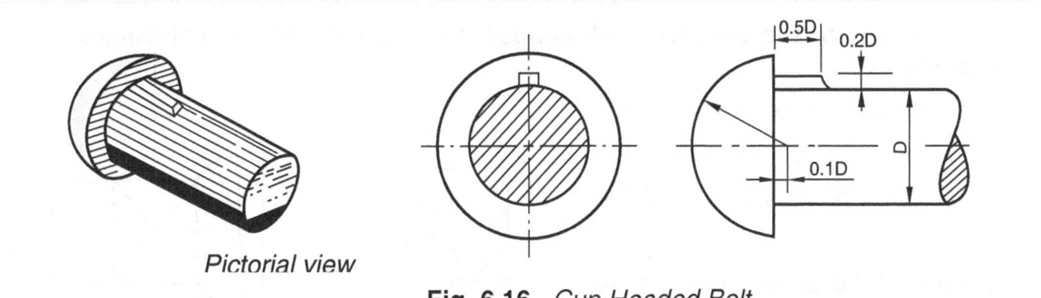

Fig. 6.16. *Cup Headed Bolt*

Proportions of the cup headed or round headed bolt : Let D, the nominal diameter of the bolt, then the

Diameter of the cup head=1.6D

Length of the sung = 0.5D (Approx.)

Thickness of the sung = 0.2D (approx.)

If, however, it is provided with square neck then the length of squareneck = 0.5D.

Note : *The square neck is always provided in place of snug.*

6.5.5. T-headed bolt : This form of bolt is just like a T and is used for securing vice, jigs, etc. to the tables of machine tools in which T-slots are cut to accomodate the T-head. The neck of the bolt is squared to prevent to rotation (see Fig. 6.17).

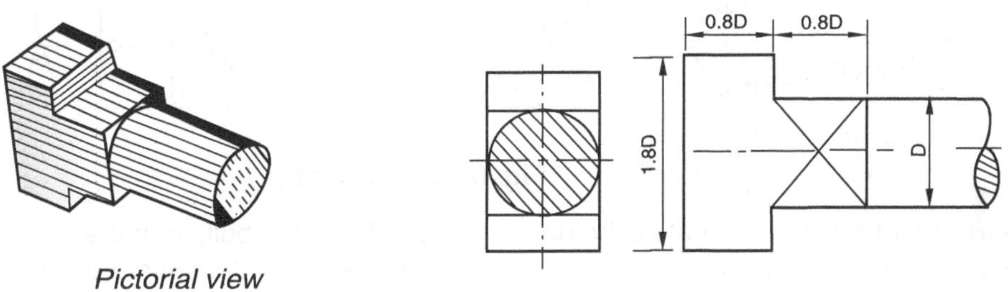

Fig. 6.17 *T-headed bolt*

Proportions of the T-headed bolt : Let D, nominal diameter of the bolt, then the

Width of the T-head = 1.8 D

Thickness of the head = 0.8 D

Length of the square neck = 0.8 D

6.5.6. Counter-sunk headed bolt : These bolts are commonly used for securing metal works to wood works *i.e.* where flush surfaces are required. It is provided with a pin (see Fig. 6-18), similar to that used in the case of cylindrical headed bolt or by means of a square portion of the shank so as prevent the rotation (see Fig. 6.19).

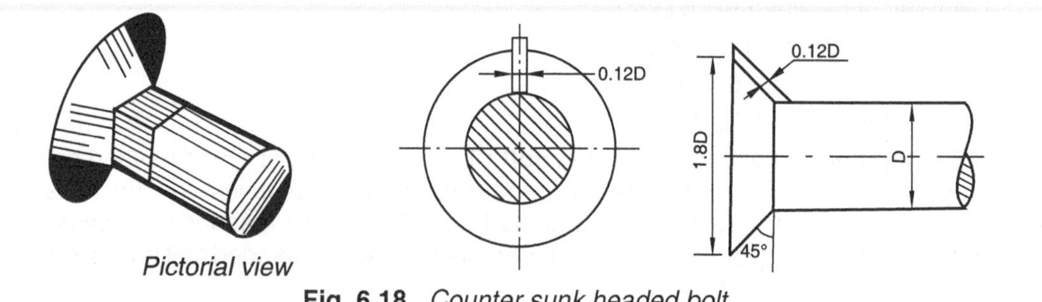

Fig. 6.18 *Counter sunk headed bolt*

Proportions of the counter-sunk headed bolt : Let D, the nominal diameter of the bolt, then the

Outside width of the head = 1.8 D

Slope of the head = 45° to base

Width of the pin or sung = 0.12 D (Approx).

Length of the square neck, if the bolt is provided with a square portion = 0.8D.

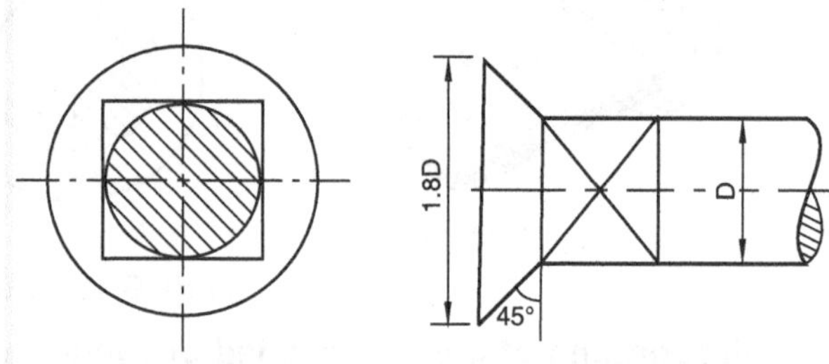

Fig. 6.19 *Counter sunk headed bolt*

6.5.7. Hook bolt : These bolts are sometimes used for clamping steel joints in position and also for fixing the electric conductor rails to the insulators on electric railway track. In this, the shank of the bolt passes through a hole in one piece only, the other piece is gripped by the hook shaped bolt head. It is usually provided with square neck (see Fig. 6.20).

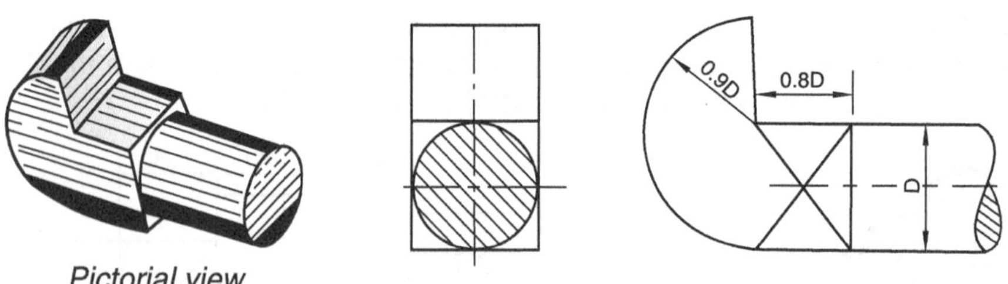

Fig. 6.20 *Hook Bolt*

Proportions of hook bolt : Let D, the nominal deameter of the bolt, then the

Diameter of the head =1.8 D

Approx Length of the square portion = 0.8 D

6.5.8. Headless tapered bolt : These are used for large marine shaft couplings, where any form of head can not be used there by involving encorachment on valuable space. Its shank is tapered and has no head as the name indicates.

Fig. 6.21 shows the headless tapered bolt with its proportions in terms of the nominal diameter of the bolt D.

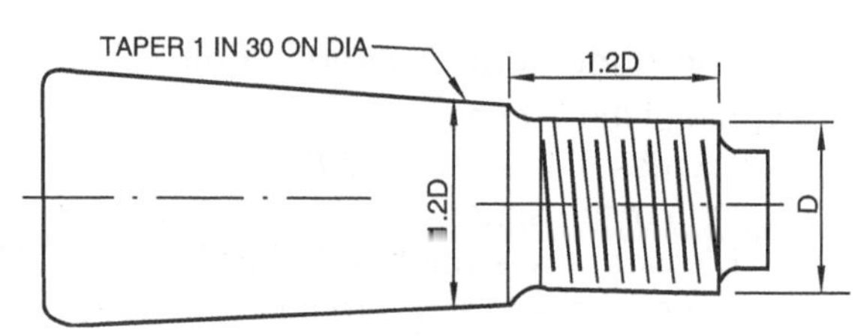

Fig. 6.21 *Counter sunk headed bolt*

6.5.9. Eye bolt : It is used as an appliance for lifting and transporting heavy machines. It consists of a ring of circular cross-section (see Fig. 6.22). This type of bolt is screwed inside a threaded hole on the top of the machine, directly above its centre of gravity. The ordinary eye bolt can have a circular ring of rectangular cross-section.

Proportions of the eye bolt : Let D, the nominal diameter of the bolt, then the

Diameter of the ring = 3 D

Diameter of bottom portion of neck = 2 D

For other dimensions, see Fig. 6.22

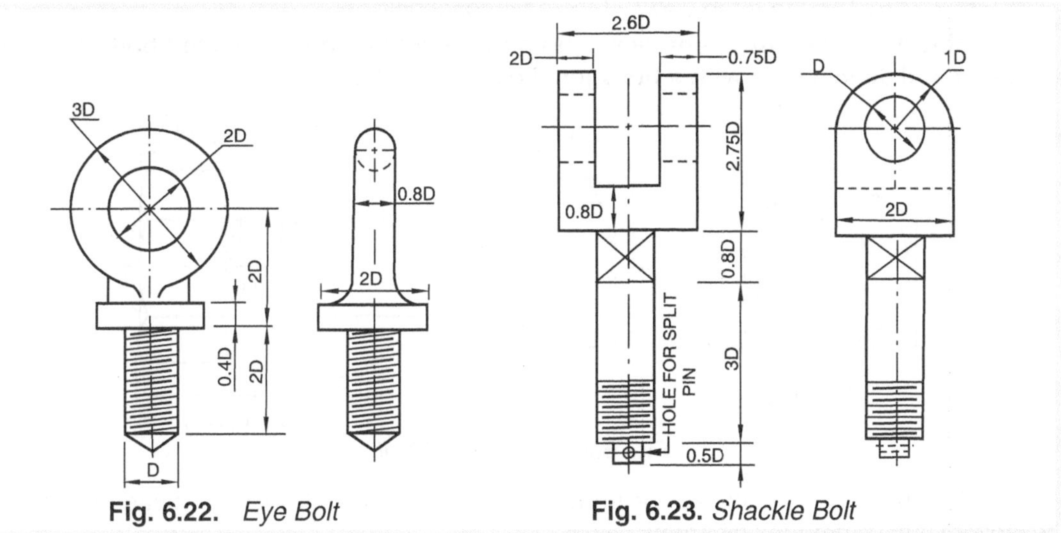

Fig. 6.22. *Eye Bolt* **Fig. 6.23.** *Shackle Bolt*

6.5.10. Shackle bolt : Fig. 6.23 shows a shackle bolt in which the head just like a fork having holes to receive a pin. A square neck is usually provided with the head to prevent the rotation.

Fig. 6.23 shows the front view and end view of the shackle bolt with its proportions in terms of nominal diameter of the bolt D.

Problem 1 : Draw the front view, top view and side view of a hexagonal bolt 24 mm diameter and 96 mm long with a hexagonal nut and a washer by following approximate proportion [Fig. 6.24].

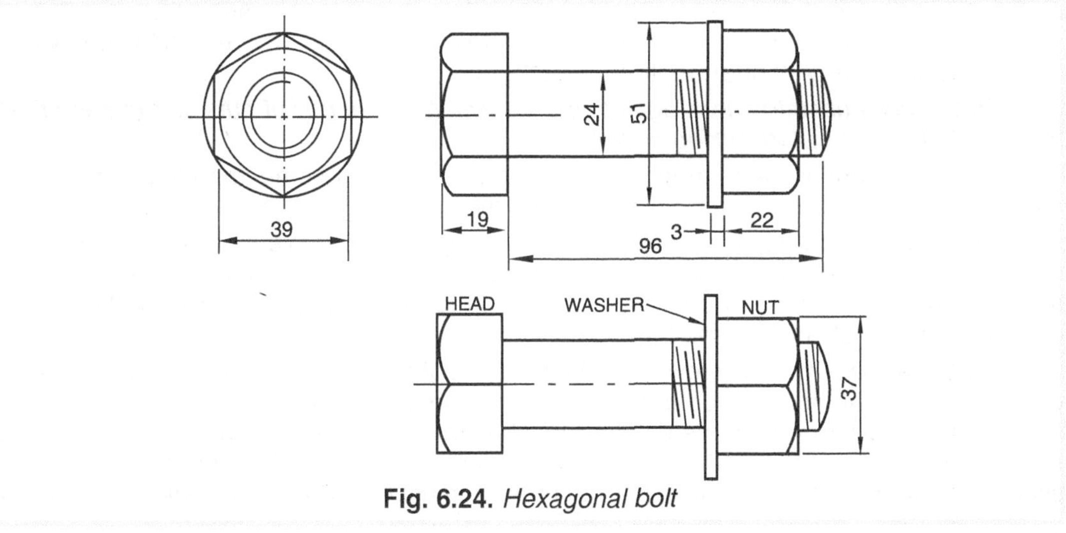

Fig. 6.24. *Hexagonal bolt*

Solution : If the diameter of bolt is 24 mm, then I.S. proportions are given below:-

Width across flat = 1.5 D + 3 = 1.5 × 24 + 3 = 39 mm

Thickness of head, T = 0.8 D = 0.7 × 24 = 19.2 mm = 19 mm (say)

Thickness of the nut = 0.9 × 24 mm = 21.6 mm = 22 mm (say)

The length of the shank of stem=4 D (Approx. for drawing purpose)=96 mm

Angle of chamfer = 30° to base of the head

Diameter of the washe r = 2 D + 3 mm = 2 x 24 + 3 = 51 mm

Thickness of washer = 0.12 D = 0.12 × 24 = 2.88 mm or say 3mm

Fig. 6.24 Shows the three views of the hexagonal bolt.

Problem 2 : Draw the front view, and side view of a square headed bolt of 24 mm diameter and 96 mm long with a hexagonal nut [Fig. 6.25].

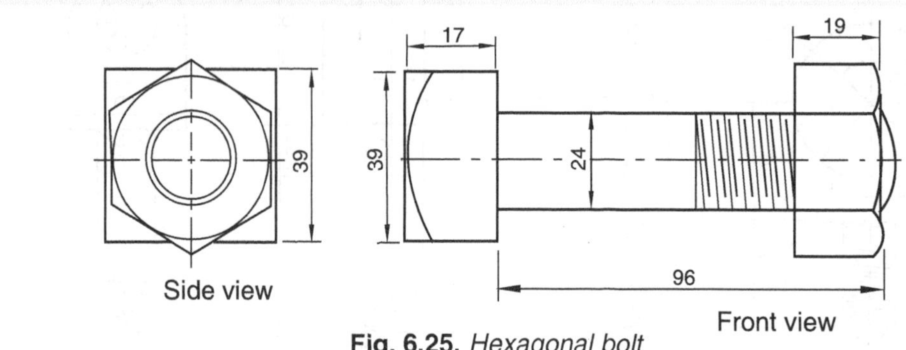

Fig. 6.25. *Hexagonal bolt*

Solution : Fig. 6.25 shows the front view and side view of a square headed bolt with hexagonal nut. Its proportions are as follow:-

Width across flat, W = 1.5 D + 3 = 1.5 × 24 + 3 = 39 mm

Thickness of head, T=0.7D × 24 =16.8 mm = say 17 mm

Angle of Chamfer=30°

Radius of chamfer=2 D = 2 × 24=48 mm

The proportions of the hexagonal nut are the same as of problem 1.

6.6. LOCKING DEVICES

As stated already that nut and bolt are extensively used to join different parts of the machine. These machines are always subject to virbrations in moving parts. Due to vibrations, there is always a tendency for the nut to get slack and to screw off the bolt which may cause to be very dangerous to the machinery.

The devices used to check against the slackening tendency of nuts and to keep them in right position are called locking devices.

There are various types of locking devices available but some important types which are generally used in engineering practice are described below:-

6.6.1. Lock nut : It is the most common locking device. In this, two nuts, one ordinary nut and the other standard nut known is *check* nut are used. The thickness of the lock nut is about 0.6 of the thickness of ordinary nut. But the top and bottom surfaces of the nut are chamfered at a angle of 30° (see Fig. 6.26).

In this device nut 'A' is screwed on the bolt, as tightly as possible. The thread in the nut presses against the bottom side of the thread in the bolt as shown in Fig. 6.26 (*i*). then the second nut 'B' is screwed on the bolt till it touches the top of the first nut. The upper nut `B' is then held by spanner while the lower nut A is turned backward as for a possible by another spanner. Now, the threads in the upper nut press against the lower side of the threads in the bolt, the lower threads in the lower nut press on the upper side as shown in Fig. 6.26 (*ii*). The two nuts are thus wedged or locked tightly against each other and against the bolt.

Figs. 6.26 *(iii), (iv)* and *(v)* show different froms of lock nut.

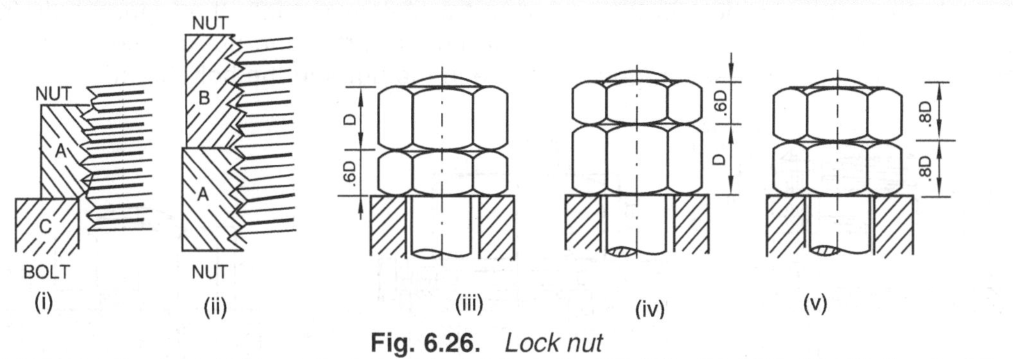

Fig. 6.26. *Lock nut*

6.6.2 Pin Nut : In this type of locking device, after tightening the nut, a hole is drilled through the nut and the bolt (see Fig. 6.27). A split pin is then inserted in the hole and the ends of the pin are opened and bent round the flats of the nut. Care should be taken while drilling the hole that nut does not loose.

6.6.3. Slotted Nut : It is hexagonal nut with slots cut in the upper and through the opposite faces (see Fig. 6.28). When the nut is tightened up, a hole previously drilled through the bolt must be in alignment with a pair of slots of the nut. A split pin is then inserted into the slots of the nut and hole. Then the legs of the split pin are opened out and bent around the nut (not shown in the figure).

Slotted nuts are largely used on jobs subject to sudden shocks and considerable virbations such as on motor cars, etc.

6.6.4. Castle Nut : It is a hexagonal nut with a cylindrical collar at the top, through which the slots are made diametrically. When the nut is in tight position, a spilt pin is inserted through the slots and the hole, as in case of slotted nut (see Fig. 6.29).

Castle nuts are widely used in automobiles and locomotive engines.

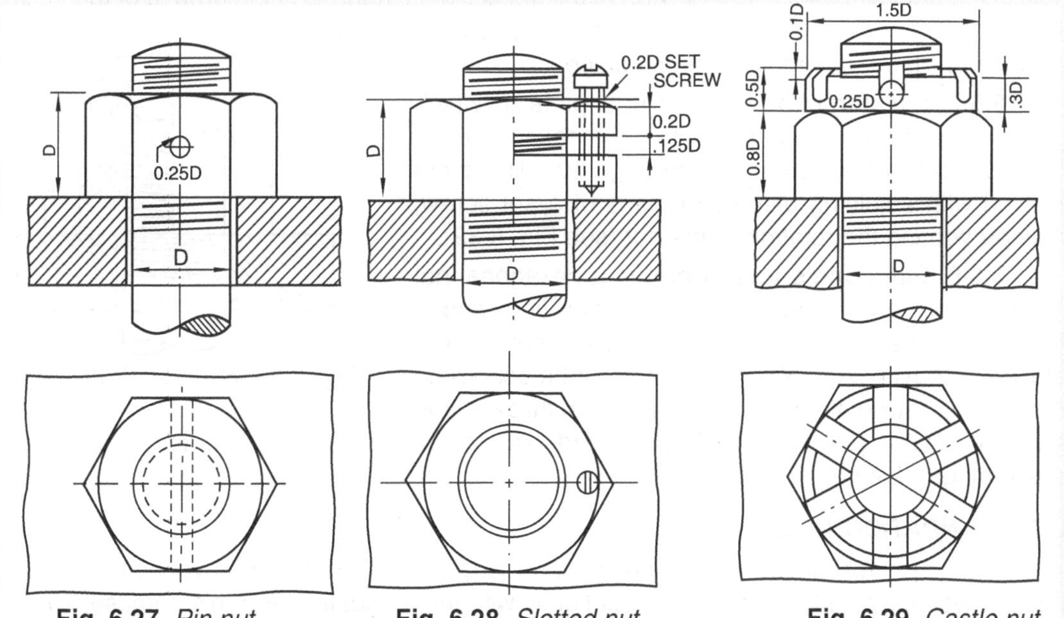

Fig. 6.27. *Pin nut* **Fig. 6.28.** *Slotted nut* **Fig. 6.29.** *Castle nut*

6.6.5. Sawn nut or Wiles Nut : It is also hexagonal nut having a saw cut, half way across it (see Fig. 6.30). The upper part of the cut portion has a clearance hole whereas in the lower part, a tapped hole is made. After tightning the nut, a set screw is passed through the clearance hole in the upper part and screwed into the tapped hole in the lower part. On tightening the set screw, the upper part of the nut presses against the lower, resulting an increase in gripping force in the nut and the bolt threads.

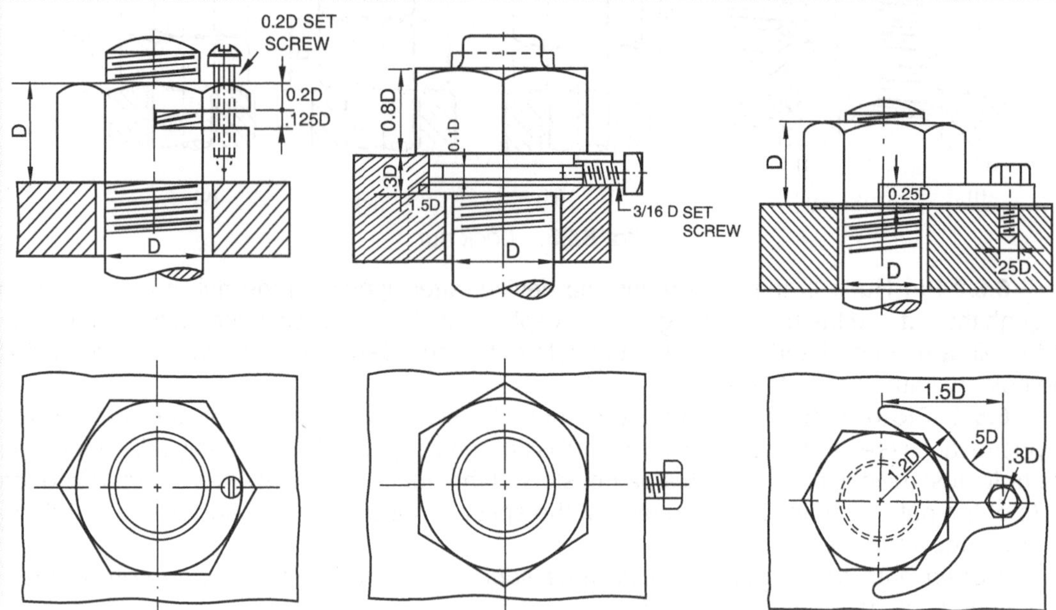

Fig. 6.30. *Sawn nut* **Fig. 6.31.** *Ring,Penn or Grooved nut* **Fig. 6.32.** *Looking plate nut*

6.6.6. Ring, Penn or Grooved nut : In this nut, a cylindrical grooved collar is made at the bottom surface of the hexagonal nut (see Fig 6.31). The collar fits into the corresponding hole in the adjoining piece and the end of the set screw is made to fit the groove of the collar. This arrangement is possible only when the bolt is placed near the face of the adjoining piece.

6.6.7. Locking plate nut : It is a plate in which grooves are made in such a way that the hexagonal nut fits into it in any position turning through 30°. This plate is fixed with adjoining piece by means of a set screw, thus preventing the rotation of the nut (see Fig. 6.32).

6.6.8. Simmond's lock nut: It is a hexagonal nut provided with a collar at its upper end and fibre ring is fitted inside the collar (see Fig. 6.33). The internal diameter of the ring is less than the core diameter of the bolt. When the nut is screwed, the bolt end cuts its own threads and gives a greater grip over the bolt threads due to high friction and thus, prevents the slackening of the nut.

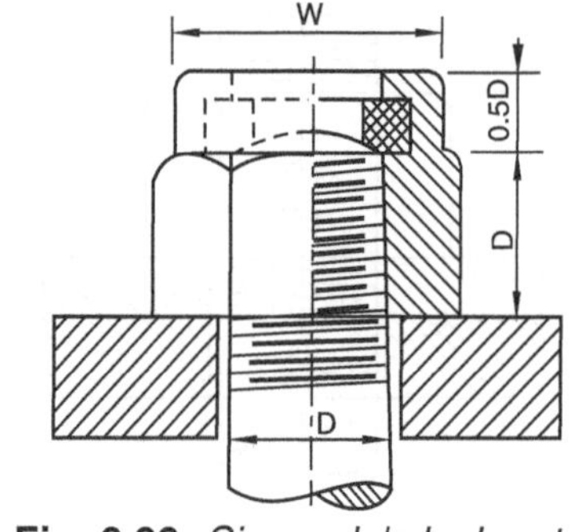

Fig. 6.33. *Simmodn's lock nut*

6.7. FOUNDATION BOLTS

The heavy machinery while running, has many sorts of vibrations in various directions. It is always desirable to fix the machines tightly with the foundation. For this purpose, special bolts of such shapes and sizes are used that they do not loosen and fail under highly vibrating loads. These special bolts are called foundation bolts.

Thus, the bolts which are used for fixing the heavy machines with foundation are called foundation bolts.

The common method for fixing the foundation bolt is to suspend it in the foundation hole which is quite large as compared to bolt. After levelling the machine, the hole is filled with a fine grout consisting of equal parts of sand and cement. When the grout is set, machine is then fixed. Sometimes, molten lead and sulphur are also used which set in the foundation within a few minutes.

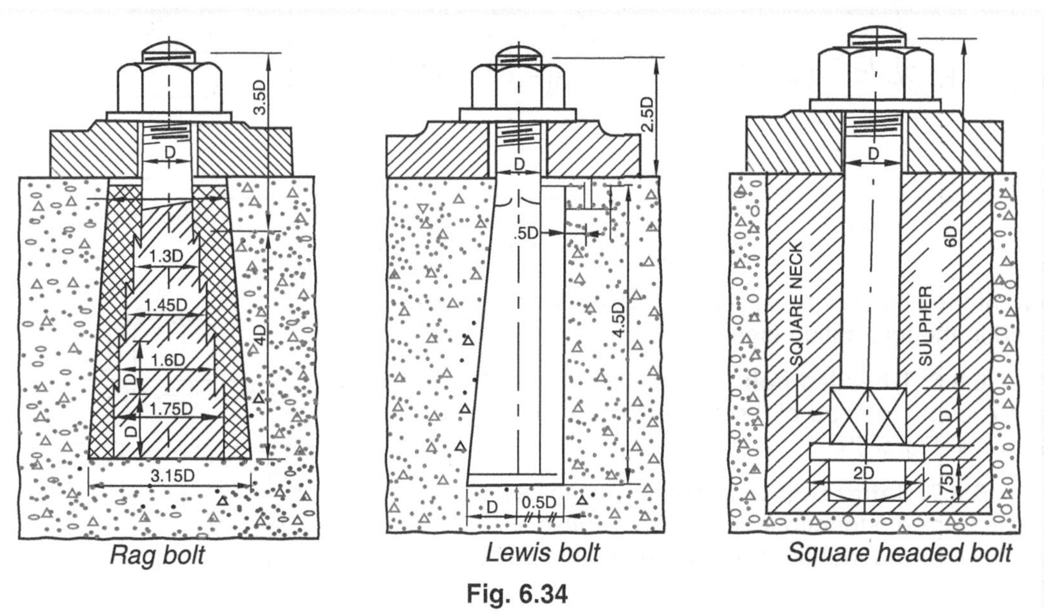

Rag bolt *Lewis bolt* *Square headed bolt*

Fig. 6.34

6.8. TYPES OF FOUNDATION BOLTS

The various types of foundation bolts used for fixing the heavy machines are as follow:-

6.8.1. Rag bolt : Rag bolt is used for fixing the heavy machines to stone or concrete foundation. It is a tapered bolt of rectangular section with edges as shown in Fig. 6.34.

5.8.2. Lewis bolt : It is used for temporary foundation. In this bolt, one side is formed straight and the other is tapered. it is fixed alongwith a key which is inserted with the straight side in the foundation. This bolt can be easily removed by simply drawing the key (see Fig. 6.34)

6.8.3. Square headed bolt : It is a simple square headed bolt with a square neck carrying a square plate as shown in Fig. 6.34. The square plate will set firmly in sulphur and will prevent the bolt from moving automatically.

6.8.4. Curved bolt : It is a simple foundation bolt made from mild steel or wrought iron bar in a curved from as shown in Fig. 6.35. The end of this bolt is rounded of twice the diameter.

6.8.5. Hoop bolt : It is also a simple form of foundation bolt which is forged from a mild steel or wrought iron bar, having an eye at the bottom. A mild steel bar passes through the

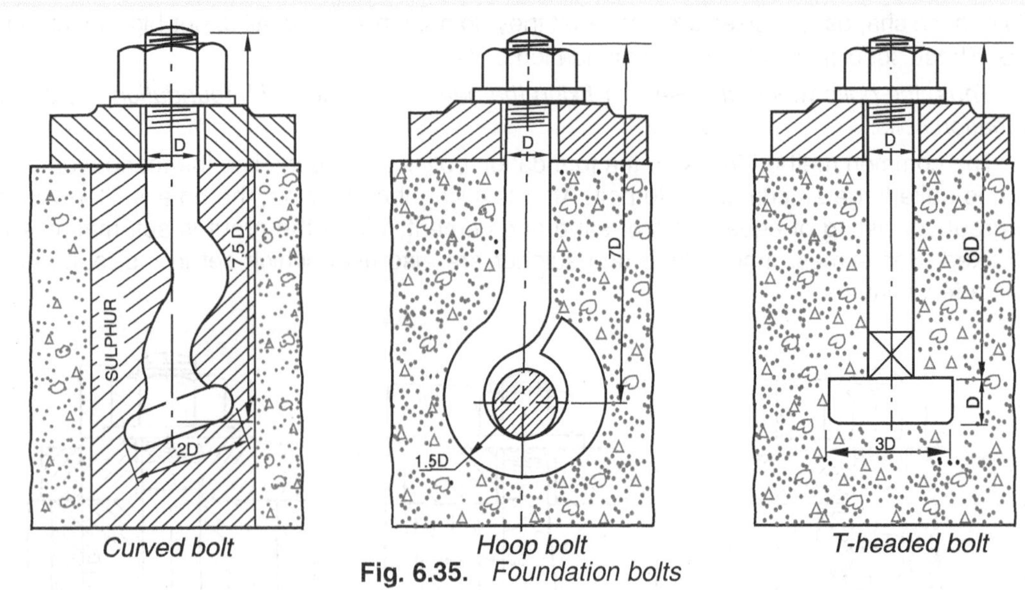

Curved bolt Hoop bolt T-headed bolt

Fig. 6.35. *Foundation bolts*

eye of the bolt at right angles to its axis by which the bolt is fixed in the foundation (see Fig. 6.35)

6.8.6. T-headed bolt : It is similar to a square headed bolt except the head of the bolt is forged into a T-shape as shown in Fig. 6.35.

6.8.7. Cotter Bolt : it is used for fixing heavy machines to the brick or stone foundation. This bolt is screwed below the floor or foundation level by means of a cotter (see Fig. 6.36).

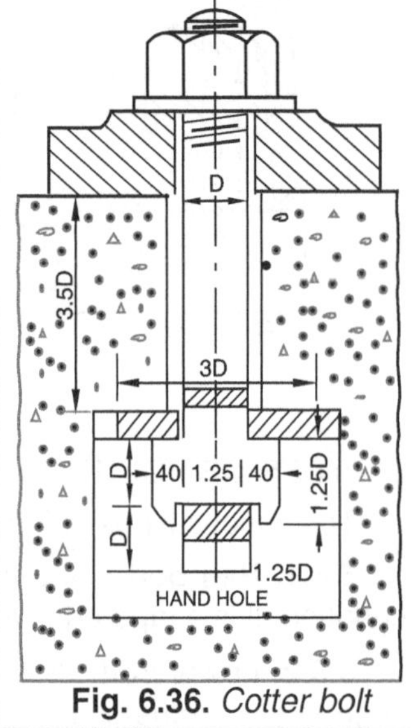

Fig. 6.36. *Cotter bolt*

6.9. STUDS

A cylindrical rod, threaded on both ends and plain in the middle is known as stud.

Stud is used where sufficient space for bolt-head and nut is not avaiable or when it is required to avoid the use of an unnecessary long bolt.

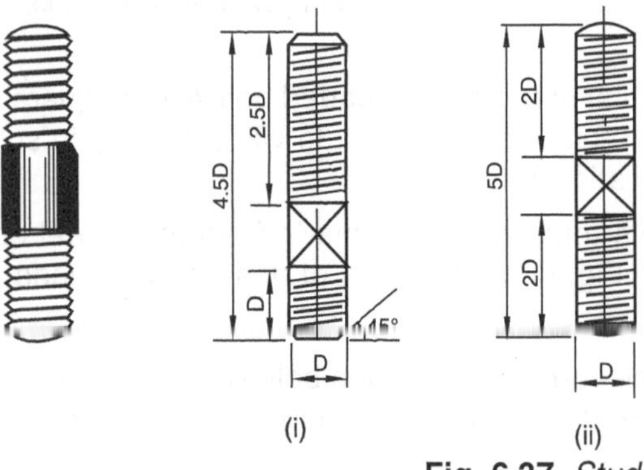

Fig. 6.37. *Studs*

Fig. 6.37*(i)* shows a stud with the middle portion plain. Fig. 6-37*(ii)* shows a stud with the middle portion made in square section. Fig. 6.37 (iii) show a stud with a Collar and is called a Collar stud. The collar gives a bearing surface to the stud and support to the adjoining part.

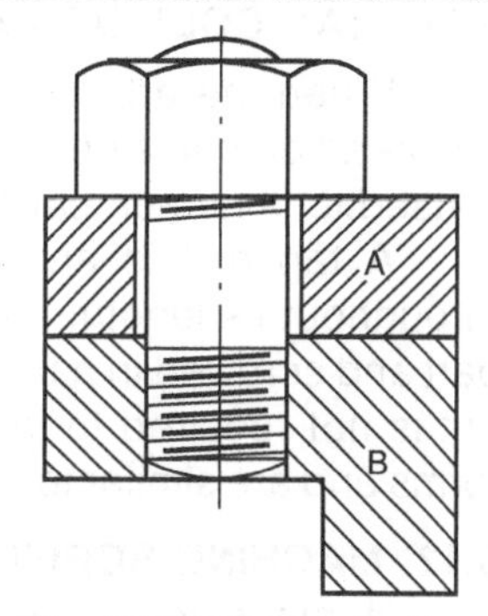

Fig. 6.38. *Cotter bolt*

Uses : *Studs are commonly used for screwing of engine cylinder cover to a cylinder casting, stop valves, etc.*

Fig. 6.38 shows a stud as a fastener for fastening two parts. The upper part has a clearance hole through which the stud passes and the lower part has a tapped hole in which the end of the stud is screwed tightly. A nut is screwed on one end of the stud which is projected over the upper part. The end of the stud which is screwed end and the opposite end in which the nut is screwed, is known as nut end. The nut end is sometimes identiified by rounding instead of chamfering. The length of the plain part between the two threaded ends depends upon the thickness of the adjoining parts.

6.10. END OF DRILLED HOLES AND SCREWED PIECE

When a stud is used for connecting one part to thick block, a hole is first drilled in thick block as shown in fig. 6.39 *(i)*. The diameter of the drilled hole is equal to the core diameter of the stud and the depth of the drilled hole is kept equal to 1.25 D.

The end of the drilled hole is conical on account of the pointed end of the drill. Fig. 6.39 *(i)* shows the end of a drill with an angle of 118°. The drilled hole in drawing is made with an angle of 120° for convenience. The hole is then tapped as shown is fig. 6.39 *(ii)*. The major diameter of the tapping hole is equal to the major diameter of the stud. The tapping cannot be done upto the end of the drilled hole, but a small length of drilled hole at the bottom is left untapped.

Note that, threads in the hole are lelf-handed for a right-handed stud as it is shown in section. Fig. 6.39 *(iii)* shows that the stud is in position in the tapped hole and two parts are connected together by means of stud and a nut.

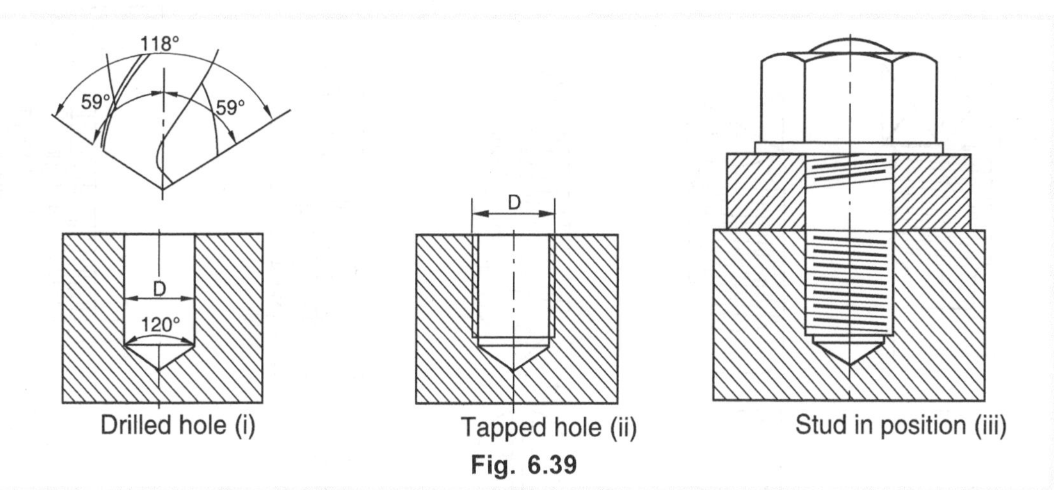

Drilled hole (i) Tapped hole (ii) Stud in position (iii)

Fig. 6.39

6.11. TAP BOLT OR CAP SCREW

When the adjoining parts do not have sufficient space for bolt head on one side and for the nut on the other, then a tap bolt is used to fasten on parts as shown in Fig. 6.40.

A tap bolt is an ordinary bolt which is threaded almost throughout its length. It passes through a clearance hole in one part and screwed in a tapped hole in the other part, so a separate nut is not required to fasten the parts. Tap bolts are of various forms and are similar to those of set-screws as shown in Fig. 6.40.

Fig. 6.40. *Tap bolt*

6.12. MACHINE SCREWS

A machine screw is a small fastner used with a nut to function in the same manner as a bolt or without a nut to function as a cap screw (see Fig. 6.41). It passes through a clearance hole in one part and is screwed into hole in the other, thus connecting the two parts. The head of the machine screws are provided with a slot for screwing on or off by a screw driver.

Uses : *Machine screws are extensively used in jigs, fixture, dies, fire-arms ete.*

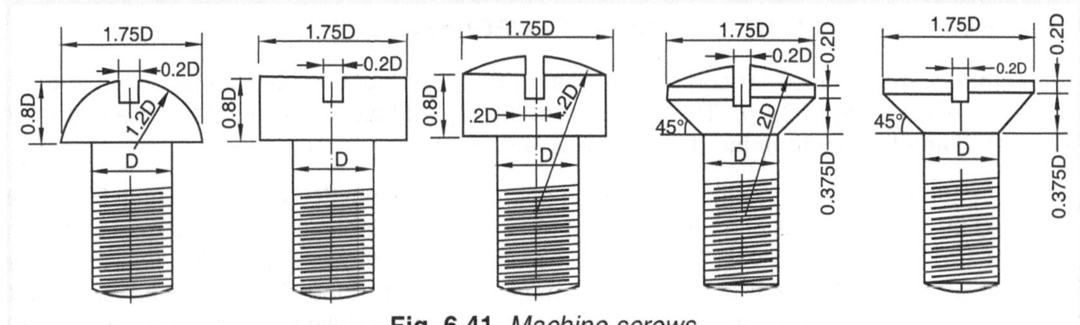

Fig. 6.41. *Machine screws*

6.13. SET SCREWS

A set screw is similar to a tap bolt, but is threaded throughout its length (see Fig. 6.42). It is used to prevent relative movement between two parts. It is screwed into a tapped hole in another part, while its point end passes on the other part, thus preventing the relative

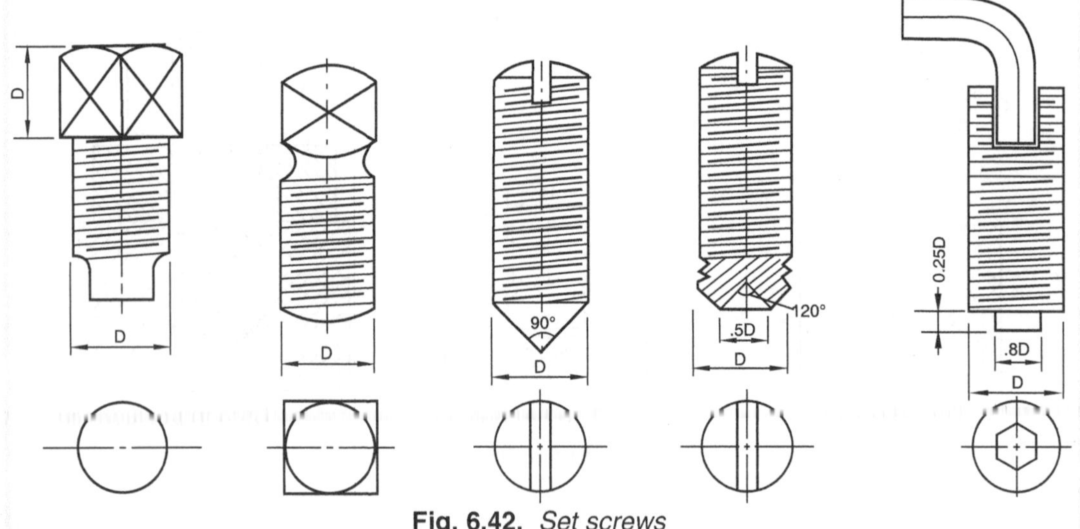

Fig. 6.42. *Set screws*

rotation and sliding movement. These are made of hardened steel.

The various types of standard set-screws, with-heads or without heads are shown in Fig. 6.42. The head types of set screws are operated by spanner. The headless types of set screws are known as grub screws and are used where the surface of the machine part is required to be free. Headless set screws are provided with a slot for screw-driver or hexagonal hole for special type of hexagonal steel bar bent at right angles, known as allen key, for screwing on and off.

Uses : *Set-screws are used in fast and loose pulleys, connecting rod ends, etc.*

6.14. WASHERS

A washer is a flat cilcular disc of metal having a hole in the centre for inserting the bolt or the stud. The hole is slightly larger in diameter than the bolt or the stud, so that the washer can be slipped easily in position. It is used for the following purposes :—

1. It is placed below the nut to give a smooth seating for the nut.
2. It distributes the force taken by the nut over a large surface area
3. It prevents the nut from cutting into the metal and thus allows the nut to be screwed more tightly.
4. It is generally used where the surface of the machine part is rough or uneven for a nut to work.

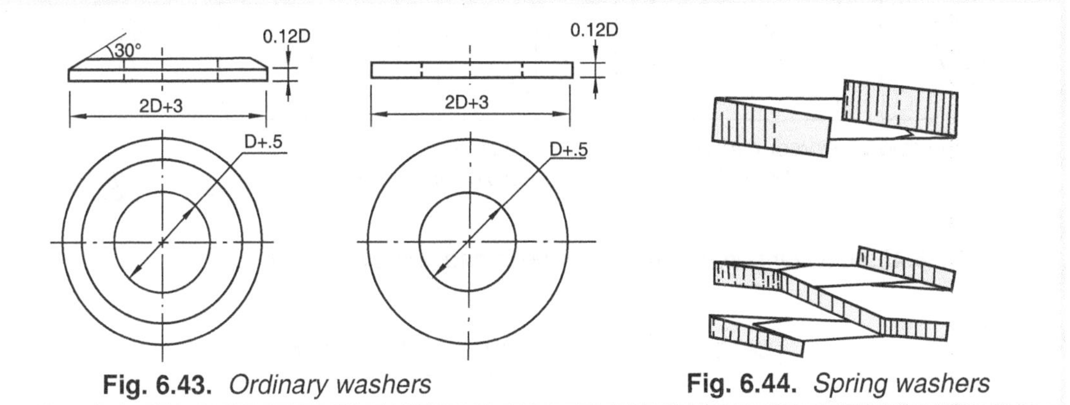

Fig. 6.43. *Ordinary washers* **Fig. 6.44.** *Spring washers*

6.15. TYPES OF WASHERS

The following are two main types of washers used in engineering practice :—

1. Ordinary washer 2. spring washers.

6.15.1 Ordinary washer : It is a simple circular piece of metal purchased out from iron sheet (see Fig. 6.43). It has hole in the centre through which a bolt or stud can pass. It is *used for general purposes.*

6.15.2. Spring Washer : Spring washer is used as a locking device for the nut. A single coiled or double coiled spring when used under a nut, the elasticity of the spring keeps the nut tight with the bolt. It is made from spring steel.

Spring washers are used in the automobiles and in moving parts where the nut may becomes losse due to vibrations. Fig. 6.44 shows standard from of spring washers.

6.16. CONVENTIONAL SYMBOLS FOR BOLTS AND NUTS

In a large drawing where numbers of nuts and bolts are used, it becomes necessary to draw all nuts and bolts by line symbols to save labour and time. Fig. 6.45 shows the conventional symbols for nuts and bolts. The application of these symbols is shown in Fig. 6.45.

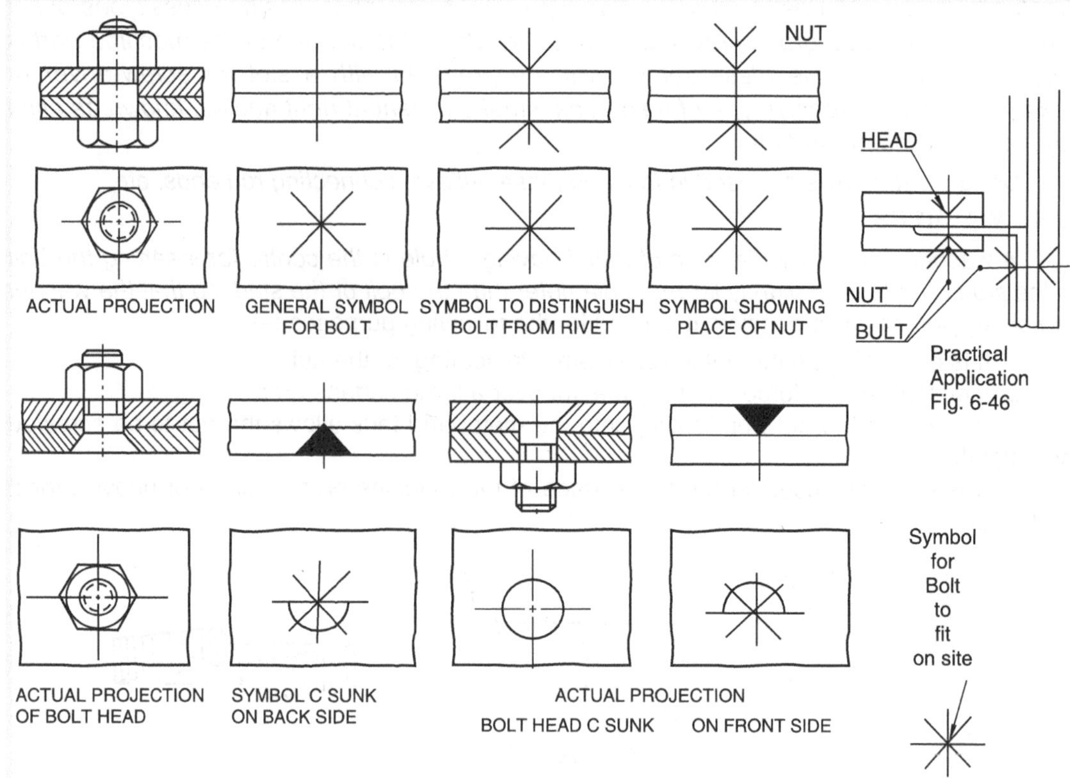

Fig. 6.45. *Conventional symbol's for bolts and nuts as per ISI*

QUESTIONS FOR SELF EXAMINATION

1. Define nut. Give the important types of nuts used in engineering practice.
2. Define bolt. Give its important types used in engineering field.
3. What are locking devices and why are they used?
4. What are the foundation bolts and where are they used?
5. What is the difference between set screw and machine screw?
6. Define washer and give its various types.

CHAPTER

7

Keys, Cotters and Joints

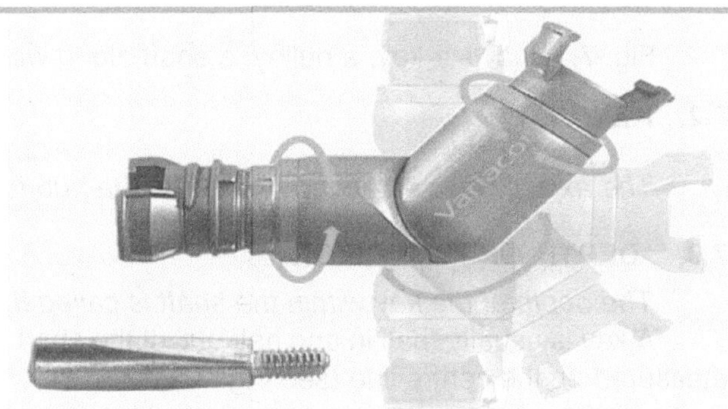

INTRODUCTION

In engineering field, we often come across with such situations where we want to connect two or more rods temporarily in order to increase their length or for transmission of motion. The length of the rods can be increased by providing joints at proper places by means of *keys, cotters, gibs, pin or other mechanical arrangements.* Therefore, the study of keys, cotters, etc. are very important for making joints temporarily.

In this chapter, we shall study key, key way, classification of keys, cotter, gib, and various important joints used in engineering practice.

7.1. KEY

A metal piece inserted between a shaft and wheel hub in an axial direction to prevent their relative movements is known as key.

Pulleys, flywheels, cranks and other similar parts are secured to the shafts by means of keys. Keys are generally made of *mild steel* as they are subjected to shearing and torsional stresses. Keys are temporary fasteners as they can be removed easily.

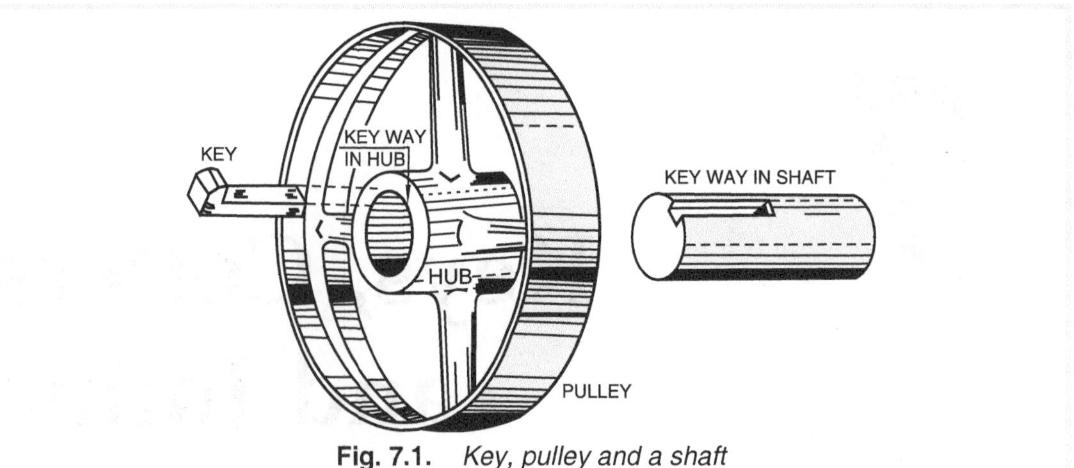

Fig. 7.1. *Key, pulley and a shaft*

Fig. 7.1 shows a key, a pulley, a shaft along with key ways in pulley and shaft

7.2. KEY WAY

The axial groove cut in the shaft or in the hub for fixing the key is known as a key-way.

7.3. DEPTH OF IMMERSION

The depth of the key within the shaft is called the depth of immersion of the key.

A key is usually "half in and half out' of the shaft when measured at the side and not when measured on the centre line (see Fig. 7.2)

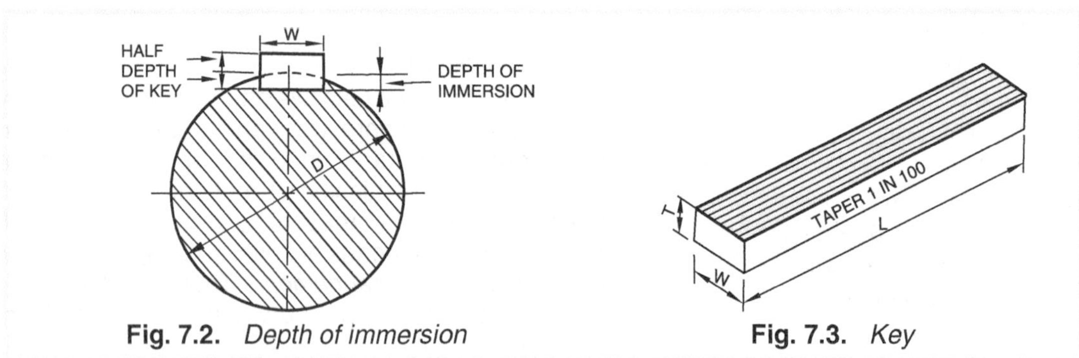

Fig. 7.2. *Depth of immersion* **Fig. 7.3.** *Key*

7.4. PROPORTIONS OF KEY

Keys are generally proportioned with relation to the diameter of the shaft on which it is to be fitted (see Fig. 7.3). The easy proportions of rectangular sunk key are as follow according to I.S : —

Let, diameter of the shaft = D and Width of the key, W = 0.25 D

Thickness of key, T=0.66 W Length of key, L=D to 1.5 D

7.5. TAPER OF KEY

A taper in key gives a tight fitting of the two parts to be fastened. It is uniform in width but tapered in thickness. The taper in the key is nominal and varies from 1 in 64 to1 in 100, but the standard taper is 1 in 100 (1 mm per 100 mm of length) and is provided on the upper surface of the key (see Fig. 7.3).

7.6. CLASSIFICATION OF KEYS

Keys are generally classified into two main categories according to the shape and the duty for which they are used :—

(a) Sunk keys (b) Saddle keys.

7.7. SUNK KEYS

When a key is inserted in the groove cut in the shaft as well as in the hub of pulley, the key is called sunk key. As the name implies, this type of keys is sunk upto a depth of half of its nominal thickness in the shaft and half into the hub of the wheel.

Types of Sunk Keys : The following are the common types of sunk keys :—

1. Parallel sunk key 2. Taper sunk key 3. Gib head key
4. Feather key 5. Wood-ruff key 6. Round or Pin key 7. Peg key.

7.7.1. Parallel Sunk Key : It can be rectangular or square in cross-section and is uniform in width and thickness throughout its length. These keys are generally used where pulleys, gears or other similar parts are secured to the shafts. The proportionate dimensions of rectangular and square parallel sunk keys are as follows :

Proportions of rectangular parallel sunk key :

Let, diameter of the shaft = D

Width of key, W = 0.25 D

Thickness of key, T=0.66 W

Length of key, L=D to 1.5 D

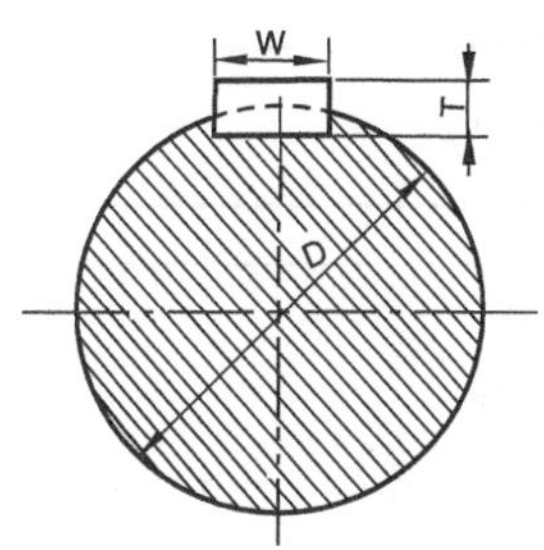

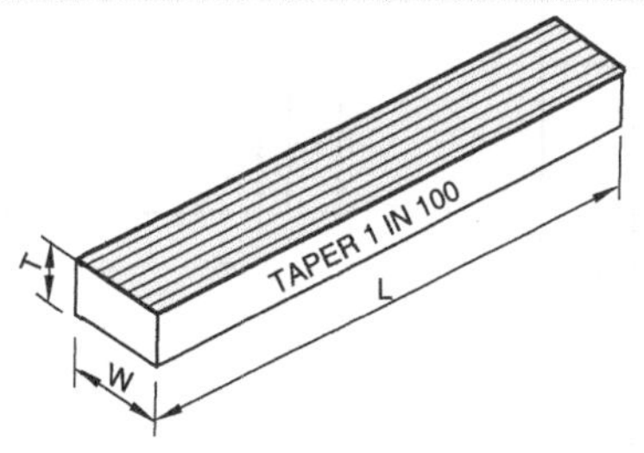

Fig. 7.4. *Rectangular sunk key*

Note : *The length of the key depends upon the length of the hub.*

Proportions of square parallel sunk key :

Let, diameter of the shaft = D

Thickness of key, T = W

Width of key, W = 0.25 D

Length of key, L = D to 1.5 D

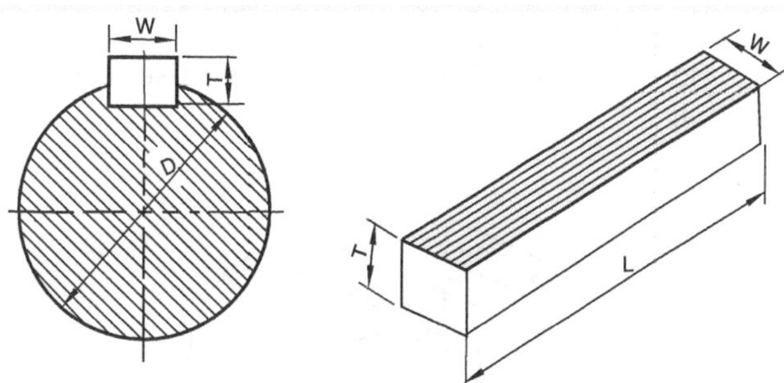

Fig. 7.5. *Square parallel sunk key*

7.7.2. Taper Sunk Key : It can be of rectangular or square in cross-section (see Fig. 7.4 and 7.5) and a standard taper 1 to 100 is given on the upper surface only. The lower surface of the key is kept straight. The thickness of the key is to be measured at the larger end.

7.7.3. Gib Head Key : The rectangular taper sunk key is often provided with a gib head (see Fig. 7.6), so that it can be easily taken out from the key way by forcing a wedge between the key head and the hub of the wheel. The head of the key when made as per given sketch is called a gib head key.

This type of key is used when the connected parts are to be separated occasionally for the purpose of repair or maintenance. Fig.7.6 shows the usual proportions of gib head key.

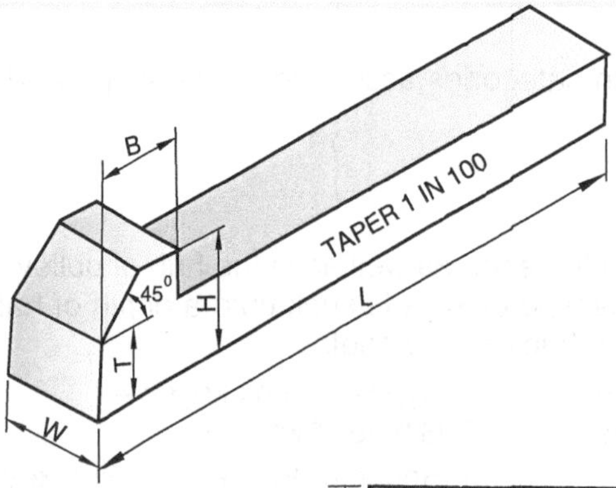

Proportions of gib head key :

Let, diameter of the shaft = D
Width of key, W = 0.25 D
Thickness of key, T = 0.66 W
Thickness of head, H = 1.75 T
Thickness of head,B = 1.5 T
Length of key = D to 1.5 D
Taper in Key 1 in 100

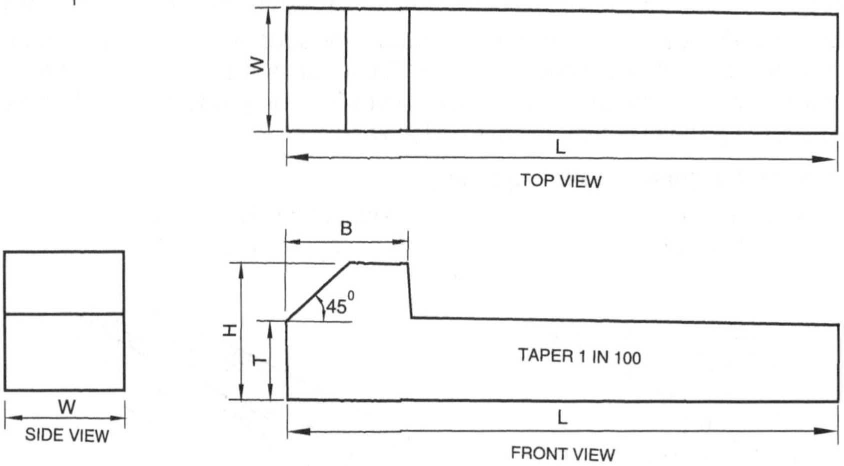

TAPER 1 IN 100

TOP VIEW

SIDE VIEW

FRONT VIEW

Fig. 7.6. *Gib Head Key*

7.7.4. Feather Key : A feather key is a particular type of parallel sunk key used to assemble such shafts and wheels which require axial movement in addition to usual rotary motion *e.g. in drilling machines, spindles and clutches.* It is fastened either to the shaft or to the hub of the wheel.

Fig. 7.7 Shows the isometric view of set screw type feather key and feather key way in the shaft.

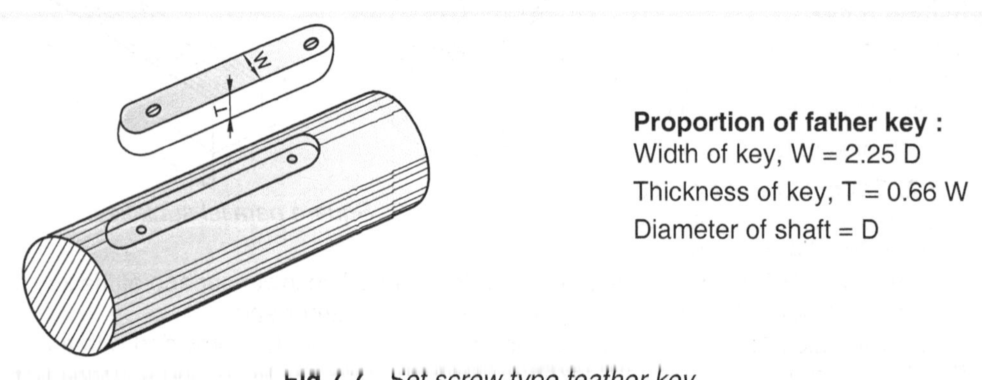

Fig 7.7. *Set screw type feather key*

Fig. 7.8. (i) shows other different kinds of the feather keys and the means by which these are fixed to the shaft and the hub.

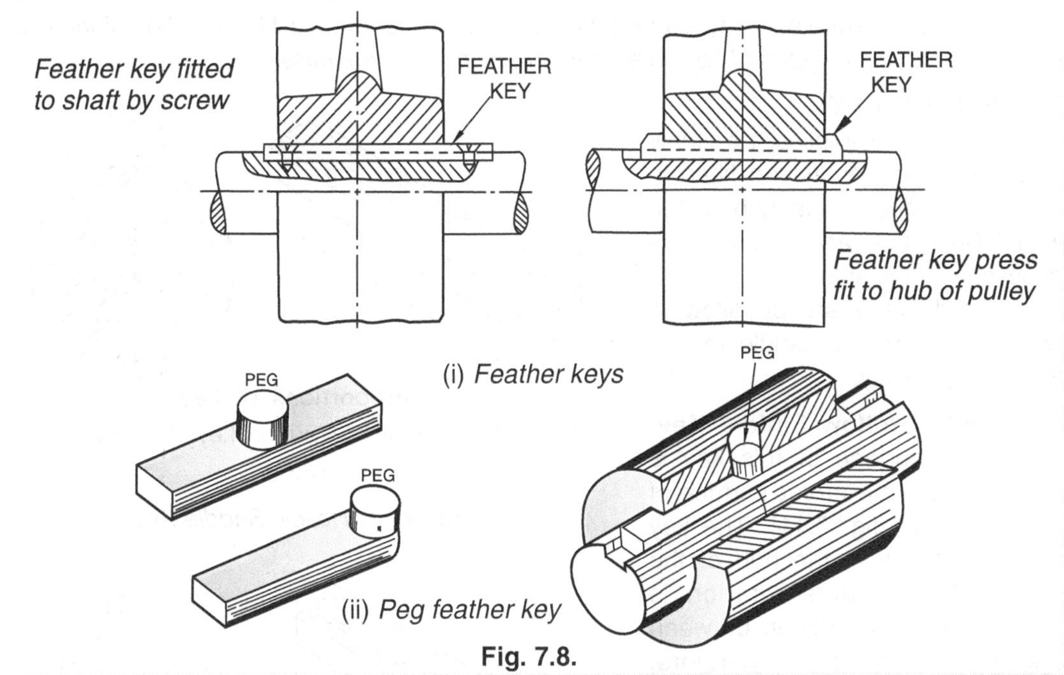

(i) *Feather keys*

(ii) *Peg feather key*

Fig. 7.8.

7.7.5. Peg Feather Key : In this type of feather keys, the pegs fit in the holes provided in the hubs of pulleys or gears. These keys allow the pulleys or gears to move axially along the shaft maddion to usual rotary motion [see Fig. 7.8 (ii)]

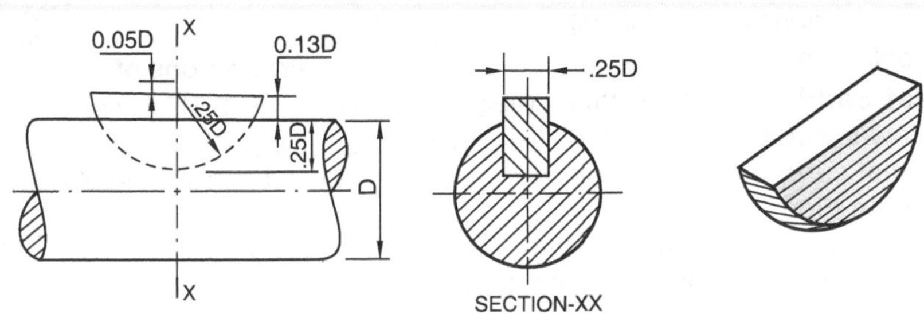

Fig. 7.9. *Wood ruff key*

7.7.6. Woodruff Key : This type of key is in the form of segment of a circular disc of uniform thickness (see Fig. 7.9). It fits into a slot of corresponding form cut in the shaft so that the flat portion projects outside the shaft. This projected portion fits into the key way cut in the hub of the wheel. Since the key and the slot in the shaft have the same radius, the key can be adjusted itself to any taper.

Woodruff key is largely used in machine tools and automobile works.

7.8. ROUND OR PIN KEY

Keys of circular cross-section are called round keys. These can be regular or tapered and are fitted in round holes drilled partly in the shaft and partly in the hub of the wheel [see Figs. 7.10 (*i*) and (*ii*)]. The diameter at the centre is the nominal diameter of the key and is usually 0.167 D of the diameter of the shaft. The ends of the round keys may be curved or chamfered.

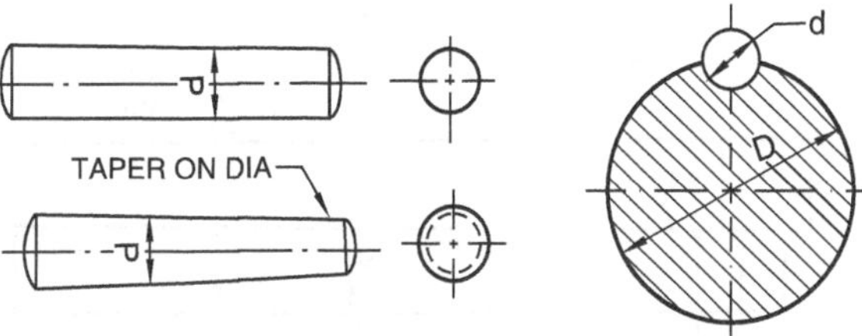

Fig. 7.10. *Round or Pin Key*

The main feature of this type of key is that, we can alter or adjust the position of the wheel easily. *Hence, these types of keys are used for low power transmission.*

7.9. SADDLE KEY

As the name implies, it is simply mounted on the shaft and fitted in recess provided in the hub of the wheel. So, in this case no key way is necessary in the shaft.

Saddle keys are of following two types : 1. Hollow saddle key 2. Flat saddle key.

Proportions of key :

If D = Dia. of Shaft Width of key, W=0.25 D

Thickness of key, T = 0.33 W

Fig. 7.11. *Hollow Saddle Key*

7.9.1. Hollow Saddle Key:

The lower surface of this key is hollow to fit on the curved surface of the shaft (see fig. 7.11). The key way is only cut in the hub of the wheel. The relative rotation is prevented by the friction between the key and the shaft. Therefore, this keys is suitable only when the power transmission is small.

Uses : *It is generally used as a temporary fastening in fixing and setting cams, eccentrics, etc.*

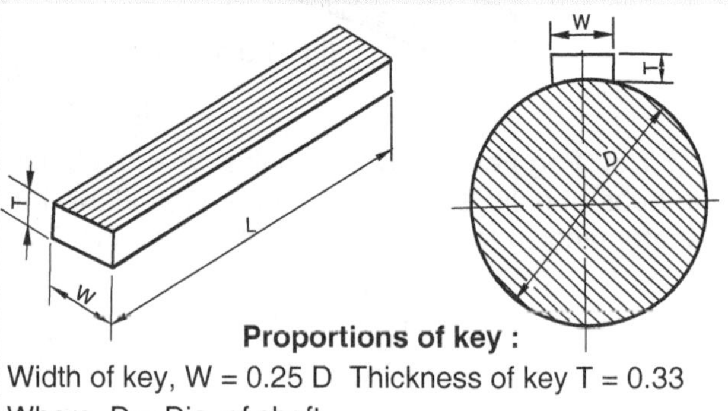

Proportions of key :

Width of key, W = 0.25 D Thickness of key T = 0.33

Where, D = Dia. of shaft

Fig. 7.12. *Flat Saddle Key*

7.9.2. Flat Saddle Key : In this key, the lower surface is flat and fits on the flat surface provided on the shaft (see Fig. 7.12). This key gives a more secured fastening than that in the hollow saddle key. Therefore, it can transmit greater power than hollow saddle key.

7.10. TANGENTIAL KEYING

In this system, keys act tangential in opposite direction as shown in Fig. 7.13. *It is used for heavy drives which may reverse direction.*

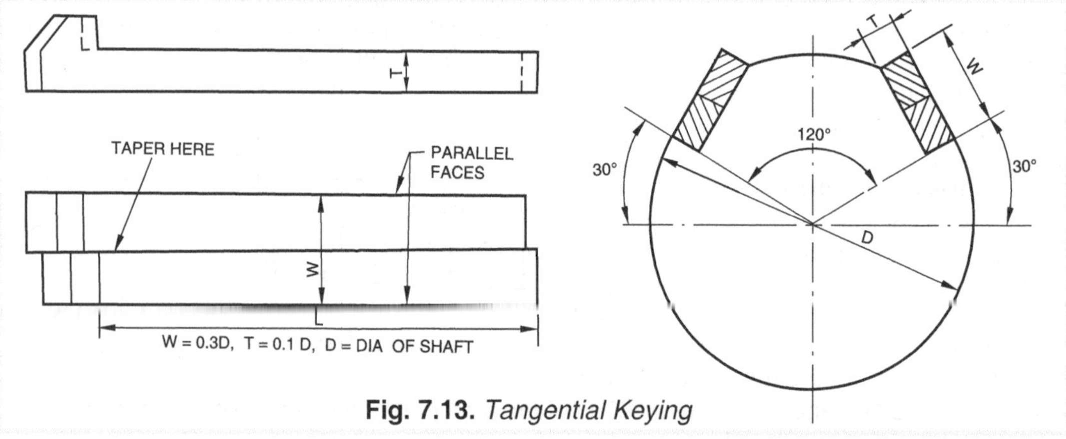

Fig. 7.13. *Tangential Keying*

7.11. COTTER

A cotter, is a flat wedge shaped metal piece of rectangular cross-section [see Fig. 7.14 (i)]. It is uniform in thickness but has taper in width throughout its length. The taper may be on one side or both sides. The usual taper is 1 in 48 to 1 in 33. Taper can be increased upto 1 in 8, if a locking device is provided. The ends of the cotter are usually like round wedges to facilitates the hammering for fixing and removing. Cotter is always inserted at right angles to the axis of the rods.

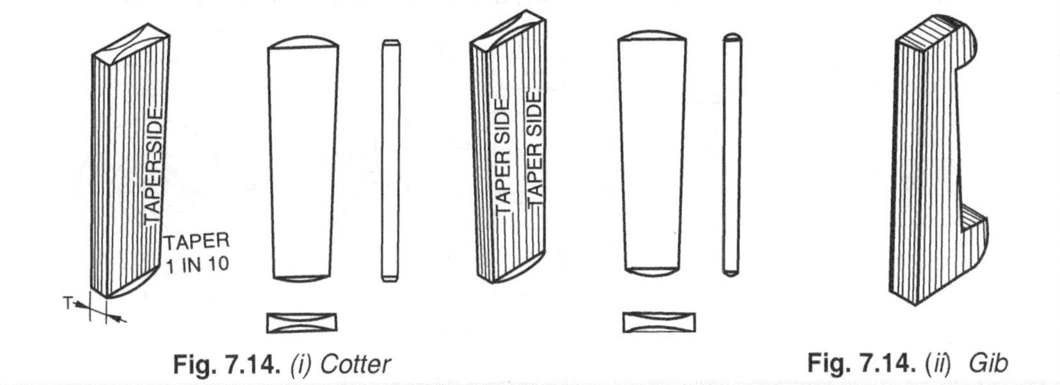

Fig. 7.14. *(i) Cotter* **Fig. 7.14.** *(ii) Gib*

The thickness of the cotter is usually 0.25 of the shaft diameter and the width is 5 times the thickness.

Uses : *Cotters are used to join two parts of bars which are subjected to tensile or compressive forces e.g. in joining of piston rod with crosshead, valve rod end and eccentric rod, foundation bolts etc.*

7.12. GIB

When two rods connected by a cotter are subjected to a pull, the friction between the cotter and the rods causes the cotter to open out. This can be prevented by use of another piece, called the gib [see Fig. 7.14 (*ii*)].

The total width of the gib and cotter is generally the same as that of a cotter. If used without a gib, the thickness of the gib is equal to the thickness of the cotter used.

7.13. SPLINE SHAFT

When the failure of a single key would cause inconvenience, it is a common practice to use a spline shaft (see Fig. 7-15). Spline shaft has splines on its see MS equally spaced grooves of uniform depth. The power transmitted is, therefore, among a number of splines.

The spline shaft is named after the number of splines on the shaft. If the number of splines are six, the shaft is called six spline shaft and so on. The hub or boss has corresponding recess for the spline shaft.

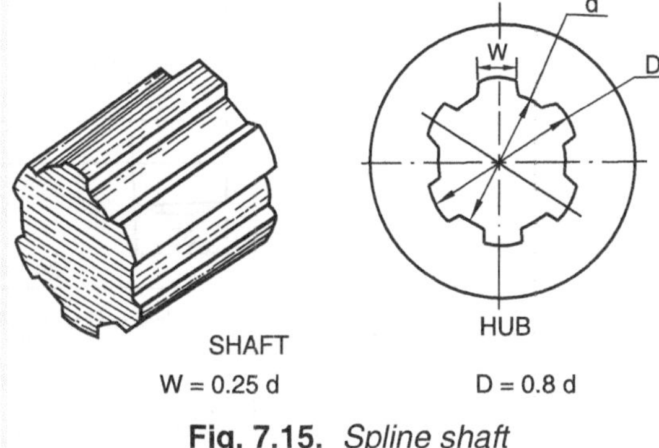

Fig. 7.15. *Spline shaft*

Use : Spline shaft is used in *driving wheel of a motor vehicle.*

7.14. JOINTS USED FOR CONNECTING RODS

The following are the various joints which are used for connecting rods subjected to axial forces in engineering practice :—

1. Sleeve and cotter joint 2. Spigot and socket joint 3. Gib and cotter joint
4. Adjustable joint 5. Knuckle joint.

7.15. SLEEVE AND COTTER JOINT

A sleeve and cotter joint is used to join two rods which are subjected to axial forces only. This joint has the following parts :—

(1) Sleeve (2) Two cotters (3) Two rods

Two slots in the sleeve and corresponding two slots near the end of the rods are provided to suit cotters. The rods are inserted through the ends of the sleeve and meet each other at the centres. Cotters are then inserted through slots to make assembly.

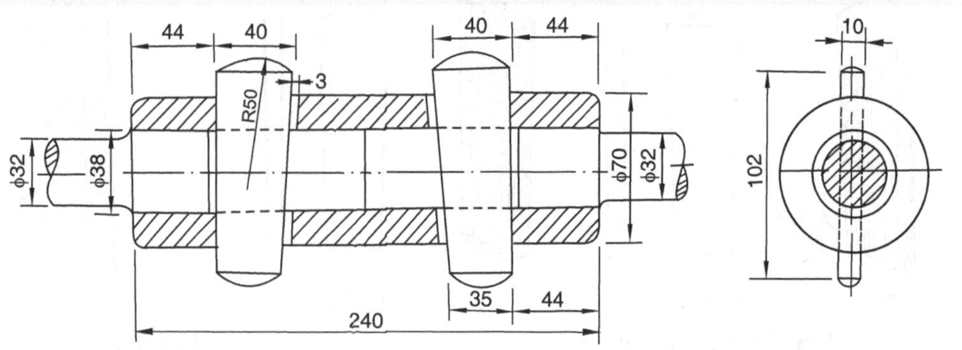

Fig. 7.16. *Sleeve and cotter joint*

Notes : (i) *Clearances between the vertical side of the cotter and sleeve are formed. The clearances are provided for better tightness and adjustment of rods.*

(ii) *The outer edges of the cotters are kept vertical and the inner edges tapered to suit the sides of cotter.*

Problem 1 : Two view of a cotter joint are shown in Fig. 7.16. Draw the following views of the joint :—

(a) Front view – upper half in section (b) Top view

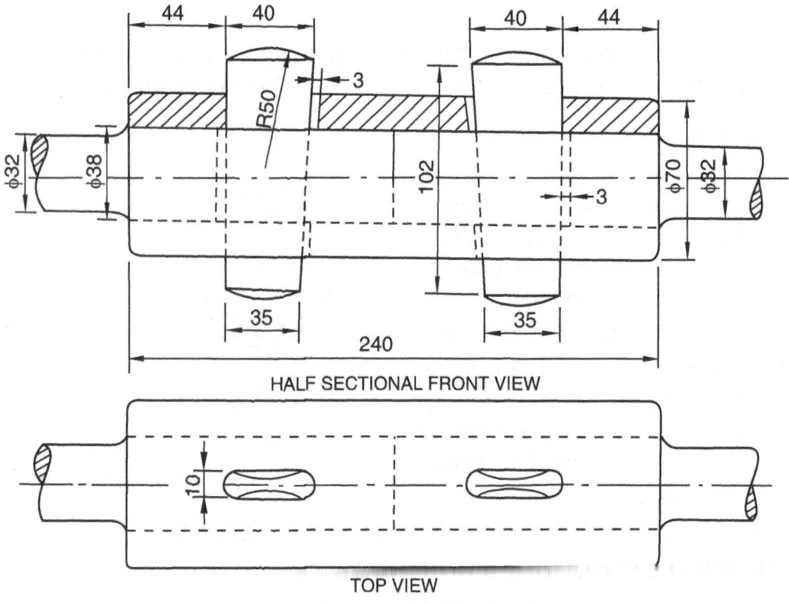

Fig. 7.17. *Cotter joint*

Solution : For its solution see Fig.7.17

7.16. SPIGOT AND SOCKET JOINT

In this joint, a socket is made by enlarging the end of one rod and the end of the other rod fitted into it. Suitable slots are made in the spigot and socket to accommodate cotter. A cotter is driven tightly through it. When the cotter is driven in, it forces the rod to move inside the socket to make the joint. The clearance in the joint varies from 1.5 to 3 mm when the cotter is driven in final position.

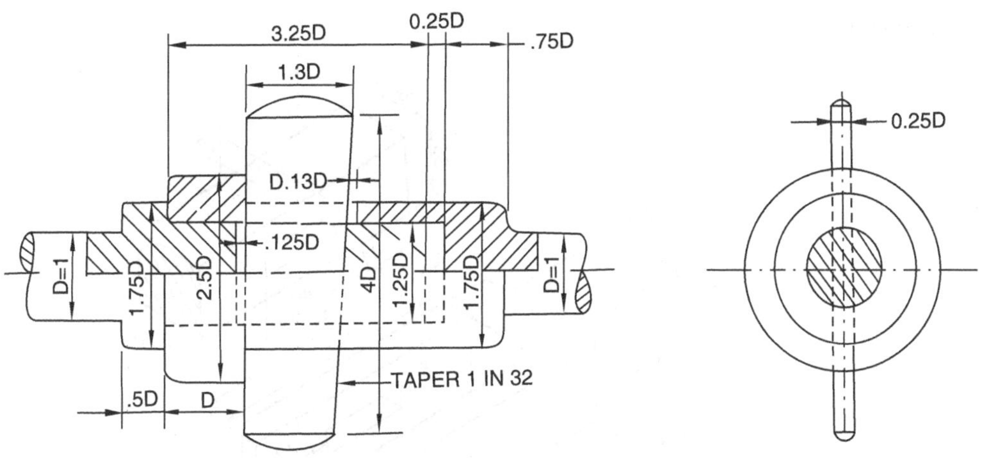

Fig. 7.18. *Spigot and socket joint*

Problem 2 : Two views of a spigot and socket joints are shown in Fig. 7.18. Draw the followings views to scale 1:1 size : —

(a) Full Sectional front view **(b) Side view**

(c) Top view, lower half in section.

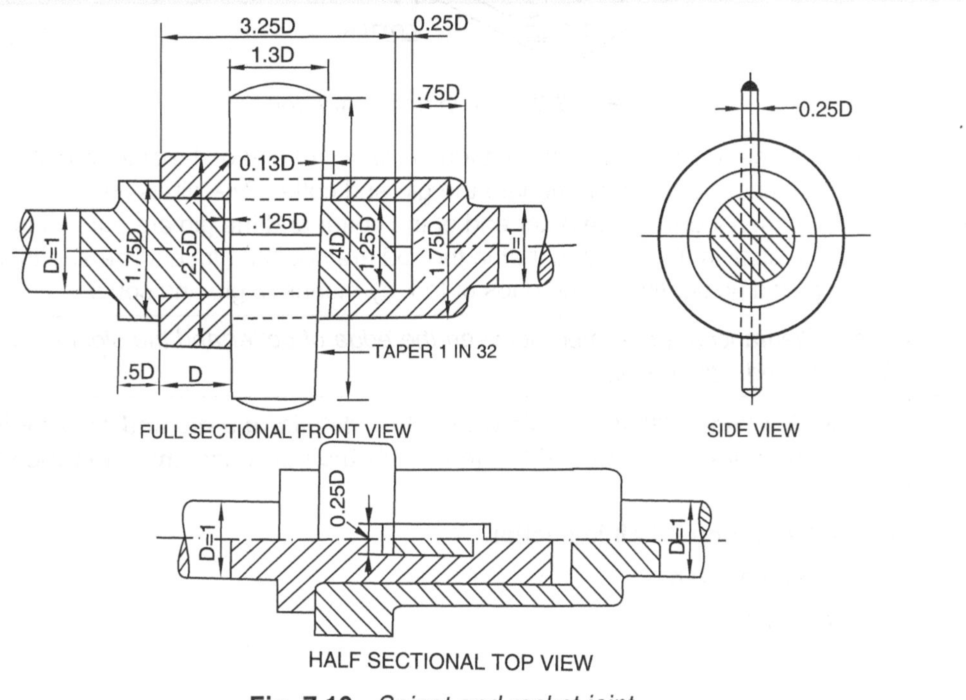

Fig. 7.19. *Spigot and rocket joint*

Solution : For its solution, see Fig. 7.19.

7.17. GIB AND COTTER JOINT

This joint is suitable for connecting rods of square or rectangular cross-section. It consists of the following four parts :—

1. Fork end rod
2. Simple rod
3. A gib
4. A cotter

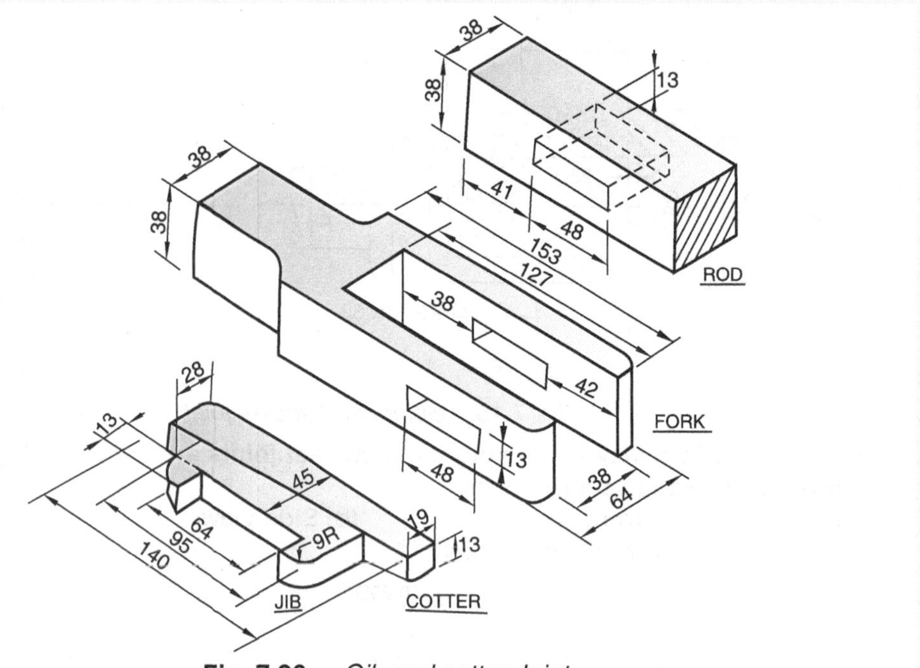

Fig. 7.20. *Gib and cotter Joint*

The end of one rod is made to form a fork, in which of end of the other rod fits. Slots are made in the fork and rod to accommodate the gib and cotter. The gib is first put into the slot and the cotter is then inerted. The outer end of the cotter and the gib are parallel to the sides of the slots. The slots in the fork and that in the rod are not in a straight line and this enables the rod to be pulled up tightly into the ends of fork by means of gib and cotter.

Note : *Clearances are formed between the edge of cotter and the slots in the fork, for better tightness and adjustment.*

Problem 3 : Fig. 7.20 shows pictorially the details of a gib and cotter joint. Draw to full size scale following views of the joint with the parts assembled in their working position :—

(a) Front view upper half in section

(b) Top view or plan

(c) Side view

Solution : For its solution, see Fig. 17.21.

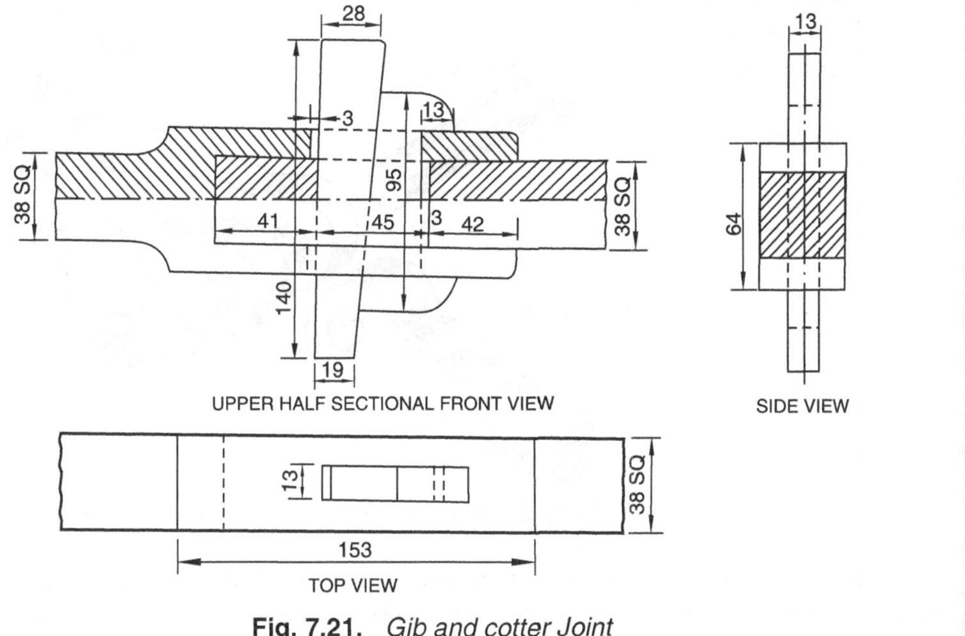

Fig. 7.21. *Gib and cotter Joint*

7.18. AJUSTABLE JOINT

This joint is used to join two round rods axially. The rods to be joined are threaded at the ends and are screwed to a coupler which has a threaded hole. The coupler is hexagonal shape in the central portion and is rounded at the ends, in order to facilitate tigntening and losening of the rods with the help of spanner (see Fig. 7.22).

For the proper adjustment and screwing action of the two rods, coupler is threaded inside its two ends in opposite direction. The coupler will thus screw off both the sides simulatneously. The rods are also threaded corresponding in opposite directions. A gap is provided inside the coupler for expansion allowances.

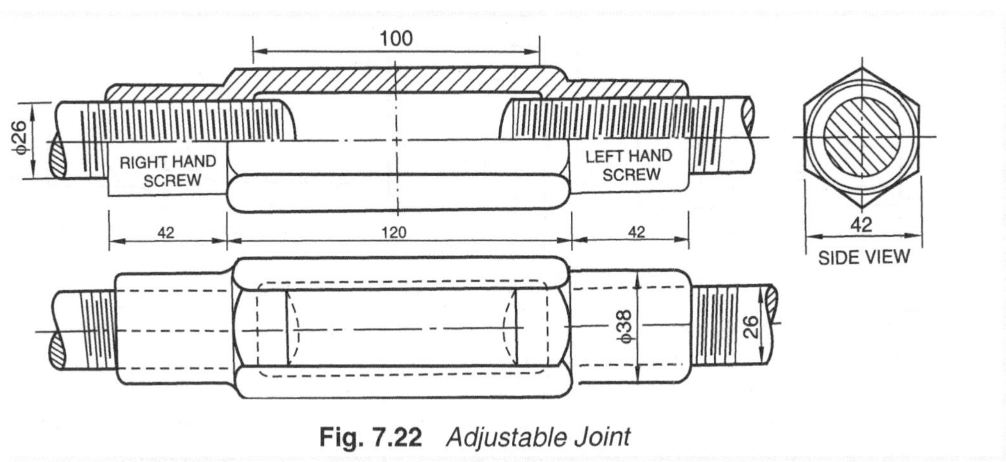

Fig. 7.22 *Adjustable Joint*

7.19. KNUCKLE JOINT

Knuckle joint is used to connect two rods, when the rods need not necessarily be lie in a straight line (see Fig. 7.23). *This joint allows a small angular movement of one rod relative to another.* It consists of following five parts :—

1. Double eye end or forked end
2. Single eye end
3. Cylindrical pin
4. Collar
5. Taper pin.

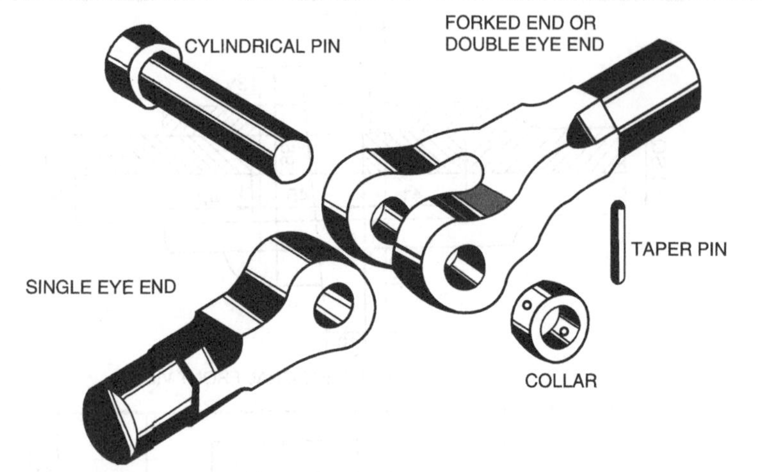

Fig. 7.23. *Details of Knuckle Joint (pictorial Views)*

In this joint, one end of the rod is forged to form a fork. Two holes are made in both the arms of the fork. The end with holes is called double eye end or forked end. The end of the other rod is also forged to form a single eye end and a hole is made through it. The single eye end is placed into the double eye end and a cylindrical pin is inserted through the holes. It is kept in position by means of a collar and taper pin. The two rods are quite free to rotate on the cylindrical pin.

Uses : *This joint is commonly used when a rotary motion is to be converted into reciprocating motion or vice versa. In case of steam engine, the eccentric rod is connected to the valve rod by means of this joint. It is also used in structure such as for connecting the bars of the roof truss, braced girders, links of a suspension chain, plate links, gearing chains, cycle chains, etc.*

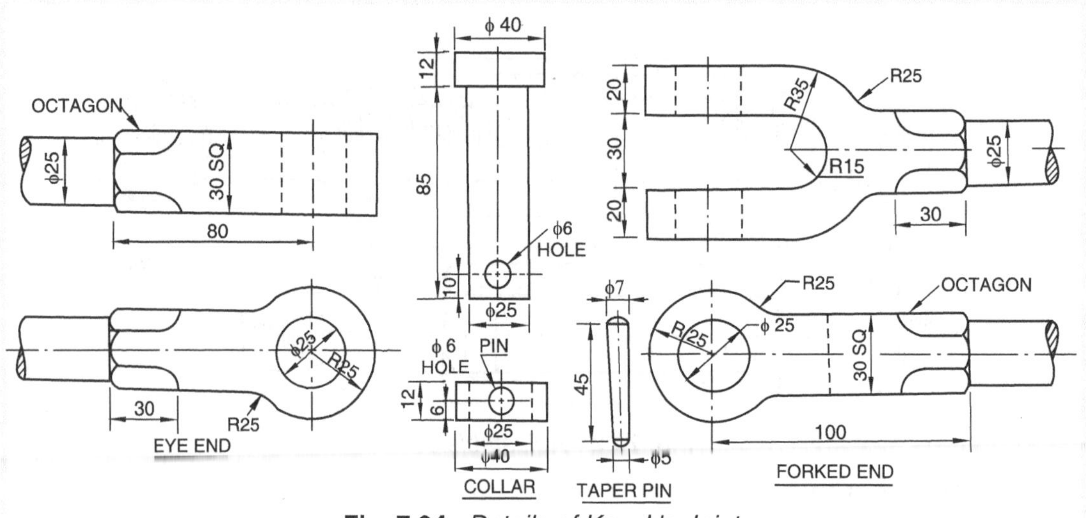

Fig. 7.24. *Details of Knuckle Joint*

Problem 5 : Fig. 7.24 shows the details of a knuckle joint. Assemble all the parts and draw the following views :—

(a) Front view

(b) Top view

Solution : For its solution, see Fig. 7.25.

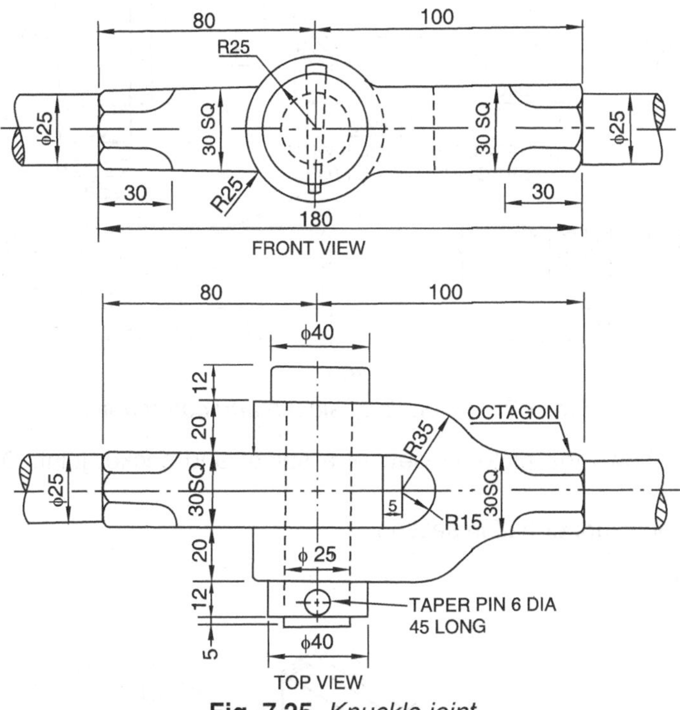

Fig. 7.25. *Knuckle joint*

QUESTIONS FOR SELF EXAMINATION

1. What are keys ? Give the main classification of keys?
2. What is the difference between sunk key and saddle key ?
3. What are cotters and where are they used?
4. What is the difference between a key and cotter ? Tell the purposes for which they are un engineering practice.
5. What do you mean by spline shaft?

PROBLEMS FOR PRACTICE

1. Draw any four types of sunk keys.
2. Sketch the following types of keys fitted on a 24 mm dia. shaft.

   (i) Flat saddle key (ii) Hollow saddle key

   (iii) Woodruff key.
3. Fig. 7.26 shows the details of sleeve and cotter joint. Assemble the parts and draw the following views:-

   (a) Half sectional front view (elevation)

   (b) End view (side view)

   (c) Top view (plan)

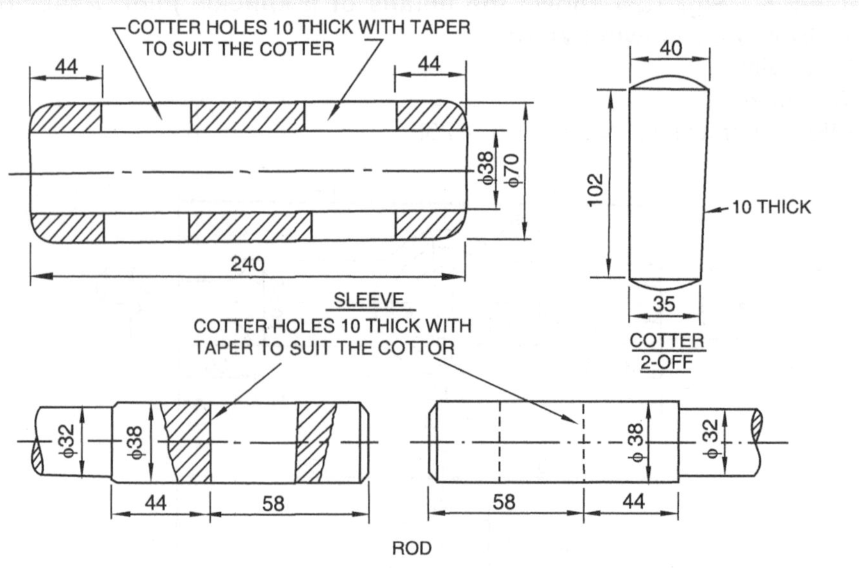

Fig. 7.26. *Details of sleeve and cotter Joint*

4. Fig. 6.27 shows the pictorial view of a spigot and socket joint. Draw the following views:-
   (a) Front view, upper half in section.
   (b) Side view
   (c) Top view

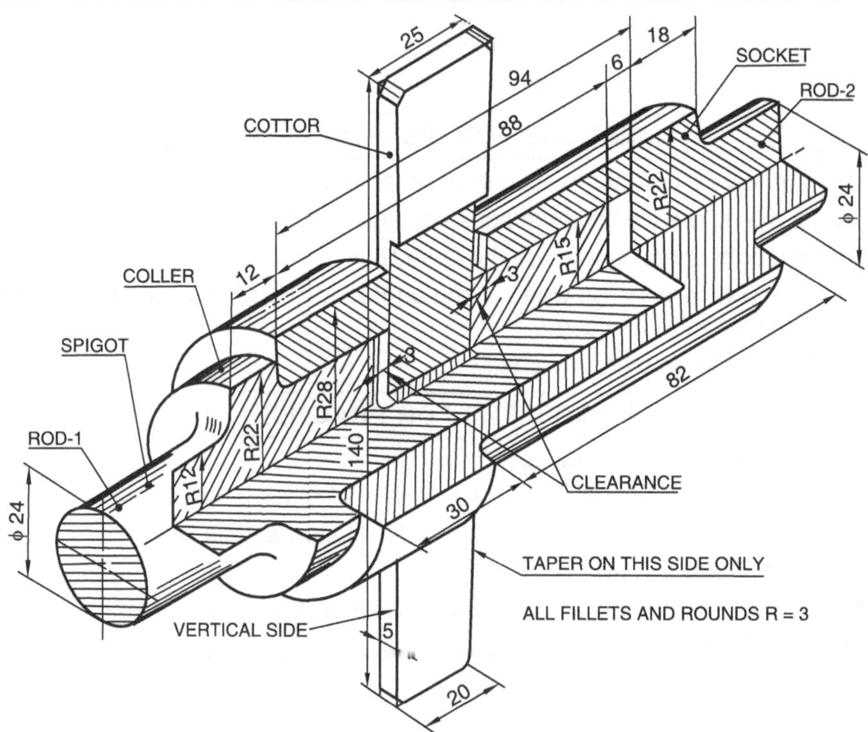

Fig. 7.27. *Pictorial view of spigot and socket joint*

5. Fig. 7.28 shows the details of a gib and cotter Joint. Assemble the parts and draw the following views :—
   (a) Half sectional front view
   (b) End view
   (c) Top view

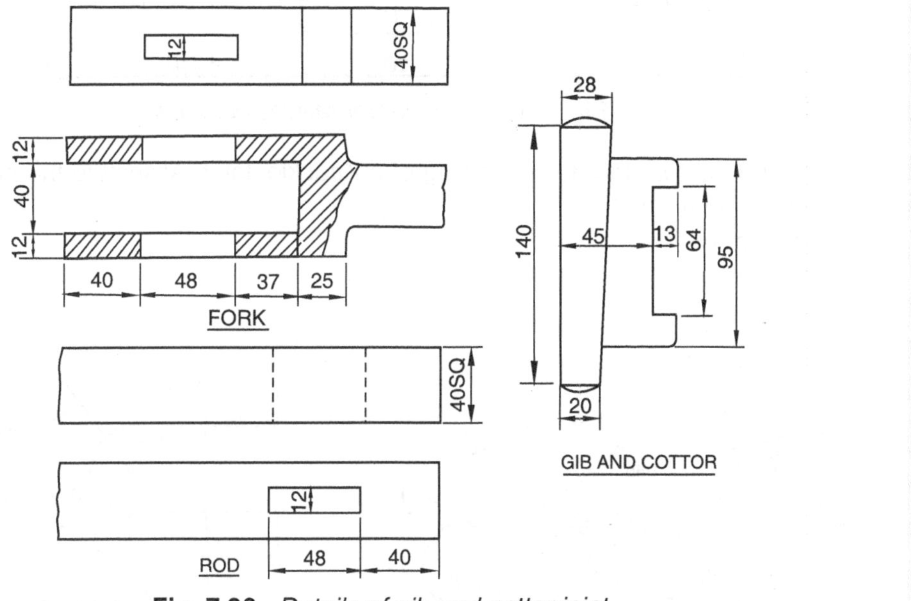

Fig. 7.28. *Details of gib and cotter joint*

6. Fig. 7.29 shows the details of spigot and socket joint. Assemble the parts and draw the following views:-
   (a) Half sectional front view (b) End view (c) Top view

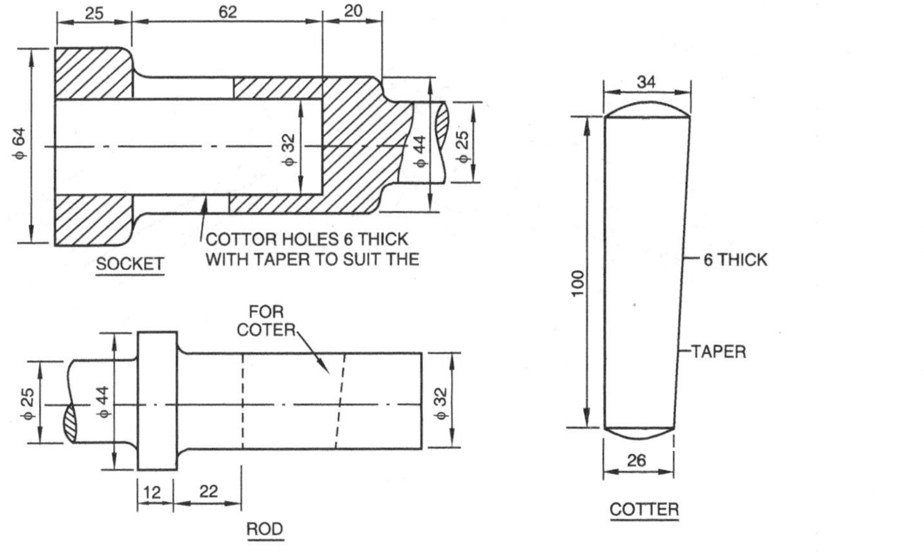

Fig. 7.29. *Details of spigot and Socket Joint*

7. Draw the front view and side view of coupler for railway coaches as show in Fig. 7.30.

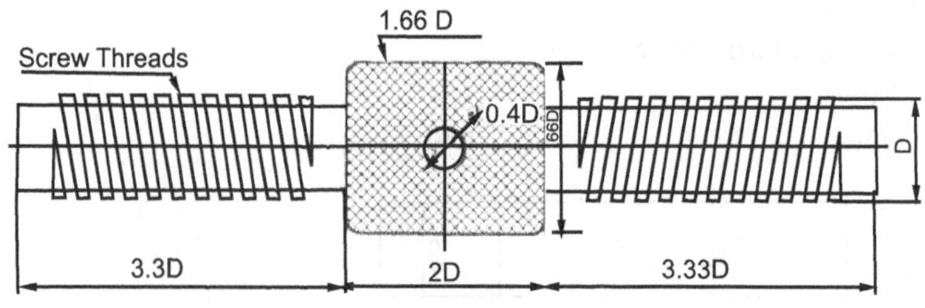

Fig. 7.30. *Coupler for railway coaches*

8. Fig. 7.31 shows the detail drawing of a Knuckle Joint. Assemble the parts and draw the following views :

   (a) Front view

   (b) Side view

   (c) Top view

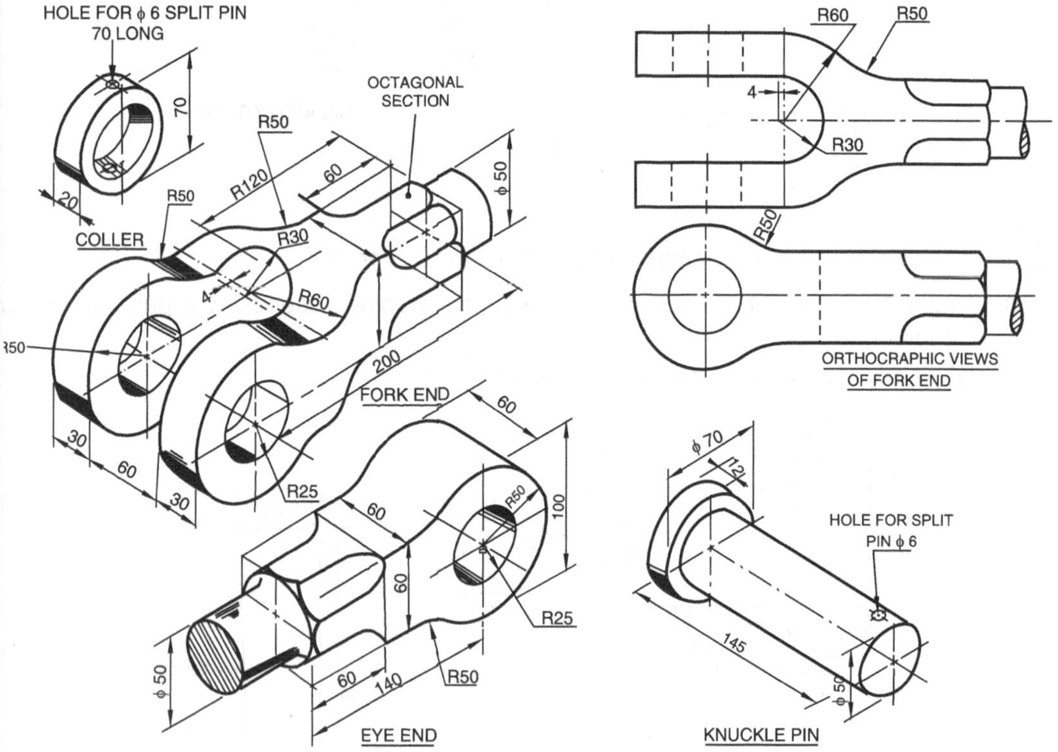

Fig. 7.31. *Details of knuckle joint*

CHAPTER

8

Shaft Couplings

INTRODUCTION

In engineering practice, mechanical power is used to run a large number of machines. One of the major problems an engineer faces is to transmit this power from one place to another. This is achieved by means of shafts, pulleys, belts, etc. Thus, the shafts play the most important role for the transmission of power.

In market, shafts are available in about 6 to 10 metres length. More lengthy shafts are not easily made due to manufacturing and transporting difficulties. But in mechanical engineering practice one requires shafts of bigger length than are available in the market. So, two for more shafts are joined axially to get the shaft of required length. The shaft coupling is achieved through sleeves, flanges, pins, cotters, bolts and nuts etc.

In this chapter, we shall study, shaft coupling, types of shaft couplings along with their actual use in engineering practice.

8.1. SHAFT COUPLING

The joining of two or more shafts by means of flanges, bolts, nuts etc. in order to increase their length is known as shaft coupling.

8.2. TYPES OF COUPLINGS

There are different types of coupling which are used for different purposes depending upon the shape and construction. But, the following two main types of couplings are used in engineering practice :—

1. Rigid Couplings
2. Non-rigid or Flexible Couplings

8.2.1. Rigid Couplings : Rigid couplings connect the two shafts rigidly in axial alignment. These types of couplings include :—

(a) Flange couplings

(b) Box or muff couplings

(c) Compression couplings

(d) Friction grip couplings.

8.2.2. Non-rigid Couplings or Flexible Couplings : Non-rigid coupling are used to connect two shafts which are not in exact alignment. But these types of couplings have important function of absorbing some shocks of the driving shaft and also prevent reversed stresses when one or both the shafts deflect under load at the coupling. These types of couplings include :—

(a) Pin type flexibe couplings

(b) Universal couplings

(d) Oldham's coupling, etc.

8.3. RIGID FLANGE COUPLING

Rigid or solid or forged flange coupling is generally used for shaftsof marine engines.

Fig. 8.1 shows a rigid coupling. Here the flanges are forged solid with the shafts and are connected together by means of a number of headless taper bolts.

Exercise 1 : Fig. 8.1 shows a rigid coupling. Draw the following views:

(a) Front view as given

(b) Side view

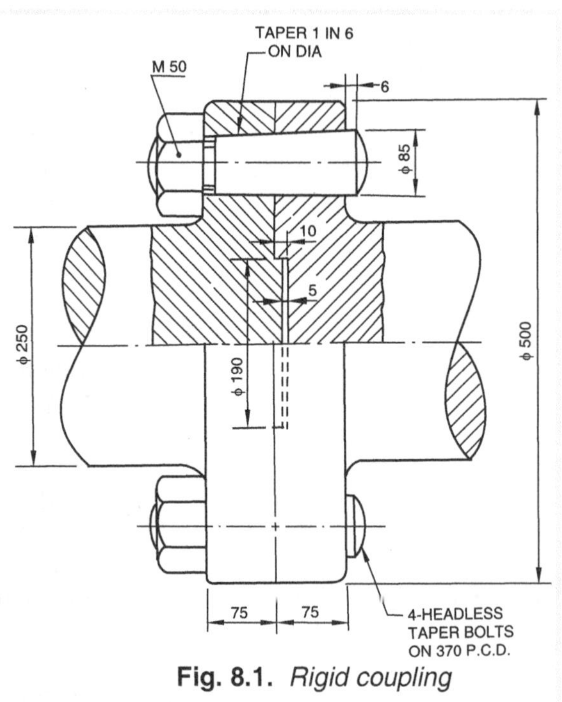

Fig. 8.1. *Rigid coupling*

8.4. FLANGE COUPLING

The flange coupling is a standard form of coupling and is extensively used in workshop to join shafts. It consists of two similar cast-iron flanges to which the shafts to be joined are keyed with the help of rectangular taper sunk keys as shown in Fig. 8.2. The flanges are then held together by means of bolts and nuts. The bolts should exactly fits in the holes of flanges and so as to transmit motion from one shaft to the other properly.

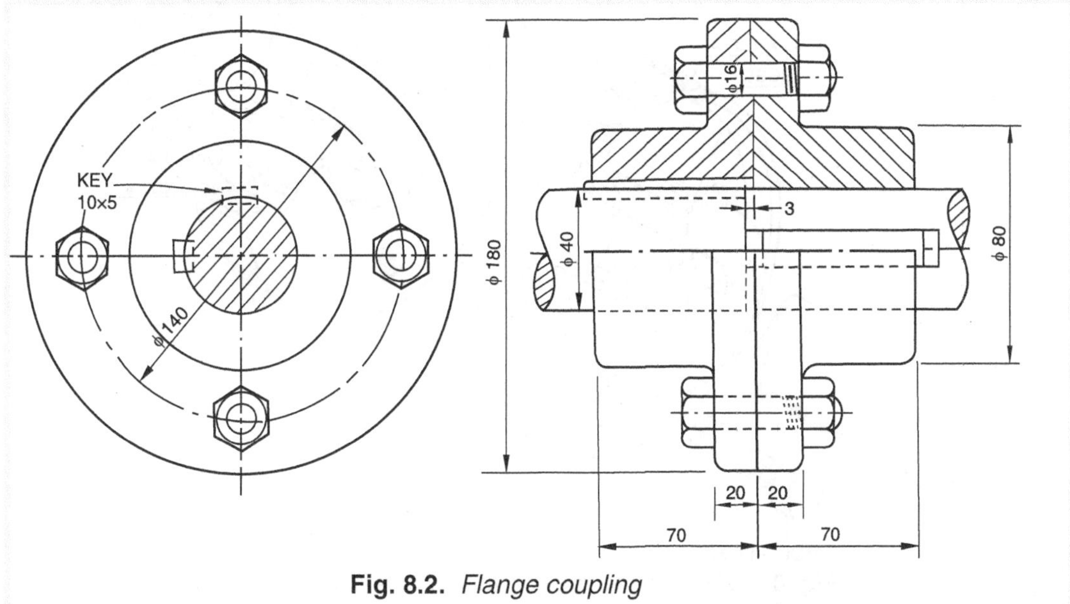

Fig. 8.2. *Flange coupling*

Note : *1. Two rectangular taper keys are inserted from inside faces of the flanges for the convenience of fitting.*

2. Two rectangular taper keys are inserted at 90° to each other so that the two shafts may not be weak at the same longitudinal section due to the presence of key ways.

Problem 1 : Fig 8.2 shows two views of a flange coupling. Draw the following views to full size scale :

(a) Full sectional front view (b) Side view

Solution : For its solution, see Fig. 8.3.

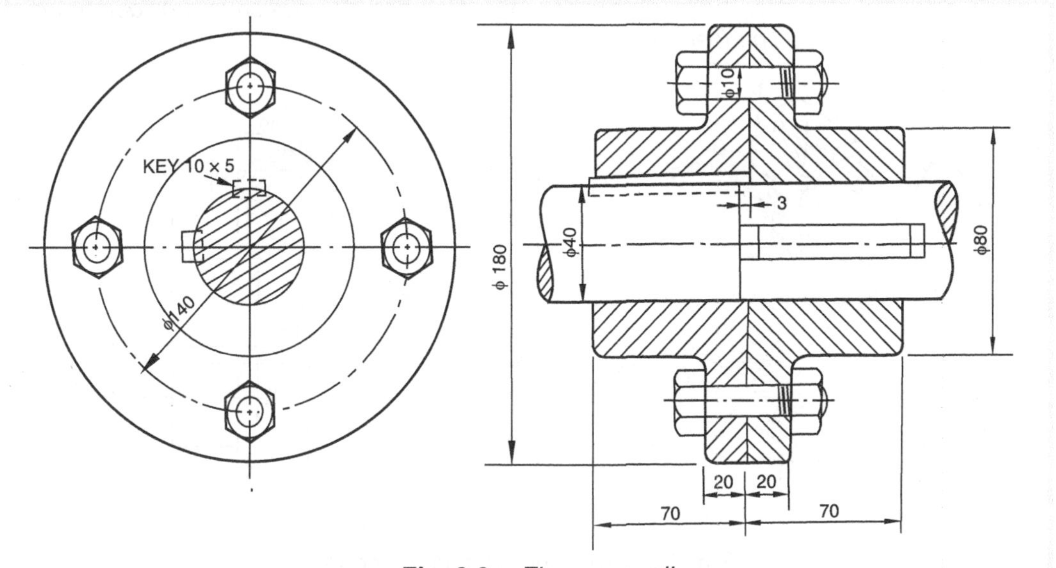

Fig. 8.3. *Flange coupling*

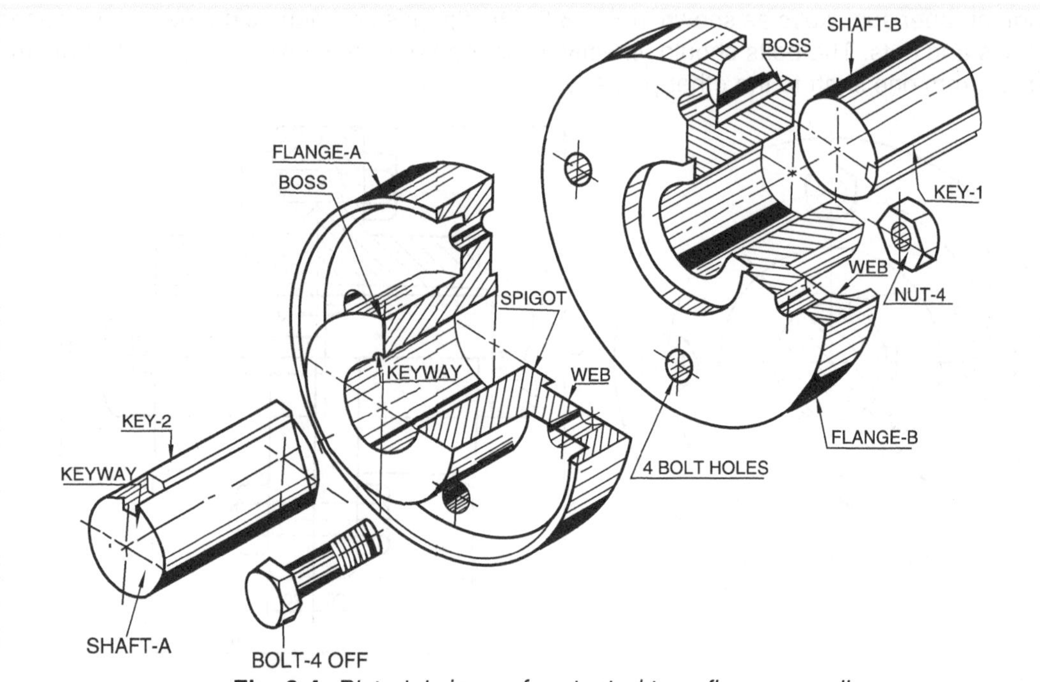

Fig. 8.4. *Pictorial views of protected type flange coupling.*

8.5. PROTECTED TYPE FLANGE COUPLING

In ordinary coupling, the bolt heads and nuts are projected beyond the flanges which are liable to cause injuries to materials, men, machinery etc. around them. To safe guard against the bolt heads and nuts, an annular space is provided on the outer surfaces of the flanges which covers them completely. The remaining arrangement is the same as explained in ordinary flange coupling. Fig. 8.4 shows a pictorial views of protected type flange coupling.

Note : *1. The alignment of the two shafts is maintained by making a circular projection on one flange which fits into the corresponding recess in the other. This arrangement is termed as spigot socket centring.*

2. An axial clearance between circular extension and depression is provided to adjust the shaft in one straight line (see Fig. 8.5).

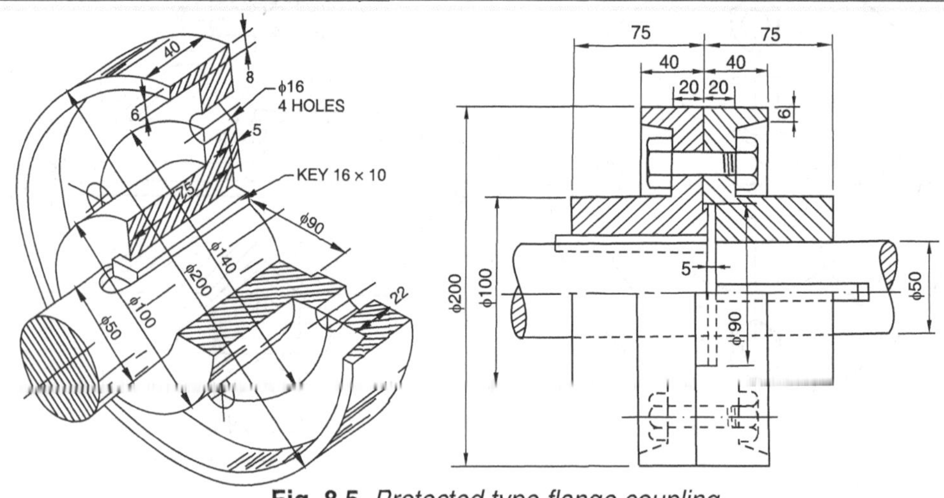

Fig. 8.5. *Protected type flange coupling*

Exercise 2 : Fig. 8.6 shows two views of a protected type flange coupling. Draw to some suitable scale the following views :

(a) Front view-full in section

(b) End view.

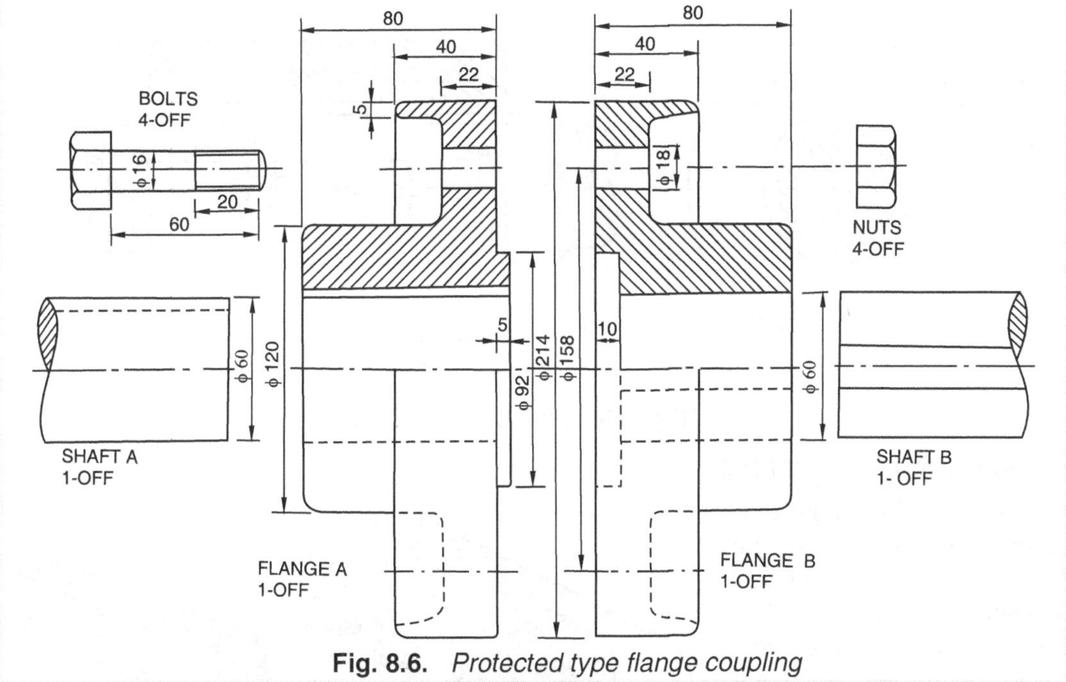

Fig. 8.6. *Protected type flange coupling*

Exercise 3 : Fig 8.6 shows details of protected type flange coupling draw with 1:1 scale the following views :

(a) Sectional front view **(b) Side view. Show the two shafts also**

8.6. MUFF COUPLING

It is the simplest form of the rigid coupling used for connecting smaller sizes of shafts. It consists of a cast iron cylindrical muff or sleeve, slides on the two ends of the shafts which are butt against each other. A long gib head key is provided on the shaft at one end of the muff to check the muff from slipping or while being inserted by hammering (see Fig. 8.7.)

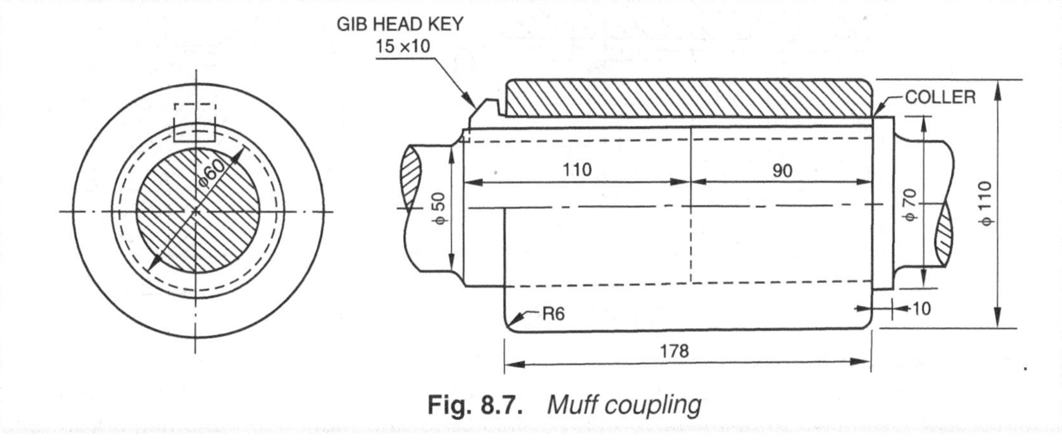

Fig. 8.7. *Muff coupling*

8.7. HALF LAG MUFF COUPLING

It is just like muff coupling, but the ends of the two shafts are made to overlap each other for a short length. A taper 1 in 12 is provided on the lag and therefore, the portion of shaft in the muff is increased. A taper key is used to connect the muff and shaft (see Fig. 8.8).

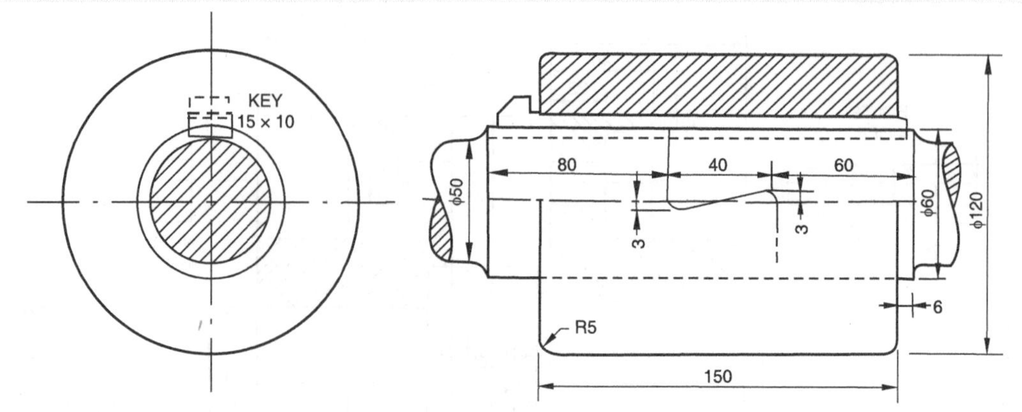

Fig. 8.8. *Half lag muff coupling*

8.8. SPLIT MUFF COUPLING

This type of coupling is used in cases where the removal of the shafts present practical difficulties. In this, the muff is split up in two semi-cylindrical halves which are joined together by means of studs and nuts. Two shafts are kept together by means of a feather key. One half of the muff is kept below and other above the shaft ends and then bolted up (see Fig. 8.9).

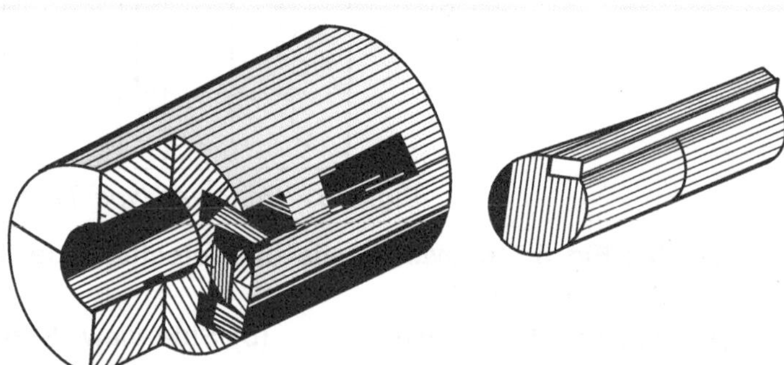

Isometric views of split muff coupling with two shafts and key

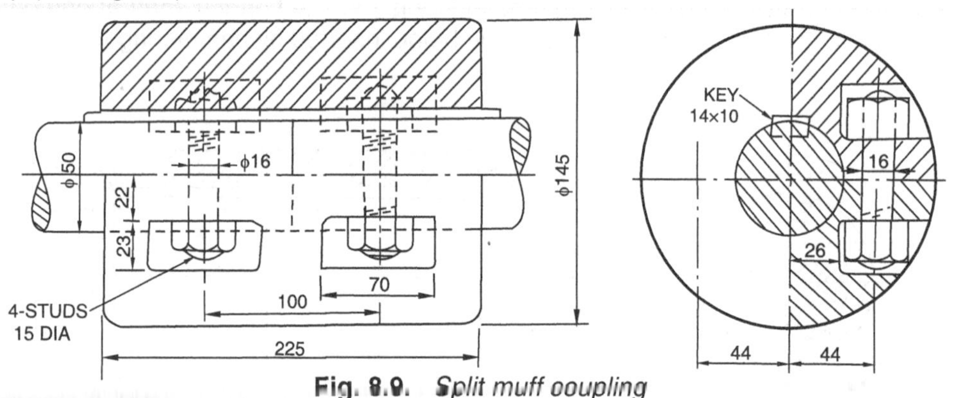

Fig. 8.9. *Split muff coupling*

8.9. FLEXIBLE COUPLING

This type of coupling is used to couple two shafts which are slightly out of line or which requires slightly relative angular movement. In this case, the motion is transmitted from one shaft to the other with the help of driving pins rigidly bolted to one flange and loosely fitted to the corresponding hole in the other flange. Compressible or flexible elements such as rubber, leather, etc. are used round the driving pins for prevention of shocks and vibrations occurring due to power transmission (see Fig. 8.10).

Flexible coupling is commonly used as a direct coupling device to minimise starting shocks *e.g. when an electric motor is directly coupled to machine.*

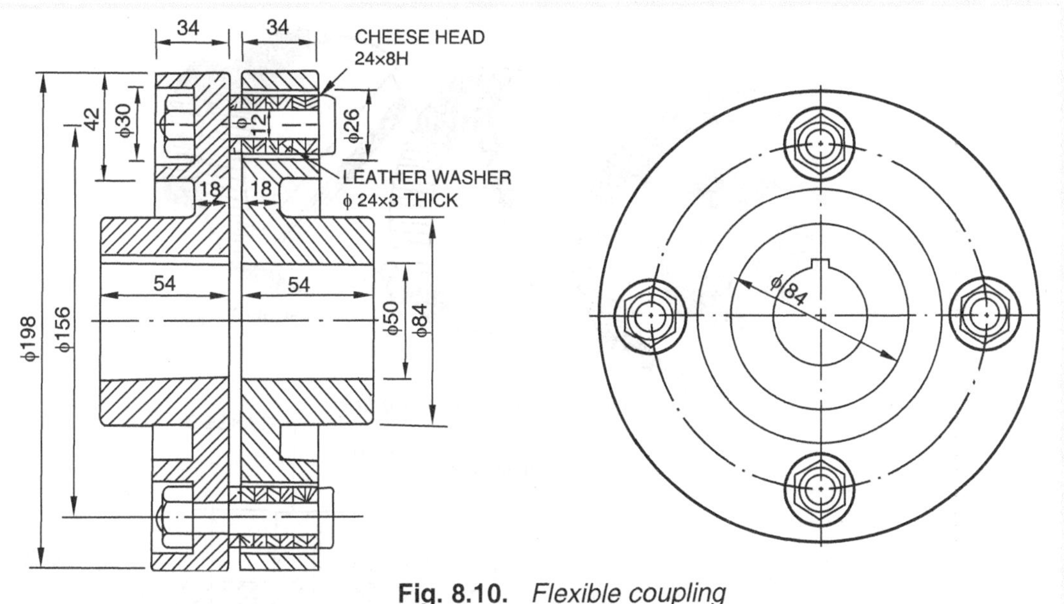

Fig. 8.10. *Flexible coupling*

Problem 2 : Fig. 8.10 shows two views of flexible coupling. Draw the following views alongwith two shafts in their proper positions :

(a) Front view – half in section **(b) Side view.**

Solution: For its solution, see Fig. 8.11.

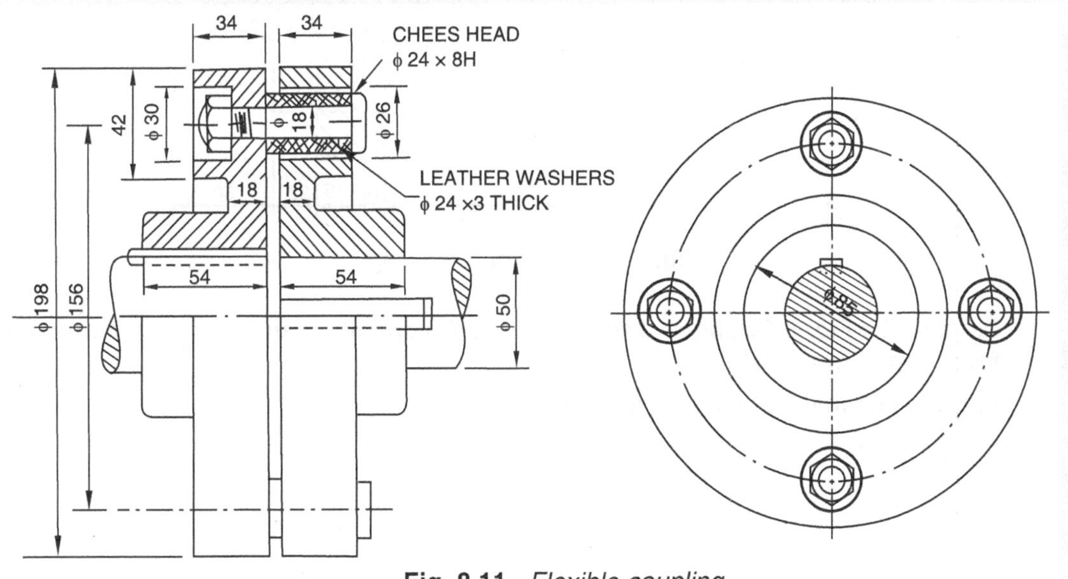

Fig. 8.11. *Flexible coupling*

8.10. OLDHAM'S COUPLING

This type of coupling is used to couple two shafts whose axes are parallel but not in alignment. It consists of two cast iron flanges (each having rectangular groove) and one centre block. The ends of shafts are keyed to the flanges. The centre block is in the form of a circular disc having two projecting parts on its opposite sides and at right angle to one another. The centre block is placed between the two flanges as shown in Fig. 8.12.

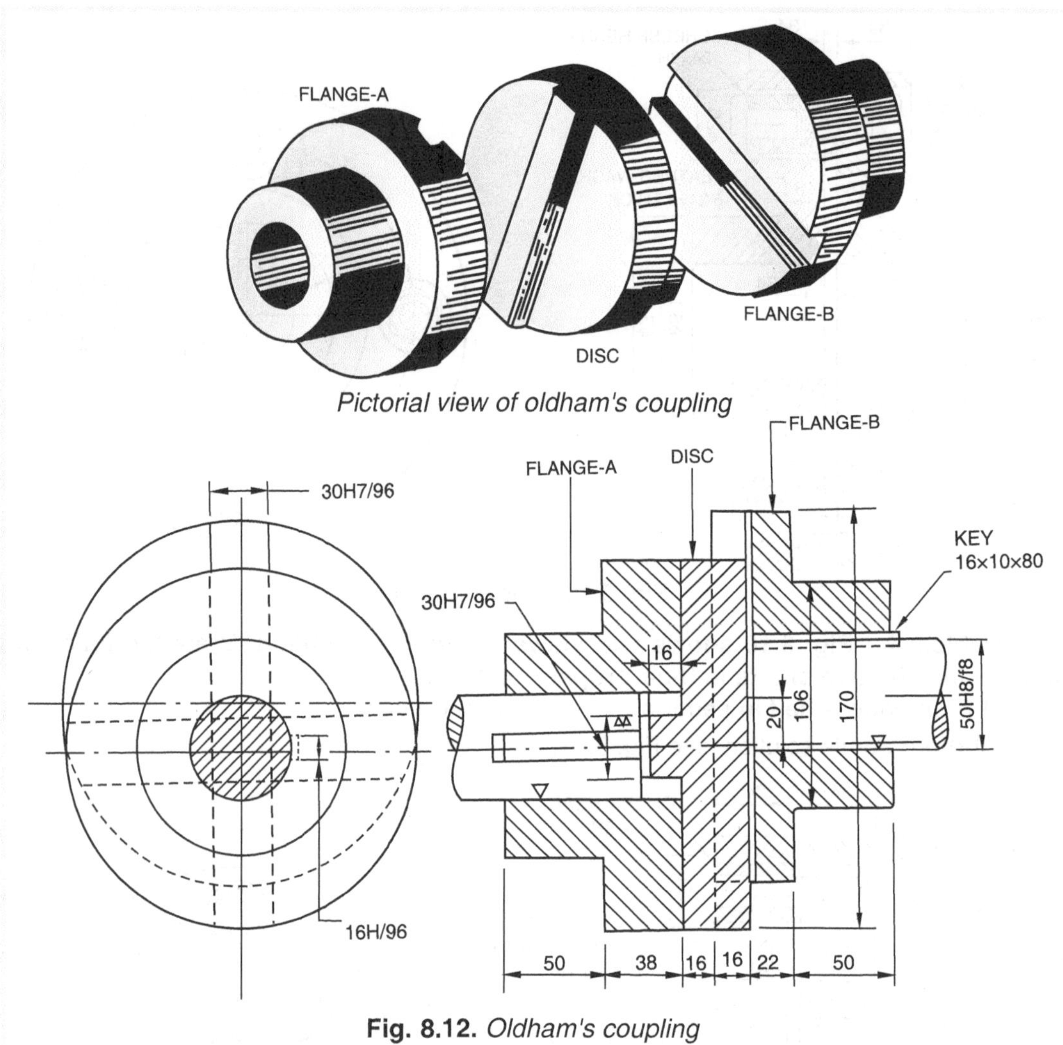

Fig. 8.12. *Oldham's coupling*

Exercise 5 : Fig. 8.12 shows two views of an oldham's coupling for 50 mm shaft. Draw following views :—

(a) Half sectional front view from centre line.

(b) End view.

8.11. UNIVERSAL COUPLING (HOOK'S JOINT)

This type of coupling is used to couple two shafts whose axes are not in line with each other, but intersect at a point. It consists of two similar forks, one centre block, two pins and two collars with lock pins. The forks are keyed on the ends of the two shafts and are then pin jointed to a centre block having two arms at right angles to each other. In this coupling the angle between the two shafts may be varied even when they are in motion.

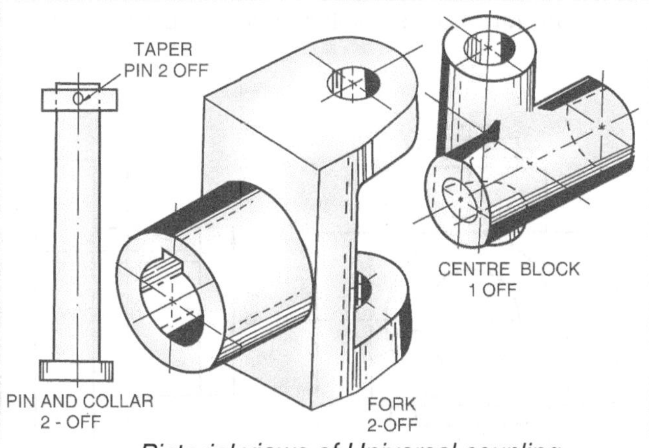

Pictorial views of Universal coupling

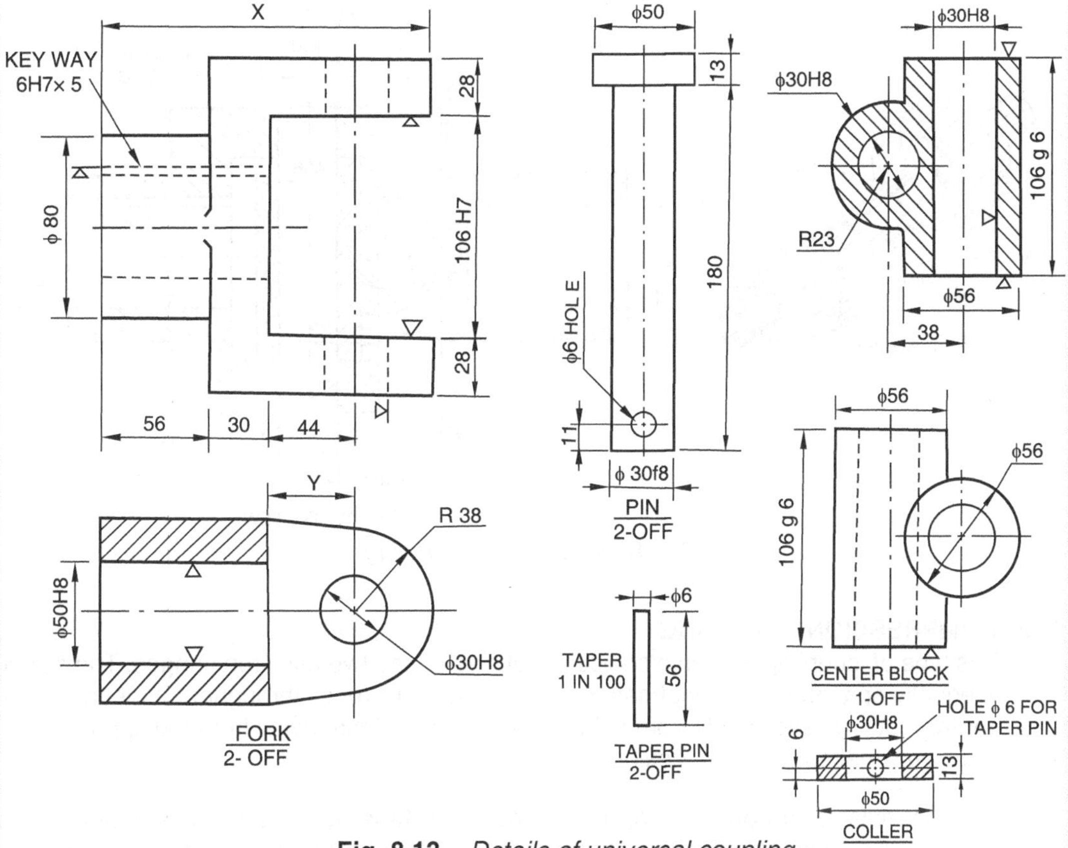

Fig. 8.13. *Details of universal coupling*

Problem 3 : Fig. 8.13 shows the details of a universal coupling. Imagine the parts assembled together and draw to a full size the following views :—

(a) Front view

(b) Side view (c) Sectional top view.

Solution: For its solution, see Fig. 8.14.

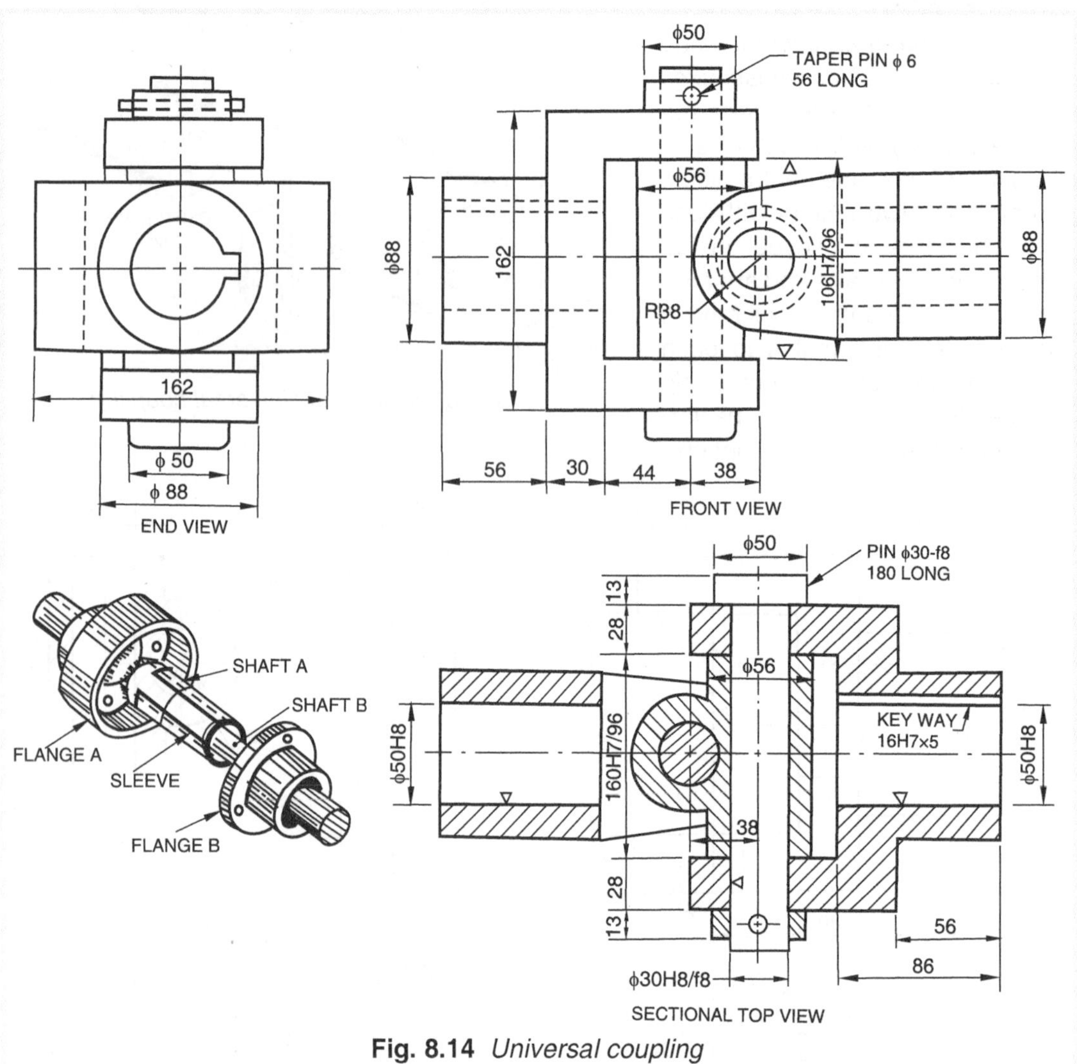

Fig. 8.14 *Universal coupling*

8.12. COMPRESSION COUPLING

This type of coupling a double conical steel sleeve and two cast iron flanges. The sleeve has six equally spaced longitudinal saw cuts. One saw cut runs throughout its length while the other five cuts run alternately from either end to within 25 mm away from the opposite end. These saw cuts make the sleeve flexible.

Two dissimilar cast iron flanges, having the same taper as that on hole of sleeve, are slipped over the sleeve. The flanges are then coupled together by means of bolts and nuts. When two lengths are tightened together, the split sleeve grips the shafts tightly due to the compression of saw cuts in the sleeve [see Fig. 8.15].

The main advantage of this coupling is that the key ways and keys are eliminated. This makes it very convenient when the shafts are to be coupled or discoupled.

Exercise 6 : Fig. 8.15 shows the details of a compression coupling. Make dimensioned drawing with all the parts assembled as follows :—

(a) Front view – Upper half in section.

(b) End view.

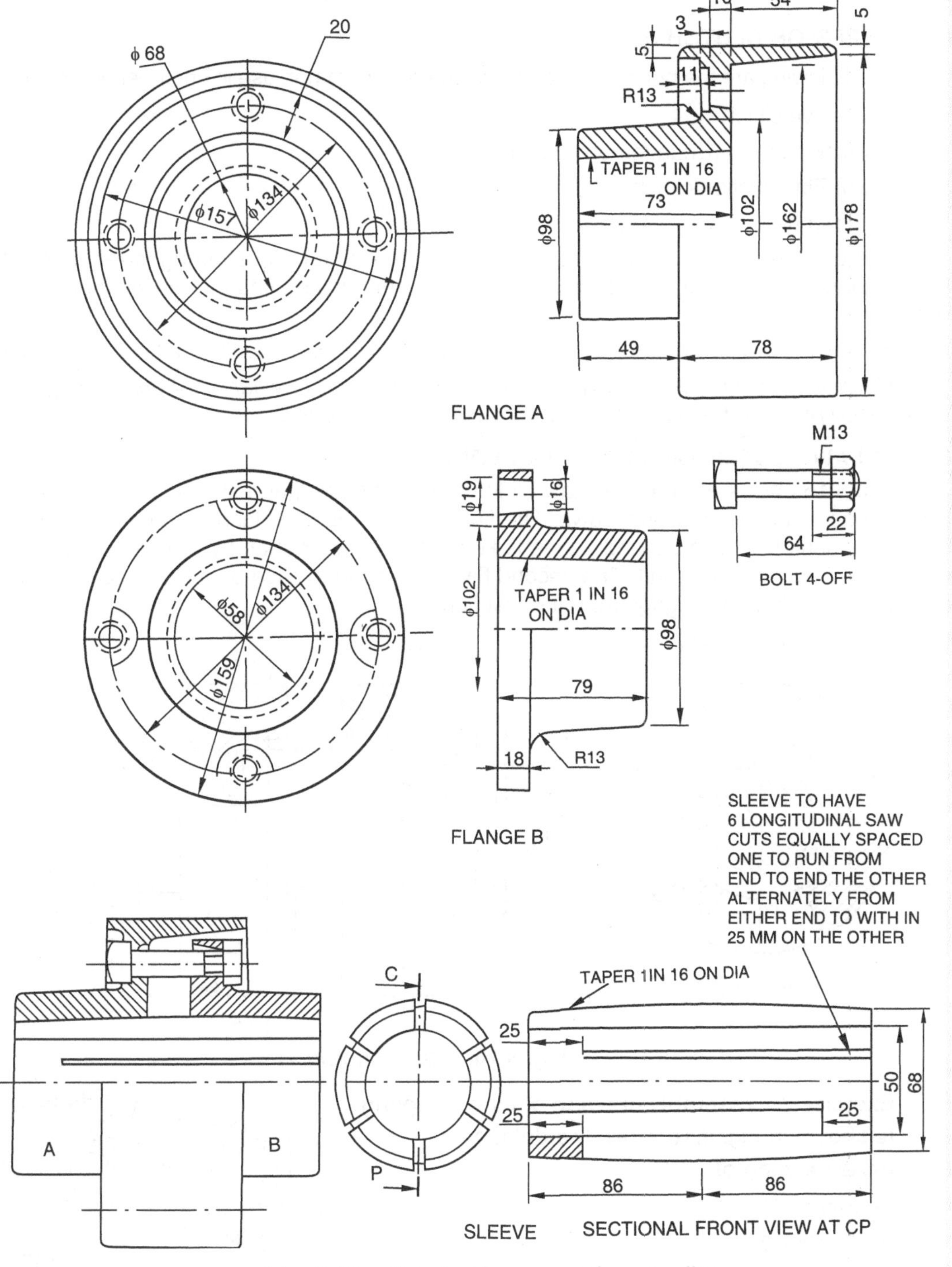

Fig. 8.15. *Details of compression coupling*

8.13. CLUTCHES

Clutch is a device used to transmit power from one shaft to another when the shafts are co-axial. The most important features of a clutch is that, the driven shaft may be cut off from the driving shaft without stopping the later at will. In case of an automobile, a clutch is provided between the engine and gear box. The transmission of power from engine shaft to gear box may be cut off at the moment while changing the gear.

8.14. TYPES OF CLUTCHES

The following are the types of clutches which are in common use in engineering practice:–

1. Positive clutch
2. Friction clutch (Engagement clutch)
   (*i*) Plate friction clutch.
   (*ii*) Conical friction clutch.

The positive clutch is used at such places where a sudden steering is required. The friction clutch is used to transmit motion gradually from driving shaft to driven shaft without shock.

Here we shall deal with two clutches (i) claw clutch or claw coupling (ii) conical clutche.

Note: *If the shaft starts slipping inside the sleeve, due to wear, tight grip can again be obtained by pressing the flanges together by tightening the bolts.*

8.15. CLAW CLUTCH OR CLAW COUPLING

Fig. 8.16 shows a three jaw claw clutch used for slow-speed shafts. It consists of two flanges having three projecting claws. The first flange is rigidly attached to end of one shaft by a taper sunk key. The second flange is keyed to the other shaft by a feather key so that it can slide freely on the shaft. The second flange is made to slide by a lever mechanism so as to engage with or disengage from the first flange.

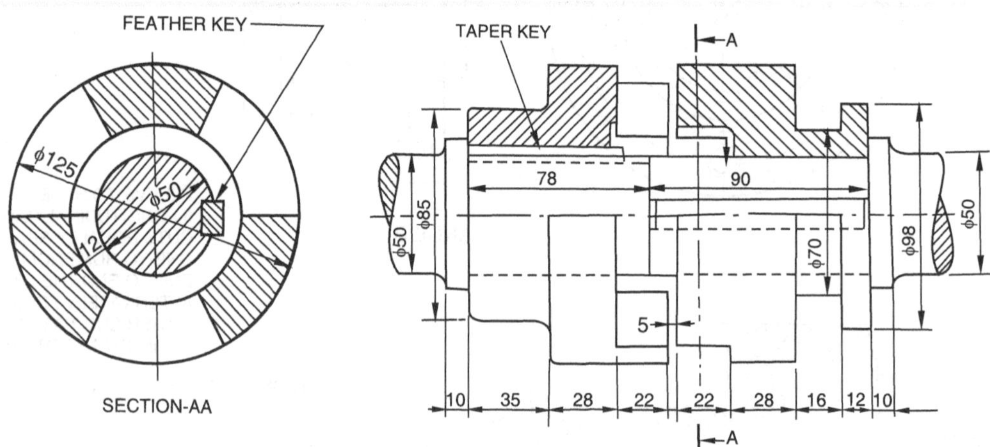

Fig. 8.16. *Claw coupling*

Exercise 7 : Assembly drawing of claw coupling is shown in Fig. 8.16. Draw the detailed parts drawing of :

(a) Flange '1'

(b) Flange '2'

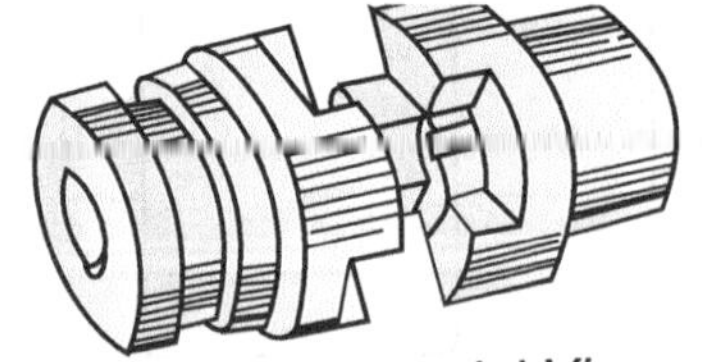

Pictorial View

8.16. CONICAL FRICTION CLUTCH

This clutch is based on the principle of friction. The motion from one shaft to another is transmitted by the friction between the engaging surfaces.

Fig. 8.17. show a simplest form of a conical friction clutch in the engaged position.

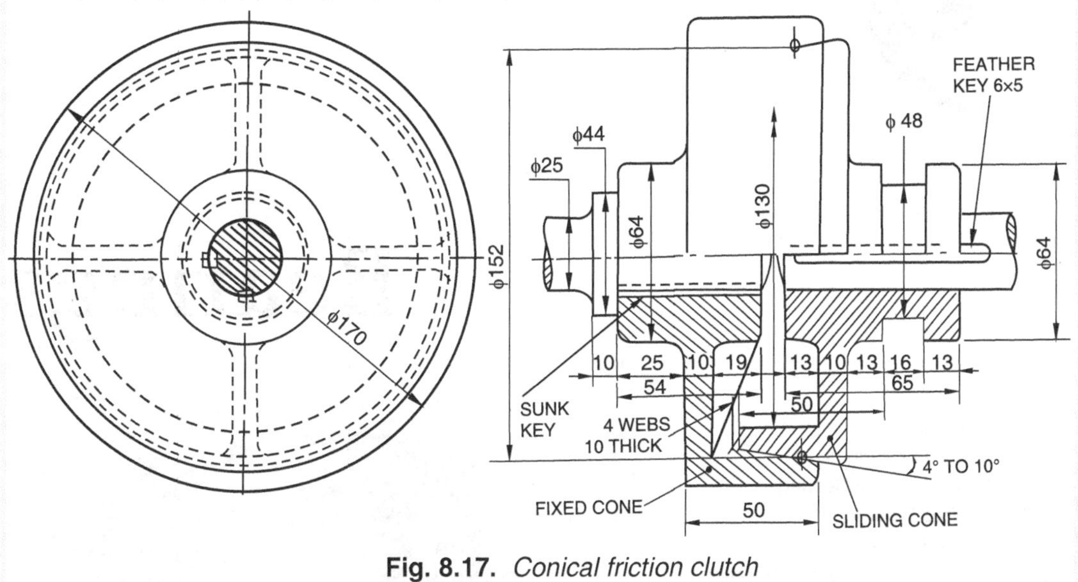

Fig. 8.17. *Conical friction clutch*

Exercise 8 : Fig. 8.17 shows the two views of a conical friction clutch. Draw the following views to full size scale :

(a) Front View – Full in section (b) Side view

QUESTIONS FOR SELF EXAMINATION

1. What do you understand by a coupling? Name the different types of couplings.
2. How do you differentiate between the following coupling :
   (*i*) a rigid and a flexible coupling;
   (*ii*) flange and box coupling; and
   (*iii*) lap type and split type of muff coupling?
3. What is flexible coupling ? What are its advantages?
4. What is the advantage of providing protective flanges?
5. What is the purpose of providing a recess in one of the flange and an extension in the other flange?
6. Match the statement of column A with corresponding ones in columns B.

Column A Type of coupling used	Column B Shafts whose axes are	Answer
(a) Flexible coupling	(i) Intersecting	(a)... (iv)
(b) Universal coupling	(ii) Parallel	(b)... (iii)
(c) Oldham coupling	(iii) not in alignment	(c)... (ii)
(d) Flange coupling	(iv) Coaxial, with slight misalignment	(d)... (i)

CHAPTER

9

Bearings

INTRODUCTION

In engineering field, we often come across such situations where we want to support shafts, generally of large length at several places. If the long shaft is supported only at the ends, the deflection at the centre due to the self weight of the shaft will be too much and the pulleys etc, if mounted on the shaft will not have a smooth circular motion. Therefore, it is necessary to have a straight long shaft throughout its length and for this, a long shaft is supported by bearings at the several places.

In this chapter, we shall study the bearings, types of bearings and their practical applications in engineering field.

9.1. BEARINGS

The devices which are used for supporting the long shafts are known as bearings.

The bearings are so named because the surface of the bearing which is in contact with the shaft is subjected to the bearing load. The contact surfaces of the bearings may wear out due to friction of the shaft. To overcome this difficulty, gun-metal bushes are provided between the contact surfaces. These bushes can be replaced when worn out too much.

Also, due to relative motion between the shafts and the bearing surfaces, there is always a frictional force and eventually the heat is produced. To minimise the frictional force and cool down the bearing, lubricant such as oil or grease is introduced between the contact surfaces and this avoids actual contact of the surfaces and prevent undue heating.

9.2. TYPES OF BEARINGS

Bearings are of different types depending upon the construction and direction of load acting upon them, but following are the three main categories of bearings :—

1. Journal Bearings
2. Foot-step Bearings
3. Thrust or collar Bearings.

9.2.1. Journal Bearings : These bearings are used for supporting the horizontal shafts and when the shafts rotate slowly. Journal bearings may be solid or split up types.

Fig. 9.1 (i) shows a simple journal bearing and Fig. 9.1 (ii) shows a simple journal bearing with a gun metal bush.

9.2.2. Foot-Step Bearing : This is also known as pivot bearing (see Fig. 9.2). Foot-step bearing supports the shaft parallel to its axis. In this bearing, the supporting pressure is parallel to the axis of the shaft. The shaft is kept in a vertical position and the end of which rests within the bearing.

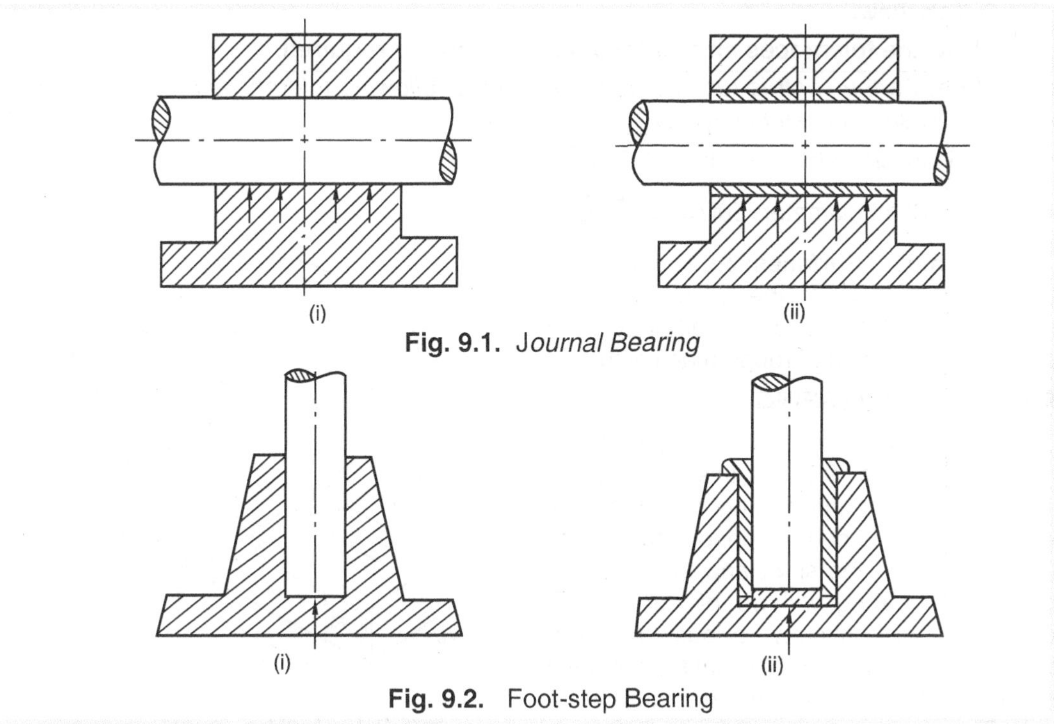

Fig. 9.1. *Journal Bearing*

Fig. 9.2. Foot-step Bearing

Fig. 9.2 *(i)* and *(ii)* show a simple and a bushed foot-step bearings.

9.2.3. Thrust or Collar bearing : The thrust bearing supports the shaft at right angle to its axis. In this bearing, the supporting pressure is also parallel to the axis of the shaft and the collars take the end pressure of the shaft. Fig. 9.3 shows a thrust bearing.

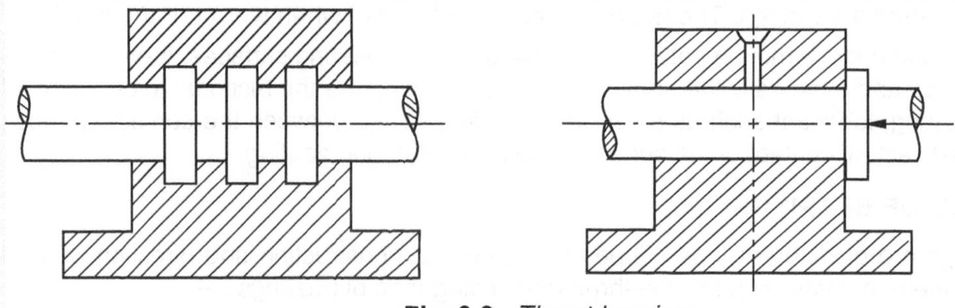

Fig. 9.3. *Thrust bearing*

9.3. SIMPLE JOURNAL BEARING

The simplest form of journal bearing is shown in Fig. 9.4. It consists of a solid casting. The upper portion of the casting is of circular shape, in which a hole is provided, for supporting and for giving free movement to the shaft. The lower portion of the casting is of rectangular shape in which two holes are made for bolting down the bearing. An oil hole is also provided at the top of the casting for lubrication.

Exercise 1 : An isometric view of a simple journal bearing is shown in Fig. 9.4. Draw to a scale 1:1 the following views :—

1. Front view, right half in section.

2. End view, left half in section.

3. Top view.

Show the tolerances and surface finish wherever required.

Note: *As the shaft is running fit, therefore the following tolerances are suggested, for 50 hole and shaft assembly =H8/f7.*

i.e. Hole diameter = 50 H8= $50\,^{+0.39}_{+0.00}$

Shaft diameter=50f7= $50\,^{+0.39}_{+0.00}$

Machining Surfaces: ϕ 50 hole surface and lower surface of the base should be machined to H8 tolerance. All other surfaces should be machined to rough machining.

9.4. BUSHED BEARING

It is modified form of a simple journal bearing (see Fig. 9.5). In this bearing, the circular hole for supporting the shaft will wear out after a certain period due to the friction between the shaft and the hole, and the whole bearing will be of no use. To overcome this difficulty, a bush made of brass, gun-metal or other antifriction alloy metal is provided between the shaft and the hole. The inside diameter of the bush is equal

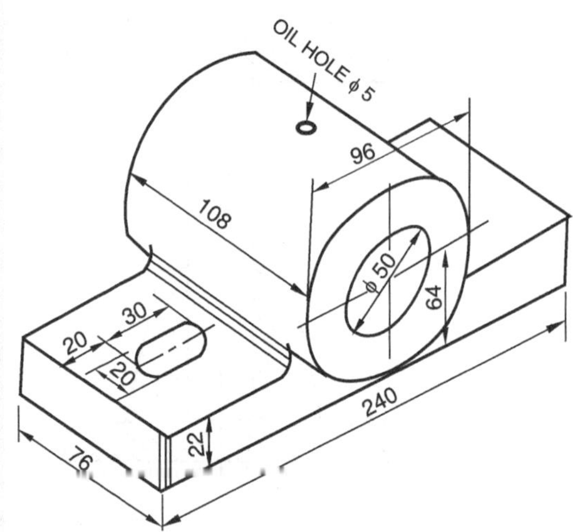

Fig. 9.4. *Simple Journal Bearing*

to the shaft diameter, having clearance for running fit. When the bush gets worn out, it is replaced by a new one.

A countersunk oil hole is drilled in the top of the body as well as through the bush thickness which leads up to the shaft for lubrication. The base of the bearing is kept hollow at the bottom to reduce the machining surface area. Two elliptical holes are made in the base for bolting down the bearing.

Notes: *1. In this bearing, the shaft is inserted endwise only. Hence, this beasing is generally placed only at or near the ends of the shafts.*

2. The bolt-holes are made in elliptical shape for adjusting the position of the bearing when bolting down.

3. Care should be taken that bush should not rotate with the shaft. The rotation of the bush is prevented by a set-screw which locks the bush in position or the bush is force fit in the body (Not shown).

Problem 1 : Fig. 9.5 shows an isometric view of a Bushed Bearing. Draw the following views :–

1. Front view, right half in section.

2. End view, left half in section.

3. Top view.

Use full size scale and show the tolerance finish wherever required.

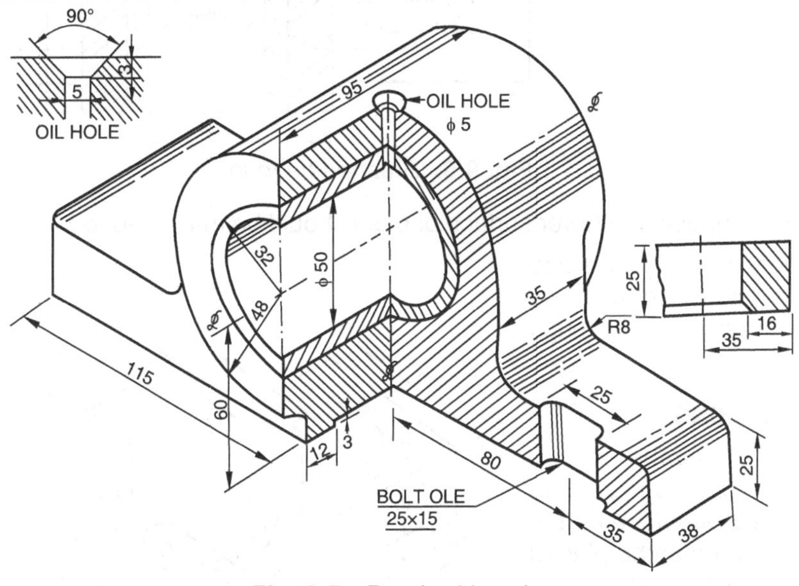

Fig. 9.5. *Brushed bearing*

Solution : For its solution, see Fig. 9.6. The following tolerances are suggested :—

For ϕ 50 bush and shaft assembly: H8/f7 and for 64 hole of body and bush assembly: 64H7/p6

Tolerance values obtained from table 2.4.

On bush diameter=50 H8= $50^{+.039}_{+.000}$ Hole of body=64 H7=64 $^{+.030}_{+.000}$

On shaft diameter=50 f7= $50^{-.025}_{-.050}$ Diameter of bush=64 p6=64 $^{+.051}_{+.032}$

The surface finish mark are shown in Fig. 9.6.

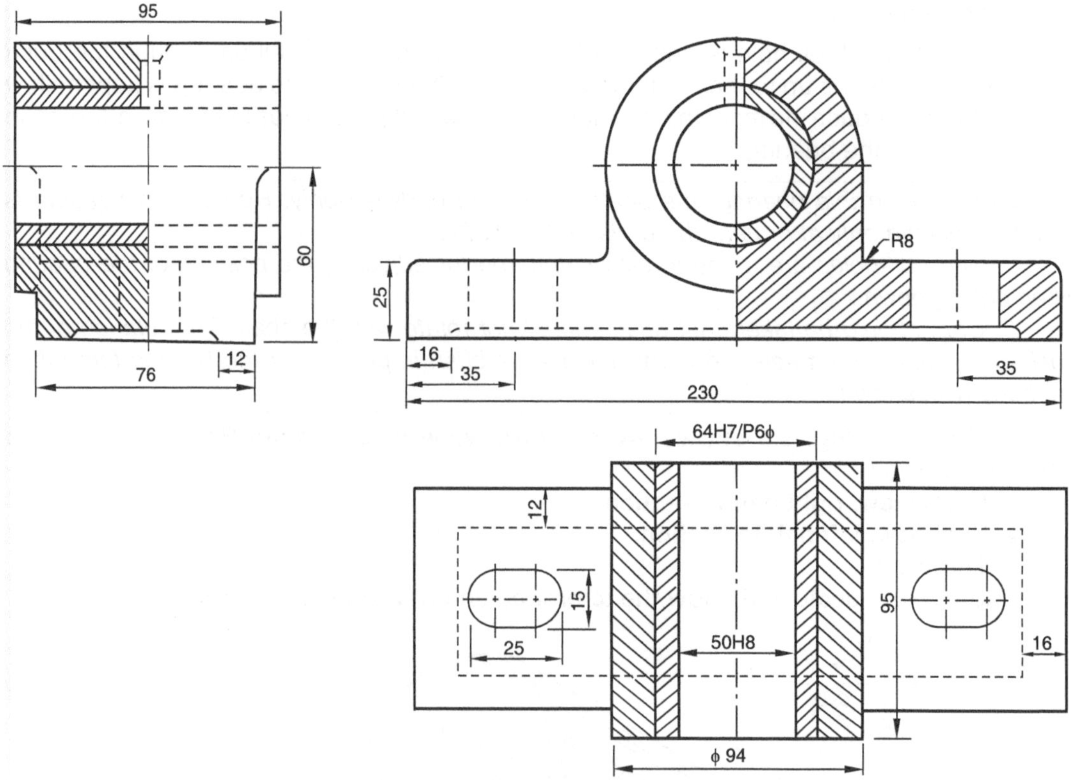

Fig. 9.6. *Brushed bearing*

i.e. 50 hole surface and lower surface of base should be machined to ∇.

64 surfaces for body hole and bush should be machined to ∇ and all other surfaces required rough machining.

9.5. OPEN BEARING

This type of bearing is used for very long or zig-zag types of shafts, e.g., crank shafts, etc. It is opened at the top and can be placed directly below the shaft at any desired position. This is provided with one half bush as shown in Fig. 9.7. The base of the bearing is provided with rectangular holes for adjustment and tightening the bearing. The sides of such bearing should be high enough so that the shaft does not fall out at high speed.

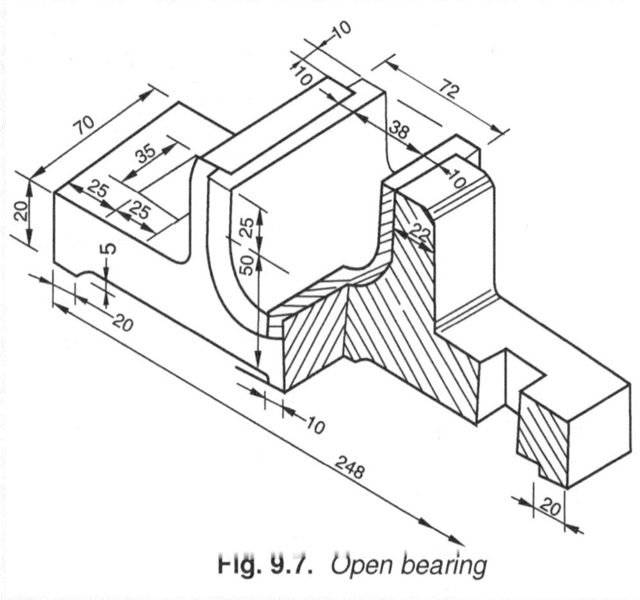

Fig. 9.7. *Open bearing*

Exercise 2 : Fig. 9.7 shows an isometric view of an Open Bearing. Draw the following views :—

1. Half sectional front view 2. Half sectional side view 3. Top view.

Show the tolerance and surface finish wherever required.

Hint : The following tolerances are suggested:

(1) For bush and shaft assembly : 38 H8/f7 [Normal running fit refer table 2-4]

(2) For bush and body assembly

On diameter: 58 H7/g6 [Sliding and location fit refer table 2-4]

On width side: 70 H7/g6

Exercise 3 : Fig. 9.8 shows two views of a simple bearing. Draw to a half full size scale the following views :—

(a) Half sectional front view (right half in section)

(b) End view projected from (a)

(c) Top view.

Show the tolerances and mark the surface finish wherever required.

Hint: *Tolerance values obtained from table 2.4 :*

Shaft and bush assembly=100 H8/f7 Bush and body assembly=140 H7/p6

$100\ H8 = 100^{+.054}_{+.000}$ $140\ H7 = 140^{+.040}_{+.000}$

$100\ f7 = 100^{-.036}_{-.071}$ $140\ p6 = 140^{+.043}_{+.068}$

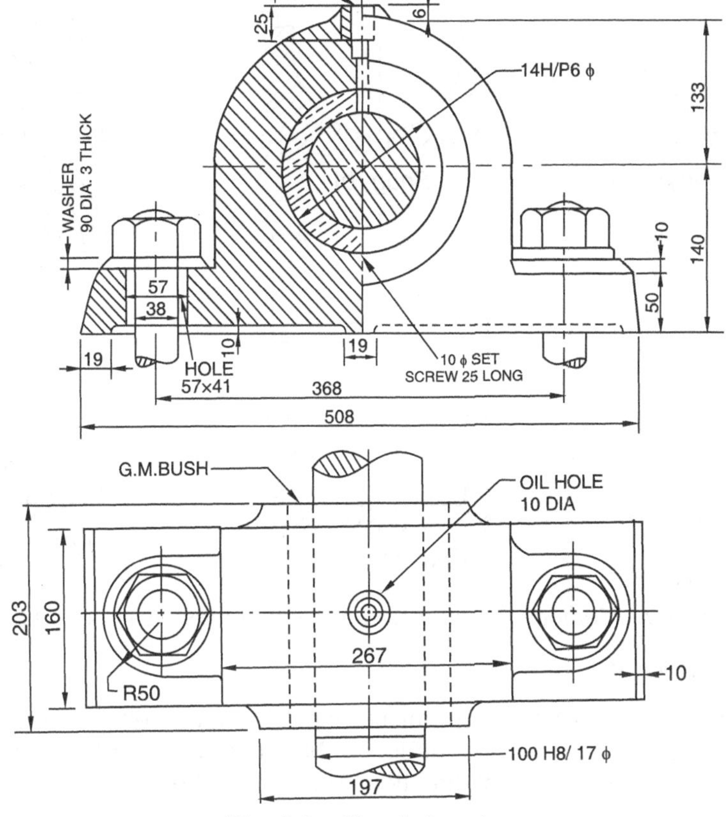

Fig. 9.8. *Simple bearing*

9.6. METHODS OF PREVENTING ROTATION OF BRASSES IN A BEARING

Following methods are used in practice to present the rotation of brasses along with the shaft in the bearing :—

1. By making the steps *octagonal near the flanges on the brasses,* which fits in the corresponding holes made in the bearing [see Fig. 9.9 (a)].

2. By providing a *snug on the outer surface of the brasses* which fits into a corresponding hole in the casting [see Fig. 9.9 (b)].

3. By providing *lugs or strips at the sides of the brasses* which fit into corresponding gaps made in the bearing [see Fig. 9.9 (c)].

4. By making the *steps rectangular near the flanges* which go into the corresponding shape made in the inner side of the block of the bearing [see Fig. 9.9 (d)].

Sometimes a set screw, half in the bush and half in the block of the bearing is used to prevent the rotation of brasses.

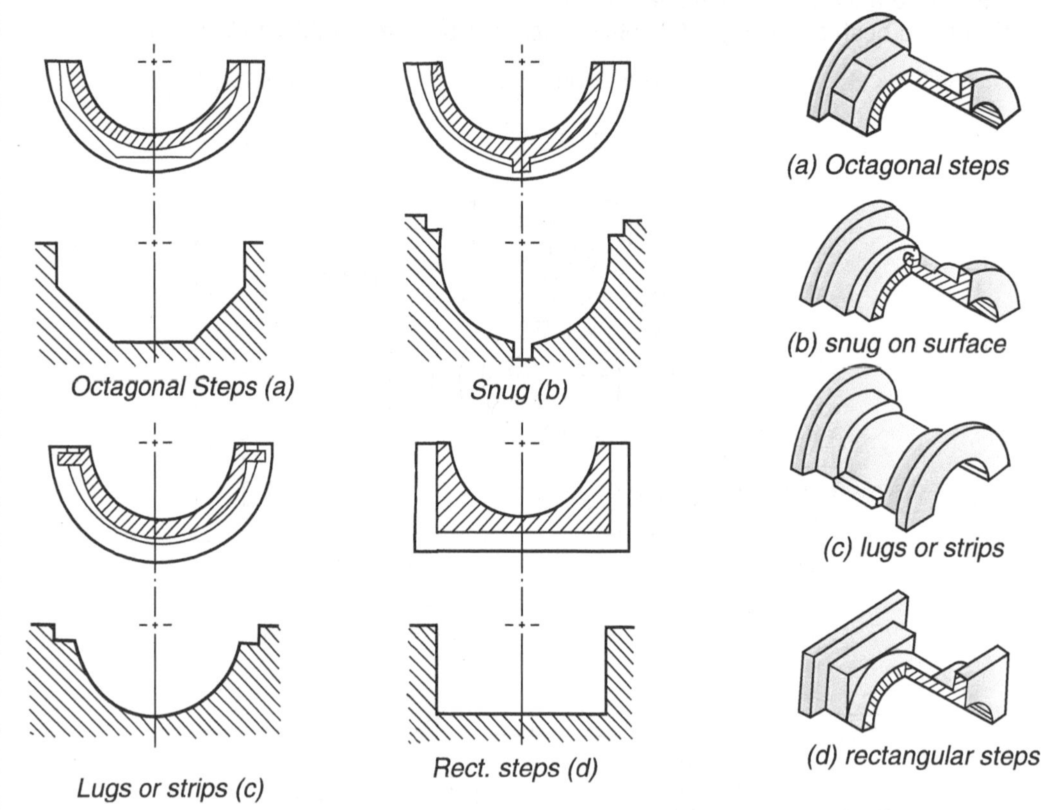

Fig. 9.9. *Methods of preventing rotation of brasses in bearings*

9.7. PLUMMER BLOCK

Plummer block is a split type of journal bearing. It is used when the speed of the rotating shaft is too high and the load varies in direction. Moreover it is installed where it is difficult to introduce the shaft from the axial direction. This type of bearing consists of :

1. A cast iron block with a base (called body)
2. Gun-metal brasses, made in two halves
3. A cast iron cap

4. Two mild steel square headed bolts
5. Two sets of hexagonal lock nuts

The body of plummer block has a semi-circular horizontal hole in which lower part of the brasses is placed. There are two elliptical holes in the base for bolting down the bearing with the support. Also, there are two holes for the square headed bolts and a small hole for fixing the snug of the lower brass. The cap has also almost semi-circular horizontal hole for the upper brass, two holes and an oil hole for lubrication purposes. The upper brass also provides an oil hole, which coincide with the cap oil hole.

In this bearing, the two halves brasses are held firmly between the body and the cap by means of square headed bolts and nuts.

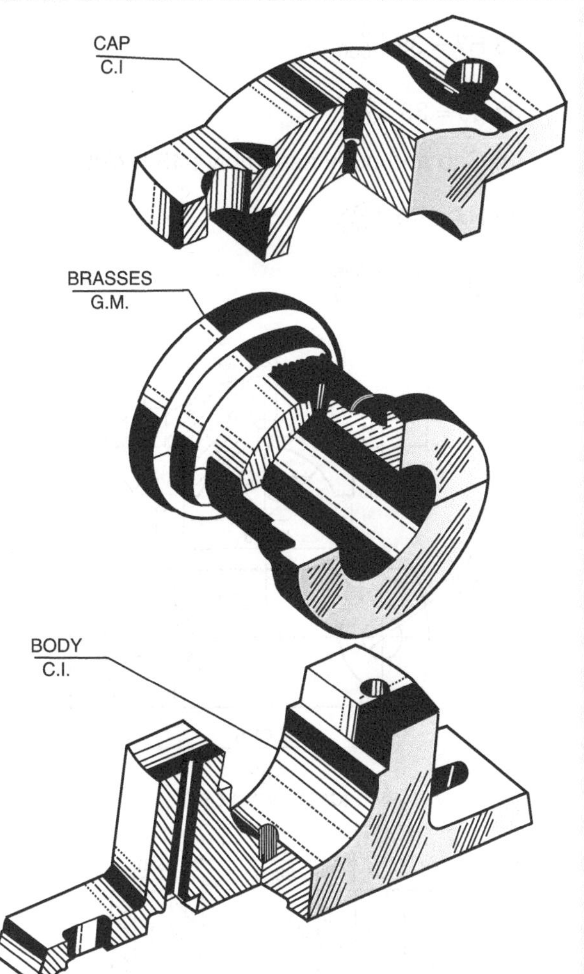

Fig. 9.10. *Isometric views of plummer block parts*

Notes: *(i) Brasses are made in two halves for easy fitting, removal and replacing.*

(ii) A gap is left between the cap for future adjustment of brasses in assembly (see Fig. 9.12). After a certain period of running the shaft, the brasses becomes elliptical in shape from inside. The elliptical shape will increase the vibration of the shaft. Then, the two brasses can be cut off at their edges to give them again a round shape from inside when assembled again. Thus, the clearance (or gap) decreases and ultimately reduces to zero. After this, the brasses are replaced by new ones.

(iii) Axial movement of the brasses is prevented by providing flanges (collars) on both sides of the brasses.

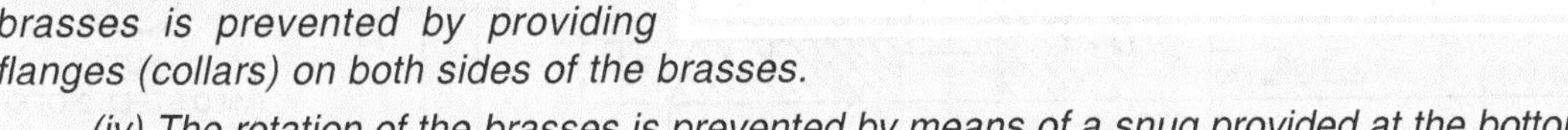

(iv) The rotation of the brasses is prevented by means of a snug provided at the bottom of lower brass, which fits into the corresponding hole in the body.

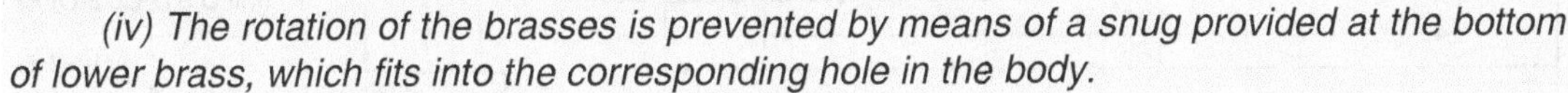

(v) Curved spaces are cut on the cap for proper replacement and rotation of the nuts.

Problem 2 : Fig. 9.11 (i) shows the details of a Plummer Block. Imagine the parts assembled together and draw to a full size the following views :—

(a) Half sectional front view. (b) Half sectional end view. (C) Top view.

Show the tolerances and mark surface finish wherever required. Give the Material List also.

Solution : For its solution, refer Fig. 9.11. (ii).

The tolerance values obtained from the table 2.4 are as under:–

(1) Bush and shaft assembly: H8/f7

Inside diameter of bush: $25\ H8 = 25^{+.033}_{+.000}$ Shaft diameter: $25f7 = 25^{-.021}_{-.041}$

(2) Bush and body assembly : 38 H 7/g6 on diameter 36 H7/g6 on width

$38\ H7 = 38^{+.025}_{+.000}$ $36H7 = 36^{+.025}_{+.000}$

$38\ g6 = 38^{+.009}_{+.025}$ $36g6 = 36^{+.009}_{+.025}$

(3) Cap and body assembly : 45 H7/g6 (4) Snug and body assembly : 6H7/h6

$45\ H7 = 45^{+.009}_{+.000}$ $6H7 = 6^{+.012}_{+.000}$

$45\ g6 = 45^{+.005}_{+.025}$ $6h\ 6 = 6^{+.000}_{+.008}$

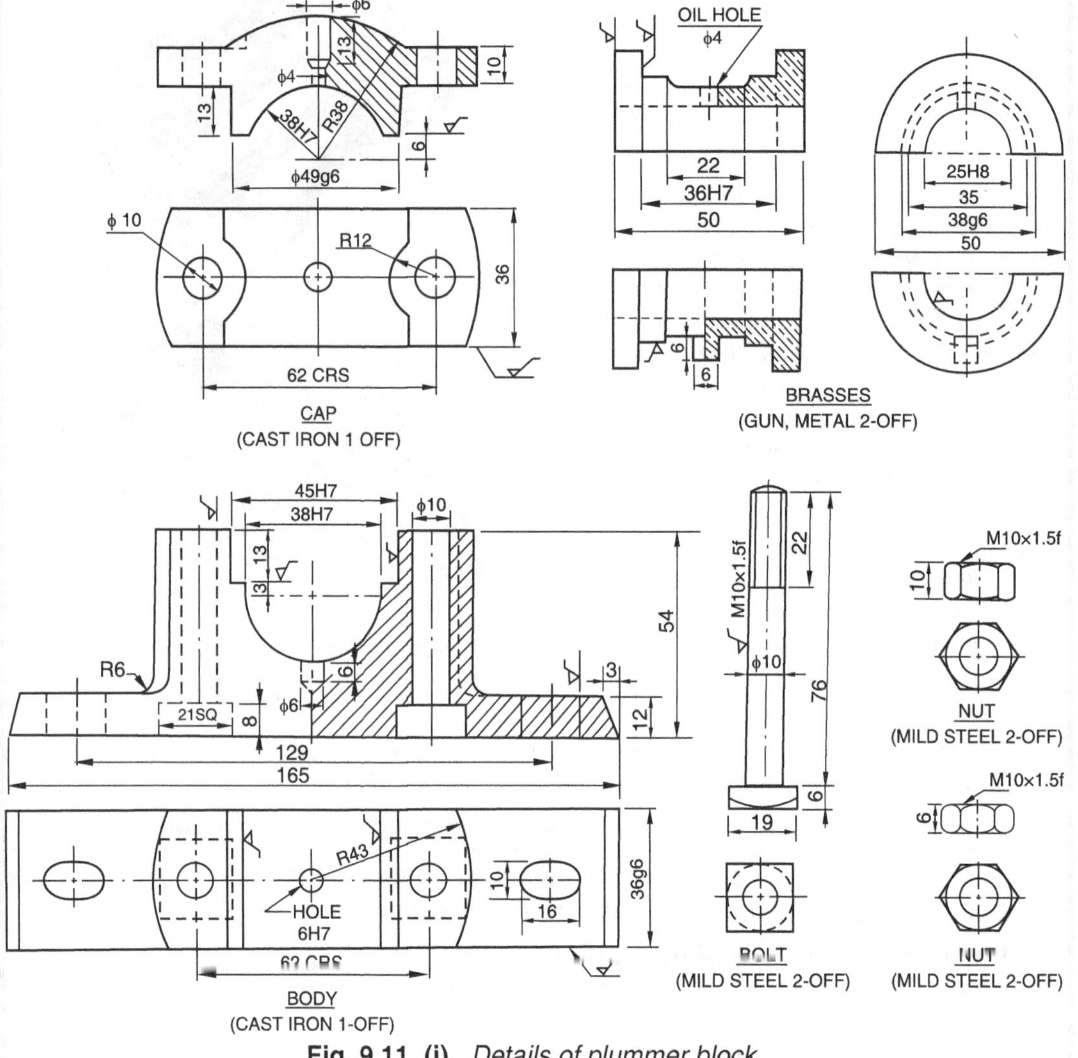

Fig. 9.11. (i) *Details of plummer block*

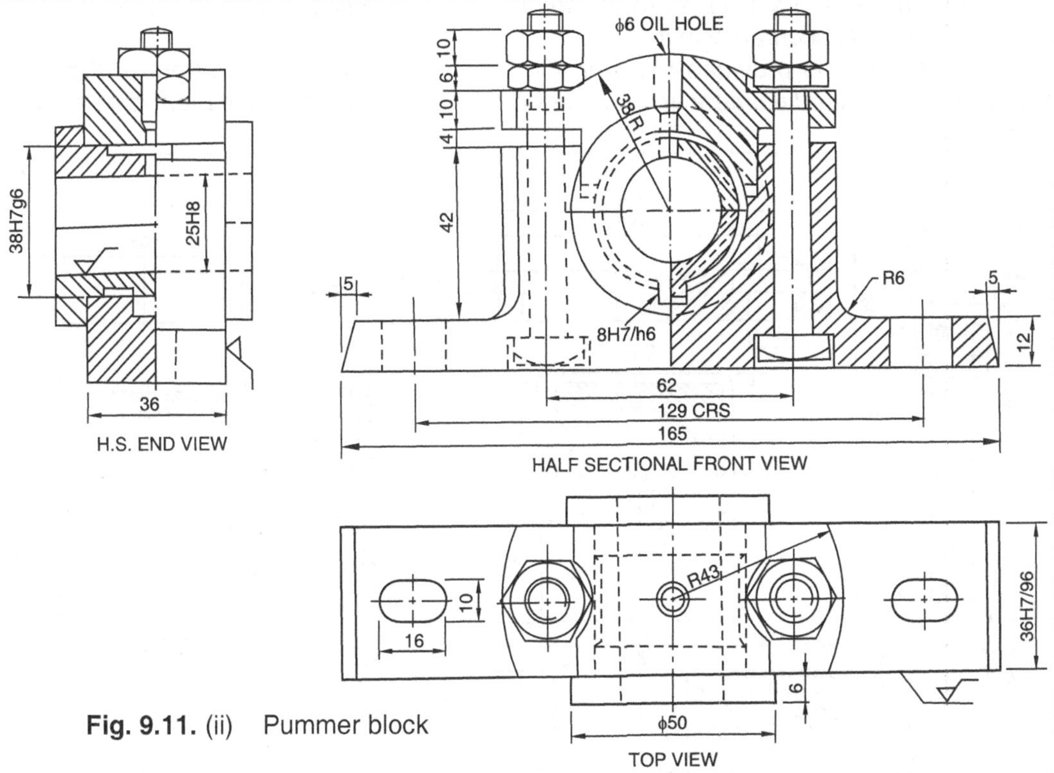

H.S. END VIEW

HALF SECTIONAL FRONT VIEW

TOP VIEW

Fig. 9.11. (ii) Pummer block

9.8. MARINE ENGINE SHAFT BEARING

This type of bearing is used for supporting the marine engine shaft. Its function is similar to the plummer block.

In this bearing the casting is a part of the engine frame and is of U-shape. The top and bottom brasses are held firmly in between the cap and casting with the help of stud bolts and nuts (see Fig. 9.12).

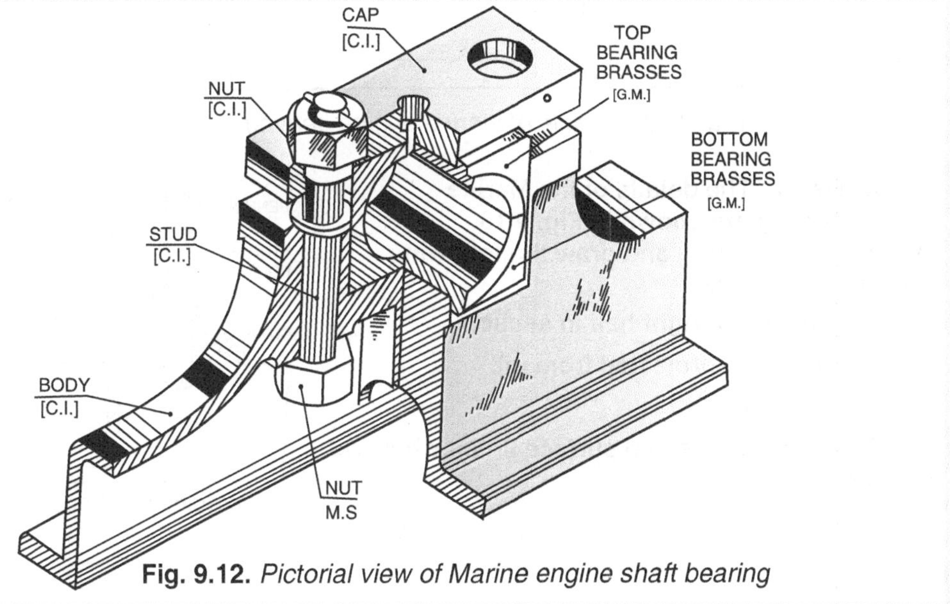

Fig. 9.12. *Pictorial view of Marine engine shaft bearing*

Note : *(i) Square saw cuts are provided in the rectangular hole of the casting so that the brasses may fit accurately in it.*

(ii) Clearance is provided between the top of the casting and bottom of the cap in assembly for future adjustment and repairs of the brasses so as to give them always round surface inside.

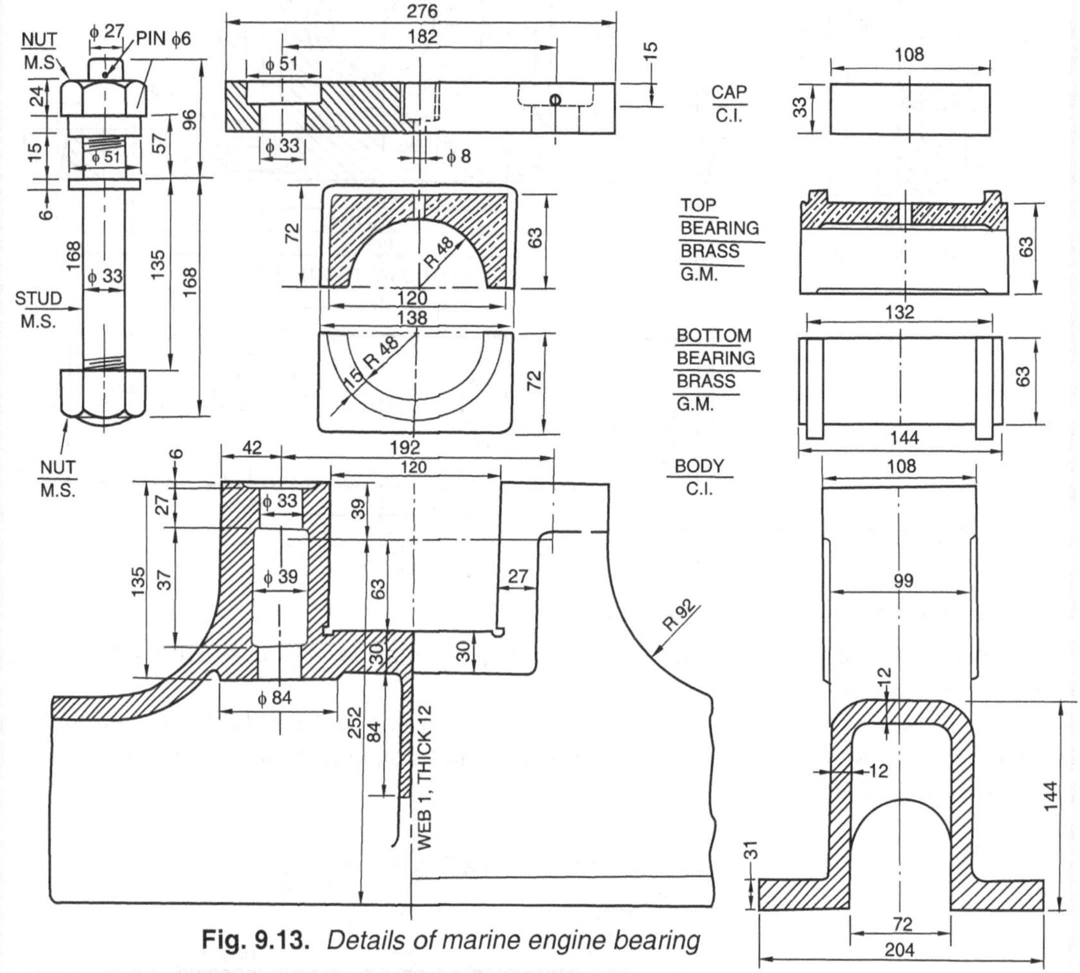

Fig. 9.13. *Details of marine engine bearing*

Exercise 4 : The details of the Marine Engine Bearing are shown in Fig. 9.13. Assemble the bearing and draw the following views to half full size:

(a) Front view, right half in section.

(b) End view, projected from (a).

(c) Top view.

Show tolerances and surface finish wherever required.

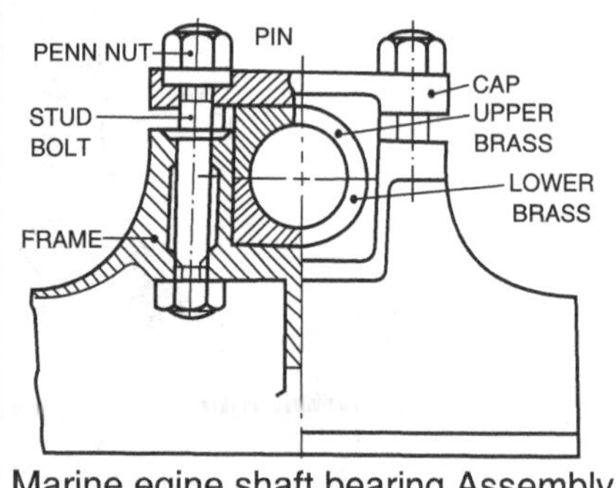

Marine egine shaft bearing Assembly

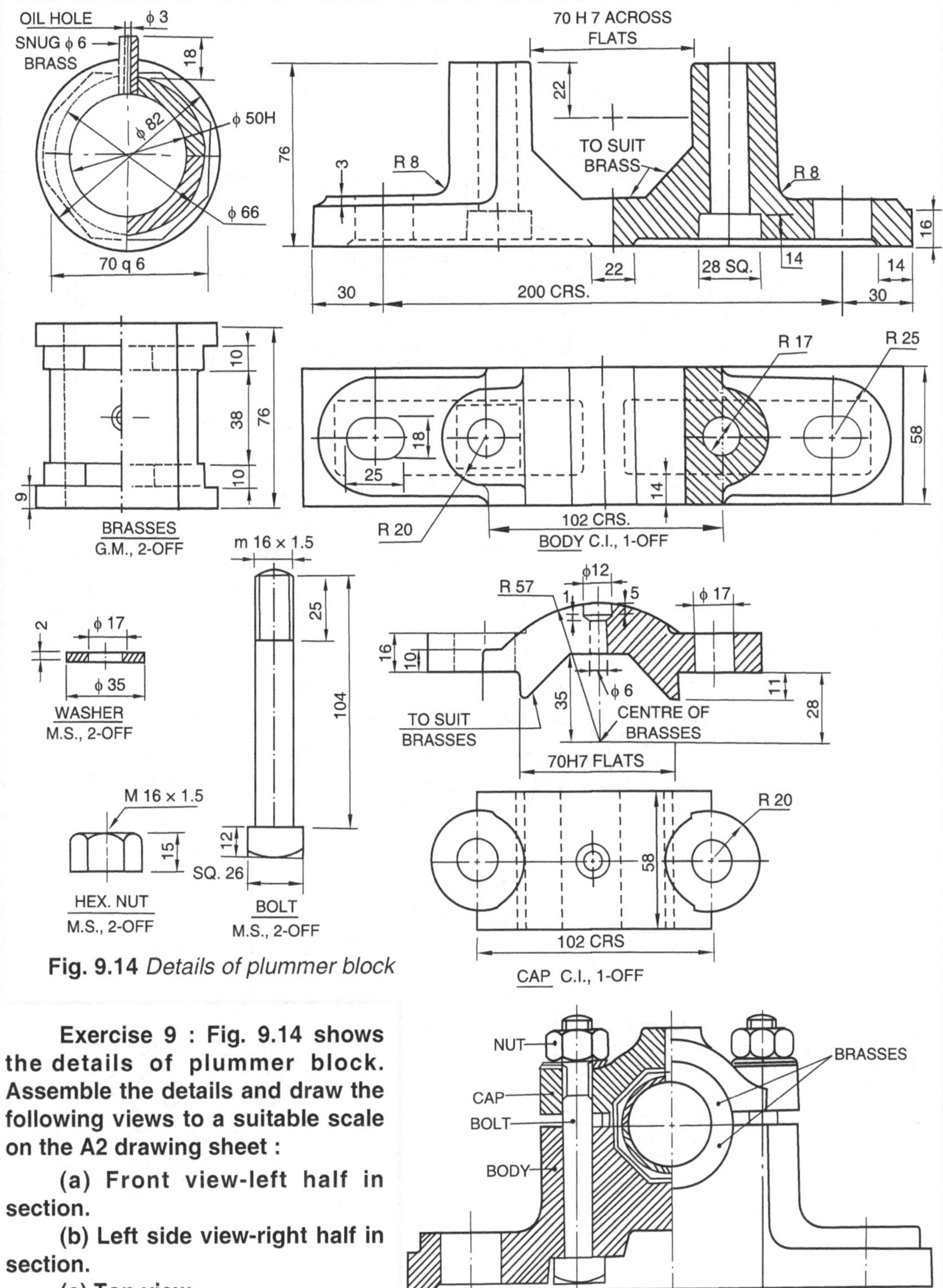

Fig. 9.14 *Details of plummer block*

Exercise 9 : Fig. 9.14 shows the details of plummer block. Assemble the details and draw the following views to a suitable scale on the A2 drawing sheet :

(a) Front view-left half in section.

(b) Left side view-right half in section.

(c) Top view.

Show the tolerance and mark surface finish wherever required. Give materials list also.

General arrangement of the assembled pedestal bearing for Fig. 9.15.

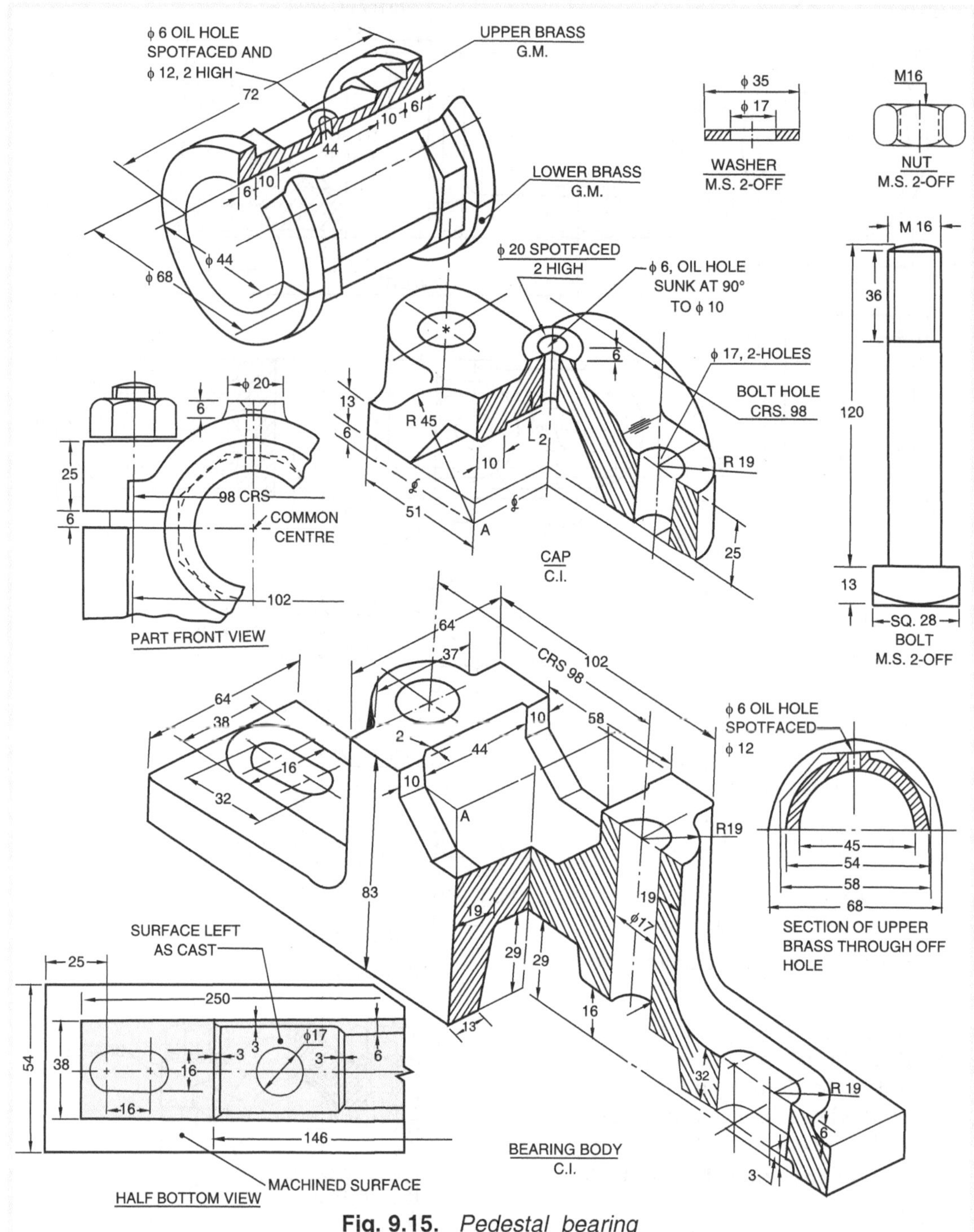

Fig. 9.15. *Pedestal bearing*

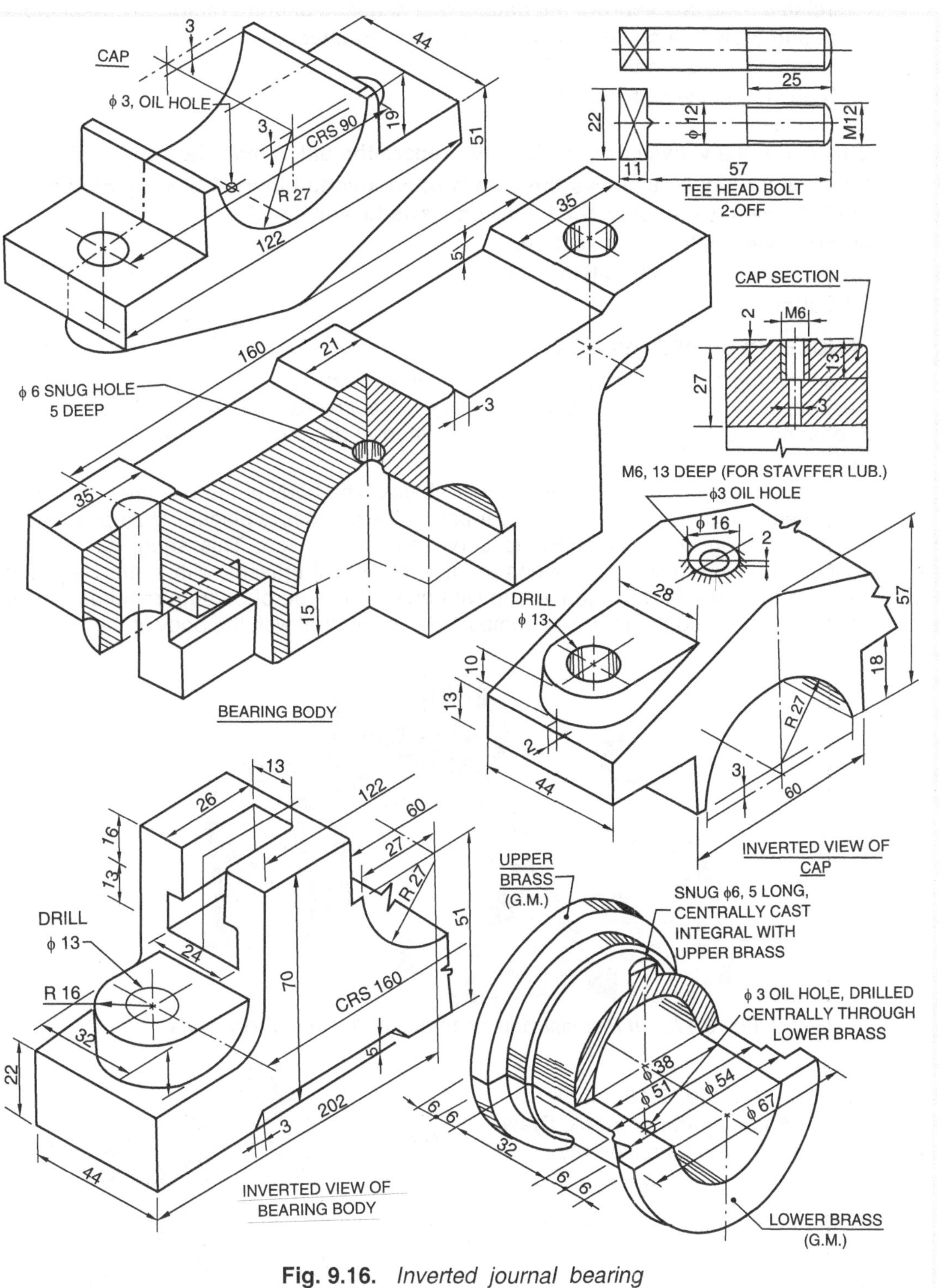

Fig. 9.16. *Inverted journal bearing*

Exercise 8 : Fig. 9.15 shows the details of a pedestal bearing pictorially. Assemble the details and draw to a suitable scale on A2 drawing sheet :—

(a) Front view–right half in section.

(b) Side view projected from (a).

(c) Top view.

Dimension the views completely, and prepare the bill of materials and parts list.

Exercise 9 : Fig. 9.16 shows the details of an inverted journal bearing pictorially. Assemble the details and draw the following views to a scale 1:2:–

(a) Front view–left half in section.

(b) Left side view–right half in section.

(c) Bottom view.

9.9. ANGLE PLUMMER BLOCK

When the top surface of the body of a plummer block is made inclined to horizontal plane (usually at 30°) the block is known as angle plummer block. The main advantage of this type of bearing is that the power can be transmitted at an angle and shaft can be easily removed.

Fig. 9.17 (i) and (ii) show angle plummer block. The body of the plummer block has an inclined semi-circular hole in which lower part of the brasses is placed. The upper brass has an integral snug which is drilled to serve an oil hole too. This snug fits into the corresponding cavity provided in the cap to stop rotation of brasses with shaft. Here, the studs replace the bolts for holding the cap with nuts. This type of bearing requires simple amount of lubrication, so the waste oil sumps are cast integral with the body of the bearing.

Fig. 9.17. *(i) Pictorial view of an Angle Plummer Block parts*

Exercise 10 : Fig. 9.18 show the details of an Angle Plummer block. Draw the following views of its assembly :

(a) Front view–right half in section.

(b) Side view–right half in section.

(c) Top view.

Prepare a parts list and bill of materials.

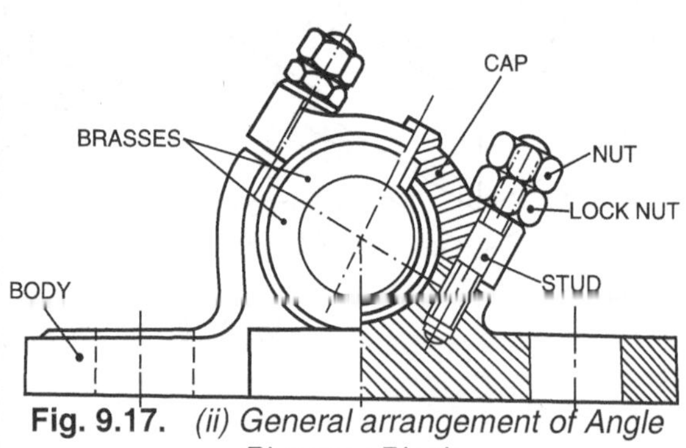

Fig. 9.17. *(ii) General arrangement of Angle Plummer Block*

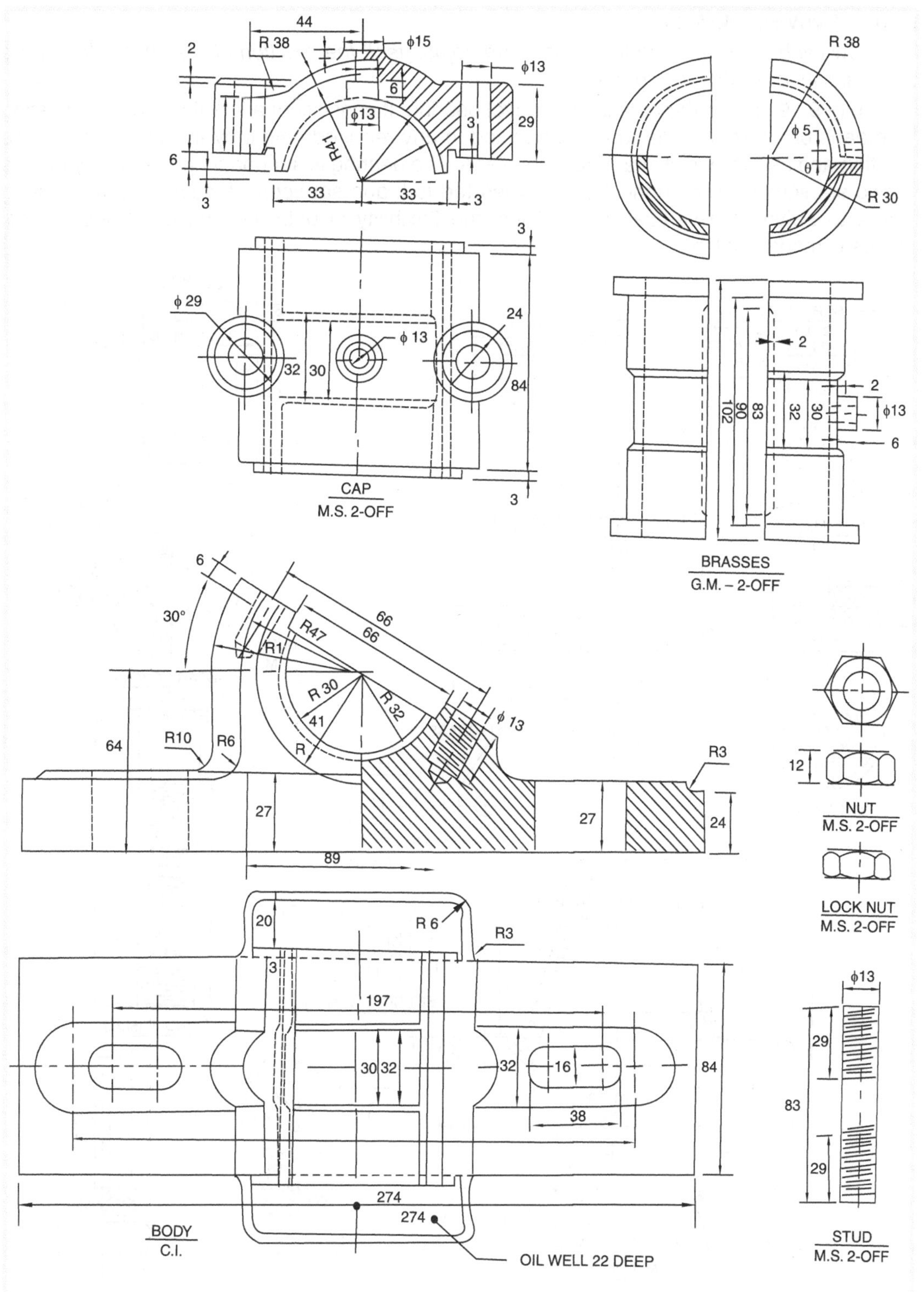

Fig. 9.18. *Details of an angle plummer block*

9.10. SWIVEL BEARING

Swivel bearing is used to support and adjust itself the position of the shaft when there is possibility of slight misalignment.

Fig. 9.19. shows the details of a swivel bearing. It consists of a fork fitted into the casting by means of a spindle. The fork can be fixed to any required height by adjusting the spindle position and free to swivel in a horizontal plane. The bearing is supported in the fork by means of two set screws. The connection between the fork and set screw is such that the bearing is free to swivel in a vertical plane. Thus, the flexibility in both the horizontal and vertical planes is made available.

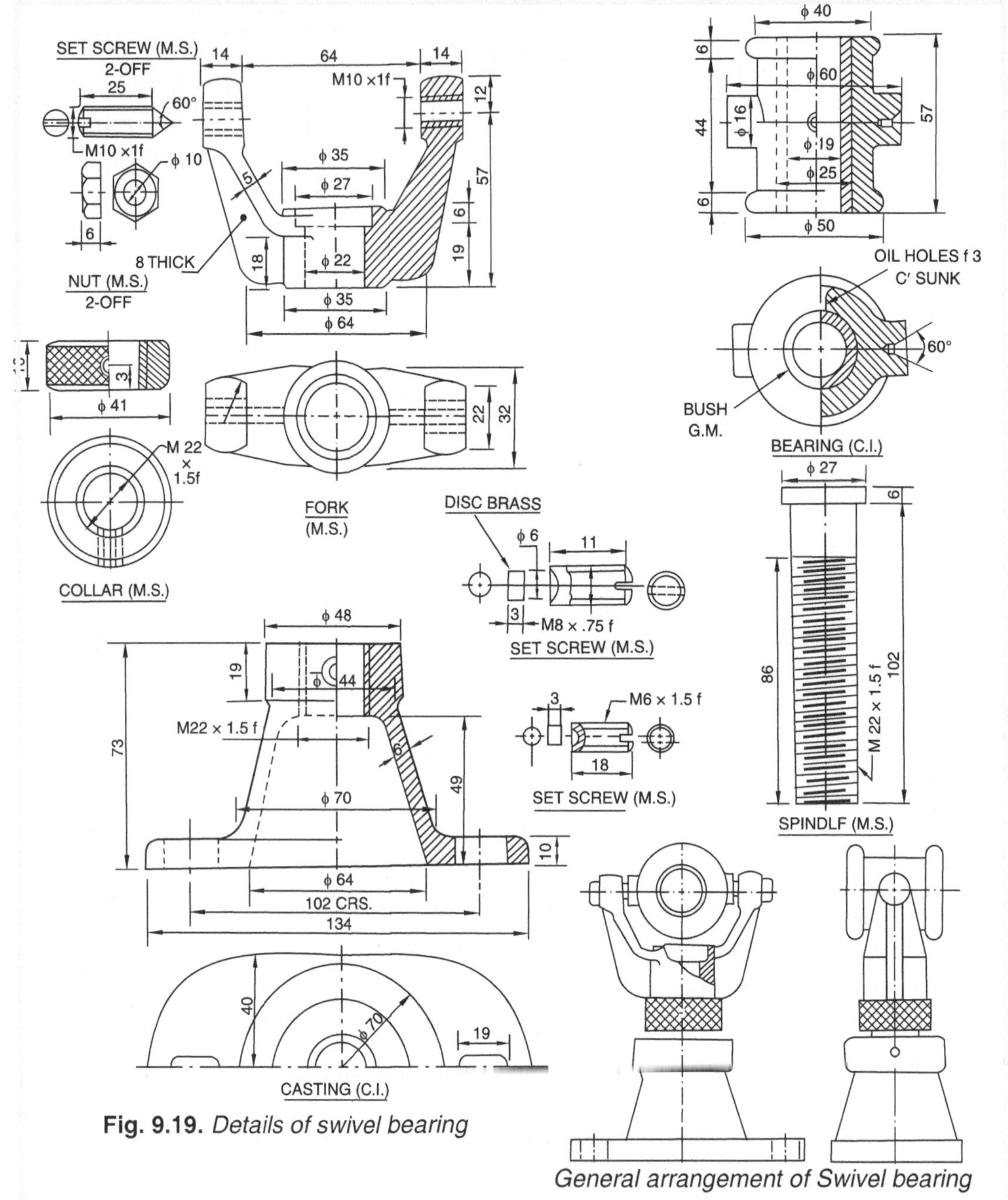

Fig. 9.19. *Details of swivel bearing*

General arrangement of Swivel bearing

Perfect alignment of shaft can be obtained by screw height and side adjustment. After the required adjustments have been made to suit the shaft position, these are locked. A lock nut with a set screw passing through it and another set screw passing through the casting are used to lock the height adjustment.

Note **:** (i) *Hexagonal nuts are used on the side adjustment set screws to lock them in position.*

(ii) *The tips of set screws tighten upon brass disc to prevent damage to the screw spindle.*

Exercise 12: Fig. 9.19 shows the details of a swivel bearing. Assemble the parts and draw the following views to a scale full size :—

(a) Front view-right half in section.

(b) End view-left half in section.

(c) Top view-outside.

Prepare a parts list and bill of materials

9.11. FOOT-STEP BEARING

Foot-step bearing is used for supporting the lower end of vertical shaft. It consists of a cast iron circular block with a base, in which a gun metal collared bush and disc are fitted. The base of the block is of rectangular shape having four elliptical holes for bolting down the bearing with the support. Circular facings are provided on the top of the holes which serve the purpose of washers.

The rotation of the disc may be prevented by fitting a pin, half in the disc and half in the body, away from the centre (see Fig. 9.23). The bush is prevented from rotation by providing a snug at the neck just below collar (not shown in the figure) or it is force fitted in the circular body. The bush and the disc can be easily replaced by new one's when worn out due to wear.

Notes: *(i) The edges of the bush where the shaft just enters it, are rounded so that the shaft can be easily inserted and removed in the bush without wearing its edges. The round edges of the bush also serve the purpose of an oil cup for lubricating the bearing.*

(ii) A gap between the upper surface of the base and lower edge of the bush is provided for an expansion due to heat and to save the material.

(iii) A gap is provided between the body and the bush, in a small length in order to reduce the machining surface area and also serve as a cup for lubrication in the case of big size bearings.

(iv) The base of the bearing is recessed at the bottom to reduce the machine surface area.

Practical Application : Vertical shafts which are supported by such bearings are not used for transmission of power, but they often occur in machine practice for paper making and pulping, in swing bridges, in textile machinery etc.

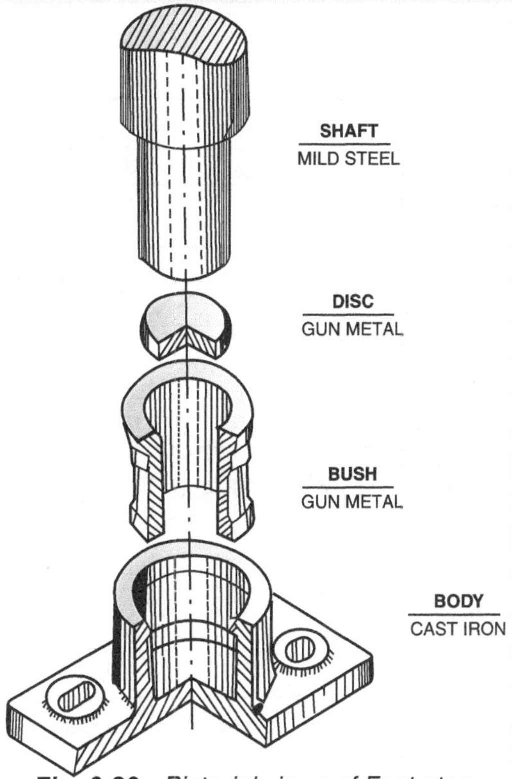

Fig. 9.20. *Pictorial views of Foot-step bearing parts*

Problem 3: Fig. 9.21 shows the details of foot-step bearing. Imagine the parts assembled together and draw to a full size scale the following views :—

(a) Front view–full section.

(b) Top view.

Show the tolerances and prepare the material list.

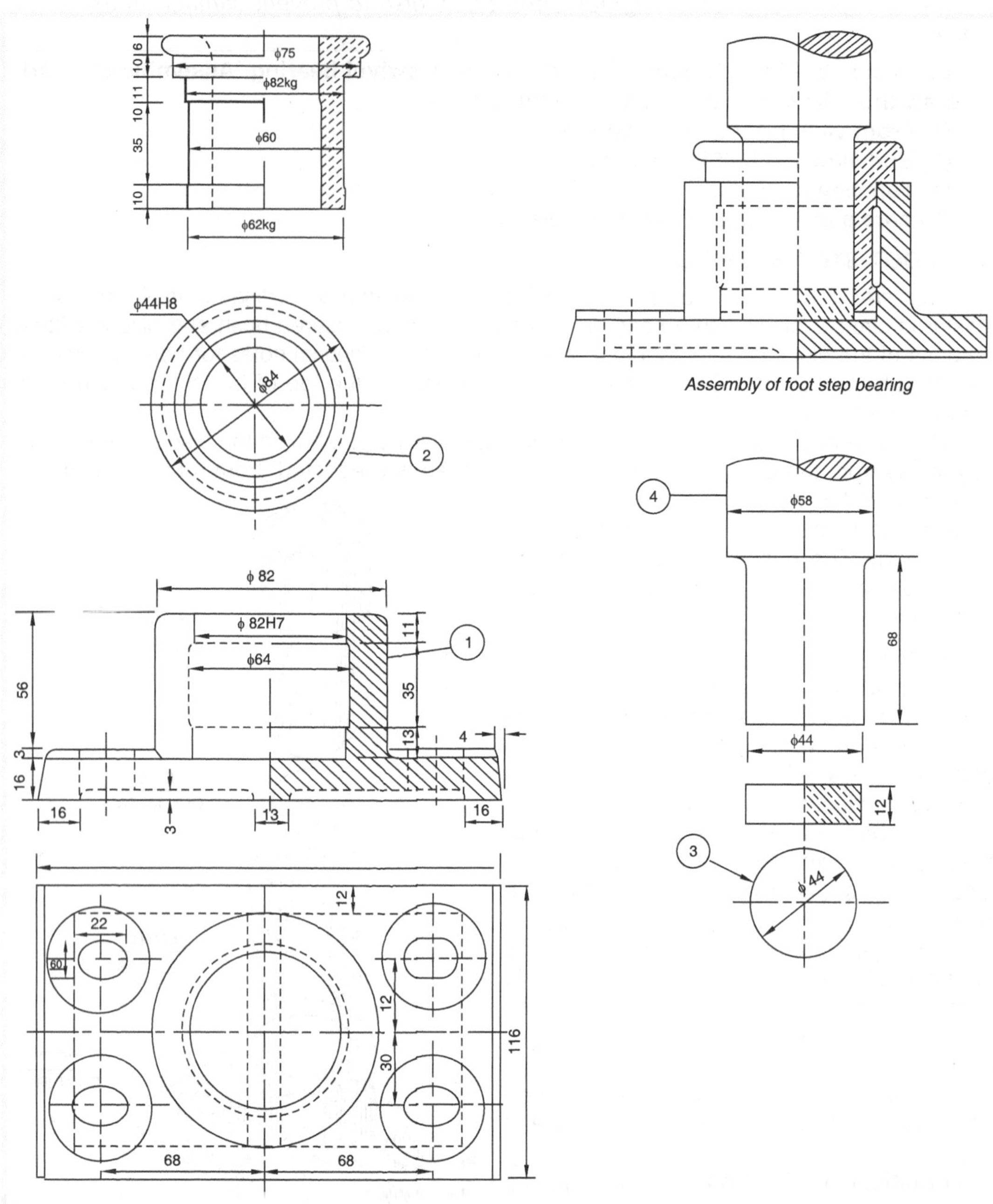

Fig. 9.21. *Details of foot step bearing*

Solution : For its solution, refer Fig. 9.22.

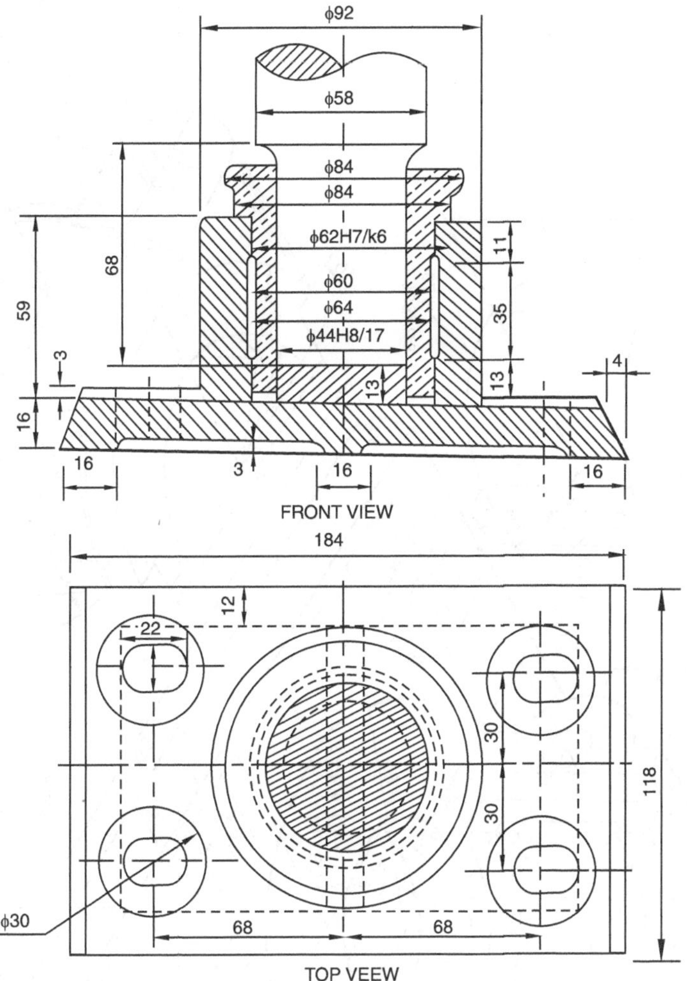

Fig. 9.22. *Foot-step Bearing*

Tolerance for body and bush assembly: 62 H7/k6 [Refer table 2–4]

$$62\ H7 = 62\ ^{+.030}_{+.000} \qquad 62\ k6 = 62\ ^{+.002}_{+.002}$$

Tolerance for bush and shaft assembly : 44 H8/f7

$$44\ H8 = 44\ ^{+.039}_{+.000} \qquad 44\ f7 = 44\ ^{-.025}_{+.050}$$

Exercise 13: Fig. 9.23 shows pictorial views of a foot-step bearing. Draw to convenient scale the following views:–

(a) Half sectional Front view.

(b) Side view.

(c) Top view.

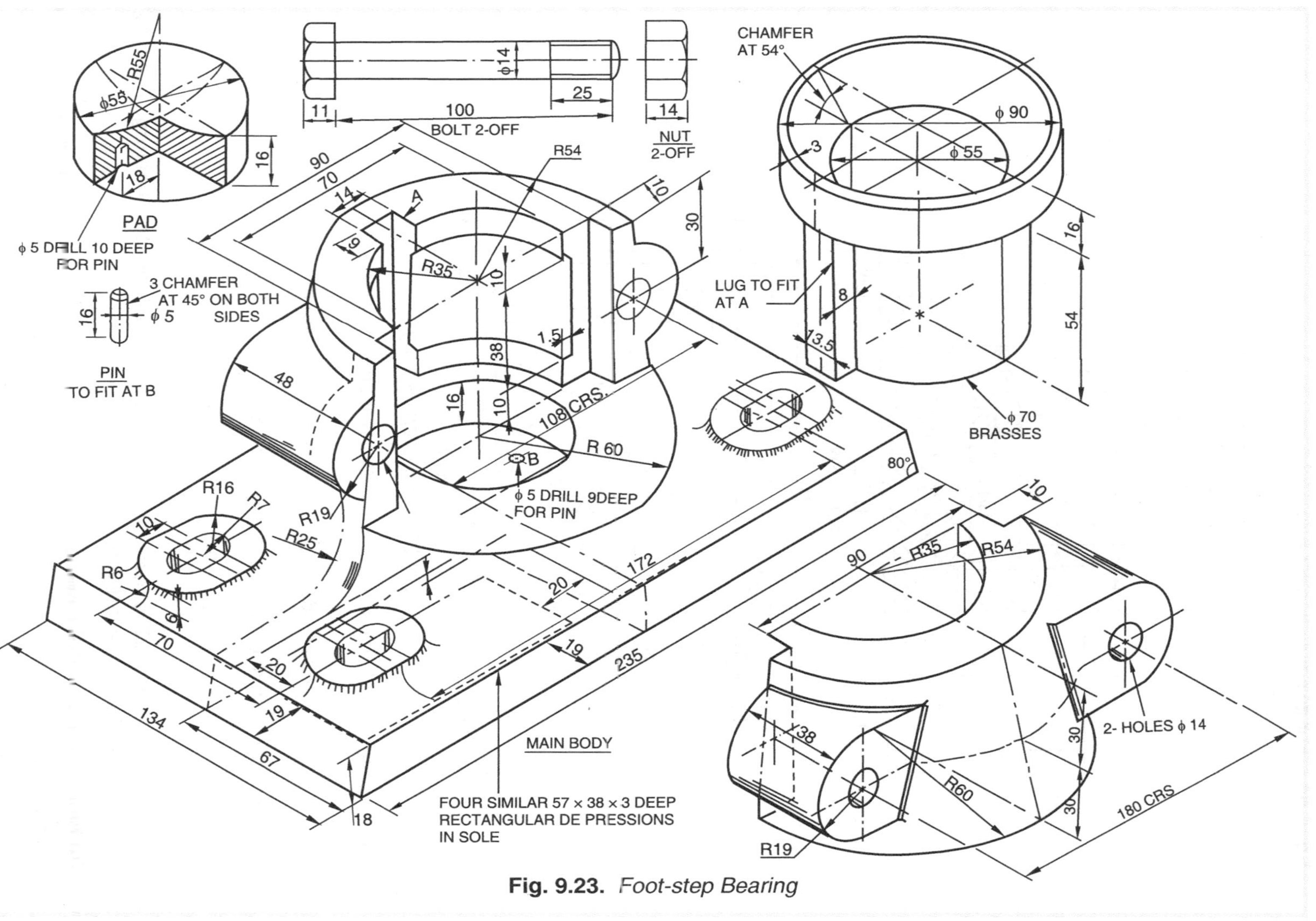

Fig. 9.23. *Foot-step Bearing*

Exercise 14: Fig. 9.24 shows pictorial view of a Foot-step bearing. Draw to a convenient scale the following views :—

(a) Half sectional front view.

(b) End view.

(c) Top view.

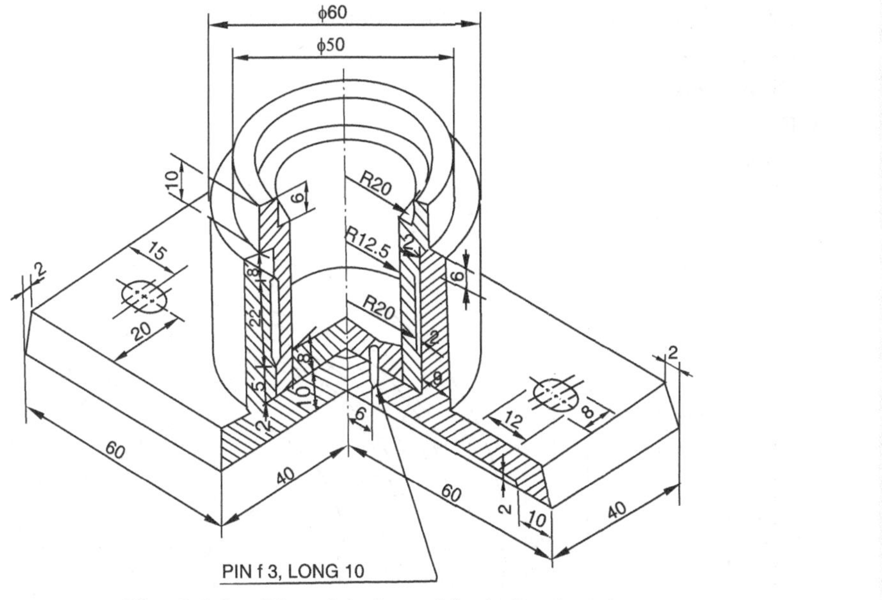

Fig. 9.24. *Pictorial view of foot step bearing*

Exercise 15: Fig. 9.25 shows two views of a thrust bearing. Draw to full size scale the following views :—

(a) Front view.

(b) End view.

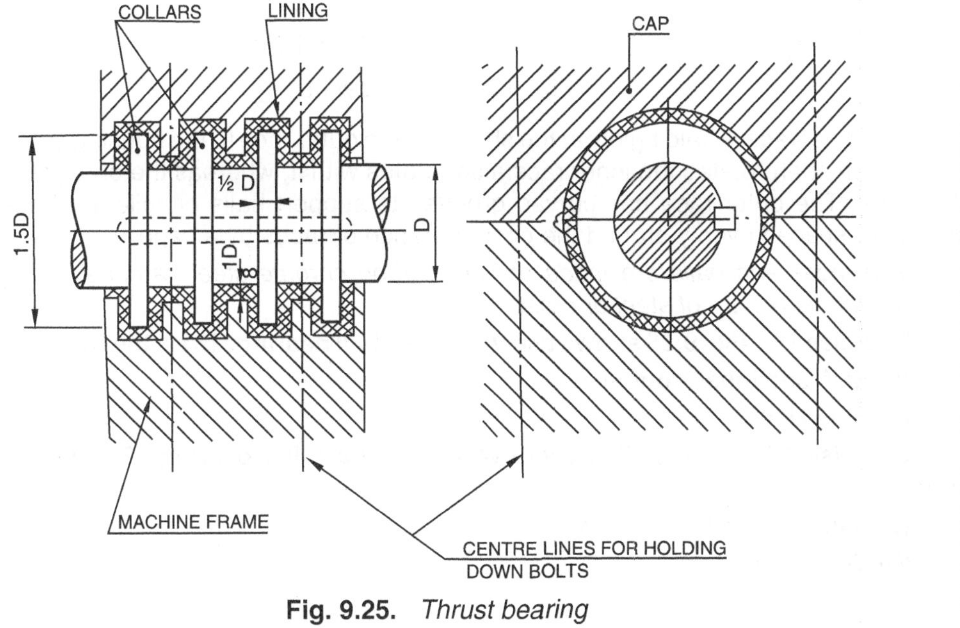

Fig. 9.25. *Thrust bearing*

9.12. BALL AND ROLLAR BEARINGS

The ball and roller bearings are used to minimise the rolling friction. These bearings require small space and much smaller quantity of lubricant as the wear and tear is practically negligible (see Fig. 9.26 and 9.27).

Ball and rollar bearings are used on large and fast moving machines such as electric motors, automobiles, workshop machines, household appliances etc.

The following are the four parts to any ball or roller bearings :—

1. Inner race
2. Outer race
3. Balls or Rollers
4. Cage.

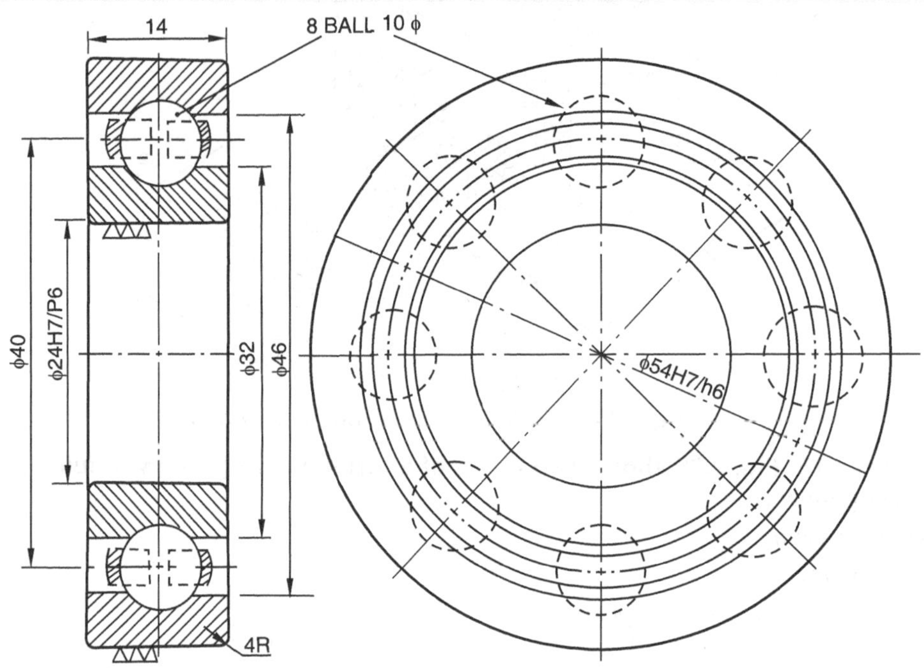

Fig. 9.26. *Ball Bearing*

The inner race is held rigidly to the shaft and the outer race to the housing of the bearing. Thus, when shaft rotates the inner race also rotates with it, whereas the outer race remains stationary. The outer race has curved grooves to support balls or rollers. These balls or rollers are spaced evenly around the circumference of races by a cage.

The races and balls are made of high carbon chrome steel hardened and polished, whereas the cages are of steel or brass.

Exercise 16 : Fig. 9.26 shows two views of a ball bearing. Draw the following views:-

(a) Half sectional front view.

(b) Side view.

Exercise 17 : Fig. 9.27 shows two views of a rollar bearing. Draw the following views:-

(a) Half sectional front view.

(b) Side view.

Show the tolerances and surface finish wherever required.

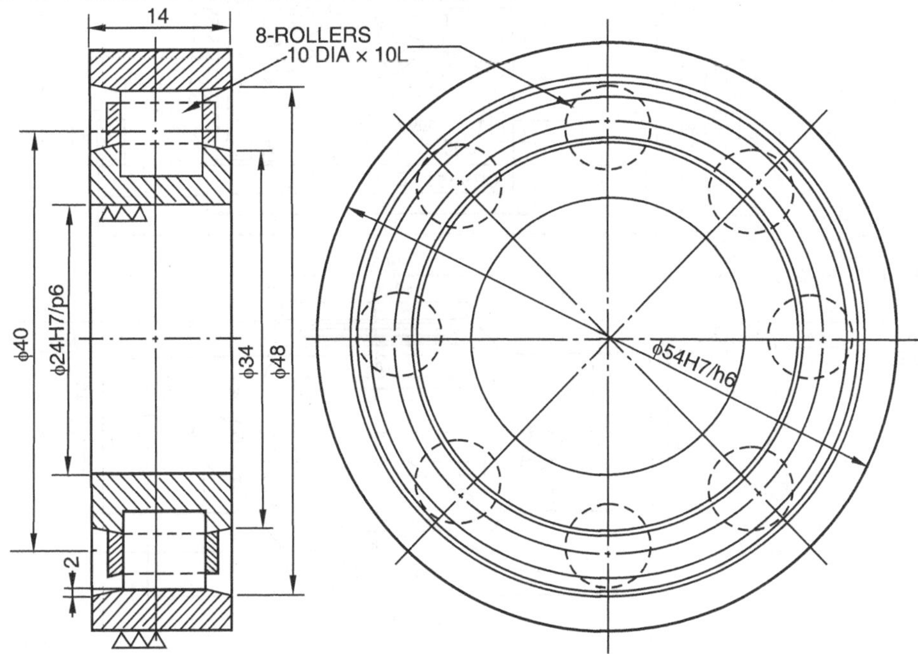

Fig. 9.27. *Roller bearing*

9.15. SELF-ALIGNING BEARING

The bearing which is used to support a horizontal shaft which gets its alignment from the bearing itself is known as self aligning bearing.

Fig. 9.28 shows details of a self aligning bearing. Here, in its body there is a double row ball bearing in which the shaft rotates. Two felt pads, one placed on each side of the ball bearing in the body to allow any fluctuation to the shaft as well as to retain the lubricant.

The tapering adopter sleeve is tightened upon the shaft by a slotted nut which is locked by a tab-locking washer.

Exercise 18 : Fig. 9.28 shows the details of a self-aligning bearing. Assemble the parts and draw to a scale 1:2 the following views :—

(1) Front view–right half in section.

(2) Side view–full in section

(3) Top view–outside.

Show the tolerances and surface finish wherever required and prepare a list of materials.

Exercise 19: Fig. 9.29 shows sectional front views and side view of φ 60 shaft ball bearing. Draw the following views in scale I : 1 :—

(a) Upper half sectional front view and side view of ball (i.e. outer race, inner race and ball cage assembled)

(b) Half sectional front view and side view of the housing

(c) One sectional view of the cover plate

Show the tolerances and surface finish wherever required.

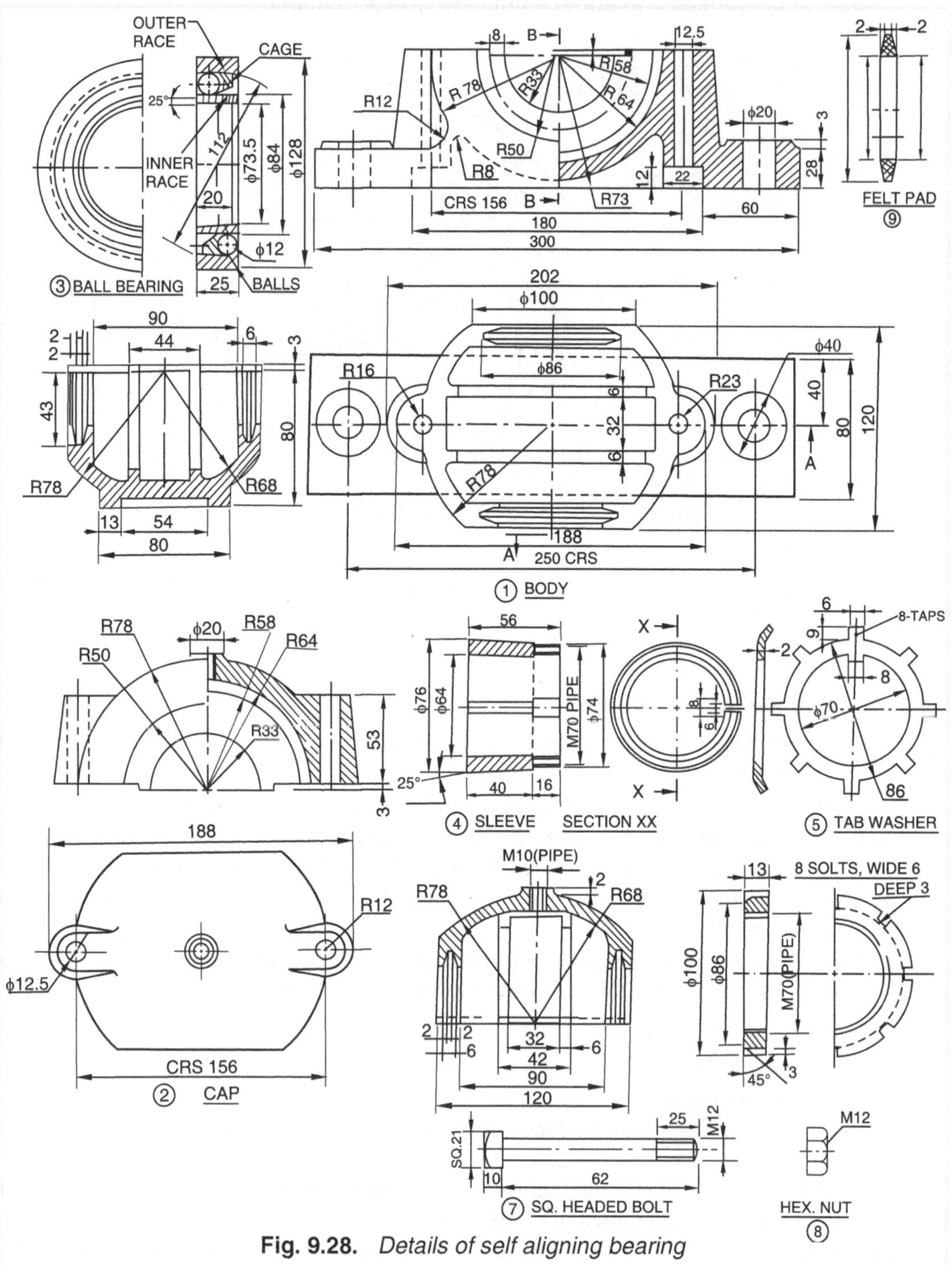

Fig. 9.28. *Details of self aligning bearing*

Exercise 20 : Fig. 9.28 shows two views of a self-aligning bearing. Draw the following views to a scale 1:2 of the following views:–

(a) Front view–right half in section

(b) Side view–full in section (c) Top view–out side view

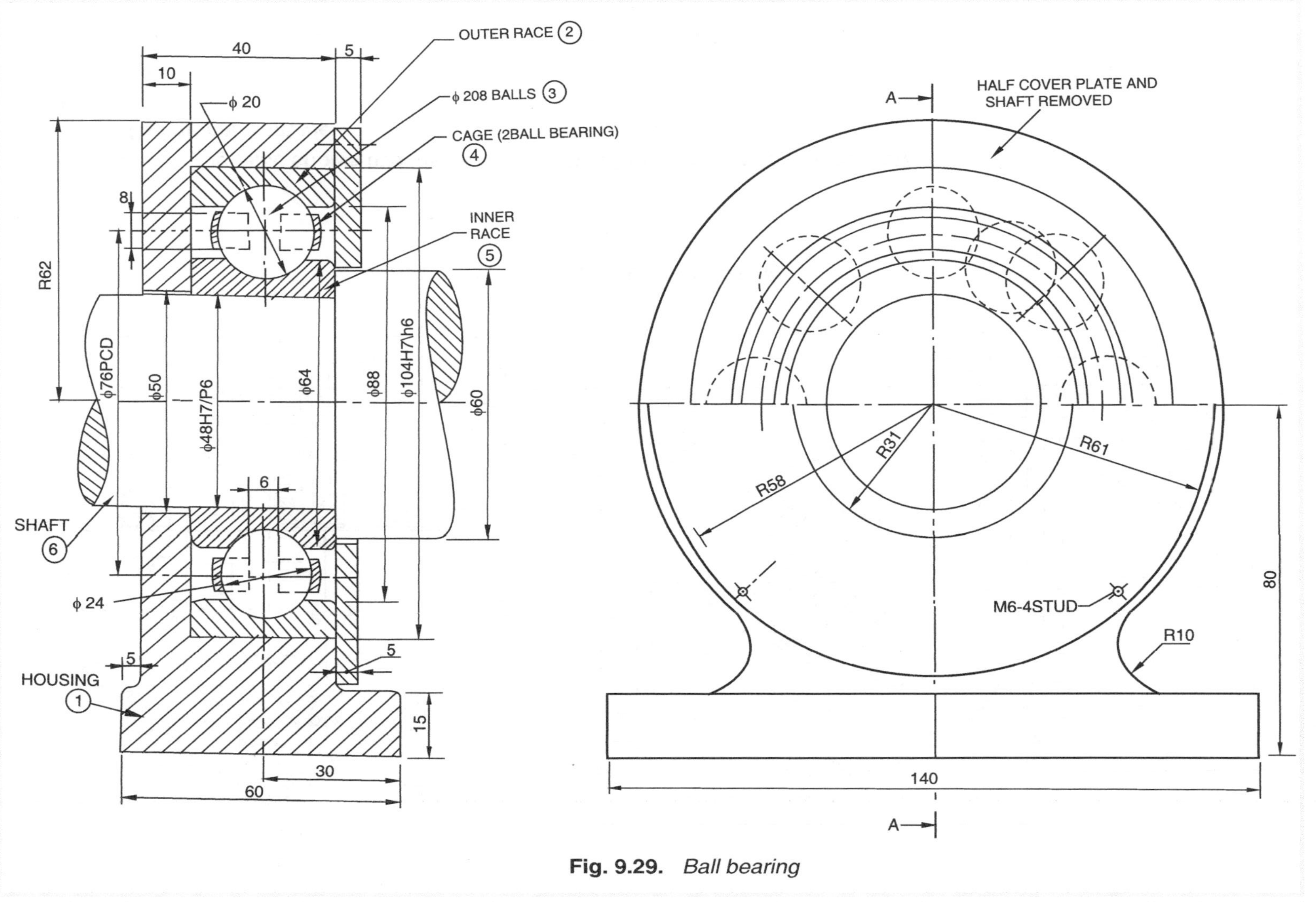

Fig. 9.29. *Ball bearing*

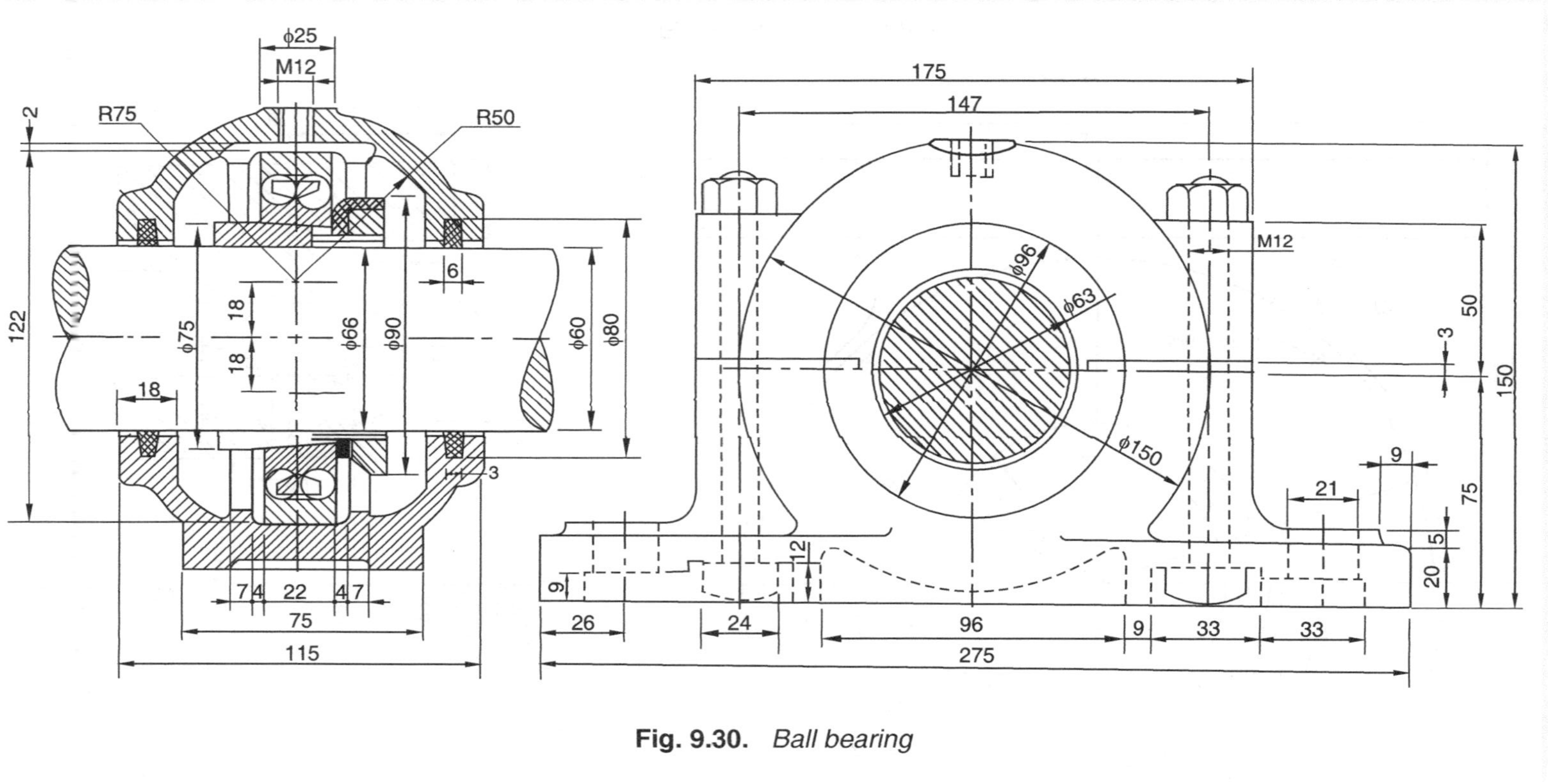

Fig. 9.30. *Ball bearing*

Exercise 21: Fig. 9.30 shows two views of a ball bearing. Draw the following views:-

(a) Half sectional front view

(b) Side

(c) Top view

Dimension the views, prepare the list of parts and material used.

QUESTIONS FOR SELF EXAMINATION

1. What do you mean by a bearing? what is the function of a bearing?
2. Name the different types of bearings?
3. What is the function of a bush in a bearing?
4. Why bearings are lubricated?
5. What are the advantages of plummer block over closed and open bearings?
6. How bushes (brasses) are prevented from rotation in bearings?
7. Why a recess is generally kept between the cap and body of a plummer block?
8. Why collars are provided on the brasses?
9. Why recess is made between the bush and body of a foot step bearing?
10. What is the function of Angle plemmer block ?
11. Where the swivel bearing is used ?
12. When the foot-step bearing is used ? Name its parts.
13. What are the advantages of ball and roller bearings upon journal bearing?
14. Match the statement of column A with corresponding ones in column B :

Column-A Type of bearing	Column-B Used for	Answer
(a) Journal bearing	(1) Vertical shafts with axial loads	(a)...(2)
(b) Coller bearing	(2) horizontal shaft with radial loads	(b)...(3)
(c) Pivot bearing	(3) horizontal shaft with axial loads	(c)...(1)

CHAPTER

10

Brackets

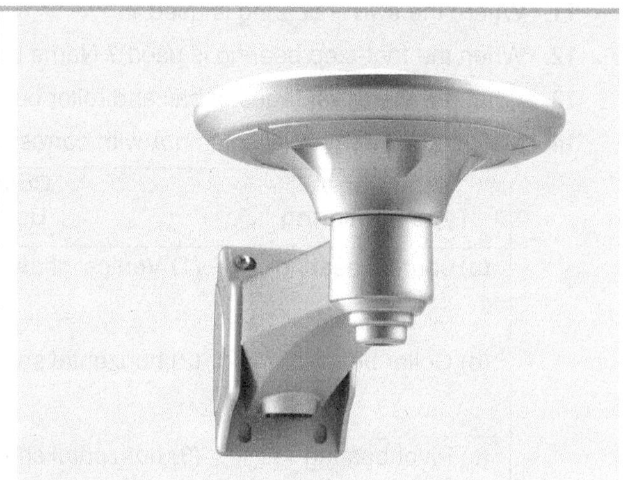

INTRODUCTION

In engineering practice, it is often desired to support the bearings which carry rotating shafts at some suitable place by some devices known as brackets. These are bolted to the walls or pillars either directly or with a wall plate between them. The bearings are supported on the horizontal shelf of the bracket. The length of the shelf depends upon the diameter of the biggest pulley mounted on the supporting shaft. Brackets are generally made of cast iron.

In this chapter, we shall deal with the study of brackets and different types of brackets alongwith their actual use in engineering practice.

10.1. BRACKETS

The devices used for supporting the bearings carrying shafts are known as brackets.

Fig. 10.1 shows a wall bracket bolted to a wall by means of bolts. For better securing to the wall, projection is provided at the lower end of vertical plate. This projection (which is in rectangular shape) fits into a slot made in the wall. The projections around the bolt-holes are provided for the washer action and give more space for tightening and loosening of the bolts (see Fig. 10.1).

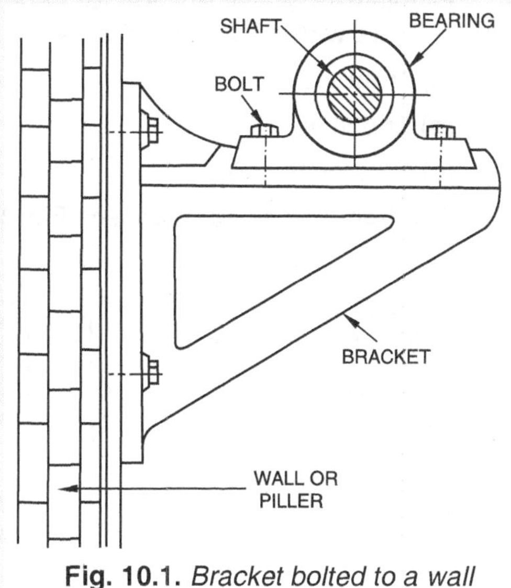

Fig. 10.1. *Bracket bolted to a wall or pillar*

The horizontal plate, having two rectangular holes is casted with the vertical plate. The holes in the sole-plate of the bearing which is to be fitted with the bracket are kept in alignment with the holes of the horizontal plate of the bracket and the two are tightly bolted. Two extensions are provided at the top of the horizontal plate to suit the outer edges of the sole plate of the bearing. These edges may be vertical, but preferably slanting. The lower web is kept hollow, because very small load is coming at this place and thus less material is required here.

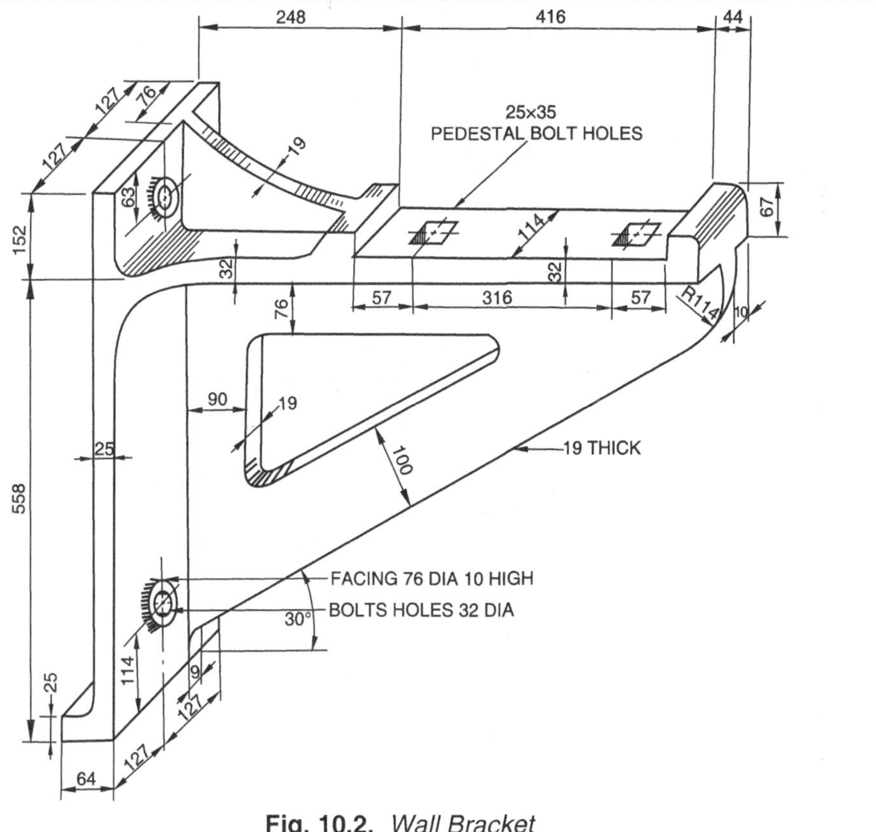

Fig. 10.2. *Wall Bracket*

Problem 1 : A wall bracket for pedestal bearing is shown in Fig. 10.2. Draw the following views with a suitable scale :—

(a) Front view

(b) End view

(c) Outside top view.

Solution: For its solution, see Fig. 10.3.

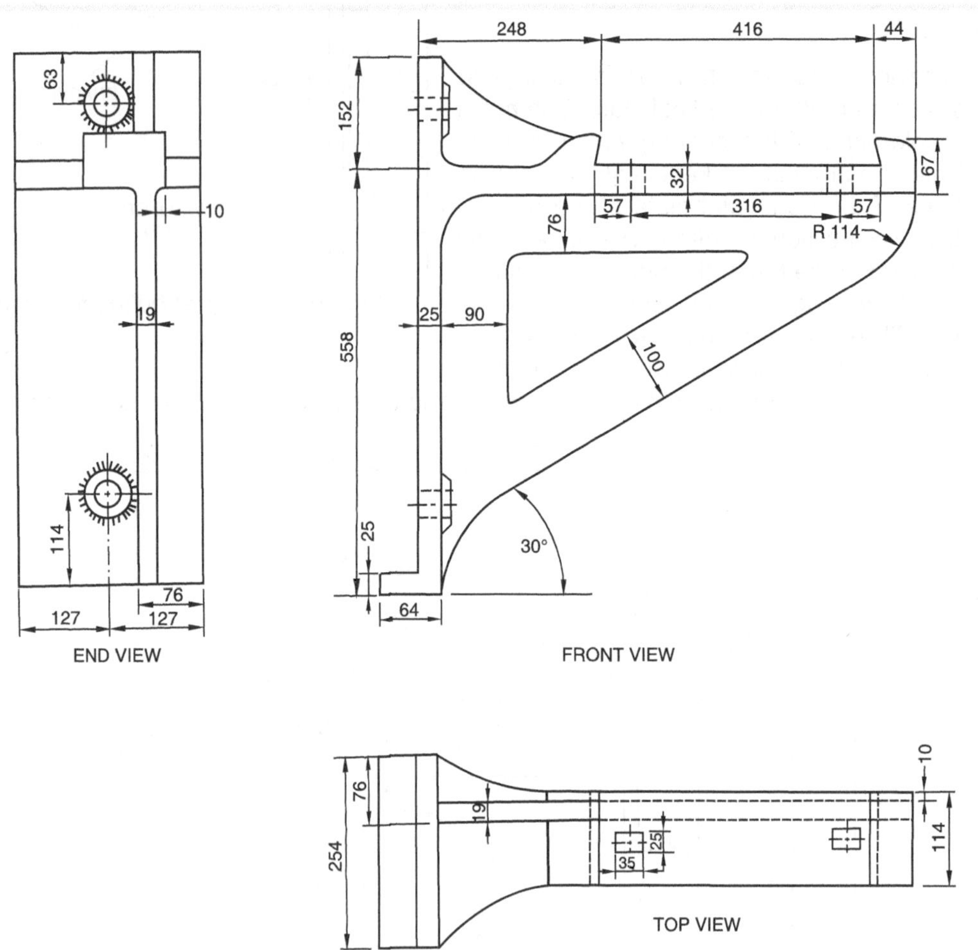

Fig. 10.3. *Wall bracket for pedestal bearing*

Problem 2 : Fig. 10.4 shows an insulator bracket. Draw the following views to a full size scale in first angle projection. :—

(a) Front view [Elevation]

(b) Side view projected from (a).

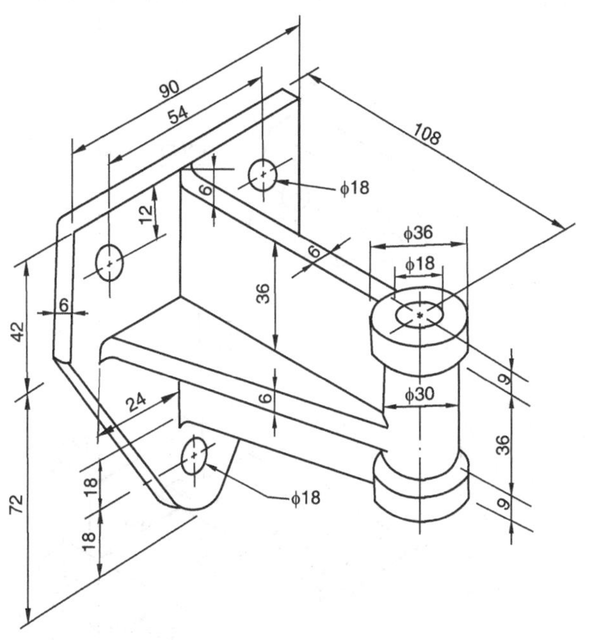

Fig. 10.4. *Insulator bracket*

Solution: For its solution, see Fig. 10.5

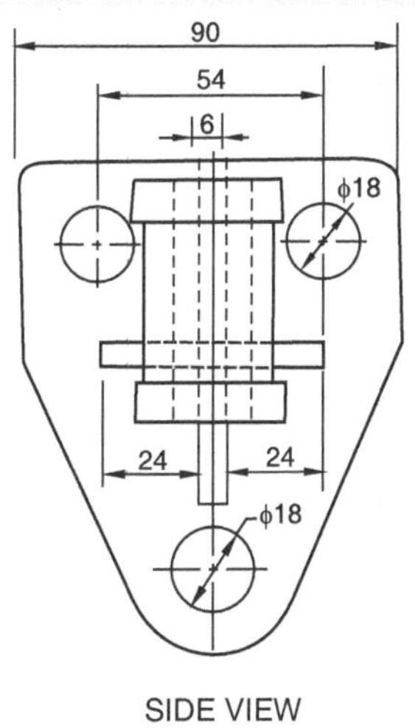

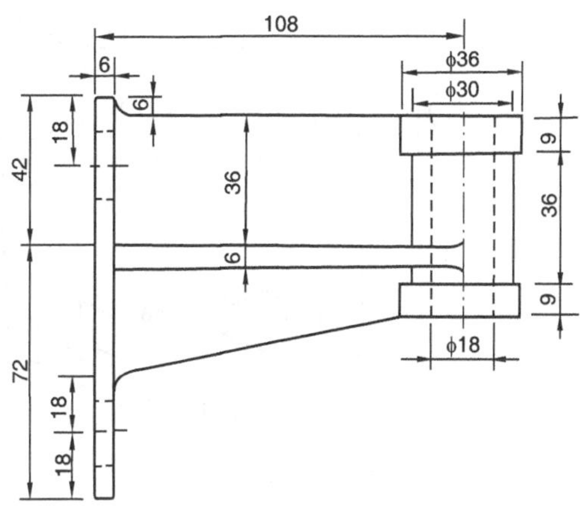

Fig. 10.5.

Exercise 1 : An isometric view of a cast iron bracket is shown in Fig. 10.6. Draw the following views to half full size scale :—

(a) Front view [Elevation]

(b) Side view [End view]

(c) Top view [Plan]

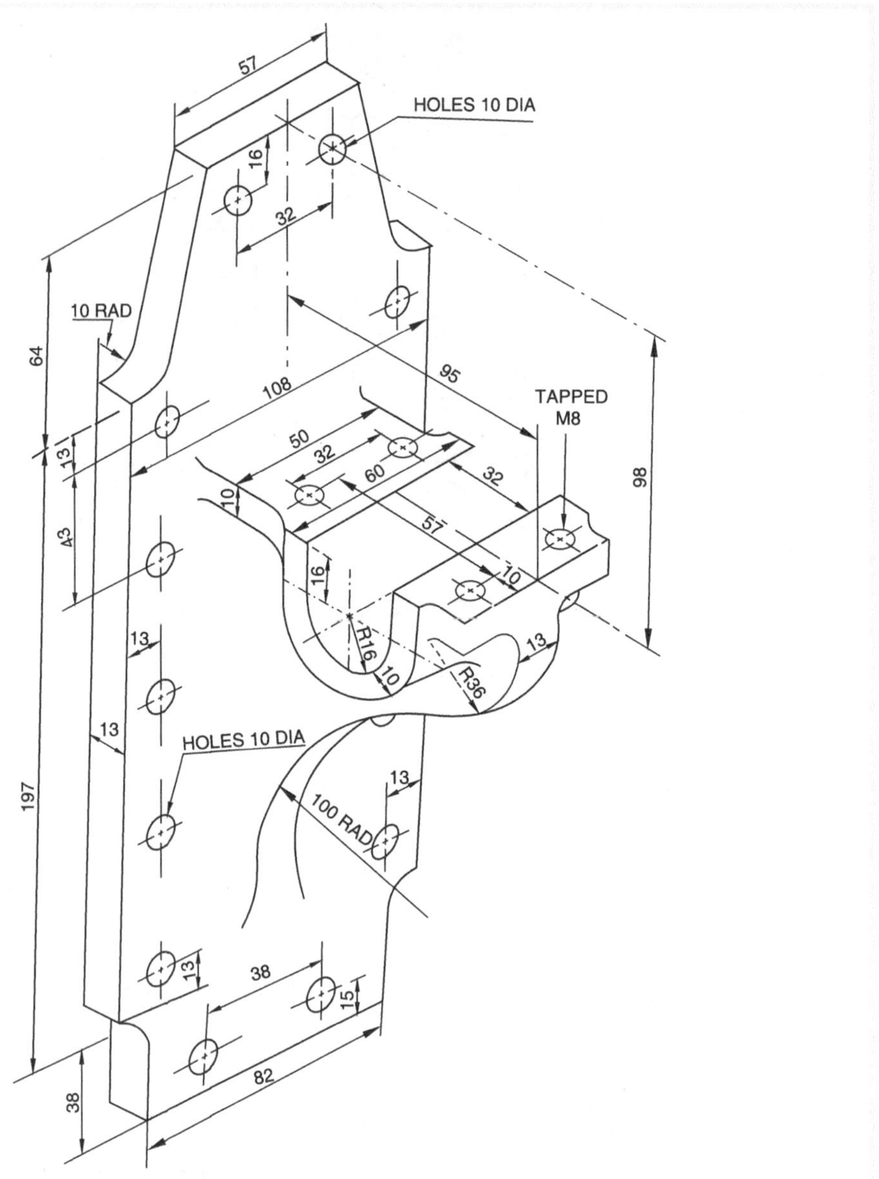

Fig. 10.6 *Cast Iron bracket*

Exercise 2 : Fig. 10.7 shows the isometric view of a wall bracket. Draw to a full size scale the following views :—

(a) Front view (b) End view (c) Top view.

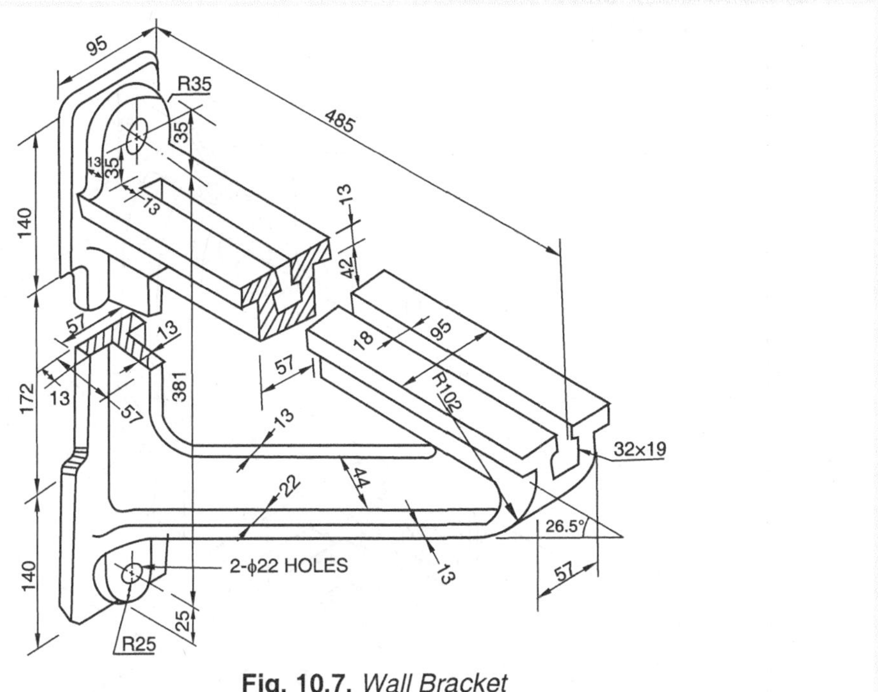

Fig. 10.7. *Wall Bracket*

Exercise 3: Fig. 10.8 shows the pictorial view of a bracket. Draw the following views to some suitable scale:–

(a) Front view (b) Side view (c) Top view.

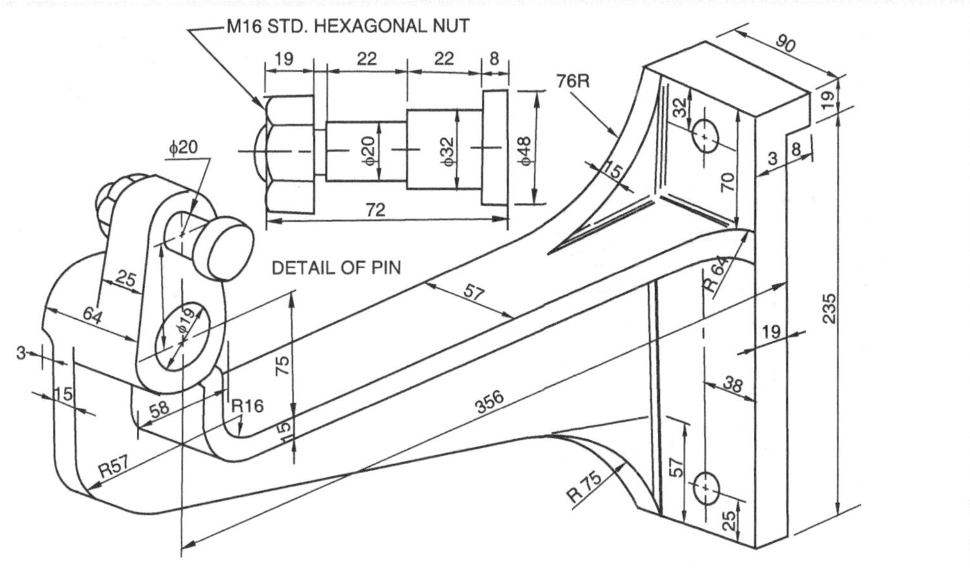

Fig. 10.8. *Wall bracket*

Exercise 4 : Fig. 10.9 shows an isometric view of a bracket. Draw the following views to a scale half full size from : —

(a) Front view [Elevation]

(b) Side view [projected from (a)]

(c) Top view [Plan].

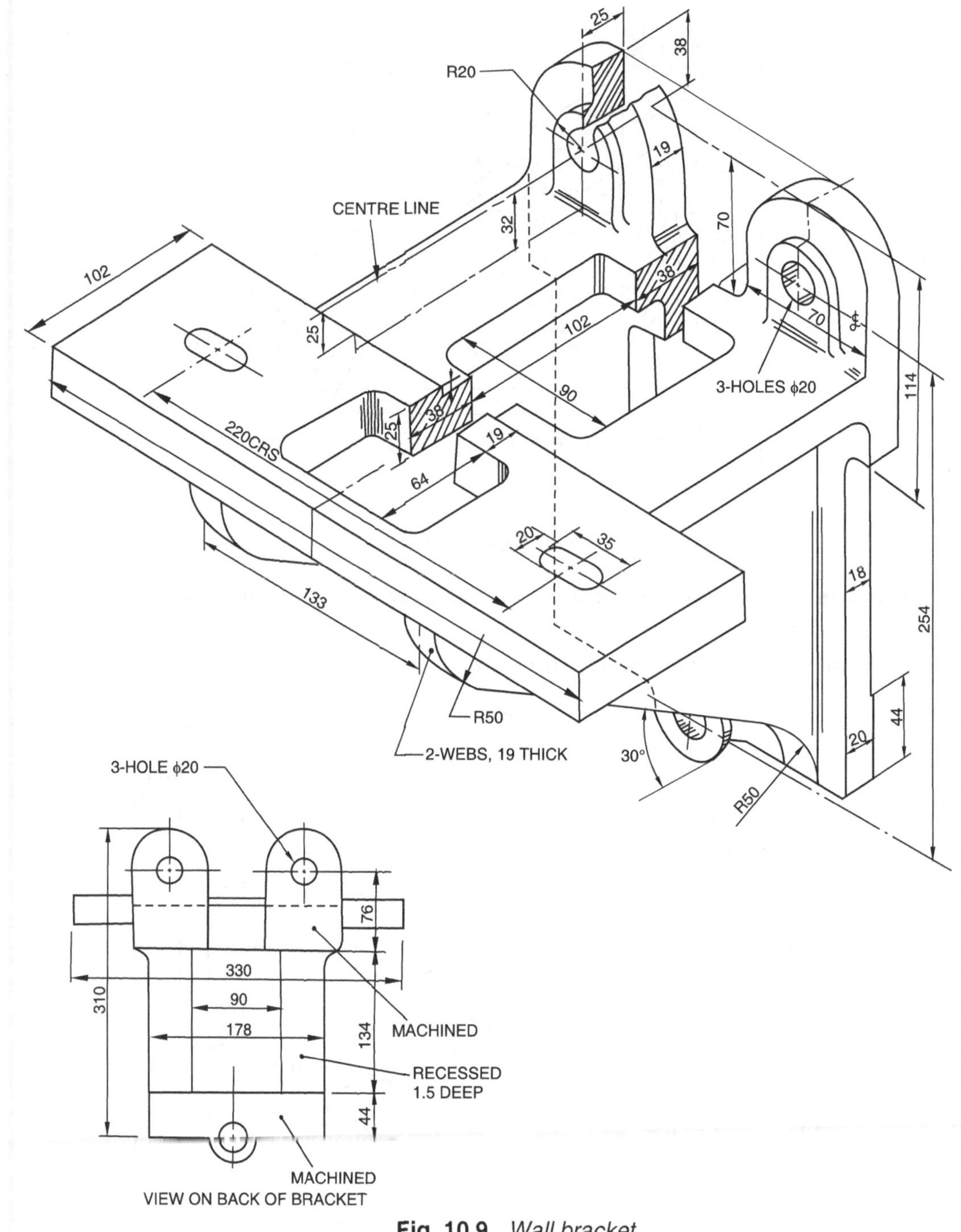

Fig. 10.9. *Wall bracket*

Exercise 5 : Fig. 10.10 shows two views of a bracket. Draw the following views:—

(a) Front view

(b) End view

(c) Top view.

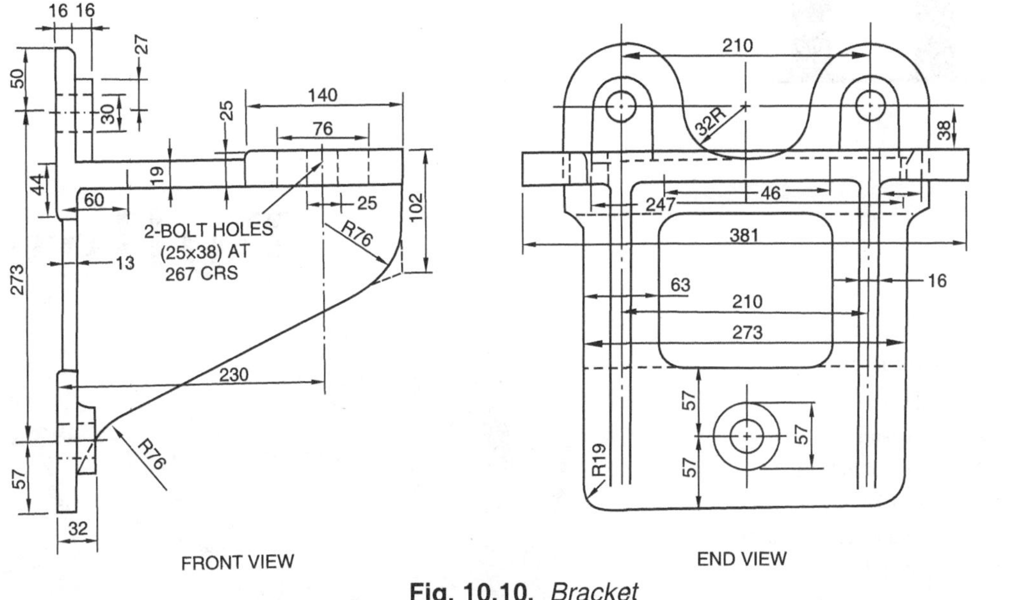

Fig. 10.10. *Bracket*

QUESTIONS FOR SELF EXAMINATION

1. What is the specific use of a bracket? How is it fixed on a wall or pillar?
2. What is the material of a bracket?
3. Why are the projections provided around the bolt holes?
4. Why is the lower web of pedestal bearing bracket kept hollow?

CHAPTER

11

Pulleys

INTRODUCTION

In engineering field, the major problem is to transmit mechanical power from one shaft to another shaft. This necessitates some devices for transmission of power. Pulleys are the devices which are used for transmission of power from one shaft to another by means of belts or ropes.

In this chapter, we shall deal with the study of pulleys, types of pulleys and their application in the engineering.

11.1. PULLEYS

The wheels which are used for transmission of power from one shaft to another by means of belts or ropes are known as pulleys.

Pulleys are made of cast iron, steel and wood in standard sizes. Generally cast iron pulleys are used in the workshop to transmit power or motion. Pulleys may be cast in one piece or bult-up from separate parts. The principal parts of all pulley are :—

1. Rim
2. Hub or Boss
3. Arms or Spokes.

The rim and hub of the pulleys are joined by arms. These arms may be straight or curved. Curved arms are better than straight arms as when the pulley contracts in the mould, there is possibilities of straight arms breaking due to the stresses developed. But these stresses can be taken up by the curved arms, as they accommodate contraction and will not break. However, with the advancement of modern foundary technology, straight arms can be casted without any fear of breakage.

The number of arms in any pulley is determined by the weight of the pulley and power to be transmitted. In general practice, if the diameter of the pulley is less than 180 mm, a solid web is used instead of arms. For diameters, 180 mm - 4 arms; 600 mm to 1500 mm - 6 arms and above 1500 mm diameter 8 arms are used.

11.2. CROWNING

Rim of the pulley is often rounded or given camber on the surface. This camber is known as crowning. Crowning is recommended for the belt to get a sound grip over the periphery of the pulley.

Due to crowing the belt will not slip off the pulley when it rotates, as it has a tendency to slide to the highest point of periphery due to centrifugal force.

Sometimes, flanges are provided on the rim in addition to crowning to keep the belt in position as in case of stepped pulley (see Fig. 11.5.)

11.3. TYPES OF PULLEYS

There are many types of pulleys used in the workshops, according to their constructions and sizes. The most commonly used pulleys are described below :—

11.4. CAST IRON PULLEYS

Cast iron pulleys are generally used in the workshop to transmit power or motion from one shaft to another.

The rim and the hub are casted with the arms. The arms of these pulleys may be straight or curved. Curved arms are better as they yield when the pulley contracts in the mould, where as the straight arms would break under such circumstances. The arms of these pulleys are made in elliptical in section (see Fig. 11.1).

Cast iron pulleys are connected to the shafts by means of sunk keys.

Notes : *1. While transmitting power, the force is least near the rim and maximum near the hub. So the cross-section of the arms near the rim is required to be least and that near the hub to be larger.*

2. Crowing is provided on the pulley to keep the belt in centre to get a sound grip.

3. The rim is slightly tapered from inside and hub on the outside (known as draw) to facilitate the removal of the pattern from the mould while casting.

Exercise 1 : Fig. 11.1 shows the pictorial view of a cast iron pulley. Draw the following views :—

(a) Front view upper half in section

(b) Side view

Give the tolerances and surface finish wherever required.

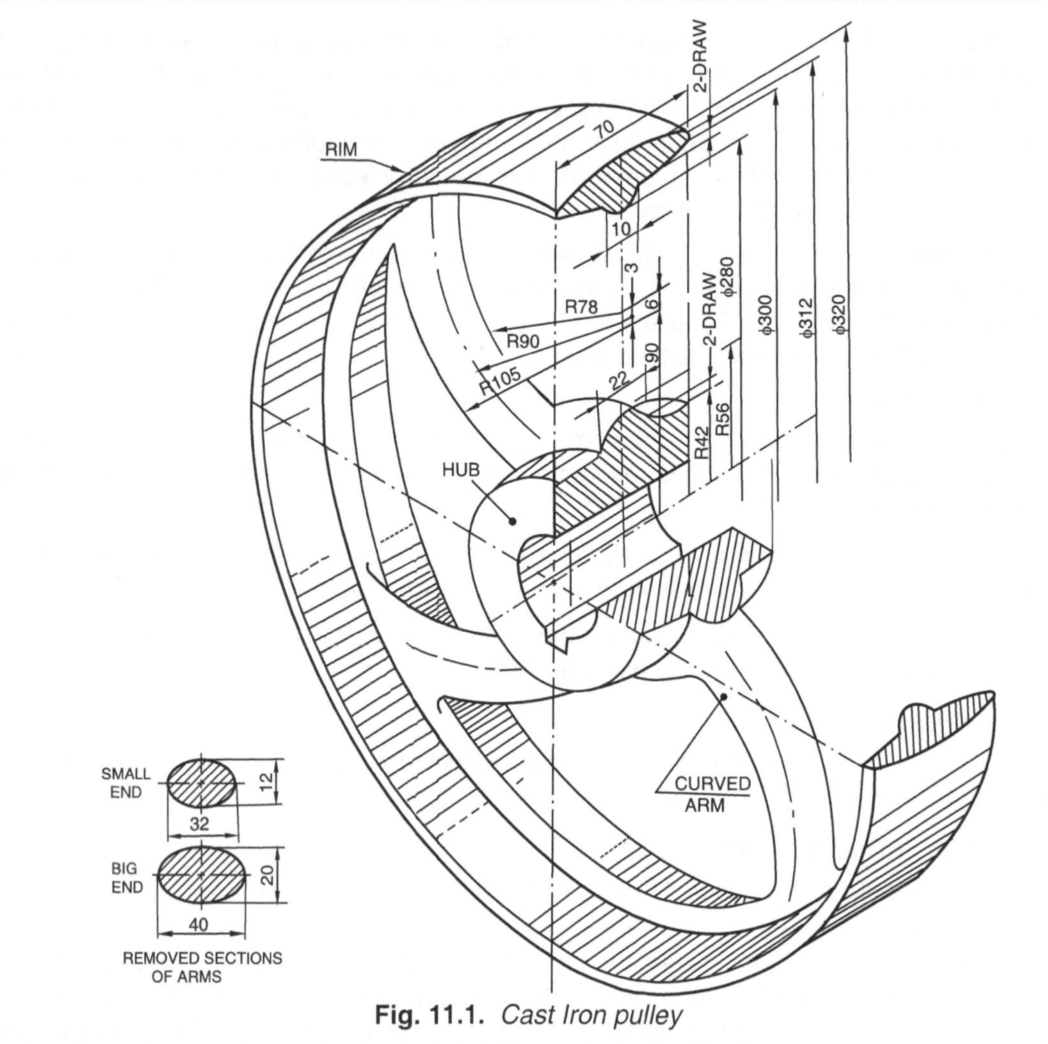

Fig. 11.1. *Cast Iron pulley*

11.5. FAST AND LOOSE PULLEYS

Fast and loose pulleys are commonly used in workshop where several machines are to be run by only one motor. The main advantage of fast and loose pulleys is that any machine can be stopped or started when required without stopping the motor (refer Fig. 11.2).

The fast pulley is fixed or keyed to the shaft while the loose pulley is mounted freely on the shaft. A gun-metal bush is provided in the loose pulley, which can be replaced by a new one when worn out due to constant friction between the pulley and the shaft. To prevent the axis movement of the loose pulley, a collar is secured to the shaft by means of a set screw (see Fig. 11.2). When it is required to run the machine, the belt is shifted to the fast pulley with the help of striking gear attached to it and when the machine is to be stopped, the belt is shifted to loose pulley.

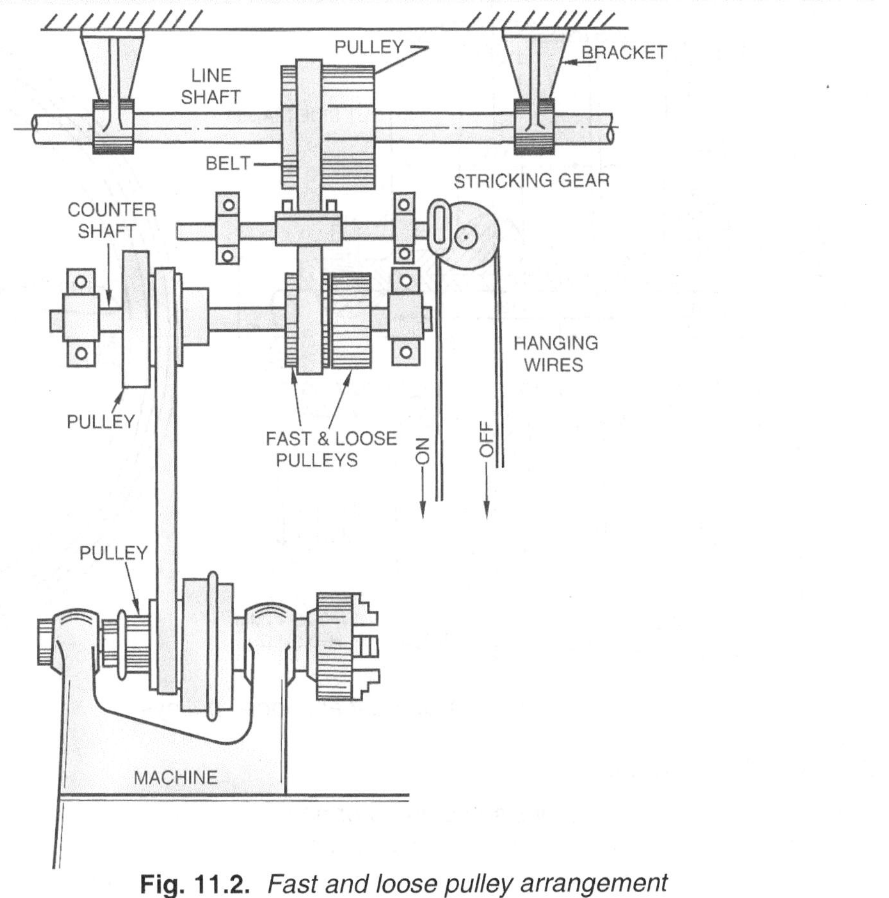

Fig. 11.2. *Fast and loose pulley arrangement*

Exercise 2 : Fig. 11.3 shows the two views of a fast and loose pulleys. Draw the following views :—

(a) Upper half sectional front view

(b) Side view.

Shows the tolerances wherever required.

Hint : The following tolerances are suggested : [Refer table 2-4]

(1) For hub and shaft assembly : 50 H7/h6

(2) For bush and hub assembly: 65h7/p6

(3) For 14 wide key way and key assembly: 14H7/n6

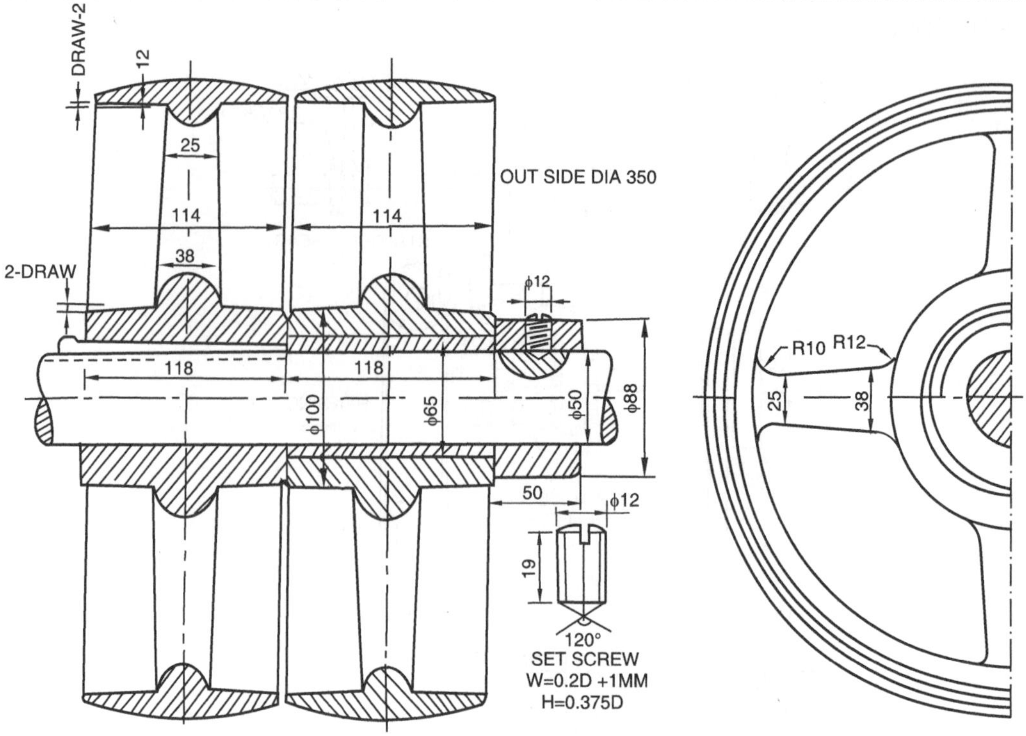

Fig. 11.3. *Fast and loose pulleys*

11.6. SPLIT-UP PULLEY

When a shaft already carries some pulley or has its end swelled and it is desired to provide a new pulley, then it will be convenient to make the pulley into two halves and joined together by means of bolts and nuts after mounting it on the shaft. *The pulleys when split up into its parts are called split or built - up pulleys. These pulleys have more strength and durability.*

The arms of these pulley are made of steel. One end of the arm is shrunk into the cast - iron hub (made in two halves) whereas the other end passes through the steel rim and riveted over. The two halves of the rim are joined together by butt joint, riveted to one half and bolted together alternately. After riveting, the rim is machined and the pulley is finally balanced.

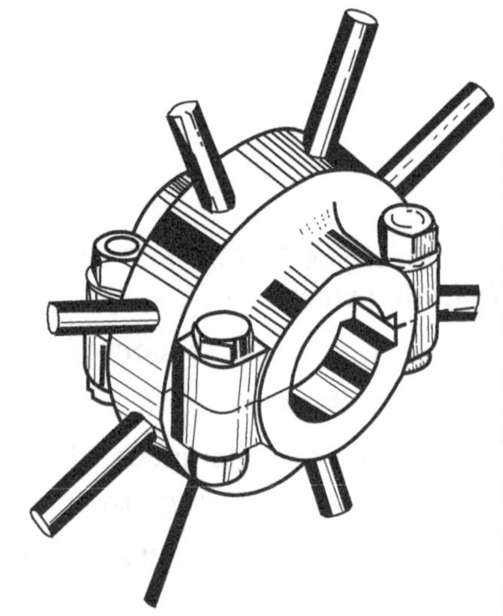

Pictorial view of hub of split pully

Exercise 3 : Fig. 11.4. shows two views of a split-up pulley. Draw the following views:

(a) Half sectional front view

(b) Side view.

Show the tolerances and surface finish wherever required.

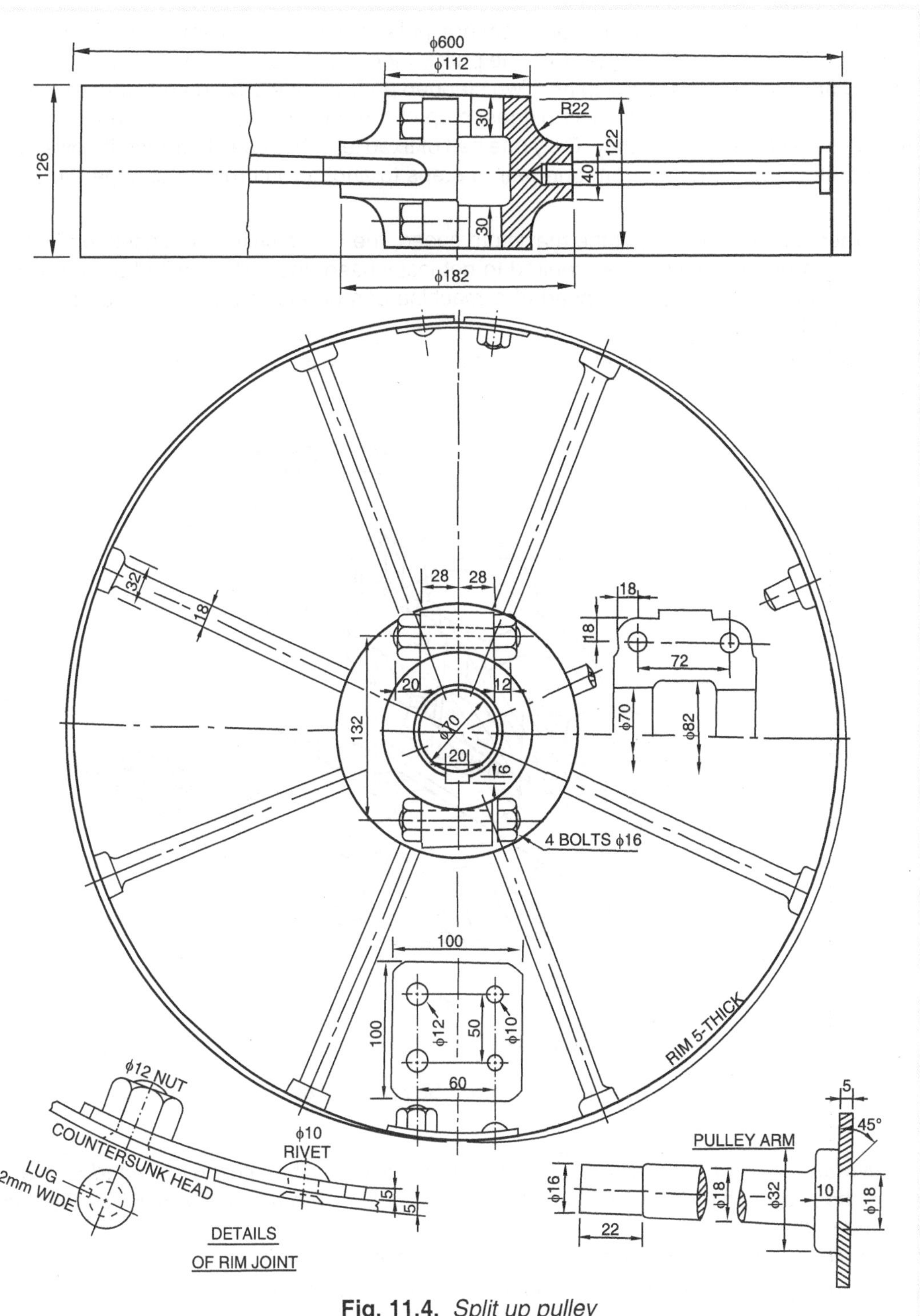

Fig. 11.4. *Split up pulley*

11.7. SPEED CONE OR STEPPED PULLEY

Stepped pulley is a combination of several pulleys of different diameters cast integrally in such a shape so as to form steps of one pulley (see Fig. 11.5). This pulley is mainly used to give different speeds to a machine. Two stepped pulleys, one driving and other driven are placed opposite to each other. The biggest step of the driving pulley is just opposite to the smallest step of the driven pulley. Only one belt of fixed length is used to transmit power from driven pulley and is shifted from one pair of steps to another in order to change the speed ratio.

When it is required to get the maximum speed, the belt is shifted to biggest diameter of the driving pulley and when it is required to get less speed, the belt is shifted on the smaller diameter step. In this way, the speed of a machine can be increased or reduced according to position of the belt.

Stepped pulley is fixed on the shaft with the help of a set screw provided at the tapped hole.

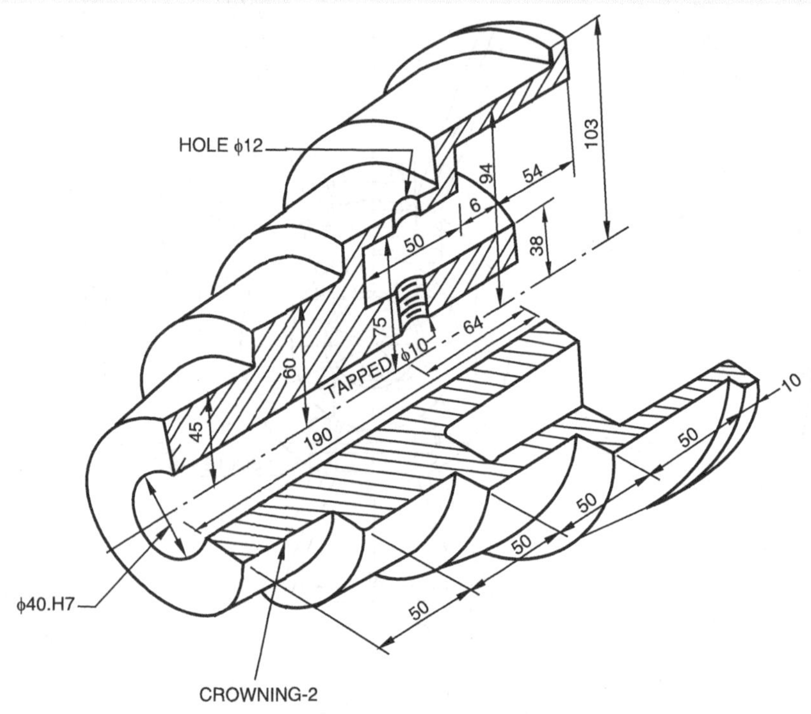

Fig. 11.5. *Stepped pulley*

Notes : *1. A clearance hole of 12 mm is provided in the second step and above the tapped hole to enable drilling, tapping and adjustment of allen set screw by which the pulley is keyed to the shaft.*

2. The pulley is made hollow to save material and to make it light.

3. The largest step is flanged so that the belt may not slip off.

Exercise 4 : Fig. 11.5 shows the pictorial view of a stepped cone pulley. Draw the following views of the cone pulley :—

(a) Front view–half in section

(b) Side view–half only.

Show the tolerances and surface finish wherever required.

11.8. ROPE PULLEY

These pulleys have V-grooves to carry ropes by means of which power is transmitted to shafts especially when the shafts are a long distance apart. Ropes made of cotton or hump of 25 mm to 50 mm in the diameters are generally employed for main driving in mills and lifting and transporting appliances.

Note : *The purpose of making the groove is to increase the frictional grip on the pulley and thus reduce the tendency to slip.*

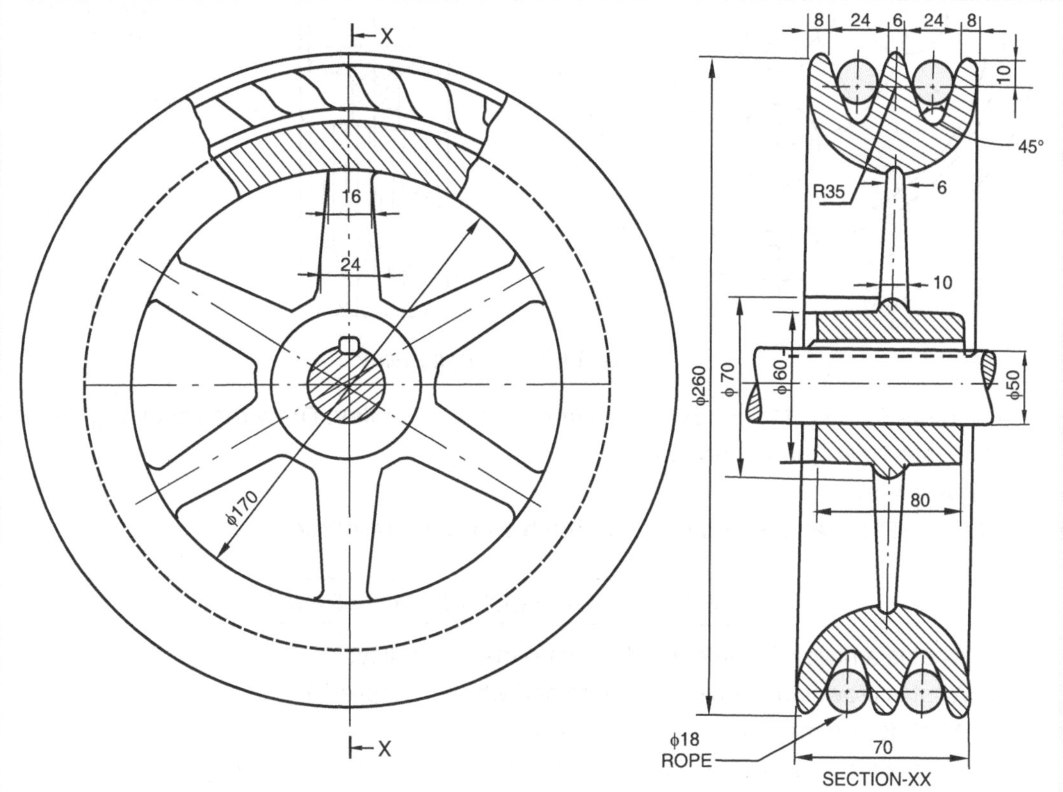

Fig. 11.6. *Rope pulley*

Exercise 5 : Fig. 11.6 shows two views of rope pulley. Draw the following views:—

(a) Front view-upper half in section

(b) Side view.

11.9. V-BELT PULLEY

V-belt pulleys have V-grooves on the rim of the pulley to carry V-belts made of rubber and fibres and are moulded as endless loops. These pulleys are widely used to transmit motion from one shaft to another when destance between them is large.

Fig. 11.7 shows two views of the grooved V-belts pulley along with details of V-groove.

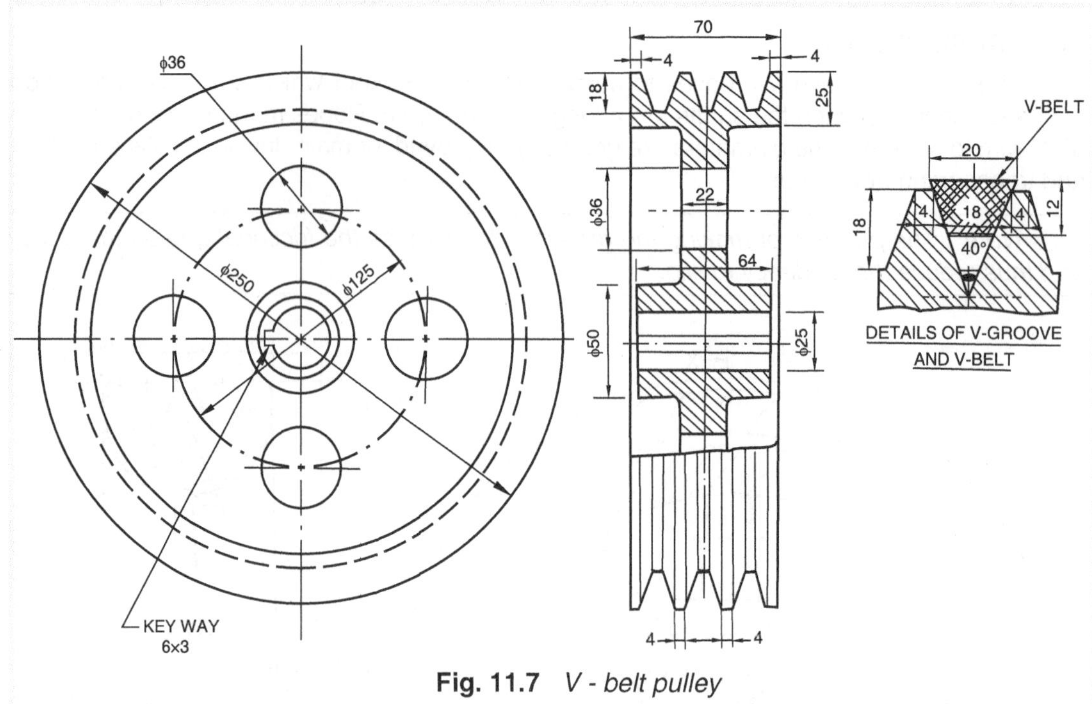

Fig. 11.7 *V - belt pulley*

Exercise 6 : Fig. 11.7 shows two views of V-belt pulley. Draw the following views:-

(a) Front view-Full in section

(b) Side view.

Show the tolerances and surface finish wherever required.

QUESTIONS FOR SELF EXAMINATION

1. What is a pulley ? Name the different parts of a pulley.
2. What is crowning of pulley ? Where and why is it applied ?
3. What do you mean by draw in a pulley ?
4. What is the special use of a fast and loose pulley ? Where and why is it used ?
5. What is the chief advantage of a stepped pulley ?
6. Where and why a stepped cone pulley is used ?
7. What is a split-up pulley ? Where and why is it used ?
8. When are the rope and V-belt pulleys used ?

CHAPTER

12

Pipe Joints

INTRODUCTION

In engineering practice, it is usual to carry fluid such as water, oil, steam, gas etc. from one place to another place by means of pipes. In market, the pipes are usually available upto 5.5 metres in length. To get the required length, a number of pipes are connected together end to end by bolts and nuts; studs and nuts, etc.

In this chapter, we shall study pipes, pipe joints, pipe meterials, size of pipe, uses of pipes and different type of pipe joints used in engineering practice.

12.1. PIPES

The hollow cylinders which are used for carrying fluid such as water, oil, steam, gas, etc from one place to another place are known as pipes.

12.2. PIPE JOINT

The *joining of two or more pipes in order to increase their length for the purpose of carrying fluid from one place to another place is known as pipe joint.*

12.3. PIPE MATERIAL

Pipes are generally made of cast-iron, wrought iron, steel, lead, copper or brass depending upon the purpose for which they are used.

12.4. SIZE OF PIPE

The size of the pipe is stated by the nominal bore of the pipe. The thickness of the pipe is not considered in it. Thus, a 50 mm pipe means the diameter of the bore is 50 mm.

TABLE 12.1 : DIFFERENT PIPES AND THEIR USES

S.NO.	NAME	SIZE IN MM	PRACTICAL USES
1.	CAST IRON PIPES	50 TO 1200	USED FOR CARRYING WATER, GAS OR STEAM
2.	WROUGHT IRON	6 TO 150	USED TO CARRY DOMESTIC WATER
			PIPES AND GAS AT LOW PRESSURE
3.	STEEL PIPES	10 TO 1800	USED FOR CARRYING WATER, STEAM, SEWAGE OR AIR AT A HIGH PRESSURE
4.	LEAD PIPES	5 TO 50	USED FOR DOMESTIC PURPOSES OR WHERE THE PIPE LINE CONTAINS FREQUENT BENDS
5.	COPPER AND	3 TO 50	USED FOR CARRYING HOT WATER IN
	BRASS PIPES		FABRICATING RADIATOR OR IN ENGINE WORK

12.5. DIFFERENT PIPES AND THEIR USES

The table 12.1 shows the different types of pipes and their uses in engineering field.

12.6. FORMS OF PIPE JOINTS

Several forms of joints are used to join pipes together in order to make a long pipe line. They are connected in various forms depending upon the material of the pipes and the purpose for which they are used. Some of the most common pipe joints are described below :–

12.7. CAST IRON FLANGED JOINT

It is the simplest form of pipe joint. In this joint, the flanges are casted at the ends of the pipes. The faces of the flanges are machined to ensure correct alignment of the pipes and bolts holes are drilled to join them by means of bolts and nuts. To prevent the leakage of fluid, a packing of iron or gasket of soft material (such as Indian rubber) is provided between the flanges.

Pictorial view of C.I. flanged joint

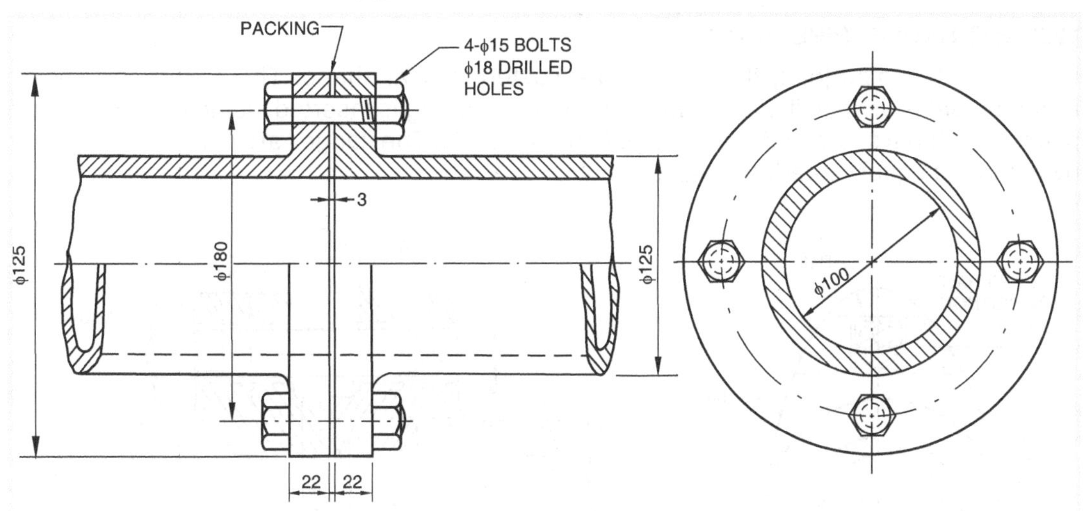

Fig. 12.1. *Cast-iron flanged joint*

Exercise 1: Fig. 12.1 shows two views of a cast-iron flanged joint. Draw the following views : —

1. Full sectional front view

2. Outside end view.

12.8. SPIGOT AND SOCKET JOINT

Spigot and socket joint is mostly employed to connect cast iron pipes which are laid underground for carrying water. Such a joint provides a better flexibility and adapts itself to small changes in level due to settlement of earth.

Pictorial View of spigot and socket joint

In this joint the spigot end of one pipe A enters into the socket end of the other pipe B(see Fig. 12.2). The space between the two ends of the pipes is filled partly by several turns of jute yarn and the remaining by pouring molten lead. The lead when solidified is caulked in tightly.

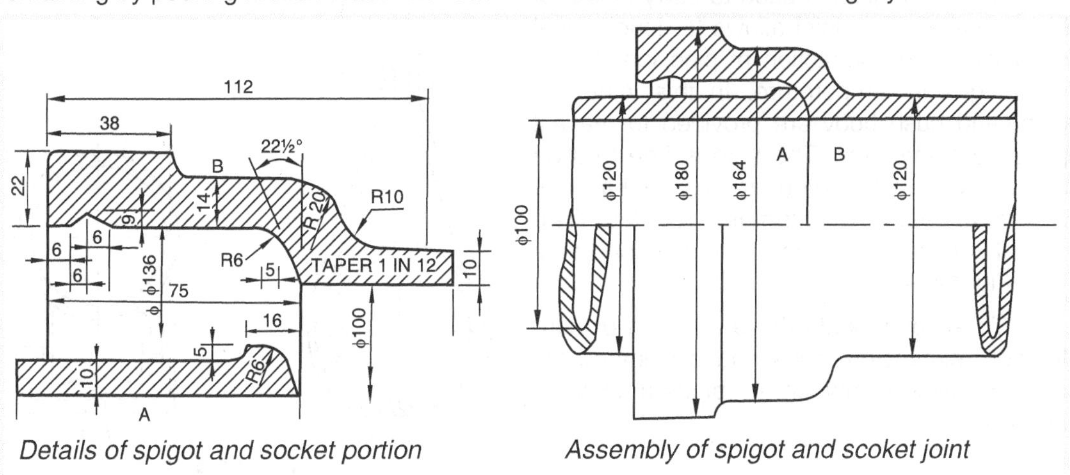

Details of spigot and socket portion *Assembly of spigot and scoket joint*

Fig. 12.2.

12.9. HYDRAULIC PIPE JOINT

In such a pipe joint, the flanges are casted in oval shape. The spigot and socket arrangement is used in it. A V-shaped gutta - percha ring is inserted inside the recess as a packing material to prevent the leakage (see Fig. 12.3). The flanges are connected together by means of square headed bolts and nuts.

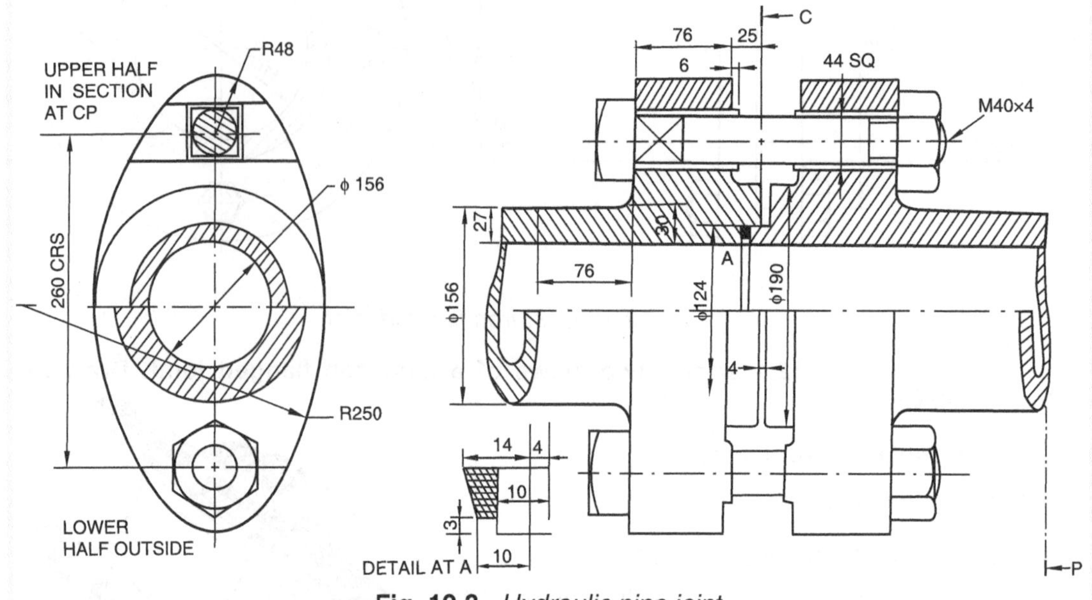

Fig. 12.3. *Hydraulic pipe joint*

Exercise 2 : Two views of a Hydraulic pipe joint are shown in the Fig. 12.3. Draw the following views :—

(a) Front view–full in section

(b) Side view.

12.10. EXPANSION JOINT

Expansion joint is used to carry steam at high pressure. This joint allows for longitudinal expansion and contraction of pipe length due to changes in temperature. In this joint, the gland and bush body are provided to make it perfectly steam tight. The pipe is free to slide in the body. For preventing the leakage of steam, asbestos packing is provided. The socket is fastened by means of bolts and nuts (see Fig. 12.5)

To accommodate the expansion or contraction, expansion loops and corrugated fillings are placed between the pipes at suitable intervals.

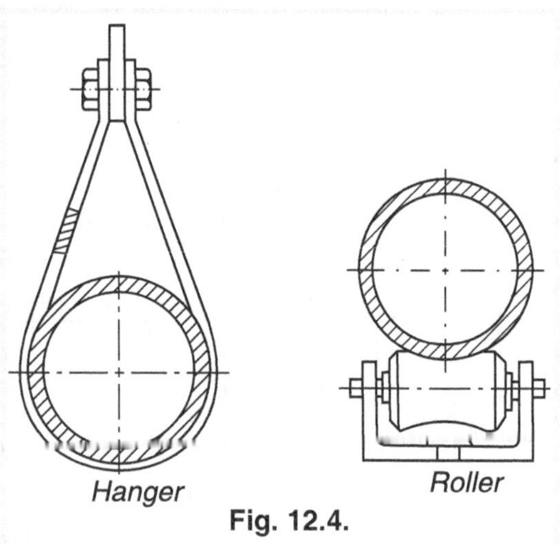

Fig. 12.4.

To allow for free expansion in length, the pipes are not clamped rigidly, but are suspended on hanger or freely supported on rollers (see Fig. 12.4).

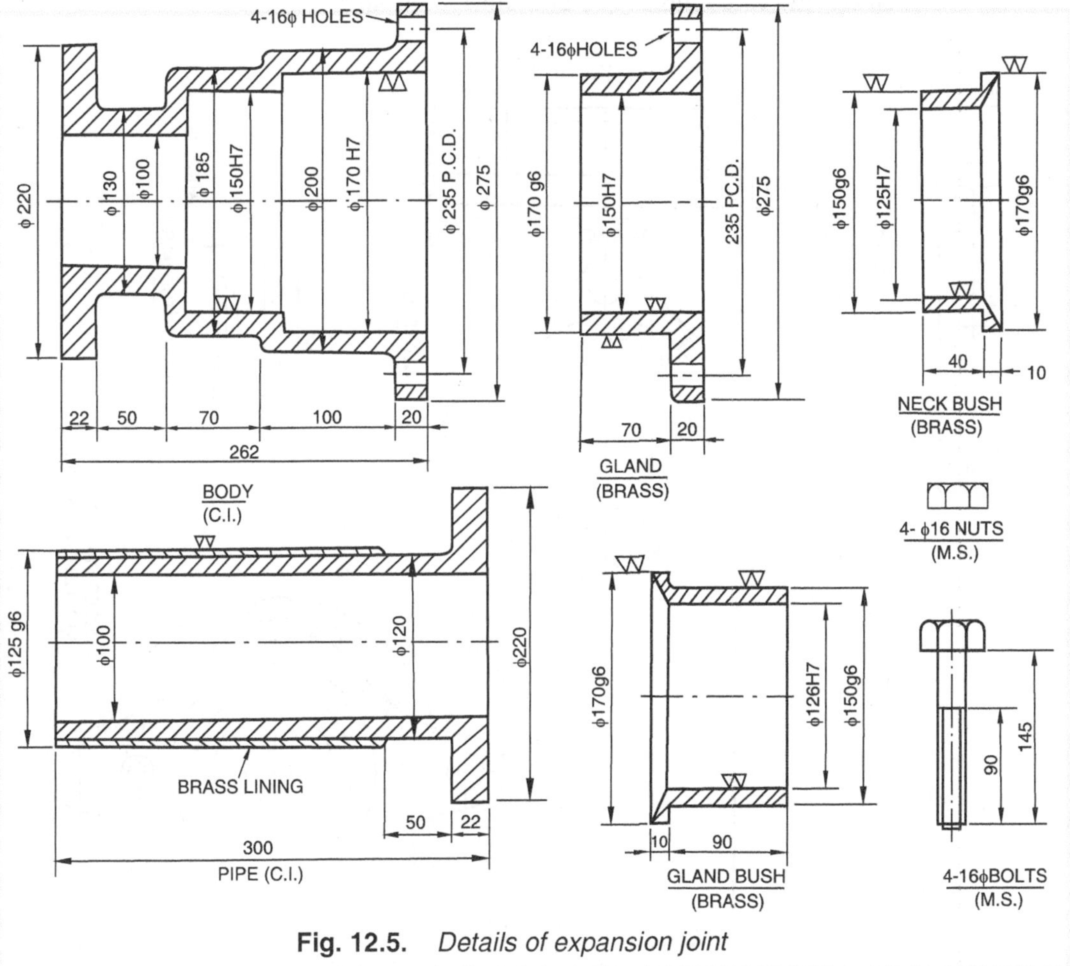

Fig. 12.5. *Details of expansion joint*

Exercise 3 : Fig 12.5 shows the details of an expansion joint. Assemble all the parts and draw the following views :—

(a) Front view – full in section.

(b) Side view.

Show the tolerances and surface finish wherever required. Give the materials list also.

Exercise 4 : Fig. 12.6 shows details of gland and stuffing box expansion joint for larger pipes. Draw the following views after assembly to some suitable scale :—

(a) Front view – full in section (b) End view – left side

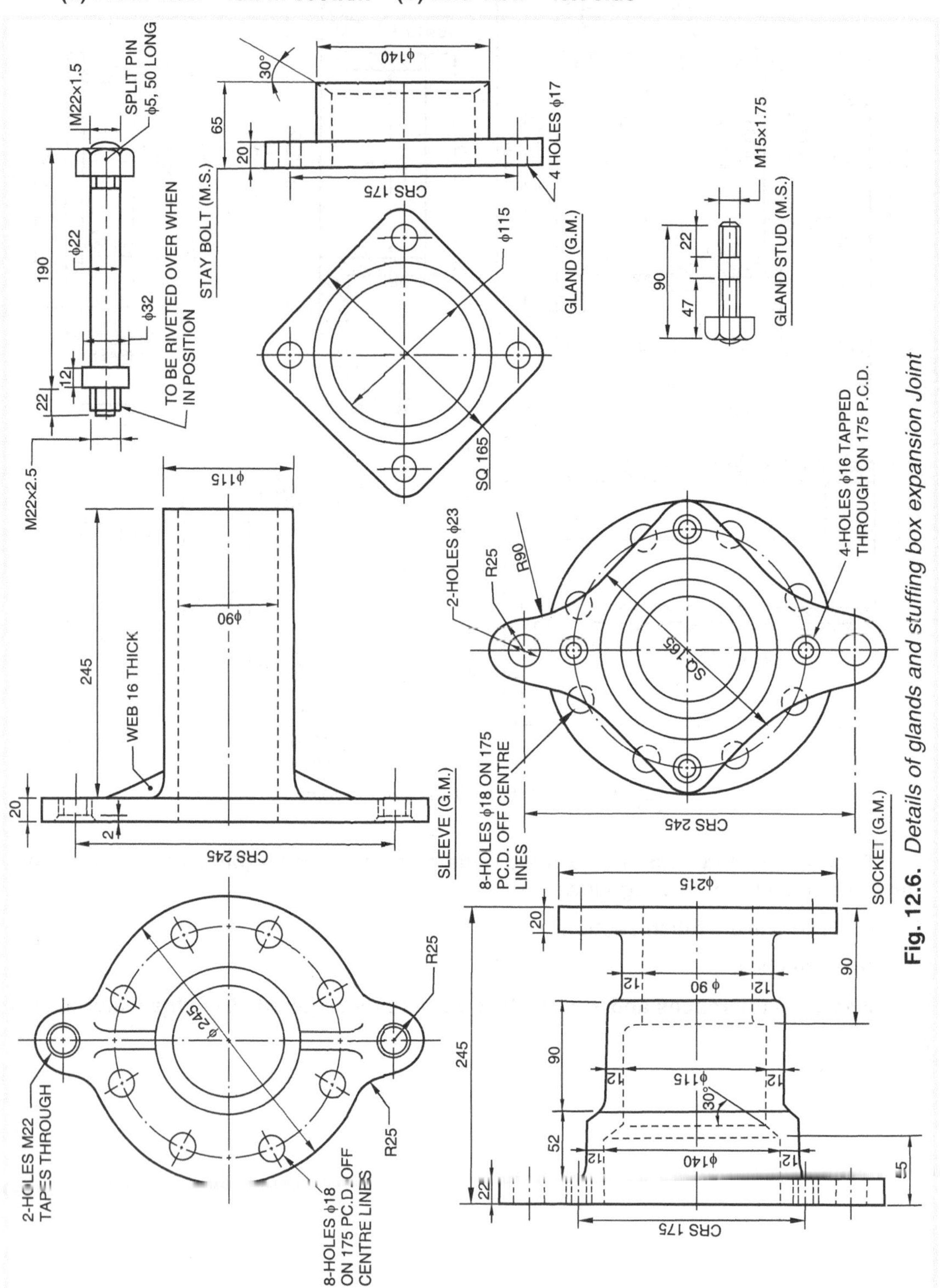

Fig. 12.6. *Details of glands and stuffing box expansion Joint*

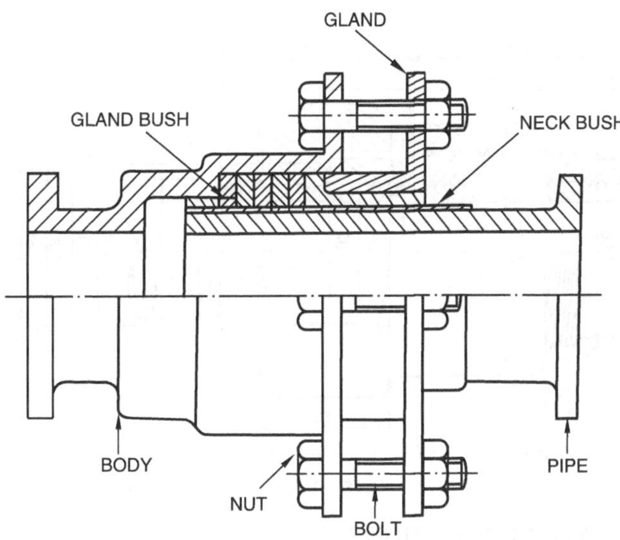

General arrangement of expansion Joint assembly

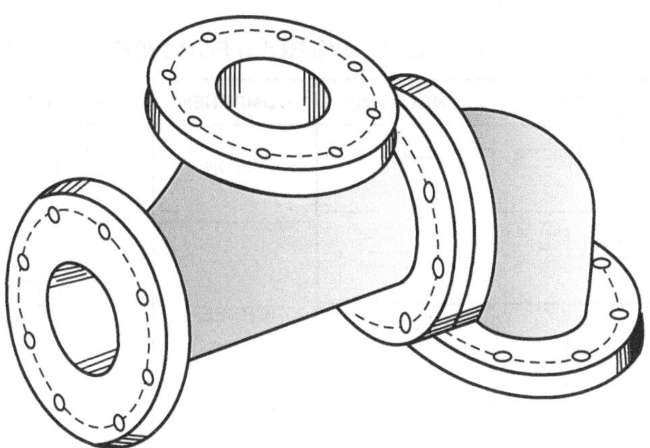

Pictorial view of right angle bend and tee pipe joint

12.8. UNION JOINT

Union joint is used for connecting pipes of small sizes. Fig 12.8 shows pipes A and B which are to be joined, have threads on the outer surfaces at their ends. The nut C which is threaded outside as well as inside is screwed on the end of pipe A. Another nut D which is also threaded is screwed to pipe B. The two nuts together with pipes A and B are drawn together by a coupler E. To prevent leakage, a packing ring is inserted between the ends of the two pipes.

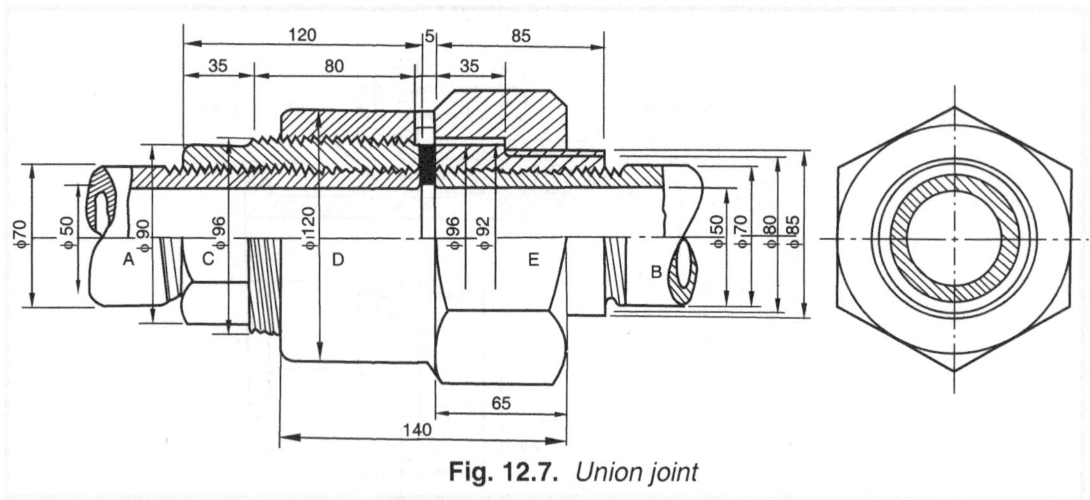

Fig. 12.7. *Union joint*

12.9. SYMBOLS FOR PIPE FITTINGS

Table 12-2 shows the symbols for standard pipe fitting in single line and double line. These symbols simplify the drawing.

TABLE 12.2. SYMBOLS FOR PIPE FITTINGS

COMPONENT	DOUBLE LINE	SINGLE LINE	COMPONENT	DOUBLE LINE	SINGLE LINE
COUPLING			UNION		
CAP			CROSS		
PLUG					
TEE			LATERAL		
90° ELBOW			GATE VALVE		
90° ELBOW TURNED DOWN					
45° ELBOW			GLOBE VALVE		
REDUCER OR REDUCING CUPLING			CHECK VALVE		

Exercise 5 : The sketch in fig. 12.8 in two views, shows a right angle bend and tee pipe joint. Draw to a scale half full size the following views :

(a) Front view - full in section

(b) Top view (out side)

(c) End view looking from the left side.

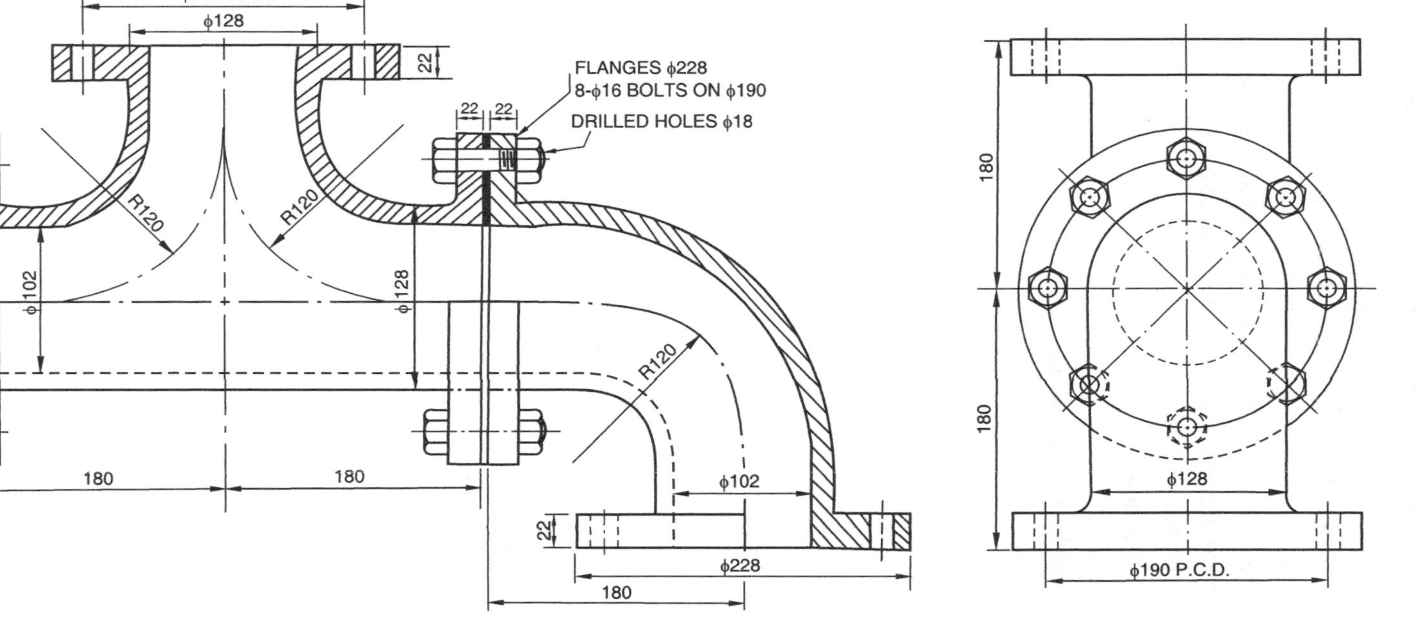

Fig. 12.8 *Right angle bend and tee pipe joint*

QUESTIONS FOR SELF EXAMINATION

1. Why are pipes jointed in common practice?
2. What are the different types of pipes ? What is the material of each?
3. Name the various types of pipe joints, you know?
4. What material is used for packing between spigot and socket pipe joint?
5. Why and where are the following pipe joints used?
   (a) Spigot and socket joint.
   (b) Hydraulic pipe joint.
   (c) Expansion joint.
   (d) Right angle bend and tee pipe joint.

CHAPTER

13

Steam Engine Parts

INTRODUCTION

In this fast developing engineering world, engines have come to stay as the most important machine to convert heat energy into mechanical energy. Heat is produced by the combustion of fuel (coal, petrol, diesel oil, or some gas, etc.). This heat is supplied to the engine for its conversion into mechanical work.

In the steam engine plant, steam is generated by heating water in the boiler. This steam is led into the cylinder where it pushes the piston to and fro in it. This to and fro motion of the piston is converted into rotary motion which is used to run most of the machines e.g. railway trains, generators of thermal power station, etc.

In this chapter, we shall deal with the study of steam engine, steam engine parts of various designs as being used in engineering practice.

13.1. STEAM ENGINE

The engine which converts heat energy of steam into mechanical energy through a link mechanism is known as steam engine.

A systematic sketch of a horizontal steam engine is shown in Fig. 13.1

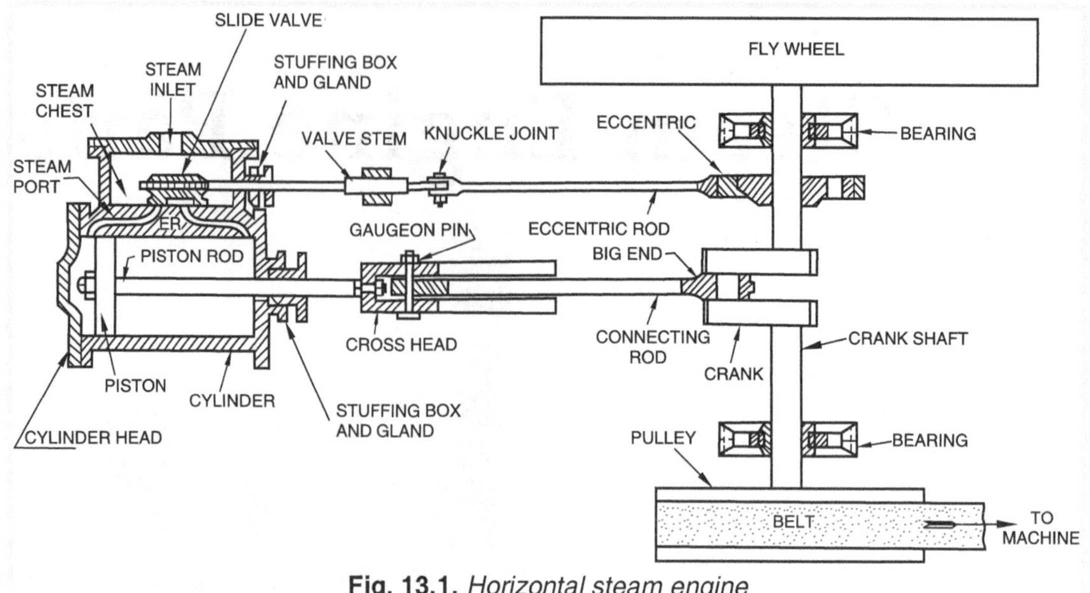

Fig. 13.1. *Horizontal steam engine*

13.2. STEAM ENGINE PARTS

A steam engine can be divided into following three parts :—

1. Stationary parts
2. Reciprocation parts
3. Revolving parts.

13.2.1. Stationary parts : *Those parts of the engine which remain stationary during the working of the engine are called stationary parts.*

In steam engine they are frame, cylinder (including steam chest and stuffing boxes), main bearings and guides.

13.2.2. Reciprocating parts : *Those parts of the engine which move linearly during the working of the engine are called recipocating parts.* In a steam engine they are piston, piston rod, crosshead, connecting rod, valve rod and slide valve.

The reciprocating parts form a complete link mechanism of the steam engine.

13.3.3. Revolving parts : *Those parts of the engine which follow circular path during the operation of the engine are called revolving parts.* In steam engine they are crank, crank shaft, eccentric and fly wheel.

In fact these are the parts which convert the linear motion of the reciprocating unit into rotary motion.

IMPORTANT PARTS OF THE STEAM ENGINE WITH THEIR DIFFERENT DESIGNS

13.3. PISTON AND PISTON ROD

A piston is a cylindrical part which reciprocates inside the engine cylinder by the action of steam. The leakage of steam from one side of the piston to the other is prevented by providing piston rings. These rings are cylindrical from of square or rectangular cross-section. The piston rings are so placed on the piston grooves that their cuts should never be in a line, otherwise the steam will leak. A small gap between the piston and ring is left for adjustment.

A rod connected to the piston is known as piston rod. One end of the piston rod is secured to piston by means of a special nut and pin. The other end of the piston rod after passing through the stuffing box is connected to the cross-head, which ensures for it a straight line motion (refer Fig. 13.1). For better gripping, usually the parts of the rod to be fitted into the piston hole is made partly tapered and partly parallel. The taper on the piston rod varies from 1 in 20 to 1 in 4 on diameter according to the length of the tapered portion.

13.4. BOX OR HOLLOW TYPE PISTON

Box type pistons are generally made of cast iron. These are casted into one piece and hollowed to reduce the weight (see Fig. 13.2). Cast iron rings are provided on the piston to prevent the leakage of steam from one side to other. Sometimes plugs are used to make the piston closed from outside, when it is used in a double acting engine.

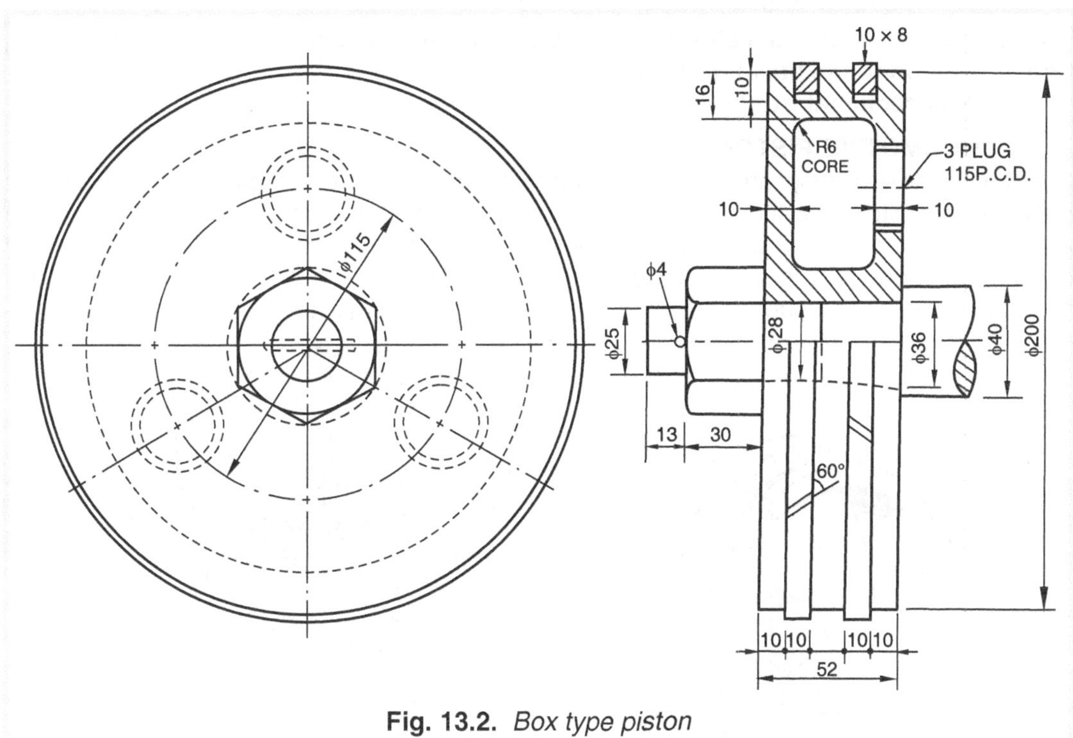

Fig. 13.2. *Box type piston*

Exercise 1: Fig. 13.2 shows two views of a box type piston.

Draw (a) Sectional elevation (b) End view.

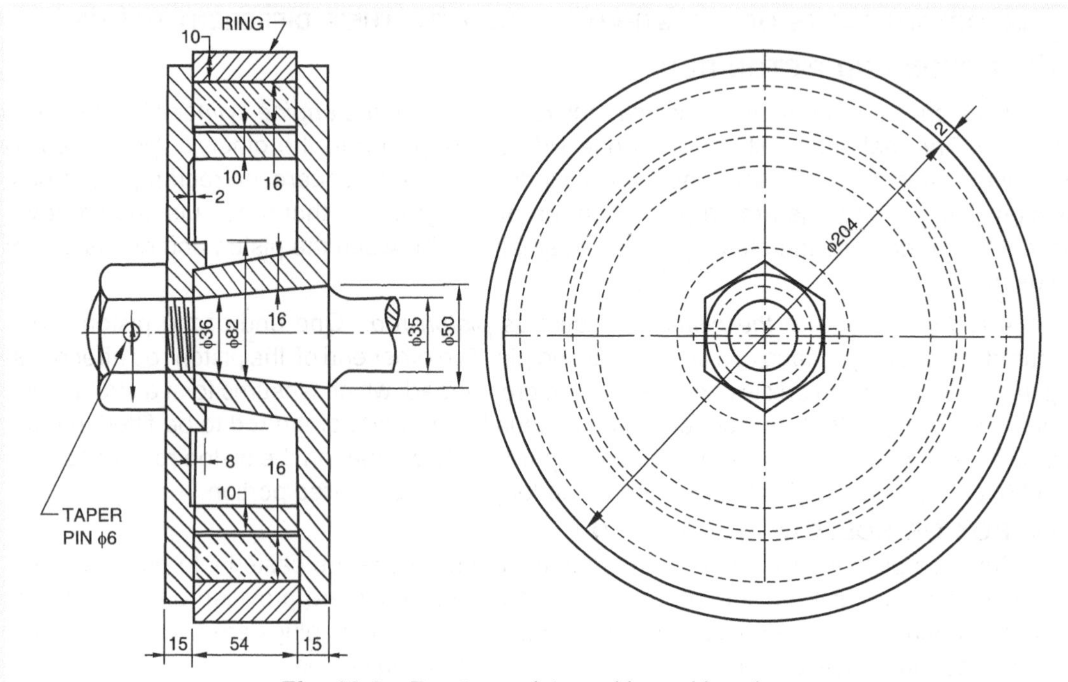

Fig. 13.3. *Box type piston with packing rings*

Exercise 2 : Fig. 16.3 shows two views of a box type piston with a brass packing ring in addition to the usual ring. Draw :

(a) Half sectional front-view (b) Side view.

13.5. LOCOMOTIVE PISTON

Locomotive pistons are generally made of cast-iron, but in these days the steel is frequently used to make the piston light. It mainly consists of a disc having grooves on the outer periphery to receive the cast iron rings. These rings are used to prevent the leakage of steam from one side to other. A tapered hole is provided in the disc in which piston rod is secured by means of a special nut and locked by a pin (see Figs. 13.4 and 13.5).

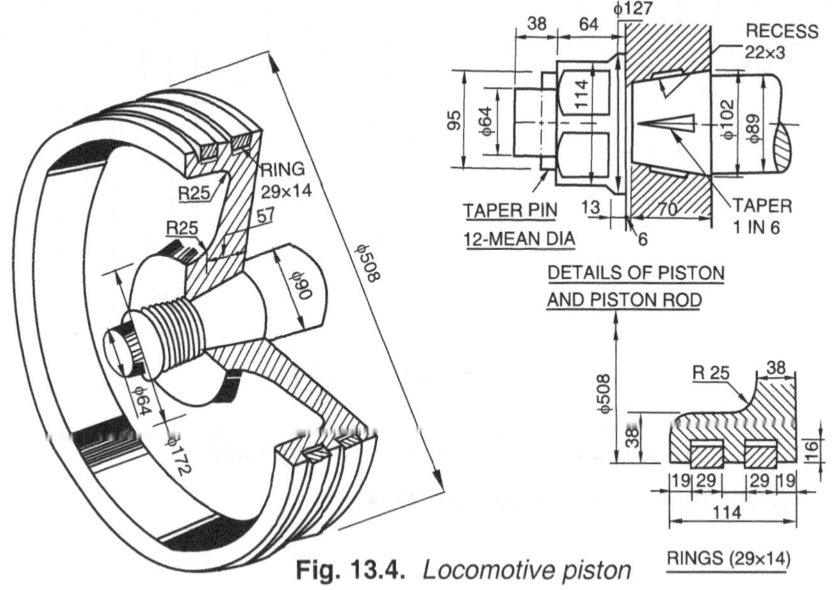

Fig. 13.4. *Locomotive piston*

Exercise 3: Fig. 13.4 shows the details of a piston and piston rod for a steam engine. Draw the following views to a scale of 1:1 :—

(a) Upper half sectional front view (b) End view.

13.6. PISTON RINGS

These are made of cast iron and produced by centrifugal casting. The molten metal is poured into a hollow cylinderical mould spinning about its axis at very high speed. The centrifugal force causes the liquid metal to flow to the mould's inner surface and settle there in the form of a hollow cylinder. The casting is removed from its mould after cooling. It is then machined all over the surface and then cut into rings of the required size. The rings are made a little larger in diameter than the bore of the cylinder.

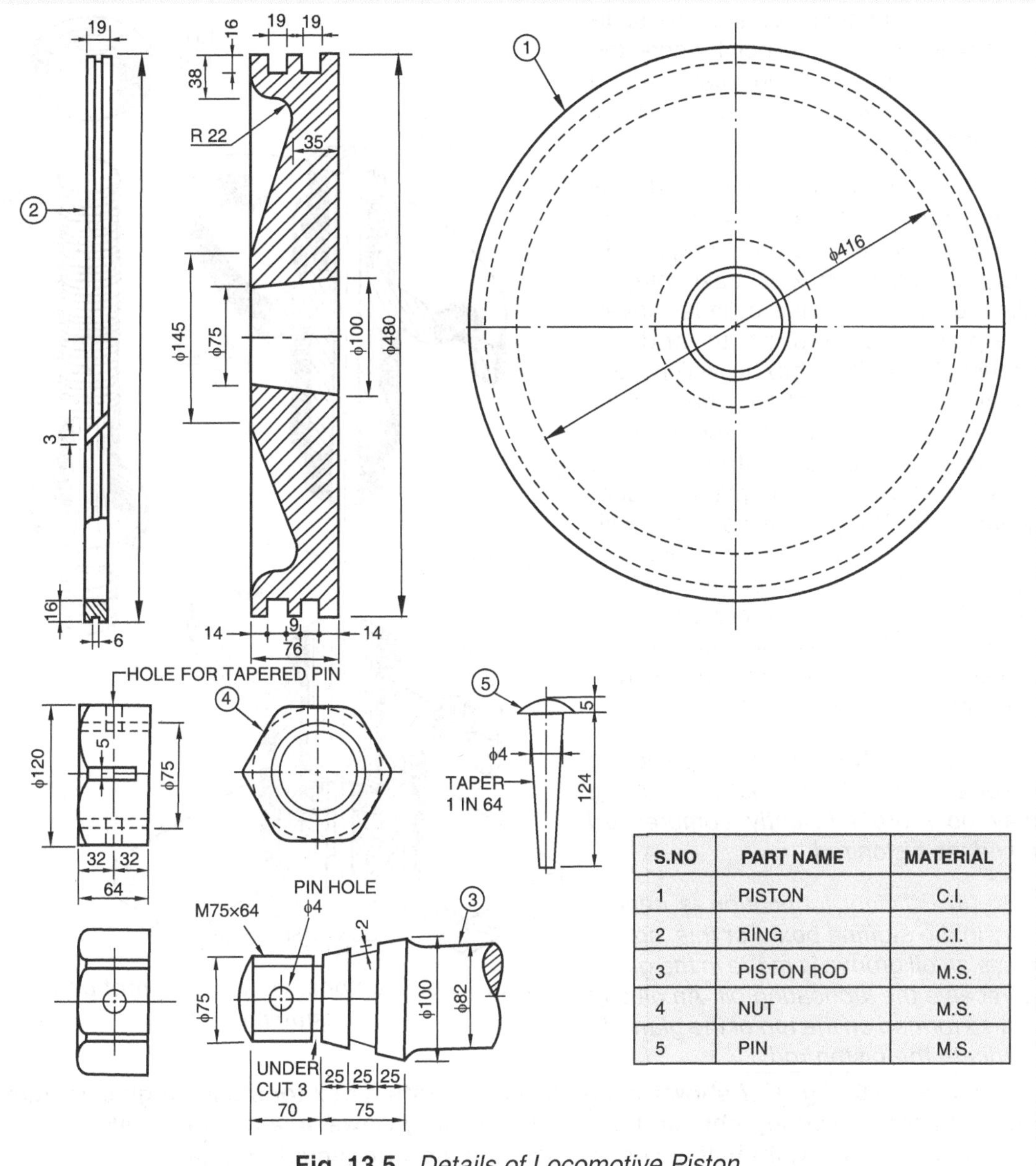

S.NO	PART NAME	MATERIAL
1	PISTON	C.I.
2	RING	C.I.
3	PISTON ROD	M.S.
4	NUT	M.S.
5	PIN	M.S.

Fig. 13.5. *Details of Locomotive Piston*

The rings are split and ends finished (see Fig. 13.5). These are then clamped on a jig, leaving a gap of 7-8 mm at the ends and turned outside diameter to its final required size. This gives the required spring action to the ring.

Exercise 4 : Fig. 13.5 shows the details of a locomotive piston. Imagine the parts assembled together and draw to a scale half full size the, following views :—

(a) Front view upper half in section (b) Side view looking from nut side

13.7. STUFFING BOX

A stuffing box is used where a sliding or rotating rod or shaft passes through the side of a vessel or cylinder containing fluid under pressure. *The function of a stuffing box is to prevent the leakage of fluid and allows a free movement of the rod.*

In case of steam engine, the stuffing box is used at the point where the piston rod and valve rod pass through the cylinder cover and the steam chest box (see Fig. 13.1).

Fig. 13.6. shows the pictorial view of different parts of a stuffing box used for a vertical steam engine. In this, the stuffing box is cast integral with the cylinder or the cylinder cover. The piston rod moves to and fro through the stuffing box in which gun-metal neck bush is fitted. The annular space around the piston rod is fitted with asbestos or any other type of packing, which is held in place by gun-metal gland. The gland is tightened with the stuffing box by means of studs and nuts. Thus, the packing around the piston rod is compressed in order to make it steam tight. When the packing becomes loose after a certain period, then it can be compressed tightly.

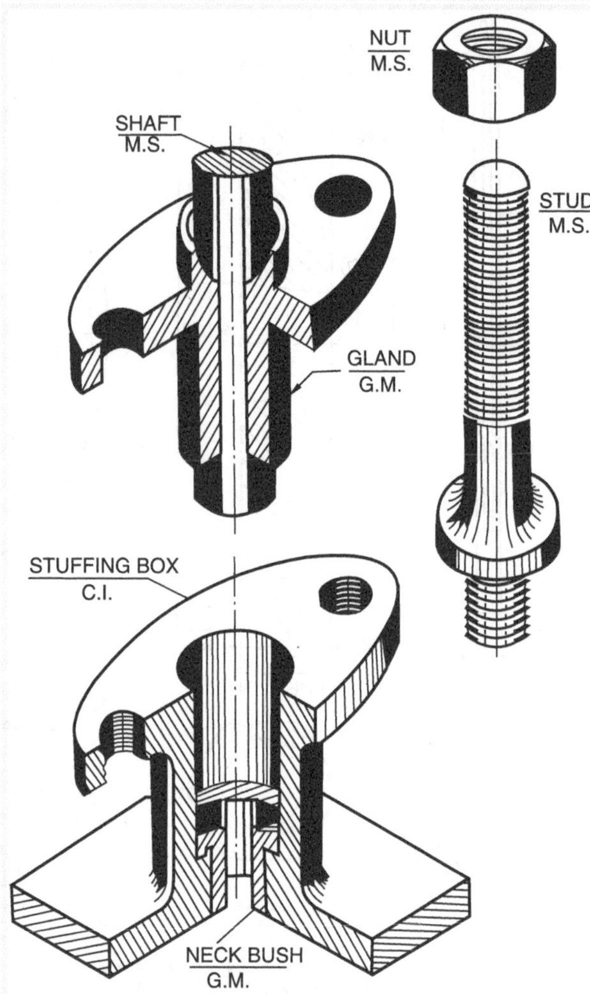

Fig. 13.6. *Poctorial view of different parts of a stuffing box*

Note : *(i) Neck bush and gland are hollowed out at 30° so that the packing may be more efficiently compressed round the piston rod.*

(ii) Efficient lubrication is necessary in the stuffing box. For this, sometimes an oil groove is made in the gland to receive the lubricating oil. An oil cup is also formed on the top of the gland to lubricate the piston rod.

Exercise 5: Fig. 13.7 shows the details of a stuffing box for steam engine. Imagine the parts assembled together and draw the following views to a full size scale :—

(a) Front view-right half in section (b) End view projected from (a)

(c) Top view.

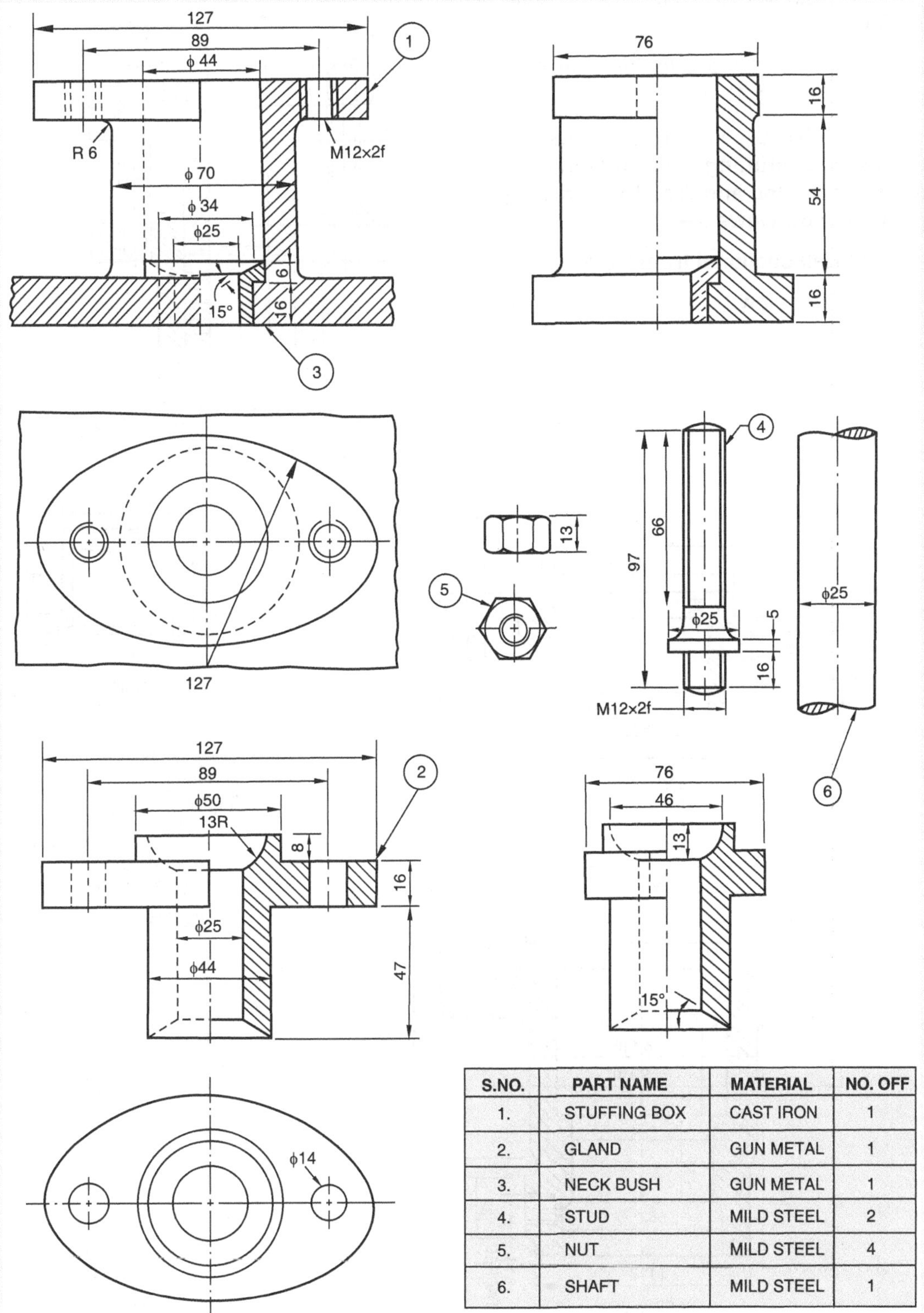

S.NO.	PART NAME	MATERIAL	NO. OFF
1.	STUFFING BOX	CAST IRON	1
2.	GLAND	GUN METAL	1
3.	NECK BUSH	GUN METAL	1
4.	STUD	MILD STEEL	2
5.	NUT	MILD STEEL	4
6.	SHAFT	MILD STEEL	1

Fig. 13.7. *Details of a stuffing box*

Fig. 13.8 shows Front view of a stuffing Box

Hint : For the solution of exercise 5 (Part a), refer Fig. 13.8.

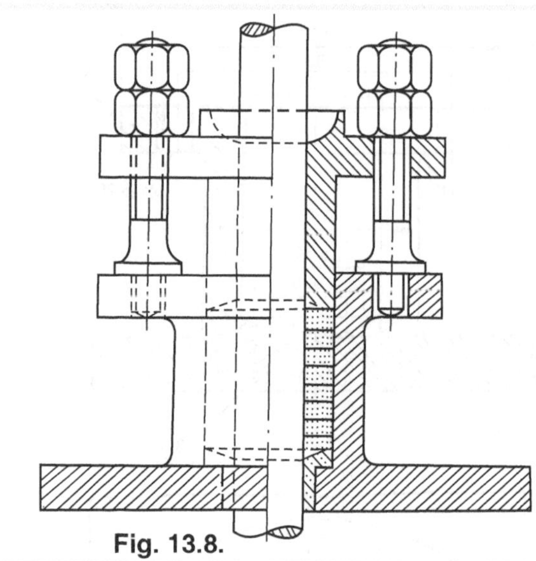

Fig. 13.8.

Exercise 6 : Fig. 13.9 shows the details of a stuffing box. Imagine the parts assembled and draw to a scale 1:1 the following views :—

(a) Half sectional front view

(b) Top view.

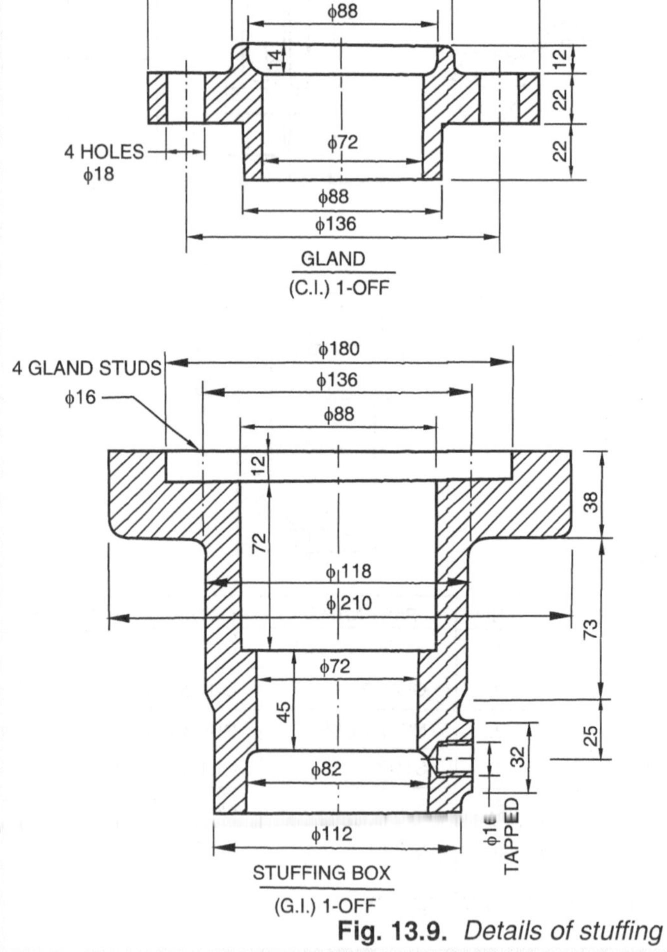

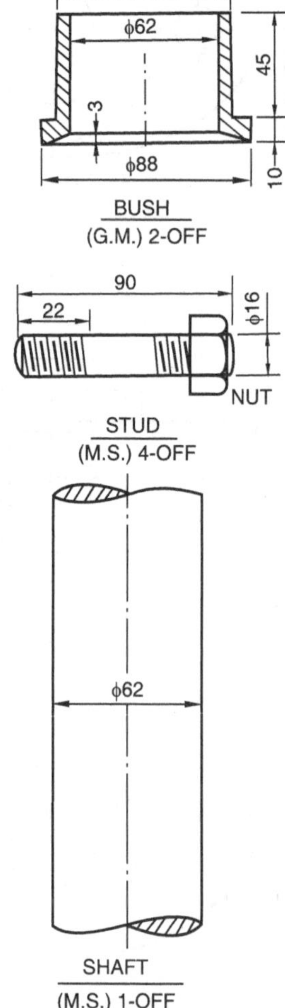

Fig. 13.9. *Details of stuffing box*

13.8. CROSS HEAD

A cross-head is a sliding block which connects piston rod to the connecting rod (see Fig. 13.10). It slides in between parallel guides and makes the piston and piston rod to move in a straight line.

On one side of the cross-head, the piston rod is secured on either by screwing or by cotter joint or it is forged with the cross-head and on the other side the small end of the connecting rod is connected by means of gudgeon pin. Fig. 13.10 shows the pictorial view of a cross-head in which the end of the piston rod is forged with cross-head.

Notes : *(i) Gun-metal bushes (brasses) are made into two halves to facilitate their fixing and removing.*

(ii) Brasses are made more thick on the two sides as these sides would be subjected to great wear and tear.

(iii) Snugs are provided on the bolt heads in order to prevent the rotation of the bolts when nuts are being screwed.

(iv) The clearance between keep plate and the body is provided for future adjustment and repair of brasses.

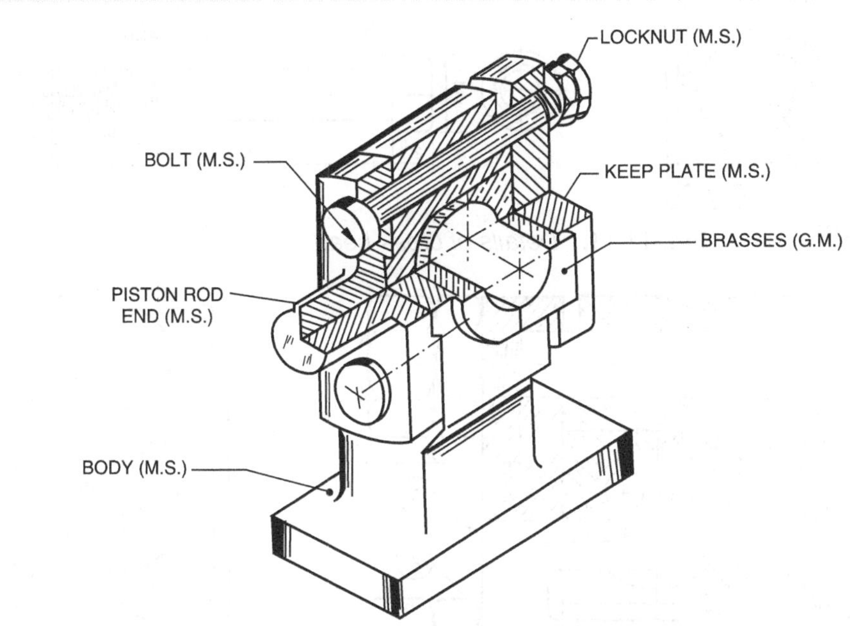

Fig. 13.10. *Pictorial view of a cross-head*

Clearance is generally left between the body and the keep plate (in assembly see Fig. 13.11). This clearance is provided for adjustment of the brasses. As after a certain period of running the pin the brasses become elliptical in shape from inside. This elliptical shape will increase the vibration of the shaft. Then, the two brasses can be cut off at their edges to give them again a round shape from inside when assembled. Then the clearance decreases and ultimately reduces to zero. After this, the brasses are replaced by new one's.

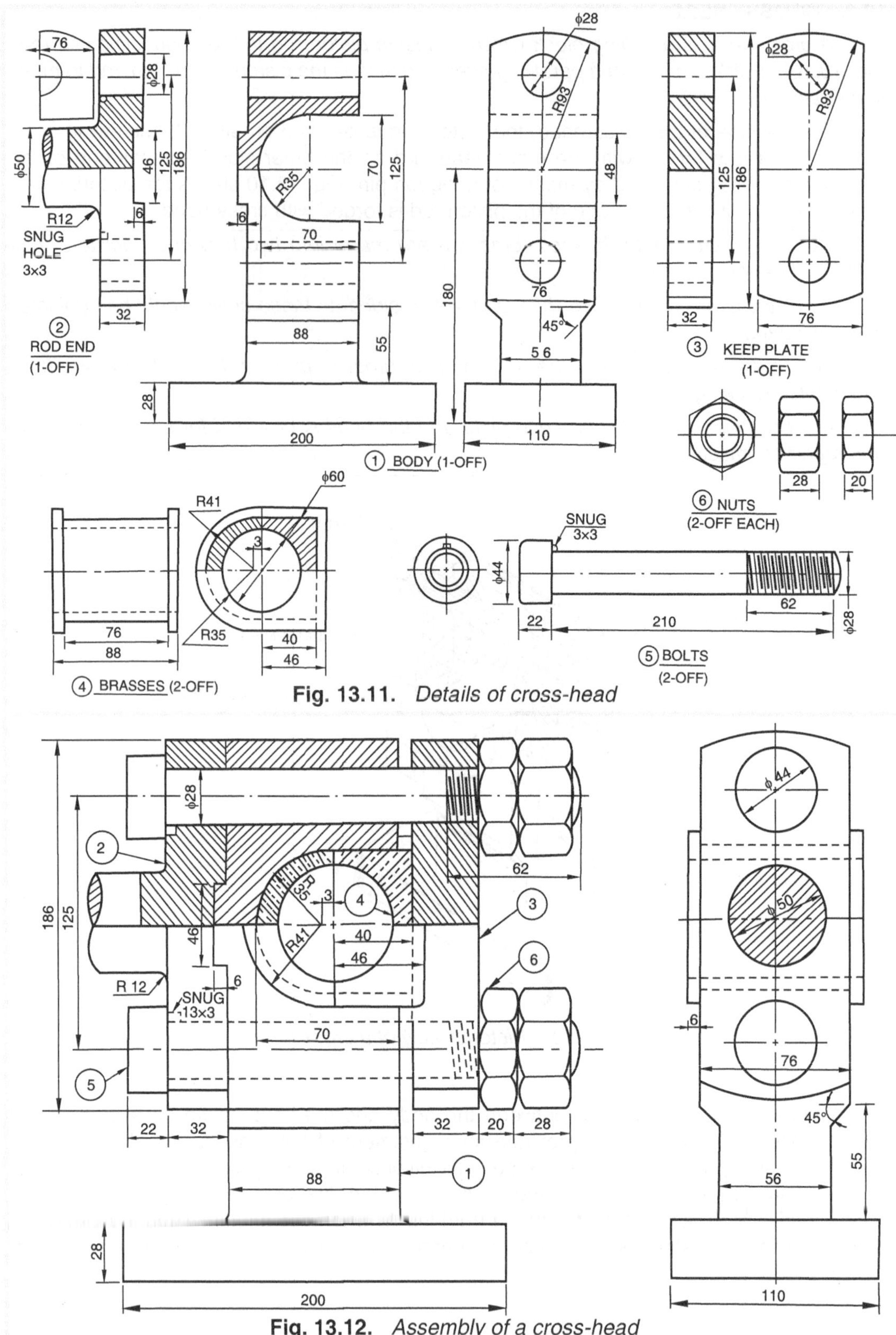

Fig. 13.11. *Details of cross-head*

Fig. 13.12. *Assembly of a cross-head*

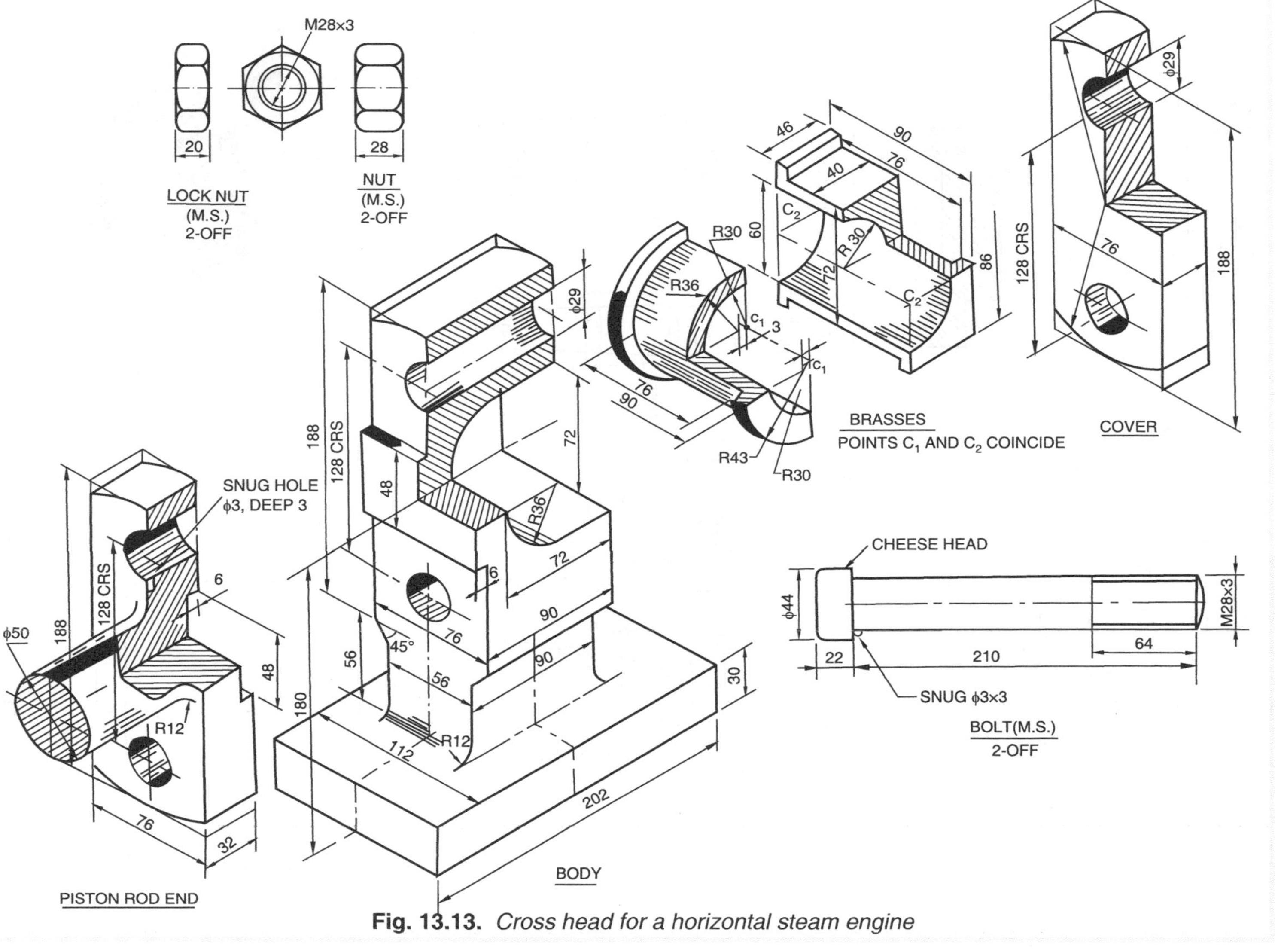

Fig. 13.13. *Cross head for a horizontal steam engine*

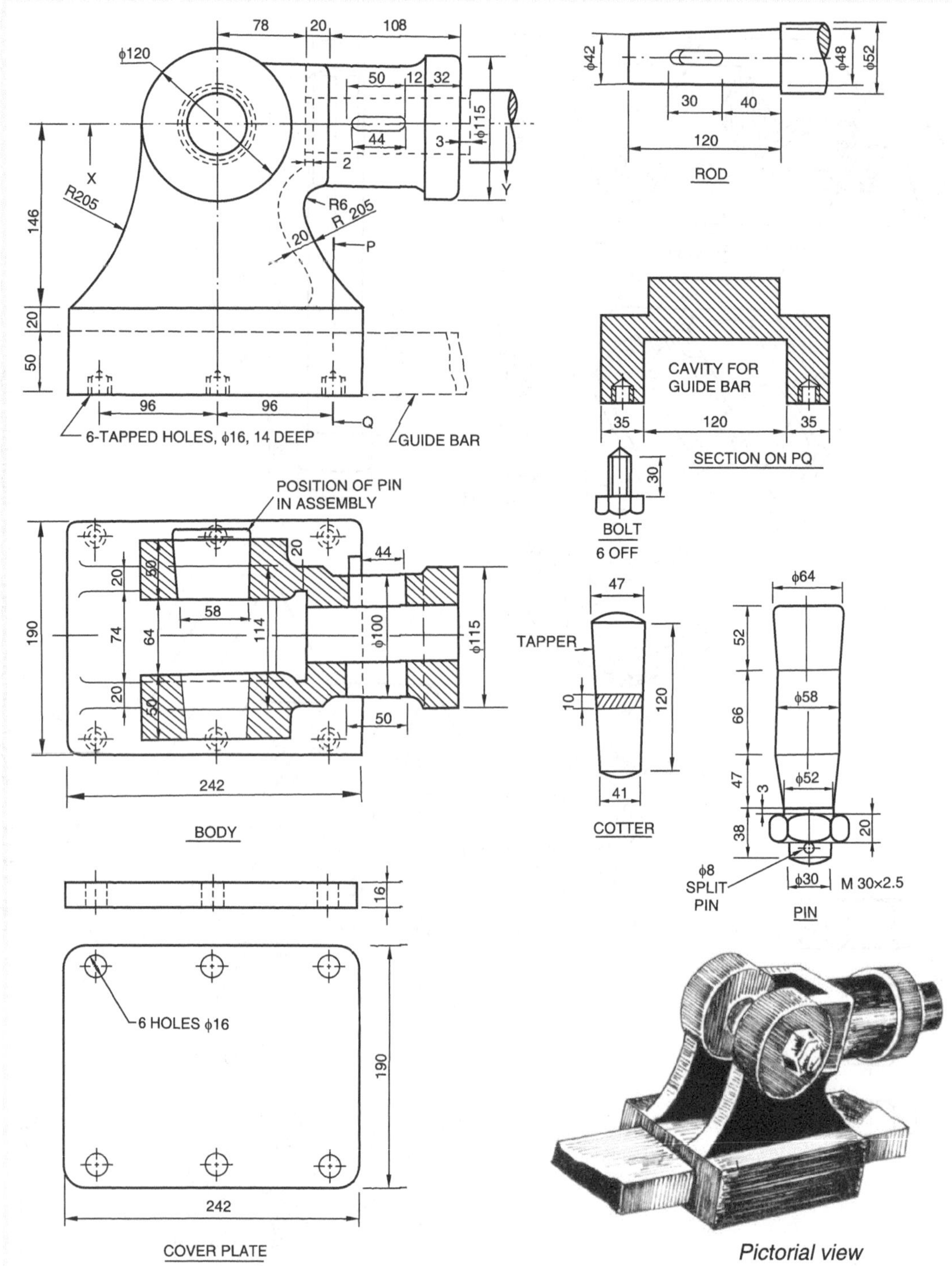

Fig. 10.14. *Cross head for a horizontal steam engine*

Problem 1 : Fig. 13.11 shows the details of a cross-head for a steam engine. Imagine the parts assembled and draw the following views :—

(a) Half sectional front view, the portion above the horizontal centre line through the brasses being shown in section.

(b) Side view.

Solution : For its solution, see Fig. 13.12.

Exercise 7 : The sketch in Fig. 13.12 shows two views of a cross-head of a steam engine. Draw to a scale full size the following views :—

(a) Front view in full section

(b) Side view upper half in section looking from the side opposite to piston rod

(c) Top view.

Exercise 8 : Fig. 13.13 shows the pictorial details of a cross-head for a horizontal steam engine. Draw the following views of the cross head assembly :—

(a) Front-view – lower half in section.

(b) R.H. side view left half in section.

(c) Top view – full in section. The section plane passing through the centre of the brasses.

Fig.13-14 shows the details of another cross-head. The design of the body of the cross-head provides for cotter joint connection with the piston and knuckle joint to the connecting rod end. To hold the cross-head against the guide bar, a rectangular cover plate attached to it at the base with the help of screws. A recess is provided on the base of cross-head body which forms a way for a rectangular guide bar.

Exercise 9 : Fig. 13.14 shows the details of a cross-head for a horizontal steam engine. Draw the following assembled views to some suitable scale;

(a) Front view (Elevation)

(b) Side view (End View)

(c) Top view (Plan).

13.9. CROSS HEAD WITH DETACHABLE SHOE

In this cross head, the piston rod is an integral part of the body. The two halves bearings are held in position by the end plate (cap). The slide block (shoe) is connected to the body with special types of screws. A projection is provided in the slide block which fits into the corresponding recess in the body. The screw which fix the slide block with the body are further locked by locking screws. The cap bolts are locked in position by locking screws.

Note : *The main advantage of detachable shoe is that it can be replaced when it wears out.*

Exercise 10 : Fig. 13.15 shows the details of the locomotive cross-head with detachable shoe. Draw the following assembled views :—

(a) Front view – right half in section

(b) R.S. end view

(c) Top view.

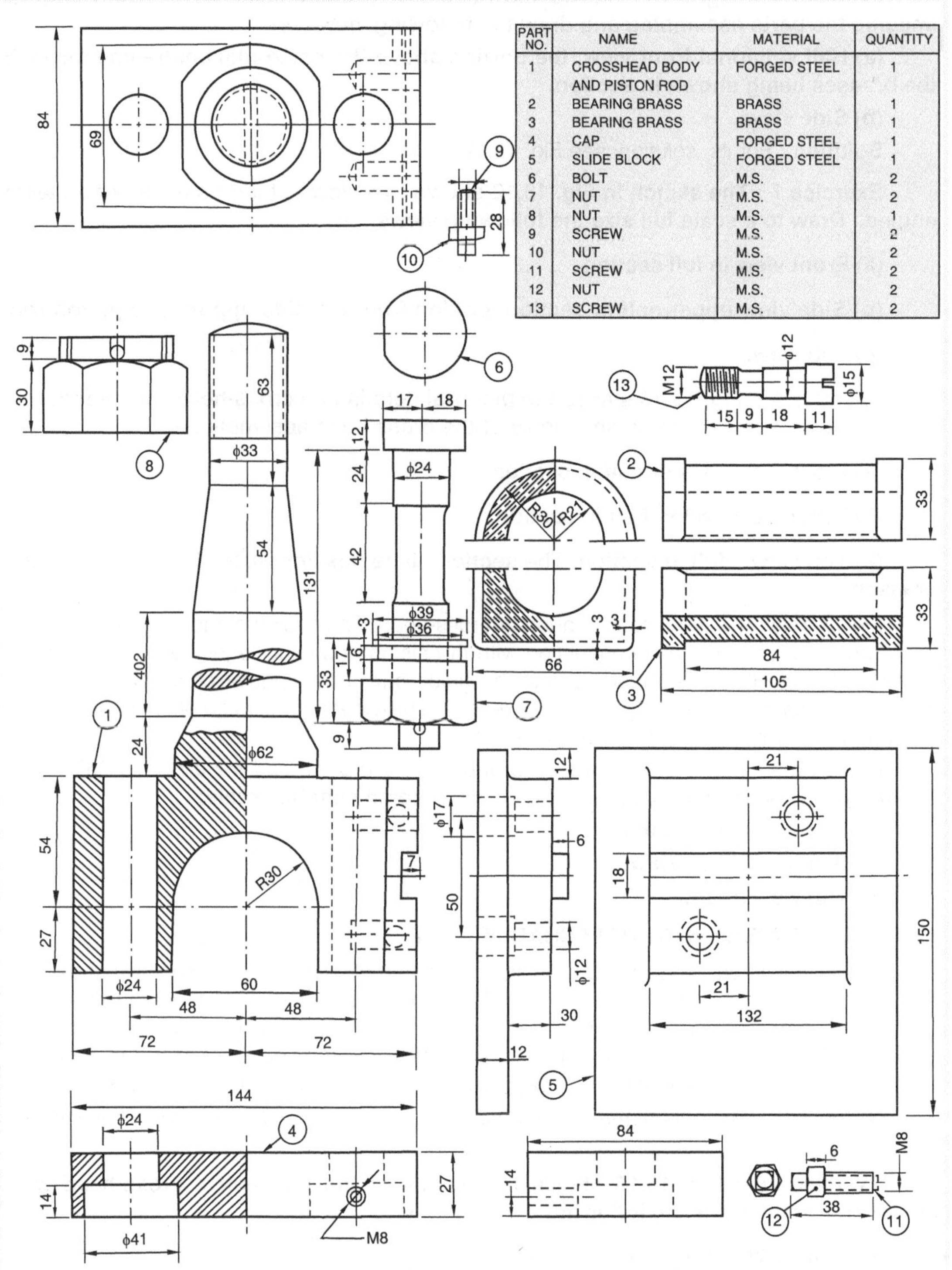

PART NO.	NAME	MATERIAL	QUANTITY
1	CROSSHEAD BODY AND PISTON ROD	FORGED STEEL	1
2	BEARING BRASS	BRASS	1
3	BEARING BRASS	BRASS	1
4	CAP	FORGED STEEL	1
5	SLIDE BLOCK	FORGED STEEL	1
6	BOLT	M.S.	2
7	NUT	M.S.	2
8	NUT	M.S.	1
9	SCREW	M.S.	2
10	NUT	M.S.	2
11	SCREW	M.S.	2
12	NUT	M.S.	2
13	SCREW	M.S.	2

Fig. 13.15. *Cross head with detachable shoe*

13.10. CONNECTING ROD END

The function of the connecting rod is to convert the reciprocating motion of the crosshead into rotary motion of the crank shaft. [see Fig. 13.1]

A connecting rod of steam engine is a link which connects the cross-head to crank. It is provided with enlargements at both ends. These enlargements are known as connecting rod ends. One end of the connecting rod which is connected to the crosshead with the help of a gudgeon pin is called the small end. The other end which is secured to the crank is called the big end.

Connecting rods are generally made of forged steel in various designs. The central portion of the connecting rod is made in circular, elliptical or I–section.

Fig. 13.16 and 13.17 show the pictorial view and detail drawings of a connecting rod used in the steam engine.

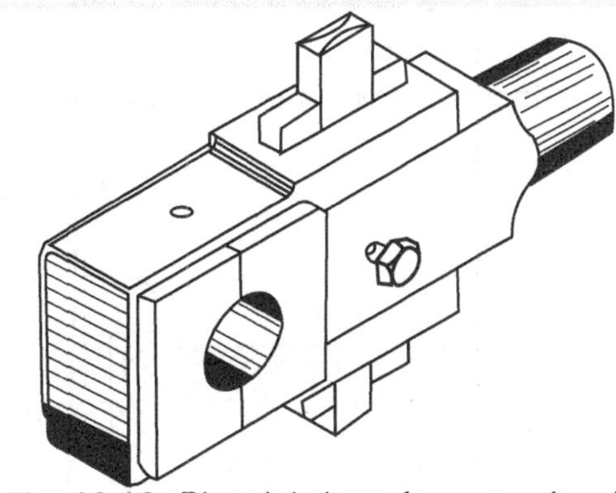

Fig. 13.16. *Pictorial view of conn. rod end*

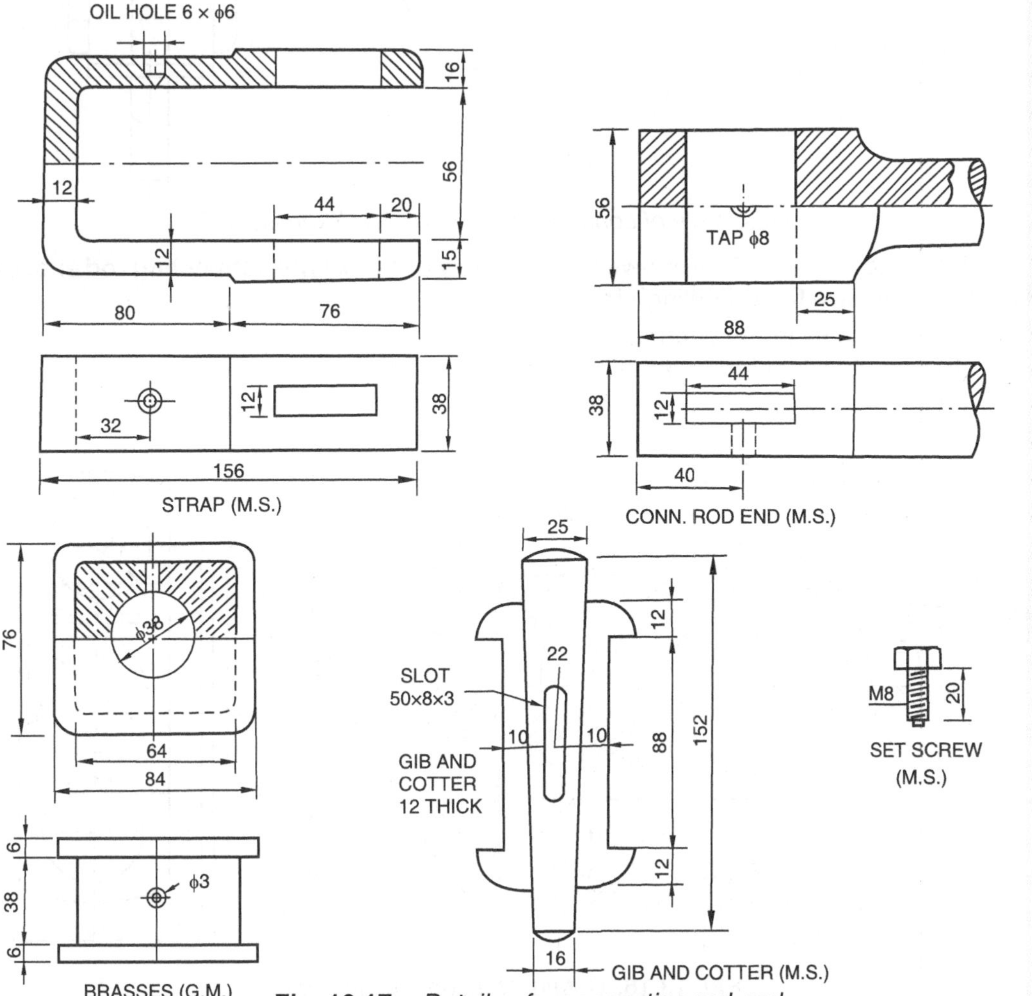

Fig. 13.17. *Details of a connecting rod end*

Exercise 10 : Fig. 13.17 shows the details of a connecting rod end. Imagine the parts assembled and draw to a scale full size the following views :—

(a) Front view upper half in section

(b) Side view looking from left

(c) Top view.

Insert crank pin in the above views.

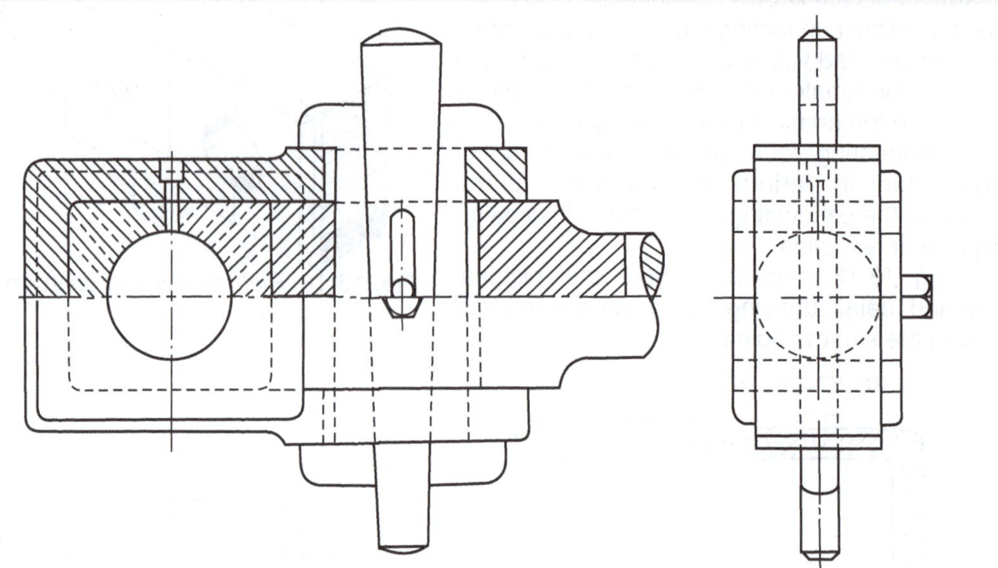

Assembly of connecting rod end [For Exercise-10]

Exercise 11 : Fig. 13.18 shows the assembly of a box type connecting rod end. Draw to a scale full size the following views :—

(a) Front view in full section

(b) End view looking in the direction of rod end

(c) Top view.

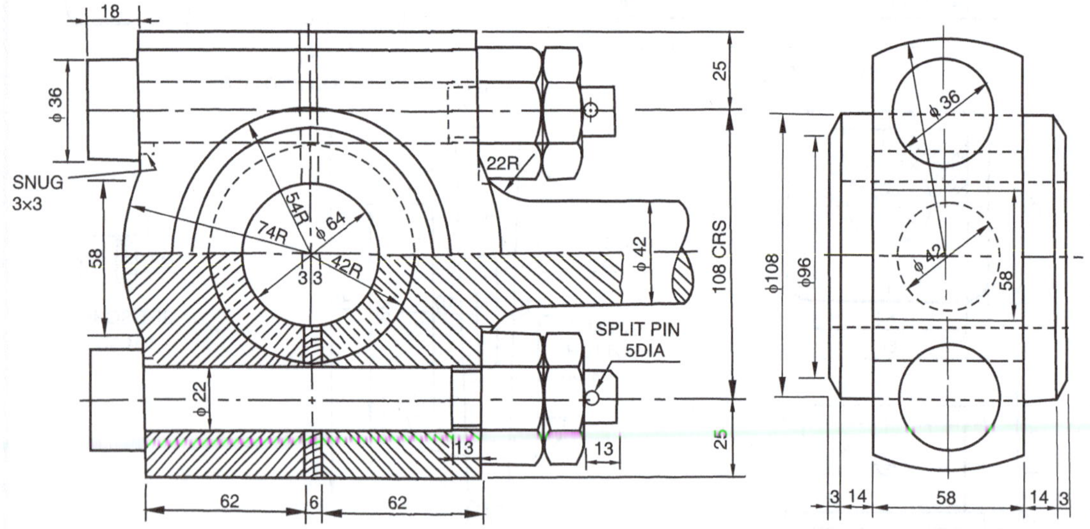

Fig. 13.18. *Details of a connecting rod end*

Exercise 12: Fig. 13.19 shows the details of big end of connecting rod. Draw the following assembled views on an A 2 size sheet :—

(a) Front view – upper half in section

(b) End view – looking from right side

(c) Top view.

Put in all necessary dimensions and centre lines. Print in a suitable lettering the necessary titles, sub-titles etc. Use your judgement for any dimensions not known.

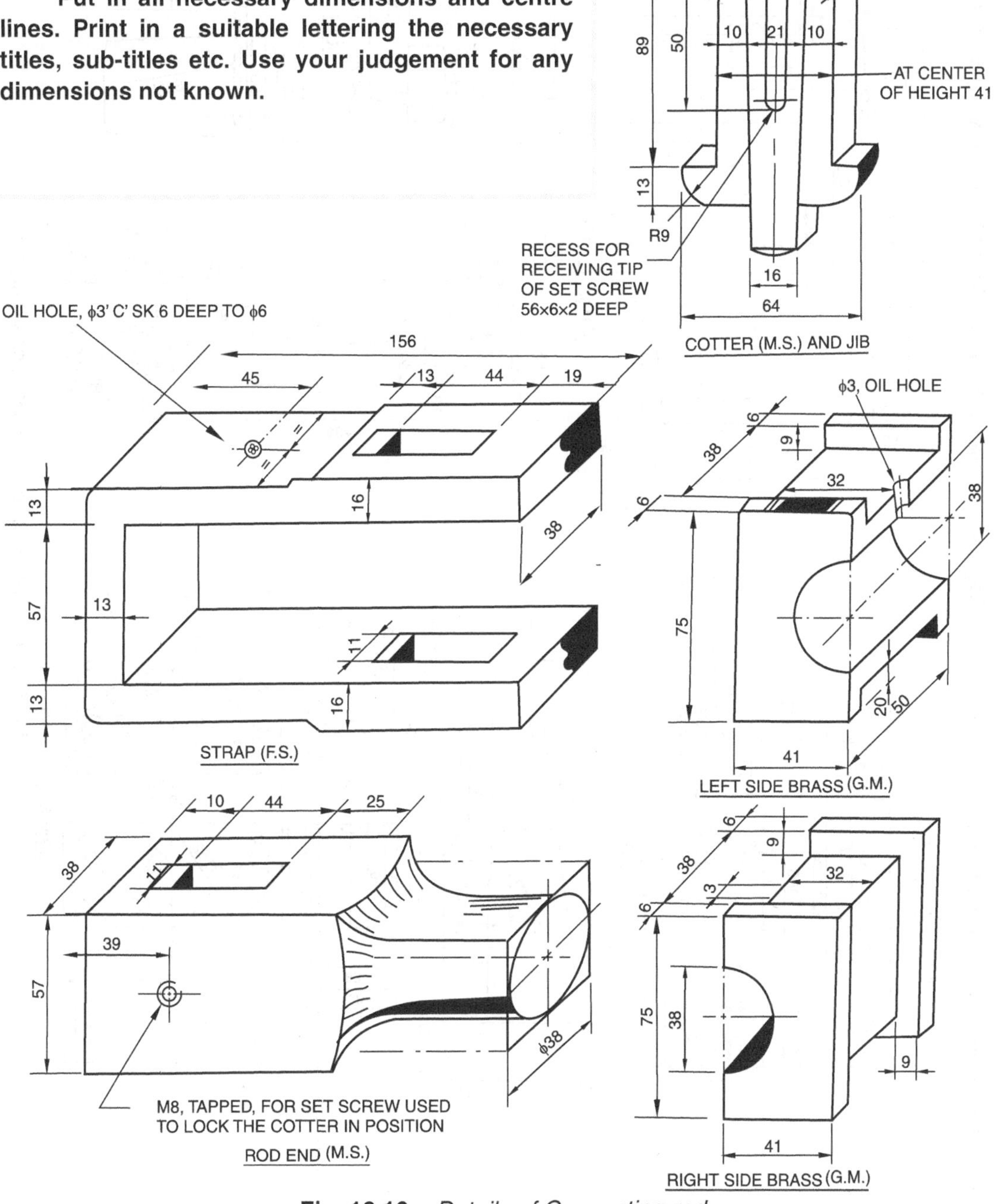

Fig. 13.19. *Details of Connecting rod*

Exercise 13: Fig. 13.20 shows the details of connecting rod. Draw (a) Half sectional front view (b) End view (c) Top view.

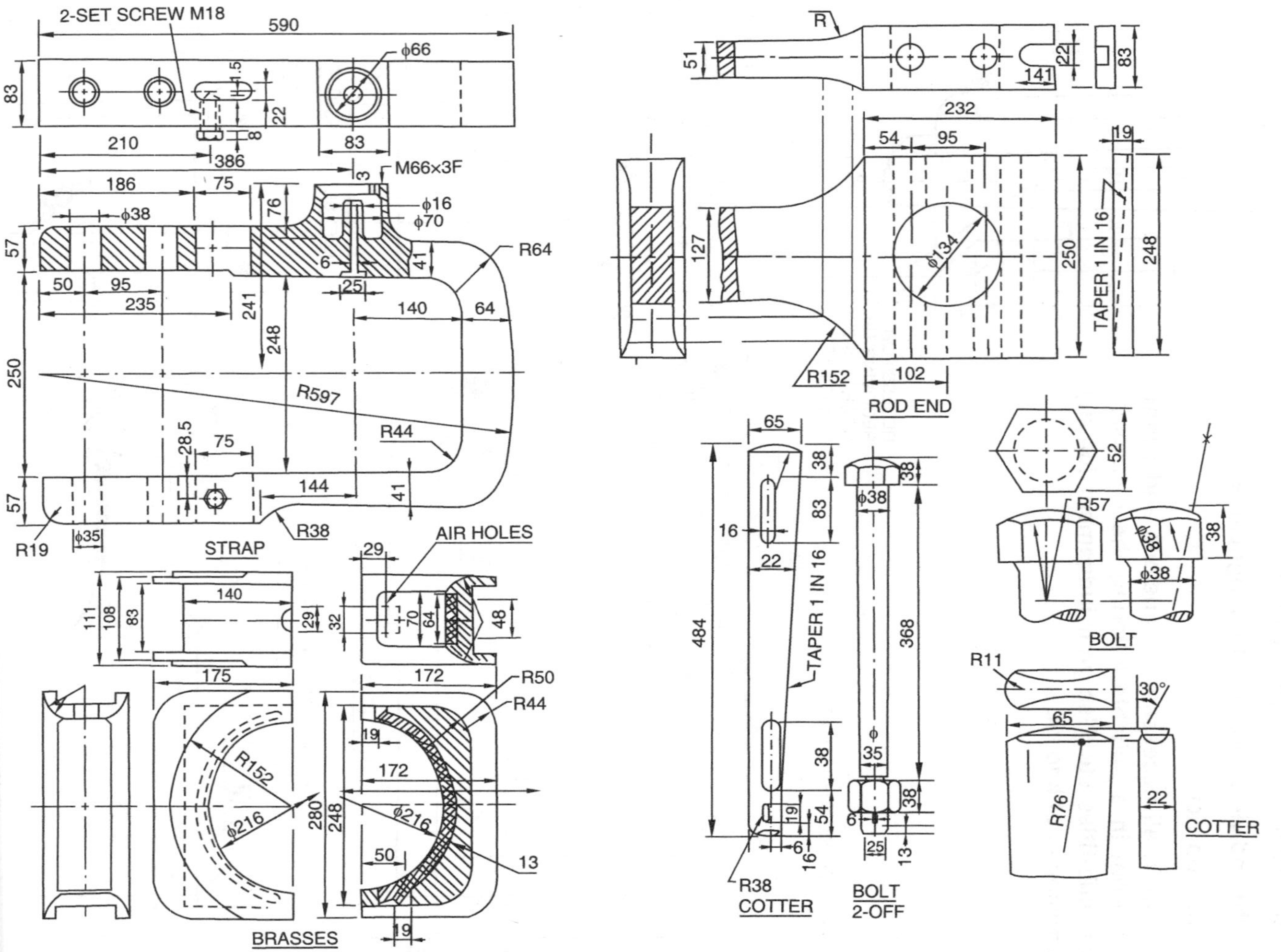

Fig. 13.20. *Details of big end of connecting rod*

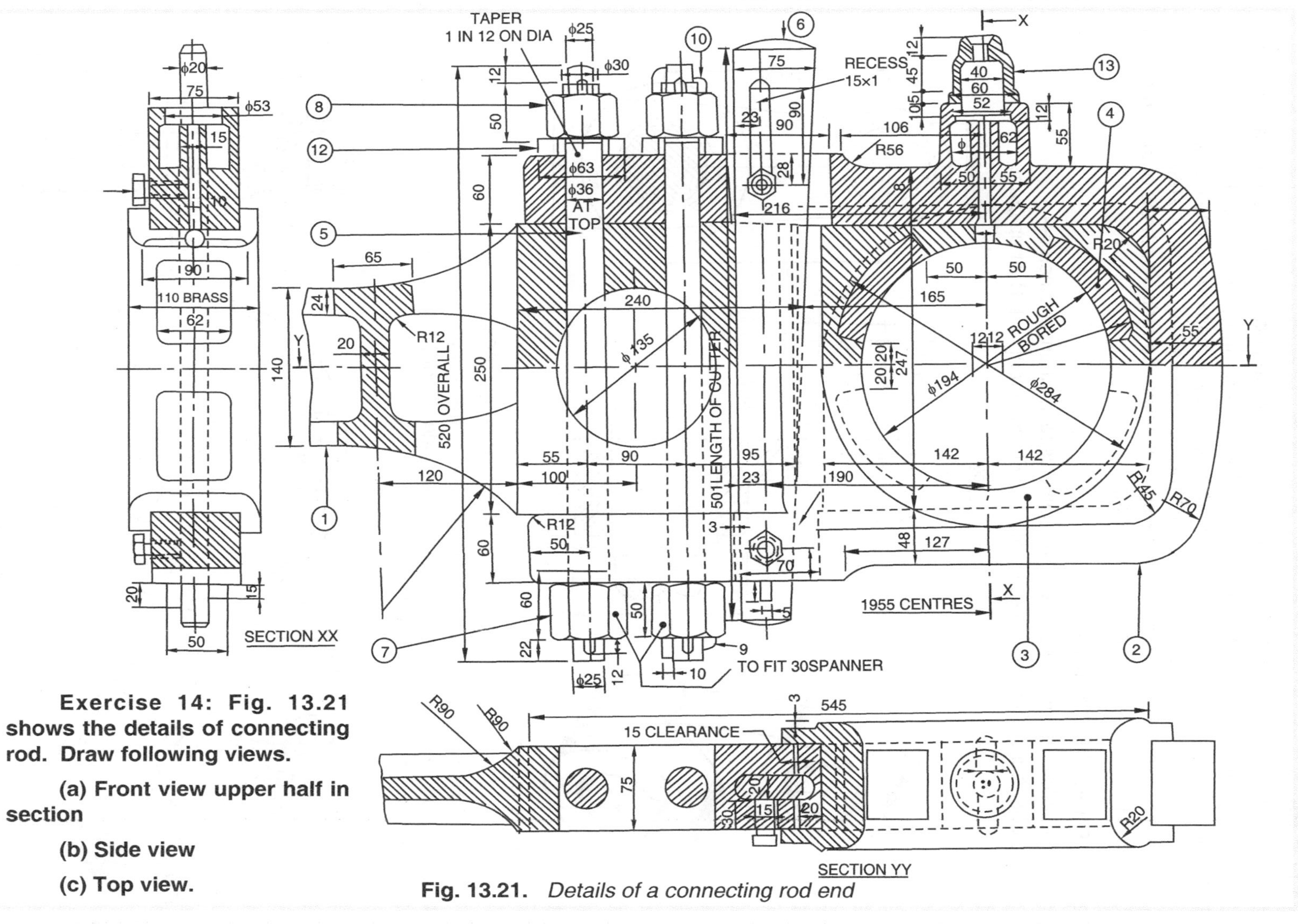

Fig. 13.21. *Details of a connecting rod end*

Exercise 14: Fig. 13.21 shows the details of connecting rod. Draw following views.

(a) Front view upper half in section

(b) Side view

(c) Top view.

13.11. CRANK

The part which convert reciprocating motion of piston into rotary motion is called crank.

The crank consists of three parts – crank pin, web and crank shaft. The crank may be forged in one piece with crank shaft or may be built up of crank webs and crank pin (see Fig. 13.24). The webs may be of steel or wrought iron while the pin is of steel. The distance between the axes of crank pin and crank shaft is called the radius or throw of the crank. The throw is equal to half the stroke of the piston.

13.12. TYPES OF CRANKS

The following types of cranks are mostly used in steam engines :—

1. Over hung crank 2. Disc crank 3. Forged crank 4. Built-up crank

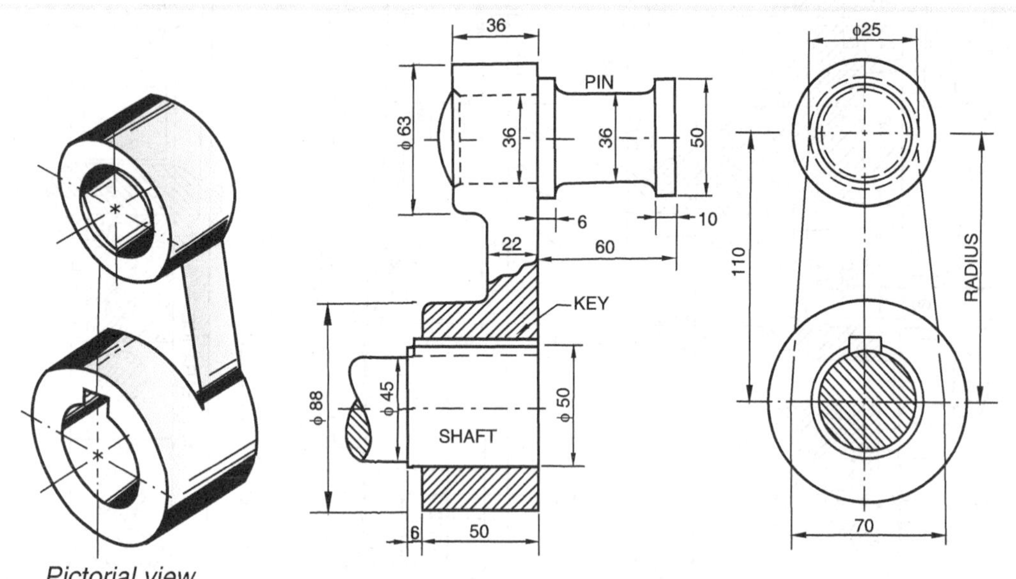

Fig. 13.22. *Overhung crank*

13.12.1. Overhung crank : This type of crank is supported by a bearing on one side only and has the pin overhanging on the otherside. It consists of two bosses-one for the pin and other for the shaft (see Fig. 13.22).

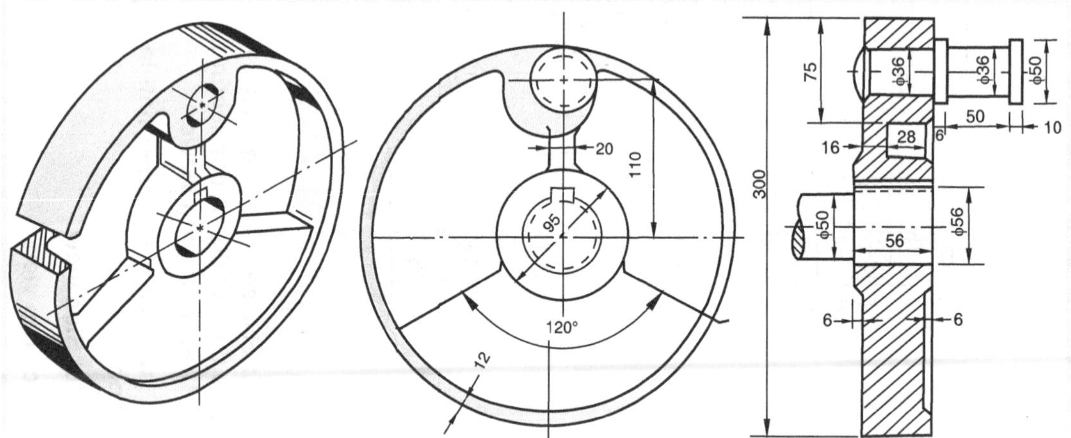

Fig. 13.23. *Disc of crank*

13.12.2. Disc crank : This type of crank consists of a cast iron disc keyed to the end of the crank shaft. The crank pin is riveted to the disc. The part of the disc on the other side on the shaft is made heavier to balance it and is used for small engines. (see Fig. 13.23).

13.12.3. Built up crank : This type of crank is built-up of separate parts and is commonly used in large marine engines. The webs are shrunk on the pin and on the shaft. A pin key is also used to secure web firmly to the shafts (see Fig. 13.24).

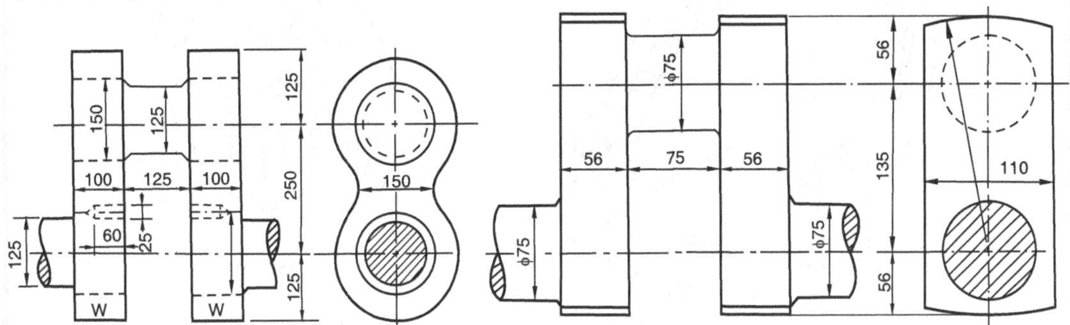

Fig. 13.24. *Built up crank* **Fig. 13.25.** *Forged crank*

13.9.4. Forged crank : In this type, the crank pin, web and shaft are forged in one piece. The webs are of rectangular shape. If there are two cranks they are spaced 90° or 180° apart. Forged cranks are commonly used in marine and locomotive engines (see Fig. 13.25).

13.13. ECCENTRICS

An eccentric is used to convert rotary motion of the crank shaft into reciprocating motion of the slide valve of a steam engine [see Fig. 13.1]

Fig. 13.26 shows details of an eccentric used in a steam engine. It consists of two main parts, sheave and strap. The sheave is a circular disc, the outer rim of which is projected in the form of a ring. The sheave (for crank shaft) is not made in the centre, but eccentric. The strap is made in two parts with annular recesses to accomodate the rim of the sheave. The sheave and straps are held together my means of bolts and nuts with packing strips placed between them. The packing strips permit adjustment for wear at a later stage. An eccentric rod is connected to one of the strips by means of studs.

When the crank shaft rotates, the sheave rotates eccenrically because of the eccentric hole and imparts reciprocating motion to the eccentric rod.

The sheave and strap are made of case iron while the bolts, nuts and studs are of mild steel.

Note : *(i)* The distance between the axes of crank shaft and the sheave is called the eccenricity *(e)* of the eccentric. It is equal to the half the travel of the slide valve.

In Fig. 13.26, $R > e + r$ where R = Radius of the sheave

e = Radius or Eccentricity of eccentric r = Radius of crank shaft.

(ii) The reverse conversion is not possible due to excessive friction between the sheave and the strap.

(iii) For balancing and to reduce the weight of eccentric, some metal is removed from the sheave.

In compound engines, the sheave is made in two parts as shown in Fig. 13.26. The larger part is made of cast iron while the other is of steel.

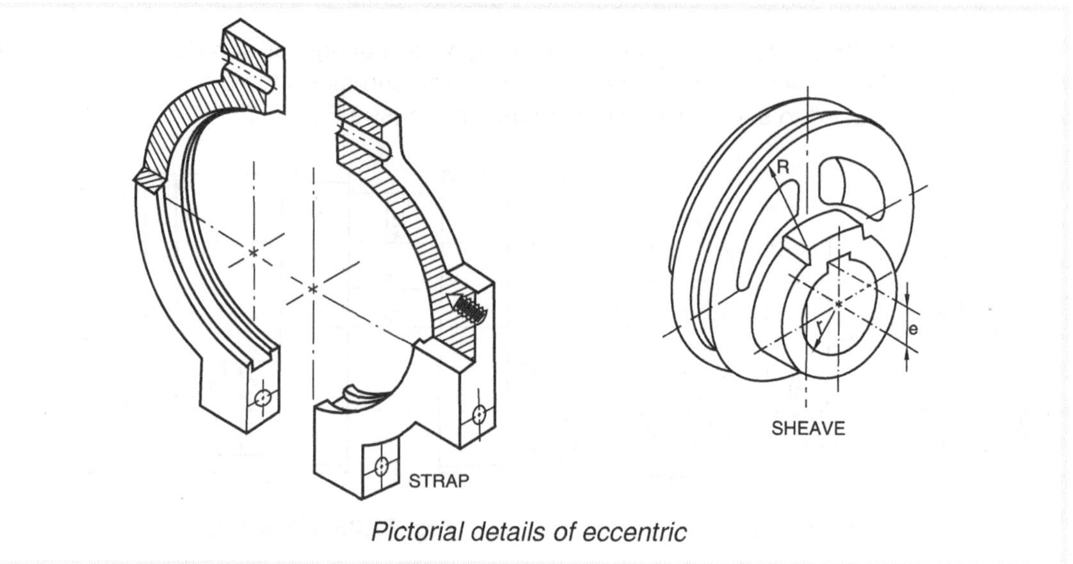

Pictorial details of eccentric

Exercise 15 : Fig. 13.26 shows the details of a simple eccentric. Draw the following assembled views to suitable scale:-

(a) Front views-upper half in section

(b) End view

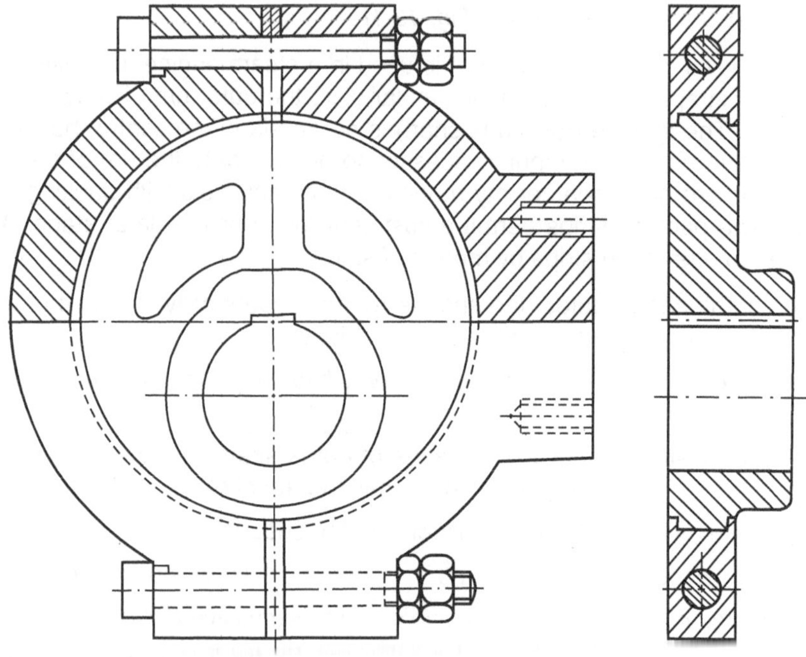

Assembly of an eccentric for Fig. 13.26

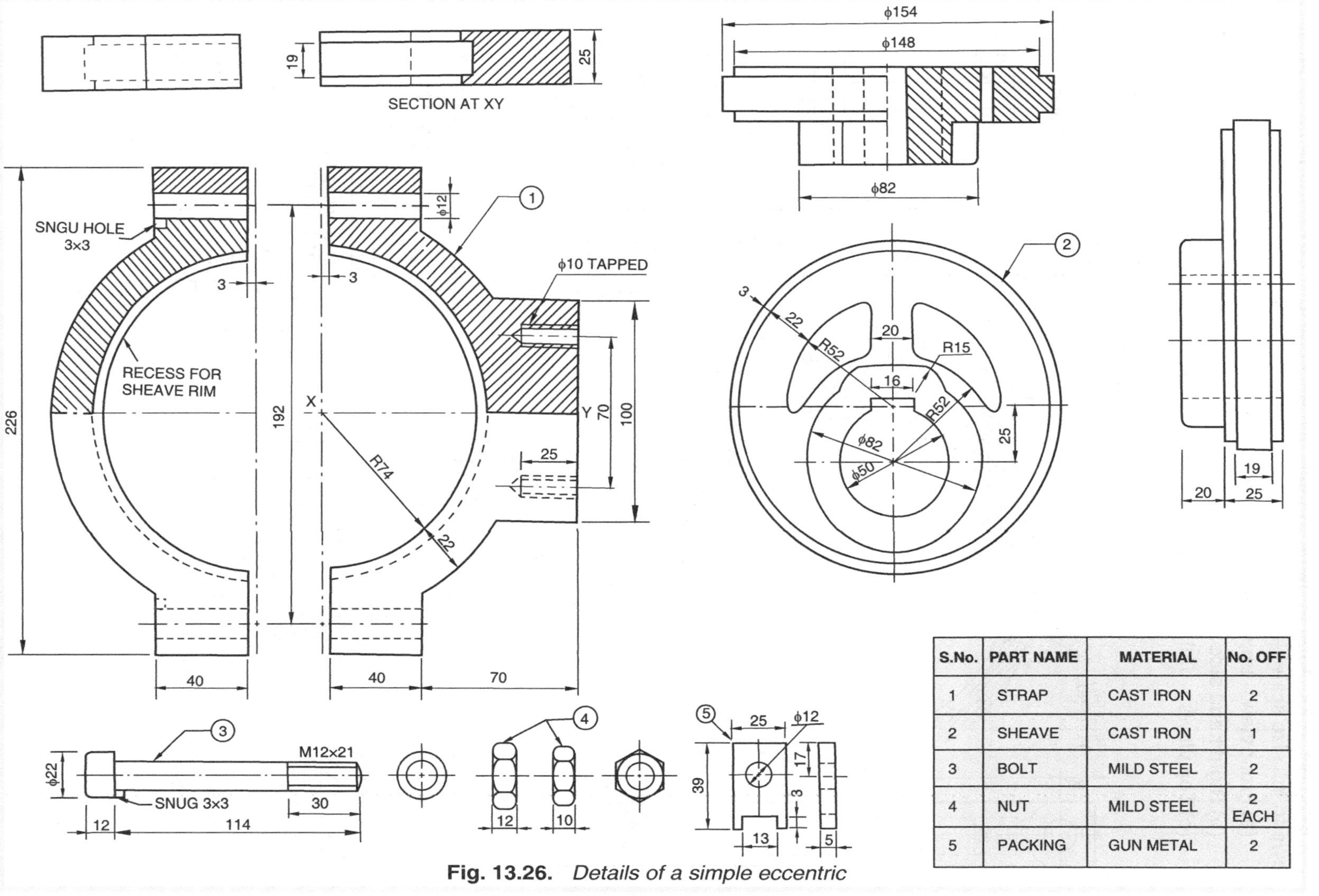

S.No.	PART NAME	MATERIAL	No. OFF
1	STRAP	CAST IRON	2
2	SHEAVE	CAST IRON	1
3	BOLT	MILD STEEL	2
4	NUT	MILD STEEL	2 EACH
5	PACKING	GUN METAL	2

Fig. 13.26. *Details of a simple eccentric*

Exercise 16. Fig. 13.27 shows the details of the eccentric. Draw the following assembled views to full size scale :—

(a) Front view – upper half in section

(b) Right hand side view

(c) Top views – full in section, through the centre of the strap.

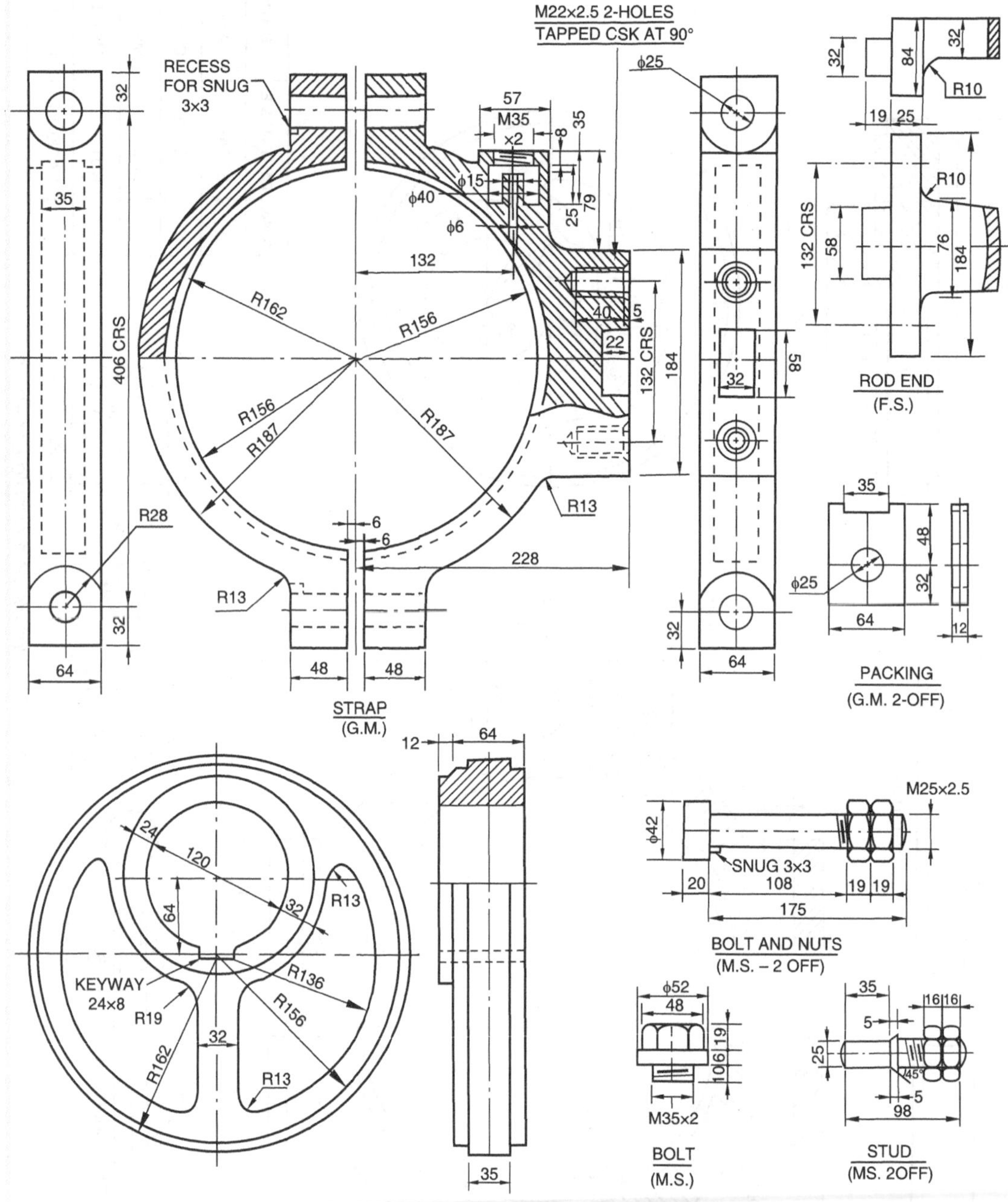

Fig. 13.27. *Details of an eccentric*

Exercise 17 : Fig 13-28 shows the details of an eccentric. Draw the following views of the eccentric assembly :—

(a) Front views-upper half in section

(b) Top views – full in section, through the centre of the straps

(c) Right side end view.

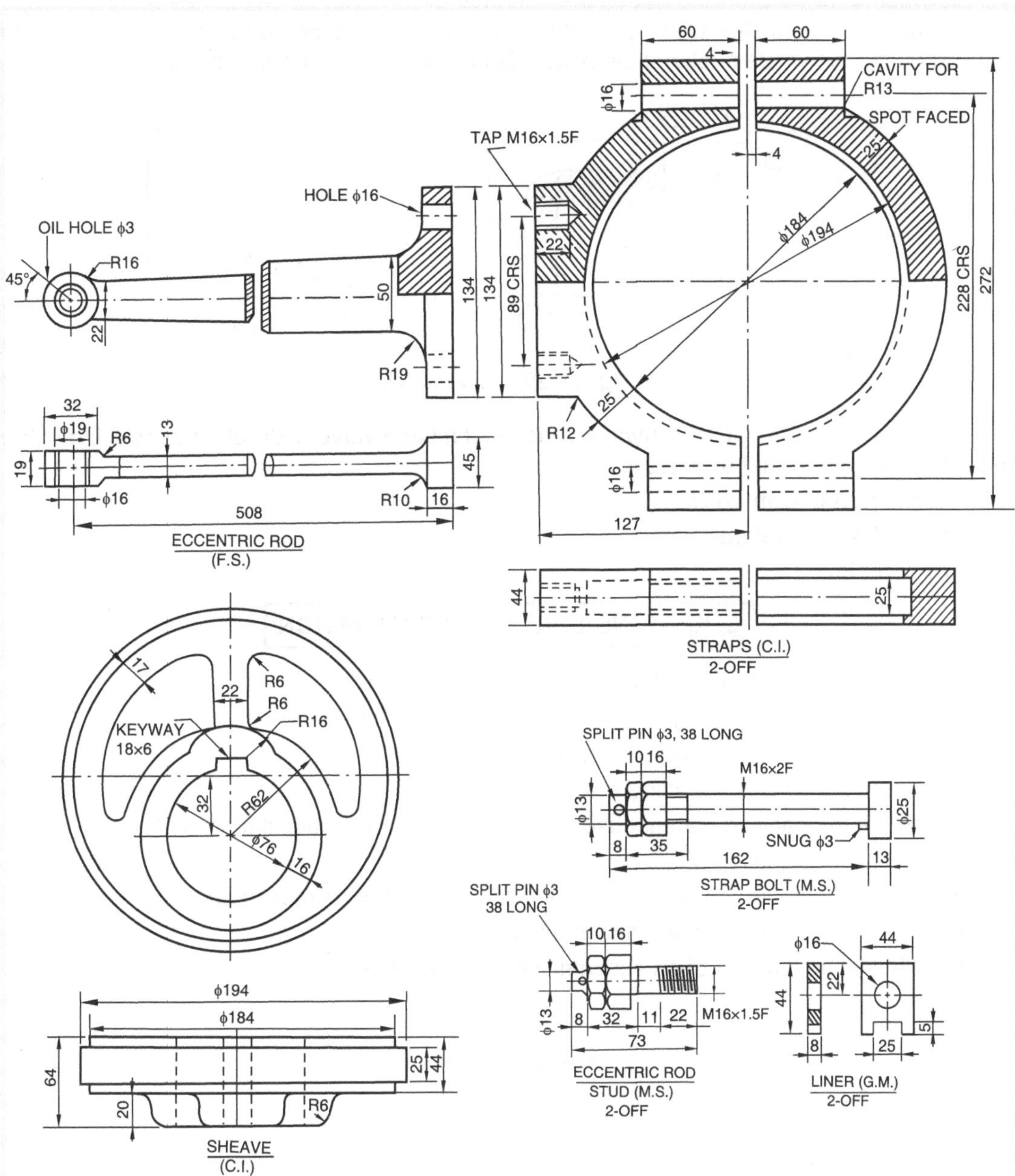

Fig. 13.28. *Details of an eccentric*

13.14. SLIDE VALVE

A slide valve is a part of steam engine which is used to control the flow of steam to and from the engine cylinder. It is also known as D-slide valve, because its cross-section is similar to the english capital letter 'D'. It is connected to valve rod by means of nuts (see Fig. 13.29).

It slides to and fro on the flat smooth surface of the cylinder which has admission and exhaust ports. When it slides, one port of the cylinder will be open to steam and the other to exhaust during the whole stroke of the piston. Slide valves are generally made of cast iron.

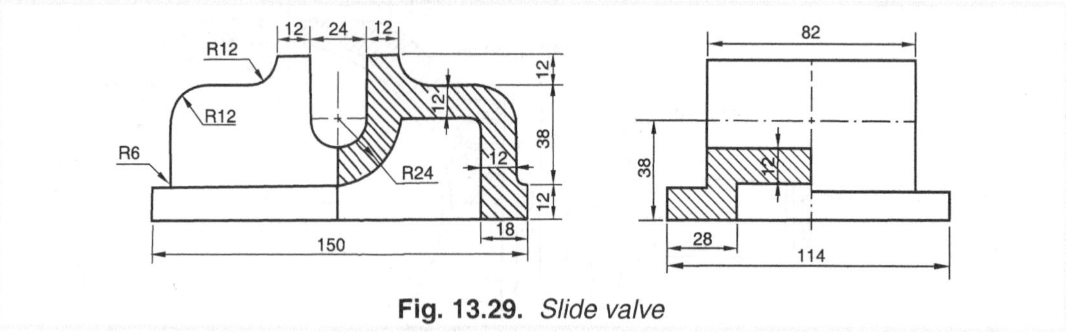

Fig. 13.29. *Slide valve*

Exercise 18 : Fig. 13.29 shows two views of a slide valve without valve rod. Draw the following views with full size scale :—

(a) Sectional front view

(b) Half sectional end view

(c) Top view.

QUESTIONS FOR SELF EXAMINATION

1. What is a steam engine? Name its reciprocating parts.
2. What is the material of piston rings? How are these fitted on the piston?
3. How is a piston rod fitted with the cross-head?
4. Where and why is a stuffing box used?
5. Why is the packing material used in stuffing box? How does it function?
6. What is the function of a cross head?
7. What is the purpose of providing detachable shoe to a cross head?
8. What is the function of a crank in an engine?
9. What is the function of an eccentric and a slide valve?

CHAPTER

14

I.C. Engine Parts

INTRODUCTION

In this modern world an internal combustion engine plays the most important role for convertion of heat energy into mechanical energy. In this, a mixture of fuel and air is burnt inside the engine cylinder. This mixture acting as the working substance, expands under combustion and in doing so it pushes the piston to and fro. This to and fro motion of the piston is converted into rotary motion through a link mechanism. It is this rotary motion which is used to run cars, scooters, diesal engine trains, generators etc.

In this chapter, we shall deal with the study of I.C. Engine and I.C. Engine parts with their different designs as used in engineering field.

14.1. I.C. ENGINE

The engine in which the combustion of fuel (petrol or diesel) takes place inside the engine cylinder and converts this heat energy into mechanical energy through a link mechanism is known as I.C. Engine.

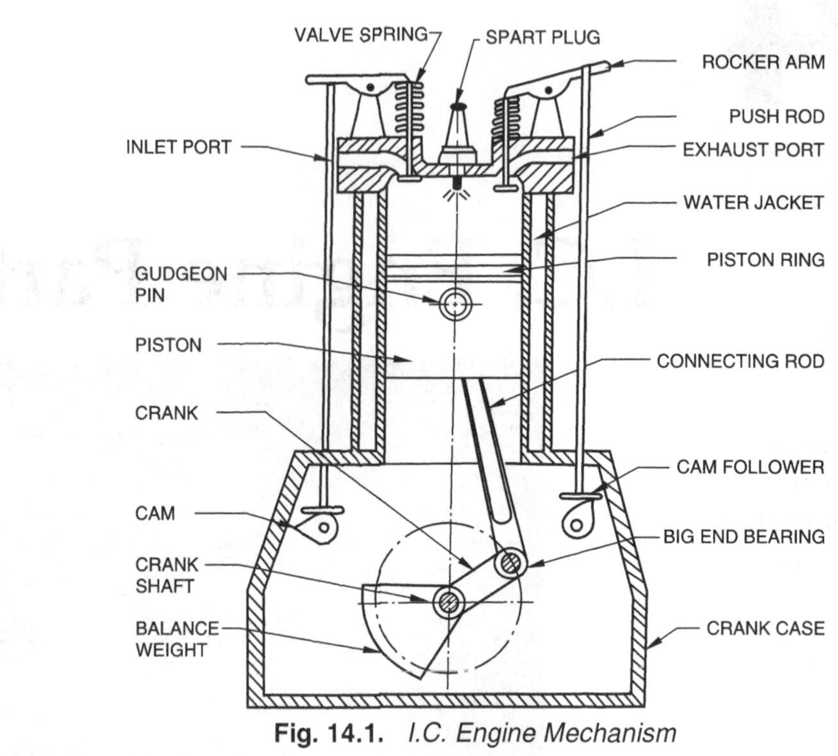

Fig. 14.1. *I.C. Engine Mechanism*

14.2. I.C. ENGINE PARTS

The internal combustion (I.C.) engine can be divided into following three parts :—

1. Stationary parts : Those parts of the engine which remain stationary during the working of the engine are called stationary parts. In I.C. engine, they are *cylinder head, cylinder block, crank case etc.*

2. Reciprocating parts : Those parts of the engine which move linerally during the working of the engine are called reciprocating parts. In I.C. engine, they are *piston, valves (inlet and outlet), push rod etc.*

3. Revolving parts : Those parts of the engine which follow circular path during the operation of the engine are called revolving parts. In I.C. engine, they are *piston rod, connecting rod, crank, crank shaft, flywheel, cam etc.*

IMPORTANT PARTS OF I.C. ENGINE WITH THEIR DIFFERENT DESIGNS

14.3. PISTON

A cylindrical piece which reciprocates inside the engine cylinder by the pressure of combustion of fuel is called piston.

Pistons are usually made of aluminium alloy, cast iron or cast steel. But the aluminium alloy pistons are commonly used due to light in construction.

The crown of the piston may be made flat, convex or concave according to the design of combustion chamber.

Exercise 1 : Fig. 14.2 shows pictorial views of a piston. Draw the following views to full size scale :

(a) Front view - Right half in section

(b) Side view projected from (a)

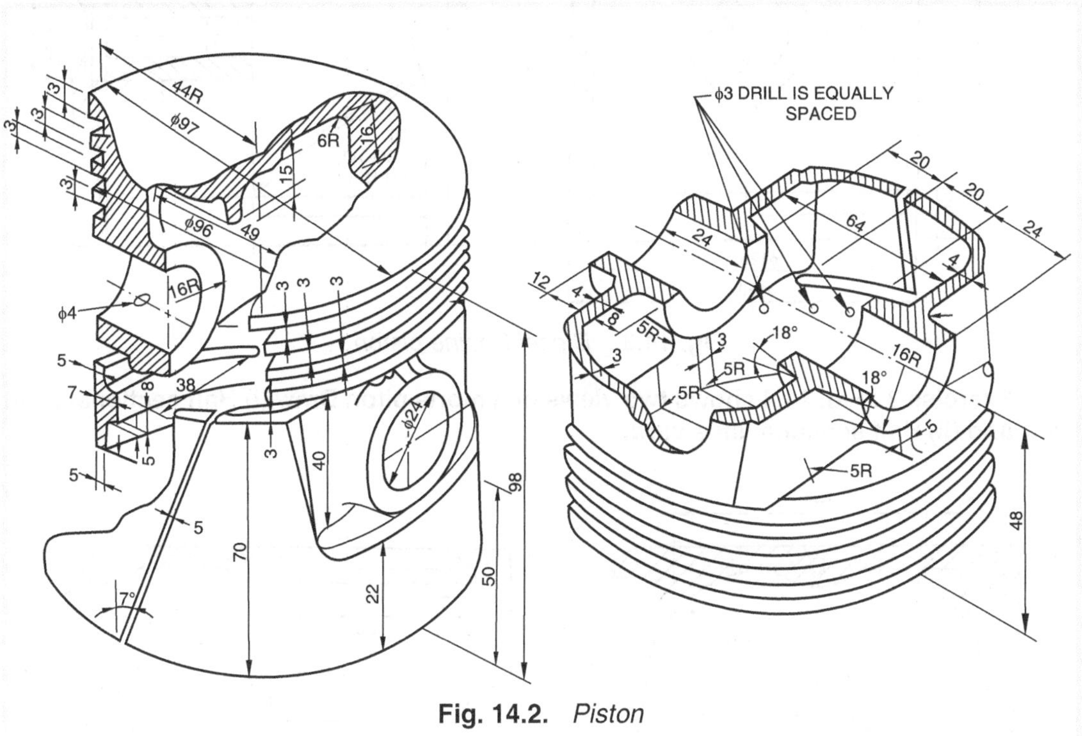

Fig. 14.2. *Piston*

Exercise 2 : Fig. 14.3 shows two views of a Diesel engine piston Draw the following views to full size scale:–

(a) Front view - full section

(b) Side view

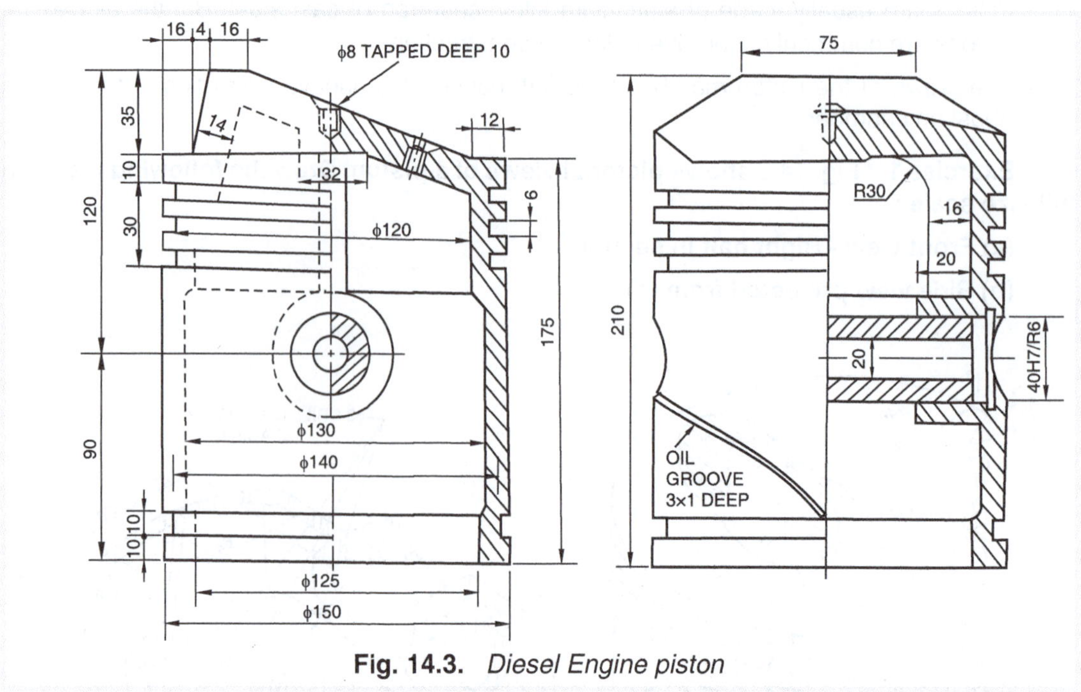

Fig. 14.3. *Diesel Engine piston*

Exercise 3 : Fig. 14.4 shows two views of a moris piton Draw (i) Half sectional front view and (ii) Half sectional side view.

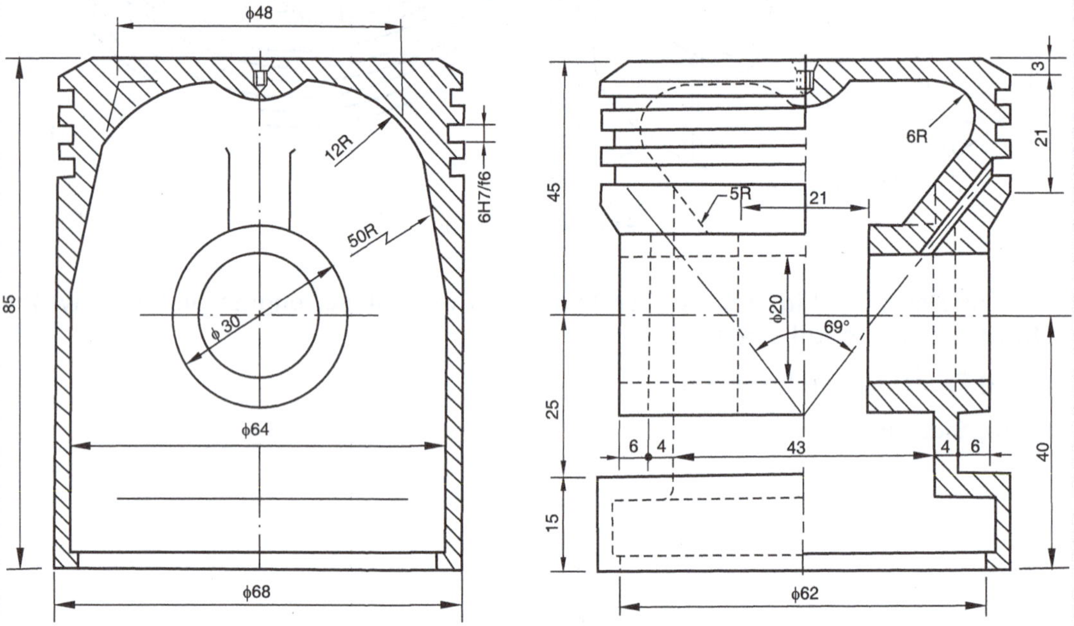

Fig. 14.4. *Moris Piston*

Exercise 4 : Fig. 14.5 show three views of a petrol engine piston. Draw following views

(a) Half sectional front view **(b) Half sectional side view projected from (a)**

(c) Top view.

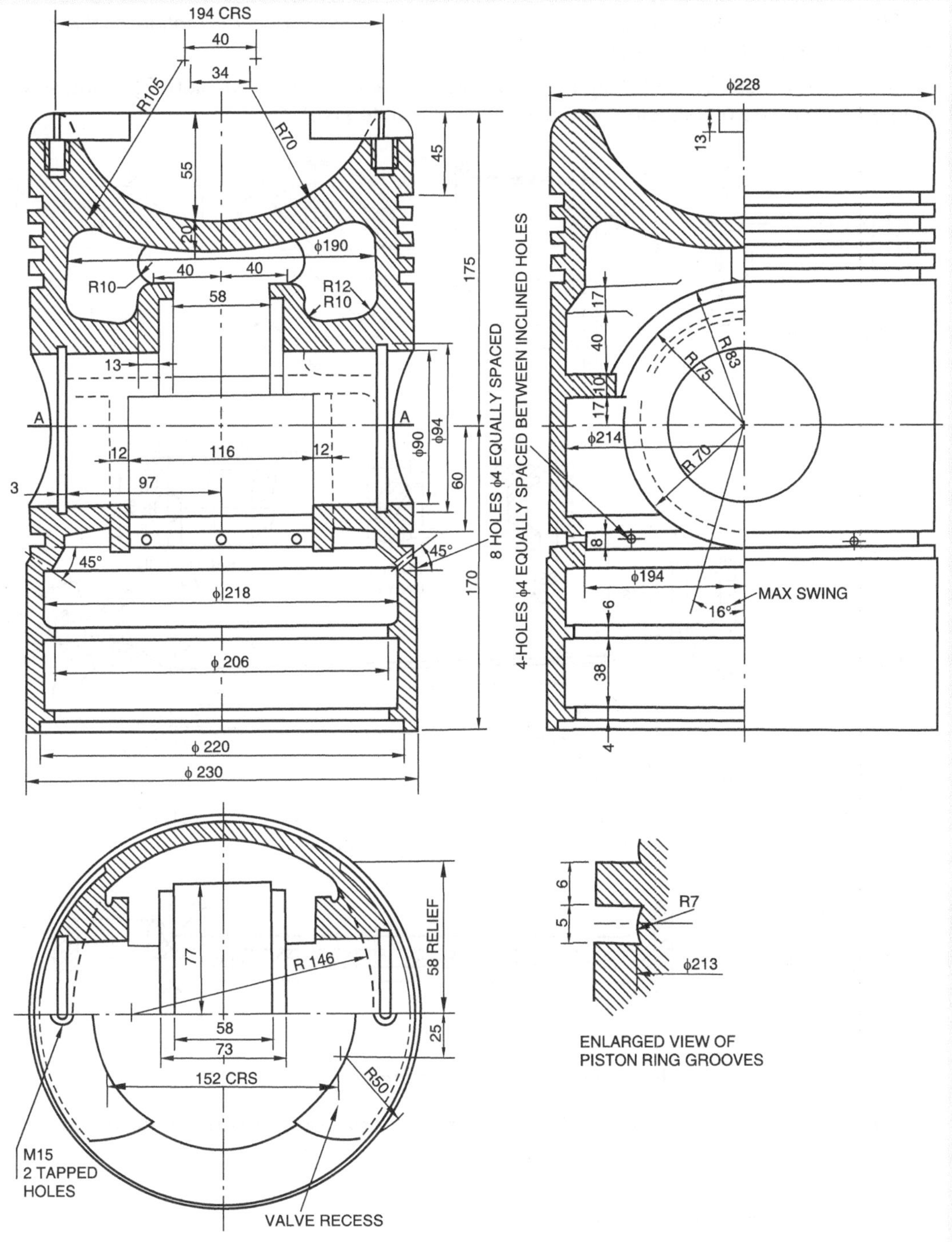

Fig. 14.5. *Petrol Engine Piston*

14.4. CONNECTING ROD

A connecting rod is a link which connects the piston to the crank. It is provided with enlargements at both the ends. These enlargements are known as connecting rod ends. One end which is connected to the piston with the help of gudgen pin is called the small end. The other end which is secured to the crank is called big end (see Figs. 14.6 & 14.7).

The function of the connecting rod is to convert the reciprocating motion of the piston into rotary motion to the crank shaft.

Connecting rods are generally made of forged steel in various design. The brasses in the big end are made of cast steel and are lined with white steel.

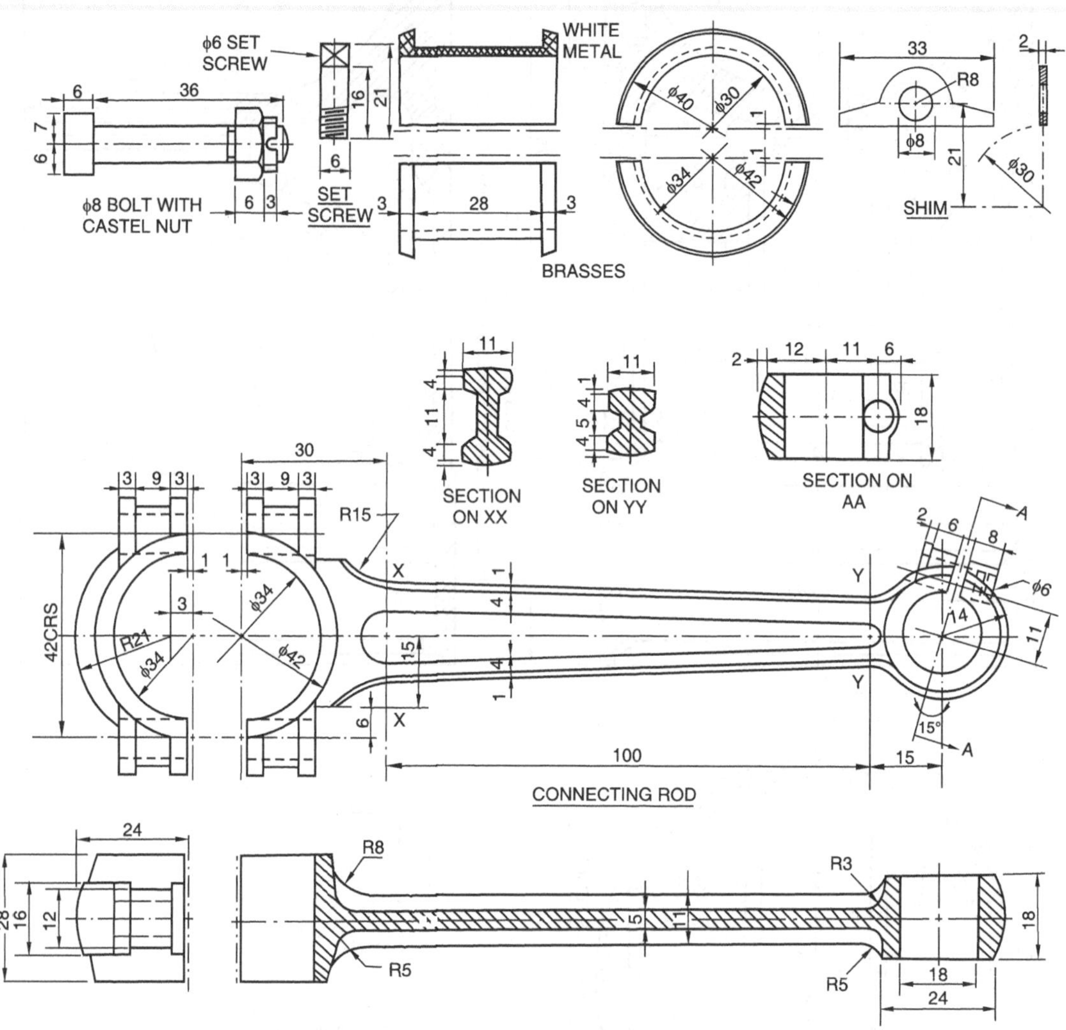

Fig. 14.6. *Details of connecting rod*

Problem 1 : Fig. 14.6 shows the details of a connecting rod for petrol engine. Assemble the parts and draw the following views :—

(a) Front view.

(b) Top view–full in section.

Show the tolerances and surface finish wherever required. Also give the material list.

Solution : For its solution, see Fig. 14.7

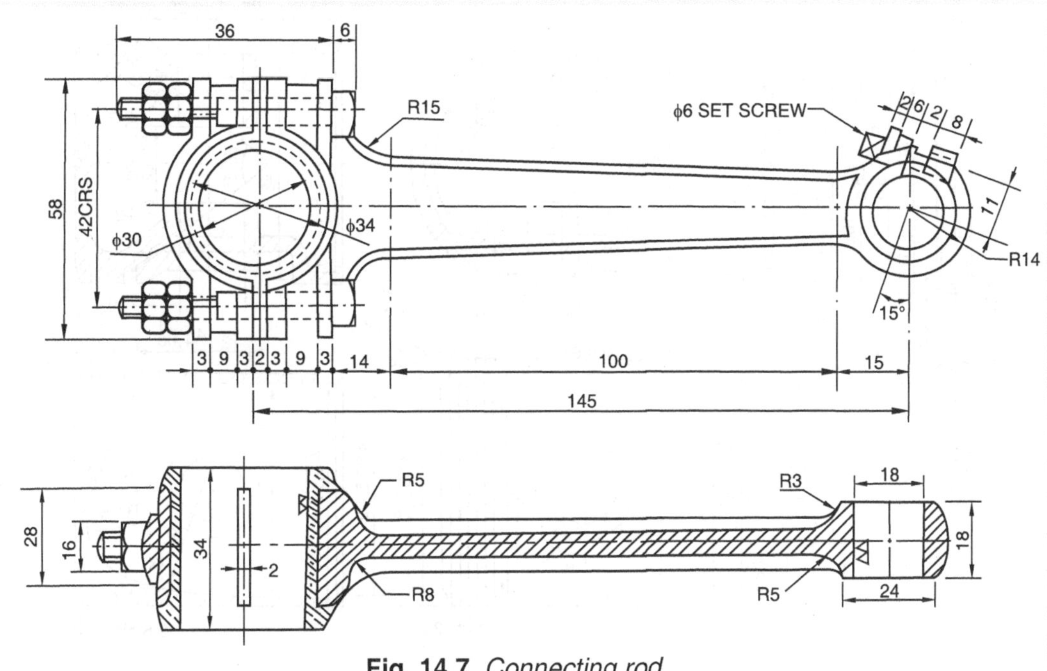

Fig. 14.7. *Connecting rod*

PARTICULARS OF PARTS

S.NO.	NAME	MATERIAL	NO. OFF
1.	CONN. ROAD	FORGED STEEL	1
2.	BRASSES	GUN-METAL	2
3.	SHIM	BRASS STEEL	2
4.	SET SCREW	MILD STEEL	1
5.	BOLTS & NUT	MILD STEEL	2

Exercise 5 : Fig. 14.8 shows the details of piston and connecting rod for an I.C. engine. Assemble all the parts and draw the following views to a scale of 1:1 :—

(a) Front view – full in section

(b) Side view

(c) Top view.

Exercise 6 : Fig. 14.9 shows the details of a connecting rod end for a gas engine. Assemble the parts and draw the following views to full size scale :—

(a) Front view upper half in section

(b) Side view

(c) Top view.

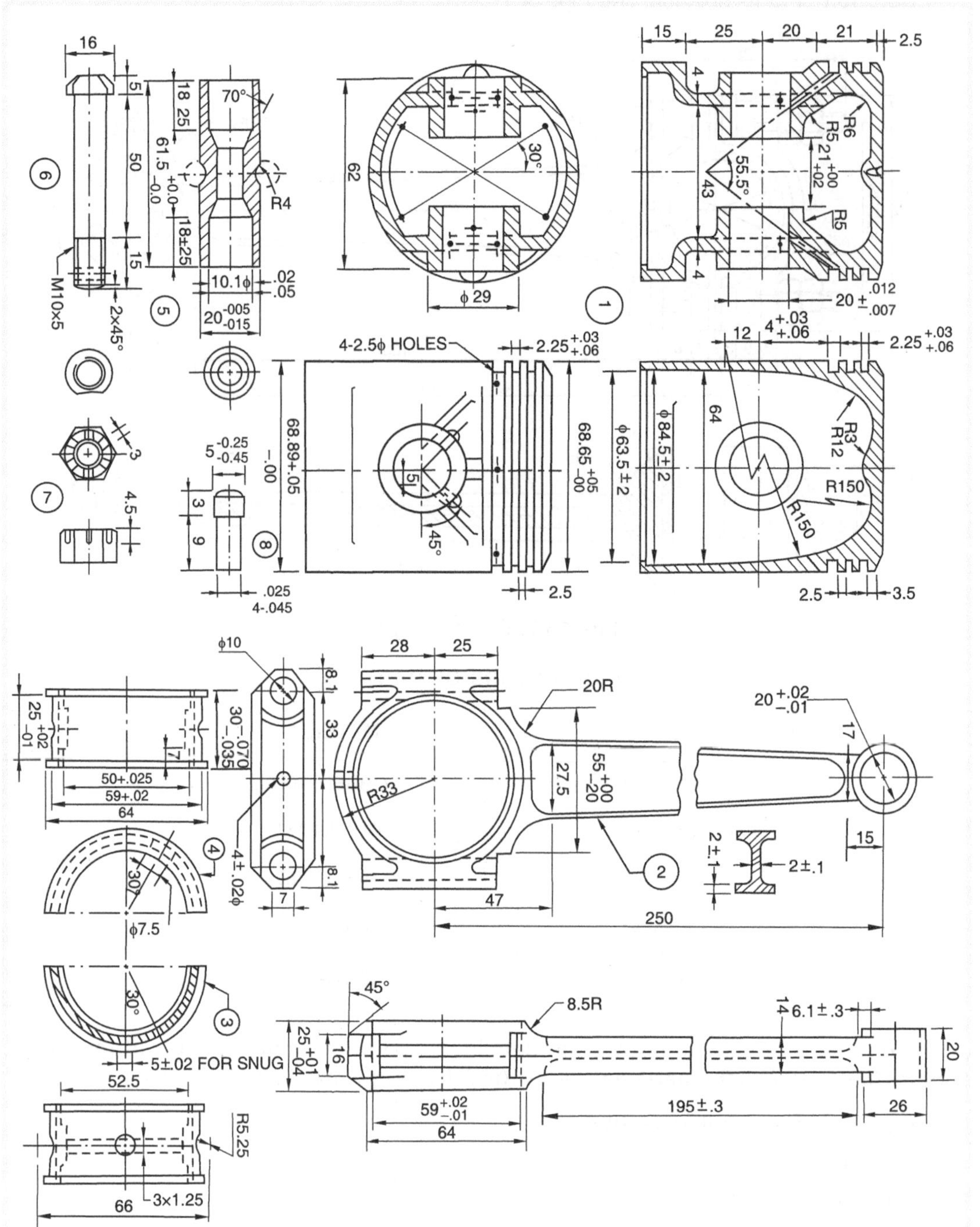

Fig. 14.8. *Details of piston and connecting rod*

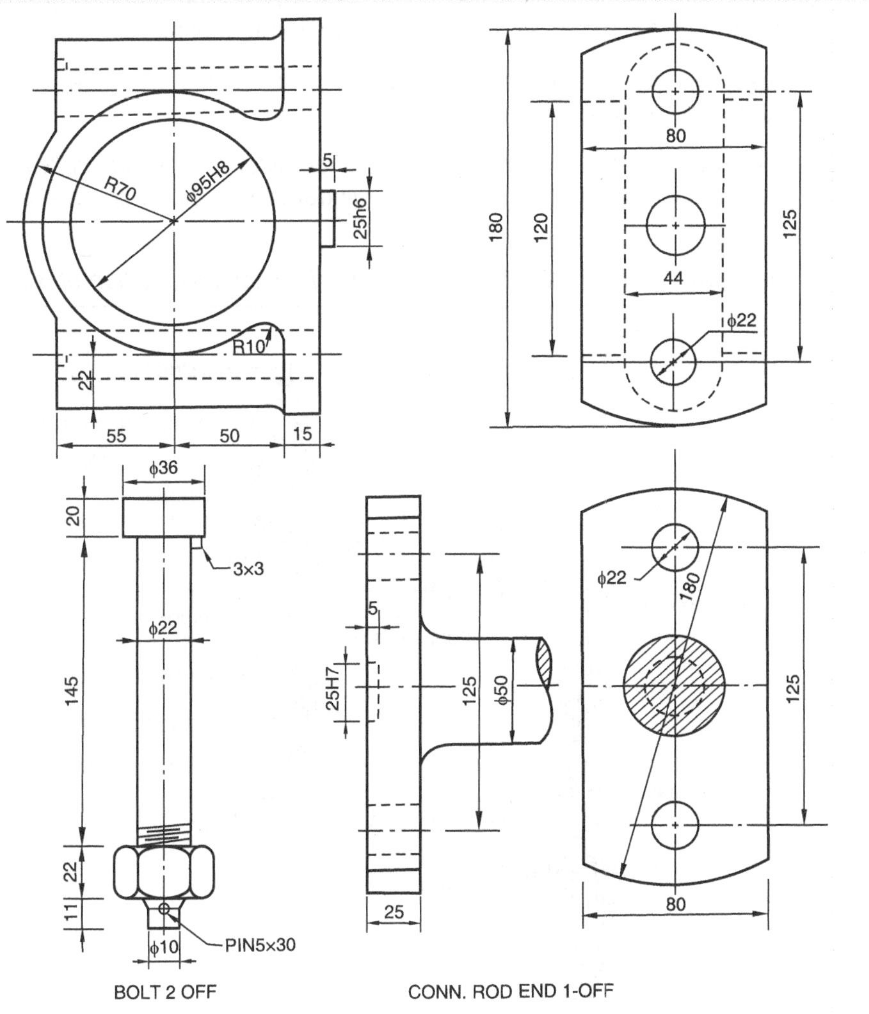

Fig. 14.9. *Details of connecting rod end*

14.5. ATOMISER OR FUEL INJECTOR

In the fast developing engineering world, I.C. engines have attained unique place as producer of mechanical power. Examples of I.C. engines are petrol engine, diesel engines etc. In a petrol engine, a mixture of air and petrol is compressed and then ignited by the spark plug in cylinder in order to obtain mechanical power.

In a diesel engine, air is compressd in the engine cylinder and at the end of compression stroke, disel fuel is injected into the cylinder. When injected fuel comes in contact with highly compressed air in the cylinder, combustion takes place and chemical energy of diesel fuel is converted into mechanical power. In order to inject the disel fuel into the engine cylinder the following two important parts are used :—

1. Fuel pump.
2. Atomiser or fuel injector.

Fig. 14.10 *shows the journey of diesel fuel from diesel tank to the engine cylinder via the fuel pump and atomiser.*

Fuel pump : *The function of the fuel pump is to deliver the required quantity of diesel fuel to the atomiser.* Fig. 14.10 shows the various parts of the fuel pump.

The atomiser of fuel injector is an important part of diesel engine and is installed between fuel pump and engine cylinder.

Atomiser or fuel injector : *The function of the atomiser to atomises or brakes the diesel fuel into fine particles and spray into the engine cylinder at the end of compression stroke for combustion.*

Note : *The spraying action of the atomiser helps in quick combustion of diesel fuel.*

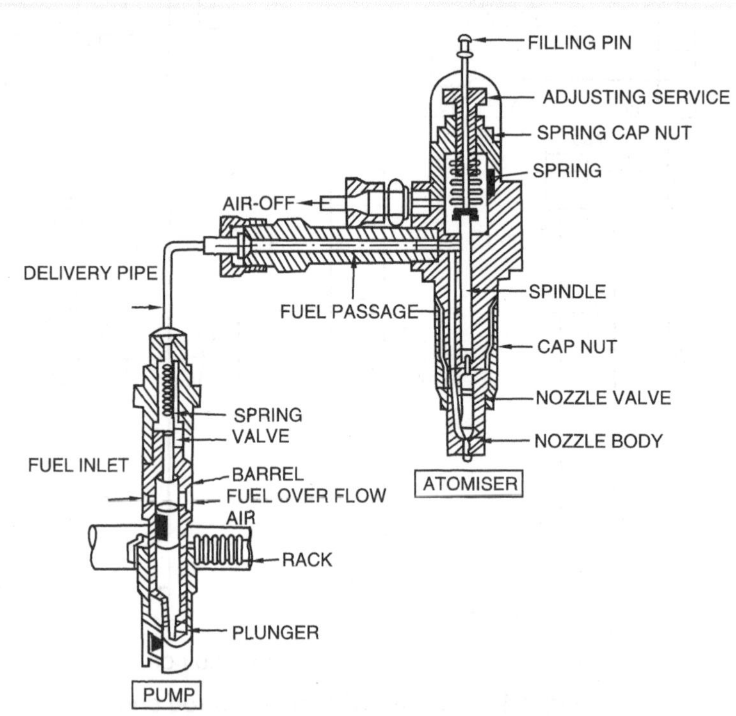

Fig. 14.10. *Details of fuel pump and atomiser*

Fig. 14.10 shows the various parts of an atomiser. In this, the diesel fuel under pressure is supplied to the injector from the fuel pump through the *fuel passages*. This high pressure fuel lifts the *nozzle* valve from its seat against the compression of the *spring* and the fuel passes down the *nozzle* and injected into the cylinder in the form of a very fine spray.

When the fuel pressure falls, the *nozzle valve* moves down under the force of the spring and rests on its seat and thus closing the inlet passage. The fuel supply is thus cut off. Any fuel leaked past the *plunger* of the *nozzzle valve,* sent to the fuel chamber through the passage. An *adjustable screw* is provided to adjust the spring pressure on the *nozzle valve*. This gives *an* adjustment to the atomisation of fuel in the cylinder. The *feeling* pin is used to determine whether the *nozzle valve* is working properly or not.

Exercise 7 : Fig. 14.11 (i) and (ii) show the details of an atomiser. Assemble all the parts properly and draw the following views :

(a) Sectional front view **(b) An end view** **(c) Top view**

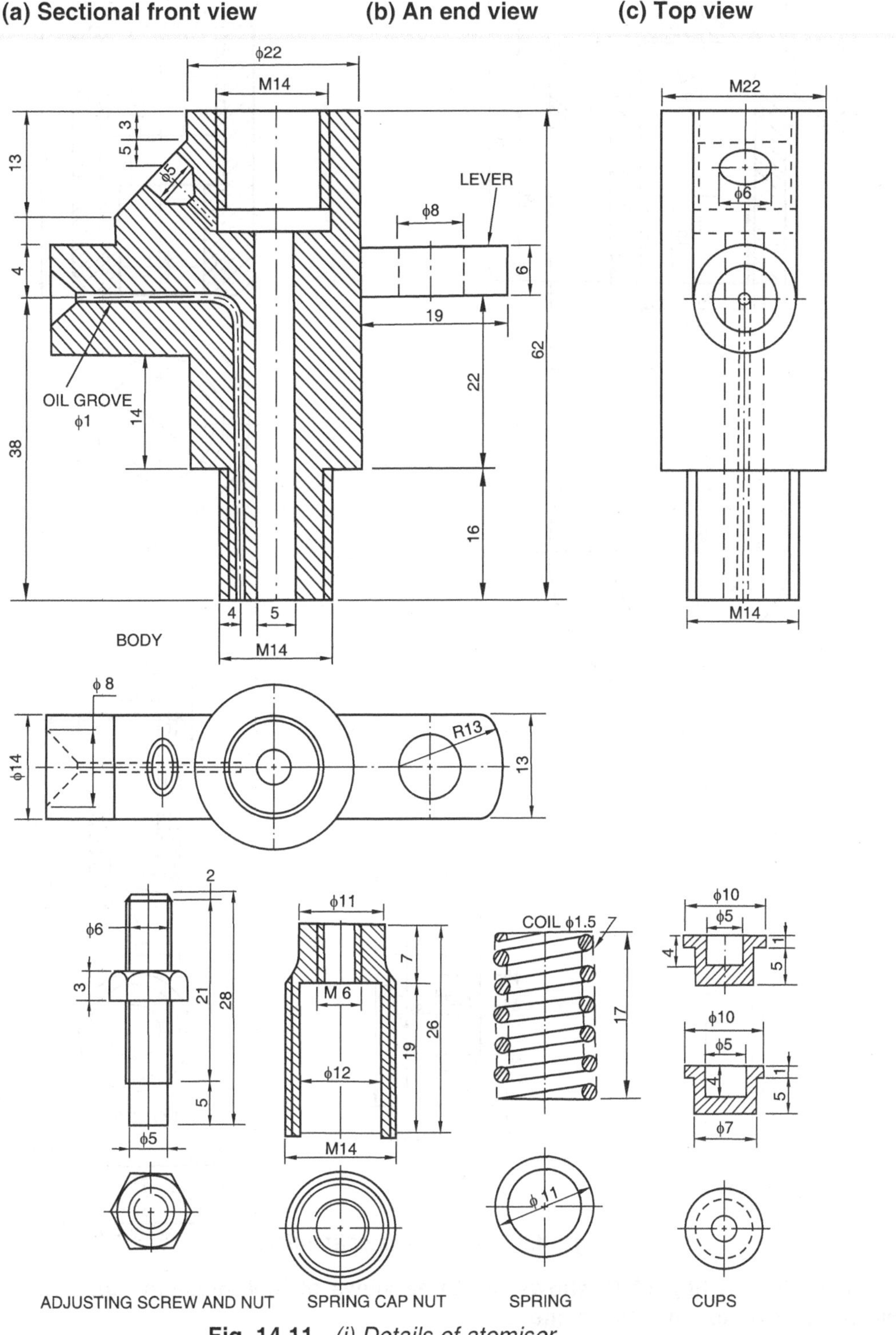

Fig. 14.11. *(i) Details of atomiser*

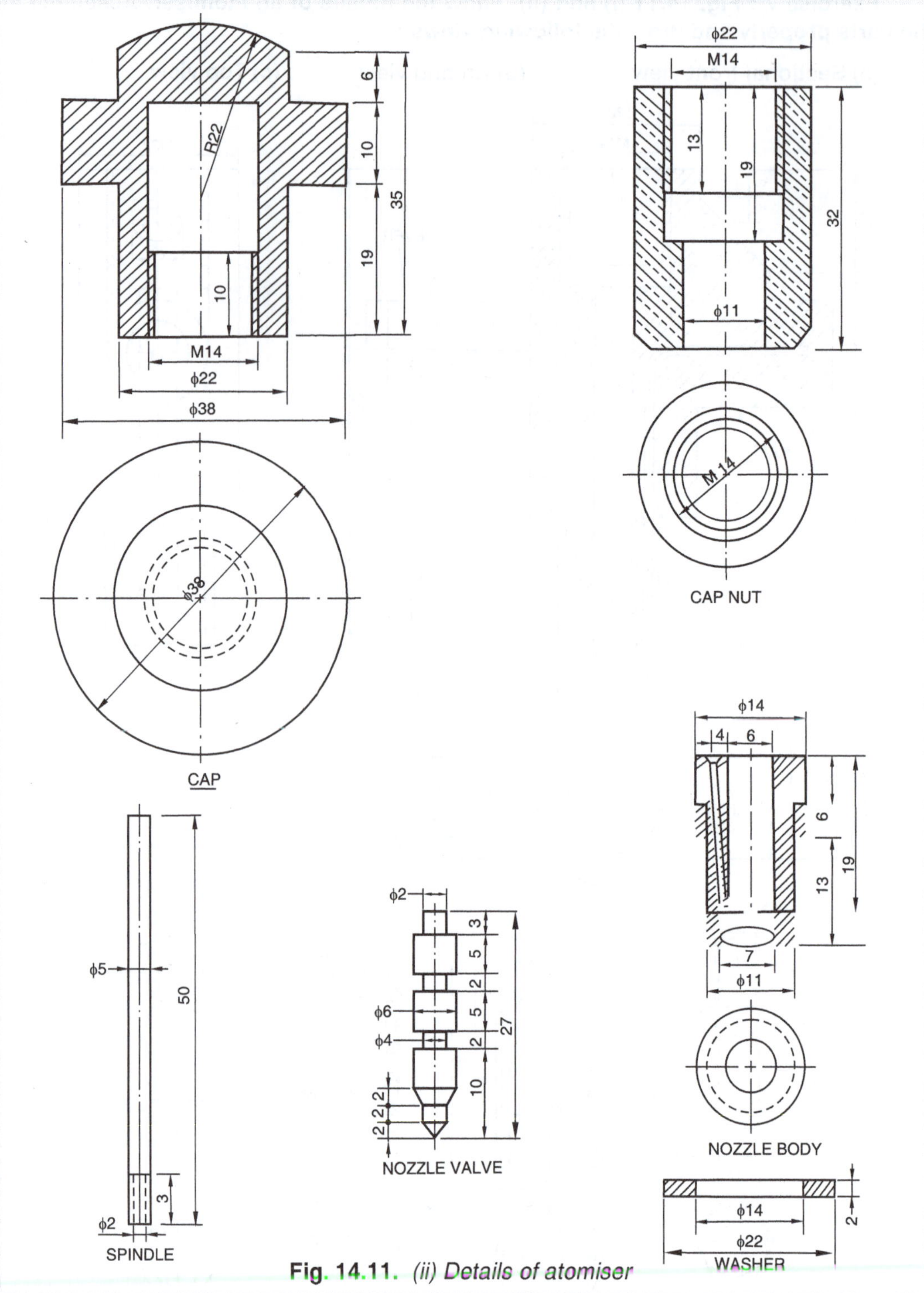

Fig. 14.11. *(ii) Details of atomiser*

Exercise 8 : Fig. 14.12 shows the three views of fuel injection pump. Draw the following views to a full size scale :

(a) Half sectional front view (b) Side view (c) Top view.

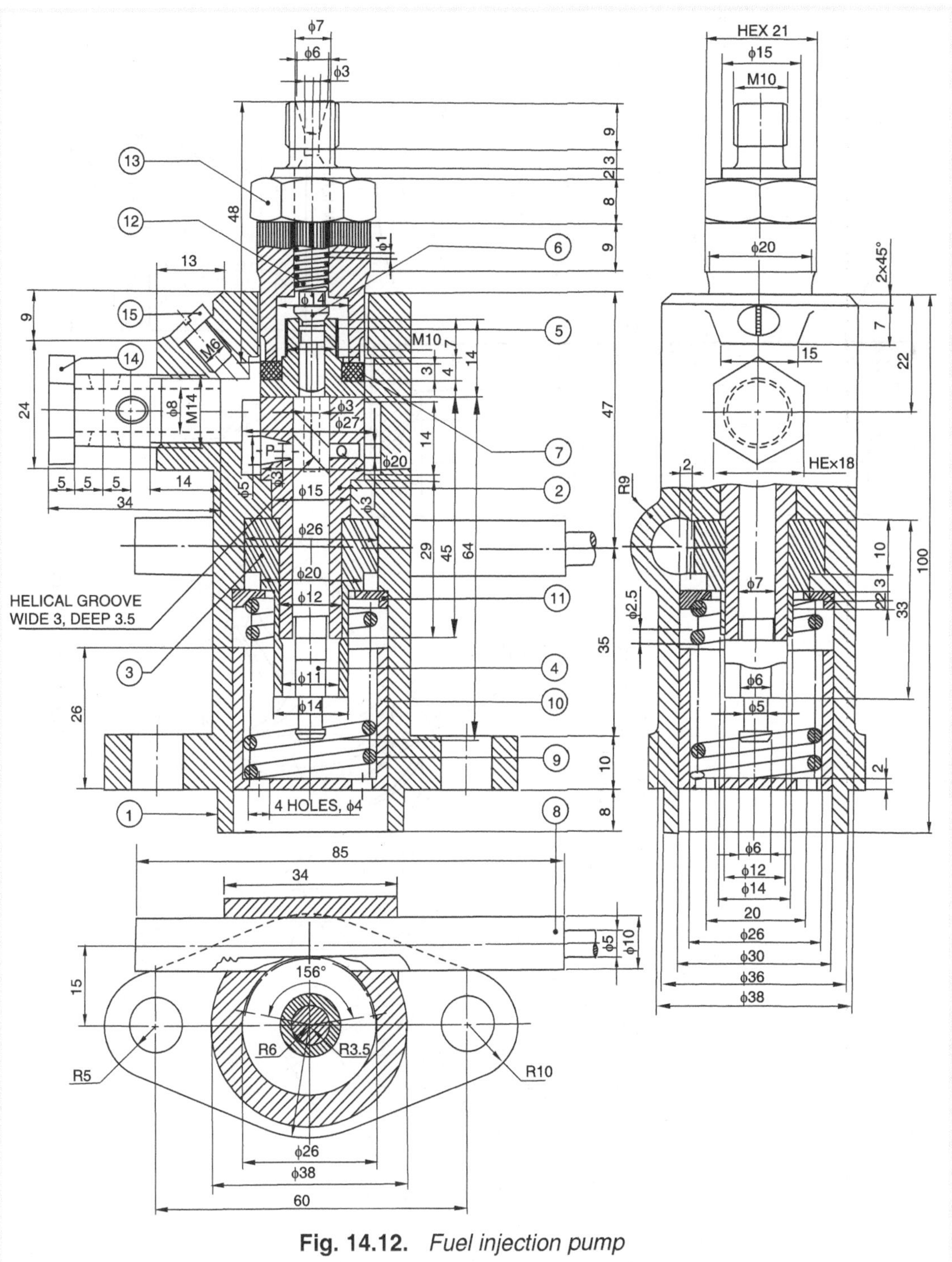

Fig. 14.12. *Fuel injection pump*

Exercise 9 : Fig. 14.13 shows details of crank shaft and fly wheel. Assemble all the parts and draw the following views:-

(a) Front view - full section **(b) Left end view**

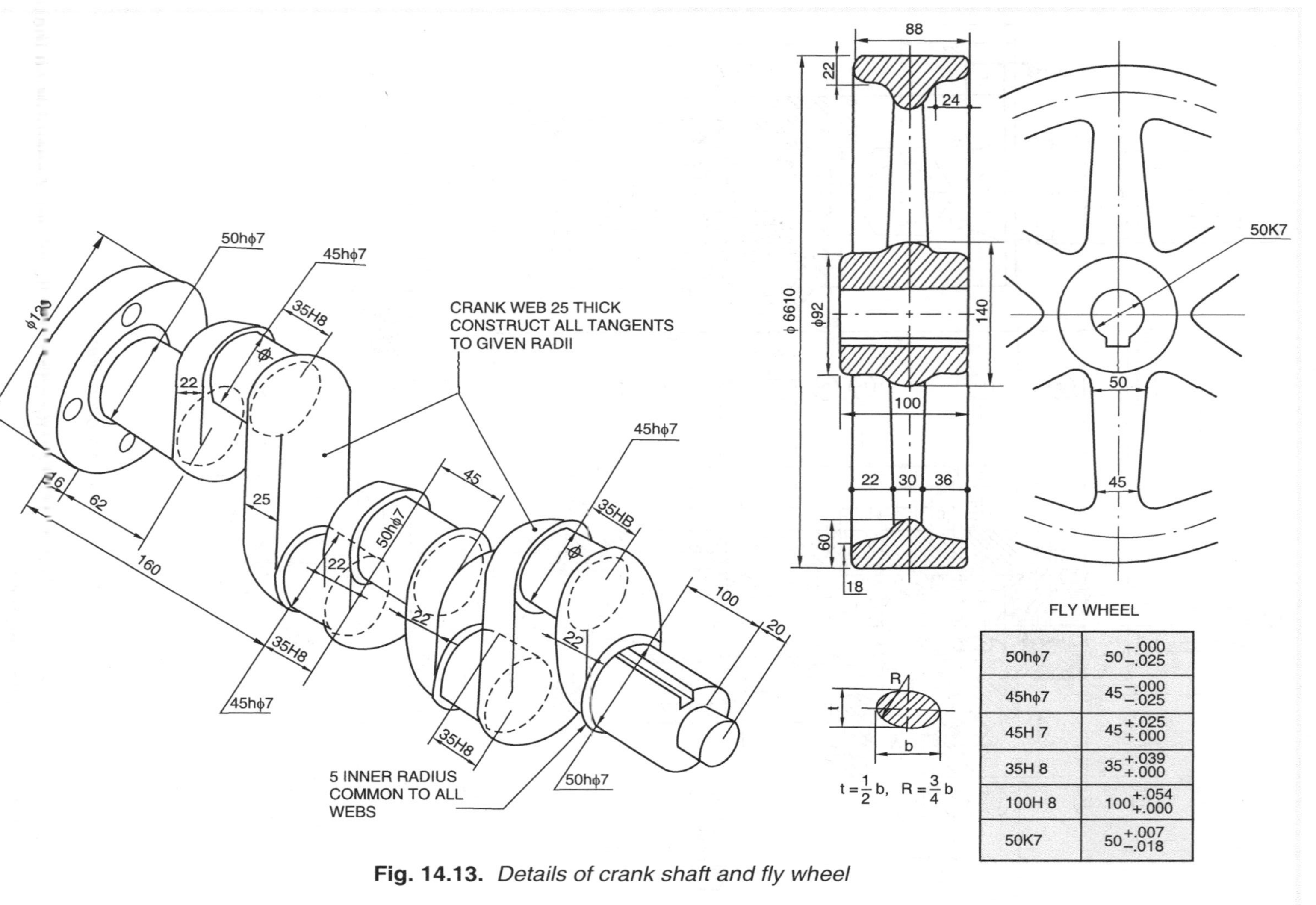

FLY WHEEL

50hϕ7	$50^{-.000}_{-.025}$
45hϕ7	$45^{-.000}_{-.025}$
45H 7	$45^{+.025}_{+.000}$
35H 8	$35^{+.039}_{+.000}$
100H 8	$100^{+.054}_{+.000}$
50K7	$50^{+.007}_{-.018}$

Fig. 14.13. *Details of crank shaft and fly wheel*

Exercise 10 : Details of a sparking plug are given in Fig. 14.14 Draw to a scale 2.5 : 1, the following assembled views :

(a) Front view – right half in section **(b) Top view**

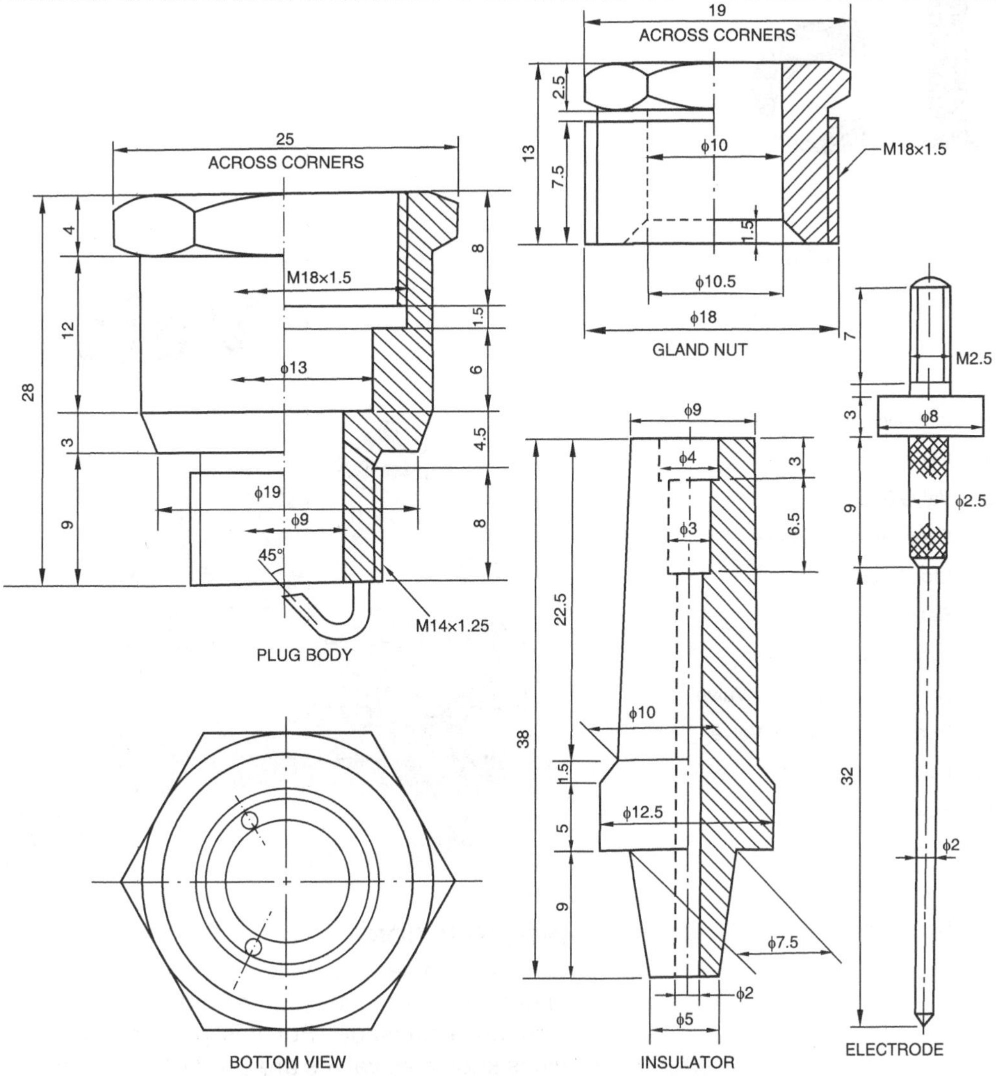

Fig. 14.14. *Details of a sparking plug*

QUESTIONS FOR SELF EXAMINATION

1. What is an I.C. Engine? Name its different parts.
2. Name the reciprocating and revolving parts of an I.C. Engine.
3. What is the difference between pertrol engine and Diesel engine?
4. What is the function of a piston?
5. Why piston rings are provided on the piston?
6. What is the function of a connecting rod?
7. How a connecting rod is fitted with the piston?
8. What is the function of atomiser and fuel injection pump?

CHAPTER

15

Valves

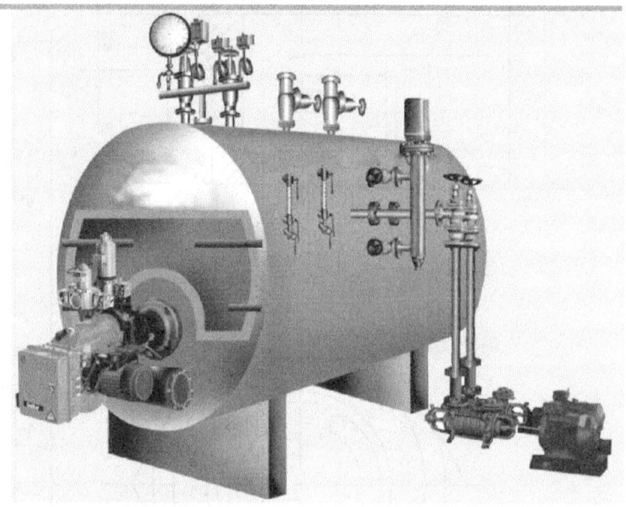

INTRODUCTION

In engineering field, we often come across situations where we want to control the flow of fluid through a passage. For this purpose various devices known as valves are used which are operated either by the pressure of fluid or by hand or by some external mechanism. Valves which are operated by the pressure of fluid are known as *non-return valves* as they automatically prevent the flow of fluid in the reverse direction. Similarly, valves which are operated by hand are known as *hand operated valves* and so on. These different valves are used at different places depending upon their operation.

In this chapter, we shall deal with the study of valves, valve seats, classification of valves and some common valves used in engineering practice.

15.1. VALVES

The devices which are used to control the flow of fluid in the pipes and pressure vessels are known as valves.

The valves are generally made of gun-metal, brass, cast iron and steel. They are bevelled at angle of 45° to its lower edges and provided with feathers, wings or guards to keep them in their position during working operation (see Fig. 15.1).

15.2. VALVE SEATS

The portion on which the valve rests is known as valve seat (see Fig. 15.1). It is generally made of gun-metal or brass in cylindrical bush form.

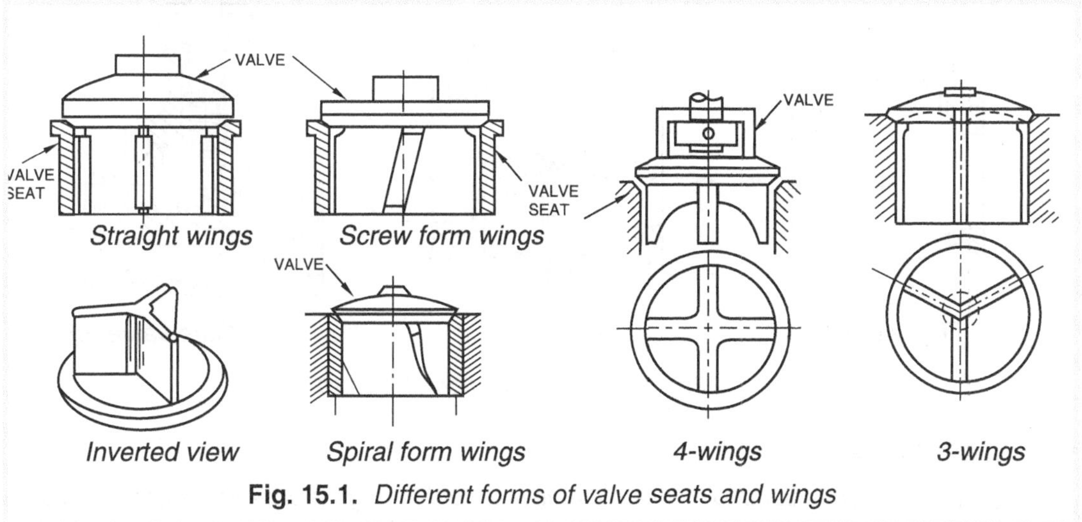

Fig. 15.1. *Different forms of valve seats and wings*

Valve seats are either screwed, pressed or secured in the main body of the valve. It is bevelled off at 45° angle at its top portion by pressing where the bevelled portion of valve rest during operation (see Fig. 15.1).

In large size valves, the valve seats are made separately and fitted to the valve body whereas in small size valves, the seats are formed in the body itself.

15.3. CLASSIFICATION OF VALVES

In engineering practice, the valves may be classified as follows depending upon operation :—

1. Flap valves
2. Lift valves
3. Gate valves

1. Flap valves : *Valves which bend or turn about a hinge is known as flap valve.* The amount of turn given to the flap controls the opening for the fluid [see Fig. 15.2 (a)]

Example : *Valve used in compressor of a refregerator.*

2. Lift valves : *Valves which move up and down perpendicular to their seats are known as lift valve [see Fig. 15.2 (b)].*

The lift of the valve is checked by the position of the lower end of the valve spindle.

Examples : *Feed check valve, non-return valves etc.*

3. Gate valves : *Valves which operate perpendicular to the fluid pressure or by rotation of the hand wheel are known as gate valves [see Fig. 15.2(c)]*

Example : *Bib cock, pillar valves, stop valves, etc.*

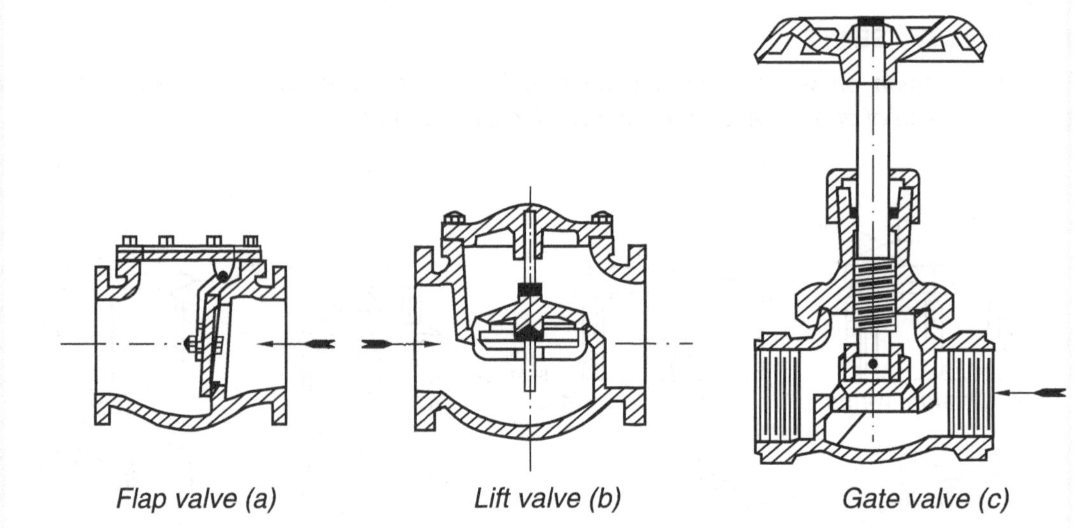

Fig. 15.2 *Classification of valves*

15.4. STOP VALVE

The valve which is used to regulate the flow of fluid through a passage is known as stop valve. The flow of fluid may be steam or water.

When the valve is mounted over the boiler to control the flow of steam from boiler to engine, it is known as *junction valve* and when the valve is placed near the engine or between the steam pipes, it is known as *stop valve.*

Fig. 15.3 shows a hand operated stop valve. It consists of a hand wheel which is fitted with the spindle to which the valve is attached. The spindle passes through a screwed portion (bridge) and then through a gland which prevents the leakage of steam. The bridge is supported by pillars screwed to the cover of the body. The valves rests over the valve seat attached to the body.

By the rotation of hand wheel, the valve moves upward or downwards direction. Thus, it opens or closes the passage for the flow of steam.

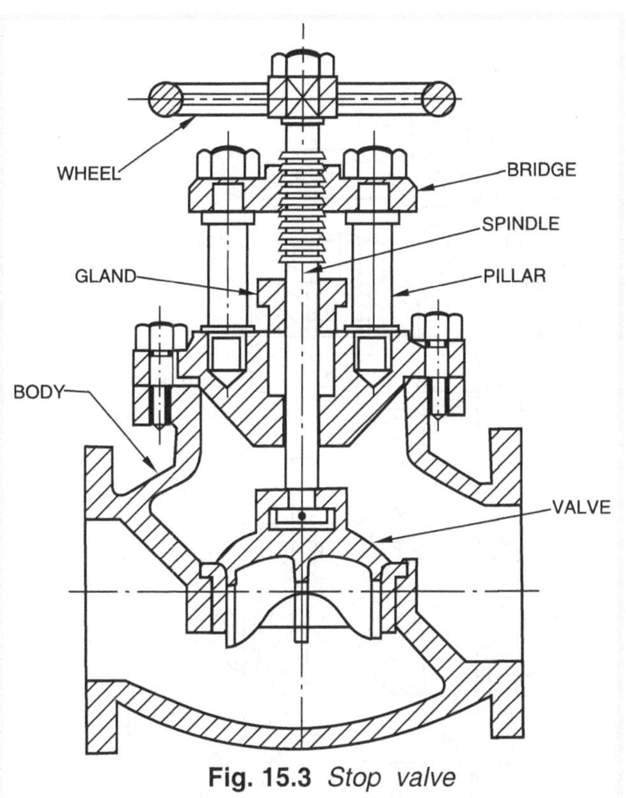

Fig. 15.3 *Stop valve*

Exercise 1: Fig. 15.4 shows the two views of a stop valve. Draw the following views to a suitable scale. (a) Front view (b) Side view

PARTICULARS OF PARTS

S.NO.	NAME	MATERIAL	NO. OFF
1.	BODY	C.I.	1
2.	COVER	C.I.	1
3.	GLAND	G.M.	1
4.	NECK BUSH	GM.	1
5.	BRIDGE	MS.	1
6.	VALVE	G.M.	1
7.	SEAT	G.M.	1
8.	SPINDLE	G.M.	1
9.	COLLAR	G.M.	1
10.	PIN	G.M.	1
11.	HAND WHEEL	C.I.	1
12.	STUD AND NUT (FOR COVER)	M.S.	1
13.	PILLAR AND NUT	M.S.	2
14.	NUT	M.S.	1
15.	STUD AND NUT (FOR GLAND)	M.S.	2

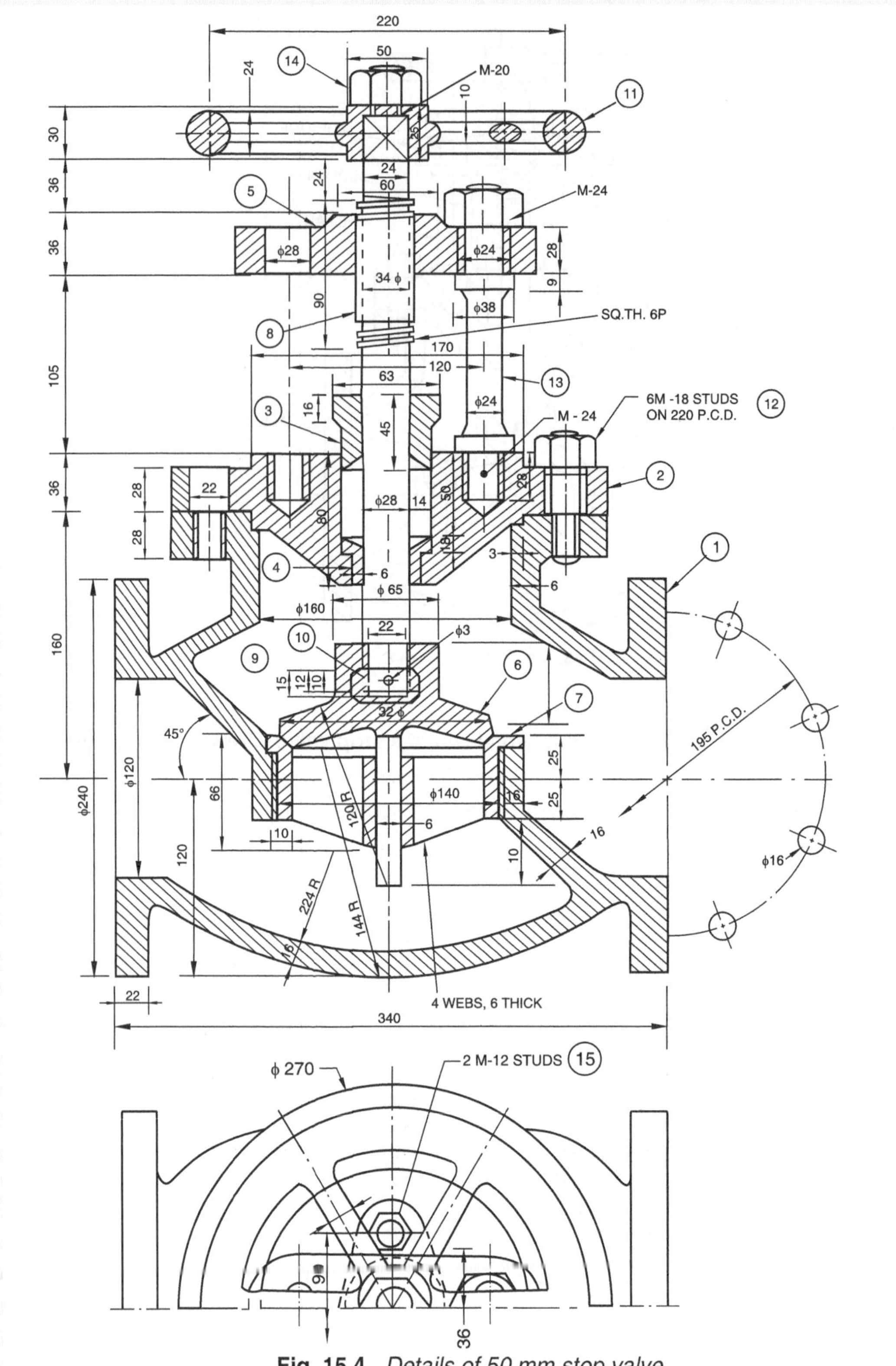

Fig. 15.4 *Details of 50 mm stop valve*

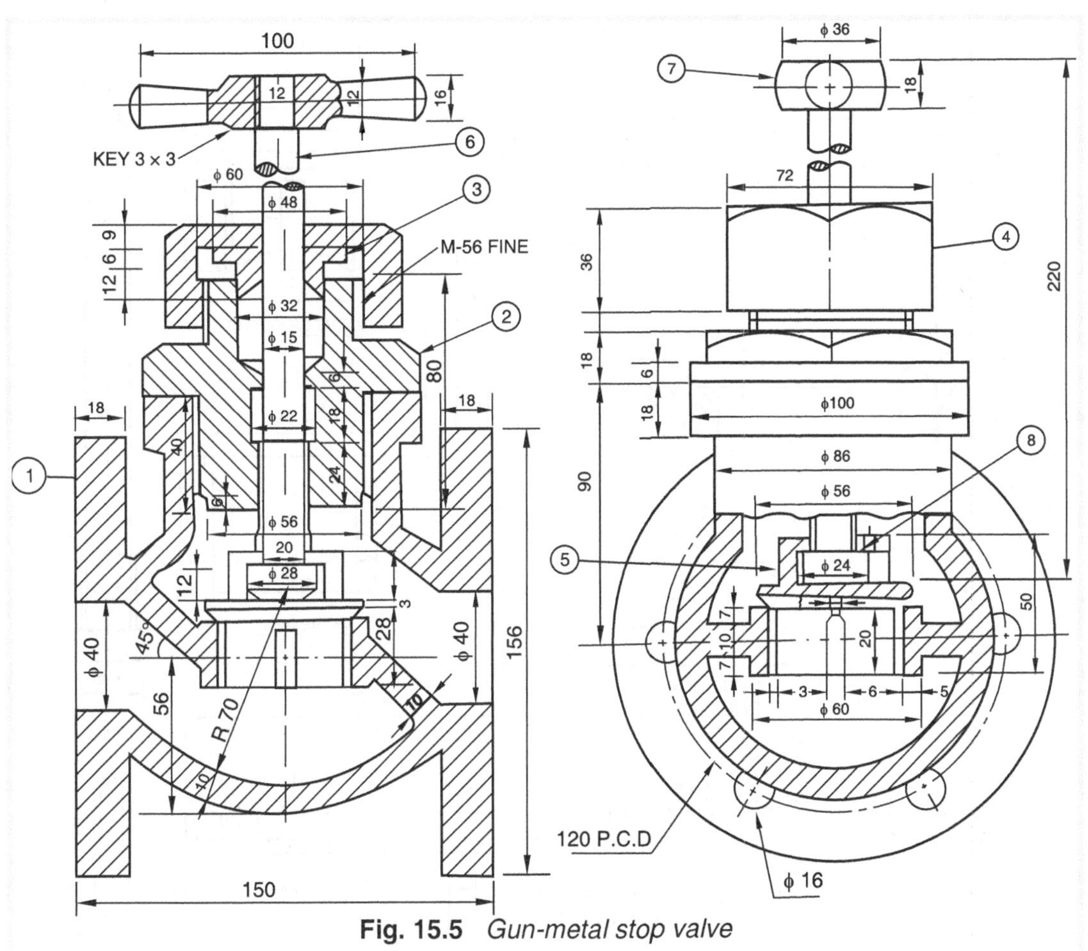

Fig. 15.5 *Gun-metal stop valve*

Exercise 2: Fig. 15.5 shows the two views of a Gun metal stop valve. Draw the following view to a full size scale:

(a) Front view **(b) End view.**

PARTICULARS OF PARTS

S.NO.	NAME	MATERIAL	NO. OFF
1.	BODY	G.M.	1
2.	COVER	G.M.	1
3.	GLAND	G.M.	1
4.	GLAND NUT	G.M.	1
5.	VALVE	G.M.	1
6.	SPINDLE	G.M.	1
7.	HANDLE	G.M.	1
8.	SPLIT PIN	M.S.	1

Exercise 3: Fig. 15.6 shows the details of 50 mm stop valve. Assemble all the parts and draw the following views to a scale half full size :—

(a) Front view full in section (b) Side view.

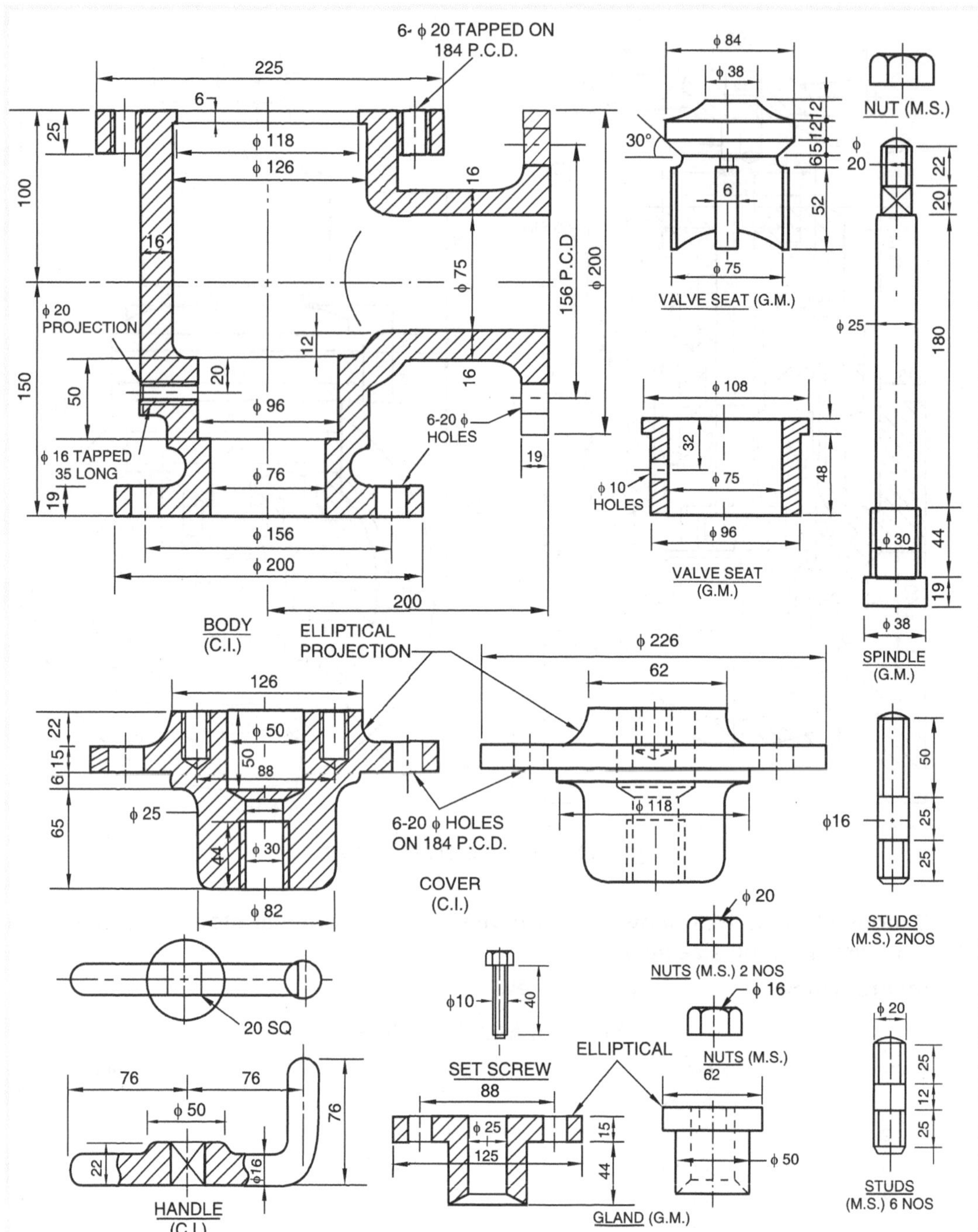

Fig. 15.6 *Details of 50 mm stop valve*

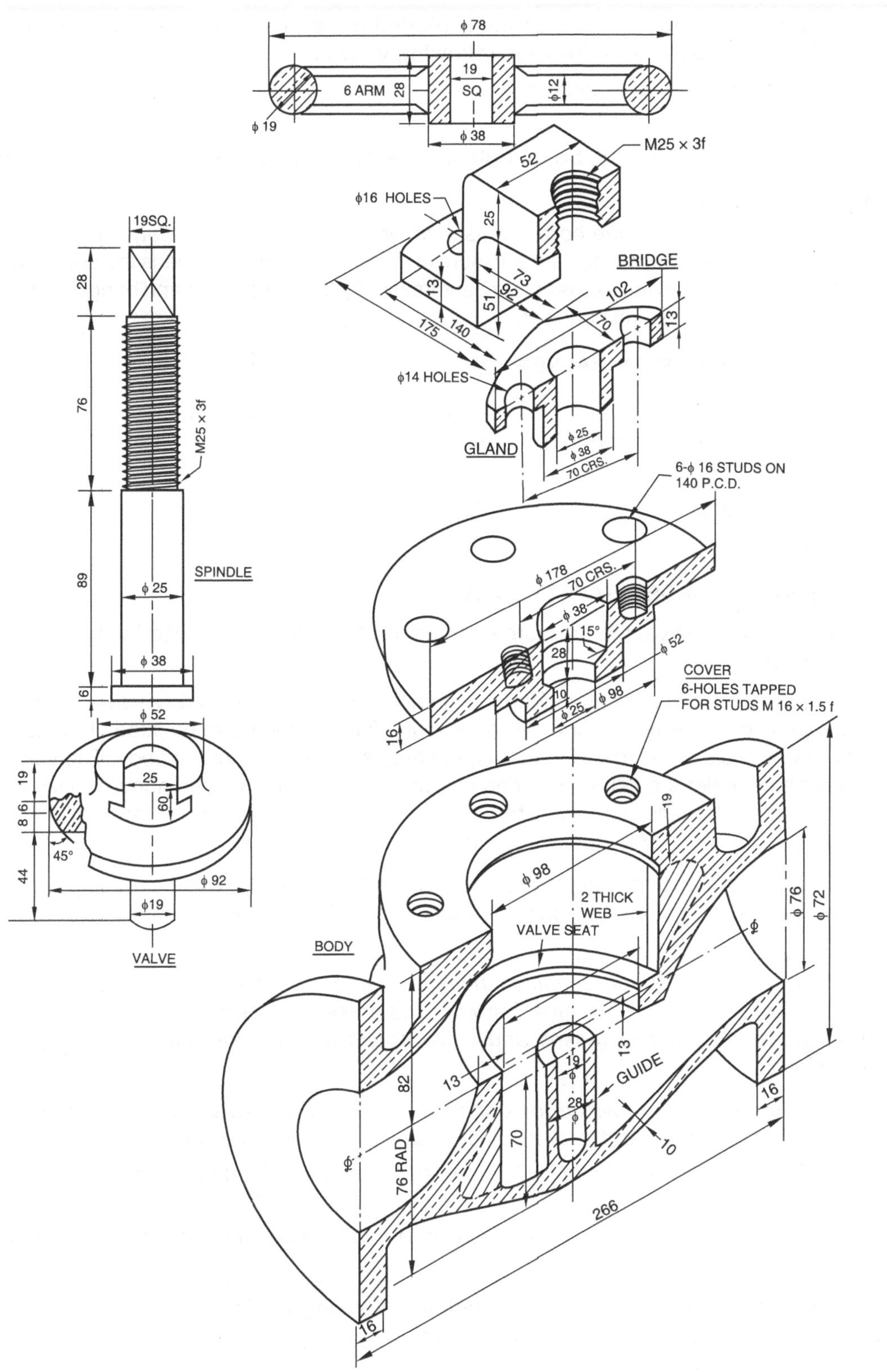

Fig. 15.7. *Gun metal steam stop valve*

Exercise 4 : Fig. 15.7 shows the exploded parts of a gun metal steam stop valve. Assemble all the parts and draw the following views :—

(a) Front view—full in section

(b) Top view

Assume suitable stud, bolts, nuts size. Prepare a part-lists and bill of materials

15.5. SAFETY VALVE

The device used on the boiler for its safe working is known as safety valve. The safety valve comes into operation when the pressure of steam in the boiler exceeds the normal working limit. It, then discharges the excess steam automatically out of the boiler and brings the pressure down to the normal working condition.

15.6. TYPES OF SAFETY VALVES

There are various types of safety valves, but in general the following types are commonly used in engineering practice :—

1. Spring loaded safety valve
2. Dead weight safety valve
3. Lever safety valve

15.7. SPRING LOADED SAFETY VALVE

In this safety valve, the pressure of steam on the valve is balanced by the force due to extended spring.

Fig. 15.8 shows a spring loaded valve called *Ram's bottom safety valve.* It consists of a U-shaped body having two limbs with a circular flange which is to be fitted at the top of the boiler. Valve seats are screwed to the upper flanges of the limbs of the body. The valves rest on their seats under the action of spring pressure. The lever is placed above valves with the help of pivots. The lever is held tight by means of helical spring, the one end of which is connected to the body and other end to the lever.

In operation, when the steam pressure exceeds the normal working pressure condition, the valve rises up against the action of the spring and causes the excess steam to escape out.

Note : *1. Under normal condition the force exerted by the spring is more than the force due to steam pressure on the valve. But when the pressure rises above the set limit, the force due to steam pressure increases and valve opens to allow the steam to escape in the atmosphere.*

2. A safety link is provided to connect the eye bolt and lever. The link is to check the blowing off of the valves and lever if, spring breakes or rise in pressure is abrupt or too much.

3. The extension on the lever is provided to operate it by hand for periodical checking of the safety valve.

4. Springs are made of nickel-chromium or high speed steel so that they can function effectively at elevated temperature.

Exercise 5: Fig. 15.8 shows the details of spring loaded safety valve. Assemble all the parts and draw the following views :—

(a) Front view (b) Top view

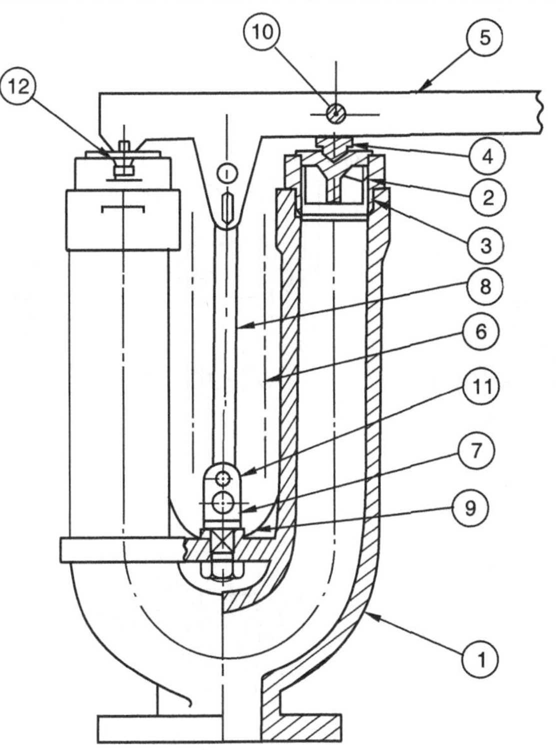

Fig. 15.8. *Assembly of a spring loaded safety valve*

PARTICULARS OF PARTS

S.NO.	NAME	MATERIAL	NO. OFF
1.	BODY	C.I.	1
2.	VALVE	G.M.	2
3.	SEAT	G.M.	2
4.	PIVOT	M.S.	1
5.	LEVER	M.S.	1
6.	SPRING	STEEL	1
7.	SHACKLE WITH NUT	M.S.	1
8.	LINK	M.S.	2
9.	COLLAR	M.S.	1
10.	PIN	M.S.	1
11.	PIN (LINK)	M.S.	1
12.	SCREW	M.S.	4

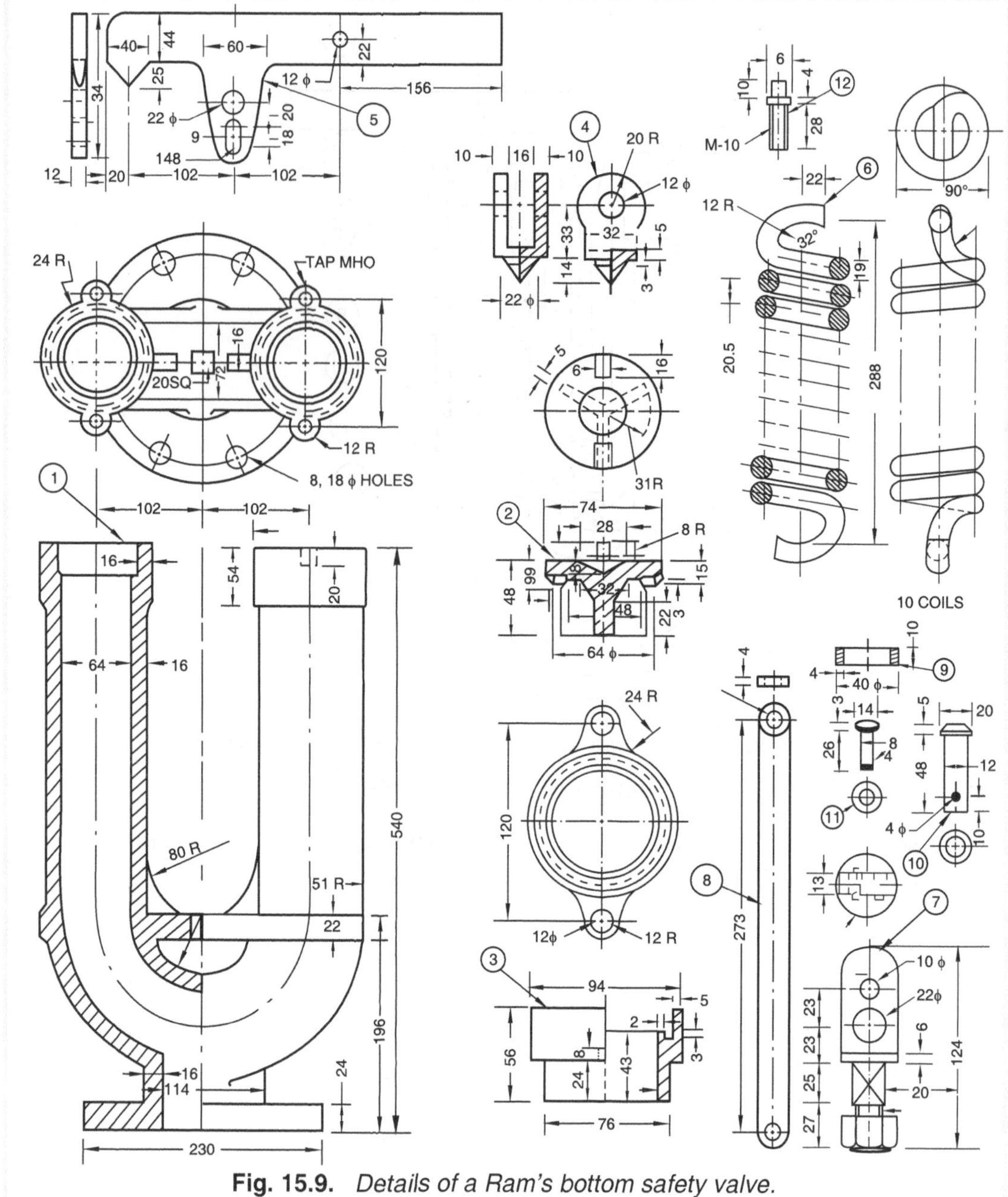

Fig. 15.9. *Details of a Ram's bottom safety valve.*

Exercise 6 : Fig. 15.10 shows the details of a spring loaded valve. Draw to some suitable scale the following views of the assembly :—

(a) Front view–left half in section (b) End view–looking from left side.

(c) Top view.

Exercise 7 : Fig. 15.11 shows the details of a Ram's bottom safety valve. Assemble all the parts and draw the following views :—

(a) Front view-left half in section (b) End view-looking from left side

(c) Top view.

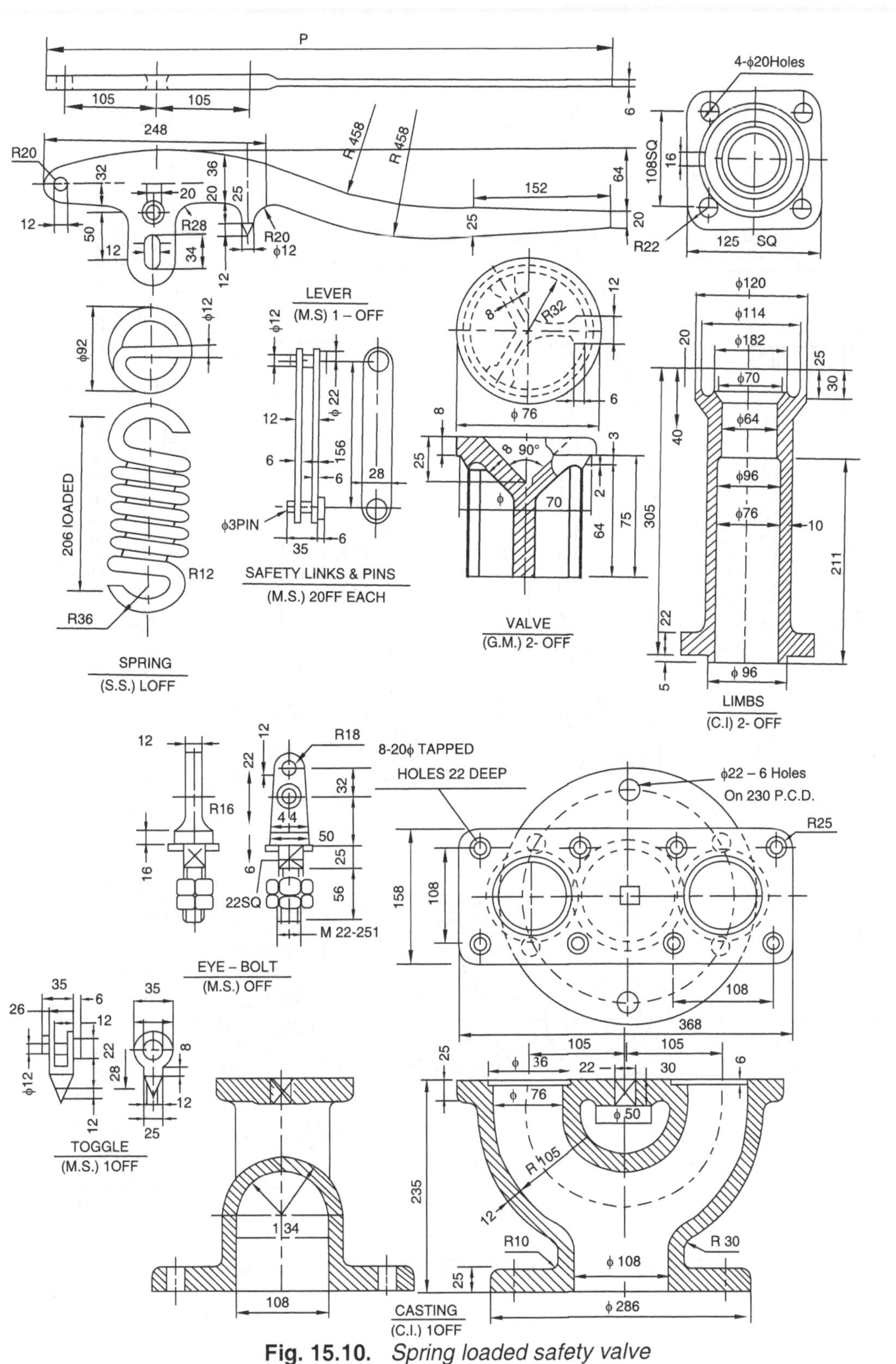

Fig. 15.10. *Spring loaded safety valve*

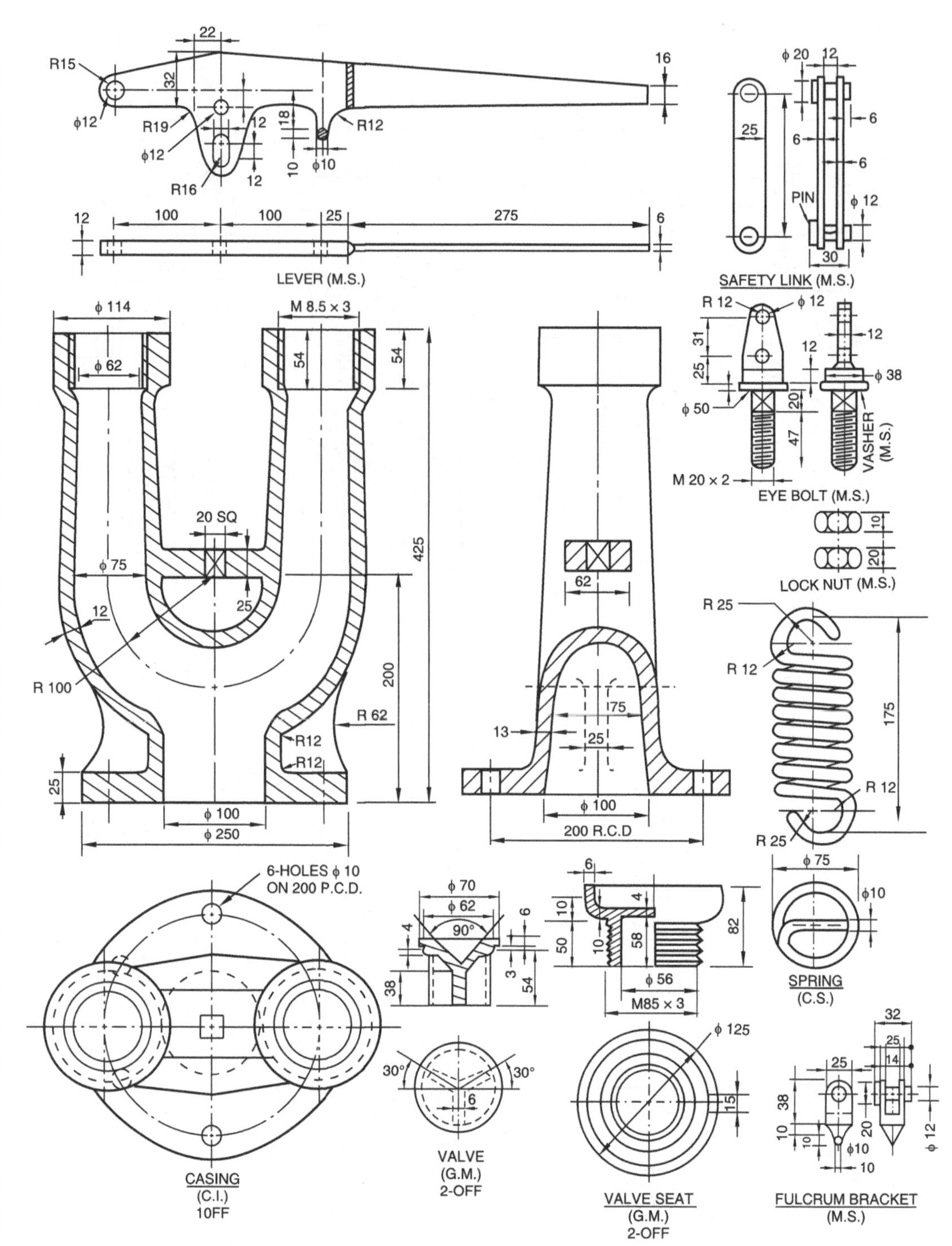

Fig. 15.11. *Details of a Ram's bottom safety valve*

Hint : Fig. 15.12. shows the front view in full section and sectional side view of safety valve.

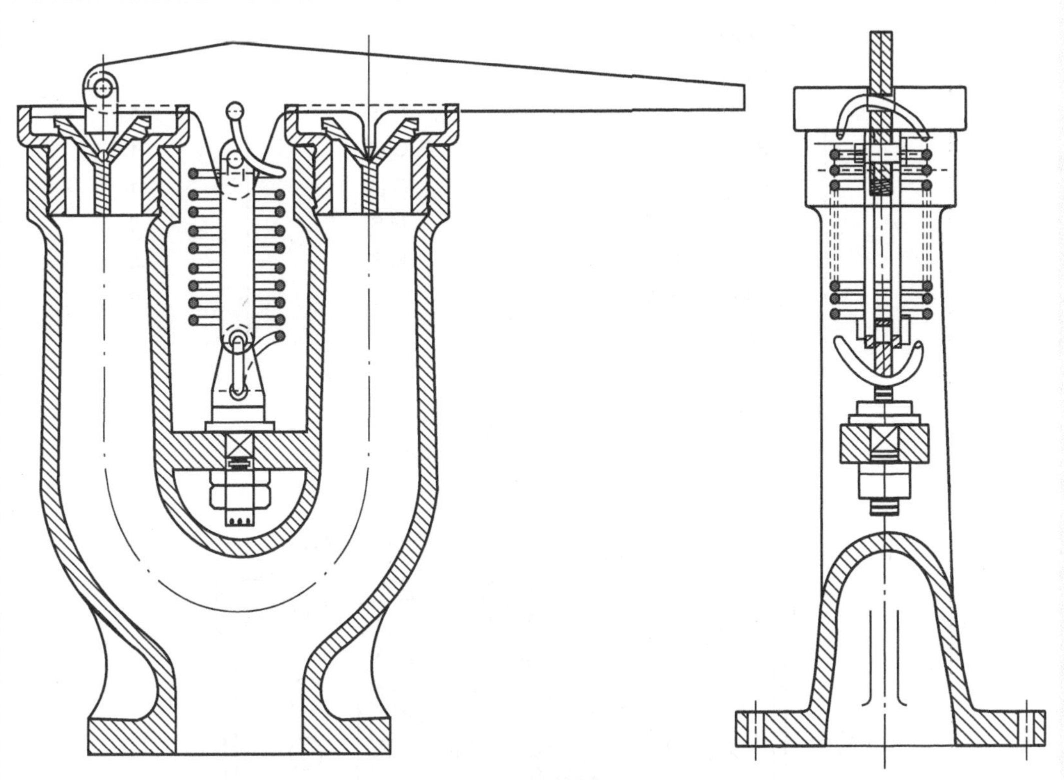

Fig. 15.12. (i) *Assembly of Ram's bottom safety valve*

15.8. DEAD WEIGHT SAFETY VALVE

In this safety valve, the pressure of steam on the valve is balanced by the force exerted due to heavy weight attached to weight carrier.

Fig. 15.12 shows a dead weight safety valve. It consists of a valve resting over the valve seat which is secured to the top end of vertical hollow body. The valve is loaded through a spindle by weights placed on the suspended weight carrier. The weights are enclosed in an outer cover.

In operation, when the steam pressure exceeds the normal condition, it lift the valve with weights and the excess steam escapes from the boiler and ultimately the steam pressure falls.

PARTICULARS OF PARTS

S.NO.	NAME	MATERIAL	NO. OFF
1.	BODY	C.I.	1
2.	CASING	C.I.	1
3.	WEIGHT-CARRIER	C.I.	1
4.	WEIGHT	C.I.	1
5.	COVER	C.I.	1

S.NO.	NAME	MATERIAL	NO. OFF
6.	RING	C.I.	1
7.	SEAT	C.S.	1
8.	VALVE	G.M.	1
9.	SPINDLE	M.S.	1
10.	BUSH	G.M.	1
11.	CAP	G.M.	1
12.	STUD AND NUT	M.S.	2
13.	BOLT AND NUT	M.S.	2
14.	CAP-SCREW	M.S.	3
15.	SET-SCREW	M.S.	2

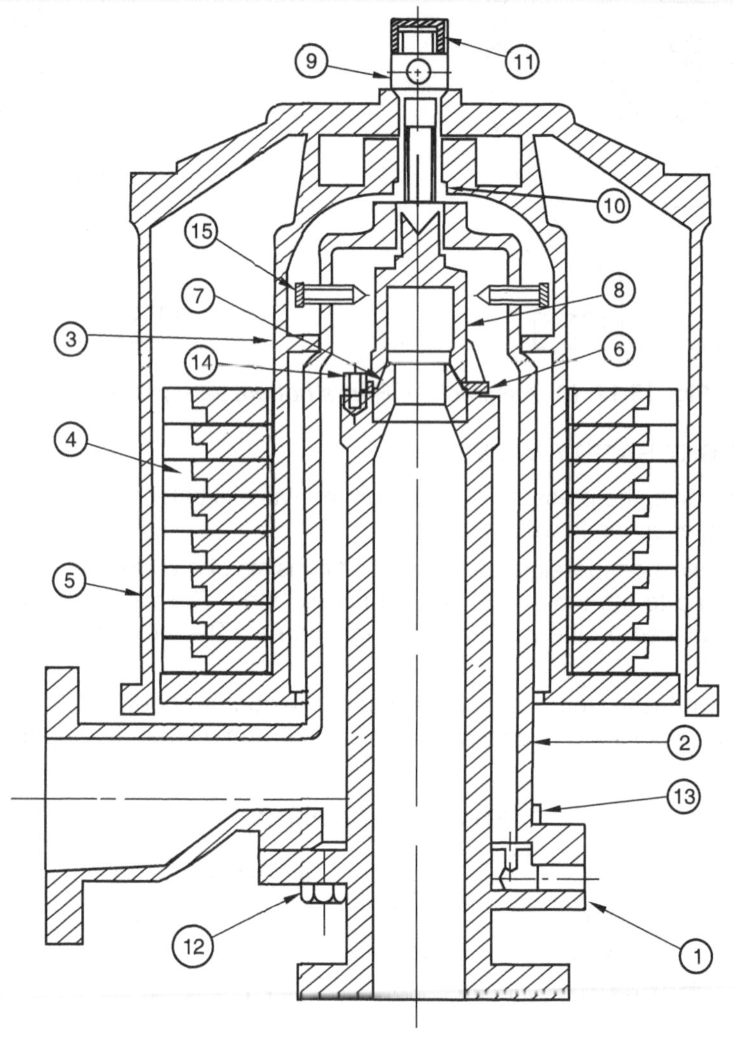

Fig. 15.12. (ii) *Details of dead weight safety valve*

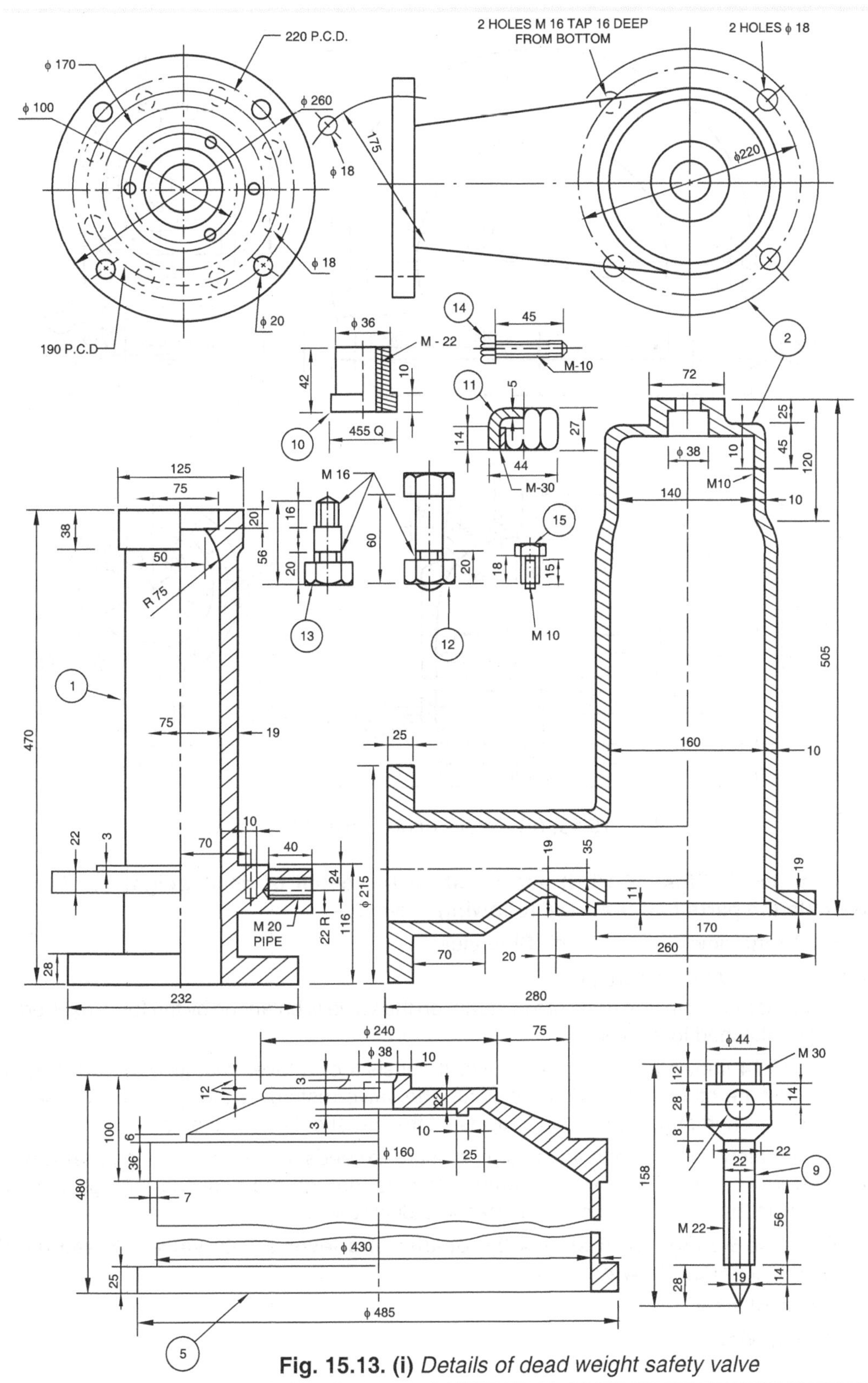

Fig. 15.13. (i) *Details of dead weight safety valve*

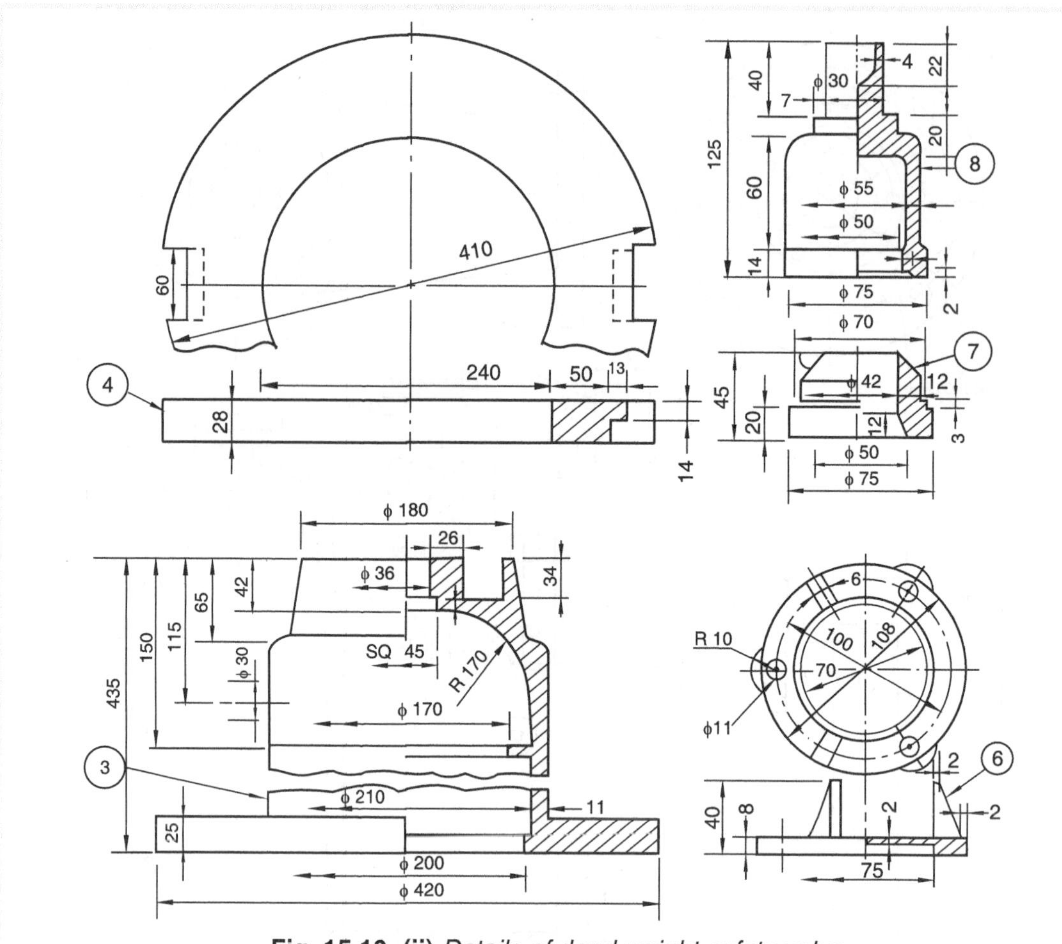

Fig. 15.13. (ii) *Details of dead weight safety valve*

Exercise 8: Figs. 15.13 (i) and (ii) show the detail of dead weight safety valve. Assemble the parts and draw the following views :—

(a) Front view **(b) Side view.**

15.9. LEVER SAFETY VALVE

In this safety valve, the pressure of steam on the valve is balanced by the force exerted due to a weight attached to a lever.

Fig. 15.14 shows a lever safety valve. It consists of a lever with its one end hinged to the body and the other end carries a weight. The valve rests over the seat which is secured to the body. A short strut is provided to transmit the thrust of lever to the valve.

In the safety valve, when the steam pressure exceeds the normal working pressure, the valve and lever lift up. The excess steam escapes through the passage through the valve and seat, with the result that the steam pressure decreases.

Exercise 9 : Fig. 15.15 shows the details of a lever safety valve. Assemble the parts and draw the following view :—

(a) Front view

(b) Top view

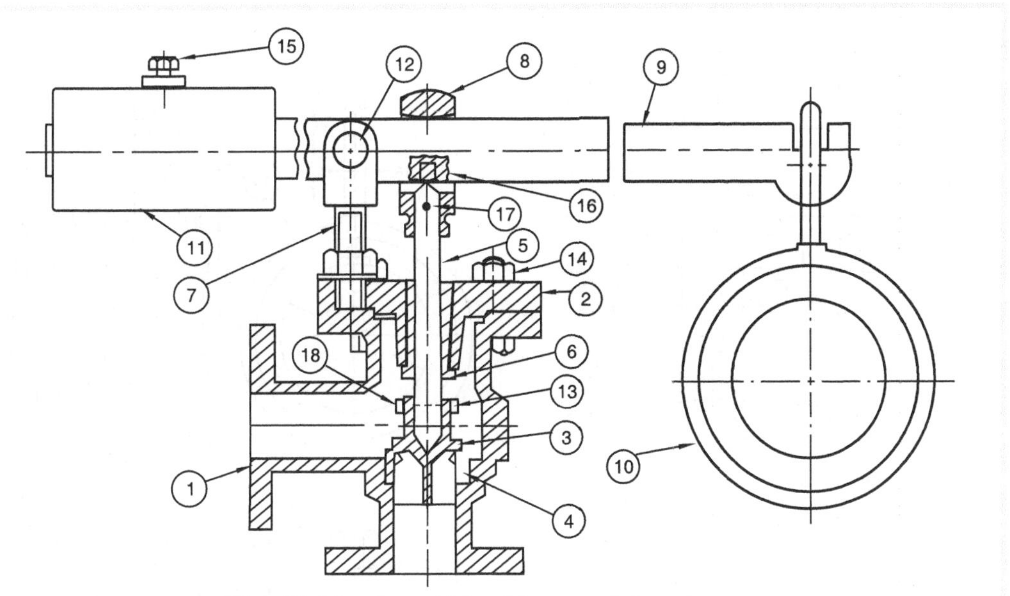

Fig. 15.14. *Assembly of lever safety valve*

PARTICULARS OF PARTS

S.NO.	NAME	MATERIAL	NO. OFF
1.	BODY	C.I.	1
2.	COVER	C.I.	1
3.	VALVE	G.M.	1
4.	SEAT	G.M.	1
5.	SPINDLE	G.M.	1
6.	BUSH BOLT	G.M.	1
7.	FULCRUM, NUT AND WASHER	G.M.	1
8.	STRUT	G.M.	1
9.	LEVER	M.S.	1
10.	WEIGHT	C.I.	1
11.	COUNTER WEIGHT	C.I.	1
12.	PIN (FULCRUM)	G.M.	1
13.	PIN	G.M.	1
14.	BOLT AND NUT	M.S.	4
15.	SET SCREW	M.S.	1
16.	BEARING	STEEL	2
17.	PIN	G.M.	1

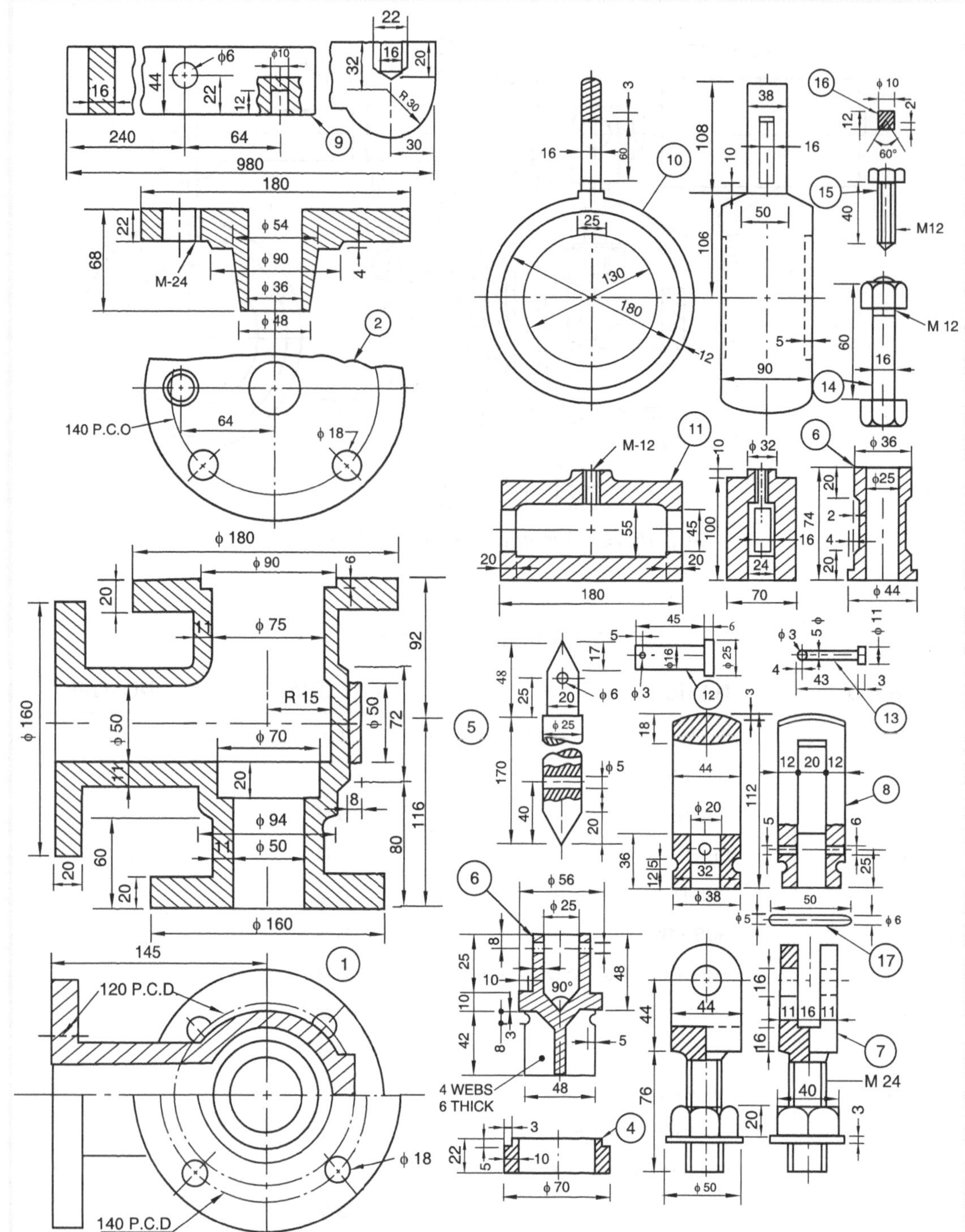

Fig. 15.15. *Details of lever safety valve*

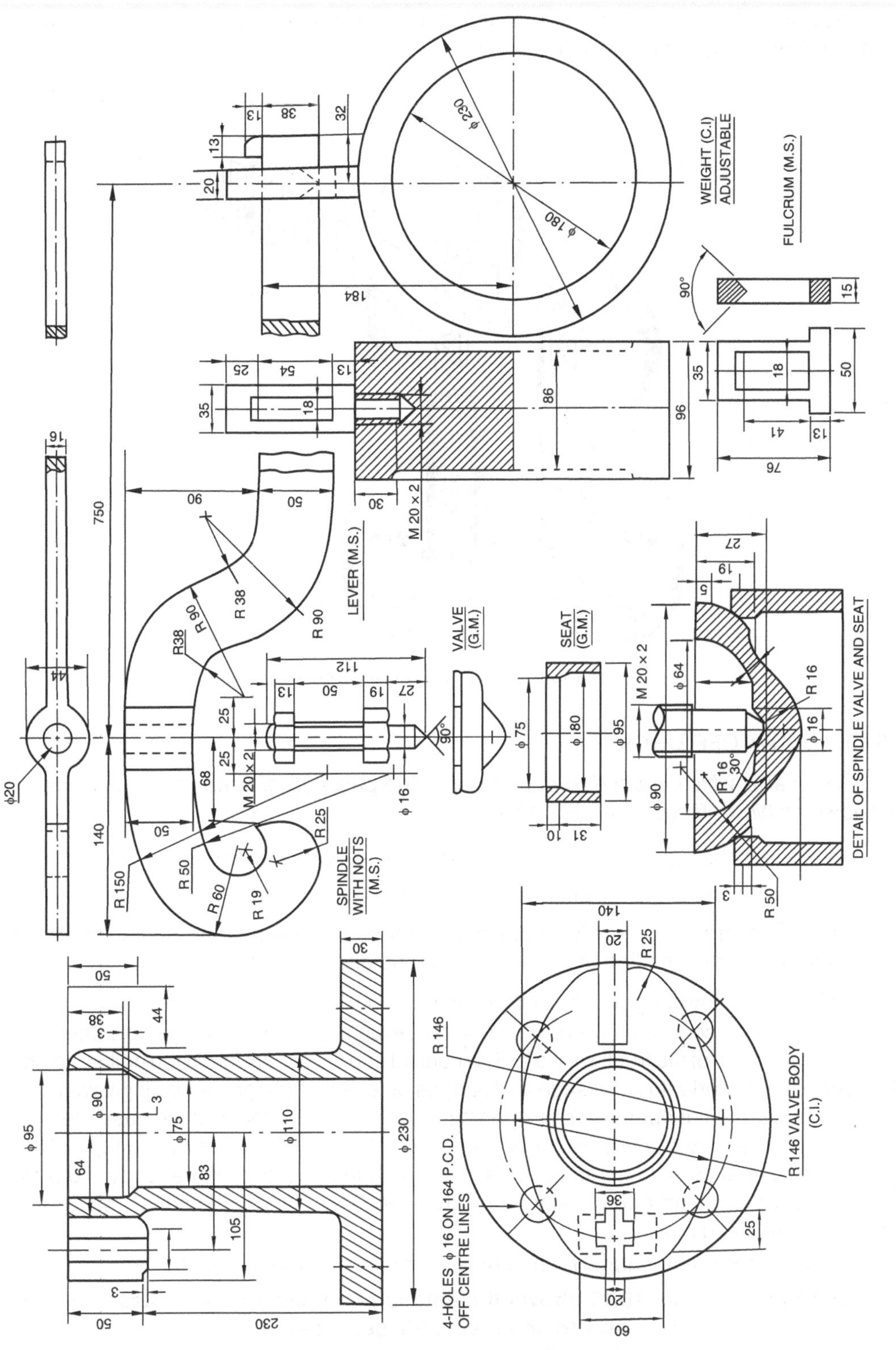

Fig. 15.16. *Details of a lever safety valve*

Exercise 10 : Fig. 15.16 shows the details of a lever safety valve. Assemble the parts and draw elevation and end view :

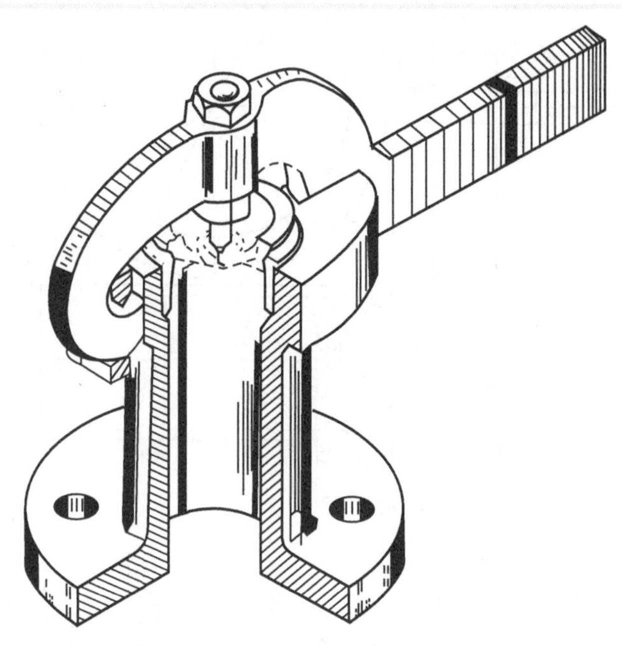

Pictorial View of a lever safety valve

15.10. BLOW OFF COCK

The valve which is used to open or shut the passage of water or other fluids in a pipe is known as blow off cock (see Fig. 15.17)

In a boiler, the blow of cock is fitted at the lowest part, in order :—

(*i*) to empty the boiler for internal cleaning and inspection.

(*ii*) to discharge periodically a portion of water so that any mud, scale and other impurities (known as sediments) settled at the bottom may be discarded.

(*iii*) for rapid lowering of the water-level in the boiler, if it becomes too high.

The blow off cock mainly consists of a tappered plug and a body (hollow shell). The plug is made in the form of the frustum of a hollow conical and fits accurately in the hollow conical shell. The body is provided with a passage and flanges on either ends for connecting it with pipes. The plug has a rectangular opening. This opening when brought in line with the passage in the body (by turning the plug) water will flow out of the boiler. The flow of water may be stopped by turning the plug such that its solid portion comes in line with the passage of the body. To prevent the leakage of water, gland and packings are provided. The gland is tigntened with the body by means of studs and nuts.

Blow off cock are generally made of cast iron, brass or gun-metal.

Problem 11 : Fig. 15.17 shows the details of a blow off cock. Assemble the parts and draw the following views to some suitable scale :—

(a) Front view full in section (b) Side view.

Solution: For its solution, see Fig. 15.18

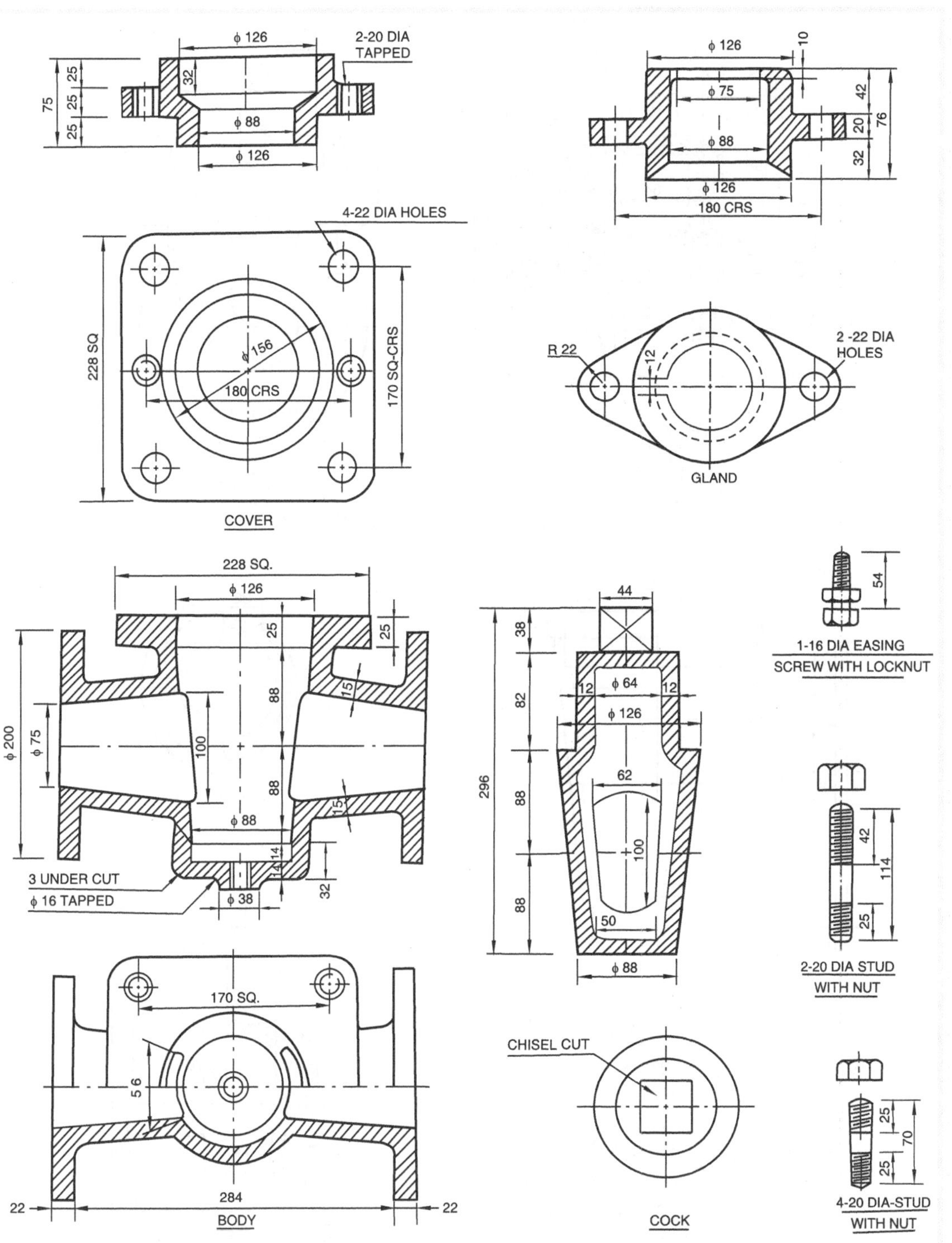

Fig. 15.17. *Blow off cock*

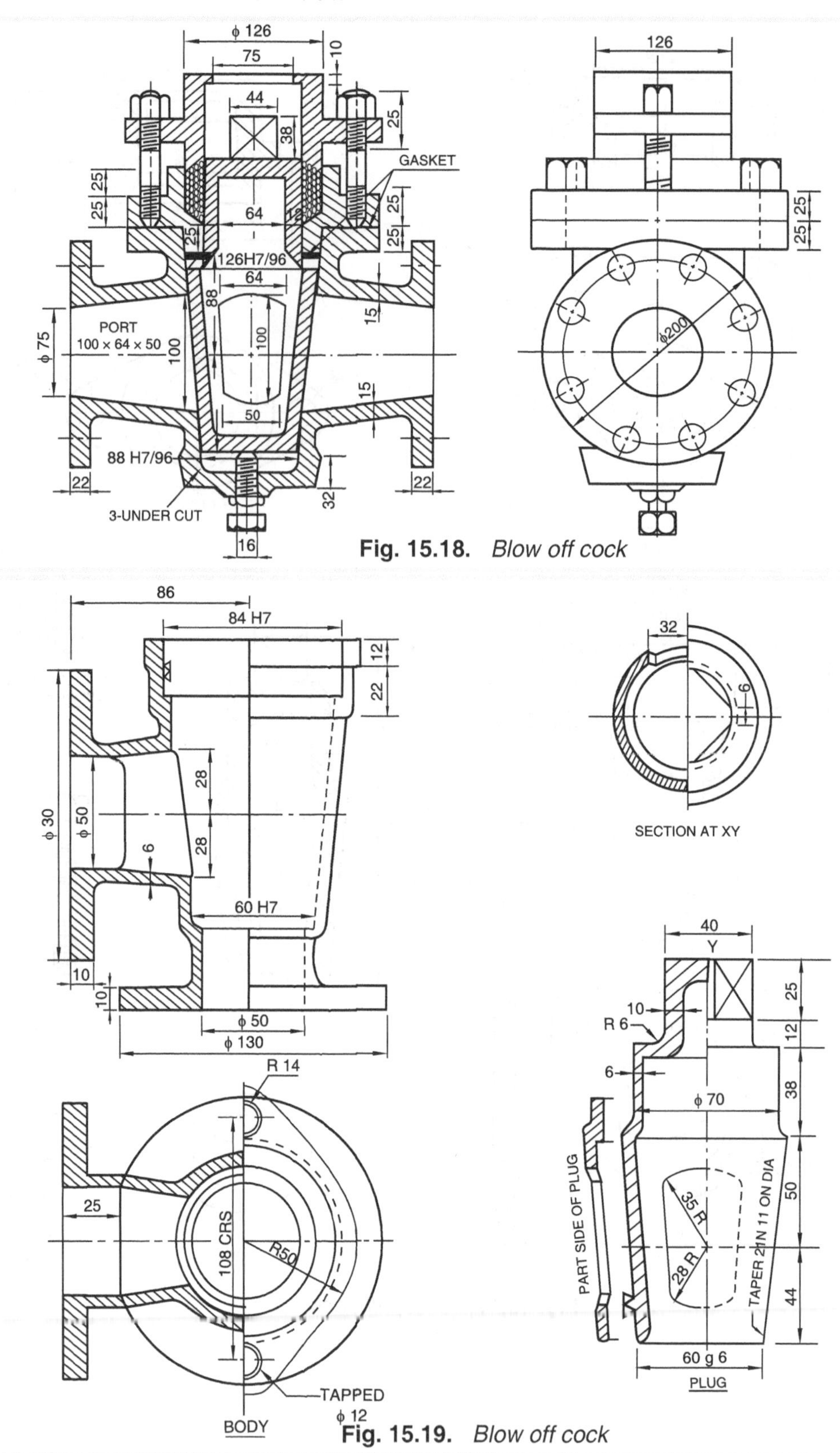

Fig. 15.18. *Blow off cock*

Fig. 15.19. *Blow off cock*

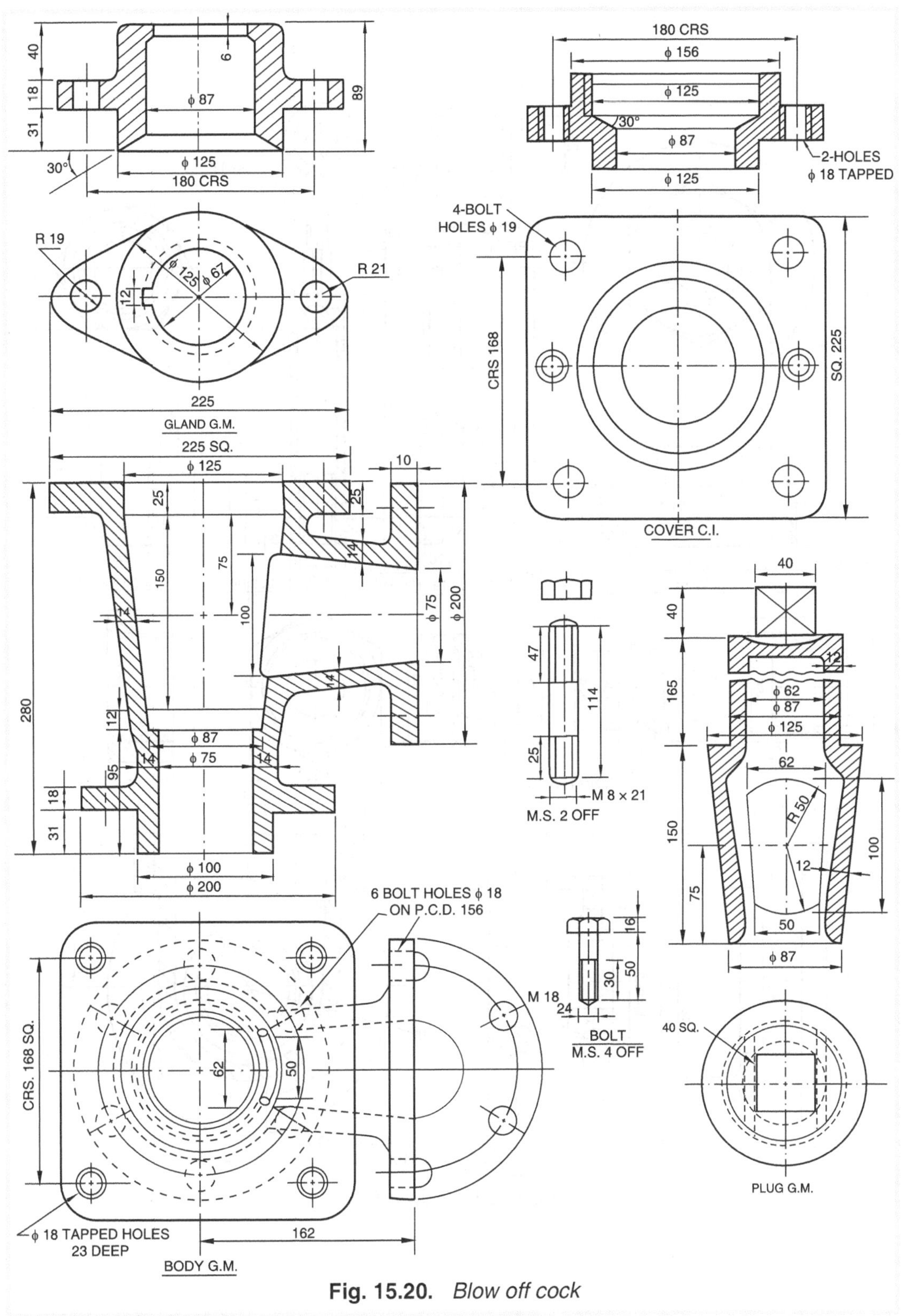

Fig. 15.20. *Blow off cock*

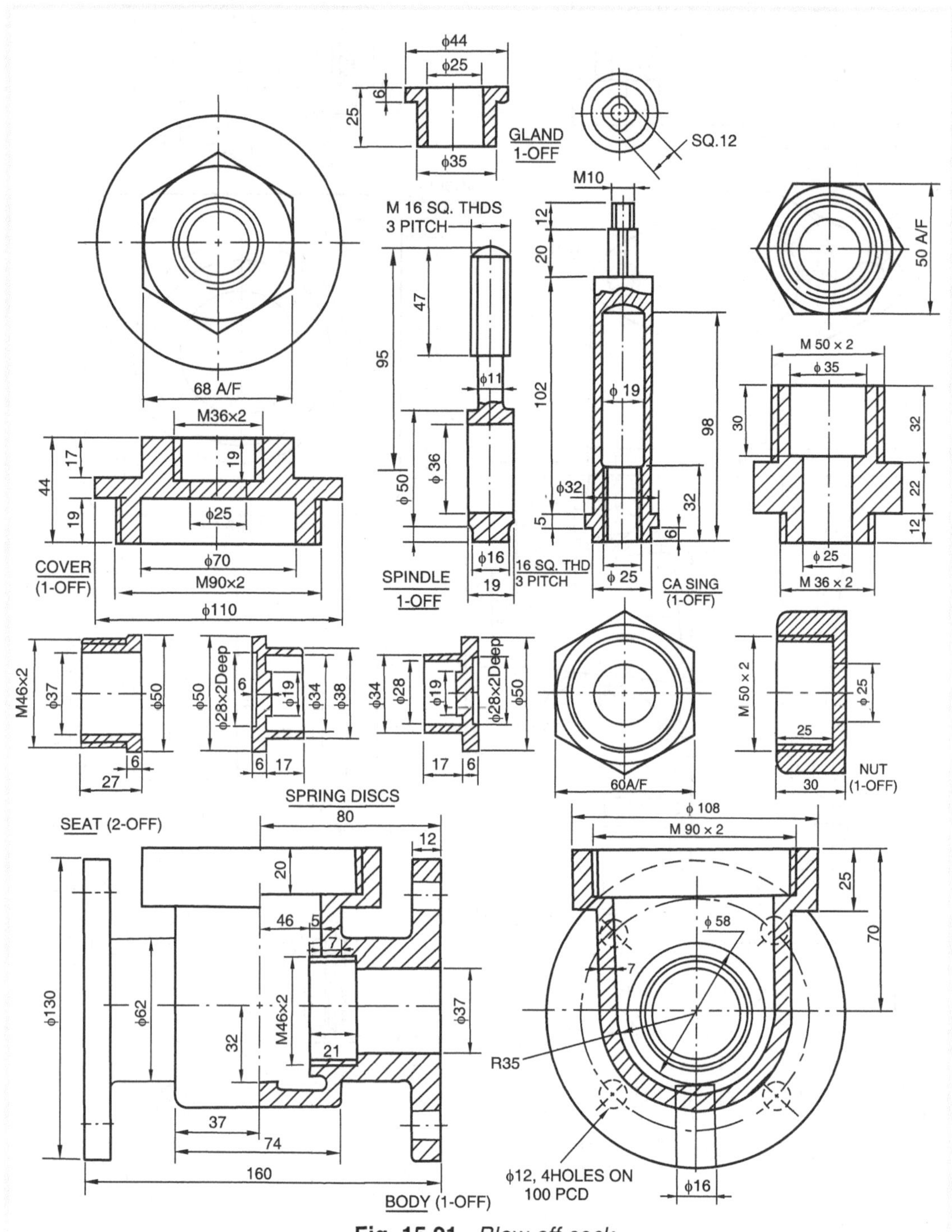

Fig. 15.21. *Blow off cock*

Exercise 12 : Fig. 15.19 shows the details of a blow off cock. Assemble the parts and draw the following views :—

(a) Elevation-full in section (b) Side view

Exercise 13 : Fig. 15.20 shows the details of a blow off cock. Assemble the parts and draw the following views:

(a) Front view (b) Top view

Exercise 14 : Fig. 15.21 shows the details of a blow off cock. Assemble the parts and draw the following views:

(a) Front view (b) Top view

15.11. FEED CHECK VALVE

A valve which control the supply of water from the pump to the boiler and prevents the back flow of water to the pump when the feed pump ceases to work is known as feed check valve. It is fitted to front end plate of the cylindrical boiler.

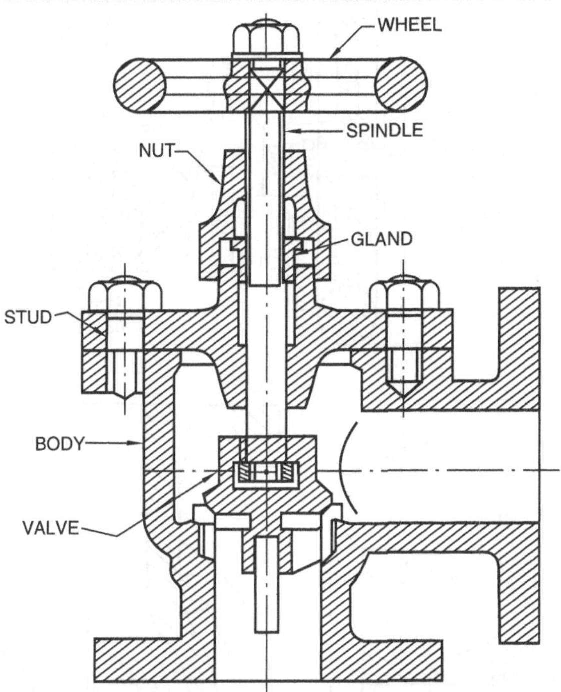

Fig. 15.22. *Assembly of feed check valve*

Fig. 15.22 shows a commonly used feed check valve. At normal condition, the valve is lifted due to the pressure of water coming from the pump and thus, the water is fed to the boiler. On the other hand when the pump stops working, the pressure of water is less than boiler pressure and valve rests on its seat thereby closing passage automatically.

Exercise 15 : Fig. 15.23 shows the details of 50 mm diametre feed check valve. Assemble all the parts and draw the following views :—

(a) Front view (b) Side view (c) Top view

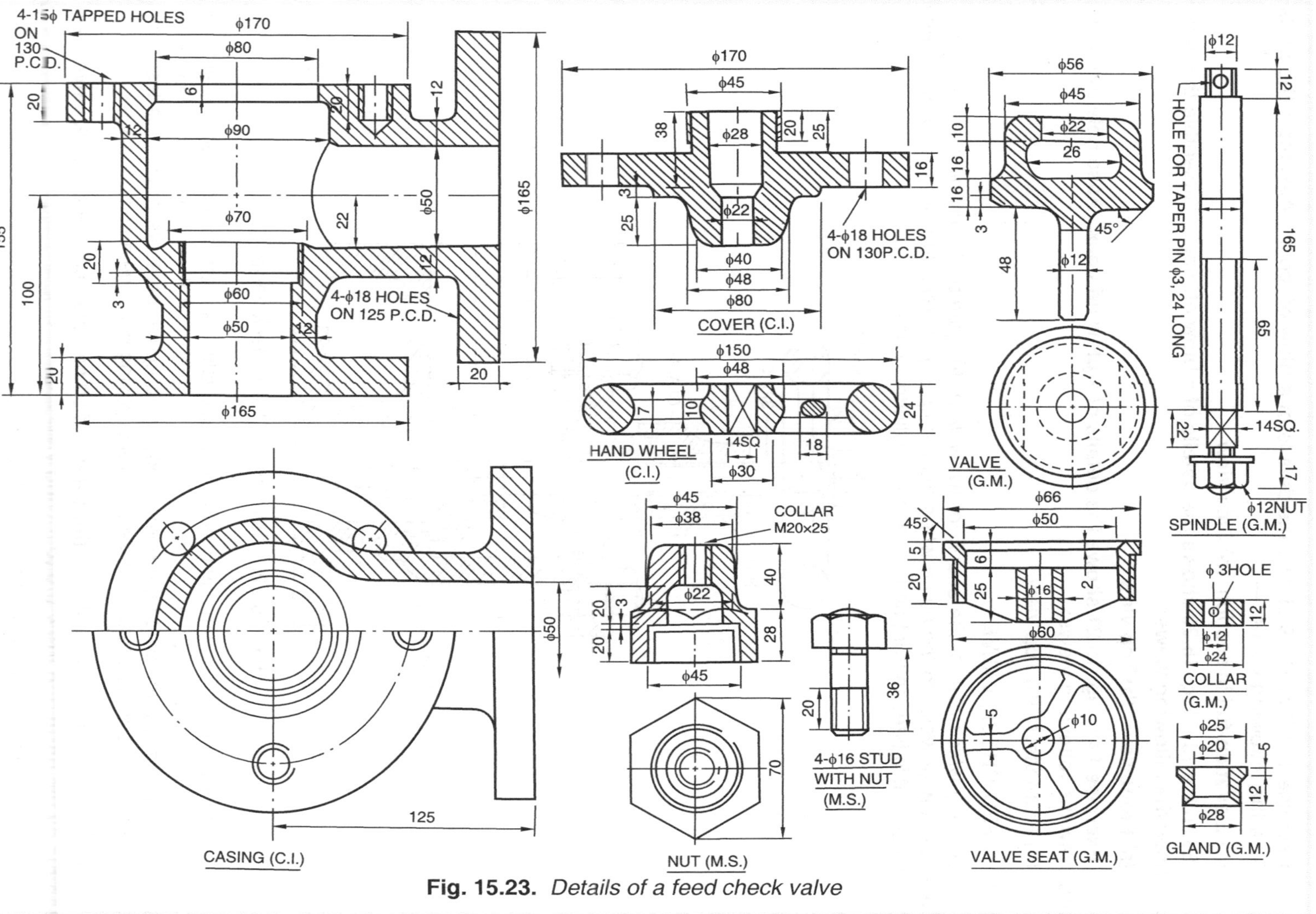

Fig. 15.23. *Details of a feed check valve*

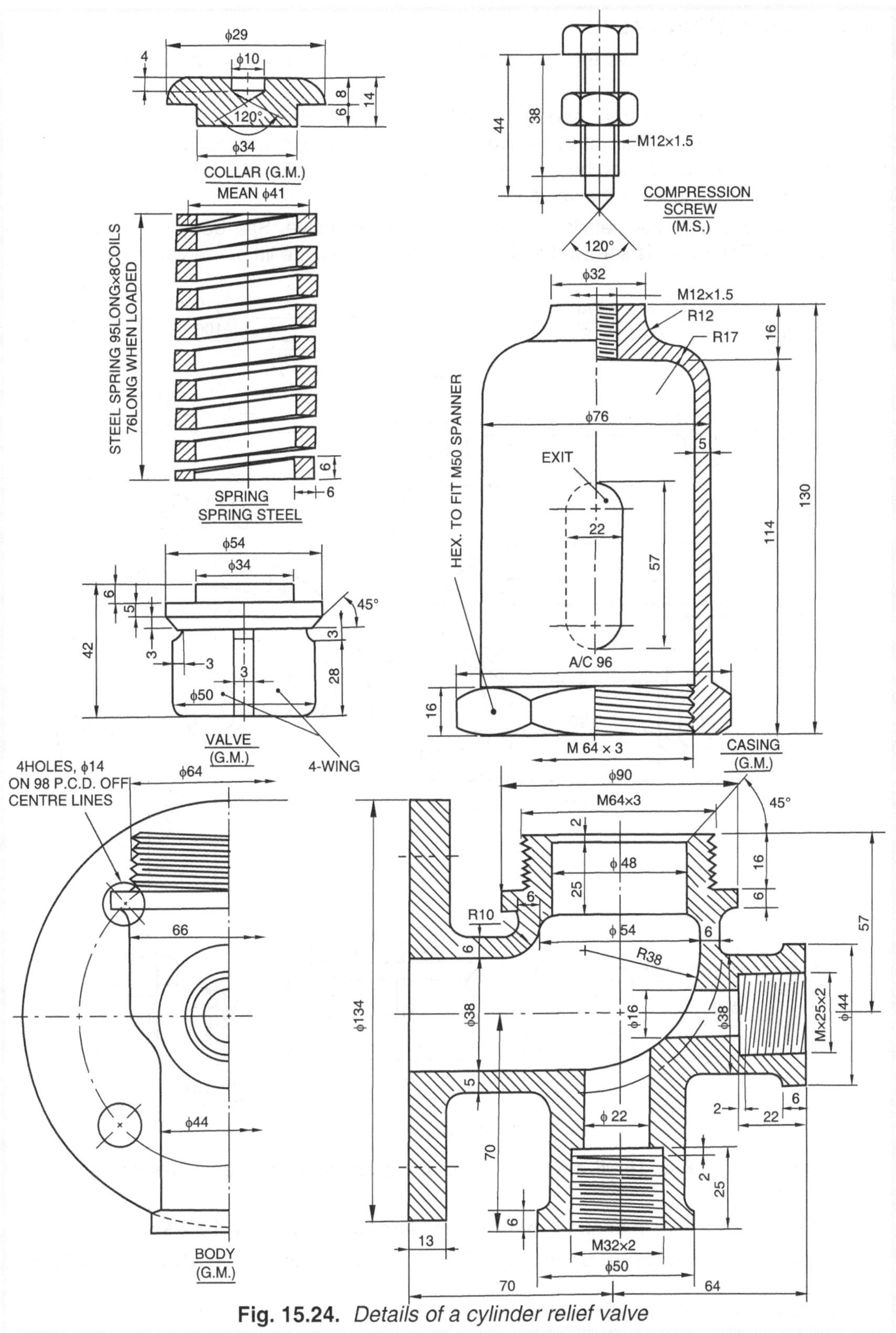

Fig. 15.24. *Details of a cylinder relief valve*

15.12. NON-RETURN VALVE

The valve which allows teh flow of fluid in one direction only is known as non-return valve.

It is mounted on the front end plate of the boiler shell and serve the same purpose as the feed check valve. The lift of the valve is limited by a projected position of the cover the cercular covar is tighted on teh body by means of 6 studs (see Fig. 15.25). A set screw is also provided in the body to keep the valve seat in its position.

Fig. 15.26 shows the details of another Non-return valve. She body of this valve is of globe pattern. The cover is screwed in the body which check the lift of the valve.

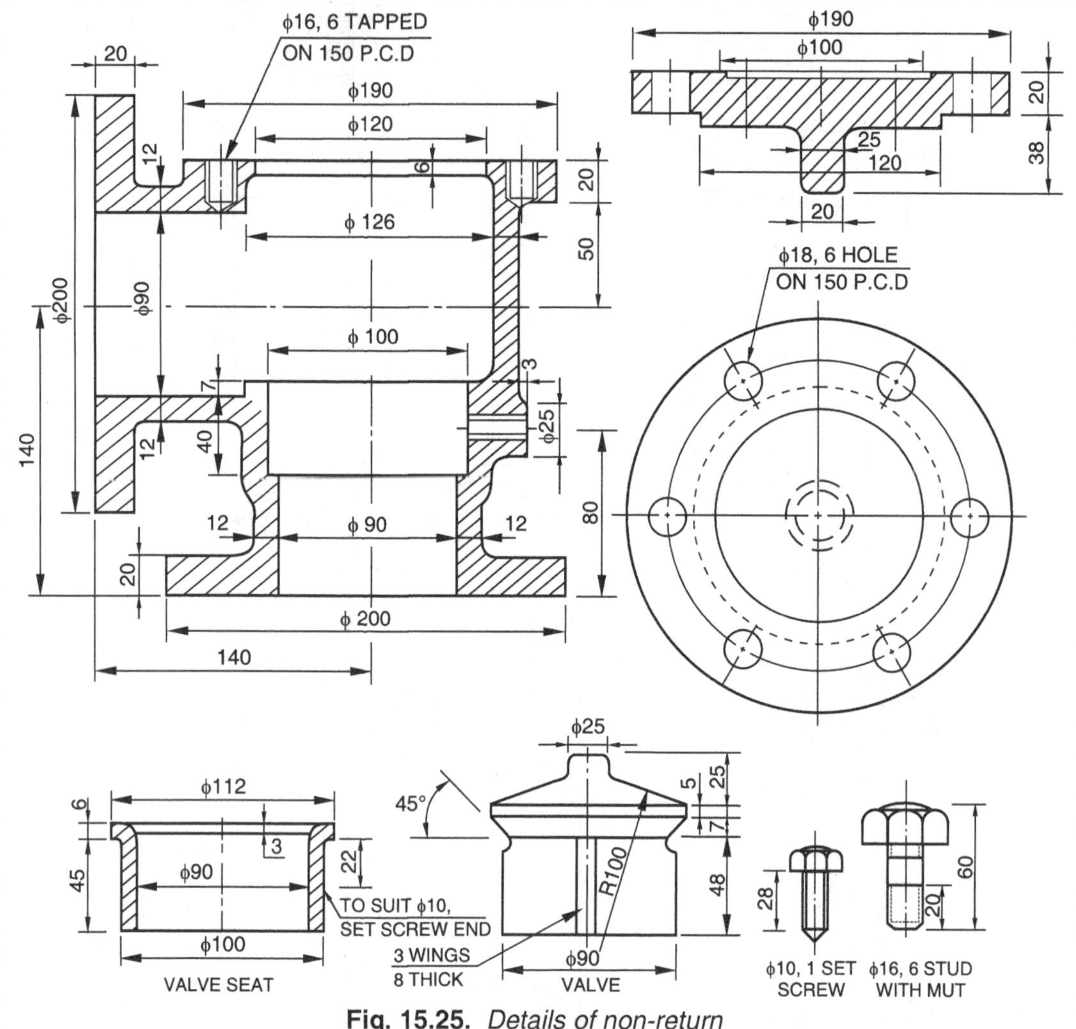

Fig. 15.25. *Details of non-return*

Exercise 16 : Fig. 15.25 shows the details of Non-return valve. Assemble all the parts and draw :

(a) Front View-In full section (b) end view and (c) top view to 1 : 2 scale.

Exercise 17 : Fig. 15.26 Shows the details of Non-return valve. Draw the following views

(a) Front view-full in section (b) Top view.

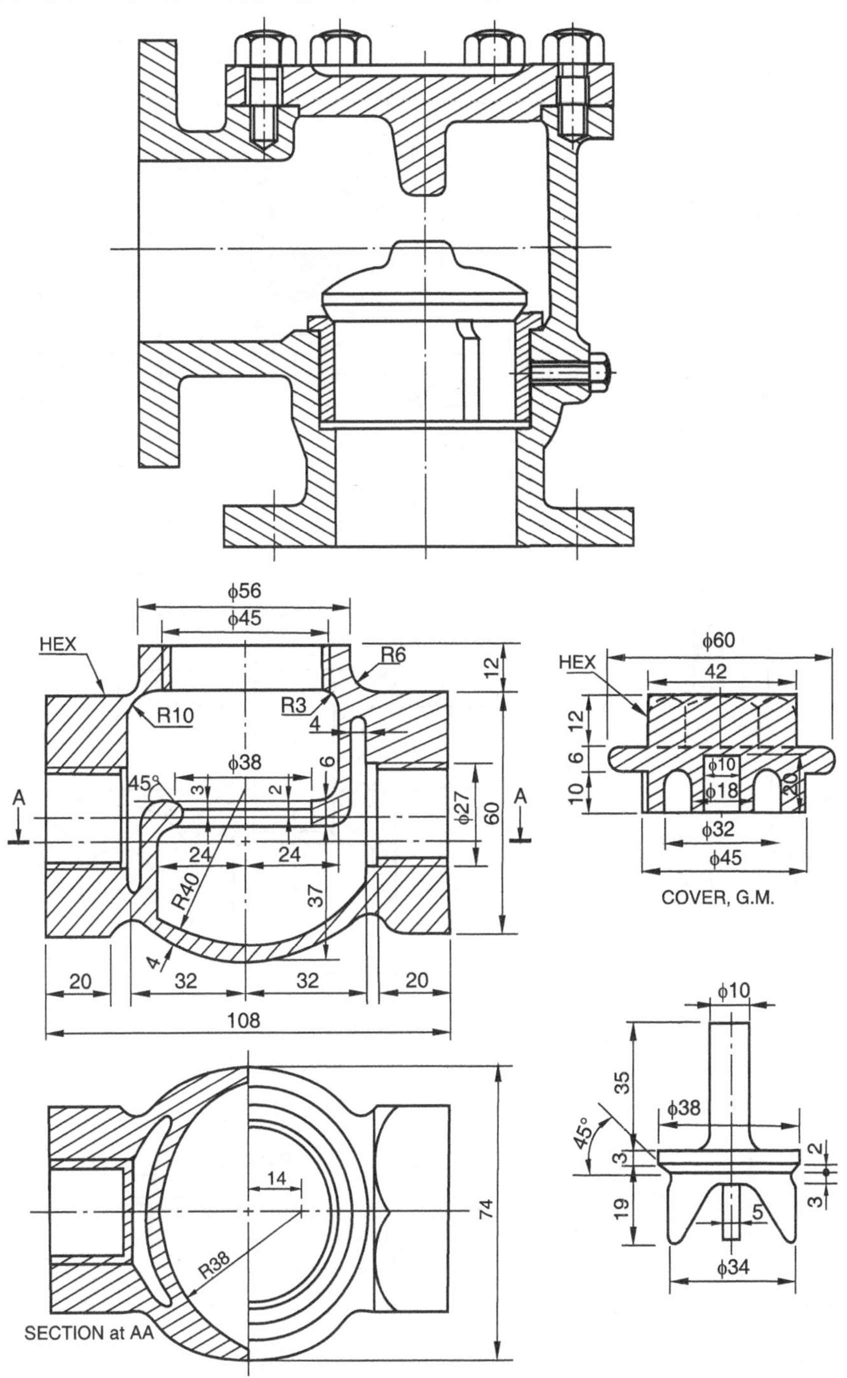

Fig. 15.26. *Details of non-return*

Exercise 18 : Fig. 15.27 shows the details of Non-return valve. Draw the following views of its assembly at scale 1 : 2.

(a) Full Sectional front view (b) End view

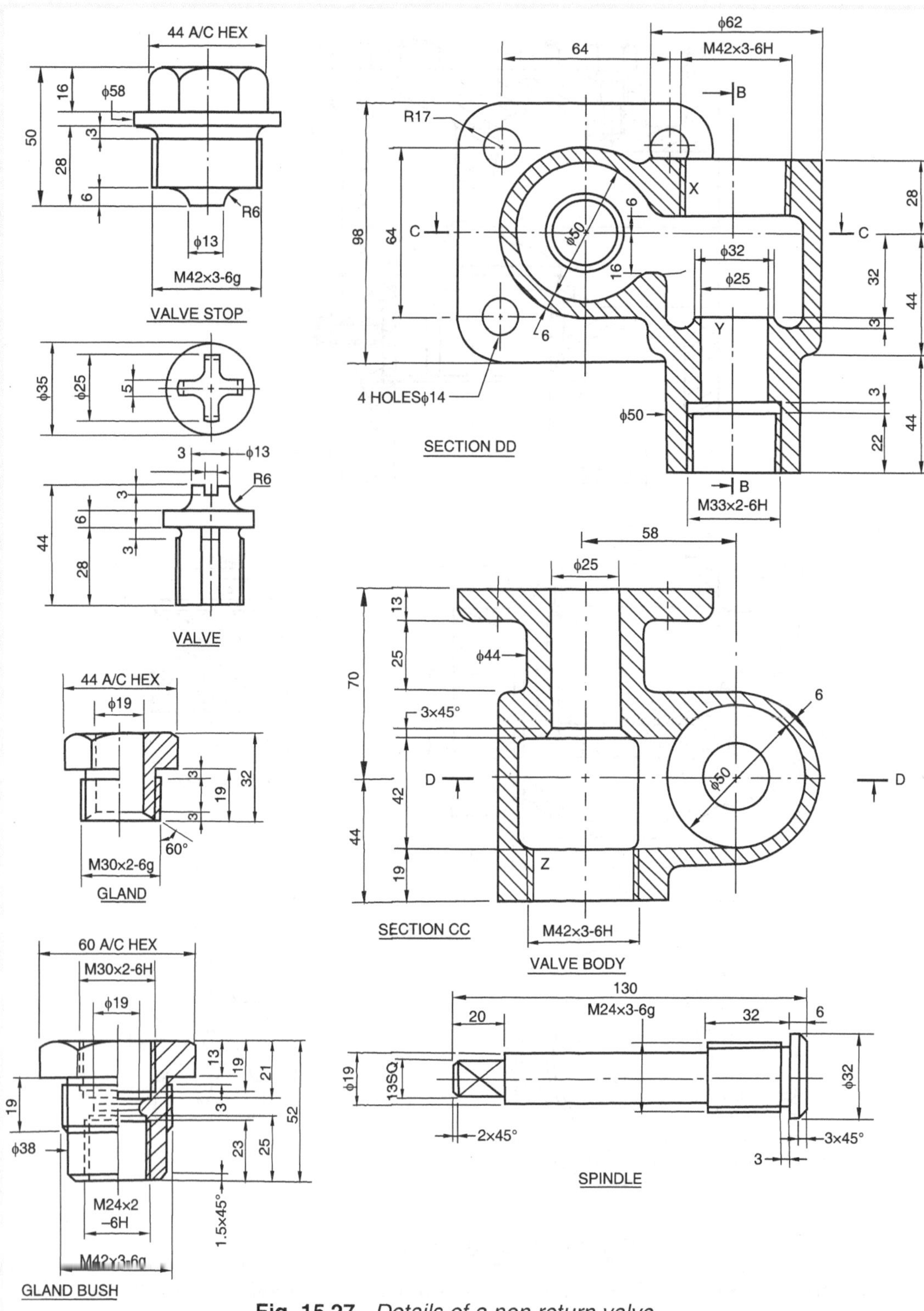

Fig. 15.27. *Details of a non return valve*

15.13. RELIEF VALVE

When a steam engine is not in working position for some time, the steam in the engine cylinder and chest condenses to form water. If this water is not removed while starting the engine, it can give rise to excessive pressure in the cylinder and hence harm the engine mechanism. As a safety measure, relief valve is fitted at end of the engine cylinder. When the pressure in the cylinder exceeds the working limit, the relief valve open to release the excessive steam pressure.

Fig. 15.24 shows the details of a relief valve. The vertical flange is fitted to the engine cylinder and the small horizontal flange at the top of the body which is machined to receive the valve. The valve is kept in closed position by a spring which is held in position by a collar and enclosed in the casing. Opening (exit) provided in the casing to release the steam. The force on the valve is adjusted by compression screw.

Exercise 19 : Fig. 15.24 shows the details of a cylinder relief valve. Draw the following assembled view on A_2 drawing sheet:

(a) Front view–Full in section

(b) End view–looking from right side.

Exercise 20 : Fig. 15.27 shows the details of a non return valve. Assemble all the parts and draw the following view:

(a) Front view-full section

(b) Side view

(c) Top view

QUESTIONS FOR SELF EXAMINATION

1. What do you mean by a valve?
2. What is a valve seat? How the valve seat is secured to the body of valve?
3. What are the different types of valves?
4. What are the functions of a stop valve?
5. What is the difference between lever safety valve and spring loaded safety valve.
6. What is the function of a blow off cock?
7. Where is the blow off cock used?
8. How is the blow off cock operated.
9. What is the function of a relief valve and where is it fitted?
10. What is a feed check valve? What is the lift of the valve in it?

CHAPTER

16

Gears

INTRODUCTION

In engineering practice, mechanical power is used to run a large number of machines. One of the major problems, which an engineer faces, is to transmit this power from one place to another e.g., in a scooter, mechanical power developed by the engine has to be transmitted to the rear wheel. The common devices which are used for transmission of power in various machines are gears, belts, ropes, chains etc.

Gears are used to transmit motion at a definite and constant velocity ratio in situations where the distance between the axes of the driving and driven shafts is short. In comparison with belt and chain drive, *gear drive is more compact, positive (without

* **Gear Drive** : Two gears meshing together for transmission of motion from one shaft to another constitute what is known as gear drive

any slip), can operate at high speed and can be used for transmission of high power.

While transmitting motion, two gears mounted on the shafts are meshed with each other. When one gear (driver) rotates the other gear (driven), and thus the motion is transmitted from one shaft to another shaft without slip.

In this chapter we shall deal with the study of gears, classification of gears, methods to draw gears, conventions for gears, etc.

16.1. GEAR

A wheel having teeth of uniform formation provided on its circumferential surface is called a gear.

Gears are manufactured by casting, machining, stamping or by power metallurgical processes. Out of all such processes, method of production of gears by machining is mostly used in engineering practice.

16.2. CLASSIFICATION OF GEARS

Gears may be classified into three groups according to the position of shaft arrangement as under :–

1. Gears which connect parallel shafts, laying in the same plane such as spur gears, helical gears, double helical or herring bone, triple helical and internal gears. In these gears, the teeth are parallel to the axis.

2. Gears which connect inclined shafts, which if produced would intersect at some angle in the same plane such as bevel gears with straight, screw or spiral teeth.

3. Gears which connect non-parallel and non-intersecting shafts, such as spiral, worm and worm wheel, hypoid and skew gears.

Fig. 16.1 to 16.4 show the important types of gears along with their practical applications.

16.3. SPUR GEARS

Gears which transmit motion between parallel shafts are called spur gears. The teeth in these gears are parallel to the axis of the shaft and are straight [see Fig. 16.1].

16.4. BEVEL GEARS

Gears which transmit power between shafts, whose axes intersect each other if produced, are called bevel gears [see Fig. 16.2]

16.5. HELICAL GEARS

Gears which transmit power between parallel shafts, but the teeth are parts of the helices, are called helical gears [see Fig. 16.3].

16.6. WORM AND WORM WHEEL OR WORM GEARING

Worm and worm wheel are used for high speed reduction, when the power is being transmitted between non parallel and non intersecting shafts. Usually the axes of the two shafts are at right angles to each other [see Fig. 16.4].

Type of Gears Along with their Practical Application

16.1. *Spur Gears*

Practical application of Spur Gears

Fig. 16.2. *Bevel Gears*

Practical application of Bevel gears

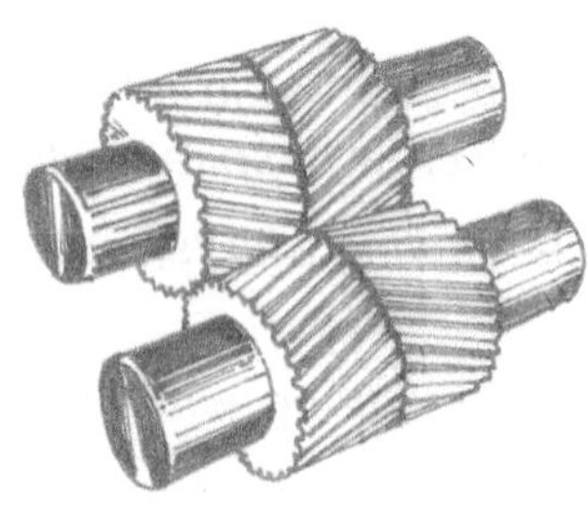

Fig. 16.3. *Helical Gears*

Practical application of Helical gears

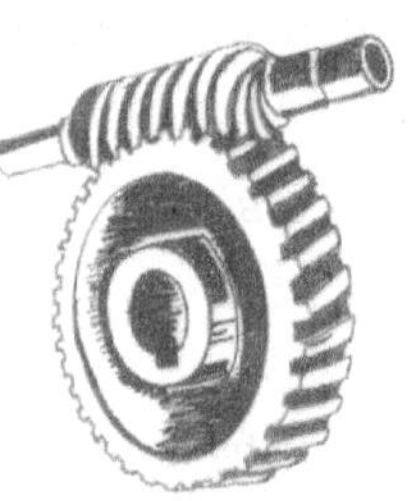

Fig. 16.4. *Worm and Worm Wheel*

Practical application of Worm and Worm Wheel

16.7. TERMS USED IN SPUR GEARS

The following are the common terms which are used in all types of spur gears [see Fig.16.5].

1. Pitch circle : it is an imaginary circle on which the teeth of the mating gears are meshed. Many important measurement are taken from the pitch circle.

2. Pitch circle diameter (P.C.D.) : It is the diameter of the pitch circle of a gear

$$\therefore \text{P.C.D.} = \frac{\text{Number of teeth}}{\text{Diameteral pitch}}$$

$$\text{or P.C.D.} = \frac{N}{\text{D.P.}}$$

3. Diameteral pitch (D.P.) : It is the number of teeth on gear per mm of pitch circle diameter.

$$\text{Diameteral pitch, D.P.} = \frac{N}{\text{P.C.D.}}$$

4. Circular pitch (C.P.): It is the arc distance on the pitch circle diameter from a point on one tooth to the corresponding point on the adjacent tooth.

$$\text{Circular pitch, C.P.} = \frac{\text{Circumference of pitch circle diameter}}{\text{Number of teeth}} = \frac{\pi \times \text{P. C. D.}}{N}$$

Relation between C.P. and D.P.

$$\text{We know, C.P.} = \frac{\pi \times \text{P. C. D.}}{N} \quad \text{---- } (i) \qquad \text{Also D.P.} = \frac{N}{\text{P.C.D.}} \quad \text{---- (ii)}$$

Multiplying (i) and (ii)

$$\text{C.P.} \times \text{D.P.} = \frac{\pi \times \text{P. C. D.}}{N} \times \frac{N}{\text{P.C.D.}}$$

$$\therefore \quad \text{C.P.} \times \text{D.P.} = \pi$$

$$\text{or} \quad \text{C.P.} = \frac{\pi}{\text{D.P.}}$$

5. Module (M) : It is the ratio of pitch circle diameter to the number of teeth in gear.

$$\text{i. e.} \quad \text{Module, } M = \frac{\text{Pitch circle diameter}}{\text{Number of teeth}} = \frac{\text{P.C.D.}}{N} \text{ (expressed in mm)}$$

Module is the main parameter in fixing the dimensions of gear teeth.

6. Chordal pitch : It is the length of the chord of the pitch circle between a point on one tooth and corresponding point on the adjacent tooth, both on the pitch circle.

7. Pitch point : It is the point of contact between the pitch circle of two gears in mesh position (see Fig. 16.5).

8. Common tangent : It is the common tangent at the pitch circle of the two mating gears at the pitch point.

9. Common Normal : It is the common normal to the two tooth profiles at the point of contact.

10. Centre distance : It is the distance between the centres of a pair of mating gears. Centre distance = Sum of the radii of the pitch circles of the two gears.

11. Tooth face : It is the side surface of the tooth above the pitch circle.

12. Tooth flank : It is the side surface of the tooth below the pitch circle.

13. Crest of tooth : It is the out side surface of tooth.

14. Root of tooth : It is the junction of tooth with material at the bottom of tooth space.

15. Tooth thickness : It is the thickness of the tooth, measured along the pitch circle (see Fig. 16.5).

$$\therefore \qquad \text{Tooth thickness} = \frac{\text{C.P.}}{2} = 0.5 \ \text{C.P.}$$

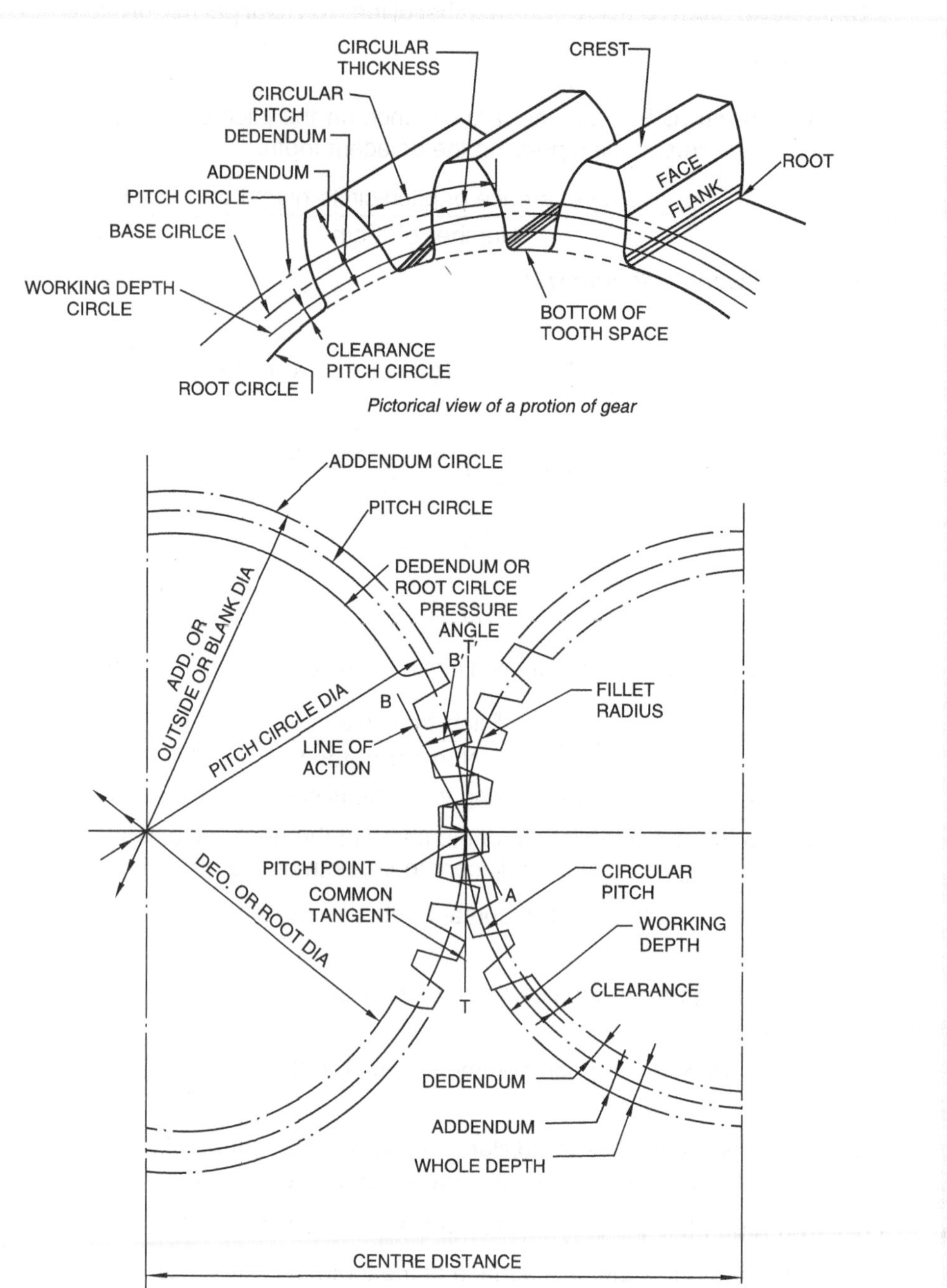

Pictorical view of a protion of gear

Fig. 16.5 *Orthographic view of mating gears.*

16. Fillet radius : It is the radius of the centre at the centre to the root of the tooth.

17. Addendum (a) : It is the radial distance between the pitch circle and the top of the tooth (see Fig. 16.5).

$$\therefore \quad \text{Addendum (a)} = \frac{1}{\text{D.P.}} = \frac{\text{C.P.}}{\pi} = 0.3183 \text{ C.P.}$$

18. Clearance (C) : It is difference between the dedendum and the addendum (see Fig. 16.5).

$$\therefore \quad \text{Clearance (C)} = \frac{\text{C.P.}}{20} = 0.05 \text{ C.P.}$$

19. Dedendum (d) : It is the radial distance between the pitch circle and the bottom of the tooth (see Fig. 16.5).

$$\therefore \quad \text{Dedendum (d)} = \text{Addendum} + \text{Clearance}$$

20. Outside or Addendum circle (D_a) : It is the diameter of the addendum or crest of the teeth.

Outside dia. or Addendum circle = P.C.D. + 2 × Addendum = P.C.D. + 2a

21. Root or dedendum circle (D_b) : It is the diameter of the dedendum or root of the teeth (see Fig. 16.5).

Root. dia. or Dedendum circle = P.C.D. – 2 × dedendum = P.C.D. – 2a

22. Whole depth : It is the sum of addendum and dedendum of a tooth.

23. Working depth: It is the distance by which a tooth extends into the space of a mating gear.

or Working depth = whole depth – clearane or twice the addendum

24. Pressurne angle or pressure of obliquity : It is the angle that determines the direction of pressure between the mating teeth. It also designates the shape of the involutee teeth and size of the base circle. *The pressure angle is generally taken as 14.5 ° or 20° (see Figs. 16.5 and 16.6).*

The larger pressure angles are advantageous as they give strong tooth form with less under cutting.

25. Base circle : It is the circle from which the involute profile is generated (see Figs. 16-5 and 16.6).

26. Line of action : It is the common tangent to the two base circles that passes though the pitch point of mating gears (see Figs. 16.5 and 16.6).

SUPER GEAR PROPORTIONS FOR 14.5° PRESSURE ANGLE

14.5°	CAST GEAR	MACHINE CUT GEAR	HIGH CLASS GEARS
ADDENDUM	0.3P	0.318P	0.318P
DEDENDUM	0.4P	0.368P	0.398P
TOOTH THICKNESS	0.48P	0.5P	0.458P
SPACE	0.52P	0.5P	0.5P
LENGTH OF TOOTH	2P TO 3P	2P TO 3P	—

P — CIRCULAR PITCH

16.8. PROFILE OR FORM OF TEETH

The outline of teeth of a gear is called the profile or form of teeth. A gear must have its tooth profile of a definite geometrical form so that it may be operated smoothly with minimum vibration and noise. The following two profiles of teeth are commonly used:–

1. Involute profile teeth 2. Cyclodial profile teeth.

16.8.1. Involute profile teeth: In involute profile teeth, the involute curve forms the profile teeth of involute gears (see Fig. 16.6).

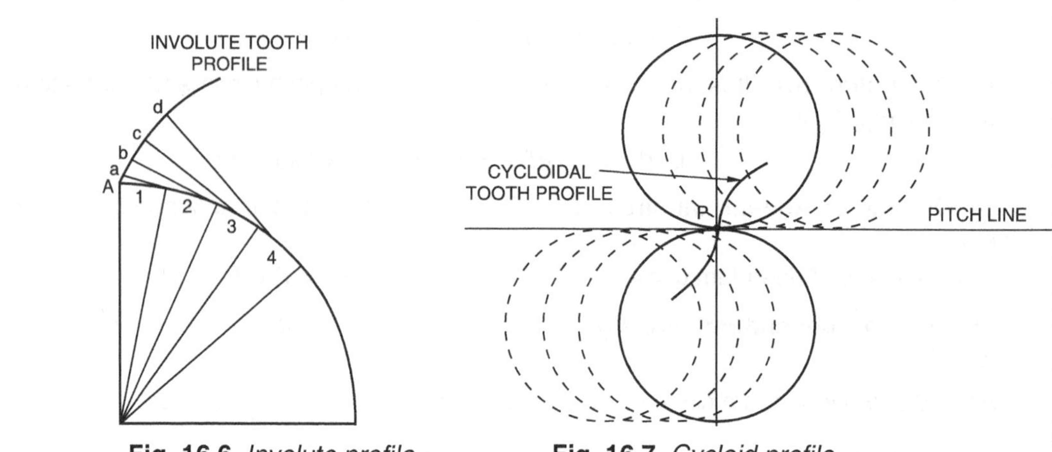

Fig. 16.6 *Involute profile* **Fig. 16.7.** *Cycloid profile*

Involute : The involute is the curve traced out by the end of a thread when unwound from the circumference of the circle.

The involute of a circle is drawn by drawing tangents at various points on the circumference of a circle and making the various points at corresponding distances along their tangents.

In Fig. 16.6, involutes generated by a point A on each of the two circles of different diameters. The lines a1,b2,c3 etc. are tangential to the circle and are equal in length to the arc lengths A1,A2, A3 etc. respectively. The curve passing through points A., a,b,c etc. is involute.

Note: 1. If the circle is of a infinitely large diameter as in case of a rock, the involute is a straight line.

2. *The involute form of teeth is generally used now a days, because it is easier to manufacture.*

16.8.2. Cycloidal teeth : In cycloidal teeth, the cycloid curve forms the profile of teeth as shown in Fig. 16.7.

Cycloid : The cycloid is the curve generated by a fixed point on the circumference of a circle which rolls without slipping along a fixed straight line.

Note : *When the circle rolls on the out side of another circle, the locus of the point on the circumference is known as epi-cycloid, conversely, when the circle rolls inside of another circle, the corresponding locus of the point on the circumference is called the hypo-cycloid.*

16.9. METHOD TO DRAW BASE CIRCLE OF A GEAR

In order the obtain base circle of a gear, for a pressure angle of 20 °, proceed as under:-

1. Draw the pitch circle of given diameter with two centre lines (see Fig. 16.8).
2. Mark the pitch point P on the pitch circle and from P draw a tangent T-T'.
3. Draw a line AB at $\theta = 20°$ pressure angle to tangent T-T'.
4. From 0 draw a line $\theta = 20°$ to AB.
5. With 0 as centre, draw a circle tangential to line AB, which is the required base circle (see Fig. 16.8).

16.10. METHOD TO DRAW BASE CIRCLES OF TWO MATING GEARS

The base circles of two mating gears for a pressure angle of $\theta = 20°$ is drawn as follows:

1. Draw the two given pitch circles touching at a point P.
2. From P, draw a common tangent T-T'.
3. Draw the line of action AB at pressure angle of $\theta = 20°$ to tangent T-T'.
4. Finally, draw the base circles tangential to the line of action AB as shown in Fig. 16.9.

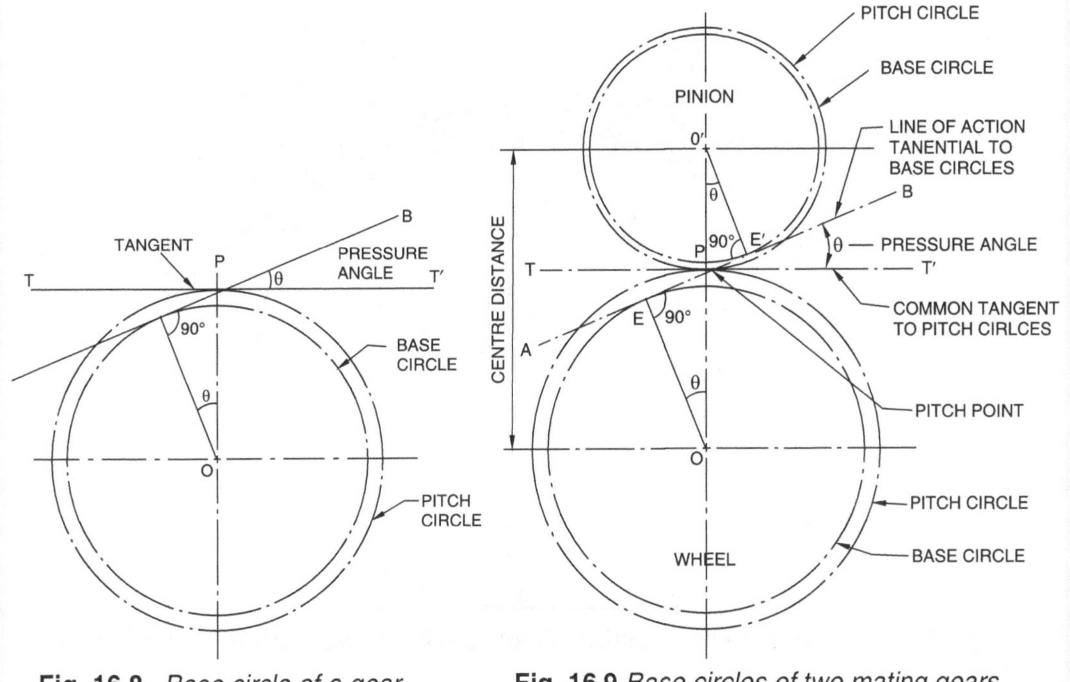

Fig. 16.8 *Base circle of a gear* **Fig. 16.9** *Base circles of two mating gears*

Problem 1: Draw the base circles for two mating gears of pitch circle diameters 250 mm and 150 mm, assuming the pressure angle to be 20°.

Solution: For its solution, see Fig. 16.9

Hint : Here $P.C.D._1 = 150$ mm (Pinion)

$P.C.D._2 = 250$ mm (wheel)

Pressure angle, $\theta = 20°$

16.11. CONSTRUCTING SPUR GEAR PROFILE (APPROX. METHOD)

The profile or shape of spur gear teeth are based on standardized formulas. These are made with either 14½° or 20° pressure angle. But most spur gears have a 20° pressure angle because they are considerably stronger.

The profile or shape of spur gear tooth is generally constructed by the following methods:—

16.11.1 Approximate Construction of Teeth Profile for gears of 30 teeth

The following steps should be followed for the construction of teeth profile for gears of 30 teeth and over :—

1. With O as centre, draw the pitch circle, addendum (outside) circle and dedendum (root) circle as determined by formulas.

2. Mark a pitch point P on the pitch circle (see Fig. 16.10).

3. With C as centre and radius equal to P.C.D./8, draw an arc to cut the semi-circle at Q.

5. Now with O as centre, draw a circle passing through Q. This is the circle on which centres of arcs for teeth profile will lie.

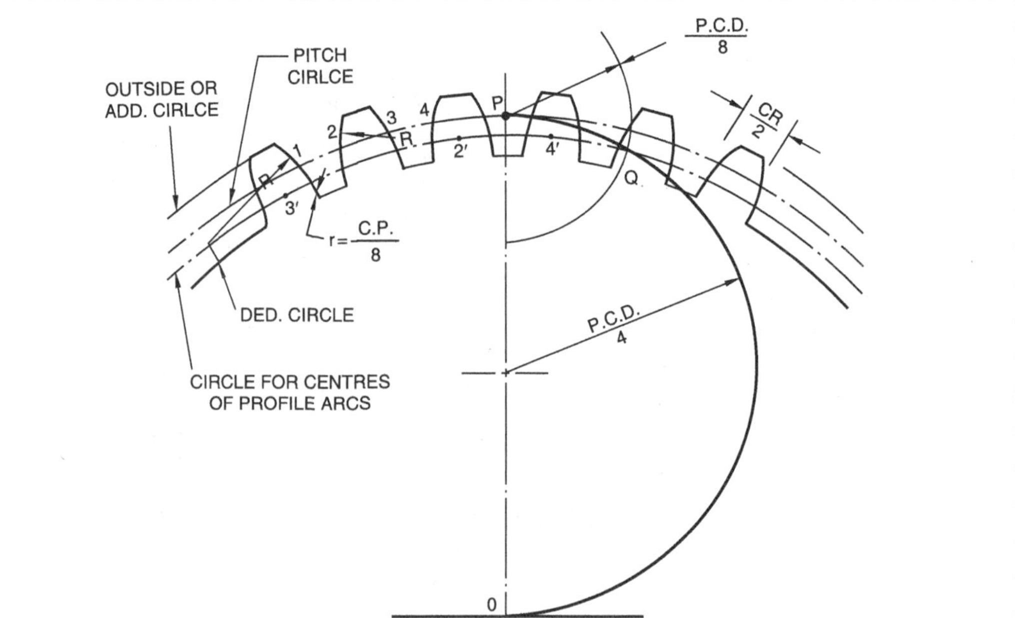

Fig. 16.10 *Approximate construction of tooth profile for gears of 30 teeth and over*

6. On the pitch circle, mark points 1,2,3,4, etc. at a distance of C.P./2 apart.

7. With 1,2,3,4 etc. as centres and radius R = P.C.D./8, locate the points 1',2',3',4' etc. on construction circle.

8. Now with 1', 2',3', 4' etc. as centres and radius R = P.C.D./8 draw arcs.

9. Complete the teeth by adding tooth fillet of radius R = C.P./8

16.11.2 Approximate construction of tooth profile for gears of less than 30 teeth:

If the number of teeth in a gear is less than 30, the Fig. 16.11 illustrated should be followed. The constructions and manner of drawing the teeth will be clear from the study of Fig. 16.11.

Problem 2 : Draw the profile of involute teeth for a gear having 25 teeth and module equal to 10 mm, assuming a pressure angle of 20°.

Solution : Here, Number of teeth, N = 25

Module, M = 10 mm

Pressure angle, θ = 20°

$$\text{We know, Module, } M = \frac{\text{P.C.D}}{N}$$

$$\therefore \text{ P.C.D.} = \text{Module} \times \text{No. of teeth} = M \times N$$

$$= 10 \times 25 = 250 \text{ mm}$$

$$\text{Circular pitch, C.P.} = \pi \times \text{Module}$$

$$= \pi \times 10 = 31.4 \text{ mm}$$

Addendum = 0.31835 × C.P. = 0.3183 × 31.4 = 10 mm.

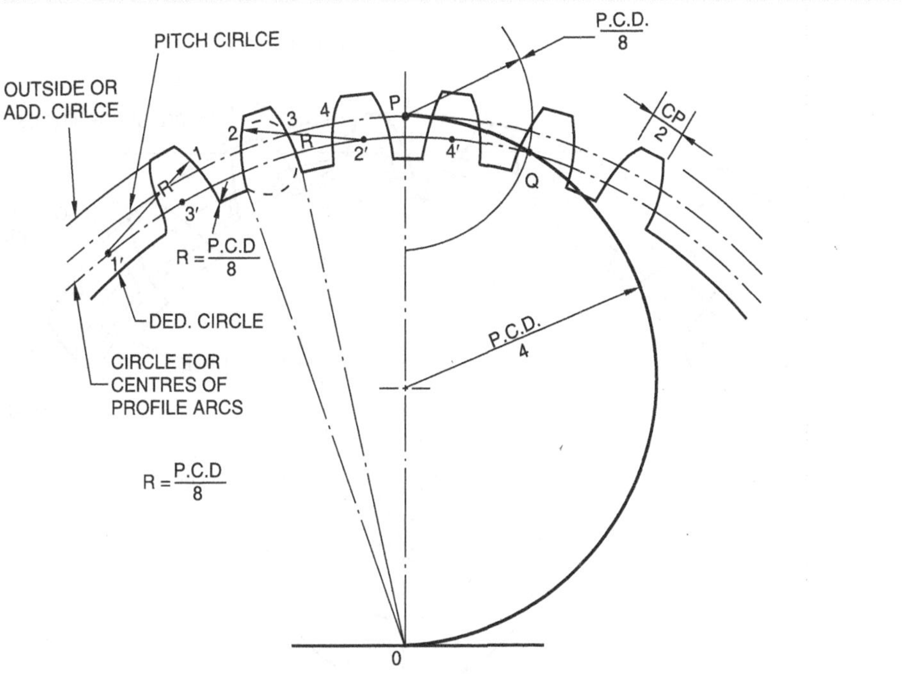

Fig. 16.11. *Approximate construction of tooth profile for gears of less than 30 teeth*

Addendum circle or outside circle diameter

$$= \text{P.C.D.} + 2 \times \text{addendum}$$

$$= 250 + 2 \times 10 = 270 \text{ mm}$$

$$\text{Clearance} = \frac{\text{C.P.}}{20} = \frac{31.4}{20} = 1.57 \text{ mm}$$

$$\text{Dedendum} = \text{Addendum} + \text{Clearance}$$

$$= 10 + 1.57 = 11.57 \text{ mm}$$

$$\text{Dedendum circle} = \text{P.C.D.} - 2 \times \text{dedendum}$$
$$\text{or (Root Circle) diameter} = 250 - 2 \times 11.57$$
$$= 226.86 \text{ mm}$$

$$\text{Tooth thickness} = \frac{\text{C.P.}}{2} = \frac{31.4}{2} = 15.7 \text{ mm}$$

Now, construct the teeth profile as shown in Fig. 16.11.

16.11.3. Approximate Construction of gear tooth by Prof. Unwin process

The following steps should be followed for construction of gear tooth by Prof. Unwin process method (see Fig. 16.12) :—

1. With O as centre, draw pitch circle, addendum circle and dedendum circle as determined by formulas.

2. Mark a pitch point (P) on the pitch circle. From P draw a tangent T-T' to the pitch circle. Draw the line of action AB at pressure angle of 14½° or 20° with the tangent T-T'.

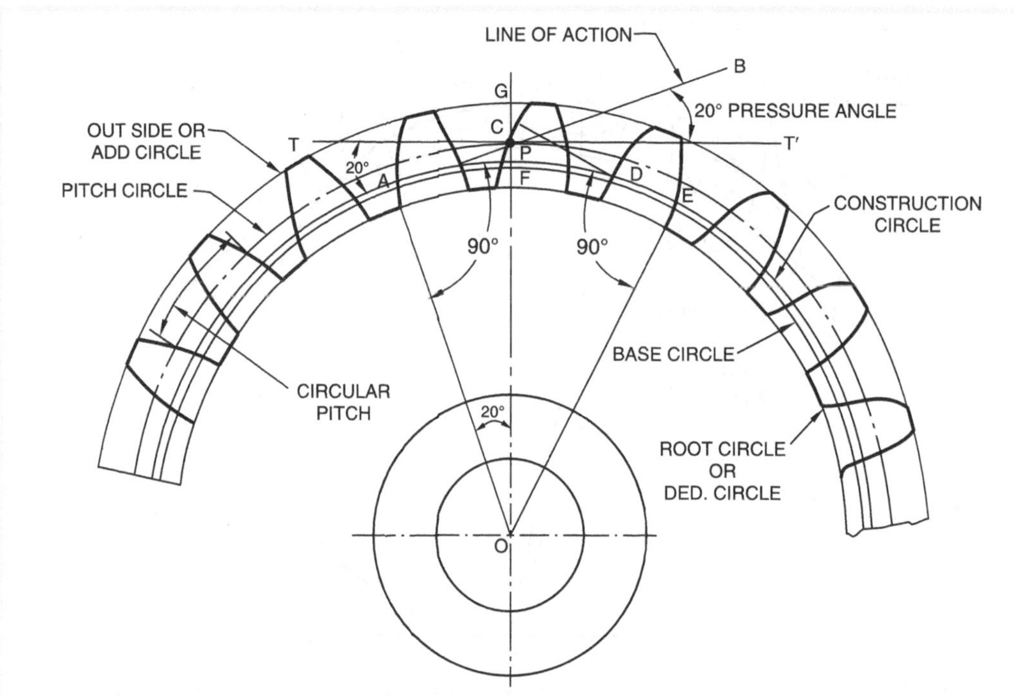

Fig. 16.12. *Constructing involute tooth profile*

F point is on base circle; $GC = \frac{GF}{3}$; CE tangent to OE; $CD = \frac{3}{4} \times CE$

3. With centre O, draw base circle tangential to the line AB. The base circle cuts the vertical line OG at F.

4. Divide GF into three parts, such that GC = GF /3.

From C draw a tangent CE on the base circle.

With D as centre and DC as radius draw an arc. This is the approximate involute tooth profile.

5. Now divide the pitch circle into arcs of lengths = C.P./2

6. With O centre and radius equal to OD draw a circle. This is the construction circle on which centres of arcs for the teeth profile will lie.

7. Now, find the centres for the teeth profiles on the construction circles by trial with DC as radius, Draw arcs through the points division.

8. Complete the teeth profile by joining the arcs with bottom of tooth space by a fillet.

16.12. EXACT METHOD OF CONSTRUCTING TOOTH PROFILE

Problem 3 : Draw the profile of involute teeth for a gear having 24 teeth and a D.P. = 1.2, taking centimetre as unit length and assuming a pressure angle of 20°.

Solution: Here, Number of teeth, N = 24

Diametral pitch, D.P. = 1.2 Pressure angle, $\theta = 20°$

We know that $\text{P.C.D} = \dfrac{N}{\text{D.P.}} = \dfrac{24}{1.2} = 20 \text{ cm} = 200 \text{ mm}$

$$\text{Circular pitch, C.P.} = \frac{\pi}{\text{D.P.}} = \frac{3.141}{1.2} = 26.18 \text{ mm}$$

$$\text{Addention} = \frac{\text{C.P.}}{\pi} = \frac{26.18}{3.141} = 8.33 \text{ mm}$$

Addendum circle or outside circle = P.C.D. + 2 addendum = 200 + 2 x 8.33 mm

Dedendum = Addendum + clearance

= 8.33 + 1.309 = 9.639 mm

$$\left[\therefore \text{Clearance} = \frac{\text{C.P.}}{20} = \frac{26.8}{2} = 1.300\right]$$

Dedendum circle = P.C.D − 2 × dedendum

= 200 − 2 × 9.639 = 180.722 mm

$$\text{Tooth thickness} = \frac{\text{C.P.}}{2} = \frac{26.18}{2} = 13.09 \text{ mm}$$

It may be necessary in development and investigation work to draw the tooth profiles of involute gears more accurately. The following steps should be followed for construction of the involute tooth profile.

1. With O as centre, draw the pitch circle, of 200 mm diameter, addendum (or outside) circle, and dedendum (or root) circle as determined by formulas.

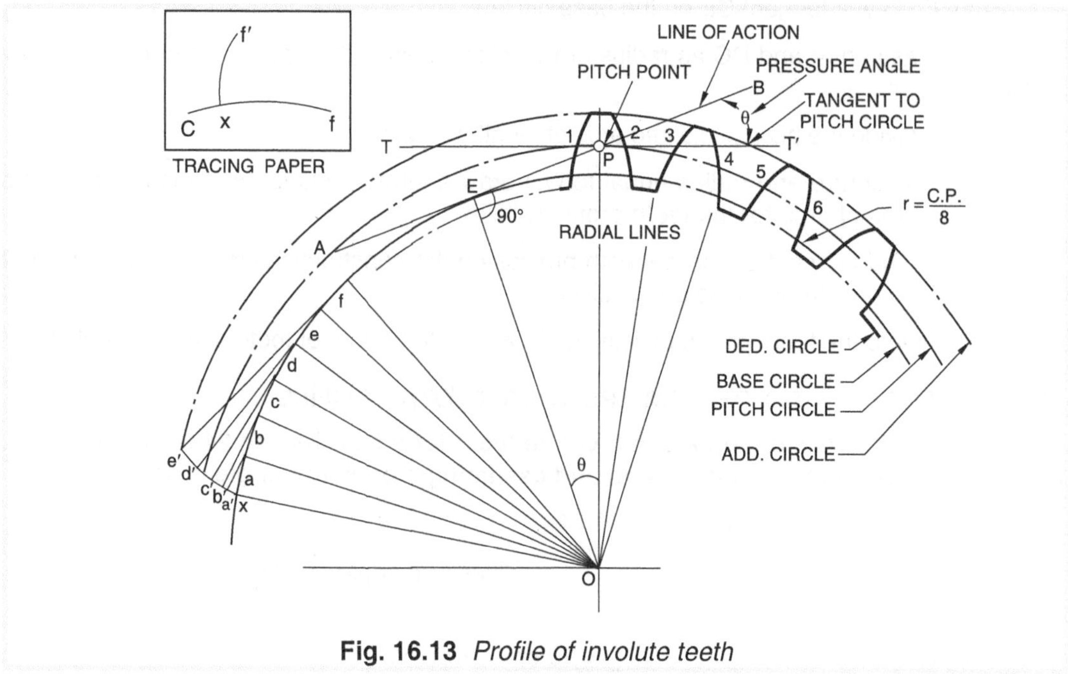

Fig. 16.13 *Profile of involute teeth*

2. Mark the pitch point P on the pitch circles. From pitch point P, draw a tangent T-T' to the pitch circle.

3. Draw the line of action AB at pressue angle of 20° with tangent T-T'

4. With centre O and radius equal to OE draw the base circle.

5. Cut off any number of equal spaces along base circle, say a,b,c,d, e etc. join these joints to the centre O.

Draw tangents to the base circle from points a, b, c, d etc.

Now with a as centre and a-x as radius draw an arc to cut the second tangent at b'. Similarly obtain the points c',d',e', etc. as shown in Fig. 16.13.

Join these points a',c',d', etc. The curve thus obtained is involute curve.

Trace out this curve on a piece of tracing paper as shown in Fig. 16.13.

6. On the pitch circle, mark points 1,2,3,4,5 etc. at distance equal to C.P./2 apart.

7. Place the tracing paper on the base circle in such a way that it coincides with it, while the curve passes through, say, point 1.

8. Pick a few points on the curve between addendum and base circle. Join these points by means of french or irregular curve. Join this curve to the bottom of the tooth space by a fillet or radius $= \frac{\text{C.P.}}{8} = \frac{26.18}{8} = 3.272$ mm

Now, reverse the tracing paper and draw the curve though the point 2. Similarly repeat the construction of other tooth.

Problem 4: Draw the front, side view and sectional side view of spur gear with the following particulars;

Module = 4 mm Number of teeth = 32 Shaft of teeth = 30 mm

Solution : We know, M = 4 and N = 32

Pitch circle diameter, P.C.D.= M × N = 4 × 32=12 mm

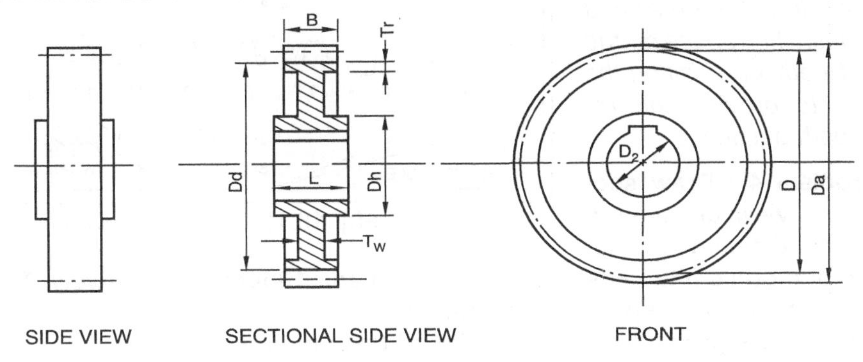

Fig. 16.14.

$$\text{Addendum circle diameter, Da} = \text{P.C.D.} + 2a = \text{P.C.D.} + 2 \times \frac{\text{P.C.D.}}{N}$$

$$= 128 + 2 \times \frac{128}{32} = 136 \text{ mm}$$

$$\text{Dedendum circle diameter, Da} = \text{P.C.D.} - 2d = \text{P.C.D.} - 2 \times \frac{1.157}{\text{D.P.}}$$

$$= \text{P.C.D.} - 2 \times 1.157 \times \frac{128}{132} = 118.74 \text{ mm}$$

$$\text{Face width, B} = 3 \times \text{P.C.D.} = 3\pi \frac{\text{P.C.D.}}{N}$$

$$= 3 \times \frac{\pi \times 128.}{32} = 37.7 \text{ mm}$$

Hub diameter, $D_h = 1.5\, D_S$ to $2\, d_s$

say, 1.75 × 35 = 52.5 mm

Hub length, h = B to 1.4B

= 1.2×37.7 = 52.5 mm [assume h = 1.2B]

$$\text{Web thickness, Tw} = \text{Pc} \times \frac{\pi \times \text{P.C.D.}}{32} = 12.6 \text{ mm}$$

Rim thickness, Tr = a+D

$$= \frac{\text{P.C.D.}}{N} + 1.157 \times \frac{\text{P.C.D.}}{N} = \frac{128}{32}\,[1 + 1.57] = 8.628 \text{ mm}$$

Views of spur gear incorporating the above dimensions can be drawn as shown Fig.16.14.

16.13. GEAR TEMPLATE

Great template is used to draw the approximate spur gear tooth profile of various sizes (see Fig. 16.15).

To use the gear template, first of all draw the outer and root circles. Then space off the required teeth and position the template to provide the required tooth out line.

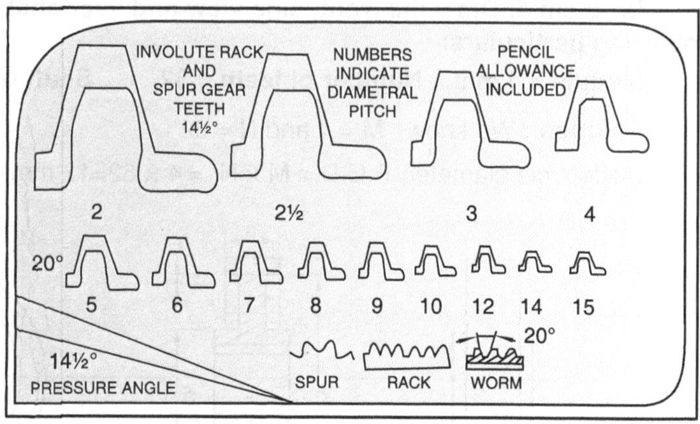

Fig. 16.15. *Gear Template*

Problem 4 : Draw sectional front view of involute spur gear and pinion in mesh position of module pitch 8mm having 30 teeth and 18 teeth respectively. The following particulars be used for spur gear and pinion.

For 30 teeth gear	For 18 teeth pinion
Face width = 36mm	Face width = 36mm
inside rim dia = 200mm	Inside rim dia. = 115mm
Boss dia.=45mm	Boss dia. = 35mm
length=48 mm	length = 48mm
Shaft dia. = 25mm	Shaft dia. = 20mm
Four arms of elliptical shape	Three lightening
With major axis = 25mm	holes of 30mm dia.
and minor axis = 20mm	in 12mm thick web
Pressure angle = 15°	Symmetrical spaced on a P.C.D of 75mm

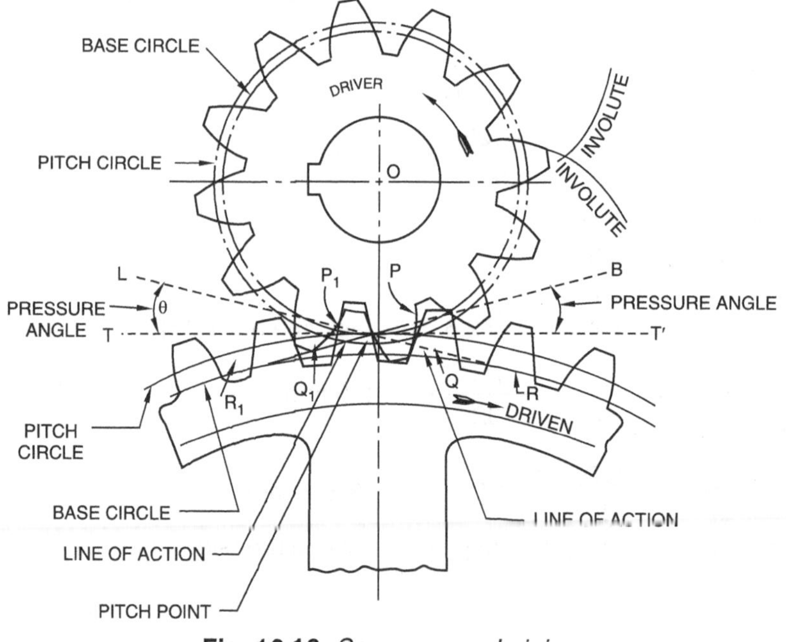

Fig. 16.16. *Spur gear and pinion*

Solution: Fig. 16.16 show the sectional front view and side view of involute spur gear and pinion.

16.14. INTERNAL GEARS

An internal gear is one in which the teeth are cut on the inside curved surface of a wheel and mesh with a small pinion as shown in Fig. 16.17.

These gears are useful when space is limited, since the centre distance between the driver and driven shafts is reduced.

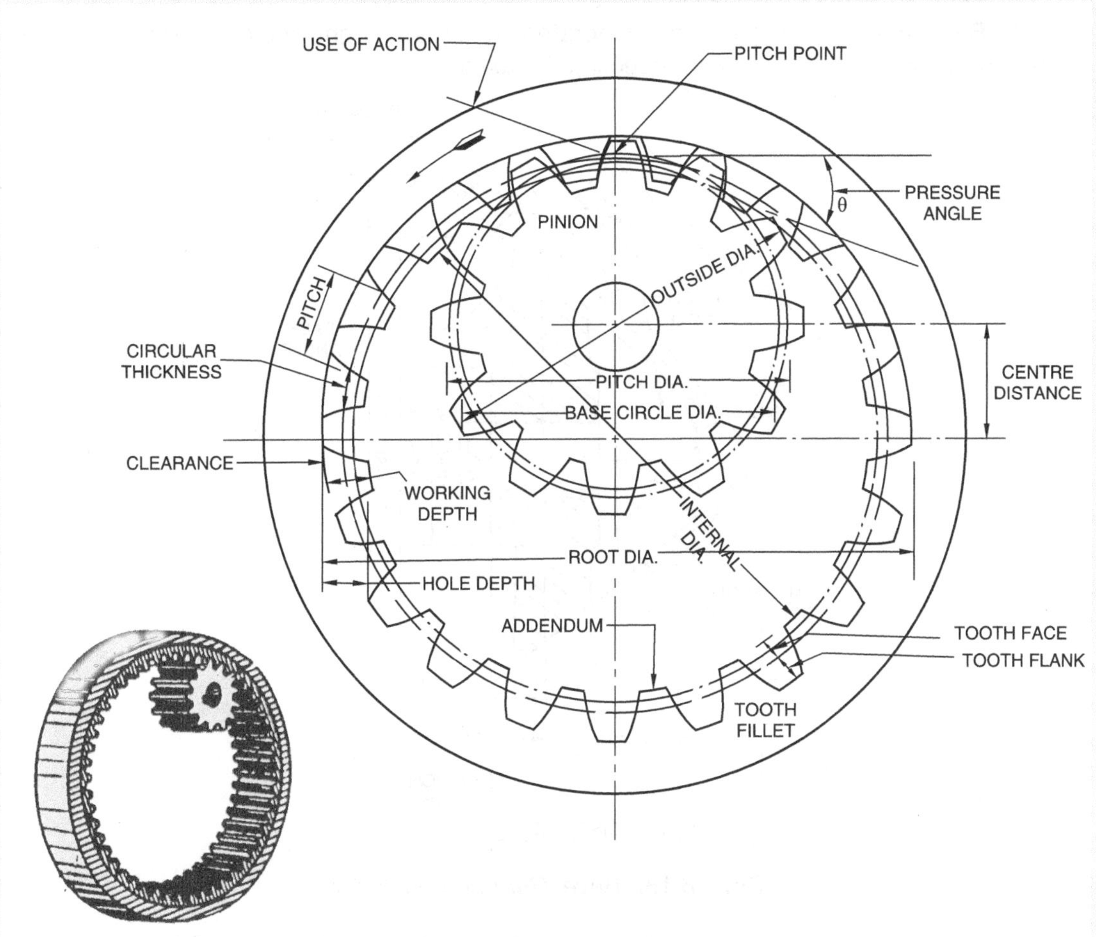

Fig. 16.17. *Internal gears*

16.15. BEVEL GEARS

Bevel gears are shaped like cone section (see Fig. 16.18). These are used for direct transmission of power between shafts whose axes intersect at an angle.

The bevel gears are always designed in pairs and used the same involute tooth form as in spur gears except that the teeth are tapered towards the cone apex.

Note : *The axes of the bevel gears may intersect at any angle, but in general the axes at right angles occurs most frequently.*

16.16. TERMS USED IN BEVEL GEARS

Most of the gear terms previously discussed for spur gears apply to bevel gears. However, some of the terms used for the bevel gears are as under :—

1. Pitch circle : It is the circle which forms the base of the pitch cone.

2. Pitch diameter : It is the diameter of the pitch circle measured at the base of the pitch cone (see Fig. 16.18).

3. Pitch angle : It is the angle between the axis and the pitch cone generator.

4. Cone distance : It is the slant length of the pitch cone.

5. Back cone : It is the cone generated by lines at right angles to the pitch cone generator at the pitch circle and intersecting the axes.

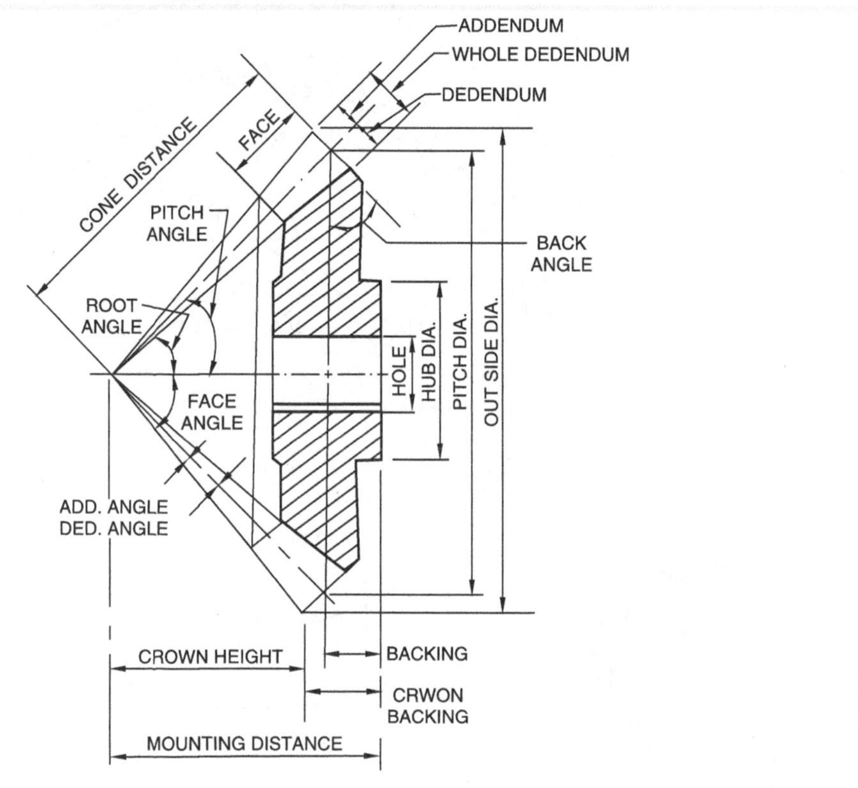

Fig. 16.18. *Bevel Gear nomenclature*

6. Addendum (a) : It is the distance between pitch circle and the tip of the tooth, measured along the back cone generator.

$$\therefore \quad \text{Addendum, (a)} = \frac{1}{\text{D.P.}}$$

7. Dedendum (d) : It is the distance between pitch and the depth of the tooth space below the pitch circle measured along back cone angle.

$$\therefore \quad \text{Dedendum, (d)} = \frac{1.157}{\text{D.P.}}$$

8. Addendum Angle (∞) : It is angle subtended by the addendum (see Fig. 16.18).

9. Dedendum Angle (δ) : It is the angle subtended by the dedendum (see Fig. 16.18).

10. Outside diameter (D_o) : It is the diameter across the tips of the teeth i.e., the diameter of the circle of intersection of the blank cone and back cone.

$$\therefore \quad \text{Outside diameter, } (D_o) = \text{Pitch Dia} + 2a \cos \pi$$

11. Face: It is the length of the tooth (see Fig. 16-18).

12. Face width (F) : It is the length of teeth measured along the pitch cone generator. It should not exceed one third of the cone distance.

13. Face angle (T_o) : It is the angle between the root of teeth and the gear axis.

14. Root angle : It is the angle between the root of the teeth and the gear axis.

15. Shaft angle : it is the angle between the intersection of shaft axes.

16. Backing: It is the distance between the base of the pitch cone to the rear of the hub.

17. Crown Backing: It is the distance between the crown of the gear to the rear of hub.

18. Crown Height : It is the distance parallel to the gear axis from the apex to the crown of the gear.

Problem 4: Draw the front view and top view of a bevel gear with the following particulars.

No of teeth = 25

Circular pitch = 25mm

Pitch cone = 70°

Face width = 50mm

Solution: Calculate the diameter of pitch cone

$$\text{Dia. of pitch cone} = \frac{\text{No of teeth} \times \text{C.P.}}{\pi} = \frac{25 \times 25}{\pi} = 200\text{mm (app.)}$$

For the construction of bevel gears follow the method given below:

16.17. METHOD TO DRAW A BEVEL GEAR

The following steps should be followed while drawing the projections of bevel gear;

For Front View :

1. Draw the pitch cone OAB (having angle $\angle OAB = 70°$) and AB equal to diameter of pitch cone.
2. Draw the back cone angle generator BC at 90° to OB and mark the addendum and dedendum distances through point B.
3. Set off BD = 36 mm on the pitch cone generator.
4. With C as centre and radius equal to CB draw pitch circle arc BB'. Mark the distance equal $\frac{\text{C.P.}}{2}$ along the arc BB' and draw the radial lines. Complete the shape of teeth by approximate method as explained already.
5. Mark the tip thickness *x* and root thickness *y* on one of the tooth.

For Top View :

6. Draw the pitch circle, addendum and dedendum circle in top view.
7. Make the space equal to circular pitch 25 mm on the pitch circle.
8. Make distances equal to *x* and *y* on addendum and dedendum circle in top view. Draw the radial centre line from O' and complete the top view as shown in Fig. 16.19.
9. Mark a,b,c,d,e and f points on addendum, pitch and dedendum circles. Project these points in front view as, a,b,c,d,e and f, on the corresponding addendum, pitch, dedendum lines.
10. Join a,b,c,d,e,f with smooth curve in top view and complete the front view as shown in Fig. 16.19.

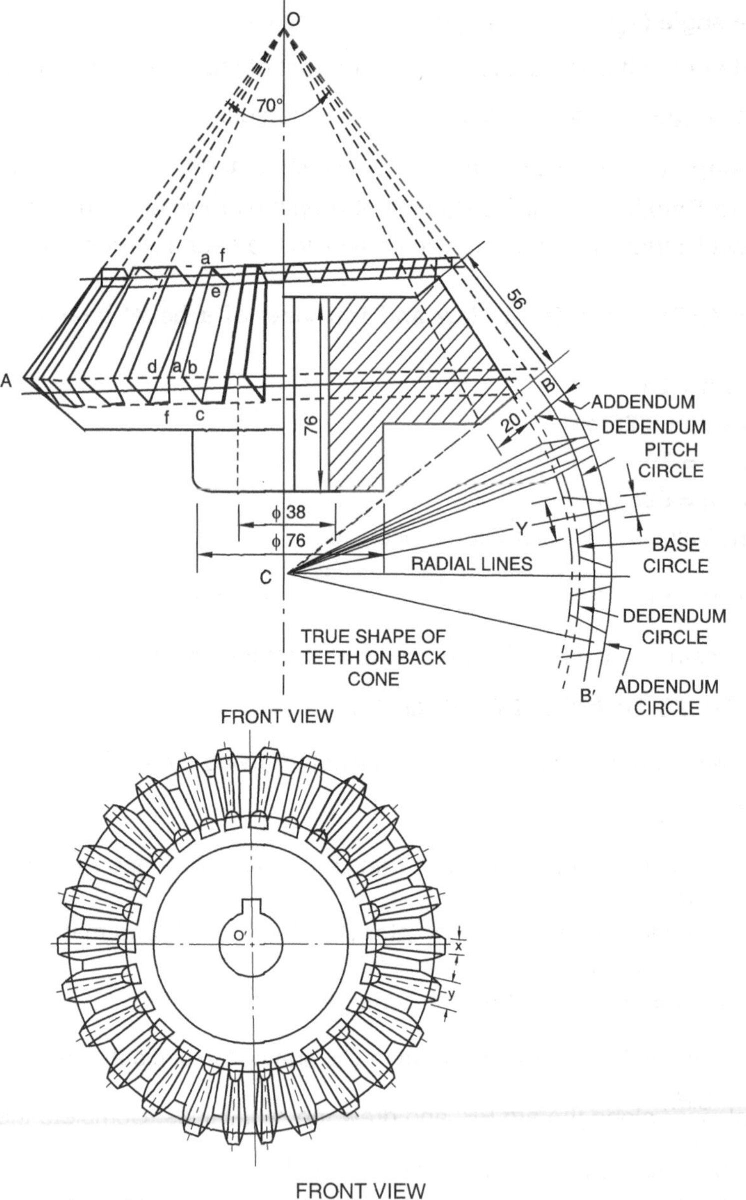

Fig. 16.19. *Construction of bevel gears*

16.18. CONSTRUCTION OF BEVEL GEAR IN MATING POSITIONS

The following steps should be taken into consideration while drawing the bevel gears in mating position:

1. Draw horizontal and vertical centre lines to intersect at O'.

2. Make the cone diameters of each gear on the centre lines and project them parallel to horizontal and vertical centre lines to meet at pitch point P.

3. Draw the pitch cones of each gear from O (see Fig. 16-20)

4. Mark the addendum and dedendum distances of each gear.

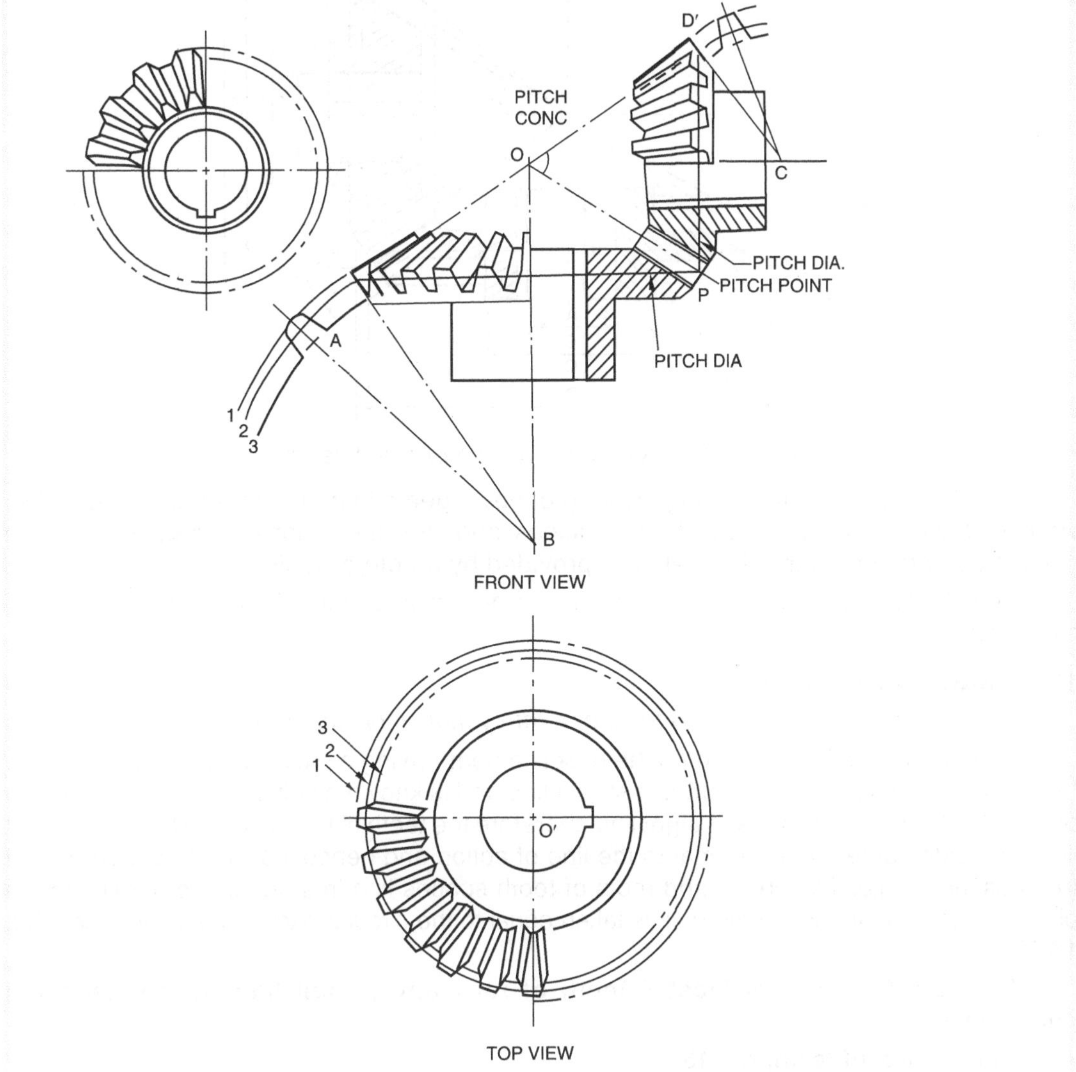

Fig. 16.20. *Bevel gears in mating positions*

5. Draw the back angles of each gear to meet the horizontal and vertical centre lines at C and B.

6. With B as centre, draw arcs 1,2, and 3 for the addendum, pitch and dedendum lines of a developed tooth.

Take radial centre lines BA and CD and draw tooth profiles of respective gear by method as discussed already.

7. Project the 1,2 and 3 points in the top view and draw the circles through 1',2', and 3' and also draw the radial centre lines from each tooth.

8. Take distances equal to circular thickness from A and transfer them to each tooth.

9. Complete the construction as discussed already.

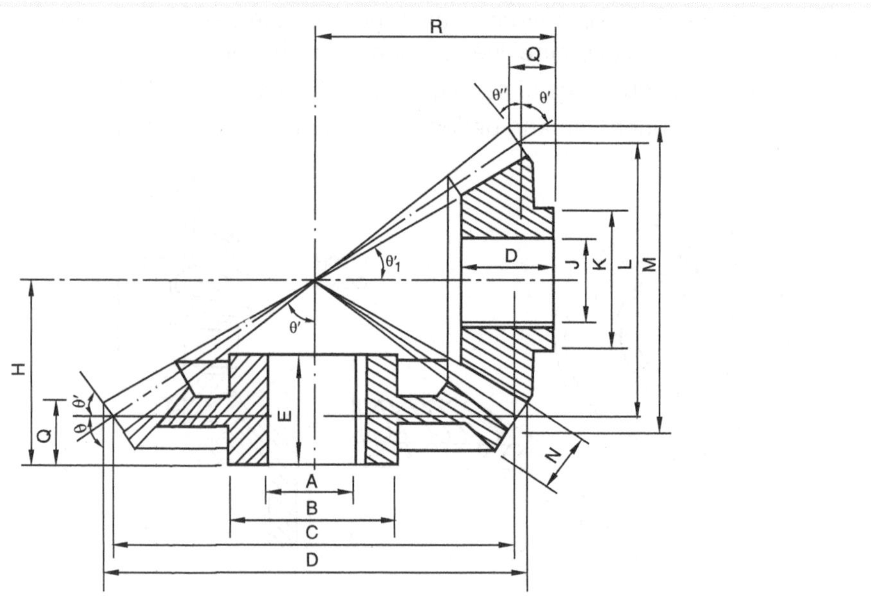

Fig. 16.21. *Working drawing of bevel gears*

Fig. 16.21 shows the working drawing of bevel gears in their operating position. The working drawing is drawn with tooth outlines and the dimensions of gear blank. The necessary data for cutting the teeth are provided by a note or table.

The Working drawing of a bevel gear can be drawn separately, if required, for detail dimensioning.

16.19. RACK AND PINION

The rack and pinion is used to transfer rotatory motion of pinion into linear motion.

Pinion is a small gear whereas the rack is a gear having a pitch circle of infinite radius. In this case the pitch circle becomes a stright line and is known as pitch line. As, the involute of a circle of infinite radius is a straight line, therefore the teeth of the rack are also in a straight line. The sides of teeth are normal to the line of action and hence inclined to the vertical at the pressure angle. The crests and roots of tooth spaces line in straight line parallel to the pitch line. The linear pitch of the rack is taken equal to the circular pitch of the pinion (see Fig. 16.22).

Problem 6 : Draw at least 5 teeth of rack and pinion having the following particulars:

Pinion : No. of teeth, N = 15

Circular pitch, C.P. = 20 mm

Pressure angle, $\theta = 20°$

Solution : Here, Pitch circle diameter, P.C.D.= $\frac{\text{C.P.} \times \text{N}}{\pi}$

$$= \frac{20 \times 15}{\pi} = 105 \text{ mm}$$

For the construction of pinion :

1. With O as centre and radius, OP = $\frac{\text{P.C.D}}{2}$ = 52.5 mm, draw the pitch circle.

2. At P draw a line MN tangential to the pitch circle. This line is known as pitch line of rack.

3. From P, draw the line of action AB inclined at angle θ = 20° to M N.

4. Draw the base circle of pinion and construct the profile of teeth.

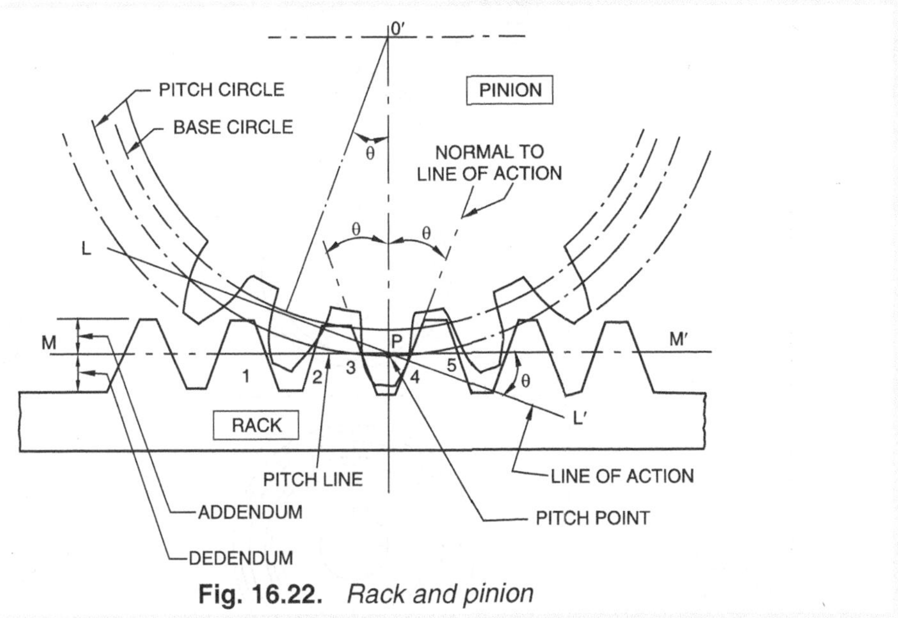

Fig. 16.22. *Rack and pinion*

For the cnstruction of rack

Draw the addendum and dedendum lines above and below the pitch line M N.

6. On M N line, mark points 1,2,3 etc. At distance C.P./2=20/2=10 mm apart.

7. Through 1,2,3 etc. draw the sides of the teeth by lines inclined at θ =20° to the vertical

8. Join each side with the bottom of the tooth space by fillet radius, r=C.P./8=20/8=2.5 mm.

Note: *In the conventional method, the angle between the sides of each tooth of the rack is generally taken as 30°.*

16.22. CONVENTIONS FOR GEARS

The conventions of gears are very important as these are used to represent various gears in most simple forms which saves time and labour.

Fig. 16.23 shows the conventional representations of spur gears, bevel gears and worm wheel in detail and assembly form according to I.S.I.

Fig. 16.24 shows the conventional representations of worm, and worm wheel in detail and assembly form according to I.S.I.

CONVENTIONAL REPRESENTATION OF GEARS

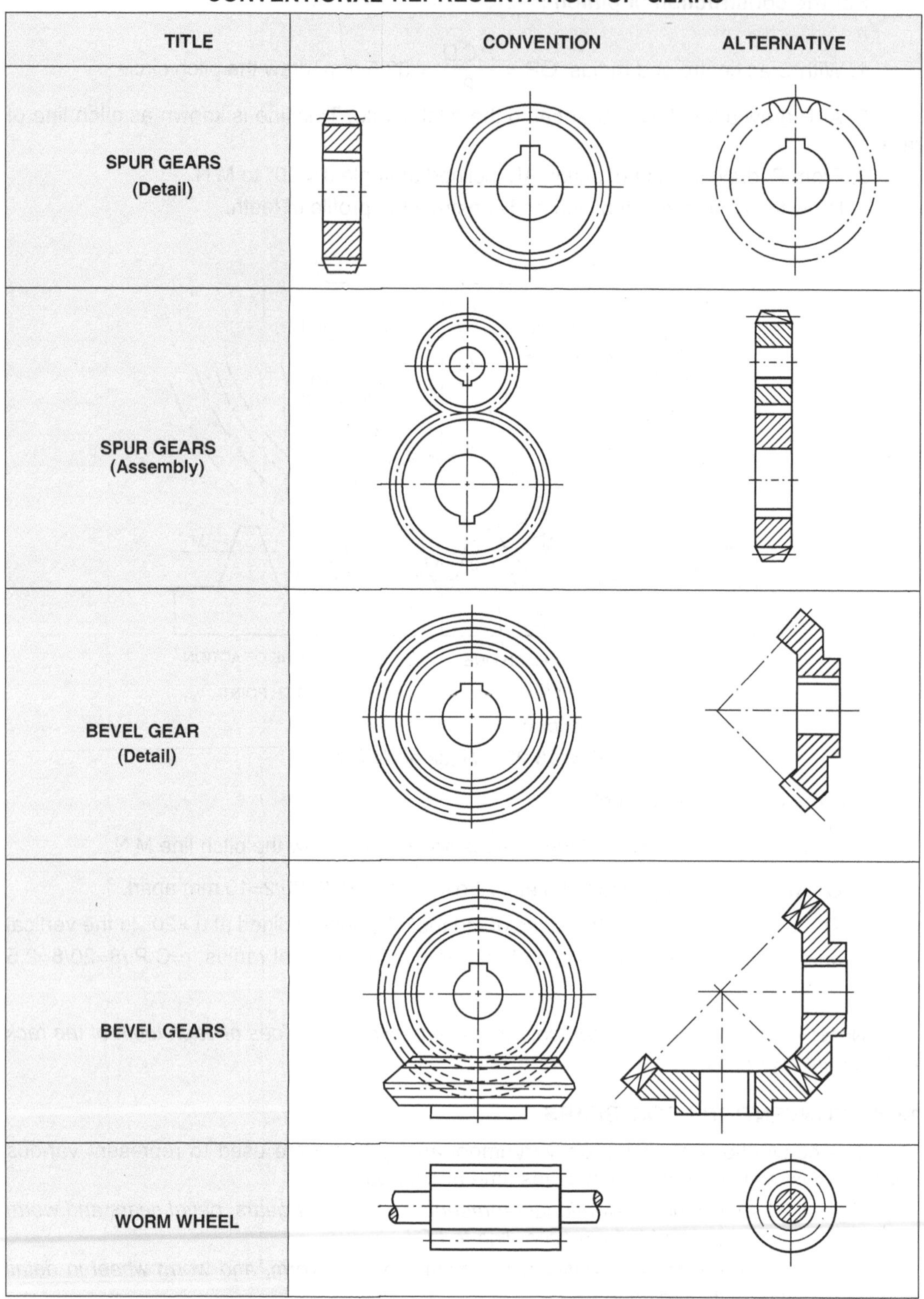

TITLE	CONVENTION	ALTERNATIVE
SPUR GEARS (Detail)		
SPUR GEARS (Assembly)		
BEVEL GEAR (Detail)		
BEVEL GEARS		
WORM WHEEL		

Fig. 16.23. *Conventional representation of gears*

CONVENTIONAL REPRESENTATION OF GEARS

TITLE	CONVENTION	ALTERNATIVE
WORM (Detail)		
WORM AND WORM WHEEL (Assembly)		

Fig. 16.24. *Conventional representation of gears*

QUESTIONS FOR SELF EXAMINATION

1. Define gear and explain the various types of gears.
2. Illustrate the following terms with neat sketches as applied to gearing; circular pitch; Module: addendum; dedendum; clearance; working depth; centre distance; crest; root; tooth; thickness and filet radius.
3. What is the relation between the module, circular pitch, number of teeth and pitch diameter in a spur gear?
4. Make an neat sketch of a gear and show the following on it:

   Pitch circle; addendum circle; root circle; construction circle.
5. Define with a neat sketch:

   (i) Pressure angle (ii) Base circle (iii) Line of action

   (iv) Pitch point (v) Common tangent.
6. What do you mean by prodile of tooth? Name the type of protiles used on gears.
7. Define with a neat sketch :

   Pitch cone, pitch diameter, pitch angle, back cone, addendum, dedendum, face, width in bevel gear.
8. Define bevel gear. Give the practical application of bevel gears.
9. Draw the conventional representations of the following gears in detail and assembly form: (i) Spur gears (ii) Bevel gears (iii) Worm and worm wheel.

CHAPTER

17

Cams

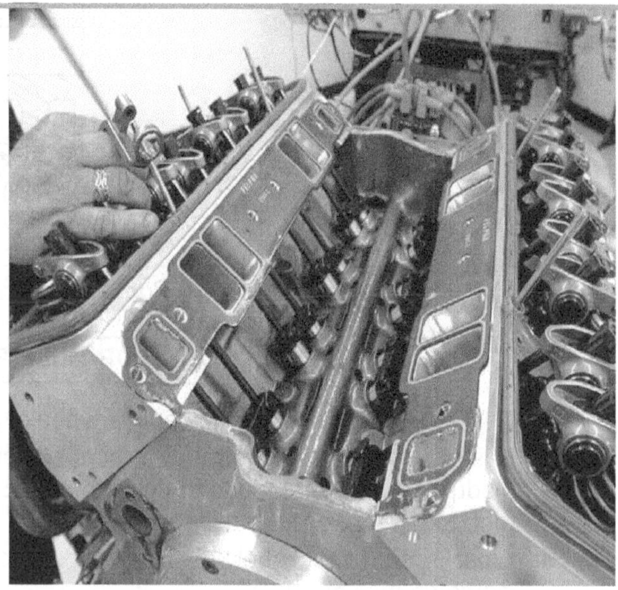

INTRODUCTION

In this engineering world, the transmission of mechanical power has become a subject of considerable importance. The present modern machines have several components which require different types of motion in different directions. Very often, it is desired to change rotary motion into reciprocating motion of complex nature e.g. operating of inlet and exhaust valves of I.C. engine. In such cases a machine part known as cam is used.

The given motion of crankshaft is supplied to cam which drives another part called the follower. The desired motion to valves is obtained from the follower. In many other machines the motion needed is of complicated type which can be obtained only by means of cam mechanism due to its simplicity in design. Thus, the use of cam in the modern machinery has become so extensive that it is of prime importance for engineering students to study its application.

In this chapter, we shall deal with study of cams, practical applications of cams, types of cams, followers and method to draw the cam profile.

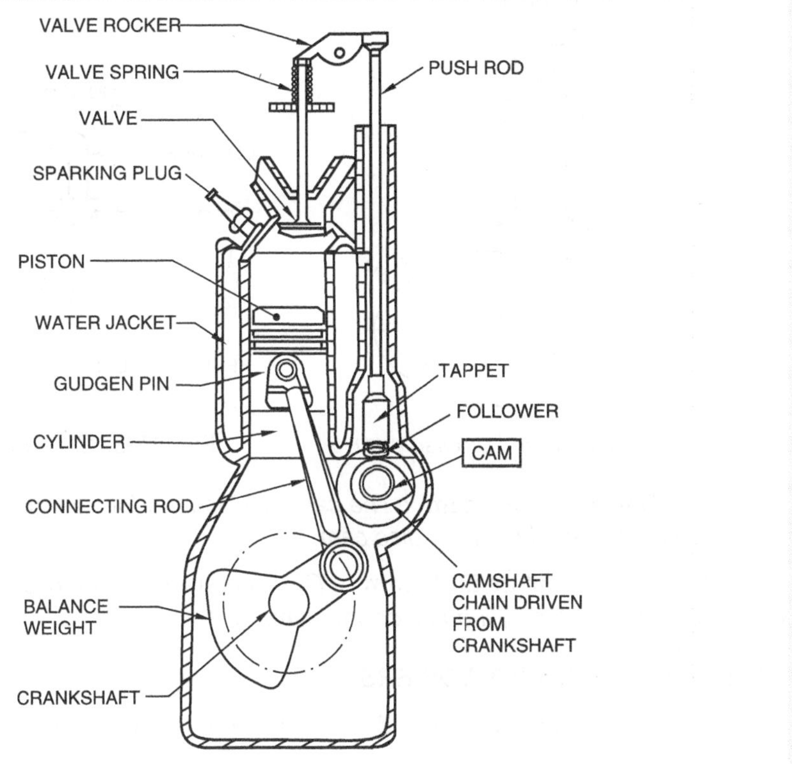

Fig. 17.1. *Cam mechanism used in I.C. engine*

17.1. CAM

A rotating device which transforms rotary motion into reciprocating motion is known as cam.

The cam drives a follower and gives it some specified type of oscillating motion. This oscillating motion of follower is transfered to receprocating motion to value. The type of motion depends upon the shape of the cam and type of the follower used. The follower is guided and remains in contact with the contour of the cam.

Cams are manufactured by a punch press, die casting or milling from the master cam.

17.2. PRACTICAL APPLICATIONS OF CAMS

Cams are extensively used in the operation of many types of machines, such as:

(i) Operating inlet and exhaust values in I.C. engines [see Fig. 17.1]

(ii) Feed mechanism of automatic lathes

(iii) Automatic screw cutting machines

(iv) Gear cutting machines

(v) Printing machinery

(vi) Spining and weaving textile machineries

(vii) Bobbing attachment of sewing machines etc.

17.3. TYPES OF CAM

The following two types of cams are used in engineering practice according to the direction of motion of follower with respect to the cam axis :—

1. Radial or disc cams
2. Cylindrical or end cams

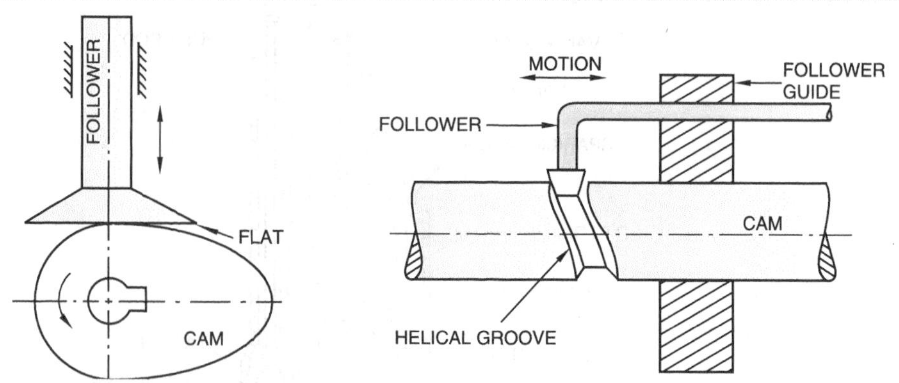

Fig. 17.2. *Radial or Disc cam* **Fig. 17.3.** *Cylinderical or End cam*

17.3.1. Radial or Disc cam : *The cam in which the follower moves perpendicular to the axis of the cam is called radial or disc cam.*

17.3.2. Cylindrical or End cam : *The cam in which the follower moves parallel to the axis of the cam is called cylindrical or end cam.*

17.4. TYPES OF CAM FOLLOWERS

Depending upon the shape and axis of motion, the following types of cam followers are used in practice :

1. Pointed or knife edge followers
2. Roller followers
3. Flat face followers and
4. Spherical followers

1. Knife edge follower : Fig. 17.4 shows a knife edge follower in which sharp knife edge is in contact with cam. This type of follower is very little in practical use due to high rate of wear at the knife edge. However cam of any shape cam work with it.

2. Roller follower : Fig. 17.4 shows a roller follower in which a roller is held by a pin to the follower assembly. This type of follower is very extensively used in practice, due to low rate of wear and perfect rolling contact with the cam.

Roller followers are used in gas and oil engines. They are also used in air craft engines due to their limited wear at high cam peripheral velocity.

3. Flat follower : Fig. 17.4 shows a flat follower in which the perfectly flat face is in contact with the cam. In this follower there will be sliding motion between the contacting surfaces but the wear can be considerably reduced by off-setting the axis of the follower. This type of follower produces high surface stresses.

Flat followers are used where the space is limited i.e. used in automobile engines to operate valves.

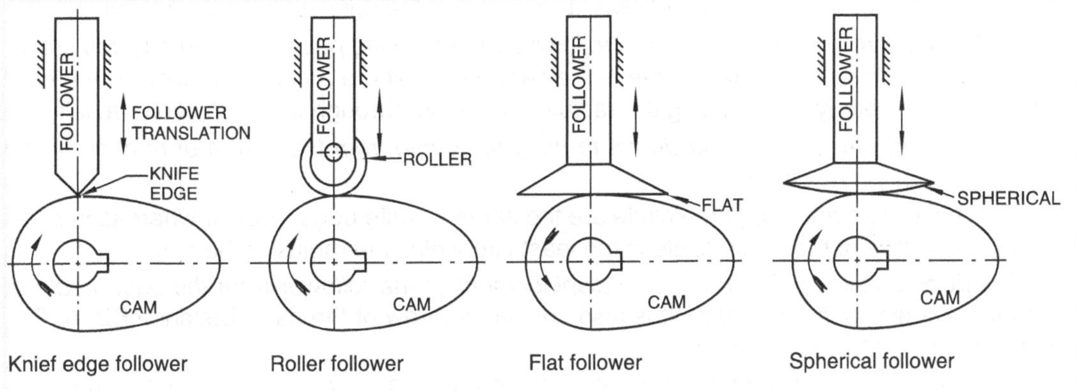

Fig. 17.4. *Different types of follower*

4. Shperical follower : Fig. 17.4 shows a spherical follower in which spherical surface is in contact with the cam.

In order to minimise the surface streses produced in the flat follower, a spherical shape having a surface of large sphere radius is given to the flat end.

The spherical followers are also used in automobile engines.

17.5. TERMS USED IN CAMS

The following are the important terms used in practice for drawing the profiles of cams:

1. Cam profile : *The actual working contour of a cam, which comes into contact with the follower to operate it, is called the cam profile.*

- *In Fig. A-B-C-D-A is the cam profile or the working contour.*

2. Base circle : *The smallest circle, drawn from the centre of rotation of a cam is known as base circle.*

- It forms the part of the cam profile and its radius is called the *least radius* of the cam (see Fig. 17.5).

The size of a particular cam depends upon the size of the base circle.

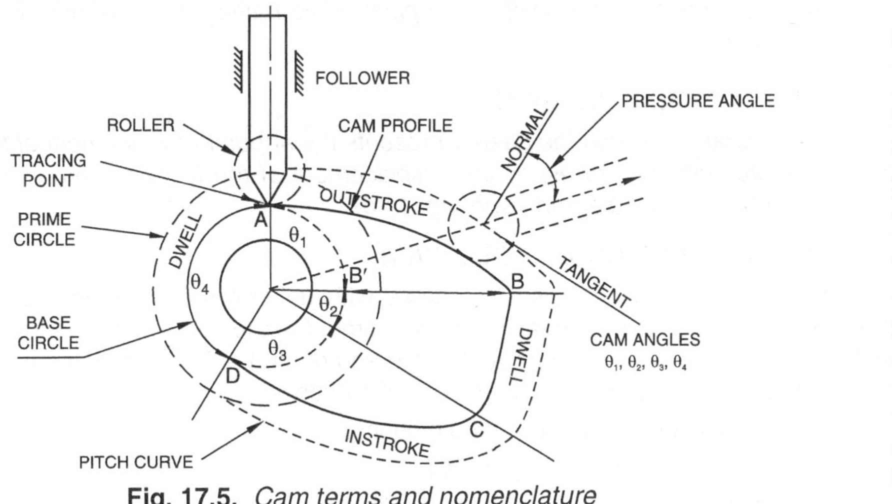

Fig. 17.5. *Cam terms and nomenclature*

3. Tracing point : *The reference point on the follower which is used in laying out the cam profile is known as tracing point.* In a knief edge follower, the tracing knife edge itself is the tracing point; where as in roller follower, the centre of the roller is the tracing point.

4. Pitch curve : *The path which is followed by the tracing point is known as pitch curve.* In a knife edge follower, the pitch curve is the cam profile where as in roller follower, the cam profile is determined by substracting the radius of the roller through the pitch curve radially.

5. Prime circle : *The circle drawn to the pitch curve from the centre of rotation of the cam is known as prime circle.*

The least circle and the prime circle are the same in knife edge follower whereas in roller follower, the radius of the prime circle is the least circle plus the radius of the roller.

6. Lift or stroke : *The maximum displacement of the follower from the base circle of the cam is known as lift or stroke.* It is also known as throw of the cam. Distance BB' or CC' is the lift of the cam (see Fig. 17.5)

7. Angle of ascent, dwell, descent, rest and action : *The angle moved by the cam for its follower to rise from its lowest position to the highest position is known as angle of ascent.* It is represented by θ_1 (see Fig.17.5)

The angle moved by the cam during which the follower remains at its highest position is called the angle of dwell. It is represented by θ_2 (see Fig. 17.5)

The angle moved by the cam for the follower to fall from its highest position to the lowest position is called the angle of descent. It is represented by θ_3 (see Fig. 17.5)

The angle moved by the cam for the follower for remaining period at uniform rest position is called the angle of rest. It is represented by θ_4.

The total angle moved by the cam for its follower from the begining of ascent to return position of rest *i.e.* after the period of ascent, dwell, descent and rest is known as total angle or angle of action.

$\therefore$ Total angle or Angle of action = $\theta_1 + \theta_2 + \theta_3 + \theta_4$

8. Pressure angle : *The angle between the line of motion of the follower and the common normal to the pitch curve at the point is called the pressure angle.* The larger the pressure angle, the larger is the side thrust.

9. Cam angle : *The angle through which a cam turns to displace the follower through a certain distance is called the cam angle.*

10. Pitch point : *The point in the pitch curve having the maximum pressure angle is known as pitch point.*

17.6. DISPLACEMENT DIAGRAM

The diagram, in which the base represents the angular displacement of the cam and the ordinate represents the corresponding displacement of the follower from its initial position is known as displacement diagram (see Fig. 17.6)

17.7. TYPES OF MOTION OF FOLLOWER

As we know, the cam usually rotates with uniform speed so that the equal angular displacement take place in equal intervals of time. On the other hand the follower starts from rest at the begining and comes to rest at the end of stroke. Thus, during the travel of follower it may follow any one of the following types of motion :—

1. Uniform velocity or Straight line motion
2. Simple harmonic motion (S.H.M.)
3. Uniformly accelerated and decelerated motion

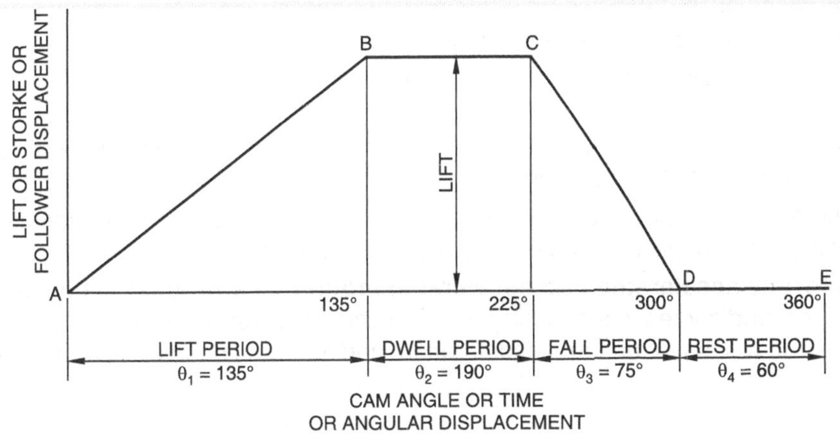

Fig 17.6. *Displacement diagram of a uniform velocity motion.*

17.7.1. Uniform velocity or Straight line motion : For uniform velocity or straight line motion of the follower, the slope of the displacement curve will be constant, as displacement is directly proportional to angle turned. Fig. 17.6 shows the displacement diagram for cam mechanism in which the follower rises with uniform velocity during 135° of cam rotation, dwells for 90°, falls with uniform velocity during 75° and finally rests for the remaining period of cam rotation.

i.e. 360° – (135°+90°+75°) = 60° (θ_4). Thus, the straight lines AB and CD indicate uniform *velocity i.e.* equal displacements in equal intervals of time for follower lift and fall periods.

17.7.2. Simple Harmonic Motion : *When the follower moves on a circular path with uniform velocity its projection on the diameter will have simple harmonic motion.* The velocity of the projection will be maximum at the centre and zero at the ends. Fig. 17.7 shows the displacement diagram for cam mechanism in which the follower rises with simple harmonic motion during 120° of cam rotation,dwells for 60° and falls with simple harmonic motion for the remaining period.

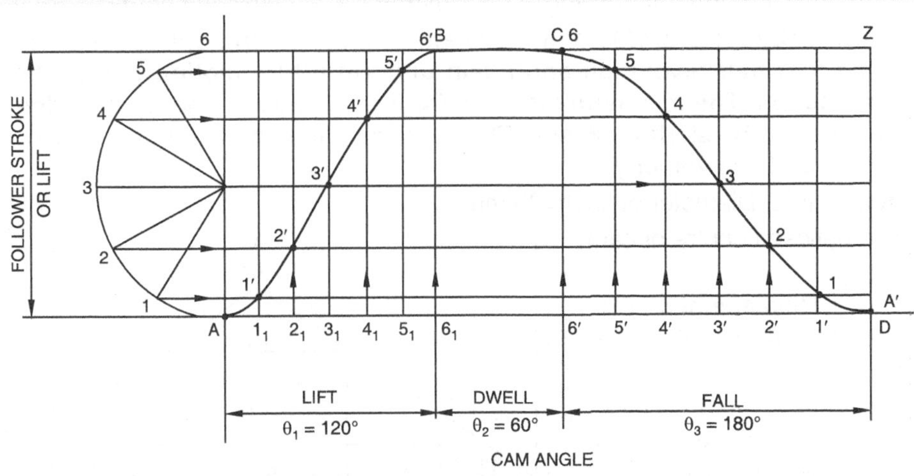

Fig. 17.7. *Displacement diagram of simple harmonic motion*

Construction of displacement diagram : *The following steps should be followed for the construction of displacement diagram for S.H.M. (see Fig. 17.7)*

1. Draw a line AA' and mark the cam angles as θ_1 =120°, θ_2=60°, and θ_3 = 180°.

2. Erect a perpendicular A Y equal to lift. Draw a semicircle on A Y of diameter equal to the follower of lift. Divide the circumference of semicircle into equal parts say 6.

3. Divide the angular displacement of cam during the lift and fall into the same number of equal parts i.e. 6.

4. Through 1,2,3,4,5 and 6 draw horizontal lines to the cut the vertical lines through 1_1, 2_1, 3_1, 4_1, 5_1 and 6_1 at 1', 2', 3', 4', 5', and 6'.

5. Draw a smooth curve passing through 1', 2', 3', 4', 5' and 6' points of intersection to get the required simple harmonic motion displacement diagram.

17.7.3. Uniformly accelerated and decelerated motion : As we know the follower of a cam starts from rest and comes to rest during outward and inward strokes, so during the first half of each stroke it will have uniform acceleration and uniform deceleration during the second half.

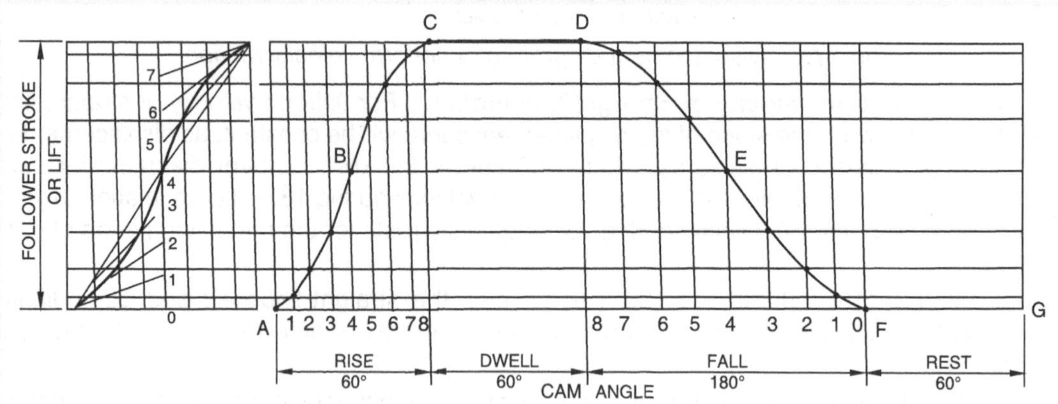

Fig . 17.8. *Displacement diagram of uniformly accelerated and decelerated motion.*

Fig. 17.8 shows the displacement time diagram for the cam mechanism in which the follower rises during outward stroke with uniform acceleration and deceleration during the 60° of cam rotation, dwells for 60° falls with acceleration and deceleration during 180° and rests for the remaining period of 60°. Fig. 17-8 illustrates the displacement-time diagram.

Problem 1 : Draw the profile of a disc cam to give uniform upward motion of 45 mm during the first half revolution and again uniform return motion during the next half of the revolution. The minimum distance from the centre of cam to the edge is 50 mm. The diameter of the shaft is 35 mm. The cam will allow the knife edge follower to reciprocate with uniform velocity.

Solution : Here, Diameter of shaft=35 mm

Distance between centre of cam to edge = 50 mm.

Part – A : Displacement Diagram–[Scale : Half full size] Fig. 17.9.

1. Take a horizontal line C D and mark the cam angles distances as θ_1 = 180° = 18 cm and θ_2 = 180° = 18 cm.

2. Erect a perpendicular DA=45 mm = lift and divide it into 6 equal parts 1, 2, 3, 4,5 and 6.

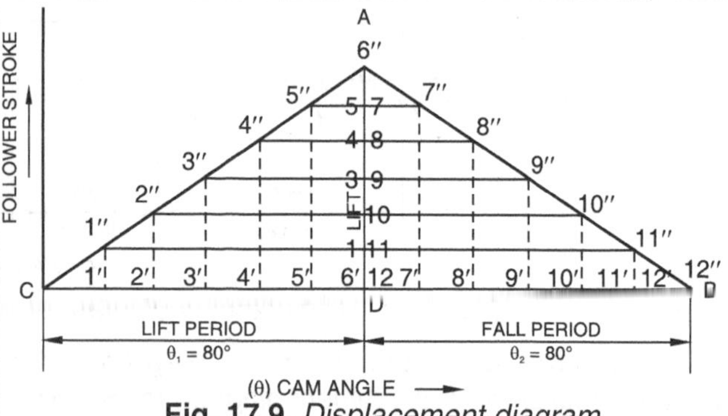

Fig. 17.9. *Displacement diagram*

3. Divide the angular displacement of cam (cam angle) during half revolution for uniform upward motion θ_1 (C D) into same number of equal part i.e. 6 as 1',2',3',4',5' and 6'.

4. Also divide the angular displacement of cam (cam angle) during next half revolution for uniform return motion (D B) into 6 parts as 7',8',9',10',11' and 12'.

5. Through 1',2',3',4' .... draw vertical lines to cut the horizontal lines drawn from 1,2,3,4 .... etc. at 1",2",3",4" ...... respectively and complete the displacement diagram as shown in Fig. 17-9.

Part – B : For Cam profile – (Fig. 17.10)

1. Draw a base circle with radius OA=50 mm

2. With O as centre and radius equal to OB=50+45 mm, draw a circle.

3. Divide the circle into 12 parts i.e. 6 parts for uniform upward motion and 6 parts for return motion.

4. Draw the radial lines from centre O.

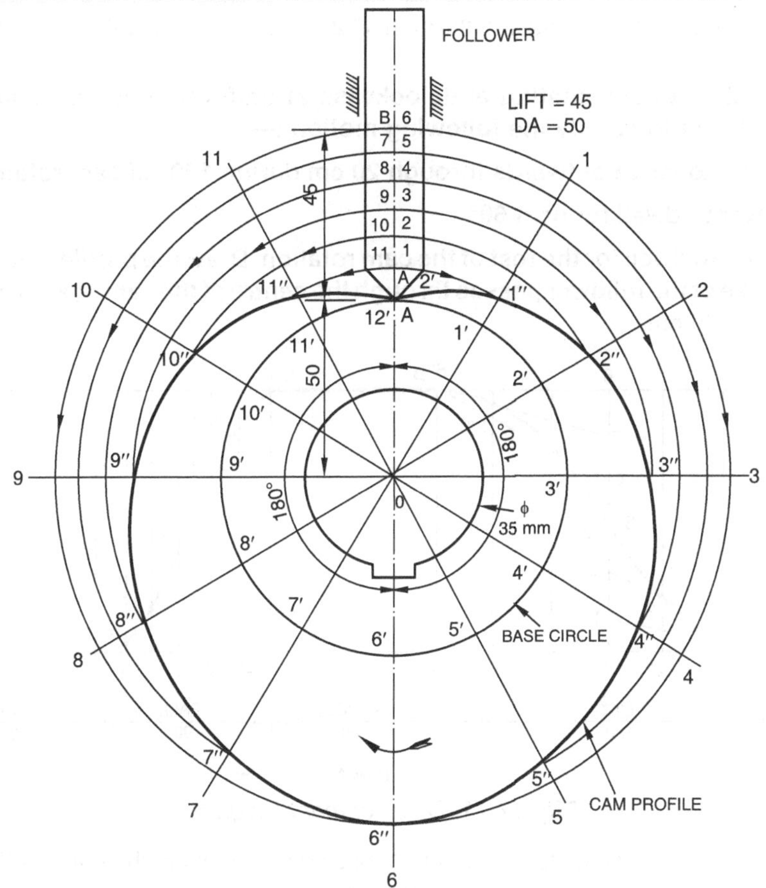

Fig. 17.10. *Cam profile*

5. On radial line 01, cut a distance 1'-1" = distance 1'-1" from displacement diagram to get point 1". Similarly on radial line 02, cut a distance 2'-2" from displacement diagram to get point 2" and so on.

6. Draw a smooth curve through points 1",2",3", ... to get the required cam profile.

This problem cam be solved without drawing the displacement diagram as discussed by the following steps:

1. Draw the base circle with radius of cam, OA=50 mm (see Fig. 17.10).

2. Mark the cam lift or stroke, AB=45 mm.

3. Divide the cam lift AB into 6 equal parts and number these parts as 1,2,3,4,5 and 6 for the out stroke and 7,8,9,10,11 and 12 for the return stroke.

4. Divide the base circle into 12 equal parts. Draw radial lines O1',O2',O3' ...... O12' from the centre O.

5. With O as centre and radius equal to OB = 50 + 45 = 90 cm draw the outer circle.

6. With O as centre and radius equal to O1 draw arc to cut the radial line O1' and O11" at 1" and 11" respectively. Similarly, with O as centre and radius equal to 02 etc. draw arcs to cut the radial lines O2' and O10' at 2" and 10" etc. respectively.

7. Draw a smooth curve passing through 1",2",3", ... 11" to get the profile of required cam.

Problem 2 : A cam rotating anticlockwise at uniform speed is required to give knife edge follower to satisfy the following motion :—

1. Follower to move outwards through 20 cm during 120° of cam rotation.

2. Follower to dwell for next 60°.

3. Follower to dwell for the rest of the cam rotation. Draw the profile of the cam, when the line of stroke of the follower passes through the centre of the cam shaft. Take minimum radius of cam = 20 cm.

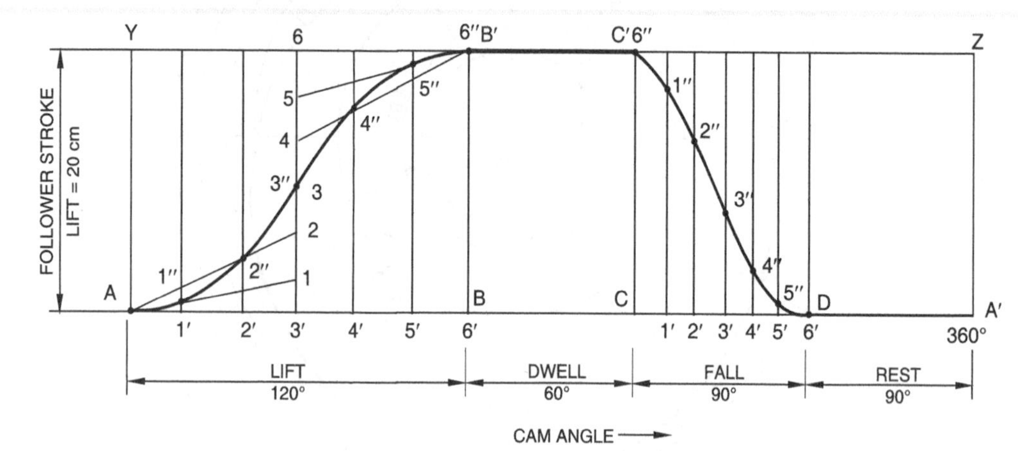

Fig. 17.11. *Displacement Diagram*

Solution : The following steps should be followed by drawing the cam profile.

Part - A : Displacement Diagram – **[Fig. 17.11]**

Draw the displacement diagram to half size scale as shown in Fig. 17.11 by steps as under:

1. Take a horizontal line AA' and mark the cam angles distances as θ_1 = 120°= 12 cm, θ_2= 60° = 6 cm., θ_3 = 90° = 9 cm and θ_4 = 90° = 9cm.

2. Erect a perpendicular A Y=20 cm=lift and divide it into 6 equal parts 1,2,3,4,5 and 6 as shown in Fig.17.11.

3. Divide the angular displacement distances A B during rise and C D during fall into the same number of equal parts i.e. 6 and 1',2',3',4',5', and 6'.

4. Through 1',2',3',4',5' and 6' draw vertical lines to cut the inclined lines at 1",2",3",4",5" and 6" and complete the displacement diagram by joining A-B-C-D-A'.

Part - B : Cam profile – [Fig. 17.12]

1. Draw a base circle with radius OA=20 cm.

2. Set off the radii O B, O C, and O D at successive angular intervals of 120°, 60° and 90° in anticlockwise direction starting from O A.

3. Divide AB and CD into the same number of equal parts as in the displacement diagram i.e. 6 parts each.

4. Draw radial lines passing through these points.

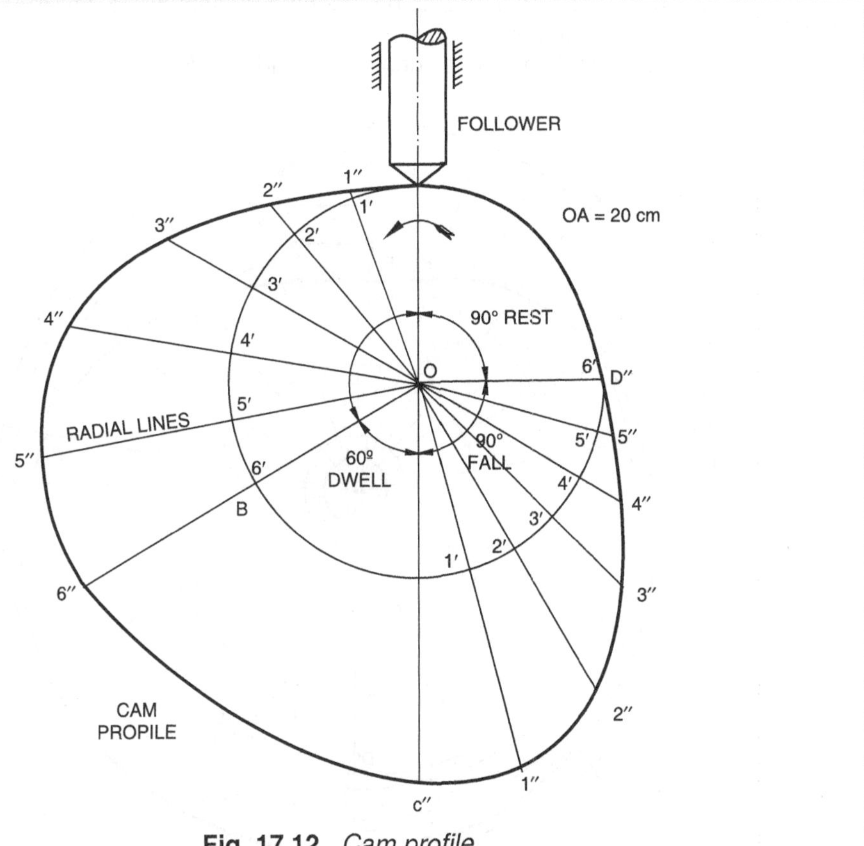

Fig. 17.12. *Cam profile*

5. Set off displacement of the follower along the corresponding radial lines beyond the base circle e.g. distance 1',-1" on the radial line=1'-1" from the displacement diagram draw curve 1"-2" and soon.

6. With O as centre and radius equal OB" draw arc B"C" for the dwell of 60°.

7. Draw the curve C" D" for fall as discussed in point 5 above.

8. With O as centre and radius equal to OD" draw arc D" A for dwell for the rest of cam rotation.

Then, AB", C", D" A is the required cam profile.

Problem 3: Draw the profile of the cam with the following specifications:

Minimum distance of cam centre to edge =35 mm

Lift of follower = 30 mm

The cam lifts the follower with simple harmonic motion during 120° of the revolution, then remains at rest for the next 60°, then falls with uniform acceleration and retardation during 120° through 15 mm and finally returns to the starting point with uniform motion.

Solution : Here,

Minimum distance of cam centre to edge, OA = 35 mm

Note: *This problem is solved without the displacement diagram as discussed below :*

The following steps should be followed for drawing the cam profile:-

1. Draw the base circle with O as centre and radius equal OA=35 mm (see Fig. 17.13)
2. Mark the cam lift, A B=30 mm.
3. Divide AB into 6 equal parts by drawing a semicircle on AB and dividing the circumference into 6 equal parts.
4. Draw the other circle with O as centre and radius OB=35+30=65 mm.

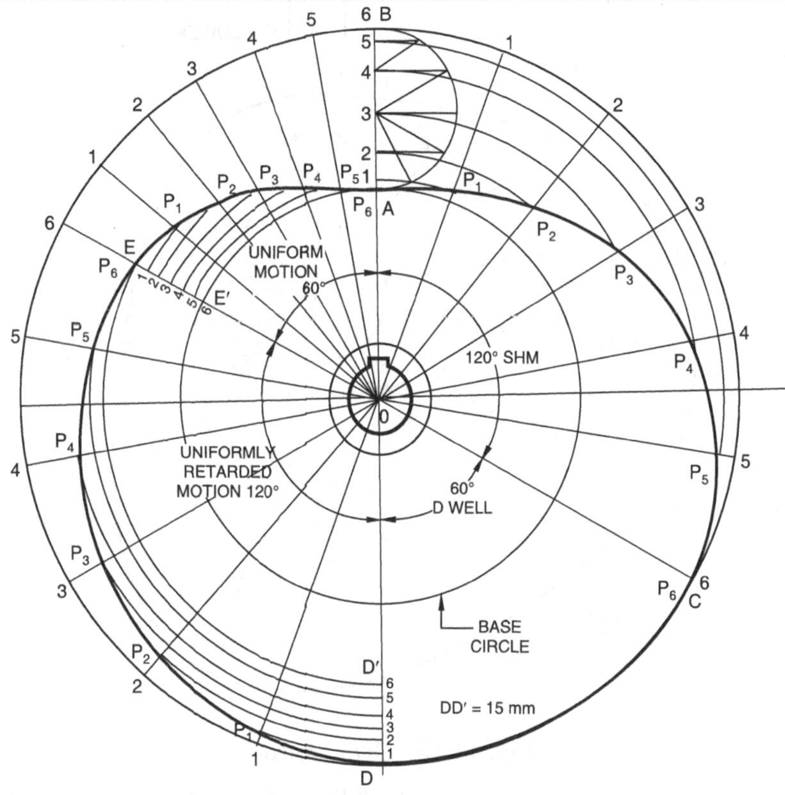

Fig. 17.13. *Cam profile*

Profile for S.H.M. during 120° clockwise (Curve AC)

5. Divide the angle of 120° into the same number of equal parts as lift A B is divided *i.e.* 6. With O as centre and radius equal 01,02,03,04,05 and 06 draw arcs to cut the radial lines at P_1, P_2, P_3, P_4, P_5 and p_6 respectively.

Draw a smooth curve A C from these points

Cam profile for rest during 60° (Curve–CD) :

6. Draw a uniform curve C D with O as centre and radius=OC

Cam profile falls with uniform acceleration and retardation during 120° through 15 mm (Curve DE) :

7. Take DD' =15 mm and divide it into 6 equal parts. Divide the angle 120° into 6 equal parts and draw the radial lines from the centre O.

With O as centre and radii equal to 01,02,03,04,05 and 06 draw arcs to cut the respective radial lines. Draw a smooth curve DE through these points.

Cam profile for return to starting with uniform motion during 60° (Curve DA)

8. Take EE'=15 mm and divide it into 6 equal parts. Divide the remaining angle of 60° into 6 equal parts and draw the radial lines from the centre O.

With O as centre and radii equal to 01, 02,03,04,05 and 06 draw arcs to cut the respective radial lines.

Draw a smooth curve E A through these points.

Thus, A C D E A is required profile of the cam.

Problem 4 : Draw the cam profile with the following particulars. Minimum distance of centre of cam from edge of the follower = 60 mm, lift = 56 mm.

The cam is to lift the follower with simple harmonic motion during one half of a revolution, then allow the follower to drop suddenly half way and further fall with uniform velocity during the remaining half of the revolution.

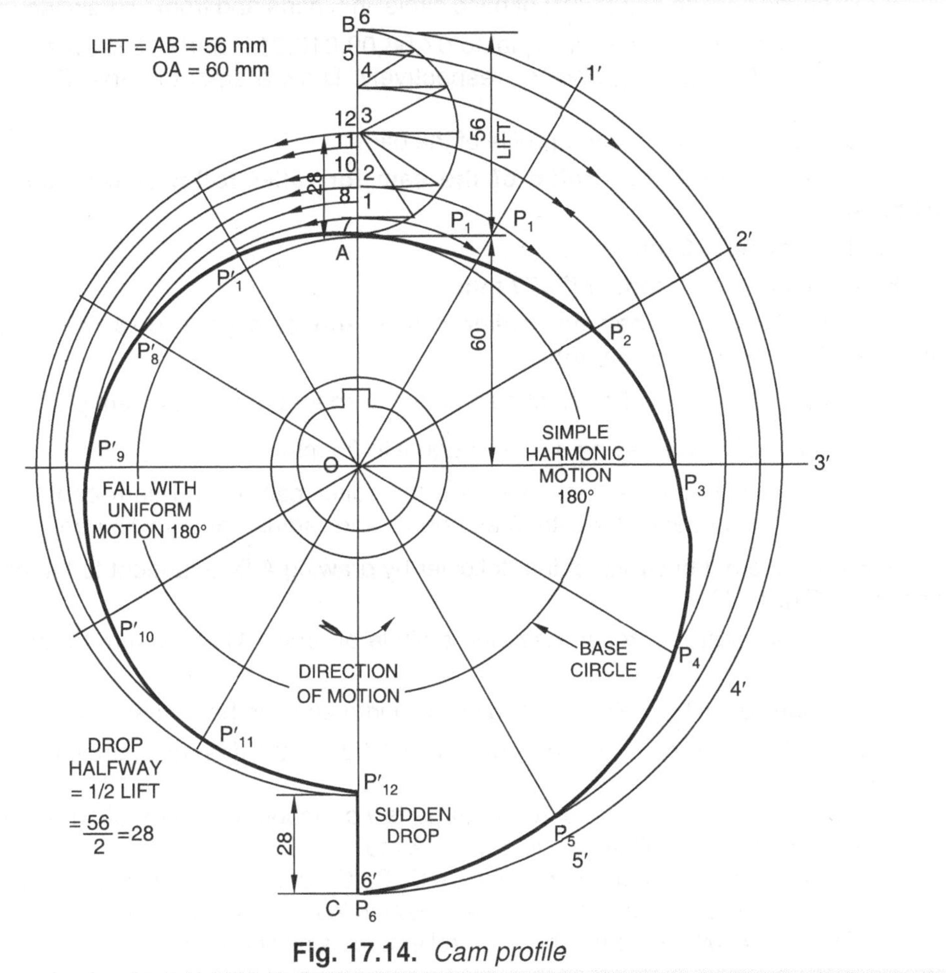

Fig. 17.14. *Cam profile*

Solution : The following steps should be followed for drawing the cam profile :—

1. Draw the base circle with radius O A=60 mm (see Fig. 17.14).

2. Mark the cam lift A B=56 mm and divided it into 6 equal parts.

(i) Cam profile for uniform S.H.M. during 180° (Curve AC).

3. Divide the first half (180°) base circle in the same number of angular as the lift A B is divided i.e. 6 parts as 1',2',3',4',5' and 6'.

Draw the radial lines 01',02',03',04',05' and 06' from the centre O.

4. With O as centre and radius equal to 01 draw an arc to cut the radial line 01' at P_1.

Similarly, with O as centre and radii equal to 02,03,04,05 and 06 to cut the radial lines at P_2,P_3,P_4,P_5 and P_6 respectively.

Draw a smooth curve AC through these points.

(ii) Cam profile for sudden drop half way (Line CD).

5. As the follower to drop suddenly half way i.e. half the lift 56/2=28 mm, therefore mark a vertical line CD equal to 28 mm for sudden drop halfway.

(iii) Cam profile with uniform motion for 180° (curve DA)

6. Divide half of the diameter of AB into 6 equal parts and number them 7,8,9,10,11 and 12.

7. Divide the second half (180°) of base circle in 6 parts and from O draw the radial lines.

8. With O as centre and radii equal to 07,08,09,010,011 and 012 draw arcs to cut the radial lines at P_7,P_8,P_9,P_{10},P_{11} and P_{12} respectively. Draw a smooth curve D A from these points.

Thus, A C D A is the required profile of the cam.

Problem 5 : Draw the profile of the cam for roller follower with the following particulars:

Shaft diameter=25 mm.

Roller diameter=12 mm. Lift=72 mm.

Horizontal distance between roller centre and cam centre=18 mm. Distance of roller from the cam centre=35 mm.

Solution : The following steps should be followed for drawing the cam profile:

1. Draw a shaft circle having diameter equal to 25 mm.

2. As the horizontal distance between the roller centre and cam centre is 18 mm, therefore draw an off set circle with O as centre and radius equal to 18 mm.

3. Fix the line of the stroke of the follower by drawing A B, a tangent to the off set circle as shown in Fig. 17.15.

4. With C as centre of roller, describe a circle of radius 12/2 = 6 mm to represents the roller.

5. Mark the cam lift C B=72 mm and divide it into any number of parts, say 8.

6. With O as centre and radii equal to 01,02,.... 08 describe circles as shown in Fig.17.15.

7. Divide the outer circle into 16 equal parts *i.e.* double the parts of lift and from each points draw tangent to the off set circle e.g. 1 - A etc.

8. Locate the various centers of roller as C_1,C_2,C_3, etc. on the respective tangent lines.

9. With C_1,C_2,C_3, etc. as centres, draw roller circles of radius=6 mm.

10. Draw a smooth curve touching the roller circles as shown in Fig. 17.15.

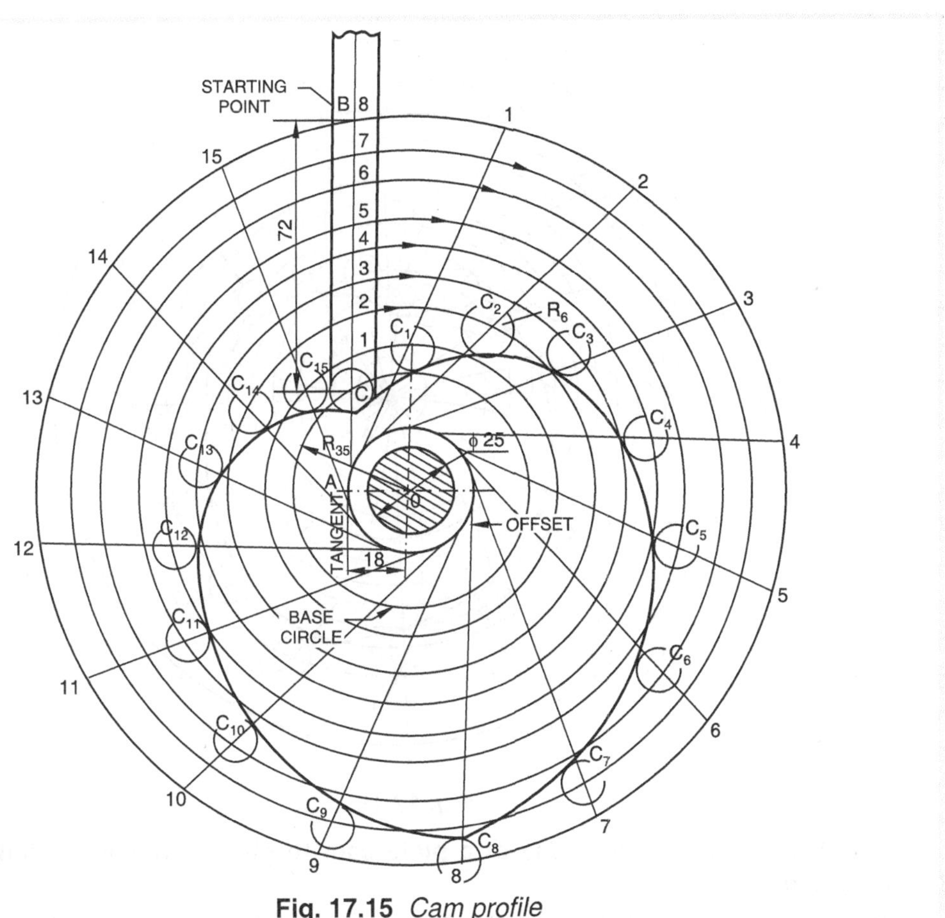

Fig. 17.15 *Cam profile*

Problem 6 : Draw the profile of a cam which allows the follower to oscillate with uniform angular velocity about a fixed centre. Every revolution of the cam completes one oscillation of the follower. Distance between the centre of cam and roller is 35 mm. Lift=75 mm, roller diameter=12 mm and length of lever=140 mm.

Solution : The following steps should be followed for drawing the profile of cam:-

1. Draw a circle, with O as centre and radius, R=OL=140 mm (see Fig. 17.16).
2. Starting from L, divide the circle into any number of equal parts, say 16.
3. From each of these parts, draw arcs passing through O e.g. with 13 as centre and radius, R=70 mm, draw arc passing through O.
4. Locate the centre C of roller such that O C=35 mm.
5. Set off the lift=75 mm and divide it into 8 equal parts. Draw the concentric circle passing through these points.

The successive points of inter section between concentric circles and arcs previously drawn are on the locus of roller centre.

6. Draw arcs, from the points of intersection circles and arcs, with radius equal to radius of roller, 12/2=6 mm.
7. Finally draw a smooth curve touching all these curved arcs, which gives the required profile of the cam.

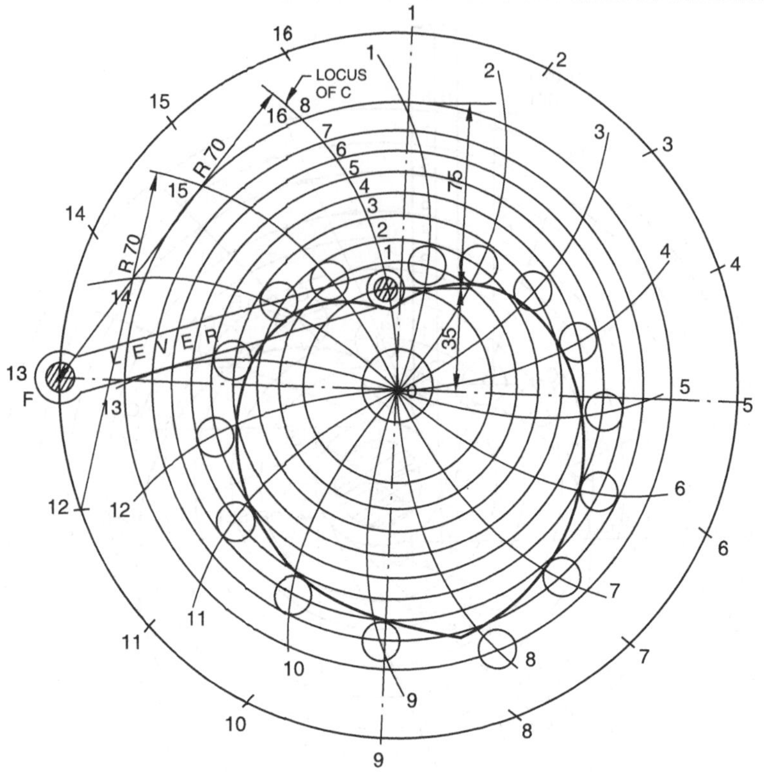

Fig. 17.16.

Problem 7 : Draw the profile of a plate cam to give an offset flat follower with following specifications:

0°-180° lift of 50 mm with uniform velocity, 180°-240° dwell, 240°-360° fall of 50 mm with uniform acceleration and retardation.

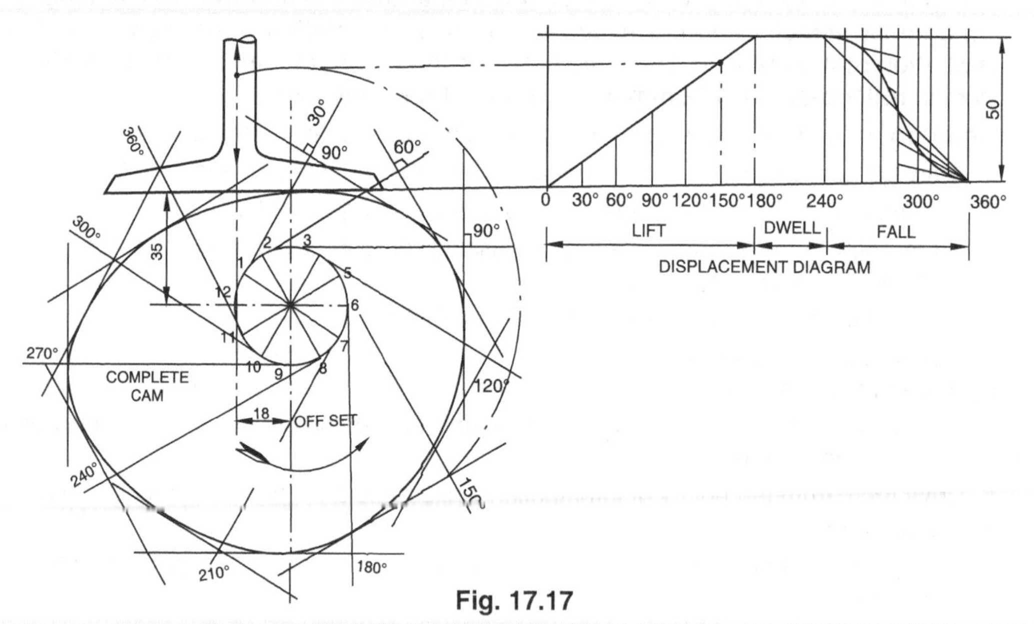

Fig. 17.17

The follower line of action is offset by 18 mm to the left of the cam shaft centre. The nearest approach of follower face to the cam shaft centre is 35 mm and the cam rotates anticlockwise.

Solution : For its solution, see Fig. 17.17.

Problem 8 : Fig. 17.18 shows the outline of a cylindrical cam. The roller follower moves through its stroke while the cam rotates once. The follower moves 75 mm to the right with uniform velocity. While the cam rotates from 0° to 120°, it then dwells from 120° to 180° and returns to starting position with S.H.M. from 180° to 360°.

Develop the cam surface and draw the front view of the cam cylinder showing the follower groove.

Solution : Fig.17.19 illustrates the method of construction of the cylindrical cam.

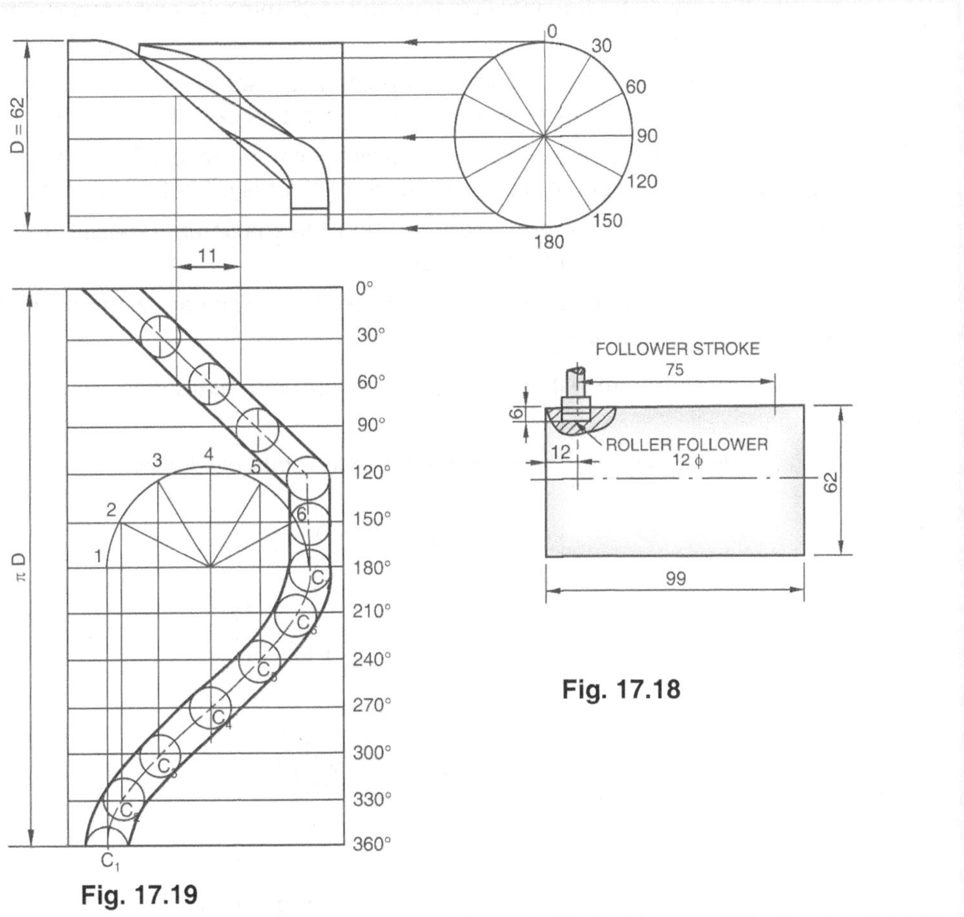

Fig. 17.19

Fig. 17.18

QUESTIONS FOR SELF EXAMINATION

1. Define a cam. Write the practical application of cams.
2. What are different types of cams and followers?
3. Define the terms: base circle, prime circle, pressure angle, lift, cam profile, angle of ascent, dwell, descent, rest and action.

4. Explain the displacement diagram. Why it is prepared?
5. What are the different types of motion of follower?
6. What is the difference between S.H.M. and U.V.M. of follower?

PROBLEMS FOR PRACTICE

1. Draw the profile of a knife edge follower for a plate cam with following specification:

   0°–90° dwell, lift of 30 mm with S.H.M. 90°–120°, 120°-240° dwell and 240°–360° fall of 30 mm with uniform velocity. The cam is to rotate clockwise and the least cam radius is to be 30 mm.

2. A radial cam rotating clockwise operates an off set roller follower and gives it the following motion. The cam lift the follower 35 mm with S.H.M. during 0°-120° of revolution then dwell for the next 60°, then fall of 35 mm with uniform acceleration and retardation during 180°

   Draw the cam profile, if the follower centre line is off-set 25 mm to the left of the cam centre of rotation. The roller radius is to be 10 mm and cam thickness is 30 mm.

3. Fig. 17.20 shows the profile of disc cam which rotates anticlockwise about centre O at a speed of 2 revolution per sec. The roller follower is off-set by 12 mm.

   Draw the follower displacement diagram, find the maximum velocity of the follower and position of the cam at which it occurs.

4. Construct the profile of line knife edge follower for a disc cam with following motion

   0°–180° lift 40 mm with S.H.M.

   180°–300° fall of 40 mm with uniform acceleration and retardation.

   300°–360° dwell.

   The least radius of the cam is 20 mm and it is to rotate clockwise.

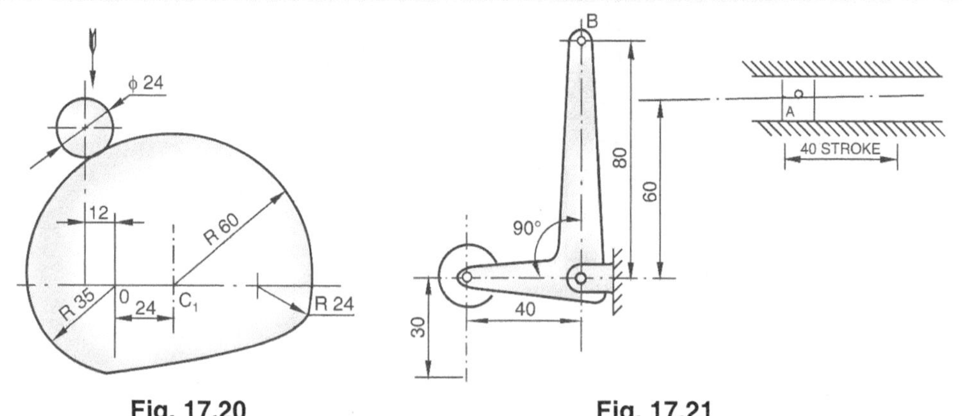

Fig. 17.20 Fig. 17.21

5. The starting position of a guide block which is operated by a plate cam through a link L M=80 mm and crank lever MNO is shown in Fig. 17.21. The block L reciprocates horizontally and is to have the motion as follows.

Movement of 45 mm to right with uniform velocity while the cam rotates through 120° remaining stationary through the next 60° of cam rotation and finally returning to its starting position with S.H.M. Draw the necessary cam profile.

CHAPTER

18

Jigs and Fixtures

INTRODUCTION

In engineering practice, the jigs and fixtures are the economical means to produce repetitive type of works by incorporating special work holding and tool guiding devices. These devices eliminate the marking out, measuring and other setting methods before machining on mass production work. In addition to this, the jigs and fixtures increases the production capacity by enabling a number of workpieces to be machined in a single set up and in some cases a number of tools may be made to operate simultaneously.

In this chapter, we shall deal with the study of jigs, fixtures, drilling jig and milling jig as used in engineeing practice.

18.1. JIG AND FIXTURE

A device which hold and locates the workpiece and also guide the cutting tool durning manufacturing process is called a jig.

It is clamped on machine table during drilling, tapping, and reaming operations.

A device which hold and locate a workpieco during manufacturing operation is called a fixture.

It is clamped on machine table, cross slide or machine bed during milling, turning, boring, grinding operations.

Note : *A fixture never guides the tool.*

18.2. DIFFERENCE BETWEEN JIG AND FIXTURE

In practice the following are the differences between a jig and a fixture :—

1. The jig holds, locates and guides the tool whereas a fixture holds and positions the workpiece but does not guide the tool.

2. The jig is made lighter for quicker handling and clamping with the table whereas the fixture is generally heavier in construction and is bolted regidly on the machine table.

3. The jig is used for holding the workpiece and guiding the tool particularly in drilling, reaming or tapping operations whereas the fixture is used for holding workpiece in milling, grinding, planing or turning operations.

18.3. DRILLING JIG

A device which is used to hold, locate the position for holes and guide the tool during drilling operation is known as drilling jig.

Drilling jig is mainly used to drill holes at the predetermined locations on the workpiece. It is also equally useful for tapping and reaming operations. The shape and size of the drilling jig varies with the shape and arrangement of their parts.

Parts of drilling jig : In general, every drilling jig consists of the following parts :—

1. Bushes or Guides – used to guide the drilling tools
2. Body – used to support the bushes
3. Locator – used to position the job for correct location
4. Clamping devices – used to lock the job in desired position
5. Base – used to support all parts of drilling jig and secure the assembly to the operating parts of the machine tool.

Fig 18.1. shows two views of a drilling jig. This type of jig is used to locate six holes spaced equally for a circular flange. This design allows for quick loading and unloading of workpieces.

For unloading, *top nut* (6) is loosened, *latch washer* (8) swivelled out of zone and then *jig plate* (3) lifted to remove the workpiece for its seating. The jig plate is designed to allow the nut to clear the centre hole. The same process is repeated in reverse to load the work piece for drilling. *Drill bushes* (7) fitted in the jig plate to guide the drill.

Note : *It should be noted that before the workpiece is used for drilling, it must be machined all over the required surfaces so that after drilling operation the part is finished and ready.*

PART NO.	NAME	MATERIALS	NO. OFF
1	BASE PLATE	C.I.	1
2	STEM	M.S.	1
3	JIG PLATE	C.I.	1
4	SCREW	M.S.	3
5	STUD	M.S.	1
6	NUT	M.S	1
7	BUSH	A.S.	6
8	LATCH WASHER	M.S	1
9	SCREW	M.S.	1

Exercise 1: Fig. 18.1. shows two views of a drilling jig. Draw the following views to same suitable scale:

(a) Front view – Half section (b) Top view.

Identify the types of fit for mating parts by using the table of tolerance. Also prepare a parts list and bill of materials.

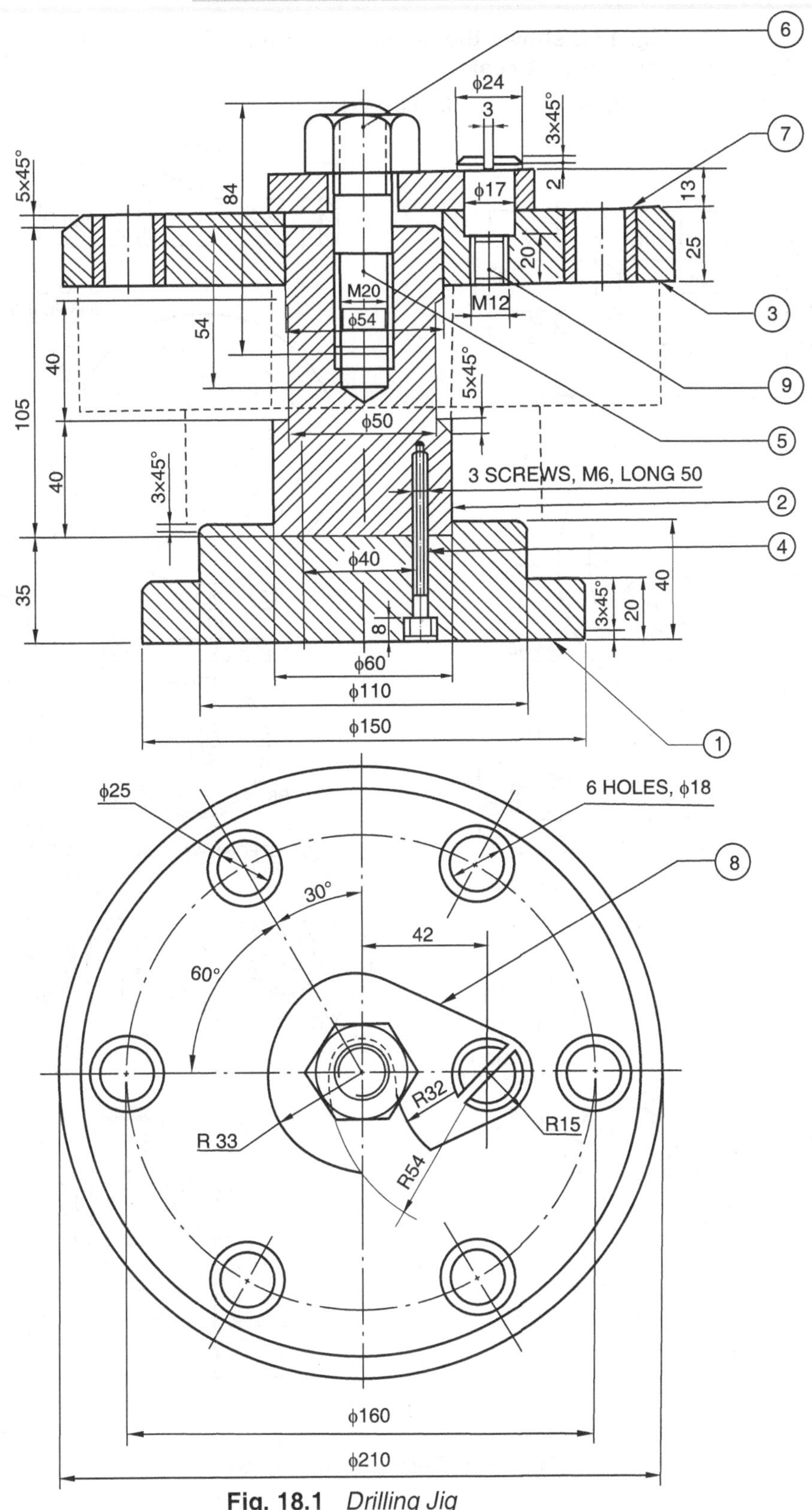

Fig. 18.1 *Drilling Jig*

Exercise 2 : **Fig. 18.2 shows the details of a drilling jig. Assemble the parts and draw the following views to 1 : 2 scale**

(a) Sectional front view **(b) Top view**

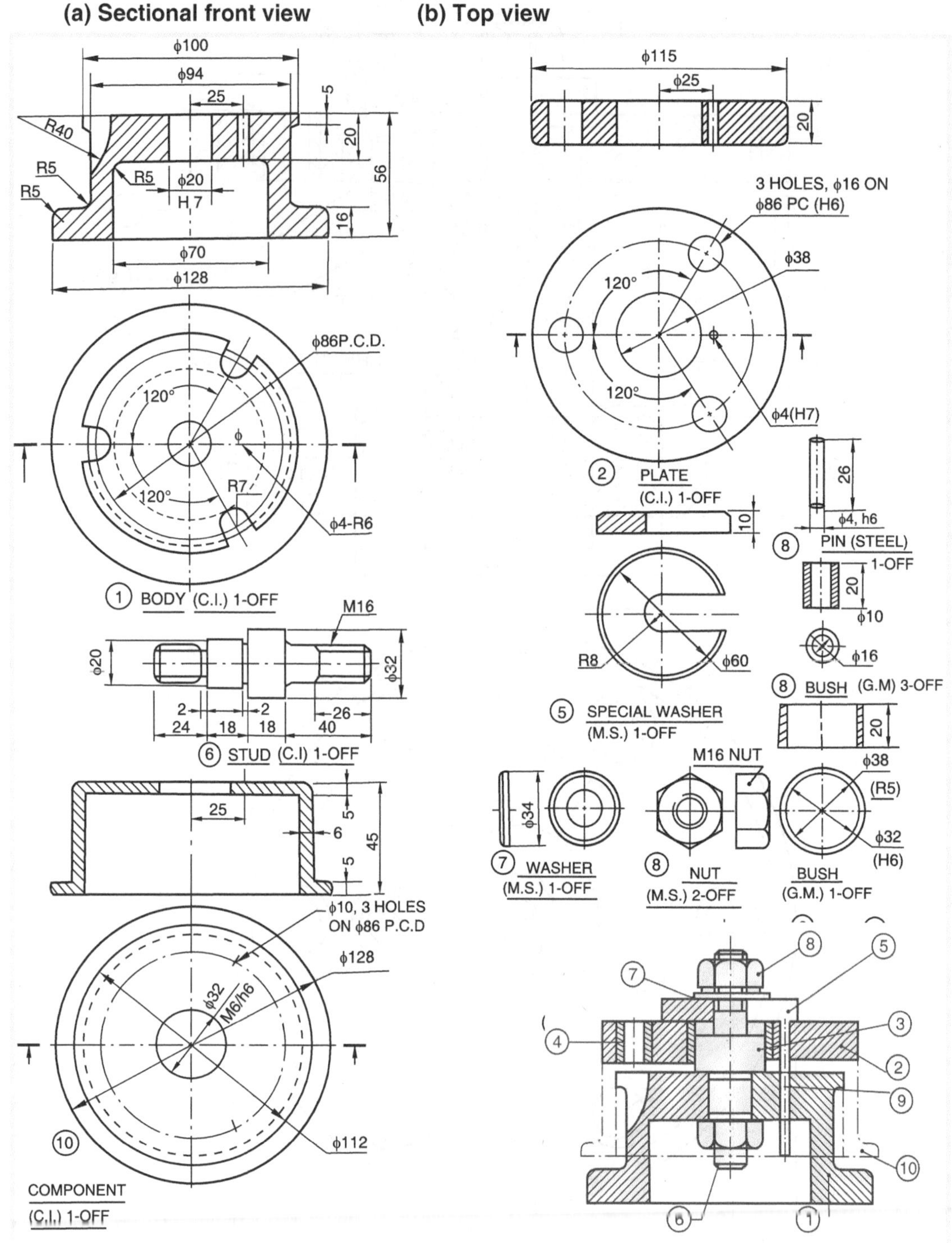

Fig. 18.2 *Details of Drilling Jig*

Exercise 3 : Detail drawings of a drilling jig are shown in Fig. 18.3. Assemble the parts and draw the following views to a convenient scale :

(a) Sectional elevation – half section **(b) Plan**

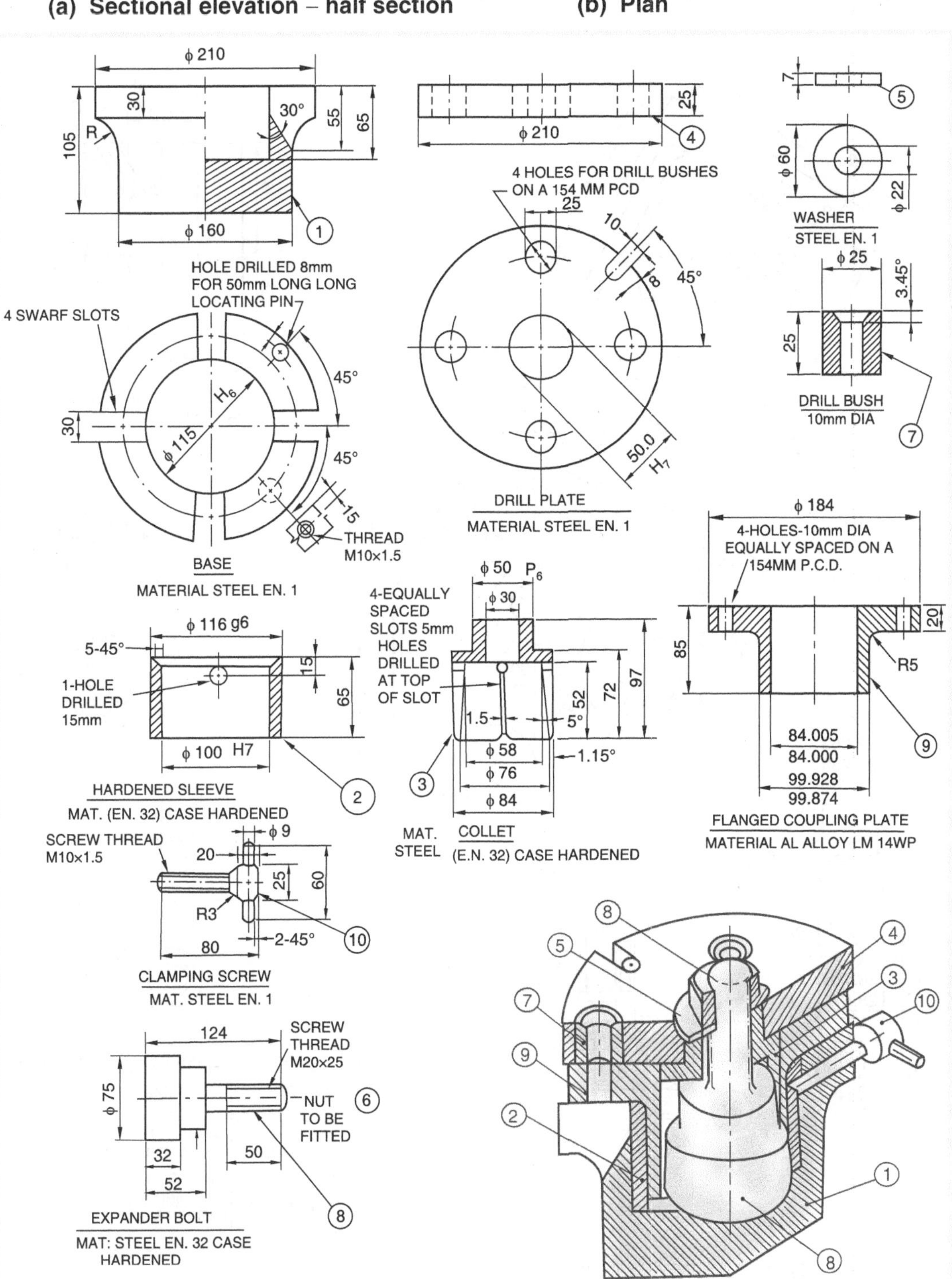

Fig. 18.3. *Details of a Drilling Jig*

Exercise 4 : Fig. 18.4 shows the details of a drilling jig. Assemble the parts and draw the following views to some suitable scale:

(a) Front view (b) End view and (c) Top view

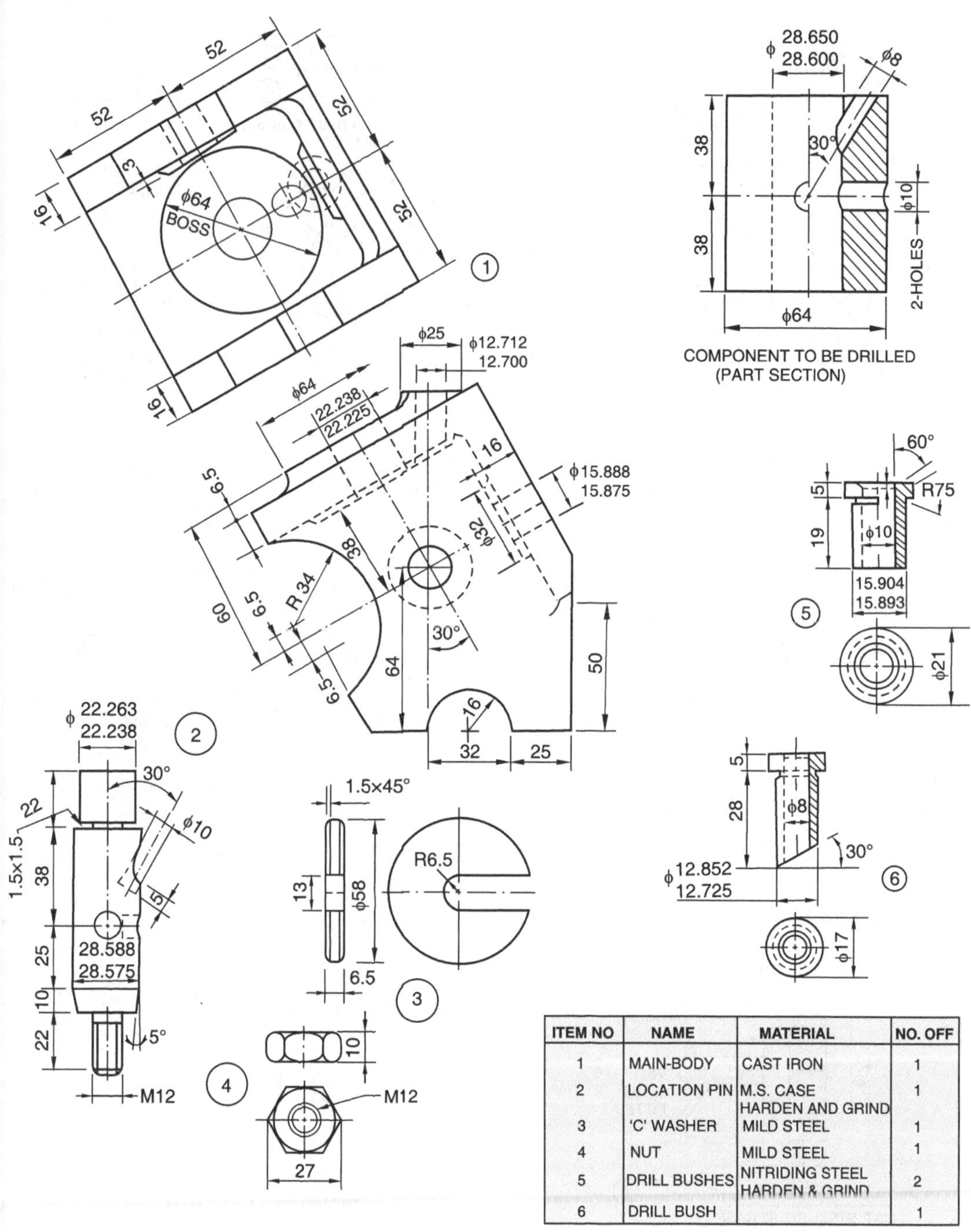

ITEM NO	NAME	MATERIAL	NO. OFF
1	MAIN-BODY	CAST IRON	1
2	LOCATION PIN	M.S. CASE HARDEN AND GRIND	1
3	'C' WASHER	MILD STEEL	1
4	NUT	MILD STEEL	1
5	DRILL BUSHES	NITRIDING STEEL HARDEN & GRIND	2
6	DRILL BUSH		1

Fig. 18.4. *Details of a Drilling jig*

18.4. DRILLING JIG FOR VALVE STEM HOLE

Fig 18.5 shows the details of a drilling jig used for a small hole of ϕ 2 in the valve stem. In this jig, drill bush is inserted into the guide hole on the top side of the base. A lock bush is used to cheek the bush against rotation. The lower step of base has a stepped vertically hole in which a locator is inserted from underneath and kept in position by means of a spring. The valve stem is inserted in the horizontal hole of the base by pressing the locator downwards against the spring. It comes upward to fit into the head hole of stem, thus locating the valve stem at predetermined distance for the hole to be drilled.

A lock handle is used to lock the locator in position while drilling the hole. When the hole is drilled, the handle is moved to release the job. The rejector is pressed inward by a button to reject the stem from the jig. The spring under the button brings the rejector back to its original position when released.

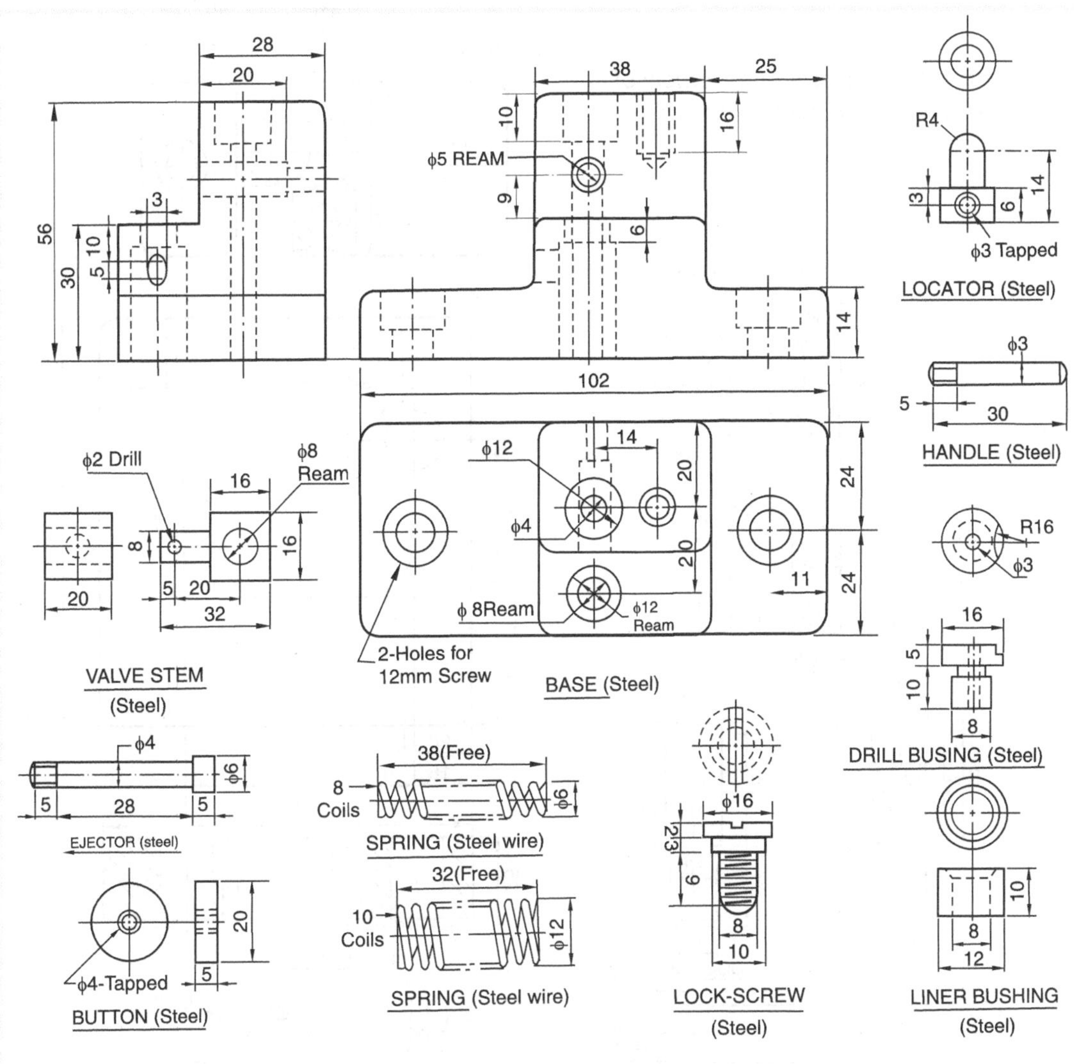

Fig. 18.5. *Details of a drilling Jig for valve stem hole*

Exercise 5 : Fig. 18.5. show the detailed drawings of a drilling jig for value stem hole. Assemble the various parts and draw the following views :—

(a) Front View (b) End View and (c) Top View

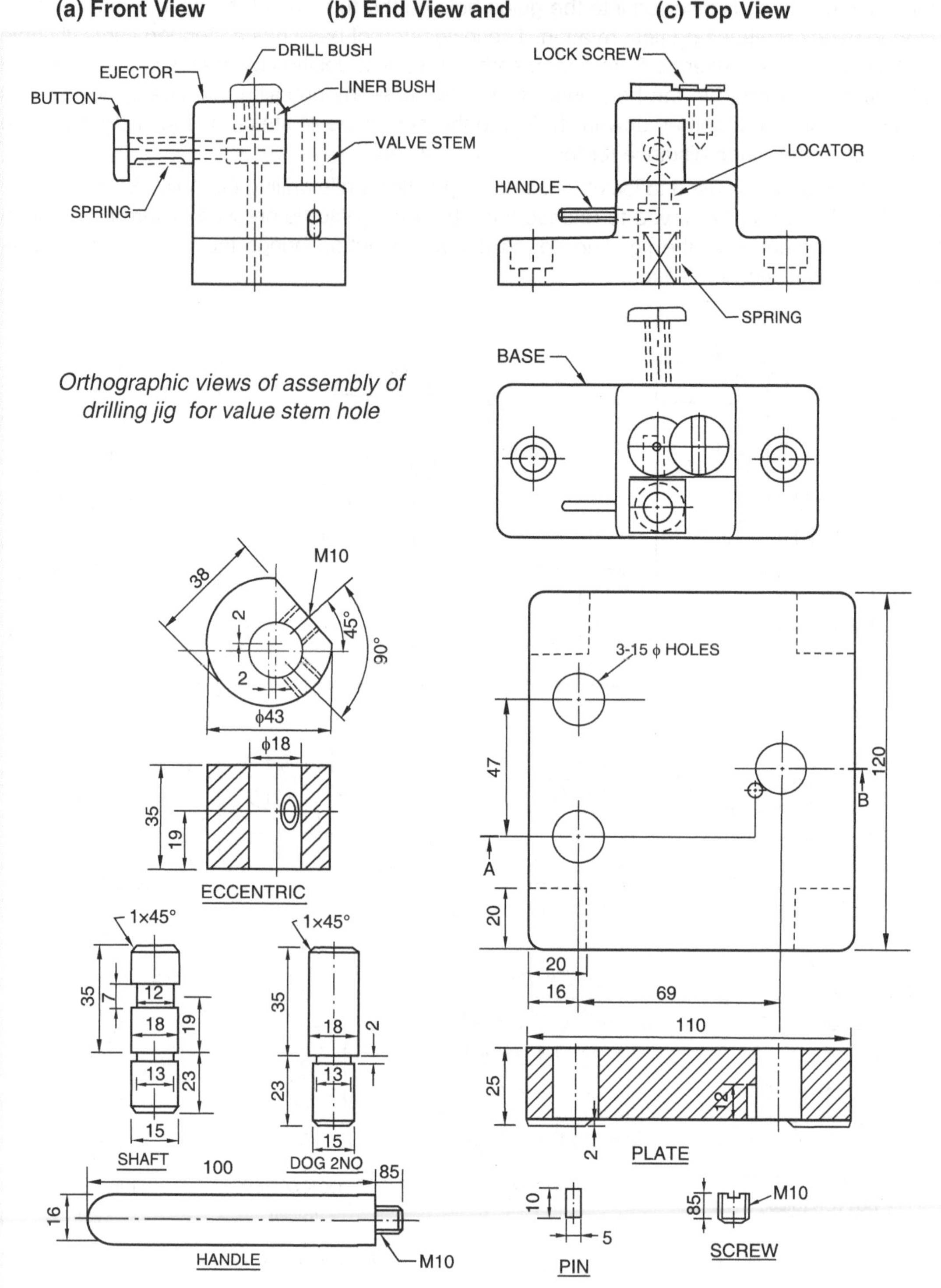

Fig. 18.6. *Details of fixture*

Problem 1 : Detail drawings of the fixture are shown in Fig. 18.6. Assemble the details and draw the following views to a scale full size:

(A) Front View – full section **(b) Top View**

Assume suitably any missing dimens. 18.7.

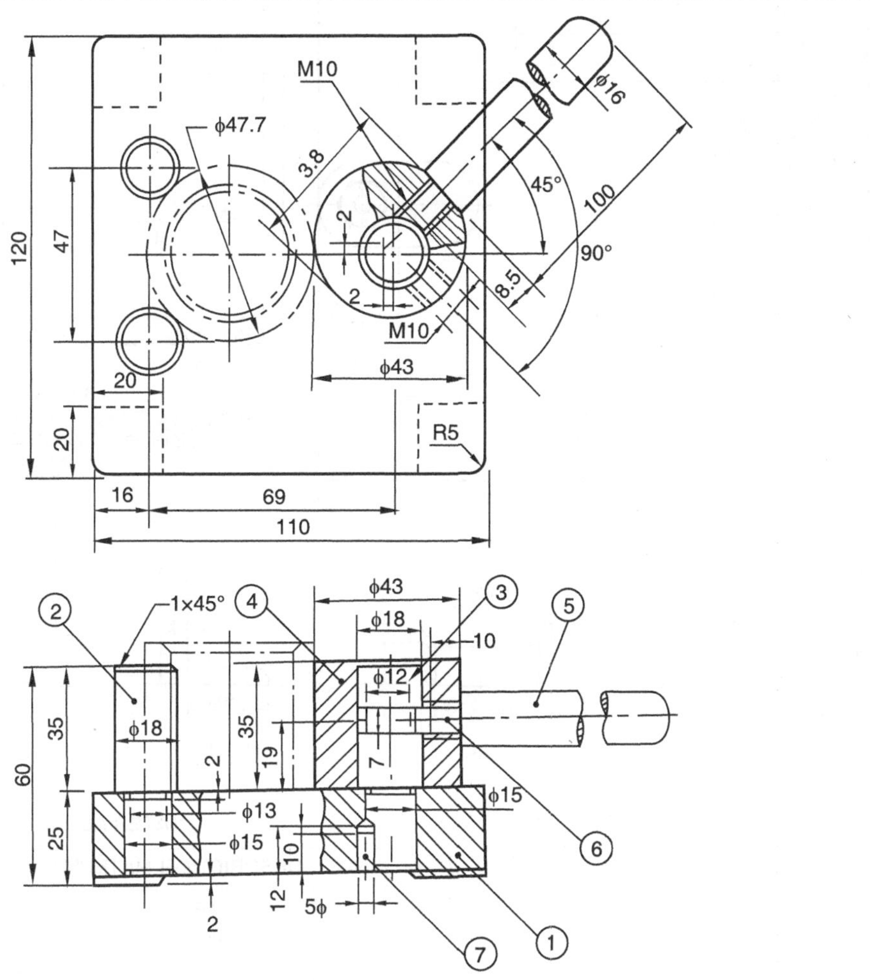

Fig. 18.7.

S. NO.	DESCRIPTION	MATERIAL	NO. OFF
1.	PLATE	M.S.	1
2.	DOG	M.S.	2
3.	SHAFT	M.S.	1
4.	ECCENTRIC	M.S.	1
5.	HANDLE	M.S.	1
6.	SCREW	M.S.	1
7.	PIN	M.S.	1

Exercise 6 : Fig. 18.8 shows the details of drilling jig and fixture. Assemble the parts and draw the following :—

(a) Sectional front view **(b) Top view**

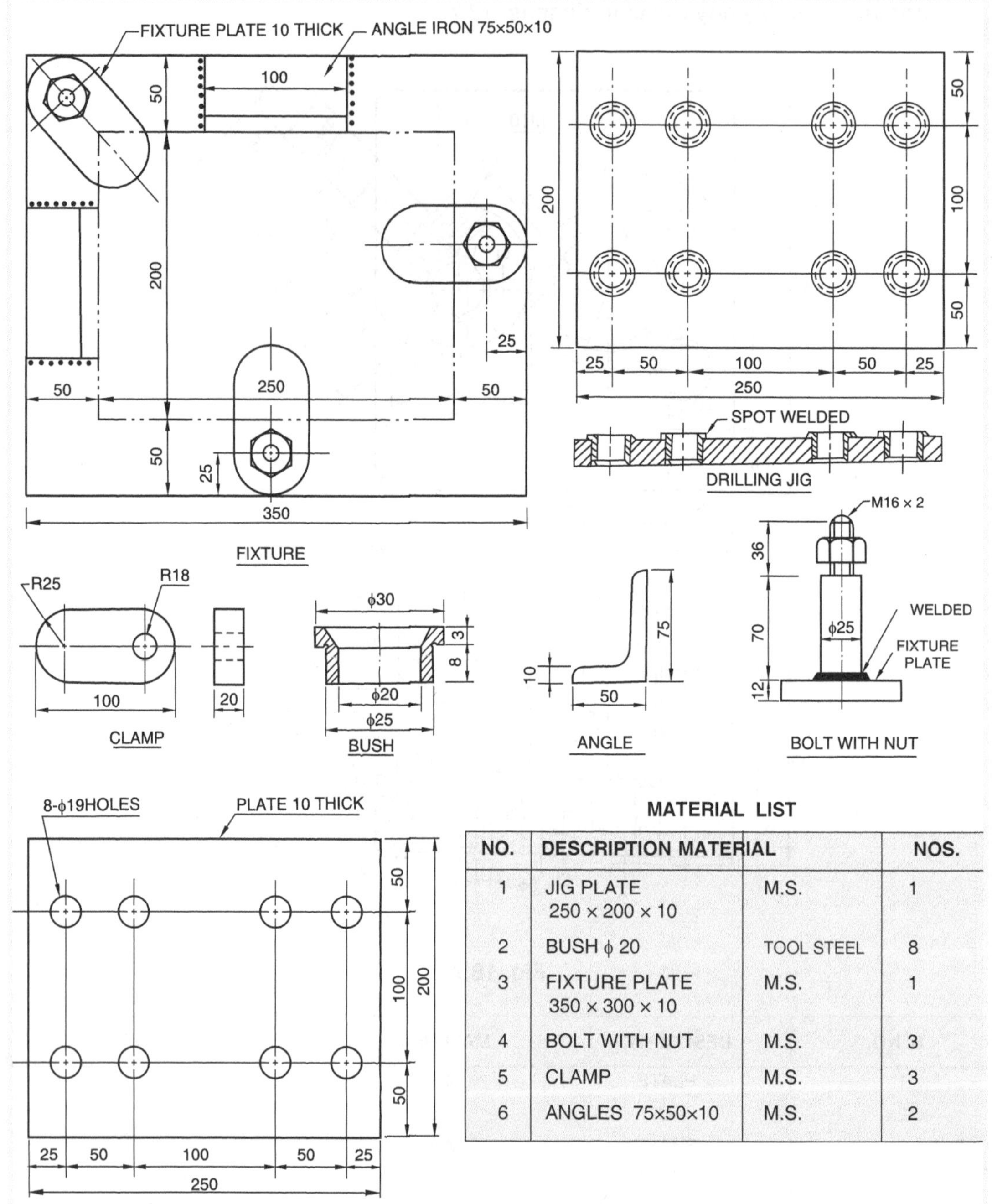

MATERIAL LIST

NO.	DESCRIPTION	MATERIAL	NOS.
1	JIG PLATE 250 × 200 × 10	M.S.	1
2	BUSH ϕ 20	TOOL STEEL	8
3	FIXTURE PLATE 350 × 300 × 10	M.S.	1
4	BOLT WITH NUT	M.S.	3
5	CLAMP	M.S.	3
6	ANGLES 75×50×10	M.S.	2

Fig. 18.8. *Details of drilling jig and fixture*

18.5. MILLING JIG

It is used to hold a special bolt during a machining operation in which a square head is milled on the end of the bolt.

Fig. 18.10 shows the details of a milling jig. Here, the component to be machined is gripped in the collet which is screwed in the body of jig. The body is held to the base by means of clamping ring, but is free to rotates into one of the four possible position. The body can be locked by a pin which engages in mating holes, drilled in the body and clamped ring. The grip of the collet is provided by the wedge effect of its outer surface, when it is forced downward side by the cap. The collet is prevented from rotation by a small key.

The four holes of ϕ 10 in the body with single hole of ϕ 10 in the clamping ring for locating the special bolt in each of the four position necessary for machining the square head.

Problem 2 : Detail drawings of a milling jig are shown in Fig. 18.10. Assemble the details and draw the following views to a scale full size:

(a) Front View - Full section

(b) Top view, with the cap removed.

Solution : For its solution, refer Fig. 18-9.

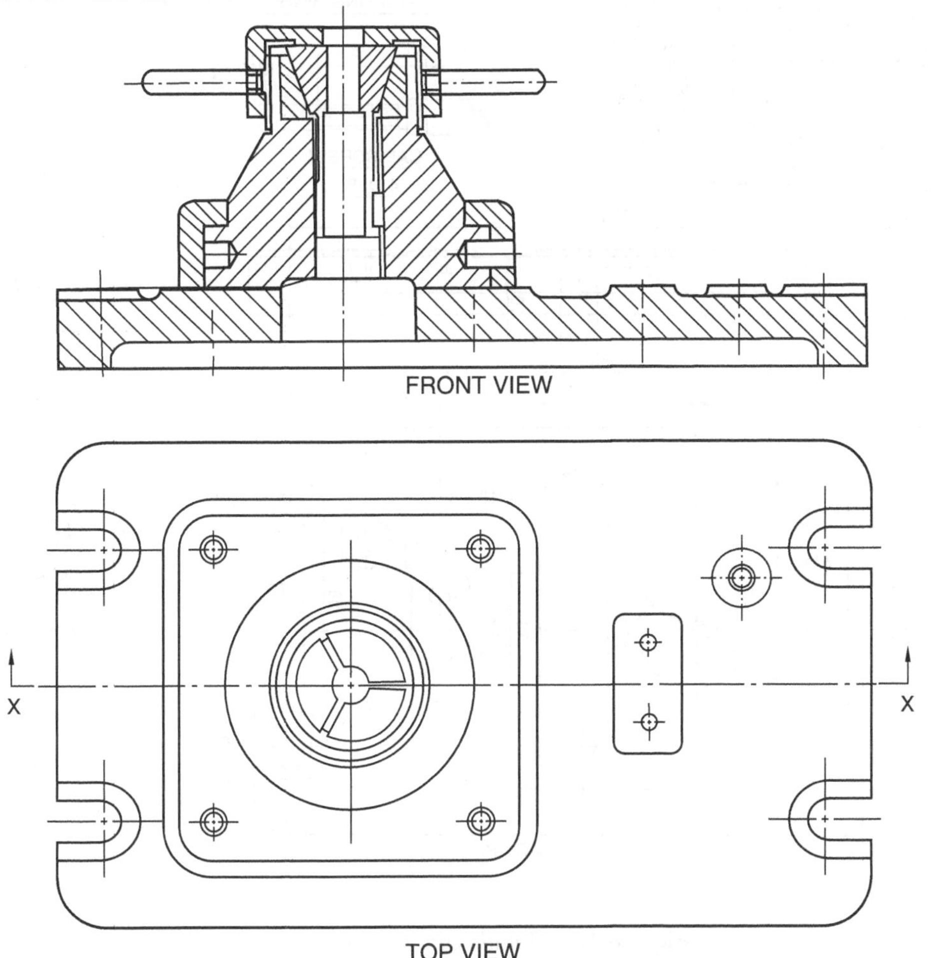

Fig. 18.9. *General arrangement, front view and top view of a milling jig.*

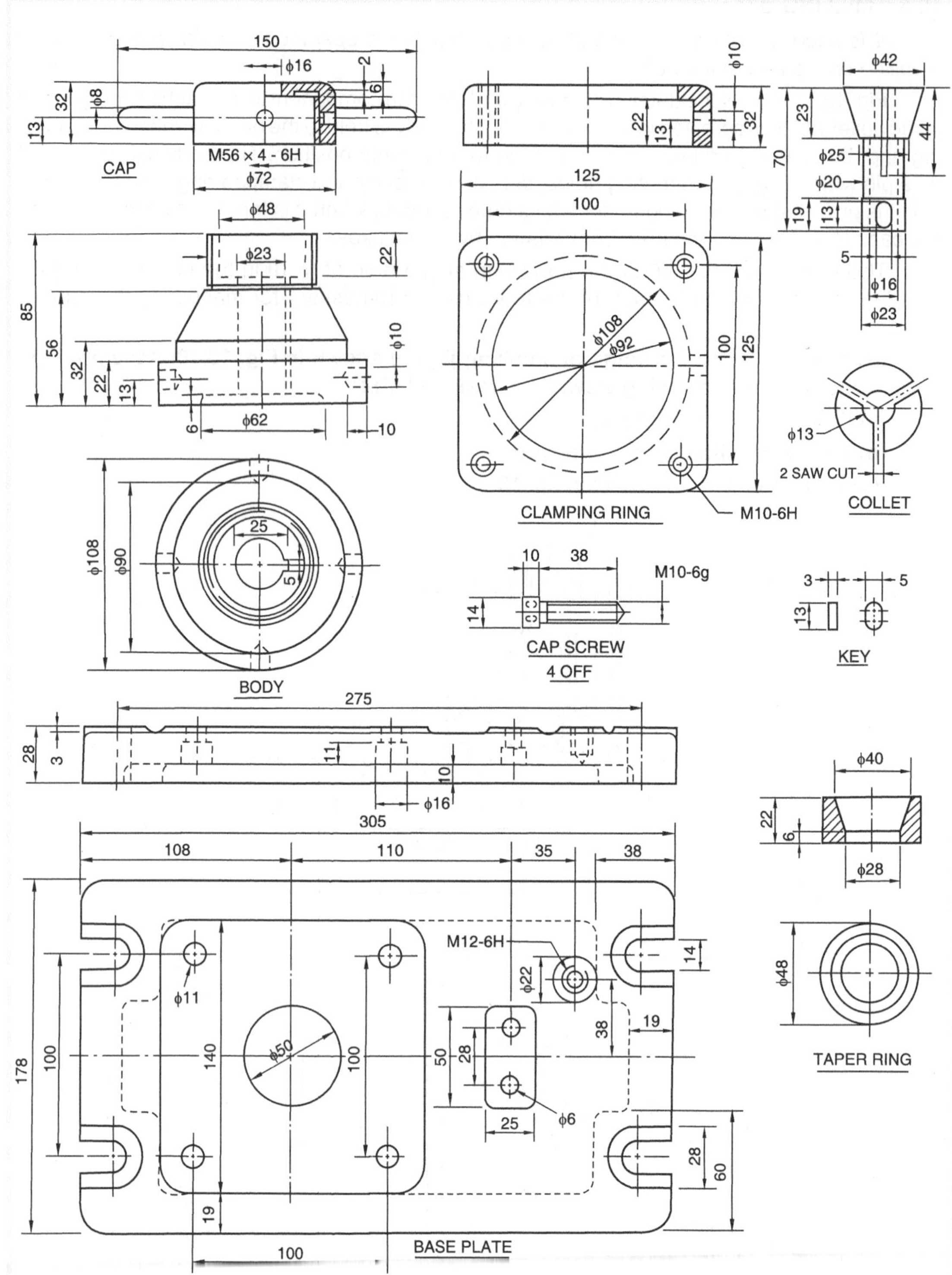

Fig. 18.10. *Details of a milling jig*

Exercise 7 : Fig. 18.11 shows the details of a tumble jig. Assemble the part and draw the following views

(a) Front view – Full section

(b) Side view

(c) Top view

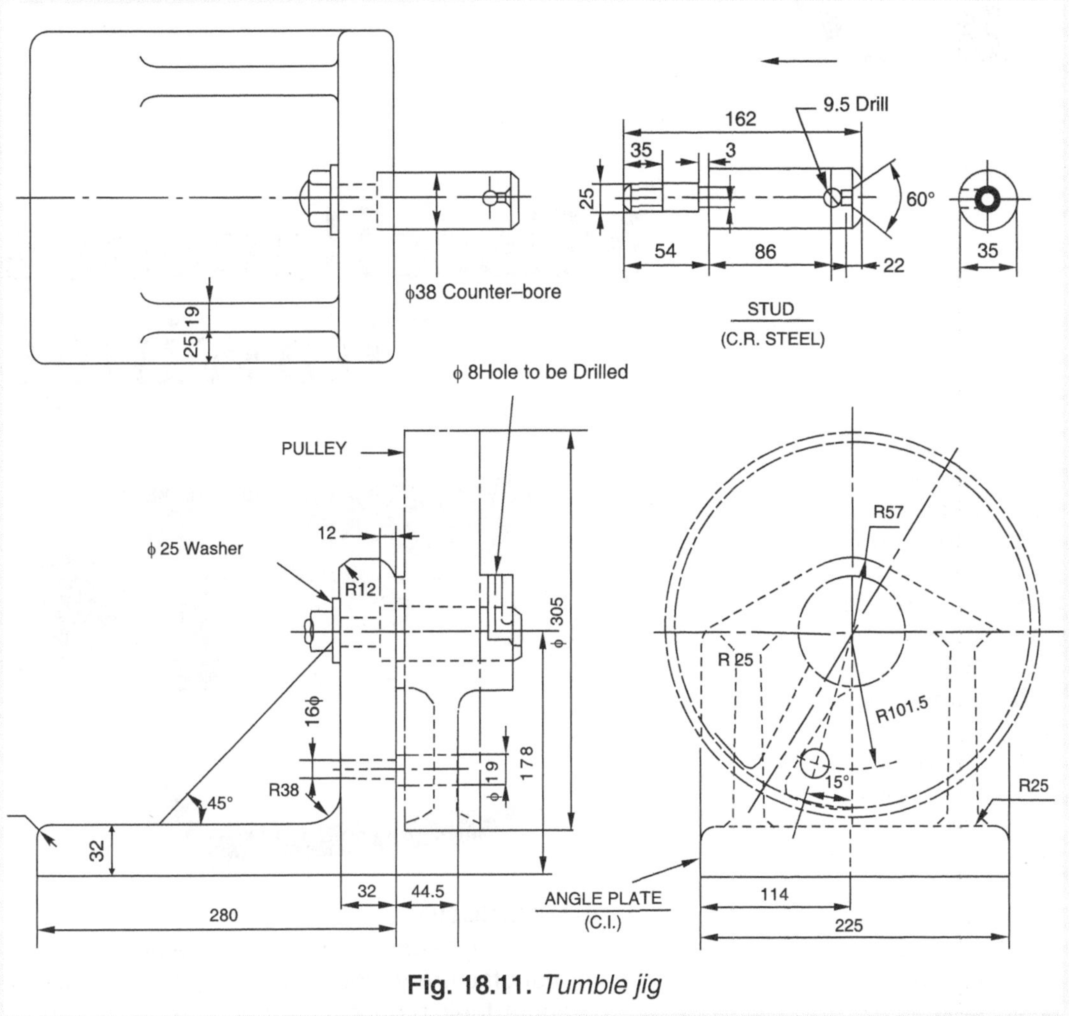

Fig. 18.11. *Tumble jig*

QUESTIONS FOR SELF EXAMINATION

1. Define jigs and fixtures. What are the difference between jig and fixture ?
2. Where a drilling jig is used ? What are the various parts of a drilling jig?
3. What is the function of a milling jig?

CHAPTER

19

Miscellaneous Drawings

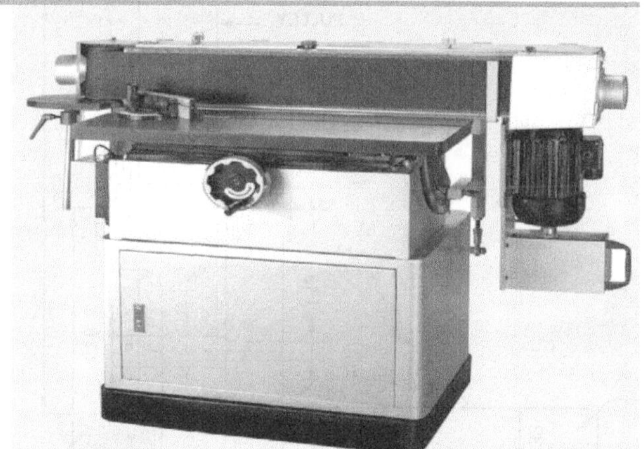

INTRODUCTION

In engineering practice, we are often required to manufacture various machine parts. In order to execute their production work, it is necessary to prepare the working drawing of these parts. With the development of new technology, a large number of machines or machine parts have come up. But we will confine our discussion to the selected machine parts *e.g. parts of a lathe machine, shaper, screw jack, vices, measuring stand, bevel gear junction box, crane hook, clutch, quick change drill holder, water pump, hand drill, governor etc.*

Although, the above mentioned machines form only a small percentage of the existing machines, yet the application of principle of these drawings may help the reader to make the drawing of any other machine or its parts.

in this chapter, we will focus our attention of the working drawing of above mentioned machine parts and machines.

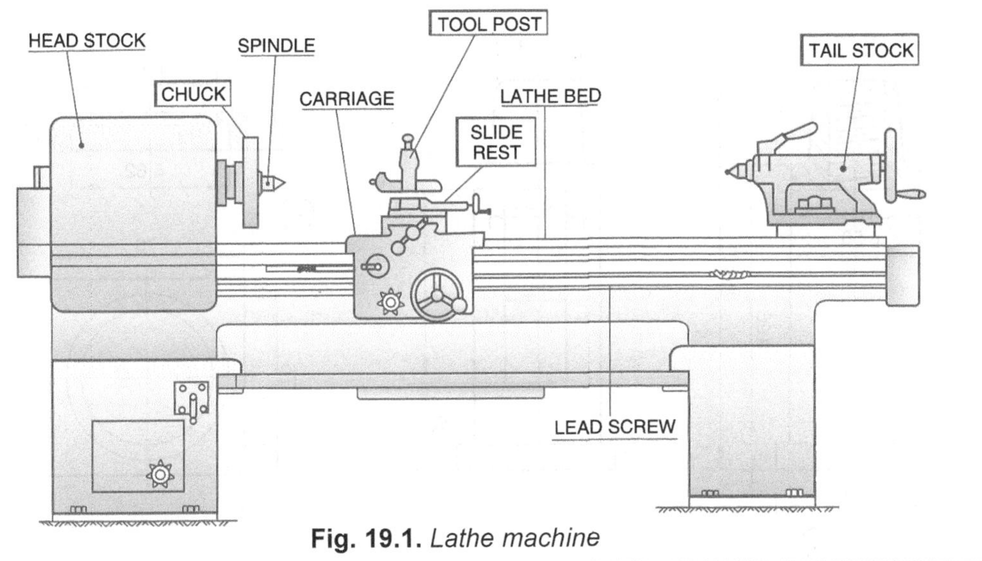

Fig. 19.1. *Lathe machine*

Fig. 19.1 shows a line sketch of a lathe. Some of its important parts are discussed as:

19.1. TOOL POST

The device which is used to hold the cutting tools on a lathe machine is called tool post (see Fig. 19.1).

Fig. 19.2 shows details of the pillar-type tool post. It is mainly used in small size lathes and only one cutting tool can be fixed in it. The entire unit is fixed on the compound rest of the lathe carriage.

The body of the tool post is of a circular pillar with a vertical slot. The tool holder (not shown) is kept in the slot and is then clamped in position by means of a square headed clamping screw placed through the pillar head.

The body can slide in T-slot, machined on the top surface of the compound rest (not shown) and can be swivelled about its vertical axes to adjust the position of the cutting tool in a horizontal plane. The wedge is a boat shaped piece placed in the vertical slot just below the tool holder. By sliding the wedge on the spherical surface ring, the height of the tool tip can be adjusted. When assembled, the square block will be resting against the inner surface of the T–slot in the compound rest and the ring will be resting on the top of the compound rest.

Exercise 1 : Fig. 19.2 shows the details of a tool post. Assemble all the parts and draw the following views:

(a) Sectional front view

(b) Right side view

(c) Top view

Prepare a parts list and materials list also.

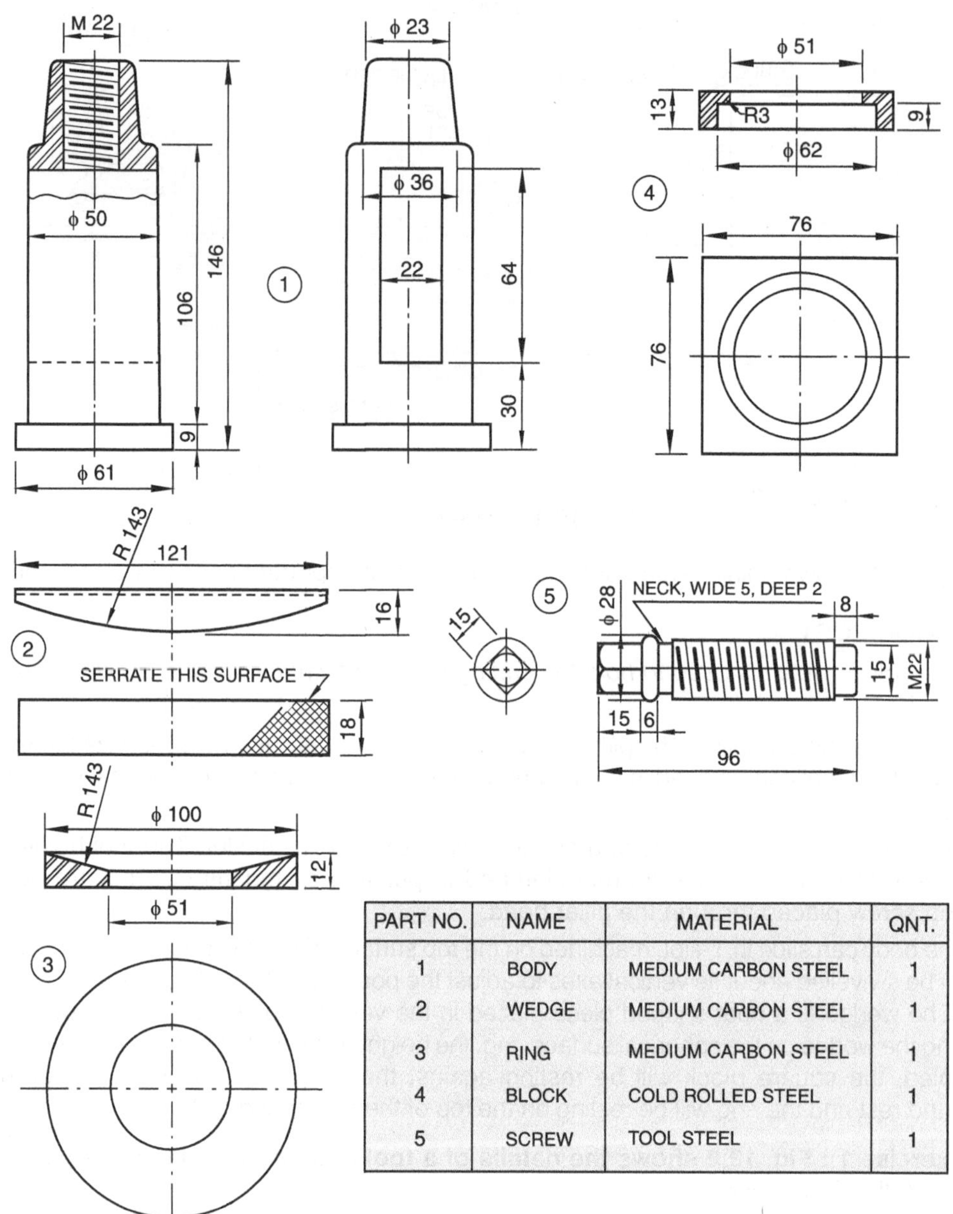

PART NO.	NAME	MATERIAL	QNT.
1	BODY	MEDIUM CARBON STEEL	1
2	WEDGE	MEDIUM CARBON STEEL	1
3	RING	MEDIUM CARBON STEEL	1
4	BLOCK	COLD ROLLED STEEL	1
5	SCREW	TOOL STEEL	1

Fig. 19.2. *Tool post for small lathe*

Exercise 2: Fig. 19.3 shows two views of a tool post. Draw the following view 1:1 size scale : —

(a) Front view

(b) Sectional side view

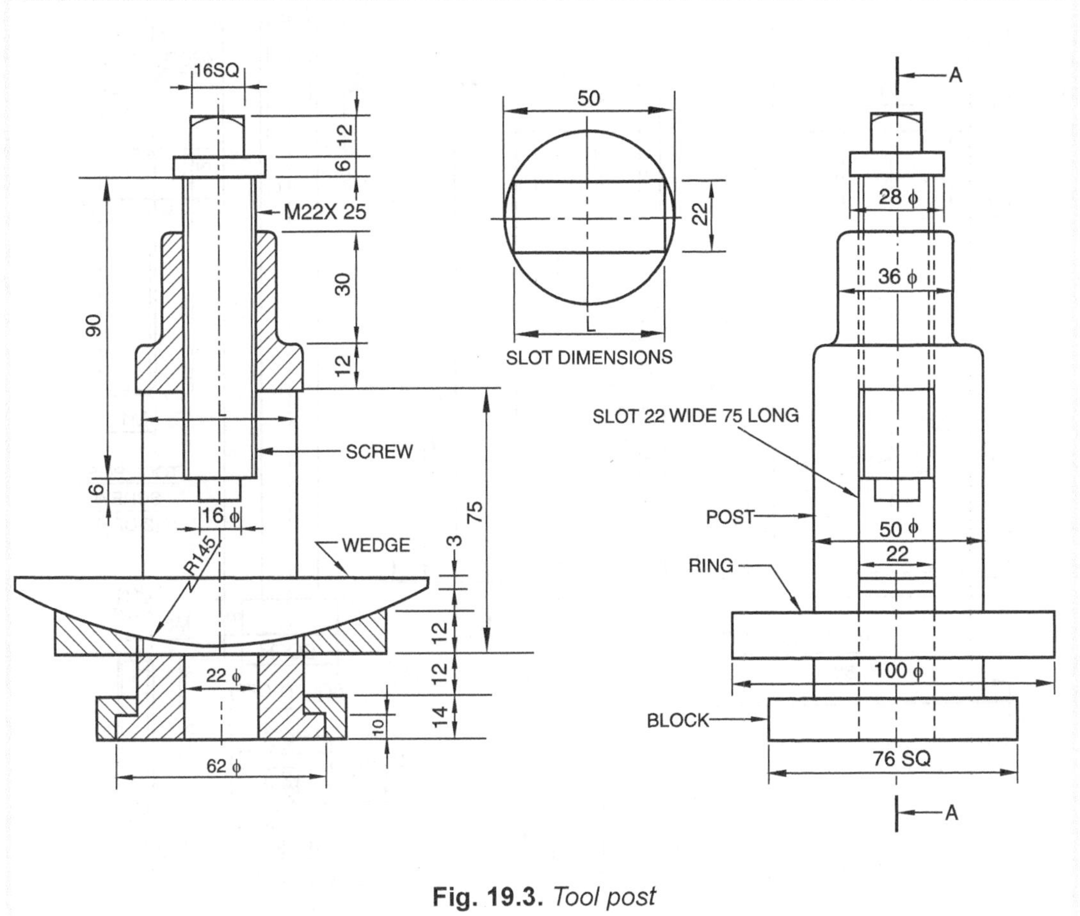

Fig. 19.3. *Tool post*

Exercise 3: Fig. 19.4 shows the details of a tool post. Assemble the parts and draw the following views to full size scale:

(a) Front view – in section

(b) Right side view, and

(c) Top view

Fig. 19.5 shows tool holder in which four different tools can be mounted at a time. The tools are held in position by the four set screws. To facilitate the use of any one of the tools at a time, the tool holder may be rotated and clamped in any position.

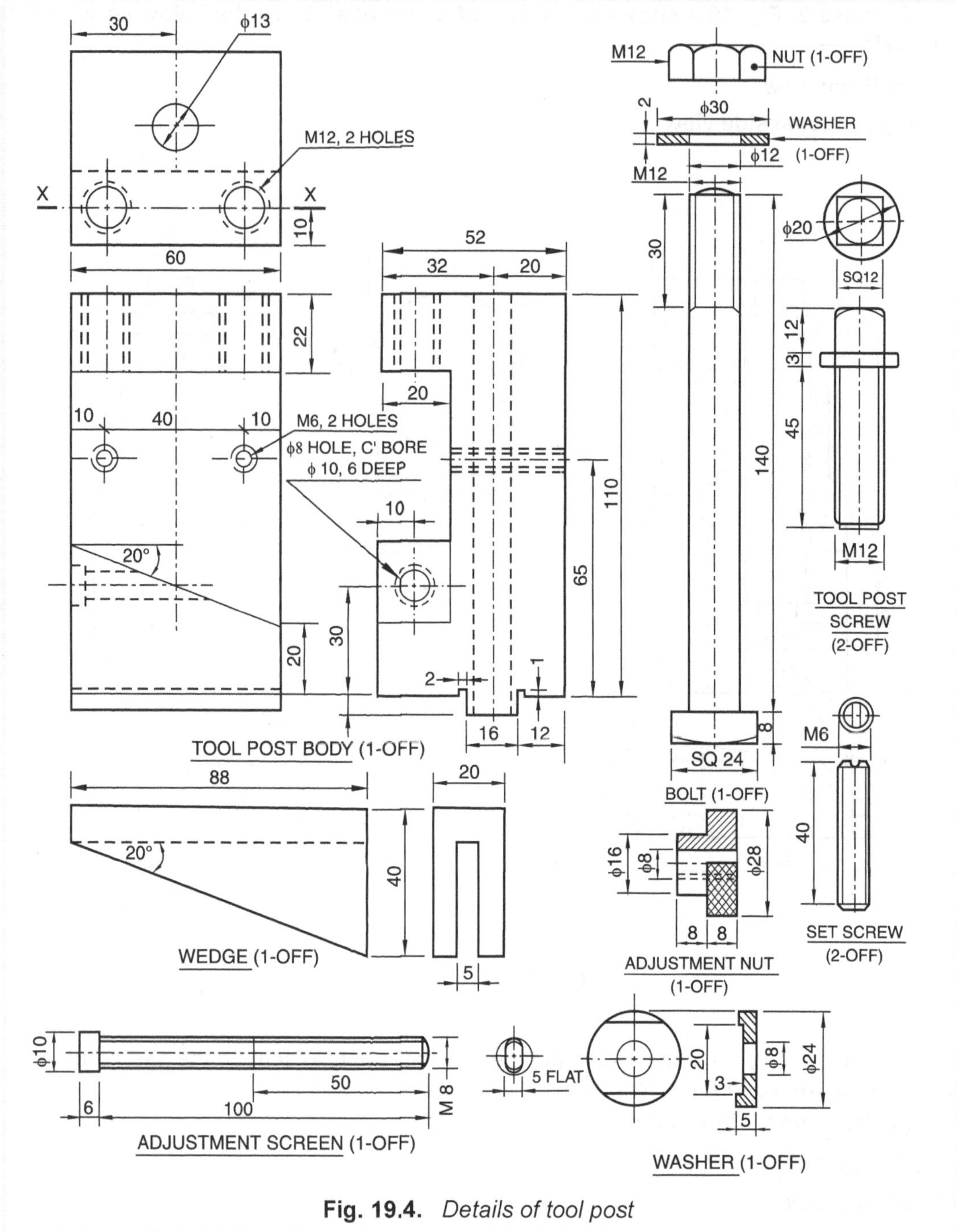

Fig. 19.4. *Details of tool post*

Exercise 4: Fig. 19.5 shows details of a tool holder. Assemble the details and draw the following views to a full size scale:

(a) Front view – full in section.

(b) top view – outside and

(c) Right side view

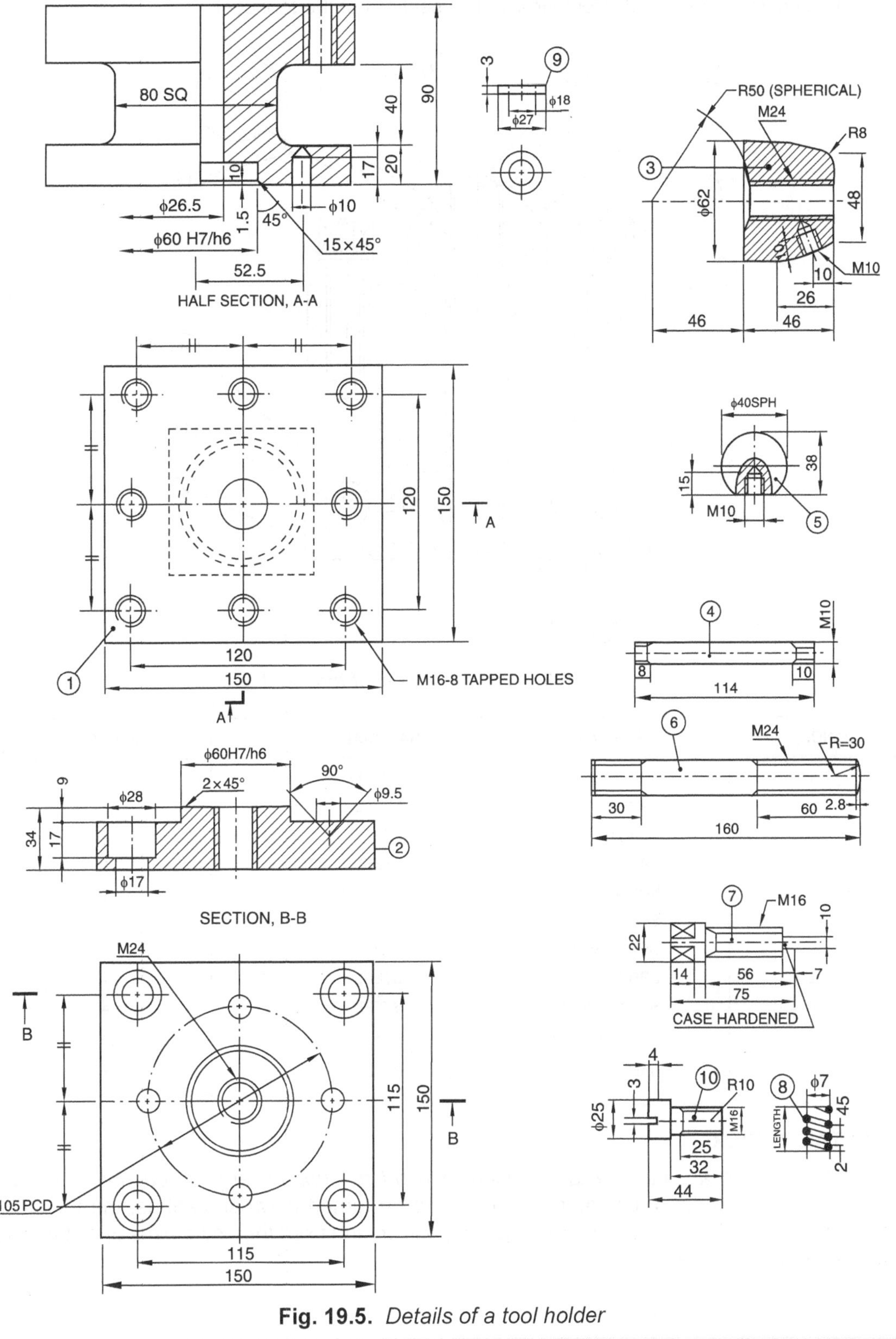

Fig. 19.5. *Details of a tool holder*

Fig. 19.6 shows the assembly front view (full in section) of the tool post.

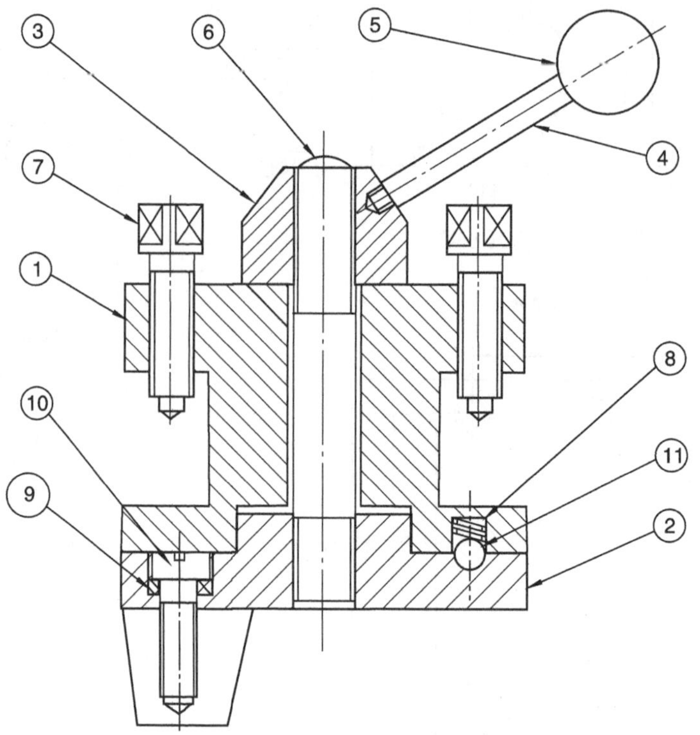

Fig. 19.6. *Assembly view of tool post*

S.NO.	PART NAME	MATERIAL	NO.OFF
1	TOOL HOLDER	STEEL	1
2	BASE PLATE	STEEL	1
3	MOVER	STEEL	1
4	HANDLE	STEEL	1
5	KNOB	EBONITE	1
6	STUD	M.S.	1
7	SET SCREW	M.S.	8
8	SPRING	STEEL	1
9	SPRING WASHER	STEEL	4
10	GROOVE SCREW	M.S.	4
11	BALL	STEEL	1

19.2. TAIL STOCK

Tail stock is a part of a lathe machine used to support lengthy job during turning operation. When works of different lengths are to be turned between centres, the tail stock can be moved on the lathe bed to the required position and fixed thereby means of a clamping bolt (see Fig. 19.1) of a lathe.

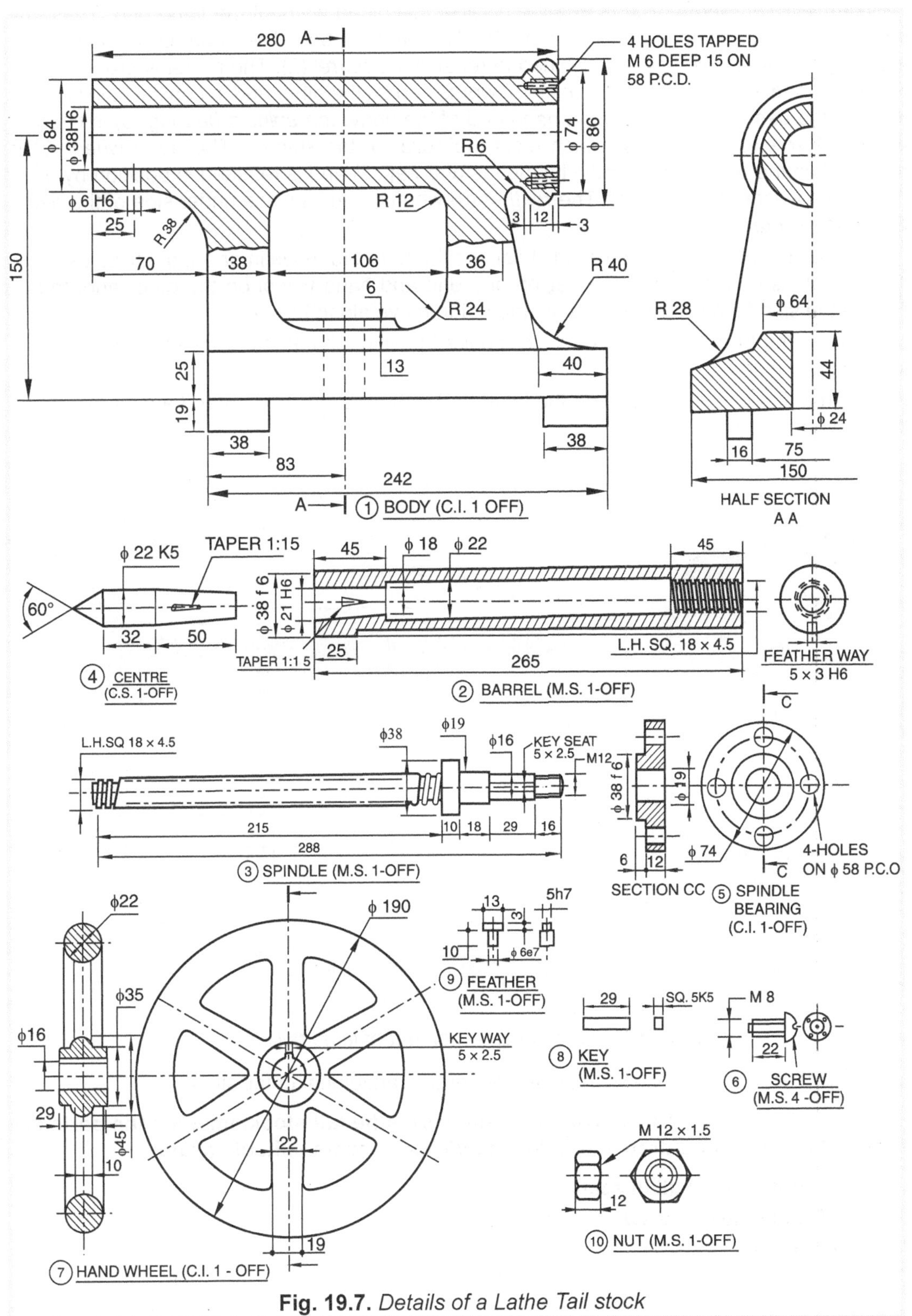

Fig. 19.7. *Details of a Lathe Tail stock*

Fig. 19.7 shows the various parts of a tail stock. The base of *body* (1) is planed and is provided with four feet which fit into the parallel machined ways of the lathe bed. Thus, the tail stock is constrained to move in a straight line on lathe bed. The upper cylindrical part of the body is cast hollow and is machined to receive the *barrel* (2). The barrel is provided with threads at one end in which the *spindle* (3) works and on the other end, a conical hole is made to carry the *centre* (4). On the right hand end of the body, the *spindle bearing* (5) is fitted by means of *screws* (6), against which rests the collar of the spindle. The *hand wheel* (7) is mounted on the end of the spindle by means of a *key* (8) and is retained in position by a *nut* (10) and a washer (not shown). The *feather* (9) when put in its place under side of the barrel prevents its rotation.

When the hand wheel is turned, it causes the barrel to move in or out of tail stock body. The spindle is confined by the collar on one end and hand wheel on the other end, thus it rotates the spindle only about the axis and not to move along the axis.

When the barrel is adjusted to the required position, it can be clamped by means of a locking lever (see Fig. 19.9).

The entire unit is clamped on the lathe bed with a clamping plate and a square headed bolt and nut.

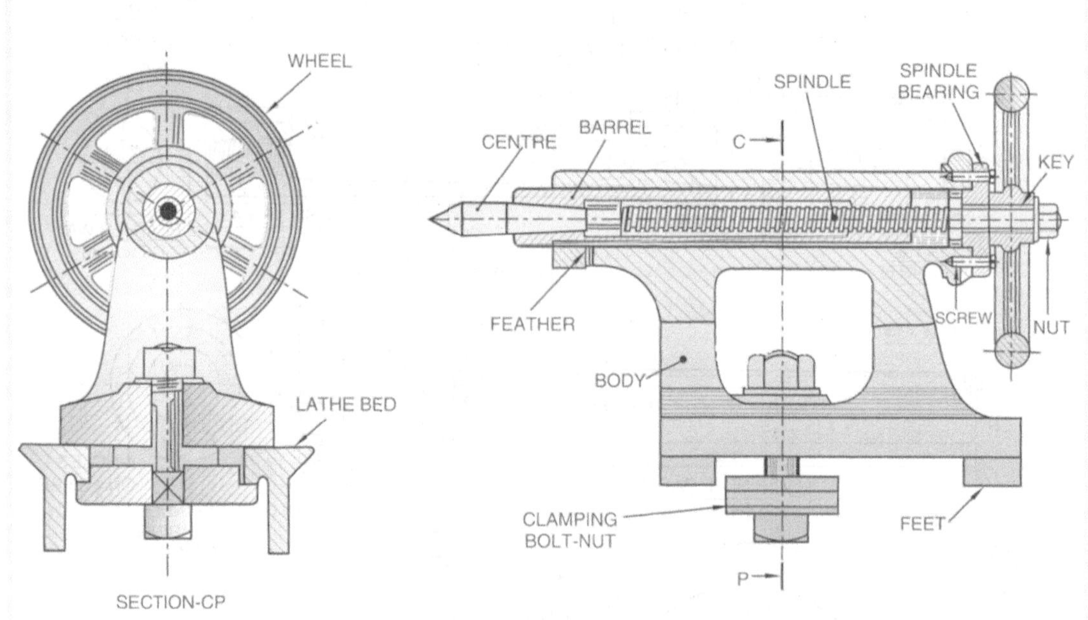

Fig. 19.8. *General arrangement of lathe tail stock*

Fig. 19.8 shows two views about the general arrangement of tail stock.

Exercise 5 : Fig. 19.7 shows the details of lathe tail stock. Imagine the parts assembled together and draw the following views to some suitable scale:

(a) Front view – Full in section

(b) Right side end view

(c) Top view

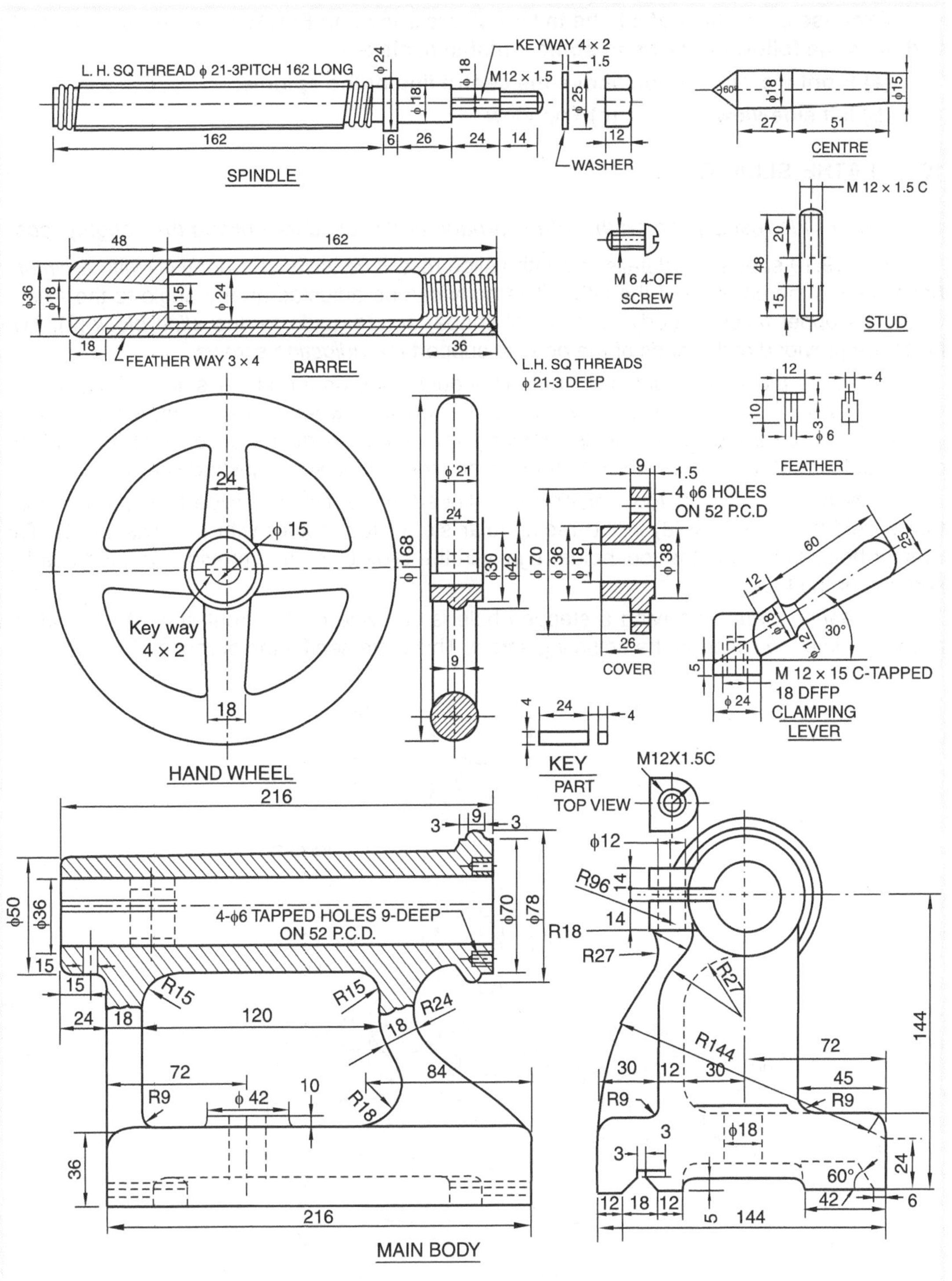

Fig. 19.9. *Details of Tail Stock*

Exercise 6 : Details of a lathe tail stock are shown in Fig. 19.9. Assemble the parts and draw the following views to some suitable scale :—

(a) Front view in section above the centre line of the spindle.

(b) left side view (c) Top view.

19.3. LATHE SLIDE REST

Lathe slide rest is fixed on the lathe carriage and is used for holding the lengthy jobs.

Fig. 19.10 shows the details of a lathe slide rest. It consists of a *body* (1) with a *circular base* fixed to the carriage of the lathe. The body can be adjusted at any angle to the axis of work. The upper faces of body (of channel shape) act as guide for the *slide block* (2). Two holes are provided at the ends of the body to support the *adjusting screws.*

The *slide block* (2) is designed to form a guide way on the under side and a T-slot on the upper side. A *wearing strip* (5) which can be replaced when it wears out, is fitted on the underside of lathe block. A hole is provided in the centre of the slide block to hold the cylindrical part of the *nut* (7) which connects the slide block with adjusting screws.

A cylindrical collar of *tool holder* (3) is slipped in the T-slot of slide block. The cutting tool is held by the *screw* (8) in the rectangular slot of tool holder with the *washer* (4). The washer helps in holding the tool holder tightly in the slide block and provides a good bearing surface for the tool.

A *bearing plate* (6) having a stepped hole is screwed on the front side of the body to accommodate the collar on the adjusting screw. This plate serves the purpose of a bearing.

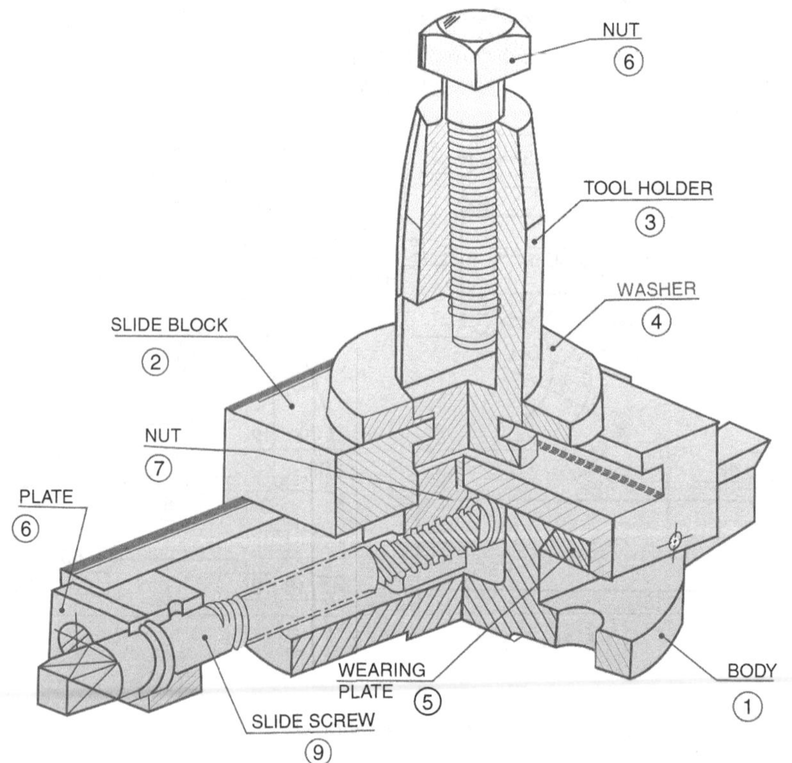

Fig. 19.11. *General arrangement (Assembly) of the lathe slide rest*

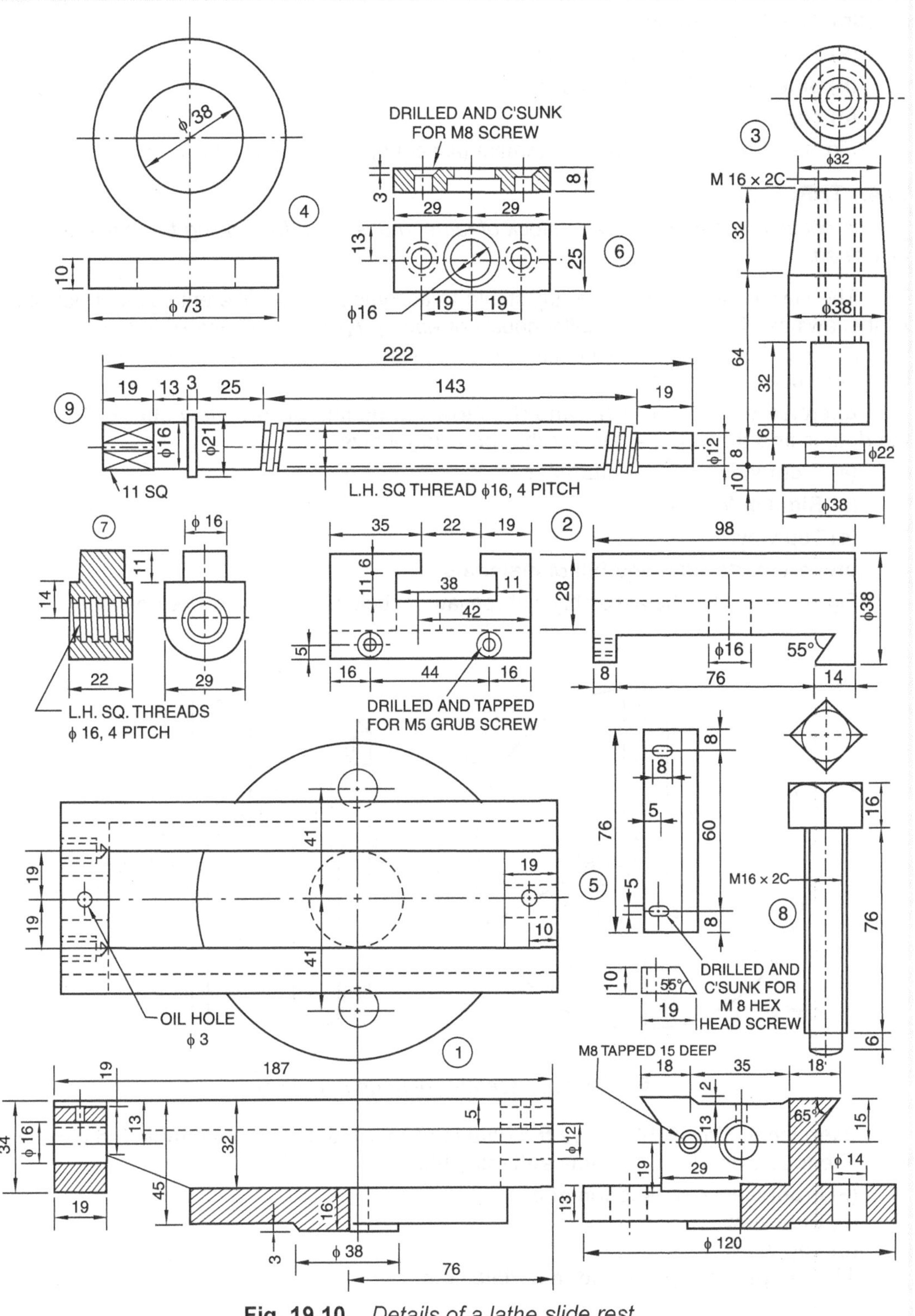

Fig. 19.10. *Details of a lathe slide rest*

Exercise 7 : Fig. 19.10 shows the details of a lathe slide rest. Assemble the parts and draw the following views:

(a) Front view - left half in section

(b) Side view - right half in section

(c) Top view.

Fig. 19.11 shows general arrangement (assembly) of the lathe slide rest

19.4. THREE JAW CHUCK

A chuck is mounted on head stock of lathe and is used for holding the bar stock to be machined.

The jaws of lathe should be adjustable to accomodate the different sizes of bar stock. All the jaws can be adjusted simultaneously or each jaw can be adjustable independently as shown in Fig. 19.12. In this chuck, jaws are adjusted so that the bar stock held in the chuck is properly centred and rotated.

Exercise 8 : Fig. 19.13 (i) and (ii) show the details of three-jaw chuck. Assemble the details and draw to a scale 1:2 the following views:

(a) Front view – in – section at c.

(b) Right side view

(c) Top view

prepare parts – list and bill of materials

Fig. 19.12 shows the assembly of three jaws chuck used on head stock of lathe.

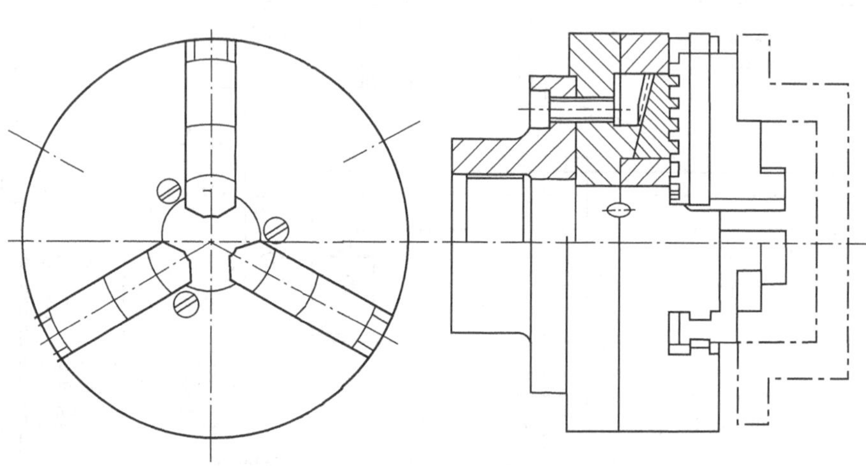

Fig. 19.12. *Assembly of three jaw chuck*

Exercise 9: Fig. 19.14 shows the details of four jaw chuck. Assemble the parts and draw the following views to some suitable side :

(a) Front view - in - section at c

(b) Side view

(c) Top view

Also prepare the parts and materials lists.

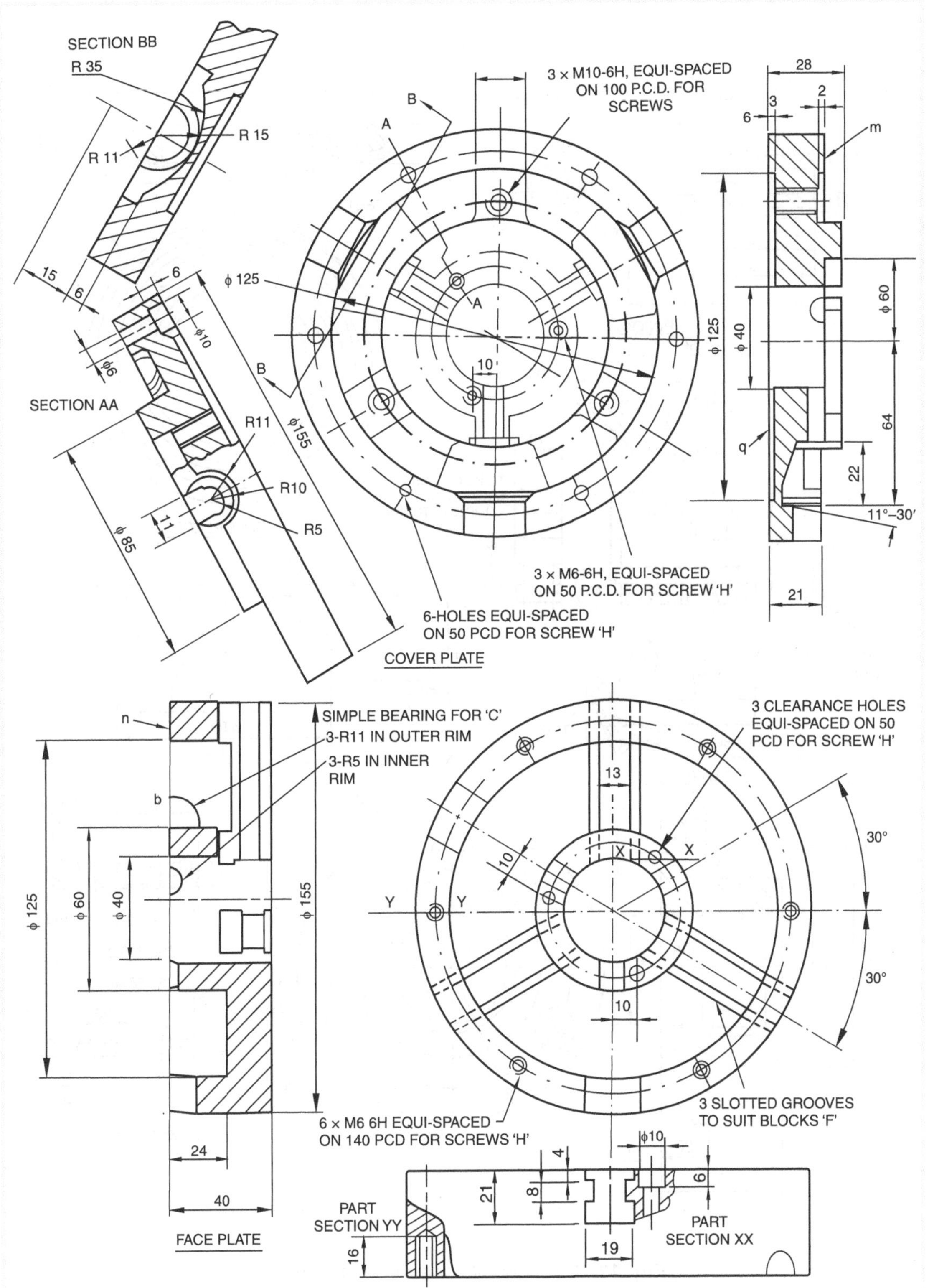

Fig. 19.13.(i) *Details of a three jaw chuck*

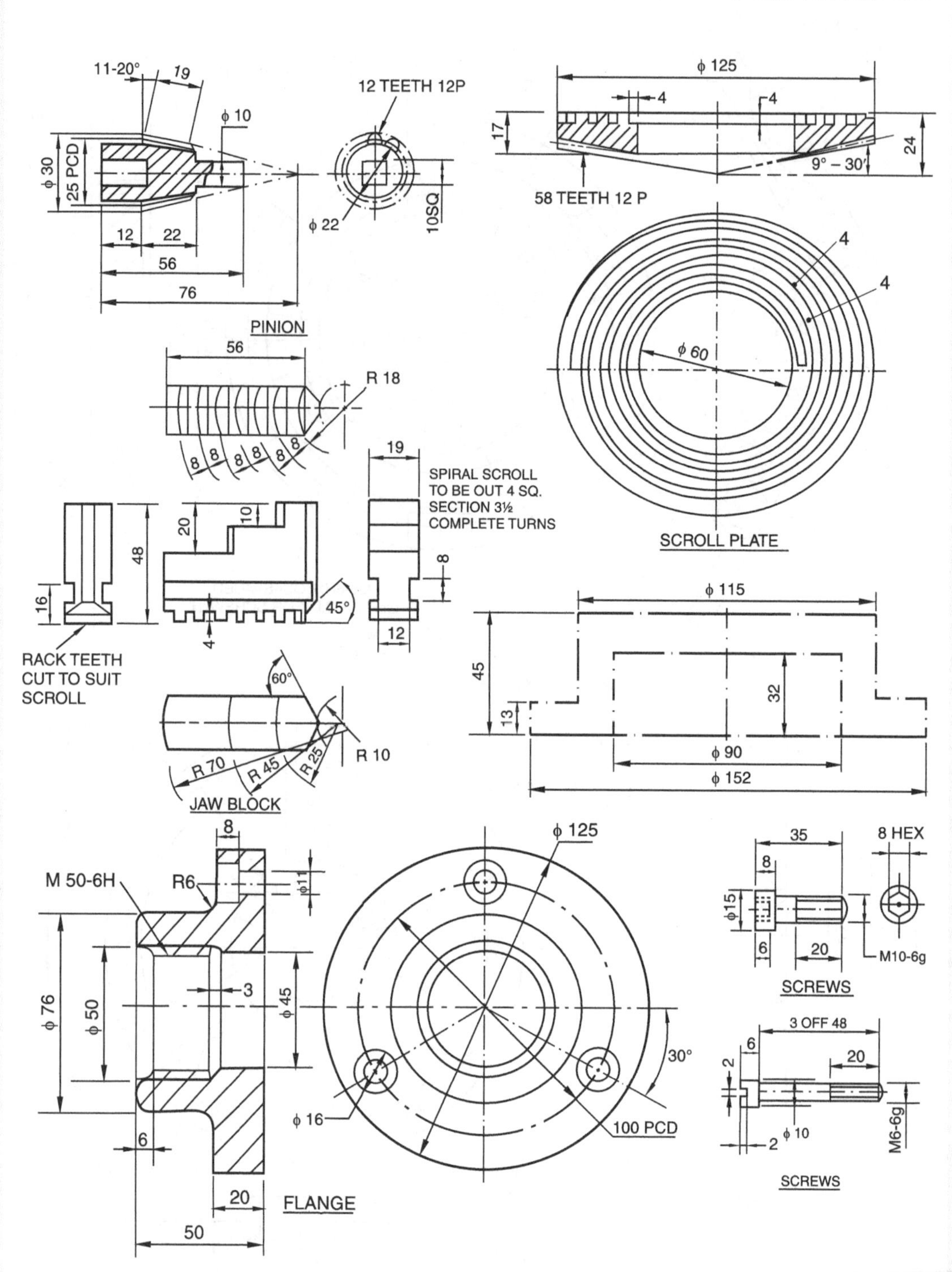

Fig. 19.13.(ii) *Details of a three jaw chuck*

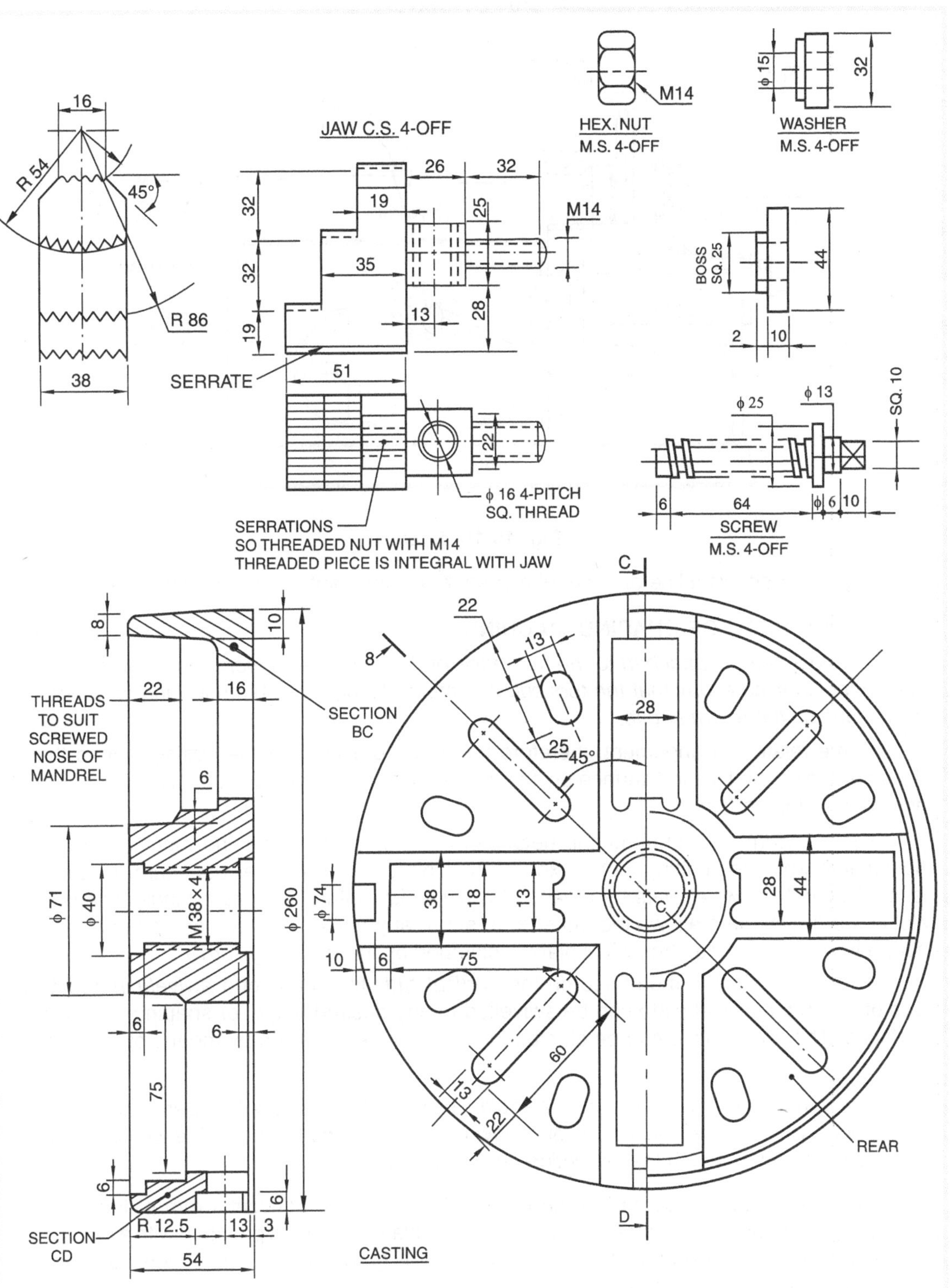

Fig. 19.14. *Details of four jaw chuck*

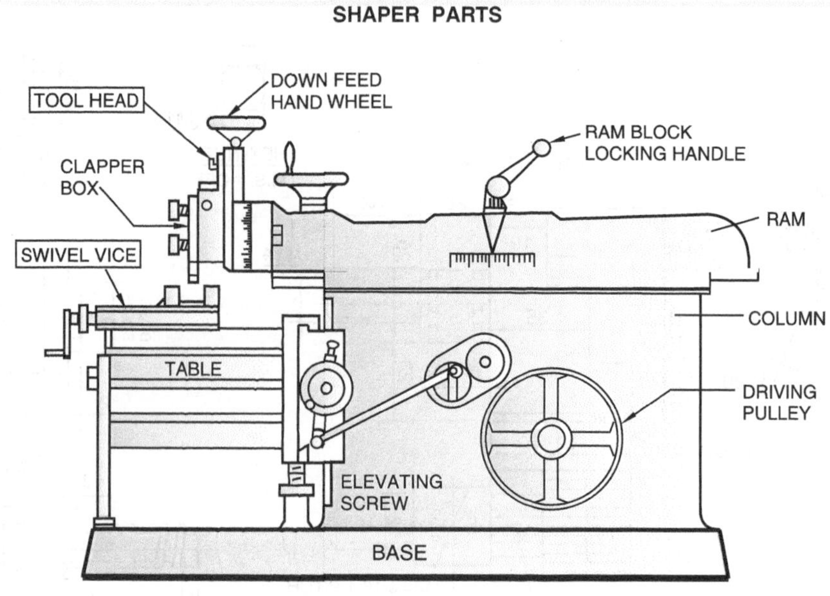

Fig. 19.15. *Shaper*

Fig. 19.15 shows a line diagram of a shaper. Its important parts are discussed below:

19.5. TOOL HEAD OF SHAPING MACHINE

The tool head is attached to the front face of the ram of a shaper. It is used to hold the cutting tool in such a way that the tool cuts the metal during forward stroke and drags on the work piece on the return stroke.

Some relative arrangement, therefore has to be provided for the cutting tool so that it does not rub on the finished surface on its return stroke. This is achieved by having a tool post hinged at its top portion.

Fig. 19.16 (i) and (ii) show the details of a tool head of a shaping machine. *The tool holder* into which the cutting tool is fixed, is positioned in the *drag release plate.* This in turn is fitted on to the *swivel plate* by means of a pin which, infact, acts as a hinge. The curved slot in the swivel plate allows the plate to be set at any angle up to 19° on each side of the centre line. After setting, the swivel plate is clamped by the *clamping screw.*

The swivel plate is then fitted on the *vertical slide* by means of a *swivel pin* about the axis of which the swivel plate can be swivelled to any desired angle for shaping the inclined surfaces. The swivel plate can be clamped in any required position by means of a clamping screw.

The vertical slide is fitted on the *back plate* by means of a *screw.* By operating the screw, the vertical slide can be moved up and down to suit the work piece. The vertical slide is provided with slot to match the projection on the back plate.

The back plate is attached to the *circular flange* (of the horizontal ram of the shaping machine) by means of a pin through the centre of the plate. The pin is provided with a hexagonal nut for tightening and adjustment. The circular flange is graduated so that the tool head can be quickly set to any desired angle.

Exercise 10 : Fig. 19.16 (i) and (ii) shows the details of a tool head of a shaping machine. Assemble the parts and draw the following views to some suitable scale:

(a) Front view - full in section (b) End view and (c) Top view

Give all necessary dimensions and prepare a bill of materials and part list.

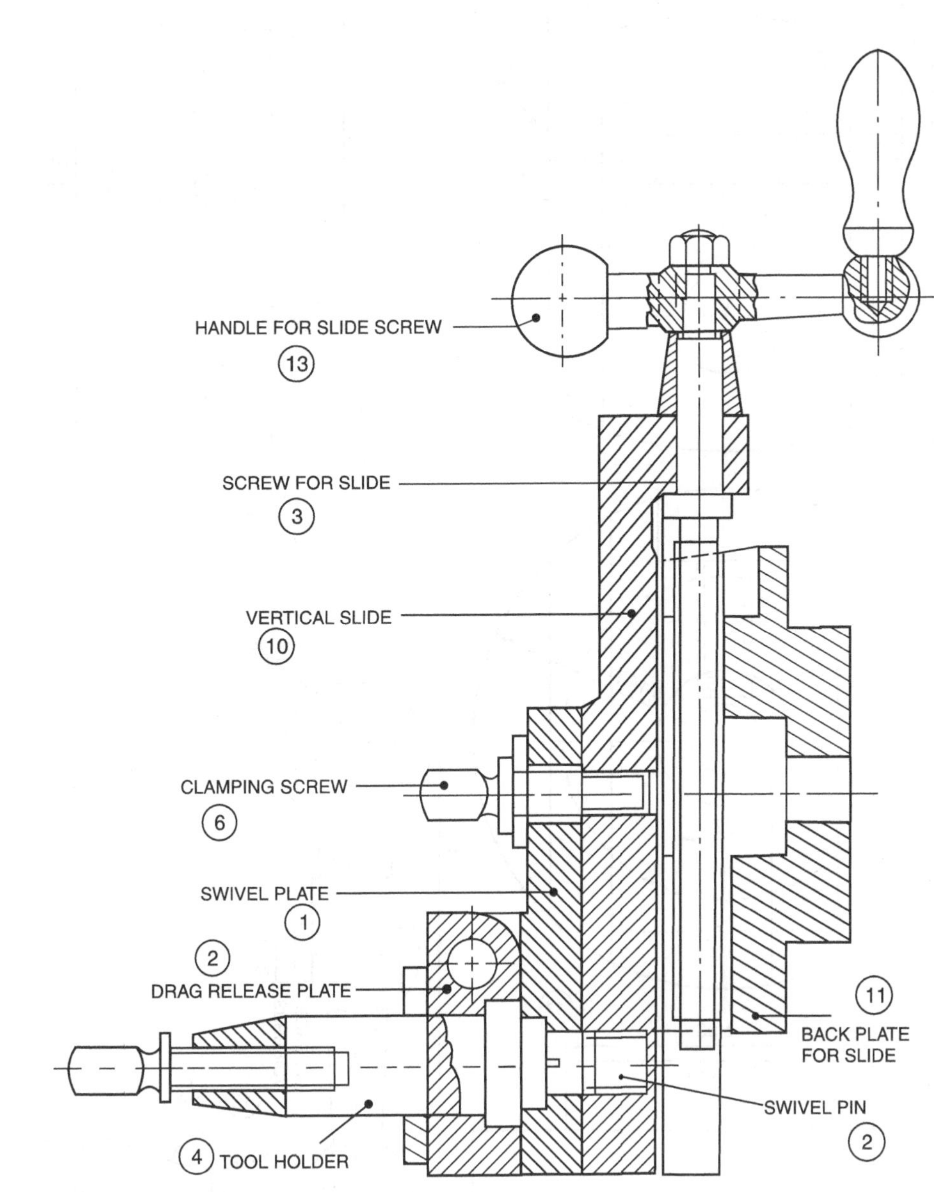

Front view-Full in Section of shaping machine for exercise 10.

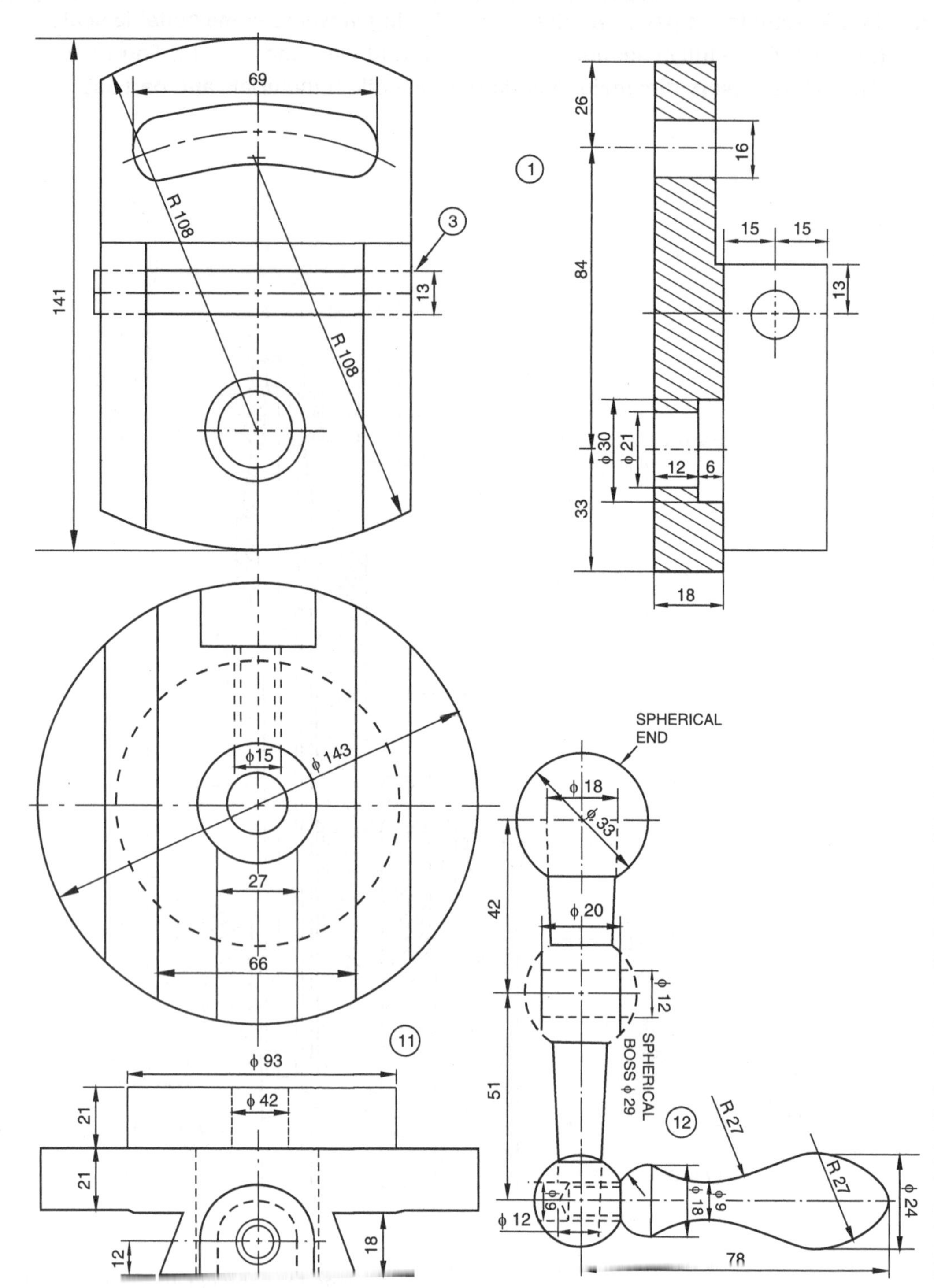

Fig. 19.16.(i) *Details of tool head of a shaping machine*

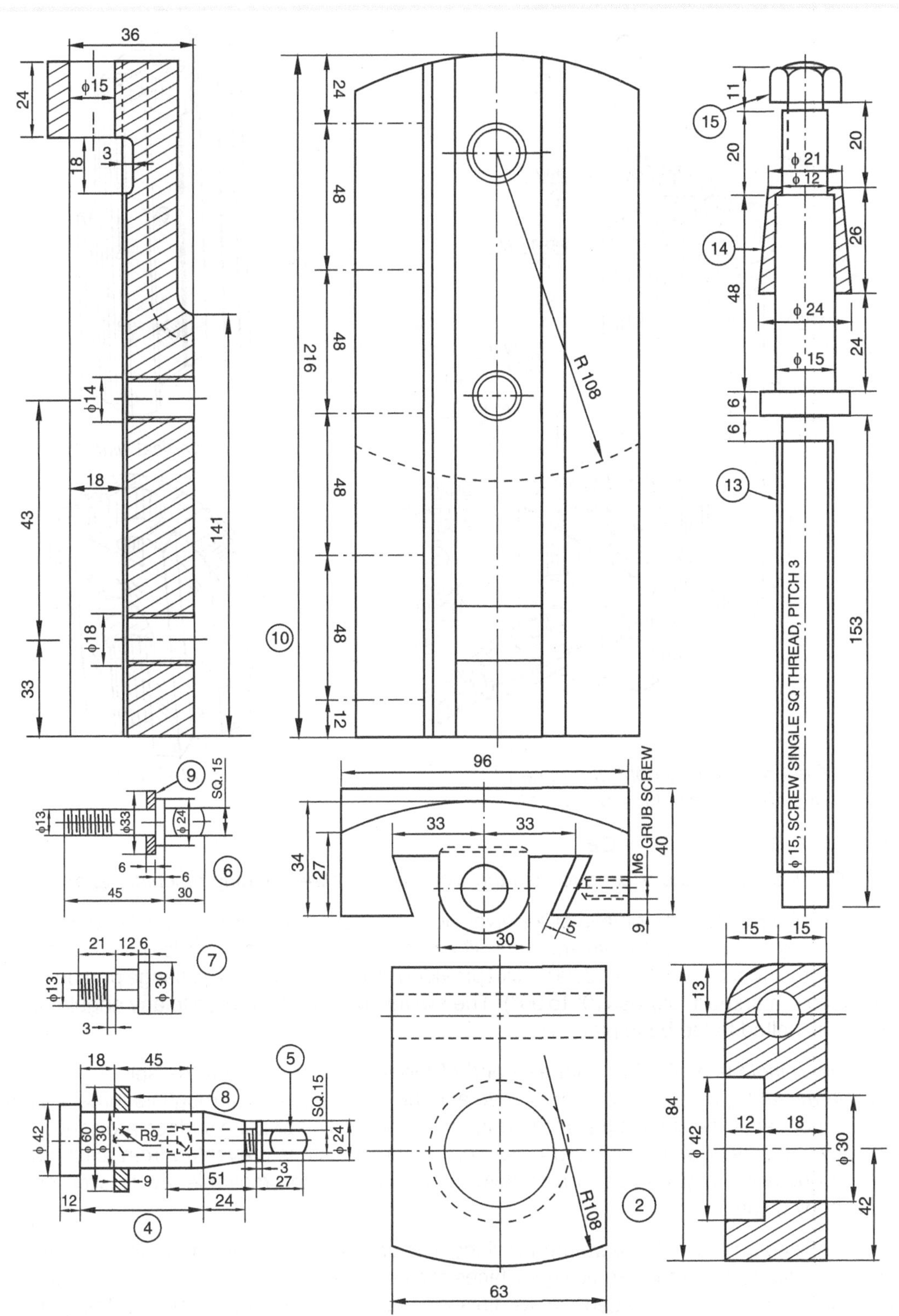

Fig. 19.16.(ii) *Details of tool head of a shaping machine*

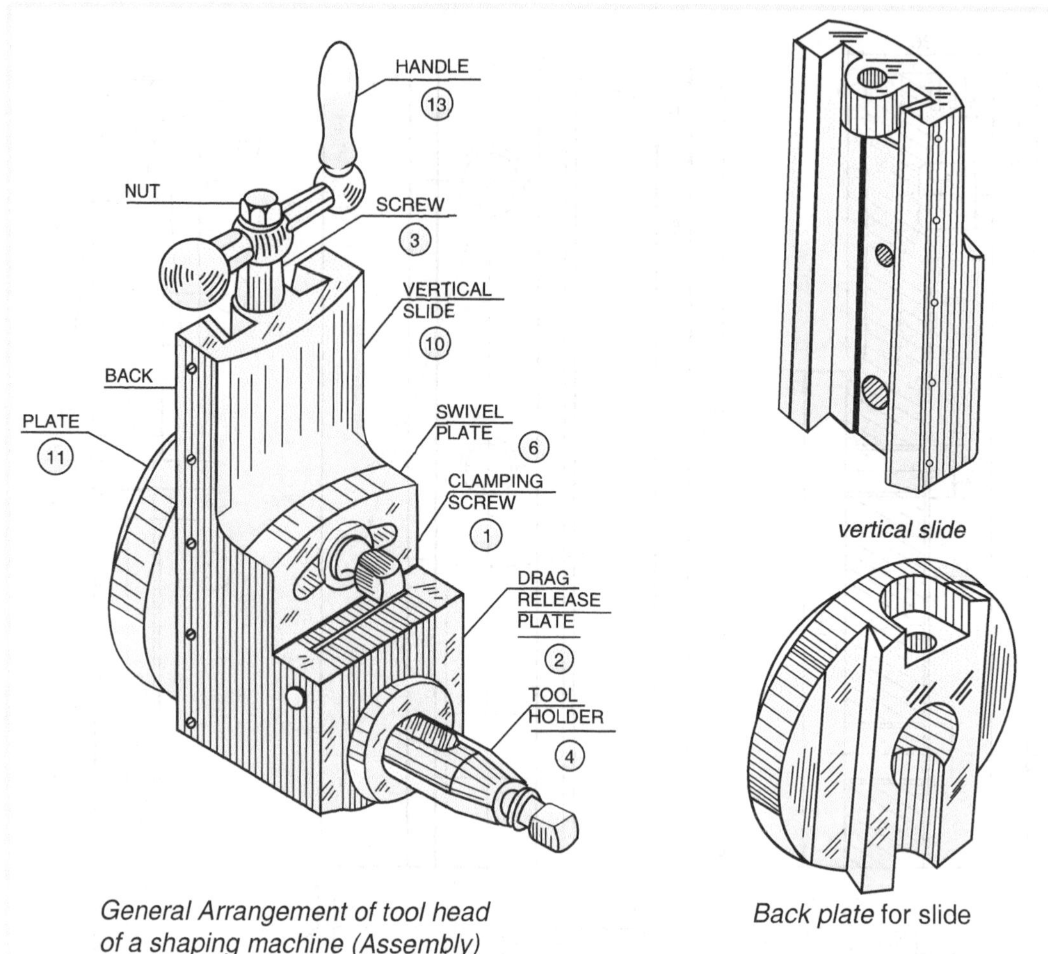

General Arrangement of tool head of a shaping machine (Assembly)

vertical slide

Back plate for slide

19.6. MACHINE SWIVEL VICE

Machine swivel vice is used to hold the work piece on the shaping machine table.

Fig. 19.18 shows the details of a swivel vice. It consists of a *swivel body* fitted on to the base plate by two *clamping bolts* and *nuts.* The heads of the bolts are made of special shape suitable for easy sliding on the circular T-slot provided in the *base plate.* The flange portion of the base plate is marked in digrees (0° to 90°). The swivel body can be turned to any angular position and is bolted to the base plate.

The *spindle* is pushed in from free end of the swivel body. When the spindle is in mid way, the *movable jaw* is inserted from above and the spindle passes through its threaded portion. Subsequently the spindle passes through a corresponding bearing surface provided in the *fixed jaw* and is fitted with a nut and a washer. As the spindle is prevented from axial motion when in operation, it gives rise to a relative motion causing the movable jaw to slide on the swivel body guide.

The fixed and movable jaws are provided with cast iron faces which are fitted by means of flat head *machine screws.* Some times, faces of aluminium or soft material are used to protect the work from getting damaged on its finished surfaces. Fig. 19.17 shows the general arrangement of swivel vice (assembly).

Exercise 11: Fig. 19.18 shows the details of a swivel vice. Assemble the parts and draw the following views to a scale half full size:

(a) Front view – full in section

(b) End view – left side

(c) Top view

Prepare the bill of materials and parts list also.

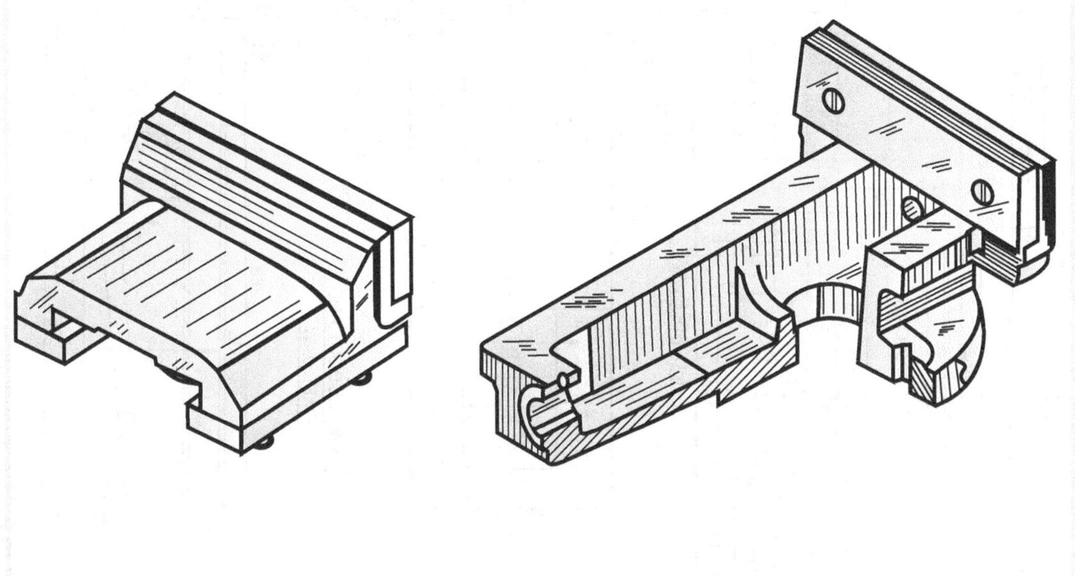

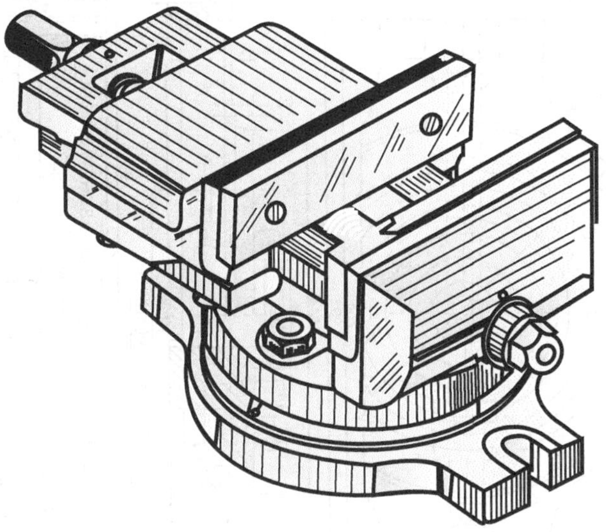

Fig. 19.17. *General arrangement of swivel vice (Assembly)*

Solution :

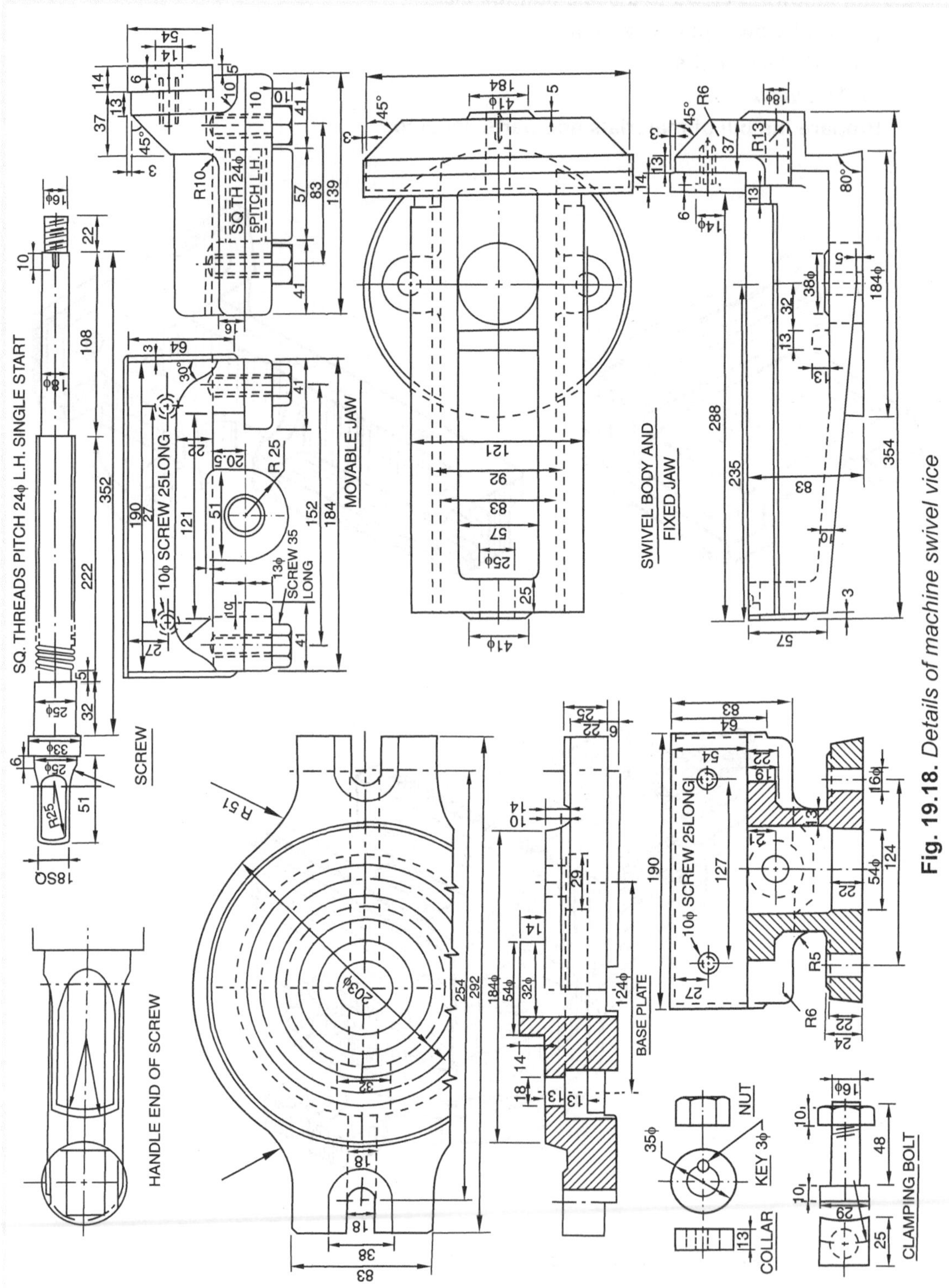

Fig. 19.18. *Details of machine swivel vice*

19.7. SCREW JACK

A device which is used to raise heavy load through some distance by applying comparatively small force is known as screw jack.

Screw jack is used to lift cars, trucks, buses, etc. for the purpose of repairs and to change the damaged wheel.

Fig. 19.19 shows the details as well as the assembly of a screw jack. It consists of a cast *iron body* in which a gun metal *nut is fitted.* A mild steel square threaded *spindle* having knurled head operates in the nut. A mild steel *tommy bar* is inserted in the hole of the spindle to give it rotation. The end of the tommy bar is knurled to have better grip. A steel cap is loosely fitted on the head of the spindle with the help of a *washer* and *set screw.*

When the spindle is turned once by the tommy bar, the load is raised vertically equal to the pitch of the thread.

Fig. 19.20 shows the general arrangement of screw jack assembly.

Important notes : *1. When the spindle is rotated to raise the load, the cup remains stationary due to loose fitting. If the cup is fixed with the spindle, the cup as well as load will rotate with the spindle.*

2. The upper face of the cup is serrated in order to achieve a better grip.

3. The spindle has square threads due to inherent property of more power transmission.

Exercise 12 : Fig. 19.19 shows the details of a screw jack. Draw the following views of the assembly to some suitable scale:

(a) Front view - right half in section (b) Top view

Also prepare a part list and bill of materials

Exercise 13 : Fig. 19.21 shows details of a screw jack. Draw the following views to some suitable scale :—

(a) Front view – half in section (b) Top view

19.8. MACHINE VICE

Machine vice is used for holding the work piece to facilitate machining operations such as drilling, shaping, milling etc.

Fig. 19.22 shows three views of a machine vice. The *jaw plate* (3) is fixed to sliding jaw (2) by means of *screw* (4). A V-cut is made in the fixed jaw to facilitate the holding of round jobs in vertical position. *Fixed jaw* (5) is attached to the base by means of two *screws* (6), but some times it can be integerated with the base of the vice. *Guide plates* (7) are attached to the underside of the sliding jaw by means of *screw* (8). Guide plates also help to prevent the lifting of the jaw during operation.

Spindle (9) is screwed at one end of the sliding jaw which carries handle *collar* (13) at the other end. The spindle is properly supported by *spindle bearing* (10) Which is fitted to the base by *screw* (6). *Collar* (11) which is fitted to the spindle by *set screw* (12), is placed between the handle collar and the spindle bearing.

The machine vice is operated by means of a rod (not shown) placed through the holes provided in the handle collar.

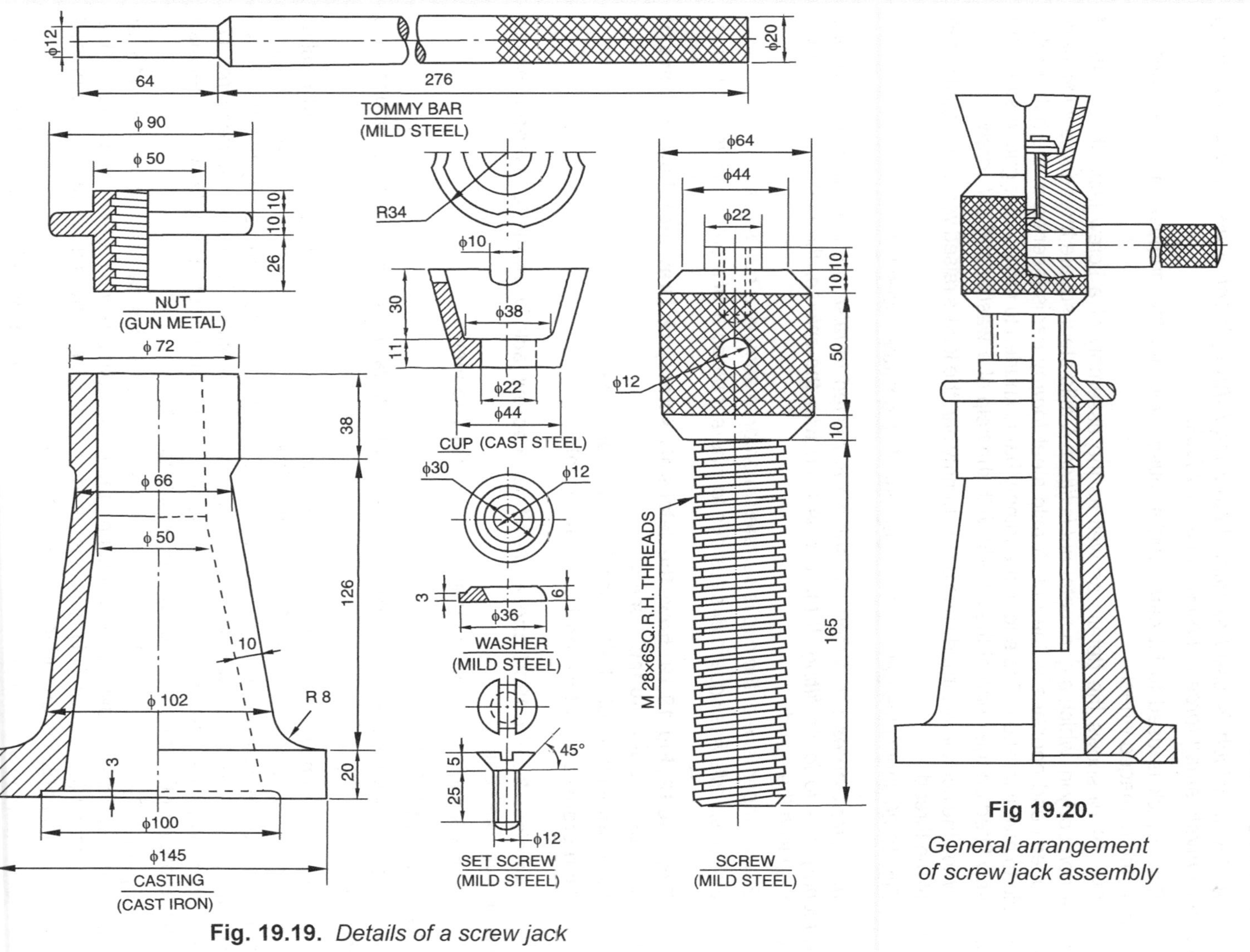

Fig. 19.19. *Details of a screw jack*

Fig 19.20.

General arrangement of screw jack assembly

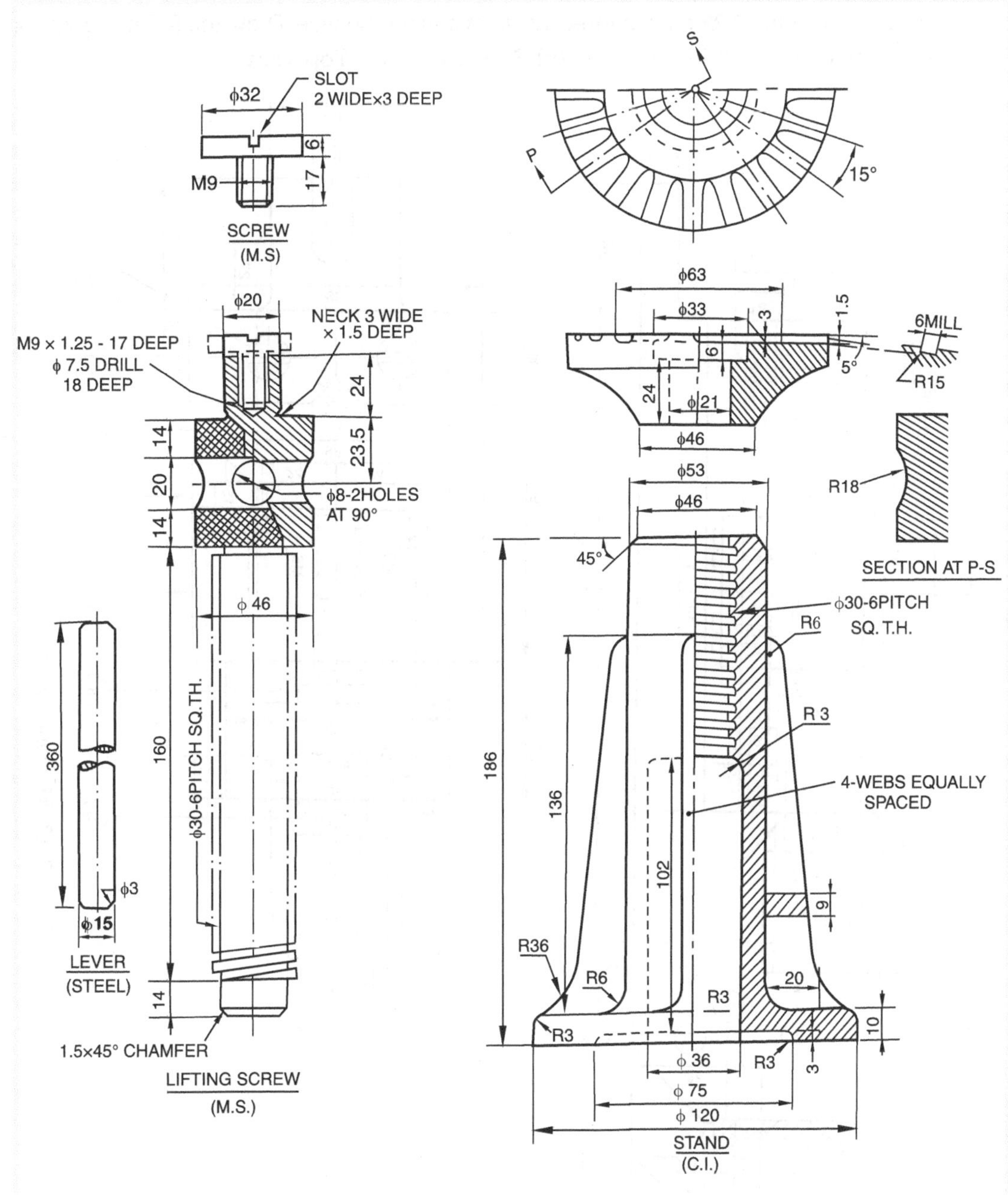

Fig. 19.21. *Details of a screw jack*

Exercise 14 : Fig. 19.23 shows the details of a machine vice. Imagine the parts assembled together and draw the following views to some suitable scale:

(a) Front view - right half in section

(b) Top view

Also prepare the parts and materials list.

Exercise 15 : Fig. 19.22 shows three views of a machine vice. Draw the following views:
(a) Front view – full in section (b) Side view (c) Top view

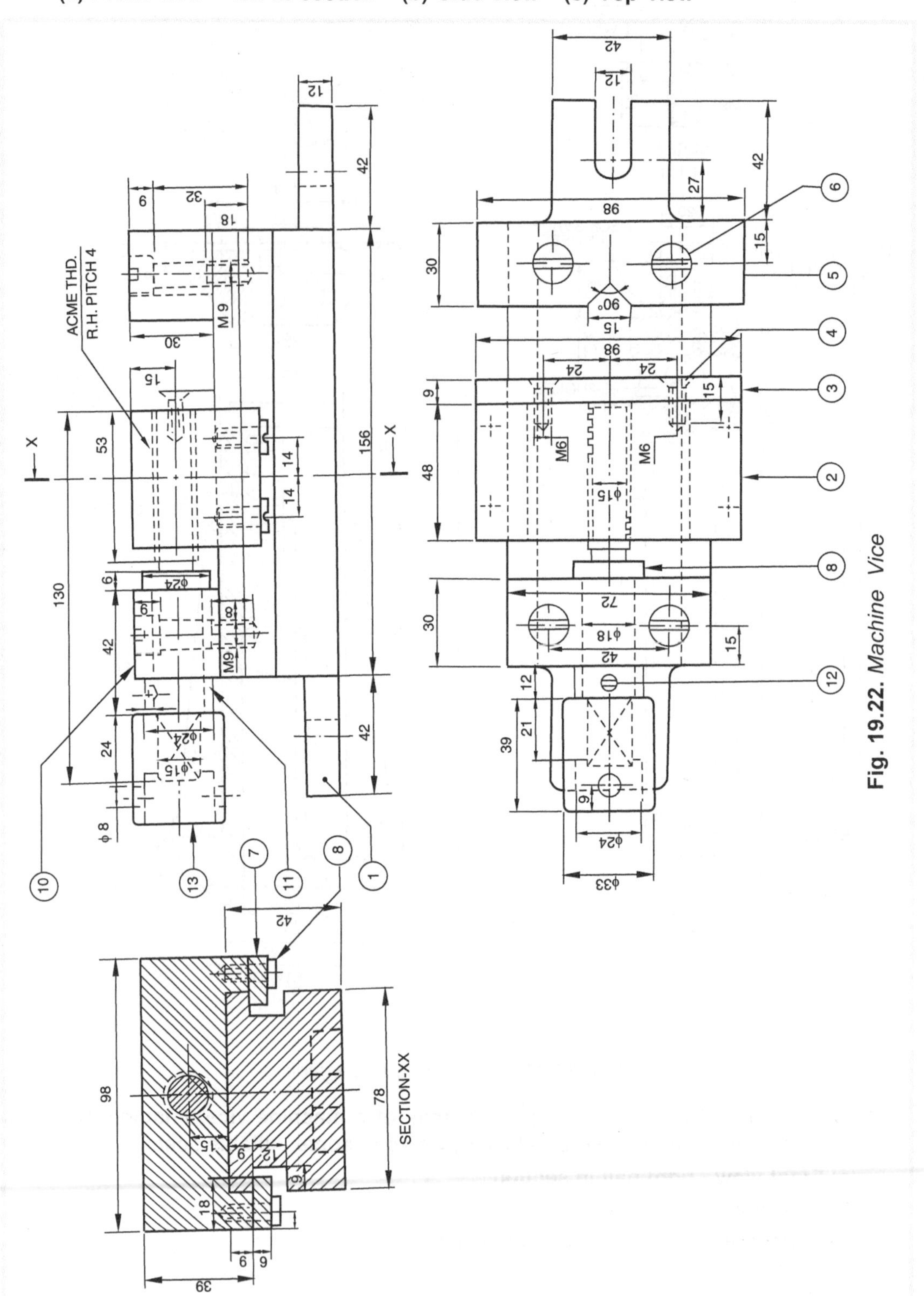

Fig. 19.22. *Machine Vice*

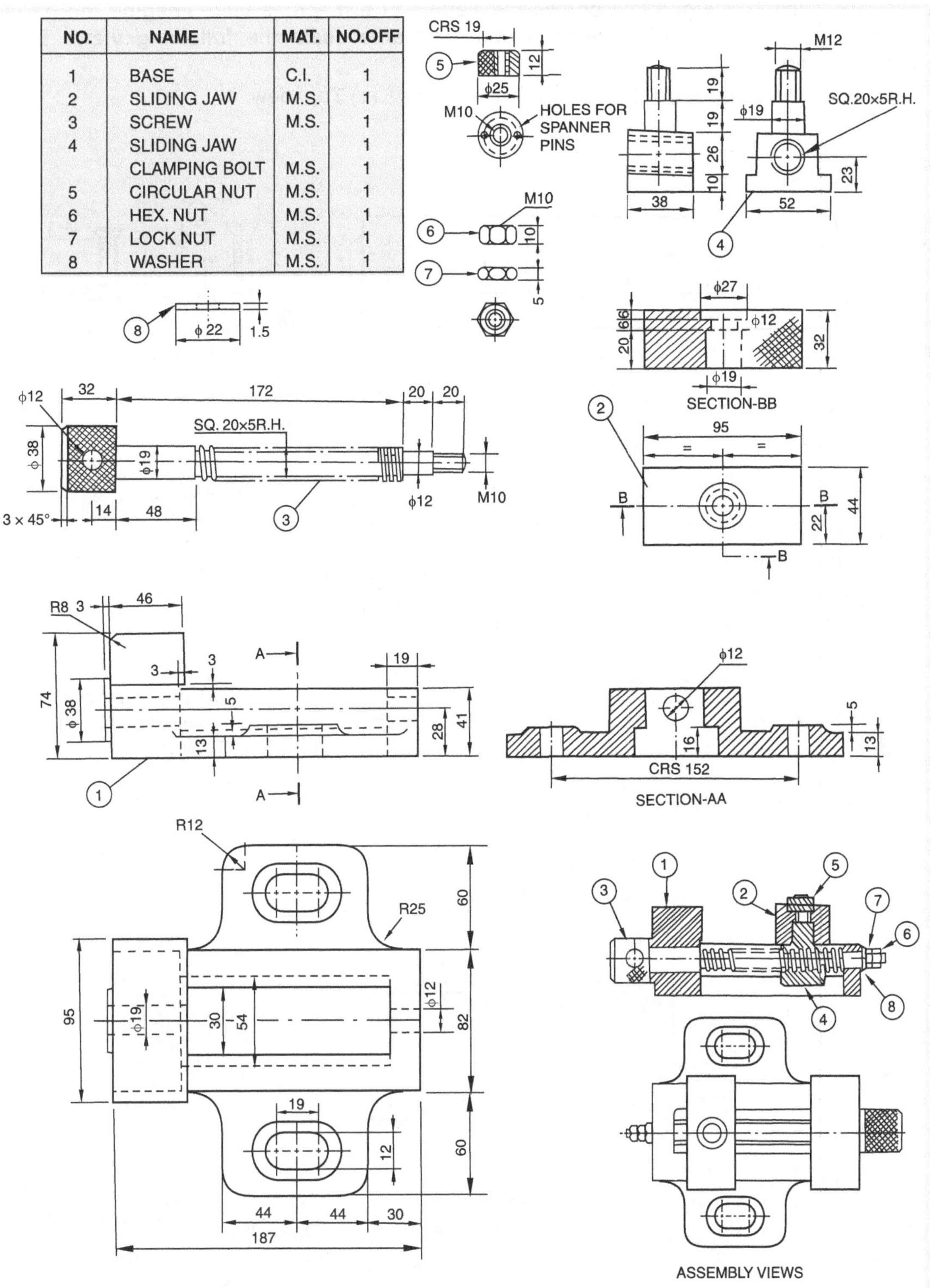

NO.	NAME	MAT.	NO.OFF
1	BASE	C.I.	1
2	SLIDING JAW	M.S.	1
3	SCREW	M.S.	1
4	SLIDING JAW		1
	CLAMPING BOLT	M.S.	1
5	CIRCULAR NUT	M.S.	1
6	HEX. NUT	M.S.	1
7	LOCK NUT	M.S.	1
8	WASHER	M.S.	1

Fig. 19.23. *Details of a machine vice*

Exercise 16 : Fig. 19.24 shows the details of a machine vice. Imagine the parts assembled together and draw to a some suitable scale the following views fully dimensioned :

(a) Half sectional front view (b) An end view (c) Top view

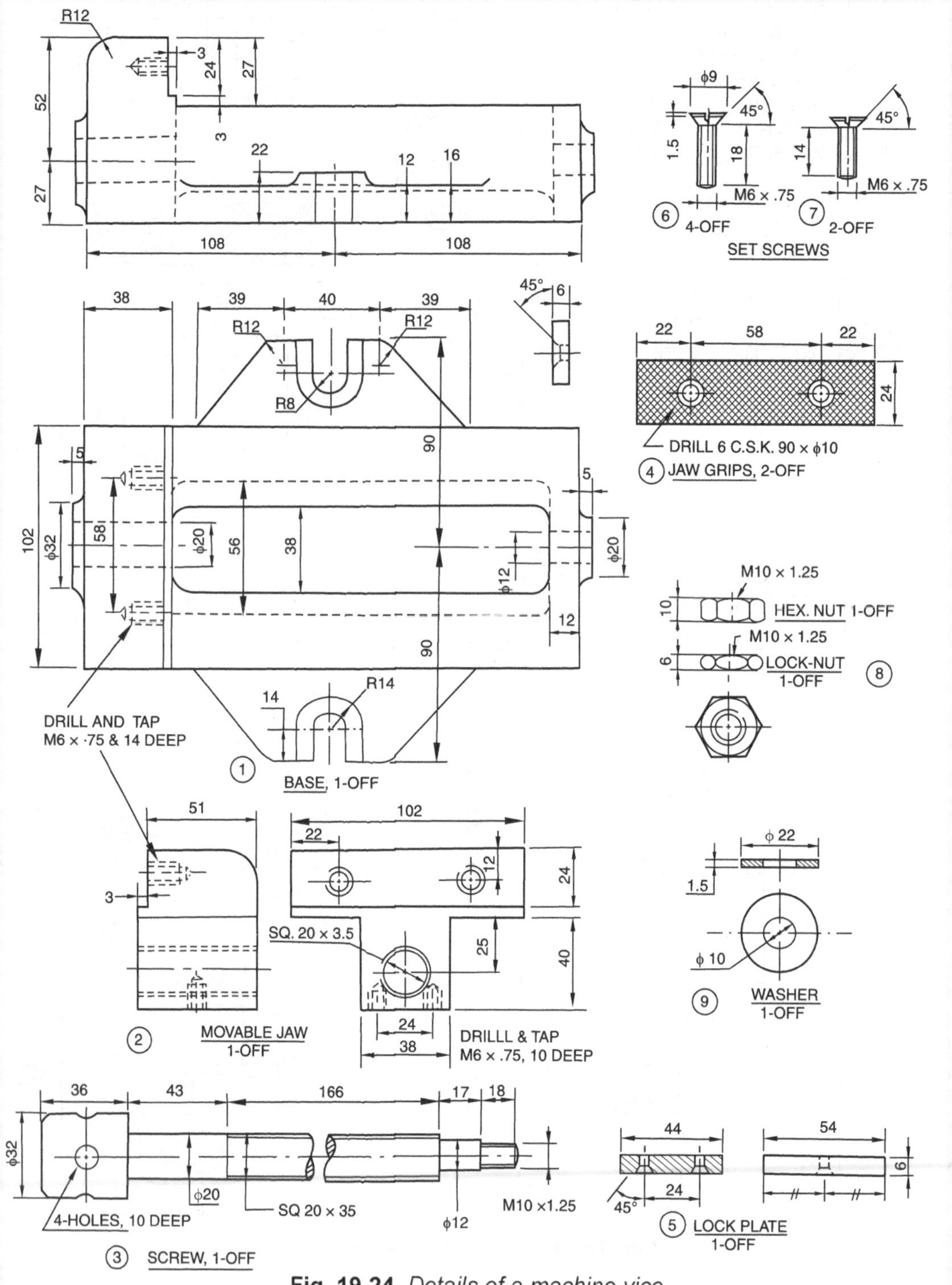

Fig. 19.24. *Details of a machine vice*

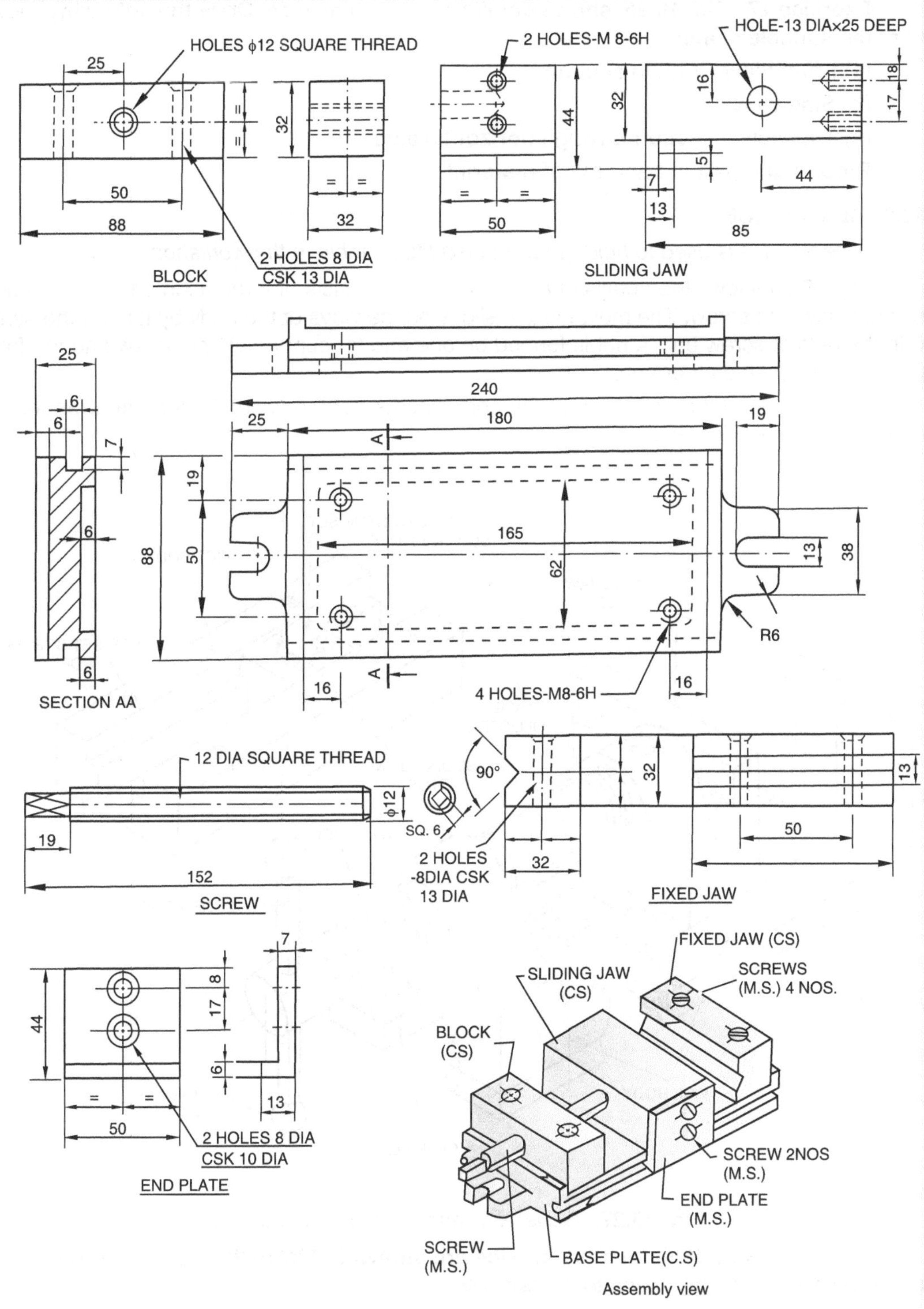

Fig. 19.25. *Details of a machine vice*

Exercise 17 : Fig. 19.25 shows details of a machine vice. Draw the following views to some suitable scale:

(a) Front view – Full in section

(b) Side view

(c) Top view – section through horizontal axis

Prepare the part list and bill of material

19.9. BENCH VICE

A bench vice is used to hold the work on a fitter's table in the workshop.

Fig. 19.26 shows the details of a bench vice. It consists of three main parts i.e., a body, movable jaw and screw. The movable jaw slides on the ways of the body by turning the screw with the help of screw bar. A nut is formed on one end for supporting the screw and the fixed jaw is formed on the other end.

To understand the assembly of the components parts of vice, its general arrangement is shown below in Fig. 19.27.

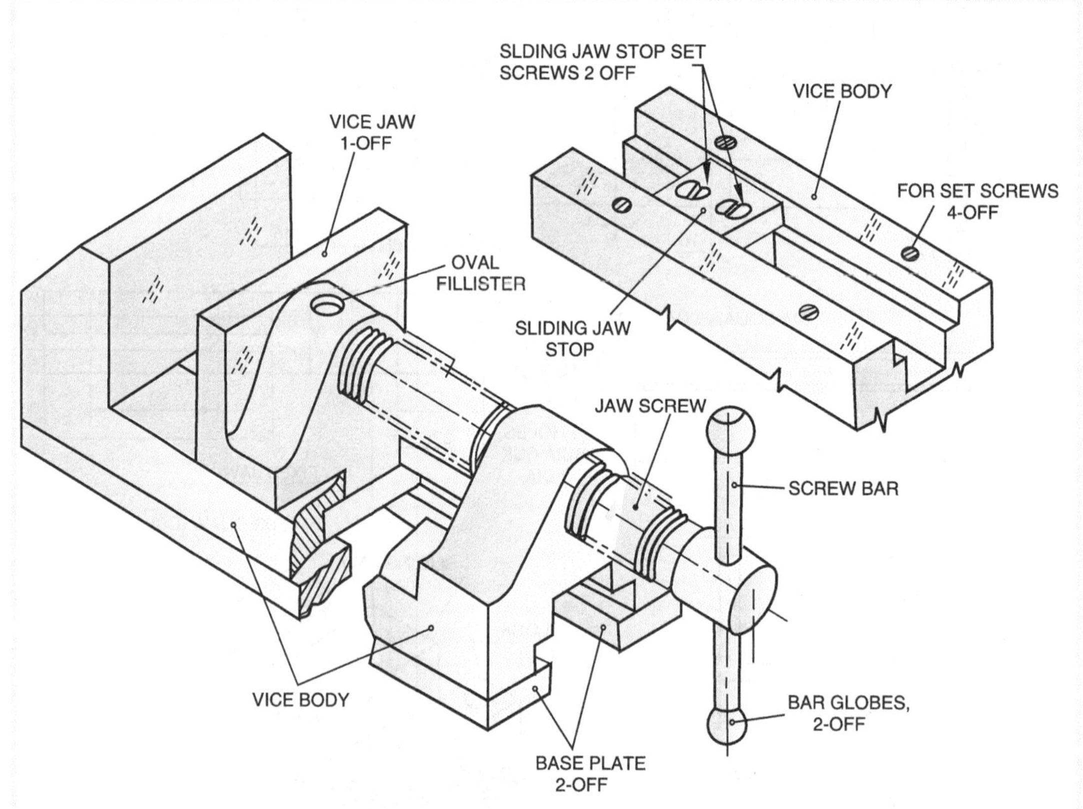

Fig. 19.27. *General arrangement of a bench vice*

Exercise 18 : Details of a bench vice are shown in Fig.19.26. Assemble the details and draw the following viewo to a scale 1:2 .

(a) Front view – full in section

(b) End view – looking from right hand side.

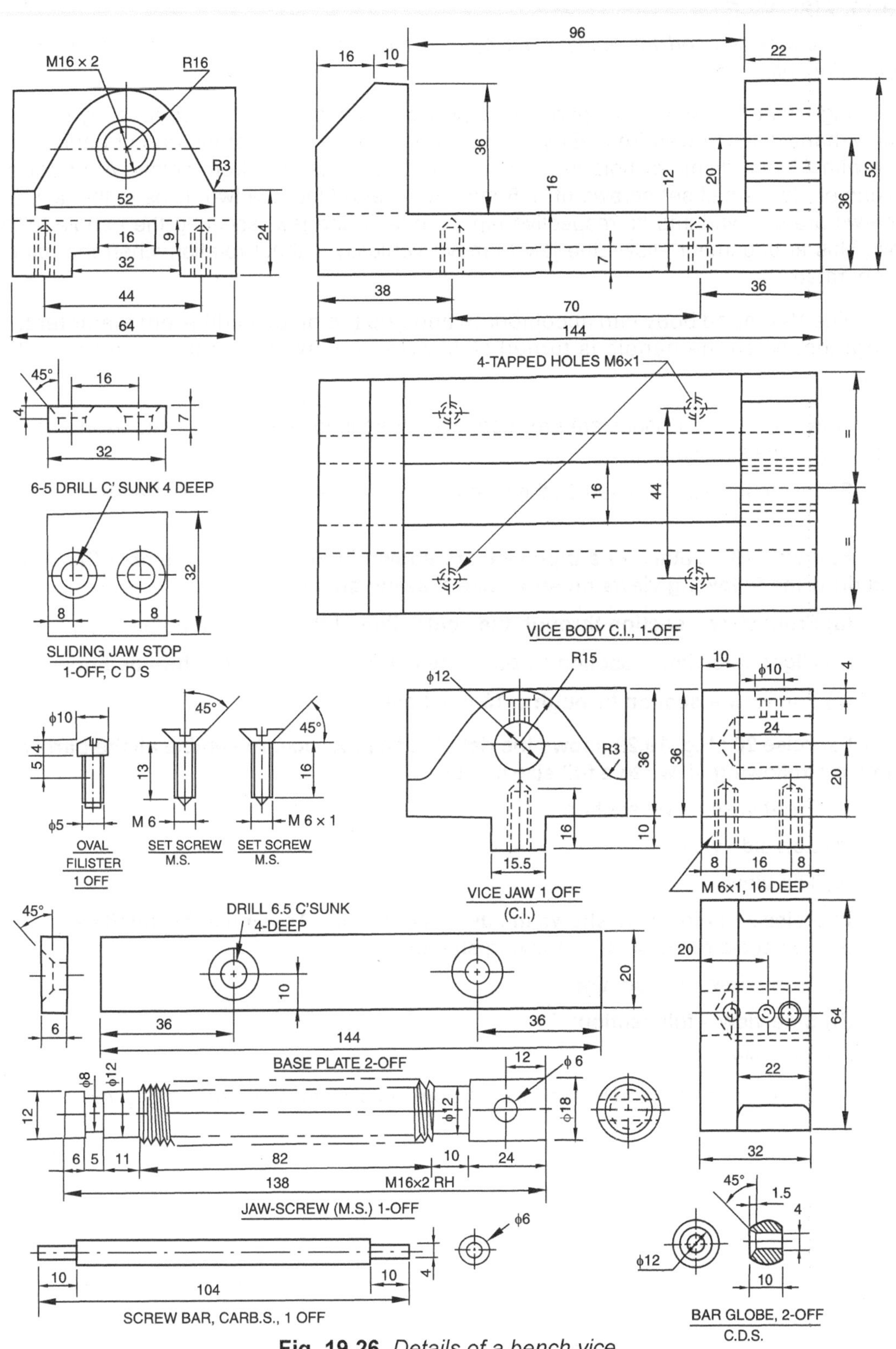

Fig. 19.26. *Details of a bench vice*

19.10. PIPE VICE

Pipe vice is used for holding the pipe to facilitate cutting it to a length, threading operation, etc.

Fig. 19.28 shows various parts of a pipe vice. It consists of V- shaped body in which screw handle is screwed from above. When the circular groove at the end of handle is in line with 15 mm diameter hole in the body the jaw is inserted from below. After correct alignment two small set screws of ϕ 5 mm are inserted into the two holes in the jaw. Set screws are screwed into its respective tapped hole in the jaw, securing the handle to the jaw. This arrangement allows the jaw to move vertically without rotation when the handle is operated.

The V-shaped body can accomodate and grip the pipes of different diameters. In operation, when the handle is turned clockwise, the jaw moves down and grips the pipe.

Note : 1. The serrations are provided in the V-shaped end of the jaw to exert effective grip on the pipe surface.

2. Square threads are used on the screw as they are stronger than V-threads.

Exercise 19 : Details of a pipe vice are shown in Fig. 19-28. Assemble the details and draw the following views on an A_2 size drawing sheet :

(a) Front view – section through the centre line of the threaded hole.

(b) Right side view – section through the centre line of square threaded hole.

(c) Top view – section through ϕ 15 mm hole.

Exercise 20 : Fig. 19.29 shows the details of a pipe vice. Assemble all the parts and draw the following views to a full size scale:

(a) Front view – Full section

(b) Left side view

(c) Top view

Exercise 21 : Fig. 19.30 shows the details of a vice. Assemble the details and draw to a full size scale on an A_2 size drawing sheet.

(a) Front view – full section

(b) Side view – full section

(c) top view

Dimension the views completely. Assume, proportionately, any dimensions not given

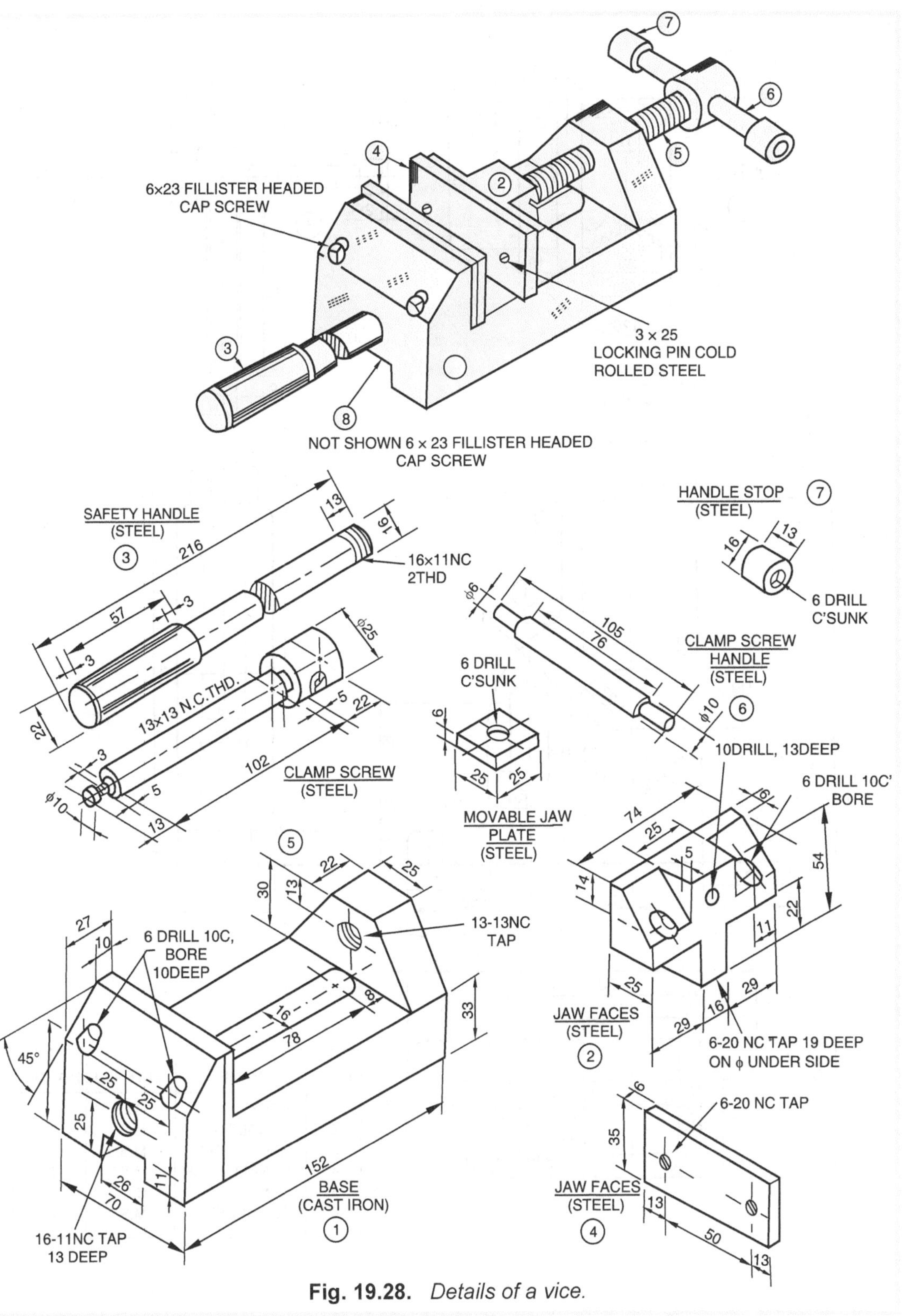

Fig. 19.28. *Details of a vice.*

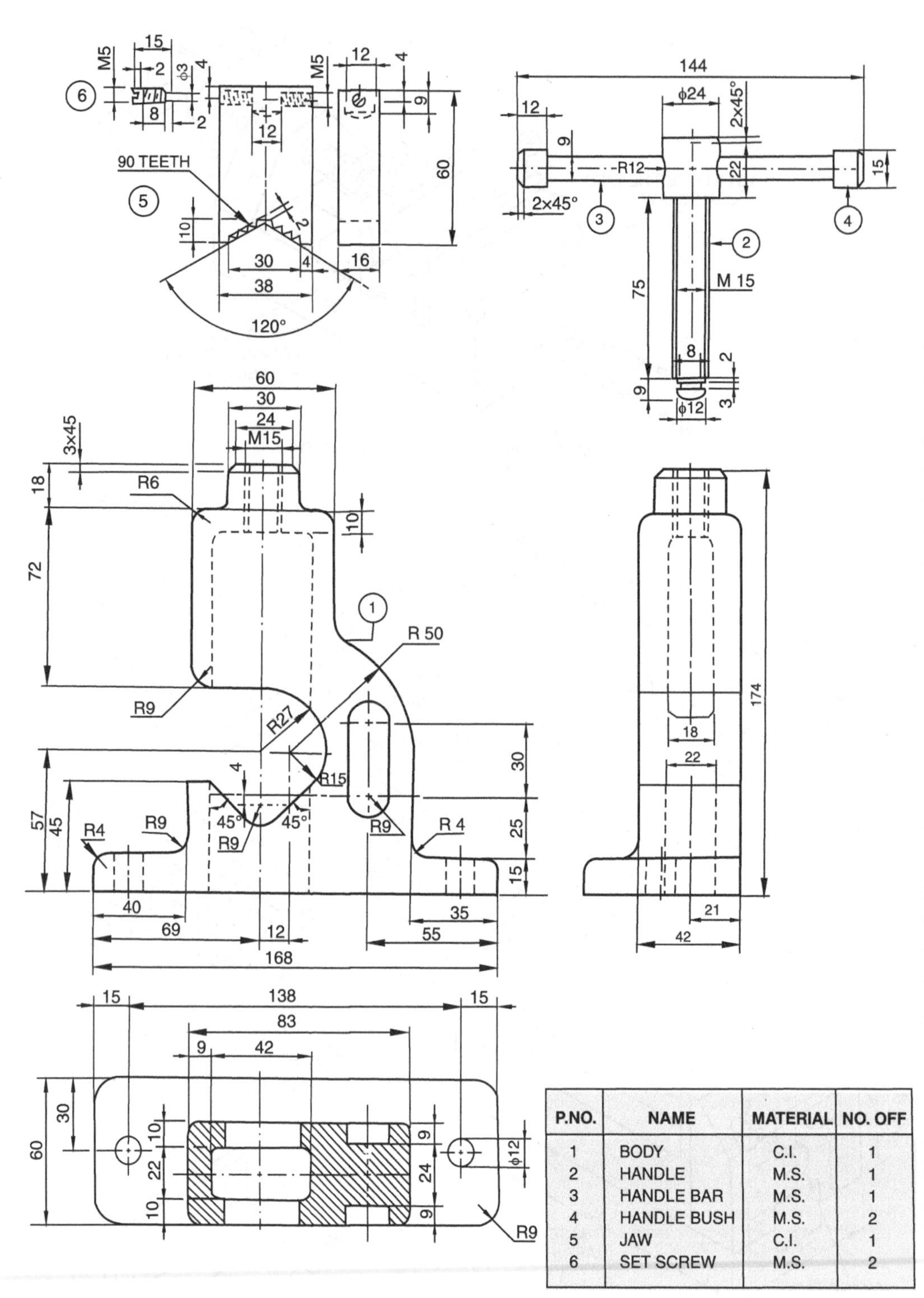

P.NO.	NAME	MATERIAL	NO. OFF
1	BODY	C.I.	1
2	HANDLE	M.S.	1
3	HANDLE BAR	M.S.	1
4	HANDLE BUSH	M.S.	2
5	JAW	C.I.	1
6	SET SCREW	M.S.	2

Fig. 19.29. *Pipe vice*

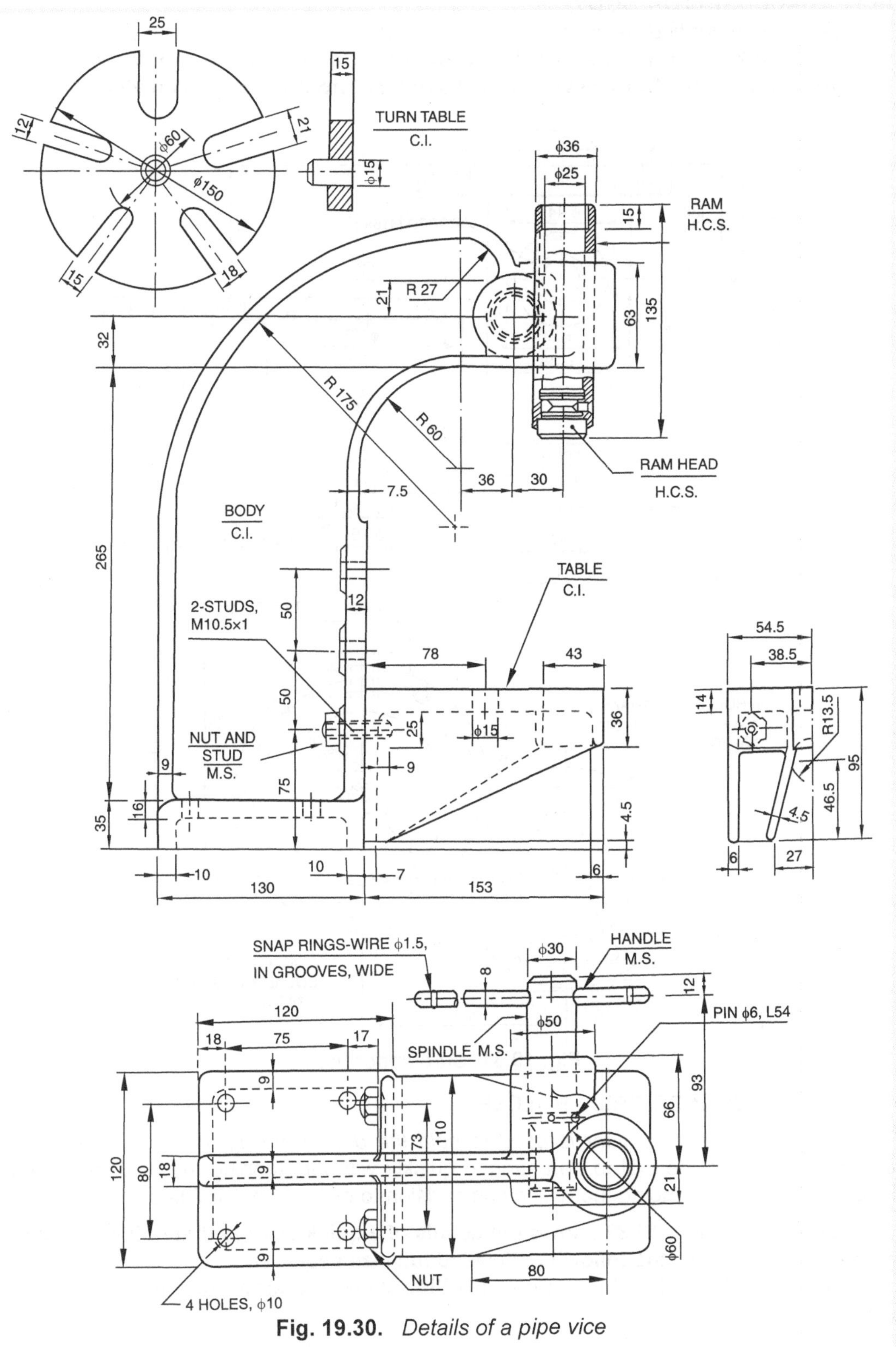

Fig. 19.30. *Details of a pipe vice*

19.11. CARBURETTOR BODY

The carburettor body is used in the petrol engine.

Exercise 22 : Fig. 19.31 shows two views of a Carburettor body. Draw the (a) Half sectional front view (b) Half sectional front view (c) Top view

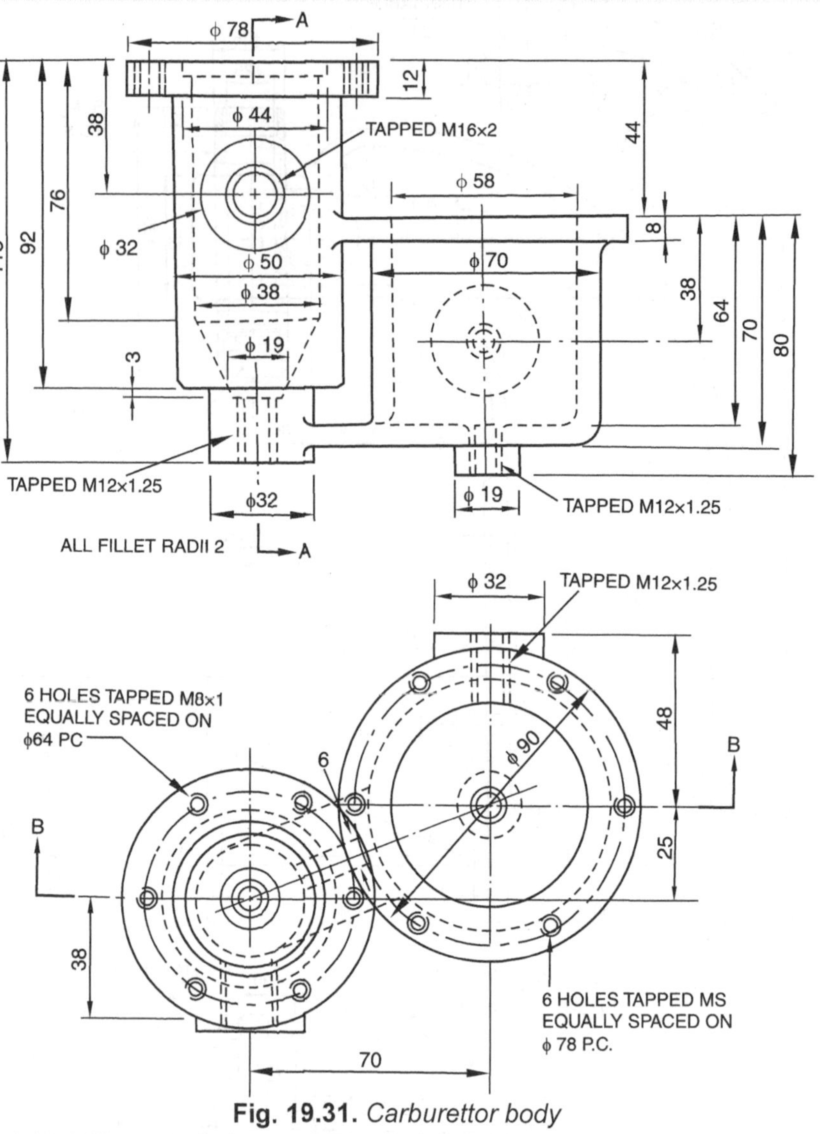

Fig. 19.31. *Carburettor body*

19.12. QUICK CHANGE DRILL HOLDER

Quick change drill holder is used to hold the drill in drilling machine during drilling operation. It allows the drill to be changed quickly without stopping the machine. Thus, it takes less time for changing the drill in comparision to ordinary drill holders.

Exercise 23 : Fig. 19.32 shows the details of a quick change drill holder. Assemble all the parts and draw the following views to full size scale :—

(a) Front View – Full in section (b) Side View - outside (c) Top View – Full in section.

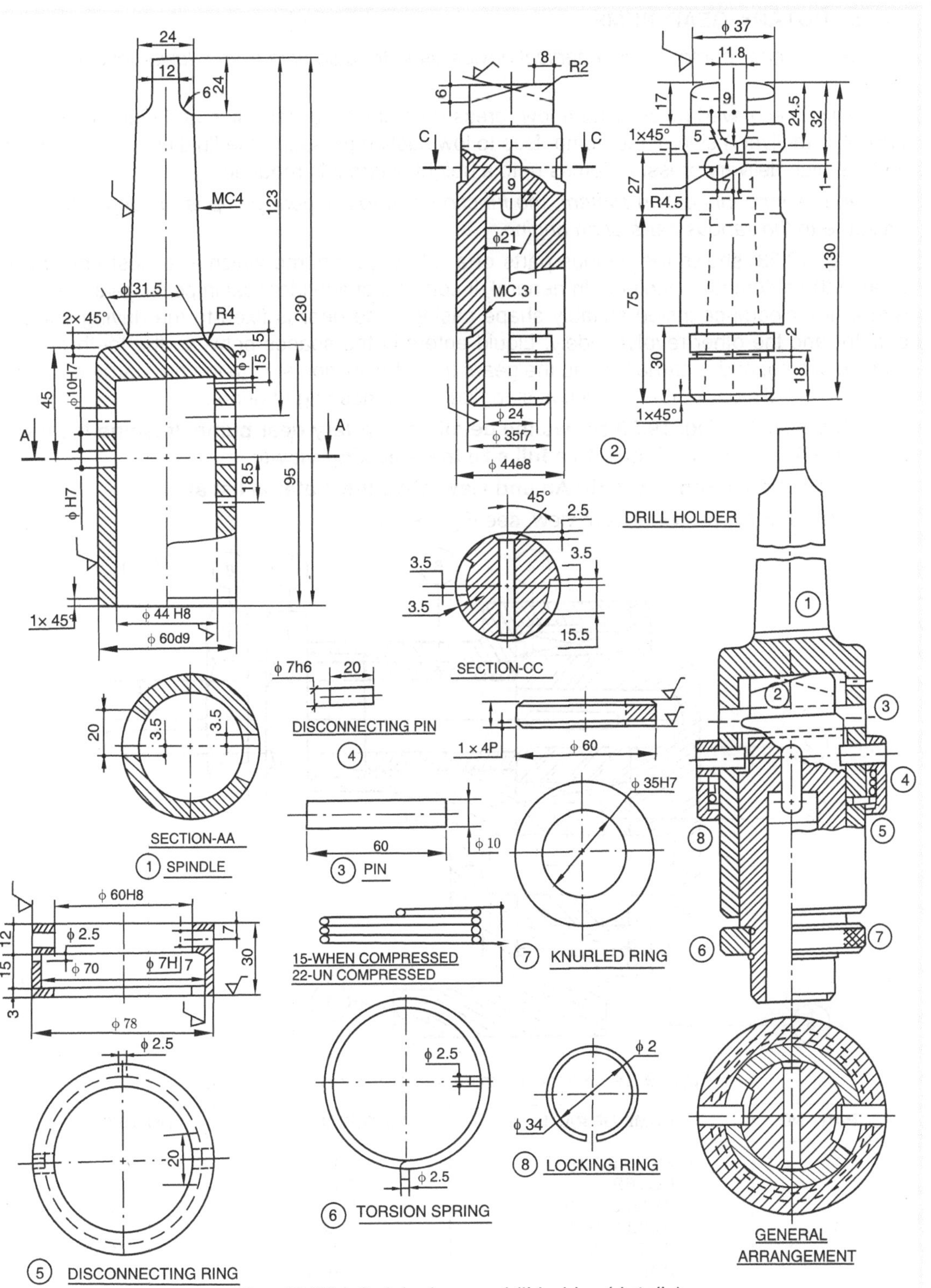

Fig. 19.32. *Quick change drill holder (details)*

19.13. ROTARY GEAR PUMP

Rotary gear pump is a mechanical device used to raise fluid from a lower level to higher level.

This is achieved by creating a low pressure at inlet (suction) side and high pressure at the outlet (delivery) side of the pump. Due to low suction pressure, the fluid rises from a depth and the high delivery pressure forces upto a height where it is required.

In pressure lubrication system, the function of a rotary gear pump is to supply oil under pressure to the various parts of an engine.

Fig. 19.33, shows the various parts of a rotary gear pump which is almost universally used in the automotive engines. In general, it consists of two identical impeller gears working with a fine clearance inside suitably shape casing. One gear is fixed to the driving shaft of a motor and the other revolves idely. Liquid enters in the spaces between the teeth of gear and the casing; and is carried round the gears from the suction side to the discharge end. The liquid cannot slip back into the suction side due to the meshing of gears.

Exercise 24 : Fig. 19.33 shows the details of a rotary gear pump. Imagine the parts assembled together and draw to a full size the following views:

(a) Sectional front view. (b) An end view. Give the material list also

Hint : For the sectional front view, see Fig. 19.34.

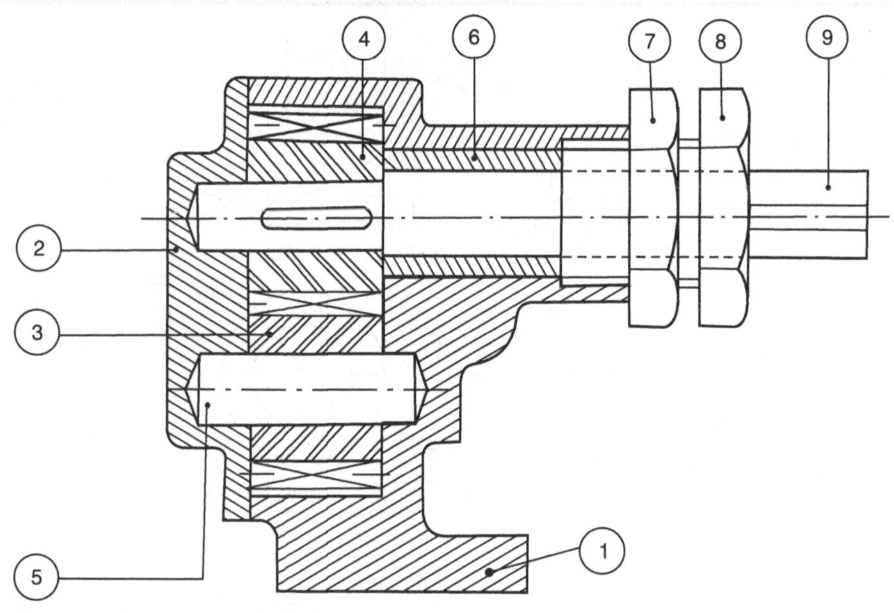

Fig. 19.34. *Sectional front view of rotary gear pump.*

PART NO.	NAME OF PART	MATERIAL	NO. OFF
1	PUMP BODY	C.I.	1
2	COVER	C.I.	1
3	IMPELLER GEAR	G.M.	1
4	IMPELLER GEAR	G.M.	1
5	FIXED SPINDLE	M.S.	1
6	BUSH	BRONZE	1
7	LOCK NUT	M.S.	1
8	GLAND	BRONZE	1
9	DRIVING SPINDLE	M.S.	1

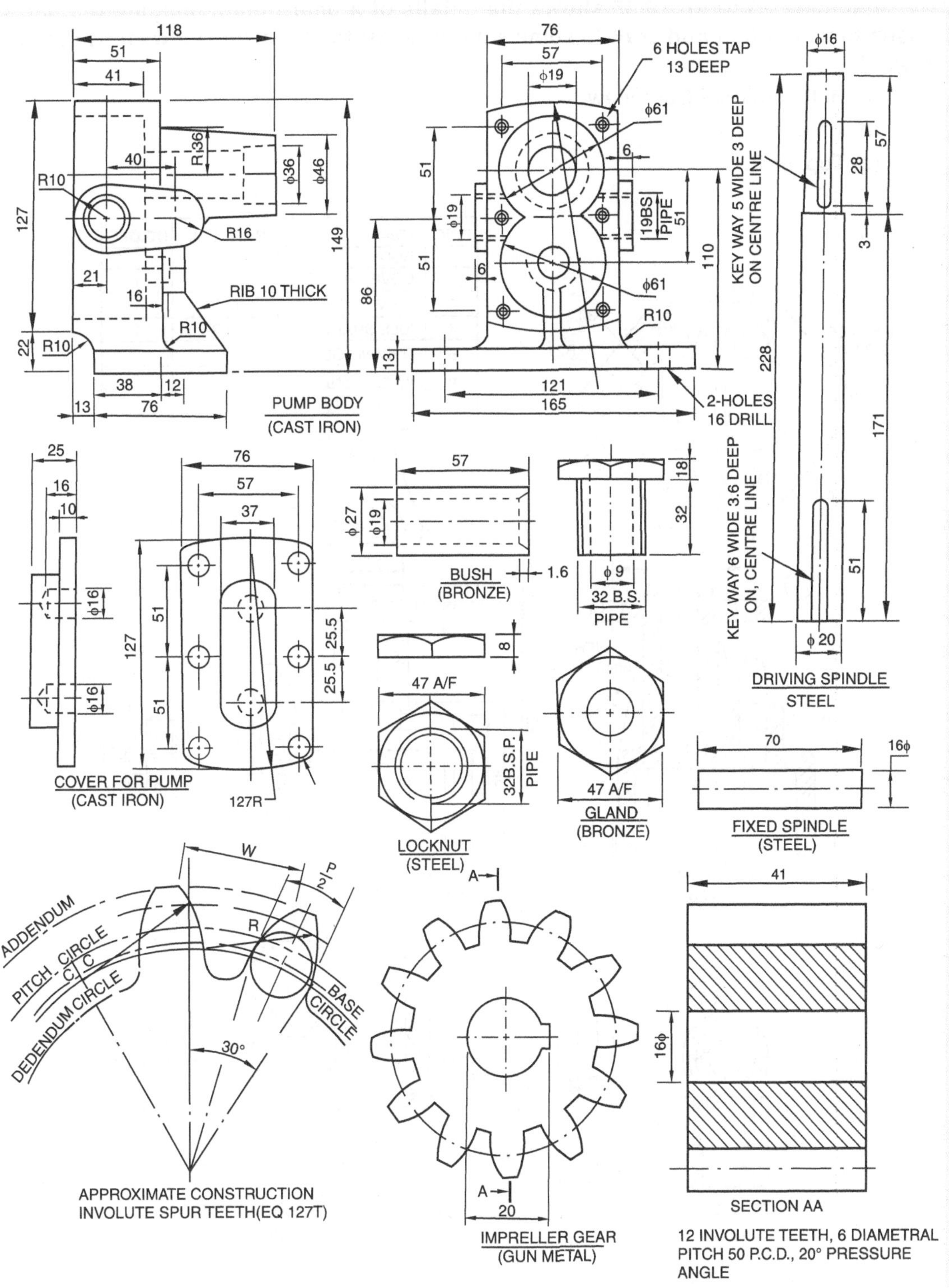

Fig. 19.33. *Details of rotary gear pump*

Exercise 25 : Fig. 19.35 shows the details of a gear pump. Imagine the parts assembled together and draw to some suitable scale the following views fully dimensioned.

(a) Half sectional front view

(b) An end view

(c) Top view.

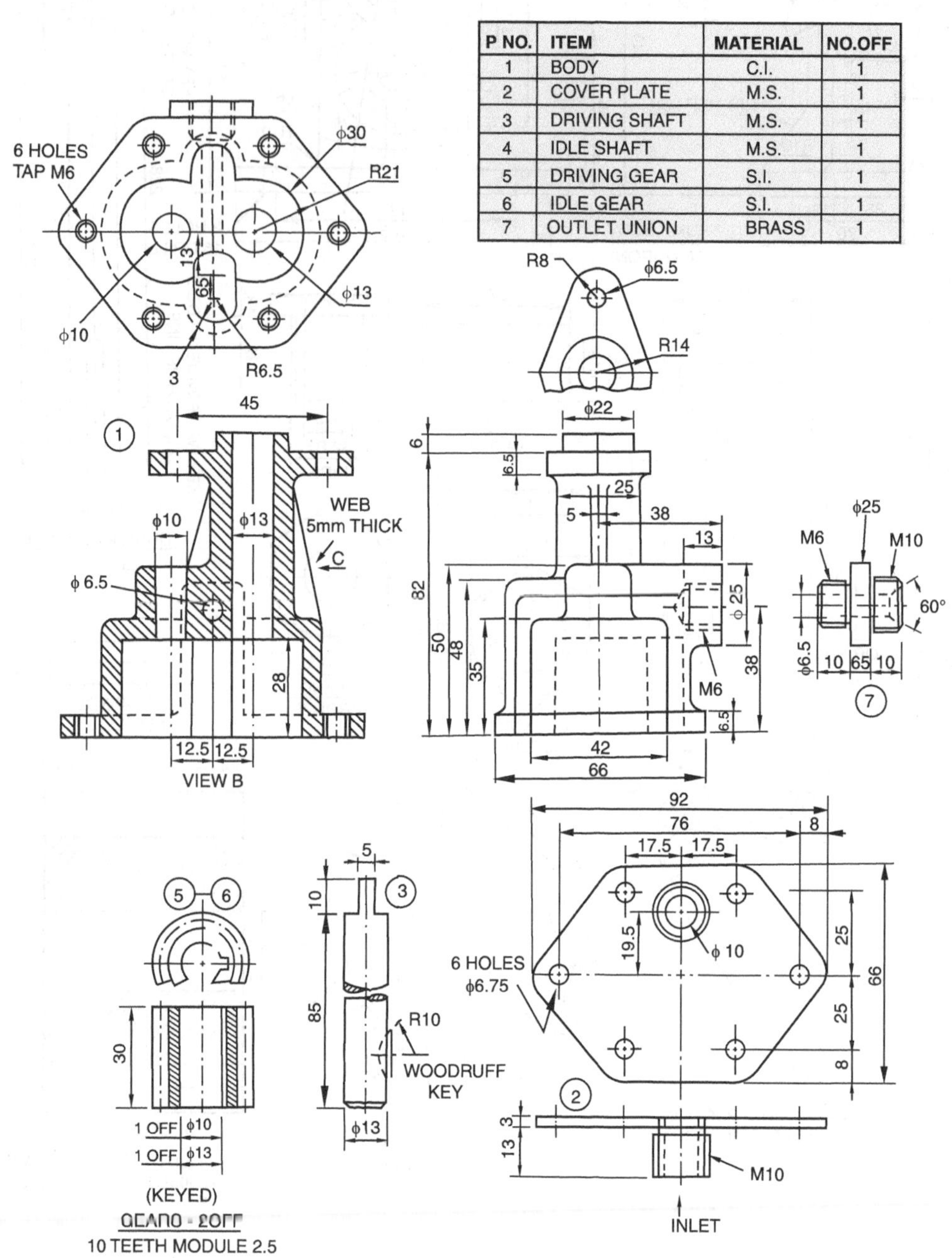

P NO.	ITEM	MATERIAL	NO.OFF
1	BODY	C.I.	1
2	COVER PLATE	M.S.	1
3	DRIVING SHAFT	M.S.	1
4	IDLE SHAFT	M.S.	1
5	DRIVING GEAR	S.I.	1
6	IDLE GEAR	S.I.	1
7	OUTLET UNION	BRASS	1

Fig. 19.35. *Gear pump*

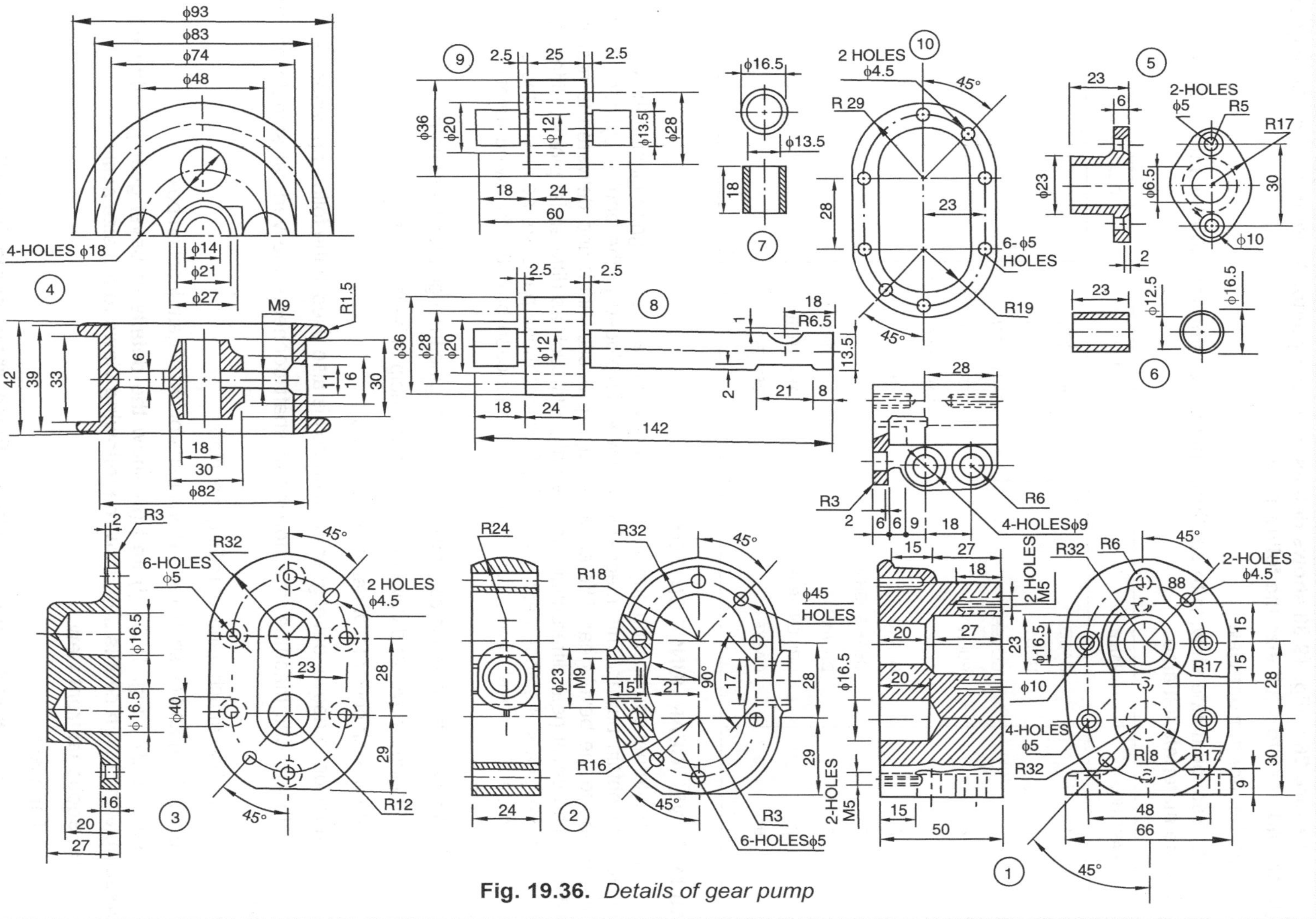

Fig. 19.36. Details of gear pump

Exercise 26 : Fig. 19.36 shows details of a gear pump. Assemble all parts of gear pump and draw to a full size the following views:

(a) Sectional front view

(b) Sectional side view, with sectional plane passing through the mid of the gears. Give the materials list also.

PART NO.	NAME	MATERIAL	QUANTITY
1	BASE	CAST IRON	1
2	BODY	CAST IRON	1
3	COVER	CAST IRON	1
4	PULLEY	CAST IRON	1
5	GLAND	CAST IRON	1
6	GLAND	BRASS	1
7	GEAR BUSHING	BRASS	1
8	DRIVING GEAR	CAST STEEL	1
9	DRIVEN GEAR	CAST STEEL	1
10	GASKET	COPPER	1

19.14. BEVEL GEAR JUNCTION BOX

In an internal combustion engine, a mixture of fuel and air is ignited inside the cylinder in order to obtain power. This mechanical power developed in the engine cylinder is first transmitted to the bevel gear differential through propeller shaft and delivered to rear two wheels. The bevel gear differential has several parts such as differential pinion, differential frame, bevel gears, instance gear, axles etc. Some of these parts are housed in the casing which is known as bevel gear junction box. In the junction box, the following functions are performed :—

(*i*) To receive the mechanical power from the engine through propeller shaft;

(*ii*) to connect driver and driven shafts with the help of bevel gears;

(*iii*) to transmit power to wheels through axles.

In addition to above, bevel gear junction box is also used in the following cases —

(*iv*) In the drill machine, the horizontal shaft connects the vertical spindle of the machine.

(*v*) In vertical attachments of milling machine.

Fig. 19.37 (i) and (ii) show the various parts of bevel gear junction box.

Exercise 27 : Fig. 19.37 (i) and (ii) show the details of bevel gear junction box. Assemble the given details and draw the following views to full size scale :—

(a) Sectional front view.

(b) Top view.

Show the tolerances and mark surface finish wherever required. Give the material list also.

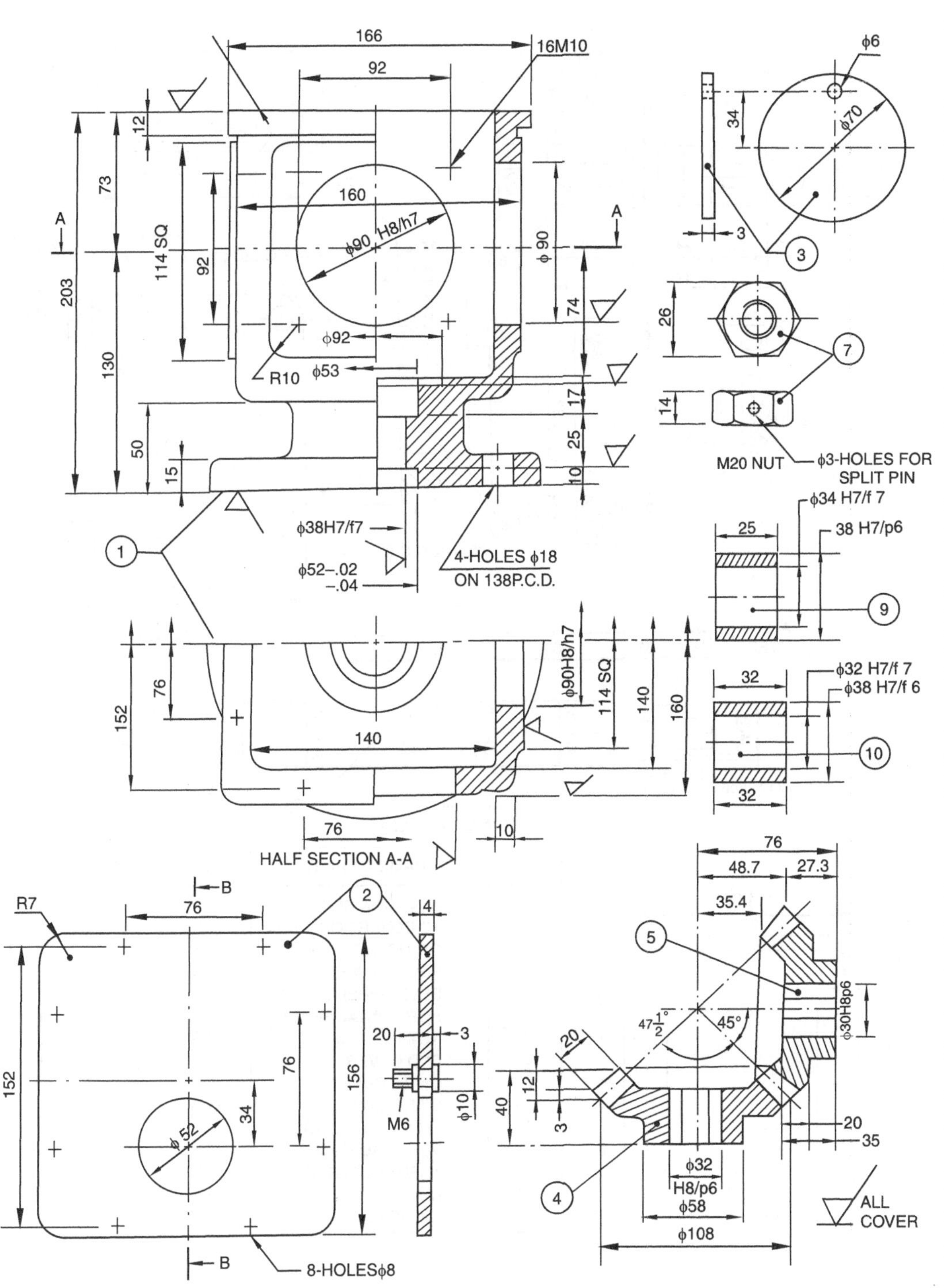

Fig. 19.37 (i) *Details of Bevel gear junction box*

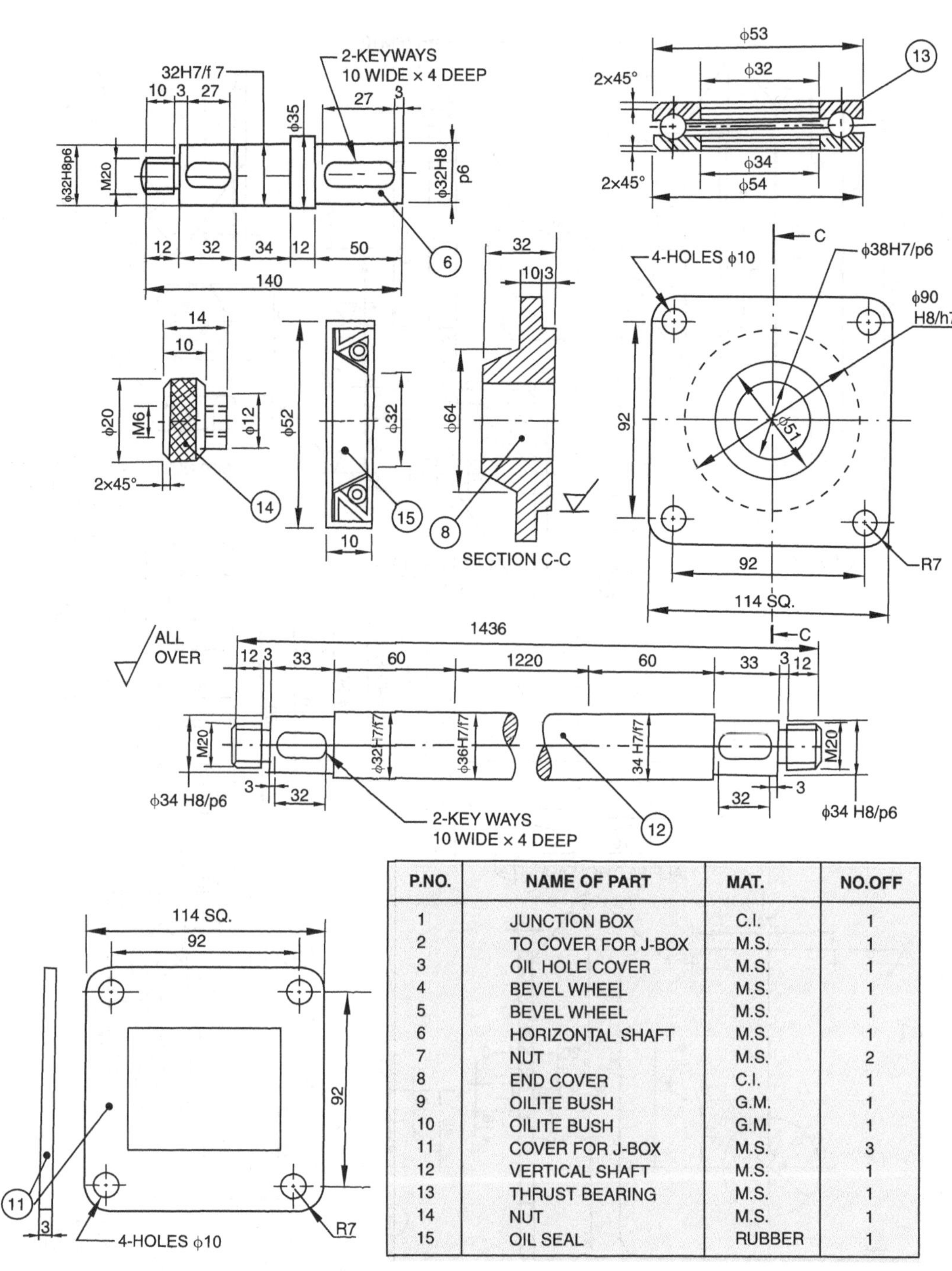

P.NO.	NAME OF PART	MAT.	NO.OFF
1	JUNCTION BOX	C.I.	1
2	TO COVER FOR J-BOX	M.S.	1
3	OIL HOLE COVER	M.S.	1
4	BEVEL WHEEL	M.S.	1
5	BEVEL WHEEL	M.S.	1
6	HORIZONTAL SHAFT	M.S.	1
7	NUT	M.S.	2
8	END COVER	C.I.	1
9	OILITE BUSH	G.M.	1
10	OILITE BUSH	G.M.	1
11	COVER FOR J-BOX	M.S.	3
12	VERTICAL SHAFT	M.S.	1
13	THRUST BEARING	M.S.	1
14	NUT	M.S.	1
15	OIL SEAL	RUBBER	1

Fig. 19.37 ***(ii)*** *Details of bevel gear junction box*

19.15. MEASURING STAND

With the modern development, accuracy of the dimensions of machine parts has become necessary. Measuring stand is an important device which can be used for measurement, inspection, testing the jobs precisely in tool room of workshop. Fig. 19.38 shows the various parts of measuring stand.

Exercise 28 : Fig. 19.38 shows the details of measuring stand. Imagine the parts assembled together and draw to a full size the following views :

(a) Sectional front view. (b) An end view. (c) Top view.

Show the machining marks wherever required and give the material list also.

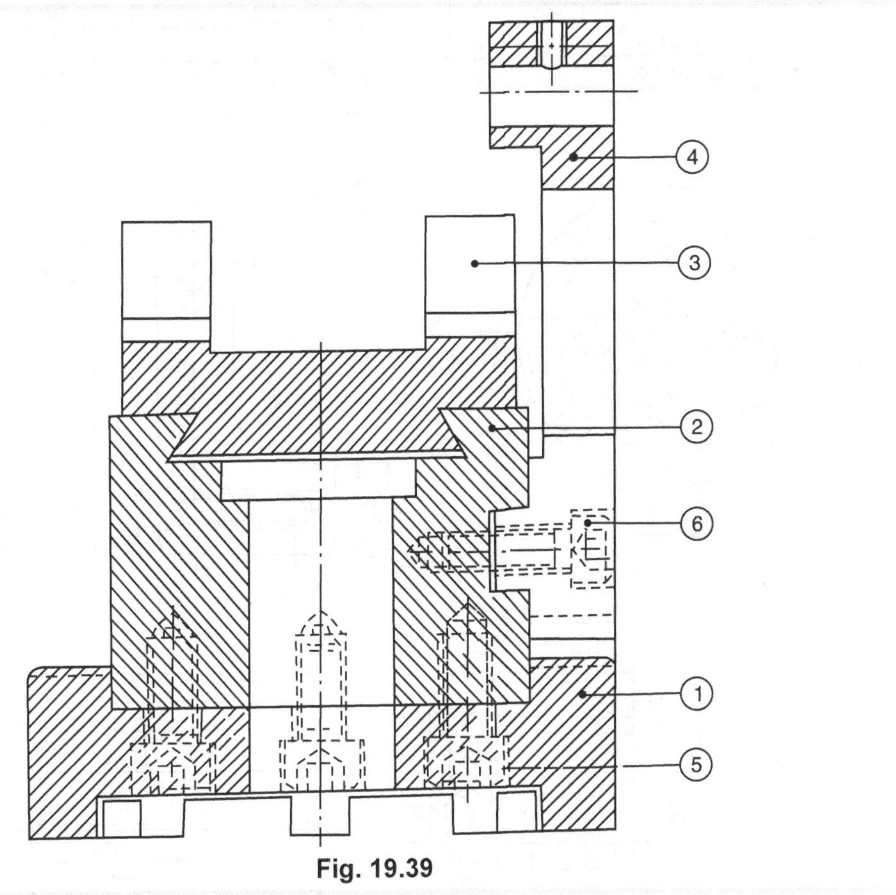

Fig. 19.39

Fig. 19.39 shows the sectional front view of measuring stand.

P.NO.	NO.OFF	NAME	MAT.
1	1	STAND	STE
2	1	SLIDE STE	"
3	1	V-BLOCK	"
4	1	BRACKET	"
5	4	ALLEN SCREW	"
6	2	SCREW	"

$12H7^{+0.018}_{0.00}$	$16f8^{-0.016}_{-0.043}$
$30H7^{+0.021}_{0}$	$66g6^{-0.010}$
$86H7^{+0.035}_{0}$	$16f7^{-0.036}_{-0.071}$
$16H8^{+0.027}_{0}$	

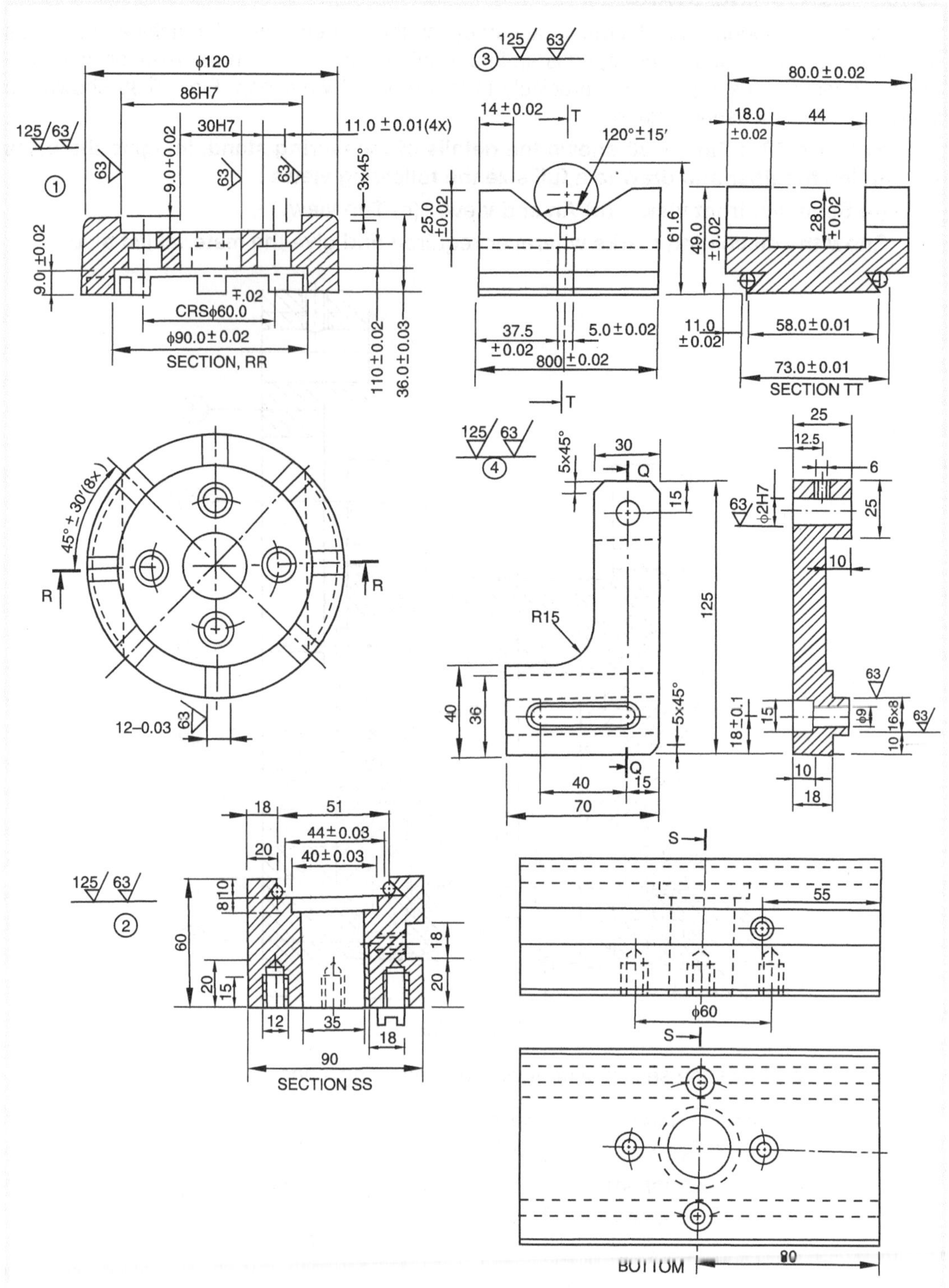

Fig. 19.38. *Details of measuring stand*

Exercise 29 : Fig. 19.40 shows the details of the measuring surface roughness stand. Imagine the parts assembled together and draw to a scale full size (a) Sectional front view (b) An end view (c) Top view.

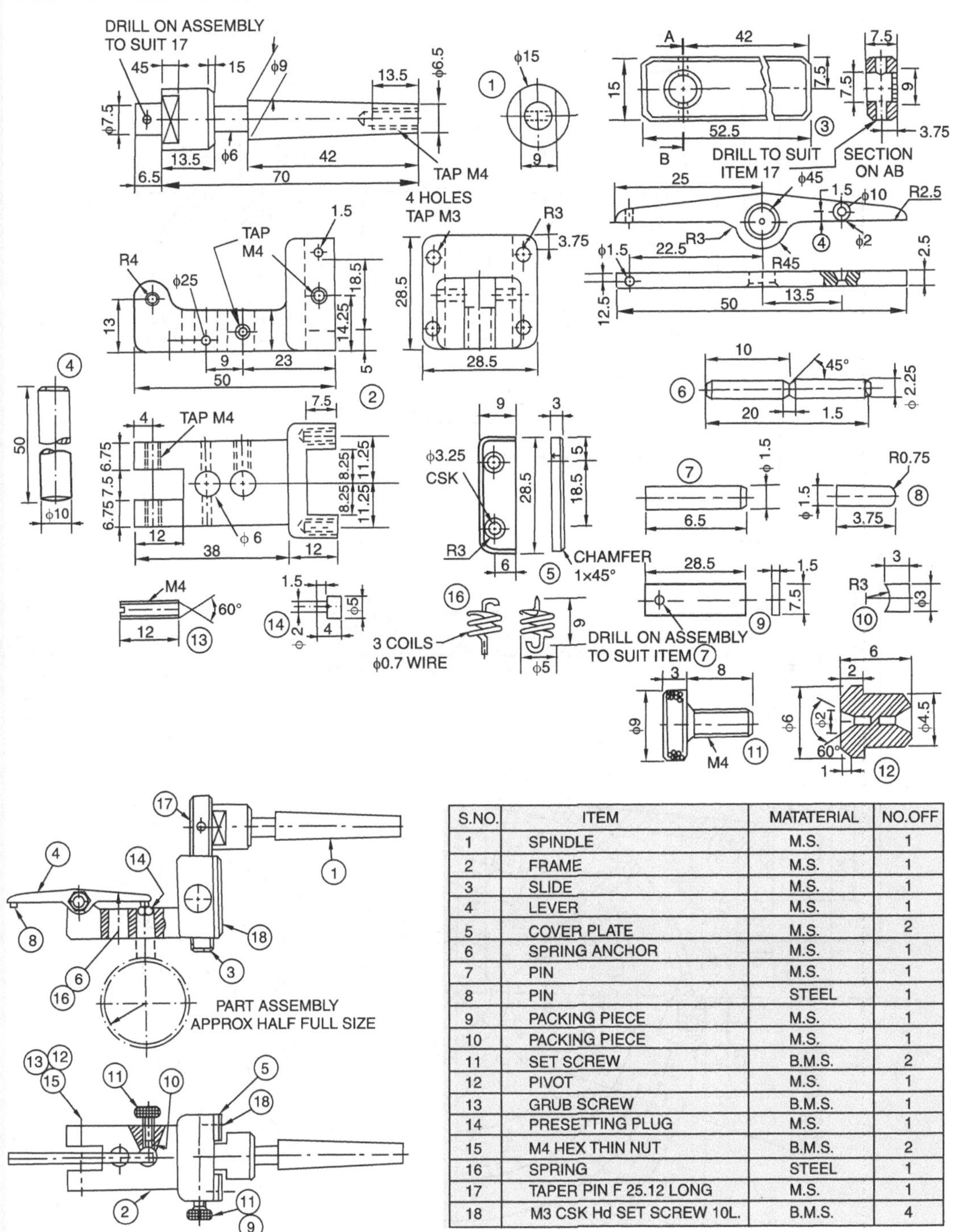

S.NO.	ITEM	MATATERIAL	NO.OFF
1	SPINDLE	M.S.	1
2	FRAME	M.S.	1
3	SLIDE	M.S.	1
4	LEVER	M.S.	1
5	COVER PLATE	M.S.	2
6	SPRING ANCHOR	M.S.	1
7	PIN	M.S.	1
8	PIN	STEEL	1
9	PACKING PIECE	M.S.	1
10	PACKING PIECE	M.S.	1
11	SET SCREW	B.M.S.	2
12	PIVOT	M.S.	1
13	GRUB SCREW	B.M.S.	1
14	PRESETTING PLUG	M.S.	1
15	M4 HEX THIN NUT	B.M.S.	2
16	SPRING	STEEL	1
17	TAPER PIN F 25.12 LONG	M.S.	1
18	M3 CSK Hd SET SCREW 10L.	B.M.S.	4

Fig. 19.40. *Measuring surface roughness stand*

Exercise 30 : Fig. 19.41 shows the a centrifugal pump body. draw the following views to some suitable scale:

(a) Front view - full section (b) Top view (c) Side view

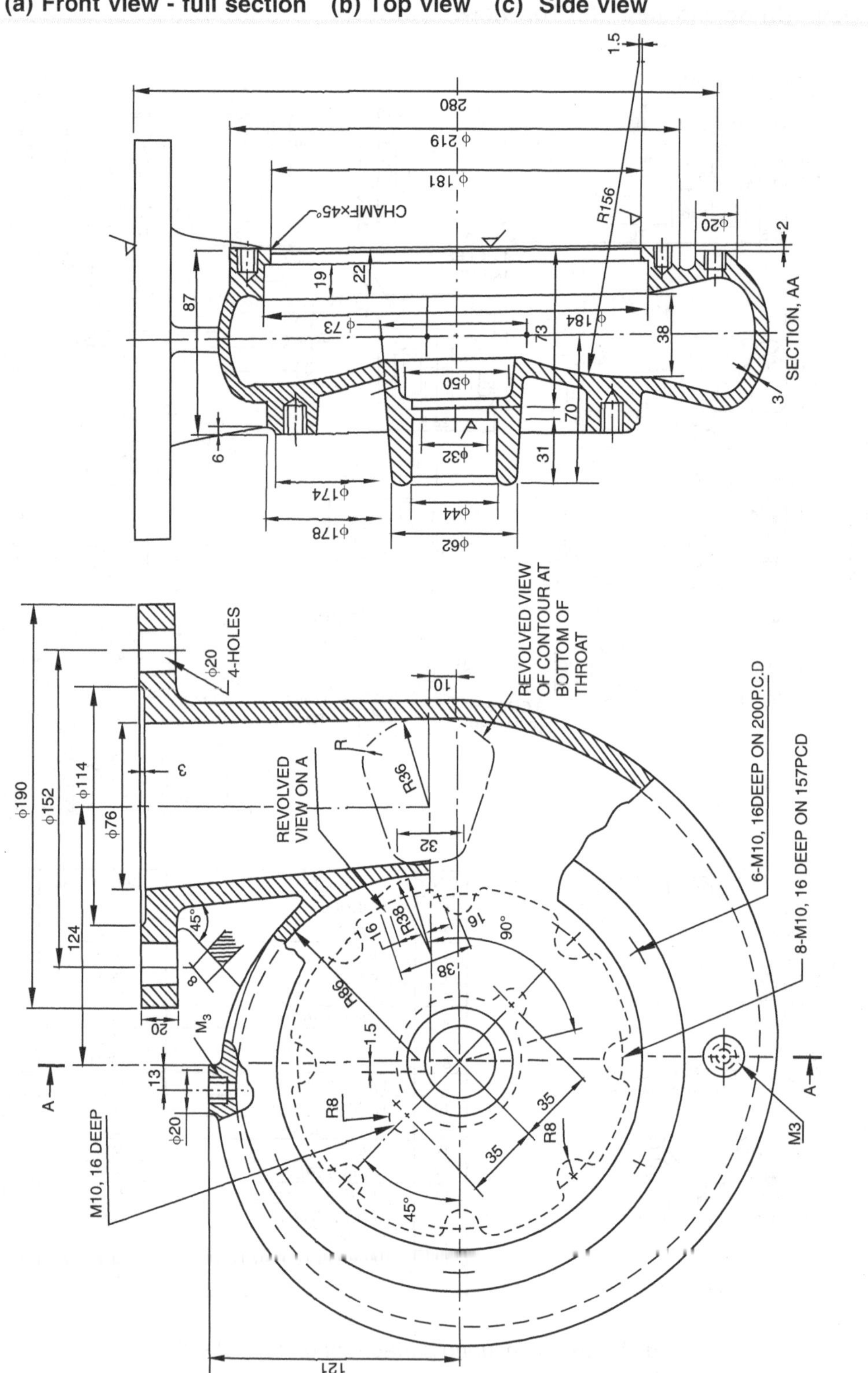

Fig. 19.41. *Centrifugal Pump*

Exercise 31 : Fig. 19.42 shows the two views of a carburettor. Draw the following views to some suitable scale :

(a) Sectional front view (b) Side view (c) Top view

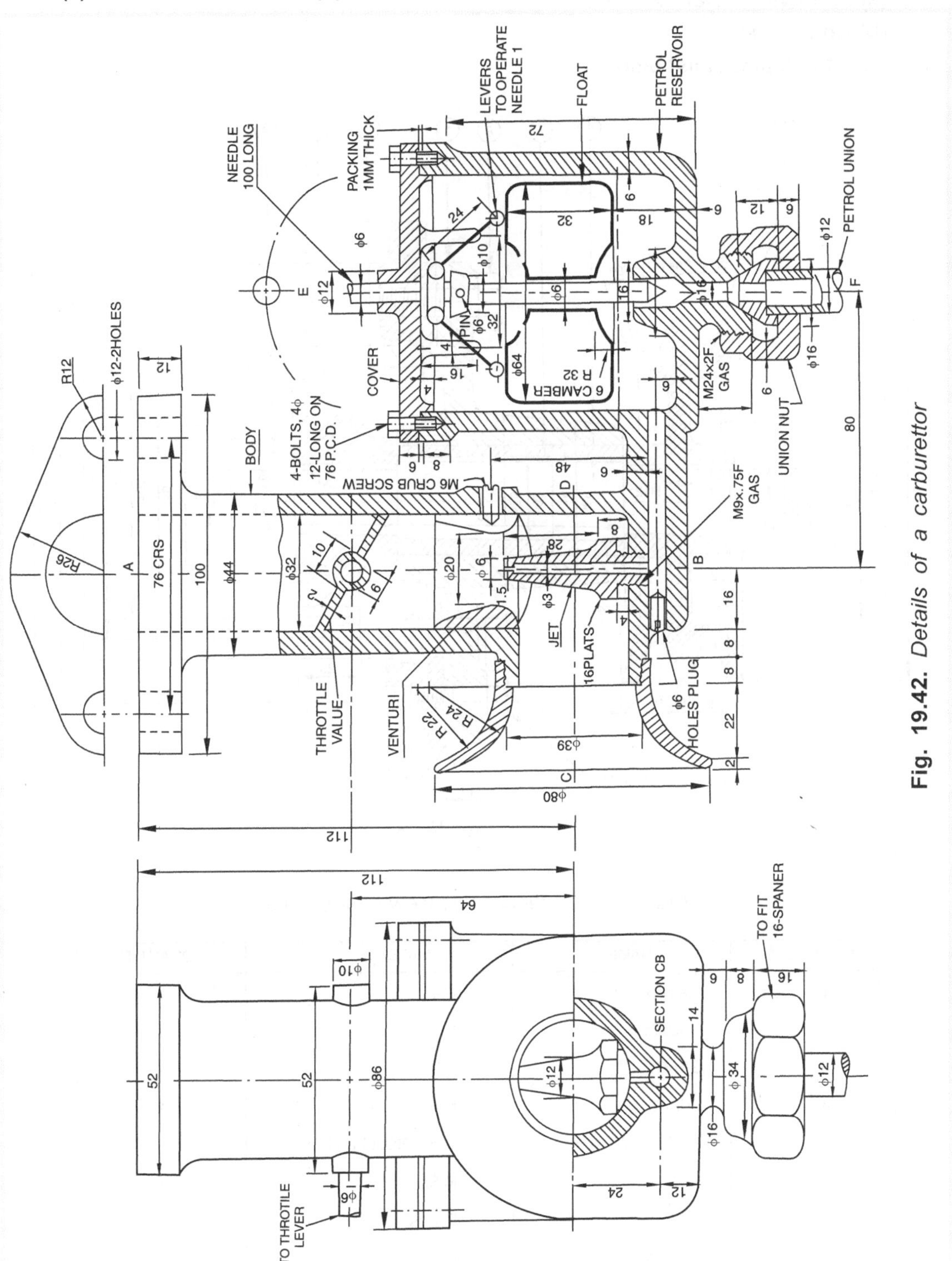

Fig. 19.42. Details of a carburettor

Exercise 32 : Fig. 19.43 shows the orthographic view of a belt drive. Draw the following views to full size scale :

(a) Front view - Half section.

(b) An end view.

Give the material list also

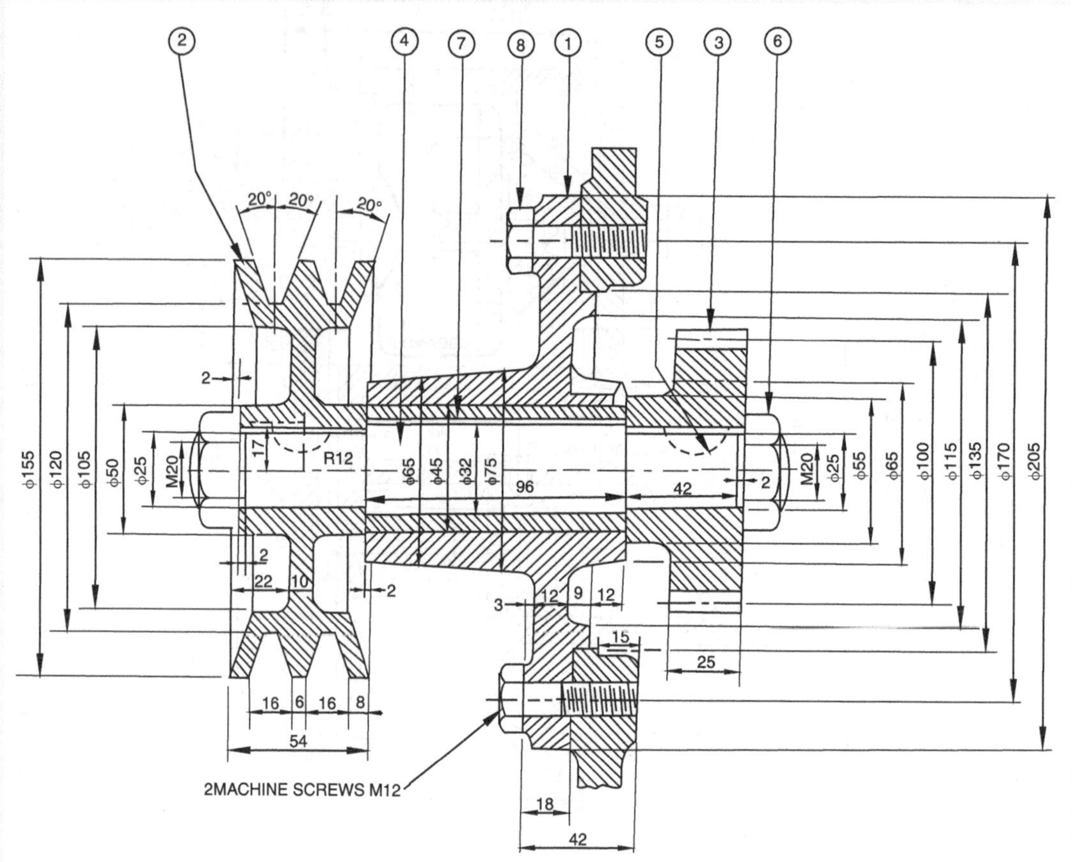

Fig. 19.43. *Orthographic view of belt drive*

PART NO.	NAME	MATERIAL	QUANTITY
1	MAIN BODY	C.I.	1
2	PULLEY	C.I.	1
3	GEAR	C.I.	1
4	SHAFT	M.S.	1
5	WOODRUFF KEY	COLD ROLLED M.S.	2
6	NUT	M.S.	2
7	BUSH	BRASS	1
8	SCREW	M.S.	1

19.16. CRANE HOOK

It is used to carry loads by means of chains or rope slings attached to the hooks.

Exercise 33 : Fig. 19.44 shows the details of a crane hook. Assemble all the parts and draw the following views:

(a) Front view - in full section

(b) Side view

(c) Top view.

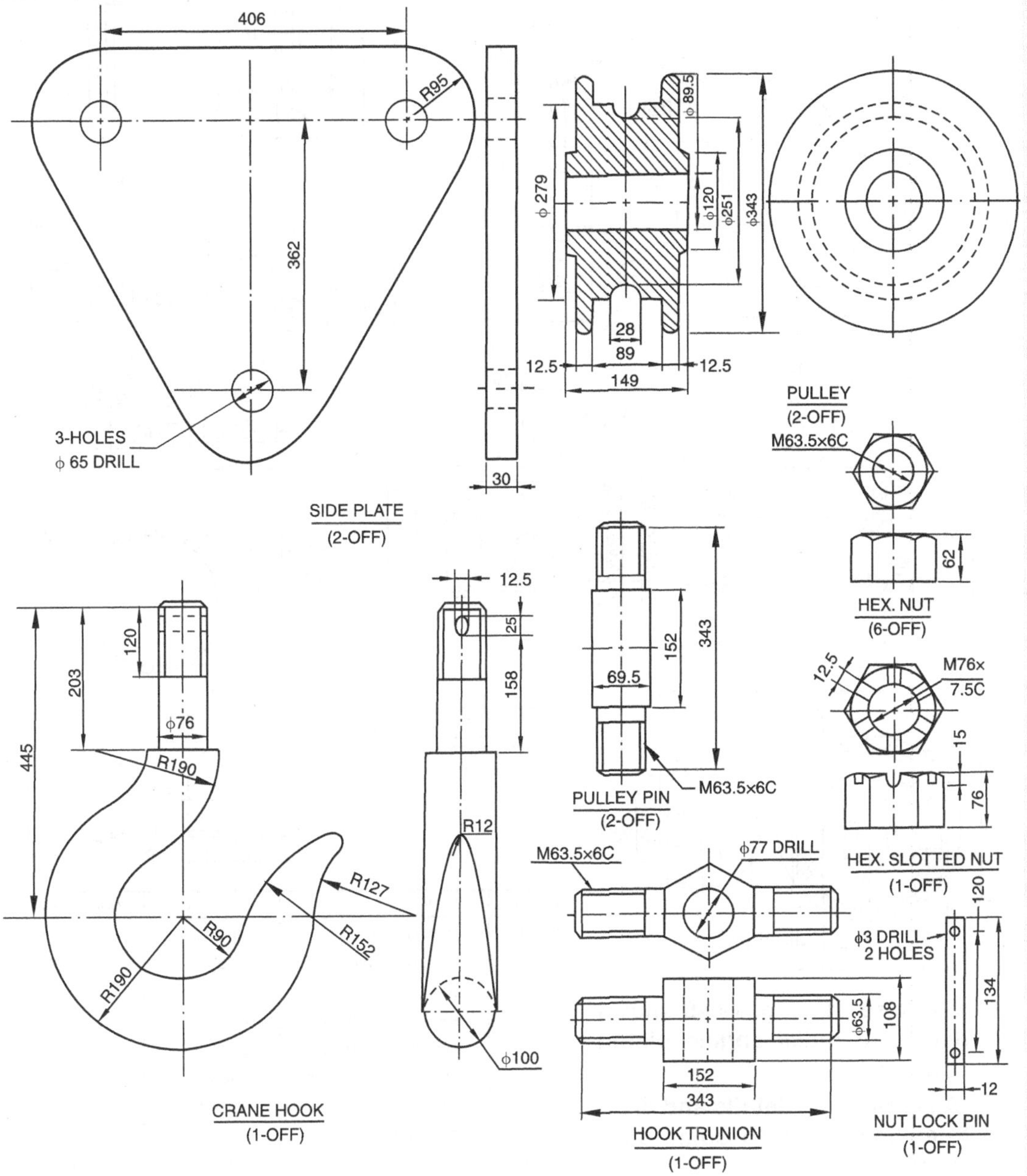

Fig. 19.44. *Details of a crane hook*

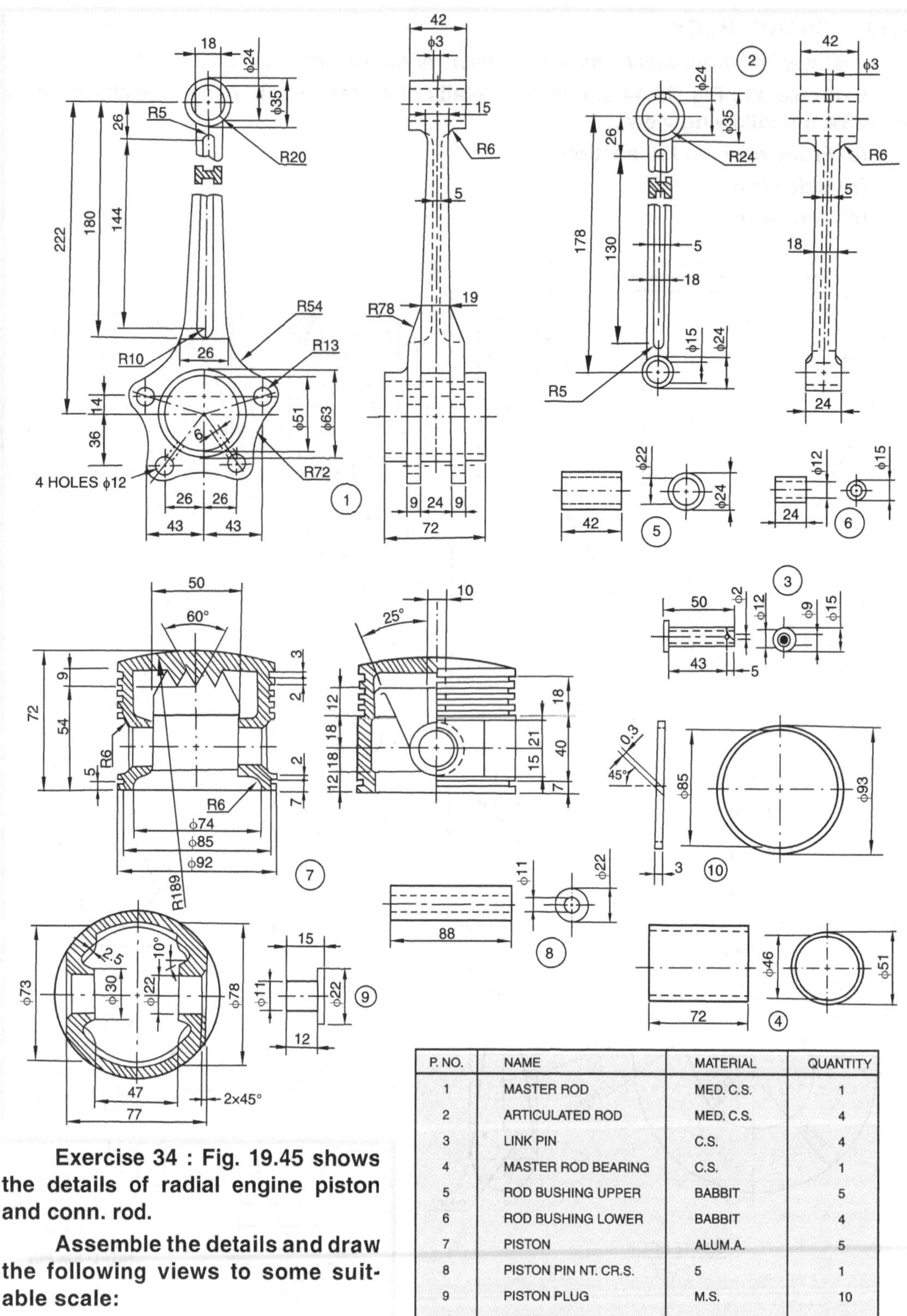

P. NO.	NAME	MATERIAL	QUANTITY
1	MASTER ROD	MED. C.S.	1
2	ARTICULATED ROD	MED. C.S.	4
3	LINK PIN	C.S.	4
4	MASTER ROD BEARING	C.S.	1
5	ROD BUSHING UPPER	BABBIT	5
6	ROD BUSHING LOWER	BABBIT	4
7	PISTON	ALUM.A.	5
8	PISTON PIN NT. CR.S.	5	1
9	PISTON PLUG	M.S.	10
10	PISTON RING	C.I. ALLOY	20

Fig. 19.45. *Details of Radial engine piston and rod*

Exercise 34 : Fig. 19.45 shows the details of radial engine piston and conn. rod.

Assemble the details and draw the following views to some suitable scale:

(a) Front view (b) Top view

Give the material list also.

Exercise 35: Fig. 19.46 shows the details of a warm gear box. Assemble the parts and draw the following views :

(a) Front view - left half in section (b) Side view - right half in section (c) Top view.

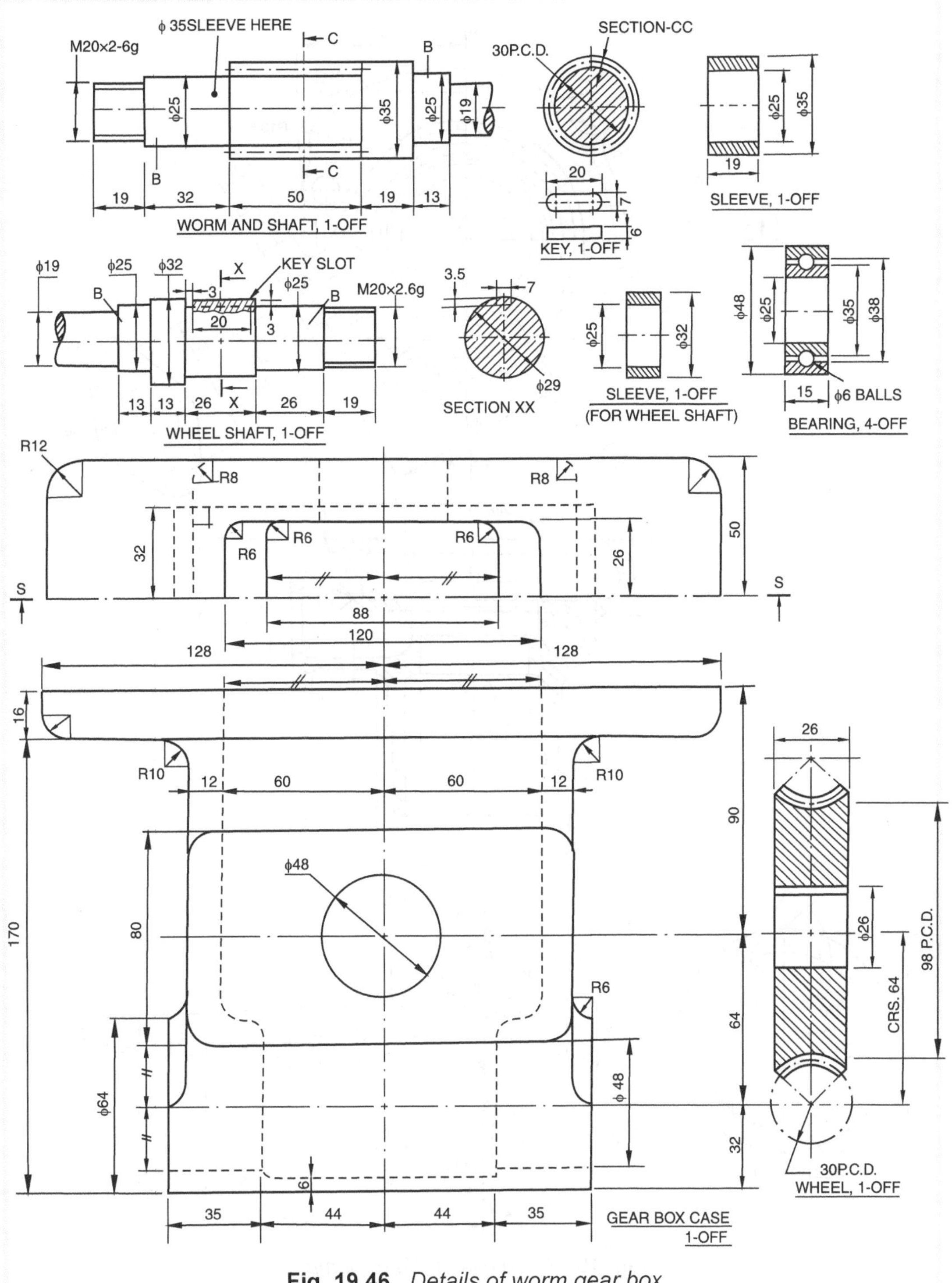

Fig. 19.46. *Details of worm gear box*

Exercise 36 : Fig. 19.47 and Fig. 19.48 show the details of a water pump. Assemble the parts and draw the following views to some suitable scale:

(a) Sectional front view (b) Side view (c) Top view

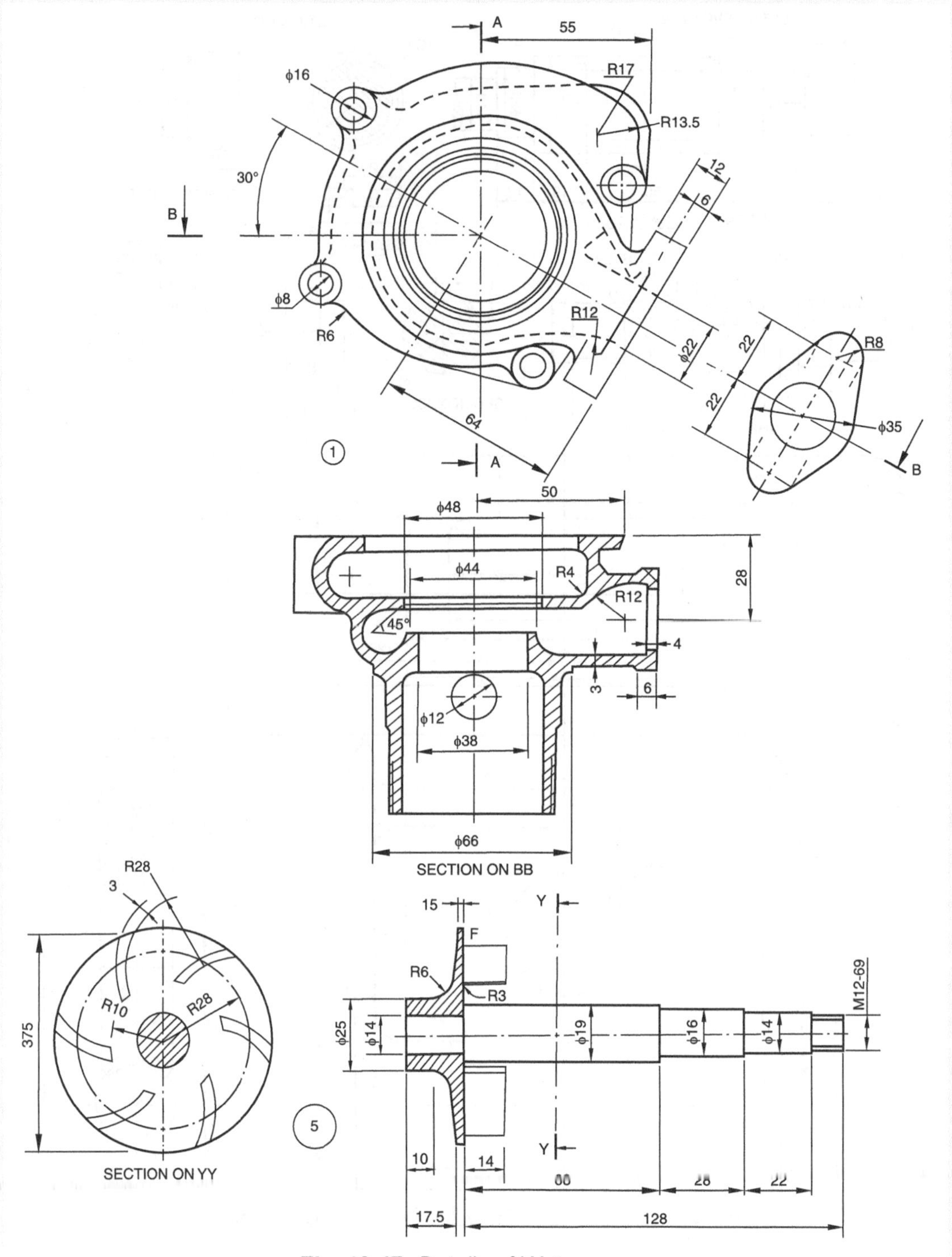

Fig. 19.47. *Details of Water pump*

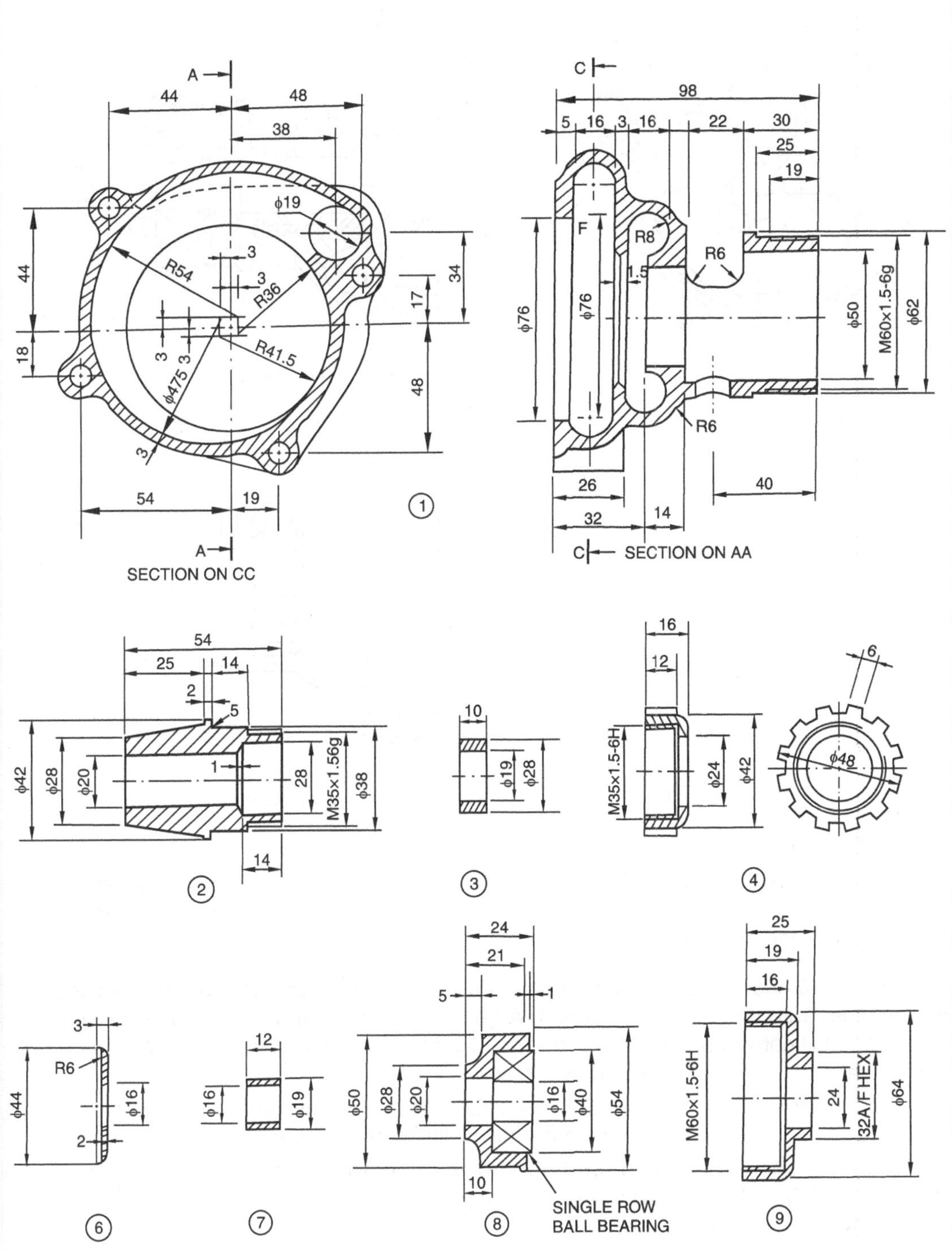

Fig. 19.48. *Details of a water pump*

Fig. 19.49. *Shows the general arrangement of water pump*

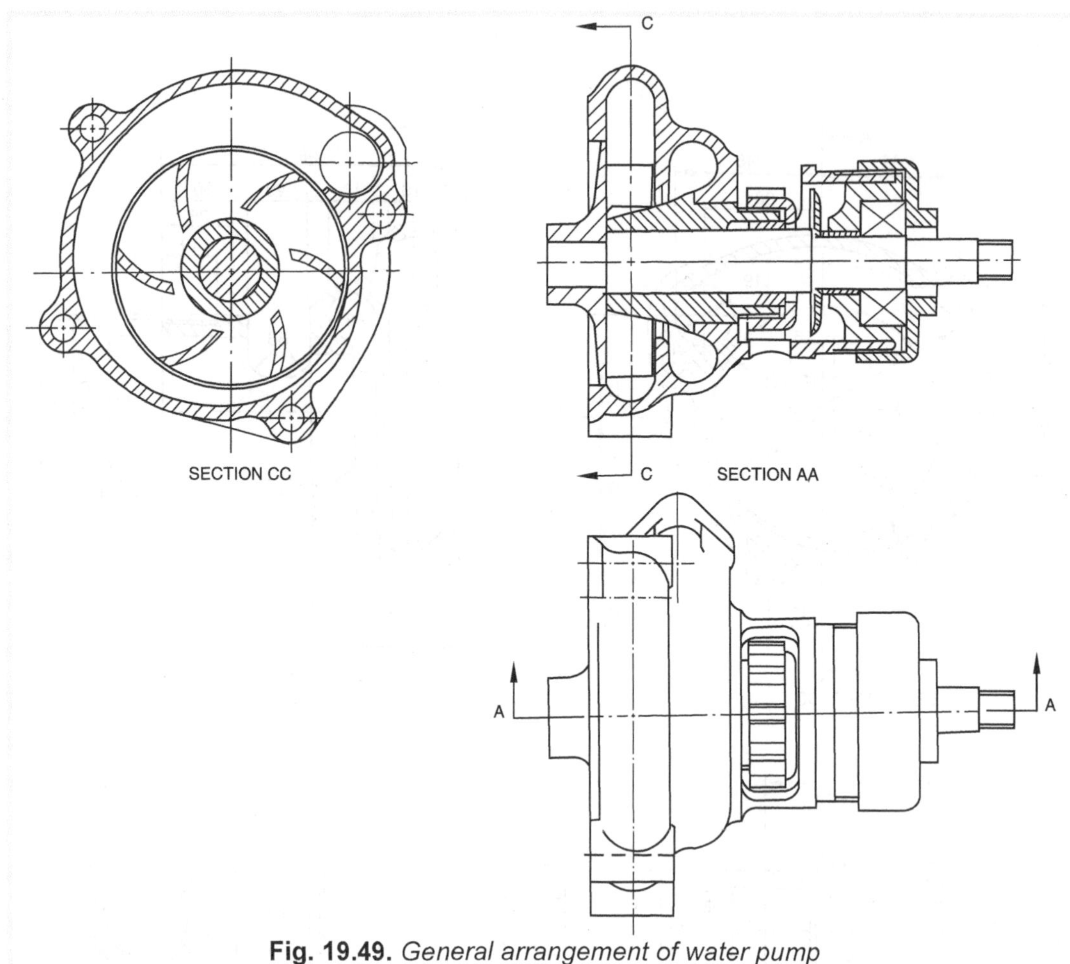

Fig. 19.49. *General arrangement of water pump*

19.17. HAND DRILL

A hand drill is used for drilling holes in wood and in sheet metal works by the pressure of hand. As regards its principle of operation, a bevel gear train is used to convert the cranking motion of the hand into rotary motion of the drill.

Fig. 19.50 shows the details of a hand drill. A housing supports two bevel pinions. The pinions mesh with a driving bevel wheel. A gear and knob shaft pass through housing. The shaft supports knob at one end freely rotating bevel pinion on the other end. The crank is fitted on the gear and the gear is retained by a screw.

A chuck, which is to hold the drill, consists of three jaws assembled together by springs inside the chuck cap. The complete chuck unit is screwed on to the lower pinion shaft.

Exercise 37 : Fig. 19.50 (i) and (ii) shows the details of a hand drill. Assemble all the parts and draw the following views to a scale full size:

(a) Front view - full section

(b) End view and

(c) Top view

Prepare the parts and material list.

Fig. 19.51 shows the general arrangement (assembly) of hand drill.

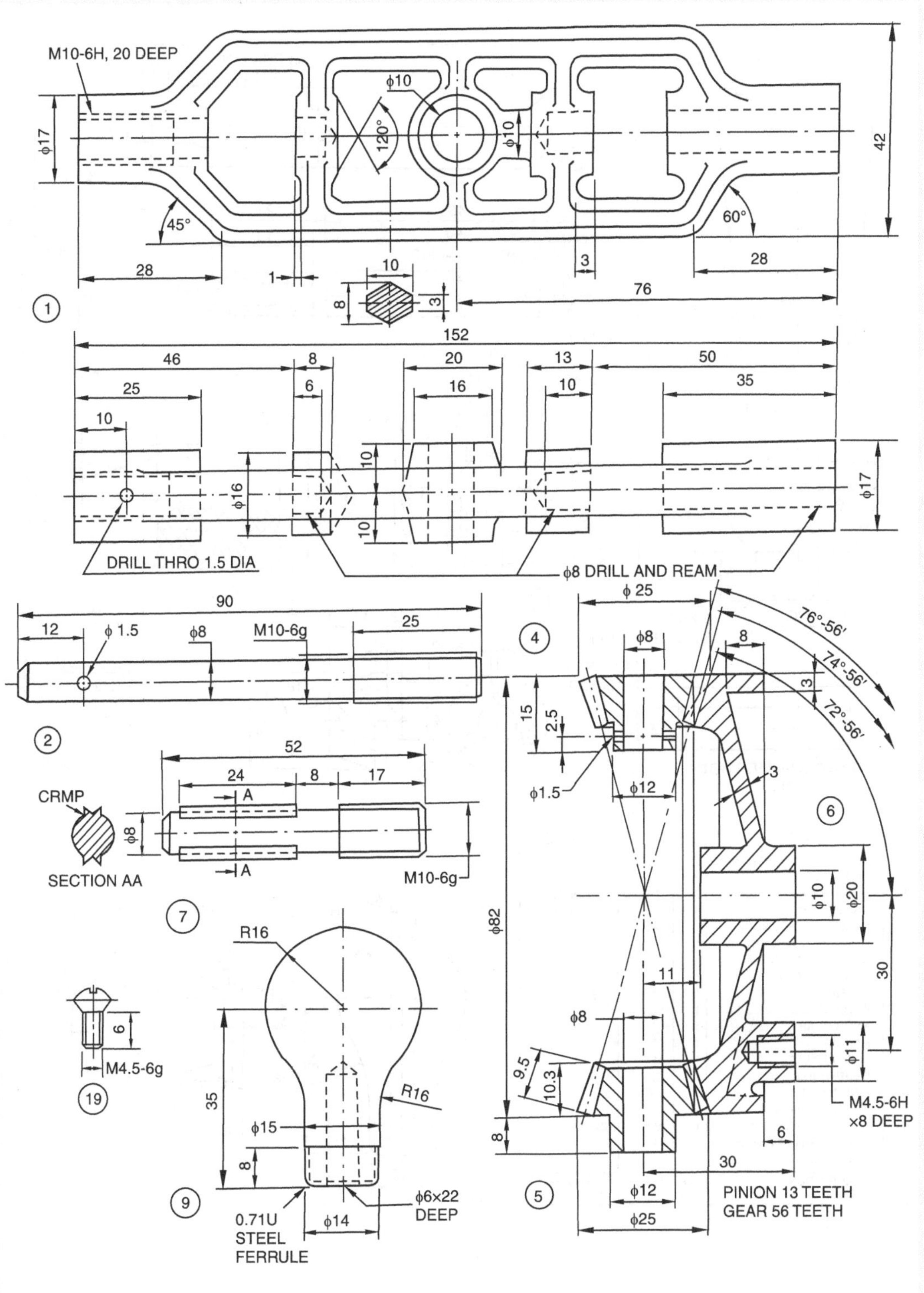

Fig. 19.50. (i) *Details of a hand drill*

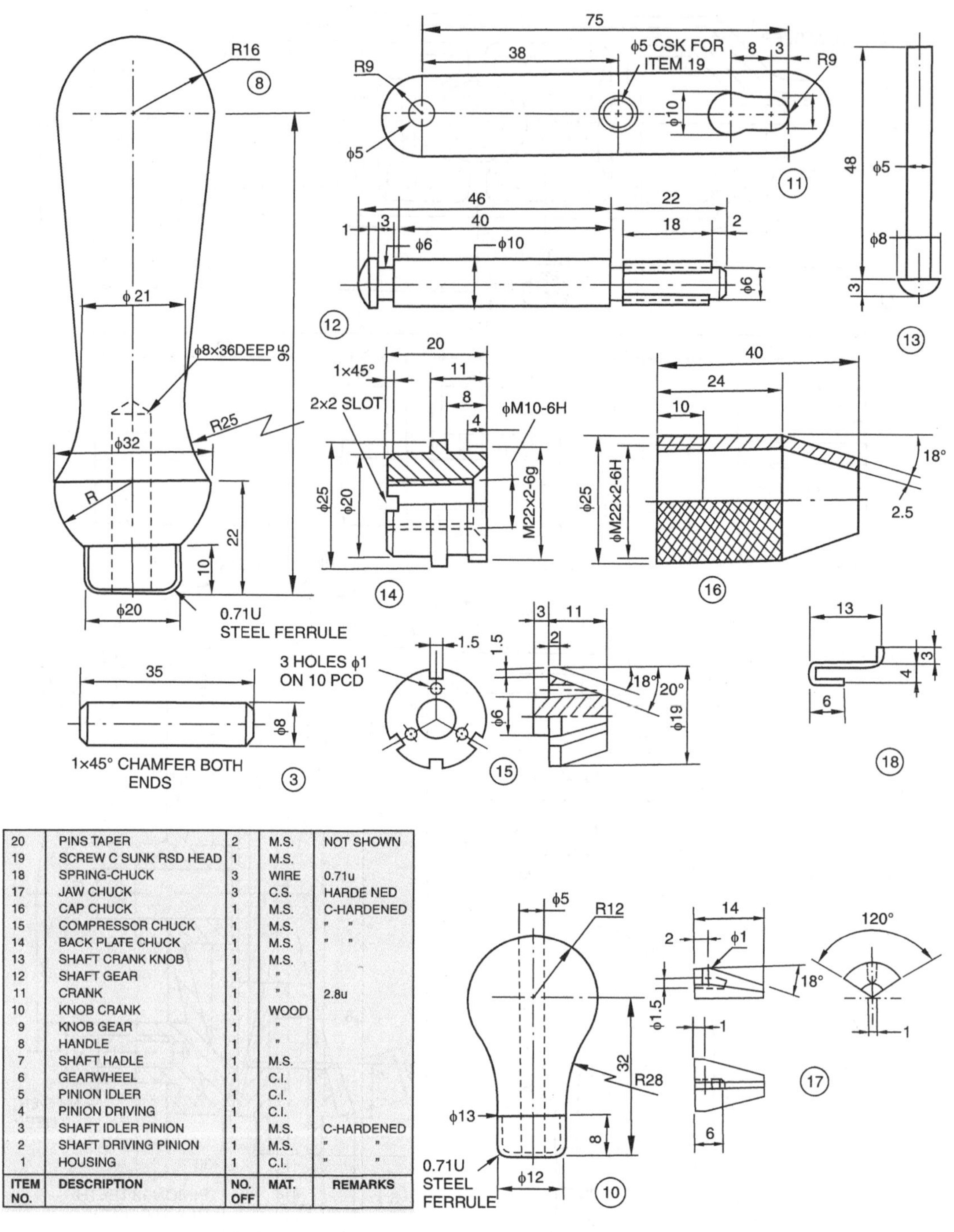

20	PINS TAPER	2	M.S.	NOT SHOWN
19	SCREW C SUNK RSD HEAD	1	M.S.	
18	SPRING-CHUCK	3	WIRE	0.71u
17	JAW CHUCK	3	C.S.	HARDE NED
16	CAP CHUCK	1	M.S.	C-HARDENED
15	COMPRESSOR CHUCK	1	M.S.	" "
14	BACK PLATE CHUCK	1	M.S.	" "
13	SHAFT CRANK KNOB	1	M.S.	
12	SHAFT GEAR	1	"	
11	CRANK	1	"	2.8u
10	KNOB CRANK	1	WOOD	
9	KNOB GEAR	1	"	
8	HANDLE	1	"	
7	SHAFT HADLE	1	M.S.	
6	GEARWHEEL	1	C.I.	
5	PINION IDLER	1	C.I.	
4	PINION DRIVING	1	C.I.	
3	SHAFT IDLER PINION	1	M.S.	C-HARDENED
2	SHAFT DRIVING PINION	1	M.S.	" "
1	HOUSING	1	C.I.	" "
ITEM NO.	DESCRIPTION	NO. OFF	MAT.	REMARKS

Fig. 19.50. (ii) *Details of a hand drill*

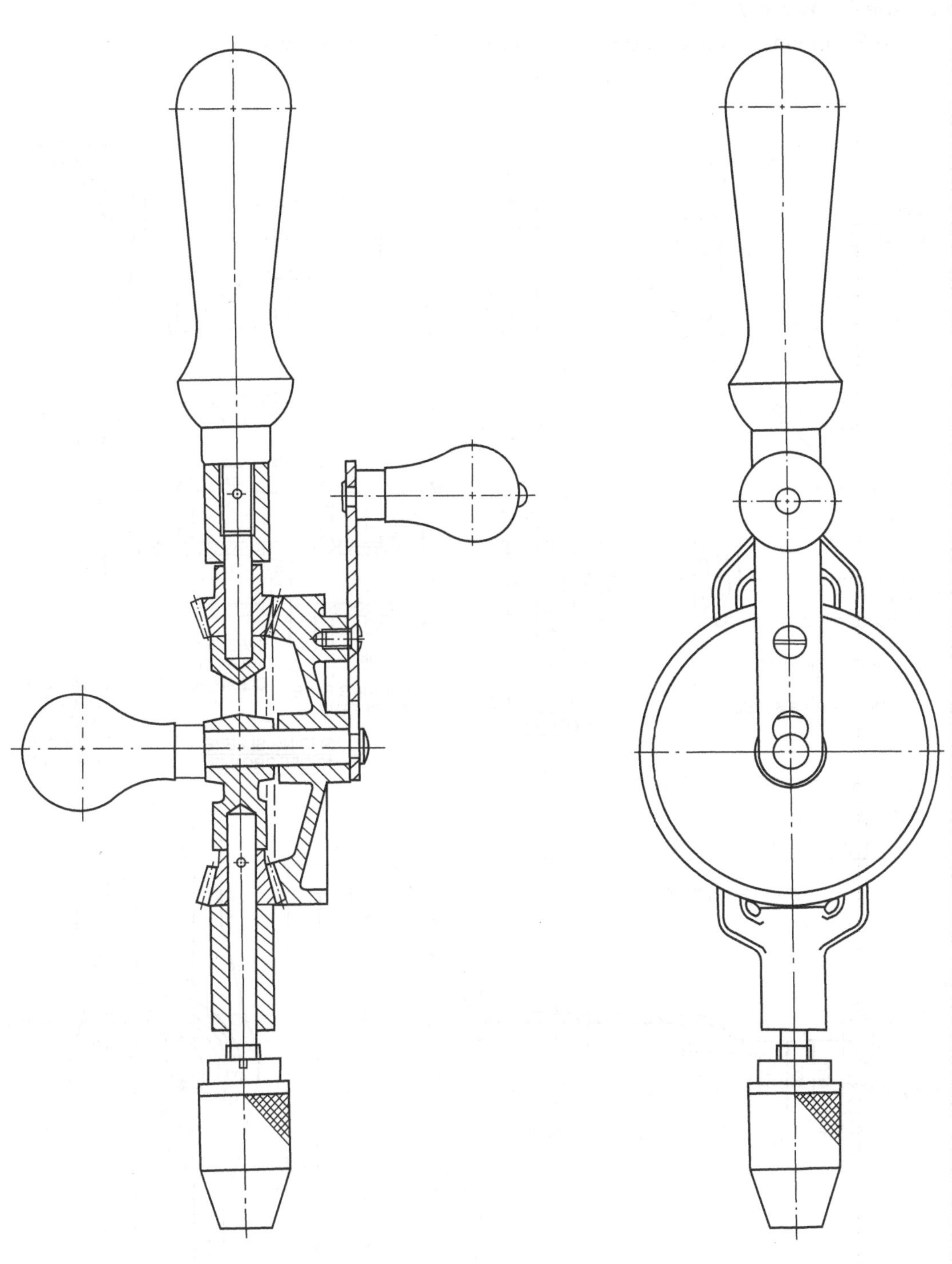

Fig. 19.51. *General arrangement (assembly) of hand drill*

Exercise 38 : Fig. 19.52 shows the details of Governor. Assemble all the parts and draw the following views.

(a) Front view - full section (b) Side view (c) Top view

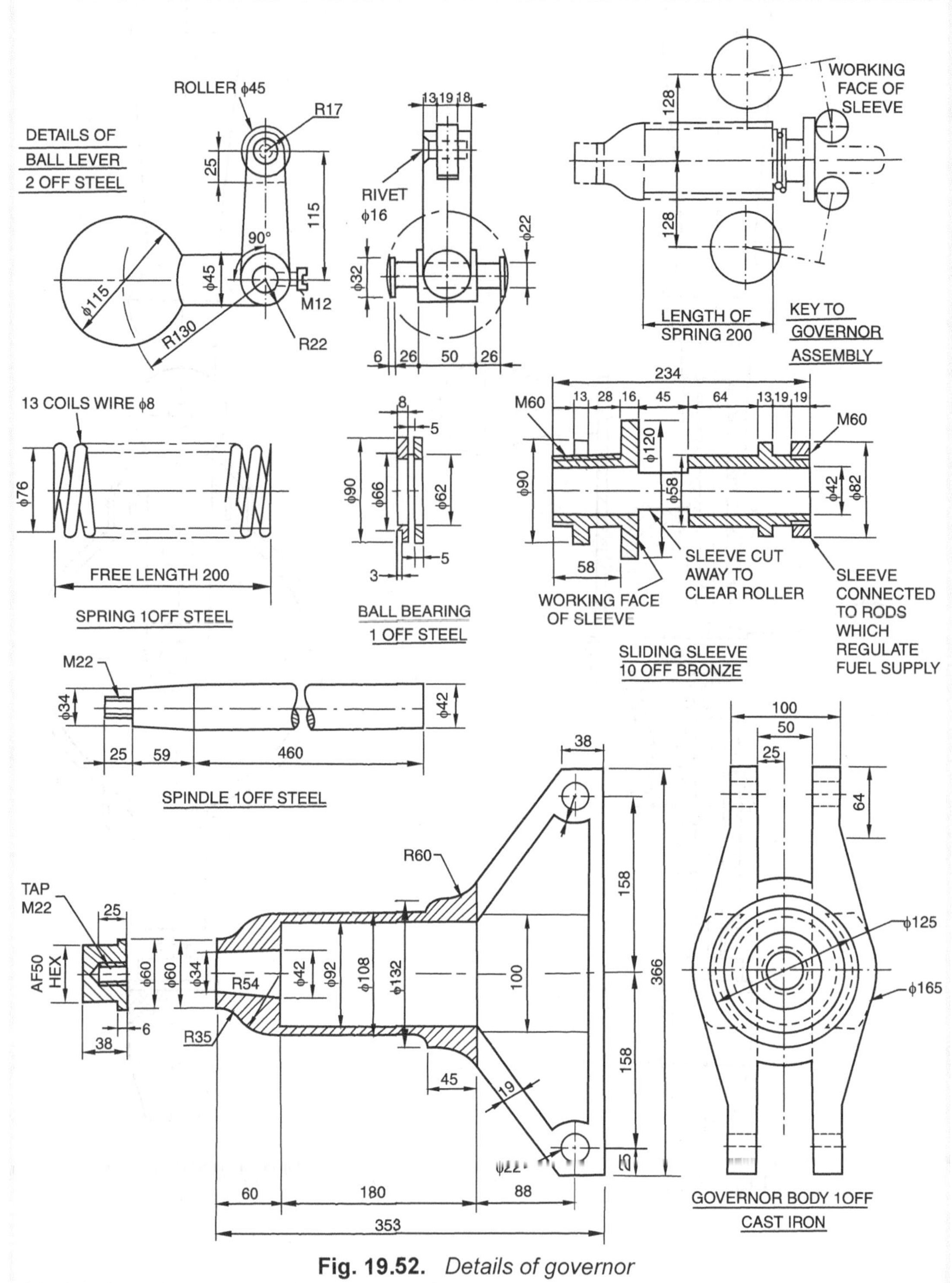

Fig. 19.52. *Details of governor*

Exercise 39 : Fig. 19.53 shows a pipe vice. Draw the following to 1 : 1 scale :

(a) Full sectional elevation

(b) End view and

(c) Top view

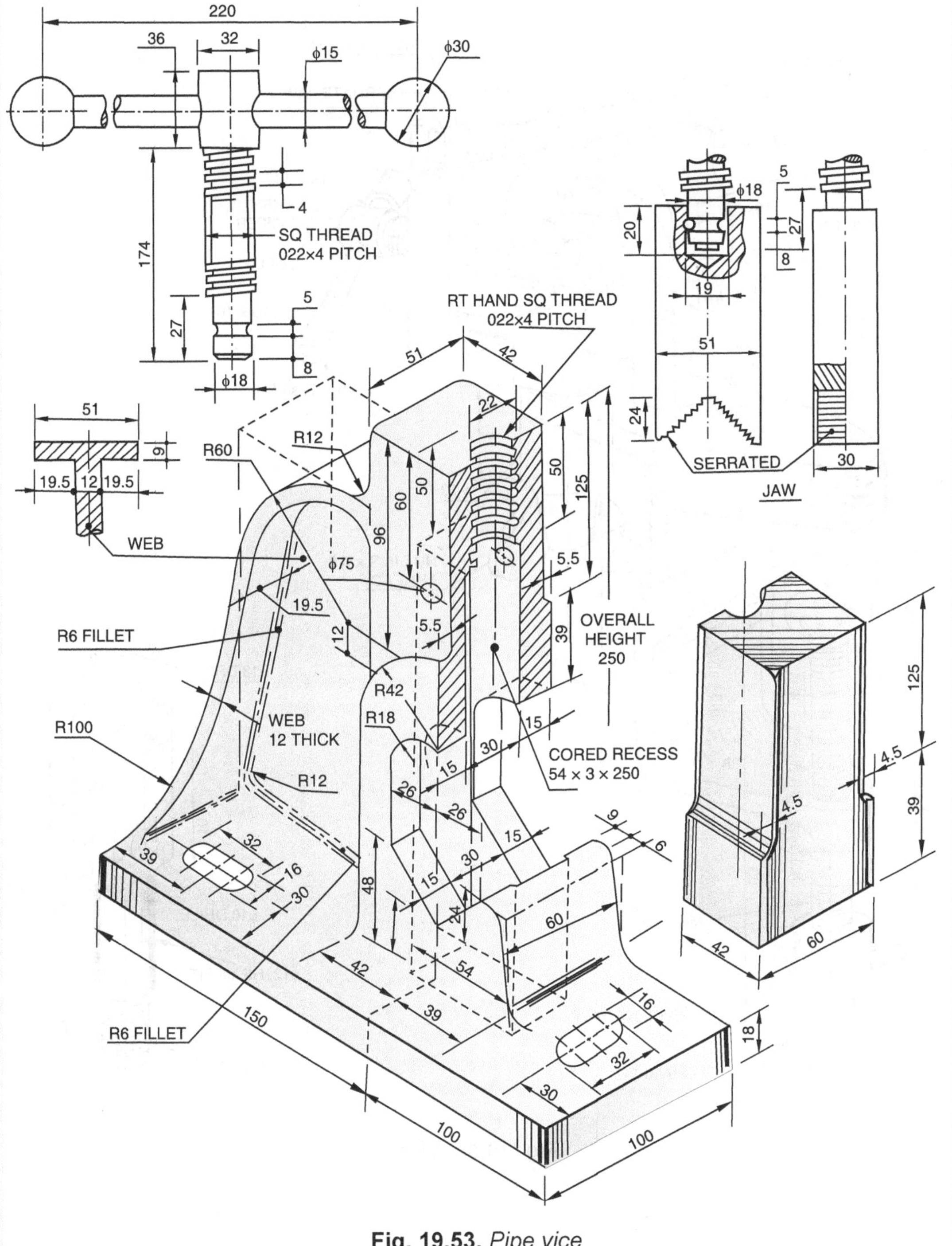

Fig. 19.53. *Pipe vice*

Exercise 40 : Fig. 19.54 shows a vice. Draw the following views to a full size scale:

(a) Sectional front view

(b) Sectional side view and

(c) Top view

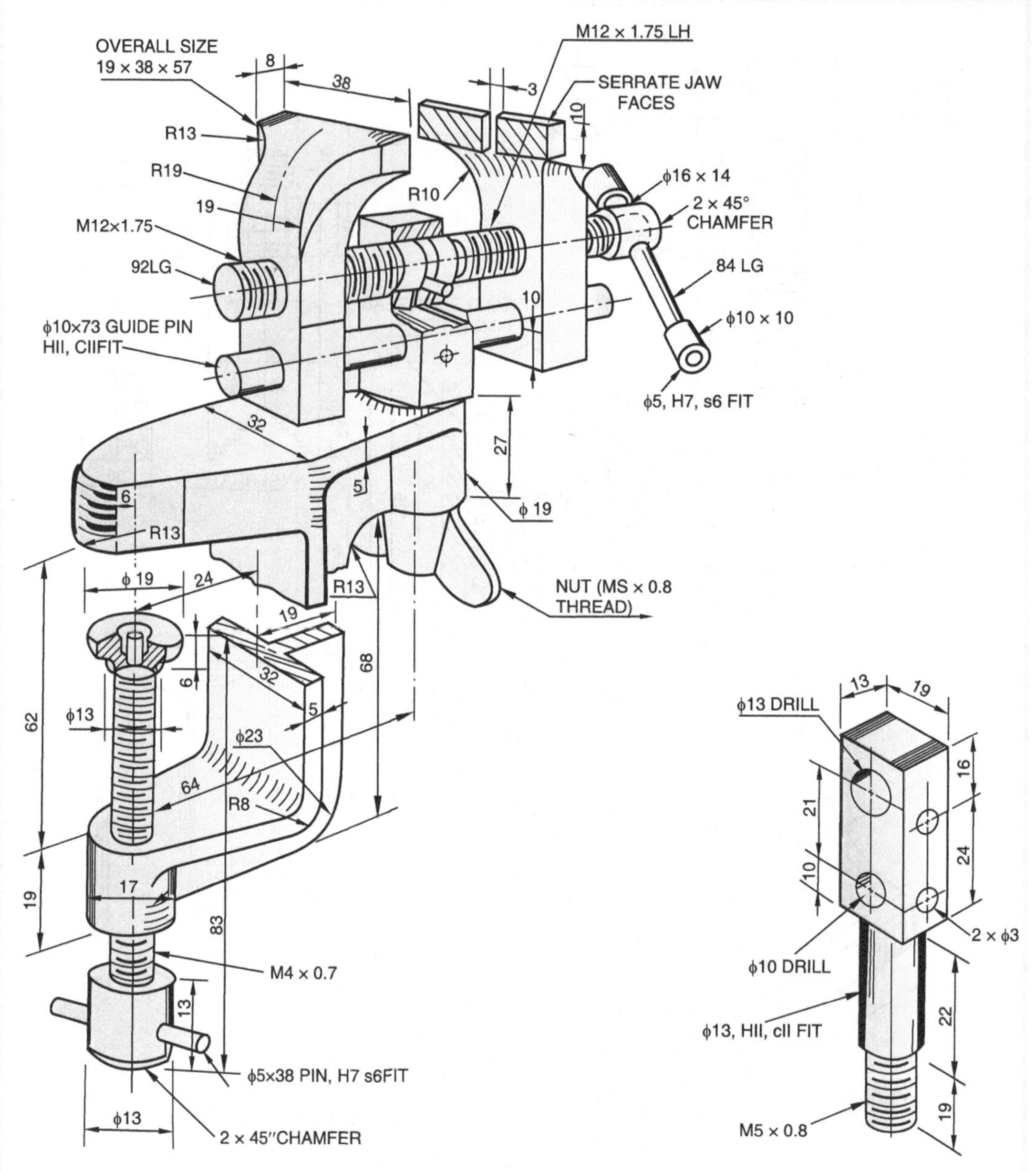

Fig. 19.54. *Vice*

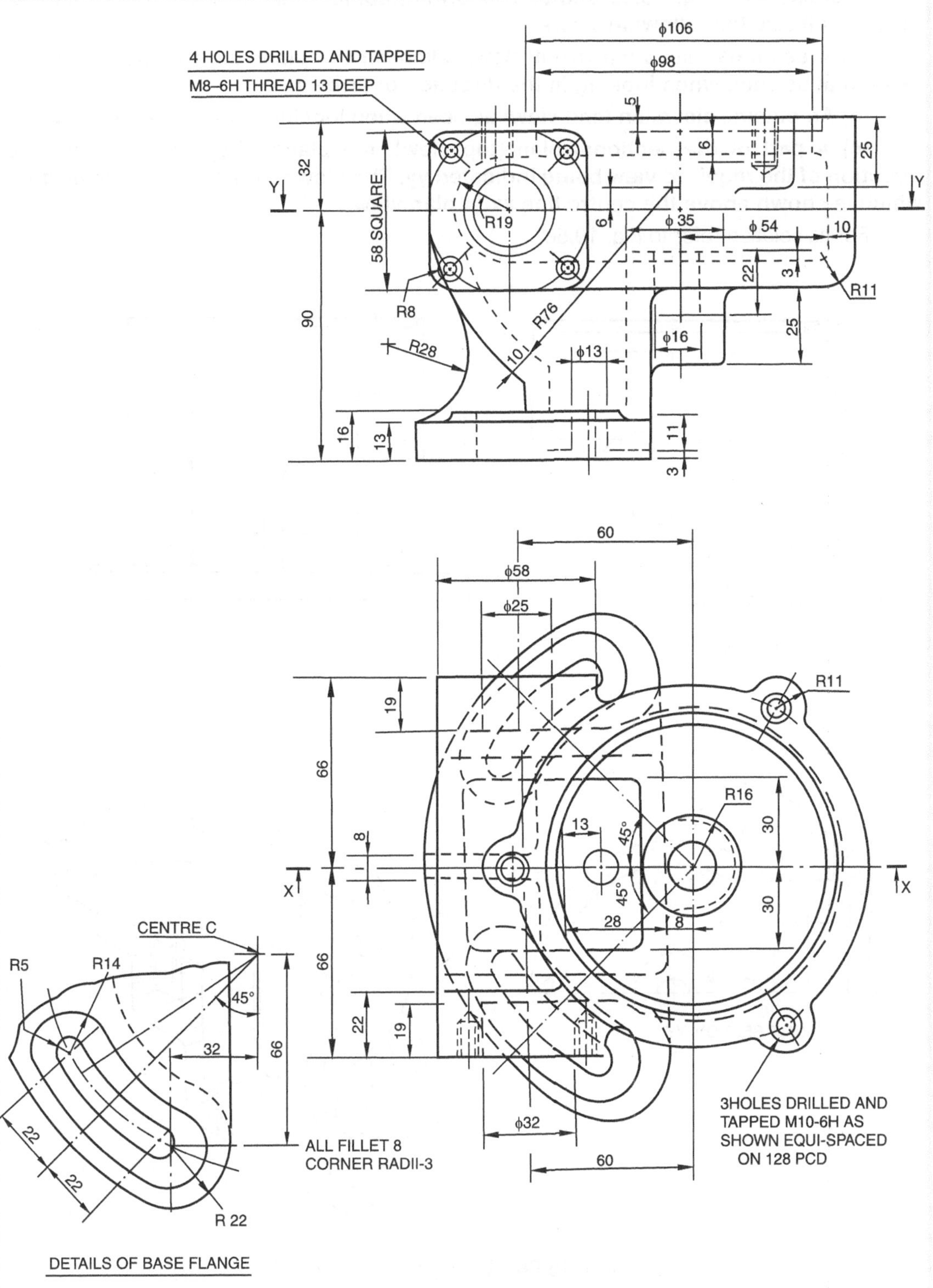

Fig. 19.55. *Orthographic views of worm gear casting*

Exercise 41 : Fig. 19.55 shows two orthographic views of worm gear casting. Draw to full size the following views :

(a) A sectional elevation (front view) taken on the plane xx, the view required being that as seen when looking in the direction of arrows.

(b) An outside elevation (end view) as seen when looking in direction of arrow E.

(c) A part of the sectional plan (top view), the plane of the section and the direction of the required view being indicated yy. The view reqruired is of that portion which is shown above the centre line xx in plan view.

For its solution refer to Fig. 19.56

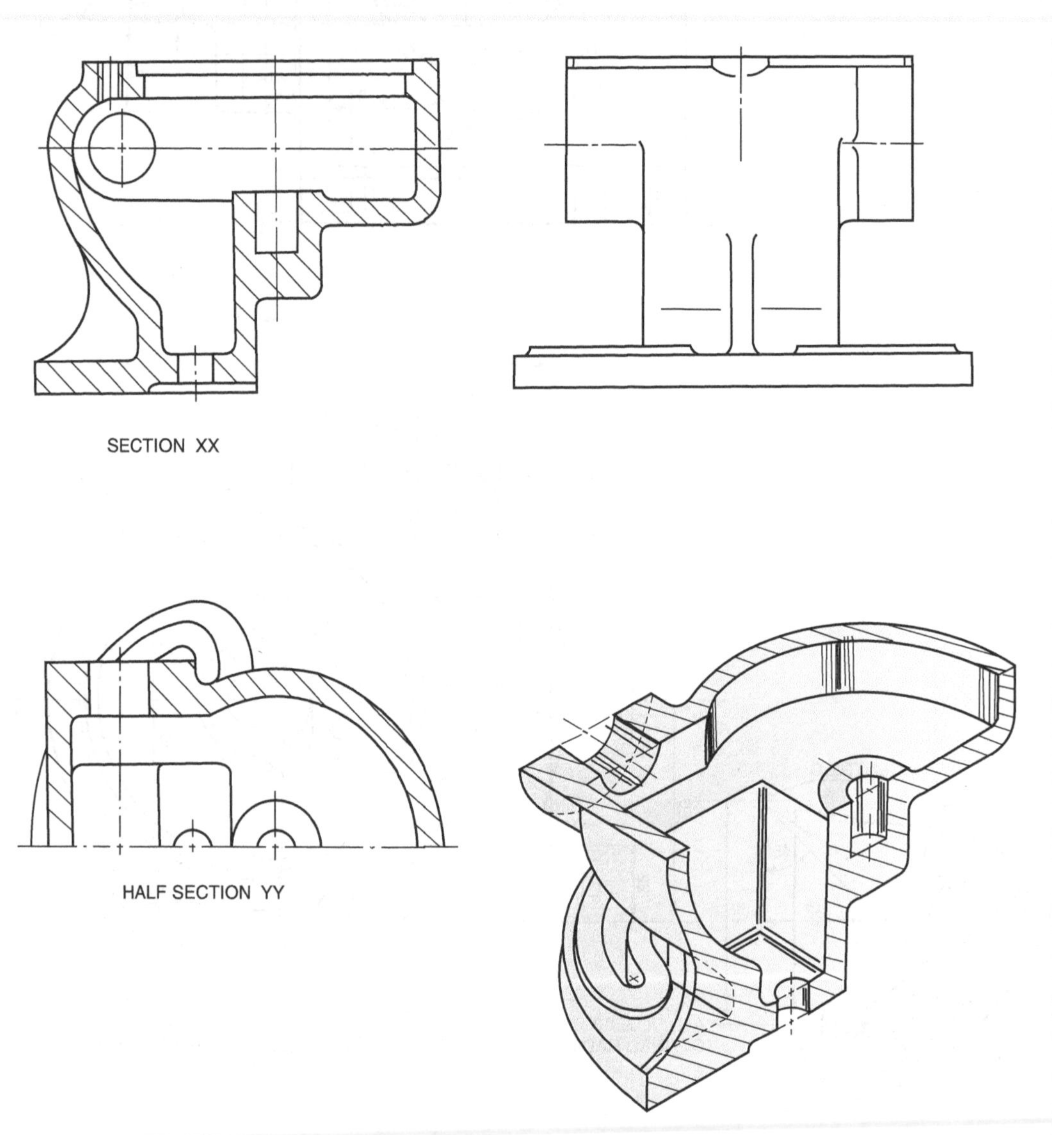

Fig. 19.56. *Worm gear casting*

19.18. SHAFT BRACKET

The shaft bracket is used for supporting the bearing carrying shaft. In this shaft bracket, the brasses, which are in the form of a split collared tube, are used to facilitate assembly and maintenance. By removing the cap and brasses, the shaft and bevel gear asserribly can be dismentled without withdrawing the shaft axially. The other gear runs in a bush of the shaft bracket.

Exercise 42 : Fig. 19.57 shows the details of a shaft bracket. Assemble the parts and draw the following views to full size scale :

(a) A full sectional front view

(b) Half outside top view and

(c) Right side view.

For part solution refer Fig. 19.58.

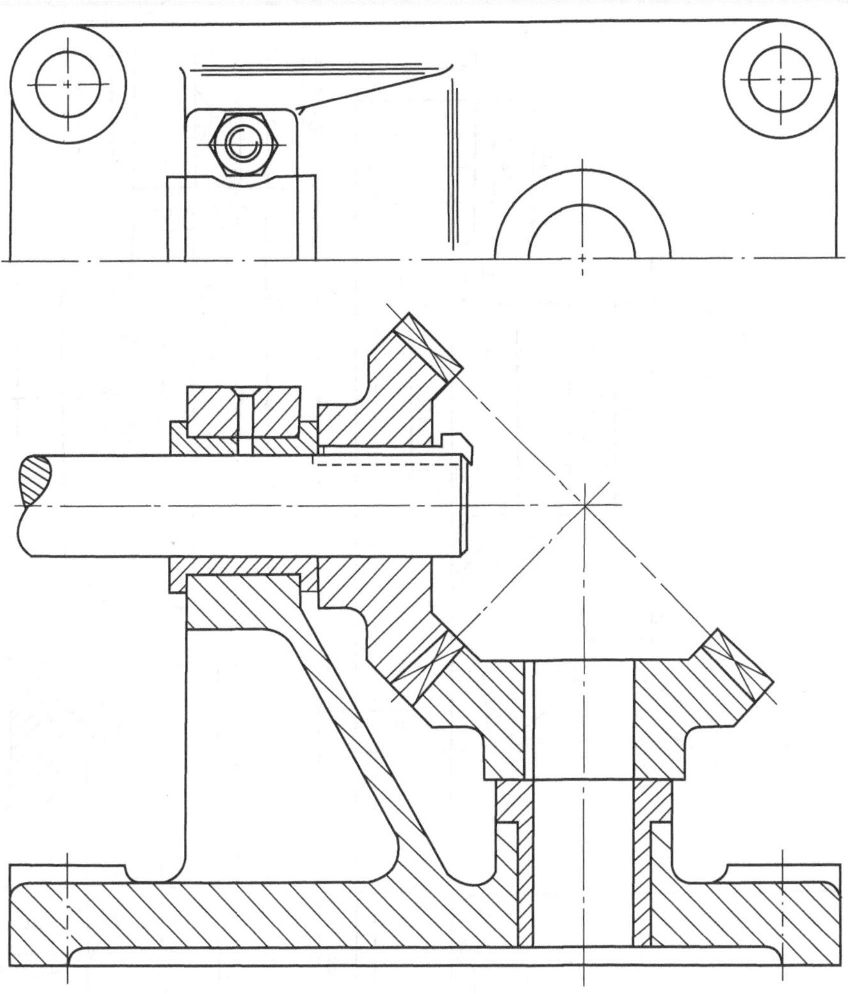

Fig. 19.58. *Part solution to exercise 41*

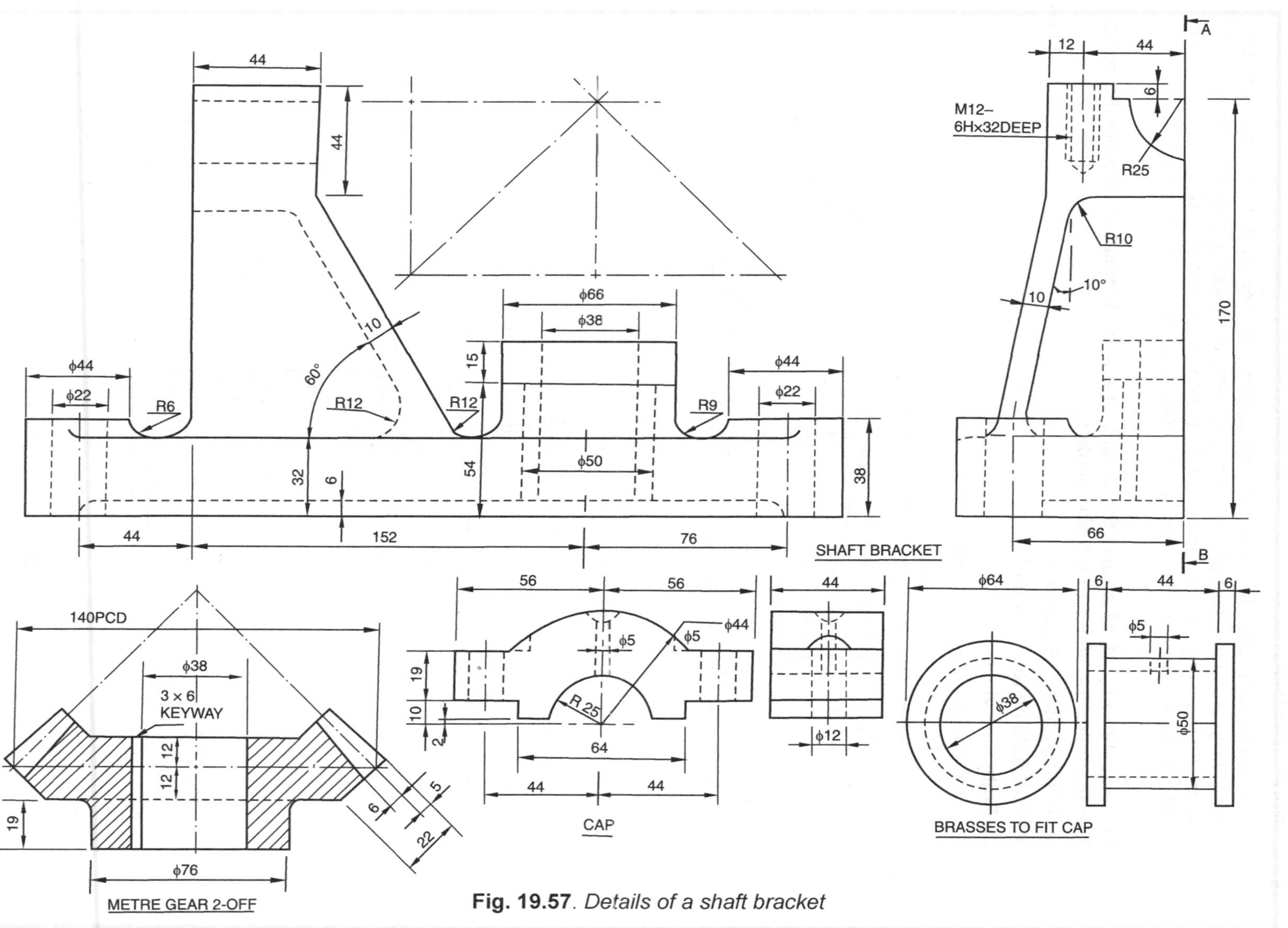

Fig. 19.57. *Details of a shaft bracket*

Exercise 43: Fig. 19.59 shows the details of a 3 way valve. Assemble the different parts and draw the following views to some suitable scale :

(a) Sectional front view (b) End view (c) Top view.

Prepare a parts list and material list also.

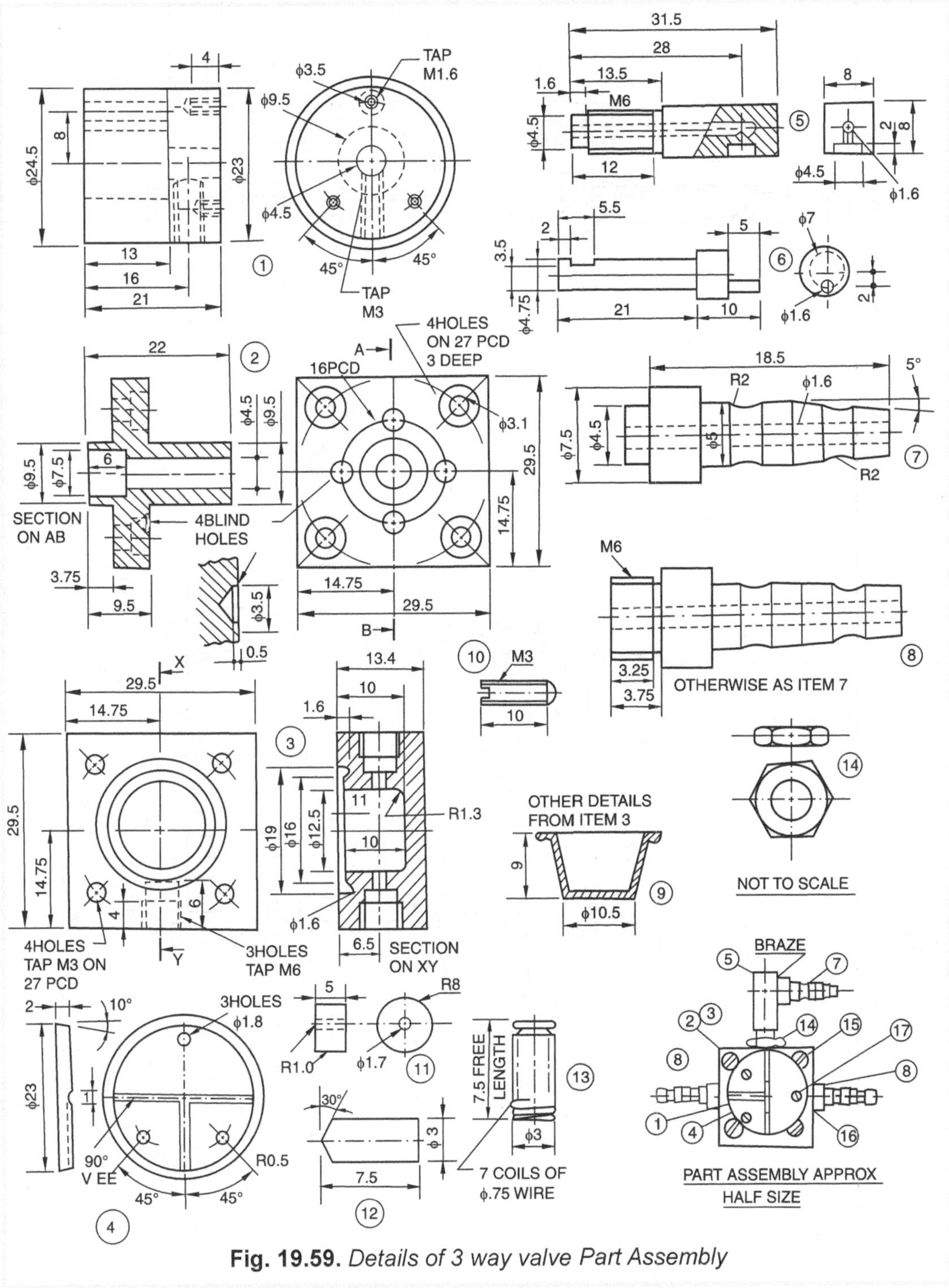

Fig. 19.59. *Details of 3 way valve Part Assembly*

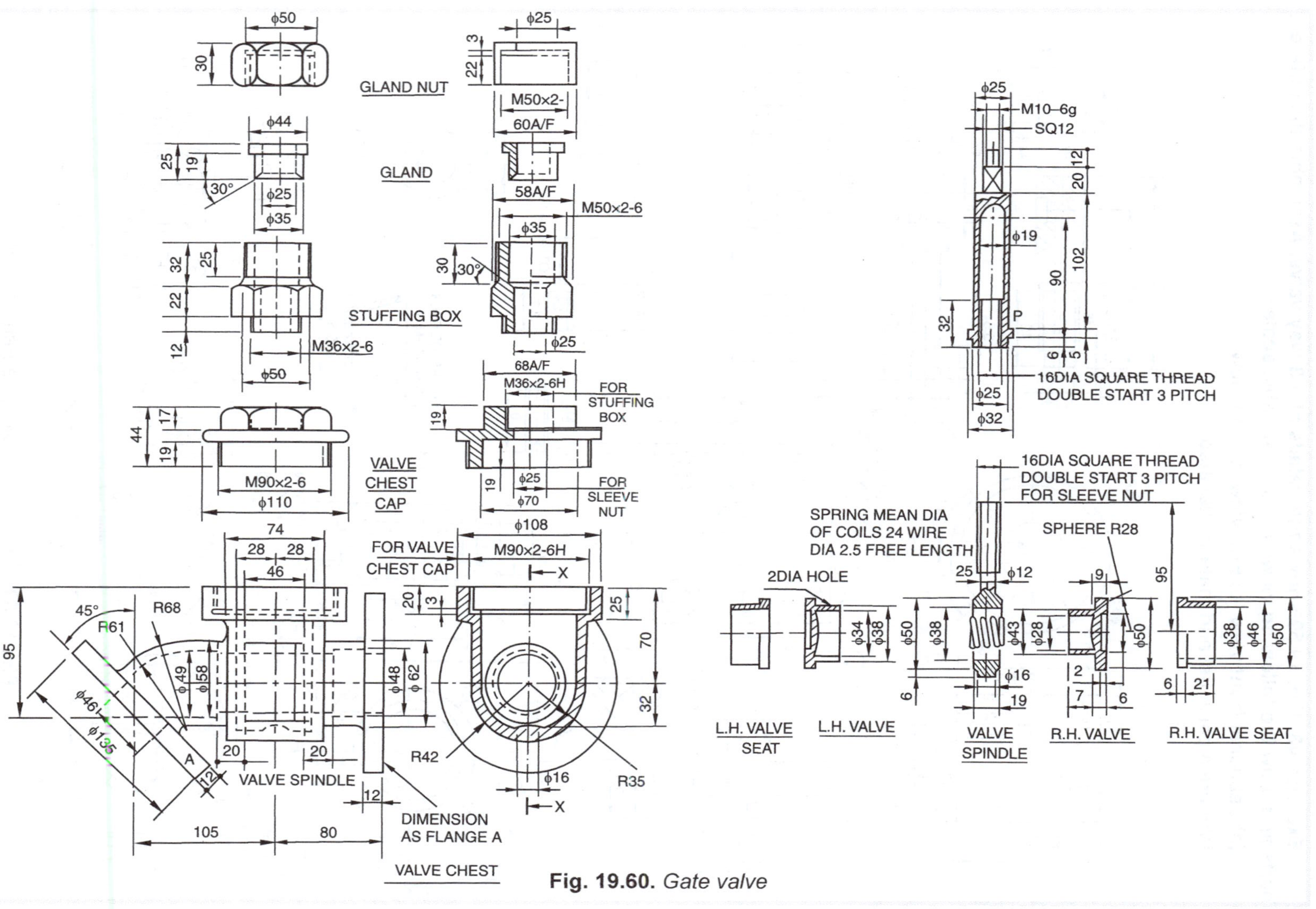

Fig. 19.60. *Gate valve*

Exercise 44 : Fig. 19.60 shows the details of the component, parts of a gate valve. Draw the following views of the assembled valve :

(a) A sectional elevation (front view) corresponding to the section plane xx shown in the given views of the valve chest.

(b) An outside elevation (end view) projected to the right hand side of view (a)

For its solution, refer Fig. 19.61.

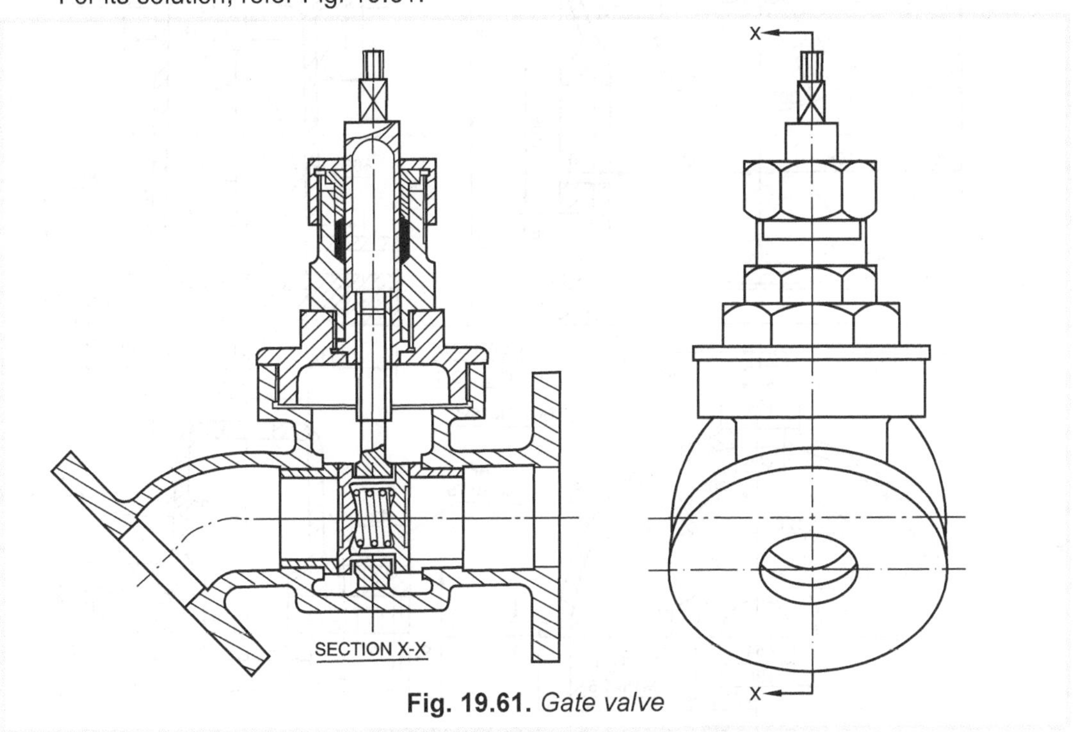

Fig. 19.61. *Gate valve*

PARTS LIST OF 3 WAY VALVE EX : 42

P.NO.	NAME	MATERIAL	NO. OFF
1.	KNOB	ALUM.	1
2.	BEARING	BRASS	1
3.	BODY	PERSPEX	1
4.	INDICATOR PLATE	ALUM	1
5.	EXTENSION PIECE	BRASS	1
6.	SPINDLE	M.S.	1
7.	NOZZLE	BRASS	1
8.	NOZZLE	BRASS	2
9.	SAC	RUBBER	1
10.	GRUB SCREW	M.S.	1
11.	ROLLER	M.S.	1
12.	PLUNGER	M.S.	1
13.	SPRING	STEEL	1
14.	M6 HEX. THIN NUT	M.S.	1
15.	M3 CH HD SCREW	M.S.	4
16.	WASHER	FIBRE	2
17.	M1-6 SCREW	M.S.	3

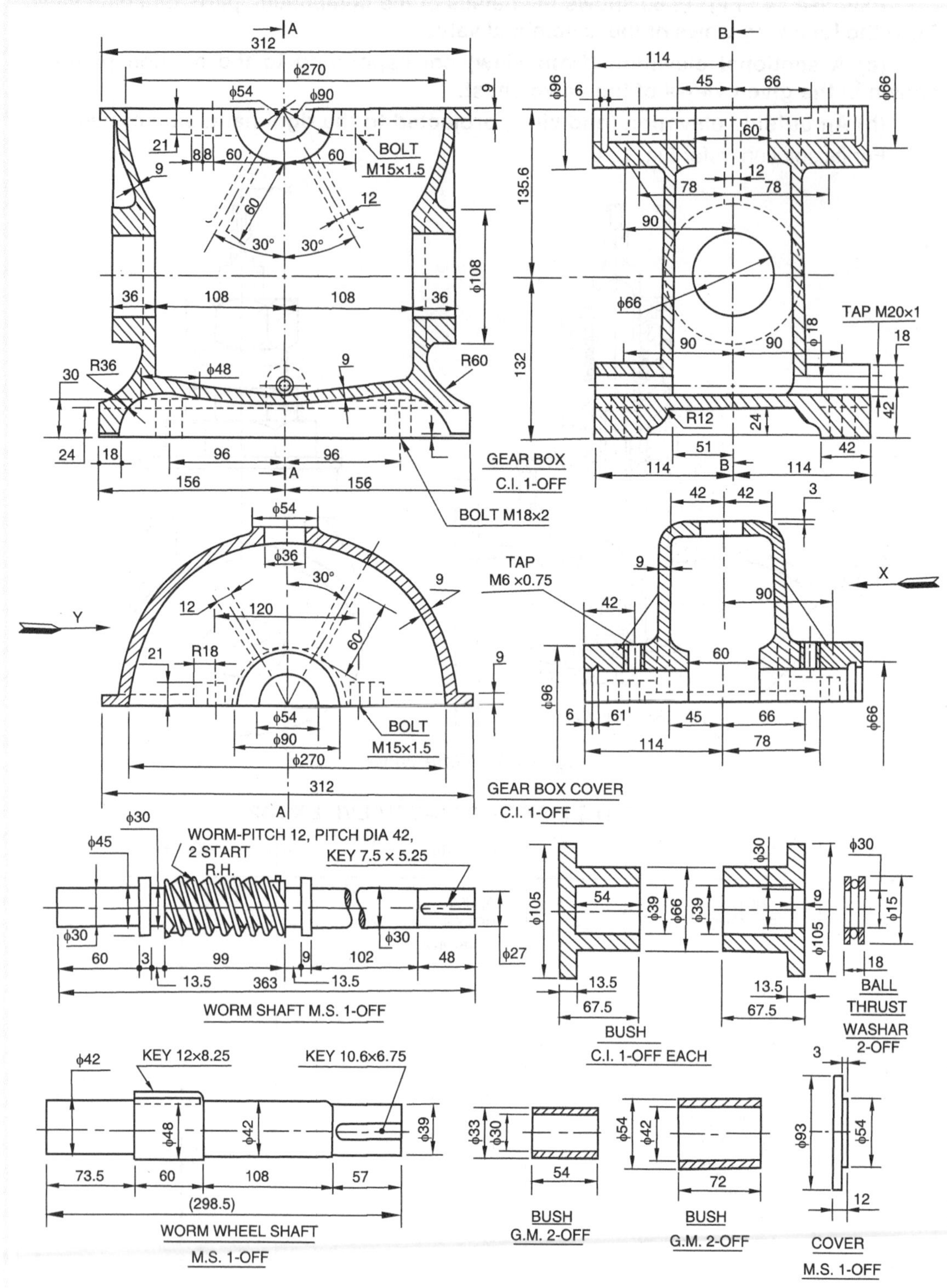

Fig. 19.62(a) *Details of worm reduction gear box*

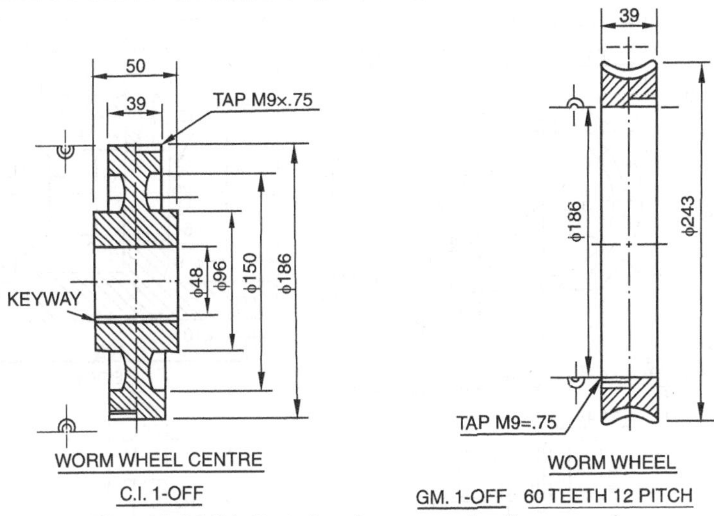

Fig. 19.62(b) *Details of worm reduction gear box*

19.19. WORM REDUCTION GEAR BOX

The worm reduction gear box of an automobile is used for changing the speed of the vehicle with the help of worm gears provided in the gear box.

Exercise 45: Figs. 19.62 (a) and (b) show the detail drawings of the parts of a worm reduction gear box. Draw the following views showing all the parts assembled in their correct positions :

(a) A half sectional front view on AA in the direction of arrow x.

(b) A half sectional end view on BB in the direction of arrow y.

(c) Top view, showing half with the cover in position, but omitting all other parts. The other half to show the cover removed, with scale 1 : 2.

19.20. INDEXING FIXTURE

It is used to drill all holes in the groove of an aluminium alloy I.C. engine piston.

Exercise 46 : Details of a hand operated indexing fixture are shown in Fig. 19.63 Assemble the details and draw the following views :

1. Sectional front view 2. End view and 3. Top view.

Fig. 19.64 shows the general arrangement of an index in a fixture.

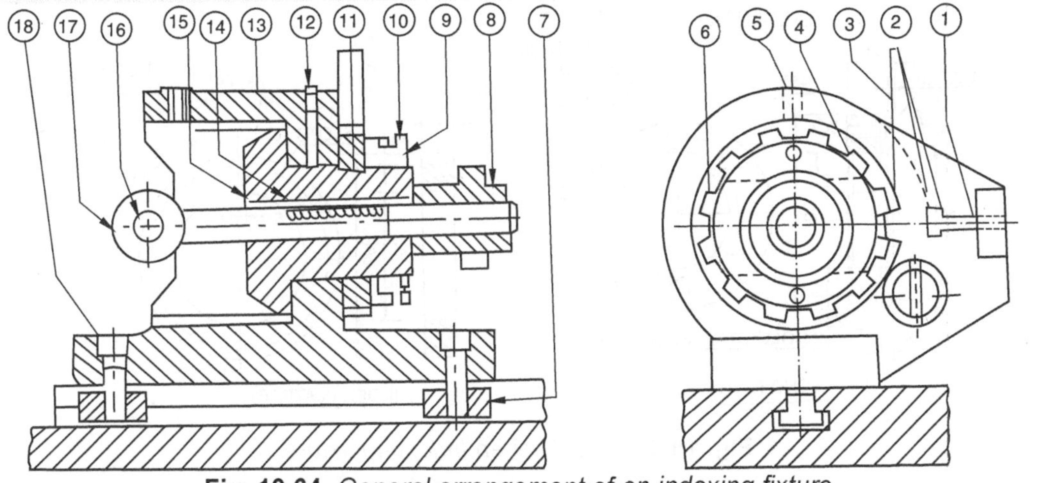

Fig. 19.64. *General arrangement of an indexing fixture.*

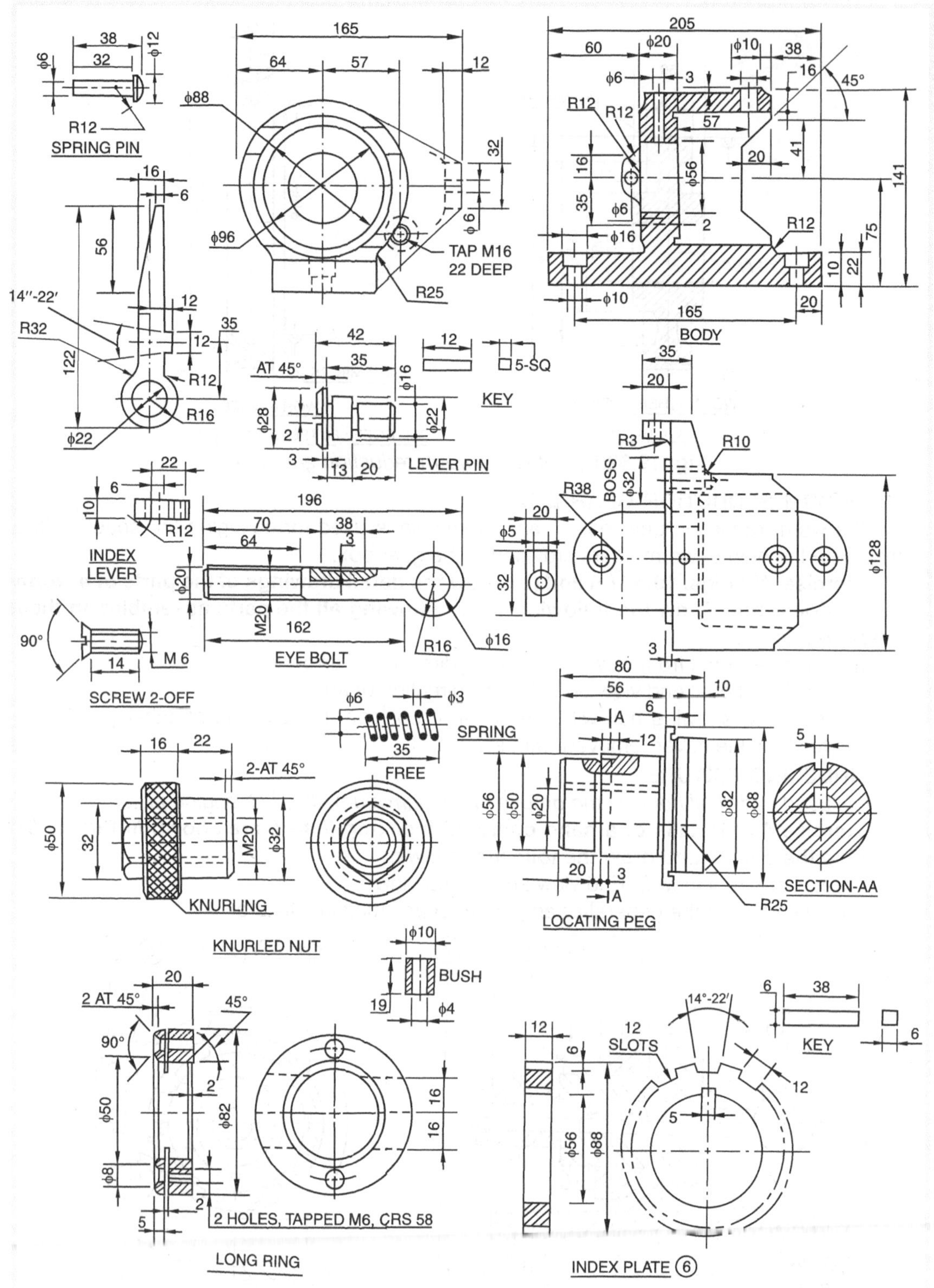

Fig. 19.63. *Details of an Indexting fixture*

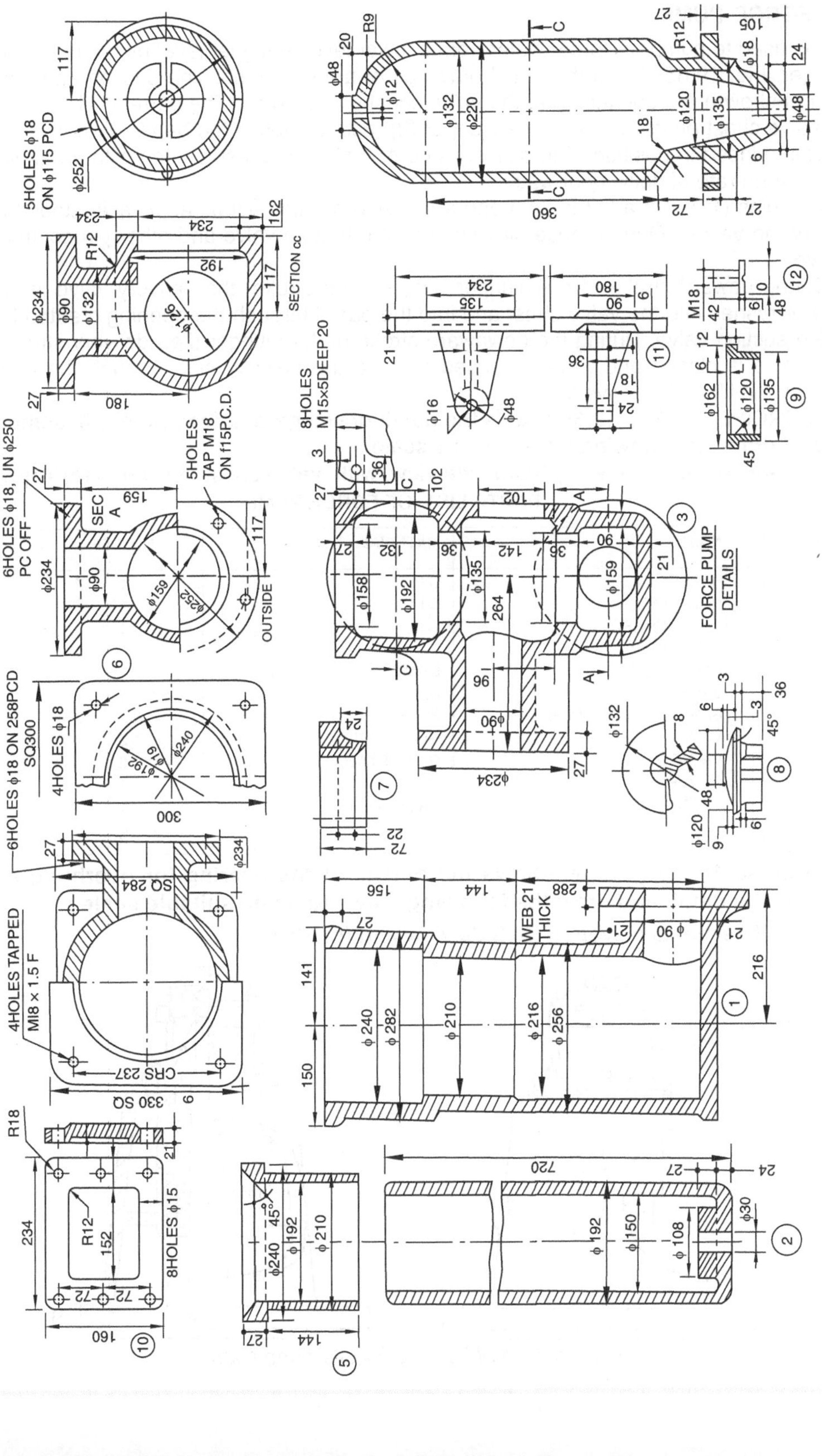

Fig. 19.65. *Details of an force pump*

19.21. FORCE PUMP

It is used to pump (or force) water from a lower level to higher level by the use of a ram or plunger. Fig. 19.65 shows the detail drawings of a force pump. The suction and delivery pipes (not shown) are connected to the valve chest of the pump. The barrel (1) is connected to the valve chest on its left side. The plunger (2) reciprocates in the barrel. Two valve seats have been placed in position. The delivery side of the valve chest is fitted with an air vessel (4) for uniform rate of delivery of water.

Two pins (12) with a flat bottom and a screwed position act as adjustable stops for the motion of the valves. One of these pins controls the bottom valve and other pin controls the top valve.

During upward stroke of the plunger (i.e., suction stroke) the delivery valve is closed and water rushes into the valve chest and into the barrel through the opening created by the lift of the suction valve. During the downward stroke of the plunger (i.e., delivery stroke) the suction valve is closed and water is forced out into the delivery side through the opening created by the lift of the delivery valve.

EXERCISE 47 : Fig. 19.65 shows the detail drawings of a foce pump. Assemble all parts and draw the following views to 1 : 1 scale :

(a) Sectional front view (b) Right side view, with valve chest in section (c) Top view.

PARTS LIST OF FORCE PUMP

P.NO.	NAME	MATERIAL	NO. OFF
1.	BARREL	C.I.	1
2.	PLUNGER	C.I.	1
3.	VALVE CHEST	C.I.	1
4.	AIR VESSEL	C.I.	1
5.	BARREL LINER	BRASS	1
6.	BARREL COVER	C.I.	1
7.	BARREL COVER LINER	BRASS	1
8.	VALVE	GUN METAL	2
9.	VALVE GLAND	GUN METAL	2
10.	COVER	C.I.	1
11.	BRACKER	CAST STEEL	1
12.	PIN	M.S.	2

Exercise 48 : Fig. 19.66 shows the details of the bearing for overhung shaft. Assemble the detail and draw the following views to some suitable scale :

(a) Half sectional elevation (b) Side view and (c) Plan

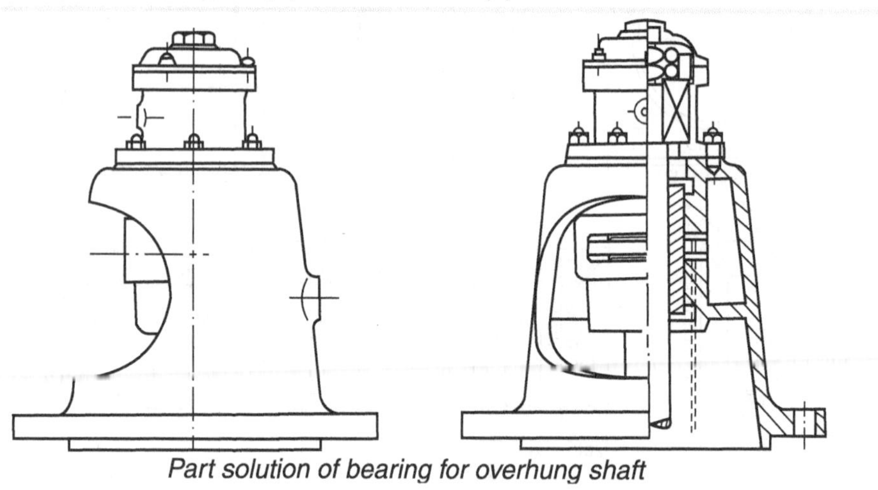

Part solution of bearing for overhung shaft

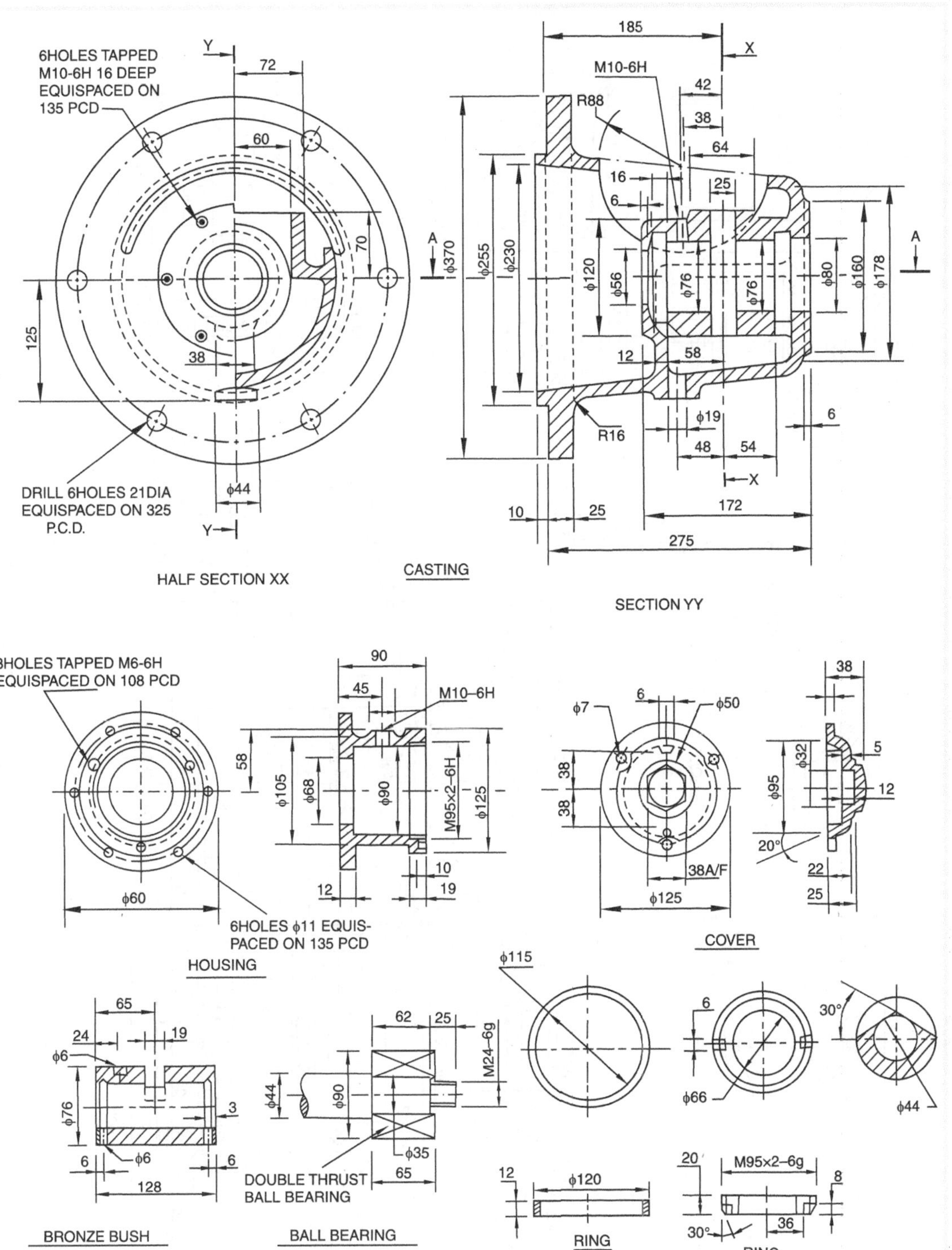

Fig. 19.66. *Details of bearing for overhung shaft*

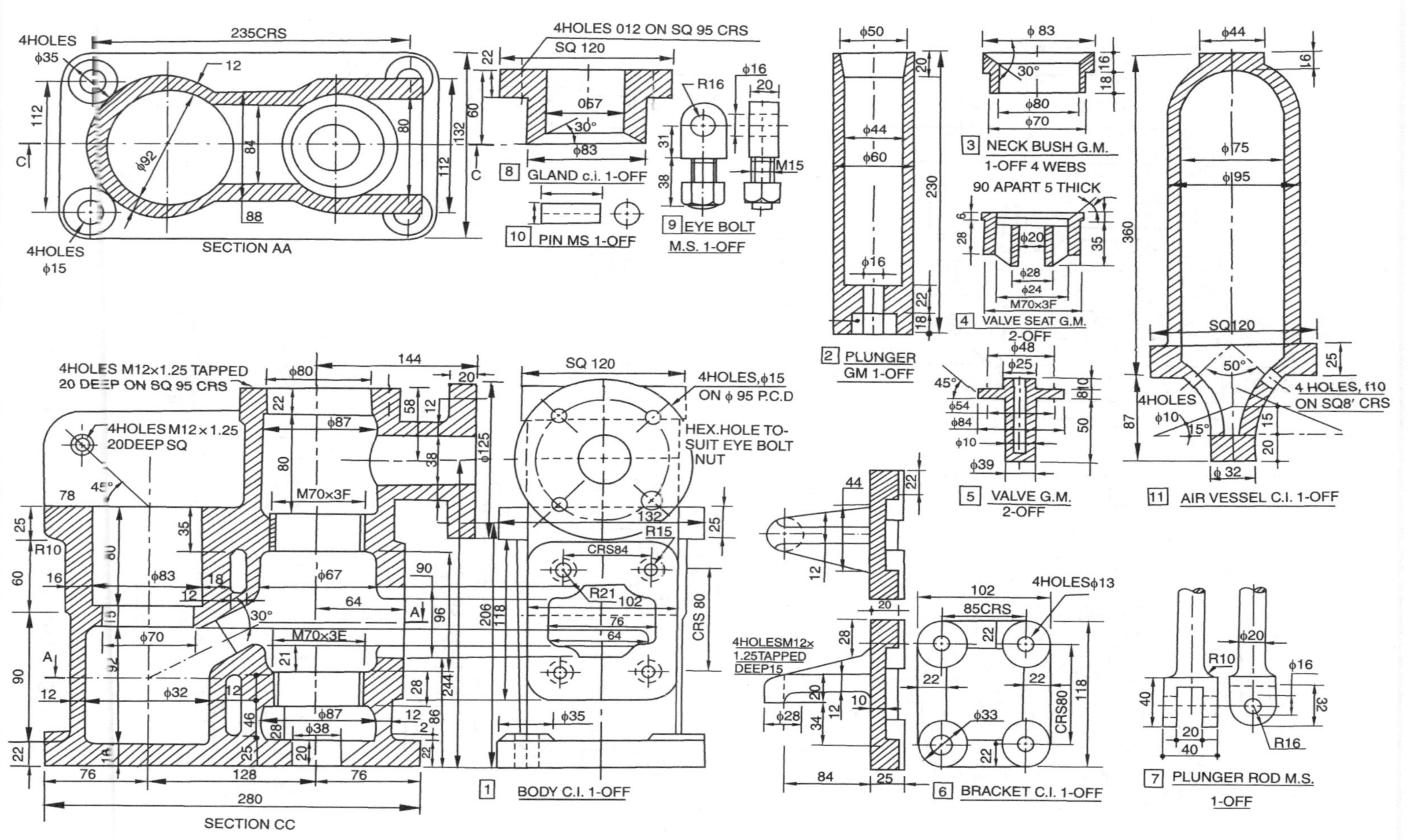

Fig. 19.67. *Ram pump*

P.NO.	NAME	MATERIAL	NO. OFF
1.	SHELL	C.I.	1
2.	FRONT COVER	M.S.	1
3.	OUTSIDE PLATE	M.S.	1
4.	INSIDER PLATE	M.S.	2
5.	CENTRE BOSS	C.I.	1
6.	STUD CARRIER	M.S.	2
7.	JAW LEVER	M.S.	3
8.	L-LEVER	M.S.	3
9.	SLIDING SLEEVE	C.I.	1
10.	STUD	M.S.	3
11.	NUT	M.S.	3
12.	SET SCREW	M.S.	6
13.	PIN FOR SLIDING SLEEVE	M.S.	3
14.	PIN FOR FORKED END	M.S.	6

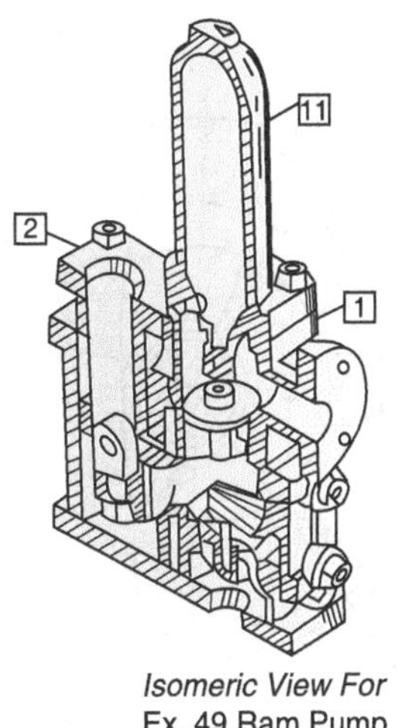

Isomeric View For
Ex. 49 Ram Pump

Exercise 49 : Fig. 19.67 shows the details of ram pump. Assemble the parts and draw (a) Section elevation (b) Side view and (c) Plan.

19.22. MULTIPLE DISC FRICTION CLUTCH

Clutch is a device used to transmit power from engine or driving shaft to the driven shaft. The most important feature of a clutch is that, the driven shaft may be cut off from the driving shaft without stopping the later.

Friction clutches which are in common use are of two types i.e. single disc and multiple disc clutches. The single plate clutch has only one plate or disc with one or two friction surfaces. In multiple disc clutch a number of plates or discs with friction surfaces on either side are used alternately on the driving and driven shafts. By bringing these discs close together by some external pressure, power can be transmitted from the driving shaft to driven shaft. The number of friction discs (plates) depends upon the amount of power to be transmitted. *Multiple disc cluches are used in heavy duty vehicles, textile and paper industries etc.*

The friction discs are made of cast iron. The contact sufraces of clutch are line with friction material e.g. leather, fabric, asbestos etc. in order to avoid slip.

Fig. 19.68 shows the details of a multiple disc clutch. The internal diameter of the shell (1) having serrations, receive the alternate friction disc known as outside plates (3). The shell whihc is mounted freely on the driven shaft has provision to receive the drive pulley.

The serrations on centre boss (5) which is keyed to the driving shaft receive the alternate friction discs known as the inside plates (4).

The front cover (2) is fixed to the shell (1) by small six set screws (12). The sliding sleeve (9) rotates freely on the driving shaft and is connected to the front cover (2) by three-L-type levers (8) and three jaw levers (7).

The last outside plate i.e. stud carrier (6) is provided with the arrangement to carry three studs (10) which are operated by sliding sleeve (9) by the L-type lever (8).

The sliding sleeve (9) roates with the shell (1) i.e. clutch body. This sleeve can be moved forwrad by some operating lever which exert pressur eon the studs (10). The plates (discs) will then get pressed together and the drive is completed.

Exercise 50 : Fig. 19.68 shows the details of multiple disc firction clutch. Assemble all the parts and draw the following views to some suitable scale :

(a) Half sectional front view, with top half in section.

(b) Right side view

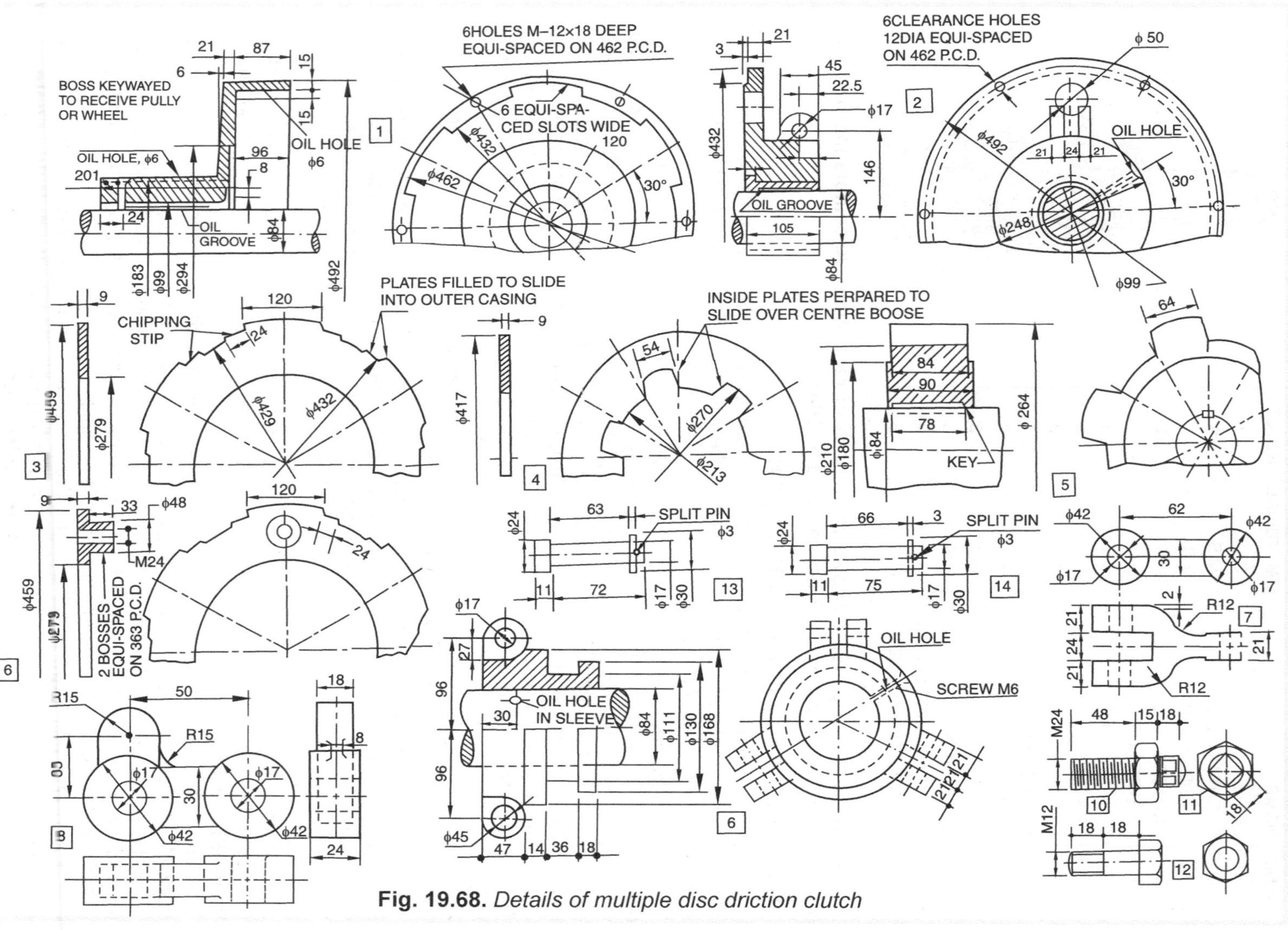

Fig. 19.68. *Details of multiple disc driction clutch*

19.23 CENTRIFUGAL CLUTCH

The centrifugal clutch unlike the plate clutch operate automatically by the action of centrigugal force. It gradually engage the driving and driven machine at the predetermined speeds and slip at overloads. Certrifugal clutches are used in belt conveyors, mopeds, squirrel cage induction motors, drums of the washing machines, etc.

Exercise 51 : Fig. 19.69 (i) and (ii) show the details of a centrifugal clutch. Draw the following views of the assembled clutch in engaged position to 1 : 1 scale :

(a) Front view with top half in section (b) Side view

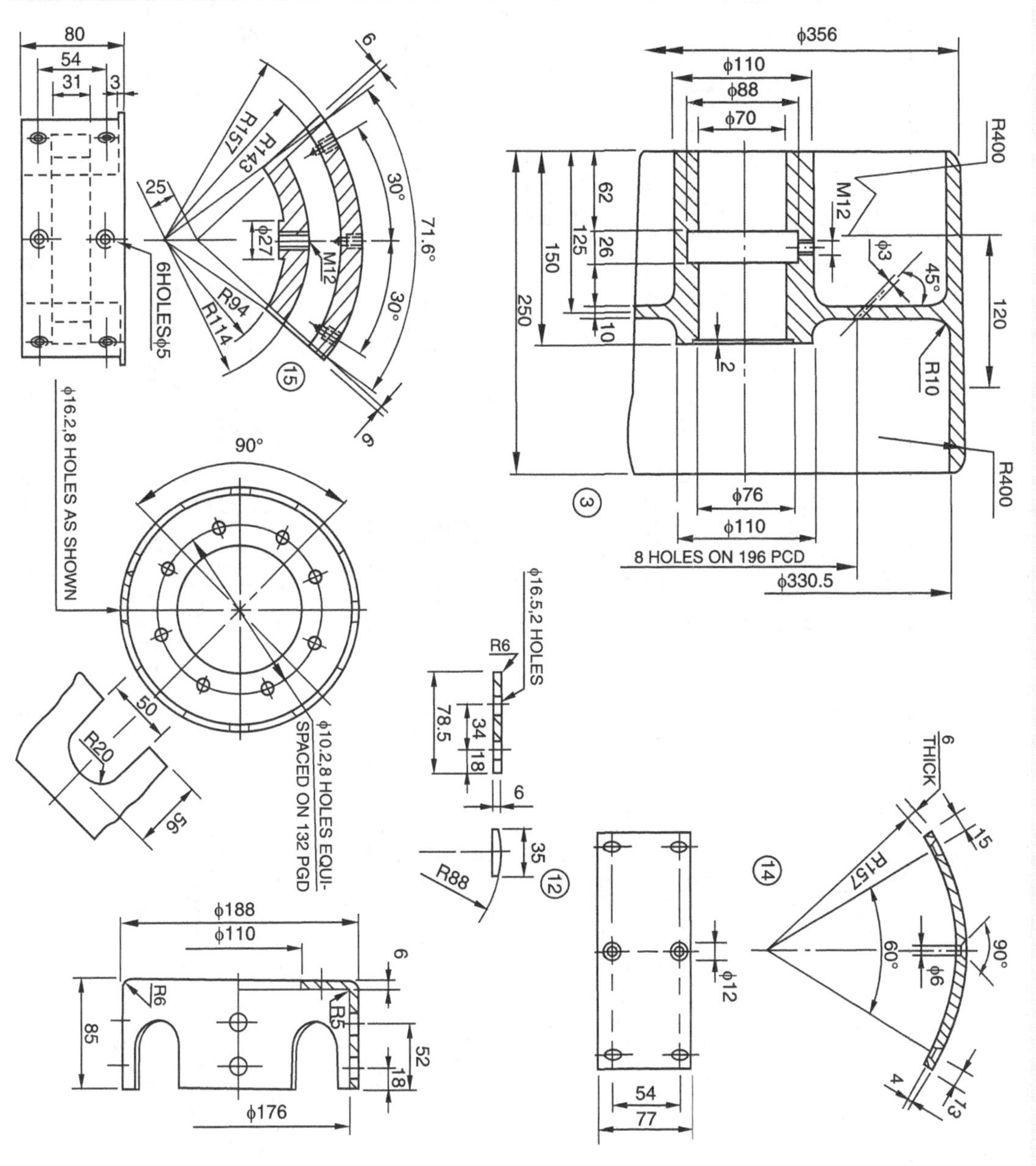

Fig. 19.69. (*i*) *Details of centrifugal clutch*

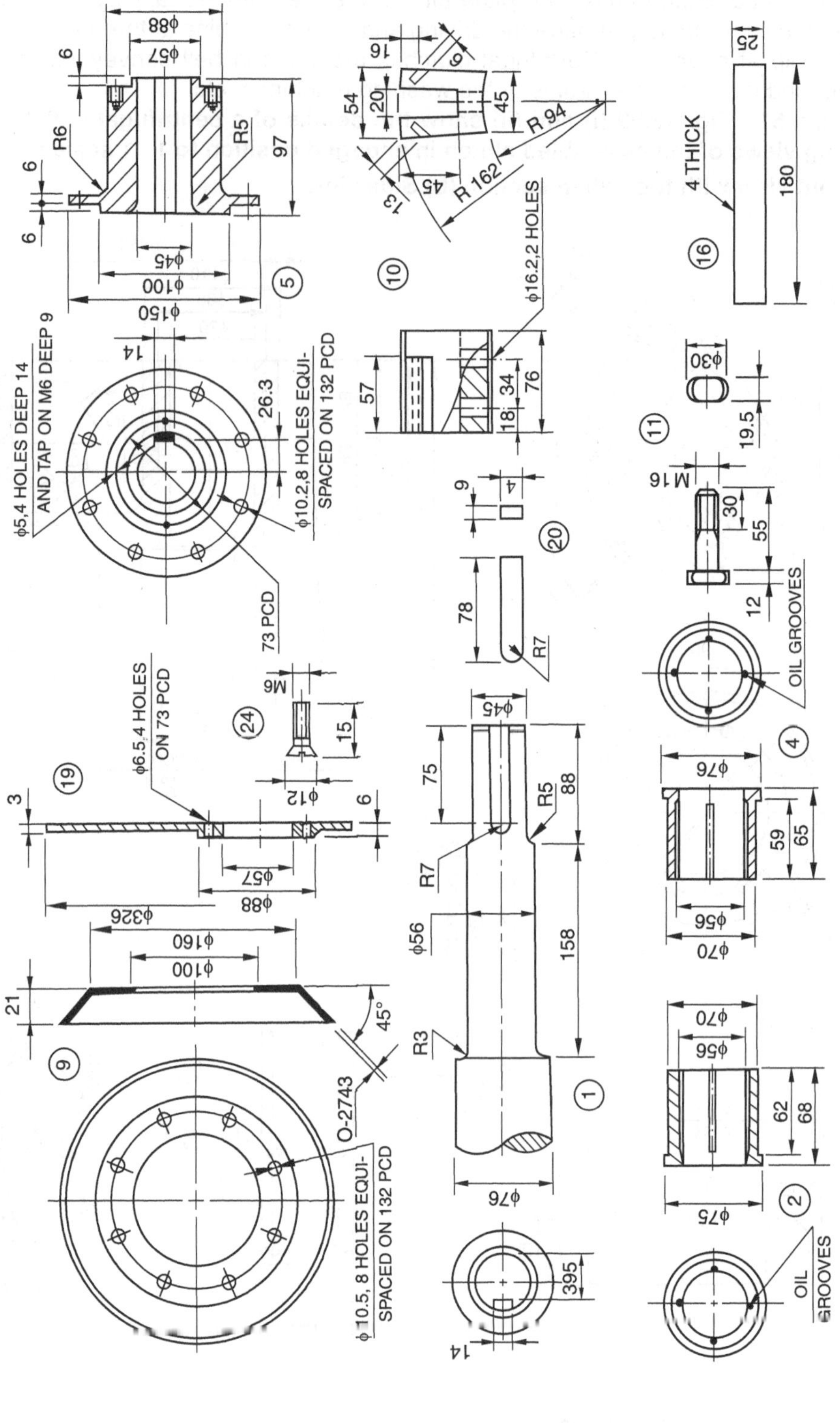

Fig. 19.69. *(ii) Details of centrifugal clutch*

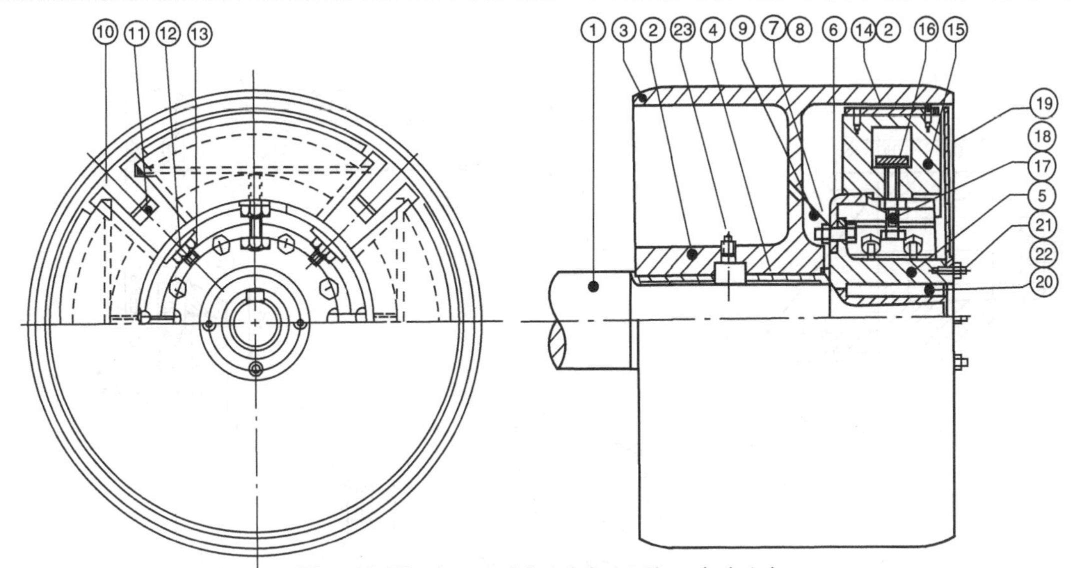

Fig. 19.70. *Assembly of Centrifugal clutch*

P.NO.	NAME	MATERIAL	NO. OFF
1.	SHAFT	MS	1
2.	BUSH BEARING NO.1	BRASS	1
3.	PULLEY	C.I.	1
4.	BUSH BEARING NO.2	BRASS	1
5.	SPIDER HUB	MS	1
6.	CARRIER RIM	MS	1
7.	HEX. SCREW M 10 × 25	MS	8
8.	NUT M10	MS	8
9.	OIL THROWER RING	MS	1
10.	ARM	MS	4
11.	SPECIAL BOLT	MS	8
12.	FACING	MS	4
13.	NUT M16	MS	8
14.	SHOE LINING	FERODE FABRIC	4
15.	SHOE	C.I.	4
16.	CONTROL LEAF SPRING	CARBON STEEL	4
17.	HEX. SCREW M12 × 50	MS	4
18.	NUT M12	MS	4
19.	COVER PLATE	MS	1
20.	KEY	MS	1
21.	STUD M6 ×16	MS	4
22.	NUT M6	MS	4
23.	END PLAY G1/8	MS	1
24.	SCREW	MS	24

Material list : Centrifugal clutch

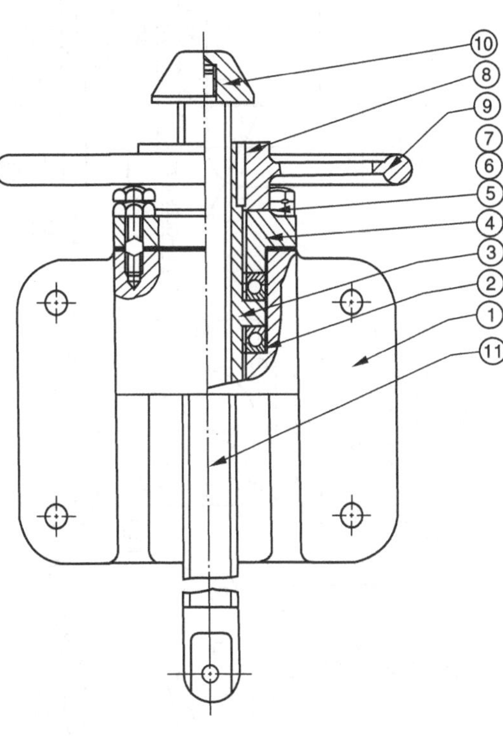

Fig. 19.72. *Assembly of Screw Hoist*

19.24 SCREW HOIST

A screw hoist is a hand operated device used for hoisting heavy parts in workshops.

Exercise 52 : Fig. 19.71 shows the details of screw hoist. Draw the following views with all the parts assembled :

(a) Front view half in section (b) End view (c) Top view

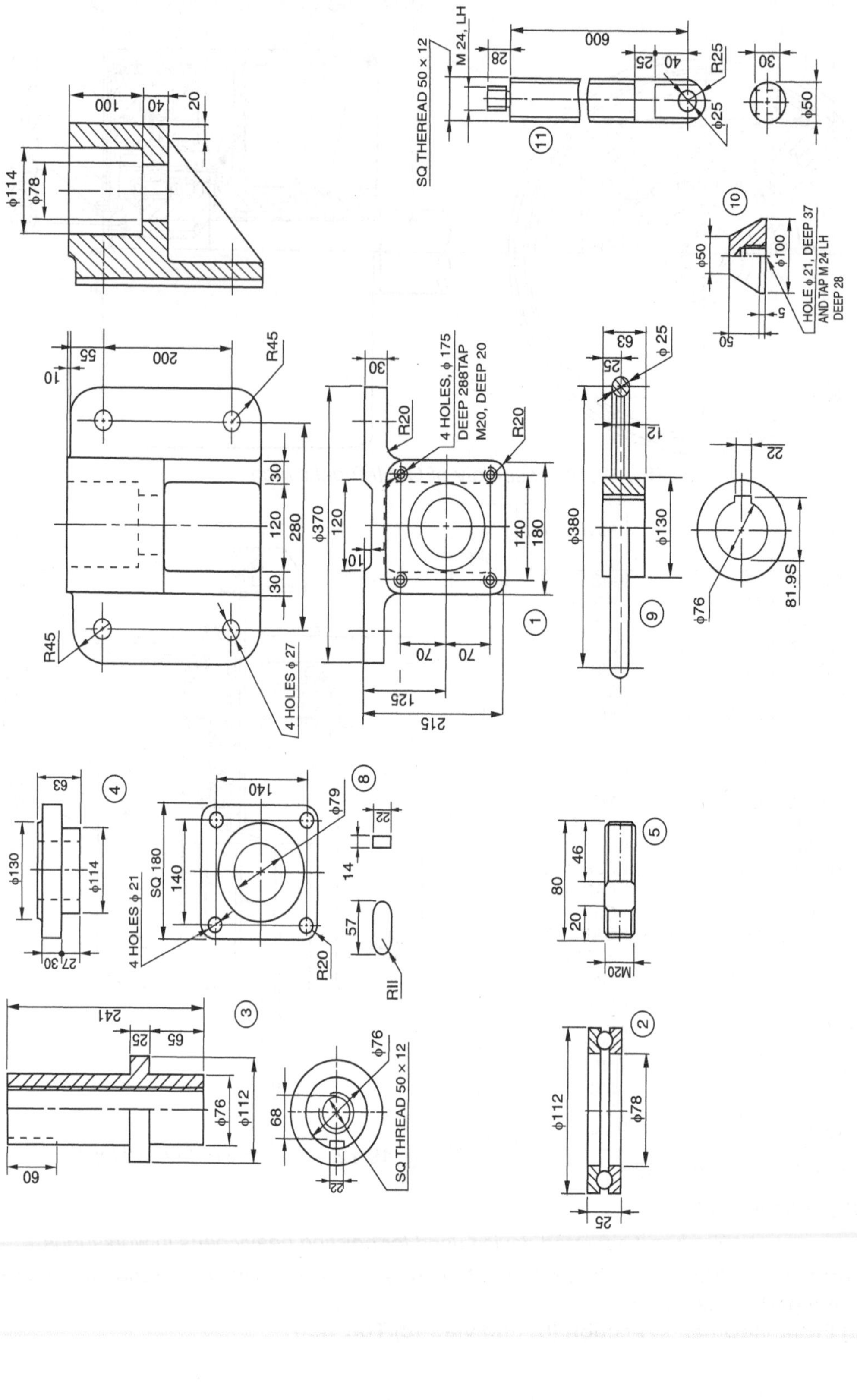

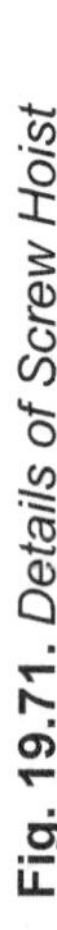

Fig. 19.71. Details of Screw Hoist

QUESTIONS FOR SELF EXAMINATION

1. What is the function of tool post of a lathe machine ?
2. Where is the tool holder used and how it is operated ?
3. What is the use of tail stock in a lathe and how it is operated ?
4. What is the function of feather in a tail stock ?
5. How the barrel moves in and out of the tail stock body of lathe ?
6. What is the function of a lathe slide rest and how it is operated ?
7. What is the use of chuck and where it is mounted ?
8. What is the function of tool head of shaping machine and where it is attached ?
9. How the cutting tool is prevented from rubbing on the finished surface on its return stroke in shaper ?
10. What is the function of machine swivel vice of a shaper ?
11. What is the use of a screw jack ?
12. How and why the cup of screw jack remains stationary when the load is raised ?
13. Why the upper face of the cup of screw jack is serrated ?
14. What is the function of machine vice and bench vice ?
15. What is use of pipe vice? Why the serrations are provided in the v-shaped end of the jaw ?
16. How the multiple disc friction clutch works ?
17. What is the use of quick change drill holder in drilling machine ?
18. What is the function of rotary gear pump ? Write the different parts of this pump.
19. What is the function of measuring stand ? Give the practical utility of measuring stand?
20. What do you mean by bevel gear junction box? Give the practical applications of bevel gear junction box ?
21. What is the use of a hand drill ?
22. What is the function of governor ? Name its different parts.
23. What is the use of shaft bracket ?
24. What is the fuction of indexing fixture ? Name its different parts
25. What is use of force pump? Give its practial utility.
26. What is the fuction of clutch ? Where the multiple clutch is used ?
27. What is centrifugl clutch ? Where the centrifugal clutch is used ?
28. Where the screw Hoist is used ?

CHAPTER

20

Computer Aided Drafting

INTRODUCTION

In this fast developing engineering world, computers find their application in all our activities, both professional and others. Computer has become the driving force for the information explosion. Jobs in fields which never heard few years, such as data communications, robotics and artificial intelligence are proving new employment opportunities every day. Due to the wide spread use and availability of computers, it is essential that future business people acquire an understanding as to what computers are and how they work in engineering. Computers have become so pervasive that virtually no business organization can function without it.

In this chapter, we will deal with the study of computer and its applications, classification, elements of computer, Hardware components, AutoCAD, Opening of AutoCAD, Computer aided drafting, Drawing Entities, Questions for self examination, Problems, etc.

20.1. COMPUTER

An electronic device, which can accept and store input data, process them, and produce output results by interpreting and executing programmed instructions is called computer.

When the computers were developed in 1947, their main application was in performing numerical calculations at a very fast rate. From 1960's onward, computers have been developed for design and drafting as well as manufacturing applications. This has given rise to new scientific disciplines of computer aided design (CAD), computer aided manufacturing (CAM) and computer aided designing and drafting (CADD). The industry has adopted CAD and CAM using the common data base for both, hence integrating the design and fabrication processes CAD and CAM utilized team of man and machine (computer) do the work better and faster. Man has analytical mind and intuition which gives him thinking capability whereas the machine (computer) has speed, accuracy, unlimited storage and instant recall capability.

All computer systems perform the five basic opertions for converting raw input data into useful information, i.e., inputting, storing, processing, outputing and controlling.

20.2. APPLICATIONS OF COMPUTER

These days computer is used in a wide range of field, such as business, banking, education, medical care, legal practice, law enforcement, military affairs, sports, personal use at home, scientific and engineering applications, etc.

20.3. CLASSIFICATION OF COMPUTER

To satisfy the various processing needs, the computers are classified as under :–

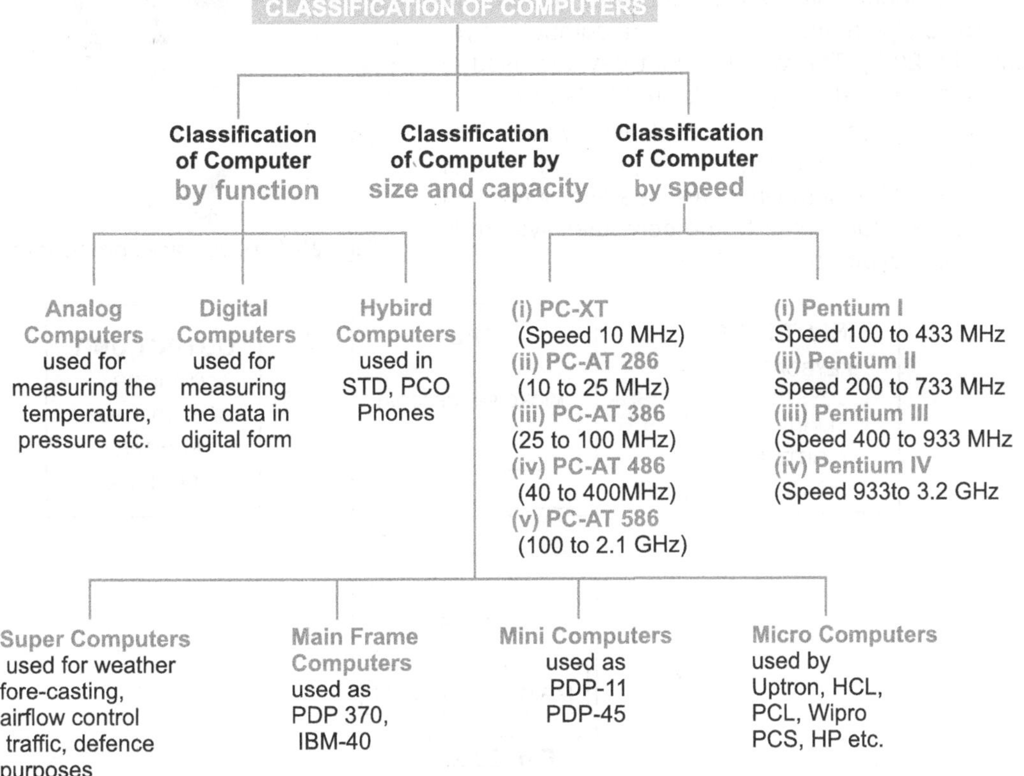

20.4. ELEMENTS OF COMPUTER

The following are the two elements which work together to make up a computer system:

1. Hardware
2. Software

20.5. HARDWARE

The physical components of a computer system is called hardware. The following are the physical components of computer termed as Hardware :

1. Input unit : (Keyboard, Joystick, Mouse and Digitizer)
2. Processor unit : (Central Processing Unit and Memory Unit)
3. Output Unit : (Monitor, Printer and Plotter)

20.6. SOFTWARE

A set of computer programs, procedures and associated documents (flowcharts, manuals etc.) related to the effective operation of a computer system is called software.

20.7. HARDWARE COMPONENTS

The hardware components (Fig. 20.1) of a computer are discussed below :

1. Input Unit : A mechanism for communication between the operator and the system is known as input unit.

The input unit converts the information into electronic signal and sends it to the computer system [Fig. 20.2]. The various input devices used in computer aided drafting are discussed below :

(i) Keyboard : The input device, which enables data entry into a computer by pressing a set of keys (labled buttons) which are nearly mounted on a key board connected to the computer system is known as keyboard.

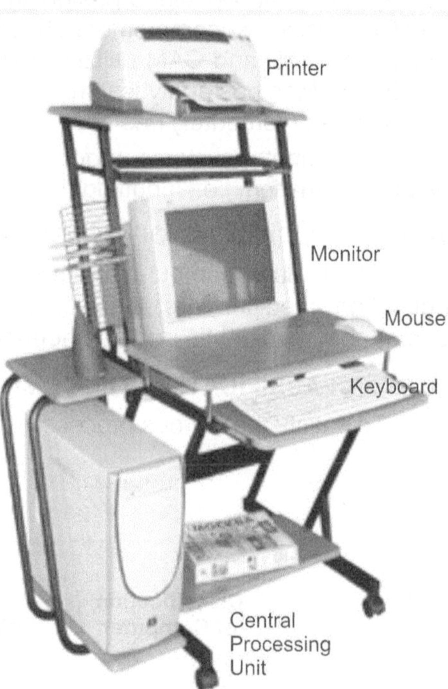

Fig. 20.1. Hardware components

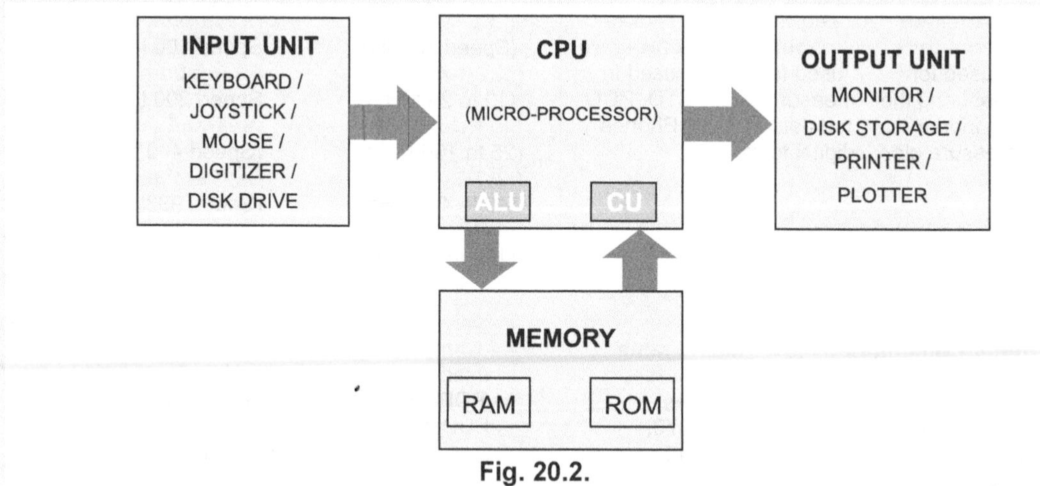

Fig. 20.2.

Fig. 20.3. *Joystick*

Fig. 20.4. *Mouse*

(ii) Joystick : An input device, which serves as an effective pointing device for applications, such as video games, flight simulators, training simulators and for controlling industrial robots is known as joystick (see Fig. 20.3). The stick is moved in the same direction as we wish the cursor to move on the screen.

(iii) Mouse : A small hand held input device, which serves as an effective point and draw device in computer is known as mouse. As the mouse is moved on a flat surface (mouse pad) instructions are sent to the processor to rapidly position the cursor on the screen.[see Fig. 20.4].

(iv) Digitizer : An input device used for converting (digitizing) pictures, maps and drawings into digital form for storage in computer is known as digitizer. This enables re-creation of drawing from the stored information whenever required and easy incorporation of change in the drawing as and when required.

It consists of flat operating surface mounted on a fine grid of waves and provided with an electronic stylus. When the tip of the stylus is at any point which as X and Y co-ordinates, the computer identifies the point and indicates it on the screen. As and when the stylus moves on the digitizer board, its movement is reflected on the screen [see Fig. 20.5]

2. Processor Unit : A unit of computer system, which interprets instructions and executes them is known as processor unit. The processor components are discussed below:

(i) Central Processing Unit (CPU) : The control unit (C.U.) and the arithmetic logic (ALU) unit of a computer system are jointly known as Central Processing Unit (CPU).

The CPU serves as the brain of the computer system and is responsible for controlling the operations of all other units of the system.

A CPU comprises of a group of processing elements often contained on a single integrated circuit chip (called microprocessor) which performs the main basic functions of a computer.

In ALU, calculations such as adding two numbers are carried out. The ALU can also carry out logical operations e.g., comparing two numbers to see which is larger.

The control unit informs the ALU to perform various operations in the sequence in which they are to be done.

(ii) Memory Unit : A device which can accept data, hold them and deliver them on demand at a later stage.

In a computer, a program may contain thousands of instructions, which cannot be carry out by the CPU. These instructions must be stored in the memory unit.There are two types of memory units–Main memory (Primary storage) and Auxiliary memory (Secondary storage). The main memory is connected directly to CPU and contained in silicon chips. The main memory are of two types : ROM and RAM.

ROM (Read Only Memory) : It stores sets of instructions needed to handle the different units of computers.

RAM (Random Access Memory) : A memory in which the time to retrieve stored information is independent of the address where it is stored.

3. Output unit : A device used in a computer system to supply information and results of computation to the outside world is known as output unit. The various output devices are discussed below :

(i) Monitor : A papular output device used for producing soft copy output is known as monitor, It displays the output on a television like screen.

(ii) Printer : An output device used to produce and output is known as printer. There are three types of printers as dot matrix printer, inject printer and laser printer.

(iii) Plotter : An ideal input device, which is often used who need to routinely generate high precision hard copy, graphic output of widely varying sizes is known as plotter. It is used by architects, engineers, city-planners and others.

Fig. 20.5. *Digitizer*

Fig. 20.6. *Plotter*

Fig. 20.6 shows a drum plotter. It has a drum with a slide which can be moved axially. The paper with perforation is attached to the drum. An ink pen is mounted on the slide. The combination of pen movement and drum rotation provides the required motion for plotting.

20.8. STORAGE DEVICE

The unit of a computer which holds the data and instructions to be processed, and the intermediate and final results of processing is known as storage unit.

The storage unit of a computer system is designed to cater to all these needs. It provides space for storing data and instructions, for intermediate results and for the final results.

The two types of storage unit are–primary storage unit and secondary storage unit. The primary storage, also known as main memory is used to hold pieces of program.

As compared to primary storage, secondary storage is slower in operation, larger in capacity, cheaper in price and can retains information even when the computer system is switched off or reset.

The secondary storage devices are in the form of disks . Two types of storage disks are hard disk and floppy disk.

Hard disk is a magnetic disk storage device made of regid metal (frequently aluminium). The hard disk come in many sizes ranging from 1 to 14 inch diameter.

Fig. 20.7. *Hard disk*

A floppy disk is a round, flat piece of flexible plastic, coated with magnetic oxide. It is encased in a square plastic or vinyl jacket cover. The jacket gives handling protection to the disk surface. The storage capacity of floppy disk ranges from 360 kB to 1.44 MB (High Density). The standard sizes are $5\frac{1}{4}''$ and $3\frac{1}{2}''$.

20.9. GRAPHIC SOFTWARE PACKAGE

A collection of pre-written programs that perform standard graphical functions required in producing 2D/3D drawings, pictures, etc. on the computer is called graphic software package.

These graphic software packages perform the following functions :

1. Generate images on the CRT (Cathode Ray Tube) screen using graphic objects, e.g. points, lines, rectangles, polygons, arcs, curves, circles, etc.
2. Accomplishes editing, printing, plotting, etc. on a screen of computer.
3. Perform certain design analysis such as finite element analysis, stress analysis, etc.
4. Produce kinematic simulation, working principle animations, prototype modeling etc.

A wide range of graphic software packages are available for engineering drawings developed by U.S.A., The most popular graphic software package used in computer aided drafting are–AutoCAD, Key CAD, Design CAD, Pro/Engineer, Pro Mechanica, etc.

20.10. COMPUTER GRAPHICS

The area of computer science, which deals with the generation, representation, manipulation and display of picture (line drawings and images) with the aid of a computer is known as computer graphics.

Computer graphics refers to generation of pictorial outputs using computer in association with other equipment like plotter and CRT (cathode ray tube). Plotter have the capability of drawing directly on paper. In CRT, a figure is displayed on a TV–like screen.

20.11. COMPUTER AIDED DRAFTING

To develop and modify the engineering drawings by a computer is known as computer aided drafting (CAD).

In traditional drafting practice, the designer express his basic ideas through free hand sketches. These drawings are followed by calculations to prove the suitability of the design. This is followed by laborious and time consuming work on the drawing board by the draftsman.

But in Computer Aided Drafting, the information related to design are communicated to CAD system by means of input devices and the drawing can be observed on the screen of computer. When the drawing work is finalised, the various functions are carried out by the CAD packages.

20.12. ADVANTAGES OF COMPUTER AIDED DRAFTING

The distinct advantages of computerised drafting over manual drafting are :

1. The engineering drawings are made easy and fast.
2. The drawings created by CAD are very accurate.
3. The scale of the drawings can be changed without redoing all over again.
4. Revisions and modifications in design can be done even at the last stage.

5. The drawings can be ported from one computer to the other.

6. Interfacing with computer aided manufacturing is a reality.

7. Editing and transformation involving the operations of objects are possible only in CAD. For e.g., copying, off setting, etc.

8. Objects can be viewed in different directions to obtains different types of view like, isometric, perspective, orthographic, etc.

9. Repeated features, e.g., bolts, screws, etc. can be drawn only once and stored in memory, then insert it wherever needed.

10. The drawings can be made more realistic by colouring and rendering help to display more distinct informations on screen.

20.13. AutoCAD

AutoCAD is a graphic software for preparing drawing with the P.C. It is the first and the foremost 'CAD' software developed in U.S.A. by Autodesk Inc.Its versions are revised periodically . Its latest version is AutoCAD 2004 that works on Windows. Now it has become very easy to prepare new drawings and make any change if so required in the drawing with the help of computer. The various benefits are achived in AutoCAD, like better engineering drawing, more standardization in the drwawing, better documentation of the design, fewer drawing errors and greater legibility.

20.14. CAPABILITIES OF AutoCAD

In addition to create a simple 2D drawing like circle or rectangle, we can design complicated 3D objects like solids, wireframes, etc. with the help of AutoCAD. The Language for its programming is bulit in the system with its name 'AutoLISP'. The 2D can be placed in 3D space as per our choice, so we can say it is a 3D modeling tool also. The Drawing Exchange File (DXF) can be used to exchange drawing among various CAD software also.

20.15. MAIN FEATURES OF AutoCAD 2004 (THE LATEST VERSION)

The following are the main features of AutoCAD 2004 :

1. More productive @ 16 % then AutoCAD-2002.

2. Very easy to handle the number of layers.

3. Very advance design centre with architecture symbols.

4. On context help features.

5. Match properties, dimensions style manager, design centre icon on interface.

6. Modification of layers properties enhanced by double click.

20.16. OPENING AutoCAD 2004

The following steps should be followed for opening AutoCAD 2004 :–

1. Click on the start button from the taskbar then start menu option will appear on the screen.

2. Choose program folder from the start menu option. Then sub menu will appear on the screen.

3. Choose AutoCAD 2004 folder then further menu will appear on the screen.

4. Then, click on AutoCAD 2004 shortcut.

20.17. COMPONENTS OF AutoCAD

The following are the main components of AutoCAD R14/2004 main window (Fig. 20.8):

1. Title bar
2. Menu bar
3. Status bar
4. Tool bar
5. Document window
6. Command window

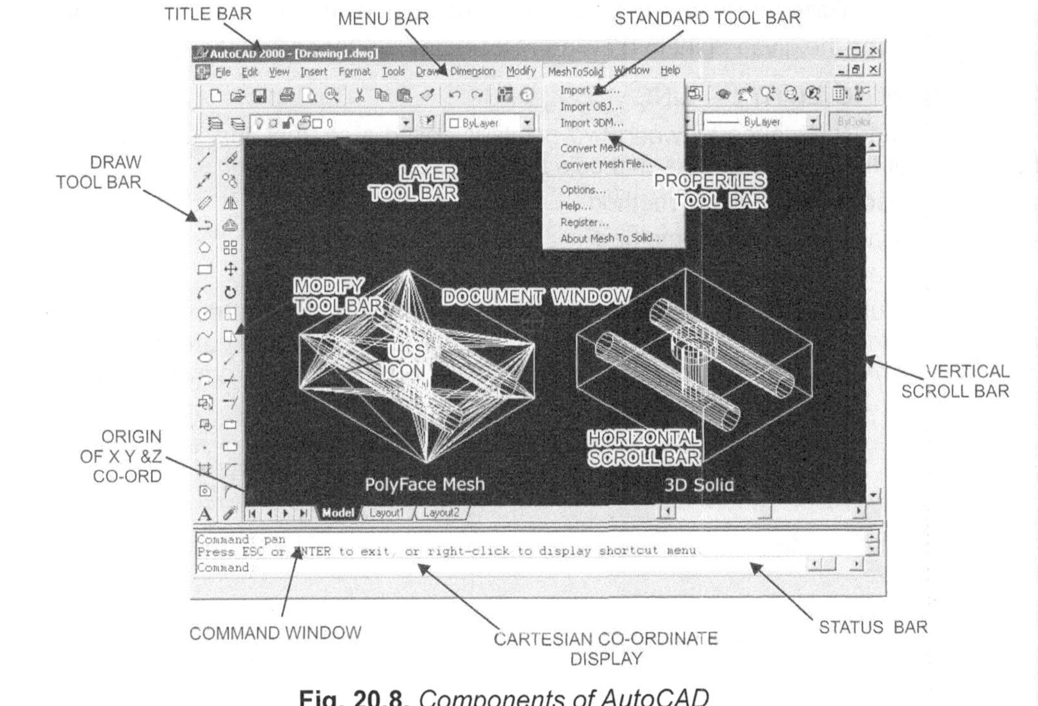

Fig. 20.8. *Components of AutoCAD*

1. Title bar : The colour strip displayed at the top of the AutoCAD widow is called the tiltle bar (Fig. 20.8). In the title bar, the file name of the drawing is indicated with in [ ].

2. Menu bar : The menu bar, located below the title bar, provides pull-down menus from which you can choose commands.

3. Tool bar : Tool Bar contains number of buttons, which you can activate commands by clicking the button on the various Tool Bar.

4. Status bar : The status bar located at the bottom of the screen shows the co-ordinates of the screen cursor as well as the current setting of various AutoCAD program mode.

5. Command window : The command is one component of AutoCAD that does not have an equivalent is most other windows programs. You can start any AutoCAD command by typing the command and then pressing enter.

Some of the components always appear in the same location. Others, such as the Tool Bars and command window, can be turned off or relocated anywhere on the windows screen.

6. Document window : The document window or drawing area, occupies most of the screen. This is the area in which you actually create your drawing [see Fig. 20.8].

20.18. DRAWING ENTITIES

An entity is a drawing element namely ; point, line, arc, circle, etc. AutoCAD provides a set of entities using draw entity command for constructing drawing.

20.19. DRAWING ENTITY – POINT

A point is that which has simply position but no magnitude. The point command places a point in the drawing. Enter the command POINT and then ask to identify the location of the point.

Problem 1 : Draw a point at the location of (60, 50). [Fig. 20.9]

Solution : Command : POINT

Point : 60, 50 (Select the point by mouse or give co-ordinates)

This places the given point in drawing at location 60,50 [see Fig. 20.9].

20.20. DRAWING ENTITY – LINE

The shortest distance between two points is called a straight line. Lines can be drawn by any one of the following three method's using the LINE command :

1. Absolute co-ordinate method,
2. Relative co-ordinate method
3. Polar co-ordinate method.

Absolute co-ordinate method : In this method, the X and Y coordination of the starting point a line (known from point) and the end point of the line (to point) are specified with left lower corner of the graphic window as the origin. The procedure of drawing a line by this method is explained in Problem 2.

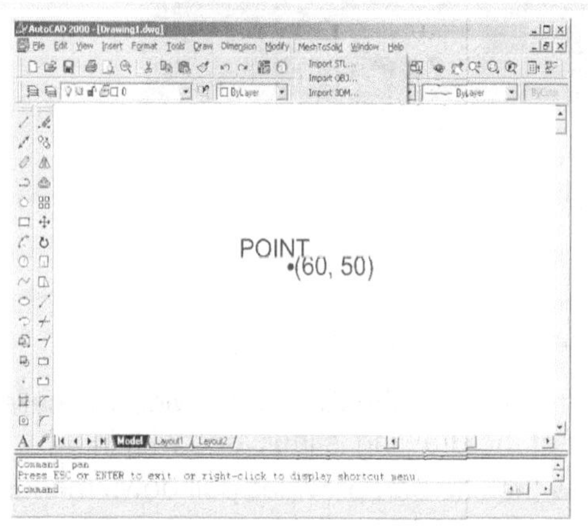

Fig. 20.9. *Point*

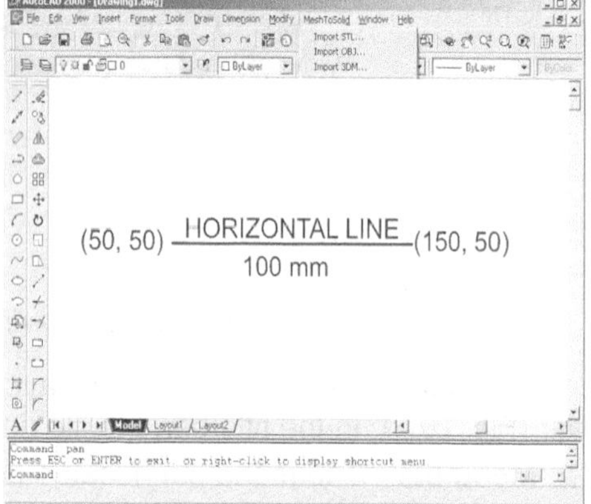

Fig. 20.10. *Horizontal line*

Problem 2 : Draw a line of 100 mm long from a point (50, 50) by using absolute co-ordinate method. [Fig. 20.10]

Solution :

Command	:	LINE
From point	:	50,50 (Select the point by mouse or enter the co-ordinates by keyboard)
To point	:	150, 50 (Enter the co-ordinates)
To point	:	Press Enter

Problem 3 : Draw a line from a point (50, 50) to point (150, 65). [Fig. 20.11]

Solution :

Command	:	LINE
From point	:	50, 50
To point	:	150, 65 at ∠90°

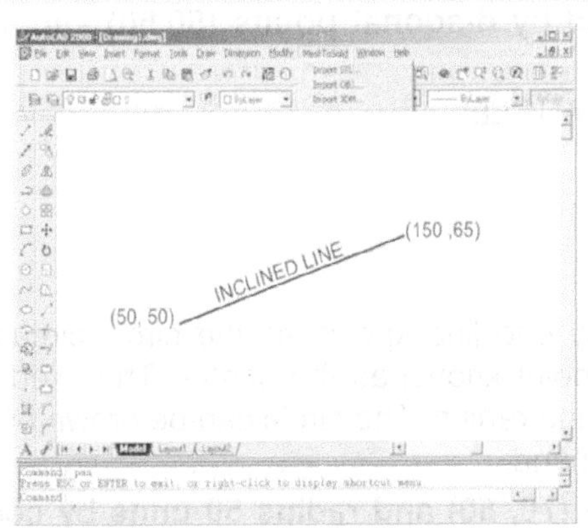

Fig. 20.11. *Inclined line*

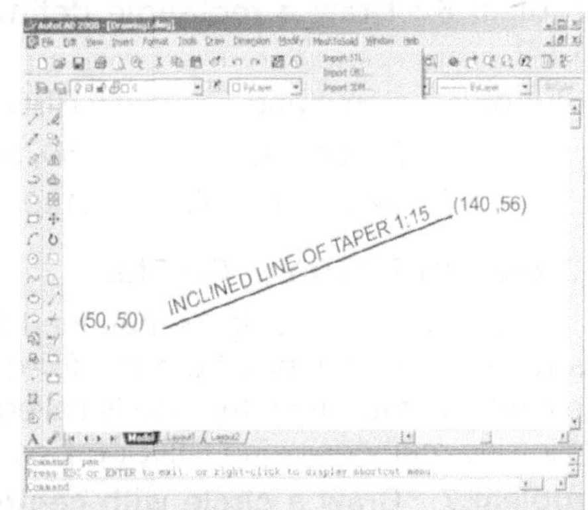

Fig. 20.12. *Inclined line of taper 1:15*

Problem 4 : Draw an inclined line from (50,50) with a taper of 1 : 15 by using relatives co-ordinates. [Fig. 20.12]

Solution :

Command	:	LINE
From point	:	50, 50
To point	:	140, 56
To point	:	Press Enter

20.21. DRAWING ENTITY – RECTANGLE

A rectangle is a quadrilateral in which the opposite sides are equal and the angles are at right angles. It is connected sequence of lines and based on two opposite corner points called diagonal points.It can be drawn by LINE command.

Problem 5 : Draw a rectangle of sides 100 mm × 50 mm, using abolute co-ordinate method. [Fig. 20.13]

Solution : Choose the starting point (50, 50).

Command	:	LINE
From point	:	50, 50
To point	:	150, 50
To point	:	150, 100
To point	:	50, 100
To point	:	C (Press Enter)

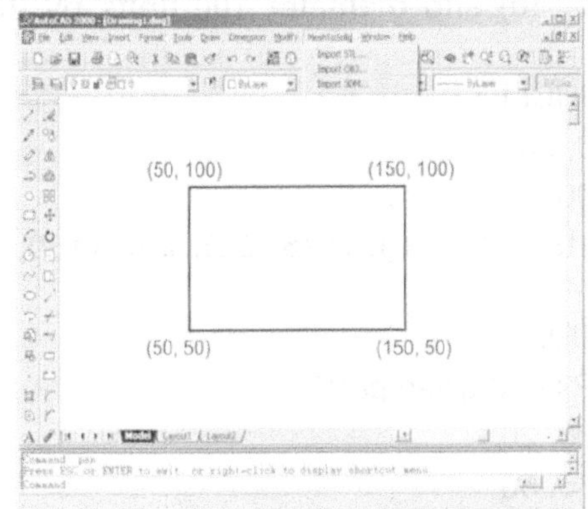

Fig. 20.13. *Rectangle*

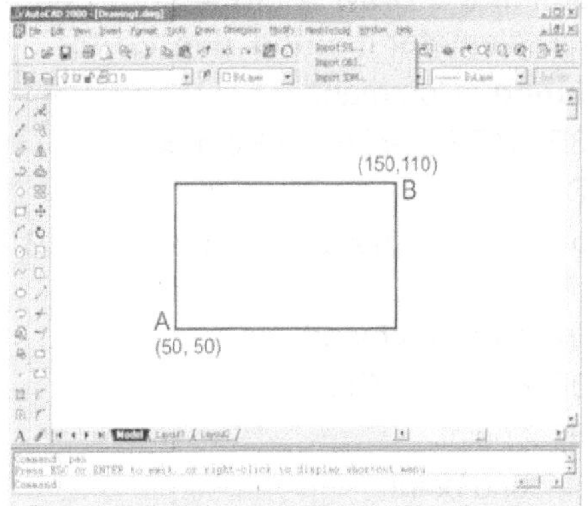

Fig. 20.14. *Rectangle*

Problem 6 : Draw a rectangle defined by diagonal points (50,50) and 150, 110. [Fig. 20.14]

Solution : Command : RECTANGLE
First corner : 50, 50
Second corner : 150, 110

20.22. DRAWING ENTITY – CIRCLE

A circle is a plane figure bounded by one line known as the circle circumference which is always equidistance from the fixed point known as the centre. The fixed distance from the centre to the circumference is called the radius. The circle can be drawn by any one of the following methods using CIRCLE Command.

Problem 7 : Draw a circle with centre (70, 50) and radius 50 units by centre and radius method. [Fig. 20.15]

Solution : Command : CIRCLE
3P/2P/TTR/ < Centre point > : 70, 50 (locate the point with mouse)
Diamter/ <Radius > : 50

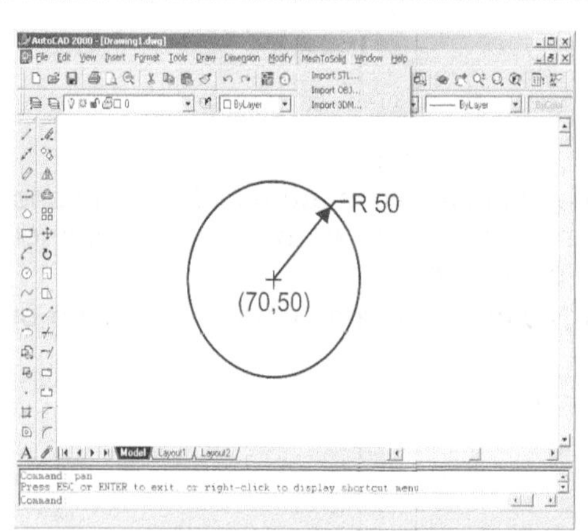

Fig. 20.15. *Circle by centre point method*

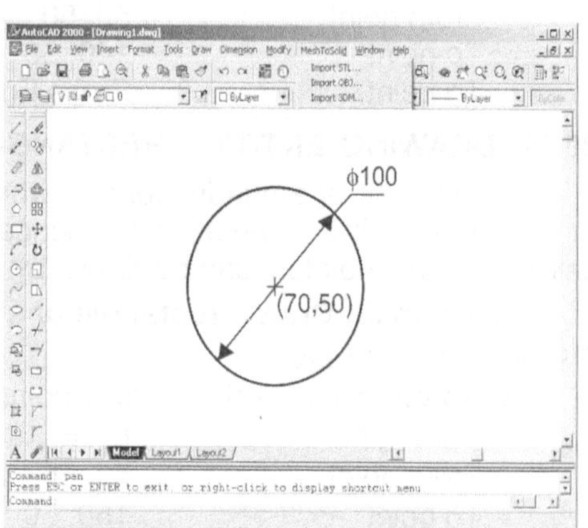

Fig. 20.16. *Circle by diameter*

Problem 8 : Draw a circle with centre (70, 50) and diameter 100 units by using centre and diameter method. [Fig. 20.16]

Solution : Command : CIRCLE
3P/2P/TTR/ <Centre Point > : 70, 50
Diameter : 100

Problem 9 : Draw a circle using 3 points (130, 20), (175, 50), and (150, 75). [Fig. 20.17]

Solution : Command : CIRCLE
3P/2P/TTR/ <Centre point > : 3P i.e, 3 given points
First point : 130, 20
Second point : 175, 50
Third point : 150, 75

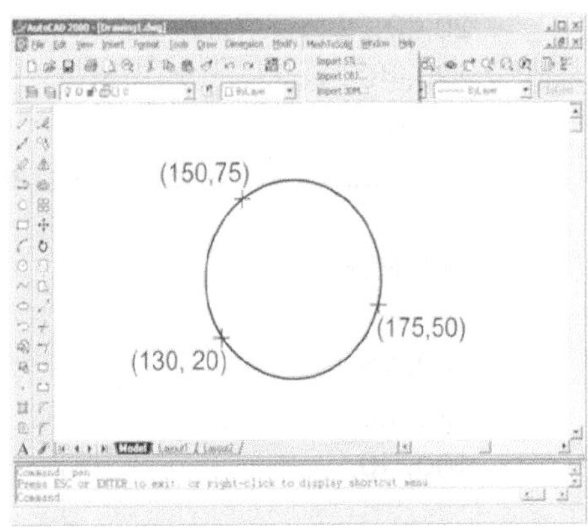

Fig. 20.17. *Circle by 3 points*

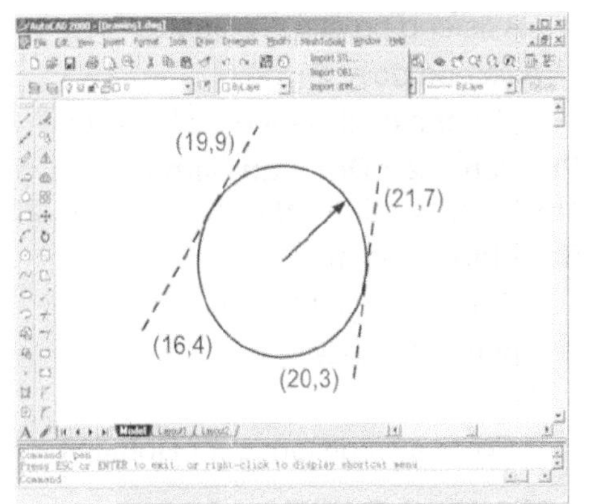

Fig. 20.18. *Circle using TTR*

Problem 10 : Draw a circle with radius 2 units and two existing lines as tangents.

Take : For line 1 – From point (16, 4) to point (19, 9), For line 2 – From point (20, 3) to point (21, 7). [Fig. 20. 18]

Solution : Command	:	CIRCLE
3P/2P/TTR/<Centre point >	:	TTR (Tangent, Tangent, Radius)
Enter Tangent Specification	:	line 1 (pick up using mouse)
Enter Second Tangent Spec.	:	line 2 (pick up using mouse)
Radius	:	2

20.23. DRAWING ENTITY – ELLIPSE

The section obtained when the section plane is inclined to the axis of cone and cut all the generators on one side of the apex is called an ellipse. The ellipse can be drawn by using ELLIPSE Command.

Problem 11 : Draw an ellipse using major end points (20, 40) (120, 40) and minor axis end point (70, 70). [Fig. 20.19]

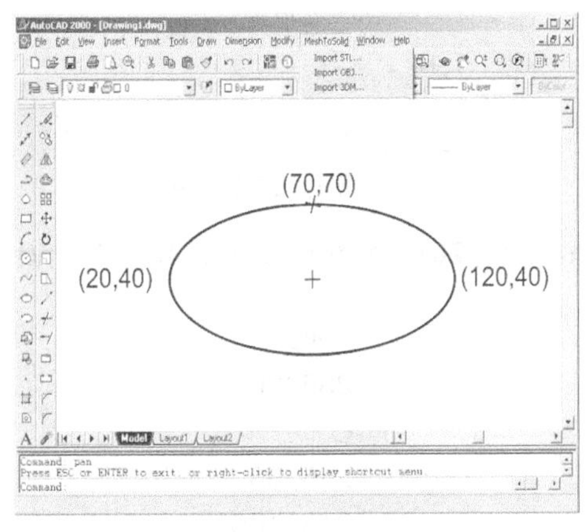

Fig. 20.19. *Ellipse*

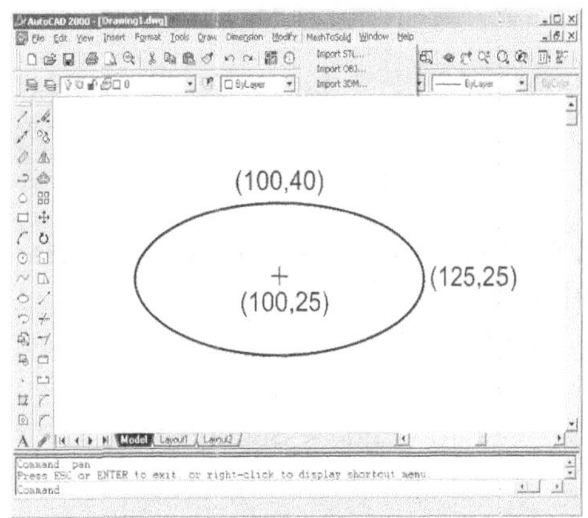

Fig. 20.20. *Ellipse*

Solution : Command : ELLIPSE

<Axis end point 1 > / Centre : 20,40

Axis end point 2 : 120, 40

< Other axis distance > / Rotation : 70,70

Problem 12 : Draw an ellipse with centre (100, 25), major axis end point (125, 25) and minor axis end point (100, 40). [Fig. 20.20]

Solution : Command : ELLIPSE

< Axis end point 1 > / Centre : C

Centre of ellipse : 100, 25

Axis end point 2 : 125, 25

< Other axis distance > / Rotation : 100, 40

20.24. DRAWING ENTITY – ARC

An arc is any part of the circumference of a circle. The arc can be drawn by using ARC command.

Problem 13 : Draw an arc using the given three points (70, 50), (50, 90) and (100, 110). [Fig. 20.21]

Solution : Command : ARC

Centre / start point > : 70, 50

Centre / End / < Second point > : 50, 90

End point : 100, 110

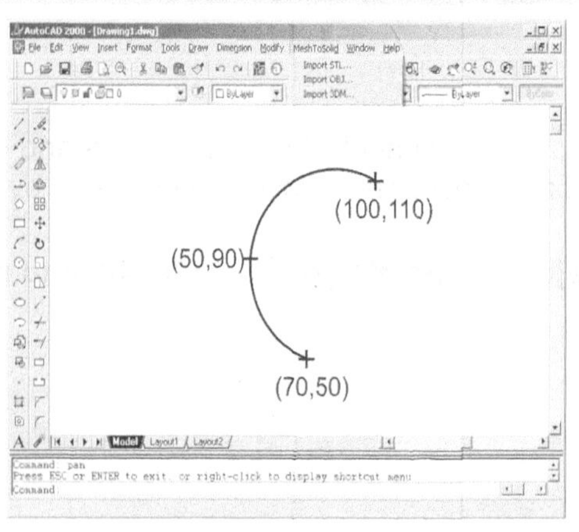

Fig. 20.21. *Arc*

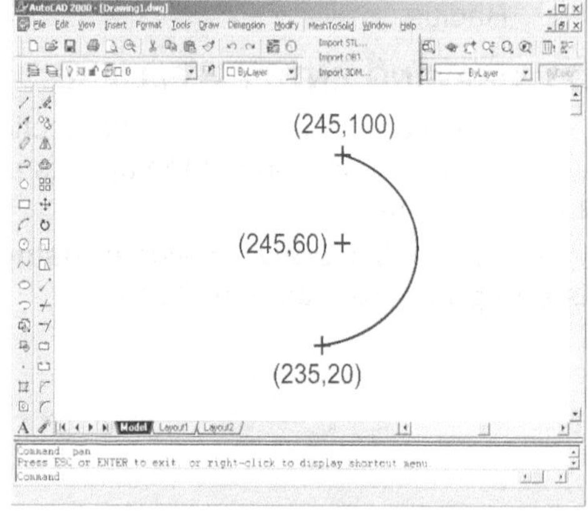

Fig. 20.22. *Arc*

Problem 14 : Draw an arc using start point (235, 20), centre point (245, 60) and end point (245, 100). [Fig. 20.22]

Solution : Command : ARC

Centre / < Start point > : 235, 20

Centre / End / < Second point > : C

Centre point : 245, 60

Angle / Length of chord / < End point > : 245, 100

20.25. DRAWING ENTITY – POLYGON

A polygon is a plane figure bounded by more than four straght lines and containing more than four angles. The polygon command draw regular 2D polygons with 3 to 1024 sides.

Problem15 : Draw a polygon of five sides inscribed in a circle of radius 50 units with centre (150, 60). [Fig. 20.23]

Solution : Command : POLYGON

Number of sides : 5

Edge/<center of polygon> : 150, 60 inscribed in circle

Circumscribed about circle (1/c) : 1

Radius of circle : 50

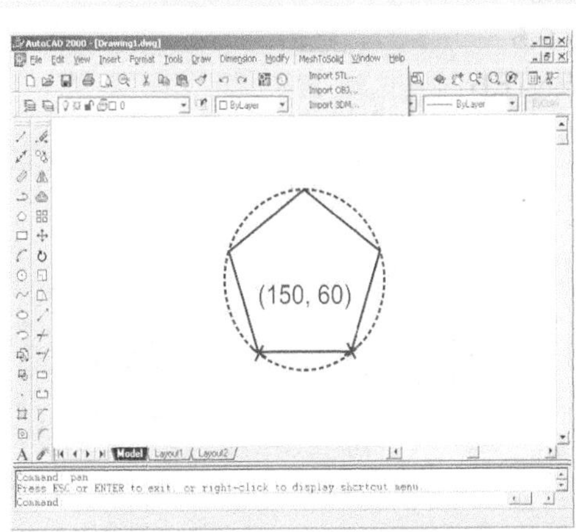

Fig. 20.23. *Polygon inscribed in a circle*

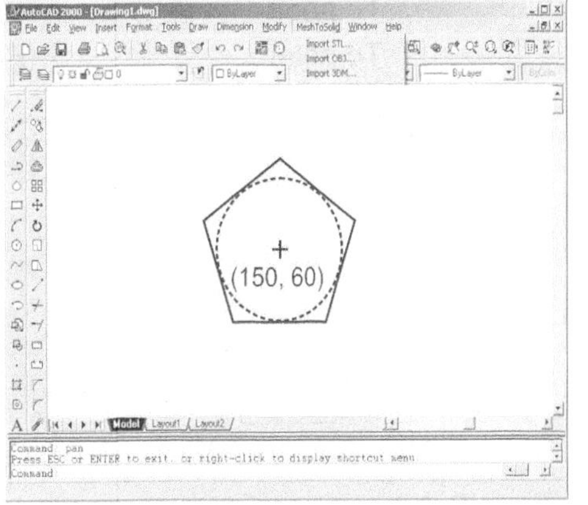

Fig. 20.24. *Polygon outside a circle*

Problem16 : Draw a pentagon with centre (150, 60) circumscribed on a circle of radius 50 unit. [Fig. 20.24]

Solution : Command : POLYGON

Number of sides : 5

Edge/<centre of polygon> : 150, 60 inscribed in circle/

Circumscribed about circle (1/c) : C

Raidus of circle : 50

Problem 17 : Draw a polygon of five sides using edge method. The first end point of the edge is (50, 50) and second end point of edge is (100, 50). [Fig. 20.25]

Solution : Command : POLYGON

Number of sides : 5

First end point of edge : 50, 50

Second end point of edge : 100, 50

Problem 18 : Draw the figure of the gauge using AutoCAD. [Fig. 20.26]

Solution:

COMMAND : LINE

From point : 10, 5

To point : @ 50 < 0

To point	:	@ 2.5, 2.5
To point	:	@ 10<90
To point	:	@ 20<180
To point	:	@ 10<90
To point	:	@ 20<0
To point	:	@ 10<90
To point	:	@ –2.5, 2.5
To point	:	@ 50<180
To point	:	@ –2.5, 2.5
To point	:	@ 30<270
To point	:	@ 2.5, –2.5
To point	:	(Press ENTER)

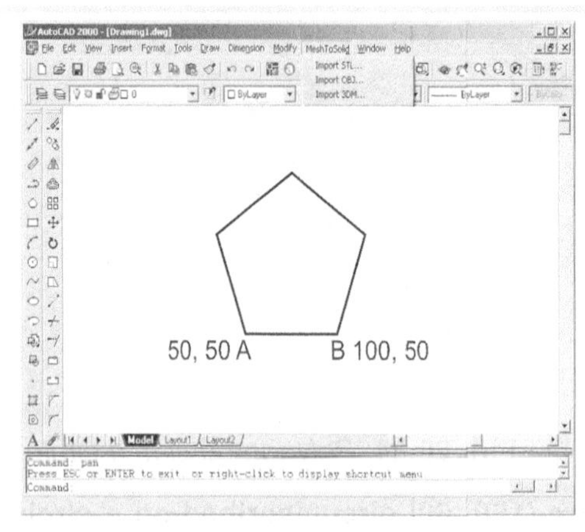

Fig. 20.25. *Pentagon*

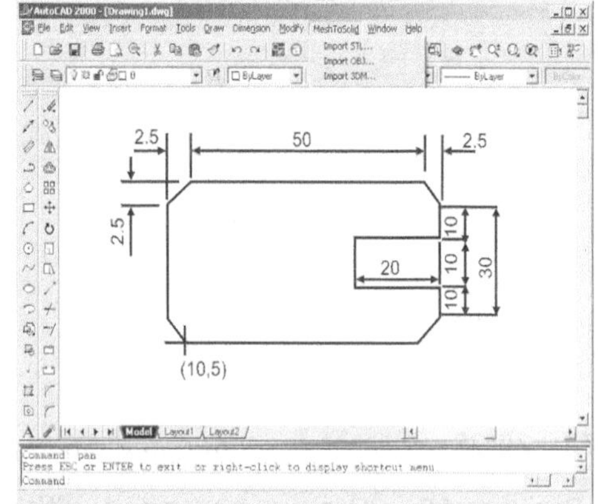

Fig. 20.26. *Gauge*

Problem 18 : Draw the figure of open bearing using AutoCAD. [Fig. 20.27]

Solution : To draw OPEN BEARING following steps should be followed :

Step (i)	Command	:	LINE
	From point	:	15, 15
	To point	:	@ 180<0
	To point	:	@ 20<90
	To point	:	@ 45<90
	To point	:	@ 55<90
	To point	:	@ 15<180
	To point	:	@ 25<270
	To point	:	(Press ENTER)

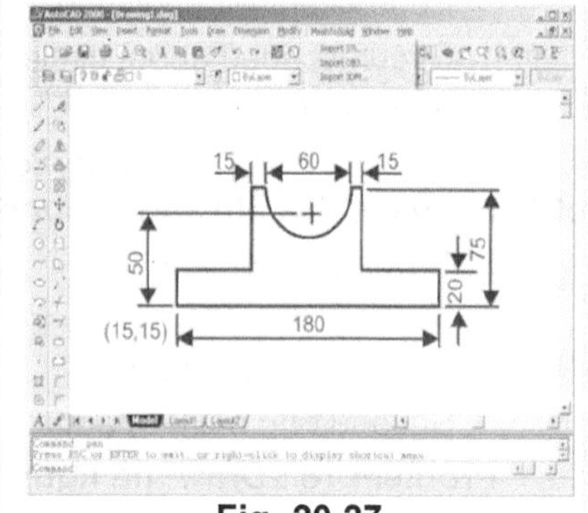

Fig. 20.27.

Steps (ii)	Command	: ARC
	Centre/<Start point>	: 130, 60
	Centre/End/<Second point>	: E
	End Point	: 70, 60
	Angle/Direction/Radius/<Centre point>	: A

Included angle : –180

Steps (iii) Command : LINE

From point : 70, 60

To point : @ 25<90

To point : @ 15<180

To point : @ 55<270

To point : @ 45<180

To point : @ 20<270

To point : (Press ENTER)

20.26. THREE DIMENSIONAL (3D) MODELLING

A model is a mathematical representation of geometric form which is stored in the computer memory.

Two dimensional model (2D) is recognised by the system as flat frame works bounded by a number of points are defined by x and y coordinates. In this, all drawings are produced only is one plane without any depth.

Three dimentional modelling (3D) is one of the most exciting features of AutoCAD. Three dimensional CAD systems store graphic data using x, y and z coordinates. 3D systems offers more capabilities but are complex and difficult to learn and use.

Three dimensional modelling takes time and practice to learn, but with time the results will be well worth.

20.27. TYPES OF MODELLING

The following are the three types of modelling supported by 3D systems.

1. Wire frame modelling
2. Surface modelling
3. Solid modelling.

20.27.1. Wire Frame Modelling : In the wireframe modelling, the objects are generated by continuing line segments. These models define the edges and surfaces of three dimensions objects. The complete object is represented by a number of lines with their end point coordinates (x, y and z) and their connectivity relationship.

Fig. 20.28 shows wire frame modelling of a cube. Wire frame models are used where the flow of energy, flow of mass and hollowness is required.

The AutoCAD packages provide the facility of preparing wire frame models of ready made and generated objects. For e.g., box, pyramid, sphere, wedge, hemisphere, etc. These may be of any irregular shape and can be fitted with wire frame surfaces by A1 MESH or 3D mesh.

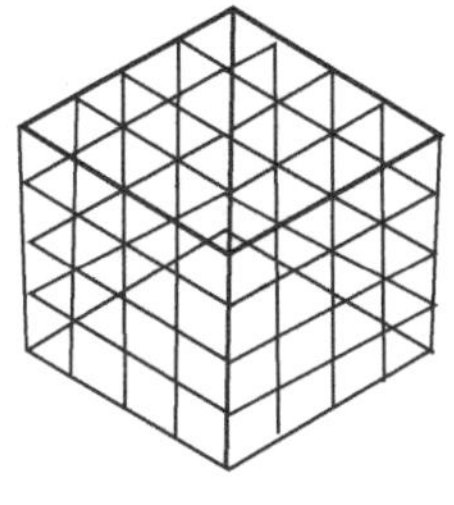

Wire frame modelling

Fig. 20.28.

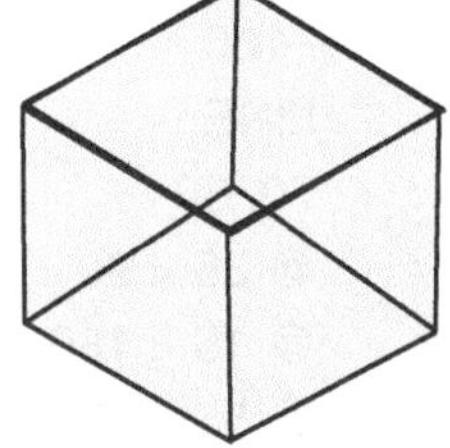

Surface modelling

Fig. 20.29.

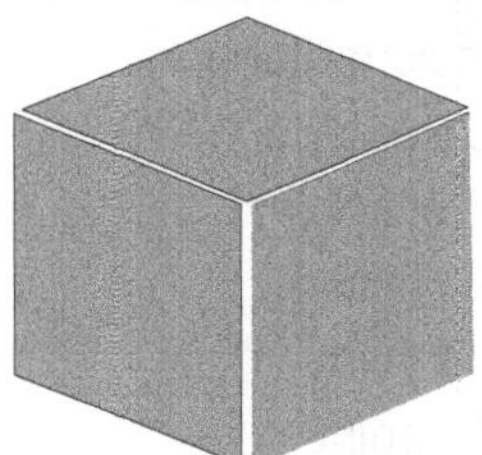

Solid modelling

Fig. 20.30.

20.27.2. Surface Modelling : In surface modelling, the objects are created by use of surfaces attached to wire frame constructions. The surfaces are very useful for preventing mass flows and allows storage of quantities because of hollowness of the object.

Fig. 20.29. shows the surface model of cube. Here, the cube is defined by twelve lines, eight points and six surfaces. In this modelling technique, a three dimensional object is created by joining of 3D surfaces and the portion behind the surface is invisible.

In AutoCAD, the surface models can be constructed by use of wire frame models attached with surfaces by use of PFACE command. The command provides the facility of creating faces of rectangular or triangular shapes.

20.27.3. Solid Modelling : In solid modelling technique, the objects are completely filled with solid material and thus the creation of solid models on the screen allow, calculation of the mass, volume, centre of gravity, etc.

Fig. 20.30 shows the solid models of cube. The AutoCAD package allows drawing of solid models like box, pyramid, cylinder, sphere, etc.

***Note** : Solid modelling may be confused with surface modelling because both can appear similar in certain cases. True solid modelling system treat the object as a real solid object.*

Problem 19 : Draw the open bearing (Fig. 20.31) by using Auto CAD.

Solution :

Command : LINE
From point : 10, 10
To point : @ 180 < 0
To point : @ 20 < 90
To point : @ 45 < 180
To point : @ 55 < 90
To point : @ 15 < 180
To point : @ 25 < 270
To point : (Press ENTER)
Command : ARC
Centre/<Start point> : 130, 60
Centre/End/<Second point> : E
End ponit : 70, 60
Angle/Direction/Radius/<centre point> : A
Included angle : –180
Command : LINE
From point : 70, 60
To point : @ 25 < 90
To point : @ 15 < 180
To point : @ 55 < 270
To point : @ 45 < 180

To point : @ 20 < 270
To point : (press ENTER)

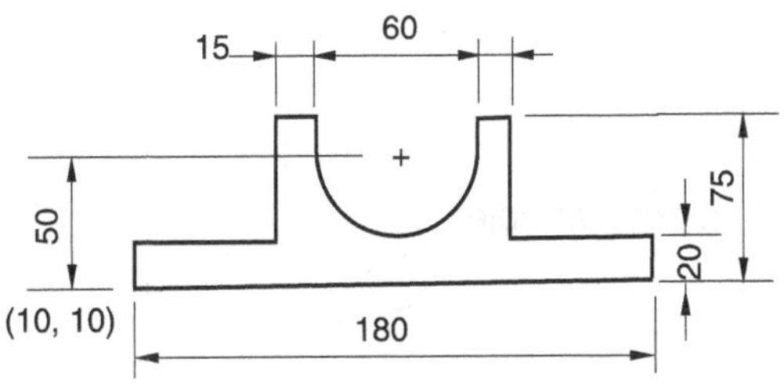

Fig. 20.31. *Open Bearing*

Problem 20 : Create 3D-V Block as shown in Fig. 20.32.

Solution : Command : limits ↵

Reset Model space limits :

Left corner ↵

Right corner : 100, 100 ↵

Command : Zoom ↵

Corner of window, enter a scale factor (nX or nXP), or

[All/ Center/Dynamic/Extents/Previous/Scale/Window] <real time> :

Command : vpoint ↵

View direction : VIEWDIR = 0.0000,0.0000,1.0000

View point : 1, –1,1 ↵

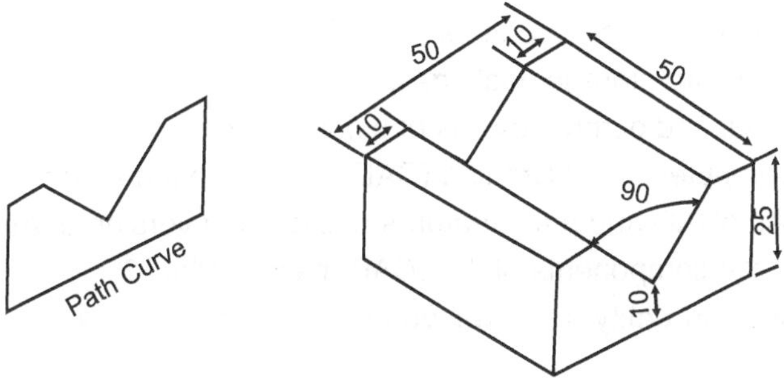

Fig. 20.32. *V-block*

Command : ucs ↵
Current ucs name : *WORLD*
An option : x ↵
X axis <907> : 90 ↵
Command : pline ↵
Start point : Pick any point

Next point : @10<270 ↵
Next point : @25<180 ↵
Next point : @25<90 ↵
Next point : @10<0 ↵
Next point : close ↵
Command : extrude ↵
Select objects : select last drawn object ↵
height : 50 ↵
Angle of taper <0> : ↵
Command : mirror ↵
Select objects : l ↵
Select objects :↵
First point : pick lower right end point
Second point : pick opposite point
Delete source objects ? [Yes/No] <n> : ↵
Command : union ↵
Select objects : all ↵
Select objects : ↵

QUESTIONS FOR SELF EXAMINATION

1. What is a computer ? What are the five basic operations performed by any computer system ?
2. What are the applications of a computer ?
3. Classify different types of computer.
4. What are the components of a computer ?
5. Distinguish between hardware and software ?
6. What are hardware components ?
7. Explain software storage packages.
8. List the input and output devices of a computer.
9. Define Computer Aided Drafting (CAD). What are the advantages of CAD ?
10. Explain AutoCAD and main features of the latest version of AutoCAD 2004.
11. What are the components of AutoCAD main window ?
12. Define drawing entity. How will you draw the enitity of a line by absolute coordinate method.
13. Define modelling, 2D modelling and 3D modelling
14. Fill in the blanks :
    (*a*) An electronic device which can accept and store input data, process them and produce output results is called ....
    (*b*) A mechanism for communication between the operator and the system is known as ....
    (*c*) The .... components of a computer system is called hardware.

(*d*) The control unit and ALU of a computer system jointly known as............. .

(*e*) AutoCAD is a............... software to produce 2D/3D drawings.

(*f*) An output device used to produce..............output is known as printer.

(*g*) .... models are true shape of 3D objects.

Ans. (*a*) computer (*b*) input (*c*) physical (*d*) CPU (*e*) graphic (*f*) hardcopy (*g*) solid

15. State whether the following statements are true or false :

(*a*) Man has mind and intuition where as the computer has speed and accuracy.

(*b*) The speed of the pentium IV computer is about 3.2 GHz.

(*c*) The two elements software of computer which work together are hardware and software.

(*d*) An output device used to produce soft copy output is known as monitor.

(*e*) AutoCAD is the first and foremost product of Autodesk Inc. UK.

(*f*) The architectural drawings created by AutoCAD are not very acurate.

(*g*) Wire frame modellings are generated by continuing line segment.

Ans. (*a*) True (*b*) True (*c*) True (*d*) True (*e*) False (*f*) False (*g*)True

PROBLEMS FOR PRACTICE

1. Draw a line from point (50, 50) to point (120, 50) by using absolute co-ordinates method.
2. Draw a rectangle defined by diagonal point (50, 50) and (160, 100).
3. Draw a circle with centre (50, 50) and radius 60 unit using AutoCAD.
4. Draw a circle with centre (65, 50) and radius 50 units by centre and radius method.
5. Draw an ellipse with centre (35, 50), major axis and end point (60, 50) and 65 degree rotation around the major axis.
6. Draw the Fig. 20.33 by using AutoCAD.
7. Create 3D object as shown in Fig. 20.34.

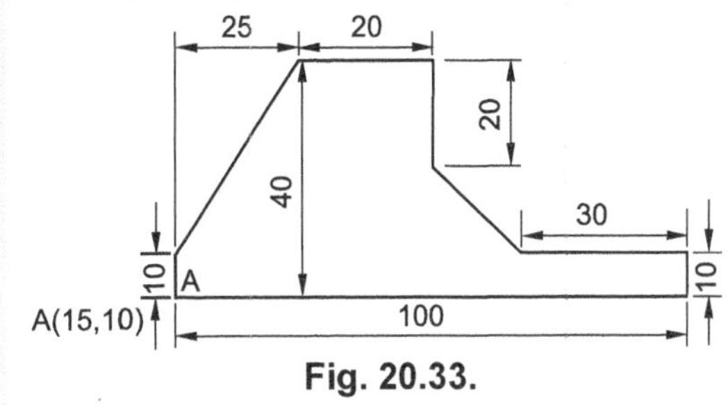

Fig. 20.33.

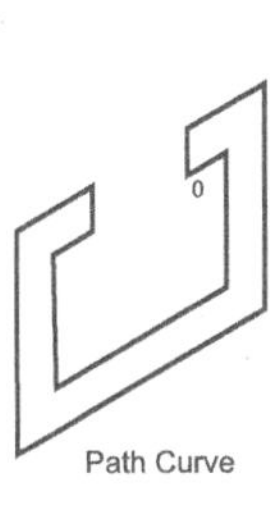

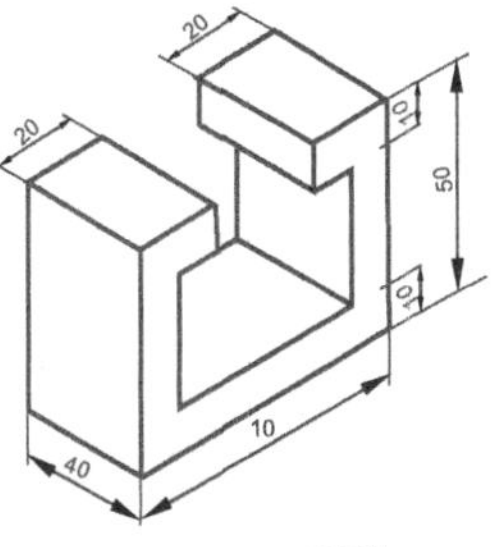

Fig. 20.34.

This Page has been intentionally left blank

Model Test Papers

MODEL TEST PAPER – 1

Time : 04 Hours *Maximum Marks : 60*

Instruction to Candidates :

(1) Section – A is ***Compulsory.***
(2) Attempt any ***Four*** *questions from Section – B.*
(3) Attempt any ***Two*** *questions from Section – C.*

Section – A

(10 × 2 = 20)

Q.1. (a) Differentiate between pitch and lead for a triple start thread.

(b) Make a free hand sketch of the Metric thread profile giving important proportions.

(c) What is use of Gib in 'Gib and Cotter Joint'?

(d) Draw the symbols along with the illustrations for the following welded joint.

(i) Fillet Weld (ii) Convex double-V Butt Weld

(e) Make a free hand sketch of the Rounded countersunk rivet head showing proportions in terms of shank dia D.

(f) Mention any two means for prevention of rotation of brasses in Plummer bearing block.

(g) What necessitates the use of expansion joint?

(h) Differentiate between "Caulking" and "Fullering" in context of rivets.

(i) Which coupling is used for connecting parallel and non-intersecting shafts? Draw free hand sketch for the same.

(j) Differentiate between Basic size and actual size.

Section – B

(4 × 5 = 20)

Q.2. Draw free hand proportionate and neat sketches for the following:

(a) Rag Foundation Bolt. (b) Use of SAWN nut as locking device.

Q.3. Draw the top view of double riveted lap joint (Chain type) for connecting two plates of thickness 9 mm. Use relevant empirical relations and show at least 3 rivet heads along each row of rivets.

Q.4. Figure 1 shows the details of Gib and cotter joint. Assemble the given parts and draw the full sectional front view of assembly.

Q.5. Figure 2 shows the details of Expansion Pipe Joint. Assemble the given parts and draw the full sectional front view of assembly.

Q.6. Figure 3 shows the details of Solid Flange Coupling. Assemble the given parts and draw the full sectional front view of assembly.

Section – C

(2 × 10 = 20)

Q.7. Figure 4 shows the details of Plummer Block. Assemble the given components and draw the right half sectional front view of assembly.

Q.8. Figure 5 shows the details of screw Jack. Assemble the given components and draw the right half sectional front view of assembly.

Q.9. Figure 6 shows the details of steam stop valve. Assemble the given components and draw the right half sectional front view of assembly.

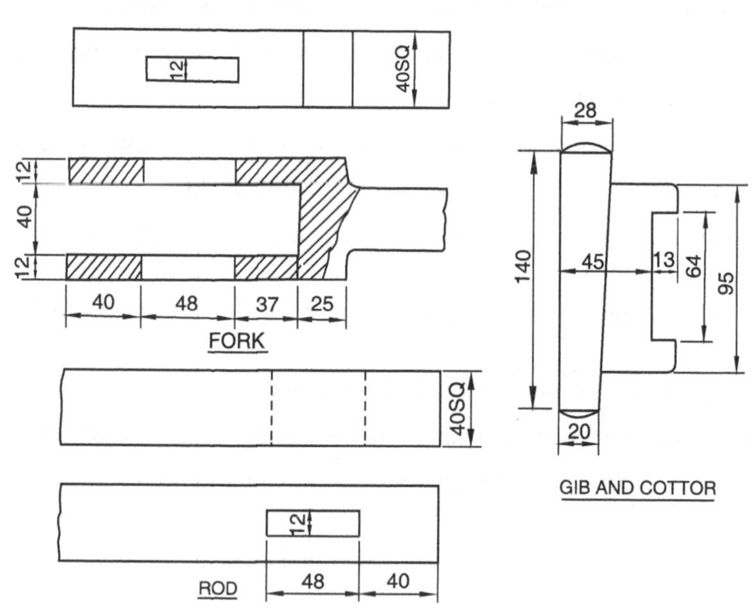

Fig. 1.

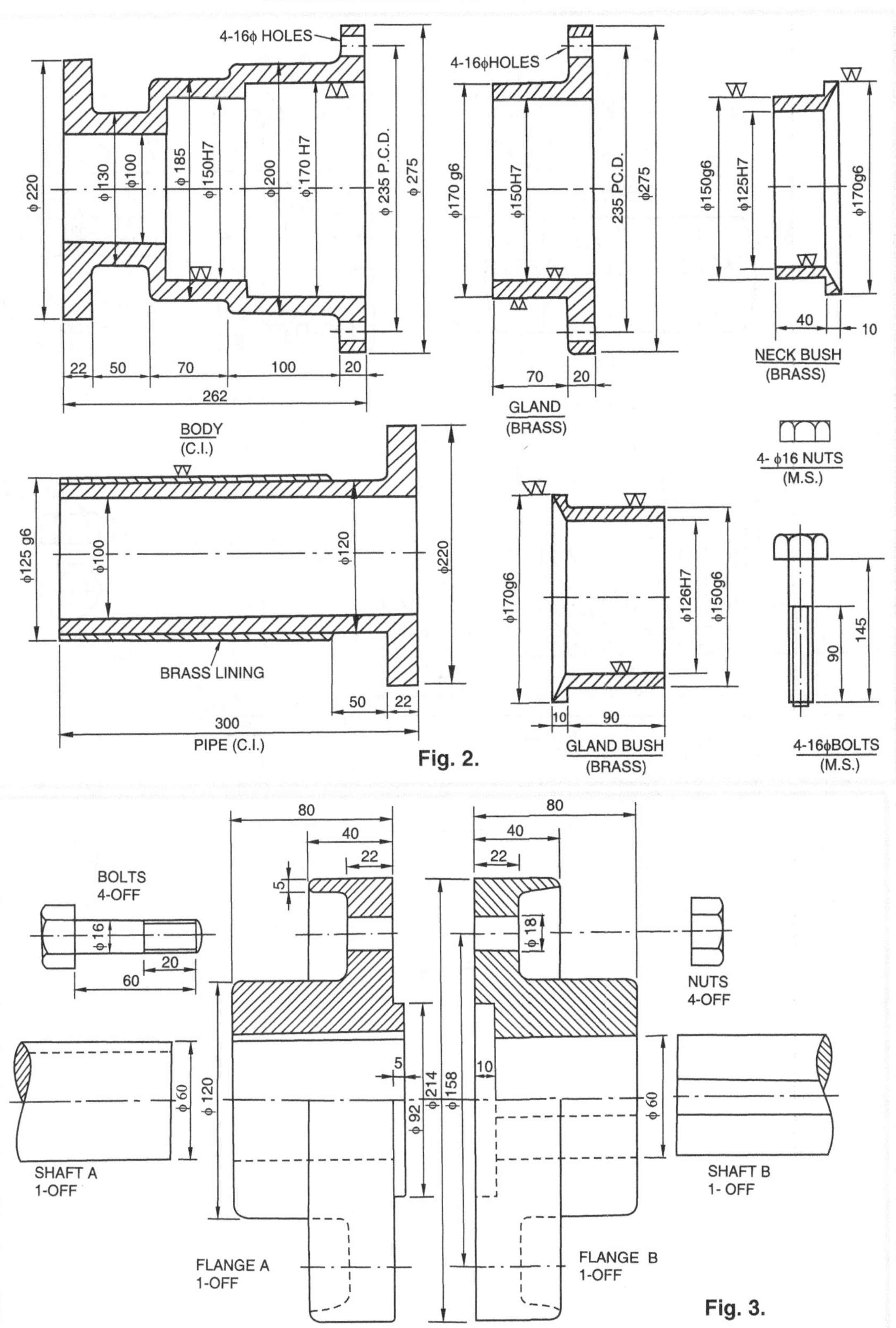
4-16ϕ HOLES
ϕ 220
ϕ130
ϕ100
ϕ 185
ϕ150H7
ϕ200
ϕ170 H7
ϕ 235 P.C.D.
ϕ 275
22
50
70
100
20
262
BODY
(C.I.)
4-16ϕHOLES
ϕ170 g6
ϕ150H7
235 PC.D.
ϕ275
70
20
GLAND
(BRASS)
ϕ150g6
ϕ125H7
ϕ170g6
40
10
NECK BUSH
(BRASS)
4- ϕ16 NUTS
(M.S.)
ϕ125 g6
ϕ100
ϕ120
ϕ220
BRASS LINING
50
22
300
PIPE (C.I.)
ϕ170g6
ϕ126H7
ϕ150g6
10
90
GLAND BUSH
(BRASS)
145
90
4-16ϕBOLTS
(M.S.)

Fig. 2.

80
40
22
5
80
40
22
BOLTS
4-OFF
ϕ16
20
60
ϕ 18
NUTS
4-OFF
ϕ 60
ϕ 120
5
ϕ 92
ϕ214
ϕ158
10
ϕ 60
SHAFT A
1-OFF
SHAFT B
1- OFF
FLANGE A
1-OFF
FLANGE B
1-OFF

Fig. 3.

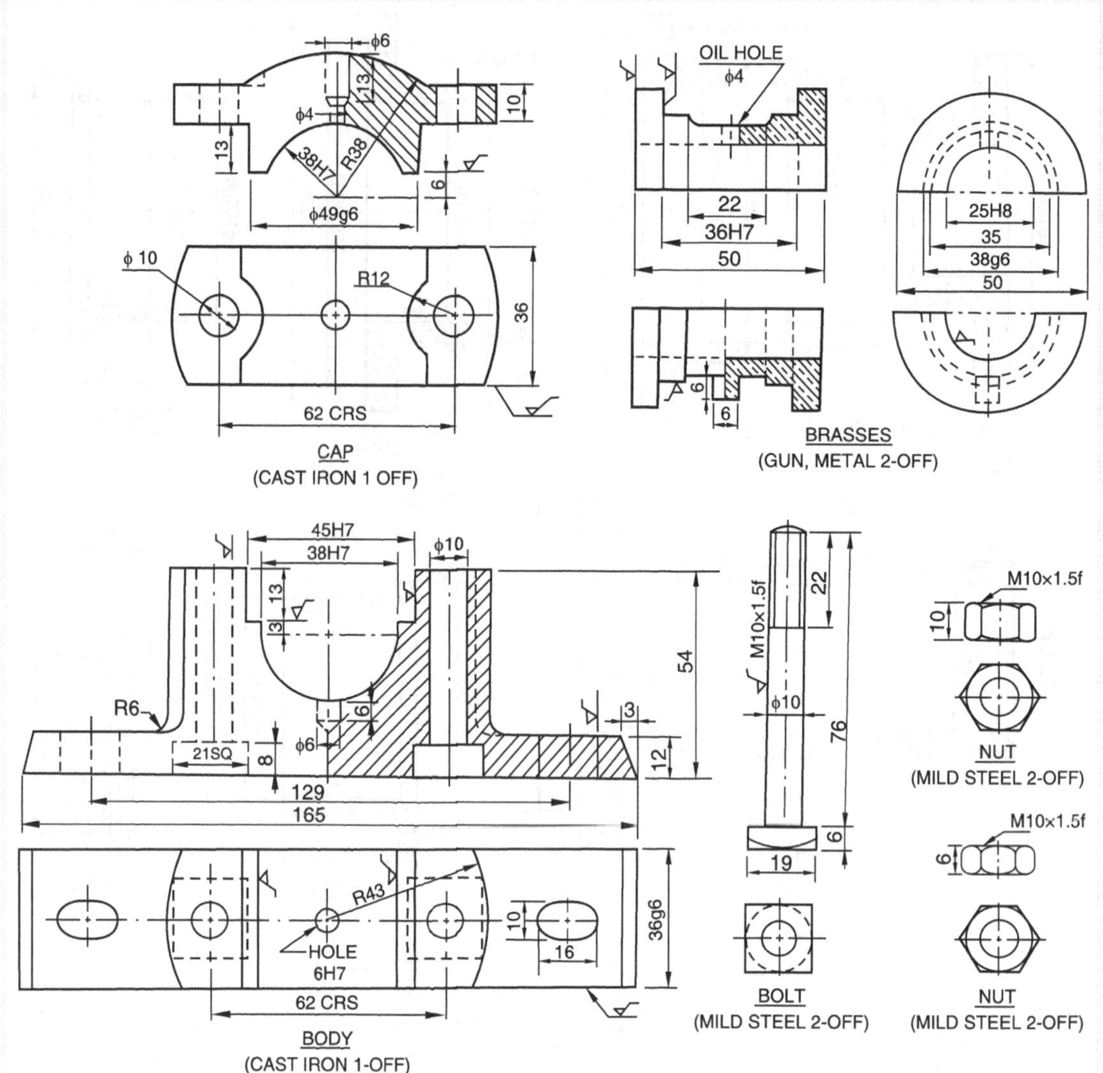

φ6
13
10
φ4
13
38H7
R38
6
φ49g6
φ 10
R12
36
62 CRS
CAP
(CAST IRON 1 OFF)
OIL HOLE
φ4
22
36H7
50
25H8
35
38g6
50
6
6
BRASSES
(GUN, METAL 2-OFF)
45H7
38H7
φ10
13
3
R6
6
φ6
21SQ
8
129
165
3
12
54
R43
HOLE
6H7
10
16
36g6
62 CRS
BODY
(CAST IRON 1-OFF)
M10×1.5f
22
φ10
76
6
19
BOLT
(MILD STEEL 2-OFF)
M10×1.5f
10
NUT
(MILD STEEL 2-OFF)
M10×1.5f
6
NUT
(MILD STEEL 2-OFF)

Fig. 4.

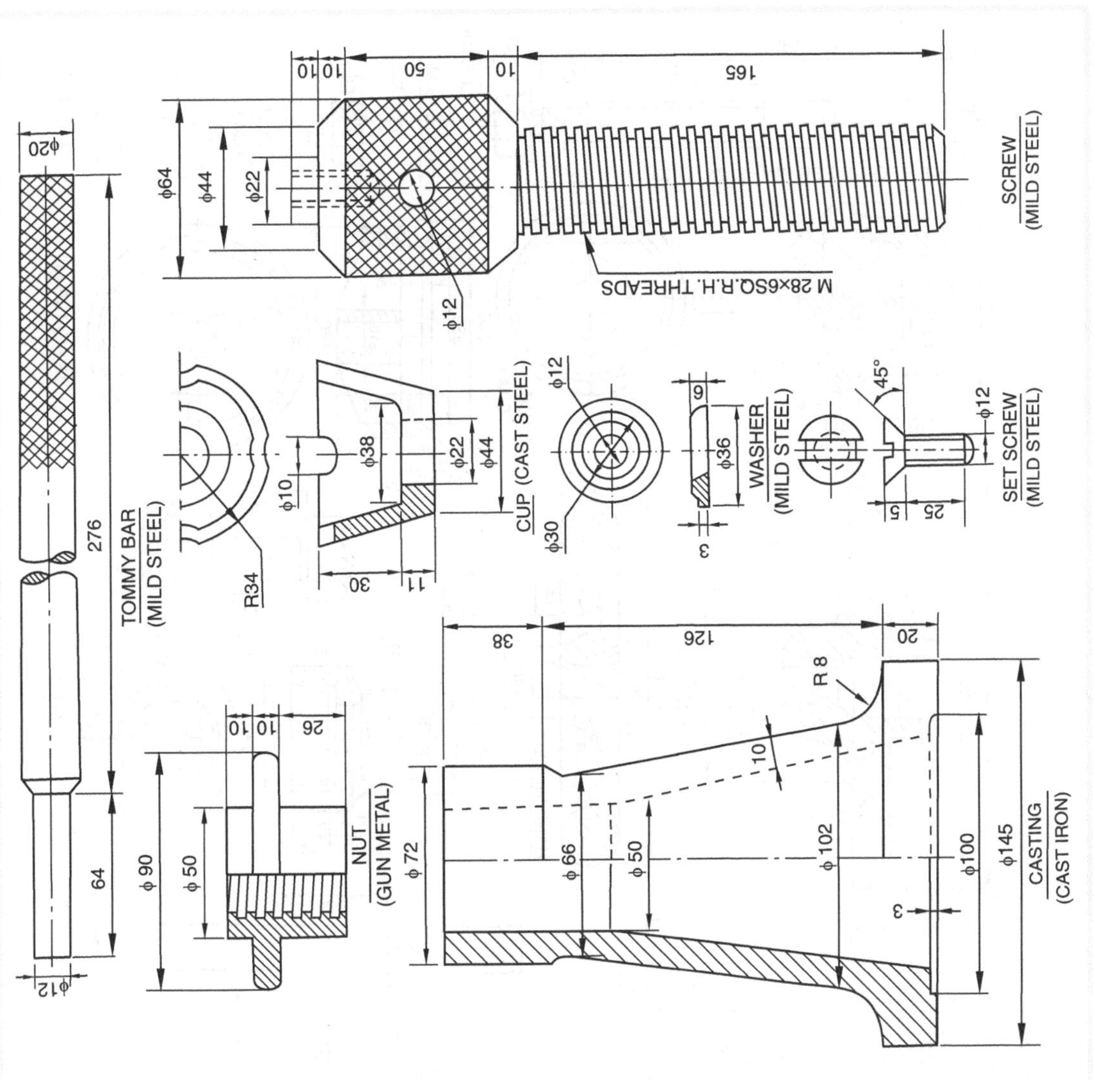
SCREW
(MILD STEEL)
M 28×6SQ.R.H. THREADS
TOMMY BAR
(MILD STEEL)
CUP (CAST STEEL)
WASHER
(MILD STEEL)
SET SCREW
(MILD STEEL)
NUT
(GUN METAL)
CASTING
(CAST IRON)

Fig. 5.

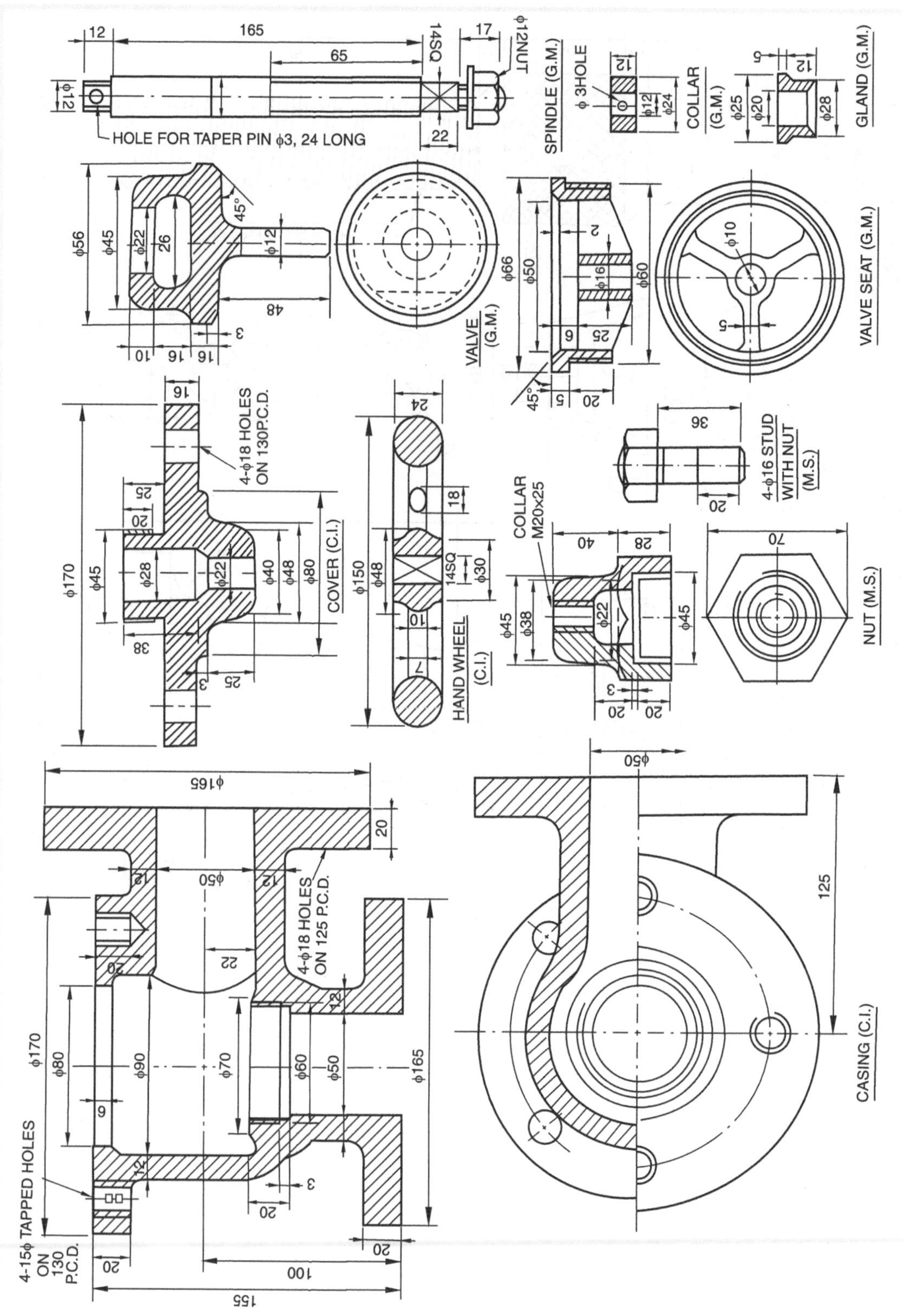

Fig. 6.

MODEL TEST PAPER – 2

Time : 04 Hours *Maximum Marks : 60*

Instruction to Candidates :

(1) *Section – A is* ***Compulsory.***
(2) *Attempt any* ***Four*** *questions from Section – B.*
(3) *Attempt any* ***Two*** *questions from Section – C.*

Section – A (10 × 2 = 20)

Q.1. (a) What are temporary and permanent fastenings?

(b) Sketch the conventional method of representing B.S.W. thread.

(c) What is the taper of a cotter and why is it provided?

(d) Sketch the basic symbols for Single and Double-V Butt welds.

(e) State the difference between rigid and non-rigid shaft couplings.

(f) Under what conditions do you use a knuckle joint?

(g) Which Joint is generally employed for pipe lines laid in the ground?

(h) Why rings are provided on the periphery of the piston?

(i) What is meant by the valve lift in a feed check valve?

(j) What is the advantage of Plummer block over a simple bushed bearing?

Section – B (4 × 5 = 20)

Q.2. Draw plan and sectional elevation of a double riveted lap joint (Chain riveting). Take diameter of rivet = 20 mm.

Q.3. Sketch (free hand) the proportionate sectional front view of a socket and spigot pipe joint.

Q.4. Draw free hand upper half sectional-front elevation of a split muff coupling on proportionate scale.

Q.5. Draw free hand sectional-front elevation and side view of a cotter joint on proportionate scale.

Q.6. Discuss the advantages of Computer Aided Drafting (CAD) over Manual Drafting.

Section – C

(2 × 10 = 20)

Q.7. Figure-1 shows the details of a Plummer block. Assemble the parts and draw to full scale the following views:

(i) Half sectional-front view and (ii) Top view. Also mark Bill of Materials.

Q.8. Figure-2 shows the details of a Rams-bottom safety valve. Assemble the parts and draw the following views:

(i) Front view- full in section, and (ii) Right side view. Use any convenient scale.

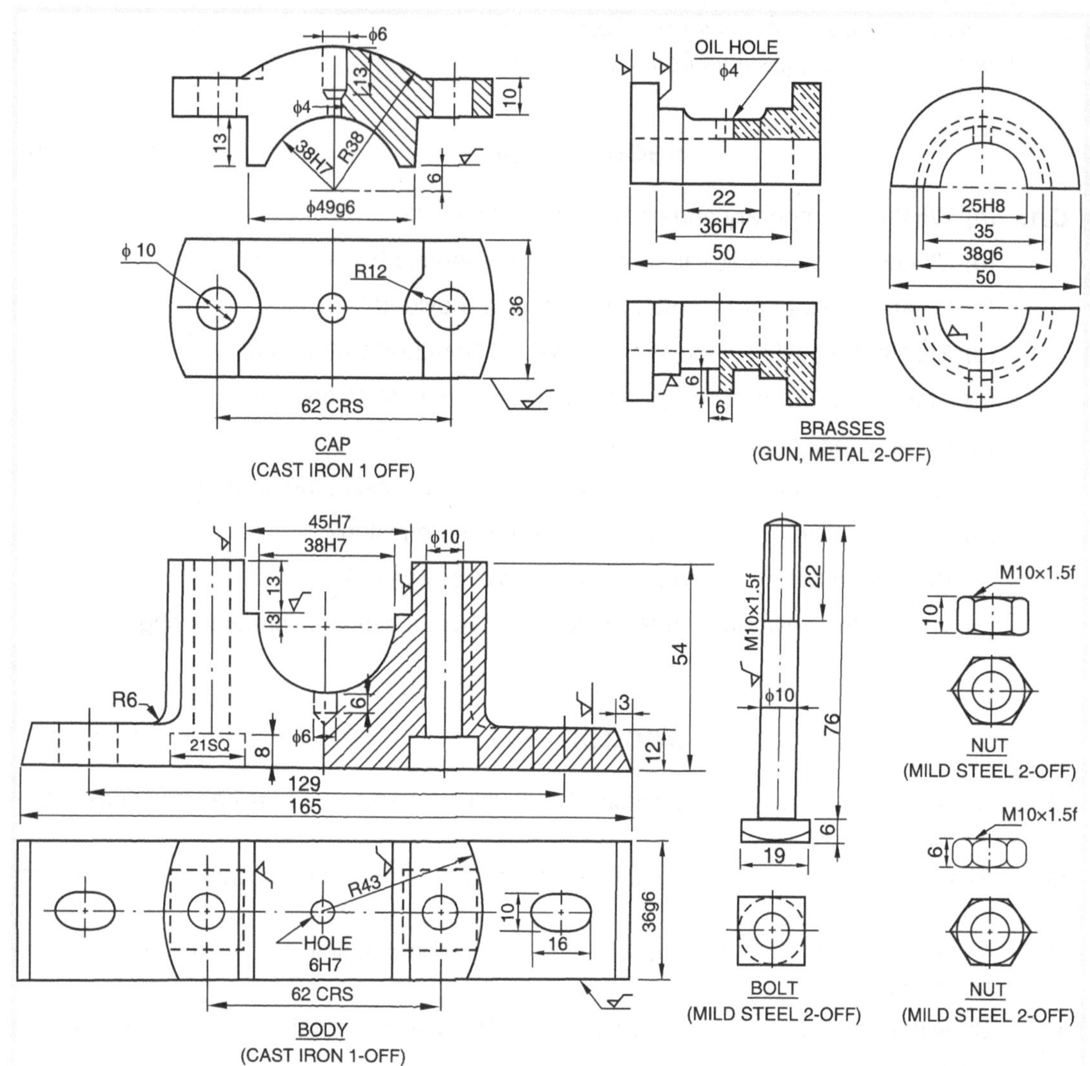

Fig. 1. *Details of plummer block*

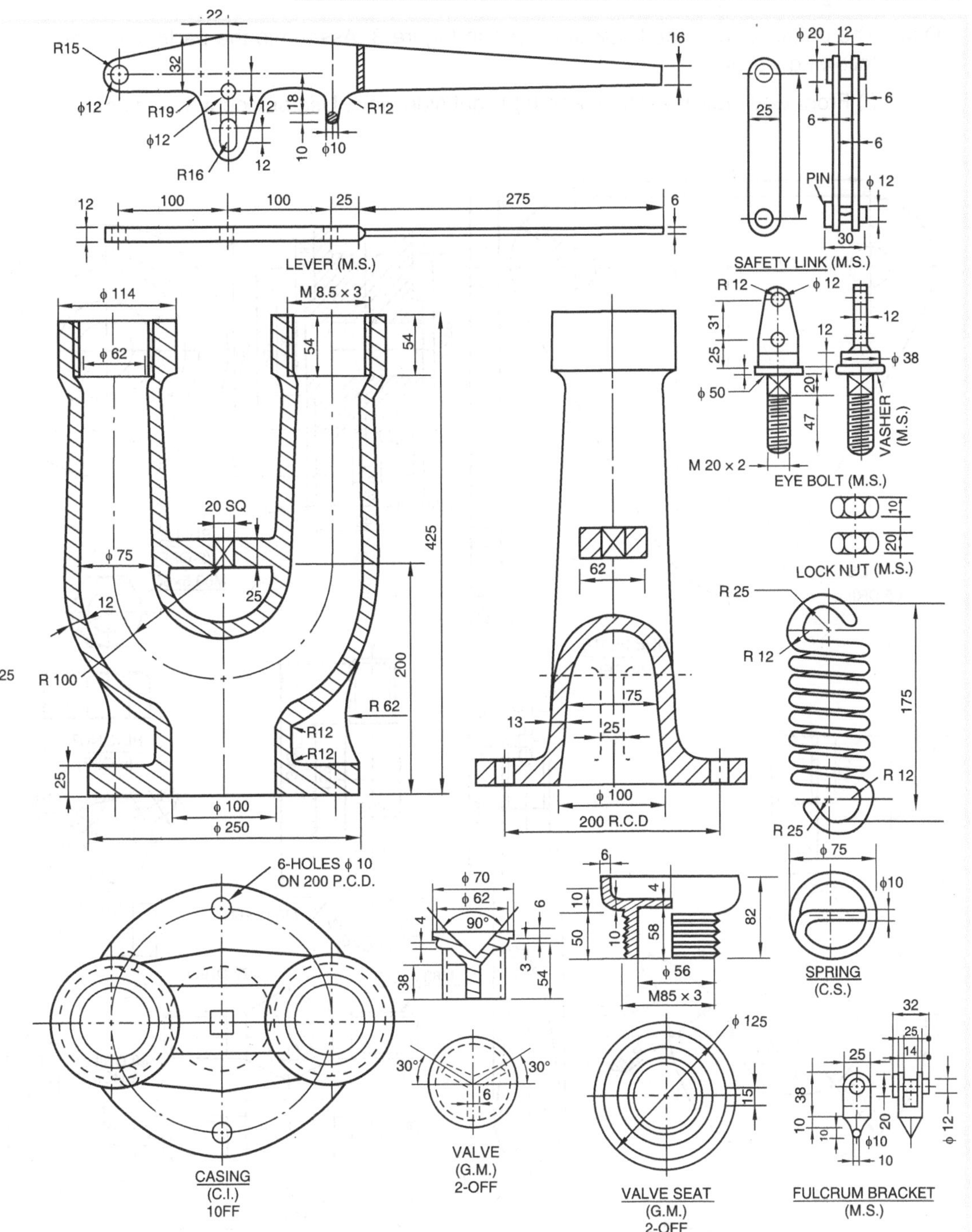

Fig. 2. *Details of a Ram's bottom safety valve*

Q.9. The details of a crane hook are given in Figure 3. Assemble the parts and draw the following views:

(i) Front view- full in section, and (ii) Right side view. Use any convenient scale.

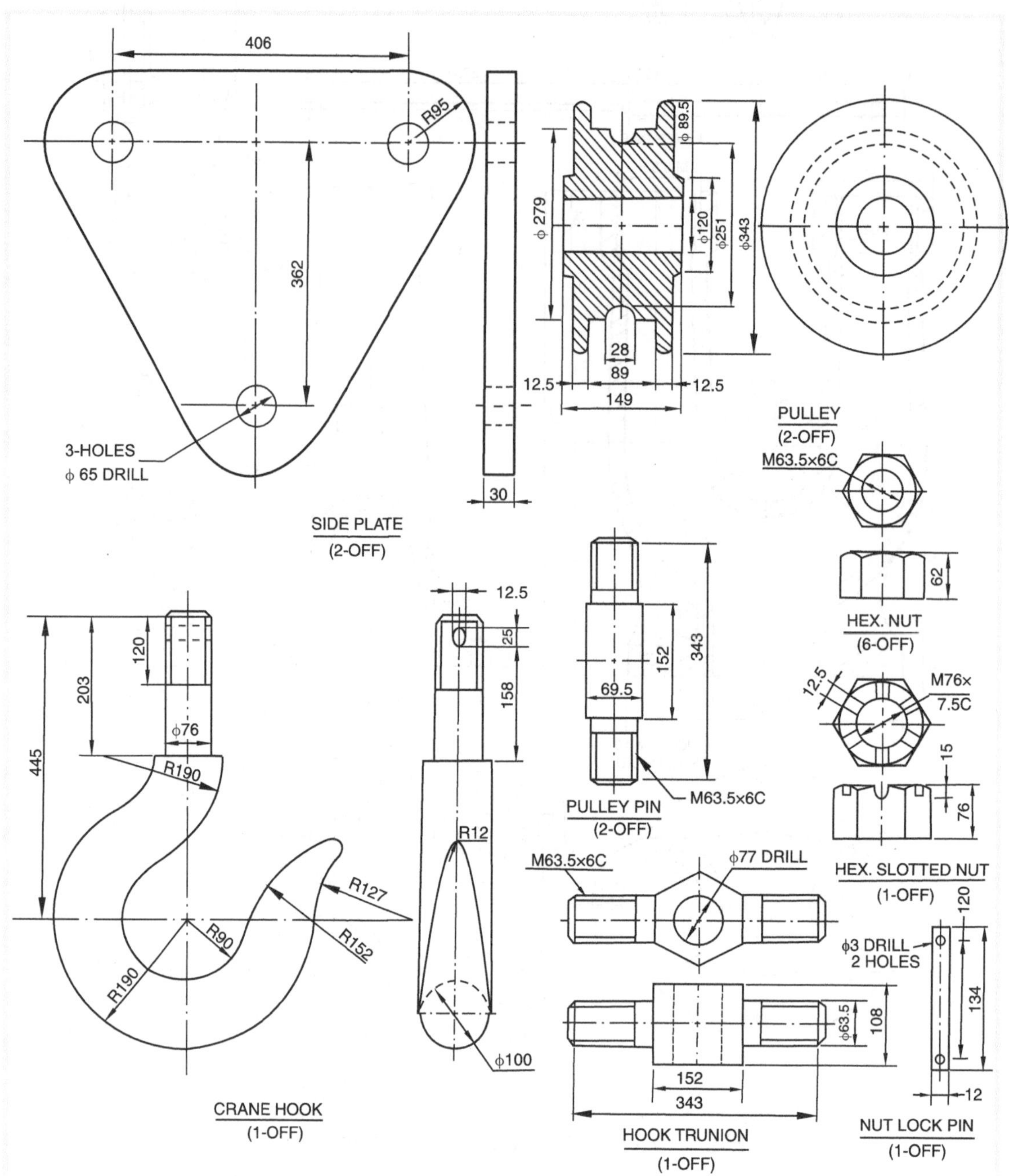

Fig. 3. *Details of a Crane Hook*

MODEL TEST PAPER – 3

Time : 04 Hours *Maximum Marks : 60*

Instruction to Candidates :

*(1) Section – A is **Compulsory.***
*(2) Attempt any **Four** questions from Section – B.*
*(3) Attempt any **Two** questions from Section – C.*

Section – A

(10 × 2 = 20)

Q.1. (a) What is the necessity of conventional representation of screw threads?
(b) What are the various type of welding?
(c) What is the difference between allowance and clearance?
(d) How do you hatch if there are more than two adjacent parts?
(e) Draw the conventional representation of Glass and Wood.
(f) Explain with a simple sketch the aligned system of dimensioning.
(g) What is a flexible coupling ? What are its advantages?
(h) What is the function of bush in a bearing?
(i) How is the blow off cock operated?
(j) What is the function of a tool post of a lathe machine?

Section – B

(4 × 5 = 20)

Q.2. Draw different types of lines used in machine drawing.

Q.3. Explain the following terms with neat sketch :
(a) Clearance
(b) Interference

Q.4. Draw the neat sketches of the following welded joints :
(a) Butt joint
(b) Lat joint
(c) T-joint
(d) Corner joint
(e) Edge joint

Q.5. Draw by conventional method a right handed square thread. Take outside dia = 64 mm, threaded length = 72 mm and pitch = 12 mm.

Q.6. Make proportionate sketch of a Union joint.

Section – C

(2 × 10 = 20)

Q.7. Draw the full sectional front view of the screw jack assembly as a shown in Fig. 1.

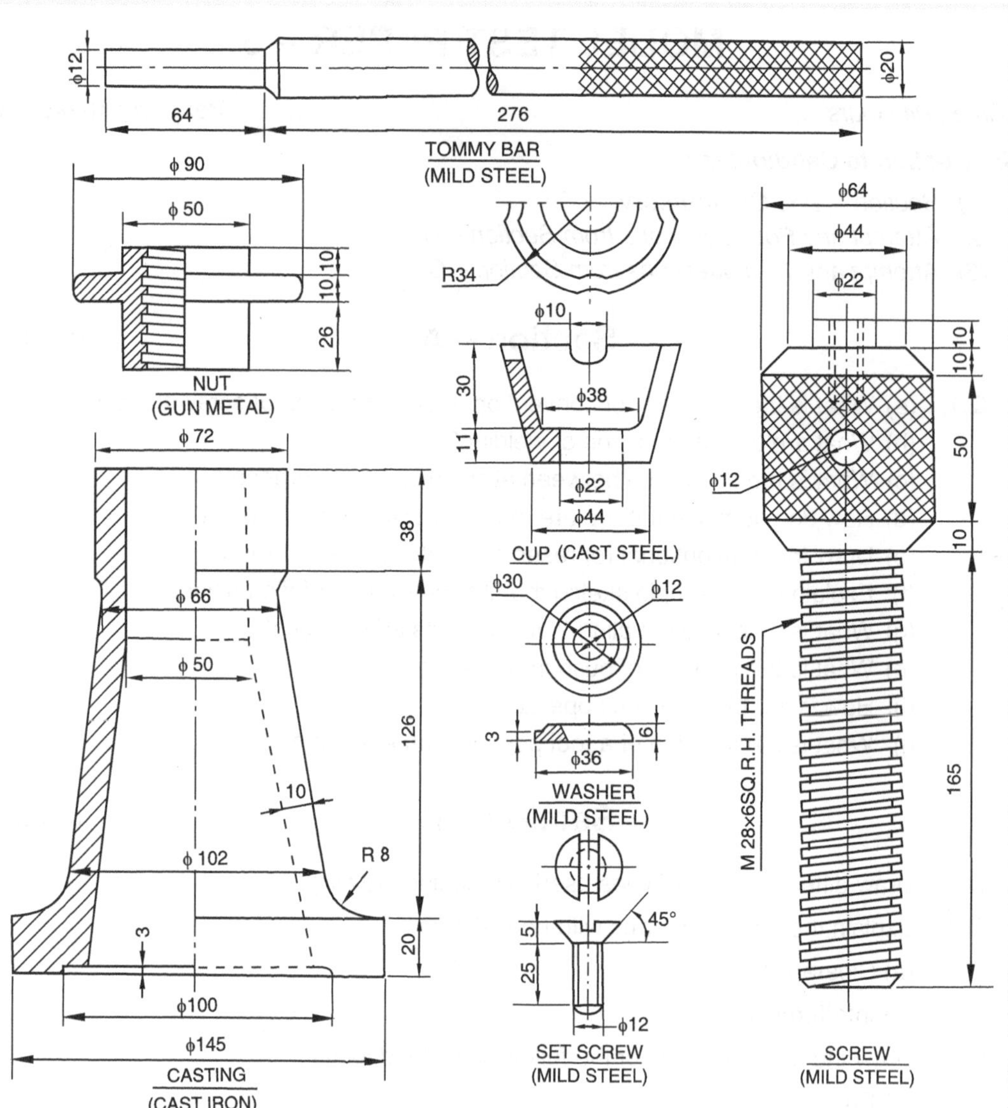

Fig. 1. *Details of screw jacks*

Q.8. Draw the full sectional front view of the I.C. Engine connecting rod assembly as shown as in Fig. 2.

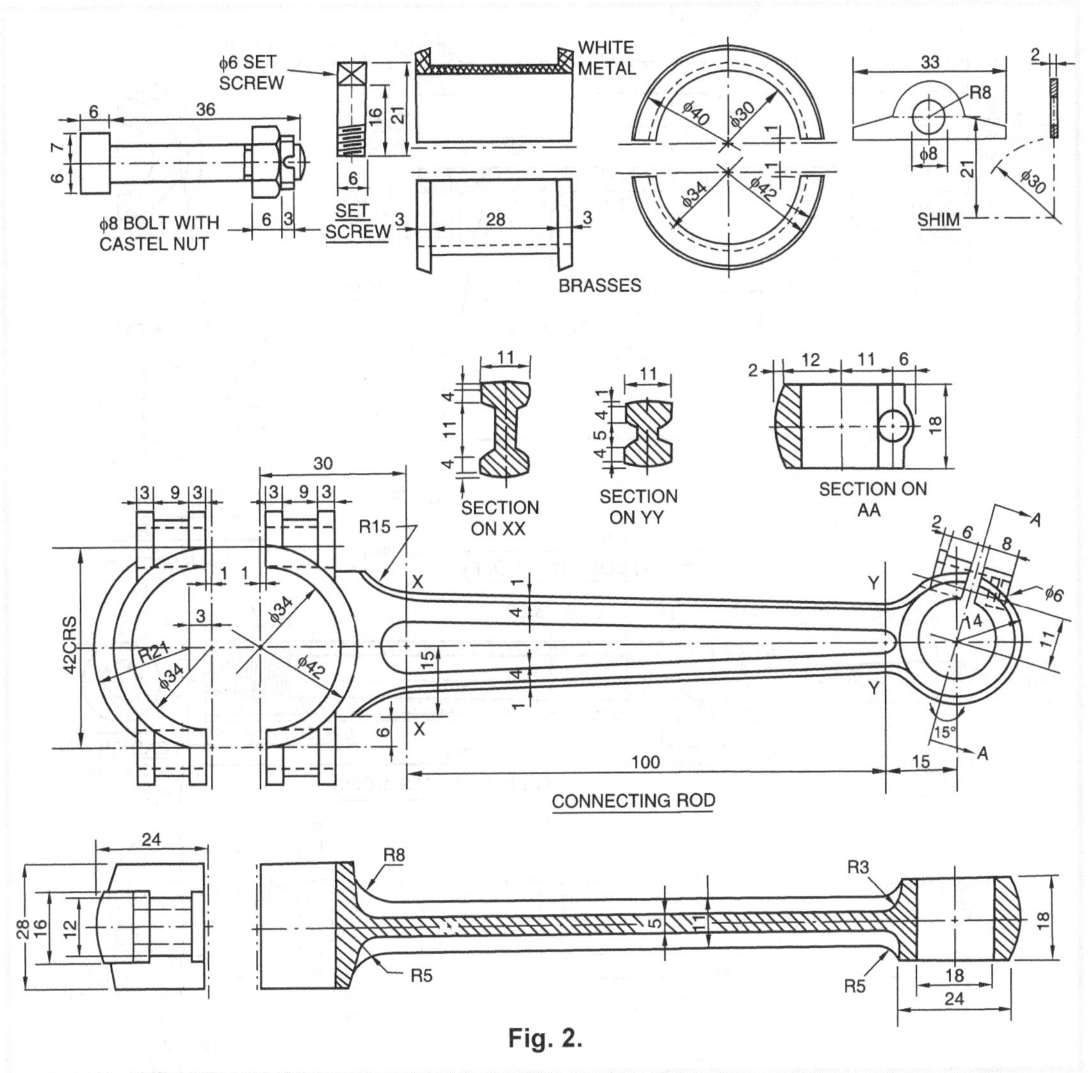

Fig. 2.

Q.9. Draw the full sectional front view of the lathe tail stock assembly as shown in Fig. 3.

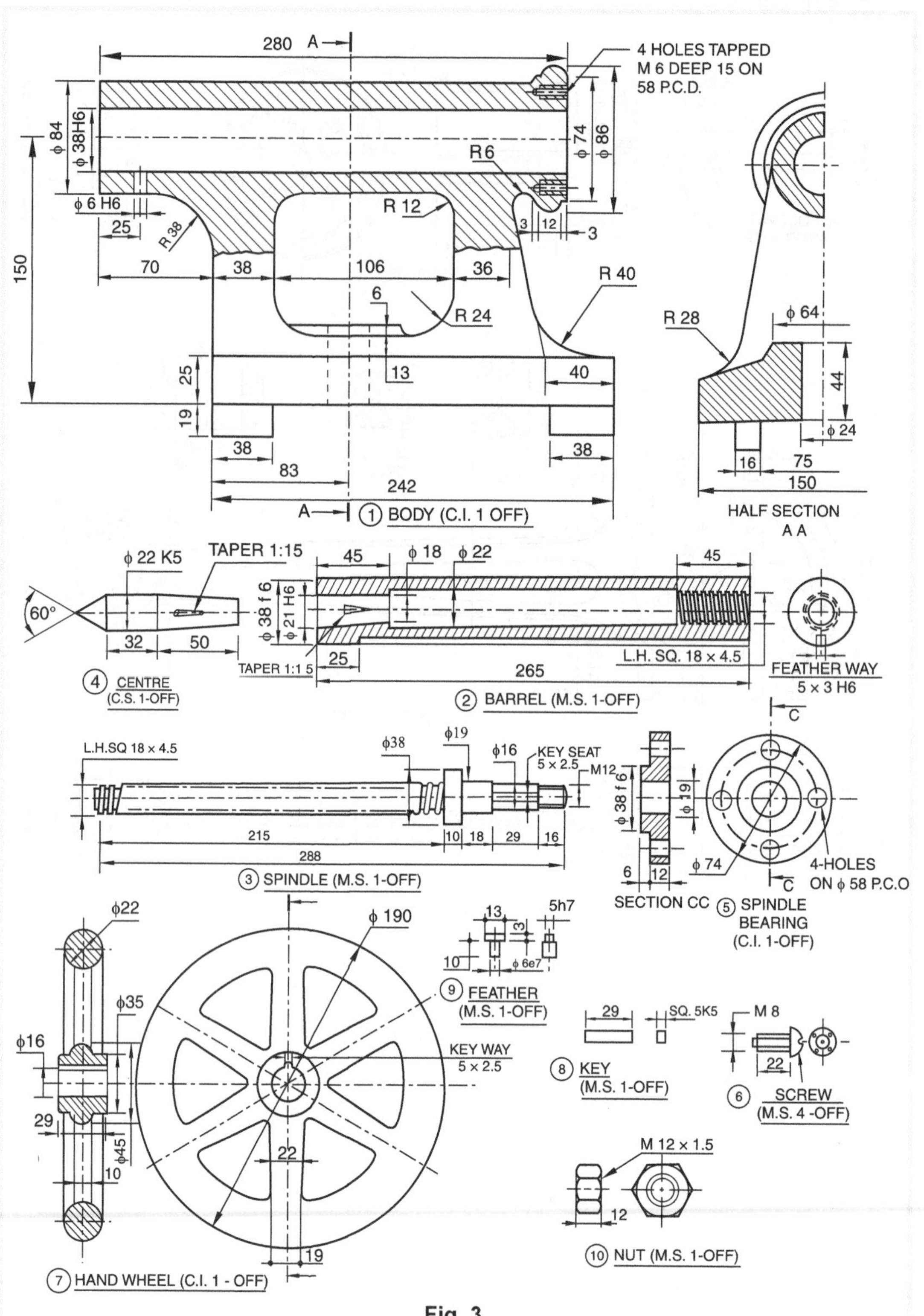

Fig. 3.

MODEL TEST PAPER – 4

Time : 04 Hours *Maximum Marks : 60*

Instruction to Candidates :

(1) Section – A is ***Compulsory.***
(2) Attempt any ***Four*** *questions from Section – B.*
(3) Attempt any ***Two*** *questions from Section – C.*

Section – A (10 × 2 = 20)

Q.1. (a) What is meant by progressive dimensioning the continuous dimensioning?

(b) How the tolerances are specified and indicated on drawings?

(c) Explain with the help of suitable sketches the method of dimensioning :
(i) Arcs and (ii) Angles.

(d) What is the use of multi-start threads?

(e) Show by sketches how the size of weld is given.

(f) What is the function of an eccentric in an engine?

(g) What is meant by the lift of the valve? What should be the maximum lift?

(h) What is the function of a tool post of a lathe machine?

(i) In a simple bushed bearing how the rotation and axial movement of the bush is prevented.

(j) What is the function of tailstock in lathe machine?

Section – B (4 × 5 = 20)

Q.2. Draw by a conventional method a right handed square thread. Take outside diameter = 64 mm, threaded length = 72 mm, and pitch = 12 mm.

Q.3. Sketch (free hand) four important pipe fittings for wrought iron pipes.

Q.4. Draw free hand upper half sectional front view of an ordinary flange coupling on proportionate scale.

Q.5. What is the function of a valve? Where a feed check valve is fitted and what are its functions?

Q.6. Discuss the use of following commands available in Auto-CAD: (a) Array (b) Offset, and (c) Mirror.

Section – C

(2 × 10 = 20)

Q.7. Figure 1 shows the details of a screw-jack. Draw the following views of the assembly to some suitable scale : (*a*) Front view-right half in section , and (*b*) Top view. Also prepare a part list and Bill of Materials.

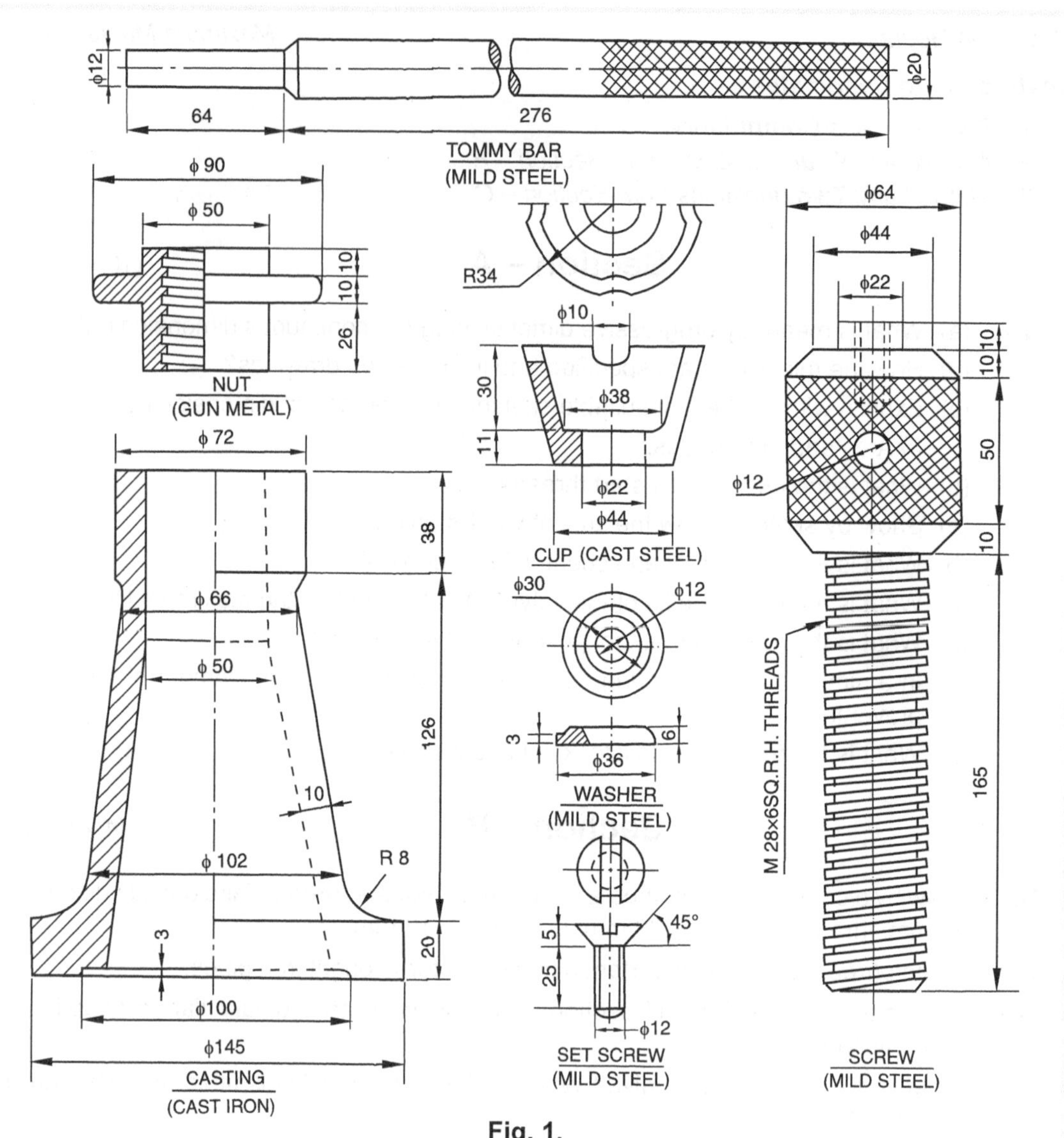

Fig. 1.

Q.8. Figure 2 shows the details of a swivel bearing. Assemble the parts and draw the following views: (a) Front view-right half in section, and (b) End view. Use any convenient scale.

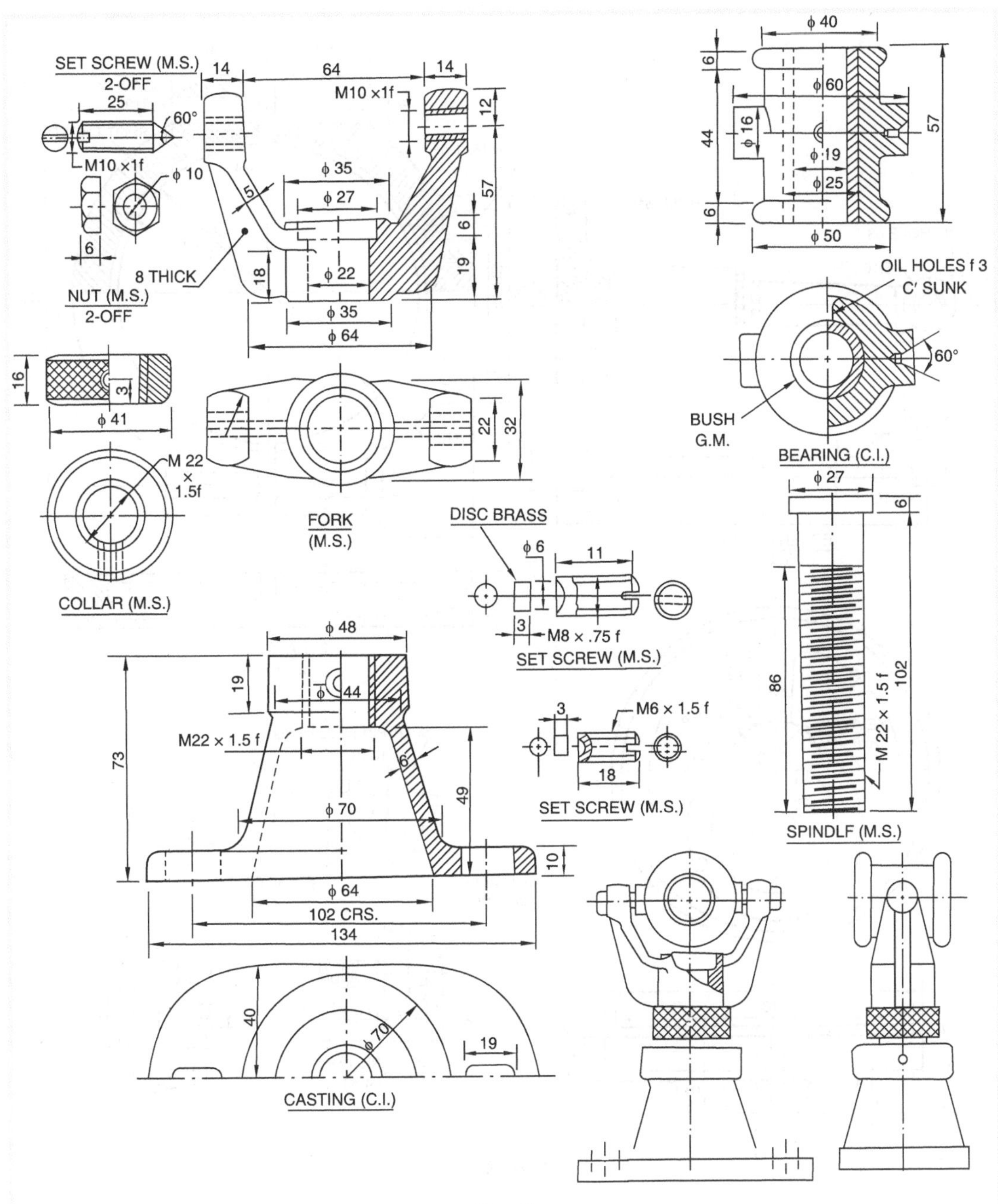

Fig. 2.

Q.9. Details of the eccentric are shown in Figure 3. Assemble the parts and draw the following views : (a) Front view upper half in section, and (b) Top view. Use any convenient scale.

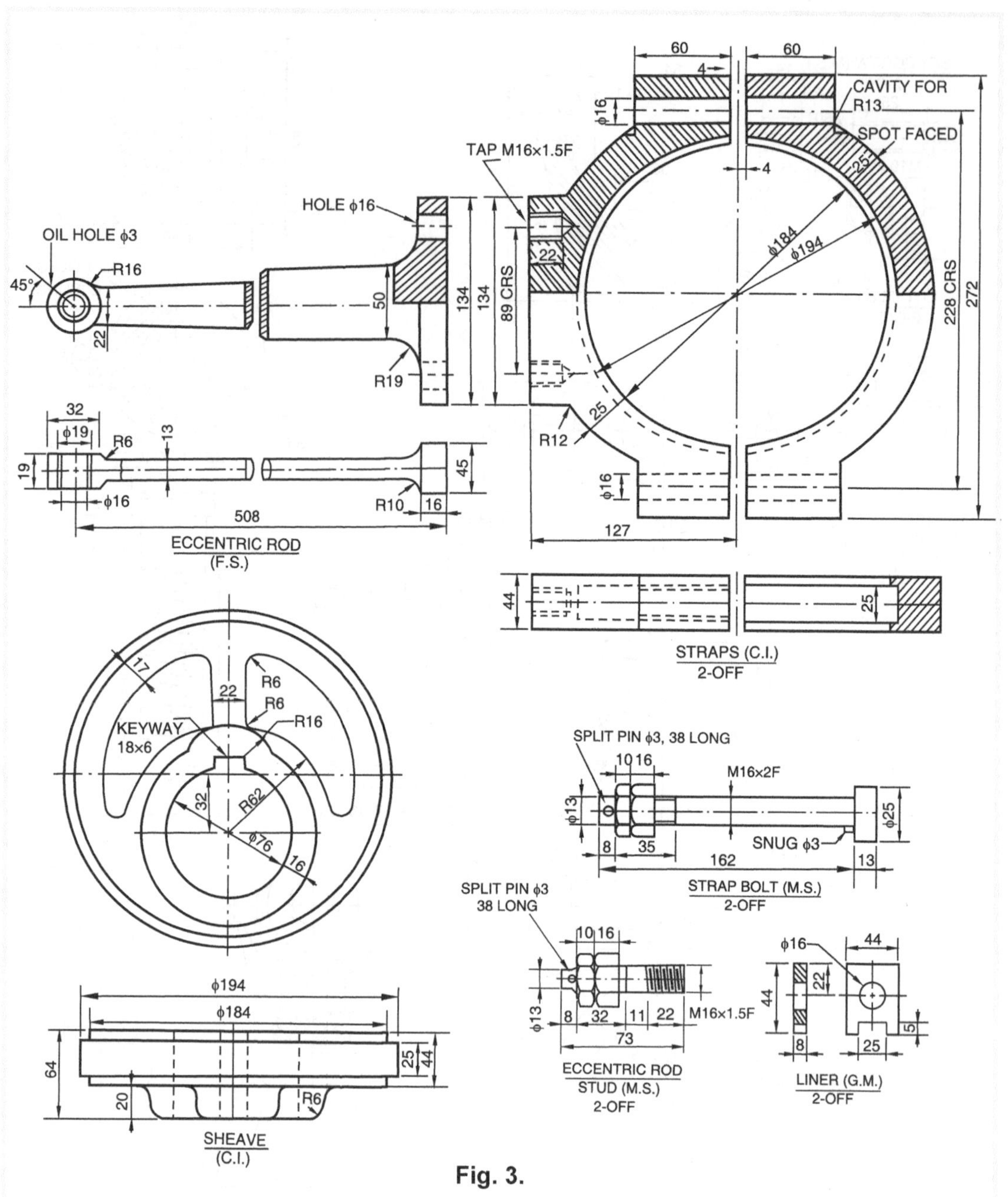

Fig. 3.

MODEL TEST PAPER – 5

Time : 04 Hours *Maximum Marks : 60*

Instructions to Candidates :

*(1) Sectrion – A is **Compulsory.***
*(2) Attempt any **Four** questions from Section – B.*
*(3) Attempt any **Two** questions from Section – C.*

Section – A

(10 × 2 = 20)

Q.1. (a) What is the difference between allowance and tolerance?

(b) What is the purpose of caulking and fullering of rivert joints?

(c) Sketch the conventioanl method of representing pipe threads (internal and external).

(d) What is the function of clearances in a cottered joint?

(e) Sketch the basic symbols for Single and Double-U Butt welds.

(f) Why are split muff couplings used in preference to solid muff?

(g) What is the specific use of an expansion pipe joint?

(h) Why brasses are used in connecting rod ends and why are these made of soft metals?

(i) What is blow-off cock and where it is used?

(j) What is the advantage of providing bush in a bearing? What is the material of brush?

Section – B

(4 × 5 = 20)

Q.2. Draw plan and sectional elevation of a double riveted butt joint (single cover and chain riveting). Take diameter of rivet = 20 mm.

Q.3. Draw free hand upper half sectional-front elevation of a protected type flange coupling on proportionate scale.

Q.4. Draw free hand sectional front elevation and plan of a knuckle joint on proportionate scale.

Q.5. Sketch any two views of the following locking devices :

(a) Slotted nut and (b) Swan nut.

Q.6. Discuss the various commands available in Auto-CAD to draw a circle.

Section – C

(2 × 10 = 20)

Q.7. Draw the full sectional Front view and Top view of the screw-jack assembly as shown in Figure 1. Also make Bill of Materials.

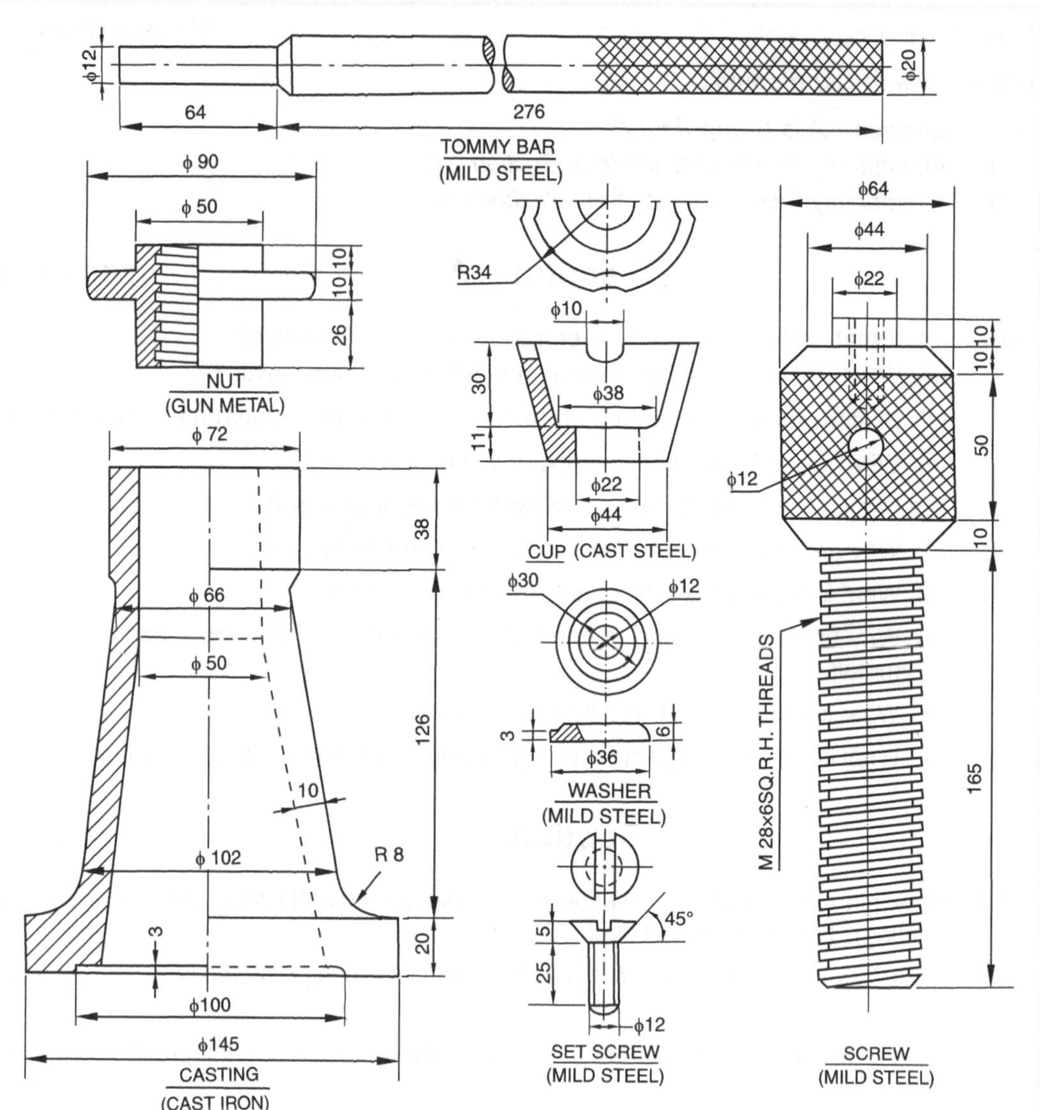

Fig. 1. *Details of a Screw-jack*

Q.8. Figure-2 shows the details of a connecting rod for Petrol engine. Assemble all the parts and draw the following views of the connecting rod :

(a) Elevation, and

(b) Plan-Full in section. Use any convenient scale.

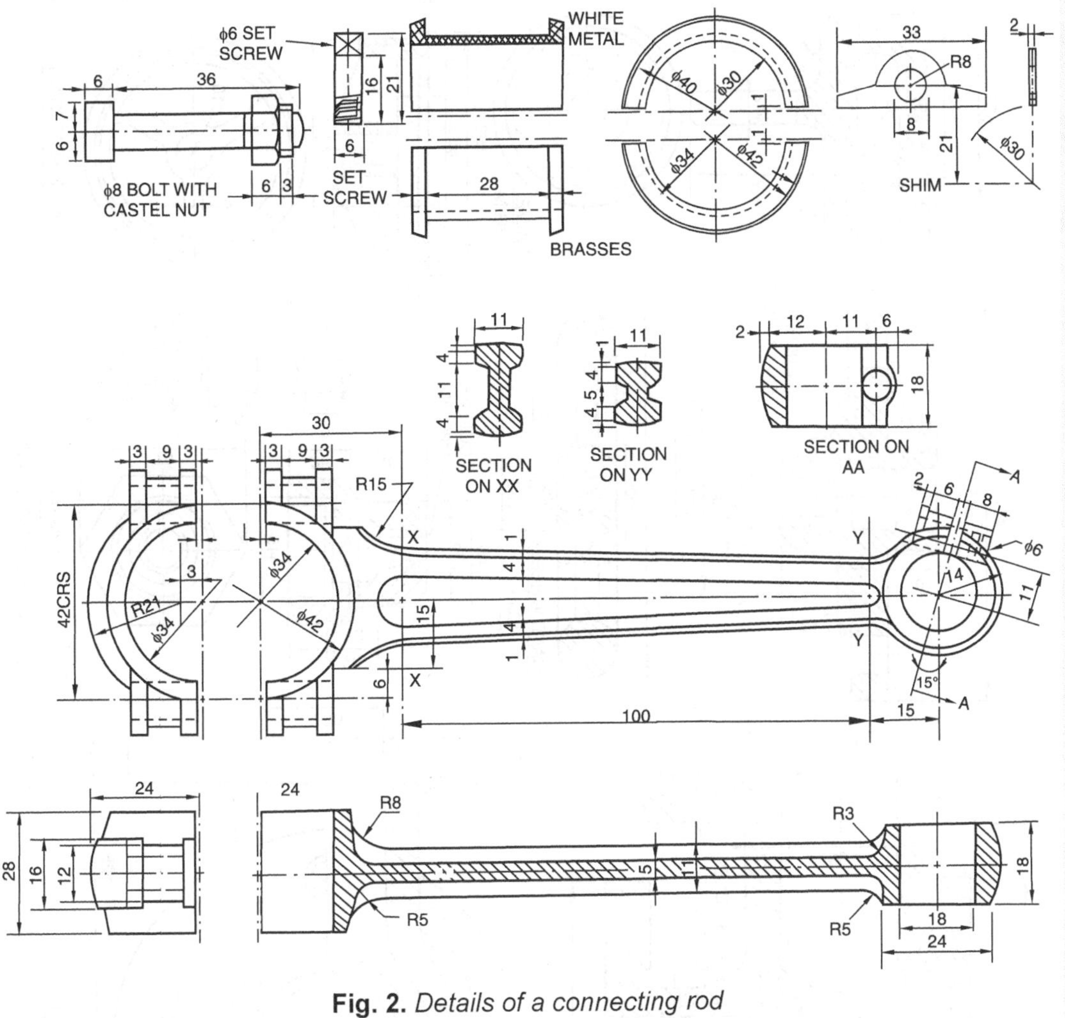

Fig. 2. *Details of a connecting rod*

Q.9. Figure 3 shows the details of a Feed check vlave. Assemble all the parts and draw the Elevation-right half in section of the assembly to half scale.

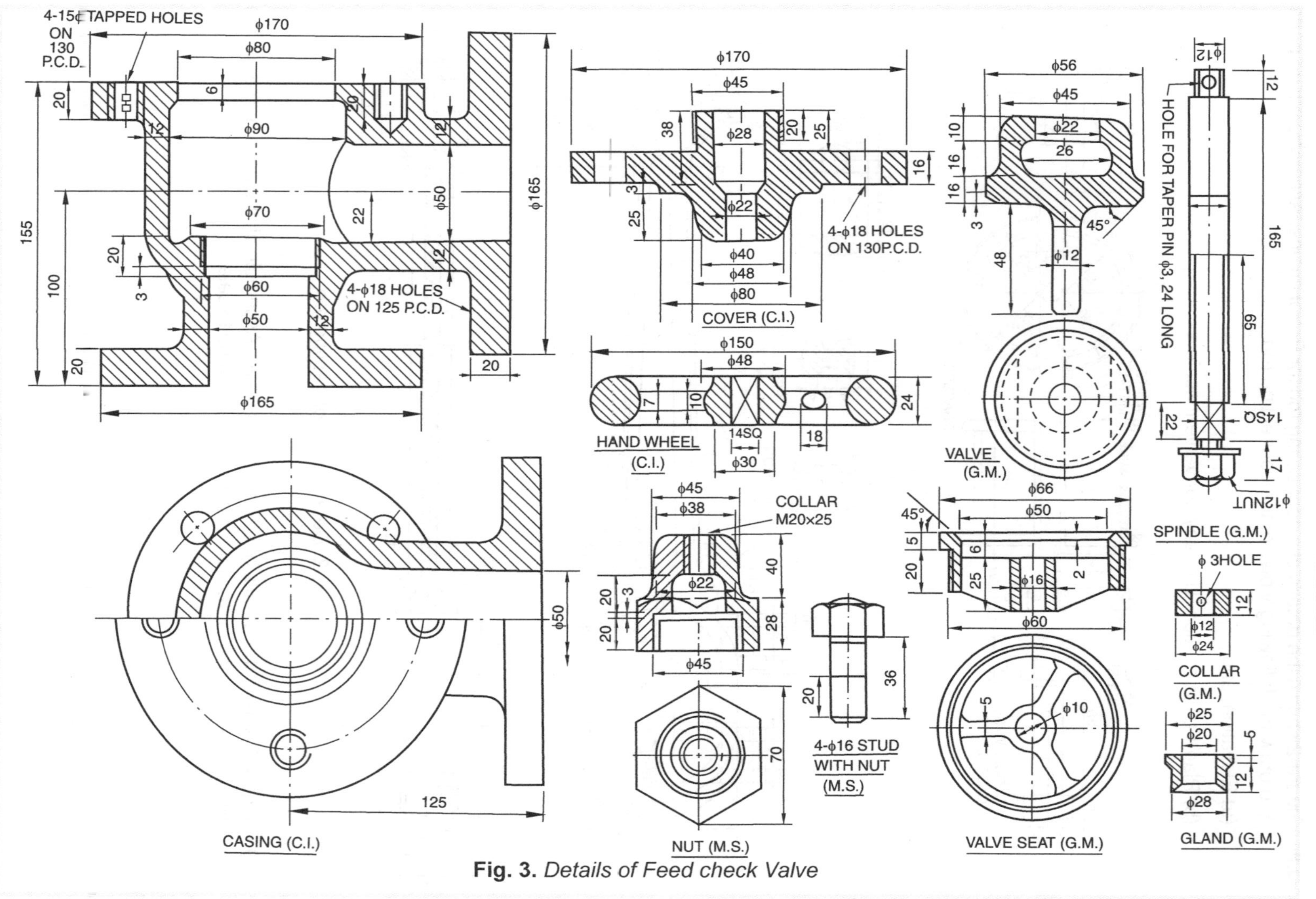

Fig. 3. *Details of Feed check Valve*

MODEL TEST PAPER – 6

Time : 04 Hours *Maximum Marks : 60*

Instructions to Candidates :

*(1) Section – A is **Compulsory.***
*(2) Attempt any **Four** questions from Section–B.*
*(3) Attempt any **Two** questions from Section–C.*

Section – A (10 × 2 = 20)

Q.1. (a) Draw the sketch of five types of lines used in machine drawing?

(b) Sketch the convention of a round section?

(c) Draw the symbol of third angle projections?

(d) The root angles in BIS metric thread and BSW threads are respectively ______ and ___________?

(e) Name two head forms of rivets?

(f) What is difference between pitch and lead?

(g) What are the functions of connecting rod in IC engine?

(h) Mention various types of bearings?

(i) Draw the free hand sketch of hexagonal bolt.

(j) Draw a symbol of fillet welding?

Section – B (4 × 5 = 20)

Q.2. What are the different types of machine drawing? Explain Production drawing in detail.

Q.3. Represent two views of hexagonal nut and square nut with proportions and dia of bolt as 30 mm.

Q.4. Draw free hand the sectional front view and right side view of the protective flanged coupling.

Q.5. Draw to scale 1:1, the standard profile of a METRIC SCREW THREAD (external), taking an enlarged pitch = 50 mm. give all the standard dimensions.

Q.6. Two steel plates, each 12 mm thick are jointed by single riveted lap joint. Draw two views to full size. Show 4 rivets and section line in plan.

Section – C

(2 × 10 = 20)

Q.7. Figure 1 shows flanges, keys and shafts to be connected in flange coupling. Assembled and draw elevation and side view in full. Note that nuts and bolts are to be added.

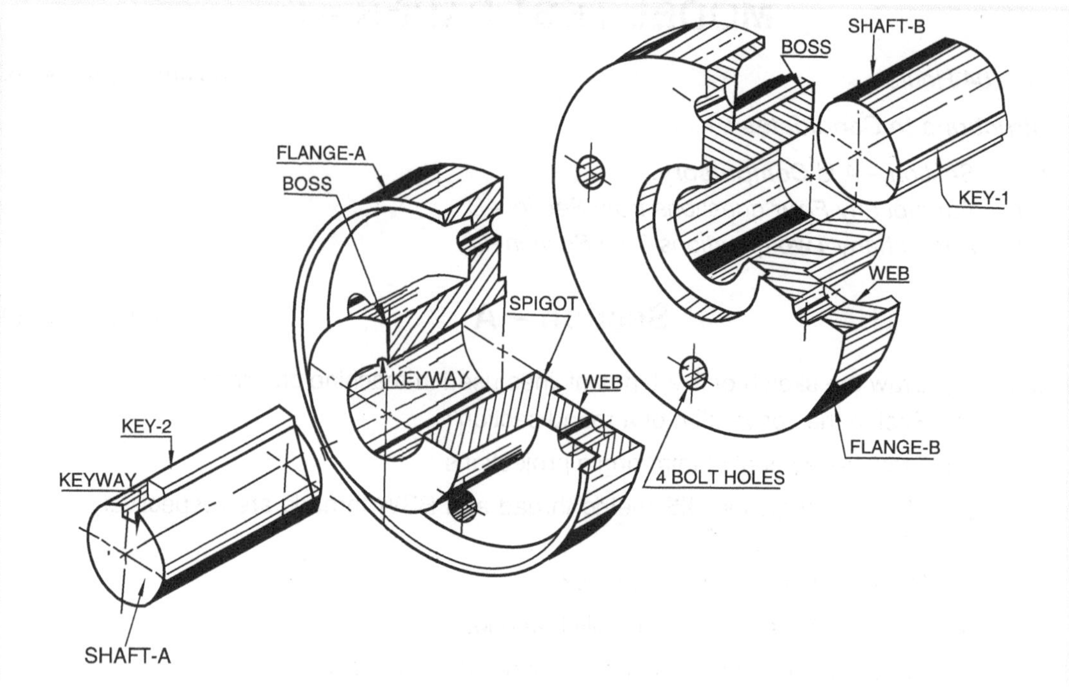

Fig. 1.

Q.8. Draw the sectional top view and front view of the petrol engine connecting rod from the given figure 2 and part list.

Part List

Part No.	**Name**	**Material**	**Quantity**
1	Rod	Forged steel	1
2	Cap	Forged steel	1
3	Bearing brass	Gum metal	2
4	Bearing bush	Phosphor bronze	1
5	Bolt	Medium carbon steel	2
6	Nut	Medium carbon steel	2

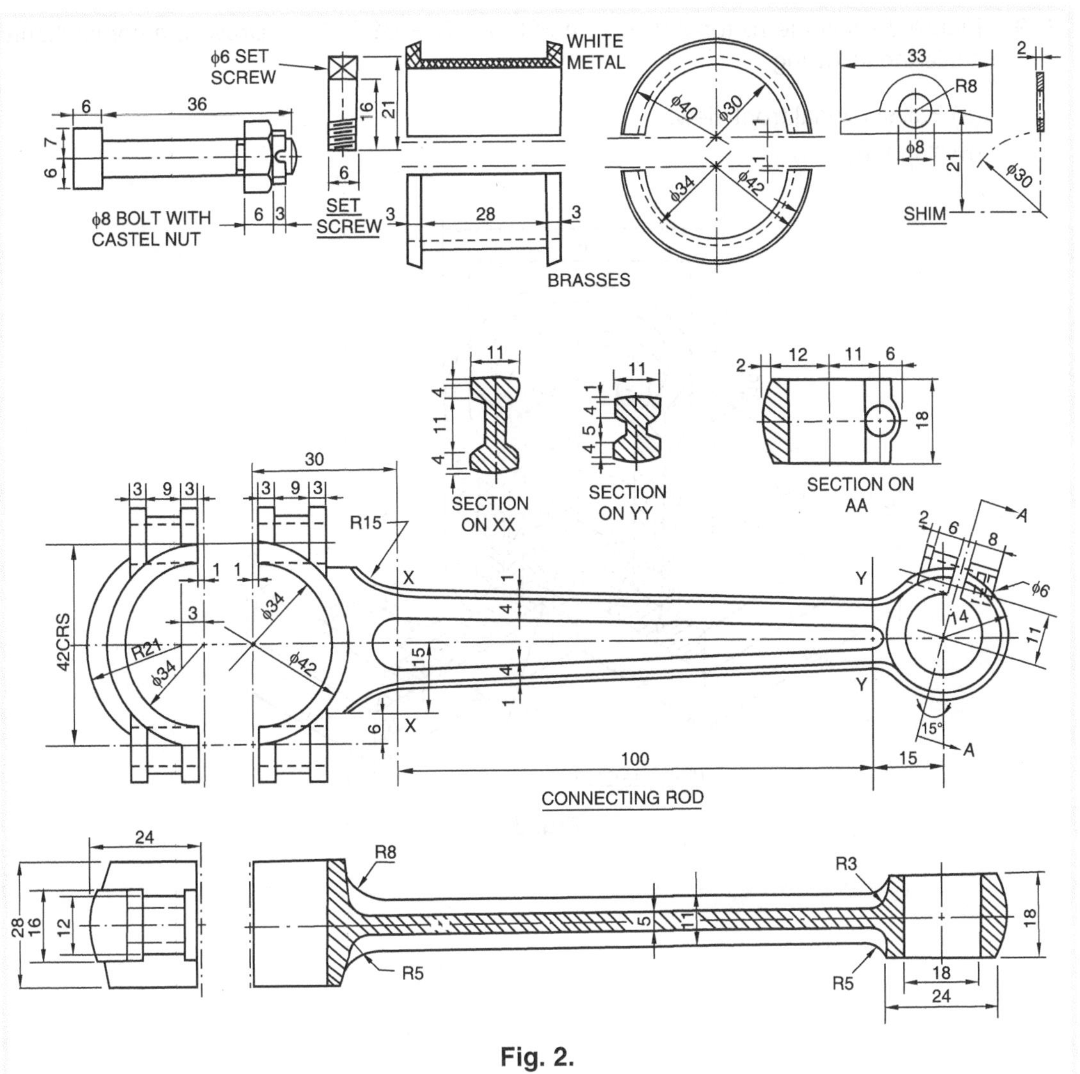

φ6 SET SCREW
WHITE METAL
φ8 BOLT WITH CASTEL NUT
SET SCREW
BRASSES
SHIM
SECTION ON XX
SECTION ON YY
SECTION ON AA
42CRS
CONNECTING ROD

Fig. 2.

Q.9. Figure 3 show the pictorial view of a FOOT STEP BEARING. Draw to a conventional scale the following :

(a) Full sectional front view.

(b) Top View.

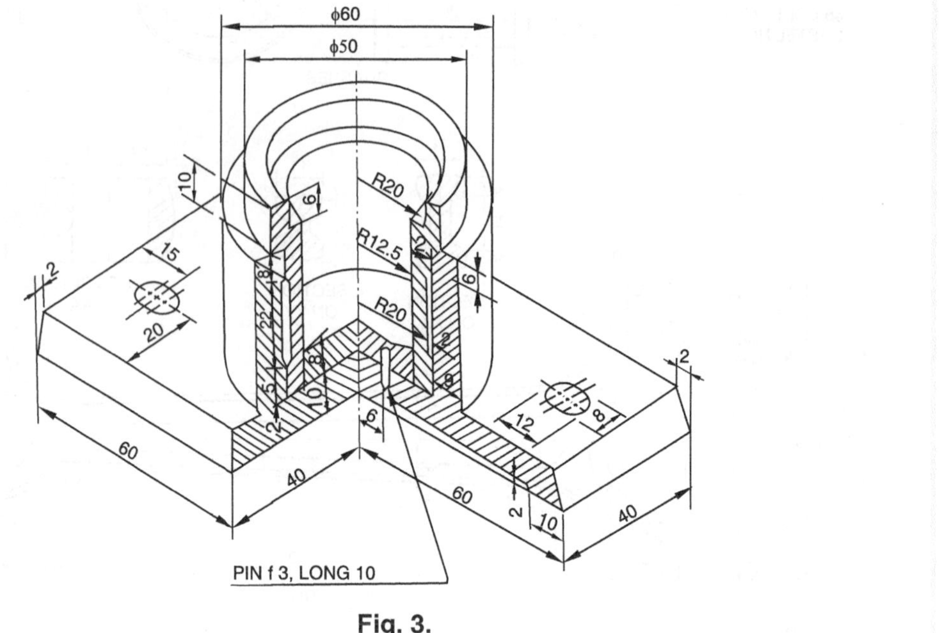

Fig. 3.

MODEL TEST PAPER – 7

Time : 04 Hours *Maximum Marks : 60*

Instructions to Candidates :

*(1) Section – A is **Compulsory.***

*(2) Attempt any **Four** questions from Section – B.*

*(3) Attempt any **Two** questions from Section – C.*

Section – A (10 × 2 = 20)

Q.1. (a) Draw the symbol of third angle projections.

(b) Explain principles of machine drawing.

(c) Write note on code IS : 296.

(d) Draw free hand sectional front view of gib head taper sunk key arrangement.

(e) Illustrate transition fit by drawing neat sketch.

(f) What is muff couping?

(g) What is flanged pipe joint?

(h) Differentiate between pith and lead?

(i) Write characteristics of IC engine piston.

(j) Name various types of welded joints.

Section – B (4 × 5 = 20)

Q.2. Draw double riveted lap joint of 16 mm thick plates using snap headed rivets. Show at least three rivets in the plan view and add a sectional elevation. Mark the dimensions in term of the rivet diameter d.

Q.3. Draw the three views of hexagonal nut of size M30. Mark the proportions in terms of the diameter d.

Q.4. Draw the free hand sectional front view and side view of knuckle joint.

Q.5. Explain in detail the different types of dimensioning.

Q.6. Draw free hand sectional front view of pipe union joint.

Section – C (2 × 10 = 20)

Q.7. From the detail of the screw jack shown in Fig. 1, draw the front view, right half in section and top view. Make part list and bill of materials.

Q.8. Fig. 2 shows the details of a 125 mm stop valve drawn in first angle. Draw the assembled front view full in section and assembled right side view. Also prepare bill of materials.

Q.9. Part drawing of a plummer block is given in Fig. 3. Assemble them and draw the elevation left in section, plan left half in section and left end view full in section.

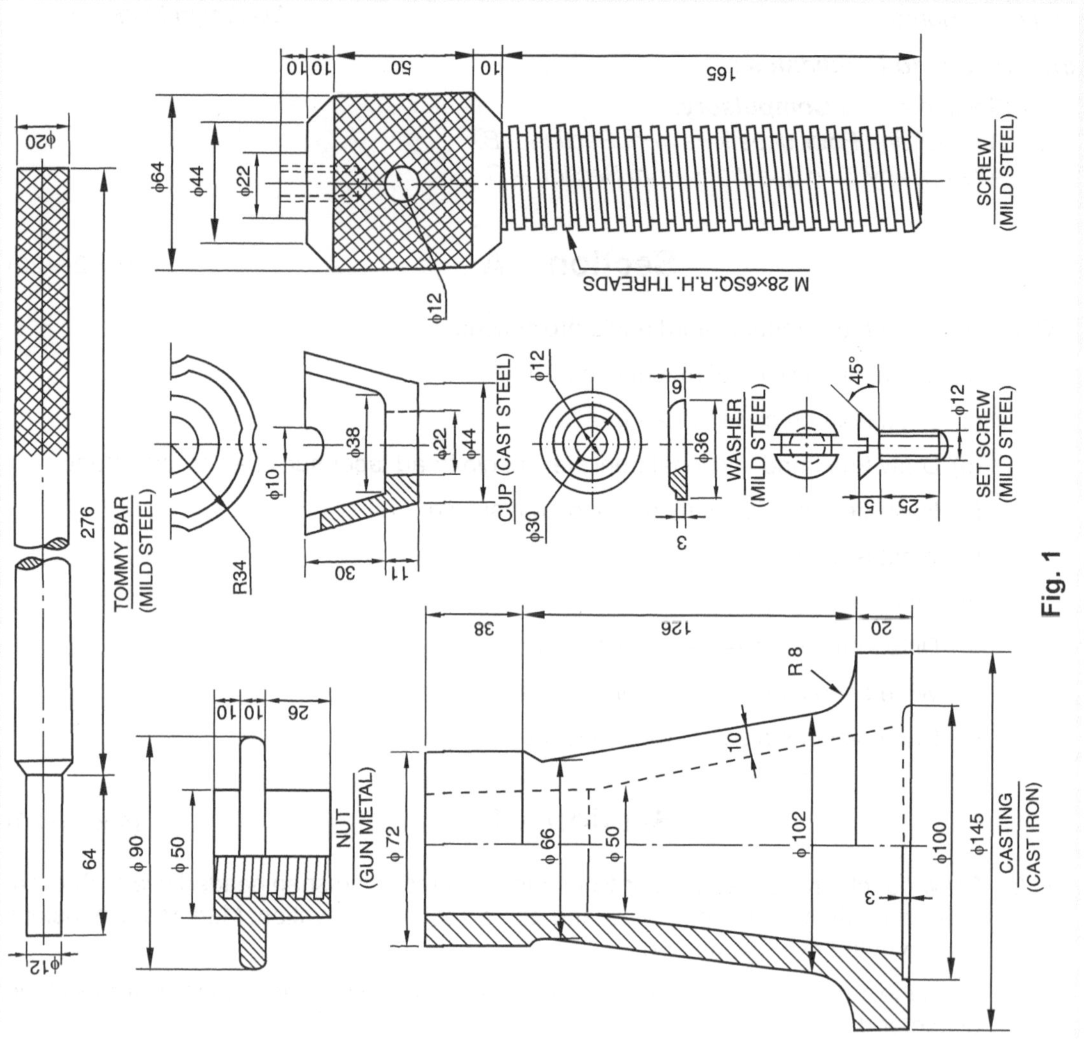

Fig. 1

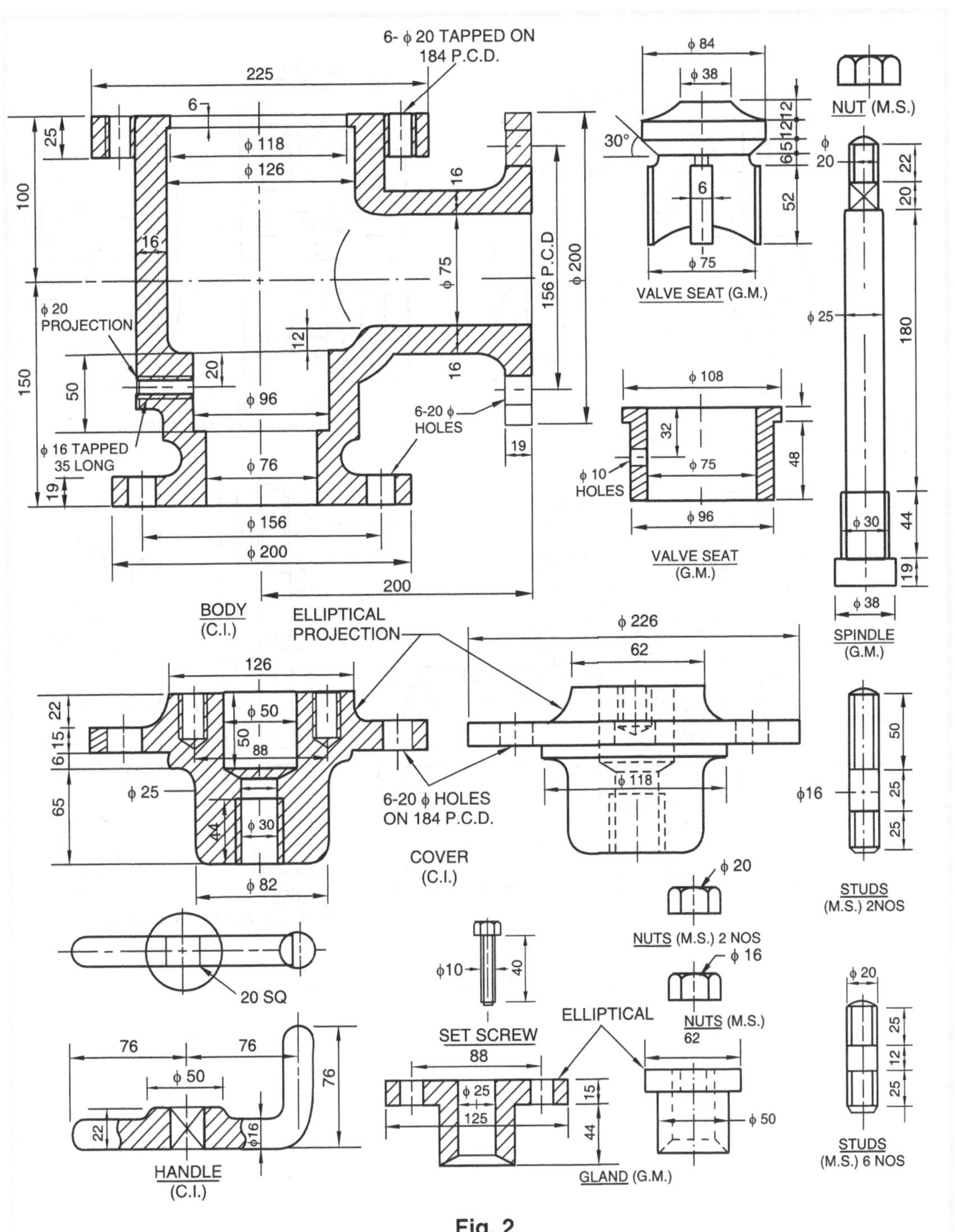
6- ϕ 20 TAPPED ON
184 P.C.D.
225
6
ϕ 118
ϕ 126
16
ϕ 75
156 P.C.D
ϕ 200
100
25
150
50
ϕ 20
PROJECTION
ϕ 96
20
12
ϕ 16 TAPPED
35 LONG
ϕ 76
6-20 ϕ
HOLES
19
ϕ 156
ϕ 200
200
BODY
(C.I.)
ϕ 84
ϕ 38
30°
6
52
ϕ 75
VALVE SEAT (G.M.)
ϕ 108
32
48
ϕ 10
HOLES
ϕ 96
VALVE SEAT
(G.M.)
NUT (M.S.)
ϕ
20
22
180
ϕ 25
ϕ 30
44
ϕ 38
SPINDLE
(G.M.)
ELLIPTICAL
PROJECTION
126
ϕ 50
88
ϕ 25
65
ϕ 30
ϕ 82
6-20 ϕ HOLES
ON 184 P.C.D.
COVER
(C.I.)
ϕ 226
62
ϕ 118
ϕ16
50
25
STUDS
(M.S.) 2NOS
ϕ 20
NUTS (M.S.) 2 NOS
ϕ 16
NUTS (M.S.)
20 SQ
ϕ10
40
SET SCREW
ELLIPTICAL
62
88
ϕ 25
125
15
44
ϕ 50
GLAND (G.M.)
76
ϕ 50
ϕ16
HANDLE
(C.I.)
STUDS
(M.S.) 6 NOS
12

Fig. 2

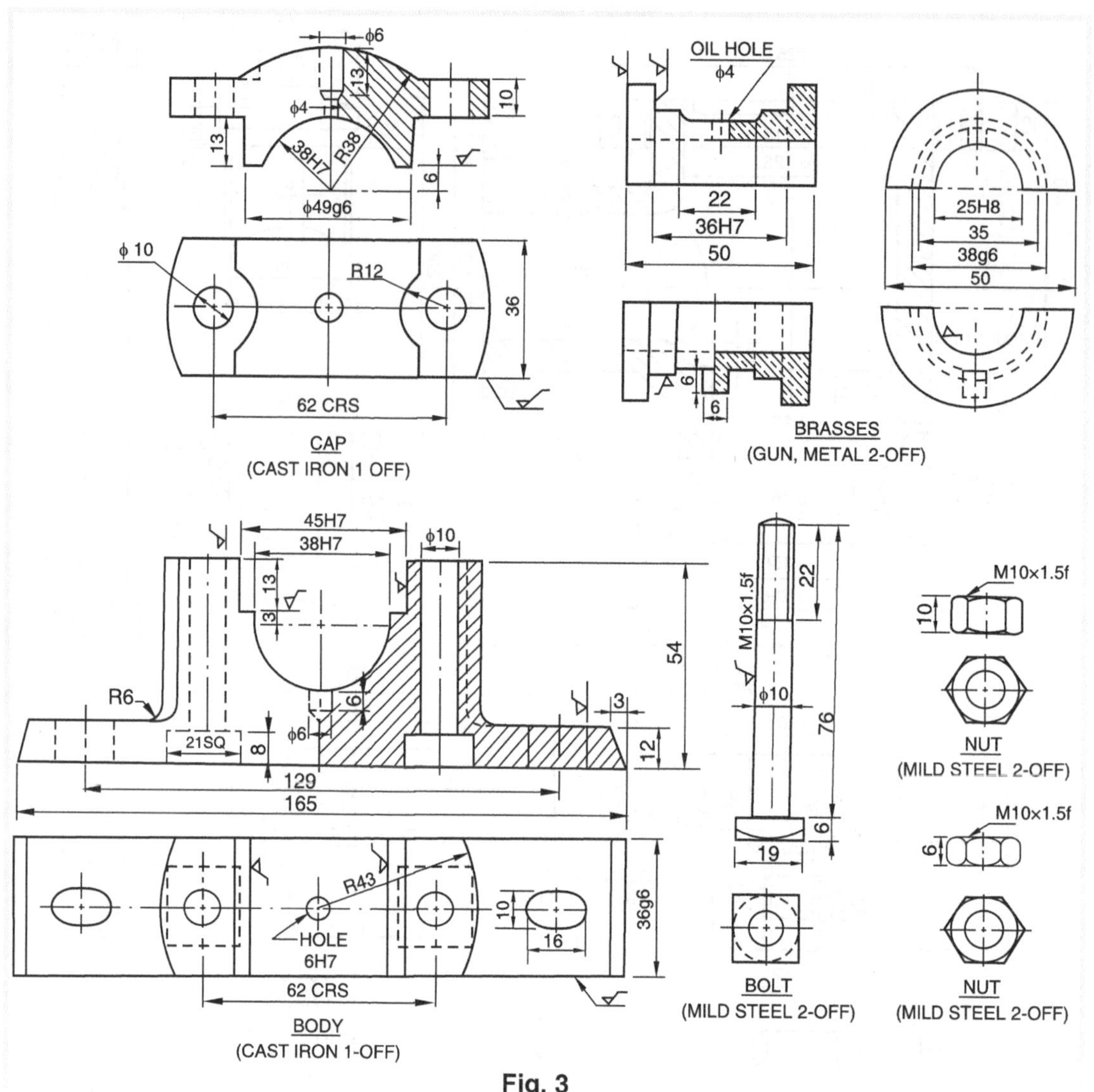

φ6
φ4
13
10
13
38H7
R38
6
φ49g6
φ 10
R12
36
62 CRS
CAP
(CAST IRON 1 OFF)
OIL HOLE
φ4
22
36H7
50
25H8
35
38g6
50
6
6
BRASSES
(GUN, METAL 2-OFF)
45H7
38H7
φ10
13
3
R6
6
φ6
21SQ
8
3
12
54
129
165
M10×1.5f
22
φ10
76
6
19
M10×1.5f
10
NUT
(MILD STEEL 2-OFF)
M10×1.5f
6
R43
HOLE
6H7
10
16
36g6
62 CRS
BODY
(CAST IRON 1-OFF)
BOLT
(MILD STEEL 2-OFF)
NUT
(MILD STEEL 2-OFF)

Fig. 3

MODEL TEST PAPER – 8

Time : 04 Hours *Maximum Marks : 60*

Instructions to Candidates :

(1) Section – A is ***Compulsory.***
(2) Attempt any ***Four*** *questions from Section – B.*
(3) Attempt any ***Two*** *questions from Section – C.*

Section – A

(10 × 2 = 20)

Q.1. (a) Name two head forms of rivets.

(b) What is pitch?

(c) What are the functions of connecting rod in IC engines?

(d) Mention various types of bearings.

(e) Sketch the convention of a round section.

(f) The root angles in BIS metric thread and BSW threads are respectively _______ and _______.

(g) What is lead?

(h) Draw the symbol of third angle projections.

(i) Draw the free hand sketch of hexagonal bolt.

(j) What is the specific use of an expansion pipe joint?

Section – B

(4 × 5 = 20)

Q.2. Draw by a conventional method a right handed square thread. Take outside diameter = 64 mm, threaded length = 72 mm and pitch = 12 mm.

Q.3. Two steel plates, each 12 mm thick are jointed by a single riveted lap joint. Draw two views to full size. Show 4 rivets and section line in plan.

Q.4. Discuss the use of following commands available in Auto-CAD:

(a) Array (b) Offset, and (c) Mirror.

Q.5. Draw free hand upper half sectional-front elevation of a protected type flange coupling on proportionate scale.

Q.6. Represent two views of hexagonal nut and square nut with proportions and dia of bolt as 30 mm.

Section – C

(2 × 10 = 20)

Q.7. Draw the sectional top view and front view of the petrol engine connecting rod from the given figure 1, and part list –

Part List

Part No	Name	Material	Qty
1	Rod	Forged steel	1
2	Cap	Forged steel	1
3	Bearing brass	Gun metal	2
4	Bearing bush	Phosphor bronze	1
5	Bolt	Medium carbon steel	2
6	Nut	Medium carbon steel	2

Q.8. Figure 2 shows the pictorial view of a FOOT STEP BEARING. Draw to a conventional scale the following:

(a) Full sectional front view.

(b) Top View

Q.9. Figure 3 shows flanges, keys and shafts to be connected in a flange coupling, Assemble and draw elevation and side view in full. Note that nuts and bolts are to be added.

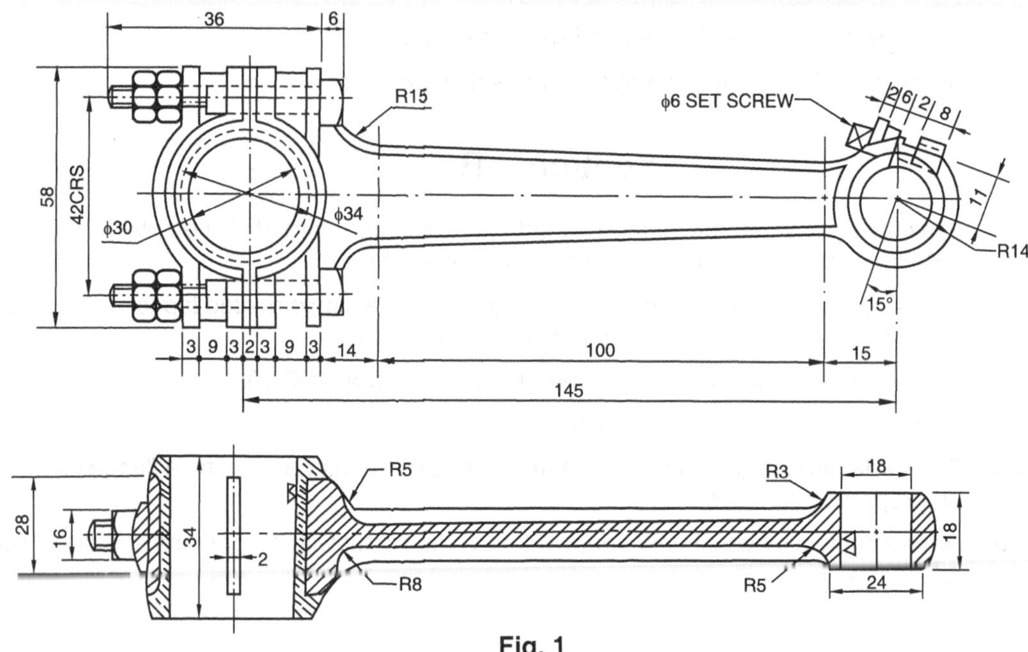

Fig. 1

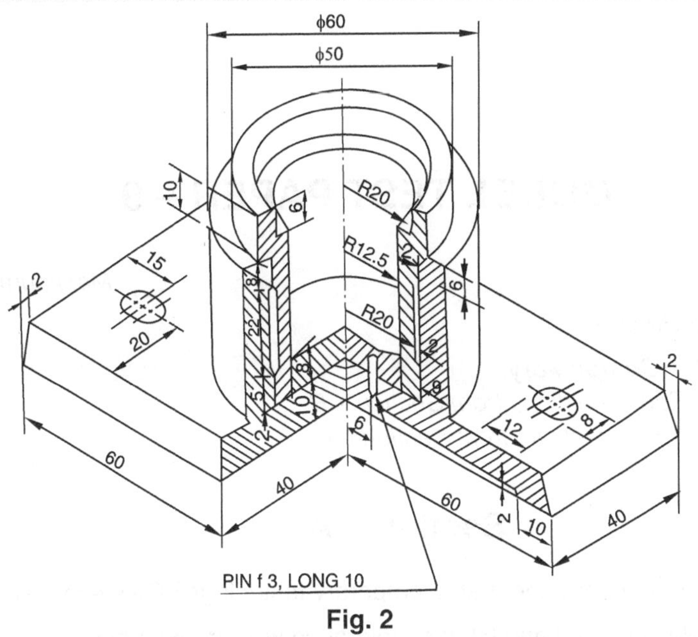

φ60
φ50
10
6
R20
R12.5
2
15
2
8
22
R20
6
2
20
8
5
10
9
2
12
8
2
6
60
40
60
2
10
40
PIN f 3, LONG 10

Fig. 2

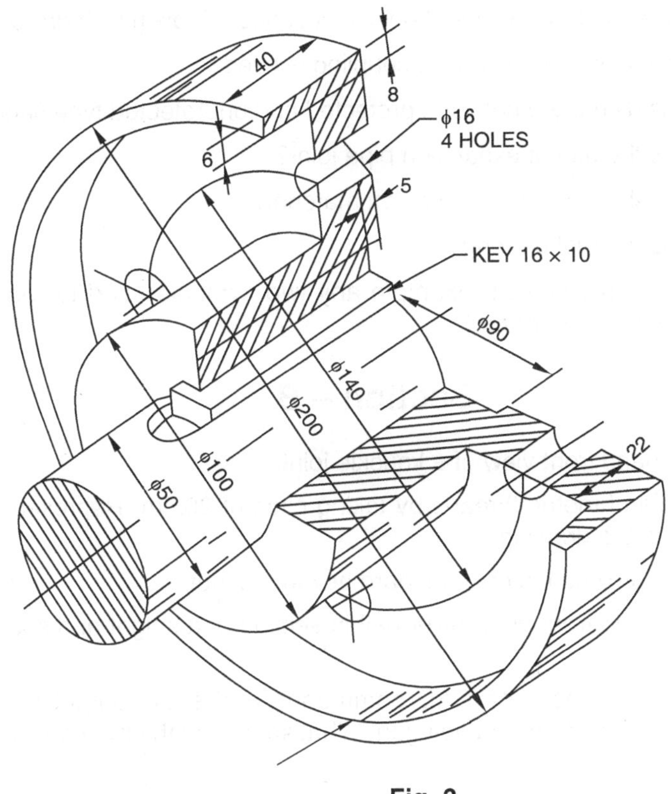

40
8
6
φ16
4 HOLES
5
KEY 16 × 10
φ90
φ140
φ200
φ100
φ50
22

Fig. 3

MODEL TEST PAPER – 9

Time : 04 Hours *Maximum Marks : 60*

Instruction to Candidates :

*(1) Section – A is **Compulsory.***
*(2) Attempt any **Four** questions from Section – B.*
*(3) Attempt any **Two** questions from Section – C.*

Section – A (10 × 2 = 20)

Q.1. (a) What are the permanent and temporary fastenings? Give example.

(b) Example unilateral and bilateral tolerance with an example.

(c) Write the procedure to tighten a lock nut.

(d) How internal threads are shown in sections? Explain with drawings.

(e) In which case the Oldham's coupling is used?

(f) What is difference between protected and unprotected type flange coupling?

(g) What is the use of expansion pipe joint?

(h) Why bushes are made from soft material?

(i) What is a thrust bearing?

(j) With suitable sketch, example any two symbols used to represent machining required on a surface.

Section – B (4 × 5 = 20)

Q.2. Draw free hand front view of a knuckle joint.

Q.3. Draw profile of metric threads by taking pitch of 20mm. Represent angle and height of thread on the drawing.

Q.4. Draw three projections of a standard hexagonal nut.

Q.5. Draw free hand front view lower half in section of a spigot and socket joint for a pipe line.

Q.6. Write step by step auto CAD commands and show output to draw I, inscribed polygon, 2, circumscribed polygon. What should be tho minimum number of sides to draw a polygon?

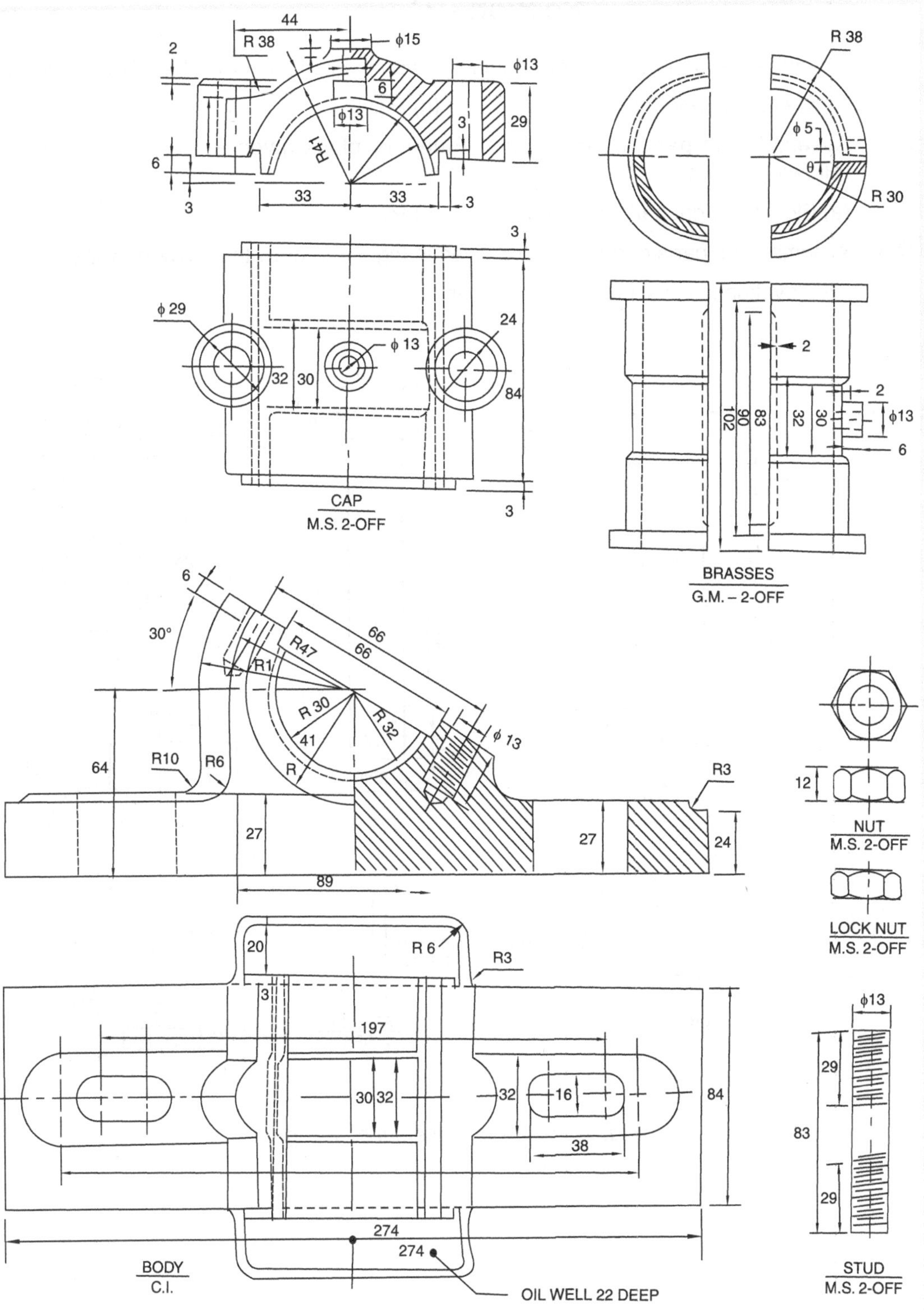

Fig. 1. *Details of an angle plummer block*

Section – C

(2 × 10 = 20)

Q.7. Assemble the parts of universal coupling given in Fig. I and draw the following views:

(a) Elevation upper half in section (b) Plan.

Q.8. Assemble the parts of an Angular Plummer block given in Fig.2 and draw the following views:

(a) Elevation right half in section (b) Plan.

Q.9. Assemble the parts of a Stop valve given in Fig.3 and draw the following views:

(a) Elevation full in section (b) Plan.

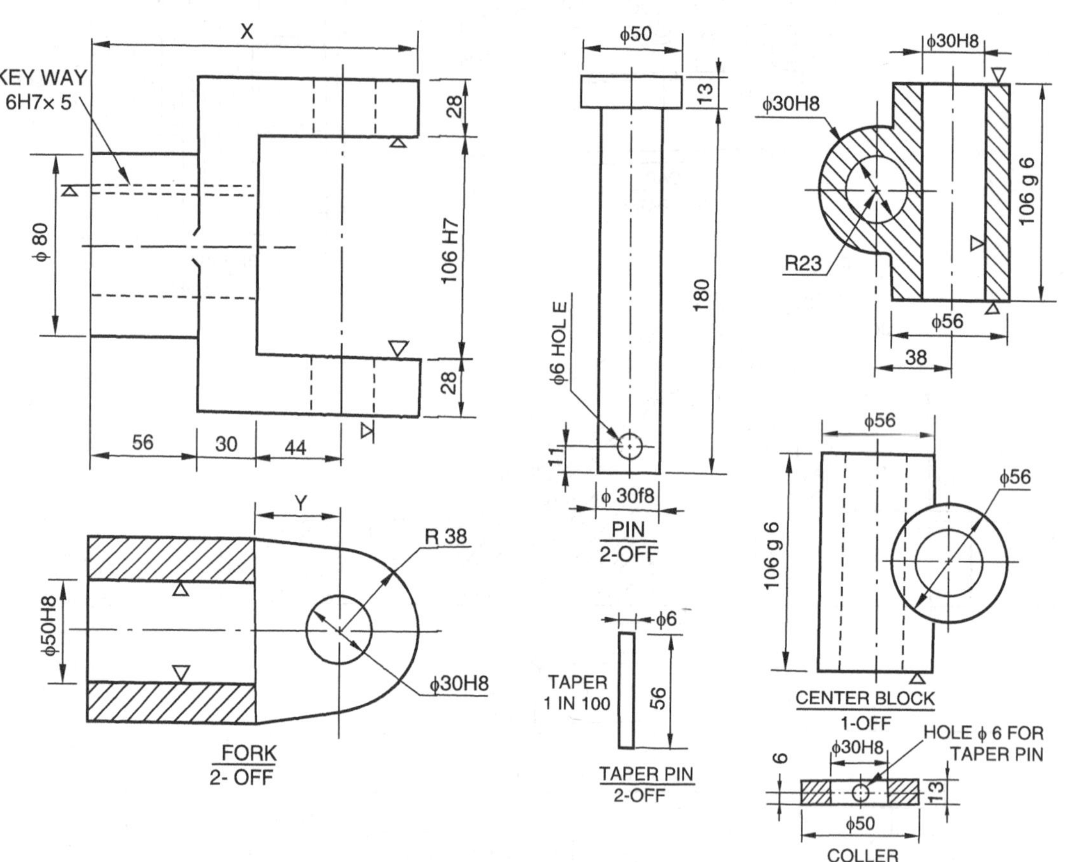

Fig. 2. *Details of universal coupling*

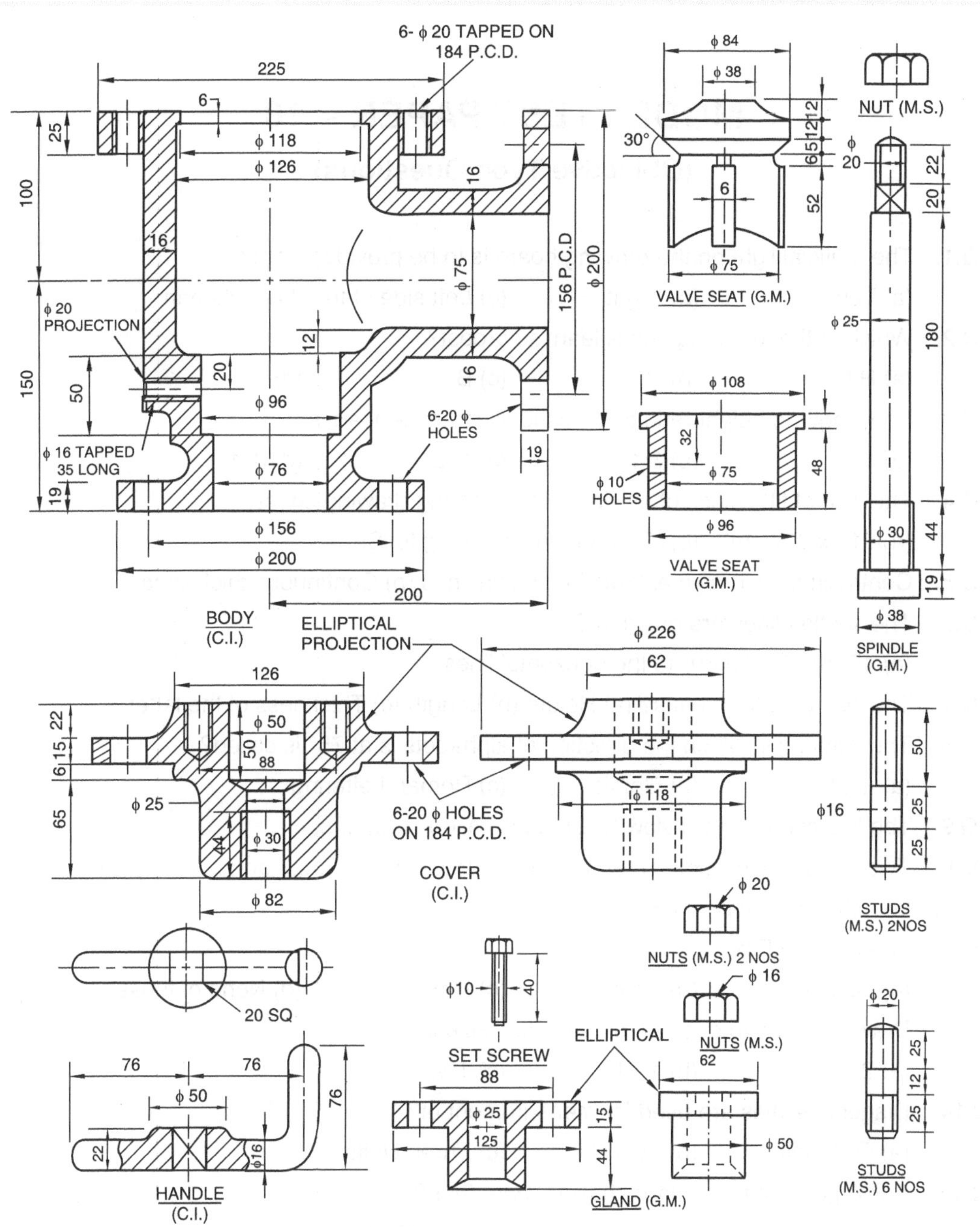

Fig. 3. *Gun metal steam stop valve*

MODEL TEST PAPER – 10

(Objective Type Questions)

Q.1. The artificial light on the drawing board is to be provided from?

(a) Top (b) Right (c) Left side of the draughtsman

Q.2. Which of the following pencil lead is hardest?

(a) H.B (b) H (c) B (d) F

Q.3. The length to height ratio of a closed filled arrow head is?

(a) 1: 3 (b) 3:1 (c) 1: 2 (d) 2: 1

Q.4. The size of title block of all size of engineering drawing sheet?

(a) 210 x 297mm (b)185 x 65mm according to ISI

Q.5. Centre lines as drawn as? (a) Chain thin lines (b) Continuous thick lines

Q.6. The section lines are drawn at?

(a) 45° (b) 30° (c) 60° to the horizontal lines

Q.7. The size of letter means? (a) Height (b) Length (c) Thickness of the letter

Q.8. The lettering in which the direction of alphabets is at 75° is called?

(a) Italic (b) Gothic (c) Roman Lettering

Q.9. The line connecting a view to a note is called ....

Q.10. Two system of placing dimensions on a drawing are .... and .... systems.

Q.11. Length of the scale = RF *x* ....

Q.12. The unit of RF is?

(a) Cu cm (b) Sq.cm (c) Cm (d) None of these

Q.13. Drawing made to one half of the actual size is?

(a) 2: 1 (b) 1 : 1 (c) 1: 2

Q.14. Diagonal scales are used for measurement of?

(a) One Unit (b) Two Units (c) Three Units

Q.15. A Polygon is a plane figure having more than?

(a) Two (b) Three (c) Four Straight Lines

MODEL TEST PAPER – 11

(Objective Type Questions)

Q.1. The projection in which the length, breath and height of an object is shown in one view is?

(a) Pictorial (b) orthographic projection

Q.2. In orthographic projection, the projectors are?

(a) Parallel (b) Perpendiculars (c) Inclined to the plane of projection.

Q.3. According to 1.5.1 symbol used for 1st angle projection is?

(a) 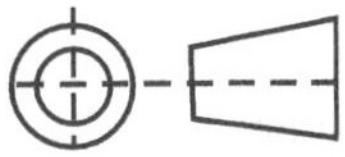 (b)

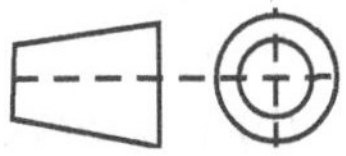

Q.4. Draw the front view and side *view* of the object as shown in Fig.1 ?

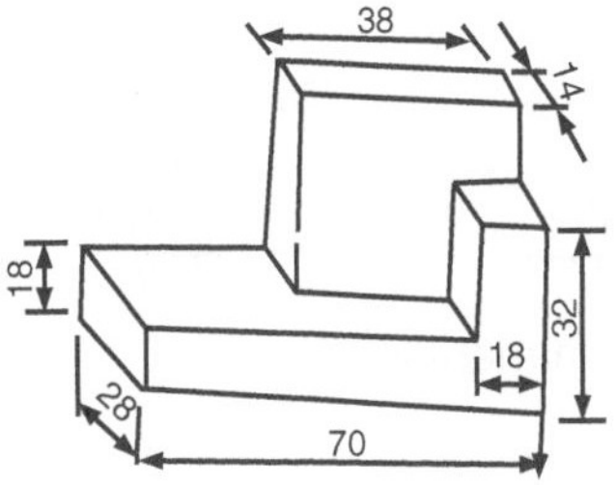

Fig. 1 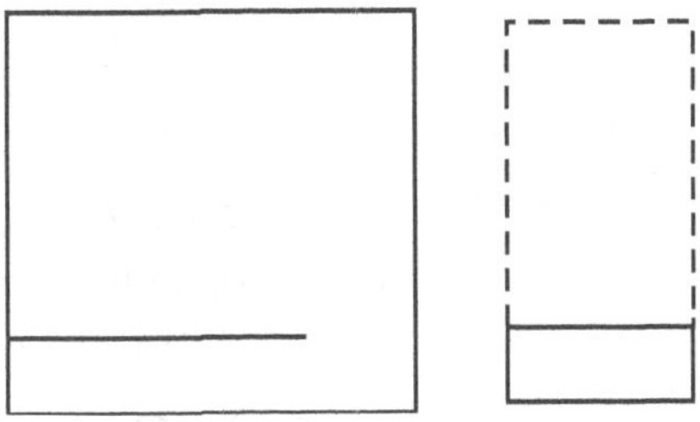 Solution

Q.5. Compare to actual diameter, isometric diameter of a sphere is?

(a) greater (b) smaller (c) equal

Q.6. A surface is represented by...........

Q.7. In a half sectional view the object is imagined to be cut off?

(a) One half (b) One fourth

Q.8. In the sectioned view, all hidden details lines are ......

Q.9. The ratio between the isometric and true length is?

(a) $2/\sqrt{3}$ (b) $\sqrt{2}/3$ (c) $\sqrt{2}/\sqrt{3}$

Q.10. In isometric projection, the receding lines are drawn with the horizontal at?

(a) 45° (b) 30° (c) 60°

Q.11. Projection on the auxiliary plane reveals the?

(a) True (b) Not true shape of the inclined surface.

Q.12. Freehand sketches are used for expressing and the recording..........

MODEL TEST PAPER – 12

(Objective Type Questions)

Q.1. A drawing which provide complete information for the production of a machine or structure is called.......drawing.

Q.2. Two extreme permissible sizes of a part between which the actual size is contained are called?

(a) Limits (b) Fits

Q.3. The difference between the dimensions of two mating parts is called?

(a) Allowance (b) Tolerance (c) Limit

Q.4. When the surface of a material is produced by any method, it is indicated as by?

(a) (b) 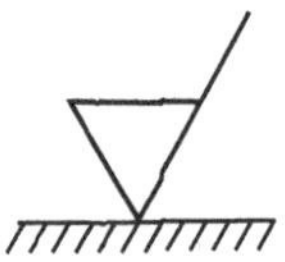 (c) 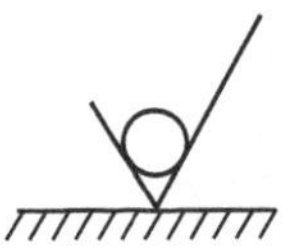 (d)

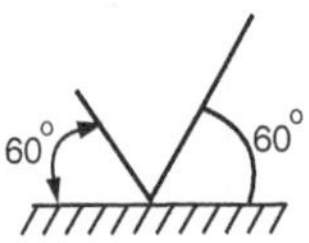

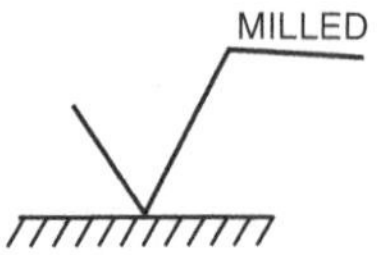

60° 60° MILLED

Q.5. The process of joining two or more metal parts permanently is?

(a) Riveting (b) Welding

Q.6. In steam engine, the eccentric rod is connected to its value rod by means of?

(a) Spigot and Socket joint (b) Knuckle joint (c) Ajustable joint

Q.7. When the two shafts are coupled whose axis are parallel but not in alignment is?

(a) Universal coupling (b) Oldham's coupling (c) Flexible coupling

Q.8. The bearing which is used to support a horizontal shaft which gets its alignment from the bearing itself is?

(a) Swivel bearing (b) Self-aligning bearing (c) Ball bearing

Q.9. The part of a steam engine which convert the rotary motion of the crank shaft into reciprocating motion of the slide valve is?

(a) Crank (b) Eccentric (c) Cross head

Q.10. The engine in which the combustion of fuel takes place inside the engine cylinder is?

(a) Steam engine (b) I.C. Engine

Q.11. The device used on the boiler for its safe working is known as?

(a) Stop valve (b) Safety valve (c) Relief valve

Q.12. Gears which are used for direct transmission of power between two shafts whose axis intersect at an angle is known as?

(a) Spur gears (b) Bevel gears (c) Worm and worm wheel

Q.13. A rotating device which transforms rotary motion into reciprocating motion in I.C. engine is known as?

(a) Gear (b) Jig (c) Cam

Q.14. A device which hold, locate and guide the work piece during manufacturing process is called?

(a) Jig (b) Fixture (c) Gear

Q.15. The device which is used to transmit power from driving shaft of engine to the driving shaft is known as?

(a) Clutch (b) Eccentric (c) Cam

Q.16. The speed of the duel core computer is about?

(a) 4.2 GHz (b) 3.8 GHz (c) 5.2 GHz

Q.17. Auto CAD is a graphic software produce?

(a) D (b) 2D